Organic Peroxides

# Organic Peroxides

## VOLUME II

DANIEL SWERN EDITOR

Fels Research Institute and Department of Chemistry
Temple University, Philadelphia, Pennsylvania

WILEY-INTERSCIENCE
A Division of John Wiley & Sons, Inc.
New York · London · Sydney · Toronto

Copyright © 1971, by John Wiley & Sons, Inc.

Library of Congress Catalog Card Number: 72-84965

ISBN 0 471 83961 2

Printed in the United States of America

10 9 8 7 6 5 4 3 2 1

# Contributors

Alwyn G. Davies
Chemistry Department
University College

R. T. Hall
Research Center
Hercules Incorporated
Wilmington, Delaware

R. Hiatt
Brock University
St. Catherines, Ontario, Canada

R. D. Mair
Research Center
Hercules Incorporated
Wilmington, Delaware

D. J. Rawlinson
Department of Chemistry
University of Wisconsin
Milwaukee. Wisconsin

Leonard S. Silbert
Eastern Regional Research
    Laboratory
U.S. Department of Agriculture
Philadelphia, Pennsylvania

G. Sosnovsky
Department of Chemistry
University of Wisconsin
Milwaukee, Wisconsin

Daniel Swern
Department of Chemistry and Fels
    Research Institute
Temple University
Philadelphia, Pennsylvania

# Preface

Volume 2 of the Organic Peroxides series continues the detailed description of many important phases of the chemistry of peroxides written by well-known authorities in their fields. The volume contains chapters describing two of the most important classes of organic peroxides, hydroperoxides and acyl peroxides. The two classes have been widely used in fundamental studies of peroxide structure and reactivity and have had many industrial applications as polymerization initiators and as starting materials in the preparation of other useful organic chemicals. The volume also contains two chapters on the determination of organic peroxides and their physical properties. To the best of my knowledge, this pair of chapters offers the most complete and detailed description of these subjects to be found anywhere. The remaining chapters in Volume 2 are devoted mainly to detailed discussions of reactions of organic and organometallic peroxides, hydrogen peroxide, and peroxydisulfate in a wide variety of important organic and inorganic reactions. As in Volume 1, extensive literature citations and tabulations of data are given. Volume 2 is about twice as long as Volume 1.

Again, it is a real pleasure to express my gratitude to the contributors for the promptness in submitting their manuscripts and for their overall excellence. Mrs. Renate Gerlach-Kyrieleis again provided invaluable help in preparing the index, typed my own contribution and, by handling the numerous details required in putting a volume of this magnitude together, played an indispensable role in the prompt completion of this work.

*Daniel Swern*
*Philadelphia, Pennsylvania*
*July, 1970*

# Contents

# Organic Peroxides

*Chapter I*

# Hydroperoxides

## R. HIATT

## I.  INTRODUCTION

Generically hydroperoxides are derivatives of hydrogen peroxide, with one hydrogen replaced by an organic radical.

$$\underset{\text{hydrogen peroxide}}{\text{H—O—O—H}} \qquad \underset{\text{hydroperoxide}}{\text{R—O—O—H}} \qquad \underset{\text{peroxy acid}}{\overset{\displaystyle \overset{\text{O}}{\|}}{\text{RC}}\text{—O—O—H}}$$

Compounds in which the carbon attached to —O—OH also bears other functional groups, as in

$$\text{RCH}\!\!\begin{array}{l}\diagup\text{OH}\\[-2pt]\diagdown\text{O}_2\text{H}\end{array}$$

are also termed hydroperoxides and are dealt with in this chapter, excepting only the case where the other functional group is carbonyl. These compounds, called peroxy acids or peracids, have a rather different chemistry and are discussed in Volume I of this series.

Hydroperoxides are known where the point of —O—OH attachment is not carbon, but Si, Ge, Sn, S, etc. These concern us only briefly here, as they are discussed in detail in Chapter IV.

Hydroperoxide chemistry had its heyday in the decade 1950–1960, following the firm establishment of these compounds as reactive intermediates in the autoxidation of olefins. Three books (1–3) published in 1961 together constitute a complete and indispensable account of the fundamental elements of hydroperoxide synthesis and reactions published in that decade and earlier.

Hydroperoxide research has not diminished in volume in the intervening years, and its most significant aspects have been threefold: the synthesis of the more recondite members of the species, particularly of nonhydrocarbons; refinements in the mechanisms of autoxidation and combustion; and attempts to elucidate more precisely the many modes of hydroperoxide decomposition. This chapter concentrates on these newer areas.

In attempting to provide a complete survey the first-named references (1–3) have been extensively consulted. More recent noteworthy articles and reviews dealing with various aspects of hydroperoxide chemistry are cited in references 4 through 26. Detailed accounts of early development of the field are contained in the classics by Rieche (27), Criegee (28, 29), and Tobolsky and Mesrobian (30).

## II.  PREPARATION OF HYDROPEROXIDES

Hydroperoxides may be prepared from organic compounds by the action of molecular oxygen, ozone, or hydrogen peroxide. Of these, molecular oxygen probably offers the greatest scope and is the most complex, at least four general types of reactions being available. It has the disadvantage, however, of frequently producing mixtures of isomers. Reactions with ozone are limited to producing α-alkoxy hydroperoxides. Displacement or addition reactions of hydrogen peroxide are most commonly used when a hydroperoxide of a particular constitution is desired, but yields are low if the compound is nontertiary and reaction conditions can be hazardous.

In most cases preparation of hydroperoxides offers severe problems in separation and purification (see Section III-C). Tables in the Appendix contain a partial list of hydroperoxides that have been prepared in sufficiently high purity to be characterized, together with references and an indication of the synthetic procedure. As a general rule, if a particular hydroperoxide can be prepared easily by one method, it can easily be prepared by other methods; if it is difficult to prepare by any one method, it will be difficult to prepare by any method.

### A.  Hydroperoxides by Reactions of Oxygen

#### 1.  Autoxidation

Free-radical-chain reactions of oxygen with organic or organometalic compounds afford a practical synthetic route for hydroperoxides from alkanes bearing tertiary hydrogens, aralkanes, olefins (including lipids, steroids, and compounds containing C=N), ethers, alcohols, and ketones. Hydroperoxides have been proposed as intermediates in the autoxidation of thioethers (31) but have not yet been isolated, probably because of facile reaction of the $-O_2H$ group with the neighboring sulfur atom (Section IV-B). Alkynes with hydrogen on the $\alpha$-carbon autoxidize more readily than the corresponding alkenes (32), but apparently there has been no attempt to isolate or characterize the resulting hydroperoxides.

The mechanism of autoxidation is well understood (at least at temperatures under 200°) and has been reviewed many times (1–4, 6, 10, 16, 24, 29, 33). The basic equations are as follows:

$$\text{initiation} \longrightarrow 2X\cdot \qquad \text{(where } X\cdot \text{ is any radical)} \tag{1}$$

$$X\cdot \ \ (\text{or } XO_2\cdot) + RH \longrightarrow R\cdot + XH \ \ (\text{or } XO_2H) \tag{2}$$

$$R\cdot + O_2 \longrightarrow RO_2\cdot \tag{3}$$

$$RO_2\cdot + RH \xrightarrow{\ k_p\ } RO_2H + R\cdot \tag{4}$$

$$2RO_2\cdot \xrightarrow{\ k_t\ } RO_2R \ \ (\text{or alcohol and carbonyl compound}) + O_2 \tag{5}$$

The rate expression* for hydroperoxide formation, and for oxygen consumption, if its formation by reaction 5 is neglected, is

$$\frac{\delta O_2}{\delta t} = \frac{\delta RO_2H}{dt} = \left(\frac{R_i}{k_t}\right)^{1/2} k_p \, [RH]$$

where $R_i$, the rate of initiation, combines reactions (1) and (2).

Janzen and co-workers (35), using electron-spin-resonance (ESR) spectra have demonstrated virtually complete dissociation of $Ph_3CO_2\cdot$ to $Ph_3C\cdot + O_2$ at 130°. The observation explains the difficulty with which triarylmethanes are autoxidized (36) and their inhibiting effect on the autoxidation of other hydrocarbons (37).

$$RO_2\cdot + Ph_3C\cdot \longrightarrow RO_2CPh_3 \tag{6}$$

The rate of autoxidation thus depends on the ratio $k_p/(k_t)^{1/2}$, about which some qualitative generalizations are possible. Experiment shows that the relative ease of autoxidation of RH roughly parallels the ease of breaking the $C-H$ bond (1) and for hydrocarbons increases in the series $n$-alkanes $<$ branched alkanes $<$ aralkanes $\cong$ alkenes $<$ alkynes (2, 32). Thus both resonance and inductive effects on the stability of the resulting radical are important (1).

*Reaction 3 is usually so fast and efficient that it can be neglected (1); it is reversible, however, at high temperatures. Using thermochemical data, Benson (34) has calculated ceiling temperatures where rates of reactions 3 and $-3$ are equal. At 1 atm of $O_2$ they are above 500° where R is alkyl but only 300 and 160° for allyl and benzyl, respectively.

More quantitative data are available for alkenes than for any other class of compounds. Bolland's rules (38) for substituted olefins of the type $\overset{1}{C}H_2=\overset{2}{C}H-\overset{3}{C}H_3$ state that alkyl substitution on C-1 or C-3 increases the rate of autoxidation by $3.3^n$ at 45°, where $n$ is the number of substituents. Substitution on C-2 has little effect. Van Sickle, Mayo, and co-workers (39) have refined these rules to include steric effects, C-3 substitution by $t$-butyl giving a rate only one-sixth as fast as substitution by methyl. They conclude that a fully formed carbon radical plus hydroperoxide is a poor model for the transition state of reaction 4. Deuterium isotope effects support this belief. Measured values of the ratio $k_H/k_D$ for reaction 4 vary from 1.05 to 5.8 (40, 41), depending on R.

For accurate predictions it is therefore best to consult tables of $k_p/(k_t)^{1/2}$ values measured experimentally, of which there are a large and growing number (3, 10, 32, 33, 39, 42–46).

Separate values for $k_p$ and $k_t$ have also now been measured for a large number of hydrocarbons (47–51), chiefly by Howard and Ingold (48, 49, 51), by rotating-sector methods. A representative sampling is presented in Table 1.

TABLE 1.  RATE CONSTANTS[a] FOR THE AUTOXIDATION OF HYDROCARBONS

| Hydrocarbon | $^b k_p/(2k_t)^{1/2} \times 10^5$ at 70° | $^b k_p/H$ at 30° | $k_t \times 10^{-6}$ at 30° | Reference |
|---|---|---|---|---|
| Propylene | 21.7 | | | 32 |
| 1-Butene | 81 | | | 32 |
| 1-Hexene | 38 | | | 32 |
| 1-Octene | 6.2[c] | 50 | 130 | 49 |
| Neopentylethylene | 13 | | | 32 |
| Isobutylene | 20 | | | 32 |
| Tetramethylethylene | 1120 | 0.14 | 0.32 | 32 |
| Cyclopentene | 220 | | | 45 |
|  | | 1.7 | 3.1 | 49 |
| Cyclohexene | 175 | | | 45 |
|  | | 1.5 | 2.8 | 49 |
| Cyclooctene | 10 | | | 45 |
| Toluene | 1.4[c] | 0.08 | 150 | 49 |
| Ethylbenzene | 21[c] | 0.65 | 20 | 49 |
| Cumene | 150[c] | 0.18 | 0.0075 | 49 |
| 1,1-Diphenylethane | 110[c] | 0.34 | 0.047 | 49 |

[a] In units of liters per mole per second.
[b] For abstraction only.
[c] At 30°.

In practical terms, however, conclusions from such data may be misleading. At 100–110° without added initiators ring-substituted cumenes autoxidize faster than cumene itself ($p$-NO$_2$ > $p$-I > $p$-Br > $p$-Cl > $p$-H) (52), although $k_p/(k_t)^{1/2}$ values predict exactly the reverse (51–53). Under these conditions the reaction is autocatalytic; the substituted $\alpha$-cumyl hydroperoxides decompose to radicals faster than the unsubstituted hydroperoxide (52).

## 2. Hydroperoxides from the Autoxidation of Alkanes and Aralkanes

### (a) THERMAL INITIATION

Most alkanes and aralkanes can be autoxidized at 100–150° without added initiators, and many aralkanes give substantial yields of hydroperoxides at lower temperatures (2, 3, 4). Typically such reactions show induction periods, followed by increasingly rapid oxygen uptake and eventual leveling off. Whatever the mechanism of original initiation (3, 10, 54) [there is now convincing evidence for direct, radical-producing reaction of tetralin with oxygen (55)], the autocatalysis clearly derives from some of the produced hydroperoxide decomposing to radicals (Section V-C). Leveling off results when the rate of decomposition of hydroperoxide equals the rate of formation (56, 57).

Since it is frequently easier to separate the hydroperoxide from unreacted starting material than from decomposition products, autoxidative syntheses are usually carried to low conversions. Typical examples are the autoxidation of *cis*-decalin at 140° to 9% by weight of hydroperoxide, a yield of 80% based on oxygen uptake (58), and the autoxidation of cyclopent-[*cd*]-indane at 50° to 10% conversion, with a 60% yield of purified hydroperoxide (59).

$$\xrightarrow[50°]{O_2} \qquad\qquad O_2H \tag{7}$$

The method is most commonly applied to syntheses of tertiary hydroperoxides—for example, from alkylcyclohexanes (1, 60) or substituted cumenes (1, 61)—but has been used by Soviet workers to prepare low concentrations of hydroperoxides in butane and *n*-decane (10). Gas-phase autoxidations of methane, ethane, and isobutane have yielded hydroperoxides, but higher homologs give mainly cleavage products (1). Addition of peroxide or azo initiators eliminates induction periods and permits use of lower temperatures, generally with improved yields. Walling (33, p. 406) lists a number of aralkyl hydroperoxides prepared in this fashion, and the method remains in frequent use (62–64).

Other initiators of hydrocarbon autoxidations include ozone (65), ionizing radiation (66–68), metal ions (Section II-B-2), and bases (Section II-B-3). Rust and Vaughan (69) found that hydrogen bromide in small amounts is a powerful catalyst for gas-phase autoxidation, acting not only as coinitiator but as a chain-transfer agent. Up to 70% yields of $t$-BuO$_2$H could be obtained.

$$HBr + O_2 \longrightarrow HO_2\cdot + Br\cdot \tag{8}$$

$$Me_3CH + Br\cdot \longrightarrow Me_3C\cdot + HBr \tag{9}$$

$$Me_3CO_2\cdot + HBr \longrightarrow Me_3CO_2H + Br\cdot \tag{10}$$

### (b) Initiation by Metal Ions

Soluble salts of transition-metal ions—chiefly cobalt, manganese, copper, and iron—are frequently added to accelerate autoxidations of hydrocarbons. Although their use is said to produce more by-products than peroxide or azo-initiated reactions (24, 70, 71), hydroperoxides of cumene (24, 72–75), *sec*-butylbenzene (22, 24), tetralin (22, 57, 76), and the like have been prepared in 15–40% yield, based on hydrocarbon. A number of references to patent literature are listed in reviews by Zavgorodnii (24), List and Kuhnen (22), and Fallab (77). Rouchaud (75) has provided a very detailed account of the effects of $Co^{2+}$ concentration on the yield of hydroperoxide from cumenes at 100°. Conversions of up to 5% gave nearly quantitative α-cumyl hydroperoxide yields with approximately $10^{-4}$ M metal ion. Higher concentrations gave faster reactions but lower yields.

The method seems most applicable to the more reactive aralkanes. Toluene gives yields of less than 1% (24). Pritzkow (78) obtained only alcohols and ketones from manganese- or cobalt-catalyzed autoxidation of heptane.

The mechanism of metal-ion catalysis is complex, but it certainly involves radical-producing reactions with hydroperoxide (Section V-D).

$$M^{n+} + RO_2H \longrightarrow M^{(n+1)+} + RO\cdot + OH^- \tag{11}$$

$$M^{(n+1)+} + RO_2H \longrightarrow M^{n+} + RO_2\cdot + H^+ \tag{12}$$

The autoxidations usually show an induction period unless a little hydroperoxide is added initially with the metal ion (24, 70, 73, 79, 80). Although $Co^{2+}$ and $Mn^{3+}$ will certainly react slowly with aralkanes to produce radicals (81–83), the much faster reaction with hydroperoxides must take over once the autoxidation begins. Too much metal ion can inhibit the autoxidation (54), presumably by destroying all the hydroperoxide as soon as it is formed.

(c) BASE-CATALYZED AUTOXIDATION

A distinction must be made between mild-base catalysis, where the base protects hydroperoxide from decomposition by acidic by-products (73, 84) and possibly produces radicals by reaction with hydroperoxide (54), and strong-base catalysis, where carbanions are involved.

Most of the Soviet quasi-industrial recipes for aralkane autoxidation (24, 80, 84–87) call for 2–3% of calcium oxide or hydroxide, or sodium hydroxide, carbonate, bicarbonate, or stearate, either as solid or in aqueous phase. The base strength does not seem high enough to create many carbanions (73), but yields are increased and lower temperatures can be used (85). Under such conditions Volkov et al. (86) claim an 86% yield at 83% conversion for 1-methyl-3-phenylindane. Yeomans and Young (88) were able to oxidize 1,4-dialkyltetralins to 80–100% conversion, with 80% of the oxygen uptake appearing as $RO_2H$.

The autoxidation of organic compounds under strongly basic conditions is proposed by Russell and Bemis (89) to proceed via carbanion formation and a radical-ion chain.

$$R_3C^- + O_2 \longrightarrow R_3C\cdot + O_2^- \tag{13}$$

$$R_3C\cdot + O_2 \longrightarrow R_3CO_2\cdot \tag{14}$$

$$R_3CO_2\cdot + R_3C^- \longrightarrow R_3CO_2^- + R_3C\cdot \tag{15}$$

Most organic hydroperoxides are too readily destroyed by base to survive these conditions (Section IV-B-4), but there are useful exceptions. Sprinzak (90) has prepared 9-alkyl or aryl-9-hydroperoxyfluorenes using Triton B in pyridine at −40 to −15°. Fluorene itself gave only fluorenone; 2,3-dimethylindene yielded a 50–50 mixture of the indenone and hydroperoxide. At higher temperatures the substituted fluorenes gave only alcohols, probably via carbanion displacement on the O—O bond.

$$R^- + RO_2H \longrightarrow RO^- + ROH \tag{16}$$

Hawthorne and Hammond (91) obtained good yields of tris(p-nitrophenyl) methyl hydroperoxide by autoxidation in alcoholic potassium hydroxide. Hydroperoxides may be obtained from hindered phenols (92, 93).

$$\tag{17}$$

Barton and co-workers (94) have obtained 60–80% yields of 17-α-hydroperoxides from 20-ketosteroids, but most simple ketones give only cleavage products through base-catalyzed decomposition of the initially formed α-ketohydroperoxides (95). α-Hydroperoxy esters have been prepared from α-phenyl acetates (96, 97).

$$Ph_2CHCO_2Me \xrightarrow[OH^-]{O_2} Ph_2\overset{\overset{\displaystyle O_2H}{|}}{C}CO_2Me \qquad (18)$$

Hydroperoxides have been synthesized from isobutyronitrile and other tertiary cyanides by the autoxidation of strongly basic solutions at $-40°$ (98).

$$Me_2CHCN \xrightarrow[O_2, -40°]{t-BuOK} Me_2\overset{\overset{\displaystyle O_2H}{|}}{C}CN \quad (15\%) \qquad (19)$$

### (d) Selectivity, Intramolecular Reactions, and Racemizations

Few quantitative studies have been made of the distribution of isomeric hydroperoxides obtained from the autoxidation of alkanes. $n$-Decane at 145° yields approximately equal concentrations of all possible secondary isomers but only traces of primary hydroperoxide (99). (Here, as in all cases cited, the hydroperoxide product was reduced to a mixture of alcohols and chromatographed.) Pritzkow and Müller (100) obtained similar results in the autoxidation of $n$-heptane with manganese or cobalt laurate at 130°, and Geiseler et al. (101) have reported random attack in the autoxidation of $n$-octadecane.

In the uncatalyzed autoxidation of cis- and trans-decalins at 140° Jaffe, Steadman, and McKinney (58) found all the possible alcohols in the LiAlH$_4$-reduced mixture, with 42% of tertiary product from the trans form, 69% from the cis form. About one-fourth as much $p$-isopropylbenzyl hydroperoxide as $p$-methylcumene hydroperoxide is obtained from the autoxidation of $p$-cymene at 60–80° (63, 102). Treibs and Schöllner (103) have demonstrated attack at both tertiary alkyl and α-aryl positions in the oxidation of dicyclohexyl phenylmethane.

$$\qquad (20)$$

Propagation reactions in doubly branched alkanes may involve intramolecular hydrogen abstraction, yielding dihydroperoxides in much greater quantity than is predictable if the two tertiary sites were oxidized indepen-

$$\underset{\overset{|}{Me_2C}-(CH_2)_n-CHMe_2}{\overset{O_2^\cdot}{}} \longrightarrow \underset{\overset{|}{Me_2C}-(CH_2)_n-\dot{C}Me_2}{\overset{O_2H}{}}$$

$$\underset{\overset{|}{Me_2C}-(CH_2)_n-CMe_2}{\overset{O_2H \qquad O_2H}{|} \quad \downarrow \quad |} \qquad (21)$$

dently (104). Rust (105) obtained 95 and 50% of dihydroperoxide, respectively, from 2,4-dimethylpentane and 2,5-dimethylhexane at approximately 10% conversion, but little or none from the analogous butane or heptane. Criegee and Ludwig (106), by developing the relevant equations for mono-hydroperoxide and dihydroperoxide formation have confirmed Rust's interpretation. They also showed intramolecular abstraction to be important for *cis*-diisopropylethylene, but not for the *trans* isomer, nor for *o*- or *p*-diethylbenzene or the dimethylcyclohexanes. Dihydroperoxides can be prepared from these compounds (106), as well as from diisopropyl benzenes (107, 108) and 1,4-dialkyltetralins (88), but only at relatively high conversions of hydrocarbon.

Gas-phase oxidations of *n*-alkanes at 200–400° show considerable evidence of intramolecular abstraction (26, 109), but the products isolated are those of cleavage and rearrangement, not hydroperoxides.

Autoxidation at an asymmetric center, as in the case of ethyl chaulmoo-grate (110) produces a racemic mixture of hydroperoxides:

$$\text{(ring)}\!\!-\!\!\overset{H}{\underset{}{C}}\!\!-\!(CH_2)_{12}CO_2Et \xrightarrow{\;O_2\;} \text{(ring)}\!\!-\!\!\overset{O_2H}{\underset{}{C}}\!\!-\!(CH_2)_{12}CO_2Et \qquad (22)$$

However, asymmetric radicals formed at the bridgehead of *cis*- and *trans*-decalin interconvert less rapidly than they react with $O_2$.

The 9-hydroperoxide fraction from the autoxidation of *cis*-decalin at 140° contains 35% *cis* and 65% *trans* isomer (58). From *trans*-decalin only 26% of the *cis*-hydroperoxide was obtained. Bartlett and co-workers (111) obtained much higher *cis/trans* or *trans/cis* hydroperoxide ratios at 50–55°, generating the 9-decalyl radicals by thermal decomposition of *cis* and *trans*-peroxy esters. Analogous results have recently been obtained by Greene and Lowry (112), who used *cis*- and *trans*-hypochlorites to generate *cis*- and *trans*-9-chlorodecalins.

### 3. *Hydroperoxides from the Autoxidation of Olefins*

Allylic hydroperoxides are most commonly prepared by the autoxidation of olefins.

$$\text{(cyclohexene)} \longrightarrow \text{(cyclohexenyl)}\cdot \xrightarrow{\;O_2\;} \text{(cyclohexenyl)}\!-\!O_2H \qquad (23)$$

The reaction may be initiated thermally (113–118) or photolytically (119), with peroxides, azo compounds (32, 38, 39, 45, 46, 120), or metal ions (121–123), at temperatures usually ranging from 25 to 80°. (Photosensitized oxidations are briefly considered in Section A-7.) Extensive lists of hydroperoxides obtained from simple cyclic and acyclic olefins can be found in the papers by Hargrave and Morris (42), Bateman (124), and Mayo and co-workers (32, 39, 45, 46) as well as in references 1, 2, 3, and 33. There is also widespread interest in hydroperoxides derived from unsaturated fats and oils (3, 125, 126) and steroids (127–129).

Yields of hydroperoxides are generally better than 70% if conversions are low (1–5%) (32, 39, 42, 45, 46); higher conversions (and higher autoxidation temperatures) produce large quantities of degradation products (39,130–133).

## (a) ADDITION VERSUS ABSTRACTION IN OLEFIN AUTOXIDATION; HYDROPEROXIDES VERSUS OTHER PRIMARY PRODUCTS

Peroxy radicals may add to the double bond of olefins rather than abstract an α-hydrogen.

$$RO_2\cdot + \quad \underset{/}{\overset{\backslash}{C}} = \underset{\backslash}{\overset{/}{C}} \quad \longleftarrow \quad RO_2 - \overset{|}{\underset{|}{C}} - \overset{|}{\underset{|}{C}}\cdot \qquad (24)$$

Thus styrene (116), α-methylstyrene (134), indene (117), and butadiene (135), as well as other similar olefins and dienes (1, 2), give long-chain polyperoxides. Some 1,5-dienes—such as squalene (124), digeranyl (124), and 1,5,9-cyclododecatriene (136) give cyclic peroxyhydroperoxides by a combination of addition and abstraction.

$$(25)$$

Mayo and co-workers have recently measured the relative rates of abstraction versus addition for simple cyclic (39, 45) and acyclic olefins (32, 39). The ratio $k_{abst}/k_{add}$ varies from about 4 for cyclopentene and cyclohexene to 0.06 for cyclooctene and from 0.5 to 2 for most straight-chain compounds (at 50–90°, initiation by azobisisobutyronitrile; the effects of temperature and solvent are small).

In most cases the major nonmonohydroperoxide product was the dimeric peroxyhydroperoxide, for example (45):

[It is undoubtedly the presence of these compounds that led earlier workers to propose cyclic peroxides,

on the basis that reduction yielded glycols (1, 2).] *Epoxides, alcohols, and carbonyl compounds can also be formed as primary products (39).

$$(26)$$

$$(27)$$

The addition-abstraction combination has led to syntheses of some unusual hydroperoxides. Olefins plus oxygen together with bromine (137, 138) or hydrogen bromide (138, 139) yield $\beta$-bromohydroperoxides.†

$$Br\cdot + H_2C{=}CHR \longrightarrow BrCH_2\dot{C}HR \xrightarrow{O_2} BrCH_2\overset{\overset{\displaystyle O_2^{\cdot}}{|}}{C}HR$$

$$(28)$$

$$BrCH_2\overset{\overset{\displaystyle O_2^{\cdot}}{|}}{C}HR + HBr \longrightarrow BrCH_2\overset{\overset{\displaystyle O_2H}{|}}{C}HR + Br\cdot$$

$$(29)$$

$\beta$-Hydroperoxy sulfides have been prepared by a similar reaction between thiols, alkenes, and oxygen (140–143).

$$RS\cdot + CH_2 = CHR' \longrightarrow RSCH_2\dot{C}HR' \xrightarrow[RSH]{O_2} RSCH_2\overset{\overset{\displaystyle O_2H}{|}}{C}HR'$$

$$(30)$$

---

* The first examples of true oxetanes (P. D. Bartlett and A. Schaap, *J. Amer. Chem. Soc.*, *92*, 3223, (1970); S. Mazur and C. S. Foote, *ibid.*, *92*, 3225 (1970) have been reported recently.

† Better yields of the same or similar compounds are obtained by electrophilic additions to double bonds (Section II-C-4).

The products are relatively unstable and yield $\beta$-hydroxy sulfoxides spontaneously at room temperature. An even more unstable hydroperoxidic material was obtained from phenylacetylene, thiophenol, and oxygen. At $-10°$ it decomposed to PhSCHOHCOPh and was supposed to be PhSCH= C(O$_2$H)Ph (144). (This appears to be the only example of a vinyl hydroperoxide.)

## (b) SELECTIVITY AND ALLYLIC ISOMERIZATION IN OLEFIN AUTOXIDATION

Olefins that give asymmetrical allyl radicals or that can give two different allyl radicals generally yield mixtures of isomeric hydroperoxides (145).

$$\text{(31)}$$

The situation is complicated by the possibility of *cis–trans* isomerization of the double bond and by the ability of allylic hydroperoxides to undergo isomerization in the presence of free radicals (146–148).

$$\text{(32*)}$$

*Reference 146.

$$(33)*$$

Thus the relative thermodynamic stability of the products probably determines the isomer ratio from any given allylic system.

In some cases the situation of the double bond is the deciding factor, particularly where formation of an allyl radical permits isolated double bonds to move into conjugation. Volger and co-workers (149) obtained only tertiary hydroperoxide from the autoxidation of either of two isomeric unsaturated ketones.

$$(34)$$

Linoleates yield 90% of a mixture of $C_9$ and $C_{13}$ hydroperoxides (150, 151).

$$(35)$$

[Methyl oleate gives nearly equal proportions of $C_8$, $C_9$, $C_{10}$, and $C_{11}$ hydroperoxides (126, 152, 153).]

Influence of double-bond stability can be inferred from the tendency of 1-hexene and 1-octene to give primary hydroperoxides (154), the 1.33:1 ratio of secondary to primary hydroperoxides obtained from methylene-cyclohexane, and the 1.88:1 ratio of tertiary to primary hydroperoxides obtained from vinylcyclohexane (45). However, tetramethylethylene yields 97% of the tertiary hydroperoxide with the double bond shifted into the less substituted position (45).

The very careful work of Van Sickle, Mayo, et al. (32) and of Balaceanu and his associates (131, 132) (Table 2) shows that for most simple olefins an approximately equal distribution of isomeric hydroperoxides results. [In

*References 147 and 148.

TABLE **2.**   ISOMER DISTRIBUTIONS IN THE AUTOXIDATION OF SIMPLE OLEFINS[a]

| | Ratio of Hydroperoxide Isomers | | | Total RO$_2$H (%) Based on O$_2$ Uptake[b] | Ref. |
|---|---|---|---|---|---|
| Alkene | Primary | Secondary | Tertiary | | |
| C=C−C−C | 1 | 1.2 | | 55 | 32 |
| C=C−C−C | 1 | 9 | | — | 132 |
| cis-C−C=C−C | 1 (34% cis) | 0.9 | | — | 132 |
| trans-C−C=C−C | 1 (90% trans) | 0.9 | | — | 132 |
| C<br>\|<br>C−C−C=C | 1 | | 1.5 | 75 | 32 |
| C<br>\|<br>C−C−C=C | 1 | | 1.1 | — | 132 |
| C<br>\|<br>C−C=C−C | 1 | 1.1 | 1.3 | 20 | 32 |
| C  C<br>\|  \|<br>C−C=C−C | 1 | | 18 | 45 | 32 |
| C=C−C−C−C−C | 1 | | 1.3 | 33 | 32 |

[a] Azobisisobutyronitrile-catalyzed at 50–70° to less than 5% conversion. Ratios by chromatography of alcohols obtained from LiAlH$_4$ or Ph$_3$P reductions of RO$_2$H.
[b] Other products are at least 90% by RO$_2$ · addition, not from RO$_2$H decomposition.

attempting to draw conclusions from this table it should be clear that monomeric hydroperoxides may account for as little as 20% of the oxygen uptake (32).]

Comfortingly, different olefins that yield the same allyl radical (as in 1- and 2-butene) give the same ratio of isomeric hydroperoxides (Table 2) (39). The result of Menguy et al. (132) for 1-butene appears to be anomalous and in direct contradiction to the work of Van Sickle et al. (32). The distribution from *cis*- and *trans*-2-butenes (132) (Table 2) suggests some transient stability of *cis*- and *trans*-allyl radicals, a conclusion also reached by Walling and Thaler (155) and demonstrated more recently by Kochi and Krusic (156).

In asymmetrical olefins that may give rise to two different allylic systems, depending on which hydrogen is abstracted—as, for example, in 1,3,5-

trimethylcyclohexene, 3,3′-bicyclohexenyl, or dicyclopentadiene—attack usually favors tertiary > secondary > primary (145, 157). The reverse, secondary > tertiary, found for such compounds as 1-isopropylcyclohexene, 1-isopropylcyclopentene, and 4-vinylcyclohexene, has been attributed to steric inhibition of resonance (157).

### 4.  *Hydroperoxides from the Autoxidation of Ethers, Alcohols, and Ketones*

Ethers autoxidize readily to α-hydroperoxy ethers [also preparable by reaction of alcohols with ozonides (Section II-B) or by addition of hydrogen peroxide to vinyl ethers (Section II-C, D)].

$$\text{Et}_2\text{O} \xrightarrow{\text{O}_2} \overset{\overset{\text{O}_2\text{H}}{|}}{\text{MeCH}}\text{—OEt} \tag{36}$$

The products are usually explosive and on standing or treatment with acid yield equally explosive dimeric and polymeric alkylidene peroxides (158, 159).

$$\tag{37}$$

Monomeric hydroperoxides from diisopropyl (160), di-*n*-butyl (161), and diisoamyl (161) ethers, tetrahydrofuran (161, 162), and dioxane (162) have been synthesized this way. Rieche and co-workers have obtained hydroperoxides from a number of substituted isochromanes (163), phthalanes (158), and 1,3-dioxolanes (159, 164, 165). Several benzylic and allylic ethers have been autoxidized by Sharp and Patrick (166), but the hydroperoxides were not isolated. The autoxidations are usually run to low conversions at room temperature or below, with ultraviolet irradiation. Using Benzophenone as a sensitizer (161), quantum yields of about 20 have been obtained.

Secondary alcohols can be autoxidized under the same conditions to give α-hydroperoxy alcohols (161, 167).

$$\text{Me}_2\text{CHOH} \xrightarrow[\substack{hv \\ \text{Ph}_2\text{CO, 12°C}}]{\text{O}_2} \text{Me}_2\text{C}\overset{\overset{\text{OH}}{\diagup}}{\underset{\underset{\text{O}_2\text{H}}{\diagdown}}{}} \tag{38}$$

[The same compounds result from reaction of hydrogen peroxide with ketones, but usually as a small fraction of a complex mixture of peroxidic materials (Section II-C-4-b).]

Yields are low, 16–25% on oxygen uptake, and quantum yields are approximately 1 (161, 168). The hydroperoxides are easily hydrolyzed to ketones and hydrogen peroxide, the only products obtained when alcohols are autoxidized at higher temperatures (1) and yield ketone derivatives when treated with 2,4-dinitrophenylhydrazine. Bis(2-hydroperoxy-2-propyl) peroxide (explosive) is a by-product from isopropanol autoxidation (167), and no monomeric hydroperoxide is obtained from cyclohexanol (169).

$$
\begin{array}{c}
\underset{\text{H}\diagup\quad\diagdown\text{OH}}{\bigcirc} \longrightarrow \underset{\text{O}_2\text{H}\diagup\quad\diagdown\text{OH}}{\bigcirc} \longrightarrow \underset{\text{HO}\quad\text{O}_2\quad\text{OH}}{\bigcirc\bigcirc}
\end{array} \quad (39)
$$

Ketones autoxidize at the α-position, but hydroperoxide products are remarkably unstable and only a few have been obtained pure. Pritzkow was able to obtain approximately 10 % solutions of the hydroperoxides of cyclohexanone (170) and 2-butanone (171) and establish their structures by reduction to the α-keto alcohols. The tertiary hydroperoxides from diisopropylketone (172, 173), some α,β unsaturated ketones (174, 175), and aryl ketones (171, 173, 176, 177) have been isolated and characterized. α-Hydroperoxy ketones have been shown by Richardson and Steed (173) to exist as internally-hydrogen-bonded structures,

$$
\underset{\text{Me}_2\text{CH}-\overset{\displaystyle\text{O}}{\overset{\|}{\text{C}}}-\text{CHMe}_2}{} \xrightarrow{\text{O}_2} \underset{\text{Me}_2\overset{\text{HO}_2}{\underset{|}{\text{C}}}-\overset{\displaystyle\text{O}}{\overset{\|}{\text{C}}}-\text{CHMe}_2}{} \qquad (40)
$$

$$
\text{Me}_2\text{CH}-\text{CH}=\text{CH}-\overset{\displaystyle\text{O}}{\overset{\|}{\text{C}}}-\text{Me} \xrightarrow{\text{O}_2} \text{Me}_2\overset{\text{O}_2\text{H}}{\underset{|}{\text{C}}}-\text{CH}=\text{CH}-\overset{\displaystyle\text{O}}{\overset{\|}{\text{C}}}-\text{Me} \quad (41)
$$

$$
\text{Ph}_2\text{CH}-\overset{\text{Me}}{\underset{|}{\text{CH}}}-\overset{\displaystyle\text{O}}{\overset{\|}{\text{C}}}-\text{Ph} \xrightarrow{\text{O}_2} \text{Ph}_2\text{CH}-\overset{\text{Me}}{\underset{\underset{\displaystyle\text{O}_2\text{H}}{|}}{\overset{|}{\text{C}}}}-\overset{\displaystyle\text{O}}{\overset{\|}{\text{C}}}-\text{Ph} \qquad (42)
$$

rather than the cyclic peroxyhemiketals originally postulated by Kohler (178).

$$
\begin{array}{c}
-\overset{|}{\underset{|}{\text{C}}}\overset{\overset{\displaystyle\text{O}-\text{O}}{\diagdown\quad|}}{\underset{\diagdown}{-\text{C}}}\overset{\text{H}}{\diagdown}\!{:}\text{O} \xrightarrow{\;\;\times\;\;} -\overset{\overset{\displaystyle\text{O}-\text{O}}{|}}{\underset{|}{\text{C}}}-\overset{|}{\underset{|}{\text{C}}}-\text{OH}
\end{array} \qquad (43)
$$

Many phenols autoxidize to ketohydroperoxides (1, 2) either under base catalysis (Section A-2-c) or in diffuse daylight (179, 180). The cyclohexadienonyl hydroperoxide (equation 45) loses oxygen spontaneously at 220° to regenerate the original phenol (180).

$$(44)$$

$$(45)$$

### 5. Hydroperoxides from the Autoxidation of Nitrogen Compounds

Compounds containing

$$\text{\textbackslash}C=X-NH-$$

where X is C or N, autoxidize easily, frequently at room temperature without added catalysts, giving hydroperoxides, $-C(O_2H)-X=N-H$. (181, 182). Busch and Dietz (183) first discovered the facile reaction of phenylhydrazones with oxygen and obtained crystalline products that they assumed to have a cyclic peroxide structure. Hydrazone hydroperoxides were also synthesized by Cook and Reed (184) and Robinson and Robinson (185) but it remained for Criegee (186), Witkop (187), Pausacker (188), and their associates to establish the structure firmly.

$$R_1R_2C=N-NH-R_3 \xrightarrow{\ O_2\ } R_1R_2\overset{\overset{\displaystyle O_2H}{\displaystyle |}}{C}-N=N-R_3 \qquad (46)$$

$$(R_1R_2 = \text{alkyl, aryl, or H}; R_3 = \text{alkyl or aryl})$$

Alkyl or aryl hydrazones of both aldehydes and ketones react (183, 186). Hydroperoxides have been prepared from the phenylhydrazones of sugars (189–191) and of more complex compounds (192).

The reaction appears to be a free-radical chain; it is catalyzed by benzoyl peroxide or light and inhibited by phenols (188). However, the rate of autoxidation is considerably slower in polar solvents than in nonpolar ones (188, 193).

2,3-Disubstituted indoles absorb oxygen readily and subsequently rearrange (181, 182, 194–196). The intermediate hydroperoxides have been isolated in a number of instances (181, 182, 194–197).

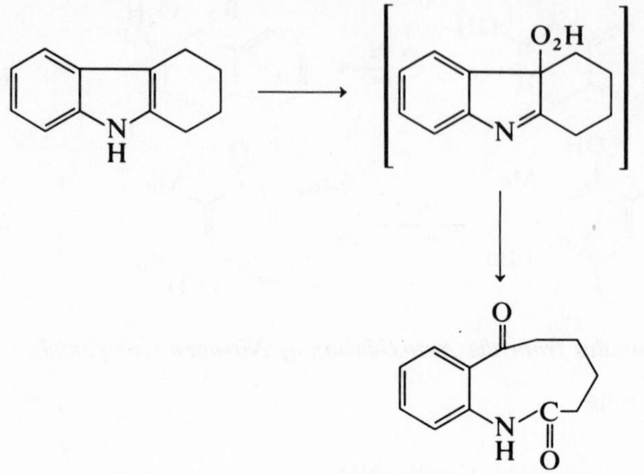

(47)

(48)

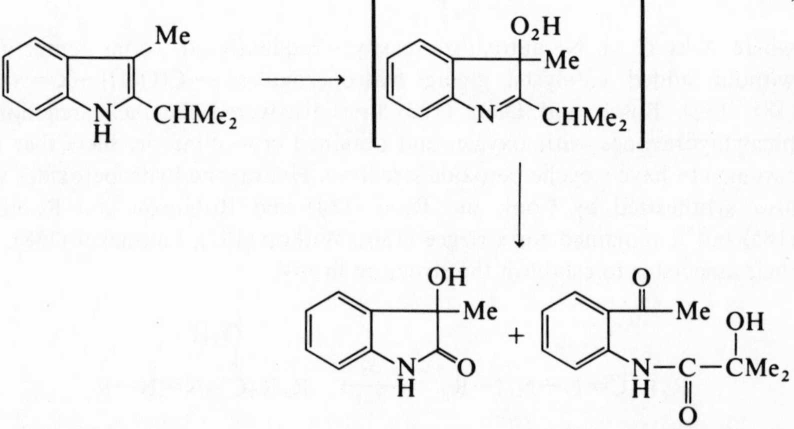

Lophine hydroperoxide has been prepared by photosensitized oxidation (196, 198–200) (Section II-A-7) or by reaction of the lophyl radical with 3% hydrogen peroxide (199). Other imidazole hydroperoxides have been synthesized similarly (200).

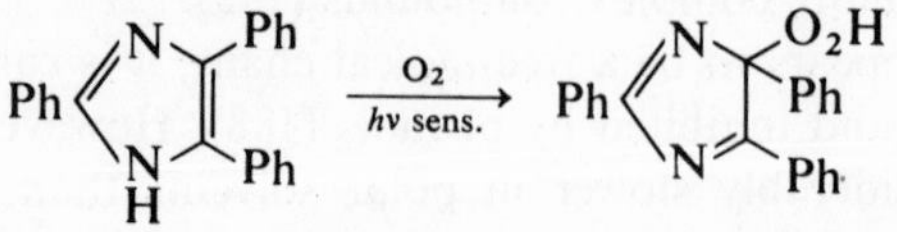

(49)

The reaction of both indoles and imidazoles has been of particular interest because of the chemiluminescence exhibited when these compounds are treated with oxygen and base (196, 197, 199–201). The hydroperoxides themselves are chemiluminescent in the presence of $OH^-$, a process now associated with the formation of the rearranged ketones in an excited state (196, 197, 199–201).

Imines also autoxidize, yielding hydroperoxides (202–205).

$$\underset{25°,\ 20hr.}{\xrightarrow{O_2}} \qquad 98\% \qquad (50)$$

Amides autoxidize much less readily than

$$\diagdown C{=}N \quad or \quad \diagdown C{=}C{-}N\diagup$$

compounds, and 1–2% conversion seems to be the maximum before extensive degradation occurs (206, 207). Rieche and Schön (206), however, have isolated a hydroperoxide from the photoautoxidation of caprolactam, and Sagar (207) has prepared hydroperoxides from several simple *N*-alkyl acetamides, using azobisisobutyronitrile or metal-ion initiation.

$$\underset{\substack{\| \\ Me-C-NH-Et}}{\overset{O}{}} \xrightarrow{O_2} \underset{\substack{\| \qquad | \\ Me-C-NH-CH-Me}}{\overset{O \qquad O_2H}{}} \qquad (51)$$

### 6. Hydroperoxides from the Autoxidation of Organometallic Compounds

Organometallic compounds autoxidize readily, sometimes inflammably (1), by what seems to be a free-radical mechanism.*

$$RO_2\cdot + \diagdown M{-}R \quad RO_2M\diagup + R\cdot \qquad (52)$$

$$R\cdot + O_2 \longrightarrow RO_2\cdot \qquad (53)$$

The resulting peroxides tend to decompose easily with loss of peroxidic oxygen, but those of lithium (210), magnesium (210–212), cadmium (213, 214), zinc (210, 213, 124), and boron (215) have been hydrolyzed to hydroperoxides in reasonable yield.

$$RO_2M\diagup \xrightarrow[H^+]{H_2O} RO_2H + HOM\diagup \qquad (54)$$

---

* Davies and Roberts (208) have shown that, at least for boron, zinc, and cadmium alkyls, autoxidation can be inhibited by galvinoxyl and that an asymmetric carbon attached to the metal atom undergoes racemization on $O_2$ insertion. Lamb and co-workers (209) found that the Grignard reagent from 6-bromo-1-hexene on oxidation and subsequent reduction of the hydroperoxides gave a 3 : 1 ratio of 5-hexen-1-ol and hydroxymethylcyclopentane. This observation is suggestive of a free-radical cyclization mechanism.

The low-temperature autoxidation and subsequent hydrolysis of Grignard reagents is a useful procedure for preparing simple primary and secondary alkyl hydroperoxides, often difficult to obtain by other methods.

$$RMgX + O_2 \longrightarrow RO_2MgX \xrightarrow{H^+} RO_2H \qquad (55)$$

The reaction, developed by Walling and Buckler (210) and used successfully by a number of others (211, 213, 216–218), may give 90% yields from tertiary Grignards and 57–66% yields from primary and secondary Grignards (210). Considerably lower yields have been reported, however (215–217).

For higher homologs Wilke and Heimbach (215) found that autoxidation of alkyl boranes gave better results. Trialkyl boranes were prepared by reaction of 2-ethylhexene, 1-octene, 1-decene, 1-undecene, styrene, α-methylstyrene, indene, cyclohexene, octene, or cyclodecene with N-triethylborazane. (Addition of B—H to the double bond was > 99% anti—arkovnikov for the simple olefins and about 88% for styrene.) Autoxidation yielded diperoxyboranes, which were further oxidized with peroxy acid, then hydrolyzed to 30–50% yields of $RO_2H$, based on the starting olefin.

$$R_3B \xrightarrow{O_2} (RO_2)_2BR \qquad (56)$$

$$(RO_2)_2BR \xrightarrow{MeCO_3H} (RO_2)_2BOR \xrightarrow[H^+]{H_2O} 2RO_2H + ROH \qquad (57)$$

Hydrolysis of the initially formed diperoxide is not a smooth reaction and yields rather more alcohol than hydroperoxide (215, 219).

### 7.   Hydroperoxides from the Photosensitized Oxidation of Olefins

A very useful synthesis of allylic hydroperoxides is the reaction of isolated double bonds with oxygen in the presence of light and photosensitizers, such as rose bengal, chlorophyll, or hematoporphyrin. Its discovery and development, by Schenck and his co-workers (22, 221), was reviewed by him in 1957 (222). More current interest in the mechanism of the reaction (223–232) has been stimulated by the finding that similar results can be obtained by nonphotochemical means (Section II-A-8).

Unlike free-radical autoxidation, the reaction invariably proceeds with migration of the double bond and is nonchain; quantum yields are 1.0 or less (223, 233).

$$MeCH=CHMe \xrightarrow[hv \text{ sens}]{O_2} Me-\overset{\displaystyle O_2H}{\overset{|}{CH}}-CH=CH_2 \qquad (58)$$

The hydroperoxidic products are fairly stable under the reaction conditions (ordinarily room temperature, neat or in alcohol); yields of 70–83% are reported (233) for the oxidations of tetramethylethylene, $\Delta^9$-octalin, and cyclohexylidenecyclohexane (nearly 100% olefin conversion). Sharp and Le Blank (234) obtained a quantitative yield in the oxidation of trimethylethylene by carrying conversion to only 42%. The reaction been used extensively to hydroperoxidize steroids (148, 235–239) and other natural products (240–242). An intriguing variant has been the recent synthesis of ozonides (243, 244) by 1,4-addition of oxygen to furans.

$$(59)$$

[Sensitized photooxidation of cyclic dienes to so-called endoperoxides had been, of course, discovered much earlier. The subject has been reviewed in detail by Davies (1) and Hawkins (2).]

$$(60)$$

For simple alkenes the ease of oxidation and yield increase with increasing alkyl substitution on the double bond: dimethylethylene < trimethylethylene < tetramethylethylene, and cyclohexene ≪ 1-methylcyclohexene < 1,2-dimethylcyclohexene (223). Limonene reacts exclusively at the triply substituted endocyclic double bond in preference to the disubstituted isopropenyl group (223). Terminal olefins, such as 1-butene (245) and 1-nonene (231), react only very slowly or not at all. Unsubstituted cyclic olefins also appear not to react.

Using competition experiments, Kopecky and Reich (231) measured relative rates of 1:450:3900:55,000 for 1-nonene, methylcyclohexene, methylcyclopentene, and tetramethylethylene, respectively, and found that the choice of photosensitizers had little effect. They suggested that the electrophilicity of the reagent was comparable to that of peroxy acids or bromine.

In addition to the invariant double-bond migration, the reaction displays steric and conformational effects that suggest a cyclic intermediate where the hydrogen is constrained to be in the plane of the $\pi$-electron cloud (223, 227).

$$(61)$$

Thus Nickon and Bagli (227) found that $\Delta^6$-chloestenes produced only $\Delta^5$-cholesten-7$\alpha$-hydroperoxides; no 7-$\beta$-isomers were obtained. 7-$\alpha$-Deuterocholesterol lost all its deuterium in being converted to 3-$\beta$-hydroxy-5-$\alpha$-hydroperoxy-$\Delta^6$-cholestene.

$$\text{(62)}$$

From the very elegant studies on the distributions of hydroperoxides obtained from the photosensitized oxidation of limonene (224, 225), carvomenthene (224, 228), and the carenes (226, 246), carried out for the most part by Schenck and his co-workers, several generalizations are possible (223). Quasi-axial hydrogens react more frequently than quasi-equatorial ones, optical activity is retained, and reaction tends to occur *trans* to bulky substituents. [In the carenes this steric control is total, and *trans*-$\Delta^2$-carene, in which the only hydrogen available for attack is *cis* to the dimethylcyclopropyl group, does not oxidize (246).]

Recent work with partially deuterated tetramethylethylenes and stilbenes has shown the deuterium-isotope rate effect to be negligible (232).

Until recently Schenck, Gollnick, and their co-workers persisted in the view that attack on olefin is by a sensitizer–oxygen complex (223), partly because the steric effects seemed too large to be accounted for otherwise.

$$\text{sens.} \xrightarrow{h\nu} \text{sens.*} \xrightarrow{O_2} \text{sens.-O}_2^* \xrightarrow{\;C=C\;} \text{sens.-O}_2$$

They showed that sensitized reactions give a product distribution different from that obtained by free-radical-initiated autoxidations (Table 3) and found that product distributions from hypochlorite–hydrogen peroxide autoxidations, which proceed via singlet oxygen (see below), appeared to simulate the latter rather than the former (223). More recent studies have

TABLE 3.   COMPARISON OF PRODUCT DISTRIBUTION[a] FROM THE
           OXIDATION OF $\alpha$-PINENE (223)

| | $CH_2OH$ | CHO | | | OH | O |
|---|---|---|---|---|---|---|
| $O_2 + h\nu + $ rose bengal $\longrightarrow$ 94% | — | — | — | — | |
| $O_2 + (PhCO)_2O_2$                         14% | 13% | 33% | 9% | 31% | |

[a] After reduction of hydroperoxides to alcohols with sodium sulfite.

shown (229, 230) that sensitized photooxidations and hypochlorite–hydrogen peroxide oxidations give identical results. The sensitizer's function is now conceded to be simply one of energy transfer (230).

## 8. Hydroperoxides by Reactions of Singlet Oxygen

In 1964 Foote and Wexler (247) published the discovery that oxygen produced by the reaction of sodium hypochlorite with hydrogen peroxide gave the same kind of products that had hitherto been obtained only from photosensitized oxidation—namely, endoperoxides from cyclic dienes and allylic hydroperoxides from alkyl-substituted monoolefins. At the same time Corey and Taylor (248) found that oxygen passed through an electric discharge would produce endoperoxides from anthracenes, although their attempts to prepare hydroperoxides by this means were unsuccessful. In both cases the active species was postulated to be an electronically excited oxygen molecule, probably in a $^1\Delta g$-state.

$$H_2O_2 + NaOCl \longrightarrow O_2^* \tag{63}$$

$$O_2^* + \ \ \diagdown C=C \diagup \diagdown_{\underset{|}{CH-}} \longrightarrow \underset{|}{-\overset{\overset{\displaystyle O_2H}{|}}{C}-\overset{|}{C}=\overset{|}{C}- \tag{64}$$

Foote and co-workers (249, 250) claim to have obtained photosensitized-type products using $H_2O_2 + OCl^-$ for a large selection of alkenes, including 1,2-dimethylcyclohexene, limonene, α-pinene, 2,5-dimethylfuran, and several steroids. Detailed procedures and yields for hydroperoxide preparation, however, are as yet available only for tetramethylethylene (249, 250). Like photosensitized oxidations, $H_2O_2 + OCl^-$ gives minuscule yields of hydroperoxides from cyclohexene (249).

2,3-Dimethyl-1-buten-2-yl hydroperoxide has been prepared by heating tetramethylethylene with anthracene endoperoxide (251). Endoperoxides are known to yield oxygen on heating (1) in now what is proposed to be an excited singlet state (251).*

---

\* Hydroperoxides have now been synthesized from singlet oxygen released from ozonides of phosphites, alcohols and ethers (R. W. Murray, W. C. Lumma and J. W.-P. Lin, *J. Amer. Chem. Soc.*, *92*, 3205 (1970)) and by direct interaction of triphenylphosphine ozonide with tetramethyethylene (P. D. Bartlett and G. D. Mendenhall, *ibid.*, *92*, 210 (1970)). Isolation of hydroperoxides from oxygen $^1\Delta g$, produced via electric discharge, has been accomplished (W. S. Gleason, A. D. Broadbent, E. Whittle and J. N. Pitts, *ibid.*, *92*, 2068 (1970)). Postulation of a perepoxide intermediate has been based on the synthesis of azido hydroperoxides from the reaction of singlet oxygen with olefins in the presence of azide ion (W. Fennical, D. R. Kearns and P. Radlick, *ibid.*, *91*, 7771 (1969)).

$$\text{C=C} + \text{(anthracene-O}_2\text{ adduct)} \xrightarrow{\Delta} \text{(anthracene)} + \underset{\underset{\text{O}_2\text{H}}{|}}{\overset{\overset{|}{}}{\text{C}-\text{C}-}} \quad (65)$$

## B. Hydroperoxides from Ozonides

The synthesis of α-alkoxy hydroperoxides by reacting ozone with olefins in the presence of alcohols derives from the work of Criegee (252, 253) and of Milas (254) on the structure of ozonides. Subsequent developments have been mainly by Rieche (255–257), Bailey (258,–262), and their co-workers. The reaction has been reviewed by Bailey (263) and more recently by Menyailo and Pospelov (264).

The mechanism involves trapping the zwitterion intermediate in the ozonide rearrangement (253), the success of which depends on the polarity of the medium and the acidity of the alcohol (265).

$$\underset{\text{Me}_2\text{C}-\text{CMe}_2}{\overset{\text{O}-\text{O}}{\underset{}{}}} \longrightarrow \text{Me}_2\overset{+}{\text{C}} + \text{O=CMe}_2 \nearrow \underset{\text{Me}_2\text{C}}{\overset{\text{O}-\text{O}}{\underset{\text{O}}{}}}\text{CMe}_2 \\ \searrow_{\text{ROH}} \text{Me}_2\text{C} \underset{\text{OR}}{\overset{\text{O}_2\text{H}}{}} \quad (66)$$

From methylenecyclohexane Sonoda et al. (265) have obtained decreasing yields of 1-alkoxycyclohexyl hydroperoxides in the series methoxy (80%), ethoxy (52%), and t-butoxy (18%). The by-products included some bis-cyclohexylidene diperoxide as well as the normal ozonide.

Normally reactions are carried out in methanol or ethanol (263, 264), but ozonization of a sulfolene in aqueous solution has yielded an α-hydroxy hydroperoxide (253).

$$\underset{\text{SO}_2}{\overset{\text{Me}\quad\quad\text{Me}}{}} \xrightarrow[\text{H}_2\text{O}]{\text{O}_3} \text{MeCOCH}_2\text{SO}_2\text{CH}_2\underset{\overset{|}{\text{OH}}}{\overset{\overset{\text{O}_2\text{H}}{|}}{\text{C}}}-\text{Me} \quad (67)$$

Ozonization of cyclic olefins may give products with the carbonyl and hydroperoxy groups intact, as in reaction (67) or in the ozonization of $\Delta^9$-octalin to 6-hydroperoxy-6-methoxycyclodecanone (252), but in some cases the isolated product is one of ring reclosure (258, 266).

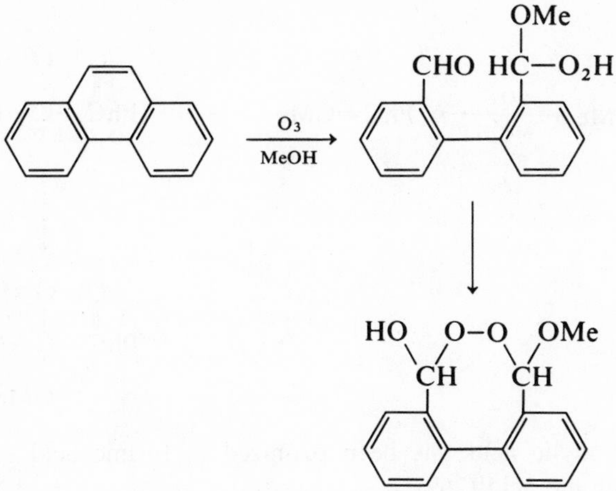

$$\text{(phenanthrene)} \xrightarrow[\text{MeOH}]{O_3} \quad \text{CHO} \quad \text{HC}(\text{OMe})\text{-O}_2\text{H} \quad \text{(biphenyl)}$$

(68)

$$\text{HO} \quad \text{O-O} \quad \text{OMe}$$
$$\text{CH} \qquad \text{CH}$$

Cyclohexene yields mainly a linear polymer by the same route (258).

For the most part, highly substituted olefins have been used in this synthesis, but 1-octene (267) and 1-hexene (264) and styrene (268) have yielded respectable amounts of the desired hydroperoxides.

Ozonolysis of olefinic alcohols has led to intramolecular trapping (255, 256).

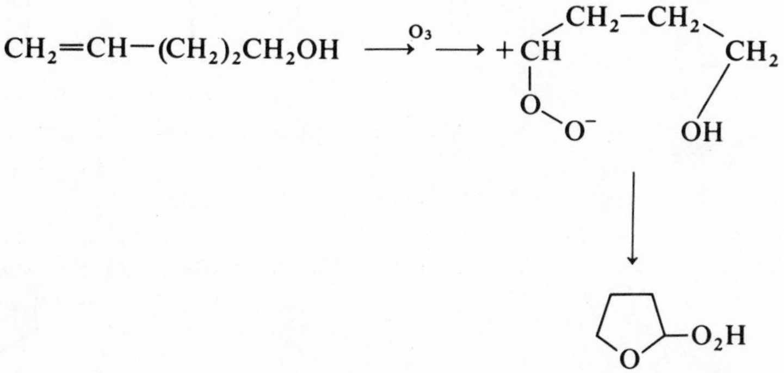

$$CH_2{=}CH{-}(CH_2)_2CH_2OH \xrightarrow{O_3} + CH \underset{O\text{-}O^-}{\overset{CH_2{-}CH_2}{\diagup}} CH_2 \diagdown OH$$

(69)

$$\text{(tetrahydrofuran ring)} {-}O_2H$$

$\alpha$-Hydroperoxy ketones can be made from diacyl-substituted alkenes (261)

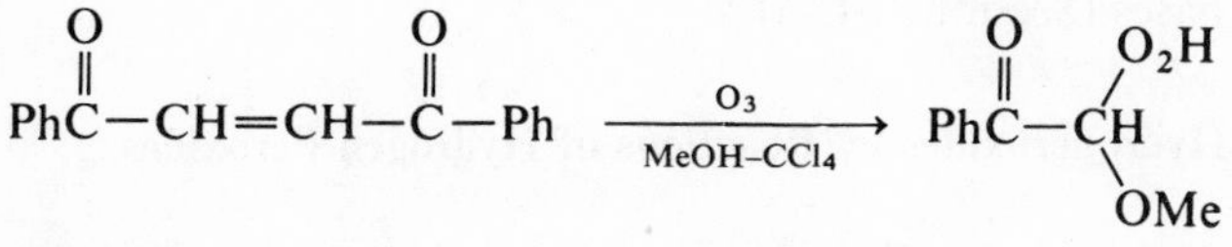

$$\underset{O}{\overset{\parallel}{Ph C}}{-}CH{=}CH{-}\underset{O}{\overset{\parallel}{C}}{-}Ph \xrightarrow[\text{MeOH–CCl}_4]{O_3} \underset{O}{\overset{\parallel}{Ph C}}{-}\underset{OMe}{\overset{O_2H}{CH}}$$

(70)

or from alkynes (262).

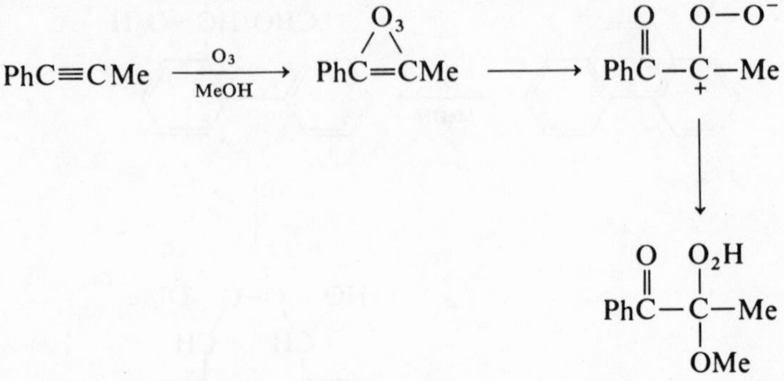

$$PhC \equiv CMe \xrightarrow[\text{MeOH}]{O_3} PhC\overset{\overset{\displaystyle O_3}{\diagup\;\diagdown}}{=}CMe \longrightarrow PhC\overset{\overset{\displaystyle O}{\|}}{-}\overset{\overset{\displaystyle O-O^-}{|}}{\underset{+}{C}}-Me$$

(71)

$$PhC\overset{\overset{\displaystyle O}{\|}}{-}\overset{\overset{\displaystyle O_2H}{|}}{\underset{\underset{\displaystyle OMe}{|}}{C}}-Me$$

Acetylenedicarboxylic acid has been ozonized in formic acid solution to hydroperoxymalonic acid (269).

Amines may be used to trap the zwitterion (257, 270). The final products are usually oxaziranes, but with care the intermediate hydroperoxides may be isolated.

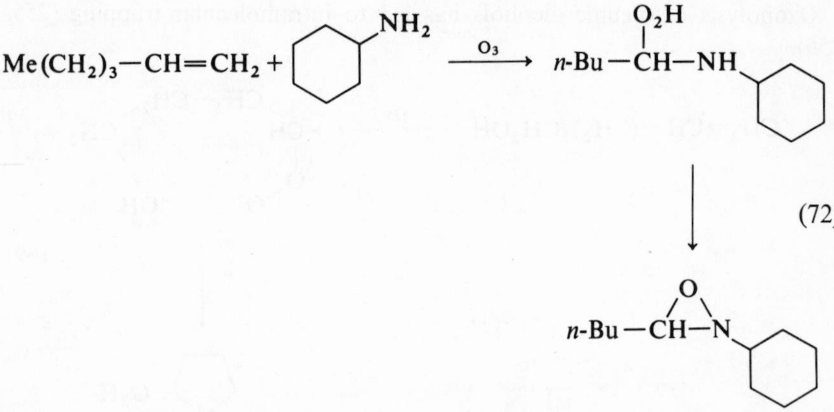

$$Me(CH_2)_3-CH=CH_2 + \underset{NH_2}{\bigcirc} \xrightarrow{O_3} n\text{-}Bu-\overset{\overset{\displaystyle O_2H}{|}}{CH}-NH\bigcirc$$

(72)

$$n\text{-}Bu-\overset{\overset{\displaystyle O}{\diagup\;\diagdown}}{CH}-N\bigcirc$$

The same hydroperoxides can be obtained by addition of hydrogen peroxide to Schiff bases (Section II-C-4).

## C.  Hydroperoxides by Reactions of Hydrogen Peroxides

Hydrogen peroxide may be reacted with organic compounds by nucleophilic displacement or by addition to unsaturated linkages to yield hydroperoxides.

$$-\overset{|}{\underset{|}{C}}-X + H_2O_2 \longrightarrow -\overset{|}{\underset{|}{C}}-O_2H + HX \qquad (73)$$

(X = OH, OR, $OSO_2R$, $OSO_2OR$, halide, etc.)

$$\overset{\diagdown}{\underset{\diagup}{C}}{=}Y + H_2O_2 - \overset{|}{\underset{\underset{Y-H}{|}}{C}}-O_2H$$

$$(74)$$

$$(Y = \overset{\diagup}{\underset{\diagdown}{C}}, N-, \text{ or } O)$$

Since introduction of the use of highly concentrated hydrogen peroxide (271) [obtained either commercially as 90–98% aqueous solutions or by drying ether extracts of the 30% material (272, 273)], the reagent has gained a high degree of popularity. The inherent dangers, however, should not be underestimated.

The explosive capabilities of 90–100% hydrogen peroxide and of solutions in organic compounds are well set out in the Becco publications (274, 275). Less publicity (276) has been given to the fact that, although addition of an alcohol to preacidified concentrated hydrogen peroxide is *usually* safe, adding concentrated acid to alcohol-concentrated hydrogen peroxide mixtures *almost always results in explosion*, particularly if the mixture is partly inhomogeneous and the alcohol a solid. Safety shields should always be employed in such reactions, and the scale of the reaction should be minimal.

### 1. Hydroperoxides from Alkyl Sulfates and Sulfonates

Ethyl hydroperoxide, the first hydroperoxide to be isolated, was synthesized by Baeyer and Villiger (277) in 1901 by using the reaction of diethyl sulfate with dilute hydrogen peroxide in the presence of aqueous potassium hydroxide. The reaction was also used to prepare methyl and propyl hydroperoxides (278, 279) and still seems to be the method of choice for these compounds (280–282). Yields are low (281) because of the concomitant base-catalyzed decomposition of the hydroperoxide. The product usually contains some dialkyl peroxides, traces of which may be responsible for the tendency of these hydroperoxides to explode in unpredictable fashion (281).

Tertiary hydroperoxides may be prepared by reaction of 30% hydrogen peroxide with alkyl hydrogen sulfates, (prepared *in situ* from equimolar quantities of alcohol—or alkene—and 70% sulfuric acid) (283, 284). *t*-Butyl and *t*-amyl hydroperoxides (283) and their higher homolog (284), as well as

acetylenic hydroperoxides (285), have been made in 50–70% yield in this
way.

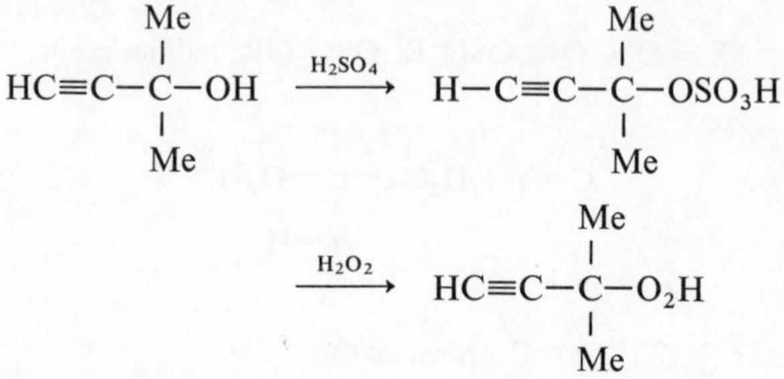

$$
\underset{\underset{\text{Me}}{|}}{\overset{\overset{\text{Me}}{|}}{HC\equiv C-C-OH}} \xrightarrow{\text{H}_2\text{SO}_4} \underset{\underset{\text{Me}}{|}}{\overset{\overset{\text{Me}}{|}}{H-C\equiv C-C-OSO_3H}} \tag{75}
$$

$$
\xrightarrow{\text{H}_2\text{O}_2} \underset{\underset{\text{Me}}{|}}{\overset{\overset{\text{Me}}{|}}{HC\equiv C-C-O_2H}}
$$

Using half the quantity of hydrogen peroxide results in good yields of bis-
alkyl peroxides (283).

Kochi (286) has provided detailed procedures for the preparation of $RO_2H$,
where R is 3-methyl-3-hexyl, 3-ethyl-3-hexyl, 4-methyl-4-octyl, and 4-methyl-
4-octyl. Synthesis of a series, $R(CH_3)_2CO_2H$, where R is $C_3$ to $C_8$, is reported
by Vilenskaya and Yurzhenko (287).

More highly branched hydroperoxides can be prepared, but they undergo
facile acid cleavage if precautions are not taken (288, 289).

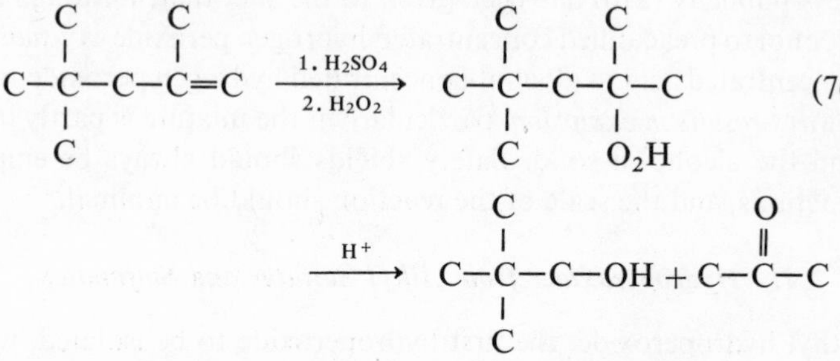

$$
\underset{\underset{\text{C}}{|}}{\overset{\overset{\text{C}\quad\text{C}}{|\quad\ |}}{C-C-C-C=C}} \xrightarrow[2.\ \text{H}_2\text{O}_2]{1.\ \text{H}_2\text{SO}_4} \underset{\underset{\text{C}\quad\ \text{O}_2\text{H}}{|\qquad|}}{\overset{\overset{\text{C}\quad\text{C}}{|\quad\ |}}{C-C-C-C-C}} \tag{76}
$$

$$
\xrightarrow{\text{H}^+} \underset{\underset{\text{C}}{|}}{\overset{\overset{\text{C}}{|}}{C-C-C-OH}} + \overset{\overset{\text{O}}{\|}}{C-C-C}
$$

[The cited example offers a convenient synthesis of neopentyl alcohol (290).]

Primary and secondary alcohols are unreactive under these conditions,
but they may be converted to hydroperoxides by solvolyzing their methane-
sulfonate esters in basic hydrogen peroxide solution (22, 291).

$$
n\text{-BuOH} \xrightarrow{\text{MeSO}_2\text{Cl}} n\text{-BuOSO}_2\text{Me} \xrightarrow[\text{KOH}]{30\%\ \text{H}_2\text{O}_2} n\text{-BuO}_2\text{H} \tag{77}
$$

This method offers what is probably the most reliable synthesis of simple
primary and secondary hydroperoxides, but yields are low, around 40 and
25%, respectively. The main by-product appears to be olefin (291). Mosher
and co-workers prepared straight-chain hydroperoxides from $C_3$ to $C_{10}$
(291), as well as allyl hydroperoxide (292). Long-chain alkyl hydroperoxides
(to $C_{18}$) have been similarly prepared (293). Tetralin hydroperoxide-$1_d$ has

recently been obtained in 40% yield by a modification in which the methane-sulfonate was not isolated (294).

Dialkyl peroxides may also be synthesized by this means, using lower ratios of hydrogen peroxide to sulfonate ester (295, 296).

## 2. Reactions of Alcohols and Ethers with Hydrogen Peroxide

### (a) MECHANISM

The alcohols that tend to undergo $S_N1$ displacements readily are converted to hydroperoxides by contact with highly concentrated hydrogen peroxide and catalytic amounts of acid. As Davies and his co-workers (1) have shown, the reaction almost certainly proceeds through a carbonium ion:

$$R_3COH \xrightarrow{H^+} R_3C^+ \xrightarrow{H_2O_2} R_3CO_2H + H^+ \qquad (78)$$

The relative ease of reaction follows the order of carbonium-ion stability (297, 298), $(CH_3)(C_2H_5)CHOH < (CH_3)_3COH < Ph(CH_3)CHOH < Ph_2CHOH < Ph_3COH$; $^{18}O$-labeled alcohols yield unlabeled hydroperoxides (299, 300); and some esters (297, 298, 301) and ethers (302) that are known to be sources of carbonium ions undergo a similar reaction.

$$(79)$$

$$Ph_2CHOEt \xrightarrow{H^+} Ph_2^+CH \xrightarrow{H_2O_2} Ph_2CHO_2H + H^+ \qquad (80)$$

The reaction of optically active alcohols with acidified hydrogen peroxide, discussed in detail by Davies (1, 303), appears to proceed largely with racemization of configuration. α-Tetralol (297, 298) and its monohydrogen phthalate (297, 298, 301) give complete racemization; 1-phenylethanol (304), its phenyl ether (305), and 1-phenylpropanol (305) give 4, 10, and 2% inversion, respectively, and 1-phenylbutanol (305), 1-(α-napthyl)ethanol (305), and 1-methyl-1-phenylethanol (299) show 3–9% retention. The claim of Choe and Tsutsumi (306) to have obtained almost complete inversion with both enantiomers of 1-methyl-1-phenylpropanol by using 30% aqueous hydrogen peroxide has been reinvestigated by Davies (303) and shown to be almost certainly in error.

The slight degree of configurational retention observed in some cases probably occurs by an $S_N i$ mechanism (1, 303):

$$
\underset{\substack{| \\ H_3C}}{\overset{\substack{C_2H_5 \\ |}}{Ph-C}} \overset{H}{\underset{\overset{O}{H}}{\underset{H}{\longrightarrow}}} \longrightarrow \underset{\substack{| \\ CH_3}}{\overset{\substack{C_2H_5 \\ |}}{Ph-C-O_2H}} \tag{81}
$$

In sharp contrast the solvolysis of methanesulfonates in base gives almost total inversion (307).

$$
H-O-\bar{O} \curvearrowright \underset{\substack{/ | \\ H \ C_5H_{11}}}{C}-O-SO_2Me \longrightarrow HO_2C \underset{\underset{C_5H_{11}}{\overset{\diagdown}{H}}}{\overset{\diagup}{=}} Me + {}^-O-SO_2Me \tag{82}
$$

No kinetic studies on the reaction of acidified hydrogen peroxide with alcohols have been published. The conditions are usually inhomogeneous; measuring small concentration changes in a mixture containing both hydrogen peroxide and a hydroperoxide is difficult, the reaction in aqueous solution is extremely fast for the water-soluble alcohols that will react at all, and in nonaqueous solvents the hydrogen peroxide tends to disappear by side reactions. Some preliminary studies in which acetonitrile was used as solvent showed second-order dependence on alcohol concentration (308).

The reverse reaction, though apparently not a problem in the last-cited work, has been encountered in studies of acid-catalyzed decomposition of hydroperoxides.

$$
RO_2H + H_2O \xrightarrow{\text{H}^+} ROH + H_2O_2 \tag{83}
$$

## (b) SYNTHESES

The reaction is usually carried out by adding alcohol to a stirred, cold ($-5$ to $0°$) ethereal or aqueous solution of hydrogen peroxide to which a drop or two of concentrated sulfuric acid has been added *previously*. [Acetic acid (309) and chloroform (310, 311) have also been used as solvents.] The mixture is then frequently allowed to come to room temperature or even heated (312), though this is probably not advisable. Typical times and temperatures are as follows: $t$-Bu(Me)$_2$CO$_2$H, 2–10 min, $0°$ (271); $n$-C$_{12}$H$_{25}$(CH$_3$)$_2$CO$_2$H, 1 hr, 25–30° (309); Ph$_2$CHOH, 6 hr, 25° (313); Ph($p$-Cl—Ph)CHOH, 66 hr, 25° (314). [The chlorine-substituted benzhydrols seem to be particularly unreactive (314).] It is advisable to keep times as short as possible to minimize acid-catalyzed decomposition of the product.

Since explosions are not infrequent, the quantities of reactants should be kept small. Davies (1) recommends a maximum of 5 ml of alcohol and not higher than 75% hydrogen peroxide. The 90% peroxide appears to be especially dangerous in contact with low-molecular-weight alcohols (271, 310), and better yields of *t*-amyl and 2,3-dimethyl-2-butyl hydroperoxides are obtained by using 50% hydrogen peroxide in refluxing chloroform (310). However, most published reports in the last 10–15 years have specified the 86–98% material. Recent examples are those of Hedaya and Winstein (310), preparation of a series of $R(CH_3)_2CO_2H$; Anderson and Smith, ring-substituted benzhydryl hydroperoxides (314) and the tertiary hydroperoxides of phenylcyclobutane, phenylcyclopentane, and phenylcyclohexane (315); Dulog and Sonner (316), a series $(CH_3)_2C(O_2H)(CH_2)_nC(O_2H)(CH_3)_2$ for $n = 1, 2, 3$, and 4.

For some alcohols—for example, 2,2,3-trimethylbutanol and 4-camphyl dimethylcarbinol (310) or 1-methylcyclobutanol (315)—good yields are obtained without acid catalysts. In the last case added acid resulted in no hydroperoxide but only products of its acid-catalyzed decomposition (315).

Generally speaking, acidified concentrated hydrogen peroxide yields hydroperoxides only from those carbinols that bear three alkyl substituents or one aryl and at least one other alkyl or aryl substituent. Foreman and Lankelma (309) were unable to prepare hydroperoxide from 2-tetradecanol. Choe and Tsutsumi (317) obtained only a 2% yield of cyclohexyl hydroperoxide from cyclohexanol.

A useful exception is a variation on the Mannich reaction discovered by Rieche and co-workers (318, 319):

$$\langle\quad\rangle NH + CH_2O + H_2O_2 \longrightarrow \langle\quad\rangle N-CH_2O_2H \qquad (84)$$

The reaction works with amides and hydrazides as well, but tends to give bisperoxides as well as hydroperoxides (318, 319).

Relatively few syntheses of hydroperoxides by cleavage of ethers have been recorded. These include the reaction of hydrogen peroxide with benzhydryl ethers (302) and the work of Rieche and his colleagues with acetals (273) and orthoesters (320).

$$H_2C\begin{matrix} OEt \\ \\ OEt \end{matrix} \xrightarrow[\substack{in\ Et_2O \\ 15hr.\ at\ 70°}]{7\%\ nonaqueous\ H_2O_2} H_2C\begin{matrix} O_2H \\ \\ OEt \end{matrix} \qquad (85)$$

$$Ph-C\begin{matrix} OMe \\ O-CH_2 \\ | \\ O-CH_2 \end{matrix} \xrightarrow{H_2O_2} Ph-C\begin{matrix} HO_2 \quad O-CH_2 \\ | \\ O-CH_2 \end{matrix} \qquad (86)$$

Similar compounds have been obtained by autoxidation (Section II-A-4) or by addition of hydrogen peroxide to vinyl ethers (Section C-4).

### 3. Reactions of Hydrogen Peroxide with Organic Halides

This reaction has been used only infrequently in hydroperoxide synthesis. For reactive halides—that is, aryl or *t*-alkyl (321)—the corresponding alcohols are more readily available and equally reactive. Less reactive ones are more easily converted to hydroperoxides via the Grignard route (Section II-A-6).

Hüttel and Ross (321, 322) prepared hydroperoxides from triphenylmethyl, α-cumyl, benzhydryl, *t*-butyl, benzyl, and *sec*-butyl chlorides and from *sec*-butyl, *sec*-amyl, and cyclohexyl bromides by direct reaction. Yields from all but the first four were, however, 5% or less. Addition of 0.2 $M$ Hg(OAc)$_2$ to the reaction medium raised yields to around 50%, but the presence of unreacted starting material and decomposition products made purification of the hydroperoxides difficult or impossible (322).

$$Ph_2CHCl + H_2O_2 \xrightarrow{\text{NaHCO}_3} Ph_2CHO_2H \quad (82\%) \qquad (87)$$

The reaction has been most useful in the synthesis of triaryl hydroperoxides (1) and of peroxides and hydroperoxides of other Group IV elements (Section II-E).

### 4. Addition of Hydrogen Peroxide to Double Bonds

(a) ADDITION OF HYDROGEN PEROXIDE TO C=C AND C=N

Olefins bearing electron-releasing substituents react with mildly acidic hydrogen peroxide to yield hydroperoxides, presumably via a carbonium-ion mechanism (1). *t*-Amyl hydroperoxide (297, 298) and a number of α-hydroperoxy ethers (323–326) have been prepared in this fashion.

$$Me_2C=CHMe \xrightarrow{\text{H}^+} [Me_2C^+-CH_2Me] \xrightarrow{\text{H}_2O_2} Me_2\overset{\overset{\displaystyle O_2H}{|}}{C}-CH_2Me$$

$$(88)$$

$$CH_2=CH-OEt \xrightarrow[\text{H}^+]{\text{H}_2O_2} Me-\overset{\overset{\displaystyle O_2H}{|}}{C}H-OEt \qquad (89)$$

$$Me-\overset{\overset{\displaystyle CH_2}{\|}}{C}-O-\overset{\overset{\displaystyle CH_2}{\|}}{C}Me \xrightarrow{\text{H}_2O_2} Me_2\overset{\overset{\displaystyle O_2H}{|}}{C}-O-O-\overset{\overset{\displaystyle O_2H}{|}}{C}-Me_2 \quad \text{(explosive)} \ (90)$$

[In the last example addition is followed by cleavage of ether linkage and displacement to form the bisperoxide (324).]

Combination of hydrogen peroxide with a source of positive halogen is a recent modification. $t$-BuOCl (327), $Br_2$ (138, 328), $N$-chloroacetamide (329), and 1,3-dibromo-5, 5-dimethylhydantoin (BMH) (329) have been used. The last two give the highest yields (55–86%) with styrene, tetramethylethylene, trimethylethylene, cyclohexene, and 1,2-dimethylcyclohexene—and the least contamination from bisperoxide by-products (327).

$$EtO-CH=CH_2 \xrightarrow[t\text{-BuOCl, }-25°]{H_2O_2} EtO-\overset{\overset{\displaystyle O_2H}{|}}{CH}-CH_2Cl \qquad (91)$$

$$CH_2=CMeCH_2Cl \xrightarrow[Br_2,\ 5-10°]{H_2O_2} CH_2Br-\overset{\overset{\displaystyle O_2H}{|}}{C}MeCH_2Cl \qquad (92)$$

$$(93)$$

Weissermel and Lederer (327) prepared $\beta$-chloroperoxides by substituting $t$-Butyl or $\alpha$-cumyl hydroperoxide for hydrogen peroxide. In what appears to be a free-radical-chain addition, using catalytic amounts of $SOCl_2$ or $SO_2Cl_2$ instead of equimolar amounts of $t$-BuOCl, they obtained nonhalogen-containing peroxides and hydroperoxides.

$$t\text{-BuO}_2H + EtO-CH=CH_2 \xrightarrow{10^{-4}\ M\ SOCl_2} EtO-\overset{\overset{\displaystyle O_2t\text{-Bu}}{|}}{CH}-CH_3\ (95\%)$$

$$(94)$$

Hydrogen peroxide gave mainly bisperoxides rather than hydroperoxides, however (327).

Enamines react with hydrogen peroxide readily (330), but isolated products are usually the bisperoxides. A number of $\alpha$-hydroperoxy amines have been prepared by Höft and Rieche (331), utilizing the low-temperature addition of hydrogen peroxide to imines.

$$(95)$$

$$(96)$$

## (b) ADDITION OF HYDROGEN PEROXIDE TO C=O

Hydrogen peroxide adds to aldehydes and ketones, yielding α-hydroxy hydroperoxides (1, 2, 332), which in turn form bis(1-hydroxyalkyl) peroxides by reaction with more carbonyl compound.

$$RCHO + H_2O_2 \longrightarrow RCH\overset{\displaystyle OH}{\underset{\displaystyle O_2H}{\big<}} \tag{97}$$

$$RCH\overset{\displaystyle OH}{\underset{\displaystyle O_2H}{\big<}} + RCHO \longrightarrow \overset{OH}{\underset{|}{R}}\!CH - O_2 - \overset{OH}{\underset{|}{C}}\!HR \tag{98}$$

Both products are in equilibrium with the reactants; at 0° for acetaldehyde in water the values of $K$ for reactions 97 and 98 are 50 l/mole and 500 $l^2$/mole$^2$, respectively (333). Equilibrium is attained in 10–20 min without added catalysts and with 0.01 $M$ $H^+$ about 10 times as fast (334).

Both α-hydroxy hydroperoxides (335, 336) and bis(1-hydroxyalkyl) peroxides (337, 338) from most straight-chain aldehydes ($C_1$–$C_{11}$) have been characterized. Equimolar mixtures of hydrogen peroxide and aldehyde in ether yield the former if the ether is evaporated at room temperature or the latter if the solvent is distilled off at higher temperature (CAUTION!). For higher members of this series the α-hydroxy hydroperoxides crystallize from the reaction medium on standing (1, 338).

For most ketones simple 1:1 addition products have eluded isolation, the exception being α-chlorocyclohexanone (339).

$$\tag{99}$$

Cyclohexanone itself gives mainly a mixture of dimers, each in turn predominating as the reaction mixture is increasingly acidified (340, 341).

$$\tag{100}$$

Solvent effects on the equilibrium, which appear to be complex, have recently been investigated by Antonovskii and Terent'ev (341, 342).

*gem*-Dihydroperoxides have been obtained from acetone (343), 2-butanone

$$Me_2C{=}O + H_2O_2 \longrightarrow Me_2C\overset{O_2H}{\underset{O_2H}{\big\langle}} \qquad (101)$$

(344), and a number of cyclic ketones (1, 345, 346). From acetone a 95% yield is obtained from neutral hydrogen peroxide.

Ledaal (345) has developed for the cyclic ketones ($C_6$–$C_{15}$) a technique that involves spraying the reaction mixture onto a porous-clay plate or spreading a thin layer on glass for rapid evaporation of water. Adding a few drops of perchloric acid to a 0° suspension or solution of the *gem*-dihydroperoxides in propionic acid produces dimeric cyclic peroxides (345).

Actually solutions of ketones with hydrogen peroxide, particularly in the presence of catalytic amounts of acid, produce a vast array of products in equilibrium. Linear or cyclic dimers, trimers, tetramers, have all been identified, with or without free hydroxy or hydroperoxy groups, and mostly explosive and detonable either by shock or heat (1).

(102)

Thus Milas and Golubovic (347) separated and identified (via chromatography) seven peroxides from the reaction of 0.2 $M$ hydrogen peroxide, 0.2 $M$ methyl ethyl ketone, and 0.05 $M$ sulfuric acid in ether at $-5°$. The cyclic trimeric peroxide (9) and dihydroperoxy dimer (5) accounted for 25 and 45%, respectively, with 10% monomer (2) and 12, 5, 2, and 1% of trimer, tetramer, pentamer, and hexamer (7, $n = 1$–4). The monomer, 2,2-dihydroperoxybutane (2), was isolated by chromatography on cellulose power. It decomposed slowly at room temperature to give oxygen and a mixture containing all the peroxides obtained from the first reaction except the cyclic trimer.

Similar results have been obtained from acidified hydrogen peroxide plus diethyl ketone (348), cyclohexanone (340, 349), acetone (324, 343), and *sym*-dichloroacetone (350).

Rieche and co-workers (351, 352) have also prepared a number of cyclic peroxides from diketones and triketones.

$$(103)$$

$$(X = H \text{ or } OH)$$

## D.  Preparation of Hydroperoxides by Miscellaneous Means

Some recent novel syntheses involve displacement or elimination reactions on halogen-substituted hydroperoxides that leave the $-O_2H$ group intact. A patent (353) claims the preparation of some cationic α-cumyl hydroperoxides.

The products are surprising in view of the relatively rapid decomposition of hydroperoxides by sulfides and amines at higher temperatures (Sections IV-B-2 and 3, respectively).

Kopecky, et al. (329) dehydrohalogenated 3-chloro-2,3-dimethyl-2-butyl hydroperoxide and 2-chloro-1,2-dimethylcyclohexyl hydroperoxide (as well as the bromo analogs) with methanolic potassium hydroxide at 0–30°. From the former, 2,3-dimethyl-3-hydroperoxy-1-butene was obtained in 70% yield.

$$
\begin{array}{cc}
\overset{\displaystyle O_2H}{\underset{\displaystyle \underset{|}{Cl}}{Me_2{-}\overset{|}{\underset{|}{C}}{-}C{-}Me_2}} & \longrightarrow \quad \overset{\displaystyle O_2H \quad Me}{Me_2{-}\overset{|}{C}{-}\!\!-\!\!-\overset{|}{C}{=}CH_2} \qquad (105)
\end{array}
$$

From the bromo analog, the hoped-for product was not the allylic hydroperoxide but the cyclic four-membered-ring peroxide, which they also synthesized (354).

$$
\underset{\displaystyle O_2H}{Me{-}\overset{\displaystyle Me \;\; Br}{\overset{|}{C}{-}\overset{|}{C}HMe}} \;\xrightarrow{\;OH^-\;}\; Me{-}\overset{\displaystyle Me \;\; Br}{\underset{\displaystyle \underset{O}{\diagdown}\!\!\diagup}{\overset{|}{C}{-}\overset{|}{C}H{-}Me}} \;\longrightarrow\; Me{-}\underset{\displaystyle O{-}O}{\overset{\displaystyle Me \;\; H}{\overset{|}{C}{-}\overset{|}{C}{-}Me}}
$$

$$(106)$$

The cyclic peroxide decomposes quantitatively at 60° to acetone and acetaldehyde, the only products of an earlier attempt by Richardson and co-workers (355) to obtain the cyclic peroxide from $ClCH_2CMe_2O_2H$.

Recently trifluoromethyl hydroperoxide has been synthesized (356) by the hydrolysis of a peroxy ester.

$$
3F\overset{\displaystyle O}{\overset{\|}{C}}{-}O{-}O{-}\overset{\displaystyle O}{\overset{\|}{C}}F + CF_2N_2 \;\xrightarrow{\;h\nu\;}\; CF_3O_2\overset{\displaystyle O}{\overset{\|}{C}}F \;\xrightarrow{\;H_2O\;}\; CF_3O_2H
$$

$$(107)$$

Hydroperoxides have been produced by the decomposition of initiators in the presence of oxygen and a source of hydrogen atoms. The synthetic utility is limited mainly to highly arylated ethanes (357), but $MeO_2H$ has resulted from the photolysis of acetone in oxygen (358, 359) and $Me_2(CN)CO_2H$ from the thermolysis of azobisisobutyronitrile (360).

$$
(Ph_2RC)_2 \;\xrightarrow{\;\Delta\;}\; 2Ph_2RC\cdot \;\xrightarrow{\;O_2\;}\; 2Ph_2RCO_2\cdot
$$
$$
\xrightarrow{\;pyrogallol\;}\; 2Ph_2RCO_2H
$$

$$(108)$$

### E. Hydroperoxides of other Elements

Although organometallic or metalloid peroxides have been synthesized from nearly every element in the periodic table (361), so far the only *hydroperoxides* to be characterized are those of C, Si, Ge, Sn, and Sb.

Claims for the preparation of $Et_3SiO_2H$, $Ph_3SiO_2H$ (362), and $Me_3SiO_2H$ (363) by reaction of concentrated hydrogen peroxide with the corresponding silanols or silyl chlorides appeared in 1955 and 1956. However, the compounds appear to be fairly impure. More recently Dannley and Jalics (364), using silyl amines as starting materials, have extended the series, improving yields and purities.

$$Ph_2MeSiNH_2 + H_2O_2 \longrightarrow Ph_2MeSiO_2H + NH_3 \qquad (109)$$

Dannley and his co-workers (365) are also responsible for the first synthesis of simple tin hydroperoxides (as well as those of germanium (365)),

$$R_3SnOH + conc. \ H_2O_2 \longrightarrow R_3SnO_2H \qquad (110)$$

$$(R = Ph \ or \ Me)$$

although Alexandrov (366, 367) had earlier obtained a hydrogen peroxide— hydroperoxide complex and found that it decomposed to a hydroxy hydroperoxide.

$$(Et_3Sn)_2O \ or \ (Et_3SnO)_2 \xrightarrow{H_2O_2} (Et_3SnO_2H)_2 \cdot H_2O_2 \longrightarrow Et_2Sn{\overset{\displaystyle OH}{\underset{\displaystyle O_2H}{\big\langle}}}$$

$$(111)$$

Dahlmann and Rieche (368) have synthesized dihydroperoxides of antimony by a reaction that also gives peroxides.

$$R_3Sb(OR')_2 + H_2O_2 \longrightarrow R_2Sb{\overset{\displaystyle O_2H}{\underset{\displaystyle O_2H}{\big\langle}}} \qquad (112)$$

## III.  PROPERTIES, SEPARATION, AND ANALYSIS OF HYDROPEROXIDES

### A.  Physical Properties of Hydroperoxides

#### 1.  General Characteristics

Hydroperoxides may be liquids or crystalline solids; the difference is usually predictable from the characteristics of the corresponding alcohol (although $RO_2H$ generally has a somewhat lower melting point and higher boiling point than ROH). Both solids and liquids are usually stable enough (if pure) to be distilled, or melted at temperatures below 70–80°; higher temperatures result in extensive degradation and, possibly, explosion.

Liquid hydroperoxides are most often characterized by densities and refractive indices, both usually higher than those of the corresponding hydro-

TABLE 4. COMPARISON OF DENSITY AND REFRACTIVE INDICES FOR RH, ROH, AND $RO_2H$

| R | RH (369) | | ROH (369) | | $RO_2H$ (1, 4, 291) | |
|---|---|---|---|---|---|---|
| | $d_4^{20}$ | $n_D^{20}$ | $d_4^{20}$ | $h_D^{20}$ | $d_4^{20}$ | $n_D^{20}$ |
| *n*-Butyl | | | 0.8098 | 1.3992 | 0.907 | 1.4057 |
| *sec*-Butyl | | | 0.8080 | 1.3954 | 0.907 | 1.4050 |
| *t*-Butyl | | | 0.7856 | 1.3838 | 0.896 | 1.4013 |
| *n*-Hexyl | 0.6594 | 1.3748 | 0.8136 | 1.4178 | 0.891 | 1.4208 |
| (cyclohexenyl ring structure) | 0.8110 | 1.4465 | $1.00_{26}^{26}$ | $1.4790^{25}$ | 1.058 | 1.4896 |
| $PhMe_2C-$ | 0.862 | 1.4915 | $0.9727_4^{19}$ | $1.5314^{19}$ | 1.0619 | 1.5242 |

carbon or alcohol (Table 4). However, the difference between ROH and $RO_2H$ are not great; both spectral characteristics and titration are better indices of purity.

## 2. Infrared Spectra

Hydroperoxides have characteristic infrared absorptions in the 820–860 $cm^{-1}$ (—O—O—), 3400–3600 $cm^{-1}$ (O—H stretch) (309, 370–376), and 6900 $cm^{-1}$ (O—H overtone) (161, 233, 377, 378) regions. Of these, the 830–860 $cm^{-1}$ bands are probably the least useful. They tend to be broad and weak, and in a region where other functional groups, notably peroxides (371–375), alcohols (371, 376), ethers (372, 373), and oxiranes (374, 375) also absorb. Thus considerable care must be exercised in using this absorption for diagnostic purposes (371–375). In cases in which the structure is known it is found that in this region hydroperoxides absorb at frequencies 30–70 $cm^{-1}$ and 20–30 $cm^{-1}$ lower than the corresponding alcohols (309, 371, 376) and dialkyl peroxides (371, 375), respectively. [The difference between ROH and $RO_2H$ diminishes with increasing molecular weight (371).]

The 838 $cm^{-1}$ absorption of $t$-$BuO_2H$ in carbon tetrachloride has been assigned to an out-of-plane deformation by Kozlov and Rabinovich (379), who find an in-plane deformation at 1335 $cm^{-1}$. In $t$-$BuO_2D$ these are shifted to 610 and 900 $cm^{-1}$, respectively. Similar assignments were made for α-cumyl hydroperoxide (379).

It has been suggested that cyclohexenyl hydroperoxide exists as two conformers on the basis of equatorial (1060 $cm^{-1}$) and axial $O_2H$ (961 $cm^{-1}$) assignments (380).

Like alcohols, hydroperoxides in solution exhibit two maxima in the O—H stretch region, a sharp peak from the monomeric species and a broad peak at higher wavelengths from the hydrogen-bonded dimer (381–383). For $t$-BuO$_2$H in carbon tetrachloride the dimer maximum occurs at 3410 cm$^{-1}$ and the monomer maximum is variously given as 3540 (383), 3555 (384), and 3562 (384) cm$^{-1}$. [Walling and Heaton (382) report maxima in 20 solvents and have used their relative intensities to determine association constants, as have others. They also measured $v_{t\text{-BuOO-D}}$ (2630 cm$^{-1}$ for the monomer in carbon tetrachloride) and found for all solvents $v_{OO-H}/v_{OO-D} \simeq 1.35$.]

Spectra that are run neat or in concentrated solution so that dimer predominates do not ordinarily permit distinction between C—O—H and O—O—H stretch (370). By using 0.001–0.01 $M$ solutions, however (cell thicknesses of 2–5 mm), so that virtually all RO$_2$H is monomeric, Soviet workers (384) have been able to estimate quantitatively mixtures of $t$-BuOH with $t$-BuO$_2$H and $\alpha$-cumyl OH with $\alpha$-cumyl O$_2$H. Under these conditions the nonhydrogen-bonded $v_{CO-H}$ is 60–70 cm$^{-1}$ higher than $v_{OO-H}$ for these and other ROH–RO$_2$H pairs (385)* and is extremely useful for estimating purity and following decomposition experiments.

A distinction between ROH and RO$_2$H can be made at higher concentrations in the near-infrared region, in which the hydrogen-bonded dimers apparently do not absorb (or at least the absorption does not interfere and has not been indentified). Alcohols show an O—H overtone at about 7040 cm$^{-1}$ (377, 378) and $\alpha$-hydroxy hydroperoxides [R$_2$C(OH)OOH] show both O—H and OO—H overtones at 7090 and 6960 cm$^{-1}$, respectively (161, 233). The intensity of the 6960 cm$^{-1}$ absorption, which decreases with increasing concentration, has been used to measure monomer–dimer association constants for the isomeric heptane hydroperoxides (385).

Kharasch and Sosnovsky (339) have published the spectra of several of the complex peroxides resulting from hydrogen peroxide–cyclohexanone reactions but have not attempted an analysis.

### 3.  Ultraviolet Spectra

Alkyl hydroperoxides and peroxides with no other functional groups absorb energy below 300 m$\mu$, but the absorptions are weak (log $\varepsilon$ for EtO$_2$H in H$_2$O = 1.0 at 250 m$\mu$) and do not exhibit maxima above 200 m$\mu$ (386, 387).

### 4.  Nuclear-Magnetic-Resonance Spectra

Surprisingly little use has been made of nuclear magnetic resonance (NMR) techniques in hydroperoxide or peroxide chemistry, considering that introduction of —O—O— changes the chemical shift of neighboring hydrogens

---

*Also unpublished work by L. Zikmund and R. Hiatt.

quite considerably, even when the precursor has been $-O-$. The most thoroughgoing* study of hydroperoxide NMR spectra has been made by Japanese workers (388) with a 40-Mc instrument and a calibration standard of benzene–cyclohexane having a peak separation of 227 cps. The spectra of *n*-propyl, *n*-butyl, *sec*-butyl, *n*-amyl, and *n*-octyl hydroperoxides and alcohols were compared and chemical shifts reported as cycles per second downfield from the $CH_3$ resonance in the same compound. The $OOH–CH_3$ separation was found to be $245 \pm 5$ cps, whereas the $OH–CH_3$ separation was only $110 \pm 10$ cps. Methylene hydrogens alpha to $-O-O-$ were $79 \pm 1$ cps downfield compared with $65 \pm 2$ cps for the corresponding alcohols.

Scattered reports in the literature of measurements on standard 60-Mc instruments confirm the very low but rather variable shielding of the hydroperoxy proton (Table 5). However, recent work by Swern and co-workers (389) on a series of hydroperoxides has shown that the signal of the hyo-dr peroxy proton is considerably downfield from that of the hydroxy proton, and its position is also concentration dependent. The upfield shift of the signal of the hydroperoxy proton with dilution is of much smaller magnitude than that in alcohols.

Although the $O_2H$ resonance may be broad or absent because of exchange, chemical shifts and splittings of other hydrogens in the molecules have been useful in several instances: structure proof for the cyclopentenylperoxy hydroperoxide dimer (45), identification of products in the reaction of hydrogen peroxide with acetylacetone (391), product analysis in styrene–tetralin cooxidation (8), and detection of impurities in di-*n*-butyl peroxide synthesis (390).

Keaveney, Berger, and Pappas (268) have published NMR data for

(113)†

---

*A more recent study is better (G. A. Ward and R. D. Mair, *Anal. Chem.*, **41**, 538 (1969)). Mixtures of ROH, $RO_2H$ and $RO_2R$ are analyzed quantitatively by comparison of the distinguishable $>CH–O–$ signals. (Note added in proof).

†Reference 391.

TABLE 5.  CHEMICAL SHIFTS[a] FOR HYDROPEROXIDES AND ANALOGOUS PEROXY COMPOUNDS

| $RO_2H$ | Solvent | OOH | $-\overset{\displaystyle \mid}{\underset{\displaystyle H}{C}}-O-O-$[b] | $-\overset{\displaystyle \mid}{\underset{\displaystyle H}{C}}-O-$[b] | Ref. |
|---|---|---|---|---|---|
| $\overset{\displaystyle O_2H}{\underset{\displaystyle \mid}{R^1R^2C}}-N=N-Ph$ | $C_6D_6$ | 0.27 | | | 193 |
| $\overset{\displaystyle O_2H}{\underset{\displaystyle \mid}{R^1R^2C}}-N=N-Ph$[c] | | 0.7 | | | 190 |
| $(CH_2)_{11}C\overset{\displaystyle O_2H}{\underset{\displaystyle O_2H}{}}$ | | −0.05 | | | 346 |
| $t$-BuO$_2$H | CDCl$_3$ | 2.03 | | | 389 |
| α-Cumyl O$_2$H | CCl$_4$ | 2.13 | | | 389 |
| α-Tetralyl O$_2$H | CDCl$_3$ | 1.57 | 5.02 | | (d) |
| $Ph-\overset{\displaystyle OMe}{\underset{\displaystyle H}{C}}-O_2H$ | CCl$_4$ | 0.0 | 4.37 | | 241 |
| Cyclopentenyl O$_2$H | CDCl$_3$ | 1.1 | 4.9 | 5.2 | 45 |
| (structure with H$^f$ H$^e$ H$^e$ O$_2$H, O$_2$) | CDCl$_3$ | ~1 | 4.9[e], 5.5[d] | 8.47–8.62 | 45 |
|  |  |  | 8.73 |  | 216 |
| $(Pr-CH_2-O)_2$ | None | | 6.06 | | 390 |

[a] 60-Mc nuclear magnetic resonance.
[b] If multiplet, taken as center.
[c] Tetra-O-benzoyl-D-glucose phenylhydrazone hydroperoxide.
[d] T. M. Mill, unpublished results.
[e] For the indicated hydrogens.
[f] For the indicated hydrogen.

several α-methoxy hydroperoxides and used differences in chemical shifts between

$$\begin{array}{ccc} & H & & H \\ & | & & | \\ -C-O_2- & & and & -C=O \\ & | & & \\ & OMe & & \end{array}$$

to distinguish quantitatively between the several products of ozonolysis reactions. It is interesting to note that the α-protons in

$$\begin{array}{ccc} & H \;\; OMe & & H \\ & | \; / & & | \\ PhC & & and & (PhC-O)_2 \\ & | & & \;\;\;\;\; OMe \\ & O_2H & & \end{array}$$

differ in chemical shift by only 0.06 ppm.

## 5. Electron-Spin-Resonance Spectra Derived from Hydroperoxides

Although hydroperoxides themselves do not exhibit an electron-spin-resonance (ESR) spectrum, peroxy radicals, which do, are very easily generated from them by action of metal ions, azo or peroxy initiators, or ultra violet radiation (392–397). The process has been most useful in determining reactivities of peroxy radicals (395, 396) and, with suitable instrumentation, might be as good a diagnostic test for the $-O_2H$ group as reaction with lead tetraacetate (398).

## 6. Bond Energies

The dissociation energy of the O—O bond has considerable relevance to many hydroperoxide reactions, particularly thermal homolysis. Since the reverse reaction would be expected to have zero activation energy, $D_{RO-OH}$ should approximately equal $E_a$ for equation 114.

$$RO-O-H \longrightarrow RO\cdot + \cdot OH \qquad (114)$$

Intuitively (399), one would expect $D_{RO-OH}$ to be the average of $D_{RO-OR}$ and $D_{HO-OH}$, or about 42 kcal/mole, but most experimental measurements of $E_a$ fall short of this by 5–12 kcal/mole. Moreover, it is questionable to calculate $D_{RO-OH}$ solely from experimental thermochemical data since the heats of combustion of hydroperoxides are difficult to obtain and are not necessarily reliable (400).

Probably the most widely accepted value is that of Benson (399), who, using group-additivity rules (401), estimated $\Delta H_f^{\circ}$ values for several peroxides and hydroperoxides; these values were in good agreement with the experimental values of Pritzkow and Müller (402). Combination of these with known values of $\Delta H^\circ_{OH}$ and $\Delta H^\circ_{RO}$ (399, 403, 404) gives $D_{RO-OH}$ of 43–44 kcal/mole for R = Me, Et, $i$-Pr, and $t$-Bu.

The RO—OH bond energy should be relatively independent of R (399, 405), but, although the latest experimental values of $E_a$ (equation 114) for $t$-BuO$_2$H and MeO$_2$H are in good agreement with the predicted values, no studies of decompositions of other hydroperoxides have as yet given comparable results.

Other semiempirical estimations of $D_{RO-OH}$ are due to Denisov (406) and Likhtenshtein (407). Both use the relationship $\Delta H^\circ_{RO_2H} = \Delta H^\circ_{ROH} + 23$ kcal/mole, which seems to be their main departure from Benson's calculations, and they obtain $D_{RO-OH}$ of about ˜35 kcal/mole. Neither Kerr (405), who has reviewed the subject, nor Benson has commented on the discrepancy.

The value of $D_{RO_2-H}$ is argued by Benson (399) to equal that of $D_{HO_2-H}$, or $90 \pm 2$ kcal/mole.

A detailed discussion of the thermochemistry of organic peroxides will be found in Volume I of this series.

## B.  Chemical Properties of Hydroperoxides

### 1.  Hydroperoxide Complexes

The tendency of hydroperoxides to dimerize in nonpolar solutions has been noted in a number of investigations (115, 381, 408–411) but has been dealt with quantitatively only recently (382, 383, 385). Hydroperoxides probably form hydrogen-bonded associations with ethers, alcohols, amines, ketones, sulfoxides, and carboxylic acids also, although the evidence derives mainly from rate studies on thermal decomposition rather than from direct physical measurements (412). Complexes, presumably of a $\pi$-type, are formed with aromatic rings and (probably) with carbon–carbon double bonds (382). Some association constants are shown in Table 6.

Walling and Heaton (382) have measured association constants for dimerization and aromatic-complex formation for $t$-BuO$_2$H using both infrared and solvent–solvent distribution techniques (Table 6). Dimerization constants have also been measured by Geiseler and Zimmerman for 1-, 2-, 3-, and 4-hydroperoxyheptanes (385) and by Zharkov and Rudnevskii for $t$-BuO$_2$H (383) (Table 6). The results are in fair accord.

There is some evidence that 1:1 complexes are not the only species. Walling and Heaton were forced to assume hydroperoxide dimer–styrene complexes in bulk styrene or in styrene–benzene to make phase-distribution results fit the infrared data. Both Geiseler and Zharkov were forced to assume hydroperoxide trimers as well as dimers in order to fit their infrared data.

Simple calculation shows that in carbon tetrachloride at 70° 0.1 $M$ $t$-BuO$_2$H

TABLE 6.   ASSOCIATION CONSTANTS FOR HYDROPEROXIDE COMPLEXES

$$RO_2H + A \; \underset{}{\overset{K}{\rightleftharpoons}} \; RO_2H \cdots A$$

| $RO_2H$ | A | Solvent | $K$ (l/mole) | $T\,(^\circ C)$ | $+\Delta H$ (kcal/mole) | Method[a] | Ref. |
|---|---|---|---|---|---|---|---|
| $t$-BuO$_2$H | $t$-BuO$_2$H | CCl$_4$ | 0.60 | 70.0 | 5.95 | A | 382 |
| $t$-BuO$_2$H | $t$-BuO$_2$H[b] | CCl$_4$ | 0.46 | 40.5 | 6.3 | A | 383 |
| $n$-Heptyl O$_2$H | $n$-Heptyl O$_2$H[c] | CCl$_4$ | 2.1 | 45.0 | 4.9 | A | 385 |
| $sec$-Heptyl O$_2$H | $sec$-Heptyl O$_2$H[d] | CCl$_4$ | 0.5 | 45.0 | 4.9 | A | 385 |
| $t$-BuO$_2$D | $t$-BuO$_2$D | CCl$_4$ | 0.64 | 70.0 | 4.9 | A | 382 |
| $t$-BuO$_2$H | Styrene | CCl$_4$ | 0.17 | 70.0 | 2.37 | A | 382 |
| $t$-BuO$_2$H | Benzene | CCl$_4$ | 0.11 | 70.0 | 1.58 | A | 382 |
| $t$-BuO$_2$H | MeCHO | CCl$_4$ | 30 | 70.0 | 4.6 | A | 413 |
| $t$-BuO$_2$H | Cyclohexanone | PhCl | 7.0[e] | 70.0 | 11.0 | B | 414 |
| $t$-BuO$_2$H | Pyridine | PhCl | 35[e] | 70.0 | 8.3 | B | 415 |
| $t$-BuO$_2$H | Me—S (cyclohexane ring), O | CCl$_4$ | 35 | 25.0 | — | A | 416 |
| $\alpha$-Cumyl O$_2$H | $\alpha$-Naphthylamine | PhCl | 300 | 74.0 | — | B | 417 |
| Decyl O$_2$H | Caproic acid | Undecane | 1830[e,f] | 70.0 | 10.2 | B | 418 |

[a] Method $A$: by infrared analysis; method $B$: from kinetics of decomposition of RO$_2$H.
[b] $K_{trimer}$ at 17.4° = 3.95 l/mole.
[c] $K_{trimer}$ at 15° = 1.4 l/mole.
[d] $K_{trimer}$ at 15° = 3.7 l/mole.
[e] Extrapolated from measurements at higher temperatures.
[f] For the complex RO$_2$H$\cdots$[R′CO$_2$H]$_2$.

is 89.4% monomer; 0.01 $M$ $t$-BuO$_2$H is 96.4% monomer. The heptyl hydroperoxides are also largely monomeric at these concentrations, but the primary
isomer shows rather different characteristics than the secondary ones at
1.0 $M$, being about 2:1 dimer:monomer, whereas the others distribute
between monomer, dimer, and trimer in roughly equal proportions (385).
Antonovskii et al. (419) have asserted that 0.7 $M$ cumene hydroperoxide
in $n$-heptane at 95° is virtually all dimer.

Kinetic evidence for complexing derives from the accelerated rates of
thermal decomposition in the presence of additives and the increased difficulty
of abstraction of the O—H hydrogen by radicals. Hydrogen bonding is
supposed to weaken the O—O bond and to make the terminal hydrogen less
readily available (412).

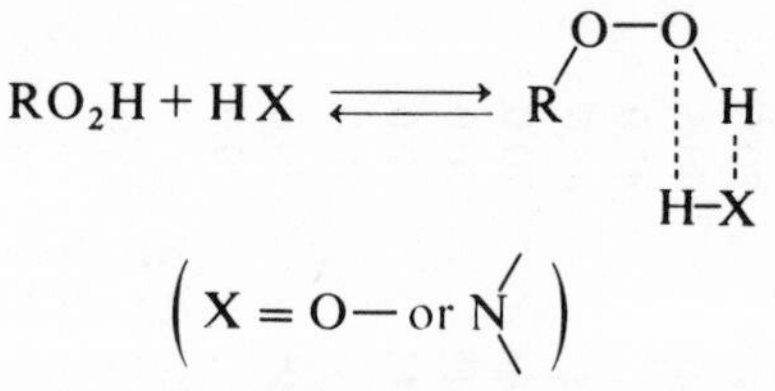

$$RO_2H + HX \; \rightleftharpoons \; R\overset{O-O}{\underset{H-X}{\diagdown \; {\vdots}\;\diagdown H}}$$ (115)

$$\left( X = O- \text{ or } N\diagup \diagdown \right)$$

The rate effects generally show saturation with increased concentration of
additive, which is indicative of complexing, and in some cases equilibrium
constants have been calculated (Table 6). The homolytic reactions following
complexing are examined in more detail in Section V, but some comments
are appropriate here.

Hydroperoxide–amine complexes postulated as intermediates to further
reaction (415, 417, 419, 420) have been isolated in some instances by Sprinzak
(90) and by Oswald and co-workers (143, 421) and patented as free-radical
initiators (422). The complexes have been called salts, although they are
apparently strongly hydrogen-bonded structures rather than ionic ones (421).

The "complex" of hydroperoxide with aldehydes is not usually hydrogen
bonded but the result of addition to the C=O bond.

$$RO_2H + \overset{\diagdown}{\underset{\diagup}{C}}=O \; \rightleftharpoons \; RO_2-\overset{|}{\underset{|}{C}}-OH$$ (116)

The rates of forward and backward reactions have been measured for a
number of hydroperoxides and aldehydes in carbon tetrachloride by
Antonovskii and Terent'ev (413) by infrared spectroscopy. The equilibrium
constants and stabilization energies calculated from their data agree reasonably well with the "complexing" constants determined by Denisov (414,
423) from the decomposition of hydroperoxides in the presence of ketones
(Table 6).

Hydrogen-bonded carbonyl–hydroperoxide complexes are possible if addition is sterically unfavorable. Richardson and Steed (173) have measured a stabilization energy of 1.86 kcal/mole for intramolecular hydrogen bonding in diisopropyl ketone hydroperoxides.

$$
\begin{array}{c}
\text{H}-\text{O} \\[2pt]
\vdots \quad | \\[2pt]
\text{O} \quad \text{O} \\[2pt]
\| \quad | \\[2pt]
\text{C}-\text{C}-\text{C}-\text{C}-\text{C} \\[2pt]
| \qquad | \\[2pt]
\text{C} \qquad \text{C}
\end{array}
$$

Cumene hydroperoxide forms a very stable 1:1 complex with its sodium salt which can be crystallized from solution (424, 425). No other hydroperoxide seems to do this—at least no other instances have been cited. The kinetics of thermal decomposition of this complex suggest that it is itself highly dimerized in solution (425).

### 2. Acidity, Reduction to Alcohols, and Summary of Other Reactions

Hydroperoxides are stronger acids than alcohols and form salts with aqueous sodium hydroxide, a property that facilitates their separation from reaction mixtures and purification (Section C). Measurements of $pK_a$ in water at 20° for several simple alkyl and aralkyl hydroperoxides give values that vary from 11.5 to 12.8 (426, 427).

$$ RO_2H \rightleftharpoons H^+ + {}^-O_2R \tag{117} $$

Hydroperoxides and their anions exhibit a number of nucleophilic reactions (Section IV) analogous to those of alcohols. Thus they form peroxy esters, peroxy acetals and ketals, and dialkyl peroxides.

$$
RO_2H + R'C\!\!\begin{array}{c}{\nearrow O}\\[-2pt]{\searrow Cl}\end{array} \longrightarrow R'C\!\!\begin{array}{c}{\nearrow O}\\[-2pt]{\searrow O_2R}\end{array} \tag{118}
$$

$$
RO_2H + {\searrow \atop \nearrow}C\!=\!O \longrightarrow {\searrow \atop \nearrow}C\!\!\begin{array}{c}{\nearrow OH}\\[-2pt]{\searrow O_2R}\end{array} \xrightarrow[RO_2H]{H^+} {\searrow \atop \nearrow}C\!\!\begin{array}{c}{\nearrow O_2R}\\[-2pt]{\searrow O_2R}\end{array} \tag{119}
$$

$$
RO_2H + R'OH \xrightarrow{H^+} RO_2R' \tag{120}
$$

$$
RO_2H + {\searrow \atop \nearrow}C\!=\!C\!\!\begin{array}{c}{\nearrow}\\[-2pt]{\searrow X}\end{array} \longrightarrow \begin{array}{c}{RO_2 \quad H}\\[-2pt]{|\qquad|}\\[-2pt]{-C-C-}\\[-2pt]{|\qquad|}\\[-2pt]{\qquad X}\end{array} \tag{121}
$$

Peroxy esters formed with *p*-nitrobenzoyl chloride (233, 271, 286) and peroxides from reaction with 1-methyl-6,8-dinitro-2-ethoxy-1,2-dihydroquinoline (428) are commonly used as derivatives.

$$\text{(122)}$$

Other compounds that react readily to produce useful derivatives include Ph₃CCl (357), Ph₃COH (298), xanthydrol (298), and dihydropyran (429).

Hydroperoxides can be reduced cleanly to alcohols by the iodide ion (430), stannous chloride (430), leuco methylene blue (431), sodium and alcohol or zinc and acetic acid (94, 299, 302, 304), aluminum isopropoxide (1), lithium aluminum hydride (432), sodium borohydride (110, 433), catalytic hydrogenation (299, 302, 304), sodium sulfite (434), sodium sulfide (76) and organic sulfides (268, 435, 436–438), phosphines (439) and phosphites (266, 440), hydrazine (233), and other amines (420).

The first three reagents of this group are most commonly used for quantitative analysis (441, 442), although iodide has been used preparatively to reduce the —O—OH function without affecting ester (96) or other halogen substituents (137) in the same molecule. Sodium sulfite (166, 224, 225, 277, 443), lithium aluminum hydride (58, 111, 132, 187, 240, 307), triphenyl phosphine (32, 444, 445), and catalytic hydrogenation (45, 105, 157, 159, 170, 269) have shared about equal popularity in recent years for the reduction of hydroperoxides in oxidation mixtures prior to gas-chromatography analysis; the reductions have been demonstrated to proceed with retention of configuration (307, 443, 445).

Triphenyl phosphine is perhaps easiest to use because of its solubility in organic liquids (439), though there is some evidence that allylic hydroperoxides may not be reduced quantitatively (32).

Hydrogenation with poisoned palladium on calcium carbonate catalyst (446) reduces allylic hydroperoxides without affecting the double bonds (45, 447), but it has yielded a substantial amount of ketone as well as alcohol in some cases (45).

$$\text{(123)}$$

Nickon and Bagli (227) have reported that Raney nickel gives ketone as the major product.

Reactions that cleave the O—O bond of hydroperoxides are complex mechanistically. The above reductions can be formulated as nucleophilic attack at the oxygen with the lower electron density:

$$B^- \quad O—O—R \quad \longrightarrow \quad BOH + {}^-OR \qquad (124)$$

$$\overset{|}{H}$$

Some of those that occur with a well-defined stoichiometry (iodide, dialkyl sulfides, phosphines and phosphites, metal hydrides) probably proceed this way. Others, such as reactions with amines (Section IV-B-3), some sulfur compounds (Section IV-B-2), phosphines and phosphites at high temperatures (Section IV-B-1) and metal ions of variable valence (Section V-D-1) appear to involve free radicals at some stage.

Nucleophiles that are also strong bases can dehydrate nontertiary hydroperoxides:

$$B^- \quad H—\overset{|}{\underset{|}{C}}— \quad \longrightarrow \quad BH + \overset{\diagdown \diagup}{\underset{\|}{C}} + OH^- \qquad (125)$$

$$O—O—H \qquad\qquad O$$

However, in most reactions of this type alcohols and C—C cleavage products are generated as well (Section IV-B-4).

Hydroperoxides are also attacked by electrophiles, most notably by $H^+$ (Section IV-A-3). The subsequent rearrangement forms the basis of the commercially important production of phenol from cumene.

$$
\begin{array}{ccccc}
\overset{Me}{\underset{|}{\underset{Me}{\overset{|}{Ph—CH}}}} & \xrightarrow{O_2} & \overset{Me}{\underset{|}{\underset{Me}{\overset{|}{PhCO_2H}}}} & \xrightarrow{H^+} & \overset{Me}{\underset{|}{\underset{O—OH_2}{\overset{|}{Ph—C—Me}}}} \\
\end{array} \qquad (126)
$$

$$Me\overset{+}{C}—O—Ph \xrightarrow{H_2O} Me_2CO + PhOH$$

*t*-Butyl hydroperoxide and primary and secondary hydroperoxides without aryl groups do not rearrange as easily and are relatively stable to acids.

The most significant practical aspect of hydroperoxide chemistry is the very facile homolytic cleavage of the O—O bond under the influence of various other substances.

$$RO_2H + A \quad \longrightarrow \quad RO\cdot + [AOH]\cdot \qquad (127)$$

(Homolytic cleavage of the O—OH bond, unless assisted in some way, is very slow at temperatures under 150°.) Thus mixtures of hydroperoxides with transition-metal salts, metal alkyls, amine salts, sulfur compounds, and the like are industrially important initiators for free-radical polymerizations and autoxidations. [The extensive patent literature is well reviewed by Mageli and Sheppard (4) and List and Kuhnen (22).] Hydroperoxides generated in, or added to, autoxidizing mixtures produce radicals and initiate chains much faster than is to be expected from a purely unimolecular process. Free-radical generation by hydroperoxide dimers and complexes with alcohols, carboxylic acids, aralkanes, alkenes, and the like has been the subject of much research by Denisov, Maizus, Emanuel, and other Soviet investigators (10).

Hydroperoxides are usually not efficient initiators; a fair fraction is wasted by induced free-radical-chain decomposition:

$$RO_2H + RO \cdot \quad (or\ R' \cdot) \quad \longrightarrow \quad RO_2 \cdot + ROH \quad (or\ R'H) \quad (128)$$

$$2RO_2 \cdot \quad \longrightarrow \quad 2RO \cdot + O_2 \tag{129}$$

The chain can be set up by any free radical or by metal ions of variable valence:

$$RO_2H + Co^{2+} \quad \longrightarrow \quad RO \cdot + [Co^{3+}OH^-] \tag{130}$$

$$[Co^{3+}OH^-] + RO_2H \quad \longrightarrow \quad Co^{2+} + RO_2 \cdot + H_2O \tag{131}$$

The reaction of metal ions with hydroperoxides usually produces oxygen, [the basis of the lead tetraacetate test for $-O-OH$ (398).] The other products, which include alcohols, carbonyl compounds and dialkyl peroxides, are difficult to predict in any quantitative fashion because of the number of possible subsequent competing reactions.

Complexes of a very few metal ions (vanadium, molybdenum, or tungsten) catalyze a very different reaction of hydroperoxides with olefins, sulfoxides, or tertiary amines. The products—epoxides, sulfones, or amine oxides, respectively—suggest that the metal complex is able to stabilize an incipient alkoxide ion in some way, giving the hydroperoxide properties of a peroxy acid.

$$RO_2H + \underset{/}{\overset{\backslash}{C}}{=}\underset{\backslash}{\overset{/}{C}} \xrightarrow{\text{metal ion}} ROH + \underset{/}{\overset{\backslash}{C}}\overset{\overset{\displaystyle O}{\overbrace{\quad}}}{-}\underset{\backslash}{\overset{/}{C}}$$

$$R_2SO \xrightarrow{\text{ROOH + metal ion}} R_2SO_2 \tag{132}$$

$$R_3N \xrightarrow{\text{ROOH + metal ion}} R_3N \rightarrow O$$

## C. Separation from Reaction Mixtures and Purification

The synthesis of hydroperoxides offers unique problems in separation and purification since conversions are frequently low and the products thermally unstable and sensitive to contaminants of many sorts. Many liquid hydroperoxides can be distilled at low pressures (Appendix); this is generally adequate for simple compounds with fewer than seven or eight carbons, but the inefficiency of distillation columns at low pressures does not permit much purification of the higher boiling ones that must be distilled at 1 torr or less to avoid thermal decomposition. Solid hydroperoxides can be recrystallized, usually from petroleum ether but this again may be ineffective if the solid contains very much of the corresponding alcohol or carbonyl compound as impurity.

Frequently hydroperoxides can be preferentially extracted from autoxidation mixtures with 80–90% aqueous methanol or ethanol (59, 126, 150, 151, 448–450). The method has been most used in lipid autoxidation where further purification of the hydroperoxide extract can be effected by countercurrent distribution methods (126, 150, 151, 449).

More commonly extractions are made with dilute or concentrated aqueous sodium hydroxide. Concentrated sodium hydroxide (30–40%) usually precipitates the sodium salt of the hydroperoxide, which can be washed with organic solvents and dried. Addition of dilute mineral acid or carbon dioxide to an aqueous solution or suspension of the salt at low temperatures regenerates the hydroperoxide. Typical are purifications of α-cumyl (61, 451) and 2-phenyl-2-butyl hydroperoxides (286, 452).

Though more susceptible to base-catalyzed decomposition than the tertiaries, primary (210) and secondary (210, 453) alkyl hydroperoxides have been extracted and purified in this manner.

The reaction of concentrated sodium hydroxide with most hydroperoxides is exothermic [though high-molecular-weight compounds, such as tertiary tetradecyl hydroperoxides, do not react at all (309)] and tends to produce some decomposition even when the solutions are kept well chilled (451, 454). The pure sodium salts of cumene and tetralin hydroperoxides can be prepared by reaction with sodium amide in toluene (451, 454) and sodium hydride in ether (455), respectively. Purification of hydroperoxides via their barium salts is also possible (456).

Alkylidene (159, 164, 255) and benzyl (322, 457) hydroperoxides, which suffer extensive decomposition on contact with concentrated base, can be extracted with 2% aqueous sodium hydroxide and regenerated with carbon dioxide or sodium bicarbonate. Simple allylic hydroperoxides appear to be unable to withstand even this mild treatment. Two percent sodium hydroxide has also been used to extract dihydroperoxides from mixtures with monohydroperoxides (106).

Chromatography on a preparative scale has rarely been used for hydroperoxide purification, though there are numerous references to analytical procedures. Milas and Golubovic (347, 348) have separated crystallizable amounts of individual peroxides from the mixtures produced from hydrogen peroxide plus ketones (Section II-C-4) by using column chromatography on cellulose powder; their procedure was adopted for purification of 2,4,6-trimethylbenzyl hydroperoxide (217).

Column chromatography on more strongly absorbing stationary phases seems limited to hydroperoxides with large nonpolar groups. Hydroperoxides of methyl oleate have been chromatographed on silicic acid (458), and alumina has been used to purify 2-alkyltetradec-2-yl hydroperoxides (309) and 5-keto-2-methyl-3-nonen-2-yl hydroperoxides (175), eluting with 2–4% acetic acid in pentane and hexane–ether, respectively.

Ordinarily chromatography on alumina is an excellent way to rid solvents of unwanted peroxidic contaminants.

Brill (146) has successfully separated usable amounts of isomeric allylic hydroperoxides by gas chromatography.

### D.   Analysis of Hydroperoxides

Fairly up-to-date treatments of quantitative peroxide analysis appear in chapters by Martin (441) and Siggia (442). (See also Chapter 6 of this volume.) This section deals briefly with methods applicable to determination of hydroperoxides.

#### 1.   Iodometric Methods

The iodide ion reacts stiochiometrically with hydroperoxides in acidic solution, producing iodine, which can be titrated with thiosulfate (459) or estimated colorimetrically (460, 461). Hydrogen peroxide, peroxy acids, diacyl peroxides, and peroxy esters react similarly (441). However, these types of compound can be distinguished from each other and from hydroperoxides by their relative rates of reaction under carefully controlled conditions (462). Dialkyl peroxides ordinarily do not react unless the medium is strongly acid (463) or contains trace amounts of iron or copper (444); heating is frequently necessary.

Although there are probably as many variations of the iodide titration as there are investigators in the field, the most convenient and commonly used are the "reflux" method (463–465) and that of Wibaut, Van Leeuwen, and Van der Wal (466). The latter employs potassium iodide in acetic acid at room temperature under a carbon dioxide atmosphere. It is somewhat more discriminating than the reflux method in that it does not titrate hydroperoxide–carbonyl adducts [peroxyhemiacetals (64)]. The "reflux" method uses sodium iodide (for greater solubility in isopropanol–acetic acid) and has the advantage of speed and lack of need for an inert atmosphere, oxygen being expelled by solvent vapor.

Tetra-n-butylammonium iodide appears to have some advantages in colorimetric determinations (461). Accuracy at 0.01 $\mu$mole/l of hydroperoxide is claimed, whereas the practical limit for titrimetric methods is about 0.0001 mole/l (461). In all cases a large excess of iodide is advisable to complex iodine as $I_3^-$ and prevent losses by volatilization or reaction with double bonds (463, 464).

## 2. Other Volumetric or Colorimetric Methods

Reduction of hydroperoxides with stannous chloride and titration of the excess stannous ion with the ferric ion or iodine solution has been used for analysis where the presence of sulfur compounds might interfere with iodide reduction (430). The method is comparable in precision to the latter, but it is lengthy and requires scrupulous purging with nitrogen to avoid high results.

Oxidation of the leucobase of methylene blue is a very sensitive method for determining hydroperoxides, ketone, and some polymeric peroxides (431, 467).

$$\text{Me}_2\text{N} \underset{\text{H}}{\overset{\text{S}}{\bigcirc}} \text{NMe}_2$$

$$\downarrow {}^{\text{H}^+ \mid \text{RO}_2\text{H}}$$

$$\text{Me}_2\text{N} \overset{\text{S}}{\bigcirc} \overset{+}{\text{N}}\text{Me}_2 + \text{ROH} + \text{H}_2\text{O} \tag{133}$$

$$\lambda_{\max} = 643 \text{ m}_\mu$$

The procedure of Sorge and Überreiter (431), which requires protection from both oxygen and light, is described in detail by Martin (441). Eiss and Giesicke (468) have reported an improvement employing benzoyl leuco methylene blue and zirconium naphthenate to speed up the reaction. [Without the additive, full-color development with $t$-BuO$_2$H takes 36 hr at room temperature (468).] It is interesting that whereas naphthenates of lead, zirconium, iron, and copper also accelerate the oxidation, cobalt naphthenate does not (468).

The mechanism of the reaction (in the absence of metal ions) has been investigated recently by Hrabak and Horacek (469), who suggest that the leucobase assists homolytic cleavage of the O—O bond:

$$\text{(structure: 4-NMe}_2\text{-substituted N-methylaniline with NH)} \quad + \; R\text{-O-O-R} \tag{134}$$

$$\downarrow$$

$$\text{(structure: radical cation } \overset{\cdot+}{N}Me_2\text{)} \quad + \; RO\cdot \; + \; RO^- \tag{135}$$

$$\text{(structure: } \overset{\cdot+}{N}Me_2 \text{ radical cation with NH)} \; + \; RO^- \longrightarrow \text{(structure: } \overset{\cdot+}{N}Me_2\text{)} \; + \; ROH \tag{136}$$

$$\text{(structure: } \overset{\cdot+}{N}Me_2\text{)} \; + \; RO\cdot \longrightarrow \text{(structure: quinonoid } \overset{+}{N}Me_2\text{)} \; + \; RO^-$$

The cupric-ion-catalyzed oxidation of phenolphthalin to phenolphthalein by peroxides is another colorimetric method of high sensitivity (441, 470, 471). However, only hydrogen peroxide reacts rapidly, and reaction with other peroxides may depend on their hydrolysis to hydrogen peroxide (441).

An analysis that certainly depends on this hydrolysis (472) is the formation of a titanium peroxysulfate complex by heating hydroperoxides with acidified titanium sulfate (472–475) or tetrachloride (476). The complex may be determined directly [$\lambda_{max}$ 407–410 m$\mu$ (472)] or further complexed with 8-quinolinol ($\lambda_{max}$ 450 m$\mu$) (474) for a threefold to fourfold increase in sensitivity. The method is best suited for minute amounts of peroxides formed in smog (473) or solvents exposed to air (472, 475). Tertiary (472) and ether (475) hydroperoxides react quantitatively.

At room temperature the reagent is quite specific for hydrogen peroxide [hydroperoxides and other peroxidic materials are not hydrolyzed fast enough to interfere (474)] and it has been used to remove hydrogen peroxide from mixtures of peroxides before polarographic analysis (477).

The oxidation of the ferrous ion to the ferric ion in the presence of the thiocyanate ion offers a rapid and sensitive technique for the detection of peroxides, a technique that has enjoyed a certain vogue as an analytical method (441, 442). Quantitative results apparently can be obtained under precisely controlled conditions (473, 478), but the approximate accuracy often noted results from a fortuitous canceling out of radical-chain decomposition of the peroxide induced by the ferrous ion and the oxidation of the ferrous ion by atmospheric oxygen (441, 442).

The reaction of hydroperoxides with diphenylpicrylhydrazyl is another complex reaction (479–481) on which an analytical method is based (480). Exactly what happens to either reactant is uncertain, though the picrylhydrazyl color [$\lambda_{max}$ 517 m$\mu$ (480)] disappears and the formation of diphenylnitroxide radical ions can be detected by ESR spectroscopy (479–482). The stiochiometry is variable and dependent on concentrations (479, 480) but at $10^{-8}$–$10^{-9}$ $M$ $RO_2H$ the disappearance of $10^{-7}$–$10^{-8}$ $M$ picrylhydrazyl, and formation of stable free radicals, is a first-order reaction with $k_1$ directly proportional to $[RO_2H]_0$ (480). The analytical procedure has been applied for the determination of oleate (480) hydroperoxides.

Other reactions that have been used for quantitative analysis of hydroperoxides include reduction by zinc and acetic acid (483) (titration of the water produced); reduction by lithium aluminum hydride (484) (measurement of hydrogen evolved); reductions by nitrite (485), bisulfite (486), sulfur dioxide (487), and triphenylphosphine (488–491) or phosphite (492). [Dulog and Burg (488) compare several gravimetric methods for the estimation of excess triphenyl phosphine; alternatively the decrease in absorption at 275 m$\mu$ can be followed (488)]. Horner and Jürgens (490) have described a procedure for analyzing mixtures of hydroperoxides, dialkyl peroxides, diacyl peroxides, and peroxy acids using reactions with iodide, diphenyl sulfide, triphenylarsine, and triphenylphosphine.

### 3. Polarographic Methods

Polarographic reduction is an excellent method for the quantitative analysis of peroxides of any type especially on a microscale [only $t$-$Bu_2O_2$ does not react (441)]. The method is amply described by Bruschweiler and Minkoff (493) and by Martin (441), and has been the subject of several more recent publications (491, 494–497), all of which give extensive tables of half-wave potentials for hydroperoxides and other peroxidic compounds. Despite the apparent enthusiasm of these workers, the technique has been used rather infrequently in hydroperoxide research, probably because of unavailability of, or unfamiliarity with, the necessary equipment. Analysis by iodide or triphenylphosphine is equally accurate (491) and just as easy.

Moreover, the expectation that individual hydroperoxides or peroxides in mixtures could be differentiated and analyzed by discrete steps in the polarographic wave has not been realized (441). The waves tend to be long and drawn out due to the nonreversibility of the reduction; a separation of about 0.4 volt in half-wave potentials is necessary before discrete steps become apparent (441). Somewhat more upright waves can be obtained by the addition of ethyl cellulose (496). The diffusion current is proportional to peroxide concentration over a $10^{-2}$–$10^{-6}$ $M$ range (496), but since the proportionality constant varies with structure, individual calibration is necessary for accurate work (441, 493).

Half-wave potentials vary with solvent; for example, for cumene hydroperoxide $E_{1/2}$ values are $+0.2$, $-0.54$, and $-0.82$ volt in 20% ethanol, 95% ethanol, and 1:1 methanol–benzene respectively (495).

In the latter solvent half-wave potentials for most hydroperoxides fall in the range of $-0.8$ to $-0.95$ volt, with that of hydrogen peroxide being $-1.14$ volts; those of dialkyl peroxides, $-1.7$ to $-1.9$ volts; and those of diacyl peroxides about 0.0 volt (495). In aqueous solution the spread between individual hydroperoxides is greater; $E_{1/2}$ values for $MeO_2H$, $EtO_2H$, $i$-$PrO_2H$, and $t$-$BuO_2H$ are reported as $-0.791$, $-0.444$, $-0.350$, and $-0.330$ volt, respectively (497).

Decomposition before polarographic analysis offers a possibility for quantitative differentiation between hydroperoxides in mixtures that probably deserves more exploration. Boardman (63) has reacted a mixture of $p$-methylcumyl hydroperoxide and $p$-isopropylbenzyl hydroperoxide, produced by the autoxidation of $p$-cymene, with ferrous sulfate. The ketone produced from the first and the aldehyde from the second were easily distinguished polarographically.

### 4. Chromatographic Analysis

A number of papers on the separation of peroxide mixtures by paper (347, 498–502) or thin-layer (503–506) chromatography have appeared in the last 10 years. All tabulate $R_f$ values for 5–25 peroxides and hydroperoxides with various adsorbents and solvents; Martin (441) describes the paper-chromatography methods of Abraham (498), Milas (347, 501), and Cartlidge (500) and their co-workers.

Paper has been used untreated (347, 501) or treated with ethylene glycol (500, 502), silicone oil (498, 502), or acetic anhydride (499) before use. Elution generally requires solvent mixtures of more or less polarity, depending on the treatment of the paper; typical are dimethylformamide–decalin (501), methanol–water (502), water–ethanol–chloroform (498), and acetic acid–dioxane–water (499). Chromatography on thin layers of alumina (506), silica (503–505), or Kieselgel (503, 504) has generally utilized toluene mixtures with carbon tetrachloride, methanol, or acetic acid.

Both types of chromatograms can be developed with acidified potassium iodide (347, 499, 501), $Fe(SCN)_2$ (498, 501), $p$-aminodimethylaniline hydrochloride (347, 499, 501), or antimony pentachloride (507). Analyses are usually qualitative, but Buxbaum (169) has obtained quantitative results good to $\pm 5\%$ by scraping previously located spots into acetic acid–potassium iodide solution and titrating the iodine produced.

Separation of a wide range of peroxide types is possible by these methods, although separation of those with closely related structures does not seem to be good. Illustrative are $R_f$ values of some aralkyl hydroperoxides chromatographed separately on silicone-treated paper with water–ethanol–chloroform (498): 1-phenylethyl, 0.86; 1-phenylpropyl, 0.80; 2-phenyl-2-propyl, 0.85; 2-phenyl-2-butyl, 0.75; α-tetralyl, 0.78.

The method has been most useful for analyzing mixtures of peroxides produced by the reaction of hydrogen peroxide with ketones (169, 347, 348, 350), for demonstrating that synthesized hydroperoxides are free of $H_2O_2$ or peroxides (158), and for following the course of reactions that have peroxides or hydroperoxides as starting materials (315, 507), intermediates (257), or products (315, 350, 358).

Gas-chromatographic analysis of oxidation mixtures after the hydroperoxides and peroxides have been reduced to alcohols (Section III-B-2) is now commonplace. Chromatographing the hydroperoxides themselves is somewhat more rare, owing to their sensitivity to heat and surfaces. From the literature it is clear that simple alkyl or alkenyl hydroperoxides with fewer than seven or eight carbons can be gas-chromatographed (109, 146, 216, 218, 219, 292, 498, 508–510); even cumene and ethylbenzene hydroperoxides are reported to be stable on treated glass beads (511) or on tricrescyl phosphate–sodium chloride (512) at 60° for short times. But, apart from the literature, it is also clear that the techniques are highly personal and subject to inexplicable difficulties.

The most comprehensive studies of optimum conditions are those of Bukata (509), Cullis (218), Öhlmann (109, 510), and their co-workers. The best columns appear to be glass (218) or Teflon (510) packed with 10–20% dinonyl or diisodecyl phthalate on Chromosorb W (109, 510), Embacel (218), or Fluoropack (146), operated at 60–80° with a rapid rate of gas flow. Cullis and Fersht (218) found it profitable to work at reduced pressure, obtained by pumping on the outlet. Even so, retention times were as long as 30 min.

Isomeric butyl (216, 219, 498, 508) and amyl (218) hydroperoxides have been separated, as well as mixtures of *t*-butyl and *t*-amyl hydroperoxides with *t*-butyl peroxy esters and peroxides (509). Öhlmann and Schirmer (109), using a polarographic cell as the detector, have analyzed hydroperoxides from heptane autoxidation.

It is difficult to say just how accurate such analyses are. Most published chromatographs appear to show some decomposition, and more thorough investigations of the technique can be expected. (As for example, T. E. Healey and P. Urone, *Anal. Chem.*, **41**, 1777 (1969) (note added in proof)).

## IV.  REACTIONS OF HYDROPEROXIDES AS
## NUCLEOPHILES AND AS ELECTROPHILES

Reactions of hydroperoxides are difficult to organize according to mechanistic type; some may start as polar reactions but end as free-radical ones and vice versa. Arbitrarily, this section covers reactions in which the *initial* bond formation or cleavage occurs by electrons moving in pairs and the subsequent steps, if any, are ionic or at least not free-radical steps.

### A.  Hydroperoxides or Their Anions as Nucleophiles

Hydroperoxides and hydroperoxide anions are generally as good or better nucleophiles than water or the hydroxide ion, respectively. Most of the quantitative comparisons available are for the hydrogen peroxide anion $OOH^-$ versus $OH^-$ (5, 12, 443, 513, 514); however, Bunton (443) estimates that relative reactivities for nucleophilic attack on acetic anhydride are as follows: water, 1; *t*-butyl hydroperoxide, 15; hydrogen peroxide, 35.

$$t\text{-}Bu\!-\!O\!-\!\underset{Me}{\overset{H}{\underset{|}{O}}}\ \ \overset{O}{\underset{Me}{C}}\ \ \overset{O}{C}\ \ O \longrightarrow t\text{-}BuO_2\!-\!\overset{O}{\overset{\|}{C}}Me + HOAc$$

$$(137)$$

Attack on *p*-nitrophenyl acetate by $O_2H^-$ or $O_2Me^-$ is $10^3$–$10^4$ times faster than attack by $OH^-$ (515).

The greater nucleophilicity for the hydroperoxide anion relative to $OH^-$ has been explained in two ways: (a) $OOR^-$ is more easily polarizable than $OH^-$ due to the unshared electrons on the neighboring oxygen (443) [the α-effect (514)]; (b) $OH^-$ is an unusually weak nucleophile because of its strong solvation in water (443). Decreased steric crowding at the reaction site must also be a factor in any comparison of $t$-$BuO_2^-$ and $t$-$BuO^-$.

The practical outcome of the high nucleophilicity of hydroperoxides and their anions is the facile synthesis of peroxides and peroxy esters by displacement or addition reactions. These reactions are quite analogous to the synthesis of hydroperoxides with hydrogen peroxide (Section II-C) and so is dicussed only briefly here.

### 1.  *Peroxides by Nucleophilic Additions or Substitutions*

Dialkyl peroxides can be formed in good yield either by direct displacement or by prior formation of a carbonium ion. Examples of the first type include reactions of sodium or potassium hydroperoxide with alkyl halides (357, 516, 517), sulfates (516) or sulfonates (295), and epoxides (518).

$$RO-O^- \quad \overset{|}{\underset{|}{C}}\text{—Br} \quad \longrightarrow \quad R-O-O-C\overset{\nearrow}{\underset{\searrow}{}} \tag{138}$$

$$RO-O^- \quad \underset{\text{H}_2\text{C—CH—Me}}{\overset{\displaystyle\overset{O}{\diagup}}{}} \quad \longrightarrow \quad R-O-O-CH_2-\overset{\overset{\textstyle O^-}{|}}{C}H-Me \tag{139}$$

The displacement of halide has been shown to proceed with inversion (517).

Syntheses via carbonium-ion intermediates have utilized triaryl halides (519), alcohols (298, 520), ethers (302), or alkenes (298).

$$\left.\begin{array}{l} Ph_3CCl \\[4pt] PhMe_2COH \quad \scriptstyle H^+ \\[4pt] Ph_2CHOMe \quad \scriptstyle H^+ \\[4pt] Me_2C{=}CH_2 \end{array}\right\} \xrightarrow{} R_1R_2R_3C^+ \xrightarrow{\text{R'O}_2\text{H}} R_1R_2R_3C-O-OR + H^+ \tag{140}$$

The reaction with diazomethane, while sometimes unproductive (521), has yielded peroxides in at least one instance (522).

$$RO_2H + CH_2N_2 \quad \longrightarrow \quad RO_2Me + N_2 \tag{141}$$

A number of cyclic peroxides have been prepared by acid-catalyzed addition of dihydroperoxides to dienes (523).

$$\tag{142}$$

$(A = CH_2-CH_2, CH{=}CH, or C{\equiv}C)$

Antonovskii and Emelin (524) have studied the kinetics of acid-catalyzed addition of $t\text{-BuO}_2\text{H}$ to isobutylene and $\alpha$-methylstyrene at 0–50°. Rates were first order in both hydroperoxide and olefin, and intermediate between first and second order in sulfuric acid. The kinetics were somewhat marred by acid-catalyzed decomposition of the hydroperoxide at the higher temperatures.

Alkylidene peroxides are very readily synthesized by this route, the $\alpha$-oxygen or nitrogen helping to stabilize the carbonium ion. The following are merely illustrative of the many examples provided by Rieche (159, 318, 320, 325), Milas (323, 343, 347, 348) and others (327, 339, 525, 526).

$$CH_2=CH-OEt + t\text{-}BuO_2H \xrightarrow{H^+} \underset{\displaystyle CH_3-\overset{\displaystyle O_2\,t\text{-}Bu}{\overset{|}{CH}}-OEt}{} \tag{143}$$

$$PrCH(OEt)_2 + PhCMeO_2H \xrightarrow{H^+} PrCH\begin{array}{l} {}^{\displaystyle OEt} \\ {}_{\displaystyle O_2CMe_2Ph} \end{array} \tag{144}$$

$$\text{(145)}$$

$$CH_2=CH-OEt + t\text{-}BuO_2H \xrightarrow{t\text{-}BuOCl} \underset{\displaystyle ClCH_2-\overset{\displaystyle O_2\,t\text{-}Bu}{\overset{|}{CH}}-OEt}{} \tag{146}$$

$$\text{(147)}$$

Nucleophilic attack by hydroperoxides on the carbonyl groups of aldehydes is quite facile even under neutral conditions (332, 413), though it can be catalyzed by acid (527).

$$t\text{-}BuO_2H + MeCHO \longrightarrow t\text{-}BuO_2\overset{\displaystyle OH}{\overset{|}{CHMe}} \tag{148}$$

The reactivity of several hydroperoxides toward acetaldehyde has been found to decrease in the order α-tetralyl $O_2H$ > $t\text{-}BuO_2H$ > α-cumyl $O_2H$ > $Ph_2MeCO_2H$ (413). A number of the peroxyhemiacetal products isolated are listed by Davies (1).

Simple peroxyhemiketals appear to be less common, although many more complex ones of the alkylidene type are known (339, 347, 348).

$$\text{(149)}$$

The products of hydroperoxide addition to keto acids have also been isolated (528).

$$\text{(150)}$$

Hydroperoxides form *gem*-diperoxides with ketones quite readily under acidic conditions (1, 345, 529, 530).

$$Me_2C{=}O + t\text{-}BuO_2H \; \rightleftharpoons \; Me_2C\!\!\begin{array}{c} \diagup OH \\[2pt] \diagdown O_2 t\text{-}Bu \end{array} \xrightarrow[t\text{-}BuO_2H]{H^+} Me_2C(O_2 t\text{-}Bu)_2$$

$$(151)$$

Aldehydes and hydroperoxides may react similarly with acid catalysis (332, 526).

Hydroperoxides react with acyl chlorides or anhydrides to produce peroxy esters. The free acid generated is neutralized by aqueous sodium hydroxide (531) or pyridine (532, 533), or removed on a rotary evaporator (534, 535).

$$t\text{-}BuO_2H + R\overset{\displaystyle O}{\overset{\|}{C}}Cl \longrightarrow t\text{-}BuO_2\overset{\displaystyle O}{\overset{\|}{C}}R + HCl \qquad (152)$$

Peroxy esters are not always obtained intact but may undergo an *in situ* Criegee-type (533) rearrangement (59, 536).

$$(153)$$

Secondary hydroperoxides, which are particularly sensitive to base, may dehydrate (536).

$$(154)$$

The kinetics of peroxy ester formation from anhydrides and hydroperoxides have recently been investigated by Antonovskii and Lyashenko (537, 538) and shown to be similar to analogous reactions of alcohols.

Peroxy esters have also been synthesized by nucleophilic addition to ketenes (539) and to isocyanates (540–543).

$$t\text{-}BuO_2H + PhNCO \longrightarrow t\text{-}BuO_2\overset{\displaystyle O}{\overset{\|}{C}}{-}NH{-}Ph \qquad (155)$$

Peroxysulfonates have been synthesized by the reaction of $t$-BuO$_2$H with sulfonyl chlorides (544, 545); a peroxysulfonate might also be expected from reaction of $t$-BuO$_2$H with benzenesulfonylazide, but instead both reactants decompose on contact (546).

Nucleophilic displacement by $RO_2H$ or $RO_2^-$ is used to produce organometallic peroxides (1, 361, 547).

$$RO_2H + R'_nMX \longrightarrow R'_nMO_2R + HX \tag{156}$$

$$(X = \text{halide, OH, OR, or } NH_2)$$

## 2. Nucleophilic Addition Followed by O—O Heterolysis. Epoxide Formation and the Baeyer–Villiger Reaction

Perhaps the simplest of the many ways in which hydroperoxides can expoxi-
dize olefins (Sections II-A-3, IV-B-5, and V-C-5) is via Michael addition to
an $\alpha$, $\beta$-unsaturated aldehyde or ketone, followed by internal carbanion
displacement on the O—O bond (548, 549). The suggested mechanism (548)
is similar to that proposed for the base-catalyzed epoxidation by hydrogen
peroxide (550).

$$\tag{157}$$

$$\tag{158}$$

Good yields of epoxide are obtained with mesityl oxide, methyl isopropenyl
ketone, chalcone, cyclohexenone (548), and cinnamaldehyde (549), using
Triton-B (548) or dilute sodium hydroxide (549) as the base catalyst. Phorone
does not react, presumably because of steric hindrance (548).

Similar addition to $\alpha,\beta$-unsaturated esters and nitriles occurs, but here
protonation of the carbanion appears to be faster than its attack on O—O,
and the products are dialkyl peroxides (548, 549, 551).

The oxidation of ketones to esters by peroxy acids (the Baeyer–Villiger
reaction) is not simulated with any facility by hydroperoxides (1, 5, 12, 13).
However, Maruyama has studied a base-catalyzed oxidation of aldehydes
(552) and of aryl ketones (553) by $t$-butyl or $\alpha$-cumyl hydroperoxides which
gives products of the Baeyer–Villiger type.

$$\tag{159}$$

Unlike the Baeyer–Villiger reaction, alkyl cleavage is preferred to aryl
(benzophenone does not react), esters are not isolated, and a $\rho\sigma$ plot for

substituted acetophenones gives a positive $\rho$ value. Alternative mechanisms are suggested (553).

$$\text{Ph}-\overset{\overset{\displaystyle O^-}{|}}{\underset{\underset{\displaystyle O \diagdown O}{|}}{C}}-\overset{|}{C}\diagdown^-\text{OH} \tag{160}$$
$$\hspace{6cm} \text{R}$$

The base-catalyzed cleavage of α-diketones and α-keto acids by hydrogen peroxide or hydroperoxides again clearly involves such intermediates (12, 13, 554) with several possibilities open for rearrangement mechanisms.

$$\text{Ph}-\overset{\overset{\displaystyle O^-}{|}}{\underset{\underset{\displaystyle O-O-\textit{t}-Bu}{|}}{C}}-\overset{\overset{\displaystyle O}{\|}}{C}-\text{Ph} \longrightarrow \text{Ph}-\overset{\overset{\displaystyle O}{\|}}{\underset{\underset{\displaystyle O^-}{|}}{C}} + \text{Ph}-\overset{\overset{\displaystyle O}{\|}}{\underset{\underset{\displaystyle O-\textit{t}-Bu}{|}}{C}} \tag{161}$$

The use of oxygen−18 isotopic labeling has not provided any definite solution (13, 554).

### 3. Acid-Catalyzed Heterolysis

Electrophilic attack by a proton may occur at either oxygen of a hydroperoxide.

$$\text{RO}_2\text{H} + \text{H}^+ \quad \overset{\displaystyle \nearrow}{\underset{\displaystyle \searrow}{\phantom{x}}} \quad \begin{array}{c} \overset{\displaystyle H}{|} \\ R-\overset{+}{O}-OH \\[1em] \overset{\displaystyle H}{|} \\ R-O-\overset{+}{O}-H \end{array} \tag{162}$$

The net result of the first possibility may be loss of hydrogen peroxide from those hydroperoxides whose R groups are sufficiently electron releasing (555).

$$\underset{\underset{+}{}}{\overset{\overset{\displaystyle Et \ \ H}{| \ \ \ |}}{\text{Ph}_2\text{C}-\text{O}-\text{OH}}} \rightleftharpoons \overset{\overset{\displaystyle Et}{|}}{\underset{+}{\text{Ph}_2\text{C}}} + \text{HO}_2\text{H} \tag{163}$$

whereas protonation of the hydroxylic oxygen leads to O−O heterolysis accompanied by rearrangement.

$$\underset{+}{\overset{\overset{\displaystyle Et \ \ \ \ H}{| \ \ \ \ \ |}}{\text{Ph}_2\text{C}-\text{O}-\text{OH}}} \longrightarrow \overset{\overset{\displaystyle Et}{|}}{\underset{+}{\text{Ph}-\text{C}-\text{O}-\text{Ph}}} + \text{H}_2\text{O} \tag{164}$$

Cleavage to hydrogen peroxide and a carbonium ion frequently tends to be obscured by the O—O heterolysis, which, unlike the former, is non-reversible. However, the equilibrium can often be forced to the right under mildly acidic conditions by trapping either product as it is formed.

$$
\underset{\substack{|\\ \text{Et}}}{\text{Ph}_2\text{C}^+} + \underset{\substack{|\\ \text{Et}}}{\text{Ph}_2\text{CO}_2\text{H}} \longrightarrow (\underset{\substack{|\\ \text{Et}}}{\text{Ph}_2\text{C}}-\text{O})_2 + \text{H}^+ \qquad (165)
$$

$$
\text{RCH}\overset{\text{O}_2\text{H}}{\underset{\text{OH}}{\diagdown}} \underset{\longleftarrow}{\overset{\text{H}^+}{\rightleftharpoons}} \text{H}_2\text{O}_2 + \overset{+}{\text{RCH}}\underset{\text{OH}}{\diagdown}
$$

$$\Big\downarrow \text{2, 4-DNP} \qquad\qquad (166)$$

$$
\text{RCH}=\text{N}-\text{NH}-\overset{\text{O}_2\text{N}}{\underset{}{\bigcirc}}-\text{NO}_2 + \text{H}^+
$$

$$
t\text{-BuO}_2\text{H} \underset{}{\overset{\text{H}^+}{\rightleftharpoons}} t\text{-Bu}^+ + \text{H}_2\text{O}_2 \xrightarrow{\text{Ti(SO}_4)_2} \text{H}_2\text{O}_2\text{-Ti(SO}_4)_2 \text{ complex}
$$

$$(167)$$

Under conditions in which O—O heterolysis predominates the hydro-peroxide–hydrogen peroxide equilibrium complicates the kinetics (556) and often results in some olefin or dialkyl peroxide being formed as by-products (1, 315, 557–559).

The acid-catalyzed O—O heterolysis of hydroperoxides, like the similar reaction of peroxides and peroxy esters (310, 560–562) depends on the ability of an alkyl or aryl group to undergo a 1,2-shift from carbon to an incipiently positive oxygen.

$$
\text{R}_3\text{C}-\text{O}-\overset{+}{\text{O}}-\text{H}_2 \longrightarrow \text{R}_3\text{C}-\overset{\delta+}{\text{O}}\cdots\overset{\delta+}{\text{OH}_2}
$$

$$
\longrightarrow \text{R}_2\overset{+}{\text{C}}-\text{O}-\text{R} + \text{H}_2\text{O} \quad (168)
$$

Two kinds of observations suggest that the alkyloxonium ion never has a free existence: (a) partial decomposition in oxygen−18 water does not yield residual hydroperoxide containing oxygen−18 (300, 559); (b) studies of relative rates for different R groups show a strong indication of anchimeric assistance in the O—O bond-breaking step (1, 314, 557, 563). Thus the rates of rearrangement of ring-substituted α-cumyl hydroperoxides (557) and benzhydryl hydroperoxides (314) follow $\rho\sigma$ plots with $\rho$ values of −4.5 and

−3.78, respectively. [The case for anchimeric assistance is more fully documented for peroxy ester rearrangements, the latest extensive study being that of Hedaya and Winstein (310).] A transition state with the migrating group partially bonded to both carbon and oxygen is postulated.

$$
\begin{array}{c}
R \quad\quad H \\
| \quad\quad | \\
Ph-C-O-OH \\
| \\
R \\
\end{array}
\quad \longrightarrow \quad
\begin{array}{c}
R \\
| \\
R-C\!-\!O\cdots\overset{\delta+}{O}-H \quad \text{products} \\
\end{array}
\tag{169}
$$

Davies (1) and Leffler (562) provide extensive tables of products from acid-catalyzed rearrangement. From these and more recent studies (Table 7) it appears that relative migratory aptitudes are cyclobutyl > aryl ≫ vinyl > hydrogen > cyclopentyl ≃ cyclohexyl ≫ alkyl. *t*-Butyl hydroperoxide resists rearrangement under the usual conditions of sulfuric or perchloric acid in acetic acid, giving instead isobutylene and hydrogen peroxide (562), but yields some acetone and methanol in the presence of boron trifluoride (564).*

The unusual effect of solvent on the competition between aryl and hydrogen migration (Table 7) has been noted by Anderson and Smith (314) and also by Van Stevenick and Kooyman (557), but no explanation has yet been advanced.

A number of recent kinetic studies of α-cumyl (556, 557, 559, 568–574) or triarylmethyl (556, 558, 574, 575) hydroperoxide decompositions have shown uniformly first-order dependence on hydroperoxide, retardation by added water, and complex dependence on acid concentration. The large solvent effects can be illustrated by pseudo-second-order rate constants for the sulfuric acid–catalyzed decomposition of α-cumyl hydroperoxide at 50° (Table 8).

The results, both kinetic and of relative migratory aptitude, can be rationalized (557) in terms of an acidity-dependent equilibrium between protonated and unprotonated hydroperoxide, followed by a rate-determining migration that is subject to specific solvation effects.

The real situation, however, is somewhat more complex due to a competing equilibrium between protonated hydroperoxide and carbonium ion + $H_2O$ (556, 575) (equation 164). Thus both phenol and *p*-cresol are produced from

---

*In 90–98% $H_2SO_4$ at 25°C, *t*-$BuO_2H$ gives the latter products exclusively (N. C. Deno, W. E. Billups, K. E. Kramer and R. R. Lastomirsky, *J. Org. Chem.*, **35**, 3080 (1970). This paper also reports products from rearrangement of propyl, other butyl, and amyl hydroperoxides and relative migratory aptitudes of Pr, Et, and H in this medium. The latter appear to depend on hydroperoxide structure. Thus for primary hydroperoxides, Pr:H:Et equals 6:2:1, whereas for secondary ones it is 2:3:1 (note added in proof).

TABLE 7.  PRODUCTS FROM ACID-CATALYZED DECOMPOSITION OF HYDROPEROXIDES

| RO₂H | Conditions | Products | Ref. |
|---|---|---|---|
| PhMe₂CO₂H | HClO₄ in 50% EtOH, 50° | PhOH + Me₂CO (100%) | 557 |
| PhCHMe<br>O₂H | HClO₄ in 50% EtOH, 50° | PhOH + MeCHO (65%), AcPh (35%) | 557 |
| Ph₂CHO₂H | H₂SO₄ in HOAc, 20° | PhOH + PhCHO (98%), Ph₂CO (2%) | 314 |
| Ph₂CHO₂H | H₂SO₄ in Et₂O, 20° | PhOH + PhCHO (68%), Ph₂CO (32%) | 314 |
| | H₂SO₄ in Me₂CO, 60° | | 88[a] |
| | | | 565 |
| | H₂SO₄ in HOAc, 20° | PhOH + (95%) | 315, 563 |
| | H₂SO₄ in HOAc, 20° | PhOH + (60–65%); PhC(CH₂)₄OAc (12%) | 315[b] |

| Hydroperoxide | Conditions | Products | Ref. |
|---|---|---|---|
| Me, $O_2H$ cyclohexane | $HClO_4$ in HOAc, 60° | cyclohexanone (0.5%), $HO(CH_2)_5Ac$ (60%) | 60[a] |
| $O_2H$ cyclohexane | $HClO_4$ in HOAc, 60° | cyclohexanone (80%) | 60[a] |
| $CH=CH_2$, $O_2H$ cyclohexene | $H_2SO_4$ in HOAc | cyclohexenone + MeCHO | 566 |
| $O_2H$ cyclohexene | $H_2SO_4$ in HOAc | $\begin{array}{c}CHO\\CHO\end{array}$ ⟶ cyclopentene–CHO | 567[a] |
| $Me(CH_2)_6CH_2O_2H$ | $HClO_4$ in HOAc, 60° | $Me(CH_2)_6CHO$ (62%), $CH_2O$ (0.1%) | 212[a] |
| $Me(CH_2)_5CHMe$ (with $O_2H$) | $HClO_4$ in HOAc, 60° | $Me(CH_2)_5CMe{=}O$ (80%), MeCHO (4%) | 212[a] |

[a] Indicated reference gives the products of the decomposition of several other closely related hydroperoxides.
[b] 1-Phenylcyclobutyl hydroperoxide is unstable under the acidic conditions of preparation but yields decomposition products of ring opening. No phenol is formed (315).

TABLE **8**. SOLVENT EFFECTS ON THE RATE OF THE ACID-CATALYZED DECOMPOSITION OF CUMENE HYDRO-PEROXIDE (559, 568)

| Solvent | $k_2(\times 10^3 \text{ l/mole-sec})^a$ |
|---------|------------------------------------------|
| EtOH–$H_2$O$(1:1)^b$ | 0.07 |
| AcOH–$H_2$O $(1:1)$ | 0.34 |
| Tetrahydrofuran | 1.7 |
| MeOH | 3.3 |
| Dioxane | 24 |
| AcOH | 94 |
| $Me_2$CO | 400 |
| PhOH–$Me_2$CO $(1:1)$ | 420,000 |

[a] At 50°; rate $= k_2[RO_2H][H_2SO_4]$ at $[H_2SO_4] = 3\text{–}7 \times 10^{-5}$ l/mole-sec.
[b] Catalyzed by $HClO_4$.

the acid-catalyzed decomposition of either $Ph_3CO_2H$ or $PhMe_2O_2H$ when the corresponding *p*-tolyl *alcohol* is added.

In the triaryl case Bissing and co-workers (556, 575) found that complete equilibration occurred before the O—O heterolysis.

$$Ph_3CO_2H \underset{H^+}{\overset{H_2O_2}{\rightleftharpoons}} Ph_3C^+ \underset{H^+}{\overset{H_2O}{\rightleftharpoons}} Ph_3COH \qquad (170)$$

$$Ph_2(p\text{-MeC}_6H_4)CO_2H \overset{H_2O}{\rightleftharpoons} Ph_2(p\text{-MeC}_6H_4)C^+$$

$$\overset{H_2O_2}{\rightleftharpoons} Ph_2(p\text{-MeC}_6H_4)COH \qquad (171)$$

The rates of $Ph_3CO_2H$ heterolysis could be duplicated starting from $H_2O_2$–$Ph_3COH$ mixtures, and a satisfactory correlation of acid dependence was obtained with $h_0$.

$$\frac{d\,[PhOH]}{dt} = Kh_0[Ph_3COH][H_2O_2] \qquad (172)$$

The equilibration of cumene hydroperoxide with alcohol $+ H_2O_2$ turned out to be somewhat slower than O—O heterolysis so that no such simple kinetic treatment was applicable (556).

## B.   O—O Heterolysis by Nucleophilic Attack

Edwards (5, 12, 13), in his discussions of nucleophilic displacements on oxygen, makes the point that as the basicity of the leaving anion increases, the rate of reaction of the peroxide should decrease.

$$\ddot{N} \quad O-O^{\diagup R} \longrightarrow \overset{+}{N}-O + \bar{O}-R \qquad (173)$$
$$\underset{R'}{|} \qquad\qquad\qquad \underset{R'}{|}$$

It is not surprising, therefore, that peroxy acids and acyl peroxides, where RO = acyloxy, and even hydrogen peroxide (R = H), should undergo such displacements readily (5, 12, 13), whereas few clean bimolecular displacements on hydroperoxides are known.*

Chief among these are the reactions of hydroperoxides with $I^-$ (1, 576, 577) and with $SO_3^{2-}$ (1), the ions that give the fastest nucleophilic displacements on hydrogen peroxide (12, p. 79; 13, p. 117). Reactions with phosphines and phosphites (Section IV-B-1) and with hydride (from metal hydrides or metals + $H^+$) (1) appear to fit in this category, as may the reactions with amines, sulfoxides, and carbon–carbon double bonds when catalyzed by certain metal complexes (Section IV-B-5). Nonmetal-catalyzed decompositions by sulfur compounds (Section IV-B-2), amines (Section IV-B-3), and double bonds (Section IV-B-5), as well as by strongly basic nucleophiles (Section IV-B-4), are complex and under some conditions yield free-radical intermediates (Section V).

Nucleophilic displacement by oxygen [either from H—O—H(R) or ⁻O—H(R)] on the O—O bond is negligibly slow even for hydrogen peroxide (1, 12). A possible exception is the decomposition by phenols, which Walling and Hodgden (578) have shown to be nonradical for diacyl peroxides. The reaction with $t$-BuO$_2$H is one-thirtieth as fast as with benzoyl peroxide (578) in qualitative conformity at least with the relative base strength of the leaving groups. The products were not reported, however, and there is some indication that phenols and alcohols can induce homolytic cleavage of hydroperoxides (Section V-C).

### 1. Reactions with Phosphines and Phosphites

Trialkyl or aryl phosphines and phosphites reduce hydroperoxides extremely rapidly† (440, 445, 579) and, in most cases, quantitatively (Section III-D-2) to alcohols.

$$PX_3 + RO_2H \longrightarrow X_3PO + ROH \qquad (174)$$

$$(X = Ar, OAr, R, \text{ or } OR)$$

---

*It is not entirely clear whether the paucity of information is due to lack of reactivity or lack of interest, nor is it clear why attack of nucleophiles must be on the oxygen atom attached to hydrogen. Bunton (12) and Davies (1) ascribe this to lower electron density; Edwards (12), to decreased steric hindrance.

†Phosphites are reported to reduce hydroperoxides "instantly" at $-40°$ in pentane solution (445).

Optically active 1-phenylethyl hydroperoxide is reduced by triphenylphosphine with complete retention of configuration (304, 305); cumene hydroperoxide and *trans*-9-decalyl hydroperoxide react with triphenyl and tri-*n*-butyl phosphines, respectively, in mixtures of ethanol and $^{18}$O water to give alcohols and phosphates containing no $^{18}$O (445); no ethers are produced, nor are alcohols found corresponding to hydrolysis of the phosphite (445).

The findings support nucleophilic attack by phosphorus on the oxygen attached to hydrogen (445) since the alternative mechanism,

$$R_3'P \quad O{-}OH \longrightarrow [R_3'\overset{+}{P}\ OR'\ OH] \longrightarrow R_3PO + ROH$$
$$\underset{R}{|}$$

(175)

could be expected to produce inversion, exchange with $^{18}$O water (and some R"OH if R' = OR"). The ability of the proton to shift readily seems to be important since dialkyl peroxides do not react similarly (see below). The rapidity of the reaction in nonpolar solvents also argues against much charge separation in the transition state, and Denney, et al. (445) have proposed a concerted transfer. No experiments with $RO_2D$ have been reported.

$$\overset{H}{\underset{}{}}$$
$$[R_3'P{\cdots}\overset{..}{O}{-}\overset{..}{O}{-}R]$$

Rate studies on a series of phosphites with 1,1-diphenylethyl hydroperoxide (580) have shown not unexpected (13) signs of steric hindrance. Bimolecular rate constants ($M^{-1}sec^{-1}$) at 20° for $(RO)_3P$ where R is Et , *i*-Pr, *p*-MeC$_6$H$_4$, and *p-t*-BuC$_6$H$_4$ were 0.56, 0.34, 0.025, and 0.0183, respectively. Actually the kinetics were somewhat higher than second order, suggesting some assistance in proton transfer similar to that postulated for hydroperoxide–dialkylsulfide reactions.

$$R_3P \quad O{\cdots}\overset{R}{\underset{|}{O}}\ H$$
$$H{\cdots}O$$
$$\underset{|}{}$$
$$OR$$

The reactions of trialkyl phosphines or phosphites with hydroperoxides or diacyl peroxides (1) contrast sharply with those with dialkyl peroxides. The latter occur only at elevated temperatures or under photoillumination and involve a fairly complex free-radical chain (581–583). There seems no reason to suppose that hydroperoxides pursue a similar path except for the suspicion that triphenylphosphine does not reduce allyl hydroperoxides cleanly (32) and is subject to some form of inhibition in the reduction of secondary alkyl hydroperoxides (584). The very rapid reaction of phosphines or phosphites with peroxy radicals is, of course, well known (583, 585–587), and a radical chain could account for the results quite nicely, assuming that RO· abstracts from hydroperoxide more rapidly than it reacts with $R_3P$(581).

$$R_3'P + RO_2 \cdot \longrightarrow R_3'PO + RO \cdot \qquad (176)$$

$$RO \cdot + RO_2H \longrightarrow ROH + RO_2 \cdot \qquad (177)$$

An opportunity for further experimentation is evident.*

### 2. Reactions with Sulfur Compounds

Edwards and his associates (13, 588) have had the latest and probably final word on the reduction of hydroperoxides by organic sulfides—a reaction first explored in the laboratories of Bateman (435, 437, 589), Overberger (590) and Modena (591–593). Their findings, using thioxane as the sulfide, merely amplify, rather than contradict, the results of the earlier workers, but the discussion leaves little room for improvement.

To summarize briefly: The reduction is rapid and quantitative for allyl (143, 435) as well as alkyl (435, 588) hydroperoxides; it is acid-catalyzed (435, 437, 588, 589) and exhibits no salt effect (588). The rate is dependent on the protic nature of the solvent, not on its polarity (588); for acetic acid, $H_2O$, $D_2O$, and alcohols the log of the relative bimolecular rate constants is proportional to $pK_a$ (588). In aprotic solvents the reaction is second order in hydroperoxide (435, 437, 589); with an added alcohol, it is first order in each (437, 588, 589). When the sulfide is diphenyl or methyl phenyl, ring substituents show a linear $\rho\sigma$ plot with $\rho = -0.98$ and $-1.13$, respectively (591–593).

Thioxane and $t$-$BuO_2H$ in water at 25° with acid catalysis give straightforward kinetics (588):

$$\text{rate} = k_2[R_2S][t\text{-}BuO_2H] + k_3[R_2S][t\text{-}BuO_2H][HClO_4] \qquad (178)$$

$$(k_2 = 1.35 \times 10^{-4} \text{ l/mole-sec}; \quad k_3 = 19.1 \text{ l}^2/\text{mole}^2\text{-sec})$$

For the uncatalyzed reaction $\Delta H\ddagger$ and $\Delta S\ddagger$ are 13.5 kcal/mole and $-27$ e.u., respectively. Thioxane reduces hydrogen peroxide 16–23 times faster than does $t$-$BuO_2H$.

---

*An extensive kinetic study of the reaction of triphenylphosphine with butyl hydroperoxides (584) has now shown this reaction to be almost certainly non-radical, and similar in most respects to the reduction of hydroperoxides by organic sulfides. Moreover, it has been shown (D. B. Denny and D. H. Jones, *J. Am. Chem. Soc.*, **91**, 5821 (1969)) that certain phosphites reduce non-tertiary dialkyl peroxides to ethers cleanly, in what appears to be a non-radical reaction (note added in proof).

All the factors point to nucleophilic attack by sulfur, facilitated by solvent-assisted proton transfer (1, 13, 588).

$$R_2'S \longrightarrow \qquad R_2'SO + AH + ROH \qquad (179)$$

The reaction has been used preparatively in a number of cases, both to reduce hydroperoxides (Section III-B-2) and to oxidize organic sulfides. A typical example is the synthesis of phenothiazine-5-oxide (594).

$$\xrightarrow[\text{dioxane, 2 hr.}]{25^\circ} \qquad (180)$$

Sulfoxides are not further oxidized by hydroperoxides except under the influence of some metal ions (Section IV-B-5). Reactions of other sulfur compounds with hydroperoxides probably involve free radicals (Section V).

### 3. Reactions with Amines

The hydroperoxide–amine reaction is complex. Although it is certainly a radical-producing reaction [ESR signals are observed (420, 595-599), thiols are oxidized to disulfides (421, 600, 601), hydrocarbon autoxidations (415, 602) and polymerizations (603) are accelerated], there is reason to believe that O—O heterolysis is the initial reaction (604).

At room temperature alkyl amines form stable complexes with hydroperoxides. Those investigated by Oswald and his associates (600, 601) (most of the possible combinations from $t$-butyl hydroperoxide, cumyl hydroperoxide, tetralin hydroperoxide, 2,5-dihydroperoxy-2,5-dimethylhexane and hexadecyl amine, piperazine, triethylenediamine), crystallized from pentane and melted sharply, without decomposition, usually below 100°. Oswald has termed these compounds "salts," although infrared and NMR spectra in nonpolar solvents indicate merely a strongly-hydrogen-bonded structure (600, 601).

$$-\!\!\!\overset{\diagdown}{\underset{\diagup}{N}}\cdots H\!-\!O_2R$$

Heating neat mixtures of amines with tertiary hydroperoxides at 100–120° produces excellent yields of alcohols (605, 606). [Secondary and primary hydroperoxides are partly dehydrated to ketones and aldehydes (Section IV-B-4).] Tertiary amines accomplish the reduction much faster than primary

TABLE 9. REDUCTION OF TERTIARY HYDROPEROXIDES BY AMINES

| Amine | $RO_2H$ | $T(^\circ C)$ | Time (hr) | ROH (%) Product | Ref. |
|---|---|---|---|---|---|
| $Et_2NOH$ | $t\text{-BuO}_2H$ | 25 | 1 | 35[a] | 604 |
| $Ph_2NNHPh$ | $t\text{-BuO}_2H$ | 20 | 2 | 30[b] | 599 |
| $Et_3N$ | $\alpha\text{-Cumyl O}_2H$ | 110–120 | 0.5 | 88 | 605 |
| $(HOCH_2CH_2)_3N$ | $\alpha\text{-Cumyl O}_2H$ | 110–120 | 0.5 | 100 | 605 |
| $Et_2NH$ | $\alpha\text{-Cumyl O}_2H$ | 110 | 11–12 | 78 | 605 |
| $i\text{-BuNH}_2$ | $\alpha\text{-Cumyl O}_2H$ | 100–110 | 16–17 | 100 | 605 |
| $Me_3CCH_2\text{--CHMeNH}_2$ | $t\text{-BuO}_2H$ | 90–110 | | 100[a] | 606 |

[a] Percentage of $t\text{-BuO}_2H$ decomposition in dilute benzene solution, $t\text{-BuOH}$ only.
[b] Percentage of $t\text{-BuO}_2H$ decomposition in dilute benzene solution, $t\text{-BuOH}$–acetone (3:1).

or secondary ones (605), and hydroxylamines (604) or hydrazines (599) faster yet (Table 9). No oxygen is evolved, and yields of water are nearly quantitative (605). The fate of the amine, in doubt for a number of years (1, 605), is now fairly well settled (604, 606). Primary and secondary amines yield imines:

$$R-CH_2NH_2 + t\text{-BuO}_2H \longrightarrow RCH=NH + t\text{-BuOH} + H_2O \quad (181)$$

Tertiary amines are cleaved, mostly to aldehydes and secondary amines (606), although low yields of hydroxylamines have been obtained under conditions under which they crystallize from solution (597); hydroxylamines usually form nitrones (604).

$$(RCH_2)_3N + t\text{-BuO}_2H \longrightarrow RCHO + (RCH_2)_2NH + t\text{-BuOH} \quad (182)$$

$$(RCH_2)_2NOH + t\text{-BuO}_2H \longrightarrow RCH=N-CH_2R + H_2O + t\text{-BuOH}$$
$$\downarrow$$
$$O$$
$$(183)$$

[The products actually isolated in most cases are those of hydrolysis and condensation (604, 606).] Aryl hydrazines yield a complex mixture of N—N cleavage products (599) but do not give clean reductions of hydroperoxide [amines that have no hydrogen on the carbon alpha to the nitrogen do not give clean reductions either (606).]

The reactions that produce these products clearly include some sort of free-radical chain. The question, so far as hydroperoxide chemistry is concerned, is whether cleavage of the O—O bond is heterolytic (nucleophilic attack by

$\underset{/}{\overset{\backslash}{-}}$N:)   or homolytic (radical displacement). There seem to be contra-

dictions in any scheme based solely on one mechanism or the other.

Thus the ESR spectra of hydroperoxide–amine reactions show *only* nitroxide radicals (420, 595–599), whereas nitryl radicals can be observed if dialkyl peroxides are thermolyzed or photolyzed in the presence of amines (420):

$$t\text{-Bu}_2\text{O}_2 \xrightarrow{\;h\nu\;} 2\;\; t\text{-BuO}\cdot \xrightarrow{\;\text{Ph}_2\text{NH}\;} \text{Ph}_2\text{N}\cdot + t\text{-BuOH} \quad (184)$$

$$t\text{-BuO}_2\text{H} + \text{Ph}_2\text{NH} \;\xrightarrow{\;\;\;\;\;} \text{Ph}_2\text{N}-\text{O}\cdot \quad\quad\quad (185)$$

De La Mare and Coppinger (604) have proposed nucleophilic displacement followed by N—O homolysis to explain this.

$$\underset{\text{H}}{\overset{\text{H}}{\text{R}_2\text{N:}}} \quad \overset{}{\text{O}-\text{O}-t\text{-Bu}} \longrightarrow \left[\underset{}{\overset{\text{H}}{\text{R}_2\text{N}-\text{OH}}}\right]^{+} \;\;^{-}\text{O}-t\text{-Bu} \quad (186)$$

$$\text{R}_2\text{NOH} + t\text{-BuOH}$$

$$\underset{\text{H}}{\overset{\text{H}\;\;\text{O}}{\text{R}_2\text{N:}}} \quad \text{O}-\text{O}-t\text{-Bu} \quad \left[\underset{}{\overset{\text{H}\;\;\text{O}}{\text{R}_2\text{N}-\text{OH}}}\right]^{+} \;^{-}\text{O}-t\text{-Bu} \quad (187)$$

$$\text{R}_2\text{NO}\cdot + \cdot\text{OH} + t\text{-BuOH}$$

However, alkoxy radical products are observed [acetone from *t*-butyl hydroperoxide (599), AcPh from α-cumyl hydroperoxide (605)] when amines and hydroperoxides are reacted in dilute solution rather than neat. The reaction is inhibited by 2,6-di-*t*-butyl-*p*-cresol [*t*-butylperoxy-substituted cyclohexadienones appear in the products (606)] and accelerated by mercaptans, which are oxidized to disulfides, whereas the amine is recovered unchanged (600, 601).

The inhibition by phenols, particularly, seems to require radical-induced hydroperoxide decomposition. Maruyama and co-workers (599) propose nitryl-radical intermediates; their nonappearance in the ESR spectra can be explained by assuming a fast reaction:

$$R_2N\cdot + t\text{-BuO}_2H \longrightarrow [R_2NOH\cdot O{-}t\text{-Bu}]$$

$$\longrightarrow R_2NO\cdot + t\text{-BuOH} \quad (188)$$

Kinetic studies (607–609) have shown very complex rate expressions for the reaction but have not materially advanced the understanding of it.

Hydroperoxides that contain a nitrogen atom in the same molecule undergo facile unimolecular decompositions, which appear to occur by simple nucleophilic displacement. An example is the smooth formation of oxaziranes from α-amino hydroperoxides (257, 331):

$$(189)$$

α-Amido hydroperoxides (207) may react similarly, although products of C—C as well as O—O cleavage are found.

For this reaction Sagar (207) has reported a unimolecular rate constant

$$k_1 = 2.3 \times 10^9 e^{-23.7/RT} \quad (77\text{--}130^\circ)$$

The reaction is unaffected by phenolic inhibitors and does not induce the polymerization of added vinyl monomers.

### 4. Base-Catalyzed Decompositions

#### (a) Intermolecular Displacements on O—O by Carbanions

Autoxidation of hydrocarbons under strongly basic conditions (73, 89–91, 97, 610, 611) or the reactions of hydroperoxides with Grignard reagents (210, 612) and other organometallic compounds (210, 613–615) yield alcohols, generally in excellent yield. Reactions with Grignard reagents (616, 617) or with organolithium compounds (1) are postulated to proceed through concerted rearrangements:

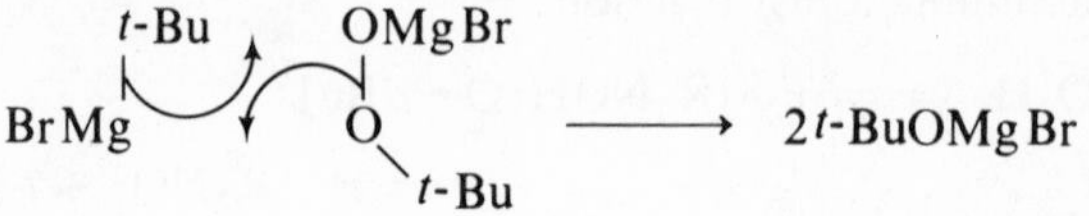

$$
\begin{array}{ccc}
t\text{-Bu} & OMgBr \\
BrMg & O & \longrightarrow \quad 2\,t\text{-BuOMgBr} \\
& t\text{-Bu}
\end{array}
\tag{191}
$$

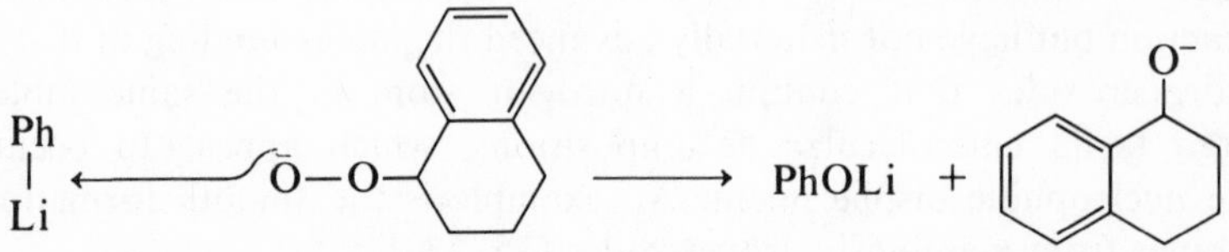

$$
\begin{array}{c}
Ph \\
| \quad \longleftarrow \quad O-O- \quad \longrightarrow \quad PhOLi \; + \\
Li
\end{array}
\tag{192}
$$

The more ionic sodio or potassio compounds may react by direct carbanion displacement (1, 90, 610).

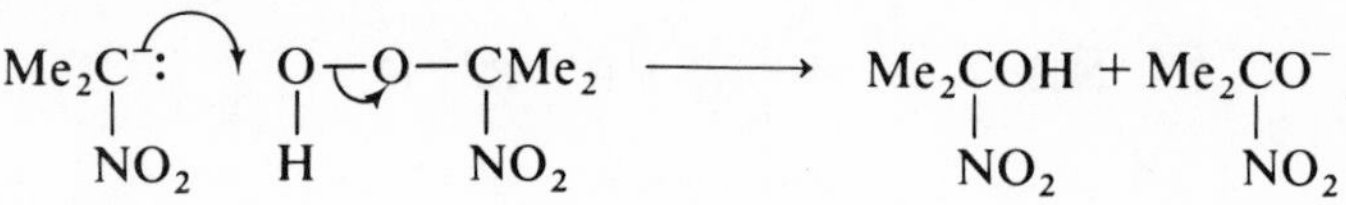

$$
\begin{array}{ccc}
Me_2C : & O-O-CMe_2 & \longrightarrow \quad Me_2COH + Me_2CO^- \\
| & | \quad | & | \qquad | \\
NO_2 & H \quad NO_2 & NO_2 \qquad NO_2
\end{array}
\tag{193}
$$

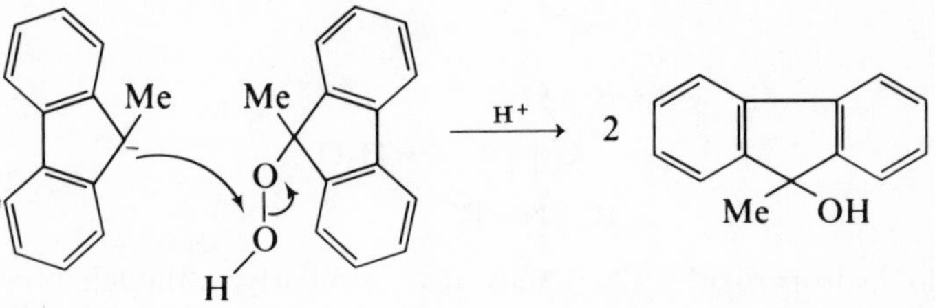

$$
\begin{array}{c}
Me \quad Me \\
\xrightarrow{\;H^+\;} \quad 2 \\
O \\
| \\
O \\
H
\end{array}
\qquad
\begin{array}{c}
\\
Me \quad OH
\end{array}
\tag{194}
$$

However in one case at least (91) triarylmethide displacements have been shown to yield free radicals.

$$(p\text{-}NO_2Ph)_3C:^- + RO_2H \longrightarrow (p\text{-}NO_2Ph)_3C\cdot + RO\cdot + OH^- \tag{195}$$

Davies (1) compares this to the reduction of hydroperoxides by the ferrous ion, the driving force being the relative stability of the triarylmethyl radical.

$$RO_2H + Fe^{2+} \longrightarrow RO\cdot + [Fe^{3+}OH^-] \tag{196}$$

Its generality, or lack of it, in the base-catalyzed autoxidations of aralkanes has not been demonstrated.

The mechanism by which α-cumyl hydroperoxide is reduced in the presence of cumene and strong bases is somewhat more ambiguous. The method, which is the subject of several patents (618–620), gives greater than 100% yields of alcohol, based on decomposed hydroperoxide, and does not produce phenol when phenyl lithium is used as the catalyst (619). The postulated ability of the ($\alpha$-cumyl $O_2H$) ($\alpha$-cumyl $O_2^- Na^+$) complex both to generate free radicals (425, 621) and to produce alcohol and oxygen by a nonradical path (621, 622) suggests several alternatives to carbanion displacement.

$$R-O-O\cdots Na^+ \longrightarrow RO\cdot + NaOH + RO_2\cdot \qquad (197)$$
$$\overset{\cdot\cdot}{H}-\overset{\cdot\cdot}{O}-O-R$$

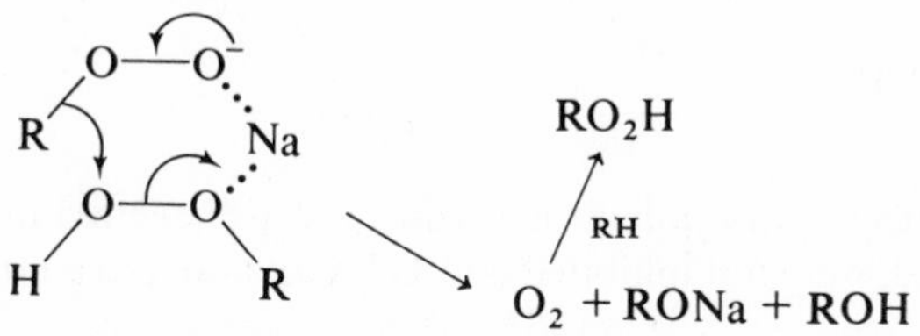

$$
\begin{array}{c}
\text{(198)}
\end{array}
$$

Although a similar mechanism has been firmly established for peroxy acids (623), the route for the base-catalyzed decomposition of tertiary hydroperoxides in inert solvents (424, 624) is much less certain. Several investigators (294, 623, 625) have now shown that the decompositions of $t$-$BuO_2H$ or $\alpha$-cumyl $O_2H$ induced by sodium hydroxide can be retarded or inhibited altogether by addition of ethylenediaminetetraacetic acid (EDTA), a metal-ion-sequestering agent. Although the reaction can be made to proceed under these conditions by raising the temperature, it does not give reproducible rates (294). Thus most probably adventitious transition-metal ions introduced with the sodium hydroxide are partly if not wholly responsible for the observed products (294, 623, 625) (Section V).

(b) INTRAMOLECULAR CARBANION DISPLACEMENTS

In 1951 Kornblum and De La Mare (626) reported the smooth room-temperature conversion of $t$-butyl-$\alpha$-phenethyl peroxide to acetophenone brought about by potassium hydroxide, sodium ethoxide, or piperidine. They cited the previously discovered base-catalyzed dehydration of isopropyl hydroperoxide to acetone (627) and generalized that all peroxides having an $\alpha$-hydrogen would be subject to such decomposition.

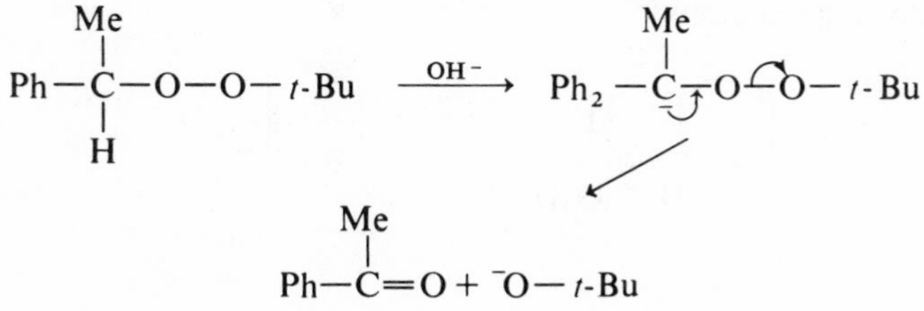

$$
\begin{array}{c}
\underset{H}{\overset{Me}{Ph-\underset{|}{\overset{|}{C}}-O-O-}}t\text{-}Bu \xrightarrow{\ OH^-\ } Ph_2-\overset{Me}{\underset{|}{C}}-O-O-t\text{-}Bu
\end{array}
\qquad (199)
$$

$$
\underset{Ph-\overset{Me}{\underset{|}{C}}=O\ +\ ^-O-t\text{-}Bu}{}
$$

This postulate has been widely accepted (1–3, 5) but only recently subjected to serious scrutiny (294). Hofmann, Pritzkow, and their associates (294) have now shown that (a) simple secondary hydroperoxides, such as cyclohexyl or 4-heptyl, yield both alcohols and ketones on treatment with sodium hydroxide and (b) such reactions can be inhibited by EDTA and are almost certainly the result of adventitious trace-metal ions.

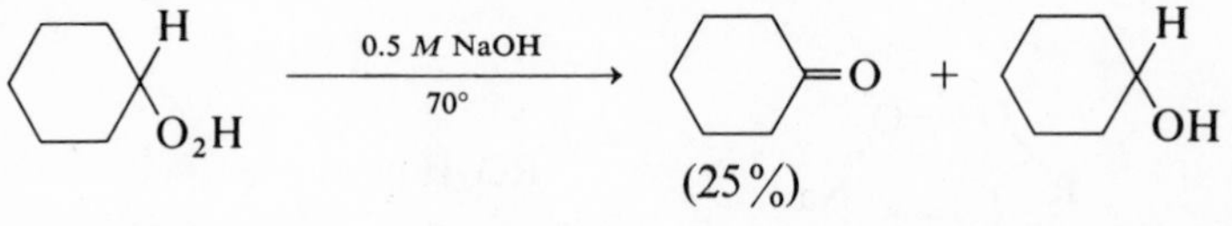

$$
\underset{O_2H}{\overset{H}{\diagdown}} \xrightarrow[70°]{0.5\ M\ \text{NaOH}} \quad \diagdown\!=\!O \ + \ \underset{OH}{\overset{H}{\diagdown}} \quad (200)
$$

$$(25\%)$$

Decompositions of tetralin hydroperoxide and α-phenethyl hydroperoxide, on the other hand, were not inhibited by EDTA and obeyed a rate expression:

$$
\frac{-\partial[RO_2H]}{dt} = \frac{k[RO_2H][OH^-]}{1 + K[OH^-]} \tag{201}
$$

(For tetralin hydroperoxide at 30° $k$ and $K$ are respectively 4.0 l/mole-sec and 4.1 l/mole; $\Delta H\ddagger$ and $\Delta S\ddagger$, are 17.3 kcal/mole and $-17$ e.u., respectively. The reaction exhibited an isotope rate effect, $k_H/k_D = 3.9$, when the α-hydrogen was replaced by α-deuterium. The results are a clear confirmation of the Kornblum–De La Mare mechanism but for only those cases in which other electron-demanding substituents alpha to the peroxide linkage facilitate carbanion formation.

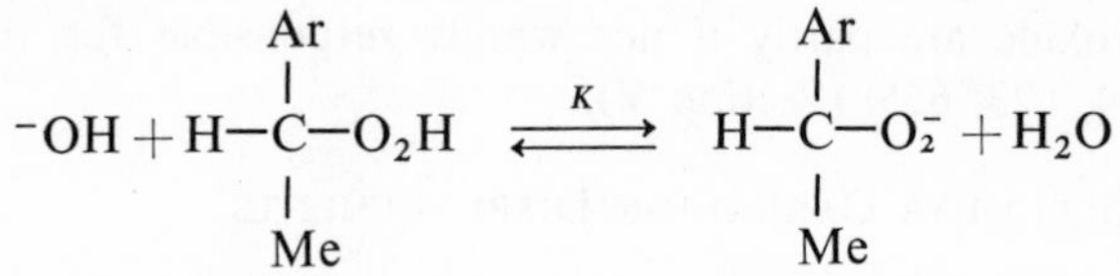

$$
{}^-OH + \underset{Me}{\overset{Ar}{H-C-O_2H}} \xrightleftharpoons{K} \underset{Me}{\overset{Ar}{H-C-O_2^-}} + H_2O \tag{202}
$$

$$
HO^- \curvearrowright H\underset{O\ \frown OH}{\overset{Me}{\frown C-Ar}} \xrightarrow{k} Me\overset{O}{\overset{\|}{C}}-Ar + H_2O + OH^- \tag{203}
$$

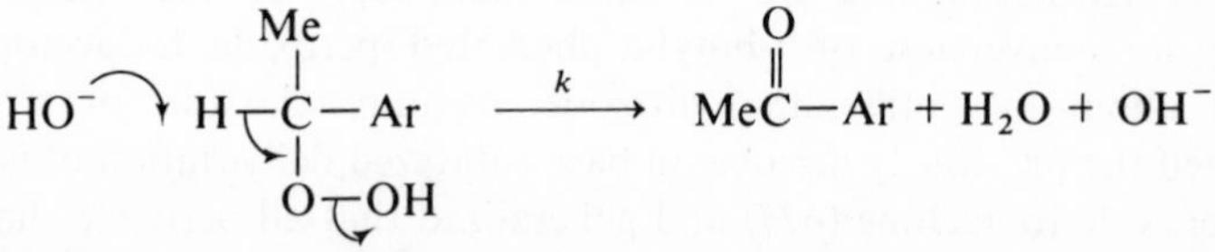

Most reported instances (90, 147, 148, 158, 268, 536, 628) of smooth base-catalyzed dehydrations of hydroperoxides fit this requirement, as, for example, in the following reactions:*

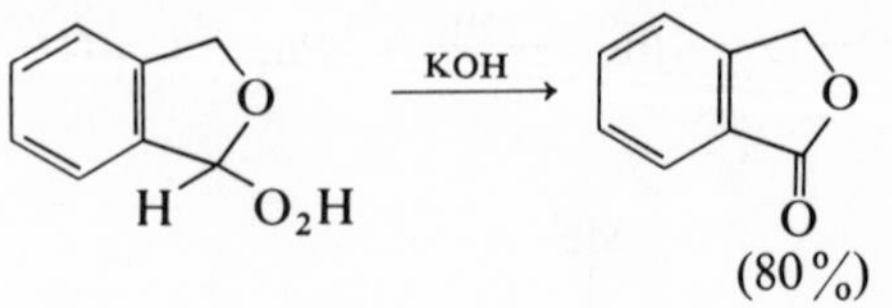

$$
\underset{H\quad O_2H}{\diagdown\!\!O} \xrightarrow{KOH} \underset{O}{\diagdown\!\!O} \tag{204}
$$

$$(80\%)$$

*Reactions 204, 205, and 206 are from references 158, 90, and 147, respectively.

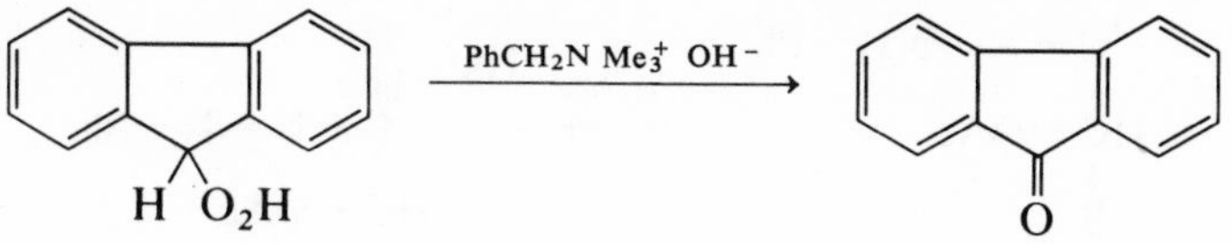

(205)

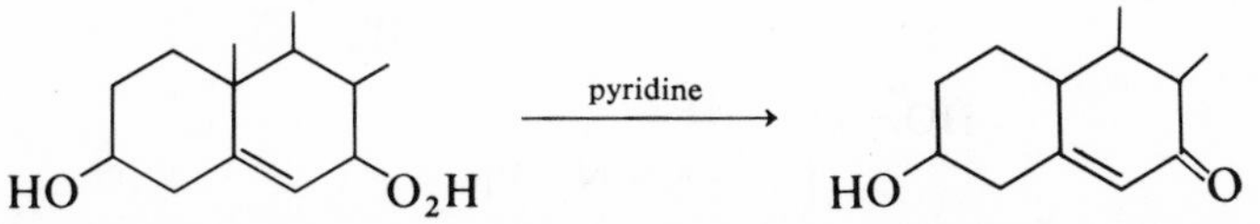

(206)

Hydroperoxides and peroxides that have nitrogen (181, 629) or oxygen (181, 319, 629) functions gamma to the peroxy linkage readily undergo concerted cleavage reactions either thermally or with mild base. No thorough kinetic studies have been reported, but the many examples extant for amine and hydrazine hydroperoxides [Schmitz, Schultz, and Rieche (189, 191, 319, 630–633)]; indole- and lophine-type hydroperoxides [Witkop (181, 182), McCapra (196, 197, 201), and others (195, 199, 200, 634)]; hydroperoxides from ozonides [Bailey (259–262)] and other diverse types (97, 635–638) leave little doubt of the generality of the ph̄ ᷈menon. Typical examples are as follows:*

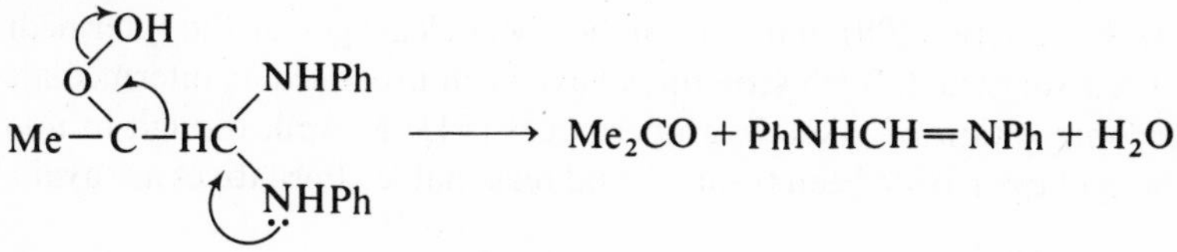

$$Me_2CO + PhNHCH{=}NPh + H_2O \quad (207)$$

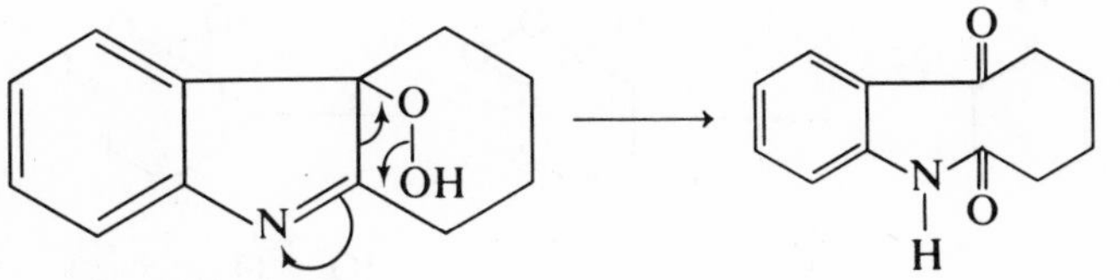

(208)

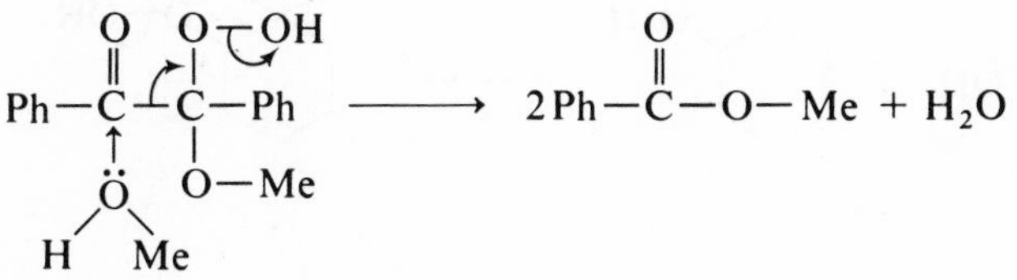

$$2\,Ph{-}C{-}O{-}Me + H_2O$$

(209)

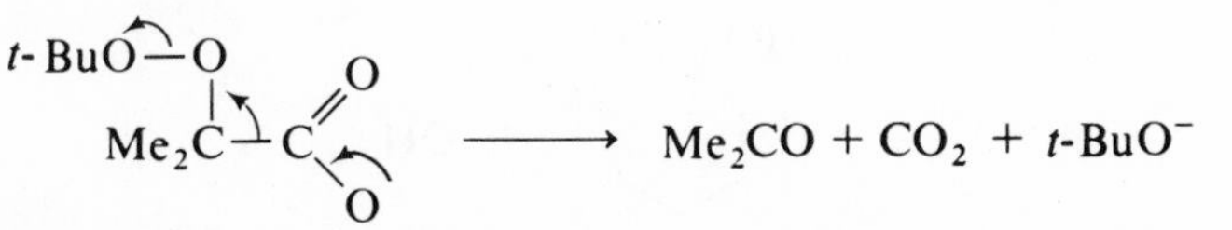

$$Me_2CO + CO_2 + t\text{-}BuO^- \quad (210)$$

A new method for the degradation of sugars (189, 191, 630–632) utilizes the principle.

*Reactions 207, 208, 209, and 210 are from references 633, 181, 260, and 635, respectively.

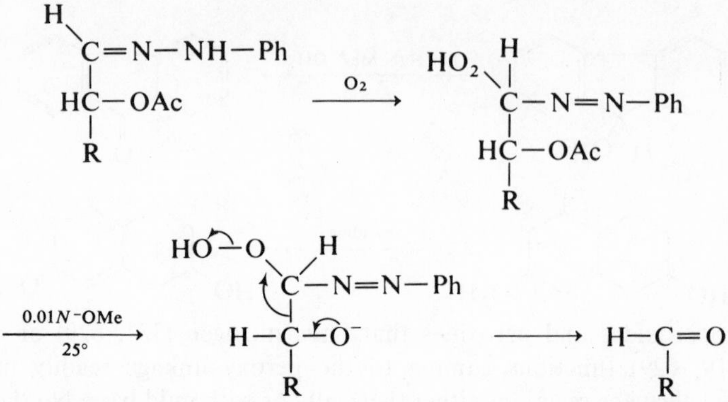

$$\underset{\underset{R}{\overset{\displaystyle H}{|}}{\overset{\displaystyle \mathrm{H}\hspace{-2pt}}{\mathrm{C}}}}{\overset{\displaystyle \mathrm{C}=\mathrm{N}-\mathrm{NH}-\mathrm{Ph}}{}} \xrightarrow{\ O_2\ } \underset{\underset{R}{\overset{\displaystyle \mathrm{H}\hspace{-2pt}}{\mathrm{C}-\mathrm{OAc}}}}{\overset{\displaystyle \mathrm{HO_2}\diagdown\overset{\displaystyle H}{|}}{\mathrm{C}-\mathrm{N}=\mathrm{N}-\mathrm{Ph}}}$$

(211)

$$\xrightarrow[25^\circ]{0.01N\,^-\mathrm{OMe}} \qquad \begin{array}{c}\mathrm{HO}\frown\mathrm{O}\quad\mathrm{H}\\ \diagup\mathrm{C}-\mathrm{N}=\mathrm{N}-\mathrm{Ph}\\ \mathrm{H}-\mathrm{C}-\mathrm{O}^-\\ |\\ \mathrm{R}\end{array} \qquad \xrightarrow{\qquad} \qquad \begin{array}{c}\mathrm{H}-\mathrm{C}=\mathrm{O}\\ |\\ \mathrm{R}\end{array}$$

A somewhat different mechanism has also been postulated for indole (196), lophine (197, 200), and acridine (201) hydroperoxides.*

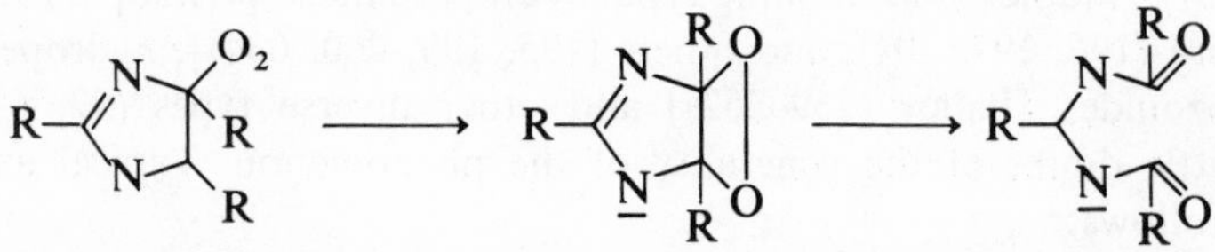

$$\begin{array}{c}\mathrm{N}\\ \mathrm{R}-\langle\quad\rangle^{\,\mathrm{O_2}}_{\,\mathrm{R}}\\ \mathrm{N}\end{array} \xrightarrow{\qquad} \begin{array}{c}\mathrm{N}\\ \mathrm{R}-\\ \underset{-}{\mathrm{N}}\end{array} \xrightarrow{\qquad} \begin{array}{c}\mathrm{R\ O}\\ \mathrm{N}\\ \mathrm{R}-\\ \underset{-}{\mathrm{N}}\end{array}$$

(212)

Both homolytic (200) and heterolytic (201) cleavages of the intermediates have been suggested. Such structures have been proposed as intermediates in the decomposition of $\alpha$-ketohydroperoxides (643) as well, though in no case has the cyclic peroxide been trapped, and reasonable alternatives are available.

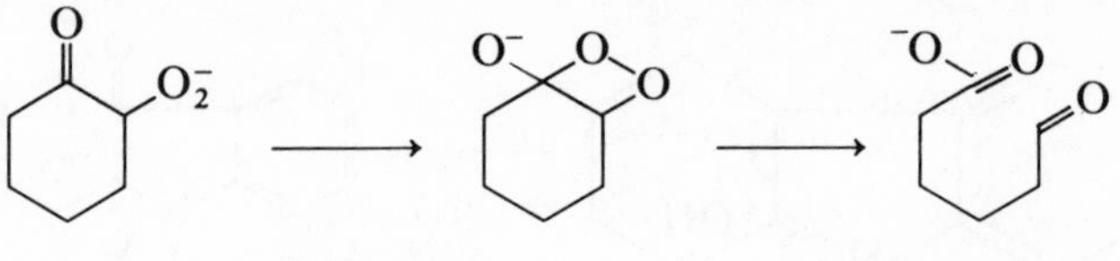

(213)

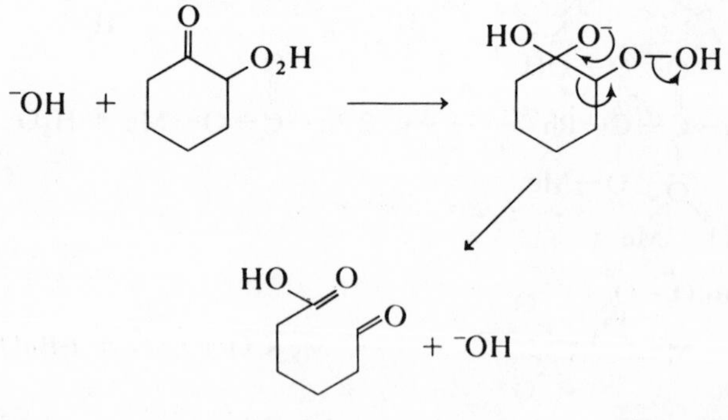

(214)

*Thermal or base-catalyzed decomposition (196, 199, 201, 637) of these hydroperoxides is photoluminescent, as result of producing carbonyl groups in an electronically excited state (196, 638–642).

## (c) Base-Catalyzed Reaction with Nitriles

The formation of amides from hydrogen peroxide and nitriles is mechanistically similar to the reactions of the preceding section (644–646).

$$R-C\equiv N + H_2O_2 \xrightarrow{\;OH^-\;} R\text{-}\text{-}\text{-}C\overset{\displaystyle HN}{\underset{\displaystyle O-OH}{\Big\langle}}$$

$$\downarrow$$

$$R-C\overset{\displaystyle HN_2}{\underset{\displaystyle O}{\Big\langle}} + OH^- \tag{215}$$

However, the base-induced decomposition of tertiary hydroperoxides catalyzed by benzonitrile (647) and dinitriles (424, 646, 647) appears to be much more complex (424, 625, 646, 647). Noting the similarity to metal-ion-catalyzed decomposition (production of $O_2$, ROH, and $R_2O_2$ from $RO_2H$) (646), Berger (647) proposed a mechanism involving both ions and radicals.

$$R'-C\equiv N + RO_2^- \longrightarrow R-C\overset{\displaystyle N^-}{\underset{\displaystyle O_2R}{\Big\langle}} \tag{216}$$

$$R'-C\overset{\displaystyle N^-}{\underset{\displaystyle O_2R}{\Big\langle}} \xrightarrow{\;RO_2H\;} R'-C\overset{\displaystyle NH}{\underset{\displaystyle O_2R}{\Big\langle}} \longrightarrow R'-C\overset{\displaystyle NH}{\underset{\displaystyle O\cdot}{\Big\langle}} + RO\cdot \tag{217}$$

In fact the presence of adventitious metal ions has now been shown to be a major factor (625). Addition of EDTA suppresses oxygen formation and retards the reaction, transforming it into simple base-catalyzed amide formation (625). The apparent synergistic effect of nitriles on metal-ion-catalyzed decomposition is intriguing but not yet explained.

### 5. Epoxidation of Olefins and Related Metal-Ion-Catalyzed Reductions of Hydroperoxides

Epoxides formed in the autoxidation of olefins result mainly from peroxy-radical addition to double bonds (39, 133, 648) (Section II-A-3). However, hydroperoxides and olefins heated in an inert atmosphere slowly produce

---

An apparently related reaction is the facile decomposition of hydroperoxides by tetranitromethane (W. F. Sager and J. C. Hoffsommer, *J. Phys. Chem.*, **73**, 4155 (1969)), in which, if Sager is correct, an intermediate trioxide, $R-O-O-O-N-C(N_2O)_3$, under-

$$\overset{\displaystyle |}{O-}$$

goes an intramolecular rearrangement (note added in proof).

epoxides (648–650) in a reaction unaffected by light or phenolic autoxidation inhibitors (648). Yields increase with increased alkyl substitution on the double bond and may be as high as 40–50% for diisobutylene, based on decomposed $t$-BuO$_2$H (649). The other products are $t$-BuOH and O$_2$; thus the reaction might still be supposed to be free radical in character but for Walling's (650) calculation from known rate constants that no more than half the epoxide formed (from $t$-BuO$_2$H and styrene) can arise from that route. Both Walling (650) and Brill (649) postulate complex formation

$$t\text{-}BuO_2\cdot + PhCH{=}CH_2 \longrightarrow Ph\overset{\cdot}{C}H{-}CH_2{-}O{-}O{-}t\text{-}Bu \tag{218}$$

$$PhH\overset{\displaystyle O}{\overset{\displaystyle \diagup\!\diagdown}{C}}{-}CH_2 + t\text{-}BuO\cdot$$

followed by carbon displacement on O—O, analogous to epoxidation by peroxy acids (651).

$$\tag{219}$$

A mixture of $t$-BuO$_2$H and 1-octene in benzene at 60° exhibits first-order rate dependence in each reactant with $k_2 = 3.8 \pm 0.3 \times 10^{-18}$ l/mole-sec (649). The much more rapid and efficient action of peroxy acids can be rationalized in terms of the less basic leaving group and/or the five-membered-ring transition state facilitating hydrogen transfer (4, 12, 13).

Epoxidation by hydroperoxides can be catalyzed by bases (Section B-4) or by boric acid (652), but the most striking catalysis is shown by complexes or salts of vanadium, chromium, molybdenum (653, 654, 655), or tungsten (656). The presence of these ions produces virtually quantitative yields of epoxide from any hydroperoxide whether produced *in situ* by autoxidation (654, 657–659) or added separately (653–661).

$$\bigcirc\!\!| \ + \ RO_2H \ \xrightarrow[\;50\text{-}80°\;]{\;VO^{2+}\;(10^{-4}M)\;} \ \bigcirc\!\!<\!O \ + \ ROH \tag{220}$$

The reaction is thus of considerable interest to the petrochemical industry and is the subject of several patents (656–659).

Several factors indicate a process of nucleophilic displacement by C=C preceded by hydroperoxide–metal ion complexing. Metal ions, such as Co$^{2+}$, Fe$^{2+}$, Mn$^{2+}$, Ni$^{2+}$, Cu$^+$, Ce$^{4+}$, that induce free-radical-chain decompositions of hydroperoxide (Section V-A) give little or no epoxides.

Rates are increased both by electron-donating substituents on the double bond (653, 655, 660) and electron-withdrawing substituents on the hydroperoxide. Activation energies are low (653, 661) and entropies, large and negative [$\Delta H\ddagger$ and $\Delta S\ddagger$ = 12.7 kcal/mole and $-19.8$ e.u., respectively, for $t$-BuO$_2$H–vanadium acetylacetonate in cyclohexene (661)]. Retardation or inhibition is produced by alcohols (660, 661) and other substances (660) that might reasonably compete with hydroperoxide for complexing sites on the metal ion (660, 661).

Recent kinetic studies (660, 661) have shown a first-order dependence on metal ion and on olefin concentration. Dependence on [RO$_2$H] is approximately first order, but in neat olefin takes the form (661)

$$\text{rate} = \frac{k[M]}{1 + (1/K[\text{RO}_2\text{H}])} \tag{221}$$

(M = metal ion)

Mechanisms proposed by several workers (653, 660, 661) are similar, the one of Gould et al. (661) being most detailed.

$$-\overset{|}{\underset{|}{V}}{}^{V}- \; + \; RO_2H \;\rightleftharpoons^{K}\; -\overset{|}{\underset{|}{V}}{}^{V}-O\overset{R}{\underset{OH}{\diagup}}$$

$$-\overset{|}{\underset{|}{V}}-O\overset{R}{\underset{OH}{\diagup}} \; + \; \bigcirc\!\!\parallel \;\xrightarrow{k}\; \left[ \;\bigcirc\!\!\!<_{O-O-\overset{|}{\underset{|}{V}}{}^{V}-}^{\overset{+}{\underset{H}{}}\;R} \; \right] \tag{222}$$

$$\bigcirc\!\!\!<O \; + \; -\overset{|}{\underset{|}{V}}{}^{V}-OH \overset{R}{} \tag{223}$$

$$-\overset{|}{\underset{|}{V}}{}^{V}-O\overset{R}{\underset{H}{\diagdown}} \; + \; RO_2H \;\rightleftharpoons\; -\overset{|}{\underset{|}{V}}{}^{V}-O\overset{R}{\underset{OH}{\diagdown}} \; + \; ROH \tag{224}$$

It seems likely that proton transfer is concerted and facilitated by other ligands of the complex, but a straightforward rationalization of their influence (660) is difficult.

The same metal ions catalyze the quantitative conversion by hydroperoxides of amines to amine oxides (22, 662, 663) and sulfoxides to sulfones (22, 664), reactions typical of peroxy acids.

$$R_3'N + RO_2H \xrightarrow{\text{V or Mo}} R_3'N \rightarrow O + ROH \tag{225}$$

$$R_2'S + 2RO_2H \xrightarrow{\text{V or Mo}} R_2'SO_2 + 2ROH \tag{226}$$

Few mechanistic details are as yet available for these reactions,* but, as in epoxidation, alcohols exert a retarding influence (663). Sulfides react considerably faster than carbon–carbon double bonds so that allyl sulfides can be oxidized cleanly to allyl sulfones (664).

## V. HOMOLYTIC REACTIONS OF HYDROPEROXIDES

Hydroperoxides undergo four general types of homolytic reaction:
  1. Unimolecular homolysis:

$$RO-OH \xrightarrow[\text{or } h\nu]{\Delta} RO\cdot + \cdot OH \tag{227}$$

  2. Assisted homolysis:

$$RO-OH + A \longrightarrow RO\cdot + [AOH]\cdot \tag{228}$$

  3. Free-radical abstraction:

$$RO-O-H + R'\cdot \longrightarrow RO_2\cdot + R'H \tag{229}$$

  4. Free-radical displacement on O—O:

$$RO-OH + R'\cdot \left\langle \begin{array}{l} ROR' + \cdot OH \\[1em] R'OH + RO\cdot \end{array} \right. \tag{230}$$

Of these, only radical abstraction is unique for hydroperoxides; the other reactions are shared with all peroxidic compounds. Compared with dialkyl peroxides and acyl peroxides, hydroperoxides give much less unimolecular homolysis and undergo assisted homolysis much more readily. Too little quantitative information is available to make a similar comparison of susceptibilities toward radical displacement on O—O.

The availability of this number of possible reactions has made a detailed

---

*For more recent discussions see E. S. Gould and M. Rado, *J. Calalysis*, **13**, 238 (1969); M. N. Sheng and J. G. Zajacek, *J. Org. Chem.*, **35**, 1839 (1970); G. R. Howe and R. Hiatt, *J. Org. Chem.*, **35**, 4007 (1970) (note added in proof).

understanding of hydroperoxide homolysis exceedingly difficult.* Despite progress in recent years, this discussion would be short indeed if it were to exclude all results that still defy clear interpretation. However, it is frequently useful to know not what is going on, exactly, but simply how fast hydroperoxides can be expected to decompose, how fast free radicals are produced, and what sort of products are formed in a given situation.

### A.  Free-Radical Abstraction

### 1.  Chain Decomposition of Tertiary Hydroperoxides

Thermal decomposition of $t$-BuO$_2$H, either neat at 100° (283) or in chlorobenzene solution at 140° (665), or photolytic decomposition at lower temperatures (666) yields $t$-BuOH and oxygen in nearly quantitative amounts. Cumene hydroperoxide behaves somewhat similarly (667–669), though substantial amounts of dicumyl peroxide and acetophenone are formed as well and yields of oxygen are lower (667–670).

The kinetics of thermal decomposition of both $t$-butyl hydroperoxide (465, 665) and $\alpha$-cumyl hydroperoxide (668, 669) have shown $1 + 3/2$ order [or $1 + 4/3$ order (399)] dependence on hydroperoxide concentration, indicating a combination of unimolecular homolysis and radical-induced decomposition (671).

Decompositions initiated by peroxy esters (672, 673), azo compounds (395), or hypochlorites (625) at 20–60° (thus uncomplicated by thermal homolysis of the hydroperoxide) have amply verified earlier conjectures (665–667) about the nature of the chain reaction.

$$\text{initiator} \xrightarrow{\ k_i\ } 2\text{X} \cdot \tag{231}$$

where X is any initiator fragment.

$$\text{X} \cdot \ (\text{or RO} \cdot) + \text{RO}_2\text{H} \xrightarrow{\ k_p\ } \text{XH} \ (\text{or ROH}) + \text{RO}_2 \cdot \tag{232}$$

where R is tertiary.

---

*The full implications of the importance of adventitious metal ions in "base-catalyzed" decompositions (Section IV-8-4) still do not seem generally recognized. Any hydroperoxide prepared via a base-catalyzed reaction or purified by means of its sodium salt is subject to contamination by trace-metal ions, whose influence can be overwhelming in concentrations as low as $10^{-8}$ $M$ (465). Likewise the discovery of W. F. Taylor, *J. Phys. Chem.*, **74**, 2250 (1970); *J. Catalysis*, **16**, 20 (1970) that finely ground Teflon accelerates hydroperoxide decomposition ignores the high probability that metal fragments introduced in the grinding process are responsible (note added in proof).

$$2RO\cdot + O_2$$

$$2\,RO_2\cdot \quad \overset{k_c}{\underset{k_t}{\diagup\diagdown}} \tag{233}$$

$$RO_2R + O_2$$

Assuming a steady state, it is possible to calculate a rate of hydroperoxide disappearance that corresponds to the experimental results (672, 673). (The reason it does not agree with high-temperature results will become apparent.)

$$\frac{\partial\,[RO_2H]}{dt} = k_i[\text{initiator}]\left[1 + \frac{k_c}{k_t}\right] \tag{234}$$

The critical factor is $k_c/k_t$, which determines the chain length. Three lines of evidence support a tetroxide mechanism for reaction (233), at least when the peroxy radicals are tertiary:

$$2RO_2\cdot \rightleftharpoons R-O-O-O-O-R$$

$$2RO\cdot + O_2$$

$$[RO\cdot\ O_2\ \cdot OR] \tag{235}$$

$$RO_2R + O_2$$

1. Interaction of $R-^{18}O-^{18}O\cdot + R-^{16}O-^{16}O\cdot$ yields $^{16}O-^{18}O$ (674).

2. Existence of tetroxides as quasi-stable species at temperatures below $-90°$ has been demonstrated (675–678).

3. Relative rates of alkoxy radical versus dialkyl peroxide formation is dependent on solvent viscosity (444, 673) in a manner indicative of solvent cage control (679).

$t$-Butyl hydroperoxide decomposed by di-$t$-butyl peroxalate (DBPO) in benzene (444, 672) or chlorobenzene (673) at 35–45° yields more than 99% $O_2$, 88–89% $t$-BuOH, and 11–12% $t$-Bu$_2$O$_2$. Calculations from both the amount of hydroperoxide versus peroxalate decomposed and the yield of $t$-Bu$_2$O$_2$ show a chain length of about 10 (444, 672, 673). Lower chain lengths obtain in more viscous solvents (673) (Table 10) and higher ones at elevated temperatures and in the gas phase (444).

Two factors tend to complicate this simple picture of induced decomposition (444, 680). The first is the ability of the alkoxy radical to cleave, which under some circumstances competes with hydrogen abstraction (444, 680) (equation 229).

TABLE **10.**   CHAIN LENGTHS FOR INDUCED DECOMPOSITION OF
*t*-BUTYL HYDROPEROXIDE[a]

| Solvent | Viscosity (cp at 25°) | Chain Length[b] |
|---|---|---|
| PhH[c] | 2.8[c] | 22 |
| PhH | 5.6 | 10.5 |
| PhCl | 7 | 10.5 |
| PhCl–cumene | 16 | 7.8[d] |
| *t*-BuO$_2$H | 33 | 7.5 |
| *t*-BuOH | 46 | 3.8 |

[a] At 45°, initiated by DBPO (673).
[b] $-\Delta_{[RO_2H]}/-2\Delta_{[DBPO]}$.
[c] At 100°, initiated by *t*-Bu$_2$O$_2$ (444).
[d] For cumene hydroperoxide.

$$R_3CO\cdot \longrightarrow R_2C{=}O + R\cdot \tag{236}$$

Although the alkyl radical formed (R·) is able to abstract hydrogen from
R′O$_2$H, (681, 682), the more likely course in the presence of

$$t\text{-BuO}_2D + CH_3\cdot \longrightarrow t\text{-BuO}_2\cdot + CH_3D \tag{237}$$

oxygen (produced from decomposing hydroperoxide) or of high concentrations of RO$_2$· is rapid termination (444, 680, 683, 684).

$$CH_3\cdot + RO_2\cdot \longrightarrow CH_3O_2R \tag{238}$$

$$CH_3\cdot \xrightarrow{O_2} CH_3O_2\cdot \xrightarrow{RO_2} ROH + CH_2O + O_2 \tag{239}$$

The consequences of alkoxy cleavage are thus lower chain lengths and more
complex kinetics since hydroperoxide concentration now becomes a factor
(444, 683). In the limiting case that *all* termination occurs via the cleavage
route, the rate expression becomes

$$\frac{-\partial[RO_2H]}{dt} = k[\text{initiator}]^{1/2}[RO_2H] \tag{240}$$

which is in fact the observed rate dependence observed for high-temperature
thermal decompositions where RO$_2$H is the initiator (465, 665, 668).

Similar results are found for low-temperature peroxalate-initiated decompositions of *t*-BuO$_2$H in solvents, such as *t*-BuOH or acetic acid (444),
that both promote alkoxy cleavage (685) and make hydrogen abstraction
more difficult by hydrogen bonding with the hydroperoxides (682). Cumene

hydroperoxide is more subject to this complication than $t$-BuO$_2$H since the cumyloxy radical cleaves more easily than $t$-BuO$\cdot$ (683, 684).

The decomposition of hydroperoxides in autoxidizable solvents introduces the second complication, which again arises from alkyl-radical formation. Radical-induced decompositions of $t$-BuO$_2$H in alkanes, refluxed to expel oxygen as it is formed, appear to terminate mainly via reactions (241) and (242).

$$RO\cdot \quad (\text{or } RO_2\cdot) + XH \quad \longrightarrow \quad X\cdot + ROH \quad (\text{or } RO_2H) \qquad (241)$$

$$(XH = \text{solvent})$$

$$X\cdot + RO_2\cdot \quad \longrightarrow \quad RO_2X \qquad (242)$$

The chain length goes through a minimum at about 100°. Below this temperature abstraction from solvent competes less favorably with abstraction from hydroperoxide; above, the dialkyl peroxides formed via equation 242 pyrolyze to radicals that continue the chain.

$$t\text{-BuO}_2X \quad \xrightarrow{\Delta} \quad t\text{-BuO}\cdot + XO\cdot \qquad (243)$$

Since XO$\cdot$ will in general be a secondary alkoxy radical, another possibility for termination exists:

$$\overset{\displaystyle O\cdot}{\underset{\displaystyle (=XO\cdot)}{\underset{\displaystyle |}{RCHR'}}} + t\text{-BuO}_2\cdot \quad \longrightarrow \quad t\text{-BuO}_2H + \overset{\displaystyle O}{\overset{\displaystyle \|}{RCR'}} \qquad (244)$$

Initiation by RO$_2$H homolysis coupled with termination by RO$\cdot$—R'O$_2\cdot$ interaction leads to a 4/3 rate-dependence on [RO$_2$H] (399).

## 2. Radical-Induced Decomposition of Primary and Secondary Hydroperoxides

Thermal decompositions of primary and secondary hydroperoxides, such as benzyl (457), tetralyl (686), or $sec$-decyl (668), exhibit greater than first-order kinetics and solvent oxidation, both typical of induced decompositions. However, such hydroperoxides do not undergo *chain* decomposition when exposed to free-radical initiators at 45° (687). Results of DBPO induced decomposition at this temperature suggest a sequence:

$$RR'HCO_2H + t\text{-BuO}\cdot \quad \longrightarrow \quad RR'HCO_2\cdot + t\text{-BuOH} \qquad (245)$$

$$2RR'HCO_2\cdot \quad \longrightarrow \quad RR'C{=}O + RR'HCOH + O_2 \qquad (246)$$

Termination on *every* interaction of primary or secondary peroxy radicals conforms to the cyclic mechanism, proposed originally by Russell

(688) and evidence by deuterium-isotope rate effects (688, 689); $k_H/k_D$ (equation 247) = 1.3–1.7.

$$2\,RR'HCO_2\cdot \longrightarrow RRC\underset{H\,\downarrow O}{\overset{O-O}{\diagdown}}O \longrightarrow products \qquad (247)$$

$$\underset{HCRR'}{|}$$

At higher temperatures (100° and above) peroxide-induced decompositions of secondary hydroperoxides exhibit short chains,* possibly due to some C—H abstraction, though this route is not yet proved.

$$RR'HCO_2H \xrightarrow{\;t\text{-BuO}\cdot\;} RR'\dot{C}O_2H \longrightarrow RR'C{=}O + \cdot OH \qquad (248)$$

### 3. Hydroperoxide–Oxy Radical Exchange Reactions

The facility with which peroxy and phenoxy radicals exchange hydrogen with hydroperoxides has important consequences both for autoxidation and its inhibition.

$$RO_2H + R'O_2\cdot \;\rightleftharpoons\; RO_2\cdot + R'O_2H \qquad (249)$$

$$RO_2H + PhO\cdot \;\rightleftharpoons\; RO_2\cdot + PhOH \qquad (250)$$

The rate constant for reaction (249) has been variously estimated at 10 (64, 690) to 100 (684) l/mole-sec. Thus the addition of hydroperoxide $RO_2H$ to autoxidizing substance R'H ensures that (at low conversion) the chain carrying radical will be $RO_2\cdot$ rather than $R'O_2\cdot$. The rate of autoxidation is affected markedly if, as in the case of tertiary versus primary or secondary peroxy compounds, the termination rate for two molecules of $RO_2\cdot$ differs substantially from that for two molecules of $R'O_2\cdot$ (684, 690). The technique has been used most effectively by Ingold and co-workers (689) for measuring absolute rate constants in autoxidations.

The last 10 years have seen considerable interest in the reactions of phenoxy (and substituted phenoxy) radicals with hydroperoxides which, together with the competing reactions, form the basis for the inhibition of autoxidation by phenolic substances. Excepting the more recent studies of Mahoney (691, 692), the results are covered in two major reviews (693, 694).

$$RO_2\cdot + PhO\cdot \longrightarrow products \qquad (251)$$

$$RO_2\cdot + RH \longrightarrow RO_2H + R\cdot \qquad (252)$$

---

*R. Hiatt and T. Mill, unpublished results.

Here we might consider two questions: (a) the effect of phenols on the radical-chain decomposition of hydroperoxides, and (b) the effect of hydroperoxide buildup on inhibition efficiency.

Phenols, and their nitrogen analogs, have been used to suppress the "induced" component of thermal decompositions (695, 696) as well as to count the numbers of free radicals produced by the homolysis (Section V-C-6). But, although decomposition rates are undoubtedly lowered, they still exceed the predicted value for unimolecular homolysis (Section V-C-1).

The reasons for this are not at all clear. The ability of a phenol to suppress radical-induced decomposition should depend on the rapidity of the phenoxy radical–peroxy radical termination (equation 251) compared with the readiness with which phenoxy reabstracts from hydroperoxide (equation 250). Inhibition experiments (694) show that whereas for phenol itself the ratio of the rate constants for equations 251 and 250 is less than 0.001, for highly hindered phenols it is greater than 1. Mahoney (692) states that the hydrogen exchange (equation 250) between hydroperoxides and the 2,4,6-tri-$t$-butyl-phenoxy radical is not reversible because of the very rapid peroxy–phenoxy termination. Reactions of this phenol with $Ph_3CO_2H$ consume only 0.5 mole of hydroperoxide for each mole of phenol (697).

The effect of phenols on the peroxide-induced decomposition of hydroperoxides at low temperatures has not been explored, but the very clean decomposition of $t$-$BuO_2H$ by galvinoxyl (698) seems to include no chain element. On this assumption the rate constant for hydrogen abstraction by galvinoxyl is 1.5 l/mole-sec at 20°, in good agreement with Mahoney's value for abstraction by the tri-$t$-butylphenoxy radical (0.7 l/mole-sec at 24°) (692). However, the reaction of galvinoxyl with polypropylene hydroperoxide dissolved in benzene is reported to be less than one-hundredth as fast (699).

Hydroperoxide added to a phenol-inhibited substrate may produce accelerated autoxidation, as when tetralin hydroperoxide is added to tetralin inhibited by phenol or by 4-methoxyphenol (700), or have no effect, as when the inhibitor is tri-$t$-butylphenol (701). The critical factor is the ratio $k_-/k$ of the phenoxy radical–hydroperoxide equilibrium (equation 250), estimated to be less than 1 for phenol (702), greater than 50 for $p$-methoxyphenol (692), and presumably higher for more highly substituted compounds (700).

The products of phenol–hydroperoxide reactions are generally said to be those of peroxy–phenoxy coupling (694, 703), that is,

Careful analysis, however, shows a complex mixture, including products from phenoxy–phenoxy coupling of both C—O and C—C types, as well as quinones if the phenol is unsubstituted at $C_4$ (704).

### 4. Abstractions by Nitrogen and Carbon Radicals

In contrast to the phenol, the effectiveness of 2,4,6-tri-*t*-butylaniline as an autoxidation inhibitor is diminished by the addition of hydroperoxide (701). This suggests that the equilibrium favors peroxy radicals to a greater extent

$$R\overset{\cdot}{N}H + R'O_2H \rightleftharpoons RNH_2 + RO_2\cdot \qquad (253)$$

than the similar equilibrium with phenols. The reaction of stable nitryl radicals, such as picrylhydrazyl, with hydroperoxides is slow, however, and rather complex (Section III-D-2).

Abstraction by carbon radicals tends to be obscured, partly by the reverse reaction, partly by their tendency to couple with peroxy radicals or to pick up oxygen in the system. Reaction of $CH_3\cdot$, however, with mixtures of *t*-BuO$_2$H and *t*-BuO$_2$D (681) yields a $CH_3D/CH_4$ ratio interpretable as a kinetic isotope effect of about 4 if it can be assumed that all the $CH_4$ results from O—H abstraction. This assumption is not entirely valid since reactions of $CH_3\cdot$ with *t*-BuO$_2$T have shown that abstraction from C—H is not negligible (705).

### B. Radical Displacements on Oxygen–Oxygen

Thermodynamically it would seem preferable for alkyl radicals to effect a displacement on the relatively weak O—O bond rather than to abstract hydrogen.

$$R\cdot + R'-O-O-H \longrightarrow R'O\cdot + ROH \qquad (254)$$

The reaction is not facile, as shown by Pryor's measurements of chain-transfer constants for dialkyl peroxides in polymerizing styrene (296), but it offers a logical explanation for the production of cyclic ethers in cool-flame autoxidations at 275° and above (706).

$$
\begin{array}{c}
\overset{\displaystyle O_2\cdot}{\underset{\displaystyle |}{\phantom{.}}} \qquad\qquad\qquad \overset{\displaystyle O_2H}{\underset{\displaystyle |}{\phantom{.}}} \\
Me\overset{|}{C}H(CH_2)_3Me \longrightarrow Me\overset{|}{C}HCH_2CH_2\overset{\cdot}{C}HMe
\end{array}
$$

$$(255)$$

$$
\begin{array}{c}
O \\
MeCH \diagup \quad \diagdown CHMe + \cdot OH \\
\diagdown CH_2CH_2 \diagup
\end{array}
$$

Similar proposals have been advanced to explain the formation of alcohols from the solvent when hydroperoxides undergo induced decomposition at moderate temperatures (669, 707) but are not entirely convincing since other explanations are possible (465).

$$R\cdot + O_2 \longrightarrow RO_2\cdot \tag{256}$$

$$2RO_2\cdot \longrightarrow 2RO\cdot + O_2 \xrightarrow{\text{RH}} 2ROH + O_2 \tag{257}$$

$$R\cdot + R'O_2\cdot \longrightarrow RO_2R' \longrightarrow RO\cdot + R'O\cdot \tag{258}$$

The only case in which alcohol formation has been *shown* not to proceed from alkoxy radicals is in the decomposition of very low concentrations of $t$-BuO$_2$H in cumene, where the high yield of cumyl alcohol with respect to acetophenone stands in contradiction to the known cleavage propensity of the cumyloxy radical (465).

In a truly remarkable article Shushunov and co-workers (708) claim alkyl attack on oxygen, to cleave not the O—O but the C—O bond.

$$Me_2PhC\cdot + Me_2PhC^*O_2H \longrightarrow Me_2PhCO_2H + Me_2PhC^*\cdot \tag{259}$$

The results—namely, the decomposition of labeled cumyl hydroperoxide in cumene to produce labeled cumene—seem to offer only one alternative: a mobile equilibrium between cumylperoxy and the cumyl radical plus $O_2$. However, the logical extension from either argument is that the decomposition of cumyl hydroperoxide in toluene should produce cumene. It does not (465).

Shelton and Kopczewski (709) have suggested that the reaction of nitric oxide with hydroperoxides, which produces a mixture of alcohols, alkyl nitrites and nitrates, involves displacement on O—O.

$$NO + t\text{-BuO}_2H \longrightarrow t\text{-BuO}\cdot + HONO \tag{260}$$

Peroxy acids are attacked as well, but dialkyl peroxides are not.

### C.  Thermal Decomposition of Hydroperoxides

#### 1.  Unimolecular Homolysis

Despite the many attempts to measure rate constants for unimolecular homolysis,

$$RO_2H \longrightarrow RO\cdot + HO\cdot \tag{261}$$

it was only recently (281, 312, 465, 710, 711) that rates nearly as low as those predicted on theoretical grounds [i.e., $E_a = 42$–43 kcal/mole; log $A = 14$–15 (399)] have been obtained. This measure of success has been achieved by using very low concentrations in benzene (312) or toluene (465) in the liquid or gas phase (281, 710). Homolysis, as opposed to induced decomposition, is determined by the yield of bibenzyl (or biphenyl). The method appears to work at high temperatures (180–220° in solution, $R = t$-Bu, $k$ for reaction

261 at 180° is $1 \times 10^{-5}$ sec$^{-1}$; 360–380° in the gas phase, R = Me) where the bibenzyl yield is nearly quantitative. At lower temperatures (100° in solution, 290° in the gas phase) the rates are 10–20 times as fast as predicted, a result perhaps of surface (465) or solvent-assisted (Section C-2) homolysis.

Two sets of decompositions *in vacuo* should be mentioned: Fisher and Tipper's (282) decompositions of $MeO_2H$ in treated vessels at 395° and Benson and Spokes' (711) decomposition of $t$-$BuO_2H$ in a very-low-pressure reactor at 500–1000°K. Assuming normal $A$ factors, $E_a$ values of 45–48 (281) and 43 kcal/mole are found, respectively.

### 2. *Effects of Structure and Solvent on Rates of Thermal Decomposition*

Table 11 shows the kinetic parameters for the thermal decompositions of a number of hydroperoxides in several solvents. The reactions are all first order in $RO_2H$, induced decomposition being suppressed by added inhibitor

TABLE 11. THERMAL DECOMPOSITIONS OF $RO_2H$ IN SOLUTION

| R | Solvent | $T$(°C) | $k_1 \times 10^5$ sec$^{-1}$ | $E_a$ (kcal/mole) | $A$ | Ref. |
|---|---|---|---|---|---|---|
| $t$-Butyl | Toluene | 180 | ~ 1.0 | 42 | | 465 |
| $t$-Butyl | $n$-Octane | 150 | 0.9 | 39 | $1.2 \times 10^{15}$ | 665 |
| $t$-Butyl | Water | | | 34 | $4 \times 10^{12}$ | 716 |
| $t$-Butyl | Styrene | 73.5 | 2.6[a] | | | 714 |
| $t$-Butyl | Aniline | 73.5 | 2.9[a] | — | — | 714 |
| $t$-Butyl | Benzyl alcohol | 73.5 | 14.7[a] | — | — | 714 |
| Cumyl | Cumene | 150 | 1.24 | 26.6 | $7.2 \times 10^8$ | 669 |
| Cumyl | Mineral oil[b] | 150 | 13.5 | 29.0 | $1.3 \times 10^{11}$ | 695 |
| Cumyl | 2 $M$ Styrene in xylene | 110 | 2.4 | — | — | 713 |
| Cumyl | $\alpha$-Methylstyrene | 85 | 2.9 | 18.9 | $1.7 \times 10^5$ | 61 |
| $\alpha$-Tetralyl | Mineral oil[b] | 150 | 13.5 | 29.0 | $1.3 \times 10^{11}$ | 695 |
| $\alpha$-Tetralyl | Dioctyl ether[b] | 150 | 31 | 29.0 | $2.9 \times 10^{11}$ | 410 |
| $\alpha$-Tetralyl | 2-Ethylhexene-1[b] | 150 | 43 | 19.8 | $7.4 \times 10^6$ | 410 |
| $n$-Octyl | Mineral oil[b] | 150 | 9.3 | 26.9 | $1 \times 10^{10}$ | 695 |
| $sec$-Decyl | $n$-Decane | 150 | 4.0 | 31.7 | $1 \times 10^{12}$ | 668 |
| Cyclohexyl | Cyclohexane | 150 | 3.1 | 34 | $1.2 \times 10^{13}$ | 453 |
| 9-Decalyl | Decalin | 150 | 9.9 | 32.1 | $8.5 \times 10^{13}$ | 715 |
| 9-Methyl-9-fluorenyl | Cumene | 150 | 8.8 | 27.7 | $1.8 \times 10^{10}$ | 712 |

[a] Calculated from the percentage of decomposition in 1 hr, assuming first-order reaction.

[b] Phenyl-$\alpha$-naphthylamine added to prevent induced decomposition.

(410, 419, 695) or styrene (61, 712–714), or analyzed for the first-order component of mixed kinetics (465, 665, 668, 669, 715).

Clearly, however, these rates are considerably greater and the activation energies consistently lower than predicted for unimolecular homolysis (Section C-1). The rate of decomposition in any particular environment is relatively independent of hydroperoxide structure, as might be expected (399, 692), although Japanese workers (713) claim a $\rho\sigma$ relationship for ring-substituted cumyl hydroperoxides.

Thomas (410) noted that decomposition rates roughly paralleled the ease of autoxidation of the solvent and suggested that concerted C—H and O—O bond breaking might be responsible.

$$RH + R'O_2H \longrightarrow R\cdot + R'O\cdot + H_2O \tag{262}$$

This proposal has been supported by measurements of the rates of free-radical production (see below), but it must be stressed that *induced* decomposition is also facilitated by autoxidizable solvents, probably through the formation of thermally labile mixed peroxides (465).

$$R'O_2 + R\cdot \longrightarrow R'O_2R \longrightarrow RO\cdot + R'O\cdot \tag{263}$$

### 3.  Molecularly Assisted Homolysis

Hydroperoxides unquestionably catalyze autoxidations (3, 46, 409, 717) and vinyl polymerizations (718–721), initiating chains at rates that are many times as fast as those that would be permitted by purely unimolecular homolysis (Table 12). Added inhibitors are used up with corresponding rapidity (410). Rate expressions for such reactions show a dependence on substrate in reasonable agreement with the known tendency of hydroperoxides to form complexes (Section III-B-1). Taken altogether, the evidence appears to offer incontrovertible proof that the homolysis of $RO_2H$ can be assisted by olefins, aldehydes and ketones, carboxylic acids, alcohols, amines and amine salts, sulfur compounds, bromide ions, aralkanes, or another molecule of hydroperoxide; in other words, by practically anything.*

---

*As a hedge against the future, two points should be noted:

1. Hydroperoxide decompositions are notoriously sensitive to minute impurities, particularly of metal ions. The substantial influence of purification procedures, not only of the hydroperoxide but of solvents as well (465), necessitates a somewhat skeptical attitude toward the following discussion.

2. Diacyl and dialkyl peroxides, which are somewhat less sensitive to metal-ion-catalyzed or induced decomposition, do not appear to exhibit the same kind of assisted homolysis that hydroperoxides do. The effect of olefins on the homolysis of the diacyl peroxide linkage now appears to be negligible unless the double bond is conveniently situated within the peroxide molecule itself (722–724).

TABLE 12. RATES OF FREE-RADICAL PRODUCTION[a,b,c] BY $RO_2H$ IN SOLUTION

| R | $[A]$[d] | $T$ (°C) | $k_2$ ($\times 10^6$ l/mole-sec)[d] | $E_a$ (kcal/mole) | $A$ | Ref. |
|---|---|---|---|---|---|---|
| Butenyl | $RO_2H$[e] | 70 | 5.5[(a)] | 25.8 | $1.5 \times 10^{11}$ | 131 |
| Cyclo-pentenyl | $RO_2H$[f] | 50 | ~1[(a)] | | | 46 |
| t-Butyl | $RO_2H$[g] | 70 | 0.13[(c)] | 23 | $5.7 \times 10^7$ | 725 |
| t-Butyl | Styrene | 70 | 0.36[b] | — | — | 118 |
| t-Butyl | Styrene[h] | 70 | 0.008[b] | — | — | 650 |
| t-Butyl | Styrene | 70 | 0.08[c] | 17.2 | $1.2 \times 10^4$ | 726 |
| t-Butyl | Pyridine[g] | 70 | 1.2[c] | 21 | $1.9 \times 10^7$ | 415 |
| t-Butyl | Cyclo-hexanone[g] | 100 | 3.0[c] | 26 | $3.6 \times 10^9$ | 414 |
| t-Butyl | n-Butanol | 70 | 0.55[c] | 20 | $5 \times 10^6$ | 725 |
| t-Butyl | Bromide ion | 70 | 153[c] | 19.5 | $4 \times 10^8$ | 727 |
| Cumyl | Cumene | 70 | 0.0014 | 26 | $5 \times 10^7$ | 419 |
| Cumyl | Styrene[h] | 73.5 | 0.45[b] | | | 721 |
| Cumyl | Styrene[i] | 73.5 | 6.2[b] | | | 721 |
| Cumyl | Pyridine | 73.5 | 3200[b] | | | 721 |
| Cumyl | Benzyl alcohol | 73.5 | 3500[b] | | | 721 |
| Cumyl | Benzoic acid[g] | 90 | 148[c] | 20.6 | $3.7 \times 10^8$ | 728 729 |

[a] From rates of autoxidation.
[b] From rates of polymerization.
[c] From rates of inhibitor consumption.
[d] $R = 2k_2[RO_2H][A]$.
[e] In 1-butene-benzene.
[f] In cyclopentene.
[g] In PhCl.
[h] In benzene.
[i] In pyridine.

#### 4. *Initiation of Autoxidation—Bimolecular Homolysis*

Hydroperoxides, either accumulated or added, exert an accelerating influence on hydrocarbon autoxidations that in many instances is in direct proportion to hydroperoxide concentration (46, 409, 717, 720). Initiation is thus proportional to $[RO_2H]^2$, and comparison with reactions initiated by azobisisobutyronitrile gives rates of radical production (Table 12). Most of these observations pertain to olefin oxidations (and hence to allylic hydroperoxides), but similar results have been obtained with tetralin hydroperoxide in tetralin

(717) and for the inhibitor-monitored production of radicals by $t$-BuO$_2$H (725).

Rates of hydroperoxide disappearance in an oxygen-free environment may also show second-order dependence if the concentration is high or the solvent nonpolar (409, 419, 668, 730), conditions that promote dimerization (Section III-B-1). The reaction may be supposed to proceed via the dimer, although the nature of the first-formed radicals is not established.

$$
\begin{array}{c}
R-O-O-H \\
\vdots \quad \vdots \\
H-O-O-R
\end{array}
\longrightarrow \quad R-O-O\cdot + H_2O + \cdot OR \quad (264)
$$

Dulog and Sonner (316), examining the thermal decomposition of a series of dihydroperoxides,

$$
\begin{array}{c}
\quad O_2H \qquad O_2H \\
\quad | \qquad\quad | \\
Me_2C-(CH_2)_nCMe_2
\end{array}
$$

have found values of $\Delta S\ddagger$ where $n = 1$ or 2 to be indicative of a high degree of constraint in the transition state, that is, interaction of the two groups. For $n = 3$ or 4 $\Delta S\ddagger$ values are normal.

### 5.  Initiation of Polymerization. Olefin-Assisted Homolysis

Decompositions in styrene (713, 718–721), α-methylstyrene (61, 712) or methyl methacrylate (721) are first order in [RO$_2$H] and initiate polymerization at a rate nearly independent of hydroperoxide structure (718) (Table 12). Estimation of the true rate of chain initiation, of course, requires assumptions about chain-transfer and chain-termination constants, but the results of different investigators are in fair accord (Table 12) and in line with the rates measured by inhibitor consumption (726).

Determination of the rate dependence on monomer concentration is complicated by solvent effects. Using pyridine, dimethylaniline, benzyl alcohol, and benzene as diluents, Tobolsky and Matlack (721) found that in the first three diluents hydroperoxide–monomer interactions were largely swamped by hydroperoxide–solvent radical–producing reactions, but were able to interpret their data as bimolecular reactions of several types (Table 12).

$$
R_i = k_2[RO_2H][X] + k_2'[RO_2H][M] \qquad (265)
$$

($R_i$ = rate of initiation, X = solvent, M = monomer)

Both $k_2$ and $k_2'$ depended on the nature of the solvent.

A more exhaustive study of styrene polymerizations in benzene and in carbon tetrachloride by Walling and Heaton, combined with their independently measured stability constants of hydroperoxide complexes (650), also led to multiterm rate expressions.

$$R_i = 2k_1[RO_2H] + 2k_2[RO_2H][M] + 2k_3[RO_2H][X][M] \qquad (266)$$

$$(R_i = \text{rate of initiation}, \text{M} = \text{monomer}, \text{X} = \text{solvent})$$

The values of $k_1$ and $k_2$ were approximately $10^{-8}$ l/mole-sec at 70°. In carbon tetrachloride the third-order term was negligible, but in styrene–benzene (1:1) it accounted for 80% of the initiation.

The mechanism of radical formation is not known but probably (726) involves O—O homolysis of the type demonstrated by Martin and Koenig (723).

$$(267)$$

[Walling and Heaton (650) showed that $t\text{-BuO}_2\text{D}$ produced an inverse isotope effect; $k_2\text{H}/k_2\text{D}$ (equation 266) $= 0.35$, thus disproving an earlier hypothesis (719) that hydrogen transfer from $RO_2H$ is involved.]

The acceleration caused by some solvents (650, 721) [and even by $O_2$ (64)] seems related to complexation which may weaken the O—O bond. Hydroperoxides are not efficient initiators in these systems. Comparisons between Tables 11 and 12 show that hydroperoxide decomposition is at least 10 times as fast as free-radical production. Walling and Heaton (650) calculated that in styrene at 70° only 1.6% of the decomposing $t\text{-BuO}_2\text{H}$ yields radicals; 7.9% is destroyed by radical-induced decomposition, and the remainder is converted by an apparently nonradical route to $t\text{-BuOH}$ and styrene oxide.

Radical-producing reactions between hydroperoxides and simple olefins appear to be largely obscured by the bimolecular homolysis of two molecules of $RO_2H$ (Section V-C-4). However, a small unimolecular component is sometimes observed and may be due to the former process (717). Brill and Indicator (649) have found activation energies of about 20 kcal/mole for the decomposition of $t\text{-BuO}_2\text{H}$ in several olefins. In 1-octene–benzene mixture $k_2$ for alkene–hydroperoxide interaction was $3.8 \times 10^{-8}$ l/mole-sec at 60°.

## 6. The Inhibitor Method

Measuring the rates of radical production from hydroperoxides by the rates of inhibitor consumption was pioneered by Thomas and Harle (410), who showed that the autoxidation of solvents containing both hydroperoxides and phenyl-$\alpha$-naphthylamine was delayed for periods inversely proportional to the rates of hydroperoxide decomposition. The method has since been very popular with Soviet workers, chiefly Denisov and his associates, who by spectroscopic measurements of inhibitor disappearance have obtained rate parameters for hydroperoxide homolysis assisted by carboxylic acids (418, 728, 729), alcohols (725), ketones (414, 423), hydroperoxide itself (419, 725), pyridine (415), styrene (726), cumene (419), and the bromide ion (727).

The validity of the technique rests with (a) the inhibitor's efficiency in radical trapping and (b) its inability to promote homolysis itself. These conditions seem to be met since the inhibitors (naphthylamines or naphthols at $10^{-3}$–$10^{-4}$ $M$ concentrations) disappear at rates independent of their concentration (417, 731, 732).

The rate expressions for homolysis, so determined, are in most cases first order in hydroperoxide and in assisting substance, and the rates (Table 12) are in fair agreement with those obtained by other methods (Sections C-4 and C-5). Reactions with carboxylic acids (418, 728, 729) and with ketones (414, 423), however, exhibit nonlinear dependence, so that it is necessary to postulate a series of complexes $(RO_2H)_n(A)_m$, where A is the assisting substance.

The mechanisms postulated by these workers, namely, hydrogen bonding leading to O—O weakening, hemiketal and ketal formation, or simple oxidation–reduction $(Br^-)$, are unfortunately purely speculative. The rates of inhibitor disappearance have not been correlated with the rate of hydroperoxide decomposition, nor have the products been ascertained.

### 7. Decompositions Assisted by Aldehydes and Ketones, Amines and Amine Salts, and Sulfur Compounds

A number of substances exert a destructive influence on hydroperoxides by diverse mechanisms that may be labeled "homolytic," although the efficiency of radical production is not known. The best understood of these, through the work of Mosher and co-workers (733, 734), is the reaction of primary hydroperoxides with aldehydes.

The decomposition of primary alkyl hydroperoxides is autocatalytic and yields molecular hydrogen (733). The effect of added aldehydes, deuterium-isotope rate effects, and product studies (734) are consistent with the decomposition of the peroxyhemiacetal adduct by a cyclic mechanism.

$$RCHO + RCH_2OH \longrightarrow R-\underset{H}{\overset{O-O}{\underset{|}{C}}}\underset{H\ H}{\underset{\phantom{|}}{}}\overset{O-O}{\underset{OH}{\overset{|}{C}}}-R$$

$$\downarrow$$

$$RCHO + H_2 + RCO_2H$$

(268)

The thermal decomposition of ring-substituted benzyl hydroperoxides produces less hydrogen than the primary alkyl compounds (217, 457). The yield is dependent on solvent (457) and temperature (217) and greatly enhanced by the addition of aliphatic aldehydes. The homolysis of the peroxyhemiacetal probably competes with the cyclic route—the mechanisms appear to be competitive in the case of alkylidene hydroperoxides and di-*sec*-butyl peroxide (444)—though it is not known to what extent.*

Acetone (465, 735) and methyl ethyl ketone (735) slightly catalyze the decomposition of $t$-BuO$_2$H. Fluorenone, diketones, and triketones are much more effective and cause an enhanced rate of polymerization when added to solutions of $t$-BuO$_2$H in styrene. Neither acetophenone nor benzophenone shows this effect, which occurs in the dark (735); that is, it is not simply photosensitization. Infrared studies suggest that hydrogen-bonded complexes, rather than peroxy ketals, are involved (735).

The complexity of amine–hydroperoxide reactions has been mentioned (Section IV-B-3). The combination serves to initiate polymerizations (422, 603) and autoxidations (423, 600, 602) at a faster rate than hydroperoxide alone. Amine hydrochlorides (736–739) and hydrobromides (738–740) exert a similar effect; Dulog (737) has shown that $10^{-3}$ $M$ di-butylamine hydrochloride $+2.5 \times 10^{-4}$ $M$ tetralin hydroperoxide at 50° initiates methyl methacrylate polymerization at a rate equivalent to that of 0.01 $M$ azobisisobutyronitrile.

The influence of the halide ion cannot be ignored but is not clear. Amine perchlorates or fluoroborates are not effective catalysts (738, 739), and Denisov has claimed that the effect of amine hydrobromide is due solely to Br$^-$ (727). However, in some cases the chlorides are equally good (738, 739) or better (740) as catalysts.

The reactions of hydroperoxides with organic sulfur compounds other than the sulfides (Section IV-B-2) are not well understood. Sulfoxides (741), disulfides (742–745), diselenides (746, 747), thiolsulfinates and sulfonates (742–745), and dialkyl dithiophosphates (748) all serve as inhibitors of autoxidation, primarily by destroying hydroperoxides. The products, insofar as these are known (743–747) and the fact that inhibition is occasionally preceded by a short burst of activity (747, 748), suggest a free-radical mechanism.

Typically such compounds exert no influence at all on the rate of autoxidation for a period and then suddenly stop it dead (742–748); similar induction periods are observed for the reactions of disulfides (743) and diselenides

---

*Subsequent explorations of $s$-Bu$_2$O$_2$ decompositions (R. Hiatt and S. Szilagyi, *Can. J. Chem.*, **48**, 615 (1970)) have ascertained the degree of competition, but have shed some doubt on the concerted cyclic mechanism for H$_2$ production (note added in proof).

(747) with hydroperoxides. Apparently oxidation produces compounds of higher activity (742, 743). Barnard and co-workers (744, 745) have examined the reaction of thiosulfinates with hydroperoxides at moderate temperatures, showing the products to be a mixture of thiosulfonates and disulfides in which the sulfurs in the original material have become scrambled. They suggest

$$
\begin{array}{ccc}
\overset{\displaystyle O}{\underset{(\uparrow)}{\phantom{.}}} & & \overset{\displaystyle O}{\underset{(\uparrow)}{\phantom{.}}} \\
R-S-SR & \xrightarrow{\;R'O_2H\;} & RS\cdot + RS\cdot
\end{array}
\qquad (269)
$$

Diselenides react to form seleninic anhydrides, and thence $SeO_2$, decomposing in the process about 50 equivalents of $t\text{-BuO}_2H$ to $t\text{-BuOH} + O_2$ (747). Peroxy esters are proposed as intermediates.

$$
t\text{-BuO}_2H + (RSe)_2\overset{\displaystyle O}{\underset{(\uparrow)}{\phantom{.}}} \longrightarrow RSeO_2\overset{\displaystyle O}{\underset{(\uparrow)}{\phantom{.}}}-t\text{-Bu} \longrightarrow t\text{-BuO}\cdot + RSeO\overset{\displaystyle O}{\underset{(\uparrow)}{\phantom{.}}}\cdot
$$

$$
(270)
$$

Sulfur dioxide catalyzes hydroperoxide decomposition rapidly at $0°$ (743), initiating vinyl polymerization (749–752), which is partly cationic and partly of the free-radical type (749, 750). The reaction of the thiosulfate ion with hydroperoxides under mildly acidic conditions also catalyzes polymerizations (443, 753) and autoxidation (754), perhaps via peroxy ester formation (443, 754).

### D.  Reactions of Hydroperoxides with Metal Ions

#### 1.  Transition-Metal Ions of Variable Valence

The review by Denisov and Emanuel (25) is a useful account of pre-1961 work, and papers by Hiatt, Mayo, et al. (680, 755) survey the field to 1967. The decomposition of hydroperoxides by millimolar amounts of transition-metal ions, such as $Co^{2+}$, $Mn^{2+}$, and $Fe^{2+}$, is quite rapid even at room temperature (25, 680, 755); the products in unreactive solvents can be rationalized by

$$
RO_2H + M^{n+} \longrightarrow RO\cdot + M^{(n+1)+}(OH^-) \qquad (271)
$$

$$
RO_2H + M^{(n+1)+} \longrightarrow RO_2\cdot + M^{n+}(H^+) \qquad (272)
$$

followed in turn by radical-chain decomposition in competition with

$$
RO\cdot + M^{n+} \longrightarrow RO^- + M^{(n+1)} \qquad (273)
$$

Thus $t\text{-BuO}_2H$ with catalytic amounts of $Co^{2+}$ or $Mn^{2+}$ yields about $95\%$ $O_2$ plus $t\text{-BuOH}$ and $t\text{-Bu}_2O_2$ by induced decomposition, for which the cycle of equations (271) and (272) provides an unflagging source of initiation (755).

Metal ions in large excess, however, compete effectively for $RO\cdot$ (equation 273), resulting in reduction to $ROH + H_2O$ (286). This reaction, together with a similar reduction of peroxy radicals (756, 757), presumably explains why relatively low concentrations of metal ions are optimal (54) for catalyzing autoxidations.

$$RO_2^{\cdot} + Mn^{2+} \longrightarrow RO_2^{-} + Mn^{3+} \tag{274}$$

Metal ions differ in their abilities to perform reactions 271, 272, and 273; reactivity can, as well, be influenced by the gegen-ion or by complexing ligands (25, 286, 755, 758). Thus the cobaltic EDTA complex is not reduced by $RO_2H$ (759), although cobaltic carboxylates are reduced very rapidly (755, 760). Most ferric salts do not reduce $RO_2H$ readily, though the iron–phthalocyanine complex does (755), but $Fe^{2+}$ appears to reduce $RO\cdot$ much faster than $Co^{2+}$ does (286, 755). No satisfactory correlation between activity and other seemingly relevant property, such as oxidation–reduction potential (286), has yet been adduced.*

Cleavage of the alkoxy radical is another product-forming reaction of consequence. Though useful in judging relative rates of alkoxy reduction by various metal ions (286, 762), the main application is perhaps to synthesis. The following are typical:†

$$\text{(cyclohexane ring with } R, O_2H) \xrightarrow{Fe^{2+}} \text{(cyclohexane ring with } R, O\cdot) \longrightarrow R-\overset{\overset{\displaystyle O}{\|}}{C}-(CH_2)_4CH_2\cdot \tag{275}$$

$$\text{(ring with } O, O_2H) \xrightarrow[\text{or } Cu^{1+}]{Fe^{2+}} \text{(ring with } O, O\cdot) \longrightarrow H\overset{\overset{\displaystyle O}{\|}}{C}-O(CH_2)_3CH_2\cdot \tag{276}$$

$$PhMe_2CO_2H \xrightarrow{Fe^{2+}} PhMe_2CO\cdot \longrightarrow Ph\overset{\overset{\displaystyle O}{\|}}{C}Me + Me\cdot \tag{277}$$

Final products may result by dimerization or abstraction of hydrogen (764, 767), reaction with anions in the system (763, 765, 766), or telomerization with added olefins (765). An unusual adaptation by Kochi and co-workers (768, 769) is the synthesis of benzyl chromium.

---

*Even considering known capabilities, it is frequently difficult to explain results. Wallace and Skomoroski (79, 761) have found $Fe^{2+}$ or $Ni^{2+}$ in combination with hydroperoxides to be much more effective catalysts for cumene oxidation than similar combinations with $Co^{2+}$ or $Cu^+$. Since all four ions decompose $RO_2H$ to free radicals, the difference must lie in subsequent reactions that have not been clearly elucidated.

†Reactions 275, 276, and 277 are from references 763–765, 766 and 767, and 63 respectively.

$$\text{PhCH}_2\text{CMe}_2\text{O}_2\text{H} \xrightarrow{\text{Cr}^{2+}} \text{PhCH}_2\cdot \xrightarrow{\text{Cr}^{2+}} \text{PhCH}_2\text{Cr(L)}_5 \quad (278)$$

where L represents other ligands.

Products of dehydration are frequently reported for primary and secondary hydroperoxides (63, 166, 171, 207, 767), but yields, where reported, are 50% or less. Hence it is not certain whether these represent results of radical–radical interactions or of reduction by the metal ion (755).*

$$\underset{\text{PhCH}-\text{OR}}{\overset{\text{O}_2\text{H}}{|}} \xrightarrow{\text{Fe}^{2+}} \underset{\text{PhC}-\text{OR}}{\overset{\text{O}}{\parallel}} \quad (279)$$

$$\underset{\text{PhCH}-\text{OR} + \text{Fe}^{3+}}{\overset{\text{O}\cdot}{|}} \longrightarrow \underset{\text{PhC}-\text{OR} + \text{H}^+ + \text{Fe}^{2+}}{\overset{\text{O}}{\parallel}} \quad (280)$$

Decompositions by metal ions in solvents that readily undergo hydrogen abstraction can produce good yields of dialkyl peroxides,

$$\text{RH} + \text{R}'\text{O}_2\text{H} \xrightarrow{\text{M}^{n+}} \text{RO}_2\text{R}' \quad (281)$$

where RH = cumene (770, 771), alkenes (770, 772), tetrahydrofuran (773), nitriles (774), or even cyclohexane (755). The carbonium-ion mechanism advanced by Kochi (20, 21) is based on the concurrent formation of RX from anions [$\text{Cl}^-$ or $\text{Br}^-$ (20, 763), $^-\text{OAc}$ (20, 775), $\text{N}_3^-$ (772)] present in the system. However, the steady-state concentrations of $\text{R}'\text{O}_2\cdot$ [as measured by ESR spectra (755, 776)] are large enough to make coupling with $\text{R}\cdot$ a feasible route (755).

$$\text{R}'\text{O}\cdot \xrightarrow{\text{RH}} \text{R}\cdot \xrightarrow{\text{M}^{n+}} \text{R}^+ \xrightarrow{\text{R}'\text{O}_2\text{H}} \text{RO}_2\text{R}' \quad (282)$$

Specific circumstances seem to determine whether alkoxy or peroxy radicals, both derived from metal ion–hydroperoxide interactions, are the predominant factor in subsequent reactions. Walling and Zavitsas (289) have shown that the selectivity in abstraction from RH versus R'H by a $t\text{-BuO}_2\text{H}$–$\text{Cu}^+$ combination is more in line with the activity of $t\text{-BuO}_2\cdot$ than of $t\text{-BuO}\cdot$. Similarly Bigot (773) claims that the decomposition of $t\text{-BuO}_2\text{H}$ by $\text{Co}^{2+}$ or $\text{Cu}^+$ in refluxing pyrrole or furan leads to $t\text{-BuO}_2\cdot$ addition.

$$\text{(furan)} + t\text{-BuO}_2\text{H} \xrightarrow{\text{Co}^{2+}} t\text{-BuO}_2\text{(ring)O}_2-t\text{-Bu} \quad (283)$$

On the other hand, decomposition in the presence of butadiene results exclusively in $t\text{-BuO}\cdot$ addition (777, 778).

---

*The substantially greater than 50% yields of benzaldehyde from cobalt ion-catalysed decomposition of benzyl hydroperoxide (E. J. Y. Scott, *J. Phys. Chem.*, **74**, 1174 (1970)) would appear to favor the latter (note added in proof).

$$t\text{-BuO}_2\text{H} + \text{C}_4\text{H}_6 \xrightarrow{\text{CuCl}} t\text{-BuOCH}_2\text{CH}=\text{CHCH}_2\text{Cl}$$

$$+ \ t\text{-BuOCH}_2\text{CHCH}=\text{CH}_2 \qquad (284)$$
$$\underset{\displaystyle \text{Cl}}{\overset{\displaystyle |}{\phantom{.}}}$$

It is generally agreed (392–397, 779) that the only radical in evidence in the ESR spectra of metal ion–hydroperoxide mixtures is $\text{RO}_2^{\cdot}$.

The kinetics of metal-ion-catalyzed decompositions of hydroperoxides are complex (25, 755). A consistent feature is autoretardation, which may be more or less extreme, depending on the temperature, the particular metal catalysts, the hydroperoxides, and the solvent (755). Apparently the products that are formed complex with the metal ion, deactivating it and causing its eventual precipitation. The precise nature of these substances is still unknown, though formic acid is suspected to be the chief offender in cases in which alkoxy cleavage can yield methyl radicals (755).

The initial rates of decomposition in most instances are first order or nearly so in $[\text{RO}_2\text{H}]$ (755) but show considerable variation in the dependence on metal-ion concentration, less than first order and greater than first order all having been reported for catalysis by complexes of cobalt alone, depending on the solvent and temperature (780).

$$\frac{-\partial[\text{RO}_2\text{H}]}{\partial t} = k[\text{RO}_2\text{H}][\text{Co}]^{0-3} \qquad (285)$$

First-order dependence on total metal-ion concentration is consistent with kinetically significant activity of the metal ion being limited to rapid oxidation–reduction cycling via equations 271 and 272 (755). Valiant attempts have been made to rationalize lower and higher orders in terms of metal ion–hydroperoxide complexes (759, 760, 781, 782). Independent efforts to identify such complexes by spectroscopic means have not been fruitful, however (122, 783), and the irregular kinetics can alternatively be explained, though not in detail, by the metal catalysts' lack of true solubility in the medium (755).

### 2. Decomposition by Other Metal Ions

The ability of lead tetraacetate to produce cyclic peroxides from 1,3- or 1,4-dihydroperoxides, discovered by Criegee and his co-workers (324, 784), has been frequently exploited by Rieche's group in their peroxide syntheses (351). The reagent produces dicumyl peroxide from cumene hydroperoxide but in low yield (785).

$$\underset{\displaystyle \text{Me}_2\overset{\displaystyle \overset{\textstyle \text{O}_2\text{H}}{|}}{\text{C}}-\text{CH}_2-\overset{\displaystyle \overset{\textstyle \text{O}_2\text{H}}{|}}{\text{C}}-\text{Me}_2} \xrightarrow{\text{Pb(OAc)}_4} \underset{\displaystyle \text{CH}_2}{\text{Me}_2\text{C}\overset{\displaystyle \text{O}-\text{O}}{\diagdown\diagup}\text{CMe}_2} \qquad (286)$$

Acetophenone and cumyl alcohol form the greatest share of the products, the relative amounts being suggestive of free-radical-induced decomposition. No detailed mechanistic studies seem to have been carried out, although an ionic mechanism has been proposed (785).

Stearates of $Pb^{2+}$ (755), $Mg^{2+}$, $Zn^{2+}$, $Ca^{2+}$, and $Sr^{2+}$ (786) have been reported to catalyze the decomposition of hydroperoxides, though much less rapidly than compounds of cobalt, manganese, or iron. A mechanism involving one-electron oxidation or reduction of these ions seem unlikely, and their mode of operation is not clear. It has been conjectured that $Zn^{2+}$ acts as a Lewis acid and that $Mg^{2+}$ in some way facilitates radical-chain decomposition, though not starting chains itself (786).

The dual role of vanadium deserves comment. As the acetylacetonate it decomposes hydroperoxides in a manner similar to that of manganese or cobalt (397, 755), which in the presence of methyl methacrylate initiates its polymerization efficiently (787, 788). In the presence of *simple* olefins this free-radical type of decomposition is negligible; instead the nonradical epoxidation (Section IV-B-5) is catalyzed. This influence of olefin structure on the mechanism has not been explained.

# APPENDIX

## Hydroperoxides of Simple Organic Compounds.

The following tables comprise a listing of hydroperoxides of simple organic compounds that have been synthesized and characterized. Some compounds are included for which melting or boiling points are not given but from the citation seem reasonably certain to have been obtained.

The key to the method of synthesis is as follows:

$A$ = autoxidation
$A_s$ = photosensitized autoxidation
$A_1$ = autoxidation by singlet oxygen, chemical or photolytic
$B$ = trialkylborane autoxidation
$G$ = Grignard autoxidation
$H_a$ = addition of $H_2O_2$
$H_d$ = displacement by $H_2O_2$
$O$ = ozone

The listed physical properties are those of the first-cited reference unless otherwise indicated. For refractive indices at temperatures other than 20° the temperature is shown in parentheses after the value.

TABLE **A-1.**   ALKYL HYDROPEROXIDES

| $C_n$ | Hydroperoxide | bp/mm Hg | $n_D^{20}$ | mp | Method and Ref. |
|---|---|---|---|---|---|
| 1 | $MeO_2H$ | 45.5–46.5/184 | 1.3654 (21°) | | $H_d$ (281) |
| 2 | $EtO_2H$ | 43–44/50 | | | $H_d$ (280) |
| 3 | $PrO_2H$ | 35/20 | 1.3890 | | $H_d$ (280) |
| | $i\text{-}PrO_2H$ | 38–38.5/20 | (25°) | | $H_d$ (1, p. 9) |
| 4 | $BuO_2H$ | 40–42/8 | 1.4057 | | $H_d$ (291) |
| | $sec\text{-}BuO_2H$ | 41–42/11 | 1.4050 | | $H_d$ (291) |
| | $i\text{-}BuO_2H$ | 39–40/15 | 1.4023 | | $H_d$ (789), $G$ (216) |
| | $t\text{-}BuO_2H$ | 33–34/17 | 1.3983 (25°) | 4.0–4.5 | $Hd$ (283), $G$ (210) |
| | $Me_2C(CN)O_2H$ | 37/1 | 1.4138 | 4–5 (360) | $A$ (790, 98, 360) |
| 5 | $n\text{-}BuCH_2O_2H$ | 41–42/4 | 1.4146 | | $H_d$ (291) |
| | $PrCHMeO_2H$ | 46–47/7 | 1.4140 | | $H_d$ (291) |
| | $Et_2CHO_2H$ | 51–52/10 | 1.4155 | | $H_d$ (291) |
| | $i\text{-}BuCH_2O_2H$ | 47–48/6 | 1.4118 | | $H_d$ (291) |
| | $sec\text{-}BuCH_2O_2H$ | — | — | | $H_d$ (218) |
| | $i\text{-}PrCHMeO_2H$ | 42.1–43/1 | | | $G$ (218) |
| | $EtCMe_2O_2H$ | 34–35/7 | 1.4120 (25°) | | $H_d$ (284, 310), $G$ (210) |
| 6 | $n\text{-}C_6H_{13}O_2H$ | 42–43/2 | 1.4208 | | $H_d$ (291) |
| | $BuCHMeO_2H$ | 29–30/1 | 1.4186 | | $H_d$ (291) |
| | $PrCHEtO_2H$ | 48–49/5 | 1.4204 | | $H_d$ (291) |
| | $PrCMe_2O_2H$ | 44/3 | 1.4198 | | $H_d$ (287) |
| | $i\text{-}PrCMe_2O_2H$ | 56–57/10 | 1.4226 (25°) | | $H_d$ (310) |
| 7 | $n\text{-}C_7H_{15}O_2H$ | 42–43/0.06 | 1.4269 | 35.5 | $G$ (385), $H_d$ (291) |
| | $n\text{-}C_5H_{11}CHMeO_2H$ | 33/0.01 | 1.4262 | | $G$ (385), $H_d$ (291) |
| | $BuCHEtO_2H$ | 35/0.08 | 1.4268 | | $G$ (385) |
| | $Pr_2CHO_2H$ | 36/0.08 | 1.4272 | | $G$ (385) |
| | $PrCMeEtO_2H$ | 43–44/1 | 1.4302 | | $H_d$(286, 298) |
| | $BuCMe_2O_2H$ | 59/3.5 | 1.4270 | | $H_d$ (287) |
| | $t\text{-}BuCMe_2O_2H$ | | | 113–114 | $H_d$ (271, 310) |
| | $Et_3CO_2H$ | 70–71/17 | | | $H_d$ (271) |

TABLE **A-1**—*Continued*

| $C_n$ | Hydroperoxide | bp/mm Hg | $n_D^{20}$ | mp | Method and Ref. |
|---|---|---|---|---|---|
| 8 | $n$-$C_8H_{17}O_2H$ | 40–44/0.01 | 1.4311 | | $G$ (212), $B$ (215), $H_d$ (291) |
| | $n$-$C_6H_{13}CHMeO_2H$ | 39–42/0.01 | 1.4229 | | $G$ (212), $H_d$ (292) |
| | $n$-$C_5H_{11}CHEtO_2H$ | 32–34/0.01 | 1.4308 | | $G$ (212) |
| | $BuCHPrO_2H$ | 35–37/0.01 | 1.4297 | | $G$ (212) |
| | $BuCHEtCH_2O_2H$ | 60/1.7 | 1.4330 | | $B$ (215) |
| | $i$-$BuCH_2CMe_2O_2H$ | 29–30/0.02 | | | $A$ (106) |
| | $t$-$BuCMeEtO_2H$ | | | 113–114 | $H_d$ (284) |
| | $n$-$C_5H_{11}CMe_2O_2H$ | 58/1.8 | 1.4310 | | $H_d$ (287) |
| | $PrCEt_2O_2H$ | 47–48/1 | 1.4371 | | $H_d$ (286) |
| | $t$-$BuCH_2CMe_2O_2H$ | 44–45/0.9 | | | $H_a$ (289) |
| 9 | $n$-$C_9H_{19}O_2H$ | 53–55/0.3 | 1.4330 | | $H_d$ (291) |
| | $n$-$C_6H_{13}CMe_2O_2H$ | 40/0.04 | 1.4332 | | $H_d$ (287) |
| | $BuCMePrO_2H$ | 55–60/1 | 1.4334 | | $H_d$ (286) |
| 10 | $n$-$C_{10}H_{21}O_2H$ | 61–63/0.3 | 1.4378 | | $H_d$ (291) |
| | $n$-$C_7H_{15}CMe_2O_2H$ | 51/0.03 | 1.4367 | | $H_d$ (287) |
| | $BuCEtPrO_2H$ | 62–63/1 | 1.4409 | | $H_d$ (286) |
| | $i$-$Pr(CH_2)_4CMe_2O_2H$ | — | — | | $H_d$ (316) |
| 11 | $n$-$C_8H_{17}CMe_2O_2H$ | 59/0.03 | 1.4400 | | $H_d$ (287) |
| | $n$-$C_{11}H_{23}O_2H$ | 69/0.005 | 1.4415 | | $B$( 215) |
| 12 | $n$-$C_{12}H_{25}O_2H$ | | | 12–13 | $H_d$ (791) |
| 13 | (4-camphyl)$CMe_2O_2H$ | | | 37–39 | $H_d$ (310) |
| 14 | $n$-$C_{14}H_{29}O_2H$ | | | 29–30.5 | $H_d$ (791) |
| 15 | $n$-$C_{12}H_{25}CMe_2O_2H$ | | 1.4474 | 13–14 | $H_d$ (309) |
| 16 | $n$-$C_{16}H_{44}O_2H$ | | | 42–44 | $H_d$ (791) |
| 18 | $n$-$C_{18}H_{37}O_2H$ | | | 49–50 | $H_d$ (791) |
| 20 | $n$-$C_{12}H_{25}CMe(C_6H_{11})O_2H$ | | 1.4715 | 21–23 | $H_d$ (309) |
| 7 | $HO_2CMe_2CH_2CMe_2O_2H$ | Oil | | | $H_d$ (784) |
| 8 | $HO_2CMe_2(CH_2)_2CMe_2O_2H$ | | | 105 | $H_d$ (784, 519) |
| 9 | $HO_2CMe_2(CH_2)_3CMe_2O_2H$ | | | | $H_d$ (316) |
| 10 | $HO_2CMe_2(CH_2)_4CMe_2O_2H$ | | | 66 | $H_d$ (271, 316) |
| 4 | $BuO_2D$ | | | | $H_d$ (789) |

Table **A-2.** cycloalkyl hydroperoxides

| $C_n$ | Hydroperoxide | bp/mm Hg | $n_D^{20}$ | mp | Method and Ref. |
|---|---|---|---|---|---|

| | | | | | |
|---|---|---|---|---|---|
| 10 | | — | — | | $H_d$ (315) |

| $C_n$ | R | | | | |
|---|---|---|---|---|---|
| 5 | H | 46–47/4 | 1.4533 | | $H_d$ (291) |
| 6 | Me | 73.5/19 | 1.4526 | | A (792) |
| | CN | — | — | | A (98) |
| 8 | Pr | 34.5/0.003 | — | | A (1, p. 23) |
| 11 | Ph | Oil | 1.5452 | | $H_d$ (315, 557) |

| | R | | | | |
|---|---|---|---|---|---|
| 6 | H | 57/1.2 (453) | 1.4622 (453) | | |
| | | 29/0.03 | 1.4646 | | G (60), A (453), B (215) |
| 7 | Me | ·38/0.03 | 1.4652 | | A (60) |
| 8 | Et | 39/0.01 | 1.4667 | | A (60) |
| 9 | i-Pr | 52/0.01 | 1.4720 | | A (60) |
| 10 | t-Bu | 61/0.01 | 1.4757 | | A (60) |
| 12 | $C_6H_{11}$ | Oil | 1.4921 | | A (60) |
| | Ph | | | 59–60 | $H_d$ (315, 557) |

TABLE A-2—*Continued*

| $C_n$ | Hydroperoxide | bp/mm Hg | $n_D^{20}$ | mp | Method and Ref. |
|-------|---------------|----------|------------|-----|-----------------|
| | R | | | | |
| 8 | 2-Me | 28–29/0.01 | 1.4672 | | $A$(106) |
| | 2-Me-2(O$_2$H) | — | — | | $A$ (106) |
| | 3-Me | 26–27/0.01 | 1.4602 | | $A$ (106) |
| | 3-Me-3(O$_2$H) | — | — | | $A$ (106) |
| | 4-Me (*trans*) | | | 22 | $A$ (106) |
| | 4-Me-4(O$_2$H) (*trans*) | | | 147 | $A$ (106) |
| 8 | (CH$_2$)$_7$CHO$_2$H | 48–49/10$^{-4}$ | 1.4830 | | $B$ (215) |
| 12 | (CH$_2$)$_{11}$CH$_2$OH | | | 27–28 | $B$ (215) |
| 10 | | | | 94–95 | $H_d$ (271) |
| 9 | | 64–65/0.01 | | | $A$ (793) |
| 17 | | | | 70 | $A$ (86) |

| $C_n$ | R$^1$ | R$^2$ | R$^3$ | bp/mm Hg | $n_D^{20}$ | mp | Method and Ref. |
|-------|-------|-------|-------|----------|------------|-----|-----------------|
| 10 | H | H | H | | | 56 | $A$ (794, 795) |
| 11 | Me | H | H | 99–100/0.1 | | | $A$ (796) |
| 12 | Me | Me | H | Oil | | | $A$ (88) |
| | Me | Me | O$_2$H | | | 149–150 | $A$ (88) |
| 13 | Me | Me | Me | | | 38.5–40 | $A$ (797) |
| 14 | Et | Et | H | | | 63–64 | $A$ (88) |
| | Et | Et | O$_2$H | | | 120 | $A$ (88) |

TABLE A-2—*Continued*

| $C_n$ | Hydroperoxide | bp/mm Hg | $n_D^{20}$ | mp | Method and Ref. |
|---|---|---|---|---|---|
| 12 | | | | 42 | *A* (59) |

| | R | | | | |
|---|---|---|---|---|---|
| 13 | H | | | 53–55 | $H_d$ (798) |
| 14 | Me | | | 150[a] | *A* (90, 712) |
| 15 | Et | | | 108–109 | *A* (90) |
| 17 | Bu | | | 56–58 | *A* (90) |
| 19 | Ph | | | 67–69 | *A* (90) $H_d$ (798) |
| 20 | PhCH$_2$ | | | 70–72 | *A* (90) |

| 13 | | | | 123–124 | H*d* (298) |

| 14 | | | | 108.5–109 | *A* (1, p. 17) |

| 14 | | | | 69 | *A* (59) |

| 26 | | | | 240 | $H_d$ (799) |

[a] Decomposes.

TABLE A-3.   ARALKYL HYDROPEROXIDES

| $C_n$ | Hydroperoxide | bp/mm Hg | $n_D^{20}$ | mp | Method and Ref. |
|---|---|---|---|---|---|

| $C_n$ | R | | | | |
|---|---|---|---|---|---|
| 7 | H | — | — | | $H_d$ (322), G (210) |
| 8 | p-Me | 51/0.05 | 1.5322 | | A (800, 457) |
| 10 | 2,4,6-Me$_3$ | | | 86–89 | G (217) |

| $C_n$ | R | | | | |
|---|---|---|---|---|---|
| 8 | H | 48.2/0.2 | 1.5265 | | $H_d$ (557), A (800) |
| 10 | p-Et | 57–58/0.01 | 1.5192 | | A (106) |
| | o-Et | 60–61/0.06 | 1.5248 | | A (106) |
| | p-(HO$_2$CHMe) | — | — | | A (106) |
| | PhCH$_2$CH$_2$O$_2$H | 55/10$^{-4}$ | 1.5290 | | B (215) |

| $C_n$ | R | | | | |
|---|---|---|---|---|---|
| 9 | H | 60/0.2 | | | A (794) |
| | p-Br | | | 43.5 | A (61) |
| | o-Cl | 73/0.05 | 1.5452 | | $H_d$ (557) |
| | m-Cl | | 1.5377 | 25–26 | $H_d$ (557) |
| | p-Cl | 85/0.4 | 1.5389 | 34 | A (61, 557) |
| | p-I | | | 56.0 | A (61) |
| | p-NO$_2$ | | | 40 | A (801) |
| | o-Me | 60/0.04 | 1.5312 | | $H_d$ (557) |
| | p-Me | 66.5/0.1 | 1.5202 | | A (557) |
| | p-CO$_2$H | | | 152 | A (102) |
| | m-i-Pr | Oil | 1.5121 | | A (557) |
| 12 | p-i-Pr | | 1.5134 | 33–34 | A (557) |
| | p-(HO$_2$CMe$_2$) | | | 140–141.5 | A (272) |

T A B L E A-3—*Continued*

| $C_n$ | Hydroperoxide | bp/mm Hg | $n_D^{20}$ | mp | Method and Ref. |
|---|---|---|---|---|---|
| 13 | p-(MeOCMe$_2$) | Oil | 1.5510 | | A (272) |
| | p-t-Bu | | | 77–78 | A (557) |
| 14 | p-(Me$_3$CCO$_2$) | | | 83–83.5 | A (802) |
| 10 | PhCMeEtO$_2$H | 68–71/0.1 | 1.5222 | | A (286, 452), H$_d$ (298, 303) |
| | PhCHPrO$_2$H | 56/0.006 | 1.5103 | | A (803) |

| | R | | | | |
|---|---|---|---|---|---|
| 10 | H | | | 38–41 | H$_d$ (310) |
| 11 | m-MeO— | Oil | | | H$_d$ (310) |
| | p-MeO— | Oil | | | H$_d$ (310) |
| 11 | Ph(CH$_2$)$_2$CMe$_2$O$_2$H | Oil | | | H$_d$ (310) |

| | X | R | | | mp | Method and Ref. |
|---|---|---|---|---|---|---|
| 13 | H | H | | | 48–49 | H$_d$ (313, 314), A (794) |
| | o-Cl | H | | | 48–51.5 | H$_d$ (314) |
| | m-Cl | H | | | 38.5–40.5 | H$_d$ (314) |
| | p-Cl | H | | | 56–58 | H$_d$ (314) |
| 14 | o-Me | H | | | 62–63.5 | H$_d$ (314) |
| | m-Me | H | Oil | | | H$_d$ (314) |
| | p-Me | H | | | 50.5–52 | H$_d$ (314) |
| | p-MeO | H | | | 50–51 | H$_d$ (298) |
| | H | Me | | | 86 | A (357), |
| 15 | p-Me | Me | Oil | | | A (804) |
| | H | Et | | | 81–82 | A (357), H$_d$ (798) |
| 16 | H | Pr | | | 39 | A (804) |
| 18 | t-Bu | Me | | | 47 | A (804) |
| | p-Ph | H | | | 162–163 | H$_d$ (298) |
| 19 | H | Ph | | | 81–82 | H$_d$ (271) |
| | (p-NO$_2$C$_6$H$_4$)$_3$CO$_2$H | | | | 190–191 | A (91) |

TABLE A-3—*Continued*

| $C_n$ | Hydroperoxide | bp/mm Hg | $n_D^{20}$ | mp | Method and Ref. |
|---|---|---|---|---|---|
| 13 | $\alpha$-$C_{10}H_7CMe_2O_2H$ | | | 66–67 | A (805) |
| | $\beta$-$C_{10}H_7CMe_2O_2H$ | | | | A (805) |
| 18 | $PhCMe(O_2H)CH_2S(\beta$-$C_{10}H_7)$ | | | 70 | A (140, 141) |
| 19 | (indanyl)—$S(\beta$-$C_{10}H_7)$, $O_2H$ | | | 10 | A (140, 141) |
| 20 | $C_{12}H_{25}CMePhO_2H$ | | 1.4902 | 34 | $H_d$ (309) |

TABLE A-4.   ALKENYL HYDROPEROXIDES

| $C_n$ | Hydroperoxide | bp/mm Hg | $n_D^{20}$ | mp | Method and Ref. |
|---|---|---|---|---|---|
| 3 | $CH_2=CHCH_2O_2H$ | | 1.4200 | | $H_d$ (292) |
| 5 | (cyclopentenyl)—$O_2H$ | 35/0.01 | | | A (434) |
| 6 | $Me_2C=CH_2CHMeO_2H$ | — | — | | A (146) |
| | $MeCH=CHCMe_2O_2H$ | — | — | | A (146) |
| | $CH_2=CMeCMe_2O_2H$ | 54–55/12 | | | $A_1$ (233) |
| | (cyclohexenyl)—$O_2H$ | 51/0.3 | 1.0588 | | A (434, 430, 806) |
| 7 | (Me-cyclohexenyl)—$O_2H$ | 47–51/0.01 | | | A (434) |
| | $AcCH=CHCMe_2O_2H$ | | | 34.5 | A (174, 149) |
| 8 | $i$-$PrCH=CHCMe_2O_2H$ | 28–29/0.5 | 1.4438 | | A (106) |
| | $HO_2CMe_2CH=CHCMe_2O_2H$ | — | — | | A (106) |

TABLE A-4—*Continued*

| $C_n$ | Hydroperoxide | bp/mm Hg | $n_D^{20}$ | mp | Method and Ref. |
|---|---|---|---|---|---|
| 9 | | 73–75/0.2 | | | A (807) |
| 0 | | | | 60 | $A_1$ (233) |
| 2 | | | | 38–40 | $A_1$ (233) |
| | S—CMe$_2$CH=CHCMe$_2$O$_2$H | | | | |
| | **R** | | | | |
| 4 | H | | | 38.5–40 | A (143) |
| | p-Cl | | | 44–45 | A (143) |
| 5 | p-Me | | | 40–41.5 | A (143) |
| 8 | p-t-Bu | | | 64–66 | A (143) |
| 7 | | | | 148–149[a] | $A_1$ (235, 237) |
| 7 | | | | 154–156.5 | $A_1$ (148) |

[a] Decomposes.

TABLE A-5. ALKYNYL HYDROPEROXIDES[a]

| $C_n$ | Hydroperoxide | bp/mm Hg | $n_D^{20}$ | mp |
|---|---|---|---|---|
| 5 | $HC \equiv CCMe_2O_2H$ | 42/17 | | |
| 6 | $HC \equiv CCMeEtO_2H$ | 38–40/5 | | |
| 8 | $HO_2CMe_2C \equiv CCMe_2O_2H$ | | | 107–109 |
| 10 | $HO_2CMe_2(C \equiv C)_2CMe_2O_2H$ | | | 95–96.5 |
| 12 | $HO_2CMeEt(C \equiv C)_2CMeEtO_2H$ | | | 44–44.5 |

| 14 | $n = 1$ | | | 95 |
| 16 | $n = 2$ | | | 96–97 |

[a] Method: displacement by hydrogen peroxide. Data from reference 285.


TABLE A-6. HYDROPEROXIDES OF KETONES, ENOLS, ESTERS, AND CARBOXYLIC ACIDS

| $C_n$ | Hydroperoxide | $b_p$/mm Hg | $n_D^{20}$ | mp | Method and Ref. |
|---|---|---|---|---|---|

| | $R^1$ | $R^2$ | $R^3$ | | | | |
|---|---|---|---|---|---|---|---|
| 5 | Me | Me | Me | | | 28–30 | A (808) |
| 7 | $-(CH_2)_5-$ | | Me | Oil | | | A (808) |
| 7 | Me | Me | i-Pr | 41–43/11 | | 43–44 | A (173, 808) |
| 21 | $Ph_2CH$ | Me | Ph | | | 127 | G (171) |
| 23 | $Ph_2CH$ | H | Mesityl | | | 117 | G (171) |
| | PhEtCH | Ph | Ph | | | 152–53 | G (173) |
| 26 | $Ph_2CH$ | Ph | Ph | | | 127 | G (536) |

TABLE **A-6**—*Continued*

| $C_n$ | Hydroperoxide | | | | | bp/mm Hg | $n_D^{20}$ | mp | Method and Ref. |
|---|---|---|---|---|---|---|---|---|---|
| | $R^1$ | $R^2$ | $R^3$ | $R^4$ | $R^5$ | | | | |
| 11 | Me | Me | Me | Me | Me | | | 108–110 | *A* (93) |
| 14 | H | OH | *t*-Bu | H | *t*-Bu | | | 137.5–138.5 | *A* (92) |
| 15 | H | OMe | *t*-Bu | H | *t*-Bu | | | 163.5–165 | *A* (92) |
| | *t*-Bu | H | Me | H | *t*-Bu | | | 115 | *A* (93) |
| | *t*-Bu | H | OMe | H | *t*-Bu | | | 84–86 | *A* (93) |
| 17 | Me | *t*-Bu | CH=CH₂ | H | *t*-Bu | | | 123–124.5 | *A* (180) |
| 18 | *t*-Bu | H | *t*-Bu | H | *t*-Bu | | | — | *A* (93) |

| 13 | | | | 135–137 | *A* (179) |
|---|---|---|---|---|---|

| 16 | | | | 179–181 | *A* (179) |
|---|---|---|---|---|---|

| | R | | | | |
|---|---|---|---|---|---|
| 14 | Me | | | 160–162 | *A* (809) |
| 15 | CH=CH₂ | | | 112–117[a] | *A* (809) |
| | Et | | | 104–107 | *A* (809, 810) |
| 16 | Pr | | | 73 | *A* (810) |
| 19 | Ph | | | 179–181 | *A* (809) |

| 20 | | | | | |
|---|---|---|---|---|---|

TABLE A-6—Continued

| $C_n$ | Hydroperoxide | | | bp/mm Hg | $n_D^{20}$ | mp | Method and Ref. |
|---|---|---|---|---|---|---|---|
| | $R^1$ | $R^2$ | X | | | | |
| | $\beta$-OH | $\alpha$-H | $H_2$ | | | 169–172 | A (94) |
| | $\beta$-OH | $\alpha$-H | O | | | 161–163 | A (94) |
| | $\alpha$-OH | $\beta$-H | $H_2$ | | | 162–163 | A (94) |
| | $\alpha$-OH | $\beta$-H | O | | | 158–160 | A (94) |

$$\begin{array}{c} O_2H \\ | \\ R^1-C-CO_2Me \\ | \\ R^2 \end{array}$$

| $C_n$ | $R^1$ | $R^2$ | bp/mm Hg | $n_D^{20}$ | mp | Method and Ref. |
|---|---|---|---|---|---|---|
| 14 | Ph | Ph | | | 69.5–70.5 | A (97) |
| 15 | Ph | $p$-Cl-C$_6$H$_4$CH$_2$ | | | 94–96 | A (97) |
| | Ph | PhCH$_2$ | | | 108.5–109.5 | A (97) |
| 16 | $p$-MeC$_6$H$_4$ | $p$-MeC$_6$H$_4$ | | | 69–71 | A (97) |
| 3 | HO$_2$CH(CO$_2$H)$_2$ | | | | 120 | O (269) |

| 5 | | | | 78 | $H_a$ (808) |
|---|---|---|---|---|---|

| 7 | | Oil | | | $H_a$(808) |
|---|---|---|---|---|---|

---

TABLE **A-7**.    ALKYLIDENE HYDROPEROXIDES

| $C_n$ | Hydroperoxide | b$_p$/mm Hg | $n_D^{20}$ | mp | Method and Ref. |
|---|---|---|---|---|---|
| | OH<br>\|<br>RCHO$_2$H | | | | |
| | R | | | | |
| 1 | H | Oil | | | $H_a$ (811) |
| 2 | Me | Oil | | | $H_a$ (812) |
| | Cl$_3$C | | | 122 | $H_a$ (813) |

TABLE A-7—*Continued*

| $C_n$ | Hydroperoxide | bp/mm Hg | $n_D^{20}$ | mp | Method and Ref. |
|---|---|---|---|---|---|
| 3 | Et | Oil | | | $H_a$ (27) |
| 4 | Pr | 40/0.1 | 1.4438 | | $A_s$ (161) |
| 5 | Bu | Oil | | | $H_a$ (27) |
| 7 | $n$-$C_6H_{13}$ | | | 40 | $H_a$ (812) |
| 8 | $n$-$C_7H_{15}$ | | | 46 | $H_a$ (812) |
| 9 | $n$-$C_8H_{17}$ | | | 50–54 | $H_a$ (812) |
| 10 | $n$-$C_9H_{19}$ | | | 61 | $H_a$ (812) |
| 11 | $n$-$C_{10}H_{21}$ | | | 62 | $H_a$ (812) |
| 12 | $n$-$C_{11}H_{21}$ | | | 65–67 | $H_a$ (812) |

$$\underset{\substack{| \\ R^1CHO_2H}}{OR^2}$$

| $C_n$ | $R^1$ | $R^2$ | bp/mm Hg | $n_D^{20}$ | mp | Method and Ref. |
|---|---|---|---|---|---|---|
| 3 | H | Et | 51–52/11 | 1.4063 | | $H_d$ (273) |
| | Me | Me | — | 1.4120 | | O(814) |
| 4 | Me | Et | 62–64/14 | | | $H_d$ (273), $H_a$ (323) |
| 5 | H | $t$-Bu | 31–33/4 | 1.4131 | | O (815) |
| | Et | Et | 35–36/0.02 | 1.4158 (21°) | | $H_d$ (273) |
| | Pr | Me | 40/0.015 | 1.4198 | | $H_d$ (273) |
| 6 | Me | Bu | 48–51/2 | 1.4197 | | $H_a$ (327) |
| | Et | Pr | 41–42/0.17 | 1.4185 (23°) | | $H_d$ (273) |
| | $t$-Bu | Me | 47/0.9 | | 37–38 | O (816) |
| 7 | $t$-Bu | Et | 38/0.25 | | 18–19 | O (816) |
| 8 | Pr | Bu | 62/0.1 | 1.4311 | | $A_s$ (161) |
| 9 | $i$-Bu | $i$-BuCH$_2$ | 50–60/0.05 | 1.4402 | | $A_s$ (161) |
| 7 | $CH_2$=CH | Bu | | | | $A$ (166) |
| 8 | Ph | Me | | | | $A$ (166), O (268) |
| 9 | $n$-$C_6H_{13}$ | Et | | | | O (264) |
| | Benzoyl | Me | | | 81–82 | O (261, 262) |
| 10 | Benzoyl | Et | | | 77 | O (261, 262) |

$$\underset{\substack{| \\ R^1R^2CO_2H}}{OR^3}$$

TABLE A-7—*Continued*

| $C_n$ | Hydroperoxide | | | bp/mm Hg | $n_D^{20}$ | mp | Method and Ref. |
|---|---|---|---|---|---|---|---|
| | $R^1$ | $R^2$ | $R^3$ | | | | |
| 3 | Me | Me | H | 20–25/0.01 | 1.4183 | | $A_s$ (161) |
| 4 | Me | Me | Me | 61–63/18 | | | O(253) |
| | Me | Et | H | 59/0.8 | 1.4420 | | $A_s$ (161) |
| 5 | Me | Pr | H | 29/0.1 | 1.4402 | | $A_s$ (161) |
| | Et | Et | H | 33/0.1 | | | $A_s$ (161) |
| 7 | $-(CH_2)_5-$ | | Me | 54.5–55/0.2 | | | O (253) |
| 10 | Benzoyl | Me | Me | | | 60 | O (261, 262) |
| 11 | Benzoyl | Me | Et | | | 57 | O (261, 262) |

$$(CH_2)_n C \begin{smallmatrix} \diagup OR \\ \diagdown O_2H \end{smallmatrix}$$

| $C_n$ | $n$ | $R$ | bp/mm Hg | $n_D^{20}$ | mp | Method and Ref. |
|---|---|---|---|---|---|---|
| 7 | 5 | Me | 54.5–55/0.02 | | | O (253, 265) |
| 8 | 5 | Et | 53–58/0.01 | | | O (265) |
| 10 | 5 | *t*-Bu | 49–54/0.01 | | | O (265) |

| $C_n$ | Hydroperoxide | bp/mm Hg | $n_D^{20}$ | mp | Method and Ref. |
|---|---|---|---|---|---|
| 6 | cyclohexane ring with OH, O$_2$H, Cl substituents | | | 76 | $H_a$ (339) |
| | cyclohexane ring with OH, O$_2$H, Br substituents | | | 82–83 | $H_a$ (339) |
| 4 | tetrahydrofuran ring with O$_2$H | 45–50/0.01 | 1.4455 | | $A_s$ (161), O (256) |
| 5 | dihydrofuran ring with Me, O$_2$H | | | 80–100[a] | $H_a$ (323) |
| 6 | ring with Me, Me, HO, O$_2$H | | | 134–136 | $A_1$ (244) |
| 7 | ring with Me, Me, MeO, O$_2$H | | | 75–77 | $A_1$ (244) |
| | ring with Me, Me, Me, Me, HO$_2$, O$_2$H | | | | |

TABLE A-7—*Continued*

| $C_n$ | Hydroperoxide | bp/mm Hg | $n_D^{20}$ | mp | Method and Ref. |
|---|---|---|---|---|---|
| 8 | | | | 135–137 | $H_d$ (326) |
| 5 | | 58–59/3 | 1.1136 (25°) | | $H_a$ (323), O (256) |

|  | $R^1$ | $R^2$ | | | |
|---|---|---|---|---|---|
| 8 | H | H | | 44–45 | *A* (158) |
| 10 | Me | Me | | 63–64 | *A*( 158) |
| 14 | Ph | H | | 102–105 | *A* (158) |
| 8 | =O | | | 100[a] | $H_d$ (255) |

|  | R | | | | |
|---|---|---|---|---|---|
| 11 | MeO | | | 94–95 | O (252) |
| 12 | EtO | | | 66–68 | O (252) |
|  | AcO | | | 112–113 | O (252) |

|  | R | bp/mm Hg | | mp | Method and Ref. |
|---|---|---|---|---|---|
| 3 | H | 20/0.2 | | | *A* (159) |
| 4 | Me | 20/0.2 | | | *A* (159) |
| 5 | Et | Oil | | | *A* (159) |
| 6 | Pr | Oil | | | *A* (159) |
| 9 | $n$-$C_6H_{13}$ | Oil | | | *A* (159) |
|  | Ph | | | 38–40 | *A* (164), $H_d$ (320) |
|  | $p$-$ClC_6H_4$ | | | 76–77 | A (164) |

TABLE A-7—*Continued*

| $C_n$ | Hydroperoxide | bp/mm Hg | $n_D^{20}$ | mp | Method and Ref. |
|---|---|---|---|---|---|
| 4 | 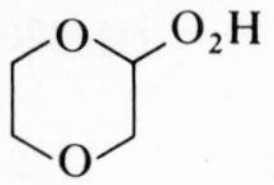 | 45.5/0.01 | 1.4390 | | $A_s$ (161) |

$R^1R^2C(O_2H)_2$

| | $R^1$ | $R^2$ | | | | |
|---|---|---|---|---|---|---|
| 3 | Me | Et | — | | | $H_a$ (344, 347) |
| 4 | Et | Et | Oil | | | $H_a$ (348) |
| 6 | $-(CH_2)_5-$ | | — | | | $H_a$ (345) |
| 7 | Me | $CH_2CO_2Et$ | — | | | $H_a$ (808) |
| 7 | Me | $(CH_2)_2CO_2Et$ | — | | | $H_a$ (808) |
| 8 | $-(CH_2)_7-$ | | — | | | $H_a$ (345) |
| 12 | $-(CH_2)_{11}-$ | | | | 140 | $H_a$ (346) |
| 15 | $-(CH_2)_{14}-$ | | — | | | $H_a$ (345) |

| 12 | | | | 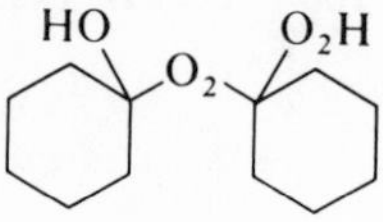 76–77 | $H_a$ (339, 817) |
|---|---|---|---|---|---|

$(R^1R^2CO)_2$ with $O_2H$

| | $R^1$ | $R^2$ | | | | |
|---|---|---|---|---|---|---|
| 6 | Me | Me | | | 37 | $A_s$ (167), $H_a$ (324) |
| | $CH_2Cl$ | $CH_2Cl$ | | | 112–114[a] | $H_d$ (350) |
| 8 | Me | Et | | | 39–42 | $H_a$ (347) |
| 10 | Et | Et | | | 30.6 | $H_a$ (348) |
| 12 | $-(CH_2)_5-$ | | | | 82–83 | $H_a$ (817) |

| 10 | | | | — | 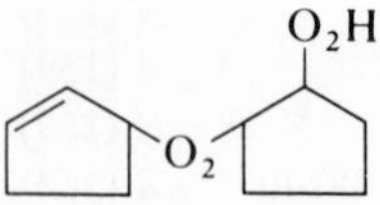 $A$ (45) |
|---|---|---|---|---|---|

T A B L E A-7—*Continued*

| $C_n$ | Hydroperoxide | bp/mm Hg | $n_D^{20}$ | mp | Method and Ref. |
|---|---|---|---|---|---|
| 7 | (structure: $O_2$ — $O_2H$ / OH / $O_2$) | | | 120–121 | $H_a$ (351) |
| | (structure: $O_2$ — $O_2H$ / $O_2H$ / $O_2$) | | | 117 | $H_a$ (351) |

T A B L E **A-8**.   HYDROPEROXIDES OF NITROGEN COMPOUNDS

| $C_n$ | Hydroperoxide | bp/mm Hg | $n_D^{20}$ | mp | Method and Ref. |
|---|---|---|---|---|---|
| | $R^1NHCR^2R^3O_2H$ | | | | |
| | $R^1$ $\quad$ $R^2$ $\quad$ $R^3$ | | | | |
| 8 | $C_6H_{11}$ $\quad$ H $\quad$ Me | | | 69–73 | $H_a$ (331) |
| 9 | $C_6H_{11}$ $\quad$ H $\quad$ Et | | | 62–65 | $H_a$ (331) |
| | $C_6H_{11}$ $\quad$ Me $\quad$ Me | | | 39–41 | $H_a$ (331) |
| 11 | $C_6H_{11}$ $\quad$ H $\quad$ Bu | | | 80 | $H_a$, O (257) |
| 12 | $C_6H_{11}$ $\quad$ $-(CH_2)_5-$ | | | 90–91 | $H_a$ (331) |
| | Ph $\quad$ $-(CH_2)_5-$ | | | 66–68 | $H_a$ (331) |
| 13 | $PhCH_2$ $\quad$ $-(CH_2)_5-$ | | | 86–89 | $H_a$ (331) |
| | $\overset{\displaystyle O}{\overset{\|}{R^2CNR^1CHR^3O_2H}}$ | | | | |
| | $R^1$ $\quad$ $R^2$ $\quad$ $R^3$ | | | | |
| 4 | H $\quad$ Me $\quad$ Me | | | 97.5 | $A$ (207) |
| 6 | H $\quad$ Et $\quad$ Et | | | 68.5 | $A$ (207) |
| | Et $\quad$ Me $\quad$ Me | | | 74 | $A$ (207) |
| 8 | Pr $\quad$ Me $\quad$ Et | | | 47.5–52 | $A$ (207) |
| 6 | H $\quad$ $-(CH_2)_4-$ | | | 139 | $A$ (206) |
| 8 | $BzNHNHCH_2O_2H$ | | | 123–124 | $H_d$ (319) |

TABLE A-8—*Continued*

| $C_n$ | Hydroperoxide | | | bp/mm Hg | $n_D^{20}$ | mp | Method and Ref. |
|---|---|---|---|---|---|---|---|
| 9 | $BzNHN(CH_2O_2H)_2$ | | | | | 110 | $H_d$ (319) |
| | $R^1N=NCR^2R^3O_2H$ | | | | | | |
| | | | | | | | |
| | $R^1$ | $R^2$ | $R^3$ | | | | |
| 8 | Ph | Me | H | Oil | | | A (193) |
| 9 | Ph | Me | Me | Oil | | | A (193) |
| | $p$-$BrC_6H_4$ | Me | Me | | | 45–47[a] | A (183) |
| 10 | Ph | Pr | H | Oil | | | A (193) |
| 12 | Ph | $-(CH_2)_5-$ | | Oil | | | A (193) |
| | | | | | | | |
| 13 | Ph | H | Ph | | | 65–66 | A (183) |
| 13 | Ph | H | $m-NO_2C_6H_4$ | | | 83–84 | A (183) |
| | $p$-$BrC_6H_4$ | H | Ph | | | 107–108 | A (183) |
| | $p$-$ClC_6H_4$ | H | Ph | | | 104–105 | A (183) |
| 14 | $p$-$MeC_6H_4$ | H | Ph | | | 77–78 | A (183) |
| | $PhCH_2$ | H | Ph | | | 70–71 | A (183) |
| | Ph | H | $p$-$MeOC_6H_4$ | | | 83–84 | A (183) |
| | $p$-$BrC_6H_4$ | Me | Ph | | | 48–49 | A (183) |

| | | | | | | | |
|---|---|---|---|---|---|---|---|
| 10 | 2,3-$Me_2$ | | | | | 113 | A (818) |
| 11 | 2,3-$Me_2$-5-MeO | | | | | 111 | A (818) |
| 12 | 2-$i$-Pr-3-Me | | | | | 101.5–102.5 | A (819) |
| 15 | 2-Ph-3-Me | | | | | 154 | A (819) |
| 16 | 2-Ph-3-Me-5-MeO | | | | | 148–150 | A (818) |

| | R | | | | | | |
|---|---|---|---|---|---|---|---|
| 12 | Me | | | | | 118–119[a] | A (202) |
| 18 | $PhCH_2$ | | | | | — | A (202) |

TABLE **A-8**—*Continued*

| $C_n$ | Hydroperoxide | bp/mm Hg | $n_D^{20}$ | mp | Method and Ref. |
|-------|---------------|----------|------------|-----|-----------------|
| 28 | 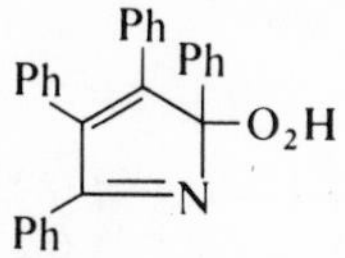 | | | 170–172[a] | $A_1$ (198) |
| 12 | 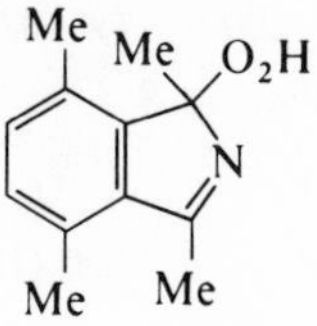 | | | 138[a] | $A$ (203) |
| | 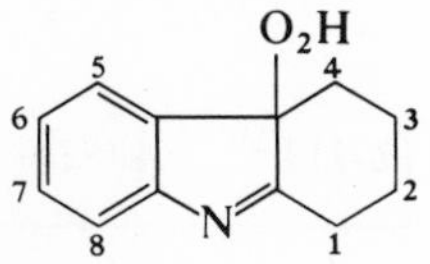 | | | 132–133 | $A$ (181, 182) |
| 12 | 6-Br | | | 148 | $A$ (820) |
| 12 | 8-Br | | | 146–150 | $A$ (820) |
| 13 | 6-MeO | | | 126 | $A$ (820) |
| 14 | 6-CONH$_2$ | | | 165 | $A$ (820) |
| | 5-CO$_2$Me | | | 149–150 | $A$ (820) |
| | 6,7-(MeO)$_2$ 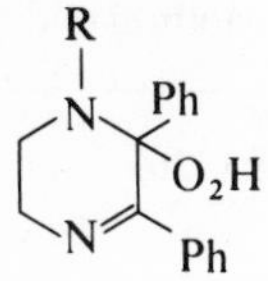 | | | 122 | $A$ (820) |

R

| | R | | | | |
|----|---------|--|--|---------|----------|
| 18 | Et | | | 103–104 | $A$ (821) |
| 22 | Ph | | | 126 | $A$ (821) |
| 23 | PhCH$_2$ 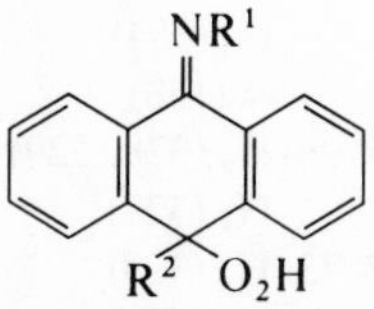 | | | 117–118 | $A$ (821) |

TABLE A-8—*Continued*

| $C_n$ | Hydroperoxide | | bp/mm Hg | $n_D^{20}$ | mp | Method and Ref. |
|---|---|---|---|---|---|---|
| | R$^1$ | R$^2$ | | | | |
| 14 | H | H | | | 130–135 | A (822) |
| 20 | H | Ph | | | 175 | A (822) |
| 27 | p-MeOC$_6$H$_4$ | Ph | | | 118–200 | A (823) |
| 21 | | | | | 110[a] | A$_s$ (199) |
| 27 | p-MeC$_6$H$_4$N=CPh—CPh$_2$O$_2$H | | | | 112–113.5 | A (824) |

[a] Decomposes.

TABLE A-9. HYDROPEROXIDES CONTAINING HALOGEN

| $C_n$ | Hydroperoxide | | | bp/mm Hg | $n_D^{20}$ | mp | Method and Ref. |
|---|---|---|---|---|---|---|---|
| 1 | CF$_3$O$_2$H | | | ~25/760 | | | S (356) |
| | R$^1$CR$^2$R$^3$O$^2$H | | | | | | |
| | R$^1$ | R$^2$ | R$^3$ | | | | |
| 3 | BrCH$_2$ | H | BrCH$_2$ | 60–62/0.02 | 1.5493/(25°) | | A (137, 138) |
| | BrCH$_2$ | H | ClCH$_2$ | 55.5–56.5/0.03 | 1.5169 (25°) | | A (138) |
| 4 | BrCH$_2$ | Me | Me | 33/0.05 | 1.4770 (25°) | | H$_a$ (328) |
| | BrCH$_2$ | Me | ClCH$_2$ | 56/0.01 | 1.5073 (25°) | | A (138) |
| | ClCH$_2$ | H | OEt | 45.5–46/1.5 | 1.4432 | | H$_a$ (327) |
| 5 | BrCHMe | Me | Me | | | | H$_a$ (354) |
| 6 | BrCH$_2$ | Me | Pr | 50–51/0.01 | — | | A (138) |
| | BrCMe$_2$ | Me | Me | | | 95–96 | H$_a$ (328, 329) |
| | ClCH$_2$ | Me | Pr | 36.5/1.5 | 1.4511 | | H$_a$ (327) |
| | ClCMe$_2$ | Me | Me | | | 72–73 | H$_a$ (329) |
| | ClCH$_2$ | H | i-BuO | 61/2 | — | | H$_a$ (327) |

TABLE A-9—*Continued*

| $C_n$ | Hydroperoxide | bp/mm Hg | $n_D^{20}$ | mp | Method and Ref. |
|---|---|---|---|---|---|
| | (cyclohexyl $O_2H$, Br) | 68–70/0.01 | — | | $A$ (138), $H_a$ (329) |
| | (cyclohexyl $O_2H$, Cl) | 61/0.01 | — | | $H_a$ (327, 328) |
| 7 | ClCHPr  H  Et | 58–60/0.01 | 1.4515 | | $H_a$ (327) |
| 8 | (cyclohexyl Me, $O_2H$, Cl, Me) | — | — | | $H_a$ (329) |
| | BrCH$_2$  H  Ph | — | — | | $H_a$ (329) |

# REFERENCES

1. A. G. Davies, *Organic Peroxides*, Butterworths, London, 1961.
2. E. G. E. Hawkins, *Organic Peroxides*, Van Nostrand, Princeton, N.J., 1961.
3. W. O. Lundberg, Ed., *Autoxidation and Antioxidants*, Vol. I, Interscience, New York, 1961.
4. O. L. Mageli and C. S. Sheppard, "Peroxides and Peroxy Compounds," in Kirk–Othmer *Encyclopedia of Chemical Technology*, 2nd ed., Wiley, New York, 1967, Vol. 14, pp. 746–820; Chapter I in *Organic Peroxides*, Vol. I, edited by D. Swern, Wiley, New York, 1970.
5. O. E. Edwards and R. Curci, "Peroxide Reaction Mechanisms," in Kirk–Othmer *Encyclopedia of Chemical Technology*, 2nd ed., Wiley, New York, 1967, Vol. I4, pp. 820–839.
6. W. A. Pryor, *Free Radicals*, McGraw-Hill, New York, 1966.
7. K. B. Wiberg, Ed., *Oxidation in Organic Chemistry*, Part A, Academic Press, New York, 1965.
8. *Oxidation of Organic Compounds*, Vols. I, II, and III, Proceedings of the International Oxidation Symposium, San Francisco, 1967. *Advances in Chemistry Series*, **75**, **76**, and **77** American Chemical Society, Washington, D.C., 1968.
9. *Free Radicals in Solution*, Proceedings of the International Symposium on Free Radicals in Solution, Ann Arbor, Mich., 1966, Butterworths, London, 1967; *Pure Appl. Chem.*, **15**, 1 (1967).

10. N. M. Emanuel, E. T. Denisov, and Z. K. Maizus, *Liquid Phase Oxidation of Hydrocarbons*, translated by B. J. Hazzard, Plenum, New York, 1967.

11. G. H. Williams, Ed., *Advances in Free Radical Chemistry*, Vol. I, Logos, London, 1965.

12. J. O. Edwards, Ed., *Peroxide Reaction Mechanisms*, Interscience, New York, 1962.

13. E. J. Behrman and J. O. Edwards, " Nucleophilic Displacement on Peroxide Oxygen," in *Progress in Physical Organic Chemistry*, Vol. IV, edited by A. Streitweiser and R. W. Taft, Interscience, New York, 1967, pp. 93–123.

14. C. Walling, " Free Radical Rearrangements," in *Molecular Rearrangements*, Part 1, edited by P. de Mayo, Interscience, New York, 1964, pp. 407–456.

15. P. A. S. Smith, "Rearrangements involving Migration to an Electron Deficient Nitrogen or Oxygen," in *Molecular Rearrangements*, Part I, edited by P. de Mayo, Interscience, New York, 1964, pp. 457–592.

16. R. Stewart, *Oxidation Mechanisms*, Benjamin, New York, 1964.

17. A. Rieche, *Angew. Chem.*, **73**, 57 (1961).

18. A. Rieche, *Kunststoffe*, **54**, 428 (1964).

19. C. Walling, *Radiation Res. Suppl.*, **3**, 3 (1963).

20. J. K. Kochi, *Tetrahedron*, **18**, 483 (1962).

21. J. K. Kochi, *Science*, **155**, 415 (1967).

22. F. List and L. Kuhnen, *Erdöl und Kohle*, **20**, 192 (1967).

23. A. H. M. Cosijn, *Chem. Weekblad*, **60**, 585 (1964).

24. S. V. Zavgorodnii, *Russian. Chem. Rev.*, **30**, 133 (1961).

25. E. T. Denisov, and N. M. Emanuel, *Russian Chem. Rev.*, **29**, 645 (1960).

26. A. Fish, *Angew. Chem. Intern. Ed.*, **7**, 45 (1965).

27. A. Rieche, *Alkylperoxide und Ozonide*, Steinkopff, Dresden, 1931.

28. R. Criegee, *Fortschr. Chem. Forsch.*, **1**, 508 (1956).

29. R. Criegee, " Herstellung und Umwandlung von Peroxiden," in *Methoden der organischen Chemie*, Vol. 8, (Houben-Weyl), Thieme, Stuttgart, 1952, p. 1.

30. A. V. Tobolsky and R. B. Mesrobian, *Organic Peroxides*, Interscience, New York, 1954.

31. T. J. Wallace, *J. Org. Chem.*, **30**, 3147, 4017 (1965).

32. D. E. Van Sickle, F. R. Mayo, R. M. Arluck, and M. Syz, *J. Amer. Chem. Soc.*, **89**, 967 (1967).

33. C. Walling, *Free Radicals in Solution*, Wiley, New York, 1957, Chapter 9.

34. S. W. Benson, *J. Amer. Chem. Soc.*, **87**, 972 (1965).

35. E. G. Janzen, F. J. Johnston, and C. A. Ayers, *J. Amer. Chem. Soc.*, **89**, 1176 (1967).

36. G. A. Russell, *J. Amer. Chem. Soc.*, **77**, 4583 (1955); *ibid.*, **78**, 1047 (1956).

37. D. G. Hendry and G. A. Russell, *J. Amer. Chem. Soc.*, **86**, 2371 (1964).

38. J. L. Bolland, *Trans. Faraday. Soc.*, **46**, 358 (1950).

39. D. E. Van Sickle, F. R. Mayo, E. S. Gould, and R. M. Arluck, *J. Amer. Chem. Soc.*, **89**, 977 (1967).

40. S. Rummel, P. Krumbiegel, and H. Huebner, *J. Prakt. Chem.*, **37**, 206 (1968).

41. G. A. Russell, *J. Amer. Chem. Soc.*, **79**, 3871 (1957).

42. K. R. Hargrave and A. L. Morris, *Trans. Faraday Soc.*, **52**, 89 (1956).

43. D. G. Hendry and G. A. Russell, *J. Amer. Chem. Soc.*, **86**, 2369 (1964).
44. W. F. Brill and B. J. Barone, *J. Org. Chem.*, **29**, 140 (1964).
45. D. E. Van Sickle, F. R. Mayo, and R. M. Arluck, *J. Amer. Chem. Soc.*, **87**, 4824 (1965).
46. D. E. Van Sickle, F. R. Mayo, and R. M. Arluck, *J. Amer. Chem. Soc.*, **87**, 4832 (1965).
47. D. G. Hendry, *J. Amer. Chem. Soc.*, **89**, 5433 (1967).
48. J. A. Howard and K. U. Ingold, *Can. J. Chem.*, **43**, 2729, 2737 (1965); *ibid.*, **44**, 1113, 1119 (1966).
49. J. A. Howard and K. U. Ingold, *Can. J. Chem.*, **45**, 793 (1967).
50. B. S. Middleton and K. U. Ingold, *Can. J. Chem.*, **45**, 191 (1967).
51. J. A. Howard, K. U. Ingold, and M. Symonds, *Can. J. Chem.*, **46**, 1017 (1965).
52. Z. Kulicki, *Zesz. Nauk Politich Slask. Chem.*, No. **36**, 3 (1964); through *Chem. Abstr.*, **67**, 116335 (1967).
53. T. I. Yurzhenko and M. A. Dikii, *Dokl. Akad. Nauk SSSR*, **137**, 1137 (1961); through *Chem. Abstr.*, **55**, 19856 (1961).
54. N. Uri, *Chem. Ind.* **1967**, 2060.
55. D. J. Carlsson and J. C. Robb, *Trans. Faraday Soc.*, **62**, 3404 (1966).
56. A. V. Tobolsky, D. J. Metz, and R. B. Mesrobian, *J. Amer. Chem. Soc.*, **72**, 1942 (1950); C. Walling, *ibid.*, *91*, 7590 (1969).
57. Y. Kamiya, S. Beaton, A. Lafortune, and K. U. Ingold, *Can. J. Chem.*, **41**, 2020 (1963).
58. F. Jaffe, T. R. Steadman, and R. W. McKinney, *J. Amer. Chem. Soc.*, **85**, 351 (1963).
59. G. Mann, *J. Prakt. Chem.*, **20**, 210 (1963).
60. W. Pritzkow and K. H. Gröbe, *Ber.*, **93**, 2156 (1960).
61. M. A. Dikii and T. I. Yurzhenko, *Zh. Obshch. Khim.*, **33**, 1360 (1963).
62. H. S. Blanchard, *J. Amer. Chem. Soc.*, **81**, 4548 (1959).
63. H. Boardman, *J. Amer. Chem. Soc.*, **84**, 1376 (1962).
64. R. Hiatt, C. W. Gould, and F. R. Mayo, *J. Org. Chem.*, **29**, 3461 (1964).
65. G. Wagner, *J. Prakt. Chem.*, **27**, 297 (1965).
66. G. Dobson and G. Hughes, *J. Phys. Chem.*, **69**, 1814 (1965).
67. M. Klang, *Oléagineaux*, **20**, 245 (1965).
68. Y. A. Ershov, A. F. Lukovnikov, and A. A. Baturina, *Kinetiki Kataliz*, **5**, 752 (1964); through *Chem. Abstr.*, **61**, 13442 (1964).
69. F. F. Rust and W. E. Vaughan, *Ind. Eng. Chem.*, **41**, 2595 (1949).
70. H. Hock and H. Kropf, *J. Prakt. Chem.*, **13**, 285 (1961).
71. A. V. Ryazanova, A. V. Bondarenko, I. A. Maizlakh, and M. I. Farberov, *Neftekhim.* **6**, 227 (1965); through *Chem. Abstr.*, **65**, 3776 (1966).
72. H. S. Blanchard, *J. Amer. Chem. Soc.*, **82**, 2014 (1960).
73. H. Hock and H. Kropf, *J. Prakt. Chem.*, **13**, 20 (1961).
74. H. Kropf and H. Hoffmann, *Tetrahedron Letters*, **1967**, 659.
75. J. Rouchaud, *Bull. Soc. Chim. Belges*, **76**, 171 (1967).
76. S. F. Naumova, V. N., Kovaleva, and K. A. Zhauverko, *Dokl. Akad. Nauk Belorusk. SSR*, **5**, 109 (1961); through *Chem. Abstr.*, **58**, 4484 (1963).
77. S. Fallab, *Angew. Chem., Intern. Ed.*, **6**, 496 (1967).

78. W. Pritzkow, *Angew. Chem.*, **67**, 399 (1955).
79. T. J. Wallace and R. M. Skomoroski, *Chem. Ind.*, **1965**, 384,
80. S. V. Zavgorodni, I. N. Novokov, V. G. Kryuchkova, and V. P. Shatalov, *Khim. Prom.*, **1962**, 185; through *Chem. Abstr.*, **58**, 5554 (1963).
81. P. J. Andrulis, M. J. S. Dewar, and R. L. Hunt, *J. Amer. Chem. Soc.*, **88**, 5473 (1966).
82. D. G. Hoare and W. A. Waters, *J. Chem. Soc.*, **1964**, 2552.
83. T. A. Cooper and W. A. Waters, *J. Chem. Soc.*, **B1967**, 455, 464.
84. V. A. Simanov and M. S. Nemtsov, *Zh. Obshch. Khim.*, **32**, 2914, 2919, 2925, 3179 (1962); through *Chem. Abstr.*, **58**, 8869 (1963).
85. A. V. Bondarenko, M. I. Farberov, A. V. Ryazanova, and N. A. Usacheva, *Neftekhim.* **6**, 423 (1966); through *Chem. Abstr.*, **65**, 13589 (1966).
86. R. N. Volkov, Yu. S. Tsybin, and L. F. Kovrizhko, *Tr. Lab. Khim. Vysokomolekul, Soedin, Voronezhsk Univ.*, 105 (1964); through *Chem. Abstr.*, **65**, 661 (1966).
87. T. I. Yurzhenko and V. A. Puchin, *Nauk Zapiski L'vov Politekh. Inst.*, No. **62**, 333 (1957); through *Chem. Abstr.*, **56**, 376 (1962).
88. B. Yeomans and D. P. Young, *J. Chem. Soc.*, **1958**, 2288.
89. G. Russell and A. G. Bemis, *J. Amer. Chem. Soc.*, **88**, 5491 (1966) and references therein.
90. Y. Sprinzak, *J. Amer. Chem. Soc.*, **80**, 5449 (1958).
91. M. F. Hawthorne and G. S. Hammond, *J. Amer. Chem. Soc.*, **77**, 2549 (1955).
92. H. Musso and D. Maassen, *Ann.*, **689**, 93 (1965).
93. H. R. Gersmann and A. F. Bickel, *J. Chem. Soc.*, **1959**, 2711; *ibid.*, **1962**, 2356.
94. E. J. Bailey, J. Eliks, and D. H. R. Barton, *Proc. Chem, Soc.*, **1960**, 214.
95. W. E. Doering and R. Haines, *J. Amer. Chem. Soc.*, **76**, 482 (1954); *ibid.*, **75**, 4372 (1953).
96. M. Avramoff and Y. Sprinzak, *Proc. Chem. Soc.*, **1962**, 150.
97. M. Avramoff and Y. Sprinzak, *J. Amer. Chem. Soc.*, **85**, 1655(1965).
98. Netherlands Patent Application 6,406,757 (1965).
99. J. L. Benton and M. M. Worth, *Nature*, **171**, 269 (1953).
100. W. Pritzkow and K. A. Müller, *Ann.*, **597**, 167 (1955).
101. G. Geiseler, F. Asinger, and H. Wein, *Ber.*, **92**, 958 (1959).
102. G. S. Serif, C. F. Hunt, and A. N. Bourns, *Can. J. Chem.*, **31**, 1229 (1953).
103. W. Treibs and R. Schöllner, *Ber.*, **91**, 2282 (1958).
104. J. P. Wibaut and A. Strang, *Proc. Nederl. Akad. Wetensch.*, **B54**, 102 (1951); *ibid.*, **B55**, 207 (1952).
105. F. F. Rust, *J. Amer. Chem. Soc.*, **79**, 4000 (1957).
106. R. Criegee and P. Ludwig, *Erdöl und Kohle*, **15**, 523 (1962).
107. British Patents 641,250 (1950); 646,102 (1950); and 727,498 (1955).
108. P. F. Ivanenko, *Neftekhim.*, **6**, 430 (1966); cf. *Chem. Abstr.*, **65**, 13589 (1966).
109. G. Öhlmann and W. Schirmer, *Kinetika Kataliz* **8**, 742 (1967); cf. *Chem. Abstr.*, **68**, 58847 (1968).
110. A. G. Davies and J. E. Packer, *Chem. Ind.* (*London*), **1960**, 1165.

111. P. D. Bartlett, R. E. Pincock, J. H. Rolston, W. G. Schindel, and L. A. Singer, *J. Amer. Chem. Soc.*, **87**, 2590 (1965).

112. F. D. Greene and N. N. Lowry, *J. Org. Chem.*, **32**, 875 (1967).

113. K. R. Hargrave and A. L. Morris, *Trans. Faraday Soc.*, **52**, 89 (1956).

114. E. H. Farmer, *Trans. Faraday Soc.*, **42**, 228 (1946).

115. J. L. Bolland and J. Gee, *Trans. Faraday Soc.*, **42**, 244 (1946).

116. A. A. Miller and F. R. Mayo, *J. Amer. Chem. Soc.*, **78**, 1017 (1956).

117. G. A. Russell, *J. Amer. Chem. Soc.*, **78**, 1035, 1041 (1956).

118. L. Dulog, *Makromol. Chem.*, **76**, 119 (1964).

119. L. Bateman and J. Gee, *Proc. Roy. Soc.*, **A195**, 376 (1948).

120. L. Bateman and A. L. Morris, *Trans. Faraday Soc.*, **48**, 1149 (1952).

121. J. Burger, C. Meyer, G. Clement, and J. C. Balaceanu, *Compt. Rend.*, **252**, 2235 (1961).

122. E. Ochial, *Tetrahedron*, **20**, 1819 (1964).

123. J. P. Collman, M. Kubota, and J. Hosking, *J. Amer. Chem. Soc.*, **89**, 4809 (1967).

124. L. Bateman, *Quart. Rev.*, **8**, 147 (1954).

125. A. Rieche, *Fette, Seifen, Anstrichmittel*, **60**, 637 (1958).

126. H. W. Schultz, Ed., *Lipids and Their Oxidations*, Second Symposium on Foods, Oregon State University, AVI Publishing Co., Westport, Conn., 1962.

127. R. Y. Kirdoni and D. S. Layne, *Biochemistry*, **4**, 330 (1965).

128. Netherlands Patent Application 6,402,851 (1965).

129. Netherlands Patent Application 6,515, 930 (1966).

130. W. F. Brill and B. J. Barone, *J. Org. Chem.*, **29**, 140 (1964).

131. A. Chauvel, G. Clement, and J. C. Balaceanu, *Bull. Soc. Chim. France*, **1962**, 1774; *ibid.*, **1963**, 2025.

132. P. Menguy, A. Chauvel, G. Clement, and J. C. Balaceanu, *Bull. Soc. Chim. France*, **1963**, 2643.

133. I. Serée de Roch, *Bull. Soc. Chim. France*, **1965**, 1979.

134. F. R. Mayo and A. A. Miller, *J. Amer. Chem. Soc.*, **80**, 2480 (1958).

135. K. Bodendorf, *Arch. Pharm.*, **271**, 1 (1933); D .G. Hendry, F. R. Mayo, and D. Scheutzle, *Ind. Eng. Chem. Prod. Res. Develop.*, **7**, 136 (1968).

136. D. E. Van Sickle, F. R. Mayo, and R. M. Arluck, *J. Org. Chem.*, **32**, 3680 (1967).

137. A. Rieche, M. Schulz, and K. Kirschke, *Angew. Chem.*, **77**, 219 (1965).

138. A. Rieche, M. Schulz, and K. Kirschke, *Ber.*, **99**, 3244 (1966).

139. German (East) Patent 55,328 (1967).

140. J. F. Ford, R. C. Pitkethly, and V. O. Young, *Tetrahedron*, **4**, 325 (1958).

141. A. A. Oswald, *J. Org. Chem.*, **24**, 443 (1959).

142. A. A. Oswald and F. Noel, *J. Org. Chem.*, **26**, 3948 (1961).

143. A. A. Oswald, B. E. Hudson, G. Rodgers, and F. Noel, *J. Org. Chem.*, **27**, 2439 (1962).

144. K. Griesbaum, A. A. Oswald, and B. E. Hudson, *J. Amer. Chem. Soc.*, **85**, 1969 (1963).

145. W. F. Brill, *Selectivity in the Liquid Phase Autoxidation of Olefins*, *Advances in Chemistry Series*, **51**, American Chemical Society, Washington, D.C., 1965.

146. W. F. Brill, *J. Amer. Chem. Soc.*, **87**, 3286 (1965).

147. G. O. Schenck, O. A. Neumüller, and W. Eisfeld, *Angew. Chem.*, **70**, 595 (1958).

148. G. O. Schenck, O. A. Neumüller, and W. Eisfeld, *Ann.*, **618**, 202 (1958).

149. H. C. Volger, W. Brackman, and J. W. Lemmers, *Rec. Trav. Chim.*, **84**, 1203 (1965).

150. O. S. Privett, W. O. Lundberg, N. A. Khan, W. E. Tolberg, and D. H. Wheeler, *J. Amer. Oil. Chem. Soc.*, **30**, 17 (1953).

151. J. A. Cannon, K. T. Zilch, S. C. Burket, and H. J. Dutton, *J. Amer. Oil Chem. Soc.*, **29**, 447 (1952).

152. A. Hautfenne, M. Loncin, and C. Paquot, *Oléagineaux*, **21**, 303 (1966).

153. R. Schöllner and R. Herzschuh, *Fette, Seifen, Anstrichmittel*, **68**, 469 (1966).

154. S. J. Moss and H. Steiner, *J. Chem. Soc.*, **1965**, 2372.

155. C. Walling and W. Thaler, *J. Amer. Chem. Soc.*, **83**, 3577 (1961).

156. J. K. Kochi and P. J. Krusic, *J. Amer. Chem. Soc.*, **90**, 7157 (1968).

157. W. J. Farrissey, *J. Org. Chem.*, **29**, 391 (1964).

158. A. Rieche and M. Schulz, *Ann.*, **653**, 32 (1962).

159. W. E. Seyfarth, A. Rieche, and A. Hesse, *Ber.*, **100**, 624 (1967).

160. K. I. Ivanov, V. K. Sovinova, and E. G. Mikhailova, *J. Gen. Chem. USSR*, **16**, 1003 (1946).

161. G. O. Schenck, H. D. Becker, K. H. Schulte-Elte, and C. H. Krauch, *Ber.*, **96**, 509 (1963).

162. R. Criegee, *Angew, Chem.* **62**, 120 (1950).

163. A. Rieche and E. Schmitz, *Ber.*, **90**, 1082 (1957).

164. A. Rieche, E. Schmitz, and E. Beyer, *Ber.*, **91**, 1935 (1958).

165. A. Rieche, H. E. Seyfarth, and A. Hesse, *Angew. Chem., Intern. Ed.*, **5**, 253 (1966).

166. D. B. Sharp and T. M. Patrick, *J. Org. Chem.*, **26**, 1389 (1961).

167. G. O. Schenck and H. D. Becker, *Angew. Chem.*, **70**, 504 (1958).

168. J. L. Bolland, and H. R. Cooper, *Nature*, **172**, 413 (1953).

169. L. H. Buxbaum, *Ann.*, **706**, 81 (1967).

170. W. Pritzkow, *Ber.*, **87**, 1668 (1954).

171. W. Pritzkow, *Ber.*, **88**, 572 (1955).

172. D. B. Sharp, L. W. Patton, and S. E. Whitcomb, *J. Amer. Chem. Soc.*, **73**, 5600 (1951).

173. W. H. Richardson and R. F. Steed, *J. Org. Chem.*, **32**, 771 (1967).

174. I. G. Tishchenko and L. S. Stanishevskii, *Zh. Obshch. Khim.*, **33**, 3751 (1963); through *Chem. Abstr.*, **60**, 7911 (1964).

175. I. G. Tishchemko and L. S. Stanishevskii, *Geterogennye Reaktsii i Reakts. Sposobnost. Sb.*, **1964**, 254; through *Chem. Abstr.*, **65**, 5357 (1966).

176. E. P. Kohler and R. B. Thompson, *J. Amer. Chem. Soc.*, **59**, 887 (1937).

177. R. C. Fuson and H. L. Jackson, *J. Amer. Chem. Soc.*, **72**, 1637 (1950).

178. E. P. Kohler, *Amer. Chem. J.*, **36**, 177 (1906).

179. J. Carnduff and D. G. Leppard, *Chem. Commun.*, **1967**, 829.

180. B. Miller, *J. Amer. Chem. Soc.*, **87**, 5515 (1965).

181. B. Witkop and J. B. Patrick, *J. Amer. Chem. Soc.*, **73**, 2196, 2188 (1951).

182. B. Witkop, J. B. Patrick, and M. Rosenbaum, *J. Amer. Chem. Soc.*, **73**, 2641 (1951).
183. W. Busch and W. Dietz, *Ber.*, **47**, 3277 (1914).
184. A. H. Cook and K. J. Reed, *J. Chem. Soc.*, **1945**, 399.
185. G. M. Robinson and R. Robinson, *J. Chem. Soc.*, **1924**, 834.
186. R. Criegee and G. Lohaus, *Ber.*, **84**, 219 (1951).
187. B. Witkop and H. M. Kissman, *J. Amer. Chem. Soc.*, **75**, 1975 (1953).
188. K. H. Pausacker, *J. Chem. Soc.*, **1950**, 3478.
189. M. Schultz and L. Somogyi, *Angew. Chem., Intern. Ed.*, **6**, 168 (1967).
190. J. Buckingham and R. D. Guthrie, *J. Chem. Soc.*, *C*, **1967**, 2268.
191. M. Schultz, H. F. Boeden, and P. Berlin, *Ann.*, **703**, 190 (1967).
192. K. Bodendorf and J. Loetzbeyer, *Ber.*, **99**, 801 (1966).
193. A. J. Bellamy and R. D. Guthrie, *J. Chem. Soc.*, **1965**, 2788.
194. A. Schönberg and R. Michaelis, *J. Chem. Soc.*, **1937**, 109.
195. H. Wasserman and M. B. Floyd, *Tetrahedron Letters*, **29**, 2009 (1963).
196. F. McCapra, D. G. Richardson, and Y. C. Chang, *Photochem. Photobiol.*, **4**, 1111 (1965).
197. F. F. McCapra and Y. C. Chang, *Chem. Commun.*, **1966**, 522.
198. C. Dufraisse, G. Rio, A. Ranjon and O. Pouchot, *Compt. Rend.*, **261**, 3133 (1965).
199. J. Sonnenberg and D. M. White, *J. Amer. Chem. Soc.*, **86**, 5685 (1964).
200. E. H. White and M. J. C. Harding, *J. Amer. Chem. Soc.*, **86**, 5866 (1964).
201. F. McCapra and D. G. Richardson, *Tetrahedron Letters*, **43**, 3167 (1964).
202. R. F. Parcell and F. P. Hauck, *J. Org. Chem.*, **28**, 3468 (1963).
203. C. O. Bender and R. Bonnett, *Chem. Commun.*, **1966**, 198.
204. E. Hoft and L. Schultze, *Z. Chem.*, **7**, 137 (1967).
205. R. J. S. Beer, T. Donavanik, and A. Robertson, *J. Chem. Soc.*, **1954**, 4139.
206. A. Rieche and W. Schön, *Ber.*, **99**, 3238 (1966); *ibid.*, **100**, 4052 (1967).
207. B. F. Sagar, *J. Chem. Soc.*, *B*, **1967**, 428, 1047.
208. A. G. Davies and B. D. Roberts, *J. Chem. Soc.*, *B*, **1967**, 17; *ibid.*, **1968**, 1074.
209. R. C. Lamb, P. W. Ayers, M. K. Toney, and J. F. Garst, *J. Amer. Chem. Soc.*, **88**, 4261 (1966).
210. C. Walling and S. A. Buckler, *J. Amer. Chem. Soc.*, **77**, 6032 (1955).
211. W. Pritzkow and K. A. Müller, *Ber.*, **89**, 2321 (1956); *Ann.*, **597**, 167 (1955).
212. W. Pritzkow and H. Schaefer, *Ber.*, **93**, 2151 (1960).
213. H. Hock and F. Ernst, *Ber.*, **92**, 2716, 2723 (1959).
214. German Patent 1,059,454 (1959).
215. G. Wilke and P. Heimbach, *Ann.*, **652**, 7 (1962).
216. A. G. Davies, D. G. Hare, and R. F. White, *J. Chem. Soc.*, **1960**, 1040.
217. W. J. Farrissey, Jr., *J. Org. Chem.*, **27**, 3065 (1962).
218. C. F. Cullis and E. Fersht, *Combust. Flame*, **7**, 185 (1963).
219. M. H. Abraham and A. G. Davies, *J. Chem. Soc.*, **1959**, 429.
220. German Patent 933,925 (1955).
221. G. O. Schenck, H. Eggert, and W. Denk, *Ann.* **584**, 177 (1953).
222. G. O. Schenck, *Angew. Chem.*, **69**, 579 (1957).
223. K. Gollnick and G. O. Schenck, *Pure Appl. Chem.*, **9**, 507 (1964).

224.  G. O. Schenck, K. Gollnick, G. Buchwald, S. Schroeter, and G. Ohloff, *Ann.*, **674**, 93 (1964).

225.  G. O. Schenck, O. A. Neumüller, G. Ohloff, and S. Schroeter, *Ann.*, **687**, 26 (1965).

226.  K. Gollnick, S. Schroeter, G. Ohloff, G. Schade and G. O. Schenck, *Ann.*, **687**, 14 (1965).

227.  A. Nickon and J. F. Bagli, *J. Amer. Chem. Soc.*, **81**, 6330 (1959); *ibid.*, **83**, 1498 (1961).

228.  R. L. Kenney and G. S. Fisher, *J. Org. Chem.*, **28**, 3509 (1963).

229.  C. S. Foote, *Vortex*, **27**, 436 (1966).

230.  Reference 8, Vol. II, articles by E. A. Ogryzlo, K. Gollnick, F. A. Litt, and A. Nickon, A. V. Khan and D. R. Kearns, R. Higgins, and C. S. Foote and H. Cheng.

231.  K. R. Kopecky and H. J. Reich, *Can. J. Chem.*, **43**, 2265 (1965).

232.  K. R. Kopecky and J. H. van de Sande, Meetings of the Chemical Institute of Canada, June 1968, Abstracts, p. 41.

233.  G. O. Schenck and K. H. Schulte-Elte, *Ann.*, **618**, 185 (1958).

234.  D. B. Sharp and J. R. Le Blank, Amer. Chem. Soc., Div. Petrol. Chem. Preprints, **5**, No. 2, C57–64 (1960); through *Chem. Abstr.*, **57**, 14923 (1962).

235.  G. O. Schenck, K. Gollnick, and O. A. Neumüller, *Ann.*, **603**, 46 (1957).

236.  G. O. Schenck, O. A. Neumüller, and W. Eisfeld, *Angew Chem.*, **70**, 595 (1958).

237.  G. O. Schenck and O. A. Neumüller, *Ann.*, **618**, 194 (1958).

238.  A. Nickon and W. L. Mendelson, *J. Amer. Chem. Soc.*, **85**, 1894 (1963).

239.  Netherlands Patent Application 6,400,153 (1965).

240.  R. A. Bell and R. E. Ireland, *Tetrahedron Letters*, No. 4, 269 (1963).

241.  S. Masamune, *J. Amer. Chem. Soc.*, **86**, 290 (1964).

242.  U.S. Patent 3,230,235 (1966).

243.  E. Koch and G. O. Schenck, *Ber.*, **99**, 1984 (1966).

244.  C. S. Foote, M. T. Wuesthoff, S. Wexler, I. G. Burstain, R. Denny, G. O. Schenck, and K. H. Schulte-Elte, *Tetrahedron*, **23**, 2583 (1967).

245.  W. F. Brill, *Oxidation of Organic Compounds*, Vol. I, *Advances in Chemistry Series*, **75**, American Chemical Society, Washington, D.C., 1968, p. 93.

246.  K. Gollnick and G. Schade, *Tetrahedron Letters*, **21**, 2335 (1966).

247.  C. S. Foote and S. Wexler, *J. Amer. Chem. Soc.*, **86**, 3879 (1964).

248.  E. J. Corey and W. C. Taylor, *J. Amer. Chem. Soc.*, **86**, 3881 (1964).

249.  C. S. Foote, S. Wexler, W. Ande, and R. Higgins, *J. Amer. Chem. Soc.*, **90**, 975 (1968).

250.  U.S. Patent 3,274,181 (1966).

251.  H. Wasserman and J. R. Scheffer, *J. Amer. Chem. Soc.*, **89**, 3173 (1967).

252.  R. Criegee and G. Wenner, *Ann.*, **654**, 9 (1949).

253.  R. Criegee and G. Lohaus, *Ann.*, **583**, 6, 12 (1953).

254.  U.S. Patent 3,022,352 (1962).

255.  A. Rieche and M. Schulz, *Ber.*, **98**, 3623 (1965).

256.  A. Rieche, M. Schulz, and D. Becker, *Ber.*, **98**, 3627 (1965).

257.  M. Schulz, A. Rieche, and D. Becker, *Ber.*, **99**, 3233 (1966).

258. P. S. Bailey, *J. Org. Chem.*, **22**, 1548 (1957).

259. P. S. Bailey and S. S. Bath, *J. Amer. Chem. Soc.*, **79**, 3120 (1957).

260. P. S. Bailey, S. B. Mainthia, and C. J. Abshire, *J. Amer. Chem. Soc.*, **82**, 6136 (1960).

261. P. S. Bailey and Y. G. Chang, *J. Org. Chem.*, **27**, 1192 (1962).

262. P. S. Bailey, Y. G. Chang, and W. W. L. Kwie, *J. Org. Chem.*, **27**, 1198 (1962).

263. P. S. Bailey, *Chem. Rev.*, **58**, 925 (1958).

264. A. T. Menyailo and M. V. Pospelov, *Russian Chem. Rev.*, **36**, 284 (1967).

265. N. Sonoda, Y. Fujiwara, and S. Tsutsumi, *Technol. Rept.*, *Osaka Univ.*, **17**, 215 (1967); through *Chem. Abstr.*, **67**, 99704 (1967).

266. J. E. Franz, W. S. Knowles, and C. Osuch, *J. Org. Chem.*, **30**, 4328 (1965).

267. S. Murai, N. Sonoda, and S. Tsutsumi, *Bull. Chem. Soc.*, *Japan* **37**, 1187 (1964).

268. W. P. Keaveney, M. G. Berger, and J. J. Pappas, *J. Org. Chem.*, **32**, 1537 (1967).

269. E. Bernatek, T. Ledaal, and S. Asen, *Acta Chem. Scand.*, **18**, 1317 (1964)

270. M. Schultz, A. Rieche, and D. Becker, *Angew. Chem.*, *Intern. Ed.*, **4**, 525 (1965).

271. R. Criegee and H. Dietrich, *Ann.*, **560**, 135 (1948).

272. S. Choe, and S. Tsutsumi, *Nippon Kagaku Zasshi*, **81**, 582, 589 (1960); through *Chem. Abstr.*, **56**, 397 (1962).

273. A. Rieche and C. Bischoff, *Ber.*, **94**, 2722 (1961).

274. Bulletin No. 46, Becco Chemical Division, Food Machinery and Chemical Corp., 1958.

275. Bulletin No. 3, Becco Chemical Division, Food Machinery and Chemical Corp.; E. S. Shanley and F. P. Greenspan, *Ind. Eng. Chem.*, **39**, 1536 (1947).

276. E. Hedaya and S. Winstein, *J. Amer. Chem. Soc.*, **89**, 5314 (1967).

277. A. Baeyer and V. Villiger, *Ber.*, **34**, 738 (1901).

278. A. Rieche and F. Hitz, *Ber.*, **62**, 2473 (1929).

279. W. Eggersglüss, *Organische Peroxyde*, *Angew. Chem.* Monograph No. 61, Verlag Chemie, Weinheim, 1951.

280. E. J. Harris, *Proc. Roy. Soc.*, **A173**, 126 (1939).

281. A. D. Kirk, *Can. J. Chem.*, **43**, 2236 (1965).

282. I. P. Fisher and C. F. H. Tipper, *Trans. Faraday Soc.*, **59**, 1174 (1963).

283. N. A. Milas and D. M. Surgenor, *J. Amer. Chem. Soc.*, **68**, 205, 643 (1946).

284. N. A. Milas and L. H. Perry, *J. Amer. Chem. Soc.*, **68**, 1938 (1946).

285. N. A. Milas and O. L. Mageli, *J. Amer. Chem. Soc.*, **74**, 1471 (1952); *ibid.*, **75**, 5970 (1953).

286. J. Kochi, *J. Amer. Chem. Soc.*, **84**, 1193 (1962).

287. M. R. Vilenskaya and T. I. Yurzhenko, *Zh. Obschch. Khim.*, **34**, 748 (1964).

288. J. Hoffman and C. E. Boord, *J. Amer. Chem. Soc.*, **77**, 3139 (1955).

289. C. Walling and A. A. Zavitsas, *J. Amer. Chem. Soc.*, **85**, 2084 (1963).

290. J. Hoffman, *Org. Syn.*, **40**, 76 (1960).

291. H. R. Williams and H. S. Mosher, *J. Amer. Chem. Soc.*, **76**, 2984, 2987 (1954).

292. S. Dykstra and H. S. Mosher, *J. Amer. Chem. Soc.*, **79**, 3474 (1957).

293. S. Wawzonek, P. D. Klimstra, and R. E. Gallio, *J. Org. Chem.*, **25**, 621 (1960).

294. R. Hofmann, H. Huebner, G. Just, L. Kratzsch, A. K. Litkowez, W. Pritz-kow, W. Rolle, and M. Wahren, *J. Prakt. Chem.*, **37**, 1022 (1968).

295. F. Welch, H. R. Williams, and H. S. Mosher, *J. Amer. Chem. Soc.*, **77**, 551 (1955).

296. W. A. Pryor, D. M. Huston, T. R. Fike, T. L. Pickering, and E. Ciuffarin, *J. Amer. Chem. Soc.*, **86**, 4237 (1964).

297. A. G. Davies and A. M. White, *Nature*, **170**, 668 (1952).

298. A. G. Davies, R. V. Foster, and A. M. White, *J. Chem. Soc.*, **1953**, 1541; *ibid.*, **1954**, 2200.

299. A. G. Davies, *J. Chem. Soc.*, **1958**, 3474.

300. M. Bassey, C. A. Bunton, A. G. Davies, T. A. Lewis, and D. R. Llewellyn, *J. Chem. Soc.*, **1955**, 2471.

301. A. G. Davies and A. M. White, *J. Chem. Soc.*, **1952**, 3300.

302. A. G. Davies and R. Feld, *J. Chem. Soc.*, **1956**, 4669.

303. A. G. Davies, *J. Chem. Soc.*, **1962**, 4288.

304. A. G. Davies and R. Feld, *J. Chem. Soc.*, **1956**, 665.

305. A. G. Davies and R. Feld, *J. Chem. Soc.*, **1958**, 4637.

306. S. Choe and S. Tustsumi, *Nippon Kagaku Zasshi*, **81**, 586 (1960); through *Chem. Abstr.*, **56**, 397 (1962).

307. H. R. Williams and H. S. Mosher, *J. Amer. Chem. Soc.*, **76**, 3495 (1954).

308. J. M. Dunstan, M.Sc., Thesis, University of Toronto, 1962.

309. R. W. Foreman and H. P. Lankelma, *J. Amer. Chem. Soc.*, **79**, 409 (1957).

310. E. Hedaya and S. Winstein, *J. Amer. Chem. Soc.*, **89**, 1661 (1967).

311. U.S. Patent 2,573,947 (1951).

312. R. Hiatt and W. J. Strachan, *J. Org. Chem.*, **28**, 1893 (1963).

313. J. I. G. Cadogan, D. Hey, and W. Sanderson, *J. Chem. Soc.*, **1958**, 4498.

314. G. H. Anderson and J. G. Smith, *Can. J. Chem.*, **46**, 1553 (1968).

315. G. H. Anderson and J. G. Smith, *Can. J. Chem.*, **46**, 1561 (1968).

316. L. Dulog and A. Sonner, *Tetrahedron Letters*, **1966**, 6353.

317. S. Choe and S. Tsutsumi, *Nippon Kagaku Zasshi*, **81**, 589 (1960); through *Chem. Abstr.*, **56**, 397 (1962).

318. A. Rieche, E. Schmitz, and E. Beyer, *Ber.*, **92**, 1206 (1959).

319. E. Schmitz, A. Rieche, and A. Stark, *Ber.*, **101**, 1035 (1968).

320. A. Rieche, E. Schmitz, and E. Beyer, *Ber.*, **91**, 1942 (1958).

321. R. Hüttel and H. Ross, *Ber.*, **89**, 2641, 2644 (1956).

322. R. Hüttel, H. Schmid, and H. Ross, *Ber.*, **92**, 699 (1959).

323. N. A. Milas, R. L. Peeler, Jr., and O. L. Mageli, *J. Amer. Chem. Soc.*, **76**, 2322 (1954).

324. R. Criegee and K. Metz, *Ber.*, **89**, 1714 (1956).

325. A. Rieche, *Angew, Chem.*, **73**, 57 (1961).

326. R. Criegee and D. Seebach, *Ber.*, **96**, 2704 (1963).

327. K. Weissermel and M. Lederer, *Ber.*, **96**, 77 (1963).

328. M. Schulz, A. Rieche, and K. Kirschke, *Ber.*, **100**, 370 (1967).

329. K. Kopecky, J. Van de Sande, and C. Mumford *Can. J. Chem.*, **46**, 25 (1968).

330. A. Rieche, E. Schmitz, and E. Beyer, *Ber.*, **92**, 1212 (1958).

331. E. Höft and A. Rieche, *Angew Chem., Intern. Ed.*, **4**, 524 (1965).

332. A. Rieche, *Angew. Chem.*, **70**, 251 (1958).
333. P. L. Kooijman and W. L. Ghijsen, *Rec. Trav. Chim.*, **66**, 205 (1947).
334. C. N. Satterfield and L. C. Case, *Ind. Eng. Chem.*, **46**, 998 (1954).
335. A. Rieche and R. Meister, *Ber.*, **68**, 1465 (1935).
336. A. Rieche, *Ber.*, **64**, 2328 (1931).
337. A. Baeyer and V. Villiger, *Ber.*, **33**, 2479 (1900).
338. Reference 27, pp. 57–60.
339. M. S. Kharasch and G. N. Sosnovsky, *J. Org. Chem.*, **23**, 1322 (1958).
340. V. L. Antonovskii, A. F. Nesterov, and O. K. Lyashenko, *Zh. Prikl. Khim.*, **40**, 2555 (1967); through *Chem. Abstr.*, **68**, 86862 (1968).
341. V. L. Antonovskii and V. A. Terent'ev, *Zh. Fiz. Khim.*, **39**, 621 (1965); through *Chem. Abstr.*, **63**, 1682 (1965).
342. V. L. Antonovskii and V. A. Terent'ev, *Zh. Fiz. Khim.*, **40**, 3078 (1966); through *Chem. Abstr.*, **66**, 88991 (1967).
343. N. A. Milas and A. Golubovic, *J. Amer. Chem. Soc.*, **81**, 6461 (1959).
344. A. I. Kirillov, *Zh. Obshch. Khim.*, **1**, 1226 (1965); E. G. Sander and W. P. Jencks, *J. Am. Chem. Soc.*, **90**, 4377 (1968).
345. T. Ledaal, *Acta Chem. Scand.*, **21**, 1656 (1967).
346. T. Ledaal and T. Solbjör, *Acta Chem. Scand.*, **21**, 1658 (1967).
347. N. A. Milas and A. Golubovic, *J. Amer. Chem. Soc.*, **81**, 5824 (1959).
348. N. A. Milas and A. Golubovic, *J. Amer. Chem. Soc.*, **81**, 3361 (1959).
349. N. A. Milas, S. A. Harris, and P. C. Panagiotakas, *J. Amer. Chem. Soc.*, **61**, 2430 (1939).
350. M. Schulz, K. Kirschke, and E. Hoehne, *Ber.*, **100**, 2242 (1967).
351. A. Rieche, C. Bischoff, and D. Prescher, *Ber.*, **97**, 3071 (1964).
352. A. Rieche, and M. Schultz, *Ber.*, **97**, 190 (1964).
353. U.S. Patent 3,262,976 (1966).
354. K. R. Kopecky and C. Mumford, abstracts of paper presented at the 51st meeting of the CIC, Vancouver, June 1968, p. 41.
355. W. H. Richardson, J. W. Peters, and W. P. Konopka, *Tetrahedron Letters*, **1966**, 5531.
356. R. L. Talbott, *J. Org. Chem.*, **33**, 2095 (1968).
357. K. Ziegler and P. Herte, *Ann.*, **551**, 206 (1942).
358. A. Cohen, *J. Chem. Phys.*, **47**, 3828 (1967).
359. J. A. Bernard and A. Cohen, *Trans. Faraday Soc.*, **64**, 396 (1968).
360. L. Dulog and W. Vogt, *Tetrahedron Letters*, **1967**, 1915.
361. G. Sosnovsky and J. H. Brown, *Chem. Rev.*, **66**, 529 (1966).
362. U.S. Patent 2,692,887 (1954).
363. W. Hahn and L. Metzinger, *Makromol. Chem.*, **21**, 113 (1956).
364. R. L. Dannley and G. Jalics, *J. Org. Chem.*, **30**, 2417, 3848 (1965).
365. R. L. Dannley, and W. A. Aue, *J. Org. Chem.*, **30**, 3845 (1965); R. L. Dannley and G. C. Ferrant, *ibid.*, **34**, 2432 (1969).
366. Y. A. Alexandrov, *Tr. po Khim. i Khim. Tekhnol.*, **2**, 485 (1962); through *Chem. Abstr.*, **59**, 8777 (1963).
367. Y. A. Alexandrov and V. A. Shushunov, *Zh. Obshch. Khim.*, **35**, 115 (1965).
368. J. Dahlmann and A. Rieche, *Ber.*, **100**, 1544 (1967).

369. *Handbook of Chemistry and Physics*, 48th ed., Chemical Rubber Co., Cleveland, 1967.

370. O. D. Shreve, M. R. Heether, H. B. Knight, and D. Swern, *Anal. Chem.*, **23**, 282 (1951).

371. A. R. Philpotts and W. Thain, *Anal. Chem.*, **24**, 638 (1952).

372. G. J. Minkoff, *Discussions Faraday Soc.*, **9**, 320 (1950).

373. G. J. Minkoff, *Proc. Roy. Soc.*, **A224**, 176 (1954).

374. L. S. Silbert, *J. Amer. Oil Chem. Soc.*, **39**, 480 (1962).

375. D. Swern and L. S. Silbert, *Anal. Chem.*, **35**, 880 (1963).

376. H. R. Williams and H. S. Mosher, *Anal. Chem.*, **27**, 517 (1955).

377. R. T. Holman and P. R. Edmonson, *Anal. Chem.*, **28**, 1533 (1956).

378. R. T. Holman, C. Nickell, O. S. Privett, and P. R. Edmonson, *J. Amer. Chem. Soc.*, **35**, 422 (1958).

379. N. A. Kozlov and I. B. Rabinovich, *Opt. Spektrosk. Akad. Nauk SSSR, Otd. Fiz. Mat. Nauk, Sb. Statei*, **3**, 192 (1967); through *Chem. Abstr.*, **67**, 77593 (1967).

380. B. V. Erofeev, I. I. Chizevskaya, and N. N. Khovratovich, *Sb. Nauchn. Robot. Akad. Nauk Belorusk. SSR, Inst. Fiz. Organ. Khim.*, **42**, (1961); through *Chem. Abstr.*, **57**, 4201 (1962).

381. H. E. De La Mare, *J. Org. Chem.*, **25**, 2114 (1960).

382. C. Walling and L. Heaton, *J. Amer. Chem. Soc.*, **87**, 48 (1965).

383. V. V. Zharkov and N. K. Rudnevskii, *Optika i Spektroskopiya*, **12**, 479 (1962); through *Chem. Abstr.*, **57**, 5477 (1962).

384. V. A. Terent'ev, N. Kh. Shtivel, and V. L. Antonovskii, *Zh. Prikl. Spectr.*, **5**, 463 (1966); through *Chem. Abstr.*, **66**, 52055 (1967).

385. G. Geiseler and H. Zimmerman, *Z. Physik. Chem. (Frankfurt)*, **46**, 84 (1964).

386. A. C. Egerton, E. J. Harris, and G. H. S. Young, *Trans. Faraday Soc.*, **44**, 745 (1948).

387. A. J. Everett and G. J. Minkoff, *Trans. Faraday Soc.*, **49**, 410 (1953).

388. S. Fujiwara, M. Katayama, and S. Kamio, *Bull. Soc. Chim. Japan*, **32**, 657 (1959).

389. D. Swern, A. H. Clements, and T. M. Luong, *Anal. Chem.*, **41**, 412 (1969).

390. W. A. Pryor and G. L. Kaplan, *J. Amer. Chem. Soc.*, **86**, 4234 (1964).

391. N. A. Milas, O. L. Mageli, A. Golubovic, R. W. Arndt, and J. C. J. Ho, *J. Amer. Chem. Soc.*, **85**, 222 (1963).

392. K. U. Ingold and J. R. Morton, *J. Amer. Chem. Soc.*, **86**, 3400 (1964).

393. J. R. Thomas, *J. Amer. Chem. Soc.*, **87**, 3935 (1965).

394. M. F. R. Mulcahy, J. R. Stevens, and J. C. Ward, *Australian J. Chem.*, **18**, 1177 (1965).

395. J. R. Thomas and K. U. Ingold, in reference 8, p. 258.

396. W. J. MaGuire, and R. C. Pink, *Trans. Faraday Soc.*, **63**, 1097 (1967).

397. R. W. Brandon and C. S. Elliot, *Tetrahedron Letters*, **1967**, 4375.

398. R. Criegee, in *Methoden der organischen Chemie*, Vol. 2, Houben-Weyl Thieme, Stuttgart, 1953, p. 1570.

399. S. W. Benson, *J. Chem. Phys.*, **40**, 1007 (1964).

400. G. J. Minkoff, *Discussions Faraday Soc.*, **10**, 329 (1951).

401. S. W. Benson and J. Buss, *J. Chem. Phys.*, **29**, 546 (1958).
402. W. Pritzkow and K. A. Müller, *Ber.*, **89**, 2318 (1956).
403. H. Hershenson, and S. W. Benson *J. Chem. Phys.*, **37**, 1889 (1962).
404. P. Gray and A. Williams, *Chem. Rev.*, **59**, 239 (1959).
405. J. A. Kerr, *Chem. Rev.*, **66**, 465 (1966).
406. E. T. Denisov, *Russ. J. Phys. Chem.*, **38**, 1 (1964).
407. G. I. Likhtenshtein, *Russ. J. Phys. Chem.*, **36**, 801 (1962).
408. R. West and R. H. Baney, *J. Phys. Chem.*, **64**, 822 (1960).
409. L. Bateman and H. Hughes, *J. Chem. Soc.*, **1952**, 4594.
410. J. R. Thomas and O. Harle, *J. Phys. Chem.*, **63**, 1027 (1959).
411. E. T. Denisov, *Zh. Fiz. Khim.*, **38**, 2085 (1964).
412. A. L. Buchachenko and O. P. Sukhanova, *Usp. Khim.*, **36**, 475 (1967).
413. V. L. Antonovskii and V. A. Terent'ev, *Zh. Organ. Khim.*, **3**, 1011 (1967); through *Chem. Abstr.*, **67**, 107884 (1967).
414. E. T. Denisov, *Dokl. Akad. Nauk SSSR*, **146**, 394 (1962); through *Chem. Abstr.*, **58**, 3284 (1963).
415. N. V. Zolotova and E. T. Denisov, *Izv. Akad. Nauk SSSR*, **1966**, 767; through *Chem. Abstr.*, **65**, 12076 (1966).
416. D. Barnard, K. R. Hargrave, and G. M. C. Higgins, *J. Chem. Soc.*, **1956**, 2845.
417. E. T. Denisov, A. L. Aleksandrov, and V. P. Shcheredin, *Isv. Akad. Nauk SSSR*, **1964**, 1583; through *Chem. Abstr.*, **62**, 2687 (1965).
418. L. G. Privalova, Z. K., Maizus and N. M. Emanuel, *Dokl. Akad. Nauk SSSR*, **161**, 1135 (1965).
419. V. L. Antonovskii, E. T. Denisov, and L. V. Solntseva, *Kinetika i Kataliz*, **6**, 815 (1965).
420. G. M. Coppinger and J. D. Swalen, *J. Amer. Chem. Soc.*, **83**, 4900 (1961).
421. A. A. Oswald, K. Griesbaum, and B. E. Hudson, *J. Org. Chem.*, **28**, 2351 (1963).
422. U.S. Patent 3,236,850 (1966).
423. E. T. Denisov, *Russ. J. Phys. Chem.*, **37**, 1029 (1963).
424. M. S. Kharasch, A. Fono, W. Nudenberg and B. Bischof, *J. Org. Chem.* **17**, 207 (1952).
425. V. A. Belyaev and M. S. Nemtsov, *Zh. Obshch, Khim.*, **31**, 3855, (1961), **32**, 3113 (1962); through *Chem. Abstr.*, **58**, 8868 (1963).
426. I. M. Kolthoff and A. I. Medalia, *J. Amer. Chem. Soc.*, **71**, 3789 (1949).
427. A. J. Everett and G. J. Minkoff, *Trans Faraday Soc.*, **49**, 410 (1953).
428. A. Rieche, E. Schmitz, and P. Dietrich, *Ber.*, **92**, 2239 (1959).
429. J. Rigaudy and E. C. Izoret, *Compt. Rend.*, **263**, 2086 (1953).
430. D. Barnard and K. R. Hargrave, *Anal. Chem. Acta*, **5**, 476 (1951).
431. G. Sorge and K. Überreiter, *Angew. Chem.*, **68**, 352, 486 (1956).
432. M. Matic and D. A. Sutton, *J. Chem. Soc.*, **1952**, 2679.
433. M. Matic and D. A. Sutton, *Chem. Ind.* (*London*), **1953**, 666.
434. R. Criegee, H. Pilz, and H. Flygare, *Ber.*, **72**, 1799 (1939).
435. L. Bateman and K. R. Hargrave, *Proc. Roy. Soc.* **A224**, 389, 399 (1954).
436. G. Ayrey, D. Barnard, and C. G. Moore, *J. Chem. Soc.*, **1953**, 3179.
437. D. Barnard, *J. Chem. Soc.*, **1956**, 489.

438. J. J. Pappas, W. P. Keaveney, E. Gaucher, and M. G. Berger, *Tetrahedron Letters*, **1966**, 4273.

439. L. Horner and W. Jurgeleit, *Ann.*, **591**, 138 (1955).

440. C. Walling and R. Rabinowitz, *J. Amer. Chem. Soc.*, **81**, 1243 (1959).

441. A. J. Martin, *Organic Analysis*, Vol. IV, Interscience, New York, 1960, pp. 1–64.

442. S. Siggia, *Quantitative Organic Analysis*, Wiley, New York, 1963, pp. 255–295.

443. C. A. Bunton, in reference 12, p. 20.

444. R. Hiatt, T. Mill, K. C. Irwin, and J. K. Castleman, *J. Org. Chem.*, **33**, 1421 (1968).

445. D. B. Denny, W. F. Goodyear, and B. Goldstein, *J. Amer. Chem. Soc.*, **82**, 1393 (1960).

446. H. Lindlar, *Helv. Chim. Acta*, **35**, 450 (1952).

447. O. S. Privett and E. C. Nickell, *Fette ,Seifen, Anstrichmittel*, **61**, 842 (1959).

448. U.S. Patents 2,430,865 (1947) and 3,256,341 (1966).

449. M. H. Brodnitz, W. W. Nawar, and I.S. Fagerson, *Lipids*, **3**, 65 (1968).

450. T. Perlstein, A. Eisner, and W. C. Ault, *J. Amer. Oil Chem. Soc.*, **43**, 380 (1966).

451. N. A. Sokolov, V. A. Shushunov, and V. A. Yablokov, *Tr. po Khim. i* Khim. Tekhnol., **58**, (1962); through *Chem. Abstr.*, **59**, 6290 (1963).

452. E. G. Hawkins, *J. Chem. Soc.*, **1949**, 2076.

453. A. Farkas and E. Passaglia, *J. Amer. Chem. Soc.*, **72**, 3333 (1950).

454. N. A. Sokolov, L. N. Chetyrbok, and V. A. Shushunov, *Zh. Obshch. Khim.*, **33**, 2027 (1963).

455. J. D'Ans and H. Gould, *Ber.*, **92**, 2559 (1959).

456. W. Pritzkow and H. Schaefer, *Ber.*, **93**, 2151 (1960).

457. W. J. Farrissey, *J. Amer. Chem. Soc.*, **84**, 1002 (1962).

458. E. N. Frankel, C. D. Evans, D. G. McConnell, E. Selka, and H. J. Dutton, *J. Org. Chem.*, **26**, 4663 (1961).

459. I. M. Kolthoff and E. B. Sandell, *Quantitative Inorganic Analysis*, Macmillan, New York, 1948, p. 614.

460. F. W. Heaton and N. Uri, *J. Sci. Food Agric.*, **5**, 476 (1958); D. K. Bannerjee and C. C. Budke, *Anal. Chem.*, **36**, 792 (1964).

461. T. R. Griffiths, *Colloq. Spectrosc. Intern.*, *12th*, Exeter, England, 1965, p. 226; through *Chem. Abstr.*, **68**, 26771 (1968).

462. T. C. Purcell and I. R. Cohen, *J. Environmental Sci.*, **1**, 431 (1967).

463. R. D. Mair and A. J. Graupner, *Anal. Chem.*, **36**, 194 (1964).

464. C. D. Wagner, R. H. Smith, and E. D. Peters, *Ind. Eng. Chem., Anal. Ed.*, **19**, 976 (1947).

465. R. Hiatt and K. C. Irwin, *J. Org. Chem.*, **33**, 1436 (1968).

466. J. P. Wibaut, H. B. van Leeuwen, and B. van der Wal, *Rec. Trav. Chim.*, **73**, 1033 (1954).

467. A. Banderet, M. Brendle, and G. Riess, *Bull. Soc. Chim. France*, **1965**, 626.

468. M. I. Eiss and P. Giesicke, *Anal. Chem.* ,**31**, 1558 (1959).

469. F. Hrabak and I. Horacek, *Coll. Czech. Chem. Commun.*, **31**, 2157 (1966).

470. E. Tikhanen, *Pharmacia*, **5**, 7 (1926).

471. E. K. Dukes and M. L. Hyder, *Anal. Chem.*, **36**, 1689 (1964).
472. H. Pobiner, *Anal. Chem.*, **33**, 1423 (1961).
473. I. R. Cohen, T. C. Purcell, and A. P. Altshuller, *J. Environmental Sci. Technol.*, **1**, 247 (1967).
474. I. R. Cohen and T. C. Purcell, *Anal. Chem.*, **39**, 131 (1967).
475. R. Strohecker, R. V. Vaubol, and A. Tenner, *Fetet und Siefen*, **44**, 246 (1937).
476. W. C. Wolfe, *Anal. Chem.*, **34**, 1328 (1962).
477. W. M. McNevin and P. K. Urone, *Anal. Chem.*, **25**, 1760 (1953).
478. I. M. Kolthoff and A. I. Medalia, *Anal. Chem.*, **23**, 595 (1951).
479. H. Ueda, *Anal. Chem.*, **35**, 2213 (1963).
480. B. G. Tarladgis, A. W. Shoenmakers, and P. Begemann, *J. Dairy Sci.*, **47**, 1011 (1964).
481. B. G. Tarladgis and A. W. Shoenmakers, *Nature*, **210**, 1151 (1966).
482. K. Moebius and F. Schneider, *Z. Naturforsch.*, **18a**, 428 (1963).
483. J. Mitchell and D. M. Smith, *Aquametry*, Interscience, New York, 1948.
484. A. P. Terent'ev, G. G. Larikova, and E. A. Bondarevskaya, *Zh. Anal. Khim.*, **21**, 355 (1966).
485. M. M. Buzlanova, L. I. Mekhryuskevo, and V. L. Antonovskii, *Zh. Anal. Khim.*, **21**, 251 (1966); through *Chem. Abstr.*, **64**, 18407 (1966); USSR Patent 172,318 (1965).
486. H. Hock and O. Schrader, *Brennstoff Chem.*, **18**, 6 (1937).
487. U.S. Patent 2,858,326 (1958); J. Mitchell and L. R. Perkins, *Appl. Polymer. Symp.*, No. 4, 167 (1967); through *Chem. Abstr.*, **68**, 30087 (1968).
488. L. Dulog and K. H. Burg, *Z. Anal. Chem.*, **203**, 184 (1964).
489. R. A. Stein and V. Slawson, *Anal. Chem.*, **35**, 1008 (1963).
490. L. Horner and E. Jürgens, *Angew. Chem.*, **70**, 268 (1958).
491. A. Niederstebruch and I. Hinsch, *Fette, Seifen, Anstrichmittel*, **69**, 637 (1967).
492. S. I. Bass, *Zh. Anal. Khim.*, **17**, 113 (1962); through *Chem. Abstr.*, **67**, 4038 (1967).
493. H. Bruschweiler and G. J. Minkhoff, *Anal. Chem. Acta*, **12**, 186 (1955).
494. I. A. Korshunov and A. I. Kalinin, *Khim. Perekisnykh Soedin.*, **1963**, 279; through *Chem. Abstr.*, **60**, 15149 (1964).
495. M. F. Romantsev and E. S. Levin, *Zh. Anal. Khim.*, **18**, 1109 (1963); through *Chem. Abstr.*, **60**, 3494 (1964).
496. L. Dulog, *Z. Anal. Chem.*, **202**, 258 (1964).
497. M. Schultz and K. H. Schwartz, *Z. Chemie*, **7**, 176 (1967); *Deutsche Akad. der Wissensch. Berlin Monatsh.*, **6**, No. 7, 515 (1964).
498. M. H. Abraham, A. G. Davies, D. R. Llewellyn, and M. E. Thain, *Anal. Chem. Acta*, **17**, 499 (1957).
499. A. Rieche and M. Schultz, *Angew. Chem.*, **70**, 694 (1958).
500. J. Cartlidge and C. F. H. Tipper, *Chem. Ind. (London)*, **1959**, 852; *Anal. Chem. Acta.* **22**, 106 (1960).
501. N. A. Milas and I. Belic, *J. Amer. Chem. Soc.*, **81**, 3358 (1959).
502. G. Dobson and G. Hughes, *J. Chromatog.*, **16**, 416 (1964).
503. E. Knappe and D. Peterli, *Z. Anal. Chem.*, **190**, 380 (1962).
504. P. Fijolka and R. Gnauck, *Plaste und Kautschuk*, **13**, 343 (1966).

505. M. M. Buzlanova, V. F. Stepanovskaya, A. F. Nesterov, and V. L. Antonovskii, *Zh. Anal. Khim.*, **21**, 506 (1966); through *Chem. Abstr.*, **65**, 4630 (1966).

506. A. N. Soroka, A. E. Batog, and M. K. Romansevich, *Zh. Obshch. Khim.*, **37**, 766 (1967).

507. R. Baranowski, Z. Gregorowicz, J. Kulicki, and Z. Kulicki, *Chem. Anal. (Warsaw)*, **11**, 135 (1966); through *Chem. Abstr.*, **64**, 18381 (1966).

508. C. F. Wurster, L. J. Durham, and H. S. Mosher, *J. Amer. Chem. Soc.*, **60**, 327 (1958).

509. S. W. Bukata, L. L. Zabrocki, and M. F. McLaughlin, *Anal. Chem.*, **35**, 885 (1963).

510. E. Ewald, G. Öhlmann, and W. Schirmer, *Z. Phys. Chem.*, **234**, 104 (1967).

511. L. Ya. Gavrilina and D. A. Vyakhirev, *Izv. Sib. Otd. Akad. Nauk SSSR, Ser. Khim. Nauk*, **1967**, 141; through *Chem. Abstr.*, **68**, 74868 (1968).

512. M. L. Vlodavets, K. A. Gol'bert, E. Ya. Chervin'skaya, N. V. Pervoskaya, and L. P. Ternovskaya, *Neftekhim.*, **5**, 613 (1965); through *Chem. Abstr.*, **64**, 1356 (1966).

513. J. O. Edwards, *Inorganic Reaction Mechanisms*, Interscience, New York, 1962, p. 82.

514. J. O. Edwards and R. G. Pearson, *J. Amer. Chem. Soc.*, **84**, 16 (1963).

515. W. P. Jencks and J. Carriuolo, *J. Amer. Chem. Soc.*, **82**, 1778 (1960).

516. F. F. Rust, F. H. Seubold, and W. E. Vaughan, *J. Amer. Chem. Soc.*, **72**, 338 (1950).

517. N. Kornblum and H. E. De La Mare, *J. Amer. Chem. Soc.*, **74**, 3079 (1952).

518. M. R. Barusch and J. Q. Payne, *J. Amer. Chem. Soc.*, **75**, 1987 (1953).

519. N. A. Milas, R. J. Klein, and D. G. Orphanos, *Chem. Ind. (London)*, **1964**, 423.

520. British Patent 896,813 (1962).

521. H. Hock and S. Lang, *Ber.*, **75**, 300 (1942).

522. A. Rieche and C. Bischoff, *Ber.*, **94**, 2457 (1961).

523. German Patent 1,222,073 (1966).

524. V. L. Antonovskii and Yu. D. Emelin, *Neftekhim.* **6**, 733 (1966); through *Chem. Abstr.*, **66**, 28165 (1967).

525. Z. F. Nazarova, A. E. Batog, and M. K. Ramantsevich, *Zh. Organ. Khim.*, **2**, 1052 (1966); through *Chem. Abstr.*, **65**, 15259 (1966).

526. A. N. Sorokina, A. E. Batog, and M. K. Ramantsevich, *Zh. Organ. Khim.*, **1**, 1881 (1965); through *Chem. Abstr.*, **64**, 3448 (1966).

527. F. H. Dickey, F. F. Rust, and R. E. Vaughan, *J. Amer. Chem. Soc.*, **71**, 1432 (1949).

528. U.S. Patent 2,455,569 (1948).

529. Belgian Patent 660,135 (1965).

530. A. Rieche, C. Bischoff, and M. Pulz, *Ber.*, **98**, 321 (1965).

531. N. A. Milas and D. M. Surgenor, *J. Amer. Chem. Soc.*, **68**, 605 (1946).

532. L. S. Silbert and D. Swern, *J. Amer. Chem. Soc.*, **81**, 2364 (1959).

533. R. Criegee, *Ber.*, **77**, 22 (1944); *ibid.*, **84**, 215 (1951).

534. N. A. Milas, D. G. Orphanos, and R. J. Klein, *J. Org. Chem.*, **29**, 3099 (1964).

535. For other references, see reference 1, p. 58, and reference 12, p. 16.

536. R. Schöllner, J. Weiland, and M. Mühlstädt, *Z. Chemie*, **3**, 390 (1963).

537. V. L. Antonovskii and O. K. Lyashenko, *Zh. Prikl. Khim.*, **40**, 243 (1967); through *Chem. Abstr.*, **66**, 85423 (1967).

538. V. L. Antonovskii and O. K. Lyashenko, *Kinetika i Kataliz*, **7**, 767 (1966); through *Chem. Abstr.*, **65**, 18438 (1966).

539. R. F. Naylor, *J. Chem. Soc.*, **1945**, 244.

540. E. L. O'Brien, F. M. Beringer, and R. B. Mesrobian, *J. Amer. Chem. Soc.*, **79**, 6238 (1957).

541. Belgian Patent 622,032 (1963).

542. U.S. Patent 2,608,570 (1952).

543. G. A. Razuvaev, N. M. Lapshin, and N. V. Balukova, *Zh. Organ. Khim.*, **4**, 73 (1968); through *Chem. Abstr.*, **68**, 86922 (1968).

544. P. D. Bartlett and B. T. Storey, *J. Amer. Chem. Soc.*, **80**, 4954 (1958).

545. P. D. Bartlett and T. Traylor, *J. Amer. Chem. Soc.*, **83**, 856 (1961).

546. J. E. Leffler and Y. Tsuno, *J. Org. Chem.*, **28**, 190 (1963).

547. A. Rieche and J. Dahlmann, *Ann.* **679**, 19 (1964).

548. N. C. Yang and R. A. Finnegan, *J. Amer. Chem. Soc.*, **80**, 5845 (1958).

549. J. Q. Payne, *J. Org. Chem.*, **24**, 2048 (1959); *ibid.*, **25**, 275 (1960).

550. C. A. Bunton and J. G. Minkoff, *J. Chem. Soc.*, **1949**, 665.

551. U.S. Patent 2,508,256 (1950).

552. O. Soga, K. Maryuama,   and R. Goto, *Nippon Kagaku Zasshi*, **81**, 668 (1960); through *Chem. Abstr.*, **56**, 310 (1962).

553. K. Maruyama, *Bull. Chem. Soc. Japan*, **33**, 1516 (1960), *ibid.*, **34**, 102, 105 (1961).

554. H. Kwart and N. J. Wegemer, *J. Amer. Chem. Soc.*, **83**, 2746 (1961); *ibid.*, **87**, 511 (1965).

555. A. G. Davies, R. V. Foster, and R. Nery, *J. Chem. Soc.*, **1954**, 2204.

556. D. E. Bissing, C. A. Matuszak, and W. E. McEwen, *Tetrahedron Letters*, **1962**, 763.

557. A. W. Van Stevenick and E. C. Kooyman, *Rec. Trav. Chim.*, **79**, 413 (1960).

558. V. P. Maslennivok, V. P. Sergeeva, and V. A. Shushunov, *Tr. po Khim. i Khim. Tekhnol.*, **1966**, 22; through *Chem. Abstr.*, **67**, 81688 (1967).

559. G. Burtzlaff, U. Felber, H. Hübner, W. Pritzkow, and W. Rolle, *J. Prakt. Chem.*, **28**, 305 (1965).

560. R. Criegee, *Ann.*, **560**, 127 (1948).

561. A. Robertson and W. A. Waters, *J. Chem. Soc.*, **1948**, 1574.

562. J. E. Leffler, *Chem. Rev.*, **45**, 385 (1949).

563. H. Kwart and R. T. Keen, *J. Amer. Chem. Soc.*, **81**, 943 (1959).

564. G. A. Razuvaev and N. A. Kartashova, *Zh. Organ. Khim.*, **3**, 993 (1967); through *Chem. Abstr.*, **67**, 107983 (1967).

565. M. S. Kharasch and J. G. Burt, *J. Org. Chem.*, **16**, 150 (1951).

566. W. F. Brill, *J. Org. Chem.*, **24**, 257 (1959).

567. E. H. Farmer and A. Sundralingam, *J. Chem. Soc.*, **1942**, 121.

568. W. Pritzkow and R. Hofmann, *J. Prakt. Chem.*, **14**, 131 (1961).

569. V. A. Shushunov, V. A. Yablokov, and N. V. Yablokova, *Tr. po Khim. i Khim. Tekhnol.*, **1964**, 441; through *Chem. Abstr.*, **57**, 1597 (1962).

570. V. P. Maslennikov and V. A. Shushunov, *Tr. po Khim. i Khim. Tekhnol.*, **1965**, 34.

571. V. P. Maslennikov, V. A. Shushunov, and V. A. Klimova, *Tr. po Khim. i Khim. Tekhnol.*, **1965**, 42.

572. V. P. Maslennikov and V. A. Shushunov, *Tr. po Khim. i Khim., Tekhnol.* **1965**, 50.

573. V. P. Maslennikov, V. P. Sergeeva, and V. A. Shushunov, *Tr. po Khim. i Khim. Tekhnol.*, **1965**, 59.

574. V. A. Yablokov, V. A. Shuchunov, and G. I. Vesnovskaya, *Tr. po Khim. i Khim. Tekhnol.*, **1966**, 198. through *Chem. Abstr.*, **68**, 12,071 (1968).

575. D. E. Bissing, C. A. Matuszak, and W. E. McEwen, *J. Amer. Chem. Soc.*, **86**, 3824 (1964).

576. R. D. Cadle and H. Huff, *J. Phys. Chem.*, **54**, 1191 (1950).

577. H. Boardmann and G. E. Hulse, *J. Amer. Chem. Soc.*, **75**, 4272 (1953).

578. C. Walling and R. B. Hogden, *J. Amer. Chem. Soc.*, **80**, 228 (1958).

579. M. S. Kharasch, R. A. Mosher, and J. S. Bengelsdorf, *J. Org. Chem.*, **25**, 1000 (1960).

580. P. A. Kirpichnikov, N. A. Mukmenova, A. N. Pudovik, and N. S. Kolyubakina, *Dokl. Akad. Nauk SSSR*, **164**, 1050 (1965); through *Chem. Abstr.*, **64**, 1926 (1966).

581. C. Walling and M. S. Pearson, *J. Amer. Chem. Soc.*, **86**, 2262 (1964).

582. W. G. Bentrude, *Tetrahedron Letters*, **1965**, 3543.

583. W. H. Starnes and N. P. Neureiter, *J. Org. Chem.*, **32**, 333 (1967).

584. R. Hiatt, R. Smythe, and C. McColeman, *Can. J. Chem.*, **49**, 000 (1971).

585. S. A. Buckler, *J. Amer. Chem. Soc.*, **84**, 3093 (1962).

586. M. B. Floyd and C. E. Boozer, *J. Amer. Chem. Soc.*, **85**, 984 (1963).

587. P. T. Levin, *Zh. Fiz. Khim.*, **38**, 672 (1964); through *Chem. Abstr.*, **61**, 1728 (1964).

588. M. A. P. Dankleff, R. Curci, J. O. Edwards, and H. W. Pyun, *J. Amer. Chem. Soc.*, **90**, 3208 (1968).

589. K. R. Hargrave, *Proc. Roy. Soc.*, **A235**, 55 (1956).

590. C. S. Overberger and R. W. Cummins, *J. Amer. Chem. Soc.*, **75**, 4783 (1953).

591. G. Modena and L. Maioli, *Gazz. Chim. Ital.*, **87**, 1306 (1957).

592. G. Modena, *Gazz. Chim. Ital.*, **89**, 834 (1959).

593. A. Cerniani and G. Modena, *Gazz. Chim. Ital.*, **89**, 843 (1959).

594. C. Bodea and I. Silberg, *Acad. Rep., Populaire Romine, Filiala Cluj, Studii Cercetari Chem.*, **14**, 317 (1963); through *Chem. Abstr.*, **62**, 14664 (1965).

595. O. L. Harle and J. R. Thomas, *J. Amer. Chem. Soc.*, **79**, 2973 (1957).

596. J. Pannell, *Mol. Phys.*, **4**, 291 (1962).

597. L. A. Harris, and H. S. Olcott, *J. Amer. Oil Chem. Soc.*, **43**, 11 (1966).

598. K. Maruyama and T. Otsuki, *Tetrahedron Letters*, **1966**, 3705.

599. K. Maruyama, T. Otsuiki, and T. Iwao, *J. Org. Chem.*, **32**, 82 (1967).

600. A. A. Oswald, F. Noel, and A. J. Stephenson, *J. Org. Chem.*, **26**, 3969 (1961).

601. A. A. Oswald, B. E. Hudson, G. Rodgers, and F. Noel, *J. Org. Chem.*, **27**, 2439 (1962).

602. E. D. Vilanskaya, K. I. Ivanov, and A. V. Koryakin, *Tr. Vse, Nauch-Tekh. Soveshch. Prisadkam Miner. Maslam*, **2**, 190 (1966); through *Chem. Abstr.*, **68**, 61309 (1968).

603. K. A. Makarov and A. F. Nikolaev, *Reaktsionnaya Sposbnost. Organ. Soedin. Tartusk. Gos. Univ.*, **2**, 86 (1965); through *Chem. Abstr.*, **65**, 9026 (1966).

604. H. E. De La Mare and G. M. Coppinger, *J. Org. Chem.*, **28**, 106 (1963).

605. C. W. Capp and E. G. E. Hawkins, *J. Chem. Soc.*, **1953**, 4106.

606. H. E. De La Mare, *J. Org. Chem.*, **25**, 2114 (1960).

607. N. M. Beileryan, S. K. Grigoryan, and O. A. Chaltykyan, *Izv. Akad. Nauk Armen. SSR, Khim. Nauk*, **17**, 245, 604 (1964); through *Chem. Abstr.*, **63**, 4125, 8147 (1965).

608. O. A. Chaltykyan, S. K. Grigoryan, and N. M. Beileryan, **18**, *Izv. Akad. Nauk Armen, SSR, Khim. Nauk*, 133 (1965); through *Chem. Abstr.*, **63**, 16156 (1965).

609. E. Beati, M. Pegoraro, and M. Kucharski, *Chem. Ind. (Milan)*, **48**, 589 (1966).

610. G. A. Russell, *J. Amer. Chem. Soc.*, **76**, 1595 (1954).

611. G. A. Russell, *J. Amer. Chem. Soc.*, **84**, 2652, (1962).

612. S. O. Lawesson and N. C. Yang, *J. Amer. Chem. Soc.*, **81**, 4230 (1959).

613. A. G. Davies and R. B. Moodie, *J. Chem. Soc.*, **1958**, 2372.

614. E. Müller and T. Töpel, *Ber.*, **72**, 273 (1939).

615. C. D. Hurd and H. J. Anderson, *J. Amer. Chem. Soc.*, **75**, 5124 (1953).

616. T. W. Campbell, W. Burney, and T. L. Jacobs, *J. Amer. Chem. Soc.*, **72**, 2735 (1950).

617. G. Roberts and C. W. Shoppee, *J. Chem. Soc.*, **1954**, 3418.

618. U.S. Patent 2,713,599 (1955).

619. U.S. Patent 2,724,729 (1955).

620. U.S. Patent 2,881,220 (1959).

621. V. A. Belyaev and M. S. Nemtsov, *Zh. Obshch. Khim.*, **32**, 3483 (1962); through *Chem. Abstr.*, **58**, 12377 (1963).

622. W. A. Waters, *Mechanisms of Oxidation of Organic Compounds*, Methuen, London, 1964, p. 46.

623. E. Koubek, M. L. Haggett, C. J. Battaglia, K. M. Ibne-Rasa, H. Y. Pyun, and J. O. Edwards, *J. Amer. Chem. Soc.*, **85**, 2263 (1963).

624. M. S. Kharasch, A. Fono, and W. Nudenberg, *J. Org. Chem.*, **16**, 113 (1951).

625. D. B. Denny and J. D. Rosen, *Tetrahedron*, **20**, 271 (1964).

626. N. Kornblum and H. E. De La Mare, *J. Amer. Chem. Soc.*, **73**, 880 (1951).

627. S. S. Medwedew and E. Alexajewa, *Ber.*, **65**, 133 (1932).

628. A. Y. Khashab, *Diss. Abstr.*, **27B**, 4310 (1967).

629. C. A. Grob and P. W. Schiess, *Angew. Chem., Intern. Ed.*, **6**, 1 (1967).

630. M. Schultz and H. Steinmaus, *Angew. Chem., Intern. Ed.*, **2**, 623 (1963).

631. M. Schultz and H. F. Boeden, *Tetrahedron Letters*, **1966**, 2843.

632. M. Schultz, H. F. Boeden, and E. Grundeman, *Z. Chemie*, **7**, 13 (1967).

633. E. Schmitz, D. Habisch, and R. Ohme, *J. Prakt. Chem.*, **37**, 252 (1968).

634. M. B. Floyd, *Diss. Abstr.*, **27B**, 3457 (1967).

635. W. H. Richardson and R. S. Smith, *J. Amer. Chem. Soc.*, **89**, 2230 (1967).

636. W. J. Linn, O. W. Webster, and R. E. Benson, *J. Amer. Chem. Soc.*, **87**, 3651 (1965); *ibid*, **85**, 2032 (1963).

637. M. M. Rauhut, D. Sheehan, R. A. Clarke, B. G. Roberts, and A. M. Semsel, *J. Org. Chem.*, **30**, 3587 (1965).

638. L. J. Bollyky, R. H. Whitman, R. A. Clarke, and M. M. Rauhut, *J. Org. Chem.*, **32**, 1663 (1967).

639. E. J. Bowen and R. A, Lloyd, *Proc. Roy. Soc.*, **A275**, 465 (1963).

640. E. J. Bowen, *Nature*, **201**, 180 (1964).

641. R. A. Lloyd, *Trans. Faraday Soc.*, **61**, 2182 (1965).

642. M. M. Rauhut, D. Sheehan, R. A. Clarke, and A. M. Semsel, *Photochem. Photobiol.*, **4**, 1097 (1965).

643. W. von E. Doering and R. M. Haines, *J. Amer. Chem. Soc.*, **76**, 482 (1954).

644. K. B. Wiberg, *J. Amer. Chem. Soc.*, **75**, 3961 (1953).

645. G. B. Payne, *J. Org. Chem.*, **26**, 651 (1961).

646. H. Berger and A. F. Bickel, *Trans. Faraday Soc.*, **57**, 1325 (1961).

647. H. Berger, *Trans. Faraday Soc.*, **58**, 1137 (1962).

648. W. F. Brill, *J. Amer. Chem. Soc.*, **85**, 141 (1963).

649. W. F. Brill and N. Indictor. *J. Org. Chem.*, **29**, 710 (1964).

650. C. Walling and L. Heaton, *J. Amer. Chem. Soc.*, **87**, 38 (1965).

651. D. Swern, *Chem. Rev.*, **45**, 1 (1949).

652. U.S. Patent 3,251,888 (1966).

653. N. Indictor and W. F. Brill, *J. Org. Chem.*, **30**, 2074 (1965).

654. K. Allison. P. Johnson, G, Foster, and M. Sparka, *Ind. Eng. Chem. Prod. Res. Develop.*, **5**, 166 (1966).

655. Canadian Patents 799,502; 799,503; 799,504 (1968).

656. Belgian Patent 641,452 (1963).

657. Belgian Patent, 657,838 (1964).

658. Belgian Patent 680,852 (1965).

659. Netherlands Patent Application 6,507,189 (1965).

660. M. N. Sheng and J. G. Zajacek, in reference 8.

661. E. S. Gould, R. Hiatt, and K. C. Irwin, *J. Am. Chem. Soc.*, **90**, 4573 (1968).

662. L. Kuhnen, *Ber.*, **99**, 3384 (1966).

663. M. N. Sheng and J. G. Zajacek, *J. Org. Chem.*, **33**, 588 (1968).

664. L. Kuhnen, *Angew. Chem., Intern. Ed.*, **5**, 893 (1966).

665. E. R. Bell, J. H. Raley, F. F. Rust, F. H. Seubold, and W. E. Vaughan, *Discussions Faraday Soc.*, **10**, 246 (1951).

666. J. T. Martin and R. G. W. Norrish, *Proc. Roy. Soc.*, **A220**, 322 (1953).

667. M. S. Kharasch, A. Fono, and W. Nudenberg, *J. Org. Chem.*, **16**, 113 (1951).

668. G. H. Twigg, *Discussions Faraday Soc.*, **14**, 240 (1953).

669. G. H. Twigg, G. W. Godin, H. C. Bailey, and J. Holden, *Erdöl und Kohle*, **15**, 74 (1962).

670. I. I. Shabalin and V. A. Simanov, *Neftekhim.*, **5**, 246 (1965).

671. P. D. Bartlett and K. Nozaki, *J. Amer. Chem. Soc.*, **69**, 2299 (1947).

672. R. Hiatt, J. R. Clipsham, and T. Visser, *Can. J. Chem.*, **42**, 2754 (1964).

673. A. Factor, C. A. Russell, and T. G. Traylor, *J. Amer. Chem. Soc.*, **87**, 3692 (1965).
674. T. G. Traylor and P. D. Bartlett, *Tetrahedron Letters*, **1960**, 30.
675. P. D. Bartlett and P. Günther, *J. Amer. Chem. Soc.*, **88**, 3288 (1966).
676. P. D. Bartlett and G. Guaraldi, *J. Amer. Chem. Soc.*, **89**, 4799 (1967).
677. T. Mill and R. Stringham, *J. Amer. Chem. Soc.*, **90**, 1062 (1968).
678. N. A. Milas and B. Plesnicar, *J. Amer. Chem. Soc.*, **90**, 4450 (1968).
679. R. Hiatt and T. G. Traylor, *J. Amer. Chem. Soc.*, **87**, 3766 (1965).
680. R. Hiatt, T. Mill, and F. R. Mayo, *J. Org. Chem.*, **33**, 1416 (1968).
681. F. H. Seubold, F. F. Rust, and W. E. Vaughan, *J. Amer. Chem. Soc.*, **73**, 18 (1951).
682. I. V. Berezin, N. F. Kazanskaya, and N. N. Ugarova, *Zh. Fiz. Khim.*, **40**, 766 (1966).
683. T. G. Traylor and C. A Russell, *J. Amer. Chem. Soc.*, **87**, 3698 (1965).
684. J. R. Thomas, *J. Amer. Chem. Soc.*, **89**, 4872 (1967).
685. C. Walling and P. J. Wagner, *J. Amer. Chem. Soc.*, **86**, 3368 (1964).
686. A. Robertson and W. A. Waters, *J. Chem. Soc.*, **1948**, 1578.
687. R. Hiatt, T, Mill, K. C. Irwin, and J. K. Castleman, *J. Org. Chem.*, **33**, 1428 (1968).
688. G. A. Russell, *J. Amer. Chem. Soc.*, **79**, 3871 (1957).
689. J. A. Howard and K. U. Ingold, *J. Amer. Chem. Soc.*, **90**, 1056, 1058 (1968).
690. J. R. Thomas and C. A. Tolman, *J. Amer. Chem. Soc.*, **84**, 2079 (1962).
691. L. R. Mahoney, *J. Amer. Chem. Soc.*, **89**, 1895 (1967).
692. L. R. Mahoney and M. A. Darooge, *J. Amer. Chem. Soc.*, **89**, 5619 (1967).
693. K. U. Ingold, *Chem. Rev.*, **61**, 563 (1961).
694. E. R. Altwicker, *Chem. Rev.*, **67**, 475 (1967).
695. J. R. Thomas, *J. Amer. Chem. Soc.*, **77**, 246 (1955).
696. V. A. Shushunov, B. A. Redoshkin, and L. V. Kotinseva, *Tr. po Khim. i Khim. Tekhnol.*, **4**, 436 (1961).
697. E. Mueller, R. Mayer, B. Narr, A. Schick, and K. Scheffler, *Ann.*, **645**, 1 (1961).
698. J. C. McGowan and T. Powell, *J. Chem. Soc.*, **1960**, 238.
699. E. G. Sklyarova, A. F. Lukovnikov, M. L. Khidekel, and V. V. Karpov, *Izv. Akad. Nauk. SSSR, Ser. Khim.* **1965**, 1093.
700. J. A. Howard and K. U. Ingold, *Can. J. Chem.*, **42**, 2324 (1964).
701. D. V. Gardner, J. A. Howard, and K. U. Ingold, *Can. J. Chem.*, **42**, 2847 (1964).
702. J. R. Thomas, *J. Amer. Chem. Soc.*, **86**, 4807 (1964).
703. T. W. Campbell and G. M. Coppinger, *J. Amer. Chem. Soc.*, **74**, 1469 (1952).
704. E. C. Harswill and K. U. Ingold, *Can. J. Chem.*, **44**, 263 (1966).
705. I. V. Berezin, N. F. Kazanskaya, and N. N. Ugarova, *Zh. Fiz. Khim.*, **40**, 766 (1966).
706. C. F. Cullis, A. Fish, A Saeed, and D. L. Trimm, *Proc. Roy. Soc.*, **A289**, 402 (1966).
707. D. J. Trecker and R. S. Foote, *Chem. Commun.*, **1967**, 841.

708. M. R. Leonov, B. A. Redoshkin, and V. A. Shushonov, *Zh. Obshch. Khim.*, **32**, 3959 (1962).

709. J. R. Shelton and R. F. Kopczewski, *J. Org. Chem.*, **32**, 2908 (1967).

710. A. D. Kirk and J. H. Knox, *Trans. Faraday Soc.*, **56**, 1296 (1960).

711. S. W. Benson and G. N. Spokes, *J. Phys. Chem.*, **72**, 1182 (1968).

712. M. A. Dikii and T. I. Yurzhenko, *Dopovidi, Akad. Nauk Ukr. SSR*, **1962**, 390; through *Chem. Abstr.*, **58**, 488 (1963).

713. Y. Ishii, A. Furuno, and S. Sumi, *Kogyo Kagaku Zasshi*, **64**, 472 (1961); through *Chem. Abstr.*, **57**, 7149 (1962).

714. V. Stannett and R. B. Mesrobian, *J. Amer. Chem. Soc.*, **72**, 4125 (1950).

715. C. H. F. Tipper, *J. Chem. Soc.*, **1953**, 1675

716. N. I. Larionova and N. N. Ugarova, *Vestn. Mosk. Univ., Ser. II, Khim.*, **1966**, 21; through *Chem. Abstr.*, **65**, 2107 (1966).

717. L. Bateman, H. Hughes, and A. L. Morris, *Discussions Faraday Soc.*, **14**, 190, (1953).

718. W. Cooper, *J. Chem. Soc.*, **1953**, 1267.

719. C. Walling and Y. W. Chang, *J. Amer. Chem. Soc.*, **76**, 4878 (1954).

720. G. Schröder, *J. Polymer Sci.*, **31**, 309 (1958).

721. A. V. Tobolsky and L. R. Matlack, *J. Polymer Sci.*, **55**, 49 (1961).

722. J. C. Martin, J. W. Taylor, and E. H. Drew, *J. Amer. Chem. Soc.*, **89**, 129 (1967).

723. J. C. Martin and T. W. Koenig, *J. Amer. Chem. Soc.*, **86**, 1771 (1964).

724. R. C. Lamb, F. F. Rogers, G. D. Dean and F. W. Voight, *J. Amer. Chem. Soc.*, **84**, 2635 (1962).

725. E. T. Denisov, *Zh. Fiz. Khim.*, **38**, 2085 (1964); through *Chem. Abstr*, **61**. 13163 (1964).

726. E. T. Denisov and L. N. Denisova, *Dokl. Akad. Nauk. SSSR*, **157**, 907 (1964).

727. E. T. Denisov, *Izv. Akad. Nauk SSSR, Ser. Khim*, **1967**, 1608.

728. V. L. Antonovskii, E. T. Denisov, and L. V. Solntseva, *Kinetika i Kataliz*, **7**, 409 (1966).

729. V. M. Solyanikov and E. T. Denisov, *Dokl. Akad. Nauk SSSR*, **173**, 1106 (1967).

730. V. Stannett and R. B. Mesrobian, *Discussions Faraday Soc.*, **14**, 242 (1953).

731. E. T. Denisov, *Kinetika i Kataliz*, **4**, 508 (1963).

732. E. T. Denisov and L. N. Denisova, *Izv. Akad. Nauk SSSR*, **1963**, 1731; through *Chem. Abstr.*, **60**, 3963 (1964).

733. H. S. Mosher and C. F. Wurster, *J. Amer. Chem. Soc.*, **77**, 5451 (1955).

734. L. J. Durham and H. S. Mosher, *J. Amer. Chem. Soc.*, **80**, 327, 332 (1958); *ibid.*, **82**, 4537 (1960); *ibid.*, **84**, 2811 (1962).

735. K. Uberreiter and W. Rabel, *Makromol. Chem.*, **68**, 12 (1963).

736. N. V. Yablokova, B. A. Redoshkin, and V. A. Shushunov, *Tr. po Khim. i Khim. Tekhnol.*, **4**, 455 (1961); through *Chem. Abstr.*, **58**, 427 (1963).

737. L. Dulog, *Makromol. Chem.*, **76**, 109 (1964).

738. H. Bredereck, A. Wagner, K. G. Kottenhahn, A. Kottenhahn, and R. Blaschke, *Angew. Chem.*, **70**, 503 (1958).

739. H. Bredereck, A. Wagner, R. Blaschke, and G. Demetriades, *Angew Chem.*, **71**, 340 (1959).

740. H. Bredereck, A. Wagner, R. Blaschke, G. Demetriades, and K. G. Kottenhahn, *Ber.*, **92**, 2628 (1959).

741. L. Bateman, M. Cain, T. Colclough, and J. L. Cuneen, *J. Chem. Soc.*, **1962**, 3570.

742. W. L. Hawkins and H. Sauter, *Chem. Ind.*, (*London*), **1962**, 1825.

743. W. L. Hawkins and H. Sauter, *J. Polymer Sci.*, Pt. **A1**, 3499 (1963).

744. D. Barnard, L. Bateman, E. R. Cole, and J. L. Cuneen, *Chem. Ind.* (*London*), **1958**, 918.

745. D. Barnard and E. J. Percy, *Chem. Ind.* (*London*), **1960**, 1332.

746. G. Ayrey, D. Barnard, and D. T. Woodbridge, *J. Chem. Soc.*, **1962**, 2089.

747. D. T. Woodbridge, *J. Chem. Soc.*, *B*, **1966**, 50.

748. S. Ivanov and D. Shopov, *Compt. Rend. Academie Bulgare de Sciences*, **18**, 845 (1965); through *Chem. Abstr.*, 7990 (1966).

749. R. C. Schulz and A. Banihaschemi, *Makromol. Chem.*, **64**, 140 (1963).

750. E. C. Eaton and K. J. Irwin, *Polymer*, **5**, 649 (1964).

751. F. Hrabak and J. Brodsky, *Coll. Czech. Chem. Commun.*, **31**, 4443 (1966).

752. German Patent, 1,220,609 (1962).

753. M. Anbor, H. Hefter, and M. L. Kremer, *Chem. Ind.* (*London*), **1962**, 1055.

754. A. Hopfinger, *Zesz. Nauk. Politeck Slask. Chem.*, No. 10, 3 (1962); through *Chem. Abstr.*, **63**, 1960 (1965).

755. R. Hiatt, K. C. Irwin, and C. W. Gould, *J. Org. Chem.*, **33**, 1430 (1968).

756. W. J. deKlein and E. C. Kooyman, *J. Catalysis*, **4**, 626 (1965).

757. H. J. Den Hartog and E. C. Kooyman, *J. Catalysis*, **6**, 347, 357 (1966).

758. K. Onuma, K. Wada, J. Yamashita, and H. Hashimoto, *Bull. Soc. Chem. Japan*, **40**, 2900 (1967).

759. W. H. Richardson, *J. Amer. Chem. Soc.*, **87**, 247 (1965).

760. W. H. Richardson, *J. Amer. Chem. Soc.*, **87**, 1096 (1965).

761. T. J. Wallace, R. M. Skomoroski, and P. J. Luccheski, *Chem. Ind.* (*London*), **1965**, 1764.

762. J. K. Kochi and F. F. Rust, *J. Amer. Chem. Soc.*, **83**, 2018 (1961).

763. F. Minisci and G. Belevedere, *Gazz. Chim. Ital.*, **90**, 1299 (1960).

764. J. B. Braunworth and G. W. Crosby, *J. Org. Chem.*, **27**, 2064 (1962).

765. J. K. Kochi and F. F. Rust, *J. Amer. Chem. Soc.*, **84**, 3946 (1962).

766. J. Kumamoto, H. E. De La Mare, and F. F. Rust, *J. Amer. Chem. Soc.*, **82**, 1935 (1960).

767. S. Murai, N. Sonoda, and S. Tsutsumi, *Bull. Chem. Soc. Japan*, **36**, 527 (1963).

768. J. K. Kochi and P. E. Mocadlo, *J. Org. Chem.*, **30**, 1134 (1965).

769. J. K. Kochi and D. Buchanan, *J. Amer. Chem. Soc.*, **87**, 853 (1965).

770. M. S. Kharasch and A. Fono, *J. Org. Chem.*, **24**, 72 (1959).

771. V. L. Antonovskii and Yu. D. Emelin, *Zh. Vses, Khim. Obshcestva im D. I. Mendeleeva*, **9**, 471 (1964); through *Chem. Abstr.*, **61**, 13222 (1964).

772. F. Minisci and R. Galli, *Tetrahedron Letters*, **1963**, 357.

773. J. A. Bigot, *Rec. Trav. Chim.*, **80**, 825 (1961).

774. M. S. Kharasch and G. Sosnovsky, *Tetrahedron*, **3**, 105 (1958).

775. D. B. Denny, R. Napier, and A. Cammerata, *J. Org. Chem.*, **30**, 3151 (1965).

776. W. H. Richardson, *J. Org. Chem.*, **30**, 2804 (1965).

777. F. Minisci, R. Galli, and U. Pallini, *Gazz. Chim. Ital.*, **91**, 1023 (1961).

778. J. K. Kochi, *J. Amer. Chem. Soc.*, **84**, 2785, 3271 (1962).

779. M. K. Shchennikova, E. A. Kuz'mina, V. A. Shushunov, and G. A. Avakumov, *Dokl. Akad. Nauk SSSR*, **164**, 868 (1965).

780. See references in reference 755.

781. H. Berger and A. F. Bickel, *Trans. Faraday Soc.*, **57**, 1325 (1961).

782. E. A. Kuz'mina and M. K. Shchennikova, *Tr. po Khim. i Khim. Tekhnol.*, **1964**, 204.

783. W. P. Griffiths and T. D. Wickins, *J. Chem. Soc.*, *A*, **1968**, 397.

784. R. Griegee and C. Paulig, *Ber.*, **88**, 712 (1955).

785. H. Hock and H. Kropf, *Ber.*, **91**, 1681 (1958).

786. H. B. Van Leeuwen, J. P. Wibaut, A. F. Bickel, and E. C. Kooyman, *Rec. Trav. Chim.*, **77**, 17 (1958); *ibid.*, **78**, 667 (1959).

787. C. H. Wang, R. McNair, and P. Levins, *J. Org. Chem.*, **30**, 3817 (1965).

788. J. A. Waterman, *J. Appl. Chem.*, **16**, 177 (1966).

789. C. Wurster, L. Durham, and W. S. Mosher, *J. Amer. Chem. Soc.*, **80**, 327 (1958).

790. M. Talät-Erben and N. Önol, *Can. J. Chem.*, **38**, 1154 (1960).

791. S. Wawzonek, P. D. Klimstra, and R. E. Gallio, *J. Org. Chem.*, **25**, 621 (1960).

792. E. J. Gasson, E. G. E. Hawkins, A. F. Millidge, and D. C. Quinn, *J. Chem. Soc.*, **1950**, 2799.

793. H. Hock and S. Lang, *Ber.*, **75**, 1051 (1942).

794. H. Hock and S. Lang, *Ber.*, **77**, 257 (1944).

795. H. Hock and W. Susemihl, *Ber.*, **66**, 61, (1933).

796. H. Hock, F. Depke, and G. Knauel, *Ber.*, **83**, 238 (1950).

797. W. Webster and D. P Young, *J. Chem. Soc.*, **1956**, 4785.

798. M. Bassey, E. Buncel, and A. G. Davies, *J. Chem. Soc.*, **1955**, 2550.

799. C. Dufraisse, A. Etienne, and J. Rigaudy, *Bull. Soc. Chim. France*, **1948**, 804.

800. H. Hock and S. Lang, *Ber.*, **76**, 169 (1943).

801. P. G. Sergeev and A. M. Sladkow, *Zh. Obshch. Khim.*, **27**, 538 (1957).

802. U.S. Patent, 2,799,695 (1957).

803. K. I. Ivanov, V. K. Savinova, and V. P. Zhakhovskaya, *Zh. Obshch. Khim.*, **22**, 781 (1952).

804. T. I. Yurzhenko, K. S. Grigor'eva, N. V. Arf'ev, and M. R. Vilenskaya, *Dokl. Akad. Nauk SSSR*, **118**, 970 (1958).

805. U.S. Patent, 2,757,207 (1956).

806. R. Criegee, *Ann.*, **522**, 75 (1936).

807. R. Criegee and H. Zogel, *Ber.*, **84**, 215 (1951).

808. R. C. P. Cubbon and C. Hewlett, *J. Chem. Soc.*, *C*, **1968**, 2978, 2983, 2986.

809. P. L. Julian, W. Cole, and G. Diemer, *J. Amer. Chem. Soc.*, **67**, 1721 (1945); P. L. Julian, W. Coleand G. Meyer, *ibid.*, **67**, 1724 (1945).

810. P. L. Julian and W. Cole, *J. Amer. Chem. Soc.*, **57**, 1724 (1935).

811. A. Rieche and R. Meister, *Ber.*, **68**, 1465 (1935).

812. A. Rieche, *Ber.*, **64**, 2328 (1931).

813. A. Baeyer and V. Villiger, *Ber.*, **33**, 2479 (1900).

814. S. D. Razumovskii and Yu. N. Yurev, *Neftekhim.*, **6**, 737 (1966); through *Chem. Abstr.*, **66**, 28232 (1967).

815. N. A. Milas, P. Davis, and J. T. Nolan, *J. Amer. Chem. Soc.*, **77**, 2536 (1955).

816. R. Criegee and G. Schröder, *Ber.*, **93**, 689 (1960).

817. R. Criegee, W. Schnorrenberg, and J. Becke, *Ann.*, **565**, 7 (1949).

818. J. S. Beer, T. Donavanik, and A. Robertson, *J. Chem. Soc.*, **1954**, 4139.

819. B. Witkop and J. B. Patrick, *J. Amer. Chem. Soc.*, **74**, 3855 (1952).

820. J. S. Beer, T. Broadhurst, and A. Robertson, *J. Chem. Soc.*, **1952**, 4351, 4946; *ibid.*, **1953**, 2440.

821. C. D. Lunsford, R. E. Lutz, and E. E. Bowden, *J. Org. Chem.*, **20**, 1513 (1955).

822. J. Rigaudy and E. G. Izoret, *Compt. Rend.*, **238**, 824 (1954).

823. J. Rigaudy and G. Cauquis, *Compt. Rend.*, **242**, 2964 (1956).

824. C. L. Stevens and R. J. Gasser, *J. Amer. Chem. Soc.*, **79**, 6057 (1957).

# Chapter II

# Chemistry of Hydroperoxides in the Presence of Metal Ions

## G. SOSNOVSKY* & D. J. RAWLINSON†

*This author was supported in part by a Public Health Service Senior Research Fellowship (1-F3-GM-31, 797-01, 1967–1968) from the National Institute of General Medical Sciences.
†Present address; Department of Chemistry, Western Illinois University, Macomb, Illinois.

**153**

## I.  INTRODUCTION

### A.  Historical Background

The chemistry of hydroperoxides in the presence of metal ions has been receiving increasing attention over the last decade because of its importance in autoxidative and polymerization processes, and because of its potential in synthetic applications. The study of interactions of hydroperoxides with metal ions commenced at the end of the last century, when Fenton (1) reported the reactions of hydrogen peroxide in the presence of ferrous ions. In the following years a large number of papers appeared dealing with the oxidative reactions of Fenton's reagent. About forty years after this discovery, Haber and Weiss (2) proposed a free-radical mechanism, which is in most parts still valid. The first step of the mechanism is as follows:

$$H_2O_2 + Fe^{2+} \longrightarrow HO\cdot + OH^- + Fe^{3+}$$

Similarly many organic peroxides, such as alkyl, cycloalkyl, and aralkyl hydroperoxides, undergo metal-ion-induced decomposition. By analogy with Fenton's reagent, the primary step of these reactions seems to be the generation of alkoxy radicals:

$$ROOH + M^n \longrightarrow RO\cdot + OH^- + M^{n+1}$$

The further fate of the alkoxy radicals depends on the nature of the alkoxy radicals, and their environment. Thus they can abstract atoms from substrates R'H.

$$RO\cdot + R'H \longrightarrow ROH + R'\cdot$$

or they can add to unsaturated compounds,

$$RO\cdot + CH_2{=}CHR' \longrightarrow ROCH_2{-}\overset{\cdot}{C}HR'$$

or they can undergo various fragmentation reactions.

Because of their great industrial, biological, and theoretical interest, the decompositions of hydroperoxides in the presence of metal ions have been extensively investigated; most of the work has been carried out since 1940. Thus Hock and co-workers in Germany studied the decomposition of hydroperoxides in the absence of unsaturated substrates. Similar investigations were carried out by Kharasch, Fono, Nudenberg, and co-workers at the University of Chicago. In addition, this group systematically investigated the interactions of *t*-alkoxy and *t*-aralkoxy radicals with polymerizable olefins. In 1952 Campbell and Coppinger in the United States, and in 1953 Bickel, Kooyman, and co-workers in the Netherlands, found the cobalt-ion-catalyzed addition reaction of *t*-butyl hydroperoxide with phenols to give unsymmetric peroxides of the type

ROOR'

(R = dienone moiety, R' = CMe<sub>3</sub>)

Also in 1953 Kharasch, Pauson, and Nudenberg, in Chicago, were able to introduce the *t*-butylperoxy group $\alpha$ to the double bond in butadiene, octene-1, and cyclohexene. In 1954 Treibs and Pellmann, in Germany, succeeded with an analogous reaction, using cumyl hydroperoxide to produce cumylperoxy cycloalkenes.

In the early 1950s the important ring-opening reaction of cyclic hydroperoxides was also discovered.

$$RC(O)(CH_2)_2(CH_2)_n CH_2\cdot$$

R = OH, alkyl, alkoxy, aryl, $n = 1, 2, 3$

In spite of the impressive volume of work and significant contributions toward establishing basic ideas on the mechanism of decomposition of hydroperoxides by metal ions, only a few useful methods were developed for preparative purposes, and it was not until about 1956 that Fono and Sosnovsky in the Institute of the late Professor Kharasch at the University of Chicago, and Chiusoli and Minisci in Italy, embarked on the preparative aspects of metal-ion-induced reactions of peroxides. As a result of these studies many useful synthetic applications have been developed. Also over the last decade De La Mare, Rust, and Kochi have made important contributions to the mechanistic aspects of metal-ion-catalyzed reactions of peroxides.

Over the years a number of books and review articles have contained sections on the chemistry of hydroperoxides in the presence of metal ions (3–11). Several articles (12, 13) are devoted primarily to mechanistic aspects. To our knowledge, there are only four readily available articles in which the emphasis is on preparative aspects (14–17). Two of these articles (14, 17) excellently cover the work of Minisci and his colleagues. In the other two articles (15, 16) an attempt has been made to cover all areas of reactions of peroxides in the presence of metal ions; however, the articles are incomplete. In continuation of the present series* an effort is made to present a comprehensive and critical review, emphasizing the preparative aspects of the reactions of hydroperoxides in the presence of metal ions. The body of this chapter is divided into two sections. Section II, entitled "In the Absence of Reactive Substrates," discusses references that are concerned with mechanistic studies or studies involving fragmentation reactions. Most of these reactions are of little current preparative importance. Section III, entitled "In the Presence of Reactive Substrates," discusses references that describe methods of more immediate potential preparative utility.

## B.  Scope and Limitations

This review describes the chemistry of hydroperoxides in the presence of metal ions that are capable of oxidation–reduction reactions. Thus reactions caused by elemental metals or by metal ions incapable of a valence change are not included. All reactions are of the homogeneous-catalysis type, and hence reactions involving surface catalysis are not included. Although most reactions involve only catalytic quantities of metal ions, reactions that involve larger than catalytic amounts of metal ions are also included, since a strict separation of these areas is impractical. Included are only those processes that seem to involve free-radical intermediates, and hence reactions of hydroperoxides in the presence of Lewis acids are not included. Not included are

*For articles on metal-ion-catalyzed reactions of symmetric peroxides and peroxyesters see Volume I of *Organic Peroxides*.

analytical procedures, polymerization studies, and most autoxidative processes. Also excluded from this review is the chemistry of hydrogen peroxide and of inorganic peroxides. The emphasis in this chapter is on synthetic methods, and hence no detailed kinetic and mechanistic discussions are presented. However, it is believed that all important references leading to these topics are included; the reaction schemes used in this chapter are based on some of these references.

### C.   Nomenclature

In most instances the nomenclature used by different authors has been retained in order to facilitate checking the original articles.

## II.   IN THE ABSENCE OF REACTIVE SUBSTRATES

### A.   Alkyl and Cycloalkyl Hydroperoxides

#### 1.   Primary

Metal-ion-catalyzed decompositions of primary hydroperoxides yield mainly alcohols, aldehydes, and carboxylic acids, plus some alkanes, and small quantities of hydrogen, carbon monoxide, and carbon dioxide gas (18, 19).

One of the simplest representatives of this group is hydroxymethyl hydroperoxide (**1**), which is decomposed by stoichiometric amounts of aqueous ferrous sulfate to formaldehyde, formic acid, and water (20).

$$2HOCH_2OOH + 2Fe^{2+} \longrightarrow 2HOCH_2O\cdot + 2Fe(OH)^{2+}$$

$$(1)$$

$$2HOCH_2O\cdot \longrightarrow CH_2O + HC(O)OH + H_2O$$

The decomposition of 1-hydroperoxy-*n*-heptane with cobalt, copper, manganese, iron, or nickel laurate yields mainly heptanol, heptanal, and heptanoic acid (18). A similar result is obtained with 1-hydroperoxy-*n*-pentane (19).

#### 2.   Secondary

Metal-ion-catalyzed decompositions of secondary hydroperoxides give mainly alcohols and carbonyl compounds, usually ketones, plus some hydrocarbons, and traces of hydrogen, carbon monoxide, and carbon dioxide gas (18, 19).

One of the simplest members of this class is 1-hydroperoxyethanol (**2**), which is decomposed by aqueous ferrous sulfate to acetaldehyde and acetic acid (21).

$$\text{MeCHOH} \xrightarrow{\text{Fe}^{2+}/\text{Fe}^{3+}} \text{MeC(O)H} + \text{MeC(O)OH}$$
$$|$$
$$\text{OOH}$$
$$(2)$$

However, if aqueous ferrous sulfate is added to an ethereal solution of compound **2**, an 86% yield of acetic acid is obtained (21).

Cobalt, copper, nickel, iron, or manganese laurate accelerates the decomposition of 3-hydroperoxyhexane (**3**) at 80° to give 23% alcohol **4** and 15% ketone **5** in 20 hr (22).

$$\text{EtCHPr} \xrightarrow{\text{M}^n/\text{M}^{n+1}} \text{EtCH(OH)Pr} + \text{EtC(O)Pr}$$
$$|$$
$$\text{OOH}$$
$$(3) \qquad\qquad (4) \qquad\qquad (5)$$

Similarly the decompositions of 2-hydroperoxy-, 3-hydroperoxy-, and 4-hydroperoxy-$n$-heptane give heptanols and heptanones in 20–40% yields (18).

The decomposition of $\alpha$-hydroperoxyethers (**6**) by ferrous sulfate at temperatures ranging from 0 to 75° gives esters (**7**) (23–26).

$$\text{RCHOR}' + \text{Fe}^{2+} \longrightarrow \text{RCHOR}' + \text{Fe(OH)}^{2+} \longrightarrow \text{RC(O)OR}'$$
$$| \qquad\qquad\qquad\qquad |$$
$$\text{OOH} \qquad\qquad\qquad \text{O}\cdot$$
$$(6) \qquad\qquad\qquad\qquad\qquad\qquad\qquad\qquad (7)$$

(a   R = Me, R' = Et)

(b   R = CH$_2$=CH, R' = Bu)

(c   R = Me(CH$_2$)$_5$, R' = Et)

In addition to compound **7**, the decomposition of compound **6c** gives ethyl formate, $n$-heptanal, and $n$-dodecane in 30, 38, and 18% yields, respectively (26).

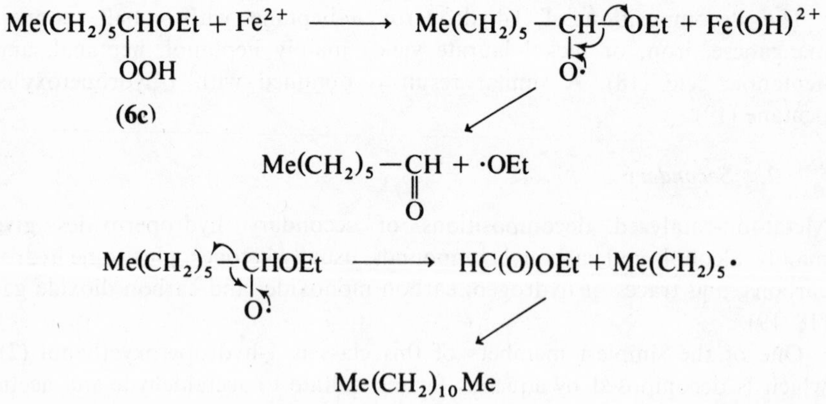

$$\text{Me(CH}_2)_5\text{CHOEt} + \text{Fe}^{2+} \longrightarrow \text{Me(CH}_2)_5-\text{CH}-\text{OEt} + \text{Fe(OH)}^{2+}$$
$$|$$
$$\text{OOH} \qquad\qquad\qquad\qquad\qquad\qquad \text{O}\cdot$$
$$(6c)$$

$$\text{Me(CH}_2)_5-\text{CH} + \cdot\text{OEt}$$
$$\|$$
$$\text{O}$$

$$\text{Me(CH}_2)_5\text{CHOEt} \longrightarrow \text{HC(O)OEt} + \text{Me(CH}_2)_5\cdot$$
$$\text{O}\cdot$$

$$\text{Me(CH}_2)_{10}\text{Me}$$

In the presence of a mixture of cuprous and cupric chlorides no *n*-dodecane is found but instead a 20% yield of *n*-hexyl chloride (**8**) is obtained (26).

$$Me(CH_2)_5 \cdot + CuCl^+ \longrightarrow Me(CH_2)_5Cl$$
$$(8)$$

The decomposition of cyclohexyl hydroperoxide (**9**) with ferrous sulfate, cupric sulfate, or manganese laurate yields mainly cyclohexanol and cyclohexanone (22, 27).

In addition, a mixture of adipic, valeric, caproic, and ε-hydroxycaproic acids is formed (22).

### 3.  Tertiary

(a)  ACYCLIC

The majority of investigations on metal-ion-catalyzed reactions of alkyl hydroperoxides have been carried out with *t*-butyl hydroperoxide, and there is an extensive literature on the mechanistic aspects of its decomposition (28–49). Recently *t*-butylperoxy radical intermediates have been detected by electron paramagnetic resonance (EPR) spectroscopy during the degradation of *t*-butyl hydroperoxide with titanium, cerium, and cobalt ions (44, 50–54). The decomposition of tertiary alkyl hydroperoxides by metal ions produces alkoxy radicals in the first decisive step.

$$RR'R''COOH + M^n \longrightarrow RR'R''CO \cdot + M(OH)^n$$

The alkoxy radicals are either reduced to alcohols or fragment to give ketones and alkyl radicals, which in turn are either reduced or oxidized to alkanes, olefins, and alcohols (19, 34, 35, 55, 56).

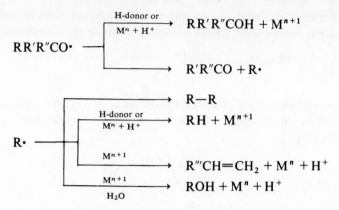

These reactions have been achieved with various organic and inorganic compounds of metals capable of oxidation–reduction reactions (e.g., copper, silver, iron, cobalt, nickel, manganese, titanium, cerium, lead, chromium, vanadium, molybdenum, and selenium). The way in which the fragmentation occurs is largely determined by the bond dissociation energies of the alkyl groups in the alkoxy radical (34). The composition of products depends also to some extent on the type, concentration, and valency state of the metal used and the solvents and temperatures employed. Thus the decomposition of $t$-butyl hydroperoxide with ferrous sulfate yields $t$-butyl alcohol, acetone, di-$t$-butyl peroxide, methane, and ethane (30, 34, 35, 41, 56). However, in the presence of iron, cobalt, copper, silver, and manganese phenanthrolines, cobalt acetates, cobalt stearates, copper 2-ethylhexanoate, dialkyl diselenide, selenic anhydride, selenic acid, cerium sulfate, or lead tetraacetate additional products—such as water, oxygen, methanol, methyl acetate, and formaldehyde—have also been observed (32, 35, 38, 43, 44, 46, 48, 55–64). In the presence of a large amount of chloride ions in the reaction mixture methyl chloride is also formed (30, 31, 35).

$$Me\cdot + M^{n+1} + Cl^- \longrightarrow MeCl + M^n$$

Reduction of alkyl radicals to alkanes can occur in the presence of ferrous, chromous, titanous, and vanadous ions (41). It has been suggested (35, 41) that the reaction proceeds through a complex (10) between the alkyl radical and the metal ion, and a carbanion-type intermediate (11), as shown by experiments with deuterium oxide (35, 41).

$$Me\cdot + M^n \longrightarrow [Me\cdot M^n] \longrightarrow [Me\!:\cdots M^{n+1}]$$

$$\text{(10)} \hspace{4cm} \text{(11)}$$

$$\xrightarrow[D_2O]{} MeD + M(DO)^n$$

$$(M = Fe, Ti, V, Cr)$$

The decomposition of *t*-butyl hydroperoxide with an equimolar quantity of ceric sulfate forms a mixture of formaldehyde, acetone, di-*t*-butyl peroxide, and oxygen (58, 59, 65). Free-radical intermediates have been proposed for this reaction; however, the mechanism is not well established (59).

The decomposition of *t*-butyl hydroperoxide with an equimolar quantity of lead tetraacetate proceeds smoothly to give mainly di-*t*-butyl peroxide and oxygen (60, 61, 62, 66). The mechanism of the lead tetraacetate reaction has been little investigated and is not well established. Ionic (64, 66) and radical (65) intermediates have both been suggested on the basis of little experimental support. More recent investigations (62, 63) at low temperatures ($-70$ to $-30°$) indicate the formation of tetroxide (**12**) and trioxide (**13**) intermediates that might form and decompose by radical processes.

$$Me_3COOH + Pb^{4+} \longrightarrow Me_3COO\cdot + Pb^{3+} + H^+$$

$$2Me_3COO\cdot \longrightarrow [Me_3CO\cdot O_2\cdot OCMe_3] \left\{ \begin{array}{l} \longrightarrow 2RO\cdot + O_2 \\ \\ \longrightarrow ROOR + O_2 \end{array} \right.$$
$$(\mathbf{12})$$

$$RO\cdot + ROOH \longrightarrow ROH + ROO\cdot$$

$$RO\cdot + ROO\cdot \underset{> -30°}{\rightleftarrows} ROOOR$$
$$(\mathbf{13})$$

To account for the major products of the decomposition of *t*-butyl hydroperoxide the following possible steps have been suggested (29, 30, 34, 35, 41, 44, 55, 56, 62, 63, 67–70):

$$Me_3COOH + M^n \longrightarrow Me_3CO\cdot + M^{n+1} + OH^-$$

$$Me_3COOH + Me_3CO\cdot \longrightarrow Me_3COH + Me_3COO\cdot$$

$$Me_3COOH + M^{n+1} \longrightarrow Me_3COO\cdot + M^n + H^+$$

$$Me_3CO\cdot + M^n \longrightarrow Me_3CO^- + M^{n+1}$$

$$Me_3CO\cdot \longrightarrow MeC(O)Me + Me\cdot \left\{ \begin{array}{l} \xrightarrow{\text{H-donor}} CH_4 \\ \\ \longrightarrow Me\!-\!Me \end{array} \right.$$

$$Me_3COO\cdot + Me_3COO\cdot \longrightarrow Me_3COOCMe_3 + O_2$$

$$Me_3COO\cdot + Me_3COOH \longrightarrow Me_3CO\cdot + Me_3COH + O_2$$

$$Me\cdot + O_2 \longrightarrow MeOO\cdot$$

$$MeOO\cdot + Me_3COO\cdot \longrightarrow CH_2O + O_2 + Me_3COH$$

$$Me_3CO\cdot + Me_3COO\cdot \longrightarrow Me_3COOMe + Me_2CO$$

$$Me_3COOMe \longrightarrow Me_3COH + CH_2O$$

$$Me\cdot + M^{n+1} + H_2O \longrightarrow MeOH + M^n + H^+$$

In addition, mechanisms involving complex formation either between the hydroperoxide or between the radicals derived from the hydroperoxide and the catalysts have been suggested (29, 32, 42, 43, 46). The mechanistic details, such as the sequence of various steps, are by no means well established, and the above steps do not necessarily indicate the correct sequence of events.

The metal-ion-catalyzed degradations of other alkyl hydroperoxides proceed by processes similar to those observed with $t$-butyl hydroperoxide. Thus the ferrous sulfate–catalyzed decomposition of $n$-butyl hydroperoxide gives a mixture of 45–55% butyraldehyde, 10–15% formaldehyde, and 10% propane (35). In the presence of cupric sulfate a 10% yield of propylene is also obtained (35). The decomposition of $t$-amyl hydroperoxide (14) with either chromous or ferrous ions results in a mixture of products consisting mainly of $t$-amyl alcohol, di-$t$-amyl peroxide, acetone, ethane, ethylene, and $n$-butane (34, 35, 41, 56). A similar composition of gas is obtained with cobalt and nickel sulfate–catalyzed reactions (31).

$$EtCMe_2OOH + M^n \longrightarrow EtCMe_2O\cdot + M(OH)^n$$

$$(14)$$

$$EtCMe_2O\cdot \longrightarrow EtCMe_2OH$$

$$EtCMe_2O\cdot \longrightarrow Me_2CO + C_2H_5\cdot$$

$$\longrightarrow CH_2{=}CH_2 + C_2H_6 + n\text{-}C_4H_{10}$$

In the presence of cupric sulfate the yield of ethylene increases (35).

$$C_2H_5\cdot + Cu^{2+} \longrightarrow CH_2{=}CH_2 + H^+ + Cu^+$$

In addition, $t$-amyl alcohol, acetone, and ethyl alcohol are formed (35).

$$Et\cdot + Cu^{2+} + H_2O \longrightarrow EtOH + Cu^+ + H^+$$

With titanous chloride the main gaseous product is ethane, whereas in the presence of halide ions the main product is ethyl chloride (35).

$$C_2H_5\cdot + M^{n+1} + Cl^- \longrightarrow C_2H_5Cl + M^n$$

The ferrous sulfate–catalyzed degradation of 2,3,3-trimethyl-2-hydroperoxy-butane (15) results in 70% ionic rearrangement to give $t$-butyl alcohol and acetone, and in 30% radical reaction to give isobutane and isobutylene (35). In the presence of cupric sulfate the decomposition gives exclusively isobutylene (35).

$$Me_3CCMe_2OOH$$

**(15)**

The cupric or ferrous ion-catalyzed decomposition of 2,4,4-trimethyl-2-pentyl hydroperoxide **(16)** (37, 40) produces 100% acetone, 55% neopentyl alcohol, 30% neopentane, 10% 2-methyl-2-butene, and a trace of 2-methyl-1-butene. In the presence of acetic acid neopentyl acetate is also obtained (37). The formation of methylbutenes is explained by the following scheme (37) :

$$Me_3CCH_2CMe_2OOH + Cu^+ \longrightarrow Me_3CCH_2CMe_2O\cdot + CuOH^+$$

**(16)**

$$Me_3CCH_2CMe_2O\cdot \longrightarrow Me_3CCH_2 + Me_2\overset{\cdot}{C}O$$

$$Me_2CCH_2 + Cu^{2+} \longrightarrow Me_3CCH_2^+ + Cu^+$$

$$\xrightarrow{\text{CH}_3:\text{ shift}} \underset{+}{Me_2CCH_2Me} \xrightarrow{-H^+} Me_2C=CHMe + \underset{\underset{Me}{|}}{CH_2=CCH_2Me}$$

However, decomposition with ferrous sulfate yields mainly 2,4,4-trimethyl-pentanol-2 and acetone (40).

The decomposition of 2-methyl-2-hexyl hydroperoxide **(17)** by copper ions gives 2-methyl-2-hexyloxy radicals, which can undergo either fragmentation to acetone and the *n*-butyl radical or 1,5-hydrogen abstraction to the 5-hydroxy-5-methyl-2-hexyl radical **(18)**. As a result the following products are obtained: acetone, butene-1, butane, 2-methyl-5-hexenol-2 **(19)**, 2-methyl-4-hexenol-2 **(20)**, and 2-methyl-2-hexanol **(21)** (36, 40).

$$Me(CH_2)_3CMe_2OOH + Cu^+ \longrightarrow$$

$$Me(CH_2)_3CMe_2O\cdot + CuOH^+$$

**(17)**

$$Me(CH_2)_3CMe_2O\cdot$$

$$\downarrow$$

$$MeCH_2CH_2CH_2\cdot + Me_2CO \quad \left[ \begin{array}{l} \xrightarrow{\text{H-donor}} MeCH_2CH_2Me \\ \\ \xrightarrow{\text{Cu(OH)}^+} MeCH_2CH=CH_2 + H^+ + Cu \end{array} \right.$$

$$Me(CH_2)_3CMe_2O\cdot \longrightarrow MeCHCH_2CH_2CMe_2OH \xrightarrow{Cu(OH)^+}$$

$$\textbf{(18)}$$

$$CH_2=CHCH_2CH_2CMe_2OH + MeCH=CHCH_2CMe_2OH + Cu^+ + H_2O$$

$$\textbf{(19)} \hspace{4cm} \textbf{(20)}$$

$$Me(CH_2)_3CMe_2O\cdot \xrightarrow[\text{or Cu}^+ + H_2O]{\text{H-donor}} Me(CH_2)_3CMe_2OH$$

$$\textbf{(21)}$$

However, decomposition with ferrous sulfate yields mainly 2-methyl-2-hexanol (36). In the presence of cupric chloride and carbon tetrachloride 5-chloro-2-methylhexan-2-ol is isolated in 26% yield (40). The degradation of 3-methylhept-3-yl hydroperoxide with ferrous sulfate gives butan-2-one (70).

The decomposition of 2,5-dimethylhexyl-2 hydroperoxide (22) with a mixture of ferrous sulfate and cupric acetate gives 44% 2,5-dimethylhexen-5-ol-2, 23% acetone, 23% 3-methylbutene, plus 2,5-dimethylhexanol-2 and 2,5 dimethylhexen-4-ol-2 (40). However, the reaction with ferrous sulfate yields 2,5-dimethylhexan-2-ol and a small amount of 2,5-dimethylhexan-2,5-diol (23) (40).

$$Me_2CHCH_2CH_2CMe_2OOH + Fe^{2+} \longrightarrow Me_2CHCH_2CH_2CMe_2O\cdot +$$

$$\textbf{(22)}$$

$$Fe(OH)^{2+} \longrightarrow Me_2CCH_2CH_2CMe_2OH \xrightarrow{Fe(OH)^{2+}}$$

$$Me_2CCH_2CH_2CMe_2OH + Fe^{2+}$$
$$|$$
$$OH$$

$$\textbf{(23)}$$

The degradation of 2-methylheptyl-2-hydroperoxide with a mixture of ferrous sulfate and cupric acetate produces 48% acetone, 48% pentene-1, 31% 2-methylhepten-5-ol-2, plus 2-methylheptanol-2 and 2-methylhepten-4-ol-2 (40).

The decomposition of 2,5-dimethyl-2,5-dihydroperoxyhexane (24) with cobaltous ions in various hydrogen-donor solvents gives mainly glycol 26 and acetone (71).

$$Me_2C(CH_2)_2CMe_2 + Co^{2+}$$
$$|\hspace{1.2cm}|$$
$$OOH \hspace{0.6cm} OOH$$

$$\downarrow$$

$$Me_2C(CH_2)_2CMe \xrightarrow{\hspace{1cm}} \begin{array}{l} \xrightarrow{\text{H-donor}} Me_2C(CH_2)_2CMe_2 \\[0.3cm] \hspace{3cm} |\hspace{1.2cm}| \\ \hspace{2.8cm} OH \hspace{0.8cm} OH \\ \hspace{3.2cm} \textbf{(26)} \\[0.4cm] \longrightarrow 2Me_2CO + [CH_2=CH_2] \end{array}$$
$$|\hspace{1cm}|$$
$$O\cdot \hspace{0.6cm} O\cdot$$
$$\textbf{(25)}$$

A similar degradation with ferrous sulfate produces a mixture of acetone, *t*-amyl alcohol, 2-methyl-2-hexanol, methanol, methyl ethyl ketone, ethyl *t*-amyl peroxide, plus ethylene, ethane, and *n*-butane (30). In the presence of cupric chloride this decomposition gives no gaseous products; the main products are 4-chloro-2-methyl-2-butanol (**28**) and 1,2-dichloroethane (30).

$$(25) \longrightarrow Me_2CO + \underset{\underset{O\cdot}{|}}{Me_2C}-(CH_2)_2\cdot \xrightarrow{CuCl_2} \underset{\underset{O\cdot}{|}}{Me_2C}(CH_2)_2Cl$$

$$(27)$$

$$(27) - \begin{cases} \xrightarrow{\text{H-donor}} & \underset{\underset{OH}{|}}{Me_2C}(CH_2)_2Cl \\ & (28) \\ \longrightarrow & Me_2CO + \cdot(CH_2)Cl \xrightarrow{CuCl_2} Cl(CH_2)_2Cl \end{cases}$$

The decomposition of mono- and dihydroperoxy-2,7-dimethyloctane (**29** and **30**, respectively) with ferrous sulfate gives acetone and a mixture of acetone and δ-methylcaproic acid, respectively (72).

$$\underset{\underset{OOH}{|}}{Me_2CH(CH_2)_4CMe_2} \qquad \underset{\underset{OOH\ \ OOH}{|\quad\ \ |}}{Me_2C(CH_2)_4CMe_2}$$

$$(29) \qquad\qquad\qquad (30)$$

(b) CYCLIC

*1-Alkyl Cycloalkyl Hydroperoxides.* The decomposition of 1-alkyl cyclo-alkyl hydroperoxides with metal ions can be utilized to prepare long-chain diketones. Thus 1-methylcyclopentyl hydroperoxide (**31**) reacts with ferrous sulfate to give a high yield of dodecane-2,11-dione (**32**) (73–75).

$$MeC(O)(CH_2)_8C(O)Me$$

$$(32)$$

The reaction of 1-methylcyclohexyl hydroperoxide (**33**) under similar conditions yields tetradecane-2,13-dione (**34**) (73, 74, 76).

This result was not observed by other investigators (77), who reported as products 10% 1-methylcyclohexanol, 34% hepten-2-one, 13% other cyclic

$$MeC(O)(CH_2)_{10}C(O)Me$$

(34)

ketones, and 43% of "higher" alcohols and ketones. In the presence of a mixture of ferrous and cupric sulfate compound 33 decomposes to give excellent yields of 6-hepten-2-one (35).

The ferrous sulfate–catalyzed decomposition of pinane hydroperoxide (35) yields 1,4-bis(3-acetyl-2,2-dimethylcyclobutyl)butane (36) (78, 79).

The decomposition of *trans*-4a-hydroperoxydecahydronaphthalene (37) with ferrous sulfate gives 2-*n*-butylcyclohexanone (38) in 80% yield (80).

The ferrous-ion-catalyzed decomposition of the lactone of hydroperoxytetrahydroabietic acid (**39**) occurs with elimination of the isopropyl group to give ketone **40** (81).

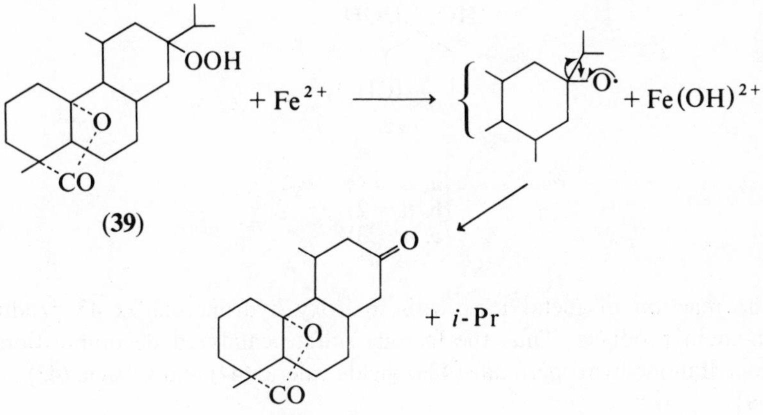

(**39**)

$+ Fe^{2+} \longrightarrow$

$+ Fe(OH)^{2+}$

(**40**)

$+ i\text{-}Pr$

*1-Aryl Cycloalkyl Hydroperoxides.* Metal-ion-catalyzed reactions of 1-aryl cycloalkyl hydroperoxides can be utilized for the preparation of long-chain diketones. Thus 1-phenylcyclohexyl hydroperoxide (**41**) is decomposed by ferrous sulfate to give 1,10-dibenzoyldecane (**42**) in 22% yield (82).

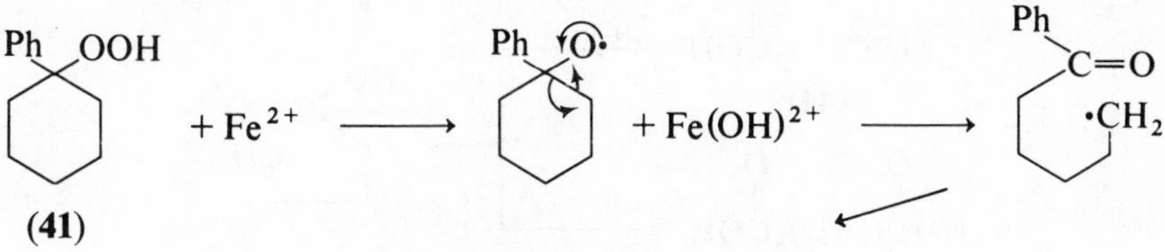

(**41**)

$+ Fe^{2+} \longrightarrow$

$+ Fe(OH)^{2+} \longrightarrow$

$PhC(O)(CH_2)_{10}C(O)Ph$

(**42**)

However, the decomposition at 50–60° is reported to yield a mixture of 1-phenylcyclohexanol, phenyl amyl ketone, cyclohexanone, and phenol (83).

*1-Hydroxy Cycloalkyl Hydroperoxides.* Cyclic ketones, such as cyclopentanone or cyclohexanone, react readily with hydrogen peroxide to give hydroxy hydroperoxides whose structures have not been fully elucidated, although Criegee (84) has made a substantial contribution in clarifying the structures of some of them. Several references (84–87) are pertinent to this subject.

Although the first product (**43**) of the reaction of an unsubstituted cyclic ketone with hydrogen peroxide has not been isolated (85), it probably exists in solution in equilibrium with the starting materials and the end products

(84, 85, 87, 88). For simplification in the schematic presentation of metal-ion-catalyzed reactions, structure **43** ($n = 1, 2, 3$, etc.) is used in the following discussion.

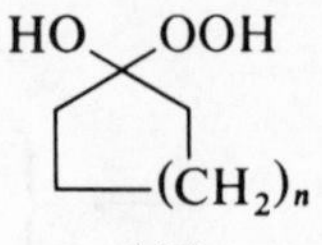

HO   OOH

—(CH$_2$)$_n$

**(43)**

(**a**, $n = 1$)
(**b**, $n = 2$)
(**c**, $n = 3$)

The reaction of metal ions with hydroxy hydroperoxides **43** produces open-chain products. Thus the ferrous sulfate–catalyzed decomposition of cyclopentanone hydroperoxide (**43a**) yields valeric (**44**) and sebacic (**45**) acids (89–91).

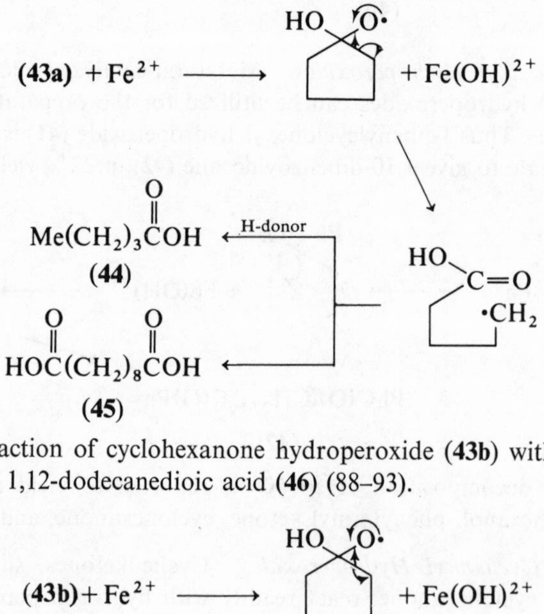

HO   O•

**(43a)** + Fe$^{2+}$ ⟶ [structure] + Fe(OH)$^{2+}$

$$\underset{\textbf{(44)}}{Me(CH_2)_3\overset{O}{\overset{\|}{C}}OH} \xleftarrow{\text{H-donor}}$$

HO
    C=O
    •CH$_2$

$$\underset{\textbf{(45)}}{HO\overset{O}{\overset{\|}{C}}(CH_2)_8\overset{O}{\overset{\|}{C}}OH} \longleftarrow$$

A similar reaction of cyclohexanone hydroperoxide (**43b**) with ferrous ions gives mainly 1,12-dodecanedioic acid (**46**) (88–93).

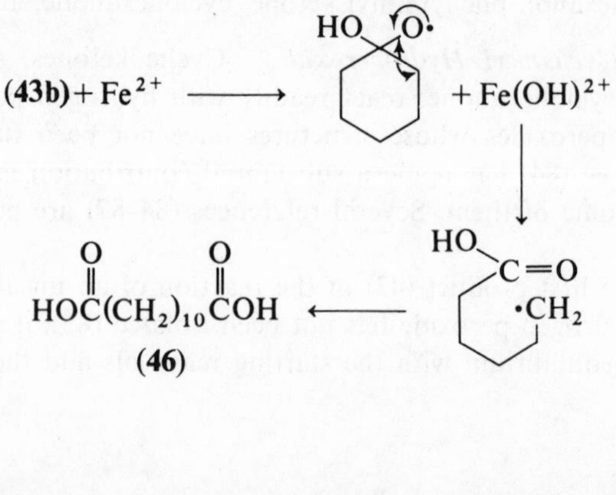

HO   O•

**(43b)** + Fe$^{2+}$ ⟶ [structure] + Fe(OH)$^{2+}$

HO
    C=O
    •CH$_2$

$$\underset{\textbf{(46)}}{HO\overset{O}{\overset{\|}{C}}(CH_2)_{10}\overset{O}{\overset{\|}{C}}OH} \longleftarrow$$

In the presence of large amounts of cupric or ferric ions 5-hexenoic acid (**47**) becomes the principal product (35, 36).

$$HOC(O)(CH_2)_4CH_2\cdot + M^{n+1}$$
$$\longrightarrow HOC(O)(CH_2)_3CH=CH_2 + M^n + H^+$$
$$\textbf{(47)}$$

When this reaction is carried out in aqueous methanol or in water, a low yield of $\omega$-hydroxycaproic acid (**48**) is also formed (35, 94).

$$HOC(O)(CH_2)_4CH_2\cdot + M^{n+1} + H_2O$$
$$\longrightarrow HOC(O)(CH_2)_4CH_2OH + M^n + H^+$$
$$\textbf{(48)}$$

If cyclohexanone hydroperoxide (**43b**) is decomposed in acidic methanol with cupric sulfate, a mixture of methyl 5-hexenoate (**49**) and methyl $\omega$-methoxyhexanoate (**50**) in 47 and 27 % yields, respectively, is obtained (35, 95), whereas decomposition with ferrous sulfate yields methyl 1,12-dodecandioate (**51**) (89). Apparently 1-methoxycyclohexyl hydroperoxide (**52**) is an intermediate.

In the presence of metal halides $\omega$-halohexanoic acids (**53**) are produced (35, 95, 96), and in the presence of acetic acid $\omega$-acetoxyhexanoic acid is obtained (95).

$$HOC(O)(CH_2)_4CH_2\cdot + M^{n+1} + X^-$$

$$\longrightarrow HOC(O)(CH_2)_4CH_2X + M^n$$

(53)

$$[X = Cl, Br, MeC(O)O]$$

The reaction of 2-chloro- and 2-bromocyclohexanone with hydrogen peroxide gives 1-hydroxy-2-halocyclohexyl hydroperoxides (**54**), which seem to be monomeric (85). Decomposition of **54a** with ice-cold aqueous methanolic ferrous sulfate produces a dichloro-1,12-dodecanedioic acid (85), which is probably composed of a mixture of two acids, **55** and **56**. Acid **56** would be expected to predominate in this mixture.

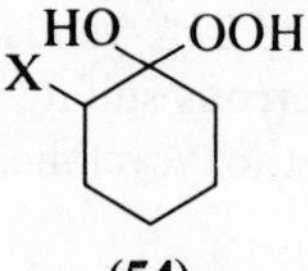

(54)

(a, X = Cl)

(b, X = Br)

$$\mathbf{54a} + Fe^{2+} \longrightarrow \quad + Fe(OH)^{2+}$$

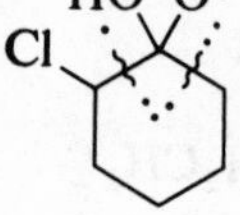

$$HOC(O)CHCl(CH_2)_4\cdot \qquad\qquad HOC(O)(CH_2)_4\underset{\cdot}{C}HCl$$

$$HOC(O)CHCl(CH_2)_8CHClC(O)OH \qquad [HOC(O)(CH_2)_4CHCl\,]_2$$

(55)                                              (56)

If this decomposition is carried out with a mixture of cupric and ferric sulfate, a low yield of chlorohexenoic acids **57** and **58** in a ratio of 4:1 is obtained (35).

$$54a + M^n \longrightarrow \quad \text{[structure]} \quad + M^{n+1}$$

$$ClCH{=}CH(CH_2)_3C(O)OH \qquad\qquad H_2C{=}CH(CH_2)_2CHClC(O)OH$$

$$(57) \qquad\qquad\qquad\qquad (58)$$

If ferrous sulfate and cupric chloride are used, saturated dichloro acids **59** and **60** are obtained in a ratio of 3:1 (35).

$$Cl_2CH(CH_2)_4C(O)OH \qquad\qquad Cl(CH_2)_4CHClC(O)OH$$

$$(59) \qquad\qquad\qquad\qquad (60)$$

The degradation of 4-methylcyclohexanone hydroperoxide (**61**) with ferrous sulfate yields a mixture of 4-methylhexanoic acid (**62**) and 4,9-dimethyl-dodecane-1,12-dioic acid (**63**) (89).

$$HOC(O)(CH_2)_2\underset{\underset{Me}{|}}{C}HCH_2Me$$

$$(62)$$

$$\text{[structure]} + M^n \longrightarrow \text{[structure]} + M^{n+1}$$

$$(61)$$

$$[HOC(O)(CH_2)_2\underset{\underset{Me}{|}}{C}H(CH_2)_2]_2$$

$$(63)$$

However, 2-methylcyclohexanone hydroperoxide (**64**) is decomposed by a mixture of ferrous and cupric sulfate to give a mixture of three heptenoic acids (**65**, **66**, and **67**) (35).

$$\underset{(64)}{\text{HO OOH}} + Fe^{2+} \qquad \underset{(65)}{H_2C=CH(CH_2)_2\overset{Me}{\underset{|}{C}}HC(O)OH}$$

$$\underset{(64)}{\text{Me}} \Big\downarrow$$

$$\underset{Me}{\text{HO O}} + Fe(OH)^{2+} \longrightarrow \begin{cases} HOC(O)\overset{Me}{\underset{|}{C}}H(CH_2)_4{}^\bullet \\[2mm] HOC(O)(CH_2)_4\overset{\bullet}{\underset{|}{C}}H \\ \phantom{HOC(O)(CH_2)_4CH}Me \end{cases}$$

$$\underset{(66)}{H_3CCH=CH(CH_2)_3C(O)OH} \qquad \underset{(67)}{H_2C=CH(CH_2)_4C(O)OH}$$

As expected, 66 and 67 predominate.

The ferrous sulfate–catalyzed decomposition of cycloheptanone hydroperoxide (43c) gives mainly n-heptanoic acid (68) and tetradecane-1,14-dioic acid (69) (89).

$$43c, + Fe^{2+} \longrightarrow \underset{}{\text{HO O}^\bullet} + Fe(OH)^{2+} \longrightarrow HOC(O)(CH_2)_6{}^\bullet$$

$$\xrightarrow{\text{H-donor}} \underset{(68)}{HOC(O)(CH_2)_5Me}$$

$$\longrightarrow \underset{(69)}{HOC(O)(CH_2)_{12}C(O)OH}$$

## B.  Peroxy Acids

The iron and cobalt-ion-catalyzed decomposition of peroxyacetic acid in acetic acid produces mainly carbon dioxide (84–86%) and carbon monoxide (10–14%) (97). The disintegration of peroxyacetic acid occurs more readily in the presence of cobaltic ions than in the presence of ferric ions (97). Several kinetic investigations have been carried out (97–99). The following fragmentation steps have been considered (97, 98):

$$MeC(O)OOH + M^{n+1} \longrightarrow MeC(O)O\cdot + H^+ + M^n$$

$$MeC(O)OOH + M^n \longrightarrow MeC(O)O\cdot + OH^- + M^{n+1}$$

$$MeC(O)O\cdot + MeC(O)OOH \longrightarrow MeC(O)OH + MeC(O)OO\cdot$$

$$MeC(O)O\cdot + M^n \longrightarrow MeC(O)O^- + M^{n+1}$$

$$MeC(O)OO\cdot + M^n \longrightarrow MeC(O)OO^- + M^{n+1}$$

$$(M = Fe, Co; n = 2)$$

The decomposition of peroxyacetic acid in acetic acid in the presence of manganese acetate is rapid and gives carbon dioxide (87–93%) and carbon monoxide (7–12%) (100). In contrast, this decomposition in water proceeds slowly to give mainly oxygen (57–64%), plus carbon dioxide (25–31%) and carbon monoxide (11–16%) (100).

The cupric chloride–catalyzed reaction of peroxyacetic acid with 3-nitro-aniline in acetic acid produces 3,3'-dinitroazobenzene (**70**), whereas in the absence of copper ions 3,3'-dinitroazoxybenzene (**71**) is formed (101).

$$ArN=NAr \qquad\qquad ArN(O)=NAr$$
$$\textbf{(70)} \qquad\qquad\qquad \textbf{(71)}$$

$$(Ar = 3-NO_2C_6H_4)$$

Similar results are obtained with 4-nitroaniline and *p*-toluidine (101). Less effective catalysts are iron, nickel, and rhodium salts, whereas in the presence of chromium, manganese, and cobalt compounds only product **71** is formed (101). The following scheme is offered to explain these results (101):

$$ArNH_2 \xrightarrow[M^n/M^{n+1}]{MeC(O)OOH} ArNHOH \xrightarrow{MeC(O)OOH} ArNO$$
$$\textbf{(72)} \qquad\qquad\qquad\qquad \textbf{(73)}$$

$$(M = Cu, Fe, Ni, Rh)$$

$$(Ar = 3\text{-}NO_2C_6H_4, 4\text{-}NO_2C_6H_4, 4\text{-}MeC_6H_4)$$

$$\textbf{72} + \textbf{73} \longrightarrow ArN(O)=NAr \qquad ArNH_2 + \textbf{73} \longrightarrow ArN=NAr$$
$$\textbf{(71)} \qquad\qquad\qquad\qquad\qquad \textbf{(70)}$$

In the uncatalyzed reaction product **72** is slowly oxidized to **73**, and **72** and **73** have time to react to give **71**, whereas in the cupric chloride–catalyzed reaction intermediate **72** is rapidly converted to **73**. As a result, **73** reacts with the starting material to give **70** (101). It is uncertain whether radical intermediates are involved in any of these steps.

The kinetics of the copper, cobalt, and nickel chloride–catalyzed decomposition of peroxybenzoic acid to benzoic acid have been studied (102, 103). The decomposition of peroxybenzoic and *p*-bromoperoxybenzoic acid with ferrous ions in the presence of styrene produces polymers. Hydroxy radicals are believed to be responsible for the initiation step (104).

$$PhC(O)OOH + Fe^{2+} \longrightarrow PhC(O)O^- + \cdot OH + Fe^{3+}$$

$$PhCH{=}CH_2 + \cdot OH \longrightarrow PhCH{-}CH_2OH$$

The cupric chloride–catalyzed decomposition of peroxybenzoic acid in a mixture of toluene and acetonitrile gives benzoic acid, benzyl chloride, and benzaldehyde. The uncatalyzed decomposition gives only a slightly different composition of products (105).

Diperoxysuccinic acid reacts with hydroquinone in the presence of copper or iron ions. In ethyl alcohol solution no reaction takes place between the peroxy acid and hydroquinone, but in the presence of cupric sulfate acetaldehyde is formed (106). Since the concentration of hydroquinone remains constant, it is believed that hydroquinone is an oxygen-transfer agent (106).

## C.  Hydroperoxides Derived from Unsaturated Hydrocarbons

The available information is too scanty to permit generalizations concerning the metal-ion-catalyzed decomposition of alkynyl and cycloalkenyl hydroperoxides.

A kinetic study has been reported on the cobaltic acetate–catalyzed decomposition of 2-methyl-2-butene hydroperoxide (107). The decomposition of 2,5-dimethylhex-3-yne-2,5-dihydroperoxide (**74**) with ferrous sulfate gives a mixture of methane, ethane, oxygen, acetone, 2,5-dimethylhex-3-yne-2,5-diol (**75**), and 5-hydroxy-5-methylhex-3-yn-2-one (**76**), plus a white solid of the possible structure **77** (108).

$$Me_2CC{\equiv}CCMe_2 \qquad Me_2C(OH)C{\equiv}C(OH)CMe_2$$
$$\quad | \qquad |$$
$$\quad OOH \ \ OOH$$
$$\quad (74) \hspace{5em} (75)$$

$$Me_2C(OH)C{\equiv}CC(O)Me$$
$$(76)$$

$$MeC(O)C{\equiv}CCMe_2{-}O{-}C{\equiv}CC(O)Me$$
$$(77)$$

The decomposition of cyclohexenyl hydroperoxide with either ferrous sulfate, ferrous sulfate plus cupric sulfate, or ferrous phthalocyanine gives cyclohexen-3-ol and cyclohexen-3-one (35, 109–111). Similar reaction of 1-methylcyclo-hexenyl hydroperoxide yields mainly 1-methylcyclohexen-3-ol and a mixture of 1-methylcyclohexen-6-one and 1-methylcyclohexen-3-one (109, 111). The decomposition of 1,2-dimethylcyclohexenyl hydroperoxide forms 1,2-di-methylcyclohexen-3-ol and 1,2-dimethylcyclohexen-3-one (109).

The hydroperoxide of 1,5-cyclooctadiene (**78**) decomposes in the presence of ferrous sulfate to give a radical (**79**) that dimerizes (112).

In the presence of a mixture of ferrous and cupric acetate in acetic acid a mixture of *cis*- and *trans*-aldehyde acetate (**80**) is obtained (113). In non-acidic solvents, such as isopropanol or acetonitrile, the *cis* isomers are formed exclusively (112). Hydrogenation of *cis* or *trans* **80** with palladium-on-carbon catalyst gives aldehyde **81** (112).

$$79 + (\overset{\overset{\displaystyle O}{\|}}{MeC})_2Cu$$

$$\underset{H_2/Pd-C}{\longrightarrow} \quad \overset{\overset{\displaystyle O}{\|}}{HC}(CH_2)_4\overset{\overset{\displaystyle OC(O)Me}{|}}{CH}CH_2Me$$

(**80**)                                        (**81**)

The hydroperoxide of 1,4-dihydronaphthalene (**82**) is decomposed by ferrous sulfate to 3-hydroxy-α-tetralone (**83**) in 41% yield (114).

           (**82**)                      (**83**)

The same reactants in aqueous methanol interact differently. A solid of unknown structure is formed (114). The decomposition of $\Delta^2$-octalin (**84**) α-hydroperoxide with iron phthalocyanine forms $\Delta^2$-α-octalone (**85**) (111).

          (**84**)                     (**85**)

The ferrous ammonium sulfate–catalyzed decomposition of 7β-hydroperoxy-dihydrolanost-8-en-3β-yl acetate (**86**) gives a mixture of 7-hydroxydihydro-lanost-8-en-3β-yl acetate (**87**), 7-oxodihydrolanost-8-en-3β-yl acetate (**88**), and lanosta-7,9-dien-3β-yl acetate (**89**) in 26, 18, and 33% yield, respectively (115).

       (**87**)               (**88**)             (**89**)

The decomposition of α-pinene hydroperoxide (**90**) with cobalt, manganese, nickel, iron, copper, or lead stearate, or with ferrous sulfate or iron phthalocyanine, yields verbenone (**91**) and verbenol (**92**) (110, 111, 115, 116).

(**90**)                                                (**91**)          (**92**)

Similarly the decomposition of terpinolene hydroperoxide (**93**) with ferrous ammonium sulfate gives a mixture of *p*-mentha-$\Delta^{1,4(8)}$-dien-3-ol (**94**) and piperitenone (**95**) (117).

(**93**)                    (**94**)                    (**95**)

The reaction of a mixture of the six hydroperoxides from (+)-limonene (**96**) with cuprous chloride in either pyridine or a mixture of pyridine and methanol at room temperature produces a mixture of products that on reduction with either lithium aluminum hydride or sodium hydrogen sulfite yields the following compounds: *cis*- and *trans*-*p*-menthadien-2,8-ol-1 (**97**), carvone (**98**), *trans*-isopiperitenol (**99**), *p*-menthadien-[1(7)8]-*trans*-ol-2 (**100**), and *trans*-carveol (**101**) (118).

(**96**)          (**97**)          (**98**)                (**99**)          (**100**)    (**101**)

The main product is **97**, consisting of 37–50% *trans* and 11–18% *cis* isomers, probably formed by the decomposition of the tertiary hydroperoxide **102** (118). Carvone (**98**) is probably formed by the decomposition of the secondary hydroperoxide **103** (118).

(**102**)                                (**97**)

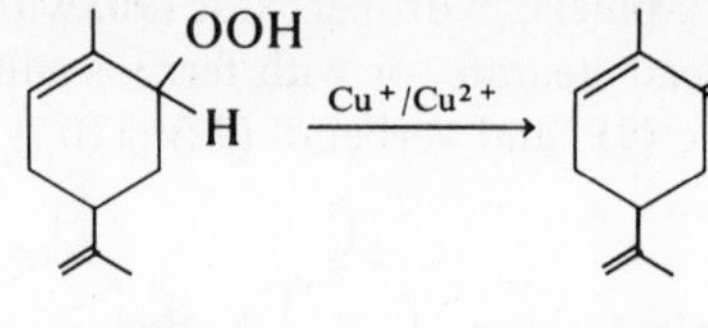

(103)                              (98)

The photosensitized autoxidation of $\Delta^5$-steroids produces $\Delta^6$-steroid-5-$\alpha$-hydroperoxides (104). In various solvents, such as chloroform, dioxane, dimethylformamide, and pyridine, peroxides 104 undergo allylic rearrangement to give $\Delta^5$-steroid-7$\alpha$-hydroperoxides 105 (119, 120). The rearrangement is accelerated by light, benzoyl peroxide, and copper chlorides, and is inhibited by hydroquinone (119, 120). In the presence of copper ions 7-keto derivatives (106) instead of hydroperoxides can be obtained (119, 120).

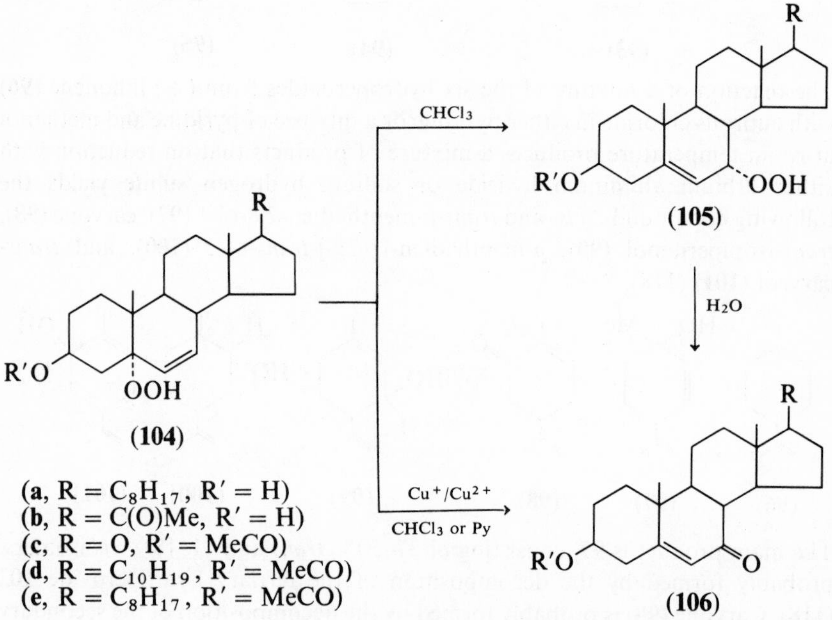

(105)

(104)

(a, R = $C_8H_{17}$, R' = H)
(b, R = C(O)Me, R' = H)
(c, R = O, R' = MeCO)
(d, R = $C_{10}H_{19}$, R' = MeCO)
(e, R = $C_8H_{17}$, R' = MeCO)

(106)

Thus the reaction of $\Delta^6$-cholesten-3$\beta$-ol-$\alpha$-hydroperoxide (104a) yields $\Delta^5$-cholesten-3$\beta$-ol-7-one (106a) (119, 120). Similarly, $\Delta^6$-allopregnen-3$\beta$-ol-20-one-5$\alpha$-hydroperoxide (104b) rearranges to $\Delta^5$-pregnen-3$\beta$-ol-20-one-7$\alpha$-hydroperoxide (105b) and $\Delta^5$-pregnen-3$\beta$-ol-7,20-dione (106b) (119, 120). Similar reactions are observed with $\Delta^6$-androsten-3$\beta$-ol-17-one-5$\alpha$-hydroperoxide-3-acetate (**104c**) and $\Delta^{6,22}$-stigmastadien-3$\beta$-ol-5$\alpha$-hydroperoxide-3-acetate (**104d**) (119, 120). A reaction of cholesteryl hydroperoxide (**104e**) with iron phthalocyanine to give 7-ketocholesteryl acetate probably belongs to the same category (111).

### D. Heterocyclic Hydroperoxides

A few metal-ion-catalyzed decompositions of heterocyclic hydroperoxides have been investigated. Thus tetrahydrofuryl hydroperoxide (**107**) is converted by ferrous sulfate to $\gamma$-butyrolactone (**108**) (121–123) and hexan-1,6-diol diformate (**109**) (123), plus a trace of 4-hydroxybutanal (**110**) (123).

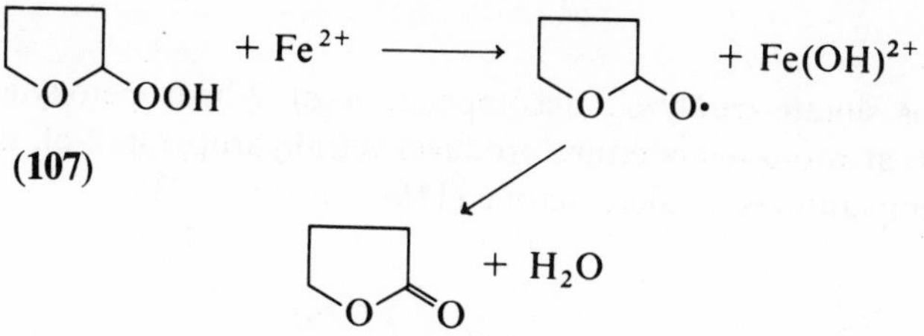

(**107**)

(**108**)

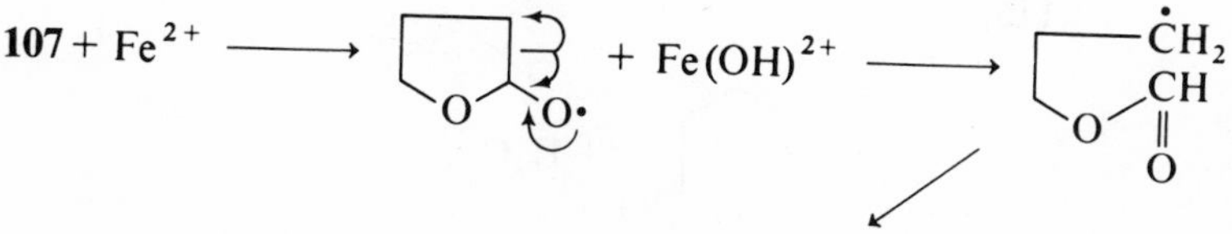

$$HC(O)O(CH_2)_6OC(O)H$$

(**109**)

$$HOCH_2CH_2CH_2CHO$$

(**110**)

At temperatures of about 50–70° in the presence of cobalt and nickel acetate **110** is also formed (124, 125). In the presence of a mixture of cuprous and cupric chloride **107** is decomposed to give a 42% yield of 3-formyloxypropyl chloride (**111**), 17% $\gamma$-butyrolactone (**108**), and a trace of 4-hydroxybutanal (**110**) (123).

$$Cl(CH_2)_3OCHO$$

(**111**)

The decomposition of 2-methyl-2-hydroperoxytetrahydrofuran by manganese acetate gives 5-hydroxypentan-2-one (**112**) (124, 125). Similarly the decomposition of 2-hydroperoxy-2,5-dimethyltetrahydrofuran (**113**) leads to hexan-2-ol-5-one (124, 125). In the presence of aqueous ferrous sulfate the decomposition of **113** gives a mixture of 46% 2,7-octanediol diacetate, 32% secondary butyl acetate (126, 127), and 19% compound **114** (126).

$MeC(O)(CH_2)_3OH$ (**112**)

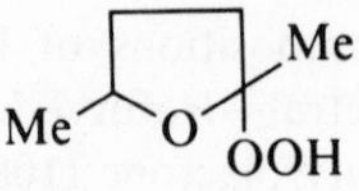

(**113**)

$ClCH_2CH_2CHMeOAc$ (**114**)

A ferrous sulfate–catalyzed decomposition of 2-hydroperoxytetrahydropyran (**115**) at room temperature produces tetrahydropyran-2-ol, whereas at elevated temperatures δ-valerolactone (**116**) is formed (23).

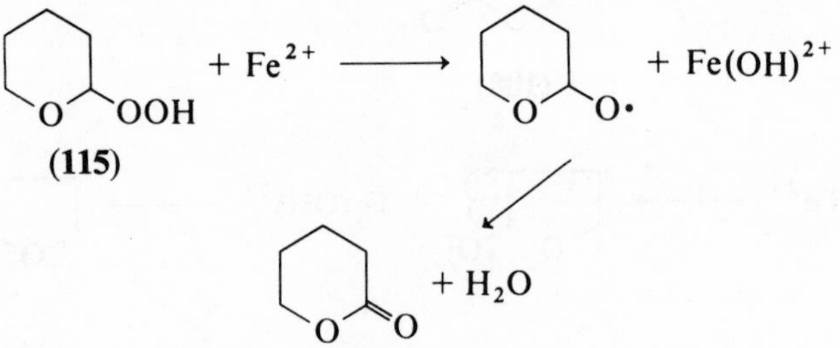

(**115**)

(**116**)

If a mixture of cupric or ferric chloride and ferrous chloride is added to an aqueous solution of **115**, a 36% yield of 4-chloro-1-butyl formate (**117**) is obtained (30).

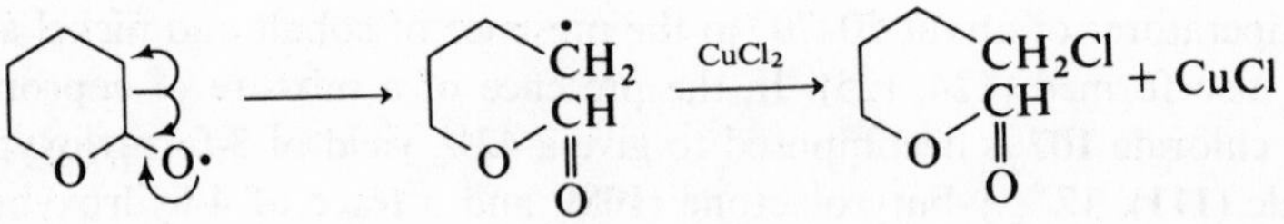

(**117**)

If the decomposition is carried out with ferrous sulfate in the absence of chlorides, 1,8-octane diformate (**118**) is produced in 50% yield (128, 129). However, if this reaction is carried out with a mixture of ferrous and cupric sulfates, the peroxide **115** is converted in 20% yield to 3-buten-1-ol (**119**) (35).

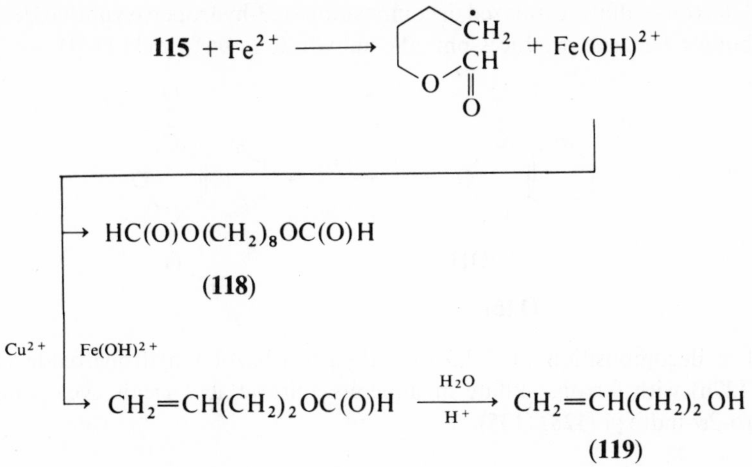

$$115 + Fe^{2+} \longrightarrow \left[ \begin{array}{c} \dot{C}H_2 \\ CH \\ O \end{array} \right] + Fe(OH)^{2+}$$

$$\rightarrow HC(O)O(CH_2)_8OC(O)H$$

**(118)**

Cu$^{2+}$     Fe(OH)$^{2+}$

$$\rightarrow CH_2{=}CH(CH_2)_2OC(O)H \xrightarrow[H^+]{H_2O} CH_2{=}CH(CH_2)_2OH$$

**(119)**

Photochemical autoxidation of dioxan yields monohydroperoxide **120** and 2,3-bis(hydroperoxide) **121** (130). The reaction of **120** with ferrous sulfate gives a mixture of acetaldehyde and formaldehyde, whereas the reaction of **121** gives only formaldehyde (130).

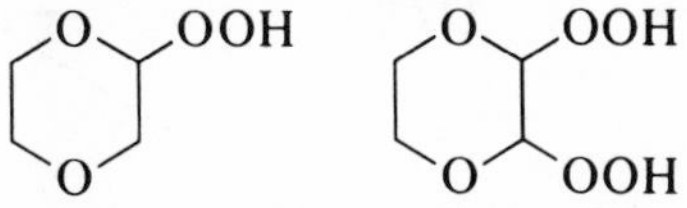

**(120)**            **(121)**

The cobaltous nitrate–catalyzed decomposition of ε-hydroperoxy-ε-capro-lactam (**122**) in boiling benzene yields 15% of adipic acid imide (**123**) in 3 hr (131, 132).

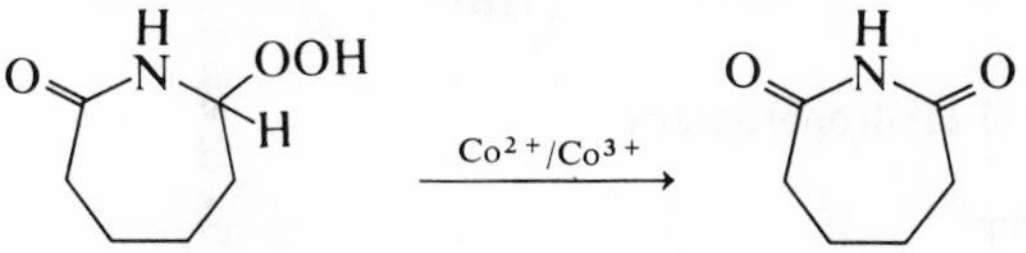

$$\xrightarrow{Co^{2+}/Co^{3+}}$$

**(122)**                              **(123)**

The reaction of phthalanyl hydroperoxide (**124**) with cobaltous nitrate yields diphthalanyl peroxide (**125**) (133).

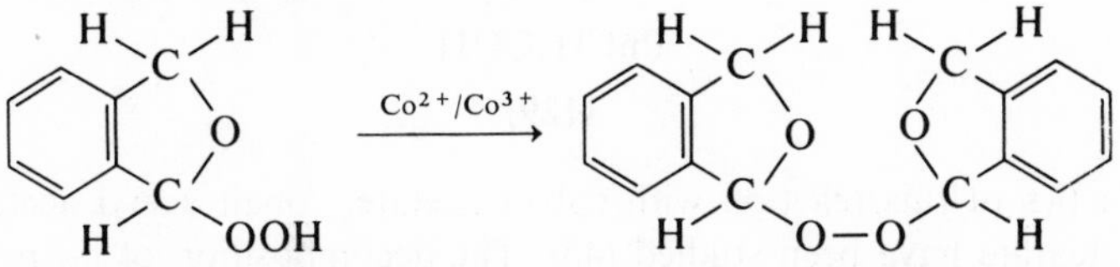

$$\xrightarrow{Co^{2+}/Co^{3+}}$$

**(124)**                              **(125)**

A ferrous sulfate–catalyzed decomposition of 3-hydroperoxyphthalide (126) in boiling benzene produces phthalic anhydride in 94% yield (134).

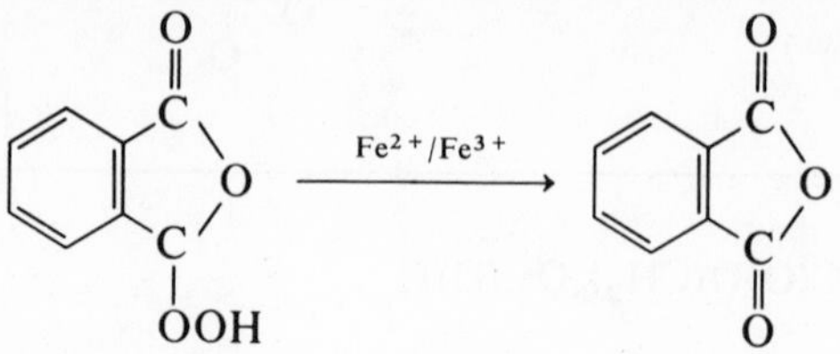

**(126)**

The decomposition of 1,2,3,4-tetrahydrocarbazolyl hydroperoxide (127a or 127b) with ferrous sulfate in aqueous sulfuric acid yields cyclopentane-spiro-2$\psi$-indoxyl (128) (135).

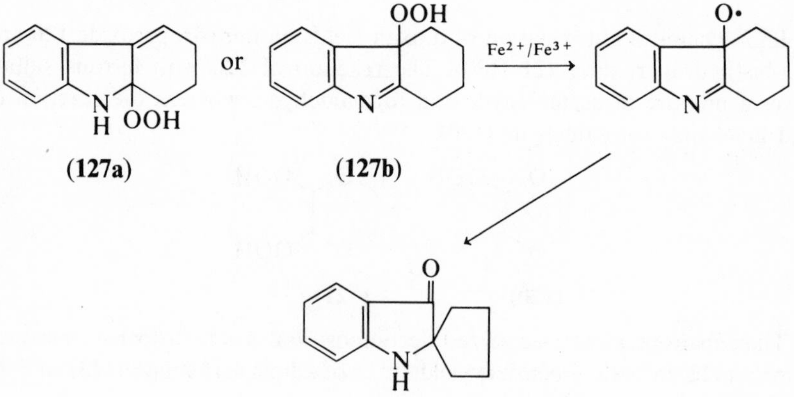

**(127a)**          **(127b)**

**(128)**

### E.  Aralkyl Hydroperoxides

#### 1.  Primary

The cobalt-ion-catalyzed decomposition of tolyl hydroperoxide (129) in acetic acid produces 60% benzaldehyde, 30% benzyl alcohol, and 10% benzyl benzoate (45). In ethylene glycol 60% benzaldehyde and 40% benzyl alcohol are formed (45).

$$PhCH_2OOH$$

**(129)**

The kinetics of this reaction with cobalt acetate, cobalt acetyl acetate, and cobalt stearate have been studied (45). The decomposition of o-, p- and m-xylene hydroperoxides with cobalt stearate in boiling xylene yields toluyl alcohols, tolualdehydes, toluic acid, carbon dioxide, carbon monoxide, and hydrogen (136).

### 2. Secondary

Hock and co-workers (137–140) showed that secondary aralkyl hydroperoxides are converted by metal ions to ketones as the major product in the absence of reactive solvents, a result later confirmed by other investigators. Thus the decomposition of $\alpha$-phenylethyl hydroperoxide (**130**) with either ferrous ammonium sulfate or cobalt stearate produces acetophenone and $\alpha$-methylbenzyl alcohol (19, 136, 138, 141), plus oxygen, hydrogen, methane, carbon monoxide, and carbon dioxide (136).

$$\begin{array}{ccc} \text{PhCHMe} + \text{M}^n & \longrightarrow & \text{PhCHMe} + \text{M(OH)}^n \\ | & & | \\ \text{OOH} & & \text{O}\cdot \\ \textbf{(130)} & & \end{array}$$

$$\begin{array}{cc} \longrightarrow & \text{PhC(O)Me} + \text{PhCH(OH)Me} \\ & (\text{M} = \text{Fe, Co}) \end{array}$$

An analogous reaction of diphenylmethyl hydroperoxide (**131**) with iron phthalocyanine gives benzophenone (111).

$$\begin{array}{ccc} \text{Ph}_2\text{CH} & \xrightarrow{\text{Fe}^{2+}=\text{Fe}^{3+}} & \text{Ph}_2\text{C}=\text{O} \\ | & & \\ \text{OOH} & & \\ \textbf{(131)} & & \end{array}$$

The decomposition of 1-phenylbut-1-yl hydroperoxide (**132**) with ferrous ions proceeds with the elimination of the propyl group to give benzaldehyde in 79 % yield (19, 142).

$$\begin{array}{ccc} \text{PhCHC}_3\text{H}_7 & \xrightarrow{\text{Fe}^{2+}/\text{Fe}^{3+}} & \text{PhCH}-\text{C}_3\text{H}_7 & \longrightarrow & \text{PhCH} + \cdot\text{C}_3\text{H}_7 \\ | & & | & & \| \\ \text{OOH} & & \text{O}\cdot & & \text{O} \\ \textbf{(132)} & & & & \end{array}$$

Secondary hydroperoxides **133** derived from alkyl aralkyl ethers can be converted with ferrous sulfate at 75° to esters **134** in 34–67 % yields (25).

$$\begin{array}{ccc} \text{RCHOR}' + \text{Fe}^{2+} & \longrightarrow & \text{RCHOR}' + \text{Fe(OH)}^{2+} \\ | & & | \\ \text{OOH} & & \text{O}\cdot \\ & \longrightarrow & \text{RC(O)OR}' + \text{H}_2\text{O} + \text{Fe}^{2+} \\ \textbf{(133)} & & \textbf{(134)} \end{array}$$

$$(\text{R} = \text{Ph, 1-naphthyl, 4-}i\text{-Pr}-\text{C}_6\text{H}_4, \text{3,4-di-Cl}-\text{C}_6\text{H}_3)$$

$$(\text{R}' = \text{Me})$$

The decomposition of tetralin hydroperoxide (135) with ferrous ions produces α-tetralol, tetralone, organic acids, water, and 1.5% oxygen (65, 111, 143, 144). If cobalt naphthenate is used for the decomposition, the amount of oxygen increases to about 16% (145). The decomposition of compound 135 in boiling benzene in the presence of cobalt stearate gives a 26% yield of bis(α-tetralyl)peroxide (146). In the presence of manganese laurate the decomposition gives, in addition to tetralol and tetralone, 7% of γ-o-hydroxyphenylbutyraldehyde (136) (22).

Mechanisms involving hydroxy radicals have also been considered (144, 145).

The decomposition of 135 has also been studied with other metal ions, such as those of copper, chromium, nickel, and lead (144, 147, 148). A number of physicochemical studies have been carried out (149–154). Equimolar quantities of lead tetraacetate decompose **135** to give α-tetralone and oxygen (60, 136, 154, 155).

$$135 + Pb[OC(O)Me]_4$$

$$+ \tfrac{1}{2}O_2 + Pb[OC(O)Me]_2 + 2MeC(O)OH$$

The mechanism of this reaction is by no means well established (61). Kharasch and co-workers (65) favor a radical mechanism, whereas Hock and Kropf (64) and Criegee (156) favor ionic intermediates.

The decomposition of the hydroperoxides of fluorene (140), indane (137), and octahydroanthracene (139) (**137, 138,** and **139,** respectively) with ferrous sulfate produces fluorenone (**140**), α-indanone (**141**), and octhracenone (**142**), respectively, in high yields.

Methyl 9-hydroperoxydehydroabietate (**143**) decomposes at 50° in the presence of ferrous sulfate to give methyl 9-oxodehydroabietate (**144**) in 82% yield (157, 158).

(143)                                        (144)

### 3. Tertiary

The metal-ion-catalyzed reactions of tertiary aralkyl hydroperoxides, in particular $\alpha$-cumyl hydroperoxide, have been studied extensively because these oxidation–reduction systems are important for the initiation of polymerization of various olefins (159). A series of polymerization studies with cumyl, substituted cumyl, and other tertiary aralkyl hydroperoxides has been reported (160–164). A series of other mechanistic and physicochemical studies has been carried out with various metal-ion catalysts—such as ferrous sulfate, copper halides, chromous perchlorate, lead dioxide, cobalt acetate, and other cobalt salts of long-chain acids; cobalt acetylacetonates; nickel acetylacetonate; nickel acetylacetonate–pyridine; nickel acetylacetonate–aniline; copper acetylacetonate; copper complexes of 2-aminopenten-2-one-4, $\alpha$-nitroso-$\beta$-naphthol, and 8-hydroxyquinoline; and copper and cobalt phthalocyanines (41, 165–178). Cumylperoxy-radical intermediates have been detected by EPR when cobalt (51), copper (179), and ceric (50) ions were used.

In the first decisive step the metal-ion-catalyzed decompositions of tertiary aralkyl hydroperoxides produce aralkoxy radicals, which undergo reduction to alcohols or fragmentation to ketones and alkyl radicals.

$$RR'R''COOH + M^n \longrightarrow RR'R''CO\cdot + M(OH)^n$$

$$RR'R''CO\cdot \begin{cases} \xrightarrow[\text{or } M^n + H^+]{\text{H-donor}} RR'R''COH \\[2ex] \longrightarrow R'R''CO + R\cdot \end{cases}$$

The alkyl radicals in turn can undergo dimerization, reduction, and oxidation reactions.

The proportion of fragmentation to reduction varies greatly with the bond-dissociation energies of the R-groups in the aralkoxy radical (19, 34). In general the decomposition of aralkyl hydroperoxides produces ketones as the major product; however, the composition of products depends also to some extent on the type of catalyst, concentration of catalyst, type of solvent,

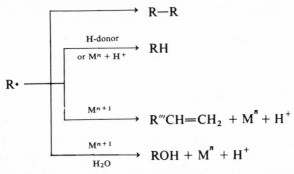

and temperature used in the reaction. Thus the decomposition of cumyl hydroperoxide with ferrous ions gives a high yield of acetophenone, plus $\alpha,\alpha$-dimethylbenzyl alcohol, methyl alcohol, methane, and ethane (35, 56). The cobalt-, manganese-, nickel-, and copper-ion-catalyzed reactions produce mainly acetophenone, dimethylbenzyl alcohol, and dicumyl peroxide (**145**) and oxygen (65, 146, 172, 180–182). Reactions catalyzed by cupric halides form acetophenone, dimethylbenzyl alcohol, acetone, and $\alpha$-methylstyrene (174, 175, 181).

$$PhCMe_2OOCMe_2Ph$$
$$\textbf{(145)}$$

The decomposition of cumyl hydroperoxide in cumene at 95° gives a 58% yield of **145** (183). The copper phenanthroline–catalyzed decomposition yields mainly cumyl alcohol, plus some oxygen and acetophenone (31). The reactions of cumyl hydroperoxide with vanadium dichloride, cobalt dichloride, or ceric sulfate yield mainly oxygen plus acetophenone (59, 65).

The reaction of cumyl hydroperoxide with vanadium pentoxide produces acetophenone, dimethylbenzyl alcohol, and $\alpha$-methylstyrene (182). The osmium tetroxide–catalyzed reaction gives a high yield of $\alpha$-methylstyrene, plus some acetophenone (182). The mechanism of these reactions has not been elucidated.

The decomposition of cumyl hydroperoxide with equimolar amounts of lead tetraacetate gives acetophenone, cumyl alcohol, compound **145**, and oxygen (64, 184). The mechanism of this reaction has not been firmly established (61). Bartlett and Günther (62) suggest the formation of trioxide and tetroxide intermediates that decompose to give cumyloxy and cumylperoxy radicals.

To account for the major products of the decomposition of cumyl hydroperoxide in the presence of metal ions the following possible steps have been suggested (19, 32, 34, 50, 56, 62, 65, 174, 176, 180, 185–188).

$$PhCMe_2OOH + M^n \longrightarrow PhCMe_2O\cdot + M(OH)^n$$
$$PhCMe_2O\cdot \longrightarrow PhC(O)Me + Me\cdot$$

$$\text{Me}\cdot \xrightarrow{\text{H-donor}} \begin{array}{c} CH_4 \\ Me-Me \end{array}$$

$PhCMe_2O\cdot + PhCMe_2OOH$   (or other H-donor)

$$\longrightarrow PhCMe_2OH + PhCMe_2OO\cdot$$

$PhCMe_2O\cdot + PhCMe_2OOH \longrightarrow PhC(O)Me + MeOH$

$$+ PhCMe_2O\cdot$$

$PhCMe_2O\cdot + M^n + H^+ \longrightarrow PhCMe_2OH + M^{n+1}$

$2PhCMe_2O\cdot \longrightarrow PhCMe_2OOCMe_2Ph$

$PhCMe_2OOH + Me\cdot \longrightarrow PhCMe_2O\cdot + MeOH$

$PhCMe_2OOH + M^{n+1} \longrightarrow PhCMe_2OO\cdot + H^+ + M^n$

$PhCMe_2OO\cdot + M^{n+1} \longrightarrow PhCMe_2OO^+ + M^n$

$$\longrightarrow PhCMe_2 + O_2$$
$$\quad\quad\quad + $$

$PhCMe_2O\cdot + PhCMe_2H \longrightarrow PhCMe_2OH + PhCMe_2$
$$\quad\quad\quad\quad\quad\quad\quad\quad\quad\quad\quad\quad\quad\quad\quad + $$

$$\xrightarrow{PhCMe_2OO\cdot} PhCMe_2OOCMe_2Ph$$

$2PhCMe_2OO\cdot \longrightarrow [PhCMe_2O\cdot O_2\cdot OCMe_2Ph]$

$PhCMe_2OOCMe_2Ph + O_2 \longleftarrow$

$2PhCMe_2O\cdot + O_2 \longleftarrow$

The mechanistic details are by no means clear; a discussion of this topic is outside the scope of this chapter.

The ferrous-ion-catalyzed decomposition of 2-phenyl-but-2-yl hydroperoxide (146) proceeds with elimination of the ethyl group to give acetophenone (19, 189).

$$\begin{array}{c} Me \\ | \\ PhC-Et \\ | \\ OOH \\ (146) \end{array} \xrightarrow{Fe^{2+}/Fe^{3+}} PhC(O)Me + \cdot Et$$

Similarly the ferrous sulfate–catalyzed decomposition of the hydroperoxide of secondary butylbenzene (**147**) (190) and of the monohydroperoxide of *p*-diisopropylbenzene (**148**) (191) yields acetophenone and *p*-isopropylaceto-phenone, respectively.

$$MeCH_2-\underset{\underset{OOH}{|}}{\overset{\overset{Me}{|}}{C}}-Ph \qquad Me_2C-\langle\bigcirc\rangle-CHMe_2$$
$$\qquad\qquad\qquad\qquad\qquad\quad \underset{OOH}{|}$$

$$\textbf{(147)} \qquad\qquad\qquad \textbf{(148)}$$

Elimination of alkyl groups also occurs in the degradation of the dihydro-peroxides of *p*-diisopropylbenzene (**149**) (190), *p-sec*-butylisopropylbenzene (**150**) (192), di-*p-sec*-pentylbenzene (**151a**) (193), and di-*p-sec*-hexylbenzene (**151b**) (192) to give *p*-diacetylbenzene.

$$Me_2C-\langle\bigcirc\rangle-CMe_2 \qquad Et(Me)C-\langle\bigcirc\rangle-CMe_2$$
$$\underset{OOH}{|}\qquad\quad\underset{OOH}{|}\qquad\qquad \underset{OOH}{|}\qquad\quad\underset{OOH}{|}$$

$$\textbf{(149)} \qquad\qquad\qquad\qquad \textbf{(150)}$$

$$R(Me)C-\langle\bigcirc\rangle-C(Me)R \qquad\qquad \textbf{(a, R} = \textit{n}\text{-Pr)}$$
$$\underset{OOH}{|}\qquad\quad\underset{OOH}{|}\qquad\qquad\qquad \textbf{(b, R} = \textit{n}\text{-Bu)}$$

$$\textbf{(151)}$$

The decomposition of α,α-dimethyl-β-phenethyl hydroperoxide (**152**) by ferrous sulfate gives a mixture of toluene, bibenzyl, benzyl alcohol, and acetone (41), whereas the decomposition by chromous ions yields acetone and the benzylchromium ion (**153**) (41, 42).

$$PhCH_2CMe_2OOH + Cr^{2+} \longrightarrow PhCH_2CMe_2O\cdot + Cr(OH)^{2+}$$
$$\textbf{(152)}$$

$$PhCH_2CMe_2O\cdot \longrightarrow PhCH_2\cdot + Me_2CO$$
$$PhCH_2\cdot + Cr^{2+} \longrightarrow PhCH_2Cr^{2+}$$
$$\textbf{(153)}$$

Triphenylmethyl hydroperoxide (**154**) reacts with ferrous ions to give radical **155**, which undergoes hydrogen abstraction and rearrangement to yield 60% of triphenylcarbinol (**156**) and 30% of benzpinacol diphenyl ether (**157**) (56, 194).

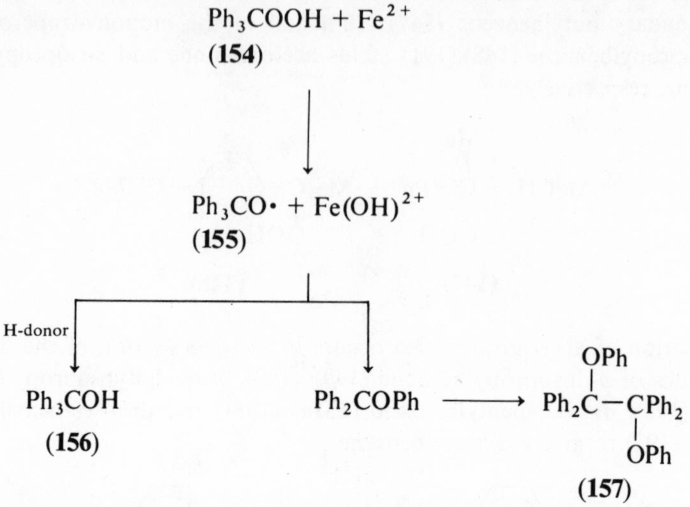

$$Ph_3COOH + Fe^{2+}$$
(154)

$$Ph_3CO\cdot + Fe(OH)^{2+}$$
(155)

H-donor

$$Ph_3COH \qquad\qquad Ph_2COPh \longrightarrow Ph_2C-CPh_2$$

(156)                         (157)

with OPh groups above and below the central carbons, shown at (157).

The ferrous sulfate–catalyzed decomposition of 1,3,3-trimethylindan-1-yl hydroperoxide (158) yields 73% of trimethylindene (159), plus some 1,3,3-trimethylindan-1-ol (160) and 3,3-dimethylindan-1-one (161) (195).

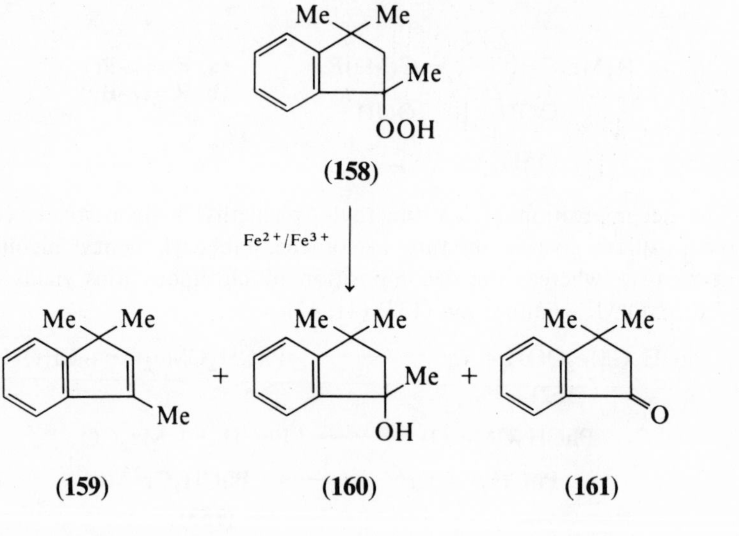

(158)

$Fe^{2+}/Fe^{3+}$

(159)            (160)            (161)

However, in the presence of ethyl alcohol the reaction produces a 33% yield of 1-ethoxy-1,3,3-trimethylindane (162), a 27% yield of compound 160, plus ketone 161 (195).

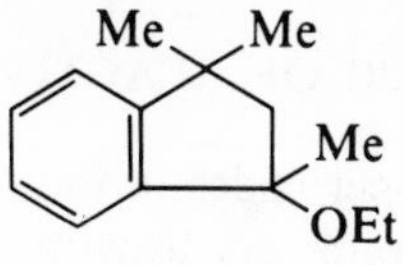

**(162)**

The tertiary hydroperoxide of methyl 9-oxodehydroabietate (**163**) decomposes in the presence of aqueous methanolic ferrous sulfate to give a low yield of compound **164** (196).

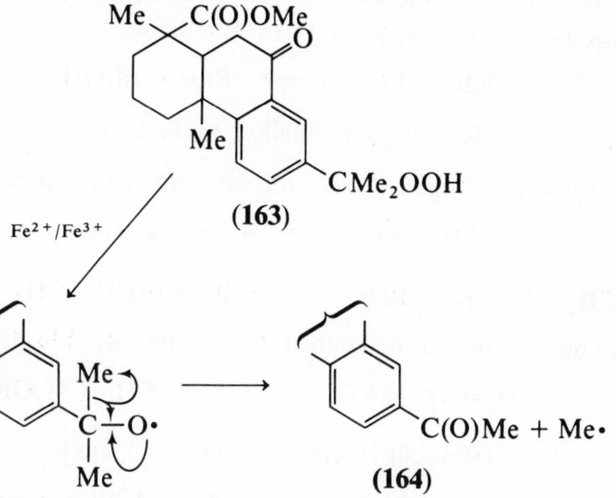

**(163)**

$Fe^{2+}/Fe^{3+}$

**(164)**

The decomposition of 1,4-diethyl-1,4-bis(hydroperoxy)-1,2,3,4-tetrahydro-naphthalene (**165**) with ferrous ions results in the formation of 1,4-naphtho-quinone (**166**) in 16% yield (197).

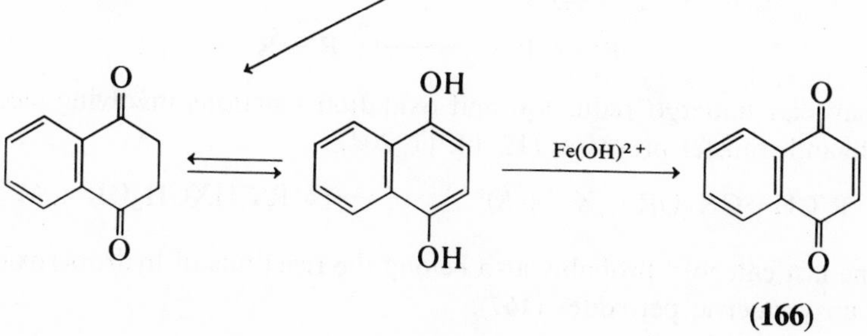

**(165)**      $+ Fe^{2+} \longrightarrow$      $+ Fe(OH)^{2+}$

$Fe(OH)^{2+}$

**(166)**

## III. IN THE PRESENCE OF REACTIVE SUBSTRATES

The decompositions of hydroperoxides in the presence of metal ions but in the absence of reactive substrates are usually of relatively little preparative interest, with noteworthy exceptions already discussed in the preceding section. However, when hydroperoxides are decomposed in the presence of various substrates, many useful derivatives can be obtained. This area of peroxide chemistry is showing rapid growth. Thus alkoxy, cycloalkoxy, aralkoxy, and other oxy radicals that are produced in the first step of the decomposition of the corresponding hydroperoxide,

$$ROOH + M^n \longrightarrow RO\cdot + M(OH)^n \tag{1}$$

$$(R = alkyl, cycloalkyl, aralkyl, etc.)$$

can either abstract hydrogen atoms from saturated or unsaturated substrates,

$$R'H + RO\cdot \longrightarrow R'\cdot + ROH \tag{2a}$$

$$R'CH_2CH{=}CH_2 + RO\cdot \longrightarrow R'CH{\cdots}CH{\cdots}CH_2 + ROH$$

or they can add to unsaturated substrates (56, 65, 141, 146, 190, 198–203).

$$R'CH{=}CH_2 + RO\cdot \longrightarrow R'CH{-}CH_2OR \tag{2b}$$

$$(R' = alkyl, vinyl, aralkyl, aryl, etc.)$$

The new radicals produced in reactions 2a and 2b can undergo various radical reactions (56, 65, 141, 146, 190, 198–203), such as abstraction of hydrogen or other atoms,

$$R'\cdot + R''X \longrightarrow R'X + R''\cdot$$

$$(R'' = hydrocarbon, polyhaloalkyl, mercapto, or other moiety)$$

$$(X = H, halogens, etc.)$$

combination, or dimerization reactions,

$$R'\cdot + R''\cdot \longrightarrow R'{-}R''$$

$$R'\cdot + R'\cdot \longrightarrow R'{-}R'$$

or they can undergo reduction and oxidation reactions involving electron- and ligand-transfer processes (12–14, 17, 204).

$$R'CH{-}CH_2OR + X^- + M^{n+1} \longrightarrow R'CHXCH_2OR + M^n$$

To the last category probably also belong the reactions of hydroperoxides to give unsymmetric peroxides (**167**).

$$ROOH + R'\cdot + M(OH)^n \longrightarrow ROOR' + M^n + H_2O \qquad (3)$$

$$(167)$$

These types of reaction were apparently observed for the first time in the early 1950s by Campbell and Coppinger (205, 206) and Bickel, Kooyman, and co-workers (201) with *t*-butyl hydroperoxide and phenols, by Kharasch, Pauson, and Nudenberg (200) with *t*-butyl hydroperoxide and olefins, and by Treibs and Pellmann (146) with cumyl hydroperoxide and olefins. However, it was not until about 1958 that Kharasch and Fono, and, in part Sosnovsky, developed this oxygenation reaction into a generally applicable synthetic method (207–210). For reaction 3 several mechanistic details have been considered (32, 201, 209, 210). Either the hydroperoxide reacts with the metal-ion oxidant to produce the peroxy radicals, which combine with the substrate radicals to give the product

$$ROOH + M(OH)^n \longrightarrow ROO\cdot + M^n + H_2O$$

$$ROO\cdot + R'\cdot \longrightarrow ROOR'$$

or the oxy radicals that are produced in reaction 1 abstract the hydrogen atoms from the hydroperoxide to give peroxy radicals, which then combine with substrate radicals to give the product

$$ROOH + RO\cdot \longrightarrow ROO\cdot + ROH$$

$$ROO\cdot + R'\cdot \longrightarrow ROOR'$$

or the substrate radicals form, with the hydroperoxide molecule, a complex (**168**), which then interacts with the metal-ion oxidant to give the products

$$R'\cdot + ROOH \longrightarrow [R'\cdot ROOH] \xrightarrow{M(OH)^n}$$

$$ROOR' + M^n + H_2O$$

$$(168)$$

Other complex formations can be also assumed. For example, it seems plausible that an association (**169**) between radical species and a paramagnetic metal ion, such as the cupric ion, can take place.

$$R'\cdot + M(OH)^n \longrightarrow [R'\cdot \cdots M(OH)^n]$$

$$(169)$$

$$ROOH + [169] \longrightarrow ROOR' + M^n + H_2O$$

$$[M = \text{paramagnetic ion (e.g., Cu}^{2+})]$$

The formation of an activated complex (**170**) between certain metal ions (e.g., cuprous ions) and unsaturated substrates is also possible (211, 212).

$$R'CH_2CH{=}CH_2 + M^n \longrightarrow R'CH_2CH{\dotequal}CH_2$$
$$\overset{\displaystyle RO\cdot}{\phantom{x}} \qquad M^n$$
$$(170)$$

$$R'\overset{\cdot}{C}HCH{\dotequal}CH_2 \; \underset{M^n}{\overset{\displaystyle \longleftrightarrow}{}} \; R'CH{\dotequal}CHCH_2\cdot \quad etc.$$
$$M^n \qquad\qquad\qquad M^n$$

$$M = Cu^+$$

This kind of complex formation can be used to explain the observed selectivity of metal-ion-catalyzed reactions of peroxides with unsaturated systems (211, 212).

In analogy with the mechanism of the peroxy ester reaction (211, 212), an interaction of substrate radicals with the metal salts of hydroperoxides involving electron and ligand transfer can also be assumed (12, 13, 204).

$$ROOH + M(OH)^n \longrightarrow ROOM^n + H_2O$$

$$R'\cdot + ROOM^n \longrightarrow ROOR' + M^n$$

The details of the individual schemes are not well established. For further information the references listed in Section II should be consulted. It is hoped that, with this background, the interpretation of the preparative methods described in this section will appear to be plausible.

### A.  Alkyl and Cycloalkyl Hydroperoxides

#### 1.  Reactions with Nonolefinic Compounds

The cuprous chloride–catalyzed decomposition of $t$-butyl hydroperoxide in cumene at 100–110° results in a 90 % yield of $t$-butyl $\alpha$-cumyl peroxide (**171**) (207, 210).

$$Me_3COOH + Cu^+ \longrightarrow Me_3CO\cdot + Cu(OH)^+$$

$$PhCHMe_2 + Me_3CO\cdot \longrightarrow Ph\overset{\cdot}{C}Me_2 + Me_3COH$$

$$\xrightarrow[\;Cu(OH)^+\;]{Me_3COOH} \underset{\substack{\displaystyle | \\ OOCMe_3}}{Ph\overset{\phantom{.}}{C}Me_2} + Cu^+ + H_2O$$
$$(171)$$

Decomposition in the presence of benzoic acid yields 20 % of $\alpha$-cumyl benzoate (**172**) (208).

$$Ph\overset{\cdot}{C}Me_2 + PhC(O)OCu^+ \longrightarrow \underset{\substack{\displaystyle | \\ OC(O)Ph}}{PhCMe_2} + Cu^+$$
$$(172)$$

Compound **171** can also be prepared by the reaction of *t*-butyl hydroperoxide with cumene in the presence of equimolar quantities of lead tetraacetate (184). Other unsymmetrical peroxides of type **171** are similarly prepared (184). The mechanism of this reaction is not fully understood (61, 63).

The cobalt acetate–catalyzed decomposition of *t*-butyl hydroperoxide in a mixture of acetic acid and acetic acid anhydride gives *t*-butyl peroxyacetate (52). The reaction can be followed by EPR spectroscopy (52).

The decomposition of *t*-butyl hydroperoxide in *p*-xylene at 50° in the presence of cuprous chloride results in an 85% yield of *p*-methylbenzyl *t*-butyl peroxide (**173**) (210).

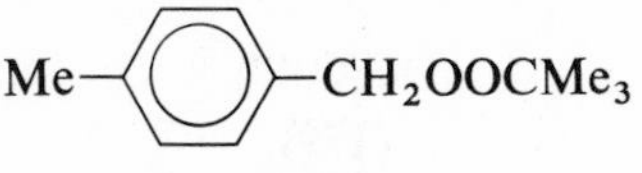

$$Me-\langle\bigcirc\rangle-CH_2OOCMe_3$$

**(173)**

The reaction of *t*-butyl hydroperoxide with ferrous sulfate, in the presence of toluene or tetralin and sodium azide, produces benzyl azide (**174**) and azidotetralin (**175**), respectively (213).

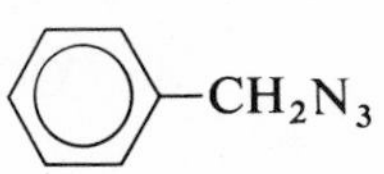

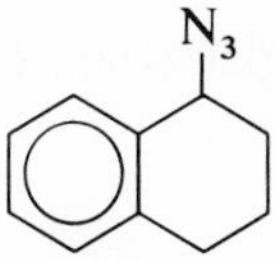

**(174)**          **(175)**

### 2. Reactions with Olefins

The reaction of *t*-butyl hydroperoxide with octene-1 in the presence of cuprous chloride at 70° gives a mixture of peroxides **176** and **177**, plus 2-octenal (**178**) (207, 210). In the presence of cobalt acetate or naphthenate this reaction was earlier reported to give compounds **176** and **178** (200). The following reaction scheme explains the formation of compounds **176** and **177**:

$$Me_3COOH + M^n \longrightarrow Me_3CO\cdot + M(OH)^n$$

$$C_5H_{11}CH_2CH=CH_2 + Me_3CO\cdot \longrightarrow Me_3COH + C_5H_{11}CH\text{---}CH\text{---}CH_2$$

$$C_5H_{11}CH\text{---}\underset{\cdot}{C}H\text{---}CH_2 + Me_3COOH + M(OH)^n$$

$$\downarrow$$

$$C_5H_{11}CH=CHCH_2OOCMe_3 + C_5H_{11}\underset{|}{\overset{}{C}}HCH=CH_2 + M^n + H_2O$$
$$OOCMe_3$$

**(176)**                    **(177)**

2-Octenal (**178**) is not a primary product but is probably formed by thermal decomposition of the primary peroxide **176** (200).

$$176 \xrightarrow{\Delta} C_5H_{11}CH=CHCHO + Me_3COH$$

**(178)**

When the decomposition of *t*-butyl hydroperoxide is carried out in a mixture of octene-1 and benzoic acid, a 50% yield of 3-benzoyloxyoctene (**179**) is obtained (208).

$$C_5H_{11}CHCH=CH_2$$
$$|$$
$$OC(O)Ph$$

**(179)**

The reaction of *t*-butyl hydroperoxide with butadiene in the presence of cobalt naphthenate at $-15$ to $-7°$ results in a mixture of 1,4-di(*t*-butyl-peroxy)butene-2 (**180**) and 3,4-di(*t*-butylperoxy)butene-1 (**181**) (200). The following pathway has been suggested (200):

$$Me_3COO\cdot + CH_2=CHCH=CH_2 \longrightarrow Me_3COOCH_2CH=CHCH_2\cdot$$
$$\updownarrow$$
$$Me_3COOCH_2CHCH=CH_2$$
$$\cdot$$

$$\xrightarrow{Me_3COO\cdot} Me_3COOCH_2CH=CHCH_2OOCMe_3 + Me_3COOCH_2CHCH=CH_2$$
$$|$$
$$OOCMe_3$$

**(180)**                                        **(181)**

The reaction of *t*-butyl hydroperoxide with cyclohexene in the presence of cuprous chloride or cobalt naphthenate produces mainly cyclohexenyl *t*-butyl peroxide (**182**) plus some cyclohexenone (**183**), which is probably formed by the decomposition of **182** (200).

$$Me_3COOH + M^n \longrightarrow Me_3CO\cdot + M(OH)^n$$

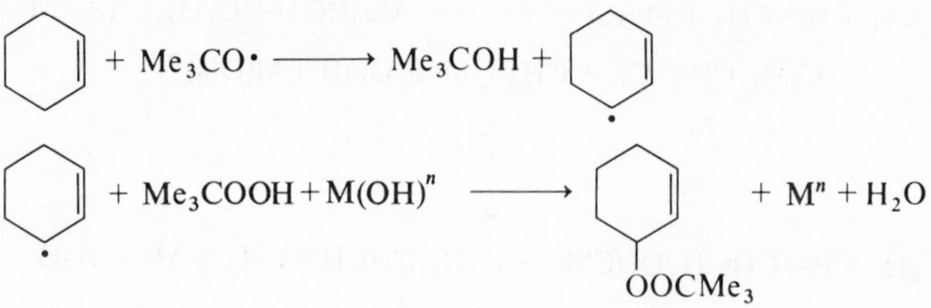

**(182)**

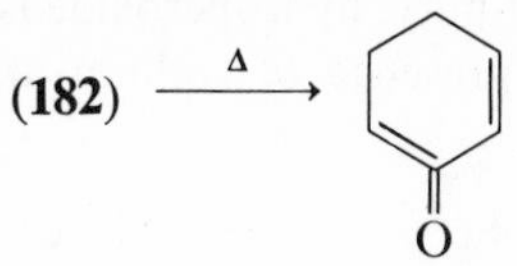

$$(182) \xrightarrow{\Delta}$$

(183)

The decomposition of *t*-butyl hydroperoxide with cuprous chloride in acetic acid containing cyclohexene gives a quantitative yield of cyclohexenyl acetate (184a) (37). Similarly in the presence of benzoic acid this decomposition produces a 90% yield of cyclohexenyl benzoate (184b) (208).

$$Me_3COOH + Cu^+ \longrightarrow Me_3CO\cdot + Cu(OH)^+$$

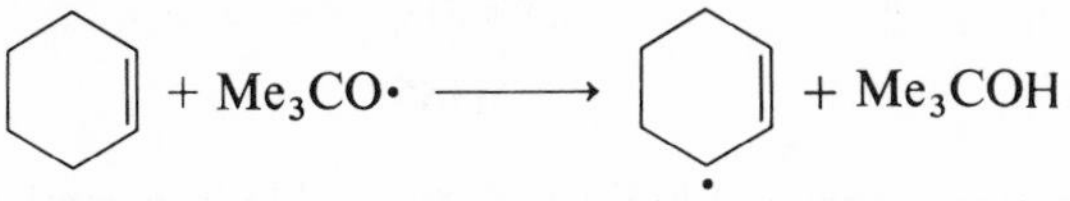

$$RC(O)OH + Cu(OH)^+ \longrightarrow RC(O)OCu^+ + H_2O$$

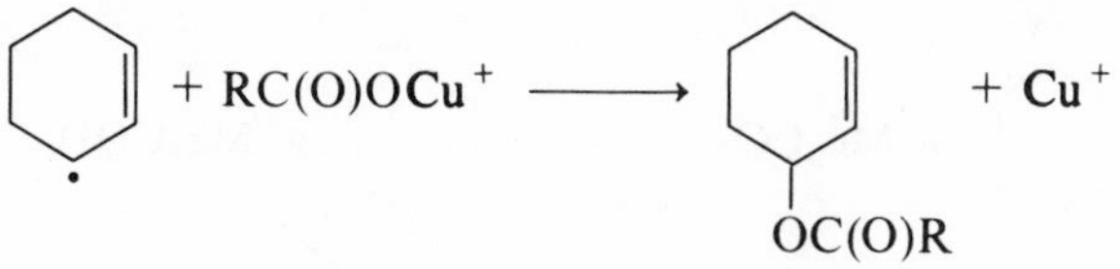

OC(O)R

(184)

(a, R = Me)
(b, R = Ph)

Compound 184a is also formed by the ferrous sulfate–catalyzed decomposition of *t*-butyl hydroperoxide in the presence of cupric acetate (214). This reaction can be applied to the preparation of optically active esters with copper salts of optically active acids. Thus cyclopentene, cyclohexene, cyclooctene, and bicyclo[3.2.1]octene-2 are converted to α-ethylcamphorates with cupric α-ethylcamphorate in 78, 84, 36, and 94% yield, respectively (215). These esters can be converted by either hydrolysis or hydrogenation with lithium aluminum hydride to $\Delta^2$-cyclopentenol, $\Delta^2$-cyclohexenol, $\Delta^2$-cyclooctenol, and bicyclo[3.2.1]octen-2-ol, respectively (215). A similar experiment with cyclohexene and cupric methyl-di-*O*-acetyl tartrate yields 61% of allylic ester, which on reduction results in 72% of $\Delta^2$-cyclohexenol (215). Thus optically active allylic alcohols are obtained. The asymmetric induction is not high, but because of the high rotation of the pure optically active materials, substantial rotation is found (215).

If the decomposition of $t$-butyl hydroperoxide is carried out in cyclohexene by ferrous sulfate in the presence of sodium azide, 3-azidocyclohex-1-ene (**185**) is obtained (214).

$$Me_3COOH + Fe^{2+} \longrightarrow Me_3CO\cdot + Fe(OH)^{2+}$$

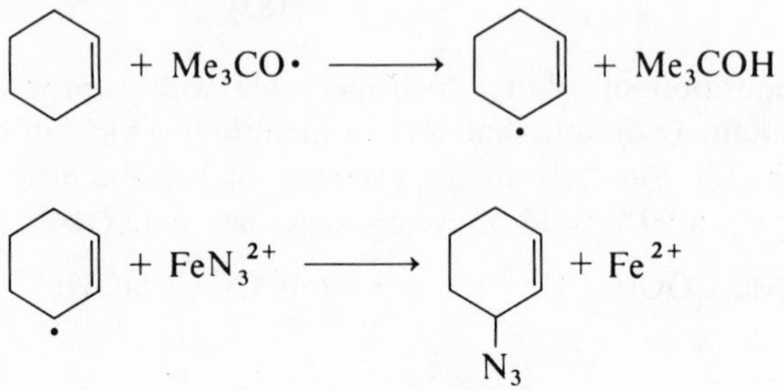

**(185)**

Similarly the decomposition of $t$-butyl hydroperoxide in a methanolic solution of cyclohexene containing ferrous sulfate and cupric nitrate at $-5°$ yields cyclohexenyl nitrate (**186**) (216).

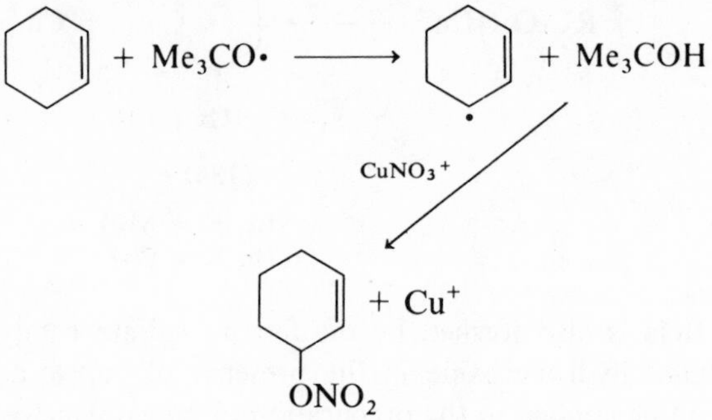

**(186)**

An analogous reaction of $t$-butyl hydroperoxide with cyclohexene in the presence of copper ions and phthalimide gives $N$-2-cyclohexenylphthalimide (**187**) (208).

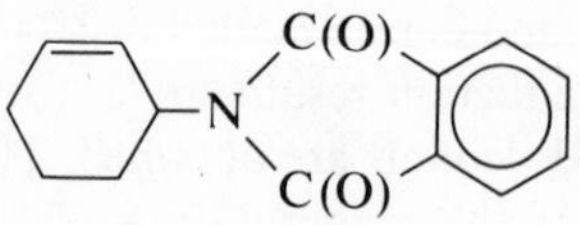

**(187)**

The reaction of 4-vinylcyclohexene with *t*-butyl hydroperoxide in the presence of cobalt naphthenate gives a mixture of peroxides (**188**), which on hydrogenation yields a mixture of ethylcyclohexanols of the following composition: 75–85% **189**, 2–15% **190**, 1–3% **191**, and 0% **192** (217).

(**189**)     (**190**)    (**191**)    (**192**)

In another investigation a 19% yield of compound **192** was obtained (218).

The reaction of *t*-butyl hydroperoxide with α-pinene at 70° for 7 hr in the presence of cuprous chloride gives a mixture of products consisting of 85% peroxides, 5% epoxides, 3% verbanone, and 5% verbanols. As a result of a radical rearrangement two peroxides (**193** and **194**) are formed (219).

(**193**)     (**194**)

A similar reaction of *t*-butyl hydroperoxide with β-pinene yields only one product (**195**) (219).

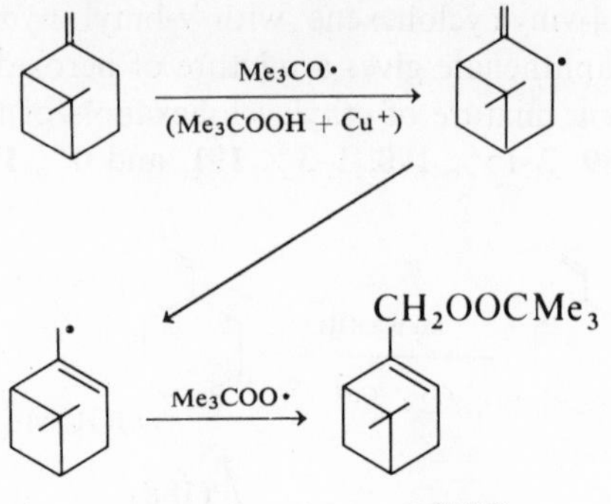

**(195)**

Analogous results are obtained with pinanyl hydroperoxide **(196)** (219).

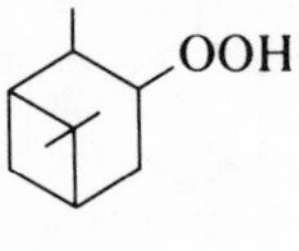

**(196)**

Under similar conditions the reaction of *d*-3-carene with *t*-butyl hydroperoxide gives a product that is probably a mixture of peroxides **197** and **198** (220).

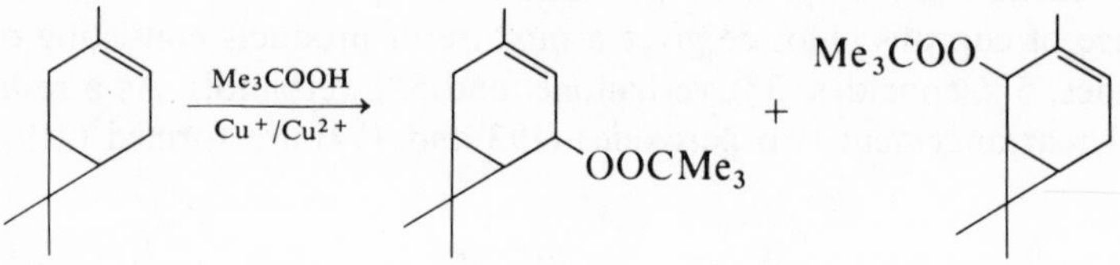

**(197)**                                      **(198)**

The metal-ion-catalyzed reactions of hydroperoxides with olefins can also give epoxides, often in excellent yield. For example, the reaction of *t*-butyl hydroperoxide with 2,4,4-trimethyl-1-pentene or cyclohexene in the presence of cobalt acetylacetonate gives the corresponding epoxides in 9–32 and 40–50% yields, respectively (221). With vanadium acetylacetonates or the octoate, yields of cyclohexene epoxide are quantitative (222). Epoxidation is strongly inhibited by *t*-butyl alcohol. Similarly the reaction of cyclohexene hydroperoxide with these olefins produces epoxides in 20–38% yields (221). Similar epoxidations of propylene, 4-methyl-2-pentene, *cis*- and *trans*-2-butene, 2-methylpentene-2, octene-1, octene-2, and 4-vinylcyclohexene are achieved with *t*-butyl hydroperoxide and cumene hydroperoxide, in the presence of acetylacetonates of copper, manganese, molybdenum, vanadium, and chromium, and molybdenum hexacarbonyl (223, 224). The reaction can be stereospecific; for example, *trans*-4-methyl-2-pentene reacts with *t*-butyl hydroperoxide in the presence of chromium acetylacetonate to give the *trans* epoxide, and the reaction of the *cis* isomer yields the *cis* epoxide (223). The yield of epoxide decreases with increasing concentrations of the catalyst, longer reaction times, and higher temperatures and pressures of oxygen (223). Nonpolar solvents are preferred, as is molybdenum hexacarbonyl catalysis (224).

2,4,4-Trimethyl-1-pentene reacts with *t*-butyl hydroperoxide and equimolar amounts of ceric sulfate to give a mixture of 55% *t*-butyl alcohol 12% oxygen gas, 10% 4,4-dimethyl-2-pentenone, and 4% 2,4,4-trimethylpentanal (221). There is evidence of a free-radical mechanism involving an epoxide intermediate (**199**) (59, 221).

$$Me_3COOH + Ce^{4+} \longrightarrow Me_3COO\cdot + Ce^{3+} + H^+$$

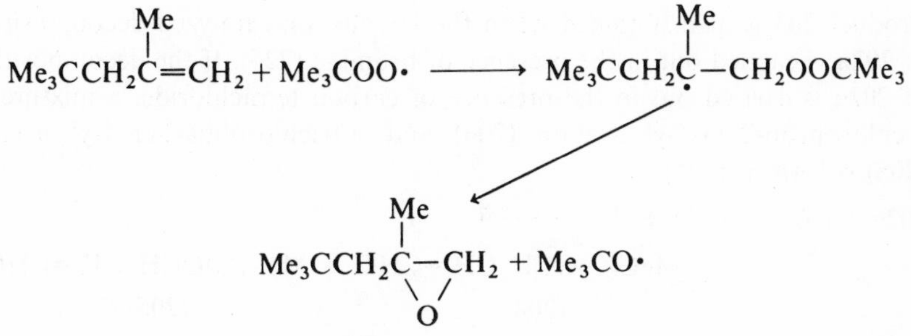

$$Me_3CCH_2\overset{\overset{\displaystyle Me}{|}}{C}=CH_2 + Me_3COO\cdot \longrightarrow Me_3CCH_2\overset{\overset{\displaystyle Me}{|}}{\underset{\displaystyle \cdot}{C}}-CH_2OOCMe_3$$

$$Me_3CCH_2\overset{\overset{\displaystyle Me}{|}}{C}\overset{\displaystyle -}{\underset{\displaystyle \diagdown_O\diagup}{}}CH_2 + Me_3CO\cdot$$

**(199)**

An interesting ring-opening reaction of hydroperoxyfuran derivatives gives straight-chain propene derivatives. Thus the ferrous sulfate–induced decomposition of 2,5-dihydrofuran 2-hydroperoxide (**200**) with bromotrichloromethane yields mainly 3-bromo-prop-2-en-1-yl formate (**201**) (225).

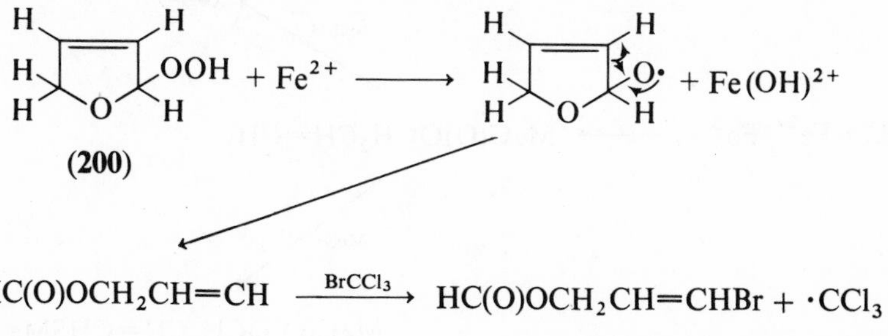

**(200)**

$$HC(O)OCH_2CH=CH \xrightarrow{\text{BrCCl}_3} HC(O)OCH_2CH=CHBr + \cdot CCl_3$$

**(201)**

The decomposition of 2-methyl-2,5-dihydrofuran 2-hydroperoxide (**202a**) with ferrous sulfate in the presence of bromotrichloromethane yields 3-bromoprop-2-en-1-yl acetate (**203**) plus hexachloroethane (225).

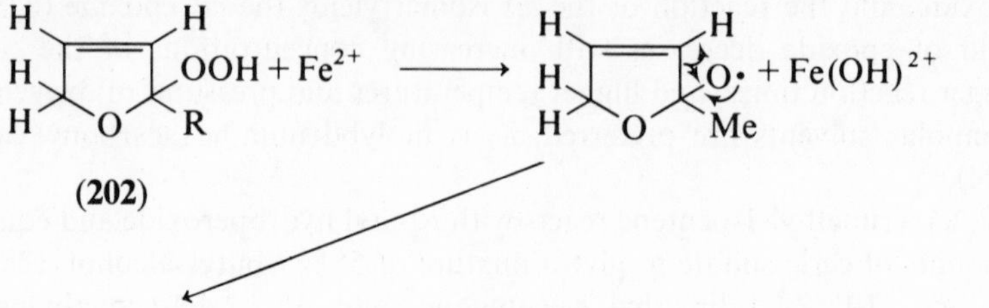

(**202**)

$$MeC(O)OCH_2CH=\dot{C}H \xrightarrow{BrCCl_3} MeC(O)OCH_2CH=CHBr + \cdot CCl_3$$

(**203**)

(**a**, R = Me)
[**b**, R = $(CH_2)_4$Me]

Product **203** is also obtained when the ferrous-ion-catalyzed decomposition of **202a** is carried out in the presence of bromine (225). If the decomposition of **202a** is carried out in the presence of carbon tetrachloride, a mixture of 3-chloroprop-2-en-1-yl acetate (**204**) and 4-trichlorobut-2-en-1-yl acetate (**205**) is formed (225).

$$202a + CCl_4 + Fe^{2+}/Fe^{3+} \longrightarrow$$

$$MeC(O)OCH_2CH=CHCl + MeC(O)OCH_2CH=CHCl_3$$
$$\qquad\qquad (204) \qquad\qquad\qquad\qquad\qquad (205)$$

The reaction of **202a** with hydrochloric acid in the presence of ferrous sulfate also produces **204** (225).

An analogous reaction with methyl mercaptan gives a mixture of prop-2-en-1-yl acetate (**206**) and 3-methylthioprop-2-en-1-yl acetate (**207**) (225).

$$MeC(O)OCH_2CH=CH_2 + MeS\cdot$$

(**206**)

MeSH

$$202 + Fe^{2+}/Fe^{3+} \longrightarrow MeC(O)OCH_2CH=\underset{\cdot}{C}H$$

MeS·

$$MeC(O)OCH_2CH=CHSMe$$

(**207**)

Similar fragmentation of 2-*n*-amyl-2,5-dihydrofuran 2-hydroperoxide (**202b**) in the presence of bromotrichloromethane yields a mixture of 3-bromoprop-2-en-1-yl caproate (**208**) and 3-trichlorobut-2-en-1-yl caproate (**209**) (225).

$$\textbf{202b} + Fe^{2+}/Fe^{3+} + BrCCl_3 \longrightarrow Me(CH_2)_4C(O)OCH_2CH=CHBr$$
$$\textbf{(208)}$$
$$+ Me(CH_2)_4C(O)OCH_2CH=CHCCl_3$$
$$\textbf{(209)}$$

The ferrous-ion-catalyzed reaction of *t*-butyl hydroperoxide with butadiene or isoprene is reported to give 1,8-bis-*t*-butoxy-2,6-octadiene and 1,8-di-*t*-butoxy-2,7-dimethyl-2,6-octadiene, respectively (198, 199). The decomposition of *t*-butyl hydroperoxide at $-10°$ in the presence of ferrous ions and thiophenol gives mainly product **210a** (203). Similarly the reaction with propan-1-thiol and *t*-butylthiol produces, among other products, **210b** and **210c**, respectively (203).

$$Me_3CO\cdot + CH_2=CHCH=CH_2 \longrightarrow Me_3COCH_2CH\text{---}\overset{\cdot}{C}H\text{---}CH_2$$

$$RSH + Me_3CO\cdot \longrightarrow RS\cdot + Me_3COH$$

$$Me_3COCH_2CH\text{---}\underset{\cdot}{C}H\text{---}CH_2 + RS\cdot \longrightarrow Me_3COC_4H_6SR$$
$$\textbf{(210)}$$
$$\textbf{(a, R = Ph)}$$
$$\textbf{(b, R = }n\text{-C}_3\text{H}_7\textbf{)}$$
$$\textbf{(c, R = Me}_3\text{C)}$$

In addition symmetrical diethers **211**, composed of two *t*-butoxy and two butadiene units, and products (**212**) of addition of thiols to butadiene, are formed (203).

$$Me_3CO(C_4H_6)_2OCMe_3 \qquad RSC_4H_6SR$$
$$\textbf{(211)} \qquad\qquad\qquad \textbf{(212)}$$

A similar reaction occurs between *t*-butyl hydroperoxide, butadiene, and hydroquinone in the presence of aqueous methanolic ferrous ammonium sulfate at $-10°$ (202). The intermediate adduct **213** is oxidized in air to compound **214** (202); the products are mixtures of 1,2- and 1,4-isomers.

$$Me_3CO\cdot + CH_2{=}CHCH{=}CH_2 \longrightarrow Me_3COCH_2CH{\dot{=}}CH{=}CH_2$$

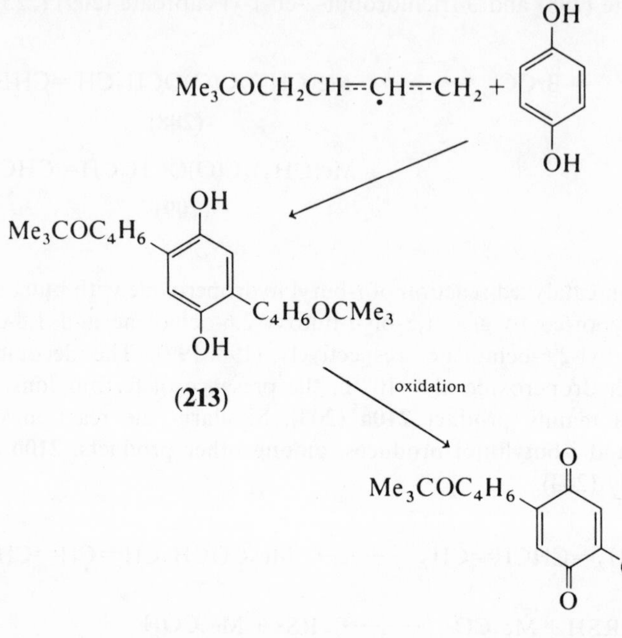

$$Me_3COCH_2CH{\dot{=}}CH{=}CH_2 +$$

(hydroquinone with OH groups)

(product 213)

$$Me_3COC_4H_6 \quad C_4H_6OCMe_3$$

**(213)**

oxidation

$$Me_3COC_4H_6 \quad C_4H_6OCMe_3$$

**(214)**

Reactions with quinone and quinhydrone give the same products, **213** and **214** (202). Analogous reaction with toluquinhydrone yields product **215** (202). However, reaction with tetrachloroquinhydrone gives compound **216** (202).

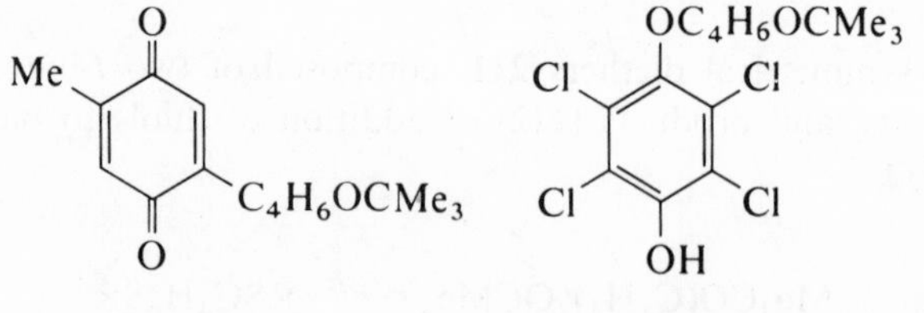

(structure with Me and $C_4H_6OCMe_3$)

**(215)**

(structure with $OC_4H_6OCMe_3$, Cl, Cl, Cl, Cl, OH)

**(216)**

The reaction of $t$-butyl hydroperoxide with butadiene at $-10°$ in the presence of both ferrous sulfate and cupric nitrate yields a mixture of $t$-butoxy nitrates **217** and **218** (216).

$$Me_3CO\cdot + CH_2{=}CHCH{=}CH_2 \longrightarrow Me_3COCH_2CH{\overset{\cdot}{=\!=}}CH{=\!=}CH_2$$

$$\Big\downarrow CuNO_3{}^+$$

$$Me_3COCH_2CHCH{=}CH_2 \;+\; Me_3COCH_2CH{=}CHCH_2ONO_2$$
$$\quad\quad\quad\quad\;\;|$$
$$\quad\quad\quad\;ONO_2$$

**(217)**                                          **(218)**

Under similar conditions the decomposition of *t*-butyl hydroperoxide in styrene produces a *t*-butoxy nitrate **(219)** (216).

$$Me_3CO\cdot + PhCH{=}CH_2 \longrightarrow PhCHCH_2OCMe_3$$

$$\xrightarrow{\;CuNO_3{}^+\;} PhCHCH_2OCMe_3$$
$$\quad\quad\quad\quad\quad\quad\;|$$
$$\quad\quad\quad\quad\quad ONO_2$$

**(219)**

The reaction of *t*-butyl hydroperoxide with butadiene and sodium azide or ammonium thiocyanate in the presence of ferrous ions results in the formation of 1-azido-4-*t*-butoxybutene-2 and 1-thiocyano-4-*t*-butoxybutene-2, respectively (226, 227).

The reaction of *t*-butyl hydroperoxide with cyclopentadiene in the presence of a mixture of ferrous and ferric sulfate and sodium azide gives a 45% yield of azido-*t*-butoxycyclopentene **(220)** (213).

$$Me_3COOH + Fe^{2+} \longrightarrow Me_3CO\cdot + Fe(OH)^{2+}$$

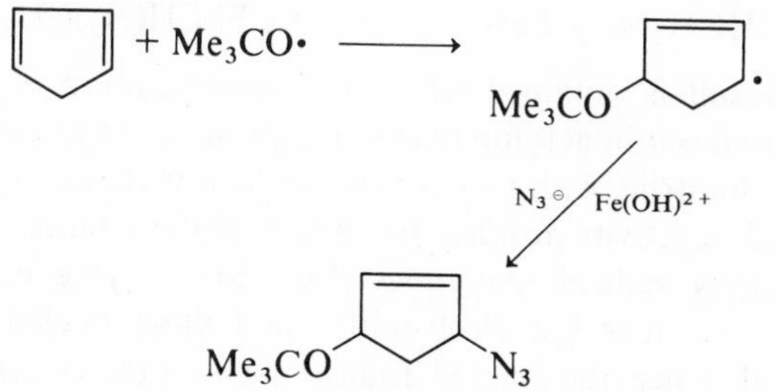

**(220)**

In the presence of large quantities of ferric chloride this reaction produces a 50% yield of the corresponding chloro-*t*-butoxycyclopentene **(221)** (213).

$$Me_3CO\underset{\phantom{x}}{\diagdown}\!\!\!\!\text{(cyclopentyl radical)}\quad\cdot\quad\xrightarrow[\text{Fe(OH)}^{2+}]{\text{Cl}^-}\quad Me_3CO\underset{\phantom{x}}{\diagdown}\!\!\!\!\text{(cyclopentyl-Cl)}\;Cl$$

(221)

The reaction of *t*-butyl hydroperoxide with styrene in the presence of ferrous and chloride ions gives α-chloro-β-*t*-butoxyethylbenzene (222) (227).

$$Me_3COOH + Fe^{2+} \longrightarrow Me_3CO\cdot + Fe(OH)^{2+}$$

$$PhCH{=}CH_2 + Me_3CO\cdot \longrightarrow Ph\overset{\cdot}{C}HCH_2OCMe_3$$

$$\xrightarrow[Fe^{3+}]{Cl^-} PhCHClCH_2OCMe_3$$

(222)

An analogous reaction in the presence of sodium azide results in a mixture of α-azido-β-*t*-butoxyethylbenzene (223) and α,β-diazidoethylbenzene (224) (213, 227, 228).

PhCH(N$_3$)CH$_2$OCMe$_3$          PhCH(N$_3$)CH$_2$N$_3$
(223)                          (224)

Because of its strongly electrophilic nature, the *t*-butoxy radical readily attacks olefinic bonds with a high electron density, and less readily olefinic bonds with electron-withdrawing groups. Thus in the case of styrene, with a double bond of high electron density, the following interactions of the *t*-butoxy radical are possible (227):

$$Me_3CO\cdot + PhCH{=}CH_2 \longrightarrow Ph\overset{\cdot}{C}HCH_2OCMe_3$$

$$\xrightarrow{FeN_3^{2+}} PhCH(N_3)CH_2OCMe_3 + Fe^{2+}$$

$$Me_3CO\cdot + FeN_3^{2+} \longrightarrow N_3\cdot + Fe^{3+} + Me_3CO^-$$

$$PhCH{=}CH_2 + N_3\cdot \longrightarrow Ph\overset{\cdot}{C}HCH_2N_3$$

$$Ph\overset{\cdot}{C}HCH_2N_3 + FeN_3^{2+} \longrightarrow PhCH(N_3)CH_2N_3$$

A different result is obtained when the decomposition of *t*-butyl hydroperoxide is carried out in acrylonitrile in the presence of ferrous and chloride ions (227). Acrylonitrile, with its strongly electron-withdrawing cyano group, has a decreased reactivity toward the electrophilic *t*-butoxy radical. As a result the *t*-butoxy radical undergoes β-scission to give acetone and the methyl radical, which is less electrophilic and more nucleophilic than the *t*-butoxy radical, since the methyl radical has a greater tendency to lose electrons than to gain them (229).

$$Me\cdot \longrightarrow Me^+ + e$$

$$Me\cdot + e \longrightarrow Me^-$$

As a result the reaction produces α-chlorobutyronitrile (225) (227).

$$Me_3CO\cdot \longrightarrow Me_2CO + Me\cdot$$

$$Me\cdot + CH_2=CHCN \longrightarrow MeCH_2\overset{\cdot}{C}HCN$$

$$\xrightarrow{\text{FeCl}^{2+}} MeCH_2CHClCN + Fe^{2+}$$

**(225)**

The reaction of *t*-butyl hydroperoxide with butadiene in acetic acid in the presence of cupric acetate produces a 50–55% yield of a mixture of 4-*t*-butoxy-3-acetoxybutene-1 **(226)** and 4-*t*-butoxy-1-acetoxybutene-2 **(227)** in 78–83 and 17–21% yield, respectively (230).

$$Me_3COOH + Cu^+ \longrightarrow Me_3CO\cdot + Cu(OH)^+$$

$$Me_3CO\cdot + CH_2=CHCH=CH_2 \longrightarrow Me_3COCH_2CH\overset{\cdot}{=\!=}CH\overset{\cdot}{=\!=}CH_2$$

MeC(O)OCu⁺

Me₃COCH₂CHCH=CH₂  +  Me₃COCH₂CH=CHCH₂
                |                                          |
           OC(O)Me                                  OC(O)Me

**(226)**                                    **(227)**

Under similar conditions allylbenzene **(228)** and *β*-methylstyrene **(229)** also yield acetoxy derivatives. However, the 3-acetoxy **(230)** and 1-acetoxy **(231)** isomers are formed in about equal quantities (37).

$$PhCH_2CH=CH_2 + Me_3CO\cdot$$

**(228)**

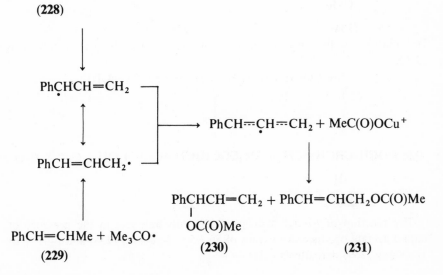

PhĊHCH=CH₂

PhCH═CH═CH₂ + MeC(O)OCu⁺

PhCH=CHCH₂·

PhCHCH=CH₂ + PhCH=CHCH₂OC(O)Me
     |
 OC(O)Me

PhCH=CHMe + Me₃CO·

**(229)**                    **(230)**                    **(231)**

The decomposition of $t$-butyl hydroperoxide in aqueous methyl alcohol in the presence of ferrous and cupric chlorides produces a 55% yield of compounds **232** (65%) and **233** (35%) (227, 230).

$$Me_3COCH_2CH \cdots \overset{\cdot}{C}H \cdots CH_2 + CuCl^+$$

$$\downarrow$$

$$Me_3COCH_2CH{=}CHCH_2Cl + Me_3CCH_2\overset{|}{\underset{Cl}{C}}HCH{=}CH_2$$

         **(232)**                              **(233)**

The same products are obtained with an aqueous mixture of $t$-butyl hydroperoxide, butadiene, ferrous sulfate, ferric chloride, and lithium chloride or hydrochloric acid (226, 228).

In the absence of chloride ions, and in the presence of large quantities of aqueous methyl alcohol, the decomposition of $t$-butyl hydroperoxide by a mixture of ferrous and cupric sulfates produces a 60–65% yield of $t$-butoxy-methoxybutenes (**234**, 72%, and **235**, 28%) (35, 230).

$$Me_3COCH_2CH \cdots \overset{\cdot}{C}H \cdots CH_2 + MeOH + Cu(OH)^+$$

$$\downarrow$$

$$Me_3COCH_2\overset{|}{\underset{OMe}{C}}HCH{=}CH_2 + Me_3COCH_2CH{=}CHCH_2OMe + Cu^+ + H^+$$

         **(234)**                            **(235)**

In addition, the hydroxy derivatives **236** and **237** are obtained in low yield (230).

$$Me_3COCH_2CH \cdots \overset{\cdot}{C}H \cdots CH_2 + H_2O + Cu(OH)^+$$

$$\downarrow$$

$$Me_3COCH_2\overset{|}{\underset{OH}{C}}HCH{=}CH_2 + Me_3COCH_2CH{=}CHCH_2OH + Cu^+ + H_2O$$

         **(236)**                            **(237)**

The reaction of $t$-butyl hydroperoxide with butene-2 in the presence of cupric acetate produces a mixture of 85–95% 3-acyloxy (**238**) and 5–15% of 1-acyloxy (**239**) derivatives (231).

$$MeCH = CHMe + Me_3CO \cdot \longrightarrow MeCH \overset{\cdot}{=} CH \overset{}{=} CH_2$$

$$\text{MeC(O)OCu}^+ \swarrow$$

$$\underset{\underset{\displaystyle OC(O)Me}{|}}{MeCHCH = CH_2} + MeCH = CHCH_2OC(O)Me$$

$$\textbf{(238)} \qquad\qquad\qquad\qquad \textbf{(239)}$$

Under similar conditions pentene-1 yields a mixture of 88–91% 3-acyloxy-pentene-1 (**240**) and some 1-acyloxypentene-2 (**241**) (231).

$$\underset{\underset{\displaystyle OC(O)Me}{|}}{MeCH_2CHCH = CH_2} \qquad MeCH_2CH = CHCH_2OC(O)Me$$

$$\textbf{(240)} \qquad\qquad\qquad\qquad\qquad \textbf{(241)}$$

If in these reactions a copper complex of phenanthroline or bipyridyl is substituted for copper acetate, the composition of the allylic isomers is distributed more evenly, indicating a change in mechanism (231). This series of reactions can be divided into two groups. To one group belong reactions that produce about equal quantities of 3- and 1-substituted allylic isomers. To the second group belong reactions that produce predominantly the less stable 3-substituted allylic isomer. The first type of reaction can be adequately explained by an electron-transfer step to give conventional carbonium-ion intermediates (**242**), which then interact with anions (X) to give the products (37).

$$Me_3COCH_2CH \overset{\cdot}{=} CH \overset{}{=} CH_2 + M^{n+1}$$

$$\downarrow$$

$$Me_3COCH_2CH \overset{}{=} \underset{+}{CH} \overset{}{=} CH_2 + M^{n}$$

$$\textbf{(242)} \qquad \downarrow X^-$$

$$\underset{\underset{\displaystyle X}{|}}{Me_3COCH_2CHCH = CH_2} + Me_3COCH_2CH = CHCH_2X$$

In the second case the predominance of 3-substituted derivatives in the products indicates that the process does not proceed through conventional radical or carbonium-ion intermediates. Originally the main proponents of this theory were Kharasch and co-workers (209, 211, 212, 232). In more recent years Kochi and colleagues have produced experimental evidence in favor of this theory (12, 13, 35, 230).

The specificity of the copper-ion-catalyzed reaction is attributed to a ligand- and electron-transfer transition state. The function of copper ions is to transfer a ligand and an electron, and to coordinate with the allylic double bond, whereby the terminal complex **243** is more stable than **244** with the internal double bond. As a result the product with the terminal double bond (**245**) predominates in the mixture of **245** and **246**.

The following minor criticisms might be advanced against this particular scheme. Since copper ions are available in the system before the formation of radicals, it might be more plausible to assume that a cuprous ion–olefin complex is first formed (211, 212). For example, an association of the type of **247** between butadiene and cuprous ions is possible.

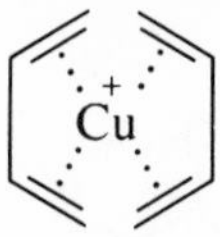
Cu

**(247)**

In the following step the addition of the *t*-butoxy radical to the complex occurs to give species **248** and **249**, which then interact with the paramagnetic cupric compound to give the products **245** and **246** (211, 212).

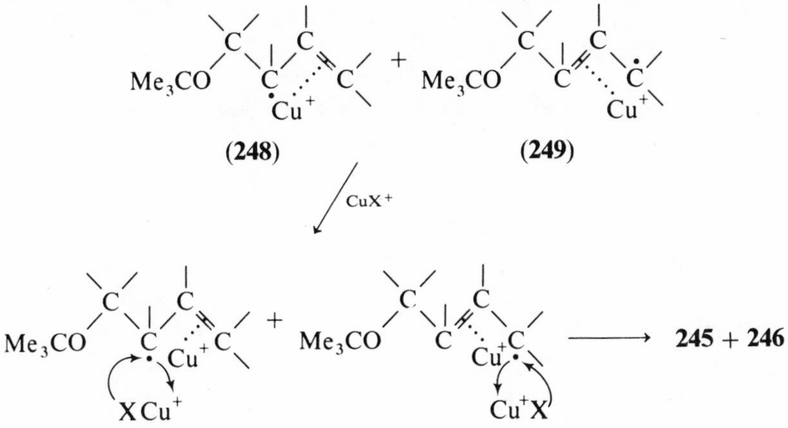
Me3CO
Cu+
(248)
Me3CO
Cu+
(249)
CuX+
Me3CO
Cu+
XCu+
Me3CO
Cu+
Cu+X
245 + 246

### 3. *Reactions with Phenols*

The oxidation of 2,6-dimethylphenol (**250**, R = Me) with *t*-butyl hydroperoxide in benzene in the presence of cobalt toluate below 30° yields 12% of **251**, R = Me, and 75% of **252**, R = Me (233). The products are separated by column chromatography with silica (233). Similarly the oxidation of 2,6-di-*t*-butylphenol (**250**, R = $CMe_3$) yields 59% of **251**, R = $CMe_3$, and 31% of **252**, R = $CMe_3$ (233).

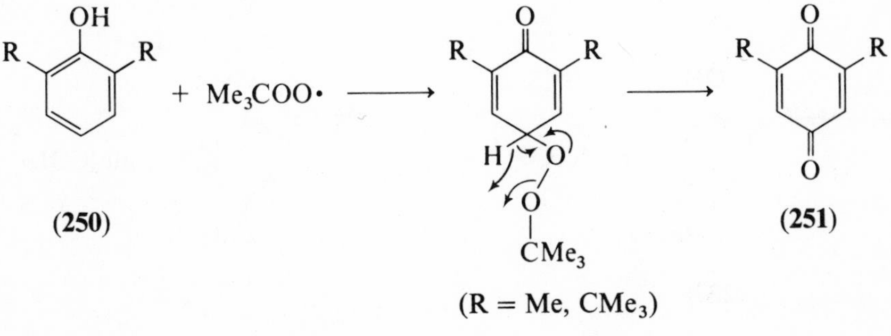
OH
R
R
+ Me3COO•
(250)
O
R
R
H
O
O
CMe3
O
R
R
O
(251)

(R = Me, $CMe_3$)

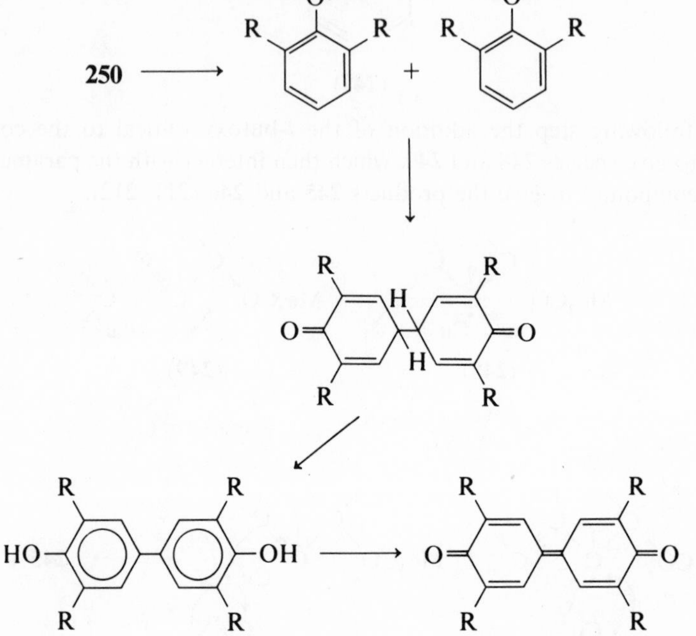

(252)

The reaction of 2,4-di-*t*-butylphenol (253) with *t*-butyl hydroperoxide in the presence of cobalt ions gives a complex mixture of products (254–264) arising from the following radical coupling reactions: (a) phenoxy–peroxy, (b) phenoxy–phenoxy at position 6, and (c) phenoxy–phenoxy through the oxygen atom of one radical and the 6-position of a second radical (234).

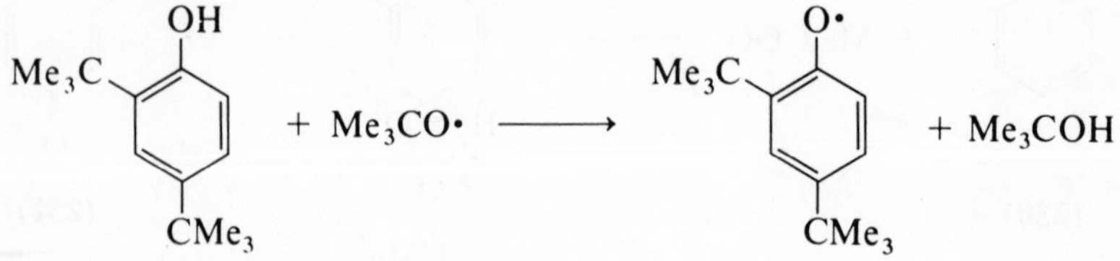

(253)

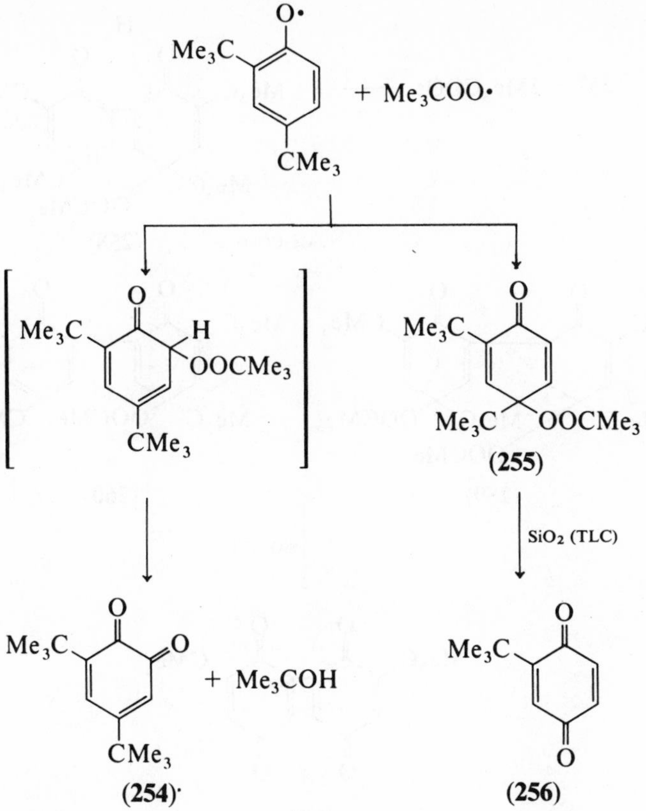
O•
Me3C
+ Me3COO•
CMe3
O
Me3C
H
OOCMe3
CMe3
O
Me3C
Me3C
OOCMe3
(255)
SiO2 (TLC)
O
Me3C
O
+ Me3COH
CMe3
(254)
O
Me3C
O
(256)

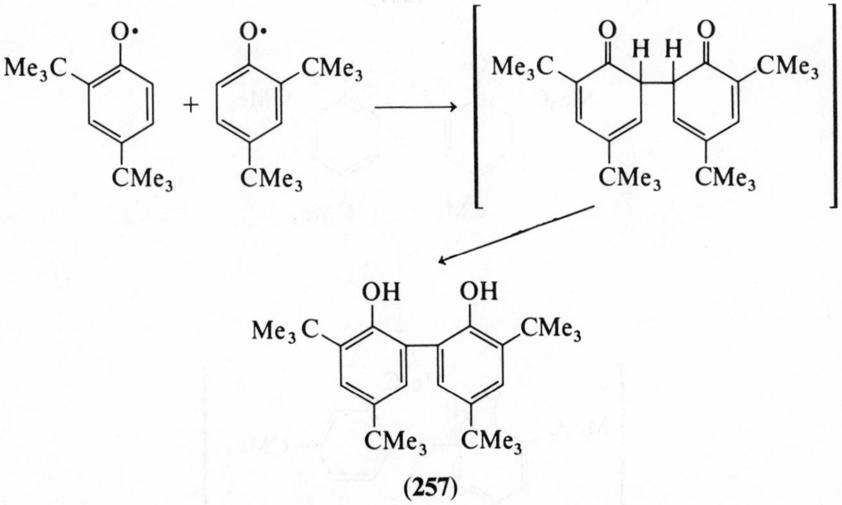
O•
Me3C
CMe3
+
O•
CMe3
CMe3
O
H H
O
Me3C
CMe3
CMe3
CMe3
OH
OH
Me3C
CMe3
CMe3
CMe3
(257)

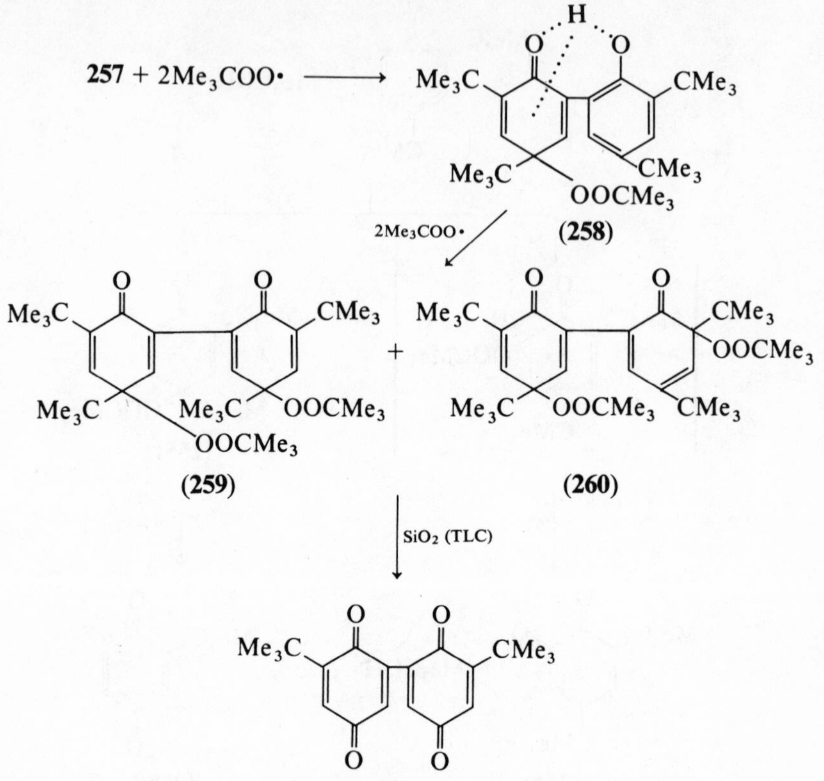

(258) (259) (260) (261)

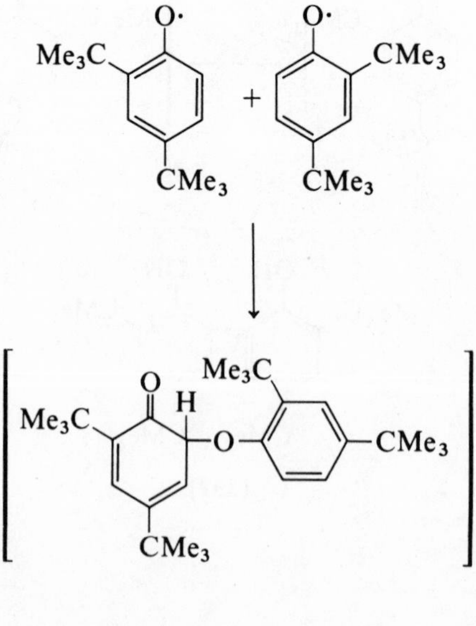

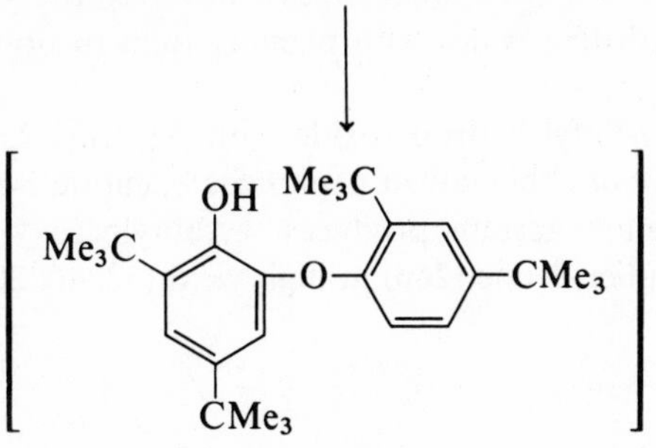

(262)

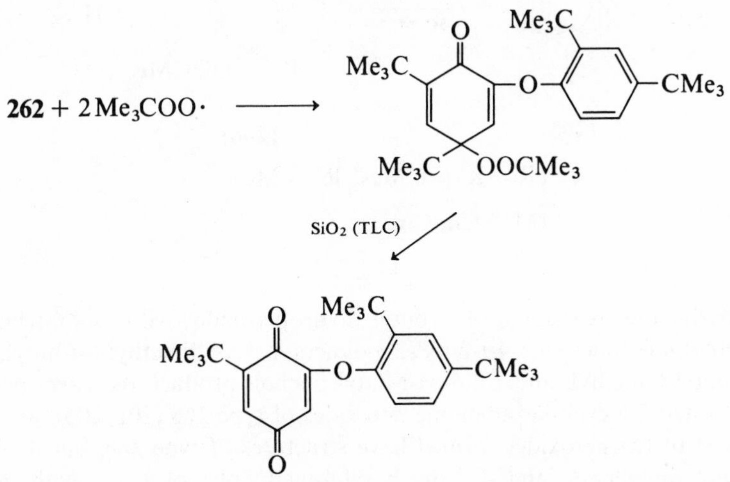

262 + 2 Me₃COO· ⟶

SiO₂ (TLC)

(264)

Product **254** amounts to about 10%, product **255** to about 5–10% (234). Products **256**, **261**, and **264** are formed on the thin-layer-chromatography (TLC) plate (234). Compound **260** is a very minor product. Compound **262** is not detected, but its oxidation product **263** is a major component (20%). The yield of compound **258** varies between 1 and 12%, depending on the ratio of *t*-butyl hydroperoxide to phenol **253**. The best yield is achieved with a ratio of 2:1 (234). Compound **257** could not be found among the products under standard conditions—namely, using a ratio of not less than 2 moles of hydroperoxide per mole of phenol. However, product **257** could be prepared in 37% yield from **253** with 1 equivalent of *t*-butyl hydroperoxide in the presence of cobalt toluate (234). Compound **259** is obtained in 60% yield from **257** with an excess of *t*-butyl hydroperoxide (234).

It is evident from these examples that the interaction of hydroperoxides with phenols can be of great complexity. However, in the following examples the reactions of hydroperoxides with phenols seem to proceed by a simpler pattern.

The reaction of $t$-butyl hydroperoxide with 4-methyl-2,6-di-$t$-butylphenol (**265**) in the presence of either cobalt naphthenate, cupric 2-ethylhexanoate, or cupric phenanthroline acetate produces 4-$t$-butylperoxy-4-methyl-2,6-di-$t$-butyl-2,5-cyclohexadien-1-one (**266**) in high yield (32, 46, 201, 205).

$$\text{(265)} \qquad\qquad\qquad \text{(266)}$$

$$(R = R' = CMe_3, \ R'' = Me$$

$$(M = Cu, Co)$$

Similarly the reactions of $t$-butyl hydroperoxide with 2,4,6-trimethyl, 2-$t$-butyl-4,6-dimethyl, 4-$t$-butyl-2,6-dimethyl, 2,4-dimethyl-6-$t$-butyl, 2,4-di-$t$-butyl-6-methyl, and 2,4,6-tri-$t$-butyl phenols produce the corresponding substituted 2,5-cyclohexadienone peroxides of type **266** (201, 205).

Most of the peroxides formed have structures of type **266**, but 4-$t$-butyl-2,6-dimethylphenol and 2,4-di-$t$-butyl-6-methylphenol react with $t$-butyl hydroperoxide to give compounds with structures of type **267** also (201).

$$\text{(267)}$$

The reaction of $t$-butyl hydroperoxide with compound **268** gives a quantitative yield of **269** (205).

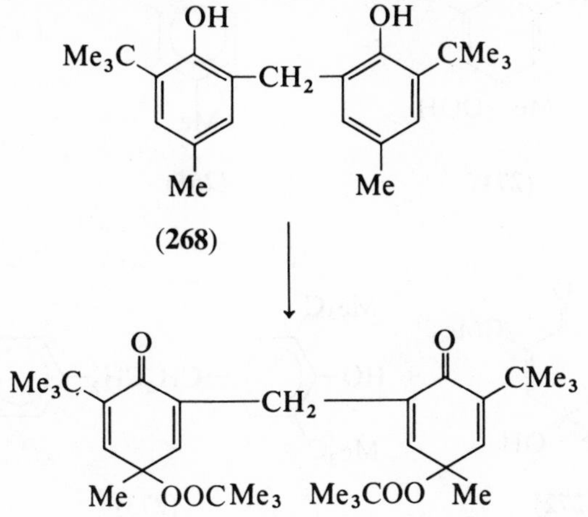

(268)

(269)

The reaction of 2,4,6-trimethylphenol and 2-*t*-butyl-4,6-dimethylphenol with *t*-butyl hydroperoxide yields stilbenequinones (270) also (201).

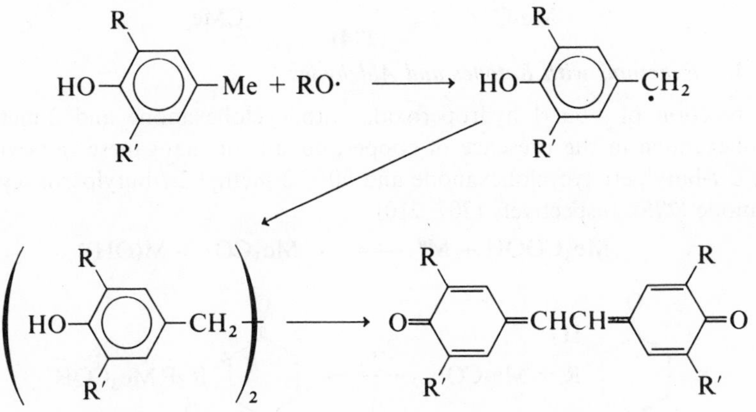

(270)

(R = Me, Me$_3$C, R' = Me)

The reaction of 4-hydroperoxy-4-methyl-2,5-di-*t*-butylcyclohexa-2,6-dien-1-one (271) with cobaltous naphthenate at room temperature in the presence of 2,6-di-*t*-butyl-4-methylphenol (265, R = R' = CMe$_3$, R" = Me) produces a mixture of the alcohol 272, 1,2-bis-3,5-di-*t*-butyl-4-hydroxyphenylethane (273) and 3,3',5,5'-tetra-*t*-butylstilbene-4,4'-quinone (274) (235).

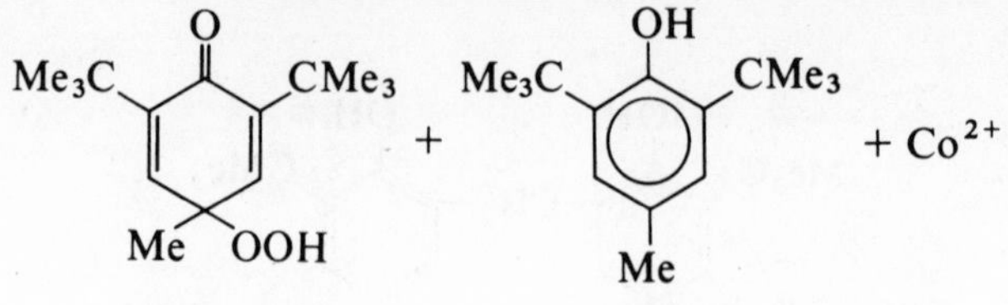

(271)                    (265)

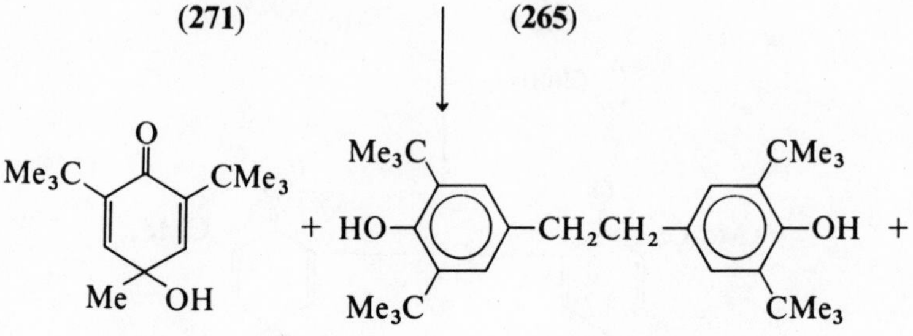

(272)                    (273)

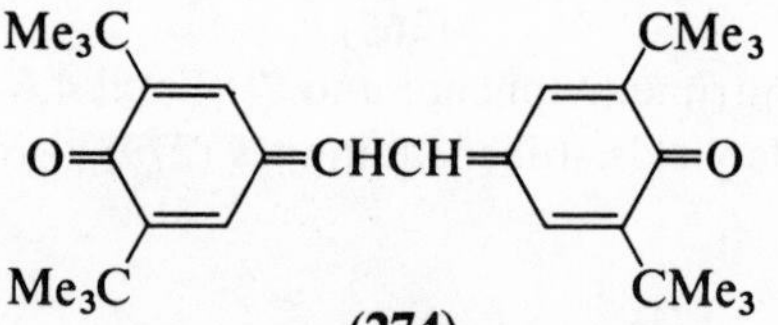

(274)

## 4. Reactions with Ketones and Aldehydes

The reaction of *t*-butyl hydroperoxide with cyclohexanone and 2-methyl-cyclohexanone in the presence of copper, cobalt, or manganese salts yields 20% 2-*t*-butylperoxycyclohexanone and 90% 2-methyl-2-*t*-butylperoxycyclo-hexanone (275), respectively (207, 210).

$$Me_3COOH + M^n \longrightarrow Me_3CO\cdot + M(OH)^n$$

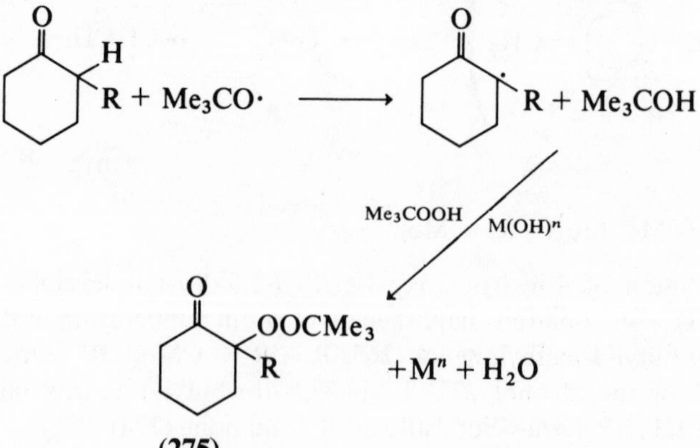

(275)

(R = H, Me, M = Cu, Co, Mn)

The reaction of *t*-butyl hydroperoxide with benzaldehyde in the presence of cuprous bromide yields *t*-butyl peroxybenzoate (209).

$$Me_3COOH + Cu^+ \longrightarrow Me_3CO\cdot + Cu(OH)^+$$

$$\underset{\displaystyle H}{PhCO} + Me_3CO\cdot \longrightarrow PhCO + Me_3COH$$

$$\xrightarrow[\text{CuOH}^+]{\text{Me}_3\text{COOH}} PhC(O)OOCMe_3 + Cu^+ + H_2O$$

### 5. Reactions with Sulfur Compounds

The reaction of *t*-butyl hydroperoxide with thiols in the presence of ferrocene gives disulfides (236).

$$Me_3COOH + Fe^{2+} \longrightarrow Me_3CO\cdot + Fe(OH)^{2+}$$

$$\xrightarrow{\text{RSH}} Me_3COH + RS\cdot \longrightarrow RSSR$$

If the ferrous-ion-catalyzed decomposition of *t*-butyl hydroperoxide is carried out in the presence of butadiene and a thiol, the following products (**276–280**) can be obtained (203, 236):

$$Me_3CO\cdot + CH_2=CHCH=CH_2 + RS\cdot$$

$$\longrightarrow \underset{(\mathbf{276})}{Me_3COC_4H_6SR} + \underset{(\mathbf{277})}{RSC_4H_6SR} + \underset{(\mathbf{278})}{RSC_4H_6C_4H_6SR} +$$

$$\underset{(\mathbf{279})}{RSC_4H_7} + \underset{(\mathbf{280})}{Me_3COC_4H_6C_4H_6OCMe_3}$$

$$(R = n\text{-}C_3H_7, Me_3C, MeCO, Ph)$$

The reaction of anthracene with thioacetic acid in the presence of *t*-butyl hydroperoxide and ferrocene produces 9-(acetylthio)anthracene (**281**, R = MeCO) and two isomers of 9, 10-di(acetylthio)-9, 10-dihydroanthracene (**282**, R = MeCO) (236). Analogous products are obtained with mercaptoacetic [**281**, **282**, R = $CH_2C(O)OH$] and thiobenzoic [**281**, **282**, R = PhC(O)] acids (236).

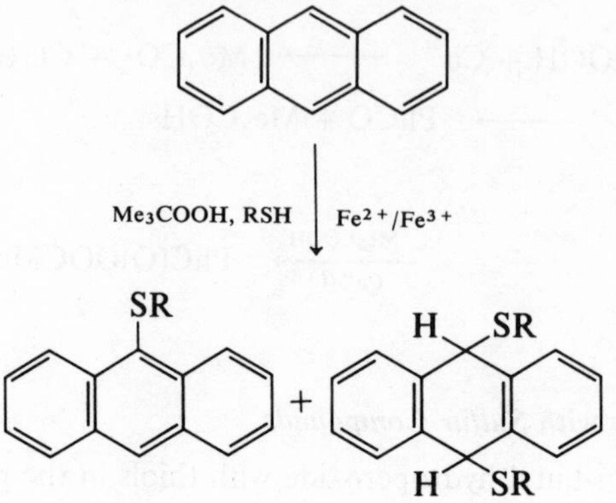

Me$_3$COOH, RSH    Fe$^{2+}$/Fe$^{3+}$

(**281**)

[R = MeC(O), PhC(O), CH$_2$C(O)OH]

Under similar conditions reaction with butane-1-thiol gives 9,10-di-($n$-butyl-thio)anthracene [**283**, R = Me(CH$_2$)$_3$] (236), and reaction with $N$-acetyl-L-cysteine gives $N$-acetyl-$S$-(9-anthryl)-L-cysteine

[**281**, R = CH$_2$CHNC(O)Me (236).
                |
                C(O)OH

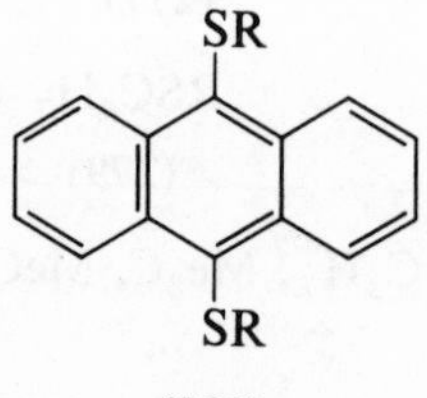

(**283**)

The reaction of 1,2-benzanthracene with thioacetic acid gives a mixture of 10-acetylthio-1,2-benzanthracene (**284**) and 9,10-di-(acetylthio)-9,10-di-hydro-1,2-benzanthracene (**285**) (236).

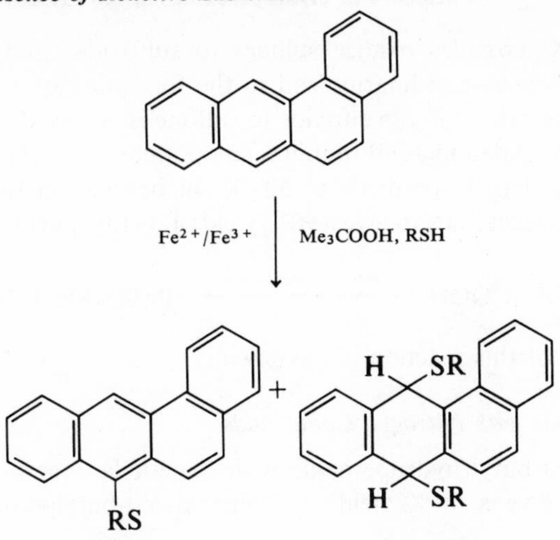

$Fe^{2+}/Fe^{3+}$   |   $Me_3COOH$, RSH

(284)   +   (285)

[R = MeC(O)]

Similar reactions of pyrene, 1,2-dibenzpyrene, and perylene with thioacetic acid, and of 1,2-dibenzpyrene with mercaptoacetic acid, yield **286**, **287**, **288**, and **289**, respectively (236).

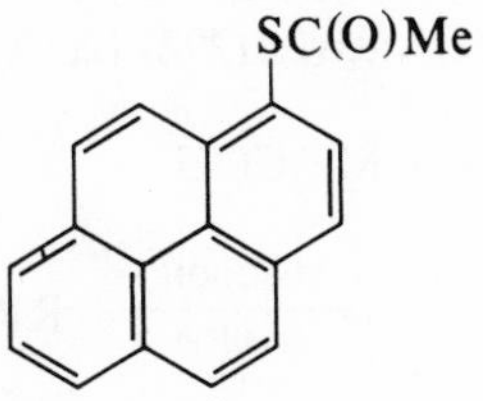

(286)

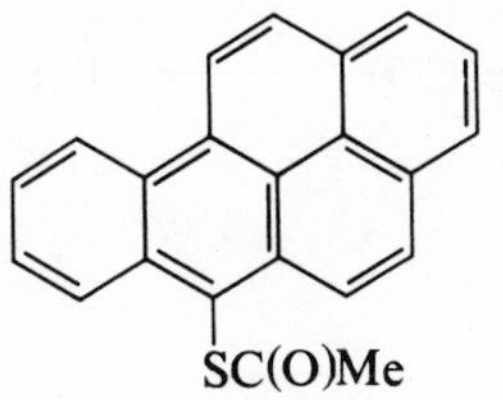

(287)

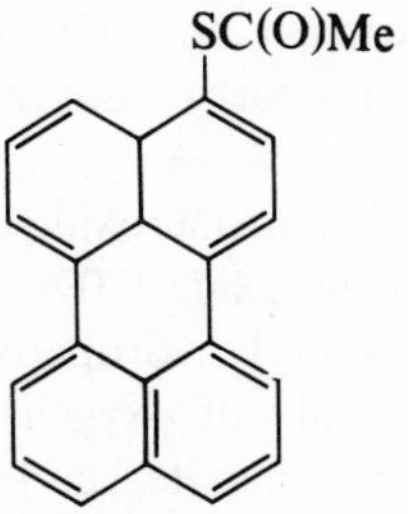

(288)

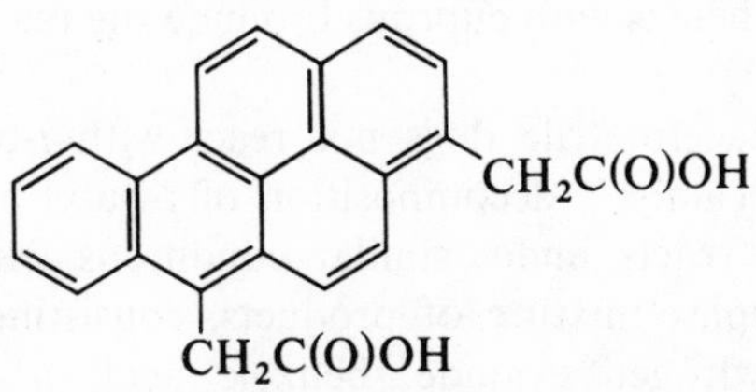

(289)

Organic hydroperoxides oxidize sulfides to sulfoxides, but an excess of hydroperoxide does not ordinarily lead to the formation of a sulfone (237). However, further oxidation of sulfoxide to sulfone is achieved in the presence of vanadium, molybdenum, and titanium compounds (237). Thus thioanisole reacts with $t$-butyl hydroperoxide at 50–70° in benzene in the presence of molybdenyl acetylacetonate to give a 90% yield of methyl phenyl sulfone (237).

$$\text{PhSMe} + 2\text{Me}_3\text{COOH} \xrightarrow{\text{Mo acetylacetonate}} \text{PhSO}_2\text{Me} + 2\text{Me}_3\text{COH}$$

The mechanism of this reaction is unknown.

### 6. Reactions with Nitrogen Compounds

The reaction of $t$-butyl hydroperoxide with dimethylaniline occurs at room temperature to give a 90% yield of $N$-methyl-$N$-$t$-butylperoxyethylaniline (**290**) (207, 210).

$$\text{Ph—NCH}_2\text{OOCMe}_3$$
$$|$$
$$\text{Me}$$
$$\textbf{(290)}$$

The reaction of $t$-butyl hydroperoxide with diphenylacetonitrile, phenylcyclohexylacetonitrile, or methylphenylacetonitrile (**291**) in the presence of cobalt or copper salt yields $t$-butylperoxy derivatives (**292**) (209).

$$\text{RR'CHCN} + \text{Me}_3\text{CO}\cdot \longrightarrow \text{RR'CCN} + \text{Me}_3\text{COH}$$
$$\textbf{(291)}$$

$$\xrightarrow[\text{M(OH)}^n]{\text{Me}_3\text{COOH}} \quad \text{RR'CCN}$$
$$|$$
$$\text{OOCMe}_3$$
$$\textbf{(292)}$$

$$(\text{R} = \text{Ph}; \text{R'} = \text{Me, cyclohexyl, Ph})$$

The reaction with cobaltous 2-ethylhexoate is carried out at room temperature, whereas with cuprous bromide the reaction is carried out in boiling benzene (209).

Dimethylacetonitrile does not react with $t$-butyl hydroperoxide but enhances the catalytic decomposition of $t$-butyl hydroperoxide (209). Phenylacetonitrile reacts under similar conditions with $t$-butyl hydroperoxide to give a complex mixture of products, consisting mainly of oxygen, $t$-butyl alcohol, hydrogen cyanide, benzoic acid, $t$-butyl benzoate, and $t$-butyl peroxybenzoate (209).

## 7.   *Reactions with Heterocyclic Compounds*

The reaction of furan with *t*-butyl hydroperoxide in methanol in the presence of ferrous sulfate and ferric chloride produces 2-methoxy-5-*t*-butoxy-2,5-di-hydrofuran (**293**) (213).

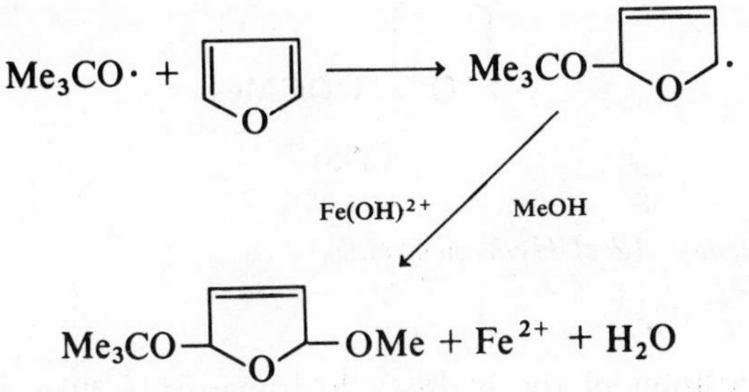

$Me_3CO \cdot \; + \; \square \longrightarrow Me_3CO-\square \cdot$

$$Fe(OH)^{2+} \qquad MeOH$$

$Me_3CO-\square-OMe + Fe^{2+} + H_2O$

**(293)**

or

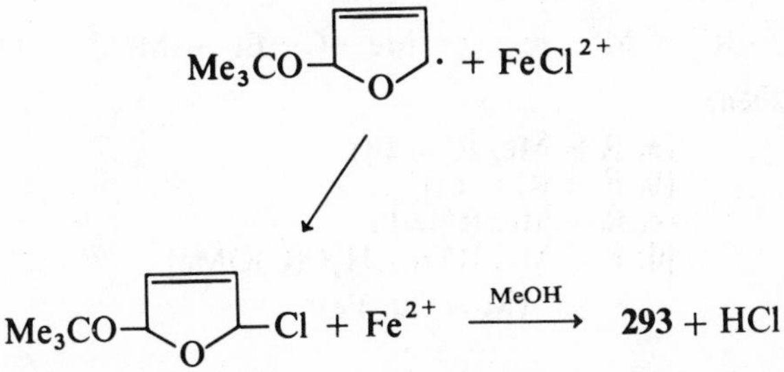

$Me_3CO-\square \cdot + FeCl^{2+}$

$Me_3CO-\square-Cl + Fe^{2+} \xrightarrow{MeOH} 293 + HCl$

If ferric chloride is replaced by sodium azide, the reaction yields 2-azido-5-*t*-butoxy-2,5-dihydrofuran (**294**) (213).

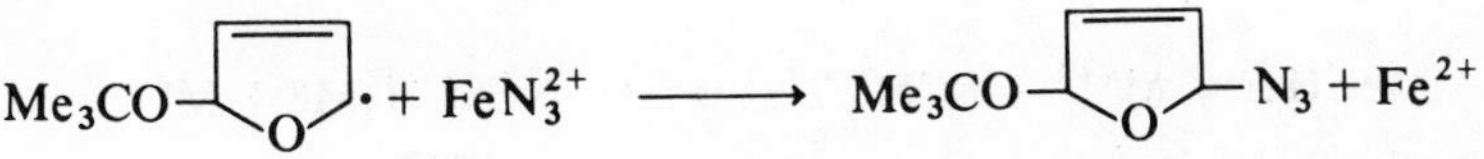

$Me_3CO-\square \cdot + FeN_3^{2+} \longrightarrow Me_3CO-\square-N_3 + Fe^{2+}$

**(294)**

However, the cobalt 2-ethylhexoate–catalyzed decomposition of *t*-butyl hydroperoxide in furan produces 2,5-di-(*t*-butylperoxy)-2,5-dihydrofuran in 55% yield (238). Similarly the decomposition in pyrrole yields presumably 2,5-di-(*t*-butylperoxy)-2,5-dihydropyrrole (238). However, the decomposition of *t*-butyl hydroperoxide in thiophene results only in the formation of tar (238). Under similar conditions the reaction of *t*-butyl hydroperoxide with

tetrahydrofuran gives 2-*t*-butylperoxytetrahydrofuran in 44% yield (238·)
The reaction of *t*-butyl hydroperoxide with dioxan in the presence of cuprous
chloride at 70° produces a 70% yield of *t*-butylperoxydioxan-1,4 (**295**) (210).

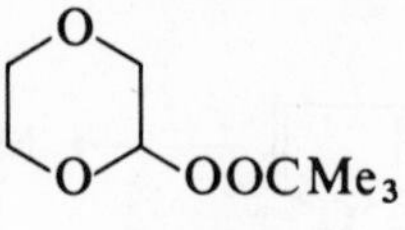

**(295)**

## 8.  1-Hydroxy Alkyl Hydroperoxides

(a)  ACYCLIC

The decomposition of the hydroxy hydroperoxide **296a**, prepared from
methyl ethyl ketone and hydrogen peroxide, with a mixture of either cuprous
and ammonium thiocyanates or ferrous sulfate and ammonium thiocyanate
in an aqueous medium at $-5$ to $5°$ forms ethyl thiocyanate (**297**) (239, 240).

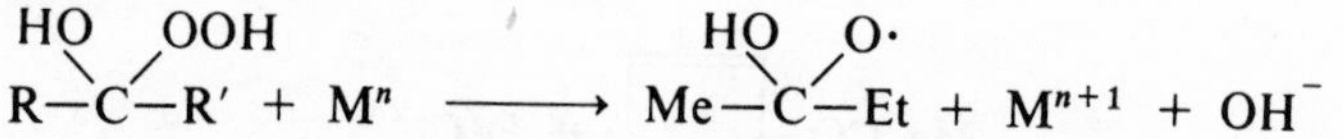

$$\underset{\substack{| \\ (\textbf{296a})}}{\overset{\substack{HO \quad OOH \\ \diagdown \quad \diagup}}{R-C-R'}} + M^n \longrightarrow \underset{}{\overset{\substack{HO \quad O\cdot \\ \diagdown \quad \diagup}}{Me-C-Et}} + M^{n+1} + OH^-$$

(**a**, R = Me, R′ = Et)
(**b**, R = R′ = Et)
(**c**, R = Me, R′ = Pr)
[**d**, R = Me, R′ = CH$_2$C(O)OMe]
(M = Cu, Fe)

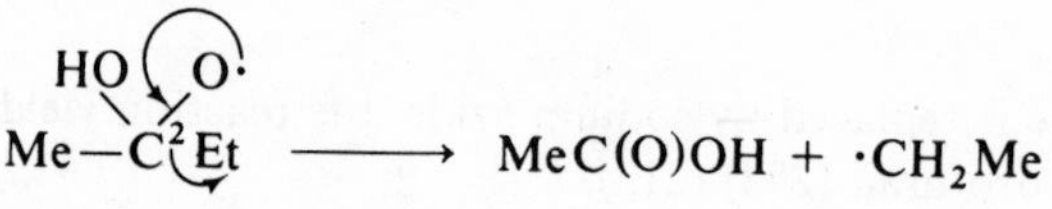

$$\overset{\substack{HO \quad O\cdot \\ \diagdown \quad \diagup}}{Me-C-Et} \longrightarrow MeC(O)OH + \cdot CH_2Me$$

$$MeCH_2\cdot + M^{n+1} + SCN^- \longrightarrow NCSCH_2Me + M^n$$
$$(\textbf{297})$$

Similarly hydroxy hydroperoxides of diethyl ketone (**296b**), methyl propyl
ketone (**296c**), and methyl acetoacetate (**296d**) undergo analogous reactions
to give ethyl thiocyanate (**297**), propyl thiocyanate (**298**), and methyl thio-
cyanoacetate (**299**), respectively (240).

$$\textbf{296b} \xrightarrow[\text{SCN}^-]{\text{M}^n/\text{M}^{n+1}} EtC(O)OH + NCSCH_2Me$$
$$(\textbf{297})$$

$$\textbf{296c} \quad \xrightarrow[\text{SCN}^-]{\text{M}^n/\text{M}^{n+1}} \quad \text{MeC(O)OH} + \text{NCSCH}_2\text{CH}_2\text{Me}$$

$$(\textbf{298})$$

$$\textbf{296d} \quad \xrightarrow[\text{SCN}^-]{\text{M}^n/\text{M}^{n+1}} \quad \text{MeC(O)OH} + \text{NCSCH}_2\text{C(O)OMe}$$

$$(\textbf{299})$$

$$(\text{M} = \text{Cu, Fe})$$

The decomposition of **296b** and **296c** with cuprous chloride or ferrous sulfate in the presence of hydrochloric or hydrobromic acid yields the corresponding ethyl and *n*-propyl halides (241).

$$\textbf{296b, c} \quad \xrightarrow[\text{X}^-]{\text{M}^n/\text{M}^{n+1}} \quad \text{MeC(O)OH} + \text{R}'\text{X}$$

$$(\text{R} = \text{Me}; \text{R}' = \text{Et, } n\text{-Pr}; \text{X} = \text{Cl, Br})$$

Similarly the decomposition of **296a** in the presence of potassium ethyl xanthogenate gives ethyl ethylxanthogenate (**300**) (242).

$$\textbf{296a} \quad \xrightarrow[-\text{SC(S)OEt}]{\text{Fe}^{2+}/\text{Fe}^{3+}} \quad \text{EtSC(O)OEt}$$

$$(\textbf{300})$$

If in addition to a halide ion an olefin is present in the reaction mixture, both
If in addition to a halide ion an olefin is present in the reaction mixture, both the olefin and the halide ion are incorporated into the product (243). Thus the reactions of peroxide **296a** in the presence of acrylonitrile or methyl acrylate yield α-halovaleronitrile (**301a**) and methyl α-halovalerate (**301b**), respectively (243).

$$\overset{\text{HO}\quad\text{OOH}}{\underset{\text{(296)}}{\text{Me}-\text{C}-\text{CH}_2\text{Me}}} + \text{M}^n \quad \xrightarrow{-5\text{to}0°} \quad \text{MeC(O)OH} + \cdot\text{CH}_2\text{Me} + \text{M}^{n+1}$$

$$\text{MeCH}_2\cdot + \text{CH}_2{=}\text{CH}{-}\text{R} \quad \longrightarrow \quad \text{MeCH}_2\text{CH}_2\overset{\cdot}{\text{CH}}{-}\text{R}$$

$$\text{MeCH}_2\text{CH}_2\text{CHR} + \text{M}^{n+1} + \text{X}^- \quad \longrightarrow \quad \underset{\text{(301)}}{\overset{}{\text{MeCH}_2\text{CH}_2\underset{\overset{|}{\text{X}}}{\text{CHR}}}} + \text{M}^n$$

$$(\text{M} = \text{Fe, Cu}; \text{X} = \text{Cl, Br}) \qquad (\textit{a}, \text{R} = \text{CN})$$
$$[\textbf{b}, \text{R} = \text{C(O)OMe}]$$

$$MeCH_2\cdot + CH_2{=}CH{-}R \longrightarrow MeCH_2CH_2\overset{\cdot}{C}H{-}R$$

$$MeCH_2CH_2\overset{\cdot}{C}HR + M^{n+1} + X^- \longrightarrow MeCH_2CH_2\underset{\underset{X}{|}}{C}HR + M^n$$

(301)

(M = Fe, Cu; X = Cl, Br)

(a, R = CN)

[b, R = C(O)OMe]

A similar reaction of various hydroxy hydroperoxides **302** with butadiene in the presence of ferrous sulfate produces dimers in accordance with the following reaction scheme (244, 245):

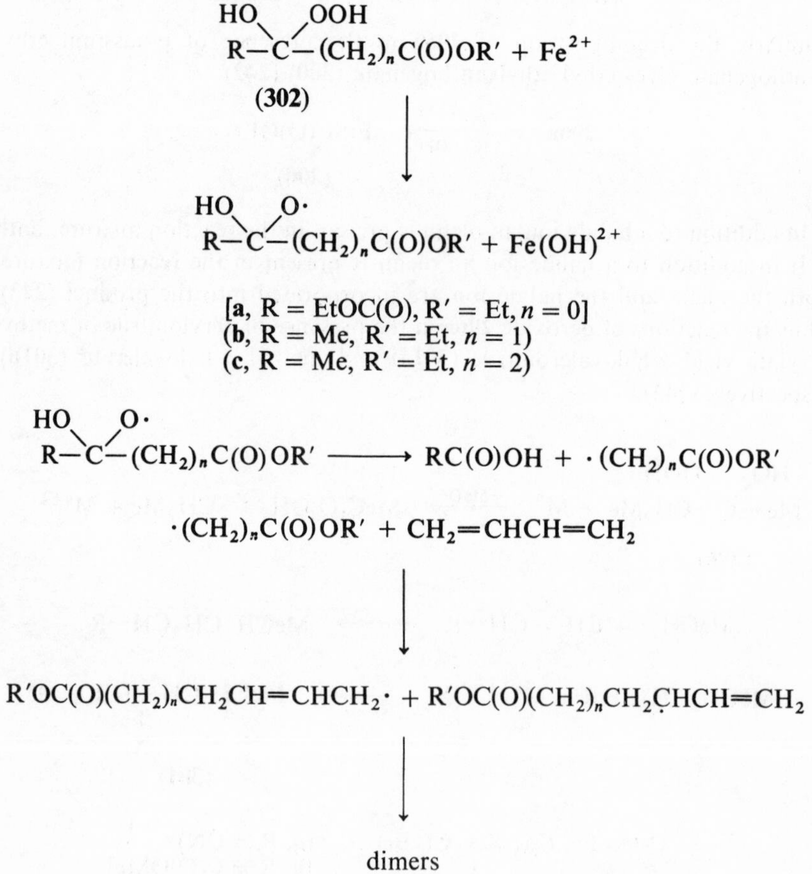

$$R{-}\overset{\overset{HO}{\diagdown}\;\overset{OOH}{\diagup}}{C}{-}(CH_2)_nC(O)OR' + Fe^{2+}$$

(302)

$$R{-}\overset{\overset{HO}{\diagdown}\;\overset{O\cdot}{\diagup}}{C}{-}(CH_2)_nC(O)OR' + Fe(OH)^{2+}$$

[a, R = EtOC(O), R' = Et, n = 0]
(b, R = Me, R' = Et, n = 1)
(c, R = Me, R' = Et, n = 2)

$$R{-}\overset{\overset{HO}{\diagdown}\;\overset{O\cdot}{\diagup}}{C}{-}(CH_2)_nC(O)OR' \longrightarrow RC(O)OH + \cdot(CH_2)_nC(O)OR'$$

$$\cdot(CH_2)_nC(O)OR' + CH_2{=}CHCH{=}CH_2$$

$$R'OC(O)(CH_2)_nCH_2CH{=}CHCH_2\cdot + R'OC(O)(CH_2)_nCH_2\overset{\cdot}{C}HCH{=}CH_2$$

dimers

Thus diethyl oxomalonate is converted to hydroxy hydroperoxide **302a**, and addition of ferrous sulfate gives the ethoxycarbonyl radical, $\cdot C(O)OEt$ ($n = 0$), which reacts with butadiene to give a mixture of esters of dicarboxylic acids. After hydrogenation and hydrolysis of this product, sebacic acid is isolated (244, 245). Similarly the hydroxy hydroperoxide of ethyl acetoacetate (**302b**) is converted to the ethoxycarbonylmethyl radical, $\cdot CH_2C(O)OEt$ ($n = 1$), which reacts with butadiene to give a mixture of esters of dicarboxylic acids, which on hydrogenation and hydrolysis gives a 28–49% yield of dodecanedioic acid (244, 245).

The hydroxy hydroperoxide of ethyl levulinate (**302c**) is converted to the 2-ethoxycarbonylethyl radical, $\cdot CH_2CH_2C(O)OEt$ ($n = 2$), which on reaction with butadiene and subsequent hydrogenation and hydrolysis of the product gives 56% of isomeric $C_{14}$ acids, from which tetradecanedioic acid can be isolated (244, 245).

When this procedure is carried out with acetonylacetone, the acetylethyl radical, $MeC(O)CH_2CH_2\cdot$ is formed (244, 245). The reaction of this radical with butadiene and subsequent hydrogenation yields a mixture of isomeric $C_{16}$ diketones (244, 245). A 30% yield of hexadecan-2,15-dione is obtained from this mixture (244, 245). Under similar conditions the reaction of ethyl 4,4,4-trifluoroacetoacetate results in only the ethoxycarbonylmethyl radical and no trifluoromethyl radical. Dodecanedioic acid is obtained from this reaction (244, 245).

## (b) CYCLIC

Cyclic ketones react readily with hydrogen peroxide to give hydroxy hydroperoxides. Although the structures of these peroxides have not been fully elucidated, their reactions can be described by using formula **303** (84–89, 92, 93).

$$\text{HO} \quad \text{OOH}$$

**(303)**

(**a**, $n = 1$)
(**b**, $n = 2$)
(**c**, $n = 3$)
(**d**, $n = 4$)

The reaction of cyclopentanone peroxide (**303a**) with ferrous ions produces a mixture of valeric (**304**) and sebacic (**305**) acids (89, 90, 91).

$$\textbf{303a} \xrightarrow{\text{Fe}^{2+}/\text{Fe}^{3+}} \quad \boxed{} \quad \longrightarrow \text{HOC(O)(CH}_2)_4\cdot$$

$$\text{Me(CH}_2)_3\text{C(O)OH} + \text{HOC(O)(CH}_2)_8\text{C(O)OH}$$

$$\textbf{(304)} \qquad\qquad\qquad \textbf{(305)}$$

However, the decomposition of **303a** with cuprous chloride or ferrous sulfate in the presence of either bromine, hydrochloric or hydrobromic acid, potassium iodide, or bromotrichloromethane gives the corresponding $\delta$-halovaleric acid (**306**) (241, 246–248). Similarly reaction in the presence of potassium cyanide, sodium azide, ammonium thiocyanate or thiosulfate gives $\delta$-cyano-, $\delta$-azido-, $\delta$-thiocyanato-, and $\delta$-thiosulfatovaleric acids (**306**), respectively (239, 240, 249). In the presence of potassium ethyl xanthogenate the corresponding valeric acid derivative [**306**, X = SC(S)OEt] is obtained,

$$\textbf{303a} + \text{M}^n \longrightarrow \text{HOC(O)(CH}_2)_4\cdot + \text{M}^{n+1}$$

$$\xrightarrow{\text{X}^-} \text{HOC(O)(CH}_2)_4\text{X} + \text{M}^n$$

$$\textbf{(306)}$$

$$[\text{X = Cl, Br, I, CN, N}_3, \text{ SCN, S}_2\text{O}_3\text{Na, SC(S)OEt; M = Cu, Fe}]$$

which on hydrolysis and oxidation gives $\delta,\delta'$-dithiodivaleric acid (**307**) (242).

$$\text{HOC(O)(CH}_2)_4\text{SC(S)OEt} \longrightarrow \text{HOC(O)(CH}_2)_4\text{S}{-}\text{S(CH}_2)_4\text{C(O)OH}$$

$$[\textbf{306}, \text{X = SC(S)OEt}] \qquad\qquad\qquad \textbf{(307)}$$

The same $\delta,\delta'$-dithiodivaleric acid (**307**) is obtained from $\delta$-thiocyanato and $\delta$-thiosulfato derivatives (**306**, X = SCN, S$_2$O$_3$Na) (40).

$$\textbf{306}, \text{X = SCN} \xrightarrow[\text{2. oxidation}]{\text{1. NaOH}}$$

$$\textbf{306}, \text{X = S}_2\text{O}_3\text{Na} \xrightarrow[\text{2. oxidation}]{\text{1. HCl}} \quad \textbf{307}$$

The cyano derivative (**306**, X = CN) can be converted to adipic (**308**) and $\varepsilon$-aminocaproic (**309**) acids or to caprolactam (**310**) (249).

$$HOC(O)(CH_2)_4CN \xrightarrow{\quad} \begin{cases} \xrightarrow{H_2O} & HOC(O)(CH_2)_4C(O)OH \quad \textbf{(308)} \\ \xrightarrow[\text{Co}]{\text{Raney}} & HOC(O)(CH_2)_5NH_2 \quad \textbf{(309)} \\ \xrightarrow{\quad} & \underline{C(O)(CH_2)_5NH} \quad \textbf{(310)} \end{cases}$$

The decomposition of **303a** in the presence of ferrous ions and carbon monoxide under 100-atm pressure at 20° gives adipic acid **(308)** (250–252).

$$\textbf{303a} + Fe^{2+} \longrightarrow \underset{\text{HO}\diagup\overset{\text{O·}}{\diagdown}}{\bigcirc} + Fe(OH)^{2+}$$

$$\downarrow$$

$$HOC(O)(CH_2)_3CH_2\cdot$$

$$\downarrow CO$$

$$HOC(O)(CH_2)_3CH_2CO\cdot \xrightarrow[\text{or 303a}]{Fe(OH)^{2+}} \textbf{308}$$

The reaction of **303a** with ferrous sulfate in the presence of sulfur dioxide yields an α-carboxy-ω-sulfonic acid **(311)** (242).

$$\textbf{303a} + Fe^{2+} \longrightarrow HOC(O)(CH_2)_4\cdot + Fe(OH)^{2+}$$

$$HOC(O)(CH_2)_4\cdot + SO_2 \longrightarrow HOC(O)(CH_2)_4\overset{\cdot}{S}O_2$$

$$\xrightarrow[H_2O]{Fe(OH)^{2+}} HOC(O)(CH_2)_4SO_3H \quad \textbf{(311)}$$

It is important that the sulfur dioxide solutions not be too concentrated to avoid decomposition of **303a** by the sulfurous acid to cyclopentanone (242).

The decomposition of **303a** by cuprous or ferrous salts in the presence of nitric oxide gives a mixture of oxime **312** and nitrosohydroxylamine **(313)**, which on hydrogenation at 100 atm and 60–100° produces δ-aminovaleric acid **(314)** (253, 254).

$$\textbf{303a} + M^n \longrightarrow HOC(O)(CH_2)_3CH_2\cdot + M^{n+1}$$

$$(M = Cu, Fe) \qquad NO$$

$$HOC(O)(CH_2)_3CH_2NO$$

$$HOC(O)(CH_2)_3CH=NOH \qquad\qquad HOC(O)(CH_2)_3CH_2NOH$$
$$\qquad\qquad\qquad\qquad\qquad\qquad\qquad\qquad\qquad\qquad\qquad\qquad NO$$

$$\textbf{(312)} \qquad\qquad\qquad\qquad\qquad\qquad\qquad\qquad \textbf{(313)}$$

$$Raney\ Ni \quad H_2$$

$$H_2N(CH_2)_4C(O)OH$$
$$\textbf{(314)}$$

The decomposition of **303a** in the presence of a mixture of cuprous chloride, aqueous hydrochloric acid, and acrylonitrile at $-10$ to $-5°$ yields compound **315**, which on hydrolysis in methanolic solution gives the methyl ester of α-chlorosuberic acid (**316**) in $18\%$ yield (243). Hydrogenolysis of **316** over Raney nickel yields suberic acid (**317**) (243).

$$\textbf{303a} + Cu^+ \longrightarrow HOC(O)(CH_2)_4\cdot + Cu(OH)^+$$

$$\xrightarrow[Cl^-]{CH_2=CHCN} HOC(O)(CH_2)_5CHCN$$
$$\qquad\qquad\qquad\qquad\qquad\qquad\qquad\qquad Cl$$
$$\qquad\qquad\qquad\qquad\qquad\qquad\qquad\quad \textbf{(315)}$$

$$\xrightarrow[MeOH]{H_3O^+} MeOC(O)(CH_2)_5CH(Cl)C(O)OMe$$
$$\qquad\qquad\qquad \textbf{(316)}$$

$$\xrightarrow[105\ atm,\ 115-120°]{H_2,\ Raney\ Ni} HOC(O)(CH_2)_6C(O)OH$$
$$\qquad\qquad\qquad\qquad\qquad\qquad\qquad\quad \textbf{(317)}$$

In addition, the methyl ester of δ-chlorovaleric acid (**306**, X = Cl) is obtained in $30\%$ yield (243).

$$HOC(O)(CH_2)_4\cdot + Fe(OH)^{2+} + Cl^- \xrightarrow{MeOH} \textbf{306}, X = Cl$$

The reaction of **303a** with ferrous sulfate in the presence of a mixture of sulfuric acid, methanol, and butadiene produces dimethyl-7,11-octadeca-diene-1,18-dioate, plus small quantities of 7-vinyl-9-hexadecene-1,16-dioate (255–257). The reaction probably involves the methyl ether **318** of **303a**.

The decomposition of cyclohexanone peroxide (**303b**) in aqueous sulfuric acid with ferrous sulfate produces a 68% yield of 1,12-dodecanedioic acid (**319**) (85–93). In addition are formed cyclohexanone, hexanoic acid (**320**), and ε-hydroxycaproic acid (**321**) (35, 88, 258). In the presence of a mixture of ferrous and cupric sulfates δ-hexenoic acid (**322**) is formed mainly (35, 55).

In the presence of hydrogen sulfide the ferrous-ion-catalyzed decomposition of **303b** yields mainly **320** (259).

$$HOC(O)(CH_2)_4CH_2 \cdot + H_2S + M^{n+1} \longrightarrow 320 + M^n + S$$

Similarly the ferrous sulfate–catalyzed reaction of **303b** with mercaptans and thiols gives, among other products, **320** (248). The reaction of **303b** with potassium ethyl xanthogenate or ammonium dithiocarbamate in the presence of ferrous ions gives **323** and **324**, respectively (242), which on hydrolysis and oxidation form $\varepsilon,\varepsilon'$-dithiodicaproic acid (**325**) (242).

$$303b + Fe^{2+} \longrightarrow HOC(O)(CH_2)_5 \cdot + Fe(OH)^{2+}$$

| $^-SC(S)OEt$ | | $^-SC(S)NH_2$ |

$$HOC(O)(CH_2)_5SC(S)OEt \qquad\qquad HOC(O)(CH_2)_5SC(S)NH_2$$

$$(323) \qquad\qquad\qquad\qquad (324)$$

$$\xrightarrow[\text{2 oxidation}]{\text{1 H}_2\text{O}} HOC(O)(CH_2)_5SS(CH_2)_5C(O)OH$$

$$(325)$$

The reaction of **303b** with sulfur dioxide yields cyclohexanone. However, in the presence of ferrous sulfate the $\alpha$-carboxylic-$\omega$-sulfonic acid (**326**) is produced (242). It is important that dilute solutions of sulfur dioxide be used to avoid the formation of cyclohexanone (239, 242).

$$303b + Fe^{2+} \longrightarrow HOC(O)(CH_2)_4CH_2 \cdot + Fe(OH)^{2+}$$

$$HOC(O)(CH_2)_4CH_2 \cdot + SO_2 \longrightarrow HOC(O)(CH_2)_4CH_2\overset{\bullet}{S}O_2$$

$$\xrightarrow[\text{Fe(OH)}^{2+}]{\text{H}_2\text{O}} HOC(O)(CH_2)_5SO_3H + Fe^{2+}$$

$$(326)$$

The reaction of **303b** with a mixture of cupric sulfate and sodium nitrite in methanol at $-10°$, followed by addition of a saturated ferrous sulfate solution, gives $\varepsilon$-nitrocaproic acid (**327**) and methyl hexenoate (**328**) (216).

$$303b + Fe^{2+} \longrightarrow HOC(O)(CH_2)_4CH_2 \cdot + Fe(OH)^{2+}$$

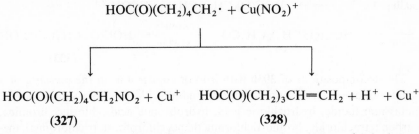

$$HOC(O)(CH_2)_4CH_2\cdot + Cu(NO_2)^+$$

$$HOC(O)(CH_2)_4CH_2NO_2 + Cu^+ \qquad HOC(O)(CH_2)_3CH{=}CH_2 + H^+ + Cu^+$$

$$\textbf{(327)} \qquad\qquad\qquad\qquad \textbf{(328)}$$

If sodium nitrite is replaced by cupric nitrate, the nitrate of ε-hydroxy-caproic acid (329) is obtained (216).

$$HOC(O)(CH_2)_4CH_2\cdot + Cu(NO_3)^+$$

$$\longrightarrow HOC(O)(CH_2)_4CH_2ONO_2 + Cu^+$$

$$\textbf{(329)}$$

The decomposition of **303b** by cuprous or ferrous salts in the presence of nitric oxide at 0–5° produces the oxime of the hemialdehyde of adipic acid (330) and the ε-nitrosohydroxylamine of caproic acid (331). Subsequent reduction of the mixture with Raney nickel produces ε-aminocaproic acid (332) (253).

$$\textbf{303b} + M^n \longrightarrow HOC(O)(CH_2)_4CH_2\cdot + M^{n+1}$$

$$\overset{NO}{\diagdown}$$

$$HOC(O)(CH_2)_4CH_2NO$$

$$HOC(O)(CH_2)_4CH{=}NOH \qquad\qquad HOC(O)(CH_2)_4CH_2NOH$$

$$\qquad\qquad\qquad\qquad\qquad\qquad\qquad\qquad | $$
$$\qquad\qquad\qquad\qquad\qquad\qquad\qquad\qquad NO$$

$$\textbf{(330)} \qquad\qquad\qquad\qquad\qquad\qquad \textbf{(331)}$$

$$\text{Raney Ni} \mid H_2$$

$$HOC(O)(CH_2)_5NH_2$$

$$\textbf{(332)}$$

$$(M{=}Cu,Fe)$$

The decomposition of **303b** in the presence of aqueous ferrous sulfate at 25° under 200 atm of carbon monoxide gives mainly pimelic acid (333) plus some caproic and ε-hydroxycaproic acids (252). At ordinary pressure less satisfactory results are obtained (252, 260, 261).

$$\textbf{303b} + Fe^{2+} \longrightarrow HOC(O)(CH_2)_4CH_2\cdot + Fe(OH)^{2+}$$

$$\xrightarrow[200\ atm]{CO} HOC(O)(CH_2)_4CH_2CO\cdot \xrightarrow[or\ \textbf{303b}]{Fe(OH)^{2+}} HOC(O)(CH_2)_5C(O)OH$$

$$(333)$$

The decompositions of **303b** with iron or copper ions in the presence of a halide donor—such as sodium chloride, indium chloride, potassium bromide, potassium iodide, hydrochloric acid, hydrobromic acid, chlorine, bromine, carbon tetrachloride, bromotrichloromethane, chloroform, trichloromethane-sulfenyl chloride, and o-nitrobenzenesulfenyl chloride—give the corresponding ε-halocaproic acids (**334**) (35, 95, 96, 214, 239, 247, 248, 252, 258, 262–266).

$$\textbf{303b} + M^n \longrightarrow HOC(O)(CH_2)_5\cdot + M^{n+1}$$

$$\xrightarrow{X^-} HOC(O)(CH_2)_5X + M^n$$

$$(334)$$

$$(M = Cu,\ Fe;\ X = Cl,\ Br,\ I)$$

Similarly this reaction in the presence of either hydrogen cyanide, potassium cyanide, sodium azide, ammonium thiocyanate, sodium thiosulfate, methyl mercaptan, dodecyl mercaptan, octadecyl mercaptan, benzyl mercaptan, or benzenethiol yields the corresponding ε-derivatives of caproic acid

$$[\textbf{334},\ X = CN,\ SCN,\ N_3,\ NaS_2O_3,\ MeS,\ Me(CH_2)_{11}S,$$
$$Me(CH_2)_{17}S,\ PhCH_2S,\ PhS]\ (214,\ 239,\ 240,\ 247-249,\ 262,\ 266,\ 267).$$

If **303b** is decomposed by copper ions in the presence of hydrogen cyanide in methanol, methyl hydrogen pimelate [**334**, X = C(O)Me] is obtained as the major acidic product (267).

The azido derivative (**334**, X = N₃) can be reduced to ε-aminocaproic acid (**309**) with Raney nickel (249).

$$HOC(O)(CH_2)_5N_3 \xrightarrow[Raney\ Ni]{H_2} HOC(O)(CH_2)_5NH_2$$

$$(309)$$

The thiocyanate and thiosulfate derivatives (**334**, X = SCN and NaS₂O₃, respectively) are readily converted to ε,ε'-dithiodicaproic acid (**325**) (239, 240).

$$\textbf{334},\ X = SCN \xrightarrow[2\ Oxidation]{1\ NaOH}$$

$$\longrightarrow HOC(O)(CH_2)_5S-S(CH_2)_5C(O)OH$$

$$\textbf{334},\ X = NaS_2O_3 \xrightarrow[2\ Oxidation]{1\ HCl} \qquad (325)$$

The decomposition of **303b** with a mixture of ferrous and cupric ions in the presence of acrylonitrile gives radical **335**, which undergoes two competitive processes: reduction to a heminitrile of azelaic acid (**336**) and oxidation to a hemiester of suberic acid (**337**) (268, 269).

$$303b + Fe^{2+} \longrightarrow HOC(O)(CH_2)_4CH_2\cdot + Fe(OH)^{2+}$$

$$HOC(O)(CH_2)_4CH_2\cdot + CH_2{=}CHCN \longrightarrow HOC(O)(CH_2)_6\overset{\cdot}{C}HCN$$

(335)

$$335 + H_2O + Fe^{2+} \longrightarrow HOC(O)(CH_2)_7CN + Fe(OH)^{2+}$$

(336)

$$335 \xrightarrow{O_2} \underset{\underset{O{-}O\cdot}{|}}{HOC(O)(CH_2)_6CHCN} \xrightarrow{Fe^{2+}} \underset{\overset{\|}{O}}{HOC(O)(CH_2)_6CCN}$$

$$\xrightarrow[+MeOH]{-HCN} HOC(O)(CH_2)_6C(O)OMe$$

(337)

The decomposition of **303b** with cuprous or ferrous ions in the presence of a mixture of acrylonitrile and chloride ions in methanol gives, after hydrolysis, a mixture of methyl esters of ε-chlorocaproic (**334**, X = Cl) and α-chloroazelaic (**338**) acids, which can be readily converted to caprolactam (**310**) and azelaic acid (**339**), respectively (243).

**303b** + Fe²⁺

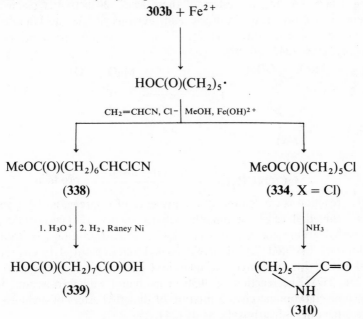

Under similar conditions the reaction of **303b** with $n$-butyl acrylate yields a mixture of $\varepsilon$-chlorocaproic (**334**, X = Cl) and $\alpha$-chloroazelaic (**338**) acid esters. Hydrolysis and subsequent hydrogenation of **338** with a Raney nickel–chromium catalyst yields azelaic acid (**339**) (243).

If the reaction of **303b** with ferrous ions is carried out in the presence of conjugated dienes—such as butadiene, isoprene, or chloroprene—long-chain unsaturated dicarboxylic acids are formed. Thus the reaction of **303b** with butadiene produces a high yield of a $C_{20}$ acid mixture (**340** and **341**), composed of about 80% of unbranched 8,12-eicosadien-1,20-dioic acid (**340**) (244, 255–257, 270).

$$\mathbf{303b} + Fe^{2+} \longrightarrow HOC(O)(CH_2)_5 \cdot$$

$$\xrightarrow{\;CH_2=CHCH=CH_2\;} HOC(O)(CH_2)_5CH_2CH=CHCH_2 \cdot$$

$$\longleftarrow HOC(O)(CH_2)_5CH_2CHCH=CH_2$$

$$\xrightarrow{\;dimerization\;} HOC(O)(CH_2)_6CH=CHCH_2CH_2CH=CH(CH_2)_6C(O)OH +$$
$$\mathbf{(340)}$$

$$HOC(O)(CH_2)_6CHCH_2CH=CH(CH_2)_6C(O)OH$$
$$|$$
$$CH=CH_2$$
$$\mathbf{(341)}$$

To prepare the dimethyl esters of these acids, **303b** is first reacted with methanol in the presence of sulfuric acid. Ferrous sulfate and butadiene are then added to the mixture (244, 255–257). The reaction probably involves the methyl ether (**342**) of **303b**.

$$MeC(O)(CH_2)_5 \cdot \xrightarrow{\;CH_2=CH-CH=CH_2\;} \text{products}$$

If this reaction is carried out in the presence of isoprene, an 80% yield of isomeric dimethyl esters of dimethyloctadecadiene-1,18-dicarboxylic acids, with methyl groups located at 7,11-, 7,12-, or 8,12-positions, is produced. In addition, 7,9- and 7,10-dimethyl-7-vinyl-9-hexadecene-1,16-dicarboxylic and 10-methyl-7-isopropenyl-9-hexadecene-1,16-dicarboxylic acids are formed (244, 256, 257). The reaction of **303b** in methanol with ferrous ions in the presence of chloroprene gives a mixture of dimethyl esters of dichloro-7,11-octadecadiene-1,18-dicarboxylic acids (244, 256, 257).

If dimethyl maleate is used instead of a diene, the reaction produces the dimer **343** and a mixture of esters **344** and **345**. Hydrolysis and hydrogenation of **344** and **345** give 3-carboxyazelaic acid (**346**) (271).

$$303b + Fe^{2+} \longrightarrow HOC(O)(CH_2)_4CH_2 \cdot + Fe(OH)^{2+}$$

$$HOC(O)(CH_2)_4CH_2 \cdot + \underset{\underset{C(O)OMe}{|}}{CH=CH} \atop C(O)OMe$$

HOC(O)(CH$_2$)$_5$CH[C(O)OMe]CHC(O)OMe

HOC(O)(CH$_2$)$_5$CH[C(O)OMe]CHC(O)OMe
|
HOC(O)(CH$_2$)$_5$CH[C(O)OMe]CHC(O)OMe
(**343**)

HOC(O)(CH$_2$)$_5$CH[C(O)OMe]CH$_2$C(O)OMe
(**344**)

1. H$_3$O$^+$
2. H$_2$

HOC(O)(CH$_2$)$_5$C[C(O)OMe]=CHC(O)OMe
(**345**)

HOC(O)(CH$_2$)$_5$CH[C(O)OH]CH$_2$C(O)OH
(**346**)

Similarly reaction with diethyl fumarate yields the ethyl ester of 7,8,9,10-tetracarbethoxyhexadecanedioic acid, which can be converted to the corresponding acid (244). The reaction of **303b** with crotonic acid in the presence of ferrous sulfate produces the dimeric 7,8-dimethyltetradecane-1,6,9,14-tetracarboxylic acid (**347**) (272).

$$303b + Fe^{2+} \longrightarrow HOC(O)(CH_2)_5 \cdot + Fe(OH)^{2+}$$

$$HOC(O)(CH_2)_5 \cdot + MeCH=CHC(O)OH$$

$$\longrightarrow \underset{\underset{C(O)OH}{|}}{HOC(O)(CH_2)_5CHCHMe}$$

$$\longrightarrow HOC(O)(CH_2)_5CH[C(O)OH]CHMe$$
|
HOC(O)(CH$_2$)$_5$CH[C(O)OH]CHMe
(**347**)

Analogous reactions of **303b** with 2-hexenoic, $\beta$-ethylacrylic, $\beta,\beta$-dimethyl-acrylic, and angelic acids yield the corresponding adducts (272).

Substituted cyclohexanone peroxides undergo ring-opening reactions similar to those of unsubstituted cyclohexanone peroxides. Thus the reaction of 4-methylcyclohexanone peroxide (**348**) with aqueous ferrous sulfate in the presence of sulfuric acid gives a mixture of 4-methylhexanoic acid (**349**) and 4,9-dimethyldodecane-1,12-dioic acid (**350**) (89).

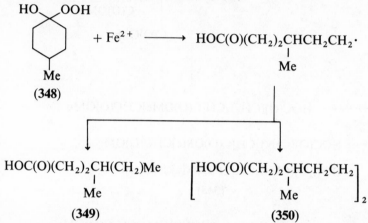

$$HO \underset{\underset{\text{Me}}{\big|}}{\overset{OOH}{\bigcirc}} + Fe^{2+} \longrightarrow HOC(O)(CH_2)_2\underset{\underset{Me}{\big|}}{CH}CH_2CH_2 \cdot$$

(348)

$$HOC(O)(CH_2)_2\underset{\underset{Me}{\big|}}{CH}(CH_2)Me \qquad \left[ HOC(O)(CH_2)_2\underset{\underset{Me}{\big|}}{CH}CH_2CH_2 \right]_2$$

(349)                                        (350)

This decomposition by either ferrous sulfate or cuprous chloride in the presence of either hydrochloric acid, hydrobromic acid, sodium azide, or potassium cyanide at $-5$ to $0°$ gives the corresponding $\varepsilon$-substituted-$\gamma$-methylcaproic acids (**351**) (241, 249). The decomposition of **348** in the presence of nitric oxide gives a product that on hydrogenation over Raney nickel yields 4-methyl-6-aminocaproic acid (**351**, X = NH$_2$) (253, 254).

$$348 + M^n \longrightarrow HOC(O)(CH_2)_2\underset{\underset{Me}{\big|}}{CH}CH_2CH_2 \cdot \ + M^{n+1}$$

$$\overset{X^-}{\longrightarrow} HOC(O)(CH_2)_2\underset{\underset{Me}{\big|}}{CH}CH_2CH_2X$$

(351)

(M = Cu, Fe; X = Cl, Br, CN, N$_3$, NH$_2$)

The decomposition of **348** by ferrous sulfate in the presence of sulfur dioxide gives the $\alpha$-carboxylic-$\omega$-sulfonic acid (**351**, X = SO$_3$H) (242).

The reaction of 4-methylcyclohexanone peroxide (**348**) with cinnamic acid in the presence of ferrous sulfate produces 3,12-dimethyl-7,8-diphenyltetra-decane-1,6,9,14-tetracarboxylic acid (**352**) (272).

$$348 + Fe^{2+} \longrightarrow HOC(O)(CH_2)_2CH(Me)(CH_2)_2 \cdot$$

$$HOC(O)(CH_2)_2CH(Me)(CH_2)_2 \cdot + PhCH{=}CHC(O)OH \longrightarrow$$

$$HOC(O)(CH_2)_2CH(Me)(CH_2)_2CH[C(O)OH]CHPh \longrightarrow \text{dimer}$$

$$(352)$$

The decomposition of a methanolic solution of **348** with ferrous sulfate in the presence of butadiene gives a mixture of methyl esters of 3,16-dimethyl-7,11-octadecadiene-1,18-dicarboxylic acid and 3,14-dimethyl-7-vinyl-9-hexadecene-1,16-dicarboxylic acid (256, 257).

In the case of 2-methyl-substituted cyclohexanone peroxide (**353**) two types of radical, **354** and **355**, might be expected. The path leading to the more stable radical **355** is preferred, since in the presence of halide, cyanide, and thiocyanate ions the corresponding $\varepsilon$-substituted heptanoic acids (**356**) are obtained (240, 241, 249–252).

$$HOC(O)CH(CH_2)_3CH_2 \cdot + M^{n+1} \qquad\qquad HOC(O)(CH_2)_4\overset{.}{C}HMe + M^{n+1}$$
$$\underset{\displaystyle Me}{|}$$

$$(354) \qquad\qquad\qquad\qquad\qquad (355)$$

$$HOC(O)(CH_2)_4CHMeX$$

$$(X = Cl, Br, CN, SCN) \qquad (356)$$

The reaction of **353** with ferrous ions and carbon monoxide under pressure gives 2-methylpimelic acid (**357**) (251, 252).

$$353 + Fe^{2+} \longrightarrow HOC(O)(CH_2)_4\overset{.}{C}HMe + Fe(OH)^{2+}$$

$$\xrightarrow{CO} HOC(O)(CH_2)_4CHMe \xrightarrow[Fe(OH)^{2+}]{353 \text{ or}} HOC(O)(CH_2)_4CHMe$$
$$\underset{\displaystyle C(O)}{|} \qquad\qquad\qquad\qquad\qquad \underset{\displaystyle C(O)OH}{|}$$

$$(357)$$

Similarly decalone hydroperoxide (358) could cleave by two routes to give radicals 359 and 360 (241).

The path leading to the more stable radical 359 is preferred, since the decomposition of 358 in the presence of halide ions results in the formation of $\gamma$-(2-halo)cyclohexylbutyric acids (361), which on hydrogenation over Raney nickel–chromium catalyst yield $\gamma$-cyclohexylbutyric acid (362) (241).

The decomposition of 2-halocyclohexanone peroxide (363) with ferrous sulfate at 0–5° in the presence of HCl or HBr produces an 85–90% yield of $\alpha,\varepsilon$-dihalocaproic acid (364) (273).

$$\underset{(363)}{\text{HO}\quad\text{OOH}} + Fe^{2+} \longrightarrow HOC(O)CHX(CH_2)_4{\cdot} + Fe^{3+}$$

$$\underset{(364)}{HOC(O)CHX(CH_2)_4Y + Fe^{2+}}$$

$$(X = Br,\ Cl,\ Y = Br,\ Cl)$$

The reaction of cycloheptanone peroxide (**303c**) with *n*-butyl mercaptan in the presence of aqueous cobaltous acetate yields butylmercaptoheptanoic acid (**365**) (266).

$$\underset{(303c)}{\text{HO}\quad\text{OOH}} + Co^{2+} \longrightarrow HOC(O)(CH_2)_6{\cdot} + Co^{3+}$$

$$\underset{(365)}{Me(CH_2)_3S(CH_2)_6C(O)OH}$$

An analogous reaction of cyclodecanone peroxide (**303d**) with methyl mercaptan in the presence of cuprous ions yields methylmercaptodecanoic acid (266).

## 9. 1-Alkoxy Cycloalkyl Hydroperoxides

In the preceding section several members (e.g., **318**, **342**) of this class of compounds were mentioned. In this section a more detailed description is presented.

The reaction of methyl and ethyl cyclopentyl ethers (**318**) with ferrous or cuprous salts in the presence of halide ions yields methyl and ethyl esters of δ-halovaleric acid (**306**, X = Cl, Br) (243, 274).

$$\underset{(318)}{\text{RO}\quad\text{OOH}} + M^n \longrightarrow ROC(O)(CH_2)_3CH_2{\cdot} + M^{n+1}$$

$$\underset{(306)}{ROC(O)(CH_2)_3CH_2X + M^n}$$

$$(R = Me,\ Et;\ M = Fe,\ Cu;\ X = Cl,\ Br)$$

A similar reaction of cyclohexanone derivatives **342** produces the corresponding esters of ε-halocaproic acid (**334**, X = Cl, Br) (274).

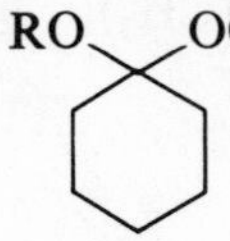

$$RO\diagdown OOH$$

(hexagon ring) $+ M^n \longrightarrow ROC(O)(CH_2)_4CH_2\cdot + M^{n+1}$

**(342)**

$X^-$

$$ROC(O)(CH_2)_4CH_2X + M^n$$
**(334)**

(R = Me, Et; M = Fe, Cu; X = Cl, Br)

However, in the absence of halide ions this reaction of **342**, R = Me, produces either the methyl ester of hexenoic acid or dodecane-1,12-dioic acid (35, 89, 95).

$$\textbf{342, R} = \text{Me} + M^n \longrightarrow MeOC(O)(CH_2)_3CH_2CH_2\cdot + M^{n+1}$$

$$MeOC(O)(CH_2)_3CH=CH_2 + M^n + H^+ \qquad MeOC(O)(CH_2)_{10}C(O)OMe$$

The reaction of **303b** at 0° in a solution containing concentrated sulfuric acid, ferrous sulfate, methyl alcohol, and an aldehyde produces keto esters **366** in 9–33% yield (275).

$$\textbf{342, R} = \text{Me} + Fe^{2+} \longrightarrow MeOC(O)(CH_2)_5\cdot + Fe^{3+}$$

$$RCHO + MeOC(O)(CH_2)_5\cdot \longrightarrow R\overset{.}{C}O + MeOC(O)(CH_2)_4Me$$

$$R\overset{.}{C}O + MeOC(O)(CH_2)_5\cdot \longrightarrow RC(O)(CH_2)_5OC(O)Me$$
**(366)**

[R = MeCH$_2$, Me(CH$_2$)$_2$, Me$_2$CH, Me$_2$CHCH$_2$, Me(CH$_2$)$_5$,

Ph, p-HO—C$_6$H$_4$, (ring figure with O).]

The reaction of **342**, R = Me, with ferrous sulfate in the presence of butadiene gives a 65–75% yield of three isomeric dimethyl eicosadienedioate dimers (**369**) (95).

**342**, R = Me + Fe$^{2+}$ $\longrightarrow$

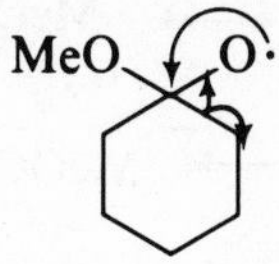

**MeOC(O)(CH$_2$)$_5\cdot$ + Fe$^{2+}$**

**(367)**

CH$_2$=CH−CH=CH$_2$

**MeOC(O)(CH$_2$)$_5$CH$_2$CH$\doteq$CH$\doteq$CH$_2$** $\longrightarrow$   dimers

**(368)**                                                              **(369)**

The reaction in the presence of a mixture of ferrous and cupric ions gives methyl 10-methoxy-8-decenoate (**370**), methyl 8-methoxy-9-decenoate (**371**), methyl $\omega$-hexenoate (**372**), and methyl $\omega$-methoxyhexanoate (**373**) in 16, 45, 6, and 2% yields, respectively (95).

**MeOC(O)(CH$_2$)$_6$CH$\doteq$CH$\doteq$CH$_2$ + Cu$^{2+}$**

**(368)**

$\downarrow$

**MeOC(O)(CH$_2$)$_6$CH$\doteq\overset{+}{CH}\doteq$CH$_2$ + Cu$^+$**

MeOH $\downarrow$

**MeOC(O)(CH$_2$)$_6$CH=CHCH$_2$OMe + MeOC(O)(CH$_2$)$_6$CHCH=CH$_2$**
$\underset{\text{OMe}}{|}$

**(370)**                                                  **(3.71)**

$\xrightarrow{\text{MeOH}}$   **MeOC(O)(CH$_2$)$_6$CH=CHCH$_2$OMe**

+ **MeOC(O)(CH$_2$)$_6$CHCH=CH$_2$**
$\underset{\text{OMe}}{|}$

**(370)**                                                  **(371)**

$$\textbf{367} + Cu^{2+} \longrightarrow MeOC(O)(CH_2)_3CH{=}CH_2 + H^+ + Cu^+$$
$$(\textbf{372})$$

$$\textbf{367} + Cu^{2+} \longrightarrow MeOC(O)(CH_2)_4CH_2^+$$
$$\xrightarrow{MeOH} MeOC(O)(CH_2)_5OMe$$
$$(\textbf{373})$$

The distribution of products **369**, **370**, **371**, **372**, and **373** is largely dependent on cupric-ion concentration and to a lesser degree on the butadiene concentration (95).

In the presence of a mixture of ferrous and cupric chlorides, butadiene, and concentrated hydrochloric acid the products are methyl $\omega$-chlorohexanoate (55%), a mixture of isomeric $C_{10}$ allylic chlorides **374** and **375** (5%), $C_{10}$ esters **370** and **371** (5%), and $C_{20}$ diesters **369** (3%) (95).

$$MeOC(O)(CH_2)_6CH{\overset{\centerdot}{{=}{=}}}CH{=}{=}CH_2 + Cu^{2+} + Cl^-$$
$$(\textbf{368})$$

$$\downarrow$$

$$\underset{\underset{(\textbf{374})}{\overset{\displaystyle |}{\underset{Cl}{}}}}{MeOC(O)(CH_2)_6CHCH{=}CH_2} + \underset{(\textbf{375})}{MeOC(O)(CH_2)_6CH{=}CHCH_2Cl} + Cu^+$$

The reaction of **342**, R = Me, with isoprene in the presence of a mixture of ferrous and cupric sulfates gives a 27% yield of methyl 8-methoxy-8-methyl-9-decenoate (**376**) and a 7% yield of methyl 10-methoxy-8-methyl-7-decenoate (**377**) (95).

$$\textbf{342}, R = Me + Fe^{2+} + \overset{\overset{\displaystyle Me}{\displaystyle |}}{CH{=}CCH{=}CH_2}$$

$$\xrightarrow[MeOH]{Cu^{2+}} MeOC(O)(CH_2)_5CH_2\underset{\underset{OMe}{\overset{\displaystyle |}{}}}{\overset{\overset{\displaystyle Me}{\displaystyle |}}{C}}CH{=}CH_2$$
$$(\textbf{376})$$

$$+ MeOC(O)(CH_2)_5CH_2\overset{\overset{\displaystyle Me}{\displaystyle |}}{C}{=}CHCH_2OMe$$
$$(\textbf{377})$$

In addition there are obtained diesters of dimethyleicosadiendioic acid (dimers of type **369**) in 20% yield and 8-methoxy-8-methyl-9-decenoic acid and esters **372** and **373** in 6.5, 10, and 5% yield, respectively (95).

The same peroxide **342**, R = Me, reacts with chloroprene in the presence of a mixture of ferrous and cupric sulfates to give a 20% yield of methyl chloromethoxydecenoate (**378**) plus a 12% yield of methyl $\omega$-chlorohexanoate (95).

$$\text{342, R = Me} + Fe^{2+} + CH_2 = \overset{\overset{\displaystyle Cl}{\displaystyle |}}{C} - CH = CH_2$$

$$\downarrow$$

$$MeOC(O)(CH_2)_5CH_2C \overset{.}{=} CH \overset{}{=} CH_2$$

$$Cu^{2+} \mid MeOH$$

$$\downarrow$$

$$[MeOC(O)(CH_2)_5CH_2\underset{\underset{\displaystyle OMe}{\displaystyle |}}{\overset{\overset{\displaystyle Cl}{\displaystyle |}}{C}}CH=CH_2] + MeOC(O)(CH_2)_5\overset{\overset{\displaystyle Cl}{\displaystyle |}}{C}=CHCH_2OMe$$

$$\text{not isolated} \qquad\qquad\qquad\qquad\qquad\qquad (378)$$

The reaction of the hydroperoxide **342**, R = Me, with acrylonitrile in the presence of a mixture of ferrous and cupric sulfates gives polymers (95). However, in the presence of ferrous and cupric chlorides a 37% yield of methyl 8-chloro-8-cyanooctanoate (**379**) and a 20% yield of methyl $\omega$-chlorohexanoate, plus compound **380** are obtained (95).

$$\text{342, R = Me} + Fe^{2+} + CH_2 = CHCN \longrightarrow$$

$$MeOC(O)(CH_2)_5CH_2\overset{.}{C}HCN \xrightarrow{\quad CuCl_2 \quad}$$

$$MeOC(O)(CH_2)_6\underset{\underset{\displaystyle Cl}{\displaystyle |}}{C}HCN + MeOC(O)(CH_2)_5(CH_2CHCN)_nCl$$

$$(379) \qquad\qquad\qquad\qquad\qquad (380)$$

$$(n = \text{predominantly 2})$$

The reaction of **342**, R = Me, with styrene in the presence of ferrous and cupric sulfates gives a 19% yield of methyl 8-methoxy-8-phenyloctanoate (**381**) and a 28% yield of methyl 10-methoxy-8,10-diphenyldecanoate (**382**) (95).

**342**, R = Me + Fe$^{2+}$ + PhCH=CH$_2$

$$\longrightarrow \text{MeOC(O)(CH}_2)_5\text{CH}_2\overset{\cdot}{\text{C}}\text{HPh} \xrightarrow{\text{Cu}^{2+}}$$

$$\text{MeOC(O)(CH}_2)_6\underset{+}{\text{C}}\text{HPh} \xrightarrow{\text{MeOH}} \text{MeOC(O)(CH}_2)_6\overset{\text{Ph}}{\underset{|}{\text{C}}}\text{HOMe} +$$

$$(381)$$

$$\text{MeOC(O)(CH}_2)_6\overset{\text{Ph}}{\underset{|}{\text{C}}}\text{HCH}_2\overset{\text{Ph}}{\underset{|}{\text{C}}}\text{HOMe}$$

$$(382)$$

The reaction of vinyl bromide with **342**, R = Me, in the presence of ferrous and cupric sulfates yields polymers (95). The reaction of vinylidene chloride and **342**, R = Me, in the presence of a mixture of ferrous chloride, ferrous sulfate, and cupric chloride gives mainly methyl $\omega$-chlorohexanoate (95). Methallyl chloride reacts with **342**, R = Me in the presence of ferrous and cupric sulfates to give a 55% yield of an adduct, believed to be mainly **383** (95).

$$\text{MeOC(O)(CH}_2)_6\overset{\text{Me}}{\underset{|}{\overset{\cdot}{\text{C}}}}\text{CH}_2\text{Cl} \xrightarrow{\text{Cu}^{2+}} \text{MeOC(O)(CH}_2)_6\overset{\text{Me}}{\underset{|}{\text{C}}}{=}\text{CHCl} + \text{H}^+ + \text{Cu}^+$$

$$(383)$$

The ferrous sulfate–initiated decomposition of **342**, R = Me in the presence of a mixture of butadiene and acrylonitrile produces a mixture of isomers **384**, which are difficult to separate (276).

**342**, R = Me + Fe$^{2+}$ $\longrightarrow$ MeOC(O)(CH$_2$)$_4$CH$_2\cdot$ + CH$_2$=CHCN

$$\longrightarrow \text{MeOC(O)(CH}_2)_4\text{CH}_2\overset{\cdot}{\text{C}}\text{HCN} \xrightarrow{\text{CH}_2=\text{CHCH}=\text{CH}_2}$$

$$[\text{MeOC(O)(CH}_2)_6\text{CH(CN)C}_4\text{H}_6]_2$$

$$(384)$$

However, the reaction of this mixture with ferrous and cupric sulfates produces compounds **385** and **386** (276).

MeOC(O)(CH$_2$)$_4$CH$_2$CH$_2$CH(CN)C$_4$H$_6\cdot$ + Cu$^{2+}$

$$\longrightarrow \text{MeOC(O)(CH}_2)_6\text{CH(CN)C}_4\text{H}_6^+$$

$$\xrightarrow{\text{MeOH}} \text{MeOC(O)(CH}_2)_6\text{CH(CN)CH}_2\text{CHCH}=\text{CH}_2$$

$$\underset{\text{OMe}}{|}$$

$$(385)$$

$$+ \text{MeOC(O)(CH}_2)_6\text{CHCNCH}_2\text{CH}=\text{CHCH}_2\text{OMe}$$

$$(386)$$

A similar reaction of **342**, R = Me, with a mixture of acrylic acid, butadiene, ferrous sulfate, and cupric sulfate gives the methyl ester of 8-carboxy-10-hydroxy-11-dodecenoic acid $\gamma$-lactone (**387**) (276).

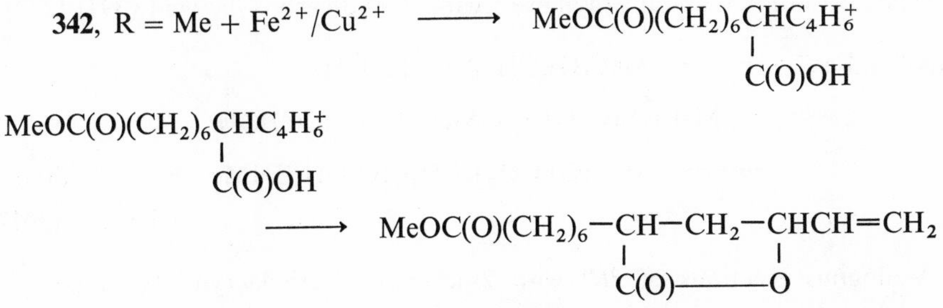

$$\textbf{342, R = Me} + Fe^{2+}/Cu^{2+} \longrightarrow MeOC(O)(CH_2)_6\underset{\underset{\displaystyle C(O)OH}{|}}{C}HC_4H_6^+$$

$$MeOC(O)(CH_2)_6\underset{\underset{\displaystyle C(O)OH}{|}}{C}HC_4H_6^+$$

$$\longrightarrow MeOC(O)(CH_2)_6{-}\underset{\underset{\displaystyle C(O)}{|}}{C}H{-}CH_2{-}\underset{\underset{\displaystyle O}{|}}{C}HCH{=}CH_2$$

**(387)**

Under similar conditions the reaction of **342**, R = Me, with a mixture of $\alpha$-methylstyrene and acrylic acid yields the methyl ester of 8-carboxy-10-hydroxy-10-phenylundecanoic acid $\gamma$-lactone (**388**) (276).

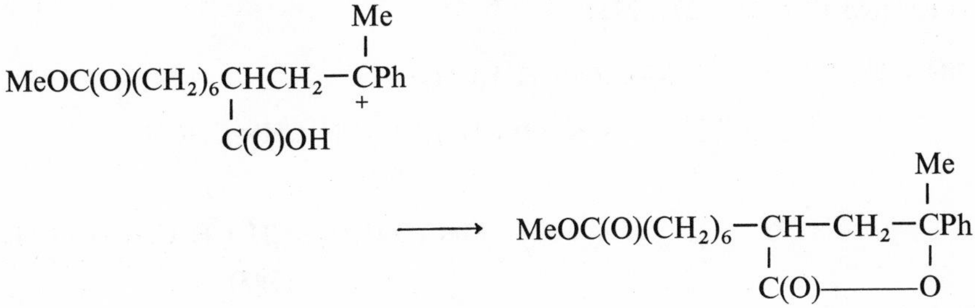

$$MeOC(O)(CH_2)_6\underset{\underset{\displaystyle C(O)OH}{|}}{C}HCH_2{-}\underset{\underset{\displaystyle +}{\overset{\overset{\displaystyle Me}{|}}{C}}}Ph$$

$$\longrightarrow MeOC(O)(CH_2)_6{-}\underset{\underset{\displaystyle C(O)}{|}}{C}H{-}CH_2{-}\underset{\underset{\displaystyle O}{|}}{\overset{\overset{\displaystyle Me}{|}}{C}}Ph$$

**(388)**

### 10.   1-Alkyl and 1-Aryl Cycloalkyl Hydroperoxides

The reactions of 1-methylcyclopentyl hydroperoxide (**389**) with aqueous ferrous sulfate or chloride in the presence of hydrochloric or hydrobromic acid at $-5$ to $0°$ produces 6-halohexanone (**390**) (274). Similarly this decomposition in the presence of cuprous cyanide and hydrogen cyanide gives 6-ketoheptonitrile (**390**) (267).

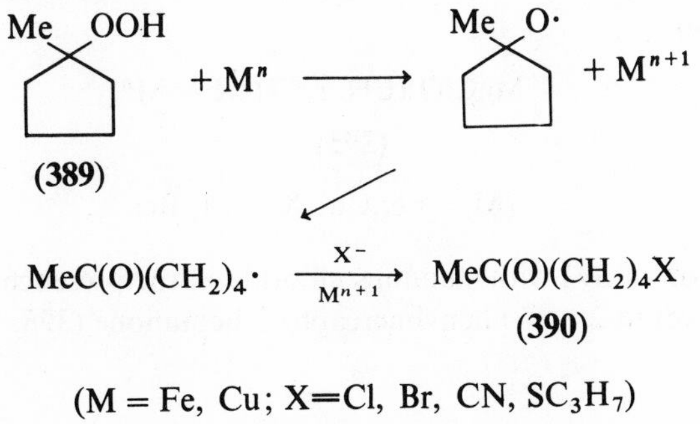

$$MeC(O)(CH_2)_4\cdot \xrightarrow[M^{n+1}]{X^-} MeC(O)(CH_2)_4X$$

**(390)**

(M = Fe, Cu; X=Cl, Br, CN, SC_3H_7)

An analogous reaction of **389** with ferrous sulfate and propyl mercaptan yields 6-propylthio-2-hexanone (**390**, $X = SC_3H_7$) (266).

The reaction of **389** with ferrous sulfate in the presence of crotonic acid yields 2,15-dioxo-8,9-dimethylhexadecane-7,10-dicarboxylic acid (**391**) (272).

$$\textbf{389} + Fe^{2+} \longrightarrow MeC(O)(CH_2)_4\cdot + Fe(OH)^{2+}$$

$$MeC(O)(CH_2)_4\cdot + MeCH{=}CHC(O)OH$$

$$\longrightarrow MeC(O)(CH_2)_4CH[C(O)OH]\dot{C}HMe \longrightarrow \text{dimer}$$

$$(391)$$

Analogous reactions of **389** with 2-hexenoic, $\beta$-ethylacrylic, $\beta,\beta$-dimethyl-acrylic, angelic, and cinnamic acids result in the formation of corresponding adducts (272).

The decomposition of **389** with ferrous sulfate in the presence of butadiene produces a mixture of dimeric diketones, resulting from the coupling of radical **392** and **393** from which a 30% yield of 8,12-eicosadiene-2,19-dione is isolated (256, 257, 277, 278).

$$\textbf{389} + Fe^{2+} \longrightarrow MeC(O)(CH_2)_3CH_2\cdot$$

$$\xrightarrow{CH_2=CHCH=CH_2} MeC(O)(CH_2)_3CH_2CH_2CH{=}CHCH_2\cdot$$

$$(392)$$

$$\longleftarrow MeC(O)(CH_2)_3CH_2CH_2\dot{C}HCH{=}CH_2$$

$$(393)$$

The reaction of 1-methylcyclohexyl hydroperoxide (**394**) with metal ions in the presence of large quantities of halide ions produces 7-haloheptan-2-one (**395**) (274, 279).

$$+ M^n \longrightarrow MeC(O)(CH_2)_4CH_2\cdot + M^{n+1} + X^-$$

(**394**)

$$MeC(O)(CH_2)_4CH_2X + M^n$$

$$(395)$$

$$(M = Fe, Cu; X = Cl, Br)$$

The reaction of **394** with stannous chloride in the presence of benzenethiol at 10° produces in 2 hr 7-phenylmercapto-2-heptanone (**395**, $X = PhS$) (266).

An analogous reaction of 1-hexylcyclohexyl hydroperoxide with 2-ethylhexyl mercaptan in the presence of chromium chloride yields 1-(2-ethylhexylmercapto)-6-dodecanone (266).

The reaction of **394** with ferrous sulfate under 100 atm of carbon monoxide gives 7-ketooctanoic acid (**396**) (250, 252).

$$\textbf{394} + Fe^{2+} \longrightarrow MeC(O)(CH_2)_4CH_2\cdot$$
$$\xrightarrow{CO} MeC(O)(CH_2)_4CH_2CO\cdot$$
$$\xrightarrow[or\ \textbf{394}]{Fe(OH)^{2+}} MeC(O)(CH_2)_5C(O)OH$$
$$(\textbf{396})$$

The reaction of **394** with ferrous sulfate in the presence of butadiene produces a mixture of diketones, arising from the coupling of radical **397** and **398**, from which a 20% yield of 9,13-docosadiene-2,21-dione is obtained (256, 257, 277, 278).

$$\textbf{394} + Fe^{2+} \longrightarrow MeC(O)(CH_2)_5\cdot + Fe(OH)^{2+}$$

$$MeC(O)(CH_2)_5\cdot + CH_2{=}CHCH{=}CH_2$$
$$\longrightarrow MeC(O)(CH_2)_5CH_2CH{=}CHCH_2\cdot$$
$$(\textbf{397})$$
$$\longleftrightarrow MeC(O)(CH_2)_5CH_2CHCH{=}CH_2 \longrightarrow dimers$$
$$(\textbf{398})$$

The decomposition of dicyanodicyclohexyl peroxide (**399**) with ferrous sulfate in the presence of butadiene gives a mixture of products from which adduct **401** is obtained (256, 257). It is possible that this reaction proceeds through a 1-cyanocyclohexyl hydroperoxide (**400**).

$$\textbf{400} + Fe^{2+} \longrightarrow CNC(O)(CH_2)_5\cdot + Fe(OH)^{2+}$$
$$\xrightarrow{CH_2{=}CHCH{=}CH_2} CNC(O)(CH_2)_5CH_2CH{=}CHCH_2\cdot$$
$$\longrightarrow [CNC(O)(CH_2)_5CH_2CH{=}CHCH_2{+}]_2$$
$$(\textbf{401})$$

The reaction of 1-phenylcyclohexyl hydroperoxide (**402**) with ferrous sulfate in the presence of butadiene gives a mixture of products from which adduct **403** is obtained (256, 257).

$$\text{Ph} \quad \text{OOH}$$

$$\text{(402)} + Fe^{2+} \longrightarrow PhC(O)(CH_2)_5\cdot + Fe(OH)^{2+}$$

$$CH_2=CH-CH=CH_2$$

$$PhC(O)(CH_2)_5CH_2CH=CHCH_2\cdot$$

$$[PhC(O)(CH_2)_5CH_2CH_2CH=CHCH_2 \tfrac{}{}]_2$$

$$\text{(403)}$$

## B.  Aralkyl Hydroperoxides

### 1.  Reactions with Hydrocarbons

In the presence of reactive substrates the decomposition of aralkyl hydroperoxides with metal ions can be used for the peroxygenation of olefinic hydrocarbons to unsymmetrical peroxides of type **167**.

$$ROOR'$$

$$\text{(167)}$$

Thus cumyl hydroperoxide reacts in the presence of either cobalt, manganese, or copper ions at 45° with cyclopentene, cyclohexene, and cycloheptene (**404**) to give the corresponding unsymmetrical peroxides (**405**) in 42, 26, and 10% yield, respectively (146).

$$PhMe_2COOH + M^n \longrightarrow PhMe_2CO\cdot + M(OH)^n$$

$$\underset{\text{(404)}}{\text{—(CH}_2)_m} + PhMe_2CO\cdot \longrightarrow \text{—(CH}_2)_m + PhMe_2COH$$

$$\text{(cyclopentenyl radical)} + PhMe_2COOH + M(OH)^n$$

$$\downarrow$$

$$\text{(cyclopentenyl—OOCMe}_2Ph) + M^n + H_2O$$
$$\textbf{(405)}$$

$$(M = Co, Mn, Cu; m = 1,2,3)$$

The cuprous chloride–catalyzed reaction of cyclohexene with cumyl hydroperoxide at 70° gives a 90% yield of cyclohexenyl cumyl peroxide (207, 210). A similar reaction of cumyl hydroperoxide with tetralin and indan in the presence of cobaltous ions produces tetralyl cumyl and indanyl cumyl peroxides in 42 and 46% yields, respectively (146). Unsymmetrical peroxides are also produced by the reaction of aralkyl hydroperoxides with equimolar quantities of lead tetraacetate in a reactive hydrocarbon (184). However, the mechanism of this reaction is not fully established (62).

If the ferrous-ion-catalyzed decomposition of cumyl hydroperoxide is carried out in the presence of a conjugated diene, such as butadiene, cumyloxy radicals, produced in the first step of the decomposition, add to the olefinic bond to give a mixture of dimers **406** containing cumyloxy groups (198, 199, 280, 281).

$$PhMe_2COOH + M^n \longrightarrow PhMe_2CO\cdot + M(OH)^n$$

$$PhMe_2CO\cdot + CH_2{=}CHCH{=}CH_2$$

$$\downarrow$$

$$PhMe_2COCH_2CH{=\!=}\overset{\cdot}{C}H{=\!=}CH_2$$

$$\downarrow$$

$$PhMe_2COC_4H_6C_4H_6OCMe_2Ph$$
$$\textbf{(406)}$$

The product consists of about 80–95% straight-chain bis($\alpha,\alpha$-dimethylbenzyl) diether of octa-2,6-diene-1,8-diol (**407**) and 5–20% of an isomer formed by 1,2-addition (198, 280).

$$(PhMe_2COCH_2—CH=CH—CH_2—)_2$$
**(407)**

Hydrogenation of **407** produces a saturated diether (281). Hydrogenation and hydrolysis produce methylstyrene and 1,8-octanediol (280).

If the decomposition of cumyl hydroperoxide is carried out in the presence of a mixture of ferrous chloride, ferrous sulfate, and sodium chloride, adducts **408** and **409** are obtained (282, 283).

$$PhMe_2COOH + Fe^{2+} \longrightarrow + PhMe_2CO\cdot + Fe(OH)^{2+}$$

$$PhMe_2CO\cdot + CH_2=CH—\overset{\bullet}{C}H=CH_2$$

$$\downarrow$$

$$PhMe_2COCH_2CH\!=\!\!=\!CH\!=\!\!=\!CH_2$$

$$\overset{\displaystyle FeCl^{2+}}{\Big\downarrow}$$

$$PhMe_2COCH_2CHClCH=CH_2 + PhMe_2COCH_2CH=CHCH_2Cl$$
$$\textbf{(408)} \hspace{5cm} \textbf{(409)}$$

A ferrous sulfate–catalyzed decomposition of tetralyl hydroperoxide in the presence of butadiene produces no adduct; instead tetralone is formed (198). The decomposition of triphenylmethyl hydroperoxide in butadiene in the presence of ferrous ions yields a mixture of adducts **411** containing the rearranged radical **410** (255).

$$Ph_3COOH + Fe^{2+} \longrightarrow Ph_3CO\cdot + Fe(OH)^{2+}$$
$$Ph_3CO\cdot \longrightarrow Ph_2\overset{\bullet}{C}OPh$$
$$\textbf{(410)}$$

$$410 + CH_2=CHCH=CH_2 \longrightarrow Ph_2(PhO)CC_4H_6\cdot \longrightarrow$$
$$Ph_2(PhO)CC_4H_6C_4H_6C(PhO)Ph_2$$
$$\textbf{(411)}$$

In the presence of vanadium pentoxide cumyl hydroperoxide reacts with olefins, with the formation of epoxides (182). Thus octene-1 and cyclohexene are transformed to 1,2-epoxyoctane and epoxycyclohexane in low yield (182). In a recent study a nonradical mechanism was proposed for an analogous reaction of cyclohexene with *t*-butyl hydroperoxide in the presence of vanadium compounds (28) (see also references 222 and 224 for more recent studies).

The reaction of a mixture of cumyl hydroperoxide, methanol, and propylene in the presence of molybdenum naphthenate at 100° produces a mixture of 1-methoxy-1-propanol, 2-methoxy-1-propanol, and propylene oxide (284). The mechanism of this reaction is unknown.

## 2. Reactions with Phenols

Metal-ion-catalyzed reactions of aralkyl hydroperoxides with phenols yield peroxycyclodienones of type **266** (201). However, some of the structures are not firmly established (201).

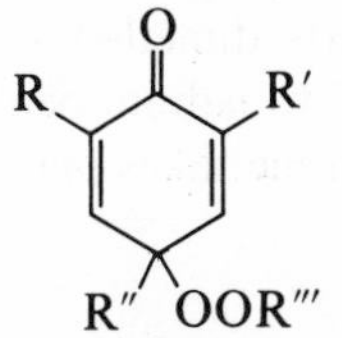

**(266)**

Hydroperoxides with the structure R'''OOH, such as 1-tetralyl, cumyl, and 1,1-diphenylethyl, react with 2,6-di-*t*-butyl-4-methylphenol in the presence of cobalt naphthenate to give the corresponding peroxy derivative (**266**, R = R' = CMe$_3$, R" = Me) (201). Similarly the reaction of 1-tetralyl hydroperoxide with 4-*t*-butyl-2,6-dimethylphenol and 2,4-di-*t*-butyl-6-methylphenol produces presumably a cyclodienone derivative of type **266** (201).

$$Ph_3COOH + Co^{2+} \longrightarrow Ph_3CO \cdot + Co(OH)^{2+}$$

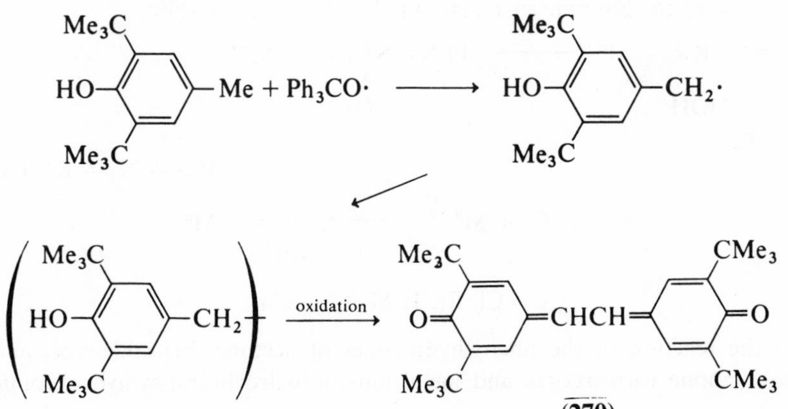

**(270)**

The reaction of 2,4,6-tri-*t*-butylphenol with cumyl hydroperoxide yields the corresponding derivatives **266**, R = R' = R'' = CMe$_3$, R''' = PhCMe$_2$ (201). However, the reaction of triphenylmethyl hydroperoxide with 2,6-di-*t*-butyl-4-methylphenol gives stilbenequinone (**270**) (201).

### 3.  Reactions with Miscellaneous Compounds

Cumyl hydroperoxide readily oxidizes isopropyl alcohol to acetone in the presence of cobalt, manganese, chromium, vanadium, tungsten, and osmium compounds (182).

The thermal decomposition of cumyl hydroperoxide with various aldehydes produces the corresponding acids (182). The reaction is facilitated by addition of osmium tetroxide, permitting the reaction to be carried out at room temperature (182). Thus acetaldehyde, butyraldehyde, 2-ethylhexanal, and benzaldehyde are converted to the corresponding acids (182).

Cumyl hydroperoxide converts dimethyl sulfoxide into dimethyl sulfone in 91 % yield in the presence of vanadium pentoxide (237). In the absence of the catalyst no oxidation to sulfone takes place (237). The mechanism of the reaction is unknown.

### C.  Hydroperoxides Derived from Nitrogen Compounds

The reaction of phenylhydrazones with oxygen gives Busch's hydroperoxides (**412**) (285–287).

$$\text{PhNHN=CRR}' + \text{O}_2 \longrightarrow \underset{\substack{| \\ \text{OOH} \\ \text{(412)}}}{\text{PhN=NCRR}'}$$

The decomposition of these peroxides with cuprous or ferrous ions in the presence of halide ions at $-5$ to $10°$ results in products (**413**) that are identical with those of the Sandmeyer reaction (239, 262, 268, 288–290).

$$\underset{\substack{| \\ \text{OOH} \\ \text{(412)}}}{\text{PhN=NCRR}'} + \text{M}^n \longrightarrow \underset{\substack{| \\ \text{O·}}}{\text{PhN=NCRR}'} + \text{M}^{n+1}$$

$$\longrightarrow \text{Ph·} + \text{N}_2 + \text{RR'CO}$$

$$\underset{\text{(413)}}{\text{Ph· + X}^- + \text{M}^{n+1}} \longrightarrow \text{PhX} + \text{M}^n$$

$$(\text{X = Cl, Br, I; M = Fe, Cu})$$

Thus the reaction of the phenylhydrazones of acetone, benzaldehyde, and cyclohexanone with oxygen and metal ions in hydrochloric or hydrobromic acids produces chloro- and bromobenzene, respectively (239, 262, 288–290). Similarly the reaction in the presence of potassium iodide yields iodobenzene (288, 289). Other anions can also be employed. Thus autoxidation of the phenylhydrazone of acetone in the presence of metal ions and ammonium thiocyanate (X = SCN), sodium azide (X = $N_3$), or potassium ethyl xanthogenate [X = SC(S)OR] produces phenyl thiocyanate, phenyl azide, and ethyl phenylxanthogenate, respectively (288, 289). If the reaction is carried out in hydrochloric acid in the presence of cupric chloride and sulfurdioxide, benzene-, sulfonyl chloride is formed (289).

$$Ph\cdot + SO_2 \longrightarrow PhSO_2\cdot$$

$$PhSO_2\cdot + Cl^- + M^{n+1} \longrightarrow PhSO_2Cl + M^n$$

If the metal-ion-catalyzed decomposition of hydroperoxides **412** is carried out in the presence of olefins and halide ions, the products (**414**) that are obtained are identical with those of the Meerwein reaction (262, 290).

$$\underset{\underset{(412)}{\underset{|}{OOH}}}{PhN{=}NCRR'} + M^n \longrightarrow \underset{\underset{|}{O\cdot}}{PhN{=}NCRR'} + M(OH)^n$$

$$\longrightarrow Ph\cdot + N_2 + RR'CO$$

$$R''CH{=}CH_2 + Ph\cdot \longrightarrow R''\overset{\cdot}{C}HCH_2Ph$$

$$R''\overset{\cdot}{C}HCH_2Ph + X^- + M^{n+1} \longrightarrow R''CHXCH_2Ph + M^n$$
$$(414)$$

Thus reactions with acrylonitrile, methacrylic acid, butyl acrylate, butadiene, styrene, and coumarin yield the corresponding products **415, 416, 417, 418, 419,** and **420** (262)

PhCH₂CHClCN          PhCH₂CMeClC(O)OH          PhCH₂CHClC(O)O—*n*-Bu
**(415)**                         **(416)**                                    **(417)**

PhCH₂CH=CHCH₂Cl      PhCH₂CHClPh
**(418)**                         **(419)**

**(420)**

In the absence of halide ions the decomposition of the phenylhydrazone hydroperoxide of acetone with ferrous sulfate in the presence of 5-hydroxy-2-hydroxymethyl-γ-pyrone (**421**), styrene, and *p*-quinone results in the formation of the corresponding phenylated products **422, 423,** and **424** (262, 290).

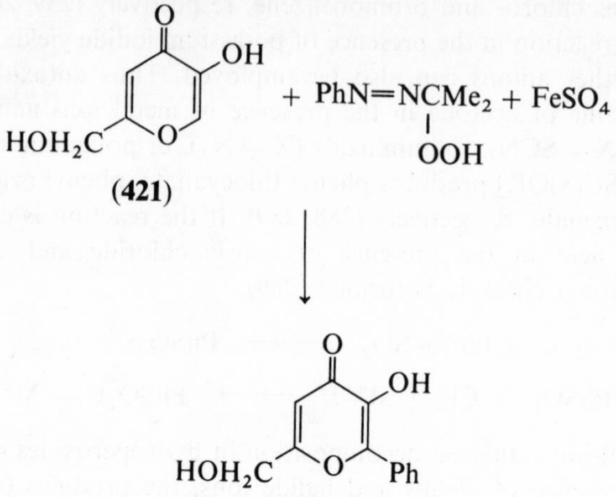

**(421)**

**(422)**

$$PhCH=CH_2 + PhN=NCMe_2 + FeSO_4 \longrightarrow PhCH_2CH_2Ph$$

$$\overset{|}{OOH}$$

**(423)**

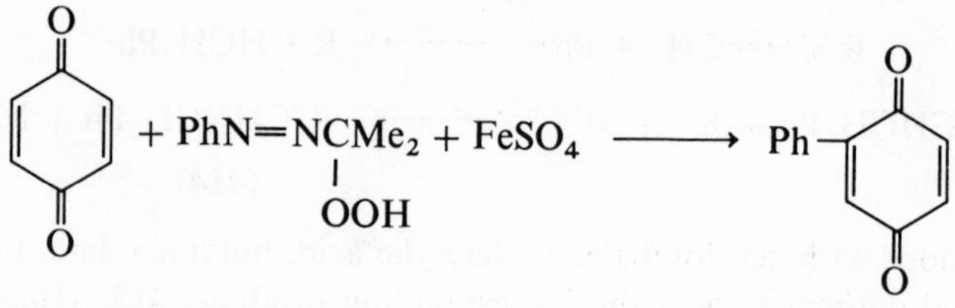

**(424)**

In the presence of large amounts of oxygen a different result is obtained. Thus autoxidation of a methanolic solution of the phenylhydrazone of acetone and acrylonitrile in the presence of copper ions at 30° yields methyl phenylacetate **(425)** (291).

$$PhCH_2\overset{\cdot}{C}HCN + O_2 \longrightarrow PhCH_2CHCN$$
$$\overset{|}{OO\cdot}$$

$$\xrightarrow{Cu^+/Cu^{2+}} PhCH_2CHCN \xrightarrow[-HCN]{MeOH} PhCH_2C(O)OMe$$
$$\overset{|}{O\cdot}$$

**(425)**

A similar result is obtained with N-acetyl-N'-phenylhydrazine **(426)**. Thus autoxidation of a methanolic solution of **426** and styrene in the presence of copper ions gives benzaldehyde and phenylbenzyl ketone (291).

$$\text{PhNHNHC(O)Me}$$
$$\textbf{(426)}$$

$$\text{Ph}\cdot + \text{CH}_2\!=\!\text{CHPh} \longrightarrow \text{PhCH}_2\text{CHPh}$$

$$\xrightarrow{\text{O}_2} \quad \underset{\substack{| \\ \text{OO}\cdot}}{\text{PhCH}_2\text{CHPh}} \quad \xrightarrow{\text{Cu}^+/\text{Cu}^{2+}} \quad \underset{\substack{| \\ \text{O}\cdot}}{\text{PhCH}_2\text{CHPh}}$$

$$\longrightarrow \text{PhCH}_2\text{C(O)Ph} + \text{PhCHO}$$

Under similar conditions the reaction in methyl acrylate forms benzaldehyde (291).

$$\text{Ph}\cdot + \text{CH}_2\!=\!\text{CHC(O)(OMe)} \longrightarrow \text{PhCH}_2\text{CHC(O)OMe}$$

$$\xrightarrow{\text{O}_2} \quad \underset{\substack{| \\ \text{OO}\cdot}}{\text{PhCH}_2\text{CHC(O)OMe}} \quad \xrightarrow{\text{Cu}^+/\text{Cu}^{2+}} \quad \underset{\substack{| \\ \text{O}\cdot}}{\text{PhCH}_2\text{CHC(O)OMe}}$$

$$\longrightarrow \text{PhCH}_2\text{C(O)C(O)OMe} \xrightarrow{\text{O}_2} \text{PhCHO}$$

The oxidation of either the phenylhydrazone of acetone or **426** in α-methylstyrene yields acetophenone (291).

$$\underset{\substack{| \\ \text{OO}\cdot}}{\text{PhC(Me)CH}_2\text{CH}_3} \xrightarrow{\text{Cu}^+/\text{Cu}^{2+}} \text{PhC(O)Me} + \text{Et}\cdot$$

Under similar conditions coumarin is converted to 3-phenyl-4-hydroxy coumarin (**427**) (291).

**(427)**

Preparation of $\alpha$-amino ketones (**428**) can be achieved by the following reaction involving olefins, $N$-chloroamines, oxygen, and ferrous salts (292). The yields vary between 20 and 75% (292).

$$R_2NCl + Fe^{2+} \longrightarrow R_2N\cdot + FeCl^{2+}$$

$$R'CH{=}CH_2 + \cdot NR_2 \longrightarrow R'\underset{\cdot}{C}HCH_2NR_2 \xrightarrow{O_2} \begin{array}{c} R'CHCH_2NR_2 \\ | \\ OO\cdot \end{array}$$

$$\xrightarrow{Fe^{2+}/Fe^{3+}} R'C(O)CH_2NR_2 + Fe(OH)^{2+}$$
$$(\mathbf{428})$$

The following amines and olefins have been successfully utilized for the preparation of $\alpha$-amino ketones: dibutylamine, piperidine, morpholine, butene-1, cyclohexene, styrene, *trans*-stilbene, and isosafrole (292). In an analogous reaction $\alpha$-azido ketones (**429**) can be prepared with an olefin, sodium azide, oxygen, and ferrous sulfate in aqueous methanol at $-5$ to $5°$ (293).

$$RCH{=}CH_2 + N_3\cdot \longrightarrow R\underset{\cdot}{C}HCH_2N_3 \xrightarrow{O_2} \begin{array}{c} RCHCH_2N_3 \\ | \\ OO\cdot \end{array}$$

$$\xrightarrow{Fe^{2+}/Fe^{3+}} RC(O)CH_2N_3 + Fe(OH)^{2+}$$
$$(\mathbf{429})$$

Thus the reaction with hexene-1, cyclohexene, styrene, and *trans*-stilbene produces the corresponding $\alpha$-azido ketones (293). However, the reaction of $\alpha$-methylstyrene yields a mixture of acetophenone, the diazide **430**, and the azido alcohol (**431**) (293).

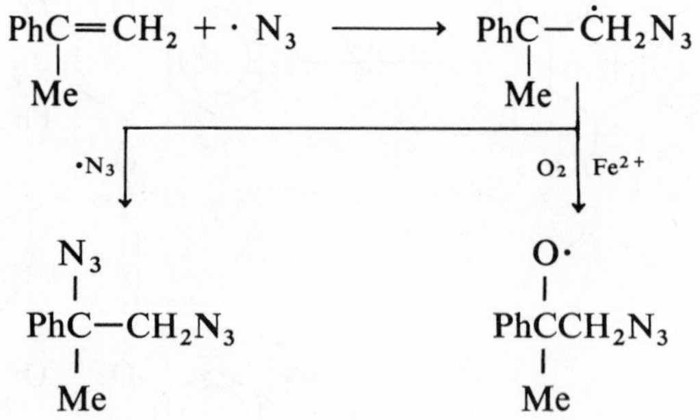

$$\begin{array}{c} \text{PhC}{=}\text{CH}_2 \\ | \\ \text{Me} \end{array} + \cdot N_3 \longrightarrow \begin{array}{c} \text{Ph}\overset{\cdot}{\text{C}}{-}\text{CH}_2N_3 \\ | \\ \text{Me} \end{array}$$

$$\xleftarrow{\cdot N_3} \qquad\qquad\qquad \xrightarrow{O_2 \mid Fe^{2+}}$$

$$\begin{array}{c} N_3 \\ | \\ \text{PhC}{-}\text{CH}_2N_3 \\ | \\ \text{Me} \\ (\mathbf{430}) \end{array} \qquad\qquad \begin{array}{c} \text{O}\cdot \\ | \\ \text{PhCCH}_2N_3 \\ | \\ \text{Me} \end{array}$$

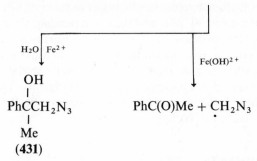

**(431)**

Under similar conditions the reaction with ethylvinyl ether produces ethyl azidoacetate (**432**) (293).

$$CH_2=CHOEt + NaN_3 + Fe^{2+} + O_2$$

$$\longrightarrow \underset{\underset{OO\cdot}{|}}{N_3CH_2CHOEt} \xrightarrow{Fe^{2+}/Fe^{3+}} N_3CH_2C(O)OEt$$

**(432)**

Similarly the reaction with acrylonitrile yields methyl azidoacetate (**433**) (293).

$$CH_2=CHCN + N_3\cdot \longrightarrow N_3CH_2\underset{\cdot}{C}HCN \xrightarrow{O_2} \underset{\underset{OO\cdot}{|}}{N_3CH_2CHCN}$$

$$\xrightarrow{Fe^{2+}/Fe^{3+}} N_3CH_2C(O)CN \xrightarrow{MeOH} N_3CH_2C(O)OMe + HCN$$

**(433)**

*Note added in November* 1968. In a recent paper Hiatt, Irwin, and Gould (294) reported the metal-catalyzed decompositions of hydroperoxides. *t*-Butyl hydroperoxide was decomposed by a variety of cobalt salts and compounds of other metals (Fe, V, Mn, Ce, and Pb). In chlorobenzene or alkanes at 25–45° half-lives for decomposition of 0.1 $M$ *t*-BuO$_2$H by $10^{-4}$ $M$ catalyst ranged from 1 to 10 min. Products included approximately 88% *t*-BuOH, 11% *t*-Bu$_2$O$_2$, 1% acetone, and 93% O$_2$. Decomposition in alcohol–chlorobenzene mixtures yielded more acetone but were only about one-hundredth as fast as in pure chlorobenzene. Reactions in all solvents were subject to autoretardation owing partly to the formation of aldehydes and carboxylic acids, and partly to changes in the catalyst, which caused its eventual precipitation. Decompositions of α-cumyl, *n*-butyl, and *sec*-butyl hydroperoxides were normally one-fourth to one-tenth as fast as those of $t$-BuO$_2$H. $n$-BuO$_2$H and $sec$-BuO$_2$H yielded 70 % O$_2$ and the corresponding alcohol and aldehyde or ketone in a ratio of about 2. The results suggest that these reactions are essentially the same as free-radical-induced decompositions but are initiated by metal ion–hydroperoxide interactions. Generally speaking the choice of metal ion, as long as it can undergo a facile one-electron oxidation-reduction reaction, has little influence on either products or rates of decomposition except in the presence of olefins.

## ACKNOWLEDGMENTS

We are grateful to all authors who sent us their valuable reprints. In particular we are indebted to Drs. A. Rieche, J. K. Kochi, F. Minisci, and M. Schulz. We would also like to thank Drs. R. Criegee. F. Minisci, J. R. Shelton, and A. Fono for their helpful letters.

# REFERENCES

1. H. J. H. Fenton, *J. Chem. Soc.*, **65**, 899 (1894).
2. F. Haber and J. Weiss, *Proc. Roy. Soc.* (*London*), **A147**, 332 (1934).
3. W. A. Waters, *The Chemistry of Free Radicals*, 2nd ed., Oxford, 1948.
4. A. V. Tobolsky and R. B. Mesrobian, *Organic Peroxides*, Interscience, New York, 1954.
5. C. Walling, *Free Radicals in Solution*, Wiley, New York, 1957.
6. B. A. Dolgoplosk, B. L. Erusalimskii, and E. I. Tiniakova, *Izv. Akad. Nauk SSSR*, **1958**, 469.
7. W. A Waters, in *Progress in Organic Chemistry*, Vol. 5, Butterworths, London, 1961, p. 1.
8. A. G. Davies, *Organic Peroxides*, Butterworths, London, 1961.
9. E. G. E. Hawkins, *Organic Peroxides*, Van Nostrand, Princeton, N.J., 1961.
10. T. Migita and K. Tokumaru, *Yuki Gosei Kagaku Kyokaishi*, **19**, 821 (1961).
11. W. A. Waters, *Mechanisms of Oxidation of Organic Compounds*, Methuen, London, 1964.
12. J. K. Kochi, *Tetrahedron*, **18**, 483 (1962).
13. J. K. Kochi, *Rec. Chem. Prog.*, **27**, 207 (1966).
14. F. Minisci, *Chim. Ind.* (*Milan*), **44**, 740 (1962).
15. S. O. Lawesson and G. Sosnovsky, *Svensk Kemisk Tidskrift*, **75**, 343 (1963); *ibid.*, **75**, 568 (1963).
16. L. S. Boguslavskaya, *Usp. Khim.*, **34**, 1199 (1965); *Russian Chem. Rev.*, **34**, 503 (1965).
17. F. Minisci, *Chim. Ind.* (*Milan*), **49**, 705 (1967).
18. W. Pritzkow and K. A. Müller, *Ann.*, **597**, 167 (1955).
19. W. Pritzkow and I. Hahn, *J. Prakt. Chem.*, **16**, 287 (1962).

20. A. Rieche and R. Meister, *Ber.*, **68**, 1465 (1935).
21. A. Rieche and R. Meister, *Ber.*, **64**, 2328 (1931).
22. W. Pritzkow and K. A. Müller, *Ber.*, **89**, 2321 (1956).
23. N. A. Milas, R. L. Peeler, and O. L. Mageli, *J. Amer. Chem. Soc.*, **76**, 2322 (1954).
24. A. Rieche and C. Bischoff, *Ber.*, **94**, 2722 (1961).
25. D. B. Sharp and T. M. Patrick, *J. Org. Chem.*, **26**, 1389 (1961).
26. S. Murai, N. Sonoda, and S. Tsutsumi, *Bull. Chem. Soc. Japan*, **37**, 1187 (1964).
27. I. V. Berezin, E. T. Denisov, and N. M. Emanuel, *The Oxidation of Cyclohexane*, Moscow University Press, 1962, translated by K. A. Allen, Pergamon Press, Oxford, 1966.
28. E. S. Gould, R. R. Hiatt, and K. C. Irwin, *J. Amer. Chem. Soc.*, **90**, 4573 (1968).
29. M. H. Dean and G. Skirrow, *Trans. Faraday Soc.*, **54**, 849 (1958).
30. J. Kumamoto, H. E. De La Mare, and F. F. Rust, *J. Amer. Chem. Soc.*, **82**, 1935 (1960).
31. J. K. Kochi and F. F. Rust, *J. Amer. Chem. Soc.*, **83**, 2017 (1961).
32. H. Berger and A. F. Bickel, *Trans. Faraday Soc.*, **57**, 1325 (1961).
33. A. N. Sergeeva, V. A. Puchin, and K. N. Mikhalevich, *Zh. Obshch. Khim.*, **31**, 871 (1961).
34. J. K. Kochi, *J. Amer. Chem. Soc.*, **84**, 1193 (1962).
35. H. E. De La Mare, J. K. Kochi, and F. F. Rust, *J. Amer. Chem. Soc.*, **85**, 1437 (1963).
36. J. K. Kochi, *J. Amer. Chem. Soc.*, **85**, 1958 (1963).
37. C. Walling and A. A. Zavitsas, *J. Amer. Chem. Soc.*, **85**, 2084 (1963).
38. E. A. Kuzmina and M. K. Shchennikova, *Tr. po Khim. i Khim. Tekhnol.*, **1964**, 194.
39. E. A. Kuzmina and M. K. Shchennikova, *Tr. po Khim. i Khim. Tekhnol.*, **1964**, 204.
40. B. Acott and A. L. J. Beckwith, *Australian J. Chem.*, **17**, 1342 (1964).
41. J. K. Kochi and P. E. Mocadlo, *J. Org. Chem.*, **30**, 1134 (1965).
42. W. H. Richardson, *J. Amer. Chem. Soc.*, **87**, 247 (1965).
43. W. H. Richardson, *J. Amer. Chem. Soc.*, **87**, 1096 (1965).
44. W. H. Richardson, *J. Org. Chem.*, **30**, 2804 (1965).
45. Y. Takegami, Y. Fujimura, H. Ishii, and T. Iwamoto, *Kogyo Kagaku Zasshi*, **68**, 196 (1965).
46. W. H. Richardson, *J. Amer. Chem. Soc.*, **88**, 975 (1966).
47. N. Indictor and T. Jochsberger, *J. Org. Chem.*, **31**, 4271 (1966).
48. K. Onuma, K. Wada, J. Yamashita, and H. Hashimoto, *Bull. Chem. Soc. Japan*, **40**, 2900 (1967).
49. E. A. Kuzmina, V. A. Shushunov, and M. K. Shchennikova, *Zh. Obshch. Khim.*, **37**, 81 (1967).
50. W. T. Dixon and R. O. C. Norman, *Nature*, **196**, 891 (1962).
51. M. K. Shchennikova, E. A. Kuzmina, V. A. Shushunov, and C. A. Abakumov, *Dokl. Akad. Nauk SSSR*, **164**, 868 (1965).

52. M. K. Shchennikova and E. A. Kuzmina, *Tr. po Khim. i Khim. Tekhnol.*, **1966**, 29.

53. M. F. R. Mulcahy, J. R. Steven, and J. C. Ward, *Australian J. Chem.*, **18**, 1177 (1965).

54. R. O. C. Norman and B. C. Gilbert, in *Advances in Physical Organic Chemistry*, Vol. 5, edited by V. Gold, Academic Press, London, 1967, p. 53.

55. H. E. De La Mare, J. K. Kochi, and F. F. Rust, *J. Amer. Chem. Soc.*, **83** 2013 (1961).

56. M. S. Kharasch, A. Fono, and W. Nudenberg, *J. Org. Chem.*, **15**, 763 (1950).

57. D. T. Woodbridge, *J. Chem. Soc.*, *B*, **1966**, 50.

58. M. S. Kharasch, U.S. Dept. of Commerce Technical Report PB 123737. Cumulative report for period July 1953 to September 1955 (Contract No. N6 ori-02040)

59. W. H. Richardson, in *Oxidation in Organic Chemistry*, Vol. 5A, edited by K. B. Wiberg, Academic Press, New York, 1965, p. 243.

60. R. Criegee, in *Methoden der organischen Chemie*, Vol. 2 (Houben-Weyl), Thieme, Stuttgart, 1953, p. 570.

61. R. Criegee, in *Oxidation in Organic Chemistry*, Vol. 5A, edited by K. B. Wiberg, Academic Press, New York, 1965, p. 277.

62. P. D. Bartlett and P. Günther, *J. Amer. Chem. Soc.*, **88**, 3288 (1966).

63. P. D. Bartlett and G. Guaraldi, *J. Amer. Chem. Soc.*, **89**, 4799 (1967).

64. H. Hock and H. Kropf, *Ber.*, **91**, 1681 (1958).

65. M. S. Kharasch, A. Fono, W. Nudenberg, and B. Bischof, *J. Org. Chem.*, **17**, 207 (1952).

66. D. Benson and L. H. Sutcliffe, *Trans. Faraday Soc.*, **55**, 2107 (1959).

67. C. E. H. Bawn, *Discussions Faraday Soc.*, **14**, 181 (1953).

68. C. E. H. Bawn, A. A. Pennington, and C. F. H. Tipper, *Discussions Faraday Soc.*, **10**, 282 (1951),

69. L. Bateman, *Quart. Rev. (London)*, **8**, 147 (1954).

70. E. G. E. Hawkins, *Organic Peroxides*, Van Nostrand, Princeton, N.J., 1961, p. 15.

71. J. P. Wibaut and A. Strang. *Proc. Koninkl. Nederland Akad. Wetenschap*, **5B**, 102 (1951).

72. K. I. Ivanov, V. K. Savinova, and V. P. Zhakhovskaya, *Dokl. Akad. Nauk SSSR*, **59**, 703 (1948).

73. E. G. E. Hawkins, *Nature*, **166**, 69 (1950).

74. E. G. E. Hawkins and D. P. Young, *J. Chem. Soc.*, **1950**, 2804.

75. J. O. Punderson, U.S. Patent 2,822,399 (1958).

76. J. O. Punderson, U.S. Patent 2,700,057 (1955).

77. A. Farkas and A. I. Smith, U.S. Patent 2,497,349 (1950).

78. R. K. Madison and G. K. Bellis, U.S. Patent 2,855,438 (1958).

79. R. K. Madison and G. K. Bellis, U.S. Patent 2,887,510 (1959).

80. H. E. Holmquist, H. S. Hothrock, C. W. Theobald, and B. E. Englund, *J. Amer. Chem. Soc.*, **78**, 5339 (1956).

81. J. Minn, T. F. Sanderson, and L. A. Subluskey, *J. Amer. Chem. Soc.*, **78**, 630 (1956).

82. D. H. Hey, C. J. M. Stirling, and G. H. Williams, *J. Chem. Soc.*, **1957**, 1054.
83. S. Akiyoshi, N. Tashiro, and S. Kanayama, *Kogyo Kagaku Zasshi*, **60**, 286 (1957).
84. R. Criegee, *Ann.*, **565**, 7 (1949).
85. M. S. Kharasch and G. Sosnovsky, *J. Org. Chem.*, **23**, 1322 (1958).
86. A. G. Davies, *Organic Peroxides*, Butterworths, London, 1961, p. 72.
87. E. G. E. Hawkins, *Organic Peroxides*, Van Nostrand, Princeton, N.J., 1961, p. 149.
88. W. Cooper and W. H. T. Davison, *J. Chem. Soc.*, **1952**, 1180.
89. E. G. E. Hawkins, *J. Chem. Soc.*, **1955**, 3463.
90. M. J. Roedel, U.S. Patent 2,601,223 (1952).
91. M. J. Roedel, British Patent 697,506 (1954).
92. N. Brown, M. J. Hartig, M. J. Roedel, A. W. Anderson, and C. E. Schweitzer, *J. Amer. Chem. Soc.*, **77**, 1756 (1955).
93. E. G. E. Hawkins, British Patent 740,747 (1956).
94. J. B. Lavigne, U.S. Patent 2,828,338 (1958).
95. J. K. Kochi and F. F. Rust, *J. Amer. Chem. Soc.*, **84**, 3946 (1962)
96. J. B. Lavigne, U.S. Patent 2,938,918 (1960).
97. S. Havel, *Sb. Ved. Praci. Vysoka Skola Chem. Technol.*, *Pardubice*, **15** (1), 95 (1967).
98. C. E. H. Bawn and J. B. Williamson, *Trans. Faraday Soc.*, **47**, 735 (1951).
99. A. V. Andrianova, E. A. Kuzmina, V. A. Shushunov, and M. K. Shchennikova, *Tr. po Khim. i Khim. Tekhnol.*, **1965**, 85.
100. M. J. Kagan and G. D. Lubarsky, *J. Phys. Chem.*, **39**, 837 (1935).
101. E. Pfeil and K. H. Schmidt, *Ann.*, **675**, 36 (1964).
102. M. K. Shchennikova, V. A. Shushunov, and A. I. Milanov, *Tr. po Khim. i Khim. Tekhnol.*, **3**, 165 (1960).
103. M. K. Tumanova, V. A. Shushunov, and V. P. Stesikov, *Tr. po Khim. i Khim. Tekhnol.*, **3**, 171 (1960).
104. W. Kern and R. Schulz, *Makromol. Chem.*, **13**, 210 (1954).
105. M. E. Kurz and P. Kovacic, *J. Org. Chem.*, **33**, 1950 (1968).
106. P. Demerseman, R. N. Feinstein, and A. Cheutin, *Bull. Soc. Chim. France*, **1961**, 268.
107. J. A. Sharp. *J. Chem. Soc.*, **1957**, 2026.
108. N. A. Milas and J. T. Nolan, *J. Amer. Chem. Soc.*, **80**, 5826 (1958).
109. E. H. Farmer and A. Sundralingam, *J. Chem. Soc.*, **1942**, 121.
110. N. P. Emelyanov, E. A. Laryutina, and A. M. Smolskii, *Dokl. Akad. Nauk Belorusk. SSR*, **10**, 245 (1966).
111. A. H. Cook, *J. Chem. Soc.*, **1938**, 1774.
112. W. J. Farrissey, *Tetrahedron Letters*, **1964**, 3641.
113. W. J. Farrissey, U. S. Patent 3,401,193 (1968).
114. H. Hock and F. Depke, *Ber.*, **83**, 327 (1950).
115. D. H. S. Horn and D. Ilse, *J. Chem. Soc.*, **1957**, 2280.
116. Y. Fushizaki and M. Saito, *Bull. Univ. Osaka, Prefect Ser. A*, **6**, 155 (1958); through *Chem. Abstr.*, **53**, 4335h (1959).

117. M. Saito, N. Takeno, M. Mita, and Y. Fushizaki, *Kogyo Kagaku Zasshi*, **61**, 326 (1958).

118. G. O. Schenck, O. A. Neumüller, G. Ohloff, and S. Schroeter, *Ann.*, **687**, 26 (1965).

119. G. O. Schenck, O. A. Neumüller, and W. Eisfeld, *Angew. Chem.*, **70**, 595 (1958).

120. G. O. Schenck, O. A. Neumüller, and W. Eisfeld, *Ann.*, **618**, 202 (1958).

121. A. Robertson, *Nature*, **162**, 153 (1948).

122. J. G. M. Bremner and D. G. Jones, British Patent 608,539 (1948).

123. S. Murai, N. Sonoda, and S. Tsutsumi, *Bull. Chem. Soc. Japan*, **36**, 527 (1963).

124. M. Mainçon and P. Chassaing, U.S. Patent 2,481,765 (1949).

125. Usines de Melle, British Patent 614,392 (1948).

126. R. V. Digman and D. F. Anderson, *J. Org. Chem.*, **28**, 239 (1963).

127. J. H. Jones and M. R. Fenske, U.S. Patent 2,989,563 (1961).

128. E. A. Youngman, G. M. Coppinger, and H. E. De La Mare, *J. Amer. Chem. Soc.*, **82**, 1935 (1960): reference 7.

129. E. A. Youngman, F. F. Rust, G. M. Coppinger, and H. E. De La Mare, *J. Org. Chem.*, **28**, 144 (1963).

130. E. K. Varfolomeeva and Z. G. Zolotova, *Ukr. Khim. Zh.*, **25**, 708 (1959).

131. A. Rieche and W. Schön, *Z. Kunststoffe*, **57**, 49 (1967).

132. A. Rieche and W. Schön, *Ber.*, **99**, 3238 (1966).

133. A. Rieche and M. Schulz, *Ann.*, **653**, 32 (1962).

134. A. Rieche and M. Schulz, *Ber.*, **98**, 3623 (1965).

135. R. J. S. Beer, L. McGrath, and A. Robertson, *J. Chem. Soc.*, **1950**, 2118.

136. W. Pritzkow and R. Hofmann, *J. Prakt. Chem.*, **12**, 11 (1960).

137. H. Hock and S. Lang, *Ber.*, **75**, 1051 (1942).

138. H. Hock and S. Lang, *Ber.*, **76B**, 169 (1943).

139. H. Hock and S. Lang, *Ber.*, **76B**, 1130 (1943).

140. H. Hock, S. Lang, and G. Knauel, *Ber.*, **83**, 227 (1950).

141. M. S. Kharasch, A. Fono, and W. Nudenberg, *J. Org. Chem.*, **16**, 128 (1951).

142. E. G. E. Hawkins, *J. Chem. Soc.*, **1949**, 2076.

143. M. Hartmann and M. Seiberth, *Helv. Chim. Acta*, **15**, 1390 (1932).

144. J. Tomiška, *Coll. Czech. Chem. Commun.*, **27**, 1549 (1962); *ibid.*, **28**, 1177 (1963).

145. A. Robertson and W. A. Waters, *J. Chem. Soc.*, **1948**, 1578.

146. W. Treibs and G. Pellmann, *Ber.*, **87**, 1201 (1954).

147. Y. Kamiya, *Bull. Chem. Soc. Japan*, **38**, 2156 (1965).

148. E. Dyer, K. R. Carle, and D. E. Weiman, *J. Org. Chem.*, **23**, 1464 (1958).

149. T. Yamada, *J. Soc. Chem. Ind. Japan*, Suppl. Binding, **40**, 422 (1937).

150. K. I. Ivanov, V. K. Savinova, and E. G. Mikhailova, *Compt. Rend. Acad. Sci. URSS*, **25**, 34 (1939).

151. A. E. Woodward and R. B. Mesrobian, *J. Amer. Chem. Soc.*, **75**, 6189 (1953).

152. O. L. Harle and J. R. Thomas, Amer. Chem. Soc., Div. Petr. Chem., Preprints, **2**, No. 1, 43 (1957).

153. Y. Kamiya, S. Beaton, A. LaFortune, and K. U. Ingold, *Can. J. Chem.*, **41**, 2020 (1963).

154. O. Cerny and Z. Gibis, *Coll. Czech. Chem. Commun.*, **29**, 2992 (1964).

155. R. Criegee, H. Pilz, and H. Flygare, *Ber.*, **72**, 1799 (1939).

156. R. Criegee, *Fortschr. Chem. Forsch.*, **1**, 509 (1950); *Angew Chem.*, **70**, 173 (1958).

157. P. F. Ritchie, T. F. Sanderson, and L. F. McBurney, *J. Amer. Chem. Soc.*, **75**, 2610 (1953).

158. P. F. Ritchie, U.S. Patent 2,656,344 (1953).

159. A. V. Tobolsky and R. B. Mesrobian, *Organic Peroxides*, Interscience, New York, 1954, p. 96.

160. J. W. L. Fordham and H. L. Williams, *Can. J. Research*, **B28**, 551 (1950).

161. J. W. L. Fordham and H. L. Williams, *J. Amer. Chem. Soc.*, **73**, 1634 (1951).

162. R. J. Orr and H. L. Williams, *J. Amer. Chem. Soc.*, **76**, 3321 (1954).

163. R. J. Orr and H. L. Williams, *J. Amer. Chem. Soc.*, **77**, 3715 (1955).

164. R. J. Orr and H. L. Williams, *J. Amer. Chem. Soc.*, **78**, 3273 (1956).

165. I. M. Kolthoff and A. I. Medalia, *J. Amer. Chem. Soc.*, **71**, 3789 (1949).

166. J. W. L. Fordham and H. L. Williams, *J. Amer. Chem. Soc.*, **72**, 4465 (1950).

167. R. J. Orr and H. L. Williams, *Can. J. Chem.*, **30**, 985 (1952).

168. R. J. Orr and H. L. Williams, *Discussions Faraday Soc.*, **1953**, 170.

169. R. J. Orr and H. L. Williams, *J. Phys. Chem.*, **57**, 925 (1953).

170. E. A. Kuzmina, V. A. Shushunov, and M. K. Shchennikova, in *The Chemistry of Peroxy Compounds*, edited by I. I. Cherniaev, G. A. Razuvaev, I. I. Volnov, and T. A. Dobrinina, Akad. Nauk SSSR, 1963, p. 231.

171. H. Kropf, *Ann.*, **637**, 111 (1960).

172. V. A. Shushunov, M. K. Shchennikova, and I. V. Volkov, *Tr. po Khim. i Khim. Tekhnol.*, **1**, 55 (1958).

173. M. K. Shchennikova, V. A. Shushunov, and E. A. Kuzmina, *Tr. po Khim. i Khim. Tekhnol.*, **4**, 449 (1961).

174. H. Hock and H. Kropf, *J. Prakt. Chem.*, **16**, 113 (1962).

175. Y. A. Aleksandrov and V. A. Shushunov, *Tr. po Khim. i. Khim. Tekhnol.*, **3**, 9 (1960).

176. R. Lombard and J. Knopf, *Bull. Soc. Chim. France*, **1966**, 3926.

177. R. Lombard and J. Knopf, *Bull. Soc. Chim. France*, **1966**, 3930.

178. J. K. Kochi and D. D. Davis, *J. Amer. Chem. Soc.*, **86**, 5264 (1964).

179. A. I. Minkov, N. P. Keier, and V. F. Anufrienko, *Kinet. Katal.*, **8** (2), 387 (1967).

180. E. A. Kuzmina, V. A. Shushunov, and M. K. Shchennikova, *Khim. Perekisnykh Soedin. Akad. Nauk SSSR, Inst. Obshch. i Neorgan. Khim.*, **1963**, 231.

181. M. K. Shchennikova, V. A. Shushunov, and A. I. Milovanov, *Tr. po Khim. i Khim. Tekhnol.*, **3**, 165 (1960).

182. E. G. E. Hawkins, *J. Chem. Soc.*, **1950**, 2169.

183. A. Farkas and R. Rosenthal, U.S. Patent 2,994,719 (1961).

184. H. Kropf and D. Goschenhofer, *Tetrahedron Letters*, **1968**, 239.

185. W. S. Wise and G. H. Twigg, *J. Chem. Soc.*, **1953**, 2172.

186.  W. Kern and H. Willersinn, *Angew. Chem.*, **67**, 573 (1955).
187.  C. Walling, *Free Radicals in Solution*, Wiley, New York, 1957, p. 505.
188.  W. A. Pryor, *Free Radicals*, McGraw-Hill, New York, 1966, p. 118.
189.  K. I. Ivanov, V. K. Savinova, and V. P. Zhakhovskaya, *Zh. Obshch. Khim.*, **22**, 781 (1952).
190.  K. I. Ivanov, V. K. Savinova, and V. P. Zhakhovskaya, *Dokl. Akad. Nauk SSSR*, **59**, 905 (1948).
191.  E. G. E. Hawkins, D. C. Quin, and F. E. Salt, British Patent 641,250 (1950).
192.  S. Tsutsumi and Y. Odaira, *Nenryô Kyôkaishi*, **36**, 535 (1957); through *Chem. Abstr.*, **52**, 3719e (1958).
193.  Y. Odaira and S. Tsutsumi, *Techn. Rep. Osaka Univ.*, **7**, 485 (1957).
194.  M. S. Kharasch, A. C. Poshkus, A. Fono, and W. Nudenberg, *J. Org. Chem.*, **16**, 1458 (1951).
195.  W. Webster and D. P. Young, *J. Chem. Soc.*, **1956**, 4785.
196.  P. F. Ritchie, T. F. Sanderson, and L. F. McBurney, *J. Amer. Chem. Soc.*, **76**, 723 (1954).
197.  B. Yeomans and D. P. Young, *J. Chem. Soc.*, **1958**, 2288.
198.  M. S. Kharasch, F. S. Arimoto, and W. Nudenberg, *J. Org. Chem.*, **16**, 1556 (1951).
199.  M. S. Kharasch, W. Nudenberg, and F. Arimoto, *Science*, **113**, 392 (1951).
200.  M. S. Kharasch, P. Pauson, and W. Nudenberg, *J. Org. Chem.*, **18**, 322 (1953).
201.  A. F. Bickel, E. C. Kooyman, W. Roest, P. Piet, and C. La Lau, *J. Chem. Soc.*, **1953**, 3211.
202.  M. S. Kharasch, F. Kawahara, and W. Nudenberg, *J. Org. Chem.* **19**, 1977 (1954).
203.  M. S. Kharasch, W. Nudenberg, and F. Kawahara, *J. Org. Chem.*, **20**, 1550 (1955).
204.  J. K. Kochi, *Science*, **155**, 415 (1967).
205.  T. W. Campbell and G. M. Coppinger, U.S. Patent 2,610,972 (1952).
206.  T. W. Campbell and G. M. Coppinger, *J. Amer. Chem. Soc.*, **74**, 1469 (1952).
207.  M. S. Kharasch and A. Fono, *J. Org. Chem.*, **23**, 324 (1958).
208.  M. S. Kharasch and A. Fono, *J. Org. Chem.*, **23**, 325 (1958).
209.  M. S. Kharasch and G. Sosnovsky, *Tetrahedron*, **3**, 105 (1958).
210.  M. S. Kharasch and A. Fono, *J. Org. Chem.*, **24**, 72 (1959).
211.  G. Sosnovsky and S. O. Lawesson, *Angew. Chem.*, **76**, 218 (1964); *Angew. Chem., Intern. Ed.*, **3**, 269 (1964).
212.  G. Sosnovsky and D. J. Rawlinson, in *Organic Peroxides*, Vol. I, edited by D. Swern, Wiley, New York, 1970, pp. 561, 585.
213.  F. Minisci, R. Galli, and M. Cecere, *Gazz. Chim. Ital.*, **94**, 67 (1964).
214.  F. Minisci and R. Galli, *Tetrahedron Letters*, **1963**, 357.
215.  D. B. Denney, R. Napier, and A. Cammarata, *J. Org. Chem.*, **30**, 3151 (1965).
216.  F. Minisci, M. Cecere, and R. Galli, *Gazz. Chim. Ital.*, **93**, 1288 (1963).
217.  J. R. Shelton and J. N. Henderson, *J. Org. Chem.*, **26**, 2185 (1961).
218.  W. J. Farrissey, *J. Org. Chem.*, **29**, 391 (1964).

219. R. LaLande and C. Filliatre, *Bull. Soc. Chim. France*, **1962**, 792.
220. L. Borowiecki, *Roczniki Chem.*, **37**, 921 (1963).
221. W. F. Brill, *J. Amer. Chem. Soc.*, **85**, 141 (1963).
222. E. S. Gould, R. R. Hiatt and K. C. Irwin, *J. Amer. Chem. Soc.*, **90**, 4573 (1968).
223. N. Indictor and W. F. Brill, *J. Org. Chem.*, **30**, 2074 (1965).
224. J. G. Zajacek and M. N. Sheng, Middle Atlantic Regional Meeting, ACS, Philadelphia, February, 1968; *Advan. Chem. Ser.*, **1968**, 418.
225. J. B. Braunwarth, U.S. Patent 3,272,856 (1966).
226. F. Minisci, R. Galli, and U. Pallini, *Gazz. Chim. Ital.*, **91**, 1023 (1961).
227. F. Minisci and R. Galli, *Tetrahedron Letters*, **1962**, 533.
228. F. Minisci and R. Galli, French Patent 1,350,360 (1964).
229. D. J. MacKinnon and W. A. Waters, *J. Chem. Soc.*, **1953**, 323.
230. J. K. Kochi, *J. Amer. Chem. Soc.*, 2785 (1962).
231. J. K. Kochi, *J. Amer. Chem. Soc.*, **84**, 3271 (1962).
232. M. S. Kharasch and A. Fono, *J. Org. Chem.*, **24**, 606 (1959).
233. E. C. Horswill and K. U. Ingold, *Can. J. Chem.*, **44**, 263 (1966).
234. E. C. Horswill and K. U. Ingold, *Can. J. Chem.*, **44**, 269 (1966).
235. G. M. Coppinger, *J. Amer. Chem. Soc.*, **79**, 2758 (1957).
236. A. L. J. Beckwith and B. S. Low, *J. Chem. Soc.*, **1964**, 2571.
237. L. Kuhnen, *Angew. Chem. Intern.*, *Ed.*, **5**, 893 (1966).
238. J. A. Bigot, *Rec. Trav. Chim.*, **80**, 825 (1961).
239. F. Minisci, *Gazz. Chim. Ital.*, **89**, 626 (1959).
240. F. Minisci, *Gazz. Chim. Ital.*, **89**, 2428 (1959).
241. F. Minisci and A Portolani, *Gazz. Chim. Ital.*, **89**, 1922 (1959).
242. F. Minisci and U. Pallini, *Gazz. Chim. Ital.*, **89**, 2438 (1959).
243. F. Minisci and U. Pallini, *Gazz. Chim. Ital.*, **91**, 1030 (1961).
244. D. D. Coffman and H. N. Cripps, U.S. Patent 2,811,551 (1957).
245. D. D. Coffman and H. N. Cripps, *J. Amer. Chem. Soc.*, **80**, 2880 (1958).
246. Montecatini, British Patent 890,141 (1962).
247. F. Minisci, U.S. Patent 3,026,334 (1962).
248. J. B. Braunwarth and G. W. Crosby, *J. Org. Chem.*, **27**, 2064 (1962).
249. F. Minisci and A. Portolani, *Gazz. Chim. Ital.*, **89**, 1941 (1959).
250. G. Chiusoli and F. Minisci, *Atti Accad. Nazl. Lincei Rend.*, **23**, 140 (1957).
251. G. Chiusoli and F. Minisci, Italian Patent 568,332 (1957).
252. G. Chiusoli and F. Minisci, *Gazz. Chim. Ital.*, **88**, 43 (1958).
253. G. Chiusoli and F. Minisci, *Gazz. Chim. Ital.*, **88**, 261 (1958).
254. G. Chiusoli and F. Minisci, Italian Patent 579,419 (1960).
255. M. S. Kharasch and W. Nudenberg, *J. Org. Chem.*, **19**, 1921 (1954).
256. M. S. Kharasch and W. Nudenberg, French Patent 1,121,551 (1954).
257. N. V. de Bataafsche Petroleum Maatschappij, British Patent 762,340 (1956).
258. F. Minisci, *Gazz. Chim. Ital.*, **89**, 1910 (1959).
259. F. Minisci, *Gazz. Chim. Ital.*, **91**, 386 (1961).
260. G. Chiusoli, Italian Patent 564,920 (1957).
261. D. D. Coffman, R. Cramer, and W. E. Mochel, *J. Amer. Chem. Soc.*, **80**, 2882 (1958).

262. F. Minisci, *Angew. Chem.*, **70**, 599 (1958).
263. F. Minisci, Italian Patent 580,012 (1957).
264. J. B. Braunwarth and G. W. Crosby, U.S. Patent 2,905,712 (1959).
265. J. B. Braunwarth, U.S. Patent 2,983,751 (1961).
266. J. B. Braunwarth and G. W. Crosby, U.S. Patent 3,061,619 (1962).
267. A. M. Hyson, U.S. Patent 2,710,302 (1955).
268. F. Minisci and R. Galli, *Chim. Ind. (Milan)*, **45**, 448 (1963).
269. F. Minisci, R. Galli, and M. Cecere, *Chim. Ind. (Milan)*, **46**, 1064 (1964).
270. G. B. Payne, C. W. Smith, and P. R. van Ess, U.S. Patent 2,957,907 (1960).
271. G. B. Payne and C. W. Smith, *J. Org. Chem.*, **23**, 1066 (1958).
272. J. B. Braunwarth and G. W. Crosby, U.S. Patent 2,933,525 (1960).
273. F. Minisci and G. Belvedere, Italian Patent 682,628 (1965).
274. F. Minisci and G. Belvedere, *Gazz. Chim. Ital.*, **90**, 1299 (1960).
275. J. B. Braunwarth and C. Rai, U.S. Patent 3,197,488 (1965).
276. F. Minisci, M. Cecere, R. Galli, and A. Selva, paper presented at the National Congress of the Italian Chemical Society, Padua, June 21, 1968.
277. D. D. Coffman and H. N. Cripps, *J. Amer. Chem. Soc.*, **80**, 2877 (1958).
278. D. D. Coffman and H. N. Cripps, U.S. Patent 2,671,810 (1954).
279. G. Cainelli and F. Minisci, *Chim. Ind. (Milan)*, **47**, 1214 (1965).
280. L. S. Abbott, British Patent 749,881 (1956).
281. E. G. E. Hawkins, British Patent 768,944 (1957).
282. F. Minisci, U. Pallini, and R. Galli, Italian Patent 648,556 (1962).
283. F. Minisci, U. Pallini, and R. Galli, French Patent 1,316,261 (1963).
284. Halcon International, Inc., Netherlands Patent Application 6,514,129; through *Chem. Abstr.*, **65**, 15230e (1966).
285. A. V. Tobolsky and R. B. Mesrobian, *Organic Peroxides*, Interscience, New York, 1954, p. 17.
286. A. G. Davies, *Organic Peroxides*, Butterworths, London, 1961, p. 27.
287. E. G. E. Hawkins, *Organic Peroxides*, Van Nostrand, Princeton, N.J., 1961, p. 118.
288. F. Minisci, *Atti Accad. Nazl. Lincei Rend.*, **25**, 538 (1958).
289. F. Minisci, *Gazz. Chim. Ital.*, **90**, 1307 (1960).
290. F. Minisci and U. Pallini, *Gazz. Chim. Ital.*, **90**, 1318 (1960).
291. F. Minisci, R. Galli, and M. Cecere, *Gazz. Chim. Ital.*, **95**, 751 (1965).
292. F. Minisci and R. Galli, *Tetrahedron Letters*, **1964**, 3197.
293. F. Minisci and R. Galli, *Chim. Ind. (Milan)*, **47**, 178 (1965).
294. R. Hiatt, K. C. Irwin, and C. W. Gould, *J. Org. Chem.*, **33**, 1430 (1968).

*Chapter III*

# Metal-Ion-Catalyzed Reactions of Hydrogen Peroxide and Peroxydisulfate

## G. SOSNOVSKY & D. J. RAWLINSON*

*Supported in part by a grant from the Research Council of Western Illinois University (Summer 1969).

## I.  INTRODUCTION

### A.  Historical Background

The chemistry of metal-ion-catalyzed reactions of hydrogen peroxide commenced at the end of the last century, when Fenton (1–5) described the reaction of tartaric acid with hydrogen peroxide in the presence of ferrous sulfate. Shortly after this discovery, Ruff (6–8) reported a modified procedure with ferric salts. This procedure has been particularly useful in carbohydrate chemistry (9).

During the next decades a large number of communications appeared dealing with the Fenton reaction and the Ruff degradation. However, it was not until 1933 that Haber and Weiss (10–14) proposed a plausible free-radical mechanism for the Fenton reaction.

$$H_2O_2 + Fe^{2+} \longrightarrow HO\cdot + Fe^{3+} + HO^-$$

In the absence of organic substrates the hydroxy radical undergoes the following series of reactions:

$$HO\cdot + Fe^{2+} \longrightarrow Fe^{3+} + OH^-$$

$$HO\cdot + H_2O_2 \longrightarrow H_2O + HOO\cdot$$

$$HO_2\cdot + H_2O_2 \longrightarrow HO\cdot + H_2O + O_2$$

$$HO_2\cdot + H_2O \longrightarrow H_3O^+ + O_2^-\cdot$$

$$O_2^-\cdot + H_2O_2 \longrightarrow OH^- + HO\cdot + O_2$$

During the following years the chemistry of the metal-ion-catalyzed decomposition of hydrogen peroxide and of peroxydisulfate received increasing attention, partly because of various biological implications (15–18) and partly because of their importance as initiators in polymerization processes (19–23).

In spite of a large volume of literature, the available review articles (9, 15–21, 24–42) that are devoted either in part or entirely to this topic are either outdated or provide information only on a specific topic. This chapter is an attempt to present a comprehensive and critical review combining various aspects of the chemistry of hydrogen peroxide and peroxydisulfate in the presence of metal ions.

### B. Scope and Limitations

This chapter describes the reactions of hydrogen peroxide and peroxydisulfate with various compounds in the presence of metal ions that are capable of oxidation–reduction processes. The discussion is restricted to transition metals of the first transition series, to silver of the second transition series, and to cerium of the lanthanide series. The discussion is further restricted to homogeneous catalysis; reactions occurring at surfaces (19, 43–54) are not included. The discussion is also restricted to the reactions that proceed, at least in part, by a free-radical mechanism. Almost all studies involving chelates (16, 55–66) and other complexes of metal ions (67–74) are omitted. Further excluded from this review are analytical methods utilizing hydrogen peroxide or peroxydisulfate and metal ions (75–82), as well as various biological problems, such as the action of metalloenzymes on hydrogen peroxide, and biological-model systems involving hydroxy radicals. The emphasis is placed mainly on the preparative side, and no discussion is made of kinetics and other mechanistic details. Nevertheless, mechanistic schemes are used, and it is believed that the most important references to theoretical aspects are included.

Although electron-spin-resonance (ESR) results (37, 83, 84) are not discussed in detail, references leading to various studies of radicals generated in the system organic substrate–metal ion–hydrogen peroxide are included throughout this chapter. However, studies of the oxidizing species generated in the absence of organic substrates (85–97) are not discussed.

### C.    Nomenclature

The nomenclature of *Chemical Abstracts* is used wherever possible. However, in many instances the original nomenclature used by authors is retained in order to avoid confusion and to facilitate cross-checking.

### II.    ALIPHATIC AND CYCLIC HYDROCARBONS

Alkanes are resistant to hydrogen peroxide. Thus no reaction takes place between $n$-pentane and 90% hydrogen peroxide in the presence of iron or manganese ions (98).

However, aromatic hydrocarbons with saturated substitutents may react at both the side chain and the aromatic nucleus. Thus the reaction of toluene with hydrogen peroxide in the presence of the ferrous ion has been reported to yield benzyl alcohol (99, 100) benzaldehyde (99–103), benzoic acid (99), hydroxy benzaldehydes (99, 100, 103), and dibenzyl (99–102), all products of side-chain reactivity, together with other products derived from attack at the benzene ring. The ESR spectrum of the benzyl radical is observable when toluene is reacted with a mixture of titanous chloride and hydrogen peroxide (104). The reaction of toluene with a mixture of ferric and ferrous sulfates and sodium azide yields benzyl azide, which may be reduced catalytically to benzylamine (105).

$$Fe^{2+} + H_2O_2 + H^+ \longrightarrow Fe^{3+} + H_2O + HO\cdot$$

$$HO\cdot + PhCH_3 \longrightarrow H_2O + PhCH_2\cdot$$

$$PhCH_2\cdot + Fe(N_3)^{2+} \longrightarrow PhCH_2N_3 + Fe^{2+}$$

$$PhCH_2N_3 \xrightarrow{\text{H}_2/\text{cat.}} PhCH_2NH_2$$

Under similar conditions the reaction of tetralin gives an azido derivative that is converted by hydrogenation to the well-defined aminotetralin derivative (105).

$$H_2O_2 + Fe^{2+} \longrightarrow HO\cdot + Fe^{3+} + OH^-$$

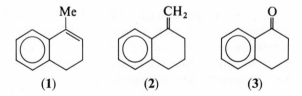

The reaction of ethylbenzene with hydrogen peroxide in the presence of acidified ferrous sulfate yields mainly the product of side-chain attack, acetophenone (101). The reaction of *p*-cymene in acetic acid with excess hydrogen peroxide in the presence of ferrous sulfate gives cuminaldehyde (106).

Unsaturated hydrocarbons undergo a variety of reactions with hydrogen peroxide in the presence of metal ions. The reactions of acetylene with hydrogen peroxide in the presence of ferrous sulfate gives mainly acetic acid, together with some acetaldehyde and ethyl alcohol (107). Under similar conditions pentene gives a mixture of acetic acid, acetone, formic acid, and carbon dioxide (108).

Oxidation of a mixture of methyl-dihydronaphthalene (**1**) and methylene-tetralin (**2**) with hydrogen peroxide in the presence of ferrous sulfate forms α-tetralone (**3**) in poor yield (109).

Me CH$_2$ O

(**1**)     (**2**)     (**3**)

Olefinic alcohols can be hydroxylated to glycols; for example, the reaction of allyl alcohol with hydrogen peroxide in the presence of a mixture of ferrous and cupric ions gives a low yield of glycerol (110).

More interesting and more useful results have been obtained from metal-ion-catalyzed reactions of hydrogen peroxide with olefins in the presence of various substrates that are either organic compounds or inorganic salts. Thus the reaction of hexene-1 with a mixture of hydrogen peroxide, ferrous sulfate, and sodium azide in methanol at $-5°$ produces a diazide, which can be reduced in the presence of a Raney nickel–chromium catalyst to 1,2-diaminohexane (105, 111). The reaction is explained by the following scheme (105, 112, 113):

$$H_2O_2 + Fe^{2+} \longrightarrow \cdot OH + Fe^{3+} + OH^-$$

$$HO\cdot + FeN_3^{2+} \longrightarrow Fe(OH)^{2+} + N_3\cdot$$

$$-CH=CH- + N_3\cdot \longrightarrow -CH-\overset{\cdot}{C}H(N_3)-$$

$$\overset{FeN_3^{2+}}{\longrightarrow} -CH(N_3)CH(N_3)-$$

Similarly the reactions of styrene (105, 111), cyclohexene (105, 111), and dehydroisoandrosterone acetate (105) produce the corresponding diazide derivatives, which can be reduced to the diamino compounds.

This type of reaction can be also applied to diolefins. Thus the reaction of cyclopentadiene produces 3,5-diazidocyclopentene, which is converted by hydrogenation to diaminocyclopentane (105, 111). A similar reaction with furan gives a highly explosive diazidodihydrofuran derivative (105).

Under similar conditions butadiene yields a mixture of 1,4-diazidobutene-2 and 3,4-diazidobutene-1 (105, 111).

Addition of chloride ions to the reaction mixture gives a product containing the azido group and the chlorine atom (105, 113).

$$\cdot OH + FeN_3^{2+} \longrightarrow Fe(OH)^{2+} + N_3\cdot$$

$$-CH=CH- + N_3\cdot \longrightarrow -\overset{\cdot}{C}HCH(N_3)-$$

$$-\overset{\cdot}{C}HCH(N_3)- + FeCl^{2+} \longrightarrow -CH(Cl)CH(N_3)- + Fe^{2+}$$

Thus the reactions of cyclohexene (105, 113), cholesterol (105), and dehydroisoandrosterone (105) with aqueous hydrogen peroxide in the presence of a mixture of ferrous sulfate, ferric chloride, and sodium azide at 25–33° result in chloroazidocyclohexane, chloroazidocholesterol, and chloroazidodehydroisoandrosterone, respectively.

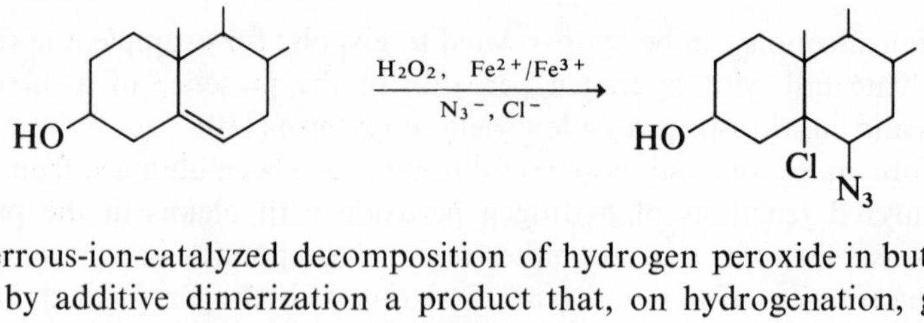

The ferrous-ion-catalyzed decomposition of hydrogen peroxide in butadiene yields by additive dimerization a product that, on hydrogenation, gives a 30% yield of octanediol (111, 114, 115).

$$
\begin{aligned}
HO\cdot + CH_2{=}CHCH{=}CH_2 &\longrightarrow HOCH_2\underset{\cdot}{C}HCH{=}CH_2 \\
&\longleftrightarrow HOCH_2CH{=}CH\underset{\cdot}{C}H_2 \\
&\longrightarrow HOC_4H_6C_4H_6OH \\
&\xrightarrow{H_2/\text{cat.}} HOC_4H_8C_4H_8OH
\end{aligned}
$$

In the presence of halide ions in this reaction mixture dihalo derivatives of dimers are obtained (116–118).

$$MX + \cdot OH \longrightarrow MOH + X\cdot$$

$$X\cdot + CH_2{=}CHCH{=}CH_2 \longrightarrow XC_4H_6\cdot \longrightarrow XC_4H_6C_4H_6X$$

$$(M = \text{Na, Fe, Ti};\ X = \text{Cl, Br})$$

Additive dimerization has been applied to a large variety of substrates, RH, such as alcohols, thiols, aldehydes, ketones, amines, acids, nitriles, and esters to give long-chain unsaturated difunctional compounds in accordance with the following scheme (119):

$$RH + \cdot OH \longrightarrow R\cdot + H_2O$$

$$R\cdot + M \longrightarrow RM\cdot \longrightarrow RMMR$$

(RH = organic compound, hydrogen atom linked to carbon)
(R· = carbon radical)
[M = butadiene, chloroprene, isoprene, acrylonitrile, maleic acid, cyclopentadiene (119), 1,1,4,4-tetrafluorobutadiene (120).]

In the case of dienes the products consist of a mixture of straight-chain compounds of type **4** and of branched-chain compounds of type **5** and **6**, which are formed by 1,2- and 1,4-addition to the diene (119).

$$R(CH_2CH{=}CHCH_2)_2R$$

**(4)**

$$\begin{array}{l} RCH_2CH{=}CHCH_2CH_2CHR \\ \phantom{RCH_2CH{=}CHCH_2CH_2C}| \\ \phantom{RCH_2CH{=}CHCH_2CH_2}CH{=}CH_2 \end{array}$$

**(5)**

$$\begin{array}{c} \phantom{RCH_2C}CH{=}CH_2 \\ \phantom{RCH_2CHC}|\phantom{HCH_2R} \\ RCH_2CHCHCH_2R \\ \phantom{RCH_2C}|\phantom{HCHCH_2R} \\ CH{=}CH_2\phantom{CH_2R} \end{array}$$

Sometimes the terminal groups R are not identical at both ends; for example, the reaction of acetaldehyde with butadiene gives products that contain R = $CH_2CHO$ and R = MeCO groups (119).

The following substrates, RH, have been used in this type of condensation: ethyl alcohol (119, 121), isopropyl alcohol (119, 121), n-butyl alcohol (119, 121), t-butyl alcohol (119–121), cyclohexanol (121), ethanethiol (123, 124), acetaldehyde (119, 122), acetone (119, 122), cyclopentanone (119, 122), cyclohexanone (122), cyclohexylamine (122), t-butylamine (120, 122), propionitrile (122), acetic acid (119, 122), propionic acid (119, 122), n-butyric acid (122), glutaric acid (119, 122), thioglycolic acid (123, 124), thiolacetic acid (123, 124), methyl formate (119), and methyl acetate (119, 122).

Autoxidation of a reaction mixture containing an olefin, hydrogen peroxide, ferrous ions, and azide ions produces a derivative containing an azide and a keto function (125).

$$HC\!=\!CH + N_3\cdot \longrightarrow -\underset{\cdot}{C}HCH(N_3)-$$

$$\xrightarrow{O_2} \underset{OO\cdot}{-CHCH(N_3)-} \xrightarrow{Fe^{2+}} \underset{O}{-CCH(N_3)-} + Fe(OH)^{2+}$$

Reactions with the following olefins produce the corresponding azidoketones: hexene-1, cyclohexene, styrene, and stilbene (125).

In some cases elimination of the azido group occurs to give ketones; for example, the reaction of α-methylstyrene yields acetophenone (125):

$$\underset{OO\cdot}{PhC(Me)CH_2N_3} + 2\,Fe^{2+} + H_2O \longrightarrow \underset{O\cdot}{PhC(Me)\!-\!CH_2N_3}$$

$$PhC(O)Me + \cdot CH_2N_3$$

The reaction of acrylonitrile gives an intermediate that readily reacts with methyl alcohol to give methyl azidoacetate (7) (125):

$$CH_2\!=\!CHCN + N_3\cdot \longrightarrow N_3CH_2\underset{\cdot}{C}HCN$$

$$\xrightarrow{O_2} \underset{OO\cdot}{N_3CH_2CHCN} \xrightarrow{Fe^{2+}} N_3CH_2C(O)CN$$

$$\xrightarrow{MeOH} N_3CH_2C(O)OMe + HCN$$
$$(7)$$

The same type of product is obtained from the reaction of ethyl vinyl ether (125):

$$CH_2\!=\!CH\!-\!OEt \xrightarrow[O_2,\,Fe^{2+}]{N_3\cdot} N_3CH_2C(O)OEt$$

Electron-spin-resonance studies have been made of radicals produced by the interaction of the titanous chloride–hydrogen peroxide system with alkenes (126–128), dienes (128), and other materials containing carbon–carbon double bonds (87, 126, 128–130).

## III. AROMATIC COMPOUNDS

The interaction between aromatic compounds and hydrogen peroxides in the presence of metal ions is described briefly in several reviews and books dealing with broader topics (17, 18, 24, 30, 39, 41, 42, 131–134). In general a variety of products may result from some combination of hydroxylation of the aromatic system, oxidative coupling, and side-chain attack; in addition, the initial products of hydroxylation, phenols, are readily oxidized further to complex products. For this reason much of the literature is of little interest in regard to synthesis. Important studies have been directed toward the determination of isomer ratios of monosubstitution products of substituted benzenes. These, however, have generally involved reaction conditions (substrate in large excess together with dilute hydrogen peroxide) that are not useful for synthetic work. Other studies dealing with mechanistic aspects, ESR spectra of transient intermediates or models for *in vivo* hydroxylation in biological systems, lie outside the scope of this chapter and are only briefly mentioned.

### A. Aromatic Hydrocarbons

An early report (107) records the isolation of catechol (32% yield), phenol (16% yield), and trace amounts of quinol from the reaction of benzene and excess dilute aqueous hydrogen peroxide at 45° in the presence of ferrous sulfate.

Later it was reported (135) that under similar conditions the yield of phenol increases with decreasing concentration of hydrogen peroxide, and with 0.55% hydrogen peroxide a yield of 18.5% is obtained; only traces of catechol are formed. Low yields of phenol also result from the reaction of 90% hydrogen peroxide with benzene in the presence of the ferrous ion (98).

The reaction of hydrogen peroxide with a mixture of benzene, acetic acid, and dilute sulfuric acid containing ferrous sulfate and a finely divided palladium catalyst yields phenol and smaller amounts of quinol (136).

The reaction of hydrogen peroxide with excess benzene in acidified aqueous medium catalyzed by ferrous sulfate yields phenol and smaller amounts of diphenyl (99, 101). Detailed studies have been made of the variation of product yield and composition with experimental variables (99, 137, 138), of the radical intermediates (17, 99, 104, 139), and of other mechanistic aspects (140–142). The following reaction scheme is accepted:

$$H_2O_2 + Fe^{2+} + H^+ \longrightarrow Fe^{3+} + H_2O + \cdot OH$$

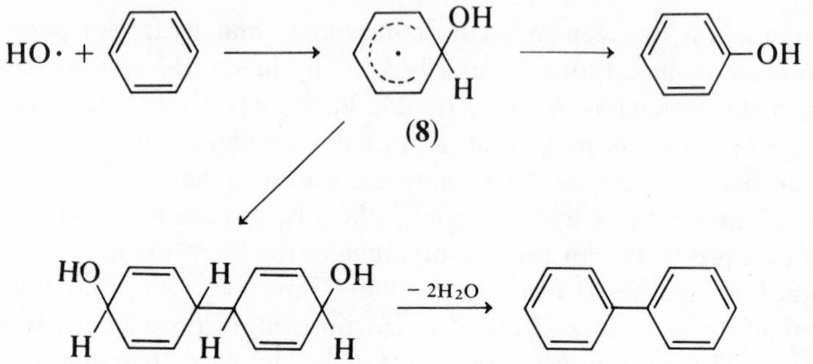

An intermediate hydroxycyclohexadienyl radical (8) is produced by the addition of the hydroxy radical to benzene. Products stem from intermediate 8 either by dimerization to 9, followed by loss of water to yield diphenyl, or by oxidation (by the ferric ion, oxygen, or radicals) to yield phenol. Intermediate 8 has been detected by ESR spectroscopy (17, 99, 104, 139) in similar systems (utilizing titanous chloride as the metal-ion catalyst). In the presence of oxygen no diphenyl is formed and phenol is the only product (137). The presence of ferric ions also favors the formation of phenol (99, 137). Conversely, the presence of phosphate (137) or fluoride (99) ions, which complex with the ferric ion, decreases the proportion of phenol.

The titanous ion (99), and to a lesser extent copper (143), ferric (99, 144), and ceric (99) ions, are also catalysts for the reaction of benzene with hydrogen peroxide. Modified systems have been investigated (145), but the work is not of interest in regard to synthesis.

The reaction of a saturated aqueous solution of naphthalene in excess (146) with hydrogen peroxide at room temperature catalyzed by ferrous sulfate produces 1- and 2-naphthol in the ratio 3.8:1. Also obtained are ring-cleavage products: salicylic acid, an unidentified o-carboxyhydroxy-cinnamic acid, and two unidentified phenols.

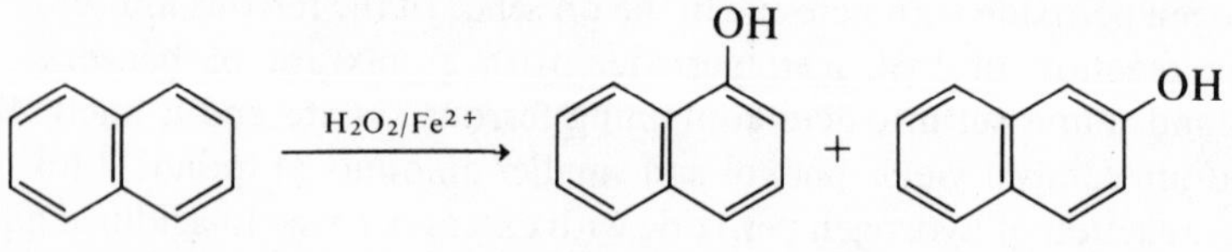

No dinaphthyls and no naphthalene diols are formed. Under the same conditions 1,1'- and 2,2'-dinaphthyl are unreactive.

The reaction of anthracene (147) with hydrogen peroxide in acetic acid at 110° catalyzed by basic ferric acetate produces mainly anthraquinone (86%) yield. Also obtained are a mixture of phenol derivatives (13% yield) containing compound **10** and a mixture of carboxylic acid derivatives (6% yield) containing the carboxylic compound **11**.

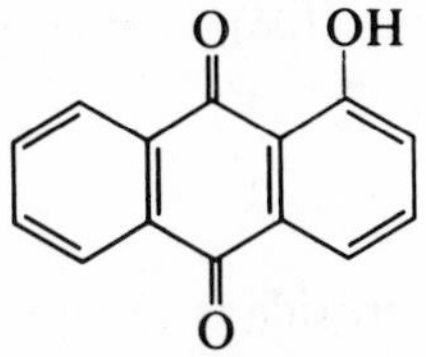

**(10)**

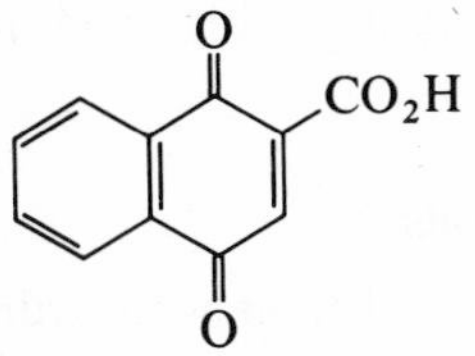

Compound **10** is suggested as an intermediate in the formation of the cleavage product **11**. The mechanism of the reaction catalyzed by the ferrous ion has been studied (148).

The ferrous-ion-catalyzed reaction of toluene with hydrogen peroxide results in hydroxylation, oxidative dimerization, and side-chain attack. Various products are reported in the literature. Merz and Waters (101) detected dibenzyl, benzaldehyde, and cresols. Boguslavskaya et al. (102) reported the formation of dibenzyl, benzaldehyde, and aromatic hydroxy compounds. Iwata, Yoshikawa, and Tzutsumi (103) found formic acid, benzaldehyde, 3,3'-bitolyl, and the isomeric hydroxybenzaldehydes; dibenzyl and cresols were not observed. Suzuki and Hotta (100) detected benzaldehyde as the major product, together with small amounts of *o*-hydroxybenzalde-hyde, benzyl alcohol, the three cresols, and dibenzyl. Dixon and Norman (104) detected the three cresols, dibenzyl, *o*-hydroxybenzaldehyde, diphenyl-methane, and 2,2'-bitolyl, benzyl alcohol, benzoic acid, and phenol.

The variety of products may be rationalized mechanistically. Formation of the nuclear-substitution products, cresols, and of bitolyls is explicable through initial addition of hydroxyl to yield hydroxycyclohexadienyl-radical intermediates of type **12**, as with benzene. Formic acid, phenol, and 2,2'-bitolyl may result from addition of the hydroxyl radical at the C-1 position of toluene, producing the cyclohexadienyl radical (**13**).

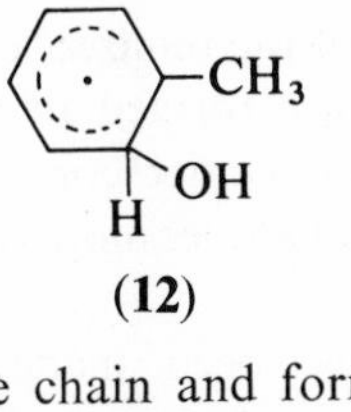

**(12)** 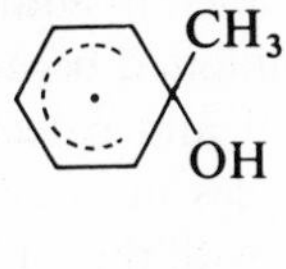        **(13)**

Oxidation of the side chain and formation of dibenzyl result from initial abstraction of hydrogen from the methyl group of toluene to yield the benzyl radical.

$$\text{⟨⟩—CH}_3 + \text{HO·} \longrightarrow \text{⟨⟩—CH}_2\text{· + H}_2\text{O}$$

The ESR spectrum of the benzyl radical in the system has been detected (104). Reactions of interest for synthesis at the side chain of toluene (105), tetralin (105), ethylbenzene (101), and *p*-cymene (106) have been described in Section II.

### B.   Phenols

The reaction of phenols with hydrogen peroxide is catalyzed by transition-metal ions. In many cases the products are complex, since further oxidation or complexing by the metal ions in their higher oxidation states may occur.

Phenol reacts vigorously in neutral or weakly acid aqueous solution with hydrogen peroxide in the presence of ferrous or ferric salts to yield complex colored products (149). The main product in the presence of ferrous salts is catechol, together with some *p*-benzoquinone, quinol, and colored materials (150). A 39% yield of dihydroxy benzenes has been recorded from the reaction in the presence of ferrous sulfate (151). In addition to catechol (3–45% yield) and traces of quinol, small amounts of pyrogallol (**14**) and purpurogallin (**15**) are formed (152).

**(14)**                    **(15)**

A yield of dihydroxy benzenes (catechol and quinol) of 72% may be obtained in the pH range 3–4 by using excess phenol (153). The catalytic activity of various metal ions on the rate of oxidation has been studied (154).

Product studies have been carried out under controlled conditions (155). The reaction of phenol with hydrogen peroxide is slower in the presence of fluoride or phosphate ions; in addition, the formation of secondary products, such as quinones, is inhibited because of the removal of the ferric ion. Under such conditions quinol and catechol are formed in a ratio of about 3:1, together with only traces of resorcinol. Under conditions where the ferric ion is not removed, or with the ferric ion alone, more catechol is formed than quinol.

Electron-spin-resonance studies have been made of transient radical intermediates formed during the reaction of phenol with the system titanous chloride–hydrogen peroxide (95, 104, 139). Different species are observed at different acidities. At pH 3.9–6.5 adduct radical **16** and at least one isomer of compound **16** are observed. In more acid solution (pH 1) the phenoxy radical (**17**) is present. In more basic solution (pH 7), the spectra of *o*-benzosemiquinone (**18**) and *p*-benzosemiquinone radicals result.

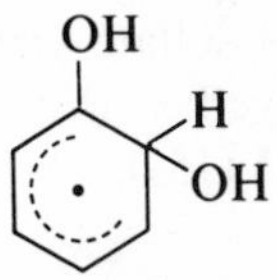

(16) 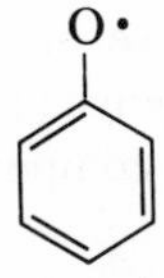       (17) 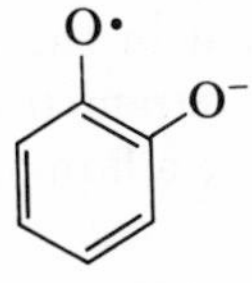       (18)

Other studies pertinent to mechanism have been made (142, 148, 156, 157).

The reaction of 1 and 2-naphthol with hydrogen peroxide in the presence of ferrous sulfate yields only polymeric material (146).

Dilute aqueous hydrogen peroxide reacts with *o*-cresol in the presence of ferrous sulfate to yield mainly uncharacterized iron complexes, together with some ditolyldiquinone (**19**) and a trace of *p*-toluhydroquinone (**20**) (158).

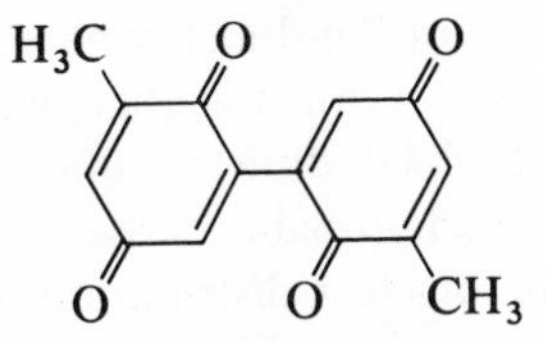

(19) 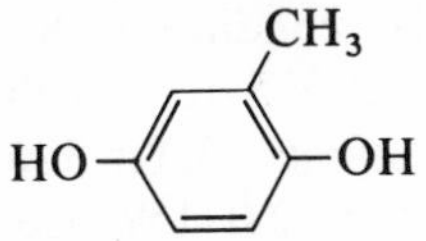           (20)

Under similar conditions the reaction of *m*-cresol yields toluhydroquinone (**20**), ditolyldiquinone (**19**), and the same iron complex (158). The reaction of *p*-cresol gives a black oxidation product that has not been characterized (158).

Under acid conditions the formation of iron complexes does not occur. The products of the reaction of *p*-cresol with hydrogen peroxide in the presence of acidified ferrous sulfate solution are 4a,8-dimethyl-4a,9b-dihydro-2(1*H*)-dibenzofuranone (**21**) and 2,2′-dihydroxy-5,5′-dimethyldiphenyl (**22**) (159).

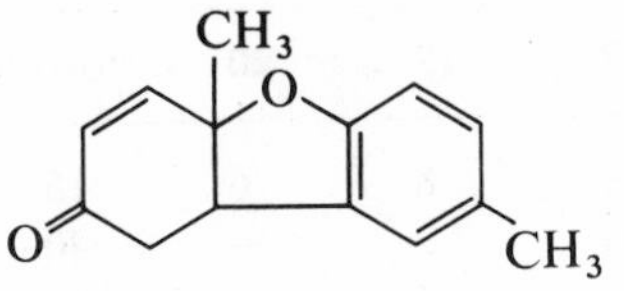

(21) 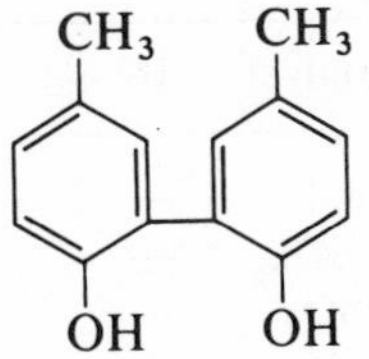           (22)

In aqueous medium the yields of compounds **21** and **22** are 13 and 20%, respectively. With 20% acetic acid present the respective yields are 18 and 11%. No homocatechol is detected. The ESR spectrum of the transient intermediate from *p*-cresol and hydrogen peroxide in acid solution in the presence of titanous chloride has been detected (104).

The reaction of 2,4-dimethylphenol (**23**) with hydrogen peroxide in acidified 20% acetic acid in the presence of ferrous sulfate yields mainly 2,2'-dihydroxy-3,5,3',5'-tetramethyldiphenyl (**24**), together with an unidentified material that may be a homolog of compound **21** (159).

(**23**)                    (**24**)

Under similar conditions the reaction of 2,6-dimethylphenol (**25**) yields 4,4'-dihydroxy-3,5,3',5'-tetramethyldiphenyl (**26**), together with the corresponding quinone, 3,3',5,5'-tetramethyl-4,4'-diphenoquinone (**27**), and 2,6-dimethylquinol (**28**) (159). Compound **28** probably derives from reduction of the quinone (**29**) during workup. The product distribution in this system has been studied with different catalysts: cuprous chloride suspension at 20°, copper sulfate at 70°, titanous chloride at 60°, and titanic chloride at 60° (160). With cuprous chloride as catalyst, a small yield of dienone dimer (**30**) is obtained in addition to compounds **27** and **29**. Yields are listed in Table 1.

TABLE 1.   REACTION OF 2,6-DIMETHYLPHENOL WITH
               HYDROGEN PEROXIDE (160)

| Catalyst | Yield (%) | | | | |
|---|---|---|---|---|---|
| | **16** | **17** | **19** | **20** | Polymer |
| CuCl | — | 9 | 8 | 6 | 15 |
| CuSO$_4$ | — | 33 | 9 | — | 40 |
| FeSO$_4$ | — | 22 | 3 | — | 55 |
| TiCl$_3$ | 9 | 20 | 51 | — | 5 |
| TiCl$_4$ | — | 50 | 38 | — | 10 |

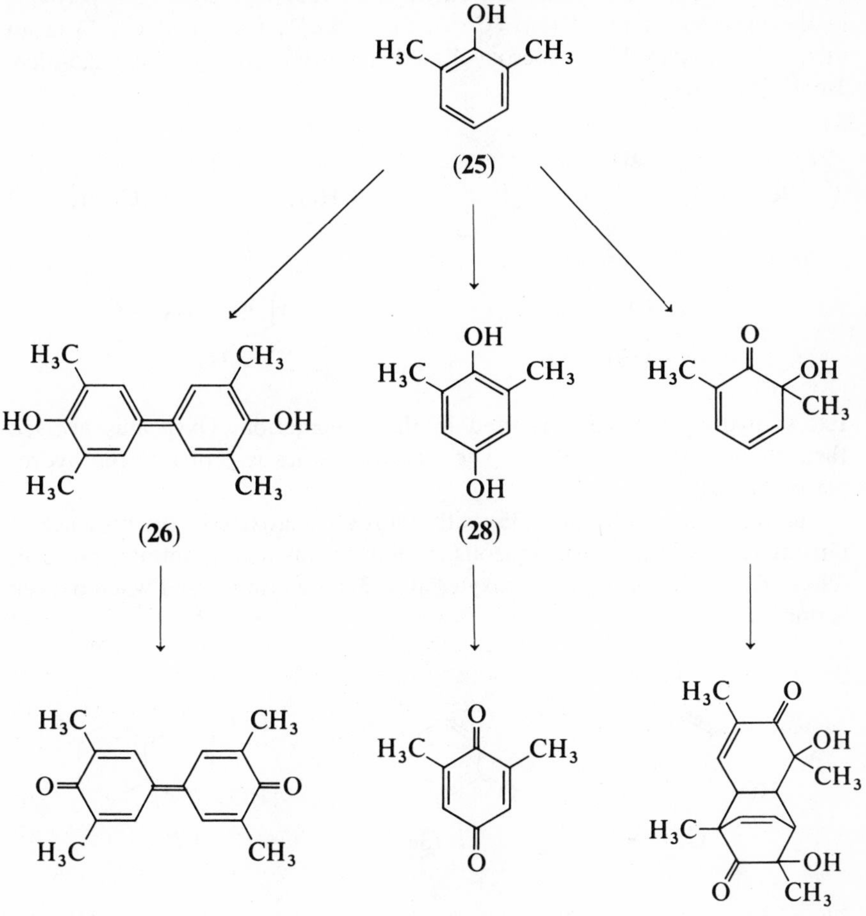

(25)

(26)    (28)

(27)    (29)    (30)

The reaction of mesitol (31) with hydrogen peroxide in the presence of ferrous sulfate in acidified 20% acetic acid yields 4,4'-dihydroxy-3,5,3',5'-tetramethyldiphenylmethane (32) and a second unidentified diphenol (159).

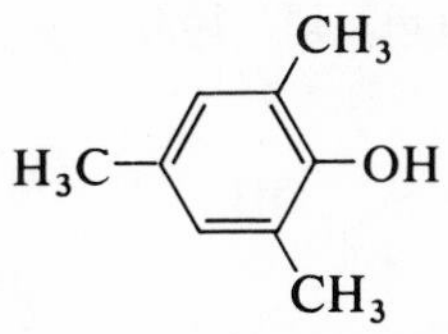

(31) 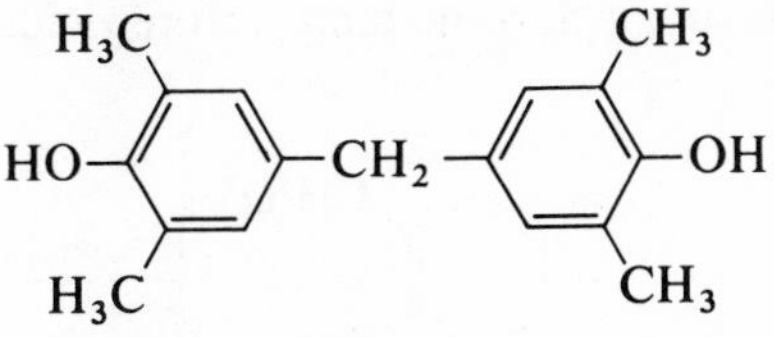    (32)

Oxidation of 4-methyl-2,6-di-*t*-butylphenol (**33**) with hydrogen peroxide in the presence of metal ions ($Co^{2+}$, $Fe^{2+}$, $Fe^{3+}$, $Cu^{+}$, and $Cu^{2+}$) in an inert solvent gives 4-hydroperoxy-4-methyl-2,5-di-*t*-butylcyclohexa-2,5-dien-1-one (**34**) (161).

Excess hydrogen proxide is used in this preparation. Oxygen is evolved throughout and has no effect. The reactive species is probably the hydroperoxy radical.

The reaction of 6-tetralol (**35**) with hydrogen peroxide in the presence of ferrous sulfate in acidified aqueous medium yields mainly tetralin *p*-quinol (**36**) (162). Traces of 6,7-dihydroxytetralin (**37**) are also formed when oxygen is present.

The yield of compound **36** is decreased in the presence of *t*-butanol or the fluoride ion.

The metal-ion-catalyzed reactions of hydrogen peroxide with other substituted monophenols have been studied, but the work is of little interest in regard to synthesis.

Reaction of salicylaldehyde with hydrogen peroxide in acidified aqueous ferrous sulfate at room temperature yields catechol (142, 157, 163).

With the hydroxybenzoic acids hydroxylation occurs *ortho* or *para* to the hydroxy group (146). Thus the reactions of salicylic acid and *p*-hydroxybenzoic acid yield, respectively, gentisic acid (**38**) and protocatechuic acid (**39**). The reaction of *m*-hydroxybenzoic acid yields a mixture of acids **38** and **39**.

A model peroxidase system composed of hydrogen peroxide, ferric, ion, ascorbic acid, and ethylenediaminetetraacetic acid (EDTA) in phosphate buffer reacts with salicylic acid to yield a mixture of 2,3- and 2,5-dihydroxybenzoic acids in a 7:2 ratio (164). Electron-spin-resonance studies have been made of the transient intermediates in the reaction of *m*- and *p*-fluorophenol and *p*-chlorophenol with the system hydrogen peroxide–titanous chloride (95).

Little work of interest for synthesis is available concerning dihydric and trihydric phenols. Thus catalyzed oxidations of catechol, resorcinol, and quinol yield uncharacterized colored products (165–167). Ferrous, ferric, and cupric ions are effective catalysts. Electron-spin-resonance studies have been made of the radical intermediates generated in the oxidation of quinol by the system titanous chloride-hydrogen peroxide (104). The reaction of pyrogallol (**14**) with hydrogen peroxide in the presence of iron and copper salts produces complex colored materials (168–170). In the presence of colloidal ferric ferrocyanide purpurogallin (**15**) is obtained (170).

## C. Baudisch Reaction

A nitrosohydroxylation reaction of benzene derivatives and phenols to give *o*-nitrosophenols was discovered by Baudisch about 30 years ago (171–177). This reaction has been attributed to the formation of nitroxy radicals, NOH, which are formed either by irradiation of nitrosopentacyanoferrate, $[Fe^{III}NO(CN)_5]^{2-}$, in the presence of hydrogen peroxide, or by reduction of nitrous acid, or by oxidation of hydroxylamine (171–178). The presence of copper or other metal ions is essential to the success of this reaction. A system consisting of aqueous hydroxylamine, 30% hydrogen peroxide, and cupric salts has been frequently used (178–181).

Phenols and substituted phenols usually react at a faster rate than that observed for hydrocarbons (180, 181). A large number of substituted benzenes and phenols have been converted to o-nitrosophenols (178).

The mechanism of the Baudisch reaction is not fully understood. Copper and other metal ions seem to be involved in the generation of hydroxy radicals, which are consumed in the formation of phenols (143, 179). In addition, metal ions are involved in the formation of stable complexes of o-nitrosophenols, thus preventing further oxidative reactions of products (178).

Strong steric effects of methyl groups located *meta* to the hydroxy group suggests that the reaction may proceed through the intermediate formation of a bulky copper(II)–hydroxylamine complex (180, 181).

A low *ortho*-to-*para* ratio of nitrosophenols obtained in the nitrosation of substituted phenols by nitrous acid, as compared with a high *ortho*-to-*para* ratio obtained with hydroxylamine–hydrogen peroxide–metal ion systems, excludes a mechanism involving the nitrosyl ion ($NO^+$) (179–181).

## D.  Hydroxylation of Azo Compounds—Pfitzner-Merkel Reaction

In 1948 Pfitzner and Merkel disclosed in a German patent (182) a remarkable hydroxylation reaction of azo dyes (183, 184). This reaction is sometimes called oxidative coppering. Since this disclosure the scope and limitations of this reaction have been established in a number of publications (185–193), including a review article (184).

The reaction makes possible the conversion of many o-hydroxyazo and o-aminazo compounds into o,o'-dihydroxyazo or o-amino-o'-hydroxyazo compounds, by using a mixture of aqueous hydrogen peroxide and copper acetate at 20–75° and pH 4.5–7.0 (184, 185, 187, 191). Other oxidizing agents—such as sodium peroxydisulfate, sodium perborate, and sodium peroxide or oxygen (air) and copper powder—are also suitable (184). This reaction is a particularly valuable one in the manufacture of azo dyes (184). It produces first a copper complex, which, on addition of hydrochloric acid, gives the free o,o'-dihydroxy or o-amino-o'-hydroxy derivative (184–193). Reduction of the azo derivative produces o-aminophenol or o-diaminophenyl derivatives. These procedures indicate the potential of the method for synthesis.

$$\text{(NH}_2) \quad \text{OH} \quad \overset{R}{\bigcirc}-N=N-\overset{R'}{\bigcirc} \quad \xrightarrow{H_2O_2,\ Cu^{2+}} \quad \overset{Cu^+}{\underset{(NH)\ O \quad O}{\overset{R}{\bigcirc}-N=N-\overset{R'}{\bigcirc}}}$$

$$\xleftarrow{HCl}$$

$$\text{(NH}_2) \quad \text{OH} \quad \text{HO} \quad \overset{R}{\bigcirc}-N=N-\overset{R'}{\bigcirc} \quad \xrightarrow{H_2/cat.} \quad \text{(NH}_2) \quad \text{OH} \quad \overset{R}{\bigcirc}-NH_2 \ + \ H_2N-\overset{R'}{\bigcirc}\ \text{HO}$$

(R,R' = substituted benzene, naphthalene, phenanthrene, acenaphthene, and chrysene rings).

If the starting material has a *o*-chloro and *o'*-hydroxy substituent, the reaction results in replacement of the chloro substituent by a hydroxy group (184).

$$\text{Cl} \quad \text{HO} \quad \overset{R}{\bigcirc}-N=N-\overset{R'}{\bigcirc} \quad \xrightarrow[2.\ HCl]{1.\ H_2O_2,\ Cu^{2+}} \quad \text{OH} \quad \text{HO} \quad \overset{R}{\bigcirc}-N=N-\overset{R'}{\bigcirc}$$

Although this reaction is applicable to a wide range of aromatic azo compounds, there are also certain limitations. For example, if the *ortho* position or other sterically important position is occupied by a strongly complexing group containing an electron pair, such as $\overset{\backslash}{\underset{/}{N}}$, $\overset{\backslash}{\underset{/}{O}}$ or $-OR$, no hydroxylation occurs (184, 192).

$$\overset{\backslash}{N}\quad \overset{R}{\bigcirc}-N=N-\overset{R'}{\bigcirc} \qquad \overset{=O}{\overset{R}{\bigcirc}-N=N-\overset{R'}{\bigcirc}} \qquad \overset{\backslash}{\underset{O}{}}\quad \overset{R}{\bigcirc}-N=N-\overset{R'}{\bigcirc}$$

In 1958 Pfitzner and Baumann (184) indicated the possibility of a free-radical mechanism involving cuprous ions in the complex together with hydroxy and hydroperoxy radical species. This conclusion was made on the basis of the following observation. The reaction requires an excess of hydrogen peroxide, and oxygen is evolved during the reaction. Furthermore, the reaction can be achieved with molecular oxygen in the presence of copper powder. It is quite certain that the metallic copper is rapidly converted into ionic species, and the reaction proceeds by a homogeneous process. It is highly unlikely that heterogeneous surface catalysis is involved. In analogy with other oxidation–reduction reactions (194) the following mechanism is now proposed:

$$H_2O_2 + Cu^+ \longrightarrow HO\cdot + Cu^{2+} + OH^-$$

Although cupric ions have always been initially added, there is little doubt that an oxidation–reduction system is immediately established to give the necessary cuprous ions to initiate the reaction. The massive amounts of cupric ions ensure a rapid conversion of the intermediate radical to the product. In detail, a concomitant electron and ligand transfer and complexing process might be involved.

### E.  Miscellaneous Compounds

Studies of product yield and isomer distribution have been made for the hydroxylation of a number of substituted benzenes, under conditions (excess of substrate interacted with dilute hydrogen peroxide) in which only mono-substitution occurs. The data are summarized in Table 2.

The relative reactivities and orientation of these substituted benzenes indicate that the oxidizing species in the ferrous-ion-catalyzed hydroxyla-tion—namely, the hydroxy radical—is reactive and electrophilic. The same orientation is obtained in the ferrous-ion-catalyzed reaction in the presence of EDTA and ascorbic acid, and also in the cuprous-or titanous-ion-catalyzed,

TABLE **2.** ISOMER DISTRIBUTIONS IN THE HYDROXYLATIONS OF
MONOSUBSTITUTED BENZENES

| Substrate | Catalyst | Isomer (%) | | | Ref. |
|---|---|---|---|---|---|
| | | *ortho* | *meta* | *para* | |
| Fluorobenzene | $Fe^{2+}$, EDTA, ascorbic acid, phosphate | 37 | 18 | 45 | 195 |
| | $Fe^{2+}$, pH 3.6 | 42 | 21 | 37 | 95 |
| Chlorobenzene | $Fe^{2+}$, EDTA, ascorbic acid, phosphate | 42 | 29 | 29 | 195 |
| | $Fe^{2+}$, pH 1 | 40–45 | 20–25 | 20–25 | 196 |
| | $Fe^{3+}$, catechol, pH 4.2 | 45 | 15 | 40 | 145 |
| Anisole | $Fe^{2+}$, EDTA, ascorbic acid, phosphate | 84 | — | 16 | 195 |
| | $Fe^{2+}$ | | | | |
| | $Cu^{+}$ | | | | |
| | $Fe^{2+}$, pH 3.6 | 85 | 3 | 12 | 95 |
| | $Fe^{3+}$, catechol, pH 4.3 | 64 | 3 | 33 | 145, 197 |
| | $Fe^{3+}$, hydroquinone, pH 4.3 | 65 | 5 | 35 | 145, 197 |
| | $Fe^{3+}$, chloroanilic acid, pH 4.3 | 74 | — | 26 | 145, 197 |
| Acetanilide | $Fe^{2+}$, EDTA, phosphate, pH 7.2 | 55–60 | 20–25 | 15–20 | 198 |
| | $Fe^{2+}$, EDTA, ascorbic acid | 47 | 3 | 51 | 199 |
| | $Fe^{2+}$, EDTA | 48 | 3 | 50 | 199 |
| | $Ti^{3+}$ | 50 | 4 | 46 | 199 |
| | $Fe^{3+}$, EDTA | 49 | 4 | 47 | 199 |
| PhC(O)OH | $Fe^{2+}$, EDTA, phosphate, pH 7.2 | 40–45 | 30–35 | 25–30 | 198 |
| | $Fe^{3+}$, EDTA, ascorbic acid, phosphate | 31 | 39 | 31 | 164 |
| $PhNO_2$ | $Fe^{2+}$, EDTA, ascorbic acid, phosphate | 24 | 30 | 46 | 195 |
| | $Fe^{3+}$, catechol, pH 4.2 | 48 | 26 | 26 | 145 |
| | $Fe^{3+}$, catechol, pH 2.9 | 47 | 26 | 27 | 145 |
| | $Fe^{2+}$, pH $\sim$ 1 | 25–30 | 50–55 | 20–25 | 200 |

reactions (195). When the catalyst is ferric ion in the presence of an enediol such as catechol (145, 197), a different isomeric distribution of products results and a different oxidizing species is involved.

In addition to hydroxylation of the benzene ring, displacement or attack of the substituent, as well as oxidative dimerization, may occur. Thus the reaction of fluorobenzene with hydrogen peroxide in the presence of acidified ferrous sulfate yields a mixture of 90% fluorophenols and 10% phenol at pH 3.6; at pH 0.8 the mixture consists of 11% fluorophenols and 89% phenol (95). The reaction of chlorobenzene with hydrogen peroxide in the presence of the ferrous ion yields, in addition to chlorophenols (101, 195, 196), some phenol and diphenyl (196). The yields of phenol and diphenyl are independent of pH but are increased in the presence of oxygen. The ferrous-ion-catalyzed oxidation of anisole by hydrogen peroxide yields hydroxylated products (95, 151, 195, 201, 202) and, in addition, formaldehyde (101), phenol (25), and quinol (201, 202). Dealkylation involves preliminary hydrogen abstraction from the side-chain methyl group, and not a direct displacement of methoxy by the hydroxy radical (201).

$$Fe^{2+} + H_2O_2 + H^+ \longrightarrow Fe^{3+} + H_2O + HO\cdot$$

$$HO\cdot + PhOCH_3 \longrightarrow H_2O + PhOCH_2\cdot$$

$$PhOCH_2\cdot + HO\cdot \longrightarrow PhOH + HCH(O)$$

Under conditions in which disubstitution does not occur the product mixture contains 93% methoxy phenols and 7% phenol at pH 3.6; at pH 0.8 the mixture consists of 2% methoxy phenols and 98% phenol (95). Similar attack of the side chain occurs in the reaction of phenacetin (**40**) (202) or phenoxyacetic acids (203, 204) with hydrogen peroxide in the presence of the ferrous ion. The reaction of phenacetin (**40**) in the presence of EDTA and ascorbic acid produces a mixture of 2- and 3-hydroxyphenacetin (**40a**) together with the dealkylated material, 4-hydroxyacetanilide (**40b**) (202).

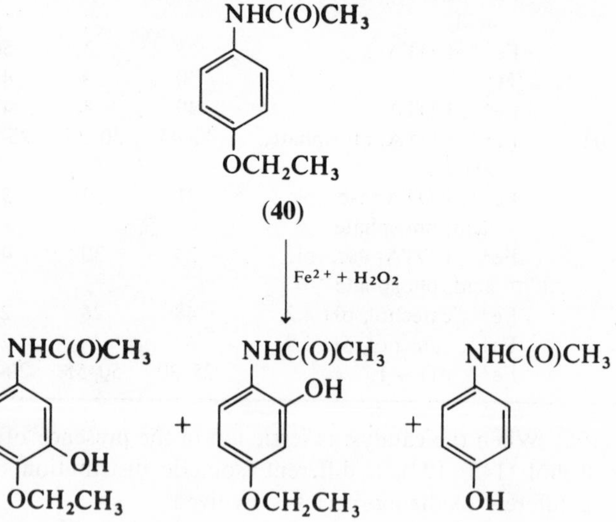

(**40**)

$Fe^{2+} + H_2O_2$

(**40a**)                    (**40b**)

The reaction of phenoxyacetic acid yields phenol (201, 204), formaldehyde (204), and carbon dioxide (204), together with 1,2-diphenoxyethane (204). Several substituted phenoxyacetic acids (2-fluoro, 4-fluoro, 2,4-difluoro, 2-chloro, and 4-chloro) react similarly (204). The reaction of benzoic acid in the presence of the ferrous ion produces concurrent hydroxylation and decarboxylation. In addition to phenolic acids (101, 143, 146, 164, 198) phenol, catechol, and resorcinol are produced (198). Similarly the oxidation of phthalic acid produces monohydroxybenzoic acids, together with 2,5- and 3,4-dihydroxybenzoic acids (146). The reaction of *o*-carboxycinnamic acid, however, yields only an unidentified *o*-carboxyhydroxycinnamic acid (146).

The side-chain reactivity of toluene and ethylbenzene has been described in Section B. Phenylacetic acid gives relatively little phenolic material; the major product is benzaldehyde (101).

The metal-ion-catalyzed reaction of hydrogen peroxide with a number of other aromatic compounds will be mentioned, although the results are often fragmentary. The reaction of acetophenone in the presence of acidified ferrous sulfate produces a mixture of *o*-, *m*-, and *p*-hydroxyacetophenones (1.9, 2.6, and 0.1% yields, respectively) together with 3.3'-diacetophenone (2.2% yield) and acetic acid (3% yield) (103). The reaction of benzaldehyde under similar conditions gives a mixture of *o*-, *m*-, and *p*-hydroxybenzaldehydes (1.3%, 1.3%, and very small yields, respectively), with some 3,3'-diphenyldialdehyde (0.5% yield) and benzoic acid (4.1% yield) (103). The reaction of benzamide yields salicylamide (101). The reaction of coumarin (**41**) with hydrogen peroxide in the presence of acidified ferrous sulfate gives a mixture of 3-, 5-, and 7-hydroxycoumarins, with traces of 6- and 8-hydroxycoumarins and *o*-coumaric acid (**42**) (146, 205).

(41)                    (42)

Benzenesulfonic acid reacts with hydrogen peroxide in the presence of ferrous sulfate to yield unidentified phenolsulfonic acids (146). Dimethylaniline in acid solution reacts to form a deep-purple solution indicative of dyes derived from *o*-aminophenol, together with formaldehyde (146).

The interaction of Fenton's reagent with heteroaromatic compounds has been reviewed by Norman and Radda (133). Individual reactions of heterocyclic nitrogen compounds, aromatic amines, amides, and amino acids are discussed in Section IX.

Some insight into the intermediates involved in the metal-ion-catalyzed reaction of aromatic compounds with hydrogen peroxide has been gained by ESR studies (83, 104, 206, 207). Most work has involved the system titanous chloride–hydrogen peroxide.

## IV. ALCOHOLS AND GLYCOLS

The oxidation of alcohols (29, 98, 208–219) by hydrogen peroxide in the presence of ferrous ions gives in most cases the corresponding aldehydes or ketones. However, depending on experimental conditions, other products can be obtained, such as carboxylic acids, formaldehyde, and carbon dioxide (Table 3). Beside iron ions other metal ions have been tried; however, they seem to be less effective (208). The following scheme explains the formation of some products:

$$H_2O_2 + Fe^{2+} \longrightarrow HO\cdot + Fe^{3+} + OH^-$$

$$MeCH_2OH + HO\cdot \longrightarrow Me\overset{\cdot}{C}HOH + H_2O$$

$$Me\overset{\cdot}{C}HOH + Fe(OH)^{2+} \longrightarrow \underset{\underset{OH}{|}}{MeCHOH} + Fe^{2+}$$

$$MeCH(OH)_2 \longrightarrow MeCHO + H_2O$$

$$MeCHO + \cdot OH \longrightarrow Me\overset{\cdot}{C}O + H_2O$$

$$Me\overset{\cdot}{C}O + Fe(OH)^{2+} \longrightarrow MeC(O)OH + Fe^{2+}$$

$$Me\overset{\cdot}{C}O \longrightarrow Me\cdot + CO$$

$$Me\cdot + Fe(OH)^{2+} \longrightarrow MeOH$$

$$MeOH + \cdot OH \longrightarrow \overset{\cdot}{C}H_2OH + H_2O$$

$$\overset{\cdot}{C}H_2OH + Fe(OH)^{2+} \longrightarrow CH_2(OH)_2 + Fe^{2+}$$

$$CH_2(OH)_2 \longrightarrow CH_2O + H_2O$$

Similar reaction schemes can be written for other systems. The presence of radical intermediates has been shown by ESR techniques (83, 86, 87, 93, 94, 130, 227–233).

An unusual elimination reaction occurs in the ferrous-ion-catalyzed oxidation of phenyl-t-butylcarbinol (43a) and p-anisyl-t-butylcarbinol (43b) with hydrogen peroxide to give 91% benzaldehyde and 78% anisaldehyde, respectively (217).

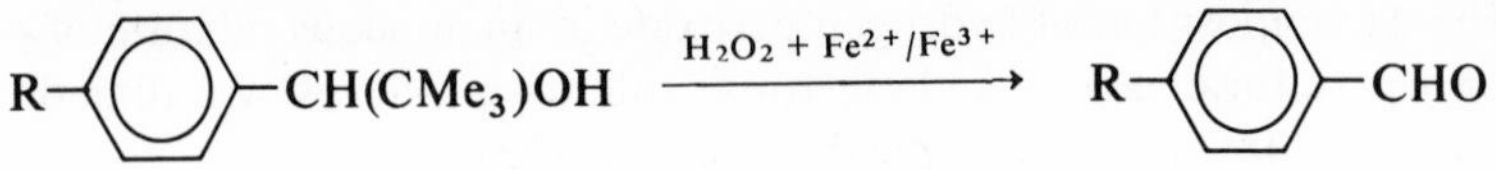

$$R-\langle\bigcirc\rangle-CH(CMe_3)OH \xrightarrow{H_2O_2 + Fe^{2+}/Fe^{3+}} R-\langle\bigcirc\rangle-CHO$$

(43)

(a, R = H)
(b, R = MeO)

TABLE 3.  OXIDATION OF ALCOHOLS AND GLYCOLS BY HYDROGEN PEROXIDE

| Alcohol/Glycol | Catalyst | Products[a] | Reference |
|---|---|---|---|
| MeOH | $Fe^{2+}$ | $CH_2O$, $HC(O)OH$, $CO_2$ | 210 211, 213 |
| EtOH | $Fe^{2+}$, $Fe^{3+}$ | MeCHO | 98, 208–211 |
| | $Co^{2+}$, $Mn^{2+}$ $Ni^{2+}$, $Cu^{2+}$ $Ce^{3+}$ | $MeC(O)OH$ (50%) $CO_2$, $H_2O$ $MeCH(OEt)_2$ | 213–215 |
| *n*-PrOH | $Fe^{2+}$ | $MeCH_2CHO$, $MeCH_2C(O)OH$ | 211, 213 |
| *i*-PrOH | $Fe^{2+}$ | $Me_2CO$, $HC(O)OH$ | 211, 213 |
| 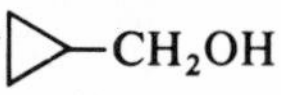 $\triangleright-CH_2OH$ | $Fe^{2+}$ | 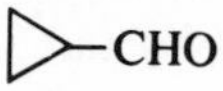 $\triangleright-CHO$ (30–37%) | 218 |
| $Me_3COH$ | $Fe^{2+}$ $Fe^{2+}$ | $Me_2CO$, $CH_2O$ $Me_2C(OH)CH_2CH_2CMe_2OH$ | 213 120, 221–223 |
| | $Fe^{2+}$ | $Me_2C(OH)CH_2OH$ | 224, 232, 234 |
| *t*-Amyl | $Fe^{2+}$ | $Me_2CO$, MeCHO, $EtC(O)Me$, $CH_2O$, | 213 |
| $PhCMe_2OH$ | $Fe^{2+}$ | $PhC(O)Me$, $CH_2O$ | 213 |
| $CH_2(OH)CH_2OH$ | $Fe^{2+}$ | $CH_2(OH)CHO$, MeCHO, $CH_2O$, $HC(O)OH$, $CO_2$, $CH(O)CH(O)$ | 213, 220, 225 |
| $Me_2C(OH)C(OH)Me_2$ | $Fe^{2+}$ | $Me_2CO$ | 213 |
| $MeCH(OH)CH(OH)Me$ | $Fe^{2+}$ | $MeC(O)C(O)Me$ | 213 |
| $CH_2(OH)CH(OH)CH_2Me$ | $Fe^{2+}$ | $EtCOC(O)OH$ | 213 |
| $MeCH(OH)CH_2CH_2OH$ | $Fe^{2+}$ | $MeC(O)CH_2C(O)OH$, $MeCH{=}CHCHO$ | 213 |
| $CH_2(OH)CH(OH)CH_2OH$ | $Fe^{2+}$ | $CH_2(OH)CH(OH)CHO$, $HC(O)OH$, $CH_2O$, $CO_2$ | 210, 213 225, 226 |

[a] Numbers in parentheses are yields.

The decomposition of hydrogen peroxide in an aqueous mixture of $t$-butyl alcohol, sulfuric acid, and ferrous sulfate produces 2,5-dimethyl-2,5-hexanediol (**44**) in 36–65% yield (120, 221–223).

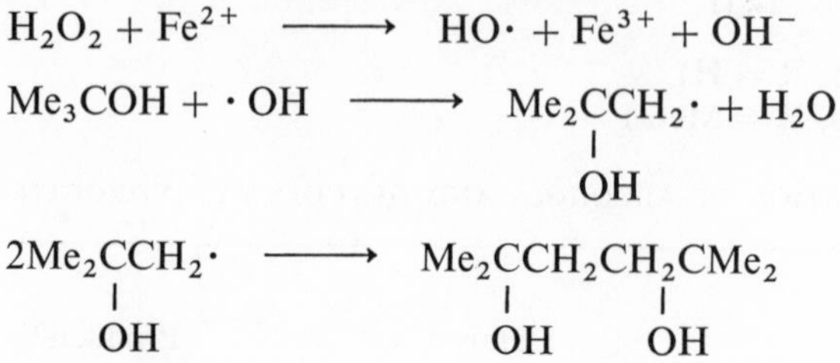

$$H_2O_2 + Fe^{2+} \longrightarrow HO\cdot + Fe^{3+} + OH^-$$

$$Me_3COH + \cdot OH \longrightarrow \underset{\underset{OH}{|}}{Me_2CCH_2}\cdot + H_2O$$

$$2\underset{\underset{OH}{|}}{Me_2CCH_2}\cdot \longrightarrow \underset{\underset{OH}{|}}{Me_2CCH_2}\underset{\underset{OH}{|}}{CH_2CMe_2}$$

(**44**)

When this reaction is carried out in the presence of a mixture of ferrous and cupric sulfates, $\alpha$-glycol (**45**) is formed (224).

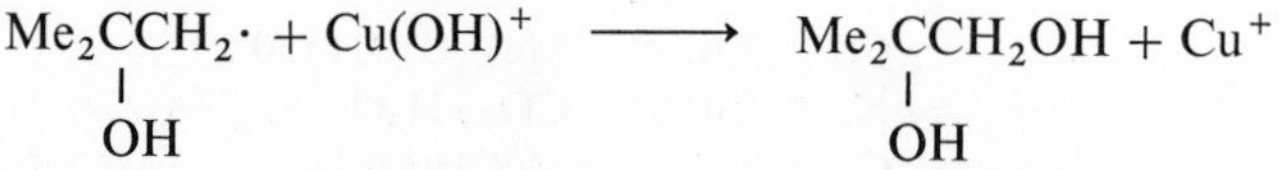

$$\underset{\underset{OH}{|}}{Me_2CCH_2}\cdot + Cu(OH)^+ \longrightarrow \underset{\underset{OH}{|}}{Me_2CCH_2OH} + Cu^+$$

(**45**)

The decomposition of hydrogen proxide in an aqueous mixture of $t$-butyl alcohol, sulfuric acid, ferrous sulfate, and carbon monoxide gives 23% $\beta$-hydroxyisovaleric acid (**46**) and 33% of dimer **44** (234, 235).

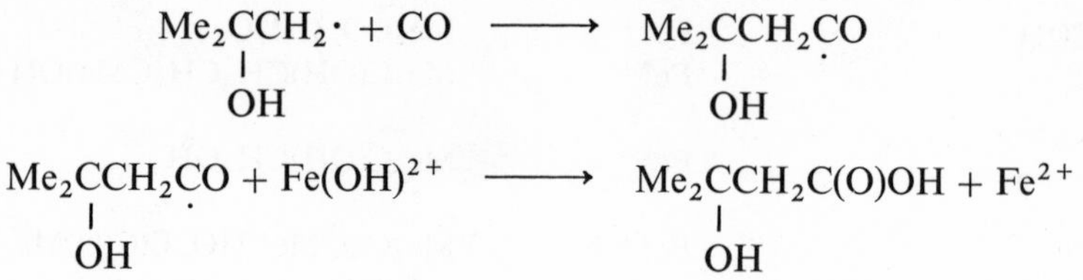

$$\underset{\underset{OH}{|}}{Me_2CCH_2}\cdot + CO \longrightarrow \underset{\underset{OH}{|}}{Me_2CCH_2}\underset{\cdot}{CO}$$

$$\underset{\underset{OH}{|}}{Me_2CCH_2}\underset{\cdot}{CO} + Fe(OH)^{2+} \longrightarrow \underset{\underset{OH}{|}}{Me_2CCH_2C(O)OH} + Fe^{2+}$$

(**46**)

If carbon monoxide is replaced by butadiene, the reaction produces 2,13-dimethyltetradeca-5,9-diene-2, 13-diol (**47**) in 60% yield (120).

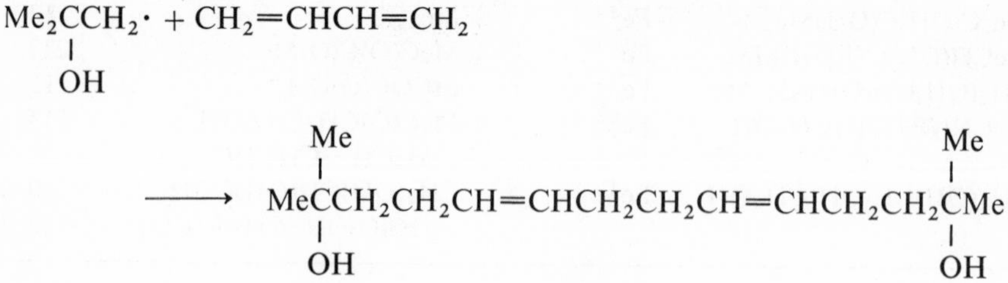

$$\underset{\underset{OH}{|}}{Me_2CCH_2}\cdot + CH_2=CHCH=CH_2$$

$$\longrightarrow \underset{\underset{OH}{|}}{\overset{\overset{Me}{|}}{Me}CCH_2CH_2CH=CHCH_2CH_2CH=CHCH_2CH_2}\underset{\underset{OH}{|}}{\overset{\overset{Me}{|}}{C}Me}$$

(**47**)

Under similar conditions the reaction with 1,1,4,4-tetrafluorobutadiene gives a 1:2 mixture of products **48** and **49** (120).

$$\underset{\quad\quad\quad\;OH\quad\quad\quad\;CH\quad\quad\quad\quad\quad\quad\;OH}{Me_2CCH_2CF_2C\!=\!CFCH\!=\!CHCF_2CH_2CMe_2}$$

$$CF_2$$

**(48)**

$$\underset{\quad\quad\quad OH\quad\quad\quad CH\;CH\quad\quad OH}{Me_2CCH_2CF\!=\!C\!-\!C\!=\!CFCH_2CMe_2}$$

$$CH_2CH_2$$

**(49)**

The ferrous-ion-catalyzed reaction of hydrogen peroxide with glycols proceeds with oxidation of either one or more hydroxy groups to give the corresponding hydroxyaldehydes, hydroxyketones, dialdehydes, diketones, and keto acids (Table 3). Also, fragmentation of glycol molecules frequently takes place, and as a result formic acid, formaldehyde, and carbon dioxide have been found on various occasions. The reaction of pinacol proceeds with rupture of the carbon–carbon bond to give acetone (213).

The ferrous-ion-catalyzed oxidation of polyhydroxy compounds, such as arabitol (236), erythritol (225), mannitol (225), dulcitol (225), sorbitol (225), and inositol (237) have also been investigated. Electron-spin-resonance spectroscopy has provided confirmation of free-radical intermediates with ethylene glycol, glycerol, propane-1,2-diol, butane-2,3-diol, dihydroxyacetone, cyclopentane-1,2-diol, sorbitol, mannitol, and hydrogen peroxide–titanium trichloride systems (228, 230, 238).

## V.   ALDEHYDES, KETONES, AND ETHERS

There appears to be no systematic account of the interaction of hydrogen peroxide with aldehydes, ketones, and ethers in the presence of metal ions, except for ESR studies. The results are fragmentary and of little preparative value (Table 4). As a consequence it is difficult to make definite conclusions. Nevertheless, it seems that the attack by the hydroxy radical occurs preferentially on the carbon–hydrogen bond located nearest to the carbonyl or ether function. The carbon radicals then undergo either substitution, dimerization, or fragmentation reactions, as shown in the examples of acetone (**50**), ethyl ether (**51**), and tetrahydrofuran (**52**) (Table 4). However, this pattern is not a

TABLE 4.    FERROUS-ION-CATALYZED REACTIONS OF HYDROGEN PEROXIDE WITH ALDEHYDES, KETONES, AND ETHERS

| Substrate | Products | References |
|---|---|---|
| $Et_2O$ | MeCHO, dimers | 213, 239 |
| $i\text{-}Pr_2O$ | $C_{12}H_{26}O_2$ dimers | 239 |
| $Me_2CO$ | $MeC(O)OH$, $MeC(O)CH_2C(O)Me$, $(MeCOCH_2)_2$ | 221, 222 |
| [cyclopentanone structure] | $Me(CH_2)_3C(O)OH$, cyclopenta-noyl cyclopentanone | 222 |
| [tetrahydrofuran structure] | $CH_2O$ + [2-hydroxytetrahydrofuran structure] + [dihydrofuran dimer with OH structure] | 213, 240 |
| [tetrahydropyran structure] | $CH_2O$, $HOC(O)(CH_2)_nCHO$ | 213 |
| [1,4-dioxane structure] | $CH_2O$, $HC(O)OH$, $[C(O)OH]_2$ | 213 |
|  | [2-hydroxy-1,4-dioxane structure] + [dioxane dimer with OH structure] | 240 |
| $CBr_3CHO \cdot H_2O$ | $CHBr_2CHO$, $CBr_3C(O)OH$, HBr | 241 |

$$H_2O_2 + Fe^{2+} \longrightarrow \cdot OH + Fe^{3+} + OH^-$$

$$MeC(O)Me + \cdot OH \longrightarrow MeC(O)CH_2\cdot + H_2O$$
(50)

$$MeC(O)\underset{\cdot}{C}H_2 + Fe(OH)^{2+} \longrightarrow MeC(O)CH_2OH + Fe^{2+}$$

$$MeC(O)CH_2OH \longrightarrow MeCHO + CH_2O$$

$$Me\underset{\cdot}{C}O + Fe(OH)^{2+} \longrightarrow MeC(O)OH + Fe^{2+}$$

$$MeCH_2OCH_2Me + \cdot OH \longrightarrow Me\underset{\cdot}{C}HOCH_2Me + H_2O$$
(51)

$$Me\underset{\cdot}{C}HOCH_2Me + Fe(OH)^{2+} \longrightarrow \underset{\underset{OH}{|}}{Me}CHOCH_2Me + Fe^{2+}$$

$$\underset{\underset{OH}{|}}{Me}CHOCH_2Me \longrightarrow MeCHO + EtOH$$

**(52)**

general one, as has been shown by ESR spectroscopy, and attack by hydroxy radicals can also occur at more remote parts of the molecule.

The type of radicals that might be involved in these reactions has been shown by ESR spectroscopy on the titanium trichloride–hydrogen peroxide system with the following compounds: formaldehyde, acetaldehyde, pyruvaldehyde, glyoxal, chloral hydrate, acetone, diethyl ketone, dihydroxyacetone, methyl ether, ethyl ether, ethyl vinyl ether, isopropyl ether, 1,1-dimethoxyethane, tetrahydrofuran, dioxan, and various acyloins (83, 93, 129, 227, 228, 231–233, 242–246).

## VI. CARBOXYLIC ACIDS

Since the discovery by Fenton of the Ferrous-ion-catalyzed oxidation of tartaric acid with hydrogen peroxide (1–5),

a large number of investigations have been performed with various carboxylic acids.

A survey of the literature indicates that organic products have been most frequently identified for the following acids: acetic (222, 247, 248), propionic (213, 221, 222, 248), *n*-butyric (222, 248, 249), isobutyric (221, 222), isovaleric (213), pivalic (120, 221, 222), palmitic (250), oxalic (251), succinic (248, 252–255), glutaric (256), adipic (256), suberic (256), crotonic (213, 248, 252), oleic (253, 257), maleic (248, 258), fumaric (248, 258), *trans*-aconitic (248), glycolic (165, 226, 248, 251, 259, 260), lactic (213, 259, 261), $\beta$-hydroxypropionic (248), $\alpha,\beta$-, and $\gamma$-hydroxybutyric (248), tartronic (259), malic (165, 248, 259, 262), glyceric (248, 258, 259), 2,3-dihydroxybutyric (248), tartaric (1–5, 165, 248, 253, 258, 263, 264, 266, 268–270), saccharic (265, 266), mucic (267), glyoxylic (251), pyruvic (165, 259, 267), and levulinic (256). Aldonic acids, in the form of soluble salts, are decarboxylated with the production of the aldehyde with one less carbon atom.

$$\underset{\underset{R}{|}}{\overset{\overset{C(O)O^-}{|}}{HCOH}} \xrightarrow{H_2O_2-Fe^{3+}} \underset{\underset{R}{|}}{HC=O} + CO_2$$

This reaction, the Ruff degradation (6–9), is described in Section VIII-B. The reactions of a number of aromatic acids [benzoic (101, 143, 146, 164, 198), phenoxyacetic (201, 203, 204), phthalic (146), phenylacetic (101), hydroxybenzoic (146, 164)] are described in Section III.

Although these studies lead to a better understanding of various biological transformations, in particular those of carbohydrates, detailed discussion of reactions of individual acids is unwarranted because mixtures of products are generally obtained. It will suffice to make some general comments.

It has been generally found that reactions of carboxylic acids with hydrogen peroxide in the presence of ferrous or copper ions proceed with decarboxylation to give products that have one carbon atom less than the initially used acid. The following scheme accounts for some of the products:

$$H_2O_2 + Fe^{2+} \longrightarrow HO\cdot + Fe^{3+} + OH^-$$

$$RCH_2C(O)OH + HO\cdot \longrightarrow RCH_2C(O)O\cdot + H_2O$$

$$RCH_2C(O)O\cdot \longrightarrow RCH_2\cdot + CO_2$$

$$RCH_2\cdot + Fe(OH)^{2+} \longrightarrow RCH_2OH + Fe^{2+}$$

$$RCH_2OH + \cdot OH \longrightarrow R\overset{\cdot}{C}HOH$$

$$R\overset{\cdot}{C}HOH + Fe(OH)^{2+} \longrightarrow RCH(OH)_2 + Fe^{2+}$$

$$RCH(OH)_2 \longrightarrow RCHO + H_2O$$

$$RCHO \xrightarrow[\text{2. } Fe(OH)^{2+}]{\text{1. } HO\cdot} RC(O)OH, \text{ etc.}$$

$$(R = alkyl)$$

A similar scheme can be used for copper ions. Although in most studies cupric ions have been employed, it is most likely that an oxidation–reduction system is rapidly established and that cuprous ions initiate this reaction sequence.

The products can be further degraded by analogous reactions, and as a result frequently such compounds as formaldehyde and formic acid are found in the product mixtures.

In a few cases dimerization of saturated carboxylic acids has been achieved

(120, 221, 222). Thus the reaction of acetic (**53a**), propionic (**53b**), isobutyric (**53c**), and pivalic (**53d**) acids with the Fenton reagent results in the formation of dimeric acids **54a, b, c,** and **d** in 4, 55, 32, and 37% yield, respectively (120, 127, 128).

$$RC(O)OH \xrightarrow{H_2O_2/Fe^{2+}} \text{[} RC(O)OH]_2$$

$$(53) \qquad\qquad\qquad (54)$$

| | |
|---|---|
| (**a** R = Me) | (**a** R = CH$_2$) |
| (**b** R = MeCH$_2$) | (**b** R = MeCH) |
| (**c** R = Me$_2$CH) | (**c** R = Me$_2$C) |
| (**d** R = Me$_3$C) | (**d** R = Me$_2$CCH$_2$) |

If the reaction of propionic acid is carried out in the presence of carbon monoxide (234), a mixture of C$_6$ dibasic acid (**54b**), α-hydroxy acid (**55**), and C$_4$ diabasic acid (**56**) is obtained.

$$53b + HO\cdot \longrightarrow Me\overset{\cdot}{C}HC(O)OH \begin{cases} \longrightarrow \textbf{54b} \\ \xrightarrow{Fe(OH)^{2+}} MeCH(OH)C(O)OH \end{cases}$$

$$(55)$$

$$Me\overset{\cdot}{C}HC(O)OH + CO \longrightarrow \underset{\underset{CO}{|}}{Me\overset{\cdot}{C}HC(O)OH} \xrightarrow{Fe(OH)^{2+}} \underset{\underset{C(O)OH}{|}}{MeCHC(O)OH}$$

$$(\textbf{56})$$

Additive dimerizations with dienes are described in Section II.

Carboxylic acids containing keto and hydroxy groups usually decarboxylate more readily than unsubstituted acids. Also in these cases mixtures of carbon dioxide and the so-called lower aldehydes, ketones, and carboxylic acids are obtained. The initial oxidative steps of tartaric acid may proceed by the following path:

$$\begin{array}{c} C(O)OH \\ | \\ HCOH \\ | \\ HCOH \\ | \\ C(O)OH \end{array} + HO\cdot \longrightarrow \begin{array}{c} C(O)OH \\ | \\ \cdot COH \\ | \\ HCOH \\ | \\ C(O)OH \end{array} \xrightarrow{Fe(OH)^{2+}} \begin{array}{c} C(O)OH \\ | \\ C(OH)_2 \\ | \\ HCOH \\ | \\ C(O)OH \end{array}$$

$$\xrightarrow{-H_2O} \begin{array}{c} C(O)OH \\ | \\ C=O \\ | \\ HCOH \\ | \\ C(O)OH \end{array} \rightleftharpoons \begin{array}{c} C(O)OH \\ | \\ COH \\ \| \\ COH \\ | \\ C(O)OH \end{array}$$

Aromatic carboxylic acids (see Section III) frequently undergo hydroxylation of the aromatic nucleus. For example, in the reaction of benzoic acid (101, 143, 146, 164, 198) with the Fenton reagent the formation of salicylic acid is formulated (206) as follows:

$$PhC(O)OH + HO\cdot \longrightarrow \underset{\text{(radical intermediate)}}{\boxed{C(O)OH,\ H,\ OH}} + H_2O \xrightarrow{Fe^{3+}} \underset{\text{(salicylic acid)}}{\boxed{C(O)OH,\ OH}} + Fe^{2+} + H$$

However, the reaction of phenylacetic acid (101) may give the benzyl radical (87, 206, 207):

$$PhCH_2C(O)OH + HO\cdot \longrightarrow \boxed{\text{cyclohexadienyl radical}} \xrightarrow[- OH^-]{- H^+,\ - CO_2} \boxed{\quad} \longleftrightarrow \underset{\dot{C}H_2}{\boxed{\text{benzyl radical}}}$$

The reaction of phenoxyacetic acid (201, 204) gives the phenoxymethyl radical (206), which reacts further to give formaldehyde, phenol, and diphenoxyethane (204).

$$PhOCH_2C(O)OH + \cdot OH \longrightarrow PhOCH_2\cdot + CO_2 + H_2O$$

Electron-spin-resonance studies (83) have been made of the interaction of the titanous chloride–hydrogen peroxide system with unsubstituted aliphatic carboxylic acids (129, 271), α-chloro acids (129), α-hydroxy acids (129), α,β-unsaturated acids (129), and phenyl-substituted aliphatic acids (87, 206, 207).

## VII. ESTERS

The reactions of esters of aliphatic carboxylic acids with aqueous hydrogen peroxide in the presence of ferrous ions (41) result in dimeric and oligomeric esters (239, 272–276) and also aldehydic, ketonic, and acidic degradation products (213, 239, 272–276). The following esters have been investigated: isoamyl formate (239), ethyl (239), n-propyl (274, 275), isopropyl (273–275), n-butyl (239, 272, 273), and isoamyl acetates (239, 274), ethyl propionate (272, 273, 275, 276), methyl butyrate (272, 273, 275, 276), methyl isobutyrate (272, 273, 275), and dimethyl adipate (272, 273). The yields of dimers and oligomers are sometimes as high as 60%, but in general the yields of well-defined products are low. Thus the reaction of methyl isobutyrate (57) gives a mixture consisting of 43% α,α'-dimethyl adipate (58), 50% α,α,α-trimethyl glutarate (59), 7% α,α,α',α'-tetramethyl succinate (60), isobutyric acid (61), acetone, monomethyl esters of $C_8$-dicarboxylic acids, plus trimers and tetramers (272, 273, 275).

$$H_2O_2 + Fe^{2+} \longrightarrow HO\cdot + Fe^{3+} + OH^-$$

MeCHMeC(O)OMe + HO·

$$\longrightarrow Me\overset{\cdot}{C}MeC(O)OMe \quad (A)$$
$$\longrightarrow \cdot CH_2CHMeC(O)OMe \quad (B)$$
$$\longrightarrow MeCHMeC(O)O\overset{\cdot}{C}H_2 \quad (C)$$

(57)

$$2A \longrightarrow \begin{array}{c} CH_2CHMeC(O)OMe \\ | \\ CH_2CHMeC(O)OMe \end{array}$$
(58)

$$A + B \longrightarrow \begin{array}{c} CMe_2C(O)OMe \\ | \\ CH_2CHMeC(O)OMe \end{array}$$
(59)

$$2B \longrightarrow \begin{array}{c} Me_2CC(O)OMe \\ | \\ Me_2CC(O)OMe \end{array}$$
(60)

$$Me_2\overset{\cdot}{C}C(O)OMe + Fe(OH)^{2+} \longrightarrow \begin{array}{c} Me_2CC(O)OMe \\ | \\ OH \end{array} \longrightarrow Me_2CO$$

$$Me_2CHC(O)O\overset{\cdot}{C}H_2 + Fe(OH)^{2+}$$
$$\longrightarrow Me_2CHC(O)OCH_2OH \longrightarrow \begin{array}{c} Me_2CHC(O)OH \\ (61) \end{array}$$

Similar reaction of methyl *n*-butyrate (272, 273, 275, 276) gives a mixture consisting of 75% dimethyl β-methyl pimelate (62), dimethyl β,β'-dimethyl adipate (63), and dimethyl suberate (64), plus propionaldehyde.

$$\begin{array}{c} CH(Me)CH_2C(O)OMe \\ | \\ CH_2CH_2CH_2C(O)OMe \end{array} \qquad \begin{array}{c} CHMeCH_2C(O)OMe \\ | \\ CHMeCH_2C(O)OMe \end{array} \qquad \begin{array}{c} CH_2CH_2CH_2C(O)OMe \\ | \\ CH_2CH_2CH_2C(O)OMe \end{array}$$
$$\quad\quad (62) \qquad\qquad\qquad\qquad (63) \qquad\qquad\qquad\qquad (64)$$

The relative quantities of the dimers indicate that in methyl isobutyrate the α carbon–hydrogen bonds of the acid part of the molecule are more reactive to attack by hydroxy radicals than the $\beta$ carbon–hydrogen bonds, whereas in methyl butyrate $\beta$ and $\gamma$ carbon–hydrogen bonds are more reactive than $\alpha$ carbon–hydrogen bonds (275).

The reactions of *n*-propyl (**65**) (274, 275) and isopropyl (**66**) (273–275) acetates give mainly 2,3-dimethylbutane-1,4-diol diacetate (**67**) and 1,4-dimethylbutane-1,4-diol diacetate (**68**), respectively.

$$MeC(O)OCH_2CH_2Me \longrightarrow \begin{array}{c} MeC(O)OCH_2CHMe \\ | \\ MeC(O)OCH_2CHMe \end{array}$$

$$(\mathbf{65}) \hspace{4cm} (\mathbf{67})$$

$$MeC(O)OCHMe_2 \longrightarrow \begin{array}{c} MeC(O)OCHMeCH_2 \\ | \\ MeC(O)OCHMeCH_2 \end{array}$$

$$(\mathbf{66}) \hspace{4cm} (\mathbf{68})$$

The $\alpha$ carbon–hydrogen bonds of the alcohol part in isopropyl acetate are eight times as reactive as $\beta$ carbon–hydrogen bonds, and in *n*-propyl acetate the $\beta$ hydrogen–carbon bonds are twice as reactive as the $\alpha$ carbon–hydrogen bonds (275). The difference in reactivity of the carbon–hydrogen bonds in the $\alpha$- and $\beta$-position of *n*-propyl and isopropyl acetate is explained by the relative stability of radicals **69** and **70**, and **71** and **72**.

$$MeCHCH_2O- \hspace{2cm} CH_2CH(Me)O-$$
$$\cdot \hspace{4cm} \cdot$$
$$(\mathbf{69}) \hspace{3cm} (\mathbf{70})$$

Radical **69** is more stable than **70**; similarly radical **71** is more favored energetically than **72**.

$$Me_2CO- \hspace{2cm} MeCH_2CHO-$$
$$\cdot \hspace{3cm} \cdot$$
$$(\mathbf{71}) \hspace{2.5cm} (\mathbf{72})$$

However, the $\alpha$-radicals, **71** and **72**, are comparatively unstable and undergo further degradation reactions, whereas radicals **69** and **70** might be further stabilized through the formation of complexes involving five- and six-membered ring systems **73** and **74** (275).

(73)                    (74)

Abstraction of hydrogen atoms from the methyl group of the acetic acid moiety either does not occur or to only a small extent (275). However, if the acid moiety has a longer aliphatic chain, such abstractions are possible (275). For example, the reaction of ethyl propionate (272, 273, 275, 276) should form the following radicals:

$$\cdot CH_2CH_2C(O)OEt \qquad Me\overset{\cdot}{C}HC(O)OEt \qquad EtC(O)OCH_2\overset{\cdot}{C}H_2$$

$$(75) \qquad\qquad (76) \qquad\qquad (77)$$

$$EtC(O)OC\overset{\cdot}{H}Me$$

$$(78)$$

The reaction gives mainly diethyl adipate, diethyl α,α′-dimethylsuccinate, and monoethyl adipate (275).

This result indicates that favored are those dimers that are derived from radicals of the acid part of the molecule. However, this result does not prove that those radicals are formed preferentially, because it is uncertain how many radicals of the alcohol part are converted to acetaldehyde (275). Studies by ESR techniques (83, 271) with ester–hydrogen peroxide–titanous chloride systems provide additional evidence for the existence of radical intermediates and the mode of hydrogen abstraction in these reactions. An ESR study dealing with esters of phosphorous, phosphoric, and methylphosphoric acids is also available (277).

## VIII.   CARBOHYDRATES

The overwhelming amount of work on the metal-ion-catalyzed interactions of hydrogen peroxide with carbohydrates has been carried out mainly with iron ions, although other metal ions, (e.g., copper, manganese, cobalt, nickel, titanium, and cerium) have also been found effective as additives in reactions of hydrogen peroxide. The iron-ion-catalyzed reactions are usually divided into two classes: (a) the Fenton reaction (1–5), involving ferrous ions, and (b) the Ruff degradation (6–8), involving ferric ions.

Since the end of the last century both methods of oxidation have been extensively used in carbohydrate chemistry and hence there is a large literature on this important subject. However, the scope and limitations of this review prevent a coverage of this area, and the reader is referred to an excellent review by Moody (9) and some additional articles (278–280) that contain the pertinent details. In this chapter the discussion is restricted to a few reactions illustrating several ideas concerning mechanistic aspects. In many instances both reactions result in extensive fragmentations of carbohydrate molecules (9, 279, 281). These results make it difficult to present a unifying concept concerning the mechanism of these reactions.

Nevertheless, it seems to be certain that both reactions proceed, at least in part, by free-radical paths; ESR studies (238, 282) support the existence of free-radical intermediates.

## A. Fenton Reaction

It seems that the reaction of carbohydrates with the Fenton reagent can proceed by several routes. Thus the reaction of L-gulono-1,4-lactone (**79**) with Fenton reagent results in the formation of a double bond to give L-ascorbic acid (**80**) (283).

$$H_2O_2 + Fe^{2+} \longrightarrow HO\cdot + Fe^{3+} + OH^-$$

This result is similar to that originally obtained by Fenton in the oxidation of tartaric acid to give a dihydroxy acid (1, 2, 4, 5).

In another type of Fenton reaction one carbon atom is eliminated as carbon dioxide to give a carbohydrate with one less carbon atom. Thus the oxidation of D-glucose (**81**) results in the formation of D-*erythro*-pentulosonic acid (**82**) (284).

$$
\begin{array}{ccccc}
\begin{array}{c}
\text{HC=O} \\
| \\
\text{HCOH} \\
| \\
\text{HOCH} \quad + \text{HO}\cdot \\
| \\
\text{HCOH} \\
| \\
\text{HCOH} \\
| \\
\text{CH}_2\text{OH} \\
\textbf{(81)}
\end{array}
&
\longrightarrow
&
\begin{array}{c}
\overset{\cdot}{\text{C}}\text{=O} \\
| \\
\text{HCOH} \\
| \\
\text{HOCH} \\
|
\end{array}
&
\xrightarrow{\text{Fe(OH)2}^{+}}
&
\begin{array}{c}
\text{C(O)OH} \\
| \\
\text{HCOH} \\
| \\
\text{HOCH} \\
|
\end{array}
\xrightarrow{\cdot\text{OH}}
\end{array}
$$

$$
\begin{array}{c}
\text{C(O)OH} \\
| \\
\cdot\text{COH} \\
| \\
\text{HOCH} \\
|
\end{array}
\xrightarrow{\text{Fe(OH)}^{2+}}
\begin{array}{c}
\text{C(O)OH} \\
| \\
\text{HOCOH} \\
| \\
\text{HOCH} \\
|
\end{array}
\xrightarrow{-\text{H}_2\text{O}}
\begin{array}{c}
\text{C(O)OH} \\
| \\
\text{C=O} \\
| \\
\text{HOCH} \\
|
\end{array}
$$

$$
\xrightarrow{-\text{CO}_2}
\begin{array}{c}
\text{HC=O} \\
| \\
\text{HOC}-\text{H} \\
|
\end{array}
\xrightarrow[\text{2. Fe(OH)}^{2+}]{\text{1. HO}\cdot}
\begin{array}{c}
\text{C(O)OH} \\
| \\
\text{HOC}-\text{H} \\
| \\
\text{HCOH} \\
| \\
\text{HCOH} \\
| \\
\text{CH}_2\text{OH}
\end{array}
\xrightarrow[\substack{\text{2. Fe(OH)}^{2+} \\ \text{3. } -\text{H}_2\text{O}}]{\text{1. HO}\cdot}
\begin{array}{c}
\text{C(O)OH} \\
| \\
\text{C=O} \\
| \\
\text{HCOH} \\
| \\
\text{HCOH} \\
| \\
\text{CH}_2\text{OH} \\
\textbf{(82)}
\end{array}
$$

An interesting demethylation reaction of methylated sugars is achieved with the Fenton reagent (9, 203, 285). However, this reaction is not specific, and further oxidation of the sugar also occurs (9, 203, 285). The reaction might be explained as follows:

$$
\begin{array}{c} | \\ -\text{C}-\text{OMe} \\ | \end{array} + \text{HO}\cdot
\longrightarrow
\begin{array}{c} | \\ -\text{C}-\text{OCH}_2\cdot \\ | \end{array}
$$

$$
\xrightarrow{\text{Fe(OH)}^{2+}}
\begin{array}{c} | \\ -\text{C}-\text{OCH}_2\text{OH} \\ | \end{array}
\longrightarrow
\begin{array}{c} | \\ -\text{C}-\text{OH} \\ | \end{array} + \text{CH}_2\text{OH}
$$

## B.   Ruff Degradation

The procedure for the degradation of a carbohydrate with Ruff reagent is usually as follows: an aldose, such as glucose, is first oxidized with bromine water to the gluconic acid. Oxidation of the calcium salt of this acid with hydrogen peroxide in the presence of ferric salts results in formation of carbonate ion and an aldose of one less carbon atom than the initial aldose.

```
        HC=O                      C(O)OCa½
         |                          |
        HCOH                      HCOH                      HC=O
         |                          |                        |
        HOCH                      HOCH                      HOCH
         |    1.  Br₂, H₂O          |    H₂O₂, Fe³⁺           |
        HCOH  ───────────→        HCOH  ───────────→        HCOH
         |    2.  CaCO₃             |                        |
        HCOH                      HCOH                      HCOH
         |                          |                        |
        CH₂OH                     CH₂OH                     CH₂OH
```

The exact mechanism of this reaction is not established; however, there are several possibilities for interpretation. In spite of the large amounts of ferric ions that are initially added, there is little doubt that an oxidation–reduction system is rapidly established also containing ferrous ions, which in turn can initiate the reaction. The massive amounts of ferric ions facilitate a concomitant electron- and ligand-transfer step to give the product. For example, the degradation of D-mannose via calcium D-mannoate (83) to arabinose (84) may proceed as follows:

$$H_2O_2 + Fe^{2+} \longrightarrow HO\cdot + Fe^{3+} + OH^-$$

```
    C(O)OCa½                          C(O)OCa½
      |                                 |
    HOCH                              HOC·
      |                                 |
    HOCH                                            Fe(OH)²⁺
      |       + HO·  ───────→            + H₂O  ───────────→
    HCOH                                 
      |                                  
    HCOH                                 
      |
    CH₂OH
    (83)
```

```
                        C(O)OCa½                    CHO
                          |                          |
                        HOCOH                       HOCH
                          |          -CO₂            |
                                   ─────────→       HCOH
                                     -H₂O             |
                                                     HCOH
                                                      |
                                                     CH₂OH
                                                     (84)
```

However, another sequence that is possible involves hydroperoxy radicals, which can be formed by one of the following routes:

$$HOOH + \cdot OH \longrightarrow HOO\cdot + H_2O$$

$$HOOH + Fe^{3+} \longrightarrow HOO\cdot + Fe^{2+} + H^+$$

Since the Ruff degradation usually proceeds at a slower rate than the Fenton reaction, the stationary hydroxy radical concentration in the Ruff reaction is probably much smaller than that in the Fenton reaction and the hydroperoxy radicals might participate preferentially in the reaction with the carbohydrate radicals to give a hydroperoxide intermediate, which in turn breaks down to give the carbonyl derivative.

$$
\begin{array}{l}
\text{C(O)OCa}_{1/2} \\
| \\
\text{HOC}\cdot \qquad\qquad + \text{HOO}\cdot \longrightarrow \\
|
\end{array}
\qquad
\begin{array}{l}
\text{C(O)OCa}_{1/2} \\
| \\
\text{HOCOOH} \\
|
\end{array}
$$

$$
\xrightarrow{\text{Fe}^{2+}}
\begin{array}{l}
\text{C(O)OCa}_{1/2} \\
| \\
\text{HOCO}\cdot \\
|
\end{array}
\xrightarrow{\text{H-donor}}
\begin{array}{l}
\text{C(O)OCa}_{1/2} \\
| \\
\text{HOCOH} \\
|
\end{array}
\xrightarrow[-H_2O]{-CO_2}
\begin{array}{l}
\text{HC=O} \\
|
\end{array}
$$

Formation of oxygen gas in the Ruff reaction by one of the following paths also indicates a high concentration of hydroperoxy radicals:

$$HOO\cdot + Fe^{3+} \longrightarrow Fe^{2+} + O_2 + H^+$$

$$HOO\cdot + H_2O_2 \longrightarrow H_2O + O_2 + HO\cdot$$

## IX  NITROGEN COMPOUNDS

### A.  Amines

The reactions of amines with hydrogen peroxide in the presence of metal ions can proceed by involving the hydrocarbon part of the molecule or the nitrogen moiety, or both. In the former case the following nonchain radical process occurs to give dimers (222):

$$H_2O_2 + Fe^{2+} \longrightarrow HO\cdot + Fe(OH)^{2+}$$

$$RH + HO\cdot \longrightarrow R\cdot + H_2O$$

$$2R\cdot \longrightarrow R-R$$

$$(RH = \text{amine})$$
$$(R\cdot = \text{carbon radical})$$

Thus the reactions of ethylamine (240), propylamine (221, 222), diethylamine (240), and *t*-butylamine (120, 221, 222) with equimolar quantities of hydrogen peroxide and ferrous sulfate form dimers **85, 86, 87,** and **88,** respectively.

$$MeCH_2NH_2 \xrightarrow{H_2O_2 + Fe^{2+}/Fe^{3+}} (H_2NCH_2CH_2)_2$$

$$(85)$$

$$MeCH_2CH_2NH_2 \xrightarrow{H_2O_2 + Fe^{2+}/Fe^{3+}} (H_2NCH_2CH_2CH_2)_2$$

$$(86)$$

$$(MeCH_2)_2NH \xrightarrow{H_2O_2 + Fe^{2+}/Fe^{3+}} EtNH(CH_2)_4NHEt$$

$$(87)$$

$$Me_3CNH_2 \xrightarrow{H_2O_2 + Fe^{2+}/Fe^{3+}} (H_2NCMe_2CH_2)_2$$

$$(88)$$

When the reaction of *t*-butylamine is carried out in the presence of carbon monoxide, a small amount of $\beta$-aminoisovaleric acid (**89**) is obtained (234, 235).

$$Me_3CNH_2 + \cdot OH \longrightarrow Me_2CCH_2 \cdot + H_2O$$
$$\underset{NH_2}{|}$$

$$\xrightarrow{CO} \underset{\underset{NH_2}{|}}{Me_2CCH_2CO} \xrightarrow{Fe(OH)^{2+}} \underset{\underset{NH_2}{|}}{Me_2CCH_2C(O)OH}$$

$$(89)$$

However, the oxidation of diethylamine and triethylamine with hydrogen peroxide in the presence of ferrous ions can also lead, by a destructive path, to acetaldehyde (213).

If the reaction of *t*-butylamine with hydrogen peroxide is carried out in the presence of butadiene, the product mixture mainly consists of compound **90**, a 12-carbon-atom diamine containing one butadiene unit (120) and a 24-carbon-atom triamine (120).

$$Me_3CNH_2 + \cdot OH \longrightarrow \underset{\underset{NH_2}{|}}{Me_2CCH_2 \cdot} \xrightarrow{CH_2=CHCH=CH_2}$$

$$\underset{\underset{NH_2}{|} \qquad\qquad\qquad \underset{NH_2}{|}}{Me_2CCH_2C_4H_6CH_2CMe_2}$$

$$(90)$$

The reaction of *t*-octylamine results in a 16-carbon-atom unit diamine, and the reaction of *t*-amylamine results in the formation of a mixture of triamines (120).

Under similar dimerizing conditions the reactions of piperidine, $\alpha$-picoline, $\beta$-picoline, and diethylaniline produce compounds **91**, **92**, **93**, and **94**, respectively (240).

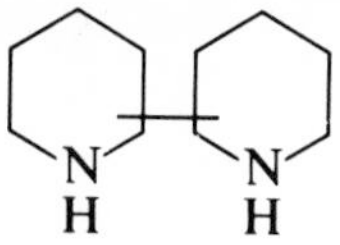

(91)

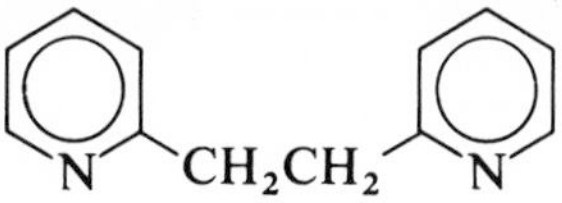

(92)

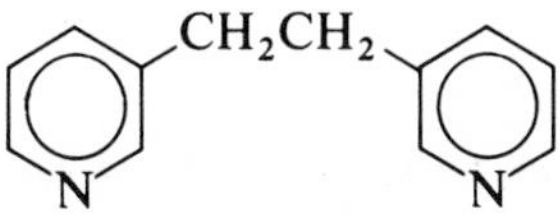

(93)

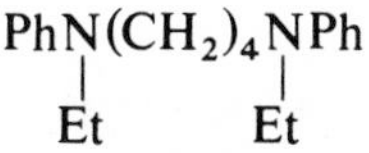

$$PhN(CH_2)_4NPh$$
$$\qquad | \qquad\quad |$$
$$\qquad Et \qquad\quad Et$$

(94)

However, cleavage of the side chain of an aromatic amine may occur without dimerization. Thus the oxidation of dimethylaniline yields formaldehyde, together with a deep-purple solution indicative of dyes from *o*-aminophenol (101). The reaction of tyramine (95) yields the hydroxylated product, 3,4-dihydroxyphenethylamine (286).

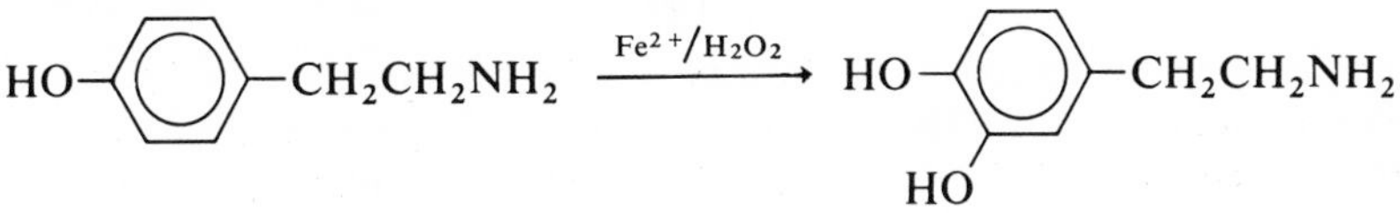

(95)

The oxidation of aniline in aqueous acetic acid with hydrogen peroxide in the presence of ferrous sulfate involves the amine group and produces a mixture of azobenzene, aminoanilinoquinone-monoanil (96), and 2,5-dianilinoquinone-monoanil (97) (287).

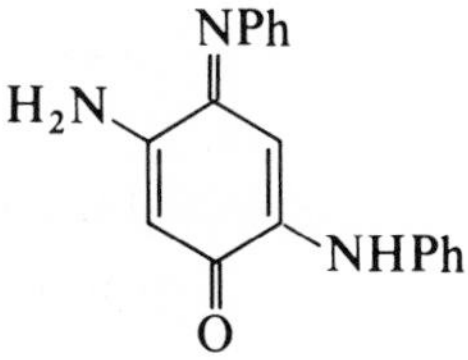

(96)

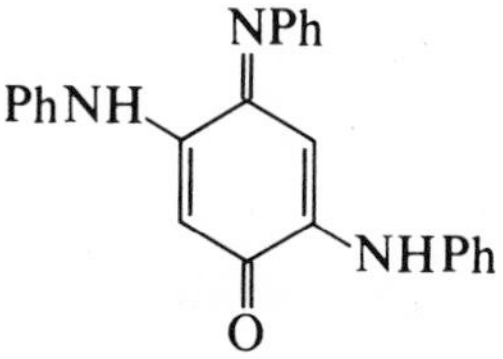

(97)

The oxidation of aniline hydrochloride in aqueous hydrochloric acid solution with hydrogen peroxide in the presence of ferrous sulfate produces emeraldine (98), which can be further oxidized to nigraniline (99) (287, 288). Similar results are obtained from reactions of *p*-aminodiphenylamine (287, 288).

$$PhNH_2 \longrightarrow PhNHOH \xrightarrow{PhNH_2} PhNHC_6H_4NH_2$$

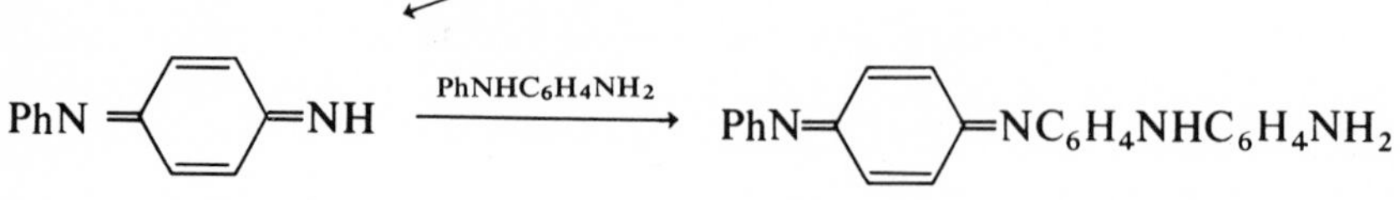

$$PhN = C_6H_4 = NH \xrightarrow{PhNHC_6H_4NH_2} PhN = C_6H_4 = NC_6H_4NHC_6H_4NH_2$$

$$PhN=C_6H_4=NC_6H_4N=C_6H_4=NC_6H_4NHC_6H_4 \; NHC_6H_4NHC_6H_4NH_2$$

$$(98)$$

oxidation

$$PhN=C_6H_4=NC_6H_4N=C_6H_4=NC_6H_4N=C_6H_4=NC_6H_4NHC_6H_4NH_2$$

$$(99)$$

Similarly the oxidation of *p*-toluidine results in a mixture of 2,7-dimethyl-3-*p*-toluidinophenazine (**100**), 4-*p*-toluidino-2,5-toluquinone di-*p*-tolylimine (**101**), 4-*p*-toluidino-2,5-toluquinone 2-*p*-tolylimine (**102**), di-*p*-tolylamine (**103**), and 4,4′-dimethylazobenzene (**104**), in low yields (15, 289).

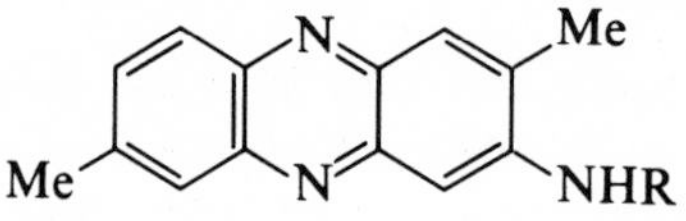

(100)

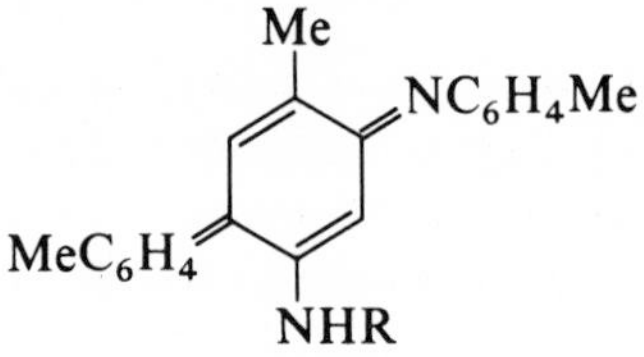

(101)

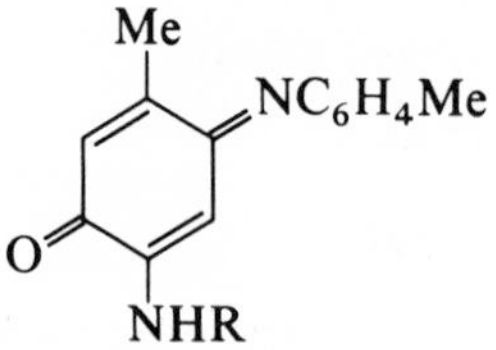

(102)

$(p\text{-}MeC_6H_4)_2NH$

(103)

$p\text{-}MeC_6H_4N=NC_6H_4\text{-}p\text{-}Me$

(104)

$(R = p - C_6H_4Me)$

A comparison of the oxidations of aromatic amines by hydrogen peroxide and the ferrous ion with that by the peroxidase system shows considerable differences (15).

The reaction of pyridine with 3% hydrogen peroxide occurs with ring cleavage to yield furfural (290). Quinoline is hydroxylated by the Fenton reagent in the presence of ascorbic acid, producing 3-hydroxyquinoline in 3–7% yield (291). The reaction of the Fenton reagent in the presence of EDTA and ascorbic acid with indole also results in hydroxylation, yielding a mixture of 4-, 5-, 6-, and 7-hydroxy indoles (292). The interaction of the Fenton reagent with heteroaromatic nitrogen compounds has been reviewed by Norman and Radda (133).

Electron-spin-resonance studies have been made of the radicals generated from the interaction of aliphatic (83, 129, 227) and cyclic (227) amines with the titanous chloride–hydrogen peroxide system.

## B. Nitriles

Reactions of aliphatic nitriles with equimolar quantities of hydrogen peroxide and ferrous sulfate result in the formation of dimers (120, 221, 222).

$$H_2O_2 + Fe^{2+} \longrightarrow HO\cdot + Fe(OH)^{2+}$$

$$RCH_2CN + \cdot OH \longrightarrow R\overset{\cdot}{C}HCN + H_2O$$

$$2R\overset{\cdot}{C}HCN \longrightarrow \underset{\underset{CN}{|}}{RCH} - \underset{\underset{CN}{|}}{CHR}$$

$$(R = alkyl)$$

For example, the dimerizations of acetonitrile (120, 221, 222) and pivalonitrile (221, 222) give 18% succinonitrile and 52% $\alpha,\alpha,\alpha',\alpha,\alpha'$-tetramethyladiponitrile, respectively.

Under similar conditions the reaction of propionitrile results in a mixture of meso-$\alpha,\alpha'$-dimethylsuccinonitrile, $\alpha$-methylglutaronitrile, and adiponitrile (222). The reaction of adiponitrile forms a tetranitrile, $C_{12}H_{14}N_4$, mp 128–132° (221, 222).

If the reaction of pivalonitrile is carried out in the presence of carbon monoxide, $\beta$-cyanoisovaleric acid (**105**) is obtained among other products (120, 234, 235).

$$CMe_3CN + \cdot OH \longrightarrow \underset{\underset{CN}{|}}{Me_2CCH_2}\cdot$$

$$\xrightarrow{CO} \underset{\underset{CN}{|}}{Me_2CCH_2CO}\cdot \xrightarrow{Fe(OH)^{2+}} \underset{\underset{CN}{|}}{Me_2CCH_2C(O)OH}$$

The interaction of aliphatic nitriles with the system titanous chloride–hydrogen peroxide has been studied by ESR spectroscopy (83, 293).

### C.  Amides, Imides, and Oximes

Dimerization of propionamide with the ferrous-ion-catalyzed decomposition of hydrogen peroxide produces adipamide in low yield, plus isomers (221, 222). The reaction of benzamide gives salicylic amide (101). The ESR spectrum of the radical generated by the interaction of benzamide with the system titanous chloride–hydrogen peroxide has, however, been attributed to the *meta* adduct (**106**) (206).

$$C(O)NH_2$$

**(106)**

The ferrous-ion-catalyzed oxidation of benzoylglycine with hydrogen peroxide proceeds slowly to give small amounts of glyoxylic acid (260).

The reactions of aromatic amides and imides—such as benzamide, toluamide, and phthalimide—in ethanolic solution with hydrogen peroxide in the presence of ferric chloride produce the corresponding hydroxamic acids (294). Only aromatic amides undergo this reaction (294). In contrast, aliphatic and aromatic oximes, (e.g., acetoxime and benzaldoxime) react under similar conditions to give hydroxamic acids (294). Electron-spin-resonance studies have been made of the radical species formed by the oxidation of oximes with the titanous chloride–hydrogen peroxide system (295, 296).

The reaction of sulfanilamide (**107**) in neutral solution with hydrogen peroxide in the presence of ferrous sulfate gives hydroxylamine, *p*-aminophenol, and a dye of the indophenol type (297). Similarly the reaction of sulfapyridine (**108**) gives hydroxylamine, *p*-aminophenol, and a dye (297).

$$H_2N-\!\!\!\bigcirc\!\!\!-SO_2NH_2 \qquad H_2N-\!\!\!\bigcirc\!\!\!-SO_2NH-\!\!\!\bigcirc$$

**(107)**                              **(108)**

The reaction of hydrogen peroxide in the presence of ferrous sulfate with cyclophosphamide (Endoxan, **109**), an important agent in cancer chemotherapy, gives hydroxylated compounds that are identical with metabolites produced enzymically from cyclophosphamide in rats. It is believed that the hydroxy group is introduced at the carbon atom of the $-NHCH_2-$ group (298).

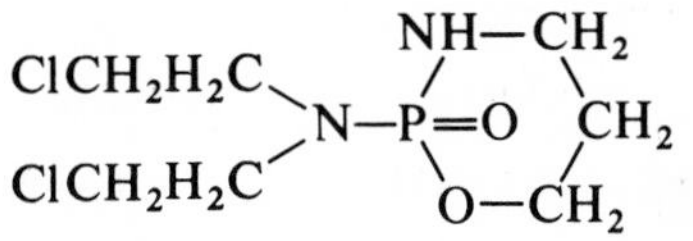

**(109)**

### D. Urea Derivatives

The condensation of methylglyoxal with urea produces an unstable product **(110)** that on treatment with aqueous hydrogen peroxide in the presence of ferrous sulfate yields a peroxide **(111)** plus a molecule of urea (299). Further oxidation of the peroxide **111** results in fragmentation to yield ammonia, carbon dioxide, and acetic acid; in this process acetylurea is an intermediate (299).

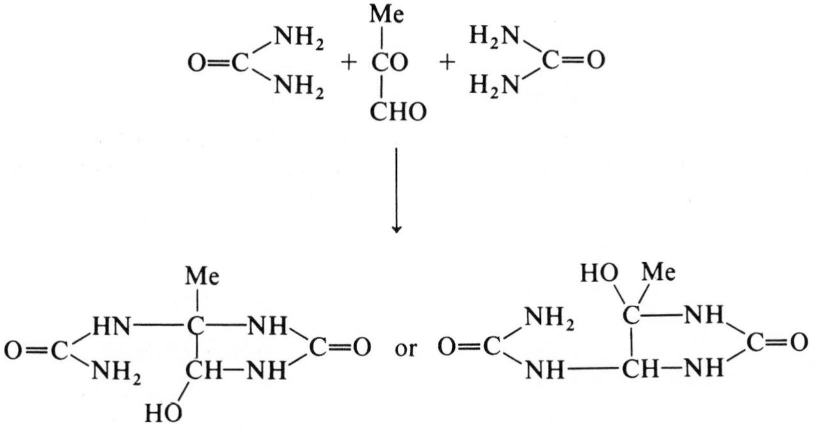

**(110)**

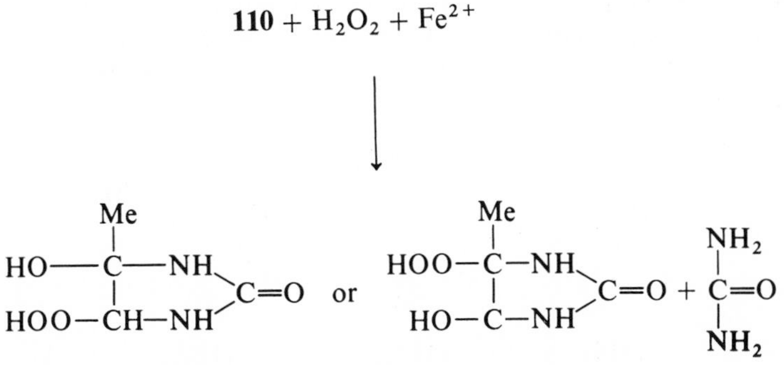

$$110 + H_2O_2 + Fe^{2+}$$

**(111)**

The reaction of creatinine (**112**) with hydrogen peroxide in the presence of ferrous sulfate forms methylguanidine (**113**), glyoxylic acid, formic acid, formaldehyde, and carbon dioxide (260).

(**112**)                    (**113**)

Under similar conditions the oxidation of thymine (5-methyluracil, **114**) forms mainly urea and pyruvic acid, plus some acetol (**115**) (300).

$$MeC(O)CH_2OH$$

(**114**)                    (**115**)

The oxidative degradation of uric acid (**116**) with 3% hydrogen peroxide in the presence of ferric chloride gives urea, carbonyldiurea (**117**), oxalic acid, and ammonia (301).

(**116**)                    (**117**)

The substitution products from the interaction of the Fenton reagent with a number of pyrimidines (302) and purines (303) have been identified. Uracil (**118**) and cytosine (**119**) are hydroxylated in the 5-position to yield, respectively, isobarbituric acid (**120**) in 6.5% yield and 5-hydroxycytosine in 4.7% yield (302).

(**118**)                (**119**)                (**120**)

With thymine (114) and 5-methylcytosine, the methyl side chain in the 5-position is attacked to form, respectively, 5-hydroxymethyluracil in 0.7% yield and 5-hydroxymethylcytosine in 0.6% yield (302). In addition to substitution, extensive fragmentation occurs. The purine bases are much less reactive (303); under similar conditions no fragmentation occurs, and only a little hydroxylation. Thus the reaction of purine (121) yields a mixture of 6- and 8-hydroxypurines. With 6-hydroxypurine, 2,6-dihydroxypurine, 2-aminopurine, and adenine (122), the products are the 8-hydroxy derivatives.

(121)          (122)

An ESR study of the reactions of nucleic acid bases with hydroxy radicals produced from hydrogen peroxide and titanium trichloride indicates that pyrimidines are more reactive than purines. Free-radical signals are detected only with thymine (114), cytosine (119), and uracil (118). The spectra are compatible with structures of addition products of hydroxy radicals to the 5,6-position double bond. Solutions of adenine (122), guanine, xanthine, and hypoxanthine show only signals derived from hydroxy radicals (304).

### E. Nitro Compounds

The reaction of a mixture of nitrobenzene and hydrogen peroxide in acetic acid with ferrous sulfate gives only small quantities of *o*-nitrophenol (101). The distribution of nitrophenol isomers obtained in the hydroxylation has been studied (145, 195, 200).

### F. Amino Acids

The oxidative degradations of amino acids with hydrogen peroxide and ferrous ions have been extensively investigated because of biological implications, such as model systems of peroxidase action and metabolism of amino acids (15, 16, 17, 260, 305–308).

Other metal ions capable of oxidation–reduction reactions—such as copper, nickel, cobalt, titanium, and cerium—have been tried; however, they are less effective than iron ions, and only copper has been used to some extent (309). Interestingly, iron ions also act as photosensitizers in degradative reactions of amino acids (310). The ferrous-ion-catalyzed reactions of hydrogen peroxide with amino acids are pH dependent, and optimum rates seem to be obtained at pH 2–4 (165).

Several patterns have been observed for the degradation of amino acids. Aliphatic amino acids can be stepwise degraded to other amino acids (308, 309, 311, 312). It seems that the hydroxy radicals attack the methylene and methyl groups (309, 312). For example, the degradation of alanine gives glycine (308, 309, 311), serine (308, 309, 311), and aminomalonic acid (123) (309).

$$
\begin{array}{c}
\text{H}_2\text{N}\quad\text{O} \\
|\qquad\| \\
\text{MeCHCOH}
\end{array}
$$

$$\text{H}_2\text{O}_2 + \text{Fe}^{2+}/\text{Fe}^{3+}$$

$$
\begin{array}{cc}
\begin{array}{c}
\text{H}_2\text{N}\quad\text{O} \\
|\qquad\| \\
\text{HOCH}_2\text{CHCOH}
\end{array}
&
\begin{array}{c}
\text{H}_2 \\
\text{ON}\quad\text{O} \\
\|\ \ |\ \ \| \\
\text{HOCCHCOH}
\end{array}
\end{array}
$$

(123)

$$- \text{CO}_2$$

$$
\begin{array}{c}
\text{H}_2\text{N}\quad\text{O} \\
|\qquad\| \\
\text{CH}_2\text{COH}
\end{array}
$$

The product distribution and mechanism of amino-acid degradation by hydrogen peroxide have been studied with a carbon-14 tracer (312).

Amino acids containing a sulfhydryl group, such as cysteine (313–315) or glutathione (313), are oxidized to the corresponding disulfide derivatives. Methionine, containing a sulfide linkage, is converted to the corresponding sulfoxide (19, 316).

Amino acids containing a ring structure may be hydroxylated in the ring; for example, phenylalanine (286, 308, 311), tyrosine (286, 308, 311), kynurenine (308, 311), tryptophan (286, 308, 311), and proline (305, 308, 311, 317–321), give tyrosine, dihydroxyphenylalanines, hydroxykynurenines, hydroxytryptophans, and hydroxyprolines, respectively. Similarly the hydroxylation of phenylglycine gives benzaldehyde, hydroxybenzaldehyde, and hydroxybenzoic acid (322). An excellent discussion of *in vitro* and *in vivo* hydroxylations and ring-opening reactions of aromatic compounds is available in a review article (17).

Frequently a more complete oxidative degradation of amino acids has been observed; it leads to a mixture of mainly ammonia, carbon dioxide, and aldehydes that contain one carbon atom less than the original amino acid (213, 260, 307, 323, 324). The intermediate formation of acids containing an $\alpha$-carbonyl function (260, 312, 324, 325) has been proposed and observed. Thus the oxidation of alanine in dilute solution with hydrogen peroxide in the presence of ferrous salt proceeds through pyruvic acid (326).

## X. PEROXYDISULFATE OXIDATION

The metal-ion-catalyzed decomposition of peroxydisulfate has been extensively investigated, and there is a large volume of literature on the mechanistic aspects of this decomposition (140, 212, 327–357). Unfortunately most of these studies either employ uninteresting substrates, as far as organic chemists are concerned, or they do not involve the isolation of products.

Although no attempt has been made to have a complete collection of these studies, it is believed that all major developments are covered by this series of references (140, 212, 327–357). Several review articles (20, 358–360) dealing with various aspects of peroxydisulfate decomposition are also available; however, most of them are several years old, and the emphasis is not on the preparative side.

### A. Aromatic Hydrocarbons, Aldehydes, Carboxylic Acids, and Esters

The information on these types is fragmentary, and therefore only tentative conclusions can be made.

The oxidation of toluene by the silver-ion-catalyzed reaction of peroxydisulfate results in a mixture of bibenzyl, benzaldehyde, and benzoic acid (361, 362). On the basis of these products, the following free-radical paths can be assumed:

$$^-O_3SO{-}OSO_3^- + M^n \longrightarrow {}^-O_3SO\cdot + SO_4^{2-} + M^{n+1}$$

$$PhMe + {}^-O_3SO\cdot \longrightarrow PhCH_2\cdot + HSO_4^-$$

$$2PhCH_2\cdot \longrightarrow PhCH_2CH_2Ph$$

$$PhCH_2\cdot + SO_4^{2-} + M^{n+1} \longrightarrow PhCH_2OSO_3^- + M^n$$

$$PhCH_2OSO_3^- + H_2O \longrightarrow PhCH_2OH + HSO_4^-$$

$$PhCH_2OH + \cdot OSO_3^- \longrightarrow Ph\overset{\cdot}{C}HOH + HSO_4^-$$

$$Ph\overset{\cdot}{C}HOH + SO_4^{2-} + M^{n+1} \longrightarrow \underset{\underset{OSO_3^-}{|}}{PhCHOH} + M^n$$

$$\underset{\underset{OSO_3^-}{|}}{PhCHOH} \longrightarrow PhCHO + HSO_4^-$$

$$(M = Ag, Fe, Cu)$$

Similarly the oxidation of ethylbenzene produces a 70% yield of aceto-phenone (363). The oxidation of isobutyric acid (**124a**)(364) gives isopropanol, acetone, and carbon dioxide, and that of trimethylacetic acid (**124b**) results in carbon dioxide and *t*-butanol (364). Similarly the oxidation of phenyl-acetic acid (**124c**) gives a mixture of carbon dioxide, benzaldehyde, and bibenzyl (361–364).

$$RR'R''CC(O)OH + \cdot OSO_3^- \longrightarrow RR'R''CC(O)O \cdot + HSO_4^-$$

$$\xrightarrow{-CO_2} RR'R''C \cdot \xrightarrow[M^{n+1}]{SO_4^{2-}} RR'R''COSO_3^-, \text{ etc.}$$

(**124**)

(**a**, R = H, R' = R" = Me)
(**b**, R = R' = R" = Me)
(**c**, R = R' = H, R" = Ph)

Under similar conditions the oxidation of acetaldehyde (365) gives acetic acid, and that of benzaldehyde (360) results in benzoic acid and oxygen. The reaction of ethyl phenylacetate yields a dimer (**125**) (361).

$$ArCHC(O)OEt$$
$$|$$
$$ArCHC(O)OEt$$

(**125**)

### B.    Alcohols and Glycols

The oxidation of alcohols with peroxydisulfate is catalyzed by silver, iron, and sometimes copper ions. The reaction results in the formation of aldehydes or ketones; for example, isopropyl (330), *n*-butyl (365), *sec*-butyl (365), and benzyl alcohols (362, 363) are converted to the corresponding carbonyls in a manner similar to that discussed in the case of toluene.

The reaction of peroxydisulfate with glycols in the presence of silver ions results in the cleavage of the carbon–carbon bond to give either aldehydes or ketones, or both, in 40–100% yields (365, 366). The reactions of primary and secondary glycols give aldehydes, whereas those of tertiary glycols yield ketones (365). In addition, small quantities of α-hydroxyketones and α-hydroxyaldehydes are obtained (365). The following glycols undergo this type of reaction (365): ethylene glycol (**126a**), 2,3-butanediol (**126b**), pinacol (**126c**), phenylethylene glycol (**126d**), 4,5-octanediol (**126e**), mesohydrobenzoin (**126f**), diethyl tartrate (**126g**), methyl 10,11-dihydroxyhendecanoate (**126h**), and *cis*- and *trans*-cyclohexanediol (**126i**).

$$RR'COH + S_2O_8{}^{2-} \xrightarrow{\text{Ag}^+} RR'{=}CO + R''R'''C{=}O + 2SO_4{}^{2-} + 2H^+$$
$$\underset{\displaystyle R''R'''COH}{|} \qquad\qquad\qquad\qquad (126)$$

(**a**, R = R′ = R″ = R‴ = H)
(**b**, R = R″ = H, R′ = R‴ = Me)
(**c**, R = R′ = R″ = R‴ = Me)
(**d**, R = R′ = R″ = H, R‴ = Ph)
(**e**, R = R″ = H, R′ = R‴ = *n*-Pr)
(**f**, R = R″ = H, R′ = R‴ = Ph)
[**g**, R = R″ = H, R′ = R‴ = C(O)OEt]
[**h**, R = R′ = R″ = H, R‴ = (CH$_2$)$_8$C(O)OMe]
[**i**, R = R″ = H, R′ = R‴ = −(CH$_2$)$_4$−]

### C.  Phenols

The oxidation of phenol and *o*- and *m*-cresol by aqueous peroxydisulfate in the presence of silver or iron ions gives ill-defined resinous polyphenols (367). However, similar reactions of *p*-cresol (361), 2,4-dimethylphenol (368), 2,6-dimethylphenol (363, 368), and 2,4,6-trimethylphenol (368) yield well-defined products (Table 5) that can be plausibly explained by free-radical paths, similar to those discussed in the case of toluene. A similar oxidation of thymol gives a dimeric product (369).

### D.  Elbs Oxidation

The oxidation of monohydric phenols to dihydric phenols by alkaline per-sulfate solution is known as the Elbs oxidation (370–373). The sulfate group enters the position *para* to the phenolic function. Subsequent hydrolysis in acid solution gives hydroquinones. If the *para* position is blocked, catechol derivatives are obtained (371–373).

Whether the Elbs oxidation proceeds by an ionic or a radical mechanism has not been convincingly resolved. Frequently Elbs oxidations yield small quantities of diphenyl derivatives, which are usually associated with free-radical coupling reactions (373, 374), and a free-radical mechanism can be assumed (373).

However, it is also possible that two concomitant reactions take place. One, an ionic reaction involving the sulfate ion, results in the formation of the main product, the dihydric phenol (375), whereas the other, involving

TABLE 5.   PEROXYDISULFATE OXIDATION OF PHENOLS

| Phenol | Catalyst | Product | Yield (%) | Reference |
|---|---|---|---|---|
| HO—⟨ ⟩—Me | Ag+, or Fe2+, or Fe3+ | | 7 | 361, 367 |
| | | | 7 | |
| | | | 15 | |
| | Ag+ | | 60 | 363, 368 |

TABLE 5. (*continued*)

| Phenol | Catalyst | Product | Yield (%) | Reference |
|--------|----------|---------|-----------|-----------|
| | | | 60 | |
| | | | 10 | |
| | Ag$^+$ | | 16 | 368 |
| | Ag$^+$ | | 22 | 368 |
| | | | 13 | |
| | | | Trace | |
| | Ag$^+$ | Dithymol | | 369 |

hydroxy radicals, results in small quantities of dimeric product (373).

$$\cdot OSO_3^- + H_2O \longrightarrow HSO_4^- + \cdot OH$$

$$^-OPh + \cdot OH \longrightarrow {}^-OC_6H_4\cdot + H_2O$$

$$^-OC_6H_4\cdot \longrightarrow dimer$$

The sulfate-ion radical could be produced by the reaction of a trace amount of metal ions (e.g., ferrous ions) with the persulfate ion.

$$^-O_3SO{-}OSO_3^- + M^n \longrightarrow {}^-O_3SO\cdot + SO_4^{2-} + M^{n+1}$$

$$(M = Fe^{2+})$$

The role of metal ions is also not entirely clear. In some synthesis work (376, 377) ferrous sulfate has been routinely used. Thus the reaction of potassium persulfate with coumarin in the presence of a mixture of potassium hydroxide and ferrous sulfate yields the hydroxycoumarin **127** (376). Similarly the reaction of methoxycoumarins **128** (376) and **129** (377) results in the formation of the methoxyhydroxycoumarins **130** and **131**, respectively.

**(127)**           **(128)**           **(129)**

**(130)**                    **(131)**

Under similar conditions dimethoxy- and diethoxycoumarin derivatives (**132**) are converted to the hydroxy derivatives (**133**) (378).

**(132)**                    **(133)**

(R = Me, Et)

The reaction of an alkaline solution of 2-hydroxypyridine with persulfate in the presence of ferrous sulfate gives a mixture of 2,3- and 2,5-dihydroxypyridines (379).

However, on the basis of these studies, it is not clear whether the addition of ferrous sulfate achieves a beneficial effect. In other investigations no effect by metal ions could be detected (373, 380).

### E. Amines and Amino Acids

The oxidation of a primary aliphatic amine with aqueous alkaline peroxydisulfate solution in the presence of silver nitrate results in a rapid exothermic reaction to yield an aldimine (363, 381, 382).

$$^-O_3SO{-}OSO_3^- + Ag^+ \longrightarrow {}^-O_3SO\cdot + SO_4^{2-} + Ag^{2+}$$

$$RCH_2NH_2 + \cdot OSO_3^- \longrightarrow R\underset{\cdot}{C}HNH_2 + HSO_4^-$$

$$R\underset{\cdot}{C}HNH_2 + SO_4^{2-} + Ag^{2+} \longrightarrow \underset{\underset{OSO_3^-}{|}}{RCHNH_2} + Ag^+$$

$$\underset{\underset{OSO_3^-}{|}}{RCHNH_2} \longrightarrow RCH{=}NH + HSO_4^-$$

$$RCH{=}NH + H_2O \longrightarrow RCHO + NH_3$$

$$RCHO + H_2NCH_2R \xrightarrow{\text{OH}^-} RCH{=}NCH_2R$$

The aldimine can be isolated, hydrolyzed to the aldehyde, or hydrogenated to the corresponding secondary amine (381, 382). The yields of aldehydes range between 14 and 96% (Table 6).

$$RCH{=}NCH_2R \quad\begin{array}{l}\xrightarrow{\;H^+\;} RCHO \\[2mm] \xrightarrow[\text{Cat.}]{\;H_2\;} (RCH_2)_2NH\end{array}$$

Under similar conditions primary amines of type **134** are converted to ketones (382) (Table 6), and primary amines of type **135** remain unchanged (382).

$$RR'CHNH_2 \xrightarrow[\text{Ag}^+,\ \text{OH}^-]{\text{S}_2\text{O}_8{}^{2-}} RR'CO \qquad\qquad RR'R''CNH_2$$
$$\textbf{(134)} \qquad\qquad\qquad\qquad\qquad\qquad \textbf{(135)}$$

$$(R = R' = R'' = Me)$$
$$(R = R' = Me,\ R'' = Me_3CCH_2)$$

Secondary amines of type **136** and **137** undergo similar reactions to give aldehydes and ketones in low yields (Table 6) (383).

$$(RCH_2)_2NH \qquad\qquad (RR'CH)_2NH$$
$$\textbf{(136)} \qquad\qquad\qquad \textbf{(137)}$$

The reactions of amino acids with peroxydisulfate in the presence of silver ions take an analogous course to give the corresponding aldehydes in low to moderate yields (383).

$$\underset{\underset{NH_2}{|}}{RCHC(O)OH} \longrightarrow RCHO + CO_2 + NH_3$$

TABLE 6. SILVER-ION-CATALYZED PEROXYDISULFATE OXIDATION OF AMINES AND AMINO ACIDS

| Amine or Amino Acid | Product | Yield (%) | Reference |
|---|---|---|---|
| $Me_2CHNH_2$ | $Me_2CO$ | 55–59 | 382 |
| $MeEtCHNH_2$ | $MeEtCO$ | 50–62 | 382 |
| cyclohexyl–$NH_2$ | cyclohexanone | 73–86 | 382 |
| $n\text{-}PrNH_2$ | $EtCHO$ | 43 | 382 |
| $n\text{-}BuNH_2$ | $n\text{-}PrCHO$ | 61–75 | 382 |
| $i\text{-}BuNH_2$ | $i\text{-}PrCHO$ | 61–71 | 382 |
| $n\text{-}C_8H_{17}NH_2$ | $n\text{-}C_7H_{15}CHO$ | 14–15 | 382 |
| $n\text{-}BuEtCHCH_2NH_2$ | $n\text{-}BuEtCHCHO$ | 66–70 | 382 |
| $t\text{-}BuCH_2CHMeCH_2CH_2NH_2$ | $t\text{-}BuCH_2CHMeCH_2CHO$ | 84–89 | 382 |
| $PhCH_2NH_2$ | $PhCHO$ | 96 | 382 |
| $(Me_2CHCH_2)_2NH$ | $Me_2CHCHO$ | 10–30 | 383 |
| $(PhCH_2)_2NH$ | $PhCHO$ | 22 | 383 |
| $(MeEt(CH_2)_2NH$ | $MeEtCO$ | 8–60 | 383 |
| (cyclohexyl)$_2NH$ | cyclohexanone | 4–11 | 383 |
| $MeCH(NH_2)C(O)OH$ | $MeCHO$ | 25 | 383 |
| $Me_2CHCH(NH_2)C(O)OH$ | $Me_2CHCHO$ | 48 | 383 |
| $EtCH(Me)CH(NH_2)C(O)OH$ | $MeEtCHCHO$ | 9 | 383 |
| $PhCH_2CH(NH_2)C(O)OH$ | $PhCHO$ | 14 | 383 |
| $HOC(O)CH_2CH_2CH(NH_2)C(O)OH$ | $HO(O)CCH_2CH_2CHO$ | 25 | 383 |
| $MeSCH_2CH_2CH(NH_2)C(O)OH$ | $Me(SO)CH_2CH_2CH(NH_2)C(O)OH$ | 25 | 383 |

# REFERENCES

1.  H. J. H. Fenton, *Proc. Chem. Soc.* (*London*), **9**, 113 (1893).
2.  H. J. H. Fenton, *J. Chem. Soc.*, **65**, 899 (1894).
3.  H. J. H. Fenton, *J. Chem. Soc.*, **67**, 48 (1895).
4.  N. J. H. Fenton, *J. Chem. Soc.*, **67**, 775 (1895).
5.  H. J. H. Fenton, *J. Chem. Soc.*, **69**, 546 (1896).
6.  O. Ruff, *Ber.*, **31**, 1573 (1898).
7.  O. Ruff, *Ber.*, **32**, 550 (1899).
8.  O. Ruff, *Ber.*, **32**, 2269 (1899).
9.  G. J. Moody, in *Advances in Carbohydrate Chemistry*, Vol. 19, edited by M. L. Wolfrom, Academic Press, New York, 1964, p. 149.
10. F. Haber and R. Willstätter, *Ber.*, **64**, 2844 (1931).
11. R. Kuhn and A. Wasserman, *Ann.*, **503**, 203 (1933).
12. F. Haber and J. Weiss, *Proc. Roy. Soc.* (*London*), **A147**, 332 (1934).
13. J. Weiss, *J. Phys. Chem.*, **41**, 1107 (1937).
14. J. Weiss, in *Advances in Catalysis*, Vol. 4, Academic Press, New York, 1952, p. 343.
15. B. C. Saunders, *Roy. Inst. Chem.* (*London*), *Lectures, Monographs, and Reports*, No. 1, 1 (1957).
16. S. Fallab, *Z. Naturwiss.-Med. Grundlagenforsch.*, **1**, 333 (1963).
17. R. O. C. Norman and J. R. L. Smith, in *Oxidases and Related Redox Systems*, Vol. 1, edited by T. S. King, H. S. Mason, and M. Morrison, Wiley, New York, 1965, p. 131.
18. H. S. Mason, in *Annual Reviews of Biochemistry*, Vol. 34, edited by J. M. Luck and P. D. Boyer, Ann. Revs. Inc., Palo Alto, Calif., 1965, p. 595.
19. E. H. E. Pietsch and A. Kotowski, and the Gmelin Institute, in *L. Gmelins Handbuch der Anorganischen Chemie*, Verlag Chemi, GMBH, Weinheim, System No. 3, Sauerstoff, 1966, pp. 2097, 2372, 2422, 2579.
20. E. H. E. Pietsch and A. Kotowski, and the Gmelin Institute, *op. cit.*, System No. 9, Schwefel, Part B, 1960, pp. 826 and 847.
21. A. I. Medalia and I. M. Kolthoff, *J. Polymer Sci.*, **4**, 377 (1949).
22. J. H. Baxendale, M. G. Evans, and G. S. Park, *Trans. Faraday Soc.*, **42**, 155 (1946).
23. M. G. Evans, *J. Chem. Soc.*, **1947**, 266.
24. W. A. Waters, *The Chemistry of Free Radicals*, Oxford University Press, New York, 1948, p. 247.
25. J. H. Baxendale, in *Advances in Catalysis*, Vol. 4, Academic Press, New York, 1952, p. 31.
26. N. Uri, *Chem. Rev.*, **50**, 375 (1952).
27. W. A. Waters, in *Organic Chemistry*, Vol. IV, edited by H. Gilman, Wiley, New York, 1953, p. 1120.
28. A. V. Tobolsky and R. B. Mesrobian, *Organic Peroxides*, Interscience, New York, 1954.

29. S. Kato and F. Masuo, *Kagaku*, **10**, 217 (1955).
30. C. Walling, *Free Radicals in Solution*, Wiley, New York, 1957, p. 564.
31. H. S. Mason, in *Advances in Enzymology*, Vol. 19, edited by F. F. Nord, Interscience, New York, 1957, p. 79.
32. A. G. Davies, *Organic Peroxides*, Butterworths, London, 1961.
33. E. G. E. Hawkins, *Organic Peroxides*, Van Nostrand, Princeton, N.J., 1961.
34. W. A. Waters, in *Progress in Organic Chemistry*, Vol. 5, edited by J. W. Cook and W. Carruthers, Butterworths, Washington, D.C., 1961, p. 1.
35. J. B. Loudon, in *Progress in Organic Chemistry*, Vol. 5, edited by J. W. Cook, and W. Carruthers, Butterworths, Washington, D.C., 1961, p. 46.
36. W. A. Waters, *Mechanisms of Oxidation of Organic Compounds*, Methuen, London, 1964.
37. W. A. Waters, Chemical Society (London), Special Publ. No.19, 71 (1964) (publ. 1965).
38. W. A. Waters, *Sci. Progr. (London)*, **53**, 413 (1965).
39. G. Tsuchihashi, *Yuki Gosei Kagaku Kyokai Shi*, **23** (2), 117 (1965).
40. R. G. R. Bacon and H. A. O. Hill, *Quart. Rev.*, **19**, 95 (1965).
41. L. S. Boguslavskaya, *Russian Chem. Rev.*, **34**, 503 (1965).
42. L. S. Boguslavskaya, *Usp. Khim.*, **34**, 1199 (1965).
43. A. Krause and E. Feret, *Roczniki Chem.*, **31**, 385 (1957).
44. A. Krause and F. Domka, *Roczniki Chem.*, **36**, 1123 (1962).
45. A. Krause and F. Domka, *Z. Anorg. Allgem. Chem.*, **315**, 110 (1962).
46. A. Krause and E. Kukielka, *Z. Physik. Chem. (Frankfurt)*, **35**, 65 (1962).
47. A. Krause and J. Orlikowska, *Naturwissenschaften*, **49**, 468 (1962).
48. A. Krause and M. Blawacka, *Z. Anorg. Allgem. Chem.*, **321**, 274 (1963).
49. A. Krause and F. Domka, *Chemiker-Ztg.*, **88**, 112 (1964).
50. A. Krause, F. Domka, and W. Sleziak, *Experientia*, **20**, 420 (1964).
51. W. Wawrzyczek, A. Jacyk, and A. Przybylski, *Kolloid- Z.*, **195**, 40 (1964).
52. A. Krause, I. Plura, and L. Lomozik, *Monatsh. Chem.*, **97**, 178 (1966).
53. J. Ghosh and S. Ghosh, *Proc. Natl. Acad. Sci. India*, **A32**, Part 2, 166 (1962).
54. J. Ghosh and S. Ghosh, *Proc. Natl. Acad. Sci. India*, **A32**, Part 2, 171 (1962).
55. H. Brintzinger and H. Erlenmeyer, *Helv. Chim. Acta*, **48**, 826 (1965).
56. H. Sigel and U. Mueller, *Helv. Chim. Acta*, **49**, 671 (1966).
57. H. Erlenmeyer, U. Mueller, and H. Sigel, *Helv. Chim. Acta*, **49**, 681 (1966).
58. R. Zell and H. Sigel, *Helv. Chim. Acta*, **49**, 870 (1966).
59. S. Petri, H. Sigel, and H. Erlenmeyer, *Helv. Chim. Acta*, **49**, 1778 (1966).
60. H. B. Jonassen and H. Thielemann, *Z. Anorg. Allgem. Chem.*, **320**, 274 (1963).
61. Z. Kovats, *Magy. Kem. Folyoirat*, **69**, 98 (1963).
62. M. T. Beck, S. Gorog, and Z. Kiss, *Magy. Kem. Folyoirat*, **69**, 550 (1963).
63. N. A. Ezerskaya and T. P. Solovykh, *Zh. Neorgan. Khim.*, **11** (9), 2179 (1966).
64. Y. Nose, M. Hatano, and S. Kambara, *Kogyo Kagaku Zasshi*, **69** (3), 566 (1966).
65. H. W. Krause and H. Mennenga, *J. Prakt. Chem.*, **32**, 283 (1966).
66. Zh. P. Kachanova and A. P. Purmal, *Zh. Fiz. Khim.*, **38**, 2483 (1964).

67.  T. Y. Lewis, D. H. Richards, and D. A. Salter, *J. Chem. Soc.*, **1963**, 2434.
68.  P. Flood, T. J. Lewis, and D. H. Richards, *J. Chem. Soc.*, **1963**, 2446.
69.  A. J. Dedman, T. J. Lewis, and D. H. Richards, *J. Chem. Soc.*, **1963**, 2456.
70.  G. A. Bogdanov, *Kataliz v Vysshei Shkole, Min. Vyshego i Srednego Spets. Obrasov SSSR, Tr. 1-go Mezhvuz. Soveshch. po Katalizu,* **1958**, No. 1, Part 1, publ. 1962, p. 299.
71.  G. A. Bogdanov, N. G. Semenova, and A. S. Chernyshev, *Izv. Vysshikh Uchebn. Zavedenii, Khim. Tekhnol.,* **7**, (3), 406 (1964).
72.  G. L. Petrova, R. I. Zaitseva, and S. A. Miroshnichenko, *Izv. Vysshikh Uchebn. Zavedenii, Khim. i Khim. Tekhnol.,* **5**, 533 (1962).
73.  A. K. Verkhovskaya and F. M. Perel'man, *Zh. Fiz. Khim.,* **38** (4) 1013 (1964).
74.  E. Wendling, *Bull. Soc. Chim. France,* **1966**, 404.
75.  E. Jasinskiene and J. Birmantas, *Izv. Vysshikh Uchebn. Zavedenii, Khim. i Khim. Tekhnol.,* **6** (6), 918 (1963).
76.  S. U. Kreingol'd, E. A. Bozhevol'nov, R. P. Lastovskii, and V. V. Sidorenko, *Zh. Anal. Khim.,* **18** (11), 1356 (1963).
77.  E. Jasinskiené and E. Bilidiené, *Lietuvos TSR Aukstųjų Mokyklų Mokslo Darbai, Chem. ir Chem. Technol.,* **4**, 33 (1964).
78.  J. Birmantas and E. Jasinskiené, *ibid.,* **6**, 5 (1965).
79.  J. Birmantas and E. Jasinskiené, *ibid.,* **6**, 13 (1965).
80.  J. Birmantas and E. Jasinskiené, *ibid.,* **6**, 31 (1965).
81.  E. Jasinskiené and E. Jankauskiené, *Lietuvos TSR Mokslu Akad. Darbai, Ser. B,* **1965** (4), 113.
82.  E. Michalski, K. Czarnecki, and K. Pietrucha, *Chem. Anal. (Warsaw),* **8** (5), 713 (1963).
83.  R. O. C. Norman and B. C. Gilbert, in *Advances in Physical Organic Chemistry,* Vol. 5, edited by V. Gold, Academic Press, London–New York, 1967, p. 53.
84.  R. O. C. Norman, *Proc. Roy. Soc. (London),* **A302**, 315 (1967).
85.  E. Saito and B. H. J. Bielski, *J. Amer. Chem. Soc.,* **83**, 4467 (1961).
86.  W. T. Dixon and R. O. C. Norman, *Nature,* **196**, 891 (1962).
87.  W. T. Dixon and R. O. C. Norman, *J. Chem. Soc.,* **1963**, 3119.
88.  L. H. Piette and G. Bulow, *Amer. Chem. Soc., Div. Petrol. Chem., Preprints,* **9** (2), C9 (1964).
89.  J. Stauff, H. J. Huster, F. Lohmann, and H. Schmidkunz, *Z. Physik. Chem. (Frankfurt),* **40**, 64 (1964).
90.  J. Stauff and H. J. Huster, *Z. Physik. Chem. (Frankfurt),* **55**, 39 (1967).
91.  F. Sicilio, R. E. Florin, and L. A. Wall, *J. Phys. Chem.,* **70**, 47 (1966).
92.  R. E. Florin, F. Sicilio, and L. A. Wall, *J. Phys. Chem.,* **72**, 3154 (1968).
93.  Y. S. Chiang, J. Craddock, D. Mickewich, and J. Turkevich, *J. Phys. Chem.,* **70**, 3509 (1966).
94.  H. Fischer, *Ber. Bunsenges. Phys. Chem.,* **71**, 685 (1967).
95.  C. R. E. Jefcoate and R. O. C. Norman, *J. Chem. Soc.,* B, **1968**, 48.
96.  K. Takakura and B. Ranby, *J. Phys. Chem.,* **72**, 164 (1968).
97.  R. O. C. Norman and P. R. West, *J. Chem. Soc.,* B, **1969**, 389.
98.  E. S. Shanley and F. P. Greenspan, *Ind. Eng. Chem.,* **39**, 1536 (1947).
99.  J. R. L. Smith and R. O. C. Norman, *J. Chem. Soc.,* **1963**, 2897.

100. N. Suzuki and H. Hotta, *Bull. Chem. Soc., Japan*, **40**, 1361 (1967).
101. J. H. Merz and W. A. Waters, *J. Chem. Soc.*, **1949**, 2427.
102. L. S. Boguslavskaya, N. A. Kartashova, V. E. Shurygin, and G. A. Razuvaev, *Zh. Obshch. Khim.*, **34**, 3081 (1964).
103. E. Iwata, S. Yoshikawa and S. Tsutsumi, *Kogyo Kagaku Zasshi*, **64**, 463 (1961).
104. W. T. Dixon and R. O. C. Norman, *J. Chem. Soc.*, **1964**, 4857.
105. F. Minisci, R. Galli, and M. Cecere, *Gazz. Chim. Ital.*, **94**, 67 (1964).
106. Y. Fujita and S. Ohashi, *J. Chem. Soc. Japan*, **63**, 93 (1942).
107. C. F. Cross, E. J. Bevan, and T. Heiberg, *Ber.*, **33**, 2015 (1900).
108. A. P. Barkhash, *Zh. Obshch. Khim.*, **5**, 254 (1935).
109. H. Hock, F. Depke, and G. Knauel, *Ber.*, **83**, 238 (1950).
110. M. Mugdan and D. P. Young, *J. Chem. Soc.*, **1949**, 2988.
111. F. Minisci and R. Galli, French Patent 1,350,360 (1964).
112. F. Minisci and R. Galli, *Tetrahedron Letters*, **1962**, 533.
113. F. Minisci and R. Galli, *Tetrahedron Letters*, **1963**, 357.
114. M. S. Kharasch, F. S. Arimoto, and W. Nudenberg, *J. Org. Chem.*, **16**, 1556 (1951).
115. M. S. Kharasch, W. Nudenberg, and F. S. Arimoto, *Science*, **113**, 392 (1951).
116. C. E. Frank and I. L. Mador, U.S. Patent 2,832,809 (1958).
117. C. M. Langkammerer, E. L. Jenner, D. D. Coffman, and B. W. Howk, *J. Amer. Chem. Soc.*, **82**, 1395 (1960).
118. E. L. Jenner, U.S. Patent 3,022,358 (1962).
119. D. D. Coffman and E. L. Jenner, *J. Amer. Chem. Soc.*, **80**, 2872 (1958).
120. W. R. McClellan, U.S. Patent 3,076,846 (1963).
121. E. L. Jenner, U.S. Patent 2,757,210 (1956).
122. E. L. Jenner, U.S. Patent 2,757,192 (1956).
123. E. L. Jenner and R. V. Lindsay, *J. Amer. Chem. Soc.*, **83**, 1911 (1961).
124. E. L. Jenner, U.S. Patent 2,842,582 (1958).
125. F. Minisci and R. Galli, *Chim. Ind.* (*Milan*), **47**, 178 (1965).
126. H. Fischer, *Z. Naturforsch.*, **19a**, 866 (1964).
127. J. Dewing, G. F. Longster, J. Myatt, and P. F. Todd, *Chem. Commun.*, **1965**, 391.
128. W. E. Griffiths G. F. Longster, J. Myatt, and P. F. Todd, *J. Chem. Soc., B*, **1967**, 530.
129. W. T. Dixon, R. O. C. Norman, and A. L. Buley, *J. Chem. Soc., B*, **1964**, 3625.
130. T. Shiga, *J. Phys. Chem.*, **69**, 3805 (1965).
131. O. C. Dermer and M. T. Edmison, *Chem. Rev.*, **57**, 77 (1957).
132. D. R. Augood and G. H. Williams, *Chem. Rev.*, **57**, 123 (1957).
133. R. O. C. Norman and G. K. Radda, in *Advances in Heterocyclic Chemistry*, Vol. 2, edited by A. R. Katritsky, Academic Press, New York, 1963, p. 164.
134. G. H. Williams, *Homolytic Aromatic Substitution*, Pergamon Press, Oxford–London, 1960, p. 110.
135. K. Ono, T. Oyamada, and H. Katsuragi, *J. Chem. Soc. Ind. Japan*, **41**, 209B (1938).

136. Imperial Chemical Industries, Ltd., Dutch Patent Appl., 5,613,970 (1967).
137. J. H. Baxendale and J. Magee, *Discussions Faraday Soc.*, **14**, 160 (1953).
138. J. H. Baxendale, *Reilly Lectures*, Vol. X, Notre Dame University Press, 1955.
139. W. T. Dixon and R. O. C. Norman, *Proc. Chem. Soc. (London)*, **1963**, 97.
140. I. P. Gragerov, A. F. Levit, Y. A. Zonov, and M. Y. Turkina, *Dokl. Akad. Nauk. SSSR.* **150**, 109 (1963).
141. N. A. Vysotskaya, *Zh. Fiz. Khim.*, **38**, 1688 (1964).
142. N. A. Vysotskaya and A. E. Brodskii, *Abh. Deut. Akad. Wiss. Berlin, Kl. Chem., Geol., Biol.*, **1964**, 653 (publ. 1965).
143. J. O. Konecny, *J. Amer. Chem. Soc.*, **76**, 4993 (1954).
144. V. S. Anderson, *Acta Chem. Scand.*, **4**, 207 (1950).
145. G. A. Hamilton, J. W. Hanifin, and J. P. Friedman, *J. Amer. Chem. Soc.*, **88**, 5269 (1966).
146. E. Boyland and P. Sims, *J. Chem. Soc.*, **1953**, 2966.
147. M. Kuroki, *Koru Taru*, **16**, 2 (1964).
148. A. I. Brodskii and N. A. Vysotskaya, *Zh. Obshch. Khim.*, **32**, 2273 (1962).
149. K. Spiro, *Z. Anal. Chem.*, **54**, 345 (1915).
150. M. Martinon, *Bull. Soc. Chim. France*, **43**, 155 (1885).
151. O. Y. Magidson and N. A. Preobrashenski, *Trans. Sci. Chem.–Pharm. Inst. (Moscow)*, **1926**, 65.
152. H. Goldhammer, *Biochem. Z.*, **189**, 81 (1927).
153. A. Chwala and M. Pailer, *J. Prakt. Chem.*, **152**, 45 (1939).
154. L. G. Gein, *Tr. Permsk. Farm. Inst.*, **1959**, 148.
155. G. Stein and J. Weiss, *J. Chem. Soc.*, **1951**, 3265.
156. A. I. Brodskii and N. A. Vysotskaya, *Khim. Perekisnykh Soedin., Akad. Nauk SSSR, Inst. Obshch. i Neorgan. Khim.*, **1963**, 249.
157. M. M. Aleksankin, N. A. Vysotskaya, and A. I. Brodskii, *Kernenergie*, **5**, 362 (1962).
158. K. Ono and T. Oyamada, *Bull. Chem. Soc. Japan*, **11**, 132 (1936).
159. S. L. Cosgrove and W. A. Waters, *J. Chem. Soc.*, **1951**, 1726.
160. R. G. R. Bacon and A. R. Izzat, *J. Chem. Soc. (C)*, **1966**, 791.
161. G. M. Coppinger, *J. Amer. Chem. Soc.*, **79**, 2758 (1957).
162. E. Hecker and G. Nowoczek, *Z. Naturforsch.*, **21b**, 153 (1966).
163. N. A. Vysotskaya, *Dokl. Akad. Nauk SSSR*, **148**, 344 (1963).
164. R. R. Grinstead, *J. Amer. Chem. Soc.*, **82**, 3472 (1960).
165. H. Wieland and W. Franke, *Ann.*, **457**, 1 (1927).
166. L. L. Ingraham, *Arch. Biochem. Biophys.*, **81**, 309 (1959).
167. P. Demerseman, R. N. Feinstein, and A. Cheutin, *Bull. Soc. Chim. France*, **1961**, 268.
168. J. Wolff, *Compt. Rend.*, **146**, 1217 (1908).
169. R. Willstätter and H. Stoll, *Ann.*, **416**, 64 (1921).
170. A. Wassermann, *J. Chem. Soc.*, **1935**, 826.
171. O. Baudisch, *Naturwissenschaften*, **27**, 768 (1939).
172. O. Baudisch and S. H. Smith, *Naturwissenschaften*, **27**, 769 (1939).
173. O. Baudisch, *Science*, **92**, 336 (1940).
174. O. Baudisch, *J. Amer. Chem. Soc.*, **63**, 622 (1941).
175. O. Baudisch and G. E. Heggen, *Arch. Biochem.*, **1**, 239 (1942).

176. O. Baudisch, *Arch. Biochem.*, **5**, 401 (1944).
177. O. Baudisch, *Science*, **108**, 443 (1948).
178. G. Cronheim, *J. Org. Chem.*, **12**, 1 (1947); *ibid.*, **12**, 7 (1947); *ibid.*, **12**, 20 (1947).
179. J. O. Konecny, *J. Amer. Chem. Soc.*, **77**, 5748 (1955).
180. K. Maruyama, I. Tanimoto, and R. Goto, *Tetrahedron Letters*, **1966**, 5889.
181. K. Maruyama, I. Tanimoto, and R. Goto, *J. Org. Chem.*, **32**, 2516 (1967).
182. H. Pfitzner and H. Merkel, German Patent 807,289 (1948).
183. H. Pfitzner, *Angew. Chem.*, **64**, 397 (1952).
184. H. Pfitzner and H. Baumann, *Angew. Chem.*, **70**, 232 (1958).
185. Z. Yoshida, K. Kazama, and R. Oda, *Kogyo Kagaku Zasshi*, **62**, 1021 (1959).
186. Z. Yoshida, K. Kazama, and R. Oda, *Kogyo Kagaku Zasshi*, **62**, 1399 (1959).
187. Z. Yoshida, S. Sagawa, and R. Oda, *Kogyo Kagaku Zasshi*, **63**, 1003 (1960).
188. Z. Yoshida, S. Sagawa, and R. Oda, *Kogyo Kagaku Zasshi*, **62**, 1402 (1959).
189. M. Kamel and S. A. Amin, *Ind. J. Chem.*, **2**, 60 (1964).
190. M. Kamel and S. A. Amin, *Ind. J. Chem.*, **2**, 62 (1964).
191. M. Kamel and S. A. Amin, *Ind. J. Chem.*, **2**, 232 (1964).
192. M. Kamel and S. A. Amin, *Ind. J. Chem.*, **2**, 282 (1964).
193. M. Kamel and S. A. Amin, *J. Prakt. Chem.*, **36**, 230 (1967).
194. G. Sosnovsky and D. J. Rawlinson, in *Organic Peroxides*, Vol. I, edited by D. Swern, Wiley, New York, **1970**, pp. 561, 585; Vol. II, 1971, pp. 272–276.
195. R. O. C. Norman and G. K. Radda, *Proc. Chem. Soc. (London)*, **1962**, 138.
196. G. R. A. Johnson, G. Stein, and J. Weiss, *J. Chem. Soc.*, **1951**, 3275.
197. G. A. Hamilton and J. P. Friedman, *J. Amer. Chem. Soc.*, **85**, 1008 (1963).
198. A. Cier, C. Nofre, M. Ranc, and A. Lefier, *Bull. Soc. Chim. France*, **1959**, 1523.
199. H. Staudinger and V. Ullrich, *Z. Naturforsch.*, **19b**, 409 (1964).
200. H. Loebl, G. Stein, and J. Weiss, *J. Chem. Soc.*, **1964**, 2074.
201. L. G. Shevchuk and N. A. Vysotskaya, *Zh. Organ. Khim.*, **2**, 1229 (1966).
202. V. Ullrich et al., *Biochem. Pharmacol.*, **1967**, 2237.
203. B. Frazer-Reid, J. K. N. Jones, and M. B. Perry, *Can. J. Chem.*, **39**, 555 (1961).
204. R. F. Brown et al., *J. Org. Chem.*, **29**, 146 (1964).
205. J. A. R. Mead, J. N. Smith, and R. T. Williams, *Biochem. J.*, **68**, 67 (1958).
206. R. O. C. Norman and R. J. Pritchett, *J. Chem. Soc.*, B, **1967**, 926.
207. H. Fischer, *Z. Naturforsch.*, **20a**, 488 (1965).
208. J. H. Walton, and C. J. Christensen, *J. Amer. Chem. Soc.*, **48**, 2083 (1926).
209. S. Goldschmidt and S. Pauncz, *Ann.*, **502**, 1 (1933).
210. A. A. Kul'tyugin and L. N. Sokolova, *Arch. Sci. Biol. Nauk SSSR*, **41**, 145 (1936).
211. L. Rosenthaler, *Arch. Pharm.*, **267**, 599 (1929).
212. J. H. Merz and W. A. Waters, *Discussions Faraday Soc.*, **2**, 179 (1947).
213. J. H. Merz and W. A. Waters, *J. Chem. Soc.*, **1949**, S15.
214. I. M. Kolthoff and A. I. Medalia, *J. Amer. Chem. Soc.*, **71**, 3777 (1949).
215. H. von Euler and H. Hasselquist, *Hoppe-Seyler's Z. Physiol. Chem.*, **303**, 176 (1956).
216. M. L. Kremer, *Nature*, **184**, Suppl. No. 10, 720 (1959).

217. W. A. Mosher, W. H. Clement, and R. L. Hillard, *Adv. Chem. Ser.*, **51**, 81 (1965).
218. C. C. Lee, A. J. Cessna, and M. K. Frost, *Can. J. Chem.*, **43**, 2924 (1965).
219. C. Heitler, D. B. Scaife, and B. W. Thompson, *J. Chem. Soc., A*, **1967**, 1409.
220. C. Neuberg and E. Schwenk, *Biol. Z.*, **71**, 114 (1915).
221. E. L. Jenner, U.S. Patent 2,700,051 (1955).
222. D. D. Coffman, E. L. Jenner, and R. D. Lipscomb, *J. Amer. Chem. Soc.*, **80**, 2864 (1958).
223. E. L. Jenner, in *Organic Synthesis*, Vol. 40, edited by M. S. Newman et al., Wiley, New York, 1960, p. 90.
224. H. E. De La Mare, K. J. Kochi, and F. F. Rust, *J. Amer. Chem. Soc.*, **85**, 1437 (1963).
225. H. J. H. Fenton and H. Jackson, *J. Chem. Soc.*, **75**, 1 (1899).
226. H. A. Spoehr, *Amer. Chem. J.*, **43**, 227 (1910).
227. W. T. Dixon and R. O. C. Norman, *J. Chem. Soc., A*, **1964**, 4850.
228. A. L. Buley, R. O. C. Norman, and R. J. Pritchett, *J. Chem. Soc., B*, **1966**, 849.
229. P. Smith, J. T. Pearson, and R. V. Tsina, *Can. J. Chem.*, **44**, 753 (1966).
230. P. Smith and P. B. Wood, *Can. J. Chem.*, **45**, 649 (1967).
231. T. Shiga, A. Boukhours, and P. Douzou, *J. Phys. Chem.*, **71**, 3559 (1967).
232. T. Shiga, A. Boukhours, and P. Douzou, *J. Phys. Chem.*, **71**, 4264 (1967).
233. P. L. Kolker, *J. Chem. Soc., Suppl.*, **1964**, 529.
234. D. D. Coffman, R. Cramer, and W. E. Mochel, *J. Amer. Chem. Soc.*, **80**, 2882 (1958).
235. D. D. Coffman, U.S. Patent, 2,687,432 (1954).
236. C. Neuberg, *Ber.*, **35**, 959 (1902).
237. J. D. Stirling, *Biochem. J.*, **28**, 1048 (1934).
238. P. J. Baugh, O. Hinojosa, and J. C. Arthur, *J. Phys. Chem.*, **71**, 1135 (1967).
239. G. A. Razuvaev and L. S. Boguslavskaya, *Zh. Obshch. Khim.*, **30**, 4094 (1960).
240. T. Shono, T. Iikuni, and R. Oda, *Nippon Kagaku Zasshi*, **81**, 1344 (1960).
241. R. J. Woods and J. W. T. Spinks, *Can. J. Chem.*, **35**, 1475 (1957).
242. A. L. Buley and R. O. C. Norman, *Proc. Chem. Soc. (London)*, **1964**, 225.
243. J. R. Steven and J. C. Ward, *Chem. Commun.*, **1965**, 273.
244. J. R. Steven and J. C. Ward, *J. Phys. Chem.*, **71**, 2367 (1967).
245. R. O. C. Norman and R. J. Pritchett, *J. Chem. Soc., B*, **1967**, 378.
246. R. O. C. Norman and R. J. Pritchett, *J. Chem. Soc., B*, **1967**, 3119.
247. F. G. Hopkins and S. W. Cole, *Proc. Roy. Soc. (London)*, **68**, 21 (1901).
248. C. Nofre, Y. Le Roux, L. Gondot, and A. Cier, *Bull. Soc. Chim. France*, **1964**, 2451.
249. R. O. Jones and I. Smedley-Maclean, *Biochem. J.*, **29**, 1690 (1935).
250. I. Smedley-Maclean and M. S. B. Pearce, *Biochem. J.*, **28**, 486 (1934).
251. S. Goldschmidt, P. Askenasy, and S. Pierros, *Ber.*, **61B**, 223 (1928).
252. C. Neuberg, *Biochem. Z.*, **67**, 71 (1914).
253. M. A. Battie and I. Smedley-Maclean, *Biochem. J.*, **23**, 593, (1929).
254. V. Subramaniam, B. H. Stent, and T. K. Walker, *J. Chem. Soc.*, **1929**, 2485.
255. H. Wieland, *Ann.*, **434**, 185 (1924).

256. A. P. Ponsford and I. Smedley-Maclean, *Biochem. J.*, **28**, 892 (1934).
257. I. Smedley-Maclean and M. S. B. Pearce, *Biochem. J.*, **25**, 1252 (1931).
258. C. Neuberg and O. Rubin, *Biochem. Z.*, **67**, 77 (1914).
259. H. J. H. Fenton and H. O. Jones, *J. Chem. Soc.*, **77**, 69 (1900).
260. H. D. Dakin, *J. Biol. Chem.*, **1**, 271 (1905).
261. G. B. Ray, *J. Gen. Physiol.*, **6**, 509 (1924).
262. H. J. H. Fenton and H. O. Jones, *J. Chem. Soc.*, **77**, 77 (1900).
263. W. H. Hatcher and W. H. Mueller, *Can. J. Research*, **3**, 291 (1930).
264. C. Neuberg and E. Schwenk, *Biochem. Z.*, **71**, 104 (1915).
265. M. Bergmann, *Ber.*, **54**, 1362 (1921).
266. F. Ferraboschi, *J. Chem. Soc.*, **95**, 1248 (1909).
267. A. Wassermann, *Biochem. Z.*, **263**, 1 (1933).
268. E. D. Pozhidaev and S. V. Gorbachev, *Zh. Fiz. Khim.*, **36**, 2512 (1962).
269. E. D. Pozhidaev and S. V. Gorbachev, *Zh. Fiz. Khim.*, **38**, 2938 (1964).
270. E. D. Pozhidaev and S. V. Gorbachev, *Zh. Fiz. Khim.*, **39**, 1678 (1965).
271. P. Smith, J. T. Pearson, P. B. Wood, and T. C. Smith, *J. Chem. Phys.*, **43**, 1535 (1965).
272. G. A. Razuvaev and L. S. Boguslavskaya, *Zh. Obshch. Khim.*, **31**, 3440 (1961).
273. L. S. Boguslavskaya and G. A. Razuvaev, *Tr. po Khim. i Khim. Tekhnol.*, **4**, 143 (1961).
274. G. A. Razuvaev and L. S. Boguslavskaya, *Zh. Obshch. Khim.*, **32**, 2320 (1962).
275. G. A. Razuvaev and L. S. Boguslavskaya, *Bul. Inst. Politekh. Iasi*, **8**, Nos. 3–4, 141 (1962).
276. L. S. Boguslavskaya and G. A. Razuvaev, *Zh. Obshch. Khim.*, **33**, 2021 (1963).
277. A. R. Metcalfe and W. A. Waters, *J. Chem. Soc.*, B, **1967**, 340.
278. W. W. Pigman and R. M. Goepp, *Chemistry of the Carbohydrates*, Academic Press, New York, 1948, p. 335.
279. G. J. Moody, *Tetrahedron*, **19**, 1705 (1963).
280. J. C. Sowden, in *Advances in Carbohydrate Chemistry*, Vol. 12, edited by M. L. Wolfrom, Academic Press, New York, 1957, pp. 47, 51, 52, 57, 61, 62, 73.
281. E. J. Bourne, D. H. Hutson, and H. Weigel, *J. Chem. Soc.*, **1960**, 5153.
282. R. O. C. Norman and R. J. Pritchett, *J. Chem. Soc.*, B, **1967**, 1329.
283. W. Berends and J. Konings, *Rec. Trav. Chim.*, **74**, 1365 (1955).
284. A. T. Küchlin, *Rec. Trav. Chim.*, **51**, 887 (1932).
285. A. N. de Belder, B. Lindberg, and O. Theander, *Acta Chem. Scand.*, **17**, 1012 (1963).
286. H. S. Raper, *Biochem. J.*, **26**, 2000 (1932).
287. P. J. G. Mann and B. C. Saunders, *Proc. Roy. Soc. (London)*, **B119**, 47 (1935).
288. A. G. Green and A. E. Woodhead, *J. Chem. Soc.*, **97**, 2388 (1910).
289. D. G. H. Daniels, F. T. Naylor, and B. C. Saunders, *J. Chem. Soc.*, **1951**, 3433.
290. C. Neuberg, *Biochem. Z.*, **20**, 526 (1909).
291. R. Breslow and L. N. Lukens, *J. Biol. Chem.*, **235**, 292 (1960).
292. E. Eich and H. Rochelmeyer, *Pharm. Acta Helv.*, **41**, 109 (1966).
293. J. T. Pearson, P. Smith, and T. C. Smith, *Can. J. Chem.*, **42**, 2022 (1964).
294. E. Oliveri-Mandala, *Gazz. Chim. Ital.*, **52** I, 107 (1922).

295. J. Q. Adams, *J. Amer. Chem. Soc.*, **89**, 6022 (1967).
296. D. Y. Edge and R. O. C. Norman, *J. Chem. Soc.*, *B*, **1969**, 182.
297. G. V. James, *Biochem. J.*, **34**, 636 (1940).
298. H. M. Rauen, K. Norpoth, and E. Golovinsky, *Naturwissenschaften*, **54**, 589 (1967).
299. L. Seekles, *Rec. Trav. Chim.*, **46**, 77 (1927).
300. O. Baudisch and L. W. Bass, *J. Amer. Chem. Soc.*, **46**, 184 (1924).
301. K. Ohta, *Biol. Z.*, **54**, 439 (1913).
302. A. Cier, A. Lefier, M. Ravier, and C. Nofre, *Compt. Rend.*, **254**, 504 (1962).
303. C. Nofre, A. Lefier, and A. Cier, *Compt. Rend.*, **253**, 687 (1961).
304. M. G. Ormerod and B. B. Singh, *Intern. J. Rad. Biol.*, **10**, 533 (1966).
305. J. Hurych and M. Chvapil, *Chem. Listy*, **60**, 312 (1966).
306. C. E. Dalgliesh, *Arch. Biochem. Biophys.*, **58**, 214 (1955).
307. H. D. Dakin, *J. Biol. Chem.*, **4**, 63 (1908).
308. C. Nofre, A. Cier, C. Michou-Saucet, and J. Parnet, *Compt. Rend.*, **251**, 811 (1960).
309. S. Goldschmidt and G. von Linde, *Erdöl und Kohle*, **15**, 81 (1962).
310. G. D. Kalyankar, C. S. Vaidyanathan, and K. V. Giri, *Experientia*, **11**, 348 (1955).
311. C. Nofre, L. Welin, J. Parnet, and A. Cier, *Bull. Soc. Chim. Biol.*, **43**, 1237 (1961).
312. C. Nofre, J. P. Charrier, and A. Cier, *Bull. Soc. Chim. Biol.*, **45**, 913 (1963).
313. N. W. Pirie, *Biochem. J.*, **25**, 1565 (1931).
314. N. Matsuura and K. Muroshima, *Sci. Papers Coll. Gen. Educ., Univ. Tokyo*, **14**, 183 (1964).
315. R. G. Neville, *J. Amer. Chem. Soc.*, **79**, 2456 (1957).
316. L. B. Jaques and H. J. Bell, *Can. J. Res.*, **24E**, 79 (1946).
317. M. Chvapil and J. Hurych, *Nature*, **184**, Suppl. No. 15, 1145 (1959).
318. M. H. Briggs, *Austral. J. Sci.*, **22**, 39 (1960).
319. J. Hurych, *Kozartstvi*, **13**, 380 (1963).
320. Y. Shibata, T. Tamaki, R. Fukada, H. Ito, and T. Tsuji, *Wakayama Igaku*, **15**, 297 (1964).
321. J. Hurych, *Hoppe-Seyler's Z. Physiol. Chem.*, **343**, 426 (1967).
322. C. Neuberg, *Biochem. Z.*, **20**, 531 (1909).
323. H. D. Dakin, *J. Biol. Chem.*, **1**, 171 (1905–6).
324. H. D. Dakin, *J. Biol. Chem.*, **5**, 409 (1909).
325. C. R. Maxwell and D. C. Peterson, *J. Amer. Chem. Soc.*, **79**, 5110 (1957).
326. G. R. A. Johnson, G. Scholes, and J. Weiss, *Science*, **114**, 412 (1951).
327. E. Bekier and S. W. Kijowski, *Roczniki Chem.*, **15**, 136 (1935).
328. C. E. H. Bawn and D. Margerison, *Trans. Faraday Soc.*, **51**, 925 (1955).
329. D. H. Irvine, *J. Chem. Soc.*, **1959**, 2977.
330. D. L. Ball, M. M. Crutchfield, and J. O. Edwards, *J. Org. Chem.*, **25**, 1599 (1960).
331. K. C. Khulbe and S. P. Srivastava, *Agra Univ. J. Res.*, **9**, Part 2, 177 (1960).
332. K. C. Khulbe and S. P. Srivastava, *Proc. Natl. Acad. Sci. India*, **30A**, 117 (1961).

333. S. P. Srivastava and G. B. Purohit, *Vijnana Parishad Anusandhan Patrika*, **4**, 199 (1961).

334. G. B. Purohit and S. P. Srivastava, *Vijnana Parishad Anusandhan Patrika*, **4**, 303 (1961).

335. K. C. Khulbe and S. P. Srivastava, *Proc. Natl. Acad. Sci. India*, **A34**, 32 (1964).

336. L. B. Subbaraman and M. Santappa, *Curr. Sci. (India)*, **33**, 208 (1964).

337. D. D. Mishra and S. Ghosh, *J. Indian Chem. Soc.*, **41**, 397 (1964).

338. D. D. Mishra and S. Ghosh, *J. Indian Chem. Soc.*, **41**, 402 (1964).

339. D. D. Mishra and S. Ghosh, *Proc. Natl. Acad. Sci. India*, **A34**, 317 (1965).

340. D. D. Mishra and S. Ghosh, *Proc. Natl. Inst. Sci. India*, **A31**, 119 (1965).

341. C. V. Bakore and S. N. Joshi, *Z. Physik. Chem. (Leipzig)*, **229**, 250 (1965).

342. M. M. Mhala and R. V. Iyer, *Indian J. Chem.*, **3**, 568 (1965).

343. L. R. Subbaraman and M. Santappa, *Proc. Indian Acad. Sci.*, **A64**, 345 (1966).

344. L. R. Subbaraman and M. Santappa, *Z. Physik. Chem. (Frankfurt)*, **48**, 163 (1966).

345. L. R. Subbaraman and M. Santappa, *Z. Physik. Chem. (Frankfurt)*, **48**, 172 (1966).

346. M. C. Agrawal and S. P. Mushran, *J. Indian Chem. Soc.*, **43**, 343 (1966).

347. S. N. Joshi and G. V. Bakore, *Curr. Sci. (India)*, **37**, 346 (1968).

348. O. A. Chaltykyan and N. M. Beileryan, *Dokl. Akad. Nauk Armen. SSR*, **31**, 275 (1960).

349. O. A. Chaltykyan and N. M. Beileryan, *Izv. Akad. Nauk Armen. SSR*, **14**, 197 (1961).

350. O. A. Chaltykyan, N. M. Beileryan, M. S. Chobanyan, and E. R. Sarukhanyan, *Izv. Akad. Nauk Armen. SSR*, **14**, 293 (1961).

351. O. A. Chaltykyan and N. M. Beileryan, *Izv. Akad. Nauk Armen. SSR*, **14**, 209 (1961).

352. N. M. Beileryan, O. A. Chaltykyan, and T. T. Gukasyan, *Dokl. Akad. Nauk Armen. SSR*, **35**, 37 (1962).

353. N. M. Beileryan and O. A. Chaltykyan, *Khim. Perekisnykh Soedin., Akad. Nauk SSSR, Inst. Obshch. i Neorgan. Khim.*, **1963**, 265.

354. O. A. Chaltykyan and N. M. Beileryan, *Khim. Perekisnykh Soedin., Akad. Nauk SSSR, Inst. Obshch. i Neorgan. Khim.*, **1963**, 260.

355. R. A. Hermens and W. H. Cone, *Trans. Illinois State Acad. Sci.*, **58**, 144 (1965).

356. B. Stehlik and F. Fiala, *Chem. Zvesti*, **20**, 97 (1966).

357. E. Ben-Zvi and T. L. Allen, *J. Amer. Chem. Soc.*, **83**, 4352 (1961).

358. A. I. Brodskii, *Magyar Tudomanyos Akad. (Budapest), Kem. Tudomanyok Osztolyanak Kozlemenyei*, **15**, 287 (1961).

359. W. K. Wilmarth and A. Haim, in *Peroxide Reaction Mechanisms*, edited by J. O. Edwards, Interscience, New York, 1960, p. 175.

360. D. A. House, *Chem. Rev.*, **62**, 185 (1962).

361. R. G. R. Bacon, R. W. Bott, J. R. Doggart, R. Grime and D. J. Munro, *Chem. Ind. (London)*, **1953**, 897.

362. R. G. R. Bacon and J. R. Doggart, *J. Chem. Soc.*, **1960**, 1332.

363. R. G. R. Bacon, *Chem. Ind. (London)*, **1962**, 19.

364. R. G. R Bacon and R. W. Bott, *Chem. Ind.* (*London*), **1953**, 1285.
365. F. P. Greenspan and H. M. Woodburn, *J. Amer. Chem. Soc.*, **76**, 6345 (1954).
366. F. P. Greenspan, University Microfilms Publ. No. 3423; *Diss. Abstr.*, **12**, 132 (1952).
367. R. G. R Bacon, R. Grime, and D. J. Munro, *J. Chem. Soc.*, **1954**, 2275.
368. R. G. R. Bacon and D. J. Munro, *J. Chem. Soc.*, **1960**, 1339.
369. P. C. Austin, *J. Chem. Soc.*, **99**, 262 (1911).
370. K. Elbs, *J. Prakt. Chem.*, **48**, 179 (1893).
371. S. M. Sethna, *Chem. Rev.*, **49**, 91 (1951).
372. H. Krauch and W. Kunz, *Reaktionen der organischen Chemie*, 3rd ed., Dr. A. Hüthig Verlag GMBH, Heidelberg, **1966**, p. 200.
373. W. Baker and N. C. Brown, *J. Chem. Soc.*, **1948**, 2303.
374. W. Baker, N. C. Brown, and J. A. Scott, *J. Chem. Soc.*, **1939**, 1922.
375. E. J. Behrmann, *J. Amer. Chem. Soc.*, **85**, 3478 (1963).
376. G. Bargellini and L. Monti, *Gazz. Chim. Ital.* **45**, 90 (1915).
377. F. Mauthner, *J. Prakt. Chem.*, **152**, 23 (1939).
378. F. Wessely and E. Demmer, *Ber.*, **62**, 120 (1929).
379. E. J. Behrmann and B. M. Pitt, *J. Amer. Chem. Soc.*, **80**, 3717 (1958).
380. E. J. Behrmann and P. P. Walker, *J. Amer. Chem. Soc.*, **84**, 3454 (1962).
381. R. G. R. Bacon, W. J. W. Hanna, D. J. Munro, and D. Stewart, *Proc. Chem. Soc.* (*London*), **1962**, 113.
382. R. G. R. Bacon and D. Stewart, *J. Chem. Soc.*, C, **1966**, 1384.
383. R. G. R. Bacon, W. J. W. Hanna, and D. Stewart, *J. Chem. Soc.*, C, **1966**, 1388.

*Chapter IV*

# The Formation of Organometallic Peroxides by Autoxidation

ALWYN G. DAVIES

## I.  INTRODUCTION

The alkylperoxy group can be introduced at a metal center by two fundamentally different types of reactions. Both start with the O—O bond already formed in one reagent and both are parallel to the two main processes by which the alkylperoxy group can be introduced at a carbon center (1).

The first of these reactions is a nucleophilic substitution by a peroxidic reagent at a metallic center:

$$ROOH + MX \longrightarrow ROOM + HX \qquad (1)$$

where M is a metal or organometallic group.

By this reaction alkylperoxy compounds of lithium, sodium, potassium, magnesium, zinc, cadmium, mercury, calcium, strontium, barium, boron, aluminum, silicon, germanium (2), tin, lead, arsenic, and antimony have been prepared.

The second reaction involves the autoxidation of an organometallic compound:

$$O_2 + RM \longrightarrow ROOM \qquad (2)$$

These reactions are more limited in scope than are those of equation 1 and as yet are restricted essentially to the preparation of organoperoxy derivatives of the nontransition metals of Groups I, II, and III. These reactions are discussed in this chapter.

There appears to be no report of the equivalent autoxidation of a metal hydride to give a metallic hydroperoxide,

$$O_2 + HM \longrightarrow HOOM \qquad (3)$$

but dimetallic peroxides have been identified in the autoxidation products of metal–metal bonded compounds:

$$O_2 + M'M \longrightarrow M'OOM \qquad (4)$$

## II.  HISTORY

The high reactivity of the organic derivatives of these metals toward air was observed in the earliest days of organometallic chemistry. In 1852 Frankland (3) heated methyl iodide with zinc and reported that the dimethylzinc that was formed has an "affinity for oxygen even more itense than potassium; in contact with atmospheric air it instantaneously ignites, burning with a beautiful blue flame, and forming white clouds of oxide of zinc; in contact with pure oxygen it burns with explosion." The lower alkyl derivatives of lithium, sodium, potassium, beryllium, magnesium, zinc, cadmium, boron, aluminum, gallium, indium, thallium, phosphorus, arsenic, antimony, and bismuth also often inflame spontaneously in air, and higher alkyl derivatives, with lower vapor pressures, may char but not ignite.

If the supply of oxygen is restricted, a smooth reaction takes place. From the alkyls of the metals from Groups I, II, and III the ultimate product is the alkoxymetallic compound ROM, but evidence was soon found for the occurrence of an oxidizing species. In 1890 Meyer and Demuth (4) treated diethylzinc in ligroin with oxygen and obtained a white solid that liberated iodine from potassium iodide; they suggested that it might be ethyl(ethylperoxy)zinc, EtZnOOEt. Wuyts (5) similarly detected traces of peroxides in the autoxidation of Grignard reagents, and Porter and Steele (6) proposed that the ultimate alkoxides were formed via intermediate peroxides (M = MgX):

$$O_2 + RM \longrightarrow ROOM \qquad (5)$$

$$ROOM + RM \longrightarrow 2ROM \qquad (6)$$

This picture was confirmed in 1953 by Walling and Buckler (7), who showed that alkyl hydroperoxides could be prepared in high yield by hydrolyzing the product obtained by adding dilute solutions of alkylmagnesium halides to ether saturated with oxygen at $-75°$ ("inverse oxidation").

$$O_2 + RMgX \longrightarrow ROOMgX \xrightarrow{H_2O} ROOH \qquad (7)$$

Shortly after this the first organoperoxymetallic compounds were isolated. In the gas phase trimethylborane was shown to react with oxygen to give dimethyl(methylperoxy)borane (**1**, R = Me) (8), and diisobutyl-*t*-butylborane in solution gave the diperoxyborane (**2**, R = *i*- or *t*-butyl) (9).

$$R_3B \xrightarrow{O_2} R_2BOOR \xrightarrow{O_2} RB(OOR)_2 \qquad (8)$$
$$\textbf{(1)} \qquad\qquad \textbf{(2)}$$

Since then peroxides have also been identified in the autoxidation of the alkyl derivatives of lithium, zinc, cadmium, mercury, aluminum, tin, and lead. It should be noted that no organic peroxides containing the alkenylperoxy, alkynylperoxy, or arylperoxy structures have yet been characterized, and, consonant with this, the autoxidation of alkenyl, alkynyl, and arylmetallic compounds gives only negligible yields of peroxides.

In Section III–VI the reactions that do give rise to identifiable organometallic peroxides are considered from a preparative point of view in the sequence of the periodic groups, and in Section VII the mechanisms of the autoxidation reactions are discussed. Previous reviews in this field are cited in references 10–15.

## III.  PEROXIDES OF THE GROUP I METALS

All the organic derivatives of the alkali metals react very rapidly with oxygen, and the lower homologs often inflame in air. Under controlled conditions a smooth reaction occurs, but the alkylmetallic compounds readily reduce the peroxides, and alkoxides rather than the peroxides are usually obtained by a combination of equations 5 and 6.

In 1955 Walling and Buckler (16) showed that this reduction could be minimized by their technique of "inverse oxidation," and, under these conditions, butyllithium (0.38 $N$ in ether) gave butylperoxylithium in 36% yield (by titration):

$$O_2 + BuLi \longrightarrow BuOOLi \qquad (9)$$

Indenyllithium under similar conditions gave indenyl hydroperoxide (3) in 75–85% yield (17), and the hydroperoxides 4 and 5 were isolated from the autoxidation of the lithium adducts of anthracene and 9,10-dimethyl-anthracene, respectively (18).

**(3)**                **(4)**                **(5)**

None of the organoperoxylithium compounds themselves, however, has been isolated from these reactions.

Tri($p$-nitrophenyl)methane reacts with alcoholic potassium hydroxide to give the potassium derivative of the carbanion, which, with oxygen, gives tri($p$-nitrophenyl)methyl hydroperoxide in 59% yield (19):

$$(4\text{-}NO_2C_6H_4)_3C^-K^+ \xrightarrow{\;O_2\;} (4\text{-}NO_2C_6H_4)_3COOH \qquad (10)$$

Triphenylmethylsodium in ether gives triphenylmethanol, along with varying amounts of bistriphenylmethyl peroxide (20, 21):

$$Ph_3CNa \xrightarrow{\;O_2\;} Ph_3COH + Ph_3COOCPh_3 \qquad (11)$$

## IV.  PEROXIDES OF THE GROUP II METALS

Alkyl derivatives of beryllium, calcium, and barium have been reported to react vigorously with air or oxygen, but no attempts to detect peroxidic products of intermediates appear to have been reported. However, it is well established that Grignard reagents and the alkyls of zinc and cadmium give alkylperoxymetallic compounds by the reaction of equation 2, and there is some evidence that alkylperoxymercury compounds are involved as intermediates in the autoxidation of dialkylmercury compounds.

Walling and Buckler's inverse oxidation of Grignard reagents provided, in 1953 (7), the first convincing evidence that organoperoxymetallic compounds were formed on autoxidation (see above). The organoperoxymagnesium compounds were not isolated, but evidence for their structure was obtained by isolating the corresponding hydroperoxides or their derivatives (e.g., equation 12):

$$t\text{-BuMgCl} \xrightarrow{\text{O}_2} t\text{-BuOOMgCl}
\begin{cases}
\xrightarrow{\text{H}_2\text{O}} & t\text{-BuOOH} \quad (82\%) \\
\xrightarrow{\text{EtBr}} & t\text{-BuOOEt} \quad (22\%) \\
\xrightarrow{\text{PhCOCl}} & t\text{-BuOOCOPh} \quad (54\%)
\end{cases}$$

Yields of peroxide increase in the sequence for R, primary < secondary < tertiary, and chlorides usually give better yields than the corresponding bromides (perhaps because the peroxide oxidizes the bromide ion to bromine). The autoxidation is very fast, and a 60% yield of $t$-butyl alcohol is formed even when $t$-butylmagnesium chloride is added to acetone saturated with oxygen.

Arylmagnesium halides give phenols in yields of about 40% and a negligible amount (<10%) of titratable peroxide; under similar conditions ethynylmagnesium halides are not autoxidized.

The autoxidation of Grignard reagents has since been utilized as a general preparative route to alkyl hydroperoxides; the results are summarized in Table 1.

The autoxidation of organozinc compounds is similar to that of Grignard reagents. The reaction has been carried out by the technique of inverse oxidation (16, 22) or by crushing an ampule of the reagent (30, 31) or injecting it from a hypodermic syringe (32), under the solvent saturated with oxygen. The reactions are summarized in Table 2.

When a dialkylzinc is autoxidized in an inert solvent, the solution stays homogeneous during the uptake of the first mole of oxygen, as the monoperoxide (6) is formed, but the diperoxide (7) separates during the uptake of the second mole. Presumably the diperoxide is coordinately polymerized through oxygen bridges, rendering the zinc four-coordinate.

$$\text{R}_2\text{Zn} \xrightarrow{\text{O}_2} \text{RZnOOR} \xrightarrow{\text{O}_2} \text{Zn(OOR)}_2 \tag{13}$$
$$\qquad\qquad\quad (6) \qquad\qquad\quad (7)$$

TABLE 1.   HYDROPEROXIDES PREPARED BY THE AUTOXIDATION OF
GRIGNARD REAGENTS

| RMgX | ROOH | Yield (%) | Ref. |
|---|---|---|---|
| EtMgCl | EtOOH | 57 | 16 |
| EtMgBr | EtOOH | 28 | 16 |
| BuMgCl | BuOOH | 57 | 16 |
| | | 80 | 22 |
| sec-BuMgCl | sec-BuOOH | — | 23 |
| t-BuMgCl | t-BuOOH | 84 | 22 |
| | | 86 | 16 |
| i-PeMgCl[a] | i-PeOOH[a] | 74 | 22 |
| i-PeMgBr[a] | i-PeOOH[a] | 10 | 22 |
| t-PeMgCl[a] | t-PeOOH[a] | 92 | 16 |
| i-PrMeCHMgBr | i-PrMeCHOOH | — | 24 |
| ClMg(CH$_2$)$_6$MgCl | HOO(CH$_2$)$_6$OOH | 44 | 17 |
| C$_7$H$_{15}$MgCl | C$_7$H$_{15}$OOH | — | 25, 26 |
| MePeCHMgCl[a] | MePeCHOOH[a] | — | 25, 26 |
| EtBuCHMgCl | EtBuCHOOH | — | 25, 26 |
| Pr$_2$CHMgCl | Pr$_2$CHOOH | — | 25, 26 |
| Me(C$_6$H$_{13}$)CHMgCl | Me(C$_6$H$_{13}$)CHOOH | 91 | 16 |
| CyclohexylMgCl | CyclohexylOOH | 83 | 22 |
| | | 66 | 16 |
| | | 79 | 17 |
| CyclohexylMgBr | CyclohexylOOH | 30 | 16 |
| 2-NorbornylMgCl | 2-NorbornylOOH | | 27 |
| 2-NorbornylMgBr | 2-NorbornylOOH | | 27 |
| PhCH$_2$MgCl | PhCH$_2$OOH | 30 | 16 |
| Bornyl MgCl | ⎰Bornyl OOH | 50 | 28 |
| | ⎱IsobornylOOH | 40 | |
| 3,5-Me$_2$C$_6$H$_3$CH$_2$MgBr | 3,5-Me$_2$C$_6$H$_3$CH$_2$OOH | 30 | 29 |

[a] Pe = pentyl.

Dimethylcadmium in cyclohexane (33) or anisole (32), absorbed about
1 mole of oxygen to give the soluble monoperoxide (**8**, R = Me) in good
yield, which was characterized by converting it into crystalline methyl
9-xanthenyl peroxide. Higher dialkylcadmiums react further to give the
insoluble diperoxides (**9**, R = Et or Bu), and the structure of di(ethylperoxy)
cadmium was confirmed by treating it with terephthaloyl chloride to give
diethyl diperoxyterephthalate.

$$R_2Cd \xrightarrow{O_2} RCdOOR \xrightarrow{O_2} Cd(OOR)_2 \qquad (14)$$
$$\textbf{(8)} \qquad\qquad \textbf{(9)}$$

TABLE 2.  THE AUTOXIDATION OF ALKYLZINC AND ALKYLCADMIUM
COMPOUNDS

| RM | Solvent | $T$ (°C) | Product | ROOM (M) | Ref. |
|---|---|---|---|---|---|
| $Et_2Zn$ | $Et_2O$ | R.T.[a] | $Zn(OOEt)_2$ | 1.93 | 31 |
| | PhOMe | R.T. | EtOZnOOEt | 0.92 | 32 |
| $Bu_2Zn$ | $Et_2O$ | R.T. | $Zn(OOBu)_2$ | 2.00 | 31 |
| $Me_2Cd$ | $C_6H_{12}$ | R.T. | MeCdOOMe | 1.13 | 33 |
| | PhOMe | R.T. | MeCdOOMe | 1.08 | 32 |
| $Et_2Cd$ | $Et_2O$ | R.T. | | | 33 |
| $Bu_2Cd$ | $Et_2O$ | R.T. | | | 33 |
| | $Et_2O$ | 0 | $Cd(OOBu)_2$ | 1.9 | 22 |
| $BuCdCl$ | $Et_2O$ | 10 | BuOOCdCl | 1.88 | 22 |
| $C_8H_{17}CdCl$ | $Et_2O$ | 0 | $C_8H_{17}OOCdCl$ | 1.80 | 22 |
| $PhCH_2CdCl$ | $Et_2O$ | $-5$ | $PhCH_2OOCdCl$ | 1.72 | 22 |

[a] R.T. = Room temperature

The autoxidation of some alkylcadmium chlorides, which gives the corresponding peroxides in high yields, has also been reported by Hock and Ernst (22). These reactions are also included in Table 2.

No compounds containing the alkylperoxymercury group have been isolated from the autoxidation of organomercury compounds, nor indeed, at the time of writing, from any other route. Primary alkylmercury compounds are usually stable toward oxygen, but secondary and tertiary alkyl-mercuries absorb oxygen slowly to give a mixture of nonperoxidic decomposition products of the intermediate peroxide [$i$-$Pr_2Hg$ (34–37), $sec$-$Bu_2Hg$ (38), $(C_6H_{11})_2Hg$ (39), $(PhCH_2)_2Hg$ (35), $t$-$Pe_2Hg$ (35)]. For example, di-$sec$-butylmercury in dimethylformamide at 70° absorbed from 0.5 to more than 2.0 moles of oxygen, depending on the pressure and rate of stirring, and gave 1-butene, butane, $cis$- and $trans$-2-butene, butan-2-one, and butan-2-ol, probably via the intermediate BuHgOOBu (38).

## V.  PEROXIDES OF THE GROUP III METALS

The autoxidation of the organic derivatives of boron has been investigated more thoroughly than that of other organometallic compounds (40).

Trialkylboranes, $R_3B$, react vigorously with air, the lower members with inflammation. Dialkylalkoxyboranes (borinic esters, $R_2BOR$) are less reactive, and alkyldialkoxyboranes [boronic esters, $RB(OR)_2$] usually show no significant reaction under normal conditions. The trialkylboroxines [boronic anhydrides, $(RBO)_3$], however, do react with oxygen; within this series, as R is varied, the reactivity follows the sequence methyl < primary < secondary < tertiary (41, 42), and the same sequence probably holds for other types of organoboranes. Thus when diisobutyl-$t$-butylborane reacts with oxygen, the relative reactivity as the first alkyl group reacts is $t$-Bu: $i$-Bu $\simeq$ 4.5: 1; as the second group reacts, this ratio is about 50: 1; thus the ultimate product consists of 99% of an isobutylboron compound (43). The introduction of phenyl substituents into the $\alpha$-position of the alkyl group increases the reactivity: 1-phenylethyldibutoxyborane [PhCH(Me)B(OBu)$_2$] in benzene reacts slowly with oxygen, and the corresponding boronic acid [PhCH(Me)B(OH)$_2$] takes up oxygen very rapidly at room temperature and atmospheric pressure (44). Vinylboranes and phenylboranes are reported to be less reactive than the corresponding alkylboranes (45), and the dimeric hydrogen-bridged alkylboron hydrides [(R$_2$BH)$_2$] are also relatively inert (46).

If the reactions are carried out at room temperature with neat liquid trialkylboranes, alkyldialkoxyboranes are formed:

$$R_3B + O_2 \longrightarrow RB(OR)_2 \qquad (15)$$

this was used as a preparative method by Frankland in 1862 (47, 48), but if air is blown into a large amount ($\sim$100 g) of tributylborane, the reaction may proceed as far as the trialkoxyborane, perhaps because the autoxidation is highly exothermic (49).

Johnson and van Campen in 1938 suggested that the reaction involved an intermediate "peroxide," $R_3B{-}O{=}O$ (50). A similar intermediate, BuBO(O$_2$), was assumed by Grummitt (41), who showed the presence of a component that oxidized potassium iodide to iodine and catalyzed the polymerization of vinyl acetate.

Following Walling and Buckler's work on the autoxidation of Grignard reagents, however, it became clear that these intermediates were more conventional peroxides containing the structure $\text{ROOB}\diagdown^{\diagup}$ In 1956 the peroxide Me$_2$BOOMe was obtained by the reaction of trimethylborane in the gas phase with oxygen (0.5 mole) at 10–15 mm Hg pressure, in a flow system with a residence time of 2–3 min (8), and higher trialkylboranes in dilute solution in inert solvents were shown to give alkyldialkylperoxyboranes, RB(OOR)$_2$ (9).

The products from the autoxidation of some boroxines have similarly been identified as the peroxyboranes (ROOBO)$_3$ (42); like trimethylborane, trimethylboroxine is less reactive than higher homologs and, in dilute solution, does not react with oxygen under normal conditions. 1-Phenylethyldihydroxyborane shows the enhanced reactivity associated with the presence of an $\alpha$-phenyl substituent and readily gives the monoperoxide PhCH(Me)OOB(OH)$_2$ (44).

These peroxides can be isolated, usually as rather unstable oils, with satisfactory analyses, nuclear magnetic resonance (NMR) spectra, mass spectra, and chemical behavior. The peroxyboranes that have been identified satisfactorily are listed in Table 3.

TABLE 3.    ORGANOPEROXYBORANES PREPARED BY AUTOXIDATION

| Alkylperoxyborane | Reference |
|---|---|
| $Me_2BOOMe$ | 8, 46 |
| $BuB(OOBu)_2$ | 51, 52 |
| $i\text{-}BuB(OO\text{-}i\text{-}Bu)_2$ | 53 |
| $sec\text{-}BuB(OO\text{-}sec\text{-}Bu)_2$ | 21 |
| $i\text{-}BuB(OO\text{-}i\text{-}Bu)(OO\text{-}t\text{-}Bu)$ | 43 |
| $PrCH(Me)CH_2B(OOCH_2CHMePr)_2$ | 51 |
| $C_8H_{17}B(OOC_8H_{17})_2$ | 54 |
| $BuCH(Et)CH_2B(OOCH_2CHEtBu)_2$ | 54 |
| $C_{10}H_{21}B(OOC_{10}H_{21})_2$ | 54 |
| $C_{11}H_{23}B(OOC_{11}H_{23})_2$ | 54 |
| $PhCH(Me)CH_2B(OOCH_2CHMePh)_2$ | 54 |
| Cyclo-$C_6H_{11}B(OO$ cyclo-$C_6H_{11})_2$ | 54 |
| Cyclo-$C_8H_{15}B(OO$ cyclo-$C_8H_{15})_2$ | 54 |
| Cyclo-$C_{12}H_{25}B(OO$ cyclo-$C_{12}H_{25})_2$ | 54 |
| $PhCH_2B(OOCH_2Ph)_2$ | 54 |
| $PhCH_2CH_2B(OOCH_2CH_2Ph)_2$ | 54 |
| $IndylB(OOindyl)_2$ | 54 |
| 2-Norbornyl $B(OO$ norbornyl$)_2$ | 55 |
| $PhCH(Me)OOB(OH)_2$ | 44 |
| $(BuOOBO)_3$ | 42 |
| $(sec\text{-}BuOOBO)_3$ | 42 |
| $(t\text{-}BuOOBO)_3$ | 42 |

Quantitative hydrogenation of the peroxy link reduces the peroxide to the corresponding hydroxyboranes and alcohols, which can be identified (43). Hydrolysis of the boron–oxygen bond may be complicated by an oxidation–reduction by reaction, perhaps by the nucleophilic 1,2-rearrangement shown in equation 16, but this can be avoided by using a peroxy acid or hydrogen peroxide, when the solvolytic reagent is reduced preferentially (equation 17) (56). Wilke and Heimbach (54) have used the latter reaction as a route to hydroperoxides via the hydroboration of olefins, followed by autoxidation of the alkylboranes and cleavage of the alkylperoxyboranes (see Table 3).

$$HO^- \ \overset{\overset{\displaystyle R}{|}}{\underset{\underset{\displaystyle OOR}{|}}{B}}{-}OOR \longrightarrow HO^-\overset{\overset{\displaystyle R}{|}}{\underset{\underset{\displaystyle OOR}{|}}{B}}{-}O{-}OR \longrightarrow HO{-}\overset{\overset{\displaystyle R}{|}}{\underset{\underset{\displaystyle OOR}{|}}{B}}{-}OR + \bar{O}R$$

$$R'C(O)OOH + \overset{\overset{\displaystyle R}{|}}{\underset{\underset{\displaystyle OOR}{|}}{B}}{-}OOR \longrightarrow R'C(O)\bar{O}{-}O{-}\overset{\overset{\displaystyle R}{|}}{\underset{\underset{\displaystyle OOR}{|}}{B}}{-}OOR$$

$$R'CO_2^- + RO\underset{\underset{\displaystyle OOR}{|}}{B}OOR$$

In the absence of any other reagent the alkyldialkylperoxyboranes slowly decompose to give principally the dialkoxyalkylperoxyboranes:

$$RB(OOR)_2 \longrightarrow (RO)_2BOOR \qquad (18)$$

Similarly dimethylmethylperoxyborane rearranges in the gas phase or in benzene by a first-order reaction with a half-life of 60 days to methyldimethoxyborane, $MeB(OMe)_2$ (8). These reactions may involve a rearrangement of the type discussed above, but, in the light of other recent work (57), a homolytic mechanism should also be considered.

1-Phenylethylperoxydihydroxyborane, on the other hand, decomposes in benzene by a first-order process (half-life 7.3 hr) to give phenol as the principal product. This reaction probably involves an oxidation–reduction rearrangement within the alkylperoxy group, facilitated by $p_\pi-p_\pi$ donation from oxygen to boron (44):

$$\underset{\underset{\displaystyle Me}{|}}{\overset{\overset{\displaystyle Ph}{|}}{CH}}{-}\underset{\underset{\displaystyle O{-}B(OH)_2}{|}}{O} \longrightarrow MeCH\overset{\displaystyle OPh}{\underset{\displaystyle OB(OH)_2}{<}}$$

It seems probable that the course of the reaction of oxygen with organo-aluminums is similar to that with organoboranes, but the products are principally trialkoxyaluminums, and only small amounts of peroxides can usually be detected. For example, the oxidation of triethylaluminum in ether at $-80°$ (58), or of butylaluminum dichloride at 0° and $-50°$ (22) gave approximately a 10% yield of peroxide, and no peroxide could be detected in the oxidation of trimethylaluminum (59, 60) or triethylaluminum (61) in the gas phase.

Again, the autoxidation of triethylindium in the gas phase is believed to involve an intermediate ethylperoxyindium compound, but only traces of the peroxide survive to be detected iodometrically (62).

## VI.  PEROXIDES OF THE GROUP IV METALS

The alkyl derivatives of silicon, germanium, tin, and lead are much less reactive toward oxygen than are the corresponding derivatives of the metals of Groups I, II, and III (except perhaps mercury), and a significant reaction occurs only under special reaction conditions (e.g., ultraviolet irradiation) or when special structural features are present (e.g., metal–metal bonding). Under these conditions the initial formation of alkylperoxymetallic compounds (ROOM) or of dimetallic peroxides (MOOM) can often be detected, but yields are usually low, and, as yet, these reactions do not constitute useful preparative methods.

Thus tetraethyltin in nonane under ultraviolet irradiation absorbed during 1 hr about 0.01 and 0.05 mole of oxygen at 20 and 80°, respectively. At very low conversions the product showed the presence of up to 16% of a peroxide that was identified by infrared spectroscopy as $Et_3SnOOEt$. Tetraethyllead was somewhat more reactive, and the presence of the peroxide $Et_3PbOOEt$ could again be detected (63). Under similar conditions triethylisopropyltin gave a mixture of the peroxides $Et_3SnOO$-$i$-$Pr$ and $Et_2$-$i$-$PrSnOOEt$ in up to 16% yield (64).

Compounds containing metal–metal bonds are more reactive. Hexaethylditin reacts with oxygen at an appreciable rate above 50° and, under ultraviolet irradiation at −30°. Under the latter conditions the presence of the unstable bistriethyltin peroxide, $Et_3SnOOSnEt_3$, was identified by infrared spectroscopy (65). There is some less direct evidence that bistriethyllead peroxide, $Et_3PbOOPbEt_3$, is an intermediate in the autoxidation of hexaethyldilead (66, 67).

Many compounds containing bonds between unlike metals react rapidly with oxygen, and it seems likely that metallic peroxides again are formed. Razuvaev (68) has investigated the products and kinetics of the autoxidation of bistriethylgermylmercury $[(Et_3Ge)_2Hg]$ in octane at −20 to 0°. Under these conditions the products contained no residual peroxide, probably because it was reduced by the reactant, but at −50° in the presence of ammonia bistriethylgermyl peroxide $[Et_3GeOOGeEt_3]$ was identified in almost quantitative yield. The part that the ammonia plays in this reaction is unresolved.

## VII.  THE MECHANISM OF THE AUTOXIDATION OF ORGANOMETALLIC COMPOUNDS

The autoxidation reactions of organometallic compounds that do give good yields of organometallic peroxides are usually very fast and heterogeneous and are not easily followed kinetically. Furthermore, until recently there were no stereospecific compounds available to permit the stereochemistry of the reaction at either the carbon or the metal center to be studied. In the absence of the powerful criteria of kinetics and of stereochemistry mechanistic interpretations are necessarily speculative.

Walling and Buckler (28) showed that diphenylamine did not inhibit the autoxidation of a Grignard reagent and that butyraldehyde was not co-oxidized when $t$-butylmagnesium chloride reacted with oxygen at $-70°$. They pointed out that, although this negative evidence did not rule out a radical mechanism, it did permit the alternative of a polar rearrangement, as shown in the following equation:

$$\begin{matrix} R \\ | \\ M \end{matrix} \; O_2 \longrightarrow \begin{matrix} R \rangle O^+ \\ | \; | \\ M-O \end{matrix} \longrightarrow \begin{matrix} R-O \\ | \\ M-O \end{matrix}$$

Similarly the autoxidation of organoboranes normally shows no induction period, and the reaction of tributylborane (56) and of tributylboroxine (41) is not inhibited by quinol, nor is that of triethylborane (69) by iodine or methyl methacrylate. The same polar mechanism (equation 20) was therefore accepted for the autoxidation of organoboranes (56, 70); the mechanism was compatible with the dependence of reactivity on structure [e.g., $R_3B >$

$R_2BOR > RB(OR)_2 ; MeB\langle\; < BuB\langle\; < sec\text{-}BuB\langle\; < t\text{-}BuB\langle\; ]$, which is

mentioned above. Grummitt's demonstration (41) that 0.1% of phenyl-$\beta$-naphthylamine did inhibit the autoxidation of tributylboroxine is now seen, in the light of subsequent knowledge, to be highly significant, but at the time its importance was overlooked, and a mechanism was proposed in which an initial oxygen complex was reduced by unreacted boroxine.

In the absence of any evidence to the contrary the polar mechanism of equation 20 was also assumed to apply to the autoxidation of the alkyl derivatives of other metals, such as zinc (31) and cadmium (33).

Subsequent demonstrations that some organometallic compounds [particularly organoboranes (71, 72)] readily undergo homolytic substitution at the metallic center rendered this interpretation less convincing: the radical scavengers that were tested might have been ineffective merely because they could not compete with the organometallic compounds in reacting with the chain-carrying species.

The optical resolution of 1-phenylethylboronic acid (73) made it possible to investigate the mechanism further (44, 74). The results are shown in equation 21:

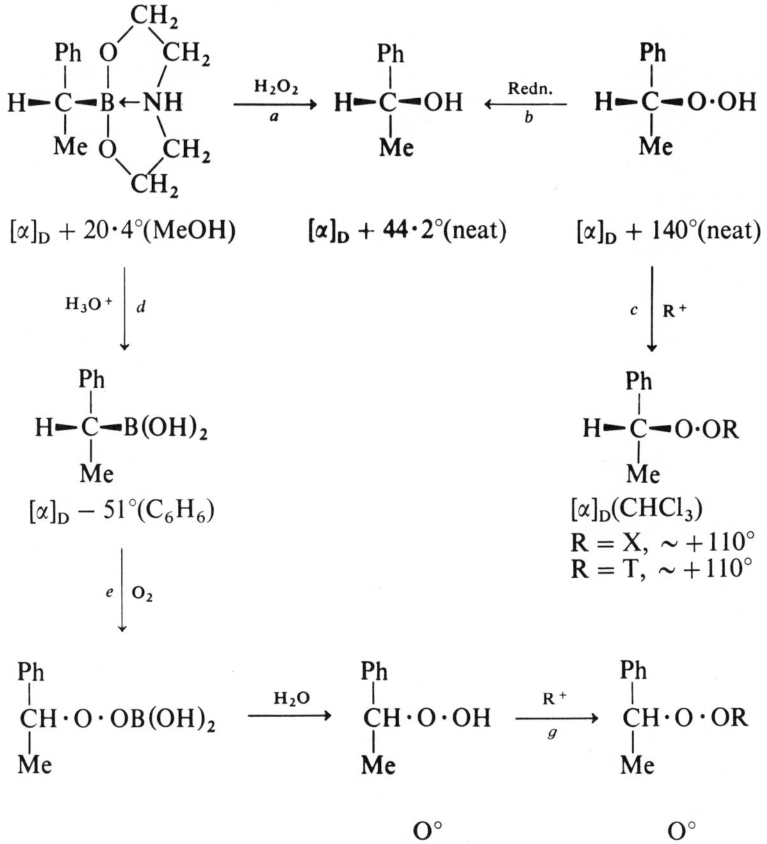

$$
\begin{array}{ccc}
\underset{\displaystyle \underset{Me}{\overset{Ph}{|}}{H-C-B\leftarrow NH}}{\overset{\displaystyle \overset{CH_2}{Ph\ O \diagup \diagdown CH_2}}{}}
& \xrightarrow[a]{H_2O_2}
& \underset{\displaystyle \underset{Me}{\overset{Ph}{|}}{H-C-OH}}{}
\xleftarrow[b]{Redn.}
& \underset{\displaystyle \underset{Me}{\overset{Ph}{|}}{H-C-O\cdot OH}}{}
\end{array}
$$

$[\alpha]_D + 20\cdot4^\circ(MeOH) \qquad\qquad [\alpha]_D + 44\cdot2^\circ(neat) \qquad\qquad [\alpha]_D + 140^\circ(neat)$

$H_3O^+ \Big| d$

$c \Big| R^+$

$$
\underset{\displaystyle Me}{\overset{\displaystyle Ph}{H-C-B(OH)_2}}
\qquad\qquad\qquad
\underset{\displaystyle Me}{\overset{\displaystyle Ph}{H-C-O\cdot OR}}
$$

$[\alpha]_D - 51^\circ(C_6H_6)$

$[\alpha]_D(CHCl_3)$
$R = X, \sim +110^\circ$
$R = T, \sim +110^\circ$

$e \Big| O_2$

$$
\underset{Me}{\overset{Ph}{CH\cdot O\cdot OB(OH)_2}}
\xrightarrow{H_2O}
\underset{Me}{\overset{Ph}{CH\cdot O\cdot OH}}
\xrightarrow[g]{R^+}
\underset{Me}{\overset{Ph}{CH\cdot O\cdot OR}}
$$

$$O^\circ \qquad\qquad\qquad O^\circ$$

(X = 9-xanthenyl, T = triphenylmethyl)

The reactions 21a–d, f, and g all involve retention of configuration. The autoxidation reaction 21e must therefore occur with at least 99% racemization, presumably through a three-coordinate carbon species, whereas the polar mechanism (equation 20) would presumably result in complete retention of configuration. The intermediate species was identified as a radical by observing the effect of more powerful inhibitors; galvinoxyl in particular, in 1 mole % amount, caused an induction period of 30 min.

The simplest mechanism compatible with the stereochemical result and with the effect of inhibitors appears to be a chain process involving the propagation steps shown in the following equations (M = boron):

$$R\cdot + O_2 \longrightarrow ROO\cdot \tag{22}$$

$$ROO\cdot + M{-}R \longrightarrow [ROO\cdot M{-}R] \longrightarrow ROOM + R\cdot \tag{23}$$

The first of these is the familiar rapid reaction of an alkyl radical with oxygen, which occurs in the autoxidation of hydrocarbons. The second assumes that alkylperoxy radicals rapidly bring about homolytic substitution at a metal center, so that only inhibitors that are very efficient at trapping the $R\cdot$ or $ROO\cdot$ radicals can interrupt the chain.

The initiation process is less clear, but it might involve reactions of the type shown in the following equations:

$$RM + O_2 \longrightarrow RM, O_2 \longrightarrow R\cdot + MOO\cdot \tag{24}$$

$$MOO\cdot + M{-}R \longrightarrow [MOO\cdot MR] \longrightarrow MOOM + R\cdot \tag{25}$$

Previously it had been suggested that a semistable borane–oxygen complex was an intermediate in the reaction (52, 75), though confirmation of this was lacking (46, 51). This appears now to be ruled out by the fact that the uptake of oxygen can be inhibited by a trace of galvinoxyl.

The termination reactions in the autoxidation presumably involve the interaction of alkyl radicals/or alkylperoxy radicals (or both), as in the autoxidation of hydrocarbons.

Similar stereochemical evidence is forthcoming from the autoxidation of the epimeric norborn-2-ylboranes (55):

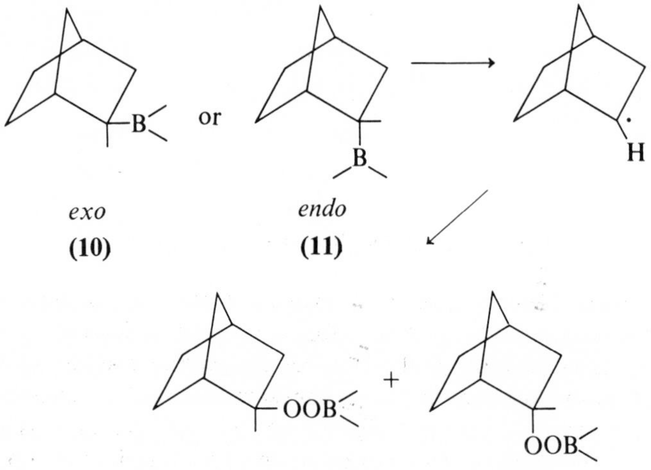

Three different mixtures of the *exo*-borane (10) and *endo*-borane (11) [>99.5% *exo*; 66% *exo*, 34% *endo*; and 14% *exo*, 86% *endo*] all gave about 76% *exo*- and 24% *endo*-norbornylperoxyboron groups (compounds 12

and **13**, respectively); this implies that all three reactions involved a common intermediate. Again, the reactions could be inhibited by galvinoxyl, implying that this intermediate is the norbornyl free radical.

Again diisopinocampheyl-*sec*-butylborane reacts with oxygen to give the *sec*-butylperoxyboron group with substantial, but now incomplete, racemization. Again, the reaction is inhibited by galvinoxyl, copper(II) dibutyldithiocarbamate and phenothiazine, substantiating a radical mechanism (76).

The test of stereochemistry can be applied to only a limited number of alkylboranes, but the effect of inhibitors is generally applicable, and the autoxidation of tri-*n*-butylboranes (77, 78), triisobutylboranes (44, 77, 78), and tri-*sec*-butylboranes (77, 78), and of the compounds (*sec*-Bu$_2$B)$_2$O and (*sec*-BuBO)$_3$ (44), are all susceptible to inhibition. The homolytic mechanism therefore appears to be general for a variety of organoborane structures.

The same mechanistic test can be applied to the autoxidation of the organic derivatives of other metals, as long as the organometallic compound does not react directly with the inhibitor and deactivate it. Galvinoxyl inhibits the autoxidation of dimethylcadmium, and the inhibited reaction can be induced to start by the addition of a copper salt. Presumably the dimethylcadmium methylates the copper, then the methylcopper compound breaks down homolytically to give initiating methyl radicals (32).

Similarly the uptake of the second mole of oxygen, though not the first, by diethylzinc is inhibited by galvinoxyl, and the inhibited reaction is again susceptible to initiation by copper salts. Presumably diethylzinc reacts too rapidly with ethylperoxy radicals to allow galvinoxyl to compete, but the molecule EtZnOOEt has a lower reactivity, permitting the galvinoxyl to divert the ethylperoxy radical (32).

The effect of inhibitors on the autoxidation of trimethylaluminum and of triphenylaluminum is less clear-cut but still significant. With the former compound, only 70% of the usual amount of oxygen is absorbed. With the latter, the autoxidation is very much slower and less complete (32).

No inhibition has yet been demonstrated in the autoxidation of Grignard reagents, probably merely because the organomagnesium halide rapidly reacts with the inhibitors to give inactive products. However, the test of stereochemistry can again be applied, by making use of the epimeric norbornylmagnesium halides:

$$\text{(27)}$$

Norbornylmagnesium chloride (43% *exo*, 57% *endo*) and norbornyl-magnesium bromide (100% *endo*, and 41% *exo*, 59% *endo*) all give norbornyl-peroxymagnesium compounds consisting of 75% *exo* and 25% *endo* ($\pm 3\%$) epimers; that is, with a composition similar to that of the isomeric mixture obtained from the norbornylboranes. This is compatible with these reactions all involving the same norbornyl-radical intermediate (27).

The photoinitiated oxidation of tetraalkyltin and lead compounds also involves, at least in part, a radical-chain process. The products that are obtained in the absence of oxygen are characteristic of alkyl radicals (63, 64), and it has been shown that the autoxidation of triethylisopropyltin can be inhibited by diphenylamine, and the radical intermediates can be detected at 77°K by ESR spectroscopy (64). Some of the oxidation may also proceed through a caged radical pair or by the direct reaction of oxygen with photo-activated molecules.

There is thus apparently no established exception to the general rule that the autoxidation of the alkyls of metals of Groups IA, IIA, IIIA, and IVA to give alkylperoxymetallic compounds proceeds at least in major part by a radical mechanism. The little work that has been carried out on the autoxida-tion of compounds that contain a metal–metal bond points to a less clear-cut situation. Hexaethylditin appears to react by a homolytic mechanism involving the propagation steps shown in equations 28 and 29, and compounds con-taining a lead–lead bond probably react by a similar mechanism (65, 66):

$$Et_3Sn\cdot + O_2 \longrightarrow Et_3SnOO\cdot \tag{28}$$

$$Et_3SnOO\cdot + Et_3Sn—SnEt_3 \longrightarrow Et_3SnOOSnEt_3 + Et_3Sn\cdot \tag{29}$$

On the other hand, the autoxidation of bis(triethylgermyl)mercury, which is believed to give initially the peroxide $Et_3GeHgOOGeEt_3$, is not inhibited by 2,6-di-*t*-butyl-4-methylphenol, 2,4,6-tri-*t*-butylphenol, or galvinoxyl (68),* and as yet none of the evidence is incompatible with a nonradical mechanism.

# REFERENCES

1.  A. G. Davies, *Organic Peroxides*, Butterworths, London, 1961.
2.  R. L. Dannley and G. C. Farrant, *J. Org. Chem.*, **34**, 2428, 2432 (1969).
3.  E. Frankland, *Phil. Trans. Roy. Soc. London*, **1852**, 417.
4.  V. Meyer and R. Demuth, *Ber.*, **23**, 394 (1890).
5.  H. Wuyts, *Compt. Rend.*, **143**, 930 (1909).
6.  G. W. Porter and C. Steele, *J. Amer. Chem. Soc.*, **42**, 2650 (1920).
7.  C. Walling and S. A. Buckler, *J. Amer. Chem. Soc.*, **75**, 4372 (1953).
8.  R. C. Petry and F. H. Verhoek, *J. Amer. Chem. Soc.*, **78**, 6416 (1956).

* Personal communication from G. A. Razuvaev.

9. M. H. Abraham and A. G. Davies, *Chem. Ind. (London)*, **1957**, 1622.
10. Reference 1, Chapter 8.
11. A. G. Davies, in *Progress in Boron Chemistry*, Vol. 1, edited by H. Steinberg and A. L. McCloskey, Pergamon, Oxford, 1964, p. 265.
12. T. G. Brilkina and V. A. Shushunov, *Reactions of Organometallic Compounds with Oxygen and Peroxides*: (a) Russian edition, Nauka, Moscow, 1966; (b) English translation, Iliffe, London, 1969.
13. A. G. Davies, Chapter 7 of reference 12b.
14. T. G. Brilkina and V. A. Shushunov, *Russian Chem. Rev.* (english translation), **35**, 613, (1966).
15. G. Sosnovsky and J. H. Brown, *Chem. Rev.*, **66**, 529 (1966).
16. C. Walling and S. A. Buckler, *J. Amer. Chem. Soc.*, **77**, 6032 (1955).
17. H. Hock and F. Ernst, *Ber.*, **92**, 2723 (1959).
18. H. Hock and F. Ernst, *Ber.*, **92**, 2732 (1959).
19. M. F. Hawthorne and G. S. Hammond, *J. Amer. Chem. Soc.*, **77**, 2549 (1955).
20. W. Schlenk and E. Marcus, *Ber.*, **47**, 1664 (1914).
21. W. E. Bachmann and F. Y. Wiselogle, *J. Amer. Chem. Soc.*, **58**, 1943 (1936).
22. H. Hock and F. Ernst, *Ber.*, **92**, 2716 (1959).
23. A. G. Davies and D. G. Hare, *J. Chem. Soc.*, **1959**, 438
24. C. F. Cullis and E. Fersht, *Combust. Flame*, **7**, 185 (1963).
25. W. Pritzkow and K. A. Müller, *Angew. Chem.*, **67**, 399 (1955).
26. W. Pritzkow and K. A. Müller, *Ann.*, **597**, 167 (1955).
27. A. G. Davies and B. P. Roberts, *J. Chem. Soc., B*, **1969**, 317.
28. C. Walling and S. A. Buckler, *J. Amer. Chem. Soc.*, **77**, 6039 (1955).
29. W. J. Farrissey, *J. Org. Chem.*, **27**, 3065 (1962).
30. M. H. Abraham, *Chem. Ind. (London)*, **1959**, 750.
31. M. H. Abraham, *J. Chem. Soc.*, **1960**, 4130.
32. A. G. Davies and P. B. Roberts, *J. Chem. Soc., B*, **1968**, 1074.
33. A. G. Davies and J. E. Packer, *J. Chem. Soc.*, **1959**, 3164.
34. G. A. Razuvaev, G. G. Putukhov, S. F. Zhil'tsov, and L. F. Kudryavtsev, *Dokl. Akad. Nauk SSSR*, **141**, 107 (1961).
35. G. A. Razuvaev, S. F. Zhil'tsov, O. N. Druzhkov, and G. G. Petukhov, *Dokl. Akad. Nauk SSSR*, **152**, 633 (1963).
36. Yu. A. Aleksandrov, O. N. Druzhkov, S. F. Zhil'tsov, and G. A. Razuvaev, *Dokl. Akad. Nauk SSSR*, **157**, 1395 (1964).
37. Yu. A. Aleksandrov, O. N. Druzhkov, S. F. Zhil'tsov, and G. G. Razuvaev, *Zh. Obshch. Khim.*, **35**, 1152, 1140 (1965).
38. F. D. Jensen and D. J. Heymann, *J. Amer. Chem. Soc.*, **88**, 3438 (1966).
39. G. A. Razuvaev, G. G. Petukhov, S. F. Zhil'tsov, and L. F. Kudryavtsev, *Dokl. Akad. Nauk SSSR*, **135**, 87 (1960).
40. Reviewed by A. G. Davies, *Progress in Boron Chemistry*, Vol. 1, edited by H. Steinberg and A. L. McCloskey, Pergamon, Oxford, 1964, Chapter 6, pp. 265–288.
41. O. Grummitt, *J. Amer. Chem. Soc.*, **64**, 1811 (1942).
42. A. G. Davies and E. C. J. Coffee, *J. Chem. Soc., C*, **1966**, 1493.
43. A. G. Davies, D. G. Hare, and R. F. M. White, *J. Chem. Soc.*, **1961**, 341.

44. A. G. Davies and B. P. Roberts, *J. Chem. Soc.*, *B*, **1967**, 17.
45. T. D. Parsons, M. B. Silverman, and D. M. Ritter, *J. Amer. Chem. Soc.*, **79**, 5091 (1957).
46. L. Parts and J. T. Miller, *Inorg. Chem.*, **3**, 1483 (1964).
47. E. Frankland, *J. Chem. Soc.*, **15**, 365 (1862).
48. E. Frankland, *Ann.*, **124**, 129 (1862).
49. S. B. Mirviss, *J. Amer. Chem. Soc.*, **83**, 3051 (1961).
50. J. R. Johnson and W. R. van Campen, *J. Amer. Chem. Soc.*, **60**, 121 (1938).
51. A. G. Davies, D. G. Hare, and O. R. Khan, *J. Chem. Soc.*, **1963**, 1125.
52. N. L. Zutty and F. J. Welch, *J. Org. Chem.*, **25**, 861 (1960).
53. A. G. Davies, D. G. Hare, and R. F. M. White, *J. Chem. Soc.*, **1960**, 1040.
54. G. Wilke and P. Heimbach, *Ann.*, **652**, 7 (1962).
55. A. G. Davies and B. P. Roberts, *J. Chem. Soc.*, *B*, **1969**, 311.
56. M. H. Abraham and A. G. Davies, *J. Chem. Soc.*, **1959**, 429.
57. D. B. Bigley and D. W. Payling, *Chem. Commun.*, **1968**, 638.
58. A. G. Davies and C. D. Hall, *J. Chem. Soc.*, **1963**, 1192.
59. C. F. Cullis, A. Fish, and R. T. Pollard, *Proc. Roy. Soc.* (*London*), **288A**, 123 (1965).
60. C. F. Cullis, A. Fish, and R. T. Pollard, *Proc. Roy. Soc.* (*London*), **289A**, 413 (1966).
61. A. Gröbler, A. Simon, T. Kada, and L. Fazakas, *J. Organometallic Chem.*, **7**, P3, (1967).
62. C. F. Cullis, A. Fish, and R. T. Pollard, *Trans. Faraday Soc.*, **60**, 2224 (1964).
63. Yu. A. Aleksandrov, B. A. Radbil, and V. A. Shushunov, *J. Gen. Chem. USSR*, (English translation), **37**, 190 (1967).
64. Yu. A. Aleksandrov and B. A. Radbil, *J. Gen. Chem. USSR*, (English translation), **37**, 2230 (1967).
65. Yu. A. Aleksandrov and B. A. Radbil, *J. Gen. Chem. USSR*, (English Translation), **36**, 562 (1966).
66. Yu. A. Aleksandrov and N. N. Vyshinskii, *Tr. po Khim. i Khim. Tekhnol.*, **4**, 626 (1962); through *Chem. Abstr.*, **58**, 3453 (1963).
67. Yu. A. Aleksandrov, N. N. Vyshinskii, and V. A. Shushunov, *Tr. po Khim. i Khim. Tekhnol.*, **4**, 3 (1961); through *Chem. Abstr.*, **56**, 492 (1962).
68. G. A. Razuvaev, Yu. A. Aleksandrov, V. N. Glushakova, and G. N. Figurova, *J. Organometallic Chem.*, **14**, 339 (1968).
69. R. L. Hansen and R. R. Hamann, *J. Chem. Phys.*, **67**, 2868 (1963).
70. T. D. Parsons, M. B. Silverman, and D. M. Ritter, *J. Amer. Chem. Soc.*, **79**, 5091 (1957).
71. J. Grotewold and E. A. Lissi, *Chem. Commun.*, **1965**, 21.
72. D. S. Matteson, *J. Org. Chem.*, **29**, 3399 (1964).
73. D. S. Matteson and R. A. Bowie, *J. Amer. Chem. Soc.*, **87**, 2587 (1965).
74. A. G. Davies and B. P. Roberts, *Chem. Commun.*, **1966**, 298.
75. S. B. Mirviss, *J. Org. Chem.*, **32**, 1713 (1967).
76. P. G. Allies and P. B. Brindley, *Chem. Ind.* (*London*). 1967, 319.
77. P. G. Allies and P. B. Bridley, unpublished data.
78. P. G. Allies, Ph. D. Thesis, London, 1968.

*Chapter V*

# Organic Peroxy Acids as Oxidizing Agents. Epoxidation.

DANIEL SWERN

---

## I.  INTRODUCTION

Organic peroxy acids are widely used oxidizing agents in organic chemistry (1–12) because of their versatility, specificity, ease of preparation, and the frequent production of excellent yields of desired product in a short time under mild conditions without the formation of difficult-to-remove by-products . For many important oxidation reactions—notably the conversion of a double bond to the oxirane or $\alpha$-glycol group, the oxidation of sulfides to sulfoxides and sulfones, the oxidation of tertiary amines to amine oxides, and the conversion of ketones to esters or lactones—organic peroxy acids are clearly the oxidizing agents of choice in the majority of cases. In these and other reactions to be described organic peroxy acids have many advantages over inorganic oxidizing agents.

Although oxidations with organic peroxy acids are usually conducted in homogeneous solution in organic solvents, such a condition is not a prerequisite for successful results. It is necessary, of course, that the substrates have sufficient solubility in the peroxy acid system for the oxidation to be completed in a reasonable time.

Oxidations with aliphatic peroxy acids, particularly peroxyformic and peroxyacetic acids, are frequently conducted in a solution or dispersion in the corresponding carboxylic acid (formic and acetic acids, respectively). Since the peroxy acid is converted to the organic acid, which may be readily separated from the reaction mixture by vacuum evaporation, the oxidation product is frequently obtained in high yield and purity as a residue.

Some of the important aliphatic peroxy acids (such as peroxyformic, peroxyacetic, and peroxypropionic acids) and the corresponding carboxylic acids are soluble both in water and in organic solvents, thus affording wide flexibility in the choice of solvents for oxidation. When the oxidation product is insoluble in water and the peroxy acid and its corresponding carboxylic acid are soluble, as is frequently the case, the oxidation product is most conveniently and simply isolated by pouring the reaction mixture into water and separating the product mechanically.

Oxidations with organic peroxy acids lend themselves readily to quantitative and kinetic study, as unconsumed peroxy acid can be rapidly and accurately determined at suitable time intervals. Thus not only are valuable kinetic data as well as the desired product obtainable from a single experiment, but a plot of the consumption of peroxy acid with time provides information on the optimum reaction time for a given temperature with a minimum of effort.

In contrast, many inorganic oxidizing agents either form voluminous and difficult-to-handle end products, such as manganese dioxide from potassium permanganate, or form complex mixtures of oxidation products, as is often the case in nitric acid or dichromate oxidations, thus rendering the isolation of the desired product tedious or impossible. In oxidations with inorganic oxidants the problem is not to effect the oxidation but to isolate the reaction products in a sufficiently pure state for identification. Also, the need for operating in aqueous media limits the usefulness of some inorganic oxidants for organic reactions. Finally, the difficulty of following the course of oxidation quantitatively requires that numerous experiments be conducted and products separated and purified to determine the time variable.

No general discussion of the use of organic peroxy acids as oxidizing agents is complete without describing the very important and widely used *in situ* oxidation technique and, of course, hydrogen peroxide, the key component in the success of that technique. Hydrogen peroxide is a stable colorless liquid available commercially in aqueous solutions containing up to 90% hydrogen peroxide, and 98% in special circumstances. Standard concentrations available commercially are 35, 50, and 70%.

The active oxygen in hydrogen peroxide is not readily available for most organic oxidation reactions, but fortunately this oxygen can be readily converted to the useful peroxy acid form by the reaction of hydrogen peroxide with organic acids or anhydrides, with little or no loss. More important, for many oxidations the peroxy acid need not be separately prepared and isolated. It is sufficient to dissolve or disperse the substance to be oxidized in the organic acid or anhydride and add the hydrogen peroxide. Temperatures above room temperature and a strong-acid catalyst may also be required. As the peroxy acid forms it is consumed, and, since its formation is an equilibrium reaction, the peroxy acid will continue to form and be consumed until very little oxidizable substance remains, provided sufficient hydrogen peroxide is used. This sequence is illustrated in the following equations:

$$RCO_2 + H_2O \underset{}{\overset{H^+}{\rightleftharpoons}} RCO_3H + H_2O$$

$$RCO_3H + \text{oxidizable substance} \longrightarrow RCO_2H + \text{oxidation product}$$

Preferably the rate of oxidation should be faster than the rate of peroxy acid formation. Such a situation contributes to convenience, speed, efficiency, and also safety of operation, as peroxy acid never accumulates.

Under appropriate conditions loss of peroxide oxygen is held to a minimum and stoichiometric utilization of the active oxygen takes place, with high or quantitative yield or product. The *in situ* technique for the preparation and consumption of peroxy acids, as illustrated, is the one most frequently used in oxidations involving peroxyformic and peroxyacetic acids. It has also been used, but not as frequently, with peroxytrifluoroacetic, peroxypropionic, and peroxybutyric acids.

Hydrogen peroxide is the most important source of active oxygen in *in situ* and preformed peroxy acid oxidations. Another *in situ* oxidation technique involves the passage of molecular oxygen or air through a solution of an oxidizable substance in an aldehyde, such as acetaldehyde, butyraldehyde, or benzaldehyde. This technique is referred to as cooxidation or coupled oxidation (2, 13–15). The peroxy acid is an important intermediate in the oxidation of aldehydes to the corresponding carboxylic acids, and in certain cases it can be used to oxidize a dissolved organic substance rather than a second molecule of aldehyde. The reactions are as follows:

$$RCHO \xrightarrow{O_2} [RCO_3H] \begin{cases} \xrightarrow{RCHO} 2\,RCO_2H \\ \xrightarrow[\text{substance}]{\text{oxidizable}} RCO_2H + \text{oxidation product} \end{cases}$$

By oxidizing aldehyde in the presence of another oxidizable substance, the peroxy acid oxygen is obtained from the atmosphere, thus providing a potentially low cost source of active oxygen. This *in situ* technique has not been utilized extensively; benzaldehyde has usually been the aldehyde of choice, although greater emphasis in the recent literature is on aliphatic aldehydes. With other aldehydes, yields may be low owing to side reactions between the oxygen and the nonaldehydic substrate and extensive conversion of aldehyde to carboxylic acid. The major product in a coupled oxidation system is the acid corresponding to the aldehyde, and with an aldehyde (e.g., benzaldehyde) the separation of large amounts of benzoic acid from the desired oxidation product is often inconvenient. With aliphatic aldehydes the separation of product is usually more convenient.

Although they are used to oxidize a broad spectrum of organic compounds, organic peroxy acids are used predominantly in the epoxidation of unsaturated compounds. This chapter is devoted largely to that subject. After extensive tabulations of compounds epoxidized since 1952, experimental details for laboratory epoxidations are given. The epoxidation of unsaturated compounds is now an important commercial reaction, and a brief description of some of those processes is also provided. The kinetics, mechanisms, and stereochemistry of epoxidation are then summarized. Unusual, abnormal, or unexplained "epoxidations" are discussed last.

## II. EPOXIDATION OF UNSATURATED COMPOUNDS

Epoxidation is the reaction in which ethylenic unsaturation is converted directly to oxirane by reaction with a peroxy acid, either preformed or generated *in situ* (1–12):

$$-\overset{|}{\underset{}{C}}=\overset{|}{\underset{}{C}}- \ + \ RCO_3H \ \longrightarrow \ -\overset{|}{\underset{}{C}}\underset{\underset{O}{\diagdown\diagup}}{\quad}\overset{|}{\underset{}{C}}- \ + \ RCO_2H$$

This reaction with preformed peroxybenzoic acid was discovered about 1908 by the Russian chemist Prileschajew (16–18). It has since been shown that peroxybenzoic acid is an extremely efficient oxidizing agent and that the reaction is general. Compounds with multiple unsaturation can also be epoxidized, either partially or completely, depending on the quantity of peroxy acid used. Many other organic peroxy acids—most commonly peroxyacetic, monoperoxyphthalic, peroxytrifluoroacetic, *m*-chloroperoxybenzoic, and peroxyformic acids—have also been widely used for epoxidation. Other peroxy acids that have been successfully employed, usually in special cases, are *p*-nitroperoxybenzoic, peroxyoctanoic, and peroxylauric acids, to name only a few.

The generality and versatility of peroxy acid epoxidation of unsaturated compounds have made it the reaction of choice in most conversions of a double bond to the oxirane in an almost limitless variety of monounsaturated or polyunsaturated systems. The main restriction is that other portions of the molecule be stable to the peroxy acid employed. Thus the presence of silicon and phosphorus(V) in unsaturated compounds does not interfere, and many expoxy monomers that contain silicon and phosphorus(V) have been prepared by direct epoxidation. Phosphorus(III), sulfur, and nitrogen in organic molecules are readily oxidized by peroxy acids, however. The controlled oxidation of organic sulfur and nitrogen compounds is another widely used oxidation reaction and is discussed separately. In such oxidations unsaturation should ordinarily be absent.

"Oxirane group" is the official *Chemical Abstracts* name for the three-membered heterocyclic ring shown above. This group, however, is frequently called the epoxide group (more properly 1,2-epoxide or $\alpha$-epoxide), thus prompting Swern, in whose laboratory considerable fundamental work on the reaction illustrated above was conducted in 1940–1950 (2, 13–15), to coin the universally accepted word "epoxidation." (The term "oxido group" is frequently found in the older literature.)

### A. Peroxybenzoic Acid

Peroxybenzoic acid is one of the most important peroxy acids for preparing oxiranes not only for historical reasons but also because of its ready availability and efficiency. In many epoxidations quantitative yields of oxiranes are obtained and frequently are in excess of 75%. The reaction can be conducted in a wide variety of inert organic solvents, most commonly benzene, chloroform, methylene chloride, ether, dioxane, acetone, and ethyl acetate. Reaction time is frequently short, but it may vary considerably, depending on the number and nature of the groups attached to the unsaturated linkage. Reaction temperature varies from 0° to room temperature usually. The reaction is especially useful for the laboratory epoxidation of nonvolatile, water-insoluble unsaturated compounds that cannot be satisfactorily converted to the oxirane by way of the chlorohydrin or by autoxidation.

Peroxybenzoic acid has been used for the epoxidation of over 600 unsaturated compounds. In his extensive reviews Swern (2, 13–15) tabulated the unsaturated compounds epoxidized to 1952, with literature references; those tables are too lengthy to be repeated here. Since 1952, however, many additional compounds have been epoxidized with peroxybenzoic acid; they are listed in Table 1. Some of the compounds listed here were also reported in earlier tabulations (2, 13–15), but it was considered desirable to provide the most recent literature references in those cases in which repetition occurs. Every effort has been made to ensure the completeness of Table 1 (and the subsequent tables listing compounds epoxidized with other peroxy acids), but it is manifestly impossible to cull from the rapidly growing and widely

TABLE 1. UNSATURATED COMPOUNDS EPOXIDIZED WITH PEROXYBENZOIC ACID SINCE 1952

| Compound | Reference |
|---|---|
| Acetaldehyde bis($\Delta^3$-tetrahydrobenzyl)acetal | 19 |
| Acetone bis($\Delta^3$-tetrahydrobenzyl)acetal | 19 |
| 2$\alpha$-Acetoxybicyclo[2.2.1]hept-5-ene | 20 |
| 21-Acetoxy-3,20-bis(ethylenedioxy)-5,9(11)-pregnadien-17$\alpha$-ol | 21 |
| 3$\beta$-Acetoxycholest-4-ene | 22 |
| 4-$\alpha$-Acetoxycholest-5-ene | 22 |
| 4-$\beta$-Acetoxycholest-5-ene | 22 |
| 2-Acetoxy-2-cyanobicyclo[2.2.1]hept-5-ene | 23 |
| 1-Acetoxycyclohexene | 24 |
| 3-Acetoxycyclohexene | 22 |
| 1-Acetoxy-3-cyclohexene | 25 |
| 3$\beta$-Acetoxy-22,23-dibromo-7 Z, 11$\alpha$-dihydroxyergost-8-ene | 26 |
| 3$\beta$-Acetoxy-22,23-dibromo-8-ergostene-7 Z, 11$\alpha$-diol | 27 |
| 3$\beta$-Acetoxy-5$\beta$,8$\beta$-epoxyergost-9-ene | 28 |
| 3$\beta$-Acetoxyergosta-7,9(11),22-triene | 29 |
| 3$\beta$-Acetoxyergost-9(11)-en-7-one 22,23-dibromide | 30 |
| 3$\beta$-Acetoxy-14,16-etioallocholadiene nitrile | 31 |
| 3-O-Acetyl-$\beta$-anhydrodigitoxigenin | 32 |
| N-Acetyl-5,6-dihydroisojervine | 33 |
| N-Acetyl-$\Delta^4$-isojervone | 33 |
| Alloaromadendrene | 34 |
| Allyl bromide | 35, 36 |
| Allyl isopropyl ether | 36 |
| Allyl-2-tosyloxy-1-hexadecyl ether | 37 |
| Allyl-2-tosyloxy-1-octadecyl ether | 37 |
| $\beta$-Anhydrodigitoxigenin | 32 |
| Benzaldehyde bis($\Delta^3$-tetrahydrobenzyl)acetal | 19 |
| 1-Benzoyl-1-decene | 38 |
| endo-2-Benzoyl-8-diphenylmethylene-3$\alpha$,4,7,7$\alpha$-tetrahydro-4,7-methanoisoindoline | 39 |
| Bicyclo[3.4.0]3-cyano-7-nonene | 40 |
| Bicyclo[2.2.1]heptan-7-one | 41 |
| Bicyclo[2.2.1]hept-5-ene-2-methyl | |
| Bicyclo[2.2.1]hept-5-ene-2-carboxylate | 42 |
| Bicyclo[2.2.1]hept-5-en-2$\alpha$-ol | 20 |
| Bicyclo[2.2.1]hept-2-en-7-ol | 41 |
| Bicyclo[2.2.1]hept-5-en-2$\alpha$-ylmethanol | 20 |
| Bicyclo[2.2.1]hept-5-en-2$\alpha$-ylmethyl acetate | 20 |
| Bicyclo[2.2.1]oct-5-ene-2-methyl | |

TABLE 1 (Continued)

| Compound | Reference |
|---|---|
| Bicyclo[2.2.2]oct-5-ene-2-carboxylate | 41 |
| Bis{2-(bicyclo[2.2.1]hept-5-ene)methyl}carbonate | 42 |
| Bis{2-(bicyclo[2.2.1]hept-5-ene)methyl}ether | 42 |
| Bis(cyclohexenyl)tetramethyl disiloxane | 43 |
| Bis(*p*-methallylphenyl)tetramethyl disiloxane | 44 |
| Butadiene | 45 |
| Buten-3-ol | 25 |
| 5-(Buten-2-yl)-2-norbornene | 46 |
| 4-*t*-Butylmethylenecyclohexane | 47 |
| Camphene | 48 |
| Carvomenthene | 49 |
| Cedrene | 50 |
| 1-Chloro-4-methylcyclohexene | 51 |
| 16α-Chloromethyl-5,17(20)-pregnadiene-3,20-diacetate | 52 |
| *trans-p*-Chlorostilbene | 53 |
| Cholest-1,4-dien-3-one | 54 |
| Cholest-4-ene | 55 |
| Cholest-5-ene-3$\beta$-carboxylic acid | 56 |
| Cholest-4-en-3α-ol | 57 |
| $\Delta^7$-Cholestenyl acetate | 58 |
| *trans*-Cinnamyl carbamate | 59 |
| Crotophorbolone | 60 |
| Cyclobutene | 61 |
| *cis*-Cyclodecene | 64 |
| *trans*-Cyclodecene | 64 |
| Cyclododecatriene | 62, 63 |
| 3,5-Cyclo-$\Delta^{8(14),22}$-ergostadiene | 65 |
| 3,5-Cyclo-$\Delta^{8(14)}$-ergostene | 65 |
| Cycloheptene | 66 |
| Cyclohexadiene | 67 |
| Cyclohexene | 68, 69 |
| $\Delta^4$-Cyclohexene-1,2-dimethanol diacetate | 70 |
| Cyclohex-2-enol | 22, 25 |
| Cyclohex-3-enol | 25 |
| Cyclohex-1-en-3-ol | 71 |
| Cyclohex-1-en-3-yl acetate | 71 |
| Cyclohex-3-enylmethanol | 20 |
| Cyclohex-3-enylmethyl acetate | 20 |
| Cyclooctatetraene | 72, 73 |
| Cyclooctatetraene-maleic anhydride adduct | 74 |

TABLE 1 (Continued)

| Compound | Reference |
|---|---|
| Cyclooctatriene | 75 |
| cis-Cyclooctene | 76 |
| cis-1-Cyclopentene-3,5-diol | 77 |
| 6-Dehydro-17-acetoxyprogesterone | 78 |
| Dehydrochlorinated polyvinyl chloride | 79, 80 |
| Dehydronorbornyl acetate | 23 |
| Dehydronorcamphor ethyleneketal | 23 |
| 2,3α-Dehydropregnan-20-one | 81 |
| 3β,20-Diacetoxy-16,20-allopregnadiene | 82 |
| 3β,17-Diacetoxy-$\Delta^{16}$-androstene | 83 |
| 1,1-Diacetoxy-2-butene | 84 |
| 3,5-Diacetoxyergosta-7,9,22-triene | 85 |
| 3,17-Diacetoxy-$\Delta^{1,3,5,16}$-estratetraene | 83 |
| 3,20-Diacetoxy-16β-methyl-5,17(20)-pregnadiene | 86 |
| 3-(2,5-Diacetoxyphenyl)propene | 87 |
| 3β-20-Diacetoxy-5,20-pregnadiene | 82 |
| 3β,20-Diacetoxy-5,16,20-pregnatriene | 82 |
| Diacetylsideridiol | 88 |
| Diallylphenylphosphine oxide | 89 |
| $\Delta^{14,16}$-Dianhydrogitoxigenin and diacetate | 90 |
| 4,5-Dibromocyclohexene | 91 |
| 2,4-Dichloro-2-butene | 92 |
| 1,1-Diethoxy-3-butene | 93 |
| Digitoxigenins | 90 |
| Dihydrocarvone | 94 |
| Dihydrocostunolide | 95 |
| exo-Dihydrodicyclopentadiene | 96 |
| Dihydroeuphyl acetate | 97 |
| 3β,6β-Dihydroxycholest-4-ene | 22 |
| 7,7-Dimethoxybicyclo[2.2.1]heptane | 98 |
| 1,1-Dimethoxy-3-butene | 93 |
| 4,4'-Dimethoxy-trans-stilbene | 69 |
| 1,2-Dimethylcyclobutene | 61 |
| Dimethyldiallylsilane | 99 |
| 4-(2,4-Dimethylphenyl)-2-chlorobutene-2 | 92 |
| 4-(2,5-Dimethylphenyl)-2-chlorobutene-2 | 92 |
| trans-1,3-Diphenyl-2-buten-1-ol | 100 |
| Diphenyldiallylsilane | 99 |
| endo-7-diphenyl-N-methyl-5-norbornene-2,3-dicarboximide | 39, 101 |
| exo-7-Diphenyl-N-methyl-5-norbornene-2,3-dicarboximide | 39, 101 |

T<small>ABLE</small> **1** (Continued)

| Compound | Reference |
| --- | --- |
| *cis*-1,3-Diphenyl-2-propen-1-ol | 100 |
| *trans*-1,3-Diphenyl-2-propen-1-ol | 100 |
| Divinyldimethylsilane | 102, 103 |
| Divinyldiphenylsilane | 103 |
| 1,5-Divinyldisiloxane | 104 |
| 1,7-Divinyltrisiloxane | 104 |
| *cis*-13-Docosenoic acid | 105 |
| *trans*-13-Docosenoic acid | 105 |
| 1,11-Dodecadiene | 106 |
| α-4,8,13-Duratriene-1,3-diol | 107 |
| Elaidic acid | 108 |
| β-Eleostearic acid–maleic anhydride adducts | 109, 110 |
| Epilumisterol | 111 |
| 13-Episclareol | 112 |
| Ergosta-7,9,22-triene-3β,5α-diol | 85 |
| Ergosteryl-D acetate | 30 |
| Ergosteryl-D acetate 22,23-dibromide | 113 |
| 4-Ethoxy-2-chlorobutene-2 | 92 |
| 2-Ethoxy-6-methyl-5,6-dihydro-2*H*-pyran | 114 |
| Ethyl crotonate | 44 |
| 1-Ethylcyclohexene | 115 |
| 3,3-Ethylenedioxy-17β-acetoxy-19-nor-5-androstene | 116 |
| 3-Ethylenedioxy-17α,20,20,21-bismethylenedioxy-16α-methyl-5-pregnen-11-one | 117 |
| Ethyl vinyl acetate | 44 |
| Euphyl acetate | 97 |
| Gibberelic acids | 118 |
| Grapeseed oil | 119 |
| 1,2,3,4,7,7-Hexachloro-5-methylene-2-norbornene | 120 |
| *trans*-2-Hexadecen-4-ol | 121 |
| 1,2,3,4,5,6-Hexamethylbicyclo[2.2.0]hexa-2,5-diene | 122 |
| *endo*-1,2,3,4,5,6-Hexamethylbicyclo[2.2.0]hex-2-ene | 122 |
| 4-Hexyloxy-2-chlorobutene-2 | 92 |
| α-Homocholest-4α-en-3α-ol | 123 |
| α-Homocholest-4α-en-3β-ol | 123 |
| Humulene | 123, 124 |
| Humulene epoxide | 124 |
| $\Delta^5$-3α-Hydroxycholestene | 125 |
| 3β-Hydroxycholest-4-ene | 22 |
| 4α-Hydroxycholest-5-ene | 55 |

TABLE 1 (Continued)

| Compound | Reference |
|---|---|
| 7α-Hydroxycholesteryl benzoate | 22 |
| 7β-Hydroxycholesteryl benzoate | 22 |
| Isodihydroeuphyl acetate | 97 |
| Isokaurene | 126 |
| Isophyllocladene | 126 |
| 1-Isopropylcyclohexene | 115 |
| Kaurene | 126 |
| Khusimol and derivatives | 127, 128 |
| Limonene | 129 |
| Lumisterol | 130 |
| Lumisteryl acetate | 130 |
| Melianotetrol triacetate | 131 |
| $\Delta^1$-Menthene | 132 |
| $\Delta^2$-Menthene | 49 |
| $\Delta^3$-Menthene | 49, 132 |
| 3β-Methoxycholest-4-ene | 22 |
| 4-Methoxy-*trans*-stilbene | 69 |
| Methyl 3α-acetoxy-$\Delta^{4,9(11)}$-choladienate | 133 |
| Methyl 3β-acetoxy-14,16-etioallocholadienate | 31 |
| Methyl 3β-acetoxy-5,14,16-etiocholatrienate | 31 |
| Methyl 3α-acetoxy-$\Delta^{9(11)}$-etiocholenate | 134 |
| Methyl bicyclo[2.2.1]hept-5-ene-2α-carboxylate | 20 |
| 3-Methylbuten-1-ol | 35 |
| Methyl cholest-5-ene-3β-carboxylate | 56 |
| Methyl crotonate | 44 |
| 1-Methylcyclobutene | 61 |
| 1-Methylcyclohexene | 115 |
| Methyl cyclohex-3-ene-1-carboxylate | 20 |
| Methyl-$\Delta^{9(10)}$-dehydroabietate | 135 |
| Methyl *trans*-2-docosenoate | 136 |
| 2-Methyl-ethoxy-1-phenylpropene | 137 |
| Methyl gibberelates | 118 |
| Methyl 10-hendecenoate | 38 |
| Methyl *trans*-2-hexadecenoate | 136 |
| Methyl methacrylate | 138 |
| Methyl *trans*-2-octadecenoate | 136 |
| 2-Methyl-2-pentene | 139 |
| α-Methylstilbene | 140 |
| 3-Methyl-*trans*-stilbene | 69 |
| 4-Methyl-*trans*-stilbene | 53, 69, 141 |

TABLE 1 (Continued)

| Compound | Reference |
|---|---|
| Methyl *trans*-3-stilbenecarboxylate | 142 |
| 4-(α-Naphthyl)-2-chlorobutene-2 | 92 |
| 4-Nitraza-1-pentene | 143 |
| Norbornadiene | 144 |
| Norbornene | 96, 145 |
| 5-Norbornenedicarboximide derivatives | 101 |
| 1,17-Octadecadiene | 106 |
| *cis*-6-Octadecenoic acid (petroselinic acid) | 105 |
| *trans*-6-Octadecenoic acid (petroselaidic acid) | 105 |
| *cis*-9-Octadecenoic acid (oleic acid) | 108 |
| 1,4,4*a*,5,6,7,8,8*a*-Octahydro-1,4,5,8-*endo,endo*-dimethanonaphthalene | 145 |
| 1,2,3,4,4*a*,5,8,8*a*-Octahydro-1,4,5,8-(*exo, endo* and *endo, exo*) dimethanonaphthalenes | 145 |
| *trans*-$\Delta^2$-Octalin | 146 |
| Octamethylcyclooctatetraene | 147 |
| Octamethyl-syn-tricyclo[4.2.0.0.$^{2,5}$]octadiene-3,7 | 147 |
| Pentachlorostyrene | 148 |
| *N*-Phenethyl-*cis*-$\Delta^4$-tetrahydrophthalimide | 149 |
| 4-Phenoxy-2-chlorobutene-2 | 92 |
| 4-Phenyl-2-chlorobutene-2 | 92 |
| 1-Phenylcyclohexene | 150 |
| Phorbol | 60 |
| Phylloeladene | 126 |
| Phytene (*trans*) | 151 |
| Phytol | 151 |
| α-Pinene | 152 |
| *cis*-Pinocarveol | 153 |
| *trans*-Pinocarveol | 153 |
| Polybutadiene | 154 |
| $\Delta^5$-Pregnen-3β-ol-20-one | 155 |
| Propyl crotonate | 44 |
| Pseudotigogenin | 156 |
| Pulegone | 157, 158 |
| (−)-*cis*-Pulegyl *p*-nitrobenzoate | 159 |
| Sclareol | 112 |
| Sideridiol | 88 |
| $\Delta^2$-Solanidene | 160 |
| Squalene | 161 |
| Stigmasterol | 162 |

TABLE 1 (Continued)

| Compound | Reference |
| --- | --- |
| cis-Stilbene | 163 |
| trans-Stilbene | 69, 141, 163 |
| cis-4-Stilbenecarboxylic acid | 142 |
| trans-4-Stilbenecarboxylic acid | 142 |
| Styrene | 69 |
| γ-Terpineol | 164 |
| 1,2,3,4-Tetrachloro-7,7-difluoro-5-methylene-2-norbornene | 120 |
| 4,5,6,7-Tetrahydroindanols | 165 |
| cis-Δ⁴-Tetrahydrophthalic anhydride | 149 |
| 1,2,3,4-Tetramethylcyclobutene | 61 |
| Tetramethylethylene | 166 |
| Tetraphenylethylene | 69 |
| Tetravinylsilane | 102 |
| trans-2-o-Tolyl-4-cyclohexenyl benzoate | 167 |
| trans-2-o-Tolyl-5-cyclohexenyl benzoate | 167 |
| Triacetyl-5,6-dihydroisojervine | 33 |
| Triacetylphorbol | 60 |
| Tribenzylallylsilane | 99 |
| Triethylallylsilane | 99 |
| Trimethylallylsilane | 99 |
| 2,4,4-Trimethylcyclohexene | 168 |
| 4-(2,4,6-Trimethylphenyl)-2-chlorobutene-2 | 92 |
| Triphenylallylsilane | 99 |
| 1,1,1-Triphenylbutene-3 | 99 |
| Triphenylethylene | 69 |
| Tripropylallylsilane | 99 |
| Vinyl acetate | 169 |
| 4-Vinylcyclohexene | 170 |
| Vinyl oleate | 171 |
| Vinyl organosiloxanes | 172 |
| Vinylpentamethyldisilane | 173 |
| Vinyltrialkylstannanes | 174 |
| Vinyltribenzylsilane | 103 |
| Vinyltriethylsilane | 103, 175 |
| Vinyltrimethylsilane | 103, 175 |
| Vinyltriphenylsilane | 103 |
| Vinyltripropylsilane | 103 |
| Withaferin A lactone | 176 |

dispersed chemical literature every compound that has been epoxidized. Patents pose a particularly difficult problem, and it is hoped that there are not too many omissions from that literature.

Most investigators have prepared a solution of peroxybenzoic acid or have isolated it by one of the numerous synthetic procedures already described (Volume I of this series). Since peroxybenzoic acid can be readily and conveniently prepared by the oxygen oxidation of benzaldehyde, several investigators have treated solutions of benzaldehyde and an unsaturated compound with air or oxygen, in the presence or absence of metal salts. The peroxy acid is consumed as it is formed, thus avoiding its separate preparation and isolation. This epoxidation method, referred to as cooxidation or coupled oxidation, has been applied to cyclohexene (177), octenes (178), oleic acid (179, 180), stilbene (179, 180), styrene (179, 180), squalene (179, 180), and α-methylstyrene (181). In general good yields of oxirane compounds are obtained, although in no case is the yield as high as that obtained with preformed peroxybenzoic acid because of competing reactions, notably hydroperoxidation of olefin, rearrangements of oxirane, and ring-opening reactions of oxirane.

## B. Monoperoxyphthalic Acid

Böhme (182, 183) was apparently the first to show that monoperoxyphthalic acid (MPA) is consumed by reaction with the ethylenic linkage, but Chakravorty and Levin (184) in 1942 were the first to isolate oxiranes by that oxidation reaction. Since then MPA has had wide use in the epoxidation of unsaturated compounds, notably sterols and other polycyclic compounds, polyenes, and many natural products. Epoxidation with MPA is conducted under the same general conditions and in the same solvents as with peroxybenzoic acid; excellent yields of oxiranes are usually obtained.

Monoperoxyphthalic acid has two advantages over peroxybenzoic acid. First, when epoxidation requires a long reaction time, its greater stability (184, 185) is an asset. Second, since epoxidations with MPA are often conducted in chloroform, in which phthalic acid (the reduction product of MPA) is insoluble, the phthalic acid precipitates and is unable to cause unwanted ring-opening reactions.

The epoxidation of unsaturated compounds with phthalic anhydride and hydrogen peroxide in benzene at moderate temperatures (25–50°) (*in situ* generation and consumption of MPA) has been patented (186). The same patent describes the use of succinic anhydride, 1,2-cyclohexanedicarboxylic acid anhydride, and α,α-dimethylbenzylsuccinic anhydride for *in situ* epoxidations.

For compounds epoxidized with MPA through 1952 the already cited reviews (2, 13–15) should be consulted. Table 2 lists unsaturated compounds epoxidized with MPA since 1952.

TABLE 2. UNSATURATED COMPOUNDS EPOXIDIZED WITH MONOPEROXYPHTHALIC ACID SINCE 1952

| Compound | Reference |
|---|---|
| 3$\beta$-Acetoxy-16$\beta$-carboxy-20$\beta$-hydroxypregn-5-ene-16,20-lactone | 187 |
| 3$\beta$-Acetoxy-16,17$\alpha$-carboxymethylenepregn-5-en-20-one | 187 |
| 3$\beta$-Acetoxy-5$\alpha$-chloro-17(20)-bisnorcholen-22-al | 188 |
| 3$\beta$-Acetoxy-4-cholestone | 189, 190 |
| 3$\beta$-Acetoxy-5-cholestene | 191 |
| 3$\beta$-Acetoxy-16$\alpha$-cyanopregn-5-en-20-one | 187 |
| 3$\beta$-Acetoxy-4,4-dimethyl-5-cholestene | 191 |
| 3$\beta$-Acetoxy-4,4-dimethyl-5-cholestene | 191 |
| 3$\beta$-Acetoxy-5$\alpha$-epoxyergosta-9,22-diene | 192 |
| 3$\beta$-Acetoxy-16,17$\alpha$-epoxy-16$\beta$-methylpregn-5-en-20-one | 187 |
| 3$\beta$-Acetoxyergosta-7,9-diene | 193 |
| 3$\beta$-Acetoxyergosta-7,9-diene-5$\alpha$-ol | 193 |
| 3$\beta$-Acetoxyergosta-7,9(11)22-trien-5$\alpha$-ol | 194, 195 |
| 3$\beta$-Acetoxyergosta-7,9,22-triene-5$\alpha$-ol | 194, 195 |
| 5-Acetoxyergosta-7,9,22-trien-3$\beta$-ol | 194, 195 |
| Acetoxylanostenone enol acetate | 196 |
| 1-Acetoxy-4-methylcyclohexene | 197 |
| 3$\beta$-Acetoxy-4$\beta$-methyl-5-cholestene | 189 |
| 3$\beta$-Acetoxy-16-exo-methylene-17$\alpha$-hydroxypregn-5-en-20-one | 187 |
| 3$\beta$-Acetoxy-16$\beta$-methylpregna-5,16-dien-20-one | 187 |
| 3$\beta$-Acetoxy-B-norcholest-4-ene | 198 |
| 3$\beta$-Acetoxy-20-oxo-14,16-allopregnadiene | 199 |
| 3$\beta$-Acetoxypregna-5,16-dien-20-one | 191 |
| 3$\beta$-Acetoxypregn-5-en-20-one | 187 |
| 3-O-Acetylcanarigenin | 200 |
| Bicyclo[2.2.2]octene | 201 |
| Bis(allylphenyl)dimethylsilane | 202 |
| Bis(p-styryl)dimethylsilane | 202 |
| Bis-(vinyldimethyl)disiloxane | 202 |
| cis-2-Butene-1,4-diol | 203 |
| trans-2-Butene-1,4-diol | 203 |
| 2-Butenes | 204 |
| 1-Buten-3-ol | 205 |
| 3-Butenyltrimethylsilane | 206 |
| 2-Butyl-2-hexenamide | 207 |
| 4-t-Butylmethylenecyclohexane | 208 |
| (+)-$\delta$-Cadinene | 209 |
| Calarene | 210 |
| Calciferol dinitrobenzoate | 211 |

Table 2 (Continued)

| Compound | Reference |
|---|---|
| Canarigenin | 200 |
| 2-Carbethoxy-7-oxabicyclo[2.2.1]hept-5-ene | 211 |
| 2-Carboxy-7-oxabicyclo[2.2.1]hept-5-ene | 212 |
| $\gamma$-Caryophyllene | 213 |
| Caryophyllenes | 214 |
| $\alpha$-Cedrene | 215 |
| Chaulmoogric acid | 216 |
| 1-Chloro-4-methylcyclohexene | 51 |
| $\Delta^3$-Cholestene | 217, 218 |
| 1,5,9-Cyclododecatriene | 219 |
| 2-Cyclohepten-1-ol | 220 |
| 1,4-Cyclohexadiene | 67 |
| Cyclohexene | 68, 221 |
| Cyclohexenyl-1-carboxamide | 207 |
| 2,8-Decadien-4,6-diyne | 222 |
| 1-Decene and decenes | 223 |
| $3\beta,20\beta$-Diacetoxy-16$\alpha$($\beta$)acetoxymethylpregn-5-ene | 187 |
| $3\beta,20\beta$-Diacetoxy-16$\alpha$-carboxypregn-5-ene | 187 |
| $3\beta,20\beta$-Diacetoxy-16$\alpha$-cyanopregn-5-ene | 187 |
| 1,10-Diacetoxy-3,8-dimethyl-2,8-decadiene-4,6-diyne | 222 |
| $3\beta,20\beta$-diacetoxy-16$\alpha$-methoxycarbonylpregn-5-ene | 187 |
| $3\beta,17\beta$-Diacetoxy-17$\alpha$-methyl-19-nor-5-androstene | 224 |
| $3\beta,17\beta$-Diacetoxy-17-nor-5-androstene | 224 |
| $3\beta,21$-Diacetoxy-20-oxoallo-14,16-pregnadiene | 199 |
| Diallyl adipate | 186 |
| Diallyl fumarate | 186 |
| Diallyl maleate | 186 |
| Diallyl phthalate | 186 |
| Dibutenyldimethylsilane | 202 |
| 2,3-Dicarboalkoxy-2,3-diazabicyclo[2.2.1]hept-5-ene | 225 |
| Dicyclopentadiene | 186 |
| Dihydroagnosteryl acetate | 196 |
| 5-Dihydroergosteryl acetate | 226 |
| $\Delta^4$-Diisododecyl tetrahydrophthalate | 186 |
| $\Delta^4$-Diisooctyl tetrahydrophthalate | 186 |
| 4,4-Dimethyl-17-(substituted)androst-5-en-3-ones | 191 |
| *dl-trans-N,N*-Dimethylcinnamide | 227 |
| 1,4-Dimethylenecyclohexane | 228 |
| 1,1-Dimethyl-7-hydroxy-9,10-octalin | 229 |
| Dimethyl-7-oxabicyclo[2.2.1]-2,5-heptadiene-2,3-dicarboxylate | 230 |

TABLE 2 (Continued)

| Compound | Reference |
|---|---|
| 4,4-Dimethyl-1-penten-3-ol | 205 |
| 3,4-Dimethylpentene-1 | 205 |
| 6,16α-Dimethyl-$\Delta^5$-pregnen-3β-ol-20-one acetate | 231 |
| 18,19-Dinor-steroids | 223 |
| Divinyldimethylsilane | 206 |
| cis-13-Docosenoic acid | 232 |
| trans-13-Docosenoic acid | 232 |
| Echinenone | 233 |
| 13-Episclareol | 112 |
| 16α,17α-Epoxypregn-5-ene-3,20-dione 3,20-bis(ethylene ketal) | 234 |
| Ergosteryl acetate | 226 |
| Estr-5(10)-ene-3,17β-diol | 235 |
| Estr-5(10)-ene-3α,17β-diol diacetate | 235 |
| Estr-5(10)-ene-3,17-dione | 235 |
| 17α-Ethinyl-$\Delta^{5(10)}$-19-norandrosten-17β-ol-3-one | 236 |
| 2-Ethylcinnamide | 207 |
| 1-Ethylcyclohexene | 68 |
| 2-Ethyl-2-hexenamide | 207 |
| 2-Ethyl-2-pentenamide | 207 |
| Euphenol acetate (isomer) | 237 |
| Fumagillin intermediates ($C_{18}H_{30}O_4$) | 238 |
| Geraniolene | 239 |
| Hexenes | 225 |
| Humulene epoxide | 240 |
| Hydnocarpic acid | 216 |
| 19-Hydroxy-5-androstene-3,17-diethyleneketal | 241 |
| 19-Hydroxy-3,3 : 17,17-bis(ethylenedioxy)-5-androstene | 242 |
| trans-4-Hydroxy-5-bromocyclohexene | 67 |
| $\Delta^5$-3α-Hydroxycholestene | 243 |
| 5α-Hydroxy-6,5-dihydroergocalciferol carbonate | 244 |
| 3β-Hydroxy-16,17α-epoxypregn-5-en-20-one | 187 |
| 19-Hydroxy-3,3-ethylenedioxy-5-cholestene | 242 |
| 2-Hydroxymethyl-7-oxabicyclo[2.2.1]hept-5-ene | 212 |
| 1-Isopropylcyclohexene | 115 |
| Isozeaxanthin | 233 |
| Limonene | 129, 152, 221 |
| Linalool | 245 |
| Linalyl esters | 246 |
| Manool | 247 |
| Manool acetate | 247 |

Tᴀʙʟᴇ **2** (Continued)

| Compound | Reference |
|---|---|
| Manool formate | 247 |
| Methyl-$\Delta^{11}$-3$\alpha$-acetoxycholenate | 217 |
| $\Delta^{5,6}$-17$\alpha$-Methyl-3$\beta$-acetoxy-17-oxyandrostene | 248 |
| 4-Methylcholest-4-en-3$\alpha$-ol | 249 |
| 4-Methylcholest-4-en-3$\beta$-ol | 249 |
| Methyl crepenynate | 250 |
| 1-Methylcyclohexene | 68, 115 |
| 4-Methylcyclohexenyl-1-carboxamide | 207 |
| cis-1-Methylcyclononene | 249 |
| cis-1-Methylcyclooctene | 249 |
| $\Delta^{5,6}$-17$\alpha$-Methyl-3$\beta$-17-diacetoxyandrostene | 248 |
| $\Delta^{5,6}$-17$\alpha$-Methyl-3$\beta$-17-dioxyandrostene | 248 |
| Methylenecyclononane | 249 |
| Methylenecyclooctane | 249 |
| trans-2-Methylenedecahydronaphthalene | 208 |
| 12-Methylenetigogenin acetate | 251 |
| Methyl linoleate | 252, 253 |
| Methyl 9,12-linoleate dimers | 254 |
| trans-9(10)-Methyl-2-methylenedecahydronaphthalene | 208 |
| Methyl endo-5-norbornene-2-carboxylate | 255 |
| Methyl exo-5-norbornene-2-carboxylate | 255 |
| Methyl trans, trans-9,11-octadecadienoate | 256 |
| 10-Methylol-$\Delta^{1(9)}$-octalin | 229 |
| 4-Methyl-1-penten-3-ol | 205 |
| Methyl sorbate | 257 |
| Monoside $\alpha$ | 200 |
| Nerolidyl formate | 258 |
| p-Nitrobenzoyloxymethyl-7-oxabicyclo[2.2.1]hept-5-ene | 212 |
| $\Delta^{5(10)}$-19-Norandrosten-17$\beta$-ol-3-one | 236 |
| Norbornadiene | 144 |
| Norbornene | 259, 260 |
| exo-5-Norbornene-2-carboxylic acid | 255 |
| B-Norcholest-4-en-3$\beta$-ol | 198 |
| 16-Norpimar-7-ene | 261 |
| 19-Nor-$\Delta^9$ steroids | 223 |
| 17-Octadecen-9,11-diynoic acid (isanic acid) | 262 |
| 1-Octadecene | 263 |
| cis-6-Octadecenoic acid (petroselinic acid) | 232, 264 |
| trans-6-Octadecenoic acid (petroselaidic acid) | 232, 264 |
| cis-9-Octadecenoic acid (oleic acid) | 232, 265 |

TABLE 2 (Continued)

| Compound | Reference |
|---|---|
| trans-9-Octadecenoic acid (elaidic acid) | 232 |
| cis-2-Octadecenol | 266 |
| trans-2-Octadecenol | 266 |
| Octamethyl-anti-tricyclo[4.2.0.0.$^{2,5}$]octadiene-3,7 | 147 |
| cis-4-Octene-4,5-$d_2$ | 267 |
| trans-4-Octene-4,5-$d_2$ | 267 |
| 1-Penten-3-ol | 205 |
| 1-Phenylcyclopentene | 268 |
| cis-2-Phenyl-2-butene | 269 |
| trans-2-Phenyl-2-butene | 269 |
| β-Pinene | 152, 221 |
| Poly(butadiene) | 270 |
| Poly(butadiene-acrylonitrile) | 270 |
| Poly(butadiene–styrene | 270 |
| Poly(diallyl phthalate) | 186 |
| Precalciferol dintrobenzoate | 211 |
| Pregnadienolone acetate | 187 |
| $\Delta^5$-Pregnen-3β,16α,20α-triol | 271 |
| cis-Propenylbenzene | 272 |
| 1-Propionoxycyclohexene | 197 |
| Propylene glycol tallate | 186 |
| 2-Propyl-2-heptenamide | 207 |
| 2-Propyl-2-hexenamide | 207 |
| 1-Propyl-4-isopropylcyclohexene | 273 |
| 2-Propyl-2-pentenamide | 207 |
| Rubber | 270 |
| Retinyl acetate | 274 |
| Sclareol | 112 |
| Soybean oil | 186, 265 |
| $\Delta^{5,20}$-Steroid diacetate | 275 |
| p-Styryltrimethylsilane | 206 |
| Tetrahydrohumulene | 240 |
| endo-Tricyclo[3.2.1.0$^{2,4}$]oct-6-ene | 276 |
| 2,2,4-Trimethylcyclohexene | 168 |
| 2,3,3-Trimethyl-4-vinylcyclopentene | 277 |
| 10-Undecen-1-ol | 278, 279 |
| Vinyl ethyldimethylsilane | 206 |
| Vinyl trianisylsilane | 280 |
| Vinyl trimethylsilane | 206 |
| Wallichiene | 281 |

## C. Peroxyacetic Acid

Peroxyacetic acid has become the most important epoxidation reagent during the last 20 years for two main reasons: first, its commercial availability, low price, and ease of preparation; second, the development of mild, high-speed reaction conditions for the preparation of a wide variety of oxiranes in high yield. Peroxyacetic acid has the added advantages that it has a low equivalent weight and can be used in both aqueous and nonaqueous solvents, and in homogeneous or heterogeneous media. The organic solvents that are employed in epoxidations with peroxybenzoic and monoperoxyphthalic acids can also be used with peroxyacetic acid. The only requirements that apply to the selection of a solvent are that it be inert under the reaction conditions and easy to dispose of.

Peroxyacetic acid is available commercially as a 40% solution in acetic acid, but if acetic acid is an undesirable diluent, it can also be prepared in a large number of other solvents and in aqueous solution. A major impetus to the widespread use and industrial study of peroxyacetic acid is its *in situ* preparation and consumption for the large-scale production of stabilizers-plasticizers for polyvinyl chloride. This development has prompted an increased study of the uses of peroxyacetic acid in a variety of potentially important commercial epoxidations and other oxidation reactions. Since *in situ* methods are used mostly in commercial epoxidations, this subject will be discussed later.

For many years, however, it was assumed that oxiranes could not be prepared by the epoxidation of olefins with peroxyacetic acid, because the products usually isolated from such reactions were α-glycols or hydroxy-acetates. Böeseken, Smit, and Gaster (282) and Smit (283), however, obtained methyl-9,10,12,13-diepoxystearate by the epoxidation of methyl linoleate with peroxyacetic acid in acetic acid solution, but yields were not reported. Repetition of this work (284), following the techniques of the earlier investigators, resulted in a low yield of the diepoxy product. Most of the methyl linoleate was converted to a polymer of unknown composition. Until the early 1950s relatively few reports suggested that oxiranes could be obtained from unsaturated compounds by epoxidation with peroxyacetic acid in acetic acid solution. Yields were either not reported or were low (30% or less). Compounds epoxidized in this manner were 3,6-diacetoxy-5-methyl-norcholestene (285), dihydrocaryophyllene (286), elaidic acid and methyl elaidate (287), ergosterol–maleic anhydride adduct (288), 5-methylnorcholestene-3,6-dione (285), β-amyrin acetate (289), and *trans*-7-octadecenoic acid (290).

On the other hand, as long ago as 1930, Arbuzow and Michailow (291) showed that *d*-Δ³-carene could be epoxidized in high yield if an ether solution of peroxyacetic acid was employed, but with an acetic acid solution they obtained only hydroxyacetates. They also showed that α-pinene could be converted to an oxirane compound in almost 90% yield by peroxyacetic acid in ether solution. Oddly, with peroxyacetic acid in chloroform, α-pinene was reported to be converted to the oxirane compound in only fair yield; the major products were hydroxyacetates. Subsequently (292) these investigators extended their work to other olefins and demonstrated that limonene, cyclohexene, anethole, and isoeugenol could also be converted to oxiranes in good yield when an ether solution of peroxyacetic acid was employed.

Arbuzow and Michailow correctly concluded therefore that the reaction of peroxyacetic acid with olefins is exactly the same as that of peroxybenzoic acid—namely, the double bond is converted to the oxirane—but when an acetic acid solution of peroxyacetic acid is employed, the oxirane is destroyed as it is formed by further reaction with acetic acid and is converted to the hydroxyacetates.

Later other investigators used a solution of peroxyacetic acid in an inert solvent to epoxidize several other olefins (293–295). The apparent need to use peroxyacetic acid in an inert solvent to obtain good yields of oxirane compounds from olefins was a serious drawback to the general applicability of peroxyacetic acid as an epoxidation reagent since peroxyacetic acid is prepared and used most conveniently in acetic acid solution and is available commercially in that solvent, whereas its isolation free (or substantially free) of acetic acid had been accomplished only with some difficulty and hazard.

In connection with a kinetic study of the reaction of peroxyacetic acid (approximately 1 $M$) in acetic acid solution with various long-chain compounds having isolated, nonterminal double bonds, Findley, Swern, and Scanlan (296) observed that if the reaction temperature is maintained between 20 and 25°, the consumption of peroxyacetic acid is substantially complete in only 2–4 hr when 1.1–1.2 moles of peroxyacetic acid per mole of double bond is employed; 85–95% of the theoretical quantity of peroxyacetic acid is consumed. The products isolated were the oxiranes, containing only small amounts of hydroxyacetates and unreacted olefin. The reaction was shown to be general and afforded a simple and convenient method for the preparation of large quantities of oxiranes; the isolation of peroxy acid and the use of inert solvents were unnecessary in most cases.

Peroxyacetic acid in acetic acid solution was then successfully applied (296) to the epoxidation of a wide variety of long-chain unsaturated compounds (acids, monoesters, triglycerides, alcohols). In a subsequent investigation Swern, Billen, and Scanlan (297) converted a series of 1-olefins to the corresponding oxiranes in fair yields by epoxidation with dilute (about 1 $M$) peroxyacetic acid in acetic acid solution. The terminally unsaturated compounds required about 24 hr of reaction time, compared with 2–4 hr for olefins with nonterminal, isolated double bonds.

Comparison of the reaction conditions employed by Swern and co-workers (296, 297) with those employed by previous investigators revealed the reason for the earlier failures to obtain high yields of oxiranes. Swern and his collaborators employed relatively short reaction times and low temperatures, whereas previous investigators employed either long reaction periods at room temperature or short reaction periods at elevated temperatures, or the solution contained sulfuric acid, which was used as the catalyst in the preparation of peroxyacetic acid.

Thus Findley, Swern, and Scanlan (296) showed that certain long-chain oxiranes dissolved in a large excess of glacial acetic acid are converted to hydroxyacetates at the rate of about 1 % per hour at 25°, but at 65° the conversion to hydroxyacetates is complete in only 4 hr, and at 100° it is complete in 1 hr. Furthermore, when a dilute solution of peroxyacetic acid in a large excess of glacial acetic acid containing 1 % of concentrated sulfuric acid is used in epoxidation reactions, no oxiranes are normally isolated and quantitative yields of hydroxyacetates can be obtained, even when mild reaction conditions are employed (296, 297). The sulfuric acid catalyzes the ring-opening reaction of the oxirane group by the excess acetic acid present. Under the proper reaction conditions, however, peroxyacetic acid in acetic acid solution can be used as a general reagent for the epoxidation of the isolated double bond, and numerous compounds have been so epoxidized, even in the presence of sulfuric acid, provided the amount of acetic acid is minimal and reaction conditions are not too severe.

Although peroxyacetic acid is most conveniently obtained in the laboratory as a solution in acetic acid, commercial procedures based on the oxidation of acetaldehyde permit the preparation of concentrated solutions of peroxyacetic acid in inert solvents, such as ethyl acetate and acetone (298–300). Peroxyacetic acid in these inert organic solvents has been used to epoxidize a wide variety of unsaturated compounds.

Azeotropic methods for preparing peroxyacetic acid (see Volume I) provide ways of obtaining it in aqueous solution (301) or in inert solvents (302, 303). Either method is readily adaptable to laboratory or commercial use. The aqueous solution contains approximately 55 % peroxyacetic acid and 45 % water, with only residual amounts of hydrogen peroxide and acetic acid. The organic solvent systems contain up to 37 % peroxyacetic acid, usually in ethyl, *n*-propyl, or isopropyl acetates. Such solutions are extremely efficient for epoxidation, as ring-opening reactions are minimal. Unsaturated compounds that cannot be epoxidized in high yield by peroxyacetic acid (preformed or prepared *in situ*) in acetic acid can ordinarily be converted to oxiranes in excellent yield by peroxyacetic acid in water or inert organic solvents.

The *in situ* method of preparing and using peroxyacetic acid for epoxidation is used mainly in commercial procedures. The preparation of peroxyacetic acid from hydrogen peroxide and acetic acid is accelerated by sulfuric acid or acid-type ion-exchange resins. The success of this method requires that the content of acetic acid be held to a minimum (about 0.5 mole or less per mole of double bond) in order for ring-opening reactions to be minor. Commercial epoxidations using the *in situ* preparation and consumption of peroxyacetic acid are discussed later.

Table 3 lists over 800 compounds that have been successfully epoxidized with peroxyacetic acid since 1952. References 2, 13, 14, and 15 should be consulted for earlier work. In this table no distinction is made with respect to solvent systems or epoxidation techniques, or, in the case of multiply unsaturated compounds, whether epoxidation is partial or complete. In a few cases acetaldehyde monoperoxyacetate has been used as the peroxyacetic acid precursor. One noteworthy feature of Table 3 is the large number of compounds partially and selectively epoxidized to obtain oxiranes still containing a polymerizable or otherwise useful double bond. Also, many of the references are to the patent literature. The only compounds listed from patents are those found in actual examples

TABLE 3. UNSATURATED COMPOUNDS EPOXIDIZED WITH PEROXYACETIC ACID
SINCE 1952

| Compound | Reference |
|---|---|
| Abietic acid | 304 |
| Acetaldehyde allyl hemiacetal acetate | 305 |
| Acetaldehyde 3-cyclohexenylmethyl hemiacetal acetate | 306 |
| Aldrin | 307, 308 |
| 2-Alkenyl imidazolines | 309 |
| Alkyd resin (soybean oil–pentaerythritol–phthalic anhydride) | 310 |
| Alloocimene | 311 |
| Allyl acetate | 312 |
| Allyl alcohol | 313, 314 |
| *m*-Allylbenzotrifluoride | 315 |
| 2-Allyl-4-*t*-butylphenyl glycidyl ether | 316 |
| Allyl chloride | 317–319 |
| Allyl crotonate | 312, 320–322 |
| Allyl crotyl ether | 312 |
| Allyl β-cyanoethoxyethyl ether | 323 |
| Allyl β-cyanoethyl ether | 323 |
| Allyl 3-cyclohexenecarboxylate | 312, 324 |
| Allyl 4-cyclohexenecarboxylate | 325 |
| Allyl 3-cyclohexenylmethyl ether | 312 |

TABLE 3 (Continued)

| Compound | Reference |
|---|---|
| *N*-Allyl-2-(2-cyclopentenyl)acetamide | 326 |
| Allyl (2-cyclopenten-1-yl)acetate | 326 |
| Allyl 2-cyclopentenyl ether | 327 |
| Allyl 3-cyclopentenyl ether | 312 |
| 2-Allyl-4,6-dichlorophenol | 316 |
| 2-Allyl-4,6-dimethylolphenyl methyl ether | 316 |
| Allyl elaidate | 328, 329 |
| Allyl 2-ethyl-2-hexenoate | 312, 320 |
| Allyl α-ethyl-β-propylacrylate | 322, 330 |
| Allyl ethyl tetrahydrophthalate | 331 |
| Allyl 5-hexenoate | 312 |
| Allyl 2-hydroxy-3-butenoate | 312 |
| Allyl 2-hydroxy-3-pentenoate | 312 |
| Allyl linoleate | 312, 328, 332 |
| Allyl methallyl ether | 312 |
| Allyl 1-methyl-3-cyclohexenecarboxylate | 333 |
| Allyl 6-methyl-3-cyclohexenecarboxylate | 312 |
| Allyl oleate | 312, 328, 329 |
| 2-Allyloxy-3-(2-propenyl)-1,4-dioxane | 324 |
| Allyl 4-pentenoate | 312 |
| *o*-Allylphenol | 335, 336 |
| *o*-Allylphenyl acetate | 336 |
| *o*-Allylphenyl allyl ether | 335, 337, 338 |
| *o*-Allylphenyl glycidyl ether | 316, 338 |
| 2-Allylphenyl methyl ether | 316 |
| Allylphenylpolyethoxy ether | 316 |
| Allyl undecylenate | 312, 328, 329 |
| 3-Benzamidocyclohexene | 339 |
| 4-Benzoylresorcinol monolinoleate | 340 |
| Bicyclic compounds | 341 |
| Bicyclo[2.2.1]hept-2-ene | 201 |
| Bicyclo[3.2.1]oct-2-ene | 342 |
| Bicyclo[4.2.0]oct-3-ene | 343 |
| Bicyclo[4.2.0]oct-7-ene | 343 |
| Bi-1-cyclopenten-3-yl | 344 |
| 1,3-Bis(allyloxy)-2,2-dimethylpropane | 345 |
| Bis(allyloxyethyl)-4-cyclohexene 1,2-dicarboxylate | 346 |
| 1,4-Bis(allyloxymethyl)cyclohexane | 345 |
| 2,5(6)-Bis(allyloxymethyl)norcamphane | 345 |
| 4,6-Bis(allyloxymethyl)-*m*-xylene | 347 |

TABLE 3 (Continued)

| Compound | Reference |
|---|---|
| 1,3-Bis(allyloxy)-2,2,4,4-tetramethylcyclobutane | 345 |
| 1,3-Bis(allyloxy)-2,2,4,4-trimethylpentane | 345 |
| Bis(endo-bicyclo[2.2.1]hept-2-en-5-yl)-1,3-butadiene | 348 |
| Bis(exo-bicyclo[2.2.1]hept-2-en-5-yl)-1,3-butadiene | 348 |
| 3,9-Bis(3-butenyl)spiro(m-dioxane) | 349, 350 |
| 2,2-Bis(4-butoxy-3-allylphenyl)propane | 351 |
| 5,6-Bis(chloromethyl)bicyclo[2.2.1]-2-heptene | 352 |
| Bis[4-(2¹-crotyl)-2,6-dichlorophenyl]-2-ethylhexyl phosphate | 353 |
| Bis(p-crotylphenyl)-2-ethylhexyl phosphate | 353 |
| 1,2-Bis(Δ'-cyclohexenyl)ethanone and monoepoxide | 354 |
| Bis(3-cyclohexenylmethyl)-2-ethylhexyl 1,2,4-butanetricarboxylate | 355 |
| Bis(3-cyclohexenylmethyl) maleate | 356 |
| Bis(3-cyclohexenylmethyl) oxalate | 356 |
| Bis(3-cyclohexenylmethyl) sebacate | 356 |
| Bis(3-cyclohexenylmethyl) terephthalate | 356 |
| Bis(3-cyclohexenylmethyl)vinyl 1,2,4-butanetricarboxylate | 355 |
| 2,3-Bis(3-cyclohexenyloxy)-1,4-dioxane | 334 |
| 3.9-Bis(3-cyclohexenyl)spirobi(m-dioxane) | 349 |
| 2,8-Bis(3-cyclohexen-2-yl)-1,3,7,9-tetraoxaspiro[4.5]decane | 357 |
| 1,3,9-Bis(3-cyclohexyl)spirobi (m-dioxane) | 350 |
| Bis(cyclopent-2-enyl) carbonate | 358 |
| Bis(2-cyclopentenyl) ether | 317, 359, 360 |
| Bis(dihydrodicyclopentadienyl) ether | 361 |
| Bis(2, 5-endomethylene-Δ³-tetrahydrobenzyl) formal | 19 |
| 2,2-Bis(4-epoxypropoxy-3-allylphenyl)propane | 351 |
| 2,2-Bis(4-epoxypropoxy-3,5-diallylphenyl)propane | 351 |
| Bis(2-ethyl-2-hexenyl) adipate | 362 |
| Bis(2-ethyl-2-hexenyl) phthalate | 362 |
| Bis(2-ethyl-2-hexenyl) tetrahydrophthalate | 362 |
| 3,9-Bis(1-ethyl-1-pentenyl)spirobi(m-dioxane) | 349, 350 |
| 1,2-Bis(isopropylidene)-3,3-dimethylcyclopropane | 363 |
| 2,8-Bis(6-methyl-3-cyclohexen-2-yl)-1,3,7,9-tetraoxaspiro[4.5]decane | 357 |
| Bis(4-methylenecyclohexylmethyl) adipate | 364 |
| Bis(4-methylenecyclohexylmethyl) 1,4-cyclohexanedicarboxylate | 364 |
| Bis(4-methylenecyclohexylmethyl) 4,5-epoxy-1,2-cyclohexanedicarboxylate | 364 |
| Bis(4-methylenecyclohexylmethyl) fumarate | 364 |
| Bis(4-methylenecyclohexylmethyl) phthalate | 364 |
| Bis(4-methylenecyclohexylmethyl) terephthalate | 364 |

TABLE 3 (Continued)

| Compound | Reference |
|---|---|
| Bis(2-methyl-2-pentenyl) phthalate | 362 |
| Bis(6-methyl-$\Delta^3$-tetrahydrobenzyl)formal | 19 |
| Bis(6-methyl-1,2,5,6-tetrahydrobenzylidene)-D-glucitol | 365 |
| Bis(6-methyl-1,2,5,6-tetrahydrobenzylidene)pentaerythritol | 365 |
| Bis[4-(2$^1$-propenyl)-2,6-dichlorophenyl]-2-ethylhexyl phosphate | 353 |
| Bis[4-(2$^1$-propenyl)-2,6-dimethylphenyl]-2-ethylhexyl phosphate | 353 |
| Bis[2-(2-propenyl)-3,4-dimethylphenyl]-2-ethylhexyl phosphate | 353 |
| 3,9-Bis(1-propenyl)spirobi(*m*-dioxane) | 349, 350 |
| N,N-Bis($\Delta^3$-tetrahydrobenzyl)acetamide | 366 |
| N,N-Bis($\Delta^3$-tetrahydrobenzyl)cyanamide | 366 |
| Bis($\Delta^3$-tetrahydrobenzyl)formal | 19 |
| N,N-Bis($\Delta^3$-tetrahydrobenzyl)formamide | 366 |
| N,N-Bis(1,2,5,6-tetrahydrobenzyl)formamide | 367 |
| N,N-Bis($\Delta^3$-tetrahydrobenzyl)hexamethylenediurethane | 366 |
| Bis(1,2,5,6-tetrahydrobenzylidene)-D-glucose | 365 |
| Bis(1,2,5,6-tetrahydrobenzylidene)pentaerythritol | 365 |
| N,N-Bis($\Delta^3$-tetrahydrobenzyl)stearamide | 366 |
| N,N-Bis($\Delta^3$-tetrahydrobenzyl)-*p*-toluenesulfonamide | 366 |
| N,N-Bis($\Delta^3$-tetrahydrobenzyl)urethane | 366 |
| Brassidic acid | 232, 368 |
| (1-Bromo-3-cyclohexenyl)methyl 1-bromo-3-cyclohexenecarboxylate | 369 |
| 5-Bromomethylbicyclo[2.2.1]hept-2-ene (*exo* and *endo*) | 344 |
| Butadiene | 45, 312, 313, 317, 319, 359, 370–374 |
| Butadiene monoxide | 317 |
| 1-Buten-3-ol | 375 |
| *cis*-2-Buten-2-ol | 376 |
| *trans*-2-Buten-1-ol | 376 |
| 3-Butenyl methacrylate | 377 |
| 2-(2-Butenyloxy)-3-(2-butenyloxy)-1,4-dioxane | 334 |
| Butenylsuccinic anhydride | 378 |
| 8(9)-(3-Butenyltricyclo[5.2.1.0$^{2,6}$]dec-3-ene | 379 |
| Butyl acetoxyricinoleate | 380 |
| Butyl crotyl phthalate | 381 |
| Butyl 1-cyclohexenecarboxylate | 382, 383 |
| 1-*t*-Butylcyclooctene | 384 |
| Butyl 1-cyclohexenecarboxylate | 382, 383 |
| Butyl 12,13-epoxy oleates | 385 |
| Butyl esters of cottonseed-oil fatty acids | 386–388 |
| Butyl esters of soybean-oil fatty acids | 387 |
| *t*-Butylethylene | 389 |

TABLE 3 (Continued)

| Compound | Reference |
|---|---|
| Butyl α-ethyl-β-propylacrylate | 383 |
| 2-t-Butyl-5,8-methano-4a,5,8,8a-tetrahydro-1,4-<br>naphthoquinone-2,3-epoxide | 390 |
| 4-t-Butylmethylenecyclohexane | 208 |
| Butyl oleate | 310, 386–388, 391–395 |
| n-Butylstyrylcarbinol | 396 |
| N-Butyltetrahydrophthlaimide | 397 |
| N-(n-Butyl)-Δ³-tetrahydrophthalimide | 313 |
| n-Butyl undecylenate | 398 |
| (−)-cis-Δ²-Carene | 399, 400 |
| (+)-Δ⁴-Carene | 399 |
| Castor oil monoglycerides | 401 |
| 3-Chloro-1-butene | 318, 402 |
| (1-Chloro-3-cyclohexenyl)methyl 1-chloro-3-cyclohexenecarboxylate | 369 |
| 2-Chloro-2,6-dimethyl-7-octene | 403 |
| 1-Chloro-2-ethyl-2-hexene | 402 |
| 5-Chloromethylbicyclo[2.2.1]-2-heptene | 352 |
| (1-Chloro-2-methyl-4-cyclohexenyl)methyl 1-chloro-2-methyl-4-<br>cyclohexenecarboxylate | 369 |
| 2-Chloro-1-octene | 402 |
| 2-Chloro-2-octene | 402 |
| 3-Chlorophenylpolyolefin | 404 |
| 3-Chloro-1-propene | 402 |
| Corn oil | 405 |
| Cottonseed oil | 387, 406 |
| Cottonseed-oil fatty acids | 407 |
| Crotyl acrylate | 312, 321 |
| Crotyl crotonate | 312, 321 |
| Crotyl methacrylate | 408 |
| o-Crotylphenol | 336 |
| o-Crotylphenyl acetate | 336 |
| p-Crotylphenyl glycidyl ether | 316 |
| 2-Cyanobicyclo[2.2.1]heptene | 323 |
| Cyano-3-cyclohexene | 323 |
| 2-Cyanophenylpolyolefin | 404 |
| 1,5-Cyclodecadiene | 409 |
| cis-1,5-Cyclododecadiene | 410 |
| trans-1,5-Cyclododecadiene | 410 |
| Cyclododecatriene | 411–413 |
| 1,3-Cycloheptadiene | 414 |

TABLE 3 (Continued)

| Compound | Reference |
|---|---|
| Cycloheptene | 415, 416 |
| 1,3-Cyclohexadiene | 414 |
| 1,4-Cyclohexadiene | 67 |
| Cyclohexane-1,2-dimethanol dilinoleate | 417 |
| 3-Cyclohexane-1,1-dimethanol dilinoleate | 418 |
| Cyclohexane-1,2-dimethanol dioleate | 417 |
| 3-Cyclohexane-1,1-dimethanol esters of linseed-oil fatty acids | 418 |
| Cyclohexene | 312, 319, 416, 419, 420 |
| 3-Cyclohexenecarbonitrile | 317, 359 |
| 3-Cyclohexenecarboxaldehyde acetals | 421 |
| 4-Cyclohexene-1,2-dicarboximide | 397 |
| 3-Cyclohexene-1,1-dimethanol dicrotonate | 422–425 |
| 3-Cyclohexene-1,1-dimethanol dilinoleate | 426 |
| 4-Cyclohexene-1,2-dimethanol dilinoleate | 417 |
| 3-Cyclohexene-1,1-dimethanol dilinoleate | 426 |
| 3-Cyclohexene-1,1-dimethanol dioleate | 423, 424, 427 |
| 4-Cyclohexene-1,2-dimethanol dioleate | 417 |
| 3-Cyclohexene-1,1-dimethanol ditallate | 423, 424 |
| 3-Cyclohexene-1,1-dimethanol diundecenoate | 423, 424 |
| 5-(3-Cyclohexenyl)bicyclo[2.2.1]-2-heptene | 352 |
| 3-Cyclohexenylmethyl acetate | 428 |
| 3-Cyclohexenylmethyl acrylate | 312, 428 |
| 3-Cyclohexenylmethyl-2-(2-butoxyethoxy)ethyl ether | 428 |
| 3-Cyclohexenylmethyl bis(decyl)1,2,4-butanetricarboxylate | 355 |
| 3-Cyclohexenylmethyl bis(2-ethylhexyl)1,2,4-butanetricarboxylate | 355 |
| 3-Cyclohexenylmethyl 3-cyclohexenecarboxylate | 429–431 |
| 3-Cyclohexenylmethyl linoleate | 432 |
| 3-Cyclohexenylmethyl methacrylate | 428 |
| Cyclohexyl 12,13-epoxyoleate | 385 |
| 1,3-Cyclooctadiene | 414 |
| 1,5-Cyclooctadiene | 414, 433–436 |
| Cyclooctatetraene dichloride | 343 |
| Cyclooctene | 437 |
| Cyclopentadiene | 414, 438–440 |
| Cyclopentene | 416 |
| 2-Cyclopentenol | 327 |
| 2-(2-Cyclopentenyl)acetamide | 326 |
| 2-Cyclopentenyl acrylate | 312 |
| 2-Cyclopentenyl bis(2-ethylhexyl)1,2,4-butanetricarboxylate | 355 |
| 2-Cyclopentenyl crotonate | 312, 327 |

TABLE 3 (Continued)

| Compound | Reference |
| --- | --- |
| N-[2-Cyclopentenyl-2-(2-cyclopentenyl)]acetamide | 326 |
| 2-Cyclopentenyl (2-cyclopenten-1-yl)acetate | 326 |
| Dehydrochlorinated polyvinyl chloride | 80 |
| 1,1-Diacetoxy-2-ethyl-2-butene | 441 |
| 1,1-Diacetoxy-2-ethyl-2-hexene | 441 |
| 3$\beta$,20-Diacetoxy-5,16,20-pregnatriene | 442 |
| 3$\alpha$,20-Diacetoxy-17(20)-pregnen-11-one | 443 |
| N,N$^1$-Diacetyl-N,N$^1$-bis($\Delta^3$-tetrahydrobenzyl)ethylenediamine | 366 |
| N,N$^1$-Diacetyl-N,N$^1$-bis($\Delta^3$-tetrahydrobenzyl)hexamethylenediamine | 366 |
| N,N$^1$-Diacetyl-N,N$^1$-bis($\Delta^3$-tetrahydrobenzyl)-p-phenylenediamine | 366 |
| Diacetylated monoglyceride esters of unsaturated fatty acids | 444–449 |
| Diacetylated monoolein | 445 |
| Diallyl p-cresyl allyl ether | 338 |
| Diallyl 4-cyclohexene-1,2-dicarboxylate | 346 |
| Diallyl dilinoleate | 346 |
| Diallyldimethyldisiloxane | 450 |
| Diallyl 8,12-eicosadienoate | 451 |
| Diallyl ether | 312 |
| Diallylidenepentaerythritol | 452 |
| Diallyl maleate | 312 |
| 2,6-Diallyl-4-methoxyphenyl glycidyl ether | 316 |
| Diallyl-p-t-octylphenyl allyl ether | 338 |
| Diallyl-o-phenylphenyl allyl ether | 338 |
| 2,6-Diallylphenyl butyl ether | 316 |
| 2,6-Diallylphenyl chloropropyl ether | 316 |
| 2,6-Diallylphenyl glycidyl ether | 316 |
| Diallylphenylphosphine oxide | 89 |
| Diallyl phenylphosphonate | 453 |
| Diallyl terephthalate | 454, 455 |
| trans-1,4-Dibromo-2-butene | 456 |
| 22,23-Dibromoergosta-7,9(11)-dien-3$\beta$-yl acetate | 457 |
| Dibutyl 2-butenylsuccinate | 458 |
| Dibutyl 4-cyclohexene-1,2-dicarboxylate | 459 |
| Dibutyl dodecenylsuccinate | 458, 460 |
| Dibutyl 4-methyl-4-cyclohexene-1,2-dicarboxylate | 459 |
| Dibutyl tetrahydrophthalate | 461 |
| Di-n-butyl tetrahydrophthalate | 331 |
| Di-n-butyl tetrahydrophthalate | 331 |
| 1,4-Dichloro-2-butene | 318, 402, 462, 463 |
| 3,4-Dichloro-1-butene | 318, 402 |

TABLE 3 (Continued)

| Compound | Reference |
|---|---|
| 6,7-Dichloro-4*a*,8*a*-epoxy-1,4,4*a*,8*a*-tetrahydro-1,4-<br>methanonaphthalene-5,8-dione | 464 |
| *trans*-5,6-Dichloronorbornene | 465 |
| Dicrotonylidenepentaerythritol | 452 |
| Dicrotylphenyl glycidyl ether | 338 |
| Dicrotyl phthalate | 381 |
| 1,2-Dicyano-4-cyclohexene | 466 |
| 1,1-Dicyano-2,2-bis(trifluoromethyl)ethylene | 467 |
| Di(2-cyclohexenyl)ether | 468 |
| Di(cyclohexenylmethyl)2-ethylhexyl 1,2,4-butanetricarboxylate | 469 |
| Di(cyclohexenylmethyl 1,2,4-butanetricarboxylate | 469 |
| Dicyclohexenylpropane | 470 |
| Dicyclopentadiene | 312, 471 |
| Dicyclopentadiene–bisphenol A adducts | 472 |
| Dicyclopentadienyl oleate | 473 |
| Di(2-cyclopenten-1-yl) ether | 474 |
| Didecyl butenylsuccinate | 458, 460 |
| Didecyl 4-cyclohexene-1,2-dicarboxylate | 459 |
| Didecyl hexenylsuccinate | 458, 460 |
| Didecyl 4-methyl-4-cyclohexene-1,2-dicarboxylate | 459 |
| Didecyl(oxo)cyclohexenylmethyl 1,2,4-butanetricarboxylate | 469 |
| Didecyl 4-tetrahydrophthalate | 475 |
| 1,3-Diepoxypropoxy-2,5-diallylbenzene | 351 |
| 1,4-Diepoxypropoxy-2,3,5,6-tetrallylbenzene | 351 |
| Diethyl allylmalonate | 476 |
| Diethyl 4-cyclohexene-1,2-dicarboxylate | 459 |
| Diethyleneglycol bis(3-cyclohexenecarboxylate) | 324, 477, 478 |
| Diethyleneglycol bis(6-methyl-3-cyclohexenecarboxylate) | 324, 477–479 |
| Di-2-ethylhexyl 2-butenylsuccinate | 458 |
| Di-2-ethylhexyl 4-cyclohexene-1,2-dicarboxylate | 459 |
| Di(2-ethylhexyl)cyclohexenylmethyl 1,2,4-butanetricarboxylate | 469 |
| Di(2-ethylhexyl) heptenylsuccinate | 458, 460 |
| Di(2-ethylhexyl) hexenylsuccinate | 458, 460 |
| Di-2-ethylhexyl 4-methyl-4-cyclohexene-1,2-dicarboxylate | 459 |
| Di(2-ethylhexyl) nonenylsuccinate | 458, 460 |
| Di(2-ethylhexyl) octenylsuccinate | 458, 460 |
| Di(2-ethylhexyl) pentenylsuccinate | 458, 460 |
| Diethyl α-ethylidene-β-methylglutarate | 382 |
| Diethyl 3,6-methylene-1,2,3,6-tetrahydrophthalate | 480 |
| 3,6-Diethyl-2,6-octadien-4-yne | 481 |

TABLE 3 (Continued)

| Compound | Reference |
|---|---|
| Diethyl tetrahydrophthalate | 331 |
| Diethyl tetramethyl-8,12-eicosadiene-1,20-dioate | 482 |
| 2,5-Dihydrothiophene-1,1-dioxide | 483 |
| 3,9-Diisopropenylspirobi(*m*-dioxane) | 349, 350 |
| Dilinoleyl 4-cyclohexene-1,2-dicarboxylate | 416 |
| Dilinoleyl hexahydrophthalate | 417 |
| 3′,6′-Dimethoxybenzonorbornadiene | 484 |
| 6,6-Dimethyl-2-acetoxymethylcibyclo[3,1.1]heptene-2 | 485 |
| Dimethyl allylmalonate | 476 |
| 2,4-Dimethyl-4-allyloxy-2-pentene | 486 |
| 2,4-Dimethyl-4-benzyloxy-2-octene | 486 |
| 6,6-Dimethyl-2-butoxymethylbicyclo[3.1.1]heptene-2 | 485 |
| 2,4-Dimethyl-4-(3-chloroethoxy)-2-hexene | 486 |
| 6,6-Dimethyl-2-chloromethylbicyclo[3.1.1]heptene-2 | 485 |
| Dimethyl 4-cyclohexene-1,2-dicarboxylate | 459 |
| 4,5-Dimethyl-4-*cis*-cyclohexene-1,2-dicarboxylic anhydride | 487 |
| 5,5-Di(6-methyl-3-cyclohexenylmethoxy)-1-pentene | 488 |
| 2,4-Dimethyl-2,4-decadiene | 489 |
| 4,7-Dimethyl-3,7-decadien-5-yne | 481 |
| Dimethyl dichloro-8,12-eicosadiene-1,20-dioate | 482 |
| Dimethyl dimethyl-8,12-eicosadiene-1,20-dioate | 482 |
| Dimethyl 8,12-eicosadiene-1,20-dioate | 482 |
| Dimethyl 1,4-endomethylene-5-cyclohexene-2,3-dicarboxylates | 490 |
| Dimethyl 1,4-endomethylene-2-methyl-5-cyclohexene-2,3-dicarboxylates | 491 |
| 2,4-Dimethyl-4-ethoxy-2-octene | 486 |
| 1,2-Dimethyl-4-formylcyclohexene | 492 |
| 2,4-Dimethyl-2,4-heptadiene | 489 |
| 2,4-Dimethylhexadiene | 489 |
| 2,4-Dimethyl-2,4-hexadiene | 493 |
| 2,5-Dimethyl-2,4-hexadiene | 414 |
| 2,5-Dimethyl-1,5-hexadiene | 312 |
| 6,6-Dimethyl-2-hydroxymethylbicyclo[3.1.1]heptene-2 | 485 |
| 6,6-Dimethyl-2-isobutoxymethylbicyclo[3.1.1]heptene-2 | 485 |
| 6,6-Dimethyl-2-isobutylbicyclo[3.1.1]heptene-2 | 485 |
| 6,6-Dimethyl-2-isopropoxymethylbicyclo[3.1.1]heptene-2 | 485 |
| 6,6-Dimethyl-2-methoxymethylbicyclo[3.1.1]heptene-2 | 485 |
| 2,3-Dimethyl-4-methoxy-2-octene | 486 |
| Di(8-methylnonyl) 3,6-endomethylenetetrahydrophthalate | 475 |
| Di(8-methylnonyl) 4-tetrahydrophthalate | 475 |
| 2,4-Dimethyl-2,4-octadiene | 493 |

TABLE 3 (Continued)

| Compound | Reference |
|---|---|
| Dimethyl 7,11-octadecadiene-1,18-dioate | 482 |
| 3,6-Dimethyl-2,6-octadien-4-yne | 494 |
| 2,6-Dimethyl-7-octen-2-ol | 403 |
| 1,1-Dimethylol-3-cyclohexene esters | 495 |
| 1,1-Dimethylol-2,5-endomethylene-3-cyclohexene esters | 495 |
| 2,4-Dimethylpentadiene | 489 |
| 2,2-Dimethyl-4-pentenyl acetate | 496 |
| 2,2-Dimethylpropanediol bis(2-cyclopenten-1-yl)acetate | 326 |
| 6,6-Dimethyl-2-propylbicyclo[3.1.1]heptene-2 | 485 |
| 6,9-Dimethyl-5,9-tetradecadien-7-yne | 481 |
| 4,4-Dimethyl-6,7,8,8a-tetrahydro-3H-2-benzopyranone | 497 |
| Dimethyl 4-tetrahydrophthalate | 475 |
| 1,4-Dioctyloxy-2,6-di(2-butenyl)benzene | 498 |
| Dioctyl tetrahydrophthalate | 333 |
| Dioleyl 4-cyclohexene-1,2-dicarboxylate | 416 |
| Dioleyl hexahydrophthalate | 417 |
| 1,3-Dioxep-5-ene (and 2-substituted derivatives) | 499 |
| Dipentaerythritol esters of soybean-oil fatty acids | 393 |
| Dipentaerythritol hexalinoleate | 500 |
| Dipentaerythritol hexaoleate | 500 |
| 4,7-Dipropyl-3,7-decadien-5-yne | 501 |
| Ditridecyl 4-cyclohexene-1,2-dicarboxylate | 459 |
| Ditridecyl 4-methyl-4-cyclohexene-1,2-dicarboxylate | 459 |
| Divinylbenzene | 317, 359 |
| Divinylbenzenes (*ortho, meta, para*) | 502 |
| Divinyldiacetoxysilane | 503 |
| 3,9-Divinylspirobi(m-dioxane) | 349, 350 |
| 1,11-Dodecadiene | 106 |
| Dodecenylsuccinic anhydride | 378 |
| Dodecyl tetradecyl 4-tetrahydrophthalate | 475 |
| 8,12-Eicosadiene-1,20-dioic acid | 482 |
| 5-Eicosenoic acid | 504 |
| 8-Eicosenyl methacrylate | 408 |
| 1,4-Endomethylene-5-cyclohexene-2,3-dicarboxylic acid anhydrides | 505 |
| 1,4-Endomethylene-2-methyl-5-cyclohexene-2,3-dicarboxylic acid | 491 |
| 1,4-Endomethylene-2-methyl-5-cyclohexene-2,3-dicarboxylic acid anhydride | 491 |
| 3,4-Epoxy-6-methylcyclohexane methylcrotonate | 320 |
| (+)-12,13-Epoxy-cis-9-octadecenoic acid | 506 |
| 2,3-Epoxypropyl methylcyclohexenyl ether | 468 |

TABLE 3 (Continued)

| Compound | Reference |
|---|---|
| 2,3-Epoxypropyl-2-cyclohexenyl ether | 468 |
| 2,3-Epoxypropylcyclopentenyl ether | 468 |
| Erucamide | 507 |
| Erucic acid | 232, 368 |
| Ethyl 2-(2-acetoxyethoxy)-12-acetoxyricinoleate | 508 |
| Ethyl acrylate | 312, 382, 383 |
| 2-Ethyl-2-butene-1,1-diol diacetate | 509 |
| 2-Ethylbutyl oleate | 510, 511 |
| Ethyl cinnamate | 382, 383, 512 |
| Ethyl crotonate | 312, 317, 359, 382, 383, 513 |
| Ethyl 3-cyclohexenecarboxylate | 312, 324 |
| Ethyl 4-cyclohexenecarboxylate | 325 |
| Ethyl $\beta,\beta$-diethylacrylate | 383 |
| Ethyl $\beta,\beta$-dimethylacrylate | 382, 383 |
| Ethylene | 419 |
| $N,N'$-Ethylene bis(2-cyclopentenylacetamide) | 326 |
| Ethylene bis(soybean-oil fatty acid) amides | 514 |
| Ethyleneglycol bis(allylphenyl) ether | 316 |
| Ethyleneglycol bis(3-cyclohexenecarboxylate) | 324, 477, 478 |
| Ethyleneglycol bis(2-cyclopenten-1-yl)acetate | 326 |
| Ethyleneglycol bis[2-(2,4-dioxaspiro[5.5]undec-9-en-3-yl]ethyl ether | 515 |
| Ethyleneglycol bis[1-methyltetrahydrobenzoate) | 333 |
| Ethyleneglycol dicrotonate | 312, 320, 383 |
| Ethyleneglycol dicyclohexenyl ether | 468 |
| Ethyleneglycol dilinoleate | 500 |
| Ethyleneglycol dioleate | 500 |
| Ethyleneglycol–epichlorohydrincyclohexenyl alcohol mixed ethers | 468 |
| Ethyleneglycol esters of tall-oil fatty acids | 393 |
| Ethyleneglycol monobutyl ether oleate | 516 |
| Ethyleneglycol monoethyl ether oleate | 516 |
| Ethyl $\beta$-ethylacrylate | 383 |
| Ethyl 2-ethyl-2-hexenoate | 312, 382 |
| Ethyl $\alpha$-ethyl-$\beta$-propylacrylate | 383 |
| Ethyl 2,4-hexadienoate (sorbate) | 383, 517 |
| 2-Ethyl-1,3-hexanediol bis(3-cyclohexenecarboxylate) | 324, 477, 478 |
| 2-Ethylhexene | 518, 519 |
| 2-Ethyl-2-hexene-1,1-diol diacetate | 509 |
| 2-Ethyl-2-hexenol | 317, 359 |
| 2-Ethyl-2-hexenyl acetate | 362 |
| 2-Ethyl-2-hexenyl acrylate | 312 |

TABLE 3 (Continued)

| Compound | Reference |
|---|---|
| 2-Ethyl-2-hexenyl benzoate | 362 |
| 2-Ethyl-2-hexenyl chloride | 318 |
| 2-Ethyl-2-hexenyl 2-ethylhexyl ether | 362 |
| 2-Ethyl-2-hexenyl bis(2-ethylhexyl)phosphate | 362 |
| 2-Ethylhexyl crotonate | 383 |
| 2-Ethylhexyl decyl nonenylsuccinate | 458, 460 |
| 2-Ethylhexyl decyl octenylsuccinate | 458, 460 |
| 2-Ethylhexyl 12,13-epoxyoleate | 385 |
| N-(2-Ethylhexyl)-1-methyl-4-cyclohexene-1,2-dicarboximide | 397 |
| 2-Ethylhexyl α-methyl-β-ethylacrylate | 383 |
| 2-Ethylhexyl oleate | 510, 511, 520 |
| Ethyl α-methylcinnamate | 382, 383 |
| Ethyl β-methyl-β-ethylacrylate | 383 |
| Ethyl 2-pentenoate | 382 |
| 2-(1-Ethyl-1-pentenyl)-1,3-dioxolane | 441 |
| 2-(1-Ethyl-1-pentenyl)-4-methyl-1,3-dioxolane | 441 |
| 3-Ethyl-3-penten-1-yne | 521 |
| Ethyl α-phenylcrotonate | 382, 383 |
| Ethylstyrylcarbinol | 396 |
| *Euphorbia lagascae* oil | 522 |
| *Euphorbia lathyris* oil | 523 |
| *trans,trans*-Farnesyl methyl ether | 524 |
| β-Formoxystyrene | 525 |
| Glyceryl monooleate | 401 |
| Glyceryl trilinoleate | 500 |
| Glyoxal tetra($\Delta^3$-tetrahydrobenzyl)acetal | 19 |
| Grapeseed oil | 119 |
| Haloalkenes | 526 |
| Hendecenamide | 408 |
| 1,6-Heptadiene | 436 |
| 1-Hepten-3-ol | 375 |
| 2-Hepten-4-ol | 527 |
| 1,2,3,4,7,7-Hexabromo-5-methylene-2-norbornene | 120 |
| Hexachlorocyclopentadiene | 528 |
| 1,2,3,4,9,9-Hexachloro-5,8-epoxy-1,4,4a,5,8,8a-hexahydro-1,4-methanonaphthalene | 529 |
| 1,2,3,4,7,7-Hexachloro-5-methylene-2-norbornene | 120 |
| 2,6,10,15-Hexadeca-2,10-diene | 524 |
| 1-Hexadecene | 530, 531 |
| 1,5-Hexadiene | 436 |

TABLE 3 (Continued)

| Compound | Reference |
|---|---|
| 1,4,5,8,9,10-Hexahydroanthracene | 532, 533 |
| 3a,4,5,6,7,7a-Hexahydro-4,7-methanoinden-5(6)-ol diesters | 534 |
| 1,2,3,4,5,6-Hexamethylbicyclo[2.2.0]hexa-2,5-diene | 122 |
| endo-1,2,3,4,5,6-Hexamethylbicyclo[2.2.0]hex-2-ene | 122 |
| Hexamethyl dimethylenecyclopropane | 535 |
| 1,6-Hexanediol bis(3-cyclohexenecarboxylate) | 324, 477, 478 |
| 1,2,6-Hexanetriol tris(2-cyclopenten-1-yl)acetate | 326 |
| 1,2,6-Hexanetriol tris(1-ethyl-2-cyclohexenecarboxylate) | 536 |
| 1,2,6-Hexanetriol tris(6-methyl-3-cyclohexenecarboxylate) | 536 |
| 3-Hexenoic acid | 497 |
| 1-Hexen-3-ol | 375 |
| 2-Hexen-4-ol | 527 |
| 5-Hexenyl methacrylate | 377 |
| 2-(2-Hexenyloxy)-3-(2-hexenyloxy)-1,4-dioxane | 334 |
| N-(Hexoxyethoxypropyl)bicyclo[2.2.1]-5-heptene-2,3-dicarboximide | 480 |
| n-Hexyl 12,13-epoxyleate | 385 |
| Hexyl oleate | 406 |
| Hexyl tallate | 392 |
| 3-Hydroxybutoxyethyl-2,4-dioxaspiro[5.5]undecene-9 | 537 |
| 2-(1-Hydroxybutyl)furan | 538 |
| 3-[2-(Hydroxy-3-chloropropoxy)ethyl-2,4-dioxaspiro[5.5]undecene-9 | 515 |
| 3-Hydroxyethoxyethyl-2,4-dioxaspiro[5.5]undecene-9 | 537 |
| 3-Hydroxyethyl-2,4-dioxaspiro[5.5]undecene-9 | 537 |
| 3-Hydroxyglyceryloxyethyl-2,4-dioxaspiro[5.5]undecene-9 | 537 |
| 2-(1-Hydroxypentyl)furan | 538 |
| 2-(1-Hydroxypropyl)furan | 538 |
| 3-Hydroxysorbitoxyethyl-2,4-dioxaspiro[5.5]undecene-9 | 537 |
| Isobutylene | 152, 317, 319, 359 |
| Isodecyl oleate | 539 |
| Isoheptene | 540 |
| (−)-trans-Isolimonene | 541 |
| Isooctyl esters of tall-oil fatty acids | 539 |
| Isooctyl oleate | 393, 539, 542 |
| Isooctyl oleate and linoleate | 539 |
| Isoprene | 319 |
| 2-Isopropenyl-1,3-dioxolane | 441 |
| Lard oil | 500 |
| d-Limonene | 543, 544 |
| (+)-Limonene 1,2-epoxide (cis and trans) | 545 |
| Linoleamide | 398 |

TABLE 3 (Continued)

| Compound | Reference |
|---|---|
| Linoleic acid | 546–548 |
| Linseed oil | 392, 404, 549 |
| Linseed oil monoesters of glycerol, ethylenediglycol, trimethylolpropane, 2,2-dimethylol-3-butanol, 1,2,4-butanetriol and pentaerythritol | 401 |
| Maleic anhydride-*trans*-1-vinyl-6-acetoxy-9-methyl-$\Delta^1$-octahydronaphthalene adduct | 549 |
| Menhaden oil | 405 |
| $\Delta^1$-*p*-Menthene | 550 |
| (−)-*trans-p*-Menth-2-en-8-ol | 540 |
| Mesityl oxide | 313, 551, 552 |
| Methallyl chloride | 313, 318, 404 |
| 2-Methallyl crotonate | 321 |
| 2-Methallyl $\beta$-cyanoethyl ether | 323 |
| Methallyl (2-cyclopent-1-yl) acetate | 326 |
| *N*-Methallylmaleimide | 553 |
| Methallyl methacrylate | 408 |
| 2-Methallyl-3-(2-methyl-2-propenoxy)-1,4-dioxane | 334 |
| 5-Methallyloxybicyclo[2.2.1]-2-heptene | 352 |
| 1,4(Methallyloxymethyl)cyclohexane | 345 |
| 2-Methoxymethyl-2,4-dimethyl-1,5-pentanediol bis(3-cyclohexenecarboxylate) | 324, 477, 478 |
| 3-Methoxy-*m*-tolylpolyolefin | 404 |
| Methyl abietate | 304 |
| 2-Methyl-1-(1-acetoxyethylidene)cyclohexane | 554 |
| Methyl acetoxyricinoleate | 555, 556 |
| Methyl brassidate | 556 |
| 2-Methyl-2-butene | 557 |
| 2-Methyl-3-buten-2-ol | 376 |
| 1-Methyl-1-butenyl formate | 525 |
| 3-Methylbicyclo[2.2.1]-5-heptene-2-carboxaldehyde di(1-methyl-3-cyclohexenylmethyl)acetal | 488 |
| Methyl crotonate | 383 |
| 1-Methylcycloheptene | 384 |
| (+)-4-Methylcyclohexene | 558 |
| 6-Methyl-3-cyclohexenealdehyde di(6-methyl-3-cyclohexenylmethyl)acetal | 488 |
| 6-Methyl-3-cyclohexenecarboxaldehyde diallyl acetal | 488 |
| 6-Methyl-3-cyclohexenecarboxylic acid | 324 |
| 6-Methylcyclohexene-4-carboxylic acid | 325 |

Table 3 (Continued)

| Compound | Reference |
|---|---|
| 6-Methyl-3-cyclohexenemethyl 2,3-epoxybutyrate | 320 |
| 6-Methyl-3-cyclohexenemethyl 2-ethyl-2-hexenoate | 320 |
| 2-(6-Methyl-3-cyclohexenylmethoxy)-5-(6-methyl-3-cyclohexenylmethoxy-1,4-dioxane) | 334 |
| 6-Methyl-3-cyclohexenylmethyl acrylate | 312 |
| 6-Methyl-3-cyclohexenylmethyl crotonate | 312, 322 |
| 6-Methyl-3-cyclohexenylmethyl $\alpha$-ethyl-$\beta$-propylacrylate | 322 |
| 1-Methylcyclohexenylmethyl 1-methylcyclohexenecarboxylate | 429 |
| 6-Methylcyclohexenylmethyl 6-methylcyclohexenecarboxylate | 429 |
| 6-Methyl-3-cyclohexenylmethyl 6-methyl-3-cyclohexenecarboxylate | 317, 359 |
| 1,2,3,4- or 6-Methyl-3-cyclohexenylmethyl 1,2,3,4, -or 6-methyl-3-cyclohexenecarboxylate | 431 |
| 6-Methyl-3-cyclohexenyl oleate | 429 |
| 1-Methylcyclooctene | 384 |
| Methylcyclopentadiene | 440 |
| 1-Methylcyclopentene | 384 |
| N-Methyl-2-(2-cyclopentenyl)acetamide | 326 |
| 2-Methyl-4,6-di($\Delta$-cyclohexenyl)trioxan | 559 |
| Methyl dihydrocyclopentadienyl ether | 560 |
| 3'-Methyldiphenylpolyolefin | 404 |
| Methyl 5-eicosenoate | 504 |
| Methyl elaidate | 561 |
| 5-Methylenebicyclo[2.2.1]-2-heptene | 352 |
| trans-2-Methylenedecahydronaphthalene | 208 |
| trans-9(10)-Methyl-2-methylenedecahydronaphthalene | 208 |
| endo-3,5-Methylene-1,2,3,6-tetrahydrophthalic anhydride | 480 |
| exo-3,5-Methylene-1,2,3,6-tetrahydrophthalic anhydride | 480 |
| Methyl 12,13-epoxyoleate | 385 |
| Methyl erucate | 507, 562 |
| Methyl esters of cottonseed-oil fatty acids | 387 |
| Methyl esters of soybean-oil fatty acids | 387 |
| Methyl 7-ethyl-3,11-dimethyl-2,6,10-tridecatrienoate | 563 |
| $\beta$-Methyl-$\beta$-formoxystyrene | 525 |
| 6-Methyl-5-hepten-2-one | 564 |
| Methyl-2,4-hexadienoate (sorbate) | 383, 517 |
| Methyl-2-ethyl-2-hexenoate | 382 |
| Methyl $\alpha$-ethyl-$\beta$-propylacrylate | 383 |
| 5-Methyl-1,3,6-heptatriene | 565 |
| 6-Methyl-2-hepten-4-ol | 527 |
| 5-Methyl-1-hexen-3-ol | 375 |

TABLE 3 (Continued)

| Compound | Reference |
|---|---|
| 5-Methyl-2-hexen-4-ol | 527 |
| 2-(4-Methyl-1-hydroxypentyl)furan | 538 |
| Methyl isopropenyl ketone | 551 |
| Methyl linoleate | 500, 566–572 |
| Methyl methacrylate | 382, 383 |
| 2-Methyl-2-nonene | 524 |
| 3-Methyl-2-nonene | 524 |
| 7-Methyl-2-octen-4-ol | 527 |
| Methyl oleate | 380, 387, 561, 566, 571, 573–575 |
| 3-Methyl-1,5-pentanediol bis(3-cyclohexenecarboxylate) | 324, 477, 478 |
| cis-2-Methyl-3-penten-2-ol | 376 |
| trans-2-Methyl-3-penten-2-ol | 376 |
| 4-Methyl-1-penten-3-ol | 375 |
| 4-Methyl-3-penten-2-one | 576 |
| Methyl petroselinate | 577 |
| N-α-Methylphenethyl-Δ⁴-cis-hexahydrophthalimide | 149 |
| 4-Methyl-2-phenyl-3-penten-2-ol | 578 |
| 2-Methyl-2-propene-1,1-diol diacetate | 509 |
| 2-Methyl-2-propenyl acrylate | 312 |
| 2-Methyl-2-propenyl crotonate | 312 |
| 1-Methyl-1-propenyl formate | 525 |
| Methyl ricinoleate | 569 |
| α-Methylstyrene | 540 |
| Methylstyrylcarbinol | 396 |
| 2-Methylsulfolene | 579 |
| 3-Methylsulfolene | 579 |
| 6-Methyl-Δ³-tetrahydrobenzaldehyde trimer | 559 |
| 3-[2-(6-Methyl-Δ³-tetrahydrobenzyloxy)-2-methylethyl]-2-4-dioxaspiro[5.5]undecene-9 | 515 |
| 10-Methyl-10-undecenyl methacrylate | 408 |
| Methyl vernolate | 570 |
| 2-Methyl-3-vinylbicyclo[2.2.1]heptane | 341 |
| Monochloroacetaldehyde bis(6-methyl-Δ³-tetrahydroobenzyl)acetal | 19 |
| Myrcene | 580, 581 |
| Neatsfoot oil | 500 |
| 4-Nitrocyclohexene | 582 |
| cis-13-Nonadecen-10-ynoic acid | 546 |
| 2-Nonen-4-ol | 527 |
| Norbornadiene | 583, 584 |
| 7,11-Octadecadiene-1,18-dicarboxylic acid | 585 |

TABLE 3 (Continued)

| Compound | Reference |
|---|---|
| *cis,trans*-Octadecadienoic acid Diels–Alder adducts | 586 |
| 17-Octadecen-9,11-diynoic acid (isanic acid) | 262 |
| *cis*-9-Octadecene | 586 |
| *trans*-6-Octadecenoic acid (petroselaidic aid) | 232, 264, 587, 588 |
| *cis*-9-Octadecenoic acid (oleic acid) | 232, 307, 313, 317, 359, 407, 419, 561, 589–592 |
| *trans*-9-Octadecenoic acid (elaidic acid) | 232, 505 |
| 9-Octadecenyl crotyl phthalate | 381 |
| *cis*-9-Octadecen-12-ynoic acid | 593 |
| *cis*-12-Octadecen-9-ynoic acid | 546 |
| Octadecyl crotyl phthalate | 381 |
| *anti*-$\Delta^4$-Octalin-1-carboxylic acid | 594 |
| *syn*-$\Delta^4$-Octalin-1-carboxylic acid | 594 |
| *cis-anti*-$\Delta^4$-Octalin-1,2-dicarboxylic acid anhydride | 594 |
| *cis-syn*-$\Delta^4$-Octalin-1,2-dicarboxylic acid anhydride | 594 |
| 1-Octene | 595 |
| *cis*-4-Octene | 267, 596 |
| *trans*-4-Octene | 267 |
| 2-Octen-4-ol | 527 |
| Octyl crotyl phthalate | 381 |
| Octyl decyl 4-tetrahydrophthalate | 597 |
| Octyl 12,13-epoxyoleate | 390 |
| Oleamide | 313 |
| 1-Olefins ($C_{11-15}$, $C_{15-18}$, $C_{15-20}$ fractions) | 531 |
| Oleonitrile | 455 |
| Oleyl acetate | 577 |
| Oleyl alcohol | 455 |
| Oleyl butyrate | 510, 511, 598 |
| Oleyl methacrylate | 408 |
| Oleyl oleate | 387, 392 |
| Olive oil | 387, 549 |
| Patchoulene | 599, 600 |
| Peanut oil | 571 |
| Pentachlorostyrene | 148 |
| *m*-Pentadecenylphenyl glycidyl ether | 85 |
| Pentaerythritol esters of unsaturated fatty acids | 601 |
| Pentaerythritol oleate–benzoate | 602 |
| Pentaerythritol tetrakis (acrylate) | 603 |
| Pentaerythritol tetrakis(crotonate) | 603 |
| Pentaerythritol tetrakis(2-ethyl-2-hexenoate) | 320 |

TABLE 3 (Continued)

| Compound | Reference |
|---|---|
| Pentaerythritol tetrakis(α-ethyl-β-propylacrylate) | 603 |
| Pentaerythritol tetralinoleate | 500 |
| Pentaerythritol tetraoleate | 500 |
| 1,5-Pentanediol bis(3-cyclohexenecarboxylate) | 324, 477, 478 |
| 1,5-Pentanediol bis(2-ethyl-2-hexenoate) | 320 |
| 1,5-Pentanediol bis(2-ethyl-3-propyl acrylate) | 383 |
| 4-Pentenyl acrylate | 377 |
| 3-Pentenyl methacrylate | 408 |
| 4-Pentenyl methacrylate | 377 |
| 4-Pentenyl 4-pentenoate | 604 |
| Perilla oil | 500 |
| Petroselinic acid | 264, 601, 605 |
| N-Phenethylbicyclo[2.2.1]hept-5-ene 2,3-dicarboximide (*exo* and *endo*) | 588 |
| N-Phenethyl-*cis*-Δ⁴-tetrahydrophthalimide | 149 |
| 2-Phenyl-2-butene | 504, 557 |
| 1-Phenyl-3-butoxy-1-butene | 606 |
| α-Phenyl-*trans*-cinnamide | 607 |
| Phenyl-2-cyclopentenyl ether | 327 |
| 1-Phenyl-3-ethoxy-1-butene | 606 |
| 1-Phenyl-3-ethoxy-1-hexene | 606 |
| 1-Phenyl-3-ethoxy-1-pentene | 606 |
| 1-(Phenylethynyl)cyclohexene | 494 |
| 4-Phenylethynyl-3-heptene | 156 |
| 1-Phenyl-1,5-hexadien-3-ol | 608 |
| 1-Phenyl-1-hexyn-3-ene | 609 |
| 1-Phenyl-3-methoxy-1-butene | 606 |
| 1-Phenyl-3-methoxy-1-hexene | 606 |
| 1-Phenyl-3-methoxy-1-pentene | 606 |
| 1-Phenyl-6-methyl-1-hepten-3-ol | 608 |
| 1-Phenyl-5-methyl-1-hexen-3-ol | 608 |
| 1-Phenyl-3-methylhex-1-yn-3-ene | 610 |
| 1-Phenyl-3-methylnon-1-yn-3-ene | 611 |
| 1-Phenyl-3-methyl-3-octen-1-yne | 612 |
| 1-Phenyl-4-methyl-1-penten-3-ol | 608 |
| 1-Phenyl-1-nonen-3-ol | 608 |
| 1-Phenyl-1-nonyn-3-ene | 613 |
| 1-Phenyl-1-octen-3-ol | 557 |
| 3-Phenyl-2-pentene | 557 |
| 1-Phenyl-1-pentyn-3-ene | 609 |

TABLE 3 (Continued)

| Compound | Reference |
| --- | --- |
| Phenylpolyolefin | 404 |
| Phenyl o-polyolefinylbenzoate | 404 |
| 3-Phenyl-2-propene-1,1-diol diacetate | 516 |
| 1-Phenyl-1-propenyl formate | 525 |
| 1-Phenyl-3-propoxy-1-butene | 606 |
| 1-Phenyl-3-propoxy-1-hexene | 606 |
| 1-Phenyl-3-propoxy-1-pentene | 606 |
| α-Pinene | 416, 419 |
| Pollack liver oil | 614 |
| Poly(acrylonitrile–isoprene–styrene) | 615, 616 |
| Poly[2,2-bis(4-hydroxy-3-allylphenyl)propane-2,2-bis(4-hydroxyphenyl)propane diglycidyl ether] | 351 |
| Polybutadiene | 154, 528, 597, 617–628 |
| Poly(butadiene–acrylonitrile) | 615, 616, 629, 630 |
| Poly(butadiene–acrylonitrile) latex | 631 |
| Poly(butadiene–α-methylstyrene) latex | 631 |
| Poly(butadiene–styrene) | 597, 615, 616, 630, 632, 633 |
| Poly(butadiene–styrene-acrylonitrile) | 622 |
| Poly(butadiene–styrene) latex | 631 |
| Poly(1,4-butanediol tetrahydrophthalate adipate) | 634 |
| Polycyclopentadienes | 635, 636 |
| Poly(2,6-dimethyl-4-bromophenol-2-methyl-6-allyl-4-bromophenol) | 637 |
| Poly(2,6-dimethyl-1,4-phenylene oxide) | 637 |
| Polyesters | 638, 639 |
| Polyesters  (adipic acid–lauric acid–tobaccoseed-oil fatty acids–propylene glycol) | 640 |
| Polyesters  (adipic acid–oleic acid–propylene glycol) | 640 |
| Polyesters  (adipic acid–palm-kernel-oil fatty acids–propylene glycol) | 640 |
| Polyesters  (1,4-butanediol–tetrahydrophthalic acid) | 461 |
| Polyesters  (butanol–ethyleneglycol–tetrahydrophthalic acid) | 461 |
| Polyesters  (tetrahydrophthalic anhydride–butanol–pentamethylene glycol) | 310 |
| Poly(hydrocarbons) | 641 |
| Polyisoprene | 329 |
| Poly(isoprene–styrene) | 615, 616 |
| Poly(2-methyl-6-allyl-1,4-phenylene oxide) | 637 |
| Poly(methylbutadiene) | 627 |
| Polymyrcene | 597 |
| Poly(styrene–diolefins) | 642 |
| Poly(styrene–methylstyrene–cyclopentadiene) | 615, 616 |
| $\Delta^{5,7,9,(11)}$-Pregnatrien-3$\beta$-ol-20-one acetate | 643 |

T ABLE **3** (Continued)

| Compound | Reference |
|---|---|
| $\Delta^{5,7,9(11)}$-Pregnatrien-3$\beta$-ol-20-one benzoate–maleic anhydride adduct | 643 |
| $\Delta^{5,7,9(11)}$-Pregnatrien-3$\beta$-ol-20-one heptanoate–maleic anhydride adduct | 643 |
| Pregnenolone acetate | 644 |
| 1,2,3-Propanetriol tris(3-cyclohexenecarboxylate) | 536 |
| 1,2,3-Propanetriol tris(2-cyclopenten-1-yl)acetate | 326 |
| 2-(1-Propenyl)-1,3-dioxolane | 440 |
| 1-Propenyl formate | 525 |
| 2-(1-Propenyl)-4-methyl-1,3-dioxolane | 441 |
| 2-$n$-Propyl-di($\Delta^3$-cyclohexenyl)trioxan | 559 |
| Propylene | 313, 317, 359, 419, 645, 646 |
| 1,2-Propyleneglycol diolate | 539 |
| Propyl 12,13-epoxyoleate | 385 |
| $n$-Propyl hendecenoate | 398 |
| Propylstyrylcarbinol | 396 |
| Pseudosarsasapogenin | 156 |
| Pseudomilagenin | 156 |
| Pseudotigogenin | 156 |
| Safflower oil | 393, 405, 406, 510, 511, 571 |
| $o$-Salicylpolyolefin | 404 |
| Shark-liver oil | 614 |
| Soybean oil   313, 386–388, 391–393, 405, 406, 455, 510, 511, 539, 542, 549, 647–650 | |
| Soybean oil fatty acids | 407 |
| Sperm oil | 614 |
| *trans*-Stilbene | 651 |
| Styrene | 312, 313, 317, 319 |
| 2-($\beta$-Styryl)-4-methyl-1,3-dioxolane | 441 |
| Styryltriacetoxysilane | 503 |
| 3-Sulfolene | 579, 652, 653 |
| Sylvestrene | 654 |
| 1,2,3,4-Tetrachloro-7,7-difluoro-5-methylene-2-norbornene | 120 |
| 1-Tetradecene | 531 |
| $\Delta^3$-Tetrahydrobenzaldehyde trimer | 559 |
| $\Delta^3$-Tetrahydrobenzonitrile | 313 |
| $\Delta^3$-Tetrahydrobenzyl alcohol acetals | 655 |
| 3-($\Delta^3$-Tetrahydrobenzylhydroxyethyl)-2,4-dioxaspiro[5.5]undecene-9 | 655 |
| 3[2-($\Delta^3$-Tetrahydrobenzyloxy)-2-methylethyl]-2,4-dioxaspiro[5.5]undecene-9 | 515 |
| Tetrahydrofurfuryl oleate | 516 |

TABLE 3 (Continued)

| Compound | Reference |
|---|---|
| Tetrahydrophthalic acid | 656 |
| Tetrahydrophthalic anhydride | 149, 331, 656 |
| Tetrahydrophthalimide | 397 |
| Tetrahydrophthalimides, $N$-substituted | 657 |
| $\alpha,\alpha,\alpha',\alpha'$-Tetrakis(4-epoxypropoxy-3-allylphenyl)-1,4-diethylbenzene | 351 |
| 1,1,2,2-Tetrakis(4-epoxypropoxy-3-allylphenyl)ethane | 351 |
| 3,3,6,6-Tetramethyl-1,4-cyclohexediene | 658 |
| Tetramethylethylene | 389, 597 |
| 4,4$^1$-Thiodibutyl bis(3-cyclohexenyl acetate) | 659 |
| 2,2$^1$-Thiodiethyl bis(3-cyclohexenecarboxylate) | 659 |
| 6,6$^1$-Thiodihexyl bis(bicyclo[2.2.1]-5-heptene-2-carboxylate) | 659 |
| 3,3$^1$-Thiodipropyl bis(3-methyl-3-butenoate) | 659 |
| $p$-Tolylpolyolefin | 404 |
| Triallyl 1,2,4-butanetricarboxylate | 469, 660 |
| 1,1,3-Tetrallyloxybutane | 488 |
| 1,1,3-Triallyloxypropane | 488 |
| Triallylphenyl allyl ether | 338 |
| 2,4,6-Triallylphenyl glycidyl ether | 316 |
| Tricyclohexenylmethyl 1,2,4-butanetricarboxylate | 469 |
| Triethyleneglycol bis(3-cyclohexenecarboxylate) | 324, 477, 478 |
| Tri(2-ethyl-2-hexenyl) 1,2,4-butanetricarboxylate | 469 |
| $m$-Trifluoromethylstyrene | 315 |
| Trimethylethylene | 389 |
| 2,4,6-Trimethyl-2,4-hexadiene | 489 |
| 2,5,5-Trimethyl-2,3-hexadiene | 75 |
| 2,4,7-Trimethyl-2,4-octadiene | 493 |
| 1,1,1-Trimethylolpropane tris(acrylate) | 603 |
| 1,1,1-Trimethylolpropane tris(3-cyclohexenecarboxylate) | 353 |
| 1,1,1-Trimethylolpropane tris(2-ethyl-2-hexenoate) | 320 |
| 1,1,1-Trimethylolpropane tris($\alpha$-ethyl-$\beta$-propylacrylate) | 603 |
| 2,4,4-Trimethyl-2-pentene | 416 |
| 2,2,4-Trimethyl-4-pentenyl acetate | 496 |
| 2,2,4-Trimethyl-4-pentenyl isobutyrate | 496 |
| Triolein | 605 |
| Trioleyl 1,2,4-butanetricarboxylate | 469, 660 |
| Trioleyl 1,2,4-cyclopentanetricarboxylate | 660 |
| Tripropylene | 518 |
| Tris(3-cyclohexenylmethyl) 1,2,4-butanetricarboxylate | 355 |
| Tris(3-cyclohexenylmethyl) 1,2,4-cyclopentanetricarboxylate | 355 |

TABLE 3 (Continued)

| Compound | Reference |
|---|---|
| Tris(2-ethyl-2-hexenyl) 1,2,4-butanetricarboxylate | 660 |
| Tris(2-ethyl-2-hexenyl) 1,2,4-cyclopentanetricarboxylate | 660 |
| Tris(1,2,5,6-Tetrahydrobenzylidene)-D-mannitol | 365 |
| Trivernolin | 522 |
| 10-Undecenyl methacrylate | 377 |
| Vernolic acid | 546 |
| *Vernonia anthelmintica* oil | 520 |
| Vinyl acetate | 312 |
| 5-Vinylbicyclo[2.2.1]-2-heptene | 352 |
| 1-Vinyl-3(4)-chloro-4(3)-ethoxymethoxycyclohexane | 662 |
| Vinyl (2-cyclopenten-1-yl)acetate | 325 |
| Vinylcyclohexane | 436 |
| Vinylcyclohexene | 317, 663 |
| 4-Vinylcyclohexene | 312, 313 |
| Vinyl cyclohexene-4-carboxylate | 325 |
| Vinyl 3-cyclohexenecarboxylate | 312, 324 |
| Vinylcyclopropane | 664 |
| 1-Vinyl-3(4)-hydroxy-4(3)-propoxycyclohexane | 662 |
| Vinyl linoleate | 332 |
| Vinyl 6-methyl-3-cyclohexenecarboxylate | 312 |
| Vinyl oleate | 171, 312, 328, 329, 665 |
| Vinylpentamethyldisiloxane | 450 |
| Vinyl-4-pentenoate | 312 |
| *m*-Vinylphenyl glycidyl ether | 316 |
| 1-Vinyl-2,2,3,3-Tetrafluorocyclobutane | 663 |
| 8(9)-Vinyltricyclo[5.2.1.0$^{2,6}$]dec-3-ene | 379 |
| Vinyl undecylenate | 312 |

In all cases listed in Table 3, the reaction conditions are sufficiently mild for the oxirane ring not to be destroyed by acetic acid originally present in the system or produced from the peroxyacetic acid. In some instances a weak base, usually excess solid sodium carbonate, is present in the reaction system. The base is essential with acid-labile epoxides. Under appropriate conditions the yields of oxiranes are high and often quantitative.

In the preparation of water-insoluble oxiranes with an acetic acid solution of peroxyacetic acid dilution of the reaction mixture with water at the termination of the reaction, as determined by iodometric titration, will precipitate the oxirane. If it is a solid, it is filtered off; if it is a liquid, it can be separated either mechanically or by solvent extraction. Since acetic acid, excess peroxyacetic acid, and hydrogen peroxide are water soluble, they are removed in the aqueous phase and the reaction product is normally obtained in clean form. The same situation applies when a water-soluble solvent (acetone or dioxane, most commonly) is used.

When an aqueous solution of peroxyacetic acid is employed, it is not essential that the unsaturated compound be soluble in the system, provided there is sufficient mutual solubility of the substrate and peroxyacetic acid. With aqueous peroxyacetic acid, ease of separation of water-insoluble reaction product by mechanical means without the necessity of dilution or distillation of solvent is an advantage.

Cooxidation of acetaldehyde and propylene with oxygen in the liquid phase at 90–130° and 80–100 atm in an inert solvent, such as ethyl acetate, and in the presence of a catalyst, such as cobalt naphthenate (about 50 ppm of Co) has been reported (666). Conversion to propylene oxide and acetic acid is 43 and 88%, respectively. Products are isolated by distillation in 57 and 70% yields, respectively. Azobisisobutyronitrile has also been used as a catalyst with the metal salt, and it is also reported that no catalyst is necessary in high-temperature cooxidations (667). The following coupled oxidations with aliphatic aldehydes have been reported (666–672) (optimum yield of epoxide in parentheses, usually at low conversions): acetaldehyde–propylene (70%); propionaldehyde–propylene (50%); methyl ethyl ketone–propylene (96%); acetaldehyde–isobutylene (82%); acetaldehyde–1-butene; crotonaldehyde–1-butene; acetaldehyde–ethylene (13%); 1-olefins–nonanal or decanal plus ozonides; and acrolein–1-hexene (74%), methacrolein–propylene (88%), and acrolein–allyl chloride (90%). The cooxidation of acetaldehyde with mixed butenes (42% isobutene, 29% 1-butene, and 18% 2-butene) gives overall yeilds of epoxides as high as 89%, or which 67% comes from isobutene, 30% from 2-butene, and only 3% from 1-butene (673).

### D.  Peroxyformic Acid

Peroxyformic acid is an unstable peroxy acid even at low temperatures and in concentrated form is a hazardous compound; it is rarely isolated. For this reason its use as an epoxidizing reagent has involved *in situ* techniques almost exclusively.

The *in situ* generation of peroxyformic acid is a rapid process requiring no added acid catalysts; equilibrium is achieved in less than 1 hr even with relatively dilute (30%) hydrogen peroxide. *In situ* oxidation with peroxyformic acid is therefore an attractive procedure both for laboratory and large-scale work since the active oxygen of hydrogen peroxide can be rapidly transferred to oxidizable substrates.

Formic acid, however, is a considerably stronger acid than acetic acid and causes opening of the oxirane ring at a much faster rate. (As is discussed later, this facile ring-opening reaction can be used to advantage in the preparation of 1,2-glycols in excellent yields from olefins.) Therefore reaction conditions for *in situ* epoxidation with peroxyformic acid must be designed to reduce ring opening while taking into account the fact that oxiranes may form more rapidly, with evolution of more heat in a shorter time.

Successful *in situ* epoxidation with peroxyformic acid requires that the quantity of formic acid be limited, usually to about 0.3 mole or less per mole of double bond to be epoxidized. Larger quantities of formic acid can be used, but then the temperature must be carefully controlled (usually at or below room temperatures). A small amount of added base or buffer is sometimes used but is not required. Dilution with an inert organic solvent (benzene, hexane) also moderates the ring-opening reaction. Epoxidation is frequently conducted at or below room temperature, but temperatures as high as 60–80° can be used with limited quantities of formic acid, both on a laboratory and commercial scale, to complete the epoxidation quickly, followed by immediate separation of the oxirane compound from the reaction mixture.

Not only must ring opening by formic acid be kept to a minimum to obtain high yields of oxiranes but, with the use of limited quantities of formic acid, the removal of any substantial portion of it in ring opening leaves an insufficient quantity for the rapid production of peroxyformic acid. Thus the overall epoxidation rate is slowed down, permitting even more ring-opening reactions to occur. Obviously, if the formic acid is largely consumed in the oxirane-ring-opening reaction, epoxidation will stop since the oxygen-carrying species (formic acid) will be depleted.

The first successful epoxidation with peroxyformic acid was reported by Byers and Hickinbottom (674), who isolated an oxirane in low yield from diisobutylene (2,4,4-trimethyl-1-penetene). Although the fact was not known to the original investigators, diisobutylene was an unfortunate choice for the olefin since its reaction is abnormal and a whole range of products is obtained, of which the oxirane is a minor one. Also, rather large volumes of formic acid were employed, a condition that is not conducive to high-yield preparation of oxiranes. With peroxybenzoic acid, on the other hand, a yield of up to 40% of the anticipated oxirane was obtained.

*In situ* epoxidation with peroxyformic acid was first shown to be a practical reaction by Niederhauser and Koroly (675), who used limited amounts of formic acid as the oxygen carrier and, in some cases, organic solvents as well to reduce ring opening. These investigators epoxidized methyl oleate, octyl oleate, propylene glycol dioleate, and soybean oil, and obtained fair yields of oxiranes. Increasing the quantity of formic acid resulted in considerably lower oxirane contents in the products, as would be expected.

Table 4 lists compounds that have been epoxidized with peroxyformic acid. No distinction has been made between partial or complete epoxidation of

TABLE 4. UNSATURATED COMPOUNDS EPOXIDIZED WITH PEROXYFORMIC ACID

| Compound | Reference |
|---|---|
| 3$\beta$-Acetoxyergosta-7,9-dien-5$\alpha$-ol | 193 |
| 3$\beta$-Acetoxyergosta-7,9(11),22-trien-5$\alpha$-ol | 194 |
| 2-(2-Acetoxyethoxy)ethyl 12-acetoxy-9-octadecenoate | 676 |
| Allopregna-7,9(11)-dien-3$\beta$,20$\beta$-diol diacetate | 677 |
| 7,9(11)-Allopregnadiene | 678 |
| Allopregna-7,9(11)-dien-3$\beta$-ol-3-one acetate | 678, 679 |
| Amellia oil | 680 |
| Bicyclo[2.2.1]heptene-2 | 201 |
| Butadiene sulfone | 681 |
| 1,4-Butanediol $\Delta^3$-tetrahydrobenzoate | 682 |
| Butyl oleate | 683 |
| Butyl oleyl phthalate | 684 |
| 3-Chloro-2-(chloromethyl)-1-propene | 685 |
| Cottonseed oil | 686 |
| Cyclooctene | 687 |
| Decyl benzyl $\Delta^4$-tetrahydrophthalate | 682 |
| 3$\beta$,5$\alpha$-Diacetoxyergosta-7,9-diene | 193 |
| 3$\beta$,5$\alpha$-Diacetoxyergosta-7,9(11),22-triene | 194 |
| 22,23-Dibromoergosteryl-D acetate | 457 |
| Dibutyl 4-tetrahydrophthalate | 682, 688 |
| 3,4-Dichloro-1-butene | 689 |
| Didecyl 4-tetrahydrophthalate | 475 |
| Di-2-ethylbutyl 3,6-endomethylene-$\Delta^4$-tetrahydrophthalate | 682 |
| Di(2-ethylhexyl) 4-tetrahydrophthalate | 475, 682 |
| 2,5-Dihydro-2,2,5,5-tetramethylfuran | 690 |
| Dilinolenyl adipate | 691 |
| Dilinoleyl adipate | 691 |
| Di(8-methylnonyl) 3,6-endomethylenetetrahydrophthalate | 475 |
| Di(8-methylnonyl) 4-tetrahydrophthalate | 475 |
| Di-$n$-octyl $\Delta^4$-tetrahydrophthalate | 682 |
| Dioleyl adipate | 691 |
| Dioleyl $\Delta^4$-tetrahydrophthalate | 682 |
| Dodecyl tetradecyl 4-tetrahydrophthalate | 475 |
| 16$\alpha$-(17$\alpha$-Epoxy-7,9(11)-allopregnadien-3$\beta$-ol-20-one acetate | 678 |
| 19$\beta$,28-Epoxy-3-nitrilo-3,4-secooleanone | 692 |
| Ergosteryl-D acetate | 30 |
| Ethylene bis(soybean-oil fatty acid) amides | 514 |
| 2 Ethylhexyloleate | 693, 694 |
| Ethyl linoleate | 686 |
| Linseed oil | 693, 694 |
| Linseed oil fatty acids | 695 |

TABLE 4 (Continued)

| Compound | Reference |
|---|---|
| Methyl elaidate | 686 |
| Methyl esters of soybean-oil fatty acids | 680, 695 |
| Methyl 12,13-*O*-isopropylidene-9-octadecenoate | 572 |
| Methyl oleate | 380, 675, 686, 696 |
| Methyl oleyl phthalate | 684 |
| Octyl decyl 4-tetrahydrophthalate | 475 |
| Octyl oleate | 675, 696 |
| Oleic acid | 697 |
| Olive oil | 697 |
| Polybutadiene | 618, 619, 698 |
| Poly(butadiene–acrylic acid) | 698 |
| Poly(butadiene–acrylonitrile) | 698 |
| Poly(butadiene–styrene) | 615, 616, 629, 633 |
| Polyesters   (adipic acid–linseed-oil acids–propylene glycol) | 640 |
| Polyesters   ($\Delta^4$-tetrahydrophthalic acid–1,6-hexanediol) | 682 |
| Polymerized soybean oil | 680 |
| Propylene glycol dioleate | 675, 696 |
| Rubber (partly depolymerized) | 698 |
| Safflower oil | 405 |
| "Sapogenin" | 699 |
| Soybean oil | 675, 680, 693, 694, 696, 697, 700, 701 |
| Soybean-oil fatty acids | 695 |
| 7,9,(11)-22$\alpha$,5$\alpha$-Spirostadien-3$\beta$-ol acetate | 678 |
| 3-Sulfolene | 652, 653 |
| 2,4,4-Trimethyl-1-pentene | 674 |
| *Vernonia anthelmintica* oil | 686 |

multiply unsaturated compounds. From the length of the table, relative to the preceding ones listing compounds epoxidized with peroxybenzoic, mono-peroxyphthalic, and peroxyacetic acids, it is evident that peroxyformic acid is not usually the epoxidation reagent of choice. However, it has found substantial utility in commercial epoxidations, and conditions have been developed for obtaining about 90% of the calculated oxirane content in the products (based on the original unsaturation).

## E. Peroxytrifluoroacetic Acid

The introduction of an electron-withdrawing group into the $\alpha$-position of an aliphatic peroxy acid or into the ring of an aromatic one not only enhances the electrophilic character of the peroxy acid but also provides a better leaving molecule in the collapse of the transition state. The result is an enhanced rate of epoxidation.

An excellent illustration of the effect of such substitution on the rate of epoxidation is given by peroxytrifluoroacetic acid, first described by Emmons and Ferris in 1953 (702). Peroxytrifluoroacetic acid is an extremely high speed oxidation reagent and therefore has found most widespread use (382, 703, 704) with negatively substituted unsaturated compounds that undergo epoxidation slowly with the usual peroxy acids (peroxybenzoic, peroxyacetic, and monoperoxyphthalic acids).

Peroxytrifluoroacetic acid is usually prepared from 90% hydrogen peroxide (CAUTION!) and trifluoroacetic anhydride, rather than from the free acid, even though equilibrium is attained very rapidly with the acid at 0°, to obtain an anhydrous solution. Methylene chloride is the solvent of choice both for the preparation and the use of the peroxy acid in oxidation reactions, although ethylene dichloride and diethyl ether have also been used. In methylene chloride peroxytrifluoroacetic acid has good stability (705).

Epoxidations with peroxytrifluoroacetic acid must be conducted with particular care because trifluoroacetic acid, the reduction product of the peroxy acid, is a strong acid and readily reacts with the oxirane group. Therefore it is not only necessary to use a buffer but the particular one chosen is important and plays a dominant role in certain cases in determining the yield of oxirane (703). Sodium carbonate, sodium bicarbonate, and disodium hydrogen phosphate have been employed. Sodium carbonate and bicarbonate are satisfactory buffers in those epoxidations in which the peroxy acid, a weak acid, reacts more rapidly with the unsaturated compound than with the buffer. Disodium hydrogen phosphate, on the other hand, is preferred with unsaturated compounds that epoxidize slowly because this buffer is able to neutralize the trifluoroacetic acid as rapidly as it is formed without making the reaction mixture so basic that the peroxy acid is destroyed before it is consumed. Under the appropriate conditions, then, excellent yields ($> 90\%$) of oxiranes can often be obtained with terminal and nonterminal olefins and with those containing electron-withdrawing substituents attached to the double bond.

Table 5 lists unsaturated compounds that have been epoxidized with peroxytrifluoroacetic acid. Two results are worth singling out. Methacrylates and acrylates can be epoxidized in 50–80% yields (703), whereas even with concentrated peroxyacetic acid in an inert solvent yields are reported to be only 20–50% and reaction times are long (382). Unsaturated organic silicon compounds are also readily epoxidized in good yields (50–80%) at 0° with peroxytrifluoroacetic acid to provide novel silicon-containing epoxy monomers, whereas peroxyacetic, peroxybenzoic, and monoperoxyphthalic acids are too slow to be effective (280).

TABLE 5. UNSATURATED COMPOUNDS EPOXIDIZED WITH
PEROXYTRIFLUOROACETIC ACID

| Compound | Reference |
|---|---|
| 1-Acetylcyclohexene | 115 |
| *N*-Benzoyl-9-azabicyclo[3.3.1]-2-nonene | 706 |
| Bicyclo[2.2.2]-5-octene-2,3-*endo*,*cis*-dicarboxylic anhydride | 707 |
| Bicyclo[2.2.2]-5-octene-2,3-*endo*,*cis*-γ-lactone | 707 |
| Cyclohepta-1,3,5-triene–maleic anhydride adduct | 708 |
| Cyclohexa-1,3-diene–maleic anhydride adduct | 708 |
| Cyclohexene | 703 |
| Cycloocta-1,3,5-triene–maleic anhydride adduct | 708 |
| Cyclopentadiene–maleic anhydride adduct | 708 |
| Cyclopentene | 703 |
| 3′,6′-Diacetoxybenzonorbornadiene | 484 |
| *trans*-1,4-Dibromo-2-butene[a] | 456, 709 |
| 7,8-Dichlorobicyclo[4.2.0]octa-2,4-diene–maleic anhydride adduct | 708 |
| 1,4-Dichloro-2-butene[a] | 709 |
| 6,7-Dichloro-1,4-dihydro-1,4-methanonaphthalene-5,8-dione | 464 |
| Diethyl cyclohex-1-enaphosphonate | 710 |
| Diethyl isopropylidenemalonate | 711 |
| 1-Dodecene | 703 |
| Ethyl acrylate | 382, 703, 704 |
| Ethyl crotonate | 382, 703 |
| Ethyl methacrylate | 703 |
| Haloalkenes[a] | 526, 709 |
| 1,5-Hexadiene | 703 |
| 1-Hexene | 703 |
| Methyl maleopimarate | 712 |
| Methyl methacrylate | 382, 703 |
| Methyl *trans*-2-octadecenoate | 713 |
| 4-Nitraza-1-alkene | 143 |
| 2-Nonen-4,6,8-triyn-1-ol (*cis* and *trans*) | 714 |
| 1-Octene | 703 |
| 1,4-Pentadiene | 715 |
| 1-Pentene | 703 |
| 2-Pentene | 703 |
| Vinyl tribenzylsilane | 280 |
| Vinyl triethylsilane | 280 |
| Vinyl tri-*p*-trifluoromethylphenylsilane | 280 |
| Vinyl tri-*p*-fluorophenylsilane | 280 |
| Vinyl triphenylsilane | 280 |
| Vinyl tri-*p*-tolylsilane | 280 |

[a] Peroxytrichloroacetic acid was also used in these epoxidations.

### F.  *m*-Chloroperoxybenzoic Acid

*m*-Chloroperoxybenzoic acid, first prepared and characterized by Lynch and Pausacker in 1955 (69), became commercially available in 1962 (716, 717). The commercial product is about 85% pure; the parent carboxylic acid, *m*-chlorobenzoic acid, is the main impurity. A 99%+ pure peroxy acid can be readily obtained by washing the commercial grade with an aqueous phosphate buffer (718). Technical data (physical and chemical properties, stability) and epoxidation procedures are presented in a bulletin (717).

*m*-Chloroperoxybenzoic acid is a stable peroxy acid, soluble in a variety of common organic solvents. The rate of epoxidation of unsaturated compounds with this peroxy acid is several times that of peroxybenzoic acid (703, 718), a decided advantage in some cases, especially where isomerizations must be avoided [see, for example, epoxidation of "Dewar" benzene (719)]. Although *m*-chloroperoxybenzoic acid has not received much attention yet as an oxidizing agent, the higher rate of epoxidation, commercial availability, and good stability suggest more widespread use of this peroxy acid for epoxidation and other oxidation reactions.

*m*-Chloroperoxybenzoic acid was first used by Lynch and Pausacker (69) for the epoxidation of *trans*-stilbene. Since their interest was primarily in the kinetics and thermodynamic parameters of epoxidation with *m*-chloroperoxybenzoic acid, they did not isolate the oxirane (a known compound) in those experiments. Similarly Schwartz and Blumbergs (718) showed that solvents that do not hydrogen bond with the peroxy acid give the highest expoxidation rates. Chlorinated solvents—notably chloroform, methylene chloride, 1,2-dichloroethane, and chlorinated benzenes—are best. A large reduction in rate (hundred fold) is experienced when *t*-butyl alcohol is the solvent.

The first published reports on the use of *m*-chloroperoxybenzoic acid for the preparation of oxiranes appeared in 1964. Table 6 lists compounds epoxidized with *m*-chloroperoxybenzoic acid.

TABLE 6. UNSATURATED COMPOUNDS EPOXIDIZED WITH
            *m*-CHLOROPEROXYBENZOIC ACID

| Compound | Reference |
|---|---|
| DL-(3,4,5)-5-Acetamido-3,4-di-*O*-acetylcyclopentenediols | 720 |
| 3-Acetoxy-2-cholestene | 721 |
| 1-Acetoxycyclohexene | 722 |
| $\Delta^6$-17$\alpha$-Acetoxy-1-dehydroprogesterone | 723 |
| *p*-Acetoxyphenyl geranyl ether | 724 |

TABLE **6** (Continued)

| Compound | Reference |
|---|---|
| $\Delta^6$-17$\alpha$-Acetoxyprogesterone | 723 |
| *p*-Acetylphenyl geranyl ether | 724 |
| *trans*-3-Benzyliden-3,4-dihydro-4-methyl-1$H$-1,4-benzodiazepine-2,5-dione | 725 |
| Bicyclo[3.2.0]hept-1-ene | 726 |
| Bicyclo[3,2,0]hept-2-ene | 727 |
| Bicyclo[2.2.0]hexa-2,5-diene (Dewar benzene) | 719 |
| Bicyclo[2.2.0]-5-octene-2,3-*endo*,*cis*-dicarboxylic anhydride | 707 |
| 17-Bromo-5$\alpha$-androst-16-ene | 728 |
| 17-Bromo-16-androsten-3$\beta$-ol acetate | 721 |
| *cis*-2-Butene | 729 |
| *trans*-2-Butene | 729 |
| 2-*t*-Butoxy-5,6-dihydro-2$H$-pyran | 730 |
| *t*-Butyl *trans*-cinnamate | 731 |
| 4-*t*-Butylcyclohexene | 732 |
| 4-*t*-Butylmethylenecyclohexane | 208 |
| (+)-3-Carene | 733, 734 |
| 2-Chlorobicyclo[2.2.1]hept-2-ene | 735 |
| (−)-(1R)-2-Chloronorbornene | 736 |
| *cis*-4-(*p*-Chlorophenyl)-5-cyanocyclohexene | 737 |
| *trans*-4-(*p*-Chlorophenyl)-5-cyanocyclohexene | 737 |
| Cholesterol | 716, 717 |
| *cis, trans*-1,5-Cyclodecadiene | 738 |
| Cycloheptene | 739 |
| 1,4-Cyclohexadiene | 67 |
| Cyclopentene | 739 |
| *cis*-1-Cyclopentene-3,5-diol | 77 |
| $\Delta^3$-Cyclopentenylmethyl brosylate | 740 |
| 1-Deuterio-10-methyl-1(9)-octalin | 741 |
| 3$\alpha$,20-Diacetoxy-$\Delta^{17,20}$-pregnene | 716, 717 |
| Diallyl isophthalate | 742 |
| Diallyl phthalate | 742 |
| Diallyl terephthalate | 742 |
| 1,3-Di-*t*-butylallene | 743 |
| 3,3-Dichloro-2-thiabicyclo[2.2.1]hept-5-ene 2,2-dioxide | 740 |
| 3,5-Dihexadecyl-6-heptadecyl-2-stearoyloxy-$\gamma$-pyrone | 744 |
| 5,6-Dihydroergosterol | 745 |
| Diisopropenyl sebacate | 743 |
| Dimethylbicyclo[2.2.2]-5-octene-2,3-*endo*,*cis*-dicarboxylate | 707 |
| 1,2-Dimethylcyclohexene | 746 |

TABLE 6 (Continued)

| Compound | Reference |
|---|---|
| 1,4-Dimethylcyclohexene | 746 |
| cis-4,5-Dimethylcyclohexene | 747 |
| 1,2α:16,17α-Dimethylenepregna-4,6-diene-3,20-dione | 748 |
| 1α,4α,β-Dimethyl-Δ$^8$-octalin | 749 |
| cis-9,10-Dimethyl-1-octal-2-yl acetate | 741 |
| cis-2,2-Dimethyl-3-pentene | 750 |
| trans-2,2-Dimethyl-3-pentene | 750 |
| cis-1,2-Diphenyl-1-dimethylphosphonatoethylene | 751 |
| trans-1,2-Diphenyl-1-dimethylphosphonatoethylene | 751 |
| Estra-4,9(10)-dien-17β-ol-3-one | 752 |
| Estra-5(10)-en-17β-ol-3-one | 752 |
| Estratetraene | 753 |
| Estrogenic steroids | 754 |
| 2-Ethoxy-6-methyl-5,6-dihydro-2H-pyran | 114 |
| 17α-Ethynylestra-4,9(10)-dien-17β-ol-3-one | 752 |
| Ethyl cinnamate | 755 |
| Ethyl crotonate | 716, 717 |
| cis-Ethyl-α-methylcinnamate | 755 |
| trans-Ethyl-α-methylcinnamate | 755 |
| cis-Ethyl-α-phenylcinnamate | 755 |
| trans-Ethyl-α-phenylcinnamate | 755 |
| m-Fluorostyrene | 756 |
| p-Fluorostyrene | 756 |
| 6-Hepten-2-one | 757 |
| 1-Hexene | 729 |
| C-nor-D-Homosapogenins | 758 |
| 4-Hydroxy-3,4-diphenyl-2-cyclopenten-1-one | 759 |
| 5-Hydroxymethylcycloheptene | 760 |
| Isopropenylcyclobutanol | 761 |
| Isopropenylcyclohexanol | 761 |
| Isopropenylcyclopentanol | 761 |
| 1-Isopropenyl-1-indanol | 762 |
| Isopropenyl stearate | 763 |
| 7-Lanosten-3β-ol | 745 |
| Levopimaric acid–acetylenedicarboxylic ester adducts | 764 |
| Mesityl oxide | 716, 717 |
| 2-Methoxy-5,6-dihydro-2H-pyran | 730 |
| Δ$^6$-16α-Methyl-17α-acetoxy-1-dehydroprogesterone | 723 |
| Δ$^6$-16α-Methyl-17α-acetoxyprogesterone | 723 |
| Methyl bisdesoxydeacetylasperulosidate tetraacetate | 735 |
| 1-Methyl-4-t-butylcyclohexene | 746 |
| 2-(N-Methylcarboxamido)-trans-cinnamanilide | 732 |

Tᴀʙʟᴇ **6** (Continued)

| Compound | Reference |
|---|---|
| Methyl cycloheptenecarboxylate | 760 |
| 1-Methylcyclohexene | 384, 739, 746 |
| 3-Methylcyclohexene | 747, 765 |
| 4-Methylcyclohexene | 747 |
| *cis*-Methyl 3,7-dimethyl-2,6-octadienoate | 563 |
| *trans*-Methyl 3,7-dimethyl-2,6-octadienoate | 562 |
| 1-Methyl-2-dimethylphosphonatocyclohexene | 751 |
| $\Delta^6$-1,2$\alpha$-Methylene-17$\alpha$-acetoxyprogesterone | 723 |
| Methylenecyclohexane | 739 |
| *trans*-2-Methylenedecahydronaphthalene | 208 |
| 3,4-Methylenedioxyphenyl geranyl ether | 724 |
| $\Delta^6$-1,2$\alpha$-Methylene-17$\alpha$-hydroxyprogesterone | 723 |
| $\Delta^6$-1,2$\alpha$-Methylene-16$\alpha$-methyl-17$\alpha$-acetoxyprogesterone | 723 |
| 17$\alpha$-Methylestra-4,9(10)-dien-17$\beta$-ol-3-one | 752 |
| Methyl *trans,trans,cis*-7-ethyl-3,11-dimethyl-2,6,10-tridecatrienoate | 766, 767 |
| *trans*-9(10)-Methyl-2-methylenedecahydronaphthalene | 208 |
| *cis*-2-Methyl-$\Delta^2$-octalin | 746 |
| *trans*-2-Methyl-$\Delta^2$-octalin | 746 |
| Methyl(+)-13-oxodeisopropylideneacetate | 768 |
| *cis*-4-Methyl-2-pentene | 769 |
| *trans*-4-Methyl-2-pentene | 769 |
| *cis*-6-Nonen-2-one | 757 |
| *trans*-6-Nonen-2-one | 757 |
| 19-Nor-5(10),9(11)-androstadienes | 770 |
| *cis*-1(9)-Octal-2-ol | 771 |
| *trans*-1(9)-Octal-2-ol | 771 |
| Octalone ketal | 772 |
| 1-Octene | 716, 717, 739 |
| *cis*-4-Octene | 773 |
| *trans*-4-Octene | 773 |
| 2-Oxabicyclo[3.2.0]hept-6-ene | 727 |
| *endo*-*N*-Phenethylbicyclo[2.2.1]hept-5-ene 2,3-dicarboximide | 774 |
| *exo*-*N*-Phenethylbicyclo[2.2.1]hept-5-ene 2,3-dicarboximide | 774 |
| Phenyl geranyl ether | 724 |
| Pregna-4,17-dien-3-one | 727 |
| *cis*-Stilbene | 773 |
| *trans*-Stilbene | 773 |
| $\Delta^6$-Testosterone acetate | 723 |
| 1,3,3-Trimethyl-5-acetoxycyclohexene | 775 |
| 2,2,3-Trimethyl-3-cyclopentene-1-ethanol acetate | 776 |
| Vinyl undecylenate | 777 |

An improved method for the preparation of volatile oxiranes with *m*-chloroperoxybenzoic acid was described by Pasto and Cumbo (729) who treated *cis*- or *trans*-2-butene with a dioxane solution of the peroxy acid at 0° for about 24 hr and then distilled the oxiranes directly from the reaction mixture. Yields were 52–60%; the epoxidations were stereospecific. Similarly 1-hexene was converted to the oxirane in 60% yield; diglyme was the solvent in this case.

### G.  *p*-Nitroperoxybenzoic Acid

*p*-Nitroperoxybenzoic acid, first reported by Medvedev and Blokh in 1933 (778) but characterized some 20 years later by Overberger and Cummins (779), is another rapid epoxidation reagent. The nitro group is a powerful electron-withdrawing group, and its substitution into the ring of an aromatic peroxy acid substantially enhances the electrophilic character of the peroxy acid (69, 780). The *p*-nitro group is more effective than chlorine; *p*-nitro-peroxybenzoic acid is 5 to 20 times more reactive than peroxybenzoic acid (69, 778, 780), whereas *m*-chloroperoxybenzoic acid is only slightly more reactive (69, 718). This greatly enhanced reactivity, coupled with outstanding stability (780, 781), often makes *p*-nitroperoxybenzoic acid the reagent of choice for olefins that epoxidize slowly. Furthermore, the low solubility of *p*-nitrobenzoic acid, the reduction product, effectively removes it from the reaction medium, thus minimizing oxirane-ring opening. The high reactivity and low solubility of *p*-nitroperoxybenzoic acid, however, limit the choice of solvents that can be used for oxidations. The solvents of choice are chloroform, ether, ether–chloroform, and tetrahydrofuran. Aromatic solvents should generally be avoided, especially those containing electron-donating groups, as they undergo facile ring attack (782).

Lynch and Pausacker (69) were the first to use *p*-nitroperoxybenzoic acid in epoxidation studies (kinetics), but they did not isolate the oxiranes. Vilkas (780) not only compared the rates of epoxidation of unsaturated compounds with *p*-nitroperoxybenzoic and peroxybenzoic acids but also was the first to isolate the oxiranes in several cases. Table 7 lists unsaturated compounds that have been epoxidized by *p*-nitroperoxybenzoic acid.

### H.  Aliphatic Peroxy Acids Containing Three or More Carbon Atoms

Although peroxyformic, peroxyacetic, and, more recently, peroxytrifluoro-acetic acids have been the aliphatic epoxy acids of choice in most epoxidations, for one reason or another longer chain aliphatic peroxy acids (peroxypro-pionic through peroxylauric acids) have found occasional use. Reasons for their selection include increased safety and stability, as well as ease of isolation of the oxidation product in certain cases. Most important, however, is the fact that $C_9$ and longer chain aliphatic peroxy acids are crystalline

TABLE 7.  UNSATURATED COMPOUNDS EPOXIDIZED WITH
p-NITROPEROXYBENZOIC ACID

| Compound | Reference |
|---|---|
| 3β-Acetoxy-4-cholestene | 25, 189 |
| 3β-Acetoxy-21-homo(5α)-17(20)-pregnene | 783 |
| 3β-Acetoxy-11β-hydroxyergost-7-ene | 784 |
| 3β-Acetoxy-13α, 14α, 17β-lanosta-8,24-diene | 780 |
| 3β-Acetoxy-5α(17)-methyleneandrostane | 783 |
| 3β-(5α)-Acetoxy-17-pregnene | 783 |
| α-Amyrine acetate | 785 |
| cis-Bicyclo[4.2.0]oct-3-ene | 786 |
| trans-Bicyclo[4.2.0]oct-3-ene | 786 |
| 2-Butene-1,4-dioleate | 417 |
| 3-t-Butylcyclohexene | 787 |
| α-Cedrene | 215 |
| Chrysanthenone | 788 |
| Citraconic anhydride–methyl resinate adduct (and methyl esters) | 789 |
| Cycloartenol acetate | 785 |
| 1,3-Cyclohexadiene | 530 |
| Cyclohexene | 780 |
| 3β,20β-Diacetoxy-4-pregnene | 25 |
| Ergocalciferol-3,5-dinitrobenzoate | 790 |
| 17-Ethylenedioxy-3β-acetoxy-4-androstene | 25 |
| Ethyl α-fenchenyl-10-carboxylate | 780 |
| Euphol acetate | 785 |
| 1,2,4-Heptatriene | 791 |
| 1,2,4-Hexatriene | 791 |
| 8,24-Lanostadiene | 785 |
| Maleic anhydride–1,3-cyclohexadiene adduct, methyl ester | 789 |
| Methyl anhydromaleopimarate | 789 |
| Methyl trans-β-methylcinnamate | 792 |
| (+)-9-Methyloctalin | 793 |
| Norbornene | 780 |
| 1,7-Octadiene | 528 |
| 1,4-Pentadiene | 436 |
| 1,2,4-Pentatriene | 791 |
| β-Pinene | 780 |
| trans-Stilbene | 794 |
| Trimethyl d-maleopimarate | 789 |

solids that are easy to prepare, store, handle, and weigh. With peroxy-pelargonic or peroxylauric acid, for example, a wide choice of inert solvents is available, whereas with peroxyformic, peroxyacetic, or peroxytrifluoro-acetic acid the corresponding aliphatic carboxylic acid is often present in large quantities unless special and sometimes dangerous preparative techniques are employed to remove it. In addition, hydrogen peroxide may also be an undesirable contaminant, especially in the epoxidation of polymerizable monomers.

Unsaturated compounds expoxidized with $C_3$ or longer chain aliphatic peroxy acids are listed in Table 8.

Peroxymaleic acid, first described in 1962 by White and Emmons (803), can be readily prepared in methylene chloride solution from maleic anhydride and 90% hydrogen peroxide (CAUTION!). Peroxymaleic acid is stable, decomposing to the extent of only 5% in 6 hr at room temperature, and can be used

TABLE 8.   UNSATURATED COMPOUNDS EPOXIDIZED WITH ALIPHATIC PEROXY
ACIDS CONTAINING THREE OR MORE CARBON ATOMS

| Acid | Compound | Reference |
|---|---|---|
| Peroxypropionic | Crotyl chloride | 318 |
| | Soybean oil | 313 |
| Peroxybutyric | Polybutadiene | 619 |
| | Poly (butadiene-styrene) | 622, 629, 633 |
| | Safflower oil | 405 |
| Peroxysuccinic | Cottonseed-oil esters | 795 |
| | Limonene | 796 |
| | Soybean oil | 186, 795, 797 |
| Peroxyoctanoic | Ethyl chaulmoograte | 798 |
| Peroxypelargonic | Butadiene | 94 |
| | Cyclopentadiene | 440 |
| | Linolenic acid | 606 |
| | Polybutadiene | 619 |
| | Poly (butadiene–styrene) | 622, 629, 633 |
| Peroxylauric | Cyanobicyclononenes | 799, 800 |
| | 4-Cyanocyclohexene | 475 |
| | 4-Cyanocyclopentenes | 799, 800 |
| | Cyanotricyclotridecene | 799, 800 |
| | Cyclopenten-3-ol | 801 |
| | cis-12-Docosene | 802 |
| | cis-12-Docosenol | 802 |
| | 4-Methylcyclopentene | 799, 800 |
| | Methyl cis-12-docosenoate | 802 |
| | Methyl cis-9-octadecenoate | 802 |
| | Methyl trans-9-octadecenoate | 802 |

in water, dioxane, or other solvents. The main advantage of peroxymaleic acid is the high rate of epoxidation, which is nearly as great as that of trifluoroperoxyacetic acid, but it requires no buffering. It has been used to convert 1-octene, methyl methacrylate, and ethyl acrylate to their corresponding oxiranes in 80, 74, and 38% yields, respectively (803), and also in the epoxidation of haloalkenes (526, 709), propylene (804), and *trans*-1,4-dibromo-2-butene (456, 709, 805). In the last-named case urea is reported to increase the epoxidation rate (805).

## I. Alkaline Hydrogen Peroxide–Nitriles ("Peroxycarboximidic Acids")

The reaction of nitriles with dilute alkaline hydrogen peroxide is an efficient, high-speed reaction for preparing the corresponding amides in excellent yields. This reaction was discovered in about 1884 and is sometimes referred to as the Radziszewski reaction (806).

Although the reaction has been widely used, its mechanism and stoichiometry were not known until 1953, when Wiberg (807) demonstrated that the rate of reaction of benzonitrile with alkaline hydrogen peroxide showed a first-order dependence on the concentration of nitrile, hydrogen peroxide, and the hydroxy ion. Further benzonitrile oxide was shown not to be an intermediate, and from oxygen-18 studies it was demonstrated that the oxygen introduced into the nitrile came from the hydrogen peroxide and not from water or the hydroxy ion. Electron-withdrawing groups in the aromatic ring caused a marked rate enhancement, a result observed in other nucleophilic reactions on the carbon atom of the nitrile group. The stoichiometry of the reaction was shown to be as follows:

$$R-C\equiv N + 2H_2O_2 \xrightarrow{-OH} R-\overset{\overset{\displaystyle O}{\|}}{C}-NH_2 + O_2 + H_2O$$

From these results Wiberg concluded that the first step in the reaction is addition of the hydroperoxide anion ($HO_2^-$) to the nitrile group, followed by a fast reaction of the addition product, the highly reactive peroxycarboximidic acid, with a second molecule of hydrogen peroxide, thus accounting for the products and stoichiometry, as follows:

$$R-C\equiv N + HO_2^- \xrightarrow[\text{step}]{\text{rate-determining}} \underset{\underset{\displaystyle O-OH}{|}}{R-C=N^-}$$

$$\xrightarrow[\text{(fast)}]{H_2O} \underset{\underset{\displaystyle O-OH}{|}}{R-C=N-H} + \bar{O}H$$

$$\underset{\underset{\displaystyle O-O-H}{|}}{R-C=N-H} + H_2O_2 \xrightarrow{\text{fast}} \underset{\underset{\displaystyle O}{\|}}{R-C-NH_2} + O_2 + H_2O$$

Peroxycarboximidic acids have not been isolated; they are unstable, highly reactive intermediates and, as already shown, oxidize hydrogen peroxide, which behaves as a reducing agent in this system. Payne, Deming, and Williams (808, 809) have shown, however, that if a more effective reducing agent, such as an olefin, is present in the system, reaction of peroxycarboximidic acids with hydrogen peroxide is largely eliminated (little or no oxygen is evolved) and the olefin is epoxidized instead:

$$\ce{>C=C< + R-\overset{\displaystyle NH}{\underset{}{C}}-OOH \longrightarrow \underset{O}{>C-C<} + R-\overset{\displaystyle O}{\underset{}{C}}-NH2}$$

By operating in this way only 1 mole of hydrogen peroxide is consumed per mole of double bond and the overall stoichiometry becomes as follows:

$$\ce{>C=C< + R-C#N + H2O2}$$

$$\downarrow {}^{-}\text{OH}$$

$$\ce{\underset{O}{>C-C<} + R-\overset{\displaystyle O}{\underset{}{C}}-NH2 + H2O}$$

(It should be noted that peroxycarboximidic acids are quite similar in structure to peroxyacids; in the former there is a $\ce{>C=NH}$ group in place of a $\ce{>C=O}$ group.) In the absence of added base, however, no reaction occurs.

Acetonitrile has been the nitrile of choice as the oxygen carrier for most of these epoxidations. Methanol is the preferred cosolvent to achieve a homogeneous solution. Reaction temperatures are in the range of 35–60°, and the $p$H is maintained at about 7.5–8 during the reaction by addition of aqueous sodium hydroxide solution (or other bases). Yields of oxiranes are frequently high (65–80%). Benzonitrile and trichloroacetonitrile have also been used as oxygen carriers; they are more reactive than acetonitrile, and epoxidation is more rapid. $t$-Butyl hydroperoxide can also be used as the oxidant instead of hydrogen peroxide (607).

Unsaturated compounds that have been epoxidized by the alkaline peroxide–nitrile method are listed in Table 9. Since the rate-determining step in the

TABLE **9.** USATURATED COMPOUNDS EPOXIDIZED BY ALKALINE
HYDROGEN PEROXIDE-NITRILES (PEROXYCARBOXIMIDIC ACIDS)

| Compound | Reference |
|---|---|
| Acrolein diethylacetal | 808, 809 |
| Acrylamide | 607 |
| Acrylonitrile | 607 |
| Allyl chloride | 808, 809 |
| 2-Allylcyclohexanone | 810 |
| Bicycloheptadiene | 808, 809 |
| Butadiene | 808, 809 |
| 4-*t*-Butylmethylenecyclohexane | 208 |
| Carbohydrates, unsaturated glycosides | 811 |
| Cinnamaldehyde | 808, 809 |
| Cyclohexene | 808, 809, 812 |
| 1,3,6-Cyclooctatriene | 813 |
| Ergocalciferol | 790 |
| Ethyl crotonate | 808, 809 |
| Ethyl isopropylidenecyanoacetate | 711 |
| 1-Hexene[a] | 808, 809 |
| Isopropylidenemalononitrile[a] | 711 |
| Mesityl oxide | 808, 809 |
| 2-Methyl-2-butene | 808, 809 |
| Methyl dihydrodicyclopentadienyl ether | 560 |
| Methylenecyclohexane | 810 |
| *trans*-2-Methylenedecahydronaphthalene | 208 |
| *trans*-9(10)-Methyl-2-methylenedecahydronaphthalene | 208 |
| Norbornadiene | 144 |
| α-Phenyl-*cis*-cinnamonitrile | 607 |
| α-Phenyl-*trans*-cinnamonitrile | 607 |
| Styrene | 808, 809, 812 |
| Vinyl *n*-butyl ether | 808, 809 |

[a] *t*-Butylhydroperoxide used also.

reaction is the addition of the hydroperoxide anion to the nitrile, differently substituted olefins, such as 1-hexene and 2-methyl-2-butene, epoxidize at the same rate. This is in contrast to the rates of the reaction of ethylenic compounds with peroxy acids, which is highly dependent on the degree of substitution at the double bond, as is discussed later under "Kinetics and Mechanisms of Epoxidation."

There is no apparent advantage in using the alkaline peroxide–nitrile reagent for the epoxidation of the vast multitude of unsaturated compounds that are epoxidized rapidly and normally by organic peroxy acids, as is the case with cyclohexene, styrene, 2-methyl-2-butene, and the like. However, if one desires to prepare an oxirane that is acid labile or if the unsaturated compound cannot be handled under acidic conditions, the alkaline peroxide–nitrile system is worth investigating. The epoxidation of acrolein diethylacetal is a striking case (107). Peroxyacetic acid epoxidation provides only a 4% yield of oxirane, whereas alkaline hydrogen peroxide–acetonitrile at 50° for 6 hr gives a 62% yield of distilled glycidaldehyde diethylacetal. The alkaline hydrogen peroxide–nitrile epoxidation method is also useful for the epoxidation of carbohydrate derivatives (811).

$\alpha,\beta$-Unsaturated nitriles are not epoxidized by organic peroxy acids owing to the rate-diminishing effect of the electron-withdrawing nitrile group. The alkaline hydrogen peroxide system just described, however, is well suited to the epoxidation of these nitriles, as the nitrile group in the molecule to be oxidized serves as the oxygen carrier without requiring additional nitrile to be added. The epoxidation is intramolecular; the double bond is converted to the oxirane and the nitrile group to the amide (607, 711).

Perhaps the most striking example of an intramolecular epoxidation of an $\alpha,\beta$-unsaturated nitrile is the preparation of glycidamide, a novel epoxy amide, in 65–70% yield from acrylonitrile with an equimolar quantity of hydrogen peroxide (607). The reaction proceeds via the intramolecular rearrangement of the nonisolable peroxyacrylimidic acid, as follows:

$$CH_2 = CH - C \equiv N + H_2O_2$$

The double bond in peroxyacrylimidic acid, no longer strongly polarized by the electron-withdrawing nitrile group, would be expected to be more receptive to electrophilic attack than the double bond of acrylonitrile, especially is the attack were intramolecular. Other examples of the epoxidation of a,$\beta$-unsaturated nitriles have been listed in Table 9.

The alkaline peroxide–nitrile technique can also be used for the selective epoxidation of a double bond in an unsaturated ketone. Thus 2-allylcyclohexanone (1) is converted to the oxirane (2) in 54% yield by benzonitrile–hydrogen peroxide, whereas peroxyacetic acid oxidation, as might have been predicted, selectively oxidizes the carbonyl group to lactone (3) in 44% yield.

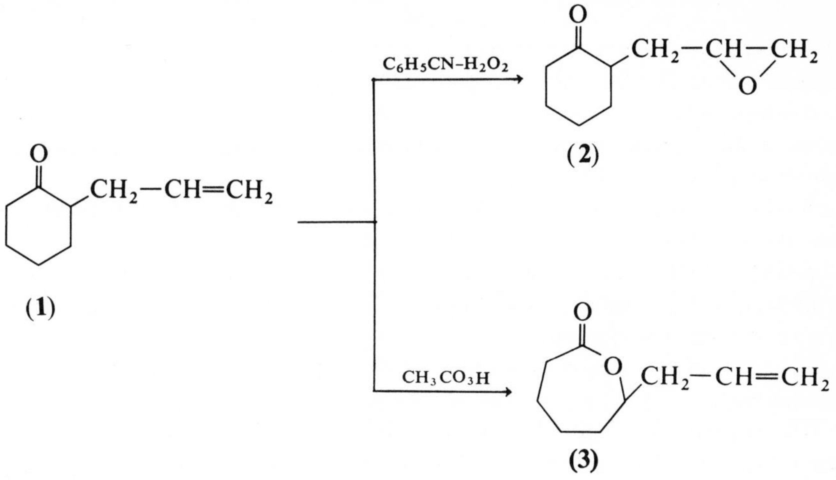

(2)

(1)

(3)

Radlick and Winstein (813) have shown a special selectivity in the epoxidation of one of the conjugated double bonds in 1,3,6-cyclooctatriene with acetonitrile–hydrogen peroxide. They obtained an equal mixture of two monooxiranes (**4** and **5**), whereas with peroxybenzoic acid the desired monooxirane (**4**) comprised only 20% of the total.

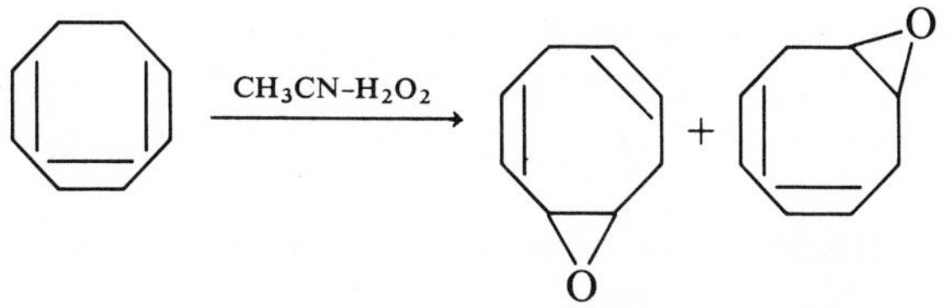

(4)          (5)

In the monoepoxidation of myrcene with hydrogen peroxide–acetonitrile the predominant position of epoxidation is also at one of the conjugated double bonds, whereas epoxidation with monoperoxyphthalic acid is, as expected, at the trisubstituted double bond (814). This special selectivity of alkaline hydrogen peroxide–nitriles in the monoepoxidation of conjugated double bonds in multiply unsaturated compounds is novel in peroxy acid chemistry and warrants further study.

Cyanamide, *N,N*-dimethylcyanamide, and salts of cyanamide have been used instead of acetonitrile or benzonitrile as oxygen carriers in the epoxidation of olefins with alkaline hydrogen peroxide (815). Cyclohexene and butenediol were epoxidized in 45–76% yields.

### J. Miscellaneous

Alkenes, cycloalkenes, triglycerides, fatty acids, and esters are reportedly epoxidized at 50–100° by hydrogen peroxide in the presence of activators consisting of strong inorganic acids and a sugar, a sugar alcohol or a sugar acid, a dicarboxylic or polycarboxylic acid, or a dehydration product of aluminum hydroxide (816, 817).

6,6-Dicarbomethoxy-2-methylhex-2-ene has been epoxidized by peroxy-naphthoic acid (818).

$\beta$-Caryophyllene (819) and styrene, 4-$t$-butyl methylenecyclohexane, and 1-methylene-3,3,5-trimethylcyclohexane (820, 821) have been epoxidized with (+)-peroxycamphoric acid.

Soybean oil has been epoxidized *in situ* with peroxyadipic and peroxycitric acids (795).

The asymmetric epoxidation of 3,3-dimethylallylquinoline or 3-(3,3-dimethylallyl)-2,4-dimethoxyquinoline with (+)- or (−)-peroxycamphoric, (+) or (−)-peroxyhydratropic, or (−)-peroxy-*endo*-norbornane-2-carboxylic acids has been used as the first step in the biosynthesis of oxygenated isoprenoids (822). The oxiranes are usually not isolated.

A series of organosilicon ene–ynes have been epoxidized at the double bond to yield silicon-containing $\alpha$-oxiranyl acetylenes (823).

The oxidation of alkenes with peroxytrifluoroacetic acid–boron trifluoride effects a one-step transformation to the corresponding ketones in generally good yield (824). The products are obtained by way of the intermediate oxiranes (not isolated), which rearrange under the influence of boron fluoride. The products are a result of H, methyl, acyl, Cl, or Br migration, as well as ring contraction.

Trichloroperoxyacetic and monochloroperoxyacetic acids have been reported to epoxidize haloolefins (709).

## III. LABORATORY EPOXIDATIONS

Epoxidation procedures are widely distributed in the literature, and the selection of a suitable one is often difficult. Typical and up-to-date laboratory epoxidation procedures are described here so that a rational choice can be made. Emphasis is on the use of the better known and readily accessible peroxy acids; unsaturated compounds have been selected to cover a wide range of olefinic structures and lability of oxirane formed. The success or failure of an epoxidation reaction is often contingent not on the ability to introduce the oxirane group but to conserve it.

### A.   Peroxybenzoic Acid

#### 1.   1,2-Epoxyethylbenzene (Styrene Oxide) (825)

To a stirred solution of peroxybenzoic acid (42 g, 0.30 mole) in chloroform (500 ml) at 0° styrene (30 g, 0.29 mole) is slowly added. The solution is maintained at 0° for 24 hr, with stirring during the first hour. At the end of 24 hr iodometric titration of an aliquot part of the solution shows that only the excess of peroxybenzoic acid remains. The chloroform solution is washed several times with 10% sodium hydroxide solution to remove benzoic acid, followed by several water washes to neutrality. The chloroform solution is dried over anhydrous sodium sulfate and filtered. Evaporation of the chloroform leaves a practically colorless liquid, which is purified by fractional distillation. 1,2-Epoxyethylbenzene boils at 188–192° at atmospheric pressure or at 101° at 40 mm Hg. The yield is about 25 g (70%).

#### 2.   cis-9,10-Epoxystearic Acid (826)

To an acetone solution (750 ml) of peroxybenzoic acid (55 g, 0.4 mole), oleic acid (85 g, 0.3 mole) is added with stirring at 0–5°. The solution is allowed to stand overnight at room temperature and then cooled to −25° and filtered. The precipitate is washed with cold acetone. The crude cis-9,10-epoxystearic acid (85 g, purity 95–99%) is crystallized twice from acetone (10 ml/g of solute) at 0 to −25° to yield analytically pure cis-9,10-epoxystearic acid, mp 59.9–59.8° (55–60 g, 62–67% yield). Its oxirane oxygen content is 5.33–5.37% (calculated 5.36%). (This procedure can also be used for the preparation of trans-9,10-epoxystearic acid, mp 55.5°, from elaidic acid.)

#### 3.   cis-Stilbene Oxide (163)

To a solution of cis-stilbene (22.2 g, 0.124 mole) in dry benzene (55 ml) a benzene solution of peroxybenzoic acid (17.8 g, 0.130 mole) is added at room temperature. After the reaction is complete (16–24 hr), as determined by iodometric titration, the yellow solution is washed several times with 5% sodium bicarbonate solution and then with water to neutrality. The benzene solution is dried over anhydrous sodium sulfate and filtered. Evaporation of the benzene leaves a yellow oil, which is dissolved in 70% aqueous ethanol and then cooled. The solid that precipitates is recrystallized from the same solvent, yielding cis-stilbene oxide (12.6 g, 52%) as fine, white needles, mp 37°. (This procedure can also be used for the preparation of trans-stilbene oxide, mp 69°, from trans-stilbene.)

### 4. exo-cis-2,3-Epoxynorbornane (96)

To solid norbornene* (65.8 g, 0.7 mole) a solution of peroxybenzoic acid (104 g, 0.75 mole) in chloroform (1340 ml) is slowly added with stirring. The reaction is strongly exothermic, and the temperature is maintained at 0–5° with an efficient cooling bath. The solution is maintained at 0–5° for 4 days, at which time iodometric titration of an aliquot part of the solution shows that the calculated quantity of peroxybenzoic acid has been consumed. The chloroform solution is washed several times with 10% sodium hydroxide solution to remove benzoic acid, followed by several water washes to neutrality. The chloroform solution is dried over anhydrous sodium sulfate and filtered. Evaporation of the chloroform through a modified Claisen flask leaves a residue, which is fractionally distilled. After a brief forerun, bp 142–146° at 630 mm Hg, exo-cis-2,3-epoxynorbornane, bp 146–148° at 630 mm Hg (59 g, 80%), is obtained as fernlike crystals, mp 121–125°, that solidify in the receiver. Recrystallization from hexane yields the analytically pure product, mp 126°.

### 5. 1,2-Epoxy-2-Methyl-3-Butene (Isoprene Monoepoxide) (827)

To a stirred solution of isoprene (16 g, 0.235 mole) in ethyl chloride (50 ml) cooled in an icebath a cold solution of peroxybenzoic acid (30 g, 0.217 mole) in ethyl chloride (150 ml) is added dropwise. (The contents of the flask and dropping funnel are protected from moisture by drying tubes.) The reaction flask is stored in the refrigerator until the oxidizing agent is completely consumed (approximately 24 hr). The solution is then shaken cautiously with double the calculated quantity of sodium bicarbonate solution (30 g/100 ml of water) in a cooled separatory funnel until evolution of carbon dioxide ceases. The aqueous layer is discarded, and the ethyl chloride solution is dried overnight in a refrigerator with anhydrous sodium sulfate. The solution is filtered, and the filtrate is fractionally distilled until unreacted isoprene begins to distill. The residual material is then fractionated twice to yield isoprene monoepoxide (7 g, 30–40%), bp 81° at 735 mm Hg, $n_D^{19}$ 1.41795 and $d_4^{19}$ 0.8616.

Butadiene is converted to the monoepoxide in the same way. Its expoxidation is considerably slower than that of isoprene. Butadiene monoepoxide has the following characteristics: bp 66.5° at 735 mm Hg, $n_D^{21}$ 1.40930 and $d_4^{21}$ 0.8712.

### B. Monoperoxyphthalic Acid

### 1. β- and α-Cholesteryl Oxide Acetates (184)

A solution of cholesteryl acetate, mp 112–114° (10 g, 0.023 mole) in ether (50 ml) is mixed with an ether solution (266 ml) containing monoperoxy-

---

* The presence of a small amount of chloroform is sometimes helpful in facilitating initial heat exchange.

phthalic acid (8.4 g, 0.046 mole). The solution is refluxed for 6 hr, and the solvent is removed by evaporation. The residue is dried under vacuum and digested with dry chloroform (250 ml). The mixture is filtered to separate the insoluble phthalic acid (6.7 g or 87% of theory), and the colorless filtrate is evaporated to dryness under vacuum. Crystallization of the residue from methanol (30 ml) gives $\beta$-cholesteryl oxide acetate (6.0 g. 58% yield), which on recrystallization gives the pure product, mp 111–112°, $[\alpha]_D^{25}$ $-21.8°$. Concentration of the filtrate gives $\alpha$-cholesteryl oxide acetate (1.55 g, 15% yield) purified by crystallization from ethanol, mp 101–103°, $[\alpha]_D^{25}$ $-44.6°$.

In a similar way cholesterol is converted to $\alpha$-cholesteryl oxide in 60% yield; the $\beta$-isomer is formed in only small amounts.

### 2. Geraniolene Monoepoxide (239)

To a stirred solution of geraniolene (64.8 g, 0.52 mole) in chloroform (1000 ml) at 0° 0.52 $N$ monoperoxyphthalic acid in ether (1000 ml) is added over a 1-hr period. Stirring is continued for 3 hr at room temperature, and the precipitated phthalic acid is then separated by filtration. The filtrate, after being washed with saturated aqueous sodium bicarbonate solution and then saturated sodium chloride solution, is dried over anhydrous sodium sulfate and filtered. The solvent is removed under reduced pressure at 60° on a water bath. The residue is then distilled through a 0.8 by 50-cm spinning-band column. The fraction boiling at 93–94° at 72 mm Hg is pure geraniolene monoepoxide (42 g, 57% yield), $n_D^{24}$ 1.4307, as shown by gas–liquid chromatography. The oxirane group in the product is exclusively at the trisubstituted double-bond position.

### C. Peroxyacetic Acid

#### 1. 1,2-Epoxydodecane (828)

To a stirred acetic acid solution of 0.9 $M$ peroxyacetic acid (408 g, 0.36 mole) at 20–25° 1-dodecene (50.5 g, 0.30 mole) is added. The reaction mixture becomes homogeneous in about 8 hr. Reaction is continued at room temperature until 85% of the calculated quantity of peroxyacetic acid has been consumed (an additional 20 hr). The reaction solution is poured into several volumes of cold water and extracted several times with ether. The ether solution is washed several times with cold water, dried over anhydrous sodium sulfate, and filtered; the ether is then evaporated. The residue is distilled through an 18- by 0.5-in. Vigreux column to yield 1,2-epoxydodecane (29 g, 52%), bp 97–98° at 3.5 mm Hg, $n_D^{20}$ 1.4356, Its oxirane oxygen content is 8.53% (calculated 8.68%).

If commercial 40% peroxyacetic acid (approximately 5 $M$) in acetic acid, in which the sulfuric acid has been neutralized with sodium acetate, is used, the reaction time is reduced considerably (4 hr or less) and yields of distilled 1,2-epoxydodecane of over 70% are obtained. Internal olefins, such as *cis*- and *trans*-4-octene (267) behave similarly with 40% peroxyacetic acid; the reaction is complete in about 1 hr.

### 2.  cis-9,10-Epoxystearic Acid (561, 829)

To a stirred acetic acid solution of peroxyacetic acid (500–2000 ml) containing 2.04 moles of peroxyacetic acid oleic acid (500 g, purity 99%, 1.8 moles) is rapidly added dropwise. The temperature is maintained at 20–25° by means of an ice bath. After about 2–4 hr, depending on the peroxyacetic acid concentration, over 96% of the calculated quantity of peroxy acid is consumed. The reaction mixture is poured into several liters of ice and water, and allowed to stand until the granular solid rises to the surface. The lower aqueous acetic acid layer is siphoned off, and the solid is washed twice with cold water and filtered. After drying, the crude 9,10-epoxystearic acid (520 g, >90% purity), mp 53–55°, is crystallized from Skellysolve B (3 ml/g) at 0° to yield pure *cis*-9,10-epoxystearic acid (400 g), mp 59.5°.

With minor modifications, the above procedure is applicable to the epoxidation of numerous other monounsaturated fatty acids, esters, and alcohols (561, 828), in particular elaidic acid, methyl oleate, methyl elaidate, oleyl alcohol, and elaidyl alcohol.

### 3.  trans-Stilbene Oxide (α,α'-Epoxybibenzyl) (651)

To a stirred solution of *trans*-stilbene (54 g, 0.3 ml) in methylene chloride (450 ml) at 20° a solution of commercial 40% peroxyacetic acid in acetic acid (65 ml, 0.425 mole), in which sodium acetate trihydrate is dissolved (5 g), is added dropwise over a 15-min period. A cooling bath is not required during the addition. The mixture is stirred for 15 hr, during which time the temperature is not allowed to rise above 35°. The contents of the flask are poured into water (500 ml), and the organic layer is separated. The aqueous phase is extracted with methylene chloride (2 × 150 ml), and the combined methylene chloride solutions are washed with 10% aqueous sodium bicarbonate (2 × 100 ml) and then with water (2 × 100 ml). The organic layer is dried over anhydrous magnesium sulfate, and the methylene chloride is removed by vacuum evaporation. The residue is recrystallized from methanol (3 ml/g of solute) to yield moderately pure *trans*-stilbene oxide (46–49 g, 78–83%), mp 66–69°. Recrystallization from hexane (3 ml/g) sharpens the melting point to 68–69° (41–44 g, 70–75%).

The above procedure is a general one for the epoxidation of relatively unreactive double bonds in which heat evolution is slow. With reactive unsaturated compounds, however, the peroxyacetic acid should be added more slowly and an efficient cooling bath is required. The fast reactions are complete in much less time, frequently 1–3 hr. The progress of the reaction is best determined by following peroxide disappearance iodometrically on aliquot portions of the reaction solution.

### 4.  Cyclooctene Oxide (437)

Commercial 40% peroxyacetic acid (125 ml), containing sodium acetate trihydrate (25 g), is added dropwise during 45 min with vigorous stirring to cyclooctene (55 g, 0.5 mole) while the reaction temperature is maintained at 27–29° with an ice bath. The reaction mixture is cooled to 0°, neutralized with 40% sodium hydroxide, and extracted with ether (3 × 200 ml). The ether extracts are dried over anhydrous magnesium sulfate and concentrated, and the residue is distilled through a 20- by 1.8-cm column packed with glass helices. The yield of cyclooctene oxide, bp 90–93° at 37 mm Hg and mp 52.5–56.5°, is 54 g (86%). Sublimation yields the pure compound, mp 56–57°, as needles.

In a similar way cycloheptene oxide, bp 83–85° at 50 mm Hg and $n_D^{25}$ 1.4615–1.4620, is prepared in 80% yield (415).

### 5.  Methyl 2,3-Epoxy-2-methylpropionate (382)

To commercial methyl methacrylate monomer (900 g, 9 moles), stabilized with 0.5% hydroquinone, a 21% solution of peroxyacetic acid in ethyl acetate (1836 g, 5.05 moles) is added at 75–85° over a $5\frac{1}{2}$-hr period, after which time about 70% of the peroxy acid has been consumed. The reaction mixture is then rapidly distilled under vacuum (about 30 mm Hg), leaving the poly (methyl methacrylate) by-product behind as a residue. Redistillation under vacuum yields methyl 2,3-epoxy-2-methylpropionate (278 g), bp 66° at 30 mm Hg, $n_D^{30}$ 1.4134, and $d^{20}$ 1.0972. The yield is 47.5 or 68.4%, based on the consumption of methyl methacrylate or peroxy acid, respectively.

Numerous other $\alpha,\beta$-unsaturated esters can be similarly epoxidized in yields based on unsaturated compound ranging from 20 to 95%, thus providing a general method for obtaining epoxy monomers of various structures (382).

### 6.  1,1,2,2-Tetramethylethylene Oxide (389, 830)

A 40% solution of peroxyacetic acid in acetic acid (372 g, 2.1 moles) is added dropwise over 6 hr to a well-stirred suspension of anhydrous sodium carbonate (300 g, 2.83 moles) in methylene chloride (300 ml) containing freshly distilled tetramethylethylene (168 g, 2.0 moles). The temperature is maintained between 5 and 10° by an ice bath. One hour after the peroxyacetic acid addition is complete, water is added (approximately 500 ml), and the aqueous phase is extracted with additional methylene chloride (3 × 200 ml). The combined methylene chloride solutions are washed with aqueous ferrous sulfate solution to destroy any residual peroxides, and the solvent is removed by distillation through a well-insulated 1- by 24-in. column packed with glass helices. The oxirane product is obtained at 89–91°; the yield is 70–80%. Redistillation gives the pure epoxide, $n_D^{20}$ 1.4016.

The procedure just described is recommended for acid-labile epoxides; it was first reported by Korach, Nielsen, and Rideout (438). Examples of other olefins epoxidized in essentially the same way are 1,1,2-trimethylethylene (389), 1,3-cycloalkadienes (414, 438), 2,4-dimethyl-2,4-hexadiene (414), 2,4,4-trimethyl-2-pentene (416), and 1-alkylcycloalkenes (384).

### D.  Peroxyformic Acid

#### *1,2-Epoxy-3,4-dichlorobutane (689)*

3,4-Dichlorobutenes (500 g, 4.0 moles) and 98% formic acid (140 g) are stirred at 60°; 85% hydrogen peroxide (CAUTION!) (185 ml) is then slowly added to maintain the temperature at about 60° (about 1 hr is required). After 4 additional hr at 60° the organic layer is separated, washed with water to remove formic acid and hydrogen peroxide, dried, and then distilled under vacuum first at 40 and then at 11 mm Hg. The fractions boiling below 42° at 40 mm Hg and below 65° at 11 mm Hg are discarded; the main fraction of crude epoxide (201 g) has bp 65–72° at 11 mm Hg. Redistillation yields the pure epoxides, bp 113–119° at 100 mm Hg, $n_D^{20}$ 1.4738–1.4768. The overall yield is about 30%.

### E.  Peroxytrifluoroacetic Acid

#### *1.  1,2-Epoxypentane (703)*

To a suspension of 90% hydrogen peroxide (8.2 ml, 0.3 mole) (CAUTION!) in methylene chloride (50 ml) cooled in an ice bath is added trifluoroacetic anhydride (50.8 ml, 0.36 mole) over a 10-min period. The solution is stirred in the cold for an additional 15 min and transferred to a dropping funnel equipped with a pressure-equalizer tube. The peroxytrifluoroacetic acid is then added over 30 min with efficient stirring to a mixture containing 1-pentene (14 g, 0.2 mole), sodium carbonate (95 g, 0.9 mole), and methylene chloride (200 ml). The solvent boils vigorously during the addition, and an ice-water-cooled condenser is necessary during the reaction to prevent escape of the 1-pentene. After the addition is complete, the mixture is refluxed for an additional 30 min and the insoluble salts are removed by centrifugation. The salt cake is triturated with 300 ml of methylene chloride and recentrifuged. The combined methylene chloride solutions are fractionated through a Todd column (90 by 1.2 cm) packed with 4-mm glass helices to remove solvent, and the residue is fractionated through a Holzman column to yield 1,2-epoxy-pentane, bp 89–90° (14 g, 81% yield).

### 2. *1,2,5,6-Diepoxyhexane (703)*

To 1,5-hexadiene (8.2 g, 0.1 mole) in methylene chloride (275 ml) containing a slurry of sodium carbonate (95 g, 0.9 mole) a solution of peroxytrifluoro-acetic acid prepared from 90% hydrogen peroxide (8.2 ml, 0.3 mole), trifluoro-acetic anhydride (50.8 ml, 0.36 mole), and methylene chloride (50 ml) is added. The addition requires 25 min, and the mixture is then refluxed for 30 min. The inorganic salts are removed by centrifugation, followed by distillation of the solvent at atmospheric pressure. The residual liquid is fractionated through a Holzman column to yield 1,2,5,6-diepoxyhexane, bp 77–80° at 22 mm Hg (8.0 g, 70% yield).

Peroxytrifluoroacetic acid is a high-speed epoxidizing reagent and is especially useful for olefins substituted with electron-withdrawing groups, such as acrylates, methacrylates, and crotonates. Yields of epoxides are as high as 85%, whereas with peroxyacetic acid yields are usually lower (382). With slowly epoxidizing olefins, disodium hydrogen phosphate is the pre-ferred buffer for use with peroxytrifluoroacetic acid. Sodium carbonate is used only with olefins that consume peroxytrifluoroacetic acid more rapidly than it reacts with the base (703).

### F. *m*-Chloroperoxybenzoic Acid

### 1. cis- *and* trans-2,3-Epoxybutenes *(729)*

In a three-neck 1000 ml flask equipped with a delivery tube, magnetic stirrer, and Dry Ice–acetone reflux condenser *m*-chloroperoxybenzoic acid (65.3 g of 85% purity, 0.322 mole) is dissolved in anhydrous dioxane (500 ml). The solution is cooled to 0°, and *cis*- or *trans*-2-butene (18.1 g, 0.322 mole) is added through the delivery tube. The contents of the flask are stirred for 10 hr under the Dry Ice-acetone condenser, at which time the butene no longer refluxes. The reaction flask is stoppered and refrigerated overnight. The mixture is then distilled, and the fractions up to 100° are collected. These are refractionated through a 2-ft glass-helix-packed column, giving either *cis*-2,3-epoxybutane (bp 58–59° at 748 mm Hg) or *trans*-2,3-epoxybutane (bp 52–53° at 748 mm Hg). Yields are 52–60% of more than 99.5% stereochemically pure epoxides.

### 2. cis- *and* trans-4,5-Epoxyoctanes *(773)*

A solution of *cis*- or *trans*-4-octene (5 g, 0.045 mole) in chloroform (50 ml) is cooled to 0°, and a solution of *m*-chloroperoxybenzoic acid (9 g of 85% purity, 0.045 mole) in chloroform (75 ml) is added, maintaining the tempera-ture at 0°. The mixture is stirred overnight at room temperature, and the *m*-chlorobenzoic acid is removed by filtration. The chloroform solution is washed with 10% sodium bicarbonate solution and dried over anhydrous sodium sulfate, followed by removal of the solvent under vacuum. Distillation of the residue yields either *cis*-4,5-epoxyoctane (bp 69° at 32 mm Hg; 3.4 g, 60% yield) or *trans*-4,5-epoxyoctane (bp 67° at 34 mm Hg; 70% yield).

### 3.   cis- *and* trans-*1-Chloro-4-methylcyclohexane Epoxides (51)*

A cold solution of *m*-chloroperoxybenzoic acid (50 g of 85% purity, 0.246 mole) in methylene chloride (675 ml) is added dropwise to a stirred solution of 1-chloro-4-methylcyclohexene (29.2 g, 0.224 mole) in methylene chloride (25 ml) at 0°. After 2 hr the precipitated *m*-chlorobenzoic acid is filtered from the cold reaction mixture and the filtrate is washed twice with saturated sodium bicarbonate solution (2 × 200 ml) and dried. Evaporation of solvent leaves a light-yellow liquid residue, which is distilled through a short-path column, giving a colorless liquid distillate (24.8 g), which is collected in a Dry Ice–cooled receiver. The crude epoxides are redistilled through a semi-micro spinning-band column to give the epoxides, a mixture of *cis*- and *trans*-isomers, bp 65.3–66° at 9 mm Hg (20.6 g, 62% yield).

### 4.   α-(+)-3,4-*Epoxycarane (733)*

To a solution of (+)-3-carene (68 g, 0.5 mole) in chloroform (750 ml) in a 3-l four-neck flask fitted with a mechanical stirrer, thermometer, condenser, and dropping funnel a solution of *m*-chloroperoxybenzoic acid (109 g of 85% purity, 0.540 mole) in chloroform (1200 ml) is added over a 75-min period at 23–27°. Stirring is continued for about 1 hr (conversion of carene to epoxide is complete within 5 min after the addition is completed), and the excess peroxy acid is destroyed by the slow addition of 10% sodium bisulfite solution until a test with starch–iodide is negative. The mixture is transferred to a separatory funnel, and the organic layer is washed successively with 10% sodium bicarbonate and saturated sodium chloride solutions, and then dried over anhydrous sodium sulfate. The solvent is evaporated, and the residue is distilled under reduced pressure through a spinning-band column to obtain α-(+)-3,4-epoxycarane (64.5 g, 84.9% yield), bp 92–93° at 23 mm Hg, $n_D^{20}$ 1.4664, and $[\alpha]^{26.5} + 13.35°$.

In a similar procedure (734) 3-carene is epoxidized in benzene solution at 40°; the yield of epoxide is about the same as above.

### 5.   *Selective Monoepoxidation of Vinyl Undecylanate (777)*

In a 1-l flask equipped with a magnetic stirrer and a stopping funnel protected with a drying tube are placed vinyl undecylenate (20.5 g, 0.095 mole), sodium bicarbonate (25 g), and benzene (200 ml). To the stirred mixture a solution of *m*-chloroperoxybenzoic acid (25 g of 85% purity, 0.145 mole) in benzene (400 ml) is added at 6–18° over 80 min. The reaction is continued at room temperature for an additional 47 hr. The excess peroxy acid is neutralized by addition of calcium hydroxide (10 g), and the mixture is stirred for several hours and filtered. The filtrate is washed successively with 2% sodium bicarbonate solution (2000 ml) and water, and then dried over anhydrous magnesium sulfate. Evaporation of the solvent under vacuum yields vinyl 10,11-epoxyundecanoate (19.1 g, 89% yield), $n_D^{30}$ 1.4640, purity 98 ± 0.5%. Distillation under vacuum through a spinning-band column causes considerable loss of the monomer as an insoluble residue owing to polymerization during thermal treatment. The pure compound is obtained in only 35% yield, bₚ 103–107° at 0.25–0.4 mm Hg, $n_D^{30}$ 1.4511.

*m*-Chloroperoxybenzoic acid, 85% minimum purity, is commercially available (716, 717). The major impurity is *m*-chlorobenzoic acid; the pure peroxy acid is relatively easy to obtain (718). Its high stability and rapid rate of epoxidation of unsaturated compounds make it a very desirable oxidant (716, 717, 831), especially for olefins with electron-withdrawing groups in the vicinity of the double bond.

### G. *p*-Nitroperoxybenzoic Acid

#### 1. (+)-trans-9-Methyldecalin-2α,3α-oxide (793)

To a solution of (+)-9-methyloctalin-2 (925 mg, 6.05 m moles) in ether (50 ml) *p*-nitroperoxybenzoic acid (780,781) (1.2 equivalents) is added. The solution is allowed to stand at room temperature for 16 hr and is then washed well with 10% sodium carbonate and dried over anhydrous magnesium sulfate. The oil obtained on evaporation of solvent is chromatographed on silica gel (66 g) to give unreacted olefin (190 mg) and the oily epoxide (747 mg, 93% based on olefin reacted); $[\alpha]_D^{29}$ + 25° (*c* 1.12, CHCl₃). The epoxide is homogeneous by thin-layer and gas–liquid chromatography.

#### 2. 3β-Acetoxy-4α,5α-epoxycholestane (189)

3β-Acetoxy-4-cholestene (10 g) is dissolved in ether (300 ml), and a solution of *p*-nitroperoxybenzoic acid (7.5 g) in tetrahydrofuran (70 ml) is added at room temperature. After 18 hr the solution is poured into saturated aqueous sodium bicarbonate solution, and the ether solution is washed with cold 1 *N* sodium hydroxide and then with saturated sodium chloride solution. After drying over anhydrous sodium sulfate, the solvents are evaporated and the oily residue is triturated with several drops of methanol to produce a crystalline product, which is then recrystallized from methanol. 3β-Acetoxy-4α,5α-epoxycholestane, mp 114–115°, is obtained in 67% yield.

### H. Miscellaneous. Alkaline Hydrogen Peroxide–Acetonitrile

#### 1. 1,2-Epoxycyclohexane (808, 809)

To a 1-l five-neck flask equipped with stirrer, thermometer, two dropping funnels, condenser, and *p*H electrodes methanol (300 ml), acetonitrile (82 g, 2.0 moles), and cyclohexene (123 g, 1.5 moles) are added. The stirred mixture is warmed to 60°, and hydrogen peroxide (1 mole of 50%) is added dropwise over 2 hr at $60 \pm 2°$ while 1 $N$ sodium hydroxide is added simultaneously to maintain an indicated $p$H of 9.5–10. After an additional 1.5 hr, the mixture is diluted with water (500 ml) and extracted with chloroform (3 × 150 ml). The combined chloroform extracts are washed with water and dried, and the solution is concentrated to about one-third its volume through a 40-tray Oldershaw column. Distillation through a 0.7- by 50-cm glass-spiral-packed column yields pure 1,2-epoxycyclohexane (68 g, 70% yield based on hydrogen peroxide), bp 54–55° at 10 mm Hg, $n_D^{20}$ 1.4525.

### 2. *Glycidaldehyde Diethylacetal (808, 809)*

Acrolein diethylacetal is epoxidized in essentially the same way (50°, about 6 hr of reaction time). Glycidaldehyde diethylacetal, bp 62–63° at 13 mm Hg, $n_D^{20}$ 1.4140, is obtained in 62% yield based on hydrogen peroxide. (Peroxyacetic acid epoxidation gives only a 4% yield owing to the lability of the acetals in acid medium.)

## IV.  COMMERCIAL EPOXIDATION PROCESSES. PLASTICIZERS-STABILIZERS

### A.  Introduction

The rapid growth in the consumption of polyvinyl chloride and its copolymers during the last decade has been paralleled by an equally rapid growth in the utilization of plasticizers and stabilizers. The most striking development in the plasticizer and stabilizer field has been the epoxidized fatty oils and fatty acid esters, whose consumption since their introduction on a commercial scale in the early 1950s has grown to over 100 million lb per year (estimated for 1970). Of this quantity, about two-thirds is epoxidized soybean oil and about one-third epoxidized esters of tall-oil fatty acids and of oleic acid. Other unsaturated oils and fatty acid esters are also epoxidized commercially, but their consumption is relatively small. The epoxidation of long-chain unsaturated compounds is an important industrial unit process and will unquestionably become more important as new uses for epoxy plasticizers-stabilizers continue to be found. Epoxy plasticizers-stabilizers now rank third in commercial production behind phthalates and phosphates (832).

The major use for epoxy plasticizers-stabilizers is the stabilization of polyvinyl chloride and its copolymers against deterioration on exposure to light and heat. These polymers, as well as other highly chlorinated substances, readily undergo deterioration when exposed to heat and/or light, resulting in degradation and discoloration, as well as cross-linking. The mechanisms by which epoxidized oils and esters stabilize chlorinated polymers and other chlorinated substances is not known, but most of the evidence suggests that stabilization occurs, not only as a consequence of scavenging hydrogen chloride, but largely because of interruption of free-radical chain degradation reactions (833). Epoxy plasticizers are used in polyvinyl chloride formulations at the secondary level, in the range of 1–10%. They are not used as the sole stabilizer in a polymeric system but are part of a complicated recipe in which metallic salts and phosphites are also used.

Although the epoxidation of unsaturated compounds has been known for about 60 years, epoxidation did not become a commercial process until *in situ* methods were developed for utilizing the active oxygen in hydrogen peroxide by means of an oxygen-carrying organic acid, usually acetic or formic acid (834–837), based on work at the Eastern Regional Research Laboratory of the U.S. Department of Agriculture (838, 839). As has already been discussed, hydrogen peroxide reacts readily with acetic acid in the presence of an acidic catalyst to form peroxyacetic acid, and it reacts with formic acid in the absence of an acid catalyst to form peroxyformic acid. In *in situ* methods the preparation of the peroxy acids is conducted in the presence of the unsaturated substance to be epoxidized so that it is not necessary to prepare and isolate the peroxy acids, which are somewhat unstable and hazardous compounds, especially on a large scale.

The successful demonstration that the oxidizing capacity of hydrogen peroxide could be utilized efficiently in organic oxidation reactions by converting the active oxygen to peroxy acid form was the key step in the development of commercial epoxidation methods. Since peroxy acid formation from hydrogen peroxide and the organic acids mentioned is an equilibrium reaction (equation 1), the equilibrium will be shifted to the right as the peroxy acid is consumed in epoxidation (equation 2), and this sequence will continue until very little unsaturation or hydrogen peroxide remains, depending on which was originally presented in excess.

$$H_2O_2 + RCO_2H \rightleftharpoons H_2O + RCO_3H$$

$$RCO_3H + {>}C{=}C{<} \longrightarrow {>}C{-}C{<} + RCO_2H$$
$$\underset{O}{\diagdown\diagup}$$

Under the proper conditions the loss of hydrogen peroxide is minimal, almost stoichiometric utilization of the active oxygen is effected, and high yields of epoxides are obtained. As the above equations show, 1 mole of hydrogen peroxide is required for each double bond epoxidized, but small excesses are used in commercial epoxidations. Although preformed peroxy acids can also be employed and are required for the epoxidation of those double bonds that are epoxidized slowly, the *in situ* preparation and consumption of peroxy acids is the currently used technique in most commercial epoxidations with hydrogen peroxide.

Rapid formation and consumption of peroxy acids must take place if *in situ* methods are to be successful. With 50–70 % hydrogen peroxide and formic acid this is no problem, even with small ratios of formic acid to hydrogen peroxide (0.25 mole per mole of hydrogen peroxide) and in the absence of added stronger acid catalyst (837). High yields of epoxide can be readily obtained from soybean oil, for example, by *in situ* epoxidation with hydrogen peroxide and formic acid. Since formic acid is a considerably stronger acid than acetic acid and cleaves epoxide rings more readily, temperatures below 60° are generally recommended for epoxidation. Peroxyformic acid is considerably less stable than peroxyacetic acid. With 50–70 % hydrogen peroxide only 0.3–0.5 mole of acetic acid is required per mole of hydrogen peroxide to achieve essentially complete epoxidation (834–836).

With hydrogen peroxide and acetic acid, peroxy acid formation is a relatively slow process, and strong-acid catalysts are needed to speed up peroxy acid formation. Strong acids ordinarily used on a commercial scale are sulfuric acid and cross-linked polystyrenesulfonic acid ion-exchange resins (8 % cross-linking) (840, 841). Strong-acid catalysts, however, also catalyze the cleavage of the epoxide ring by the carboxylic acid, which is necessarily present, thus reducing the yield of epoxide. Although sulfuric acid is the cheapest acid catalyst, its use leads to somewhat more cleavage and loss of oxirane than when sulfonic acid ion-exchange resins are employed, but the use of the latter is somewhat more costly. There is usually no difficulty in epoxidizing unsaturated compounds on a commercial scale; the problem is to retain the epoxide function.

### B.  Important Variables in Commercial Epoxidation

The important variables in reducing epoxide cleavage and thus obtaining high epoxide yields on a commercial scale are (a) the ratio of carboxylic acid to hydrogen peroxide, (b) reaction temperature and time, (c) quantity and type of strong-acid catalysts, and (d) stirring rate during epoxidation. Detailed literature is available from manufacturers of hydrogen peroxide on all of these variables and only the highlights can be discussed here (842–845).

### 1.  Carboxylic Acid–Hydrogen Peroxide Ratio

Usual molar ratios of acetic acid to 50 % hydrogen peroxide are about 0.5:1. With 70 % hydrogen peroxide the acetic acid can be approximately halved; ratios of 0.25–0.33 mole of acetic acid per mole of 70 % hydrogen peroxide are sometimes recommended (842–844). The lower acetic acid ratio is one of the advantages of using 70 % hydrogen peroxide, although its addition to the mixture of unsaturated compound and acetic acid must be more carefully controlled and is done stepwise to avoid the buildup of explosive mixtures.

With formic acid as the oxygen-carrying acid, approximately 0.25 mole of acid per mole of hydrogen peroxide is sufficient (837).

### 2.   *Reaction Temperature and Time*

Reaction temperature of about 60–80° are used commercially in *in situ* peroxyacetic and peroxyformic epoxidations. Reaction times vary from about 8–14 hr, depending on the temperature. Temperatures in excess of 80° reduce reaction time, but peroxide losses are favored. Temperatures below 60° make reaction times excessively long and also encourage the loss of oxirane and peroxide.

### 3.   *Catalysts*

Epoxidation times are related to the speed with which the peroxy acid is formed. The two most widely used catalysts for speeding up peroxyacetic acid formation are sulfuric acid (1–2% based on combined weights of acetic acid and hydrogen peroxide) and cross-linked polystyrene sulfonic acid cation-exchange resins (8% cross-linking). The ion-exchange resins are employed in two main ways (840, 841, 846–849). In most cases large quantities are employed (10–25% based on the combined weights of acetic acid and hydrogen peroxide), but the resin must be used many times to keep catalyst cost at a reasonable level. Alternatively, about 2% by weight of resin may be used and then discarded after a single run. The latter technique requires longer reaction times, epoxidation yields are somewhat lower, and costs are slightly higher.

In comparing the two methods of catalysis sulfuric acid is cheaper, but epoxide cleavage is somewhat higher. Ion-exchange resins produce less cleavage, but there have been problems in the past in handling swollen and easily fractured resins. Traces of metals, notably iron and copper, in the resin and in the reaction system have a tremendous effect on resin degradation and peroxide loss. Commercial resins are currently available with suitable purity and mechanical strength.

### 4.   *Agitation*

Agitation is required during epoxidation of water-insoluble unsaturated substances. Normally one would anticipate that increased agitation would be beneficial in a heterogeneous reaction system, since the rate of reaction is ordinarily increased. With epoxidation, however, high agitation rates result in marked increase in cleavage (850, 851). A controlled agitation rate is required for best epoxidation results. The optimum rate is that in which the oily material is just broken into discrete globules and layering is eliminated.

The rate of agitation for commercial epoxidation must always be determined experimentally and depends on the size and shape of the reactor and the kind of stirrer used. It is a compromise between using sufficiently rapid agitation to permit the best heat transfer but not so rapid that excessive cleavage and emulsion formation will result.

With appropriate control of the reaction variables just discussed, epoxidized soybean oil containing approximately 7.0% oxirane oxygen can be obtained routinely with either peroxyacetic or peroxyformic acid generated and consumed *in situ*.

### C.  Equipment (840, 841)

Both glass-lined and 300 series stainless-steel reactors can be used. Commercial preference is for the latter, which have the advantages of greater durability and better heat-transfer properties. Heat transfer is the limiting factor in the speed with which commercial epoxidations can be conducted. Stainless steel has the disadvantage of producing some contamination of the reaction mixture with iron and other metals that may promote the decomposition of hydrogen peroxide or peroxy acid (or both). This is not serious after the reactor has been run in for a period of time, because the metal surface eventually becomes passivated. In initial runs, however, metal contamination may be serious and result in low epoxide yields and higher peroxide loss. Break-in time can be reduced by treating the equipment with nitric acid. Glass-lined reactors minimize decomposition but have poor heat-transfer characteristics.

Storage tanks and other equipment with which hydrogen peroxide is in long-term contact are fabricated of high-purity aluminum (99.6%) and some synthetic polymers (852). Aluminum cannot be used with peroxy acids.

### D.  Reaction Procedure and Product Workup (840–845)

On a commercial scale, 50–70% hydrogen peroxide is added over 2–8 hr to the stirred mixture of unsaturated compound, aliphatic acid, and acid catalyst at 50–80°. The rate of addition is dependent on heat transfer. The usual design value is approximately 60 kcal/mole of epoxy oxygen introduced. Also, when 70% hydrogen peroxide is used, care must be taken to avoid the dangerous detonable region by ensuring that an exotherm has occurred before more than 25% of the hydrogen peroxide is added.

If epoxidation is stopped too soon, low oxirane and high iodine numbers will result. If the reaction is run too long, oxirane values will again be low because of excess cleavage. Periodic determinations of iodine number and oxirane oxygen are reasonably satisfactory, although the time required in preparing the sample and then analyzing it may cause slightly long epoxidation times. Graphical methods in which the drop in iodine number and increase in oxirane oxygen are plotted against time in hours are useful because an accurate estimate of the optimum time to stop the epoxidation reaction can be obtained by extrapolation.

Product workup is particularly important in commercial epoxidations. Complete removal of organic and inorganic acids from the epoxide is necessary to prevent subsequent epoxide cleavage or polymerization (or both), and to obtain a product acceptable for use with polyvinyl chloride. Residual hydrogen peroxide must also be removed to ensure a product with the best heat stability.

Initial separation of the epoxidized oil from the aqueous phase is accomplished by gravity separation. Centrifugation is employed by some manufacturers not only to speed up the separation but also to facilitate the breaking of emulsions that sometimes form. The oil phase is then thoroughly washed successively with minimum quantities of water until all acid is removed. The oil is then dried by vacuum steam stripping.

The large number of water washes that are required to remove all acid can be reduced by the use of calcium hydroxide (853). In this method the initially separated oil is washed thoroughly only once. The water is separated and a slight excess of finely divided calcium hydroxide is stirred into the wet oil. After the slurry has been stirred for a sufficient time to complete the neutralization, water is removed by vacuum stripping. The excess calcium hydroxide and other insoluble salts are then separated from the dry product by filtration, using a filter aid.

The main difficulty with the calcium hydroxide method is slow filtration of the relatively viscous product, accompanied by oil loss in the filter cake. Also, electrical properties of plasticized and stabilized polyvinyl chloride may suffer if residual calcium salts are left in the plasticizer.

### E.   Recovery of Organic Acids

In general acetic and formic acids are not recovered unless a minimum of 1 million lb/yr or more is processed. All peroxides must be destroyed before distillation.

### F.   Cost of Materials (840, 841)

Epoxidized plasticizers-stabilizers have become widely accepted not only because they perform a useful function but also because they are relatively inexpensive. Table 10, is a summary of the materials cost in the commercial epoxidation of soybean oil (iodine number 130) to an iodine number of 4 or below and an oxirane oxygen of 6.5% minimum. Hydrogen peroxide costs are based on the use of a 10% molar excess. The cost of soybean oil is arbitrarily taken as 10 cents per pound and that of hydrogen peroxide (50%) as 25.0 cents per pound.

TABLE 10.   MATERIALS COST IN THE MANUFACTURE OF
EPOXIDIZED SOYBEAN OIL

| Starting Materials | Amount Required To Obtain 100 lb of Product | Unit Cost (dollars) | Cost per 100 lb of Product (dollars) |
|---|---|---|---|
| Soybean oil (alkali refined) | 93 | 0.10 | 9.30 |
| Hydrogen peroxide (50%) | 35.7 | 0.25 | 8.89 |
| Acetic acid | 14.3 | 0.10 | 1.43 |
| Sulfuric acid[a] | 1 | 0.01 | 0.01 |
| | | | 19.63 |
| | | Materials cost per lb | 0.196 |

[a] For resin catalyst add 1–3 cents per pound to materials cost.

Other costs—such as labor, overhead, depreciation, and taxes—are variable and depend on the conditions in a given plant, its location, and the scale of the operation. For a plant producing 2 million lb or more of epoxidized soybean oil these additional costs have been estimated at about 10–15% of the cost of chemicals. Approximately 20-odd chemical companies manufacture epoxidized plasticizers-stabilizers.

### G.  Advantages of Hydrogen Peroxide (854)

Some of the advantages of hydrogen peroxide in commercial epoxidation reactions are as follows:

| | |
|---|---|
| High reaction efficiency | Absence of metallic or inorganic contamination |
| High yields | Applicable to aqueous and nonaqueous systems |
| High reaction specificity | Applicable to homogeneous and heterogeneous systems |
| Easy product isolation | Recovery of organic acids for reuse if desired |
| Minimum by-products | Applicable to standard-type process equipment. |
| Stoichiometric use of peroxide | |

### H.  Peroxyacetic Acid by Acetaldehyde Oxidation

Although the first procedures for preparing epoxidized oils commercially were *in situ* epoxidations with hydrogen peroxide and aliphatic organic acids, the preparation of peroxyacetic acid by liquid-phase oxidation of acetaldehyde was developed into a high-yield, efficient process on a plant scale in 1956

(855). By low-temperature-catalyzed oxidation with oxygen, acetaldehyde is converted into an intermediate, acetaldehyde monoperoxyacetate (AMP), a highly explosive compound not separately isolated, which by controlled thermal decomposition in solution is converted to peroxyacetic acid and regeneration of acetaldehyde for recycling. The peroxyacetic acid is obtained in solution in inert organic solvents, such as ethyl acetate or acetone.

Vapor-phase oxidation of acetaldehyde to peroxyacetic acid is also a commercially feasible way of obtaining peroxyacetic acid (856). In such processes it is necessary to limit the amount of oxygen to approximately 30% or less of that required to convert all the acetaldehyde to peroxyacetic acid. The high temperatures of oxidation necessitate the use of high-speed separation techniques and organic solvents to ensure that the peroxyacetic acid that is formed does not oxidize the acetaldehyde present.

Solutions of peroxyacetic acid containing between 20 and 40% peroxyacetic acid are obtained by liquid- or vapor-phase oxidation of acetaldehyde. Such solutions are high-speed oxidizing reagents and can be used for epoxidizing ureactive double bonds that cannot be epoxidized by *in situ* methods, such as terminal bonds and those to which electron-withdrawing groups are attached. Since the peroxyacetic acid solutions obtained by the oxidation of acetaldehyde are essentially free of water, hydrogen peroxide, and, most important, strong acids, oxirane-ring-opening reactions are minimized. The rate of epoxidation of unsaturated compounds with preformed peroxyacetic acid permits relatively short overall reaction times, thereby introducing additional economy.

## I. Manufacture of Peroxyacetic Acid from Hydrogen Peroxide and Acetic Acid

The preparation of peroxyacetic acid from hydrogen peroxide and acetic acid will probably continue to be an important commercial procedure for the preparation of peroxyacetic acid. Although the preparation of peroxyacetic acid by the liquid- or vapor-phase oxidation of acetaldehyde appears to be more economical, specialized equipment and techniques are required, and it has been suggested that it is better suited for companies that already oxidize acetaldehyde for the preparation of acetic acid and acetic anhydride. The preparation of peroxyacetic acid from hydrogen peroxide and acetic acid has considerable versatility and flexibility, as it can be conducted on a small or large scale, the process can be conducted batchwise or in a continuous manner, and, with proper precautions, the process is one that can be operated with greater safety. Peroxyacetic acid can be prepared without complicated and elaborate equipment, and, as the price of hydrogen peroxide decreases, the process becomes even more attractive.

Peroxyacetic acid can be prepared in (a) acetic acid solution (840–851), (b) organic solvents (857), or (c) aqueous solution (858). Each of these three procedures is briefly described.

### 1. Peroxyacetic Acid in Acetic Acid Solution

#### SULFURIC ACID CATALYSIS

A solution of peroxyacetic acid (40%) in acetic acid solution is produced commercially from hydrogen peroxide and acetic acid, using sulfuric acid in small quantities as the catalyst to speed up the reaction. The solution can also be prepared, if desired, at the epoxidation plant by carefully mixing 1.6 moles of glacial acetic acid with 1 mole of 90% hydrogen peroxide with approximately 2–3% of sulfuric acid; equilibrium is attained in approximately 12–16 hr at room temperature.

Such solutions contain some unconverted hydrogen peroxide, water, and large quantities of acetic acid, in addition to peroxyacetic acid. Water may be eliminated by the addition of small quantities of acetic anhydride, which also shifts the equilibrium toward the formation of additional peroxyacetic acid (859). Acetic anhydride should be added judiciously since it also reacts with peroxyacetic acid to form the highly explosive diacetyl peroxide as a by-product. Peroxyacetic acid solutions prepared by mixing acetic anhydride and 35% hydrogen peroxide at 40° have been reported by Wallace (840) to contain about 80% diacetyl peroxide; such mixtures have exploded. The latter procedure, however, is different from that in which hydrogen peroxide and acetic acid are allowed to react and the only function of the acetic anhydride is to eliminate the water. Nevertheless, the potential hazards in the use of acetic anhydride should be noted.

The reaction of hydrogen peroxide with acetic acid has been used in batch preparations only, and, in general, it is recommended that the sulfuric acid present be neutralized before the peroxyacetic acid is used for epoxidation. This is usually accomplished by adding mild bases, such as sodium acetate, sodium bicarbonate, or calcium carbonate. Alternatively, if the sulfuric acid is not neutralized, the peroxyacetic acid solution can be added to the unsaturated compound to be epoxidized in the presence of an excess of the appropriate weak base, but the former method of neutralization is preferred. If the sulfuric acid is not neutralized, oxirane yields are usually low and the major products are the ring-opened products, hydroxyacetates and small amounts of glycols from the hydrolysis of those esters. Sulfuric acid can be tolerated in *in situ* prepared peroxyacetic acid because the acetic acid is present in limited amounts.

#### SULFONIC ACID CATION-EXCHANGE-RESIN CATALYST

The use of polystyrenesulfonic acid cation-exchange resins to prepare peroxyacetic acid solutions free of strong mineral acids has been extensively described (834–836, 840, 841, 846–849). The method has been operated both continuously and batchwise. Figure 1 shows a schematic flow diagram given by

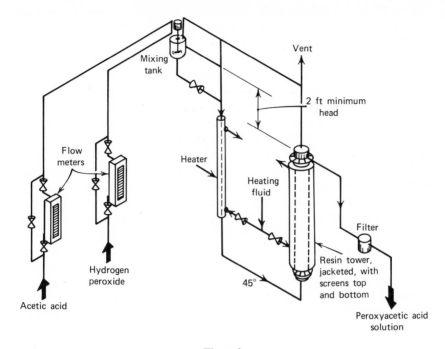

Vent

2 ft minimum head

Mixing tank

Flow meters

Heater

Heating fluid

Filter

Hydrogen peroxide

45°

Resin tower, jacketed, with screens top and bottom

Acetic acid

Peroxyacetic acid solution

**Figure 1.**

Wallace (840) for the continuous preparation of peroxyacetic acid with a resin catalyst.

In the illustrated procedure a mixture of glacial acetic acid and hydrogen peroxide is passed through a cation-exchange resin in the acid form, which has been pretreated with acetic acid to remove the water. The preparation of peroxyacetic acid is conducted at approximately 45°, and the appropriate contact time between hydrogen peroxide and acetic acid in the catalyst chamber is adjusted by controlling the flow rate of the hydrogen peroxide–acetic acid mixture.

For the preparation of 40% peroxyacetic acid solutions 90% hydrogen peroxide is required. A 15% peroxyacetic acid solution may be prepared at the rate of about 265 lb/hr by feeding a mixture containing a ratio of 6.33 gal. of glacial acetic acid to 1 gal. of 15% hydrogen peroxide to the resin tower. Adequate contact times are maintained by charging 17–18 lb of cation-exchange resin to the tower for each gallon per hour of 50% hydrogen peroxide introduced. For an effluent rate of 265 lb/hr of peroxyacetic acid a cylindrical tower 6 ft high with an 8 in. inner diameter is filled with about 75 lb of resin to a 4-ft mark. Suitable resins include Amberlite IR-120,

Dowex 50-X-8, Chempro C-20, Nalcite HCR, and Permutite Q. The tower and accessories are constructed of type 18–8 stainless steels.

When high-strength hydrogen peroxide (50% or higher concentration) is used, certain safety precautions must be taken in operating the resin process. Devices should be installed to measure the temperature and warn of an undue rise. The flow of hydrogen peroxide to the tower should be controlled by an automatic device actuated by temperature that shuts off its flow above a certain point. Automatic flooding of the tower with water and a quick opening valve to dump the charge should also be provided. When 90% hydrogen peroxide is employed, it must not come in contact with the resin undiluted with acetic acid. Contamination of the resin by heavy metals, such as copper and iron, should not be permitted to occur. Metal contamination may cause extremely rapid decomposition of hydrogen peroxide and result in explosion.

Two azeotropic processes for preparing peroxyacetic acid from acetic acid and hydrogen peroxide have recently been described (857, 858). Although these procedures are apparently not in use commercially, they offer considerable promise since they produce concentrated solutions of peroxyacetic acid in organic solvents or in water, and both solutions are essentially free of unreacted hydrogen peroxide and excess acetic acid. These peroxyacetic acid solutions are extremely efficient in the case of double bonds that cannot be epoxidized by *in situ* methods because of their slow rate of epoxidation. With preformed peroxyacetic acid solutions, however, epoxidation is a facile, rapid, and high-yield reaction.

### 2. Peroxyacetic Acid in Organic Solvents

In one procedure, described by Phillips, Starcher, and Ash (857), approximately stoichiometric quantities of acetic acid (a slight excess is used) and 35–90% hydrogen peroxide are allowed to react in the presence of a strong-acid catalyst, such as sulfuric acid or a sulfonic acid ion-exchange resin, a peroxide stabilizer, and an excess of inert organic solvent. The solvent serves not only as a diluent and moderator of the reaction but also as an azeotropic agent for the continuous removal of water by vacuum distillation, thereby continuously shifting the equilibrium to the right, as already shown in equation 1 (Section 4-A) and permitting the use of the calculated quantity of reactants.

Peroxyacetic acid remains in the distillation pot in the particular solvent used. Preferred solvent-azeotropic agents are ethyl, isopropyl, and *n*-propyl acetates, but chloroform, ethylene dichloride, and ethyl propionate have also been used. The reaction is carried out under moderate vacuum in the range 70–450 mm Hg at a reaction temperature of 40–70°. Yields of peroxyacetic acid range from about 75 to over 90%. Solutions containing up to about 37% peroxyacetic acid are obtained. Peroxypropionic acid solutions can be similarly prepared.

### 3. Aqueous Peroxyacetic Acid

In the second azeotropic method Jones (858) prepared concentrated aqueous peroxyacetic acid solutions that are virtually free of acetic acid, hydrogen peroxide, and strong acids. In this procedure no organic solvents are used, but the minimum-boiling peroxyacetic acid–water azeotrope (860), consisting of about 55% peroxyacetic acid and 45% water, is fractionally distilled continuously under vacuum from the acid-catalyzed reaction of acetic acid with 50–70% hydrogen peroxide at about 35° and 45 mm Hg. Conversion of hydrogen peroxide to peroxyacetic acid exceeds 95%.

The concentrated aqueous peroxyacetic acid is only slowly hydrolyzed in the absence of strong acids. At 5° little reduction in peroxyacetic acid occurs after 3 days, whereas at 25° significant hydrolysis takes place after about 1 day. Thus it is better to use up the aqueous peroxyacetic acid as it is formed. The aqueous peroxyacetic solution is also a very efficient epoxidizing reagent, even though unsaturated compounds may not be soluble in the aqueous system. Separation of the epoxidized products is a relatively easy process because no organic solvents are present.

## J. Manufacture of Epoxidized Oils and Esters

### 1. Epoxidation Process Notes

A variety of commercial processes have been employed for the preparation of epoxidized oils and esters. The *in situ* processes utilize hydrogen peroxide-formic acid (837, 861), resin-catalyzed hydrogen peroxide–acetic acid (840, 841, 846–849), and sulfuric acid–acetic acid–hydrogen peroxide processes (834–836, 850, 851). No information is available to the writer on commercial processes utilizing preformed peroxy acids, and it is assumed that they are modeled after laboratory procedures.

### 2. Catalysts

In general strong-acid catalysts are readily interchangeable. In the case of monounsaturated fatty acid esters equivalent results can be obtained with sulfuric acid, alkanesulfonic acids, or sulfonic acid cation-exchange resins. When polyunsaturated fatty acid esters (e.g., soybean oil) are being epoxidized, ion-exchange resins permit higher conversion to epoxide with less cleavage of the oxirane ring than is obtained with sulfuric or alkanesulfonic acid catalysts.

### 3. Solvents

Solvents are not ordinarily used on a commercial scale, but they are sometimes beneficial, particularly when sulfuric acid or alkanesulfonic catalysts are used (862). Aliphatic solvents, such as hexane or heptane, and aromatic solvents, such as benzene, toluene, or xylene, are the ones most commonly recommended. Generally 20% of solvent based on the weight of unsaturated compound is useful in repressing secondary and unwanted ring-opening reactions of epoxides. In many cases higher conversions to epoxide can be obtained when solvents are employed. When solvents are used at or near their reflux temperature, an auxiliary benefit as a heat reservoir for the exothermic reaction is provided. Furthermore, the solvent aids in the washing and separation of the epoxidized product. Epoxidation reactions utilizing ion-exchange-resin catalysts generally do not require a solvent.

### 4. Temperature

Reaction temperatures of 60–80°, depending on the catalyst system and organic oxygen-carrying acid, are often employed. Lower reaction temperatures give longer reaction times, with poor conversions to epoxide. Higher temperatures can also be utilized, but shorter times are then employed. In general greatly accelerated reactions can be achieved but at a sacrifice of epoxy yields at high temperatures. Ordinary reaction times range from about 8 to 14 hr.

### 5. Molar Ratio of Acid to Hydrogen Peroxide

Generally a molar ratio of 0.25:0.5 of acetic or formic acid to hydrogen peroxide is employed. Higher molar ratios may speed up the oxidation reactions but generally result in greater ring opening and consequently lower oxirane yields.

The acid requirements are usually sufficiently small to allow discard of the diluted acid at the end of the reaction. For large-scale applications, however, where acid recovery may be desirable, the processes described below allow recovery of the acid, with slight modification. Prior to recovery of acids by distillation, residual peroxide must be destroyed by the addition of suitable reducing agents.

### 6. Heat of Reaction

The heat of reaction for the epoxidation of a double bond is about 60 kcal/mole.

### 7. Sulfuric-Acid-Catalyzed Processes

In situ procedures for the epoxidation of soybean oil from acetic acid and hydrogen peroxide with sulfuric acid as the catalyst have been developed by numerous industrial concerns. In one procedure (two-stage epoxidation) the sulfuric acid catalyst is added last admixed with glacial acetic acid (863). Sulfuric-acid-catalyzed opening of the oxirane ring is held to a minimum because the system is a heterogeneous one. The process is operated first by adding 20% of the total requirement of 50% hydrogen peroxide to the soybean oil and heating the mixture to about 50°. The solution of acetic acid containing the acid catalyst is then added to the charge over a 4-hr period with separate and simultaneous addition of the remainder of the hydrogen peroxide. After about 13 hr at 50–60° with agitation, the aqueous and oil layers are allowed to separate. The oil layer contains the epoxidized soybean oil, and it is refined and dried. The aqueous layer contains about 5–6% of unused peroxide oxygen, 25–30% of acetic acid, and 1–2% of sulfuric acid. It is not discarded but is used in the partial epoxidation of fresh soybean oil, thus giving a greater overall utilization of hydrogen peroxide, the most expensive reagent in the process. Utilization of the aqueous layer in this way contributes to overall reaction efficiency and good economics. The aqueous layer is separated from the partially epoxidized oil, treated with a peroxide decomposition reagent, and then stripped of acetic acid. The partially epoxidized oil, which contains about 18% of its final oxirane oxygen, is then allowed to react with fresh hydrogen peroxide and acetic acid as just described to complete the epoxidation.

Another recently described procedure for the commercial epoxidation of soybean oil with hydrogen peroxide–acetic acid–sulfuric acid employs an inert solvent (e.g., benzene or hexane) to minimize opening of the oxirane ring. In a typical reaction (864) 2000 lb of soybean oil is charged to a 600-gal. stainless-steel-jacketed reactor fitted with an agitator, cooling coils, reflux condenser, vent, rupture disk, sample line, direct and recording thermometers, feed lines, and manhole. The reactor is also equipped with an automatic system for flooding the reaction with water if the reaction becomes uncontrollable. The soybean oil is heated by applying steam to the jacket, and 400 lb of hexane is added. To the solution of soybean oil and hexane 320 lb of glacial acetic acid is then added, followed by 796 lb of 50% sulfuric acid. When the reaction mixture reaches 50°, the steam is shut off, cooling water is circulated through the coils, and the addition of hydrogen peroxide is started. About 765 lb of 50% hydrogen peroxide is added over a period of 2 hr, with the temperature maintained between 50 and 60°. After completion of the peroxide addition, the temperature is allowed to rise to 60–65° and is maintained within this range until the hydrogen peroxide has been consumed. When the reaction is complete, the organic layer is washed with water and pumped to a vacuum stripping column where water and solvent are removed.

## 8. Repeated Resin Process

Successful epoxidations of soybean oil with hydrogen peroxide and acetic acid with large quantities of sulfonic acid ion-exchange resins used repeatedly have been described (840, 841, 865). Of the various processes now available for the *in situ* epoxidation of unsaturated oils and fatty acid esters, the repeated resin technique probably eliminates unsaturation most effectively and produces the highest oxirane-oxygen content. The general practice is to use 10–15% (dry weight) of the cation-exchange resin based on the weight of the unsaturated oil or ester to be epoxidized. Because such a large quantity of relatively expensive catalyst is used, it must be reused in order for the process to become economical and competitive with sulfuric acid catalysis. However, advantages include high oxirane-oxygen content, relatively small by-product formation, essentially complete elimination of unsaturation, low reaction temperatures (60°), and short reaction periods.

In one process (840, 841) the unsaturated oil or ester containing 1.0 mole of unsaturation is mixed with 0.55 mole of glacial acetic acid and 12% of dry resin, based on the weight of material to be epoxidized. Hydrogen peroxide (1.1 moles) is added slowly so that a reaction temperature of 60° is maintained. After about 4 hr at 60°, the reaction mixture is separated from the resin catalyst by decantation or filtration. The resin catalyst is allowed to remain in the reactor for succeeding runs.

Wallace (840) has pointed out that the average polystyrenesulfonic acid catalyst available commercially can be reused in approximately 6–8 runs at a 10–12% catalyst level. Factors that contribute to degradation of the resin and militate against repeated reuse include (a) the presence of heavy-metal contaminants, which cause rapid attack on the resin by hydrogen peroxide and/ or peroxyacetic acid; (b) the slow degradation of resin cross-linkage by the peroxy acid; and (c) the physical breakdown of the beads of catalyst by mechanical means. However, recent advances in the technology of preparing high-quality resin catalysts have more than tripled resin lifetimes. Improvement in the life of the catalyst also minimizes the production of fine particles, which introduce problems in filtration. The preparation of high-quality resin catalysts is a notable advance in the technology of producing epoxidized unsaturated oils and esters.

### 9.  *Minimal Resin Process*

The minimal resin process for epoxidation requires the use of about 2% resin (dry weight) based on the weight of the material to be epoxidized (846, 847, 849). Since the quantity of catalyst is approximately one-fifth that used in the repeated resin process, the normal reaction variables of epoxidation must be modified because the available surface area of the catalyst is much smaller. This can be accomplished, in part, by operating at higher temperatures (75–80°) and for longer times (7–8 hr). In spite of these changes, the minimal resin process is less effective for introducing oxirane oxygen by approximately 10–15% and in this regard is approximately equivalent to sulfuric-acid-catalyzed processes.

A schematic flow sheet for the epoxidation of unsaturated fatty oils and esters by the minimal resin process is shown in Figure 2 (841). This procedure has been extensively studied on a laboratory scale and offers interesting guidelines for commercial utilization.

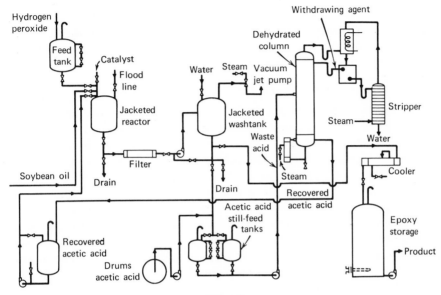

**Figure 2.**

The operation of the minimal resin process is essentially the same as that indicated for the repeated resin and sulfuric acid processes with adjustment of reaction variables as mentioned. When the reaction is complete, the resin catalyst is filtered from the reaction mixture and discarded. The reaction mixture is then pumped to a steam-jacketed washtank equipped with a vacuum jet pump. Oil and aqueous layers are allowed to separate; the lower aqueous layer is separated, and the acetic acid may be recovered by azeotropic dehydration if desired (866). The crude epoxidized product is washed successively with water or treated with alkaline reagents and then washed with water to remove residual acid. Traces of water are then removed at 60–75° under vacuum.

Continuous laboratory epoxidation procedures are known and should be amenable to commercial use (849, 867).

### 10. In Situ *Peroxyformic Acid Epoxidation of Methyl Oleate* (837)

In a reaction vessel equipped with a mechanical stirrer, thermometer, and dropping funnel 2 moles of methyl oleate and 34 g (0.66 mole) of 90% formic acid are placed. While the mixture is being stirred, 136 g (2 moles) of 50%

hydrogen peroxide is added slowly at 25–28°. The temperature is maintained in that range by the use of a cooling bath. After complete addition of the hydrogen peroxide the temperature is allowed to rise to about 40°, where it is held by cooling. The reaction mixture is stirred at 30–40° for 24 hr, after which the oil layer is separated from the aqueous layer. The oily layer is washed first with a saturated aqueous solution of sodium bicarbonate and then with water. The oil is dried at 100° under vacuum to leave 611 g of a light-yellow product containing 3.48% oxirane oxygen.

### K.  Safety Considerations (854)

In common with all peroxides, hydrogen peroxide and organic peroxy acids are thermodynamically unstable and subject to decomposition by contaminants, even when these are present in trace amounts. The decomposition reaction is highly exothermic and produces large volumes of gas. With due attention to simple safety precautions dictated by these properties little difficulty should be encounted in handling hydrogen peroxide and organic peroxy acids. These materials have been employed in large-scale organic reactions for many years. For general safety and handling considerations in the use of hydrogen peroxide and organic peroxy acids, particularly peroxyformic acid and concentrated hydrogen peroxide, a potential user should consult technical literature supplied by the manufacturers of hydrogen peroxide (836, 842–845).

When hydrogen peroxide and organic peroxy acids are to be employed as oxidizing agents for organic reactions, consideration must be given to the potential impact sensitivity of specific mixtures, especially on a large scale. Hydrogen peroxide forms potentially detonable mixtures of organic material in well-defined hydrogen peroxide concentration ranges, and such information has been published (868, 869). Generally the sensitive region is one of high hydrogen peroxide concentration. As a rough approximation, potentially detonable mixtures will be formed when hydrogen peroxide represents about one-third by volume of the final mixture—for example, 3 ml of hydrogen peroxide in 10 ml of total reactants. The line of demarcation between the explosive and safe region is not sharp. Consideration must also be given to the possibility of a phase separation or shifting of a composition into the impact-sensitive region during the course of the reaction.

Impact-sensitive mixtures of hydrogen peroxide and organic materials vary greatly in sensitivity. Many require considerable initiating energy to set them off. Nevertheless, under no circumstances should an organic oxidation reaction with hydrogen peroxide or organic peroxy acids be carried out with a mixture that falls into that range. Fortunately, from the practical standpoint, the dangerous region of hydrogen peroxide concentration is in excess of that normally encountered in epoxidation reactions. Slow and controlled addition of hydrogen peroxide to the reactants is the most effective means of avoiding impact-sensitive mixtures. The reaction should be carried out in such a way that a considerable portion of the total hydrogen peroxide employed has reacted before hydrogen peroxide addition is completed.

General safety precautions to be taken in carrying out oxidation reactions of hydrogen peroxide and peroxy acids are as follows (854):

1. Personnel should wear rubber gloves and safety glasses.

2. Reactors and storage vessels should be provided with vents.

3. Oxidizing agents should be stored in cool places.

4. All spilling should be generously flushed with water.

5. Hydrogen peroxide, peroxy acids, and reactors should be protected from contamination, especially metals.

6. Oxidants once removed should never be returned to the reactor.

7. Laboratory reactions should always be run behind a safety shield.

8. Hydrogen peroxide and peroxy acids should always be added to organic materials and not the reverse.

9. Hydrogen peroxide and peroxy acids should be added slowly.

10. Reaction mixtures should always be stirred while hydrogen peroxide or peroxy acids are being added.

11. Cooling facilities should be provided for all reactions.

12. Reaction products should never be recovered from the final reaction mixture by distillation unless residual peroxides have been destroyed.

13. Large-scale reactors should be provided with water-flooding and rapid-dumping facilities.

14. Peroxy acids and hydrogen peroxide should not be splashed on the walls of reactors in large-scale operations. A perforated pipe partially dipping under a liquid surface is suggested for feeding hydrogen peroxide into reactors.

## L. Epoxidized Oils and Esters as Plasticizers-Stabilizers (*832, 870–873*)

The major uses of epoxidized oils and esters are as plasticizers-stabilizers for polyvinyl chloride resins. Epoxidized oils and esters are incorporated into polyvinyl chloride systems to improve flexibility, elasticity, and toughness, but their major role is to impart stability to the mixture on exposure to heat and light. Epoxidized soybean oil is the plasticizer-stabilizer of choice because of its low price, but it also has the lowest volatility of the commercially available epoxy plasticizers-stabilizers. Epoxidized soybean oil cannot be used as a primary plasticizer because its compatibility with polyvinyl chloride at the levels needed for preparing highly flexible compositions is borderline.

In contrast, monomeric epoxy esters, such as epoxidized tallates and oleates, can be used both as primary plasticizers and stabilizers for polyvinyl chloride. Monomeric epoxy esters are efficient plasticizers and impart good low-temperature flexibility to polyvinyl chloride. The monomeric epoxy esters, as is the case with other efficient low-temperature plasticizers, have relatively high migration and volatility characteristics at and above room temperature. The molecular weight (chain length) of the alkyl group of the ester can be increased to reduce volatility and migration loss to more acceptable levels, but compatibility and efficiency are adversely affected. Thus, an alkyl epoxy-stearate in which the alkyl group contains 10 or more carbon atoms is incompatible with polyvinyl chloride at the 35 % level. Alkyl epoxy esters, which represent a compromise of the various factors required in a plasticizer, are epoxidized butyl, hexyl, and 2-ethylhexyl esters of oleic acid or tall-oil fatty acids.

Perhaps the most important single factor contributing to the low compatibility of an epoxidized oil or fatty acid ester with polyvinyl chloride is the presence of cleavage products containing hydroxy groups. For this reason cleavage must be kept to a minimum during epoxidation, following the recommendations already discussed.

Table 11 lists some selected properties of a commercial polyvinyl chloride–acetate copolymer (95 : 5) plasticized with various epoxy esters and epoxidized soybean oil. For comparison, data for compositions plasticized with di-2-ethylhexyl phthalate (DOP) and tricesyl phosphate (TCP) are also included (872).

The modulus column shows that epoxidized soybean oil is less efficient than DOP but more efficient than TCP. The stiffening temperature of the composition containing epoxidized soybean oil is substantially higher than that

TABLE 11.  CHARACTERISTICS OF POLYVINYL CHLORIDE PLASTICIZED WITH EPOXIDIZED OILS AND ESTERS[a]

| Epoxidized Ester | Evaluation Results on Plasticized Compositions | | Stiffening Temperature (C), Clash–Berg |
|---|---|---|---|
| | Elongation | 100% Modulus | |
| Epoxidized soybean oil | 280 | 1600 | −15 to −20 |
| Butyl epoxystearate | 450 | 1000 | −50 |
| Butyl epoxytallate | 380 | 1200 | −40 |
| Hexyl epoxystearate | 300 | 1300 | −47 |
| 2-Ethylbutyl epoxystearate | 350 | 1300 | −55 |
| DOP[b] + 1 part PbCO$_3$ | 390 | 1300 | −30 |
| TCP[c] + 1 part PbCO$_3$ | 300 | 2000 | 0 |

[a] Recipe: Polyvinyl chloride 64.5%, epoxy ester 35%, stearic acid 0.5%.
[b] Di-2-ethylhexyl phthalate.
[c] Tricresyl phosphate.

with DOP but below that with TCP. After about a month, especially on exposure to light, sheets plasticized with epoxidized soybean oil as the only plasticizer at the 35% level show marked exudation. The monomeric epoxy esters, however, are not only more efficient plasticizers than DOP, as the modulus and elongations show, but also superior low-temperature plasticizers. Furthermore, the alkyl epoxystearates show a small rate of increase of modulus with decrease in temperature. Therefore compositions containing alkyl epoxystearates stiffen gradually and not abruptly as the temperature is lowered.

The most striking attribute of all epoxy esters listed, however, and the main reason for their wide industrial acceptance, is the marked improvement they impart to the light stability of plasticized polyvinyl chloride. Thus exposure of the epoxy plasticizer–polyvinyl chloride compositions of Table 11 to ultraviolet light in a weatherometer at 70–75° produces little or no darkening in color after 400 hr. In contrast, sheets plasticized with DOP or TCP developed a pronounced amber color in 72 or 24 hr, respectively.

The usual commercial practice for stabilizing polyvinyl chloride is to use epoxidized oils and/or esters at a 1–10% level, based on the total weight of the composition. Although epoxy esters used alone improve heat stability considerably, their effect is to triple or quadruple light stability when used in synergistic combinations. The following recipe is typical for obtaining good heat and light stability (874):

Polyvinyl chloride, 100
Di-2-ethylhexyl phthalate, 45–50
Epoxidized oil or ester, 3–5
Ba-Cd laurate, 2
Phosphite chelater, 1; and substituted benzophenone, 2.

Toxicity and pharmacological studies on epoxidized soybean and linseed oils have been conducted (875, 876).

## M. Other Epoxidized Compounds and Applications

New uses for epoxidized plasticizers-stabilizers and other epoxidized substances are expected to increase the demand for these substances sharply. Promising applications in the polymer field (alkyds, polyesters, and epoxy resins) are among the more important developments. Considerable activity in these fields occurred between 1950 and 1960.

It has been reported (877) that alkyd surface coatings of low free acidity are conveniently prepared by the use of epoxidized soybean oil as a component of the reaction system. Also, use of epoxidized soybean oil as a partial replacement for glycerol in a linseed fatty acid–phthalic anhydride–glycerol alkyd of long oil length is claimed to yield surface coatings in a shorter processing time (878, 879). These air-dry faster and are lighter in colour, tougher, and better adhering than those without epoxidized oil. Use of larger quantities of epoxidized soybean oil is reported to give surface-coating alkyds containing free epoxy groups (880). Such coatings are claimed to have better adherence and anticorrosive properties. The epoxy groups permit the facile incorporation of groups containing phosphorus and chlorine for flame resistance.

Compatible epoxy-resin modifying agents are prepared from epoxidized soybean oil, phthalic anhydride, and rosin (878, 879). Epoxidized soybean oil has been added directly to epoxy resins as an inexpensive modifying agent (880). In epoxy ester surface coatings cost has also been reduced and corrosion resistance increased by replacement of part of the epoxy resin with epoxidized soybean oil (880). In view of the low price of a partially epoxidized soybean oil, such a material is worthy of consideration for cost reduction.

Emulsions of polyepoxides are reported to be of value as surface coatings and as additives for fabrics, paper, and leather to improve the shrink resistance, crease resistance, abrasion resistance, and wet strength (881, 882). Among the polyepoxides reported are epoxidized glycerol trioleate and trilinoleate. These polyepoxide films have been cured with acid or amine cross-linking agents. However, the most likely epoxides for such purposes are terminal epoxides because of their higher reactivity.

Epoxy resins have also been prepared by polymerization of various types of cycloaliphatic epoxides with epoxidized linseed oil and phthalic anhydride (883).

Uses have been proposed for unsaturated epoxides in epoxy resins. Unsaturated epoxides that have been suggested are vinyl and allyl mono- and diepoxystearates (884, 885). These monomers are initially polymerized by vinyl polymerization to yield saturated epoxy-containing prepolymers that are finally cured via the epoxy groups. Conversely, unsaturated prepolymers can be prepared by epoxy-type polymerization, followed by vinyl-type polymerization or copolymerization.

Vinyl–epoxy copolymers have also been studied as surface coatings. Films prepared from a vinyl acetate–vinyl epoxystearate latex can be cross-linked to form insoluble coatings by amines or dibasic acids (886).

Epoxy-resin prepolymers have also been prepared by the epoxidation of low-molecular-weight saturated fatty acid polyesters, such as the reaction product of dimerized linoleic acid and ethylene glycol (887), and an oil-modified polyester prepared from glycerol trioleate, pentaerythritol, and phthalic anhydride (888, 889). The epoxidized prepolymers are then cross-linked by reaction of epoxide groups with dibasic acids or polyamines.

Epoxidized glycol esters and ether alcohol esters of fatty acids have been prepared. Some of them—such as ethylene glycol monobutyl ether epoxy-stearate and tetrahydrofurfuryl epoxystearate—are claimed to be primary plasticizers as well as stabilizers for polyvinyl chloride (890). Epoxidized polymeric plasticizers, such as the epoxidized polyester from oleic acid, propylene glycol, and adipic acid, have also been reported to be primary plasticizers and stabilizers (891).

Epoxidized esters of unsaturated fatty alcohols and adipic acid, such as epoxidized dioleyl adipate, have also been prepared for plasticizer use and are claimed to be primary plasticizers (892).

Epoxy esters containing epoxy groups in both the alcohol and the acid portions of the molecule have also been investigated as plasticizers-stabilizers. Some examples are glycidyl epoxystearate (893), 3,4-epoxy-cyclohexylmethyl epoxystearate (894, 895), and 3,4-epoxycyclohexylmethyl diepoxystearate (895, 896). The last one is claimed to be an unusually effective light stabilizer for polyvinyl chloride.

Incorporation of epoxidized oils and esters into nitrocellulose lacquer formulations is reported to eliminate important deficiencies of these lacquers (897).

The epoxidation of the Diels–Alder product (aldrin) of hexachlorocyclopentadiene and bicyclo[2.2.1]-2,5-heptadiene with preformed peroxyacetic acid yields generally useful insecticides (898, 899). The epoxidized products are known commercially as dieldrin (*endo-exo* isomer) and endrin (*endo-endo* isomer).

Epoxides of the steroid series have been prepared with peroxyacetic acid, peroxybenzoic acid, and other epoxidation agents (900, 901).

The epoxidation of unsaturated polymers and copolymers to introduce the highly reactive oxirane function is another relatively new development in epoxidation technology. The subject has been reviewed (902), and only a few major points are discussed here.

Numerous unsaturated polymers are commercially available at low cost from conjugated dienes, chiefly butadiene and isoprene, or by copolymerization of such dienes with styrene or other monomers. The double bonds remaining in these polymers and copolymers can be partially or completely epoxidized either by preformed peroxy acids or by *in situ* methods. More efficient conversion of double bonds to oxirane groups with fewer unwanted side reactions is accomplished by using preformed peroxy acids. The introduction of highly reactive oxirane groups onto the backbone of a relatively unreactive hydrocarbon polymer permits facile cross-linking and other chemical reactions to be carried out (903–907).

Generally, for ease of epoxidation and subsequent handling, the unsaturated polymer should be a liquid of relatively low molecular weight (861) or a solid readily soluble in a solvent suitable for carrying out epoxidation reactions. A tabulation of polymers or copolymers that have been epoxidized by preformed peroxy acids or by hydrogen peroxide and a lower aliphatic acid can be found in references 840, 841, and 902.

The epoxidation of latexes of copolymers of butadiene and styrene, butadiene and acrylonitrile, and butadiene and α-methylstyrene with peroxyacetic acid has been reported (908). The epoxidation of polyesters of tetrahydrophthalic acid with hydroxyl-terminated unsaturated alkyd resins by a modified *in situ* hydrogen peroxide–acetic acid process using dehydrated resin has also been described (888). Isolation of the epoxidized polyesters poses a problem, however. The epoxidation of partially dehydrochlorinated polyvinyl chloride (909) and depolymerized rubber (910) has also been reported. Interaction of polyoxyethylene glycols and epoxidized soybean oil is reported to yield nonionic detergents (911).

## N.  Miscellaneous

Other references that are relevant to the subjects of this section are the following: epoxidation of oils and esters of unsaturated fatty acids (912–919); partial epoxidation of polyunsaturated fatty materials (920–924) and their use as plasticizers-stabilizers before or after hydrogenation of residual unsaturation; continuous epoxidation (925–927); epoxidation of diacetylated monoglycerides (928, 929); epoxidized fatty materials as lubricant additives (930, 931); curing epoxidized oils and epoxidized polymers (932–941); and epoxidized hexahalocyclopentadiene Diels–Alder products as herbicides (942).

## V.  KINETICS, SOLVENT EFFECTS, STEREOCHEMISTRY, AND MECHANISMS

### A.  Kinetics and Solvent Effects

Epoxidation of unsaturated compounds is the most general and widely used reaction for the introduction of the oxirane group into organic compounds. An enormous number of unsaturated compounds have been partially or completely epoxidized by this reaction (see Section II). Rates of epoxidation span a wide range and are dependent on the number and kind of substituents attached or proximal to the double bond.

Numerous investigators have examined the kinetics of epoxidation. The earliest work is that of Prileschajew (16–18), the discoverer of the epoxidation reaction, who reported that in the epoxidation of diunsaturated compounds the less "hydrogenated" (more highly alkyl substituted) of the two unsaturated linkages is epoxidized first. Prileschajew also concluded that the velocity of the epoxidation reaction and the properties of the oxirane compound obtained depend not only on the distribution of oxygen-containing groups in the vicinity of the double bond but also on their structure—namely, whether they are aldehyde, hydroxyl or carbonyl groups. Thus geraniol, on reaction with 1 mole of peroxybenzoic acid, is epoxidized almost exclusively at the more highly substituted double bond, as shown below:

$$(CH_3)_2C = CH - (CH_2)_2 - C(CH_3) = CH - CH_2OH$$

$$\downarrow \quad C_6H_5CO_3H \quad \text{(1 mole)}$$

$$(CH_3)_2C \underset{O}{\overset{}{\diagdown \diagup}} CH - (CH_2)_2 - C(CH_3) = CH - CH_2OH$$

Likewise linalyl acetate could not be converted to the diepoxide, the double bond adjacent to the acetoxy group being oxidized only with great difficulty. Citral behaves similarly, the double bond $\alpha,\beta$ to the carbonyl group being essentially unepoxidizable, whereas the terminal double bond to which two methyl groups are attached is epoxidized readily. Cinnamic acid was also reported as not epoxidizable, as was methyl cinnamate somewhat later (943). Although Prileschajew did not realize the full impact of his conclusions, which were based solely on qualitative rate studies, in modern terms it is now known that electron-donating groups in the vicinity of the double bond speed up epoxidation whereas electron-withdrawing groups slow it down.

In the period from 1920 to 1940 several groups of investigators systematically studied the effect of the structure of unsaturated compounds on the rate of reaction with peroxy acids. The primary contributions were made by Boeseken and his co-workers (943–952), who studied the rates of reaction of peroxyacetic and peroxybenzoic acids with a wide variety of unsaturated compounds over a range of temperatures. Although there are some inconsistencies in the work of Boeseken, owing largely to the difficulty in assessing the purity and structure of unsaturated compounds without modern instruments, nevertheless, general guidelines for defining the effect of structure on rate of epoxidation and the order and thermodynamic parameters of the reaction were fairly well defined by 1940. Epoxidation is a second-order reaction, first order in olefin and first order in peroxy acid. Other investigators who made significant contributions during that period were Meerwein et al. (953), Bodendorf (954), Medvedev and Blokh (778) and Heinanen (955).

In the early 1940s, however, Swern and co-workers (956) not only studied the kinetics of epoxidation of pure unsaturated compounds but also organized the large volume of widely dispersed kinetic information then available and described in detail the relationship between the structure of unsaturated compound and the rate of epoxidation. Since then numerous investigators have studied the kinetics of epoxidation and calculated activation parameters; only selected literature references are given (957–978).

Table 12 is a general summary of the effect of the structure of unsaturated compound on the rate of epoxidation by peroxyacetic acid in acetic acid.

The effect of olefin structure on rate applies to other peroxy acids as well. Although the absolute rates vary with change in peroxy acid structure and solvents used (to be discussed later), the relative rates in Table 12 generally apply.

TABLE **12.**    RELATIVE RATES OF EPOXIDATION OF UNSATURATED COMPOUNDS WITH PEROXYACETIC ACID IN ACETIC ACID

| Compound | Formula | Relative Rate of Epoxidation |
|---|---|---|
| Ethylene | $CH_2=CH_2$ | 1 |
| *A. Monoalkyl-Substituted Ethylenes*[a] | | 25 |
| Propylene | $CH_3CH=CH_2$ | |
| 1-Pentene | $CH_3(CH_2)_2CH=CH_2$ | |
| 1-Hexene | $CH_3(CH_2)_3CH=CH_2$ | |
| 1-Heptene | $CH_3(CH_2)_4CH=CH_2$ | |
| 1-Octene | $CH_3(CH_2)_5CH=CH_2$ | |
| 1-Decene | $CH_3(CH_2)_7CH=CH_2$ | |
| Methyl 10-hendecenoate | $CH_2=CH(CH_2)_8COOCH_3$ | |
| *B. Dialkyl-substituted Ethylenes*[b] | | 500–600 |
| 2-Butene | $CH_3CH=CHCH_3$ | |
| 2-Methyl-1-propene | $CH_3C(CH_3)=CH_2$ | |
| 2-Pentene | $CH_3CH_2CH=CHCH_3$ | |
| 2-Hexene | $CH_3(CH_2)_2CH=CHCH_3$ | |
| 3-Hexene | $CH_3CH_2CH=CHCH_2CH_3$ | |
| 3-Heptene | $CH_3(CH_2)_2CH=CHCH_2CH_3$ | |
| 4-Nonene | $CH_3(CH_2)_3CH=CH(CH_2)_2CH_3$ | |
| Methyl oleate | $CH_3(CH_2)_7CH=CH(CH_2)_7CO_2CH_3$ | |
| *C. Trialkylsubstituted Ethylenes*[c] | | 6000 |
| 2-Methyl-2-butene | $CH_3C(CH_3)=CHCH_3$ | |
| *D. Maleate Esters* $(RCO_2CH=CHCO_2R)$ | | ca.0[d] |
| *E. Vinyl Esters* | | ca. 0.2[d] |
| Vinyl acetate | $CH_3CO_2CH=CH_2$ | |
| Vinyl laurate | $CH_3(CH_2)_{10}CO_2CH=CH_2$ | |
| *F. Allyl Esters* | | 1–2[d] |
| Allyl acetate | $CH_3CO_2CH_2CH=CH_2$ | |

TABLE **12** (continued)

| Compound | Formula | Relative Rate of Epoxidation |
|---|---|---|
| | *G.*   *Cyclic Olefins* | |
| Cyclopentene | $(C_5H_8)$ | 1000 |
| Cyclohexene | $(C_6H_{10})$ | 675 |
| Cycloheptene | $(C_7H_{12})$ | 900 |
| | *H.*   *Aryl-Substituted Olefins* | |
| Allylbenzene | $-CH_2CH{=}CH_2$ | 10 |
| Stilbene | $-CH{=}CH-$ | 27 |
| Styrene | $-CH{=}CH_2$ | 60 |
| 1-Phenyl-1-propene | $-CH{=}CH-CH_3$ | 240 |
| 1,1-Diphenylethylene | $-C{=}CH_2$ | 250 |

[a] Specific reaction-rate constants $4.0–5.0 \times 10^{-3}$ l/mole-min at $25.8°$.
[b] Specific reaction-rate constants $90–100 \times 10^{-3}$ l/mole-min at $25.8°$.
[c] Specific reaction-rate constants about $1000 \times 10^{-3}$ l/mole-min at $25.8°$.
[d] Estimated

If the relative rate of epoxidation of ethylene is taken as 1, Table 12 shows that substitution of one hydrogen by an electron-donating alkyl group, as in propylene or methyl undecylenate, increases the rate approximately 25 times. The size and bulk of the alkyl substituent make no difference; a small alkyl group has essentially the same effect as a large one. The substitution of two electron-donating alkyl groups irrespective of size on one or both of the carbon atoms of the double bond causes a further increase in rate by a factor of approximately 20 (979). Three electron-donating alkyl groups cause a further rate increase of approximately 12 times.

In contrast, electron-withdrawing groups attached to the double bond cause a marked reduction in the rate of epoxidation. The reduction in rate is most dramatically seen with maleate esters, in which decomposition of the peroxy acid is faster than epoxidation, which is, for all practical purposes, immeasurably slow.

With cycloalkenes the epoxidation rate is on the order of that of dialkyl-substituted ethylenes, possibly somewhat higher. Aryl-substituted ethylenes cover a wide range, with the phenyl group behaving either as an electron-donating group (styrene, 1,1-diphenylethylene, 1-phenyl-1-propene) or electron-withdrawing group (stilbene).

In the epoxidation of conjugated dienes one double bond is epoxidized rapidly, but the second reacts slowly (946, 948). It is assumed that the reduction in the rate of epoxidation of the second double bond is caused by its proximity to the electron-withdrawing oxirane function. The rate difference between the two double bonds is sufficiently large for monoepoxides of conjugated dienes, such as butadiene monoepoxide and isoprene monoepoxide, to be prepared. With isoprene the more highly substituted double bond is epoxidized first (827, 980). The rate of epoxidation of conjugated dienes increases in the order butadiene, isoprene, piperylene, 2,3-dimethylbutadiene, 2,4-hexadiene (980).

$\alpha,\beta$-Unsaturated aldehydes, ketones, acids, and esters react extremely slowly with peroxybenzoic acid, an expected result (954). Cinnamyl alcohol, however, is reported to epoxidize at a normal rate (954). In the epoxidation of sorbic acid, its esters, and other derivatives the double bond distant from the carboxy group or carboxylate function epoxidizes considerably more rapidly than that conjugated with the electron-withdrawing group (955). Sorbic derivatives can therefore be selectively and cleanly monoepoxidized.

Selective epoxidation is a useful technique for determining external and internal double bonds in synthetic rubbers as the internal double bonds are epoxidized considerably more rapidly (981, 982). If the rates of epoxidation of standard unsaturated compounds containing a terminal double bond, as in undecylenic acid, and an internal double bond, as in oleic acid, are known under specific condition, it is possible to calculate the ratio of internal and external double bonds in styrene–butadiene synthetic rubbers by means of a calibration curve (981, 982).

Early in the history of epoxidation it was reported that *cis* double bonds in straight-chain fatty acids and esters are epoxidized approximately 1.5 times as rapidly as *trans* double bonds (956), and this conclusion has been repeatedly confirmed. Work on the isomeric *n*-undecenes indicates that this conclusion is general with straight-chain olefins (979). However, *trans*-cyclodecene and 1,5-cyclododecadiene are reported to epoxidize about six to eight times as fast as their *cis*- isomers (64, 410).

The first report on the effect of peroxy acid structure on the epoxidation rate is that of Medvedev and Blokh (778), who studied the rate of epoxidation of cyclohexene at 25° with peroxyacetic, peroxybenzoic, peroxyphenylacetic, *p*-methoxyperoxybenzoic, α- and β-peroxynaphthoic, and *m*- and *p*-nitro-peroxybenzoic acids. Besides confirming that epoxidation is second order and the energy of activation is about 14 kcal/mole, they noted that *m*- and *p*-nitro-peroxybenzoic acids are faster epoxidizing reagents than peroxybenzoic acid and also that that *p*-methoxyperoxybenzoic acid is slower. The conclusion in modern terms is that electron-withdrawing groups in a ring-substituted peroxybenzoic acid increase the electrophilic character of the oxidant and therefore enhance the epoxidation rate, whereas electron-donor groups diminish it. This conclusion has been confirmed many times (69, 780). Furthermore, electron-withdrawing groups in aliphatic peroxy acids also enhance the epoxidation rate; a good example is trifluoroperoxyacetic acid, which is an extremely rapid epoxidation reagent. *m*-Chloroperoxy-benzoic acid, also a rapid epoxidizer, is available commercially and has become the epoxidation reagent of choice in many cases.

The first systematic and thorough determination of activation parameters in epoxidation with aromatic peroxy acids were reported in 1955 by Lynch and Pausacker (69), although some information was in the older literature (950). These investigators showed that electron-withdrawing substituents in the *meta* and *para* position of *trans*-stilbene, as anticipated, cause a diminution in rate, whereas electron-donating groups cause rate enhancement. The reaction was shown to be second order. Table 13 lists rate constants and activation parameters in the epoxidation of variously substituted stilbenes with peroxy-benzoic acid. Table 14 presents similar information on the epoxidation of stilbene with substituted peroxybenzoic acids.

The most significant conclusion from Tables 13 and 14 is that the entropy of activation is sensibly constant regardless of the structure of the stilbene or peroxy acid. The changes in the heats ($\Delta H \pm$) and free energies ($\Delta F \pm$) of activation are small but significant; electron-donating groups in *trans*-stilbenes decrease and electron-withdrawing groups increase the values. With the substituted peroxybenzoic acids, electron-withdrawing groups decrease and

TABLE 13.   KINETIC AND THERMODYNAMIC DATA[a] FOR
EPOXIDATION OF *p*-SUBSTITUTED *trans*-STILBENES
WITH PEROXYBENZOIC ACID IN BENZENE[d]

| *p*-Substituent | $\Delta H\pm$ | $\Delta F\pm$ | $-\Delta S\pm$ | $k^c\ (\times 10^4)$ | |
|---|---|---|---|---|---|
| | | | | 35° | 30° |
| 4,4′-(OMe)₂ | 14.0 | 20.4 | 21.1 | 83 | 128 |
| OMe | 14.5 | 21.2 | 22.3 | 20 | 31 |
| Me | 15.0 | 21.7 | 22.2 | 9.7 | 15 |
| H | 15.2 | 22.2 | 22.8 | 4.3 | 6.6 |
| Cl | 15.6 | 22.4 | 22.6 | 2.7 | 4.3 |
| NO₂ | 16.6 | 23.3 | 22.0 | 0.6 | 1.0 |

[a] $\Delta H\pm$ and $\Delta F\pm$ in kilocalories per mole; $\Delta S\pm$ in entropy units.
[b] Data from reference 69.
[c] Specific reaction-rate constant, in liters per mole per second.

TABLE 14.   KINETIC AND THERMODYNAMIC DATA[a] FOR
EPOXIDATION OF *trans*-STILBENE WITH *p*-SUBSTITUTED
PEROXYBENZOIC ACIDS IN BENZENE[b]

| *p*-Substituent | $\Delta H\pm$ | $\Delta F\pm$ | $-\Delta S\pm$ | $k^c\ (\times 10^4)$ | |
|---|---|---|---|---|---|
| | | | | 25° | 30° |
| NO₂ | 13.8 | 20.8 | 23.1 | 44 | 67 |
| Cl | 14.9 | 21.6 | 22.4 | 10 | 15 |
| H | 15.2 | 22.2 | 22.8 | 4.3 | 6.6 |
| Me | 15.6 | 22.5 | 22.9 | 2.4 | 3.7 |
| OMe | 16.0 | 23.0 | 22.8 | 1.3 | 2.2 |

[a] $\Delta H\pm$ and $\Delta F\pm$ in kilocalories per mole; $\Delta S\pm$ in entropy units.
[b] Data from reference 69.
[c] Specific reaction-rate constant, in liters per mole per second.

electron-donating substituents increase the activation parameters ($\Delta H\pm$ and
and $\Delta F\pm$). Furthermore, a Hammett plot of the data gives satisfactory straight
lines with $\rho$ values of $-0.9$ and $+0.9$, respectively, for the reaction of peroxybenzoic acid with the substituted stilbenes and *trans*-stilbene with substituted
peroxybenzoic acids (69, 966). Similar results have been obtained (970) in
the peroxybenzoic and ring-substituted peroxybenzoic acid epoxidation of
α-methylstilbene and ring-substituted α-methylstilbenes in benzene solution.
In a study of the epoxidation of *meta*- and *para*-substituted styrenes with

peroxybenzoic acid in benezene solution a Hammett $\rho$ of $-1.30$ was observed in the *meta* series (983). These results are consistent with the conclusion (956) that electron-donor groups in the olefin (nucleophile) enhance the rate and electron-withdrawing groups diminish it; the converse applies with respect to the effect of substituents in the peroxy acid (electrophile).

The epoxidation reaction was reported to be uncatalyzed by acid and of second order (69). However, Berti and Bottari (984) report that the rate of epoxidation of stilbenecarboxylic acids with peroxybenzoic acid is increased (catalyzed?) by trichloroacetic acid. Sapunov and Lebedev (985) studied the kinetics of the epoxidation of allyl chloride with peroxyacetic acid and also reported catalysis by acids in accordance with the Bronsted equation.

With phenyl-substituted ethylenes Lynch and Pausacker (69) also reported that the rate of epoxidation with peroxybenzoic acid decreases in the following order: styrene, *trans*-stilbene, triphenylethylene, tetraphenylethylene. *cis*-Stilbene is epoxidized about twice as fast as *trans*-stilbene. The latter is planar, and the electron density at the double bond is diminished by delocalization into the aromatic rings. It was also noted that peroxybenzoic acid epoxidation is faster in benzene than in diethyl ether solution (69).

More recently a study of solvent and structural effects on kinetics and activation parameters in the epoxidation of unsaturated compounds with peroxyacetic acid was reported by Frostick, Phillips, and Starcher (969). Table 15 lists the actual and relative rates of reaction of various olefinic compounds with peroxyacetic acid in ethyl acetate at 25°. The results are generally consistent with the data in Table 12 and confirm the conclusions of the effect of olefin structure on the rate of reaction.

TABLE **15.** KINETICS OF REACTION OF UNSATURATED COMPOUNDS WITH PEROXYACETIC ACID IN ETHYL ACETATE AT $25°$[a]

| Compound | $k$[b] $(\times 10^6)$ | Relative Rate of Reaction |
|---|---|---|
| Vinyl acetate | $<0.1$[c] | 0[c] |
| Ethyl acrylate | $<0.1$[c] | 0[c] |
| Ethyl crotonate | 0.150 | 1 |
| Allyl acetate | 0.669 | 4.5 |
| Ethyl 2-ethyl-2-hexenoate | 4.18 | 28 |
| Allyl ether | 5.29 | 35 |
| Styrene | 34.3 | 229 |
| Ethyl 3-cyclohexenecarboxylate | 123 | 820 |
| Cyclohexene | 283 | 1885 |

[a] Data from reference 969.

[b] Specific reaction-rate constant, in liters per mole per second.

[c] Estimated value, as rate of decomposition of peroxyacetic acid is apparently greater than epoxidation rate.

Table 16 lists specific reaction-rate constants and activation parameters for the peroxyacetic acid epoxidation of unsaturated compounds in ethyl acetate, acetic acid, and benzene solution (969). As the table shows, specific reaction-rate constants are altered by change of solvent. Typically $k_{EtOAc}:k_{HOAc}:k_{C_6H_6}$ for styrene at 20° is 1:5.5:7.5; for allyl acetate at 45° the ratio is 1:2.7:3.5. Although the effect of solvent on rate is only moderate, variations in the heat of activation $\Delta H\pm$ and entropy of activation $\Delta S\pm$ are more marked. The choice of solvent therefore offers another means of controlling the epoxidation rate. From the practical point of view in achieving high yields of oxirane with minimum ring opening, especially with slowly reacting olefins, the solvent selected can mean the difference between success and failure.

TABLE 16. SPECIFIC REACTION RATE CONSTANTS AND ACTIVATION PARAMETERS IN EPOXIDATION OF UNSATURATED COMPOUNDS WITH PEROXYACETIC ACID IN VARIOUS SOLVENTS[a]

| Unsaturated Compound | Solvent | Temp. (°C) | $k^b$ ($\times 10^6$) | $\Delta H\pm$ (kcal/mole) | $-\Delta S\pm$ (e.u.) |
|---|---|---|---|---|---|
| Diallyl ether | Ethyl acetate | 20 | 3.14 | 18.0 | 22.2 |
| | | 30 | 8.56 | | |
| | | 35 | 13.9 | | |
| | | 50 | 55.3 | | |
| | Benzene | 25 | 38.3 | 15.2 | 28.1 |
| | | 35 | 88.6 | | |
| Allyl acetate | Ethyl acetate | 24 | 0.585 | 24.0 | 6.2 |
| | | 30 | 1.35 | | |
| | | 45 | 9.10 | | |
| | Acetic acid | 23 | 2.65 | 19.5 | 18.0 |
| | | 30 | 5.70 | | |
| | | 45 | 24.3 | | |
| | Benzene | 23 | 3.38 | 18.8 | 19.9 |
| | | 45 | 31.0 | | |
| Cyclohexene | Ethyl acetate | 0 | 20.2 | 16.9 | 16.2 |
| | | 15 | 104 | | |
| | | 20 | 172 | | |
| | | 25 | 283 | | |
| | Acetic acid | 15 | 82.2 | 15.6 | 18.6 |
| | | 25.8 | 2150 | | |
| | | 39.6 | 6740 | | |

TABLE 16 (Continued)

| Unsaturated Compound | Solvent | Temp. (°C) | $k^b$ (×10⁶) | $\Delta H\pm$ (kcal/mole) | $-\Delta S\pm$ (e.u.) |
|---|---|---|---|---|---|
| Ethyl crotonate | Ethyl Acetate | 30 | 0.262 | 20.1 | 22.0 |
| | | 40 | 0.793 | | |
| | | 60 | 5.27 | | |
| Ethyl 3-cyclo-hexenecar-boxylate | Ethyl acetate | 25 | 12.3 | 16.9 | 19.7 |
| | | 35 | 313 | | |
| | Acetic acid | 15 | 345 | | |
| | Benzene | 0 | 115 | 14.2 | 20.0 |
| | | 15 | 452 | | |
| Ethyl 2-ethyl-2-hexenate | Ethyl acetate | 40 | 19.2 | 18.8 | 18.5 |
| | | 60 | 117.5 | | |
| Styrene | Ethyl acetate | 20 | 22.0 | 16.2 | 22.6 |
| | | 25 | 34.3 | | |
| | | 30 | 55.3 | | |
| | | 40 | 127 | | |
| | Acetic acid | 20.5 | 124 | 13.9 | 29.3 |
| | | 25 | 186 | | |
| | | 25.8 | 188 | | |
| | | 40.8 | 574 | | |
| | Benzene | 0 | 26.8 | 14.3 | 27.0 |
| | | 20 | 162 | | |

ᵃ Data from reference 969.
ᵇ Specific reaction-rate constant, in liters per mole per second.

The first report in which the rate of epoxidation was compared in a large number of solvents is that of Renolen and Ugelstad (986), who compared the rate of epoxidation of cyclohexene at 20° with peroxybenzoic acid in 11 solvents; results are summarized in Table 17. The highest rates are with halogenated solvents, notably chloroform, methylene chloride and chlorobenzene, and the lowest rates are with solvents of higher basicity, notably ethers but also ethyl acetate, which can presumably solvate the peroxy acid effectively.

The effect of solvation on rate (Table 17) can be largely explained by the differences in the values of $\Delta H\pm$. The increase in $\Delta H\pm$ in going from carbon tetrachloride to diethyl ether is probably due to hydrogen bonding between diethyl ether and the peroxy acid, which must be ruptured in order to achieve

TABLE 17.   RATES OF EPOXIDATION OF CYCLOHEXENE WITH PEROXYBENZOIC
ACID IN VARIOUS SOLVENTS[a]

| Solvent | $k^{b}$ ($\times 10^4$) | Relative Rate of Reaction | $\Delta H\pm$ (kcal/mole) | $-\Delta S\pm$ (e.u.) |
|---|---|---|---|---|
| Chloroform | 472 | 122 | 9.6 | 31.9 |
| Methylene chloride | 225 | 58 | 9.8 | 32.7 |
| Chlorobenzene | 188 | 48 | 11.2 | 28.0 |
| Benzene | 156 | 40 | 11.0 | 29.3 |
| Carbon tetrachloride | 77.2 | 2.0 | 9.9 | 34.7 |
| Cyclohexane | 33.0 | 8.5 | 11.1 | 32.1 |
| n-Hexane | 24.0 | 6.2 | 11.9 | 29.5 |
| Ethyl acetate | 12.3 | 2.3 | 13.3 | 26.3 |
| Dioxane | 9.75 | 2.5 | 13.3 | 26.9 |
| Tetrahydrofuran | 4.33 | 1.1 | 14.4 | 24.6 |
| Diethyl ether | 3.88 | 1 | 14.4 | 24.9 |

[a] Data from reference 986.
[b] Specific reaction-rate constant, in liters per mole per second at 20°.

the transition state between olefin and peroxy acid required for epoxidation
to proceed. Entropy effects are essentially the reverse (ether more favorable
than carbon tetrachloride), but they do not have the same quantitative effect
on rate. Attempts to correlate the rate of epoxidation with the dielectric
constant of the solvent have not been especially successful; there is a rough
correlation between the dielectric constant and the epoxidation rate.

Solvent effects on the rates of epoxidation of trans-stilbene and ethyl
crotonate with m-chloroperoxybenzoic acid have been reported by Schwartz
and Blumberg (718). Tables 18 and 19 summarize the results for trans-stilbene
and ethyl crotonate, respectively. As Table 18 shows, the chlorinated solvents
and benzene provide the highest epoxidation rates, in agreement with the
earlier results of Renolen and Ugelstad (986). A large rate decrease is noted
when t-butyl alcohol is the solvent. The correlation between the epoxidation
rate and the dielectric constant of the solvent is quite rough, and it is doubtful
that any meaningful conclusion can be drawn. However, t-butyl alcohol,
which might be expected to disrupt the intramolecular hydrogen bonding
of the peroxy acid (987) and form its own hydrogen bonds with the peroxy
acid, is the solvent in which the epoxidation rate is the lowest.
$(k_{t\text{-BuOH}} : k_{CH_2Cl_2}$ $_{(CHCl_3 \text{ or } o\text{-}C_6H_4Cl_2)}$ $1:86)$.

As Table 19 shows, the rate of epoxidation of ethyl crotonate is substantially lower than that of trans-stilbene, a quite anticipated result from
previous work (Table 15) (969), and solvent effects are not as dramatic.
Nevertheless, the chlorinated solvents are again superior and acetic acid is

TABLE 18. SPECIFIC REACTION-RATE CONSTANTS FOR EPOXIDATION OF *trans*-STILBENE WITH *m*-CHLOROPEROXYBENZOIC ACID IN VARIOUS SOLVENTS AT 30°[a]

| Solvent | Dielectric Constant, 30° | $k$[b] $(\times 10^4)$ | Relative Rate of of Reaction |
|---|---|---|---|
| *o*-Dichlorobenzene | 9.45 | 36.3 | 86 |
| Methylene chloride | 8.7 | 36.1 | 86 |
| Chloroform[b] | 4.6 | 36.0 | 86 |
| 1,2-Dichloroethane[d] | 10.1 | 32.7 | 78 |
| Chlorobenzene | 5.54 | 27.3 | 65 |
| Benzene | 2.26 | 18.9[e] | 45 |
| Carbon tetrachloride | 2.2 | 14.6 | 35 |
| *t*-Butyl alcohol | 11.7 | 0.42 | 1 |

[a] Data from reference 718.
[b] Specific reaction-rate constant, in liters per mole per second. Reactants at about 0.05 mole/l.
[c] Two mole % of trifluoroacetic acid present per mole of peroxy acid.
[d] Rate unaffected by 2 mole % of trifluoroacetic acid per mole of peroxy acid.
[e] Lynch and Pausacker report 23.3 (69).

TABLE 19. SPECIFIC REACTION RATE CONSTANTS FOR EPOXIDATION OF ETHYL CROTONATE WITH *m*-CHLOROPEROXYBENZOIC ACID IN VARIOUS SOLVENTS[a]

| Solvent | Dielectric Constant | $k$[b] $(\times 10^4)$ | $T$ (°C) |
|---|---|---|---|
| 1,2-Dichloroethane | 10.1 | 1.73 | 30 |
| Benzene | 2.26 | 1.73 | 30 |
| Carbon tetrachloride | 2.23 | 1.15 | 25 |
| Chloroform[c] | 4.73 | 1.15 | 25 |
| Methylene chloride | 8.89 | 1.15 | 25 |
| 1,2-Dichloroethane[d] | 10.36 | 1.15 | 25 |
| 1,2-Dichloroethane[e] | 10.36 | 0.77 | 25 |
| Acetonitrile | — | 0.625 | 25 |
| Acetic acid (90%) Water (10%) | — | 0.53 | 30 |
| Ethyl acetate | 6.02 | 0.193 | 26 |
| Acetic acid (95%) Acetic anhydride (5%) | — | 0.075 | 30 |

[a] Data from reference 718.
[b] Specific reaction-rate constant, in liters per mole per second.
[c] Two mole % of trifluoroacetic acid present per mole of peroxy acid.
[d] Rate unaffected by 2 mole % of trifluoroacetic acid per mole of peroxy acid.
[e] Equivalent amounts of trichloroacetic acid, peroxy acid, and ethyl crotonate.

poor, especially when it contains acetic anhydride. The rates of epoxidation of *trans*-stilbene and ethyl crotonate are substantially unaffected by strong acids, such as trifluoroacetic and trichloroacetic acids.

Sapunov and Lebedev (988) have shown similar effects of solvents on the rate of epoxidation of allylic compounds with peroxyacetic acid (Table 20).

TABLE 20.    SPECIFIC REACTION-RATE CONSTANTS FOR EPOXIDATION OF ALLYL COMPOUNDS WITH PEROXYACETIC ACID IN VARIOUS SOLVENTS[a]

| Compound and Solvent | $T(°C)$ | $k^b$ $(\times 10^6)$ | Relative Rate of Reaction |
|---|---|---|---|
| Allyl alcohol: | | | |
| Carbon tetrachloride | 20 | 49 | 10 |
| Chlorobenzene | 20 | 47 | 9.6 |
| Nitrobenzene | 20 | 45 | 9.2 |
| Acetic acid | 20 | 26 | 5.3 |
| Ethyl acetate | 20 | 9.2 | 1.9 |
| Dioxane | 20 | 6.6 | 1.3 |
| Methanol | 20 | 6.3 | 1.3 |
| Tetrahydrofuran | 20 | 5.1 | 1 |
| Acetone | 20 | 4.9 | 1 |
| Allyl methyl ether: | | | |
| Chlorobenzene | 30 | 9.3 | 3.2 |
| Nitrobenzene | 30 | 8.9 | 3.0 |
| Benzene | 30 | 8.8 | 3.0 |
| Acetic acid | 30 | 7.6 | 2.6 |
| Ethyl acetate | 30 | 3.4 | 1.2 |
| Dioxane | 30 | 2.9 | 1 |
| Allyl chloride: | | | |
| Chlorobenzene | 30 | 5.9 | 3.5 |
| Nitrobenzene | 30 | 5.1 | 3 |
| Acetic acid | 30 | 3.3 | 2 |
| Ethyl acetate | 30 | 1.1 | 1 |

[a] Data from reference 988.
[b] Specific reaction-rate constant, in liters per mole per second. A fourfold to fivefold excess of unsaturated compound was used.

Allyl alcohol is epoxidized at a faster rate (about 10 times) than is either allyl methyl ether or allyl chloride. Again, the chlorinated and aromatic solvents (nitrobenzene and benzene) afford the highest rates, and acetic acid, ethyl acetate, dioxane, methanol, tetrahydrofuran, and acetone the lowest. The rate constants decrease with increasing solvent basicity.

Silbert and Konen (989) have conducted a two-pronged study: (a) to ascertain the limit to which rate enhancement in epoxidation can be extended by modifications in peroxy acid structure and (b) to find solvents with greater rate-enhancing effects than those currently known. The combination of more reactive peroxy acids and solvents with higher epoxidation rates would the extend the range of unsaturated compounds amenable to direct efficient epoxidation.

Table 21 lists the rate of epoxidation of 1-octene in two standard solvents,

TABLE **21**. SPECIFIC REACTION-RATE CONSTANTS FOR EPOXIDATION OF 1-OCTENE WITH PEROXYBENZOIC AND SUBSTITUTED PEROXYBENZOIC ACIDS[a]

| Substituted Peroxybenzoic Acid | $k^b$ ($\times 10^3$) | |
| --- | --- | --- |
| | Chloroform | Methyl Acetate |
| $p$-CH$_3$O | 1.4 | — |
| H | 2.1 | — |
| $p$-NO$_2$ | 16 | — |
| $o$-NO$_2$ | 22 | 2.53 |
| 3,5-(NO$_2$)$_2$ | $\sim$72; 18.3 (10°) | 2.03 |
| Pentafluoro | $\sim$80 | $\sim$2 |
| Peroxylauric acid | 0.63 | — |

[a] Data from reference 989.

[b] Specific reaction-rate constant, in liters per mole per second at 30°, except where otherwise indicated.

chloroform (fast) and methyl acetate (slow), with peroxybenzoic acid and a series of substituted peroxybenzoic acids (peroxylauric acid illustrates the reduction rate with an aliphatic peroxy acid). Table 21 clearly demonstrates the superiority of chloroform over methyl acetate (rate ratio 40:1). The results are almost identical with those of Renolen and Ugelstad (Table 17) (986), who compared chloroform with ethyl acetate as solvents for the epoxidation of cyclohexene.

Table 21 also shows that $p$-nitroperoxybenzoic acid is about eight times more rapid than peroxybenzoic acid for the epoxidation of 1-octene, a ratio perviously noted but with other unsaturated compounds (69,778,780). $o$-Nitro-peroxybenzoic acid, a previously uninvestigated epoxidation reagent, is even more reactive than the *para* isomer. This is explained on the basis that epoxidation diminishes steric hindrance between the $o$-nitro group and the intramolecularly-hydrogen-bonded bulky peroxycarboxy group.

Introduction of a second electron-withdrawing group into the benzene ring, as in 3,5-dinitroperoxybenzoic acid, increases the rate by another factor of about 4, but little or no additional rate enhancement is noted with pentafluoroperoxybenzoic acid.

The effect of a variety of solvents of different structural types on the rate of epoxidation of methyl oleate with peroxylauric acid at 30° is shown in Table 22 (989). The anticipated rate order is observed, demonstrating that

TABLE 22.    SPECIFIC REACTION-RATE CONSTANTS FOR EPOXIDATION OF METHYL OLEATE WITH PEROXYLAURIC ACID IN VARIOUS SOLVENTS[a]

| Solvent | $k^b$ $(\times 10^3)$ | Relative Rate of Reaction |
|---|---|---|
| Chloroform | 14.8 | 29 |
| Methylene chloride | 11.0 | 22 |
| Benzene | 4.72 | 9.4 |
| Cyclohexane | 1.55 | 3 |
| Methyl acetate | 0.633 | 1.2 |
| 1,2-Epoxybutane | 0.502 | 1 |
| Dialkyl ethers | 0.504 | 1 |

[a] Data from reference 989.
[b] Specific reaction-rate constant, in liters per mole per second at 30°.

the peroxy acid or olefin used plays only a minor role in the effect of solvent on rate. The result with 1,2-epoxybutane, which provides the lowest rate, is of interest as it suggests that specific rate constants in epoxidation should decrease as the concentration of oxirane forming increases.

Since chlorinated aliphatic solvents are, in general, the most effective, Silbert and Konen (989) undertook a detailed study of changes in the epoxidation rate with variations in the structure of the chlorinated solvent. Table 23 shows the specific reaction-rate constants for the epoxidation of methyl oleate with peroxylauric acid in various aliphatic halogenated solvents at 30°. The basic rate (relative rate 1) is obtained with $n$-alkyl halides (halogen is chlorine or bromine; with iodine oxidation of solvent occurs). Replacement of primary hydrogen atoms by chlorine or bromine causes an increase in rate in proceeding to methylene chloride or bromide (relative rate 5) and chloroform (relative rate 7), but there is a drop in rate with complete hydrogen substitution (carbon tetrachloride, relative rate 1.9) almost to the basic rate. Halogen

TABLE 23. SPECIFIC REACTION-RATE CONSTANTS FOR EPOXIDATION OF METHYL OLEATE WITH PEROXYLAURIC ACID IN HALOGENATED SOLVENTS[a]

| Solvent | $k^b$ ($\times 10^3$) | Relative Rate of Reaction |
|---|---|---|
| *n*-Alkyl chloride | 2.1 | 1 |
| Methylene chloride | 11.0 | 5.2 |
| Chloroform | 14.8 | 7.0 |
| Carbon Tetrachloride | 3.9 | 1.9 |
| 1,2-Dichloroethane | 6.0 | 2.9 |
| 1,1,2,2-Tetrachloroethane | 25.8 | 12.3 |
| 1,1,1-Trichloroethane (methyl chloroform) | 2.2 | 1 |
| 1,2-Difluorotetrachloroethane | 3.1 | 1.5 |
| *n*-Alkyl bromide | 2.2 | 1 |
| Methylene bromide | 13.2 | 6.3 |

[a] Data from reference 989.
[b] Specific reaction-rate constant, in liters per mole per second at 30°.

substitution thus increases the rate, but at least one hydrogen atom appears to be necessary.

The effect of twinning alkyl chlorides also causes an increase in rate (1,2-dichloroethane, relative rate 2.9, and 1,1,2,2-tetrachloroethane, relative rate 12.3), about two to three times that of the untwinned halides. The need for a hydrogen atom is again shown by the low rate in methyl chloroform (relative rate 1).

The electronegativity of the halogen has little or no effect on rate (cf. alkyl chlorides and bromides). Even the introduction of two fluorine atoms has little effect on the basic rate in the absence of hydrogen atoms in the molecule (1,2-difluorotetrachloroethane, relative rate 1.5). The role of hydrogen atoms in the solvent is not known.

No attempt has been made in the preceding discussion to detail all of the published work on solvent effects, kinetics, and activation parameters. Other references containing data on solvent effects are 225, 778, 780, 953, and 990. Additional information on kinetics and activation parameters can be found in references 645, 991–1001. In studies of kinetics it has usually been customary to follow the consumption of peroxy acids iodometrically. Ricciuti, Silbert, and Port (957) have determined the rate constants in the epoxidation of vinyl esters with peroxylauric acid, using a polarographic method to follow the disappearance of peroxy acid. The technique is applicable on a micro or submicro scale.

Information on the kinetics of the epoxidation of unsaturated compounds of different structures permits the design of experiments for efficient selective partial epoxidation of polyunsaturated compounds, a subject briefly mentioned earlier (827, 946, 948, 955, 980–982). For best results the double bond(s) to be preferentially epoxidized should react at least 10 times as fast as the other(s) to avoid contamination with fully epoxidized material. One of the early reports is the selective epoxidation of vinyl oleate at the internal (chain) double bond leaving the vinyl double bond unaffected (171, 665, 969). Since the chain double bond is epoxidized at least 20 times as fast as the vinyl double bond (957), a clean epoxidation to yield vinyl epoxystearate can be carried out. Numerous other selective epoxidations have been carried out since. Some typical examples are vinyl undecylenate (777, 969); internal olefins in the presence of 1-olefins (1002); manool and its esters (247); nerolidol and its formate (258); allyl undecylenate, oleate and, linoleate (969); vinyl-cyclohexene (969); alkenyl acrylates, crotonates, and other alkenoates (969); stigmasterol (990); 7-di-phenylmethylnorbornenes (36); and many other compounds dispersed in the patent literature.

The selective epoxidation of vinyl alkenoates [or other analogous compounds (969)] yields vinyl monomers readily polymerized or copolymerized by free-radical initiation, leaving the oxirane groups substantially unaffected for subsequent ionic cross-linking (171, 665). The report (998) that linoleic and linolenic acids are epoxidized faster than oleic acid with peroxyacetic acid does not appear to be correct (971). The rate of epoxidation is only slightly affected by steric hindrance (1003, 1004).

## B.  Stereochemistry

The stereochemistry of epoxidation has been of interest for a long time, but it has been only within the last two decades that the necessary instrumentation has been available to give unequivocal conclusions. Swern (1005) and Witnauer and Swern (1006) were the first to prove that unsaturated compounds are epoxidized in a stereospecific *cis* manner; *cis*-alkenes yield *cis*-epoxides and *trans*-alkenes yield *trans*-epoxides.

The epoxidation of allylic alcohols, such as cinnamyl alcohol (949), allyl alcohol (970), terpene alcohols (1007) and 3-hydroxycyclohexene (964, 965, 1008), not only proceeds at an anomalously fast rate but the hydroxy group has a directive effect. The introduction of an oxygen substituent in the vicinity of a double bond ordinarily causes a reduction in the rate because of the electron-withdrawing inductive effect (956), and when an alcohol is used as the epoxidation solvent the rate is also slowed down substantially (718). Allylic alcohols react even more rapidly than their corresponding methyl and ethyl ethers, as shown in Table 24 (964, 965). More important, allylic alcohols direct attack stereospecifically at the *cis* side of a double bond with respect to the hydroxy group even in steroids and related model compounds where $\beta$-approach of reagents is normally difficult (233, 236, 241, 594, 784, 964, 965). The hydroxy group in allylic alcohols is clearly exerting some stereo-promoting effect that may be correlated with its directive influence in giving *cis*-epoxy alcohols. Norbornene yields *exo-cis*-epoxide in excellent yield; epoxidation occurs almost exclusively from the less hindered side (96, 145, 1009). Bicyclo[3.2.1.]octene-2 also yields the *exo*-epoxide (342). The *cis* approach of the peroxy acid is also observed with 3-benzamidocyclohexene (339).

TABLE **24.**  SPECIFIC REACTION-RATE CONSTANTS FOR EPOXIDATION OF CYCLOHEXENE AND SUBSTITUTED CYCLOHEXENES WITH PEROXYBENZOIC ACID IN BENZENE[a]

| Cycloalkene | $k^b$ $(\times 10^4)$ | Relative Rate of Reaction |
|---|---|---|
| Cyclohexene | 63.3 | 23.7 |
| 3-Hydroxycyclohexene | 34.5 | 12.9 |
| 3-Methoxycyclohexene | 4.25 | 1.6 |
| 3-Ethoxycyclohexene | 4.72 | 1.8 |
| 3-Acetoxycyclohexene | 2.90 | 1.1 |
| 4-Hydroxycyclohexene | 2.67 | 1 |

[a] Data from references 964 and 965.
[b] Specific reaction-rate constant, in liters per mole per second at 5°.

In cyclohexenes with $CH_2OH$, $CH_2OCOCH_3$, or $CO_2CH_3$ substituents at the more remote 4-position approach of peroxybenzoic acid is preferentially *trans* to the substituent (968). In the bicycloheptene series the usual direction of attack by peroxy acids (*exo* or on the same side as the methylene bridge) is not altered by OH, $OCOCH_3$, $CH_2OCOCH_3$, $CO_2H$, or $CO_2CH_3$ substituents on the other side of the ring (968). With a $CH_2OH$ substituent, the primary alcohol group renders intramolecular nucleophilic assistance to the olefin–peroxy acid reaction, as indicated by a high reaction rate and the isolation of a tetrahydrofuran derivative.

The directing effects of near and remote groups on epoxidation have been reviewed by Henbest et al. (1008, 1010, 1011). The epoxidation of 4-methyl-cyclopentene with peroxylauric acid gives mainly the *trans*-epoxy compound (*trans*-to-*cis* ratio 3:1) in polar or nonpolar solvents. This suggests that the alkyl group is exerting a steric rather than a polar effect (in this context *cis* and *trans* refer to the spatial relationship between the substituent and oxirane

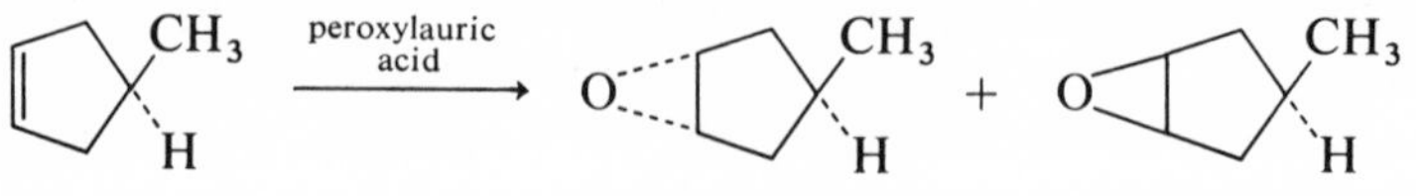

(*trans*, 76%)                    (*cis*, 24%)

groups). 4-Methylcyclopentene is a simple prototype of steroidal olefins in which axial $\beta$-methyl groups in angular positions cause reagents to attack at the $\alpha$-face of the molecule; $\Delta^2$-cholestene gives at least 85% $\alpha$-epoxide.

With unsaturated nitriles (**6**, **7**, and **8**) epoxidation with peroxylauric acid also takes place predominantly from the side of the molecule *trans* to the substituent in a hydrocarbon solvent even in a tricyclic compound (**8**). The *trans*-directive effect is strongly dependent on solvent polarity, suggesting that a polar effect is operative with the nitrile group.

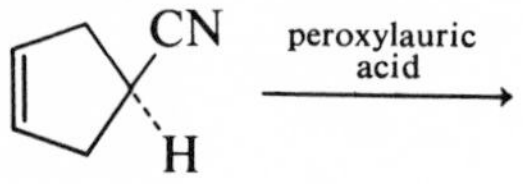

*trans*-epoxide; 95% in cyclopentane
76% in acetonitrile

(**6**)

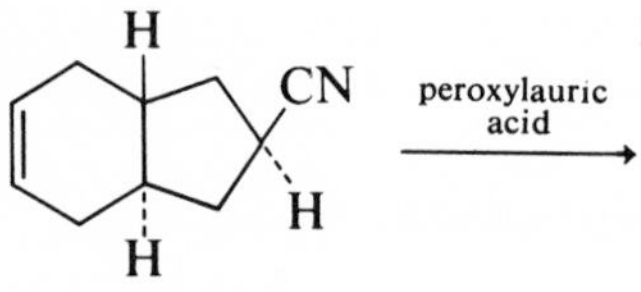

*trans*-epoxide; 69% in cyclopentane
54% in acetonitrile

(**7**)

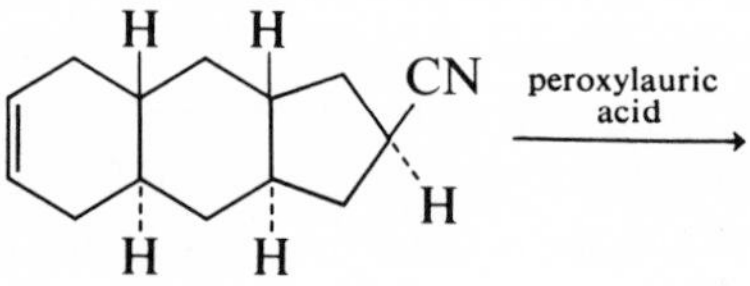

*trans*-epoxide; 57% in cyclopentane
46% in acetonitrile

(**8**)

Replacement of the hydrogen atom in compounds **6** and **7** by a methyl (**9**) or isopropyl (**10**) group substantially reduces the amount of reaction *trans* to the nitrile group, with the larger isopropyl group exerting the greater effect.

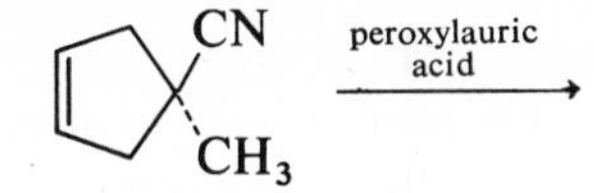

trans-epoxide; 87% in cyclopentane
53% in acetonitrile

**(9)**

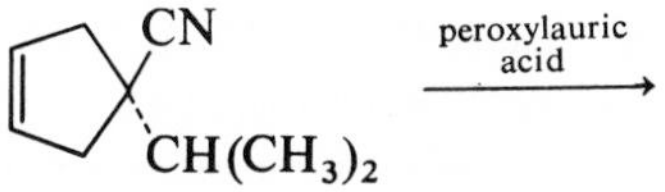

trans-epoxide; 80% in cyclopentane
26% in acetonitrile

**(10)**

Directive effects caused by steric hindrance are appreciable only when there are not more than two carbon atoms between the reaction site and the substituent (975). The following ratios of stereoisomers from epoxidation reactions appear to be the result of differential steric hindrance and not to inductive effects (direction of epoxidation shown by arrow; references in parentheses):

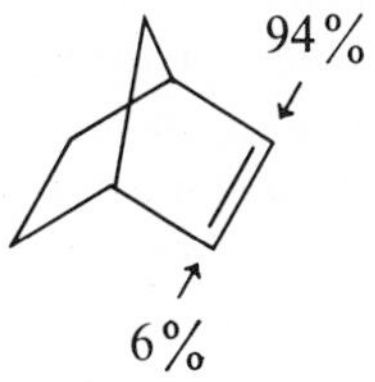

(1009)

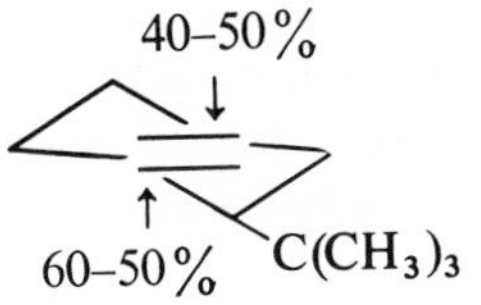

(725, 770, 1012)

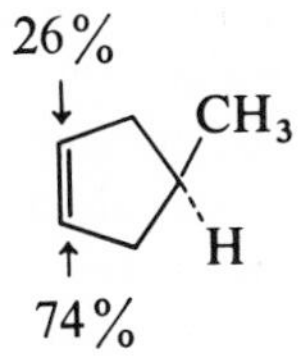

(1008, 1010, 1011)

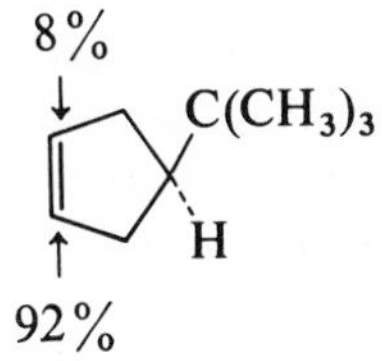

(975)

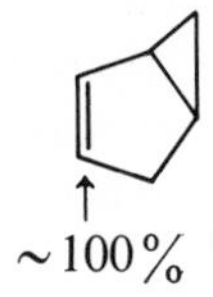

(1013)

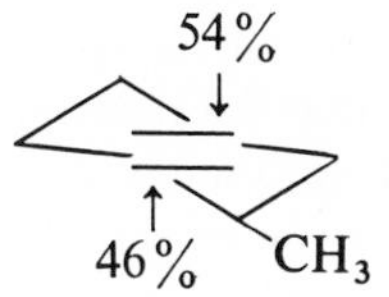

(977)

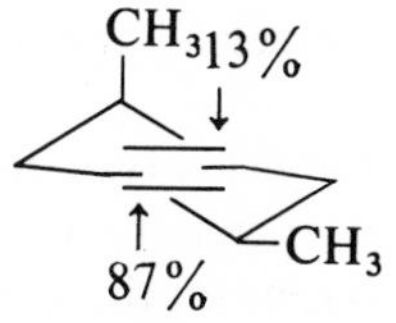

(977)

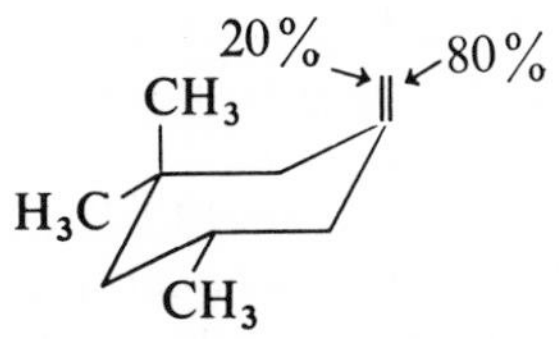

(975)

4-*t*-Butylcyclohexene (50:50 or 40:60 *trans–cis* epoxidation) appears to be an exception and suggests that the equatorial *t*-butyl group has little directive effect. A 4-methyl group also has little directive effect (*trans*-to-*cis* ratio 54:46) (977). With 3-methylcyclohexene the *trans*-to-*cis* ratio is reported to be 50:50 (765). However, with *cis*-4,5-dimethylcyclohexene epoxidation with *m*-chloroperoxybenzoic acid is predominantly *trans* (87%) (977). With the trimethylmethylenecyclohexane epoxidation is largely equatorial (80%) (975). The rate of epoxidation of 3- and 4-alkyl-substituted cyclohexenes is slower (one-half to one-fourth) than that of cyclohexene itself—as expected on the basis of steric and conformational agreements but not on the basis of inductive effects, which predict the opposite (977). 3-Methylcyclohexene, however, gives a 50:50 ratio of *cis–trans* epoxides (765).

The hindering effect in 4-alkylcyclopentenes is estimated to be about 0.6–0.8 kcal/mole (975). The strength of the directive effect of the *t*-butyl group is about 1.4 kcal/mole from the 92:8 *trans*-to-*cis* ratio in the epoxidation of 4-*t*-butylcyclopentene.

Gray and co-workers (149, 976) have shown that the $\beta$-epoxy compound is formed apparently exclusively from *cis*-$\Delta^4$-tetrahydrophthalic anhydride and as the predominant product from derived *N*-substituted imides on epoxidation. Also, tetrahydrophthalic anhydride is epoxidized with *m*-chloroperoxybenzoic acid about 25–80 times faster than either its *exo* or *endo* bicyclic anhydride or imide counterparts. *Exo* isomers are more rapidly epoxidized than their *endo* analogs and afford *exo*-epoxides exclusively. The results are the opposite of those predicted from Henbest's work. Explanation of the results of epoxidation of *exo* and *endo* isomers of the bicyclic anhydrides or imides is based on a field effect of the anhydride or imide carbonyl groups, which provide electron-deficient centers acting across space to deactivate the double bond to electrophilic attack from either direction. Deactivation should be more pronounced in the *endo* (and axial) isomers in which the electron-deficient center would be situated much closer to the double bond than in the *exo* (and equatorial) analogs.

Relatively small structural changes can cause large differences in the direction of attack of the peroxy acid in fused-ring systems. Whereas *syn*- and *anti*-$\Delta^4$-octalin-1-carboxylic acid and *cis*-*anti*-$\Delta^4$-octalin-1,2-dicarboxylic acid anhydride yield only $\alpha$-epoxide with peroxyacetic acid in chloroform, epoxidation of the *cis*-*syn* anhydride is not stereospecific and gives both $\alpha$- and $\beta$-epoxides (594).

Methyl linoleate (*cis*, *cis*-9,12-octadecadienoate) yields two stereoisomeric diepoxyoctadecanoates, a liquid and a solid, in epoxidation with peroxyacetic acid (570). The liquid has both oxirane oxygens on the same side, and the solid has them on opposite sides of the molecular plane. The monoepoxidation of methyl linoleate with monoperoxyphthalic acid yields two monoepoxides separable by chromatography (253).

## C. Miscellaneous Stereochemical Studies

Limonene, 1-menthene, and 3-menthene yield an approximately 1:1 mixture of *cis*- and *trans*-epoxides (132, 545, 1014, 1015), as expected from the earlier work of Rickborn and Lwo (977). The epoxidation of 1,4,5,8,9,10-hexahydroanthracene with peroxyacetic acid yields a 60:40 mixture of *syn* and *anti* isomeric epoxides (532). *m*-Chloroperoxybenzoic acid epoxidation of *cis*-9,10-dimethyl-1-octalin 2-acetate gives a 2:1 mixture of *trans–cis* isomers (741). The epoxidation of estra-4,9(10)-dien-3-ones at the 9(10)-double bond with *m*-chloroperoxybenzoic acid yields epoxides with 9α, 10α stereochemistry, indicating α-face attack (752). γ-Terpineol yields a 70:30 ratio of *trans–cis* epoxides on epoxidation with peroxybenzoic acid in benzene (164). The epoxidation of ($-$)-*cis*-$\Delta^2$-carene and ($+$)-$\Delta^4$-carene with peroxyacetic acid yields *cis*-caran-*trans*-2,3-epoxide and caran-*trans*-4,5-epoxide, respectively (399). 4,5-*cis*- Dimethylcyclohexene-1,2-dicarboxylic anhydride yields the *trans*-epoxide with peroxyacetic acid in chloroform (487). 2-Methoxy- and 2-*t*-butoxy-5,6-dihydro-2*H*-pyran react with *m*-chloroperoxybenzoic acid to give mixtures of epoxides in which the *trans*-to-*cis* ratio is 3:1 and 9:1, respectively; the steric control is attributed to the 2-alkoxy group (730). Other references are 77, 277, 707, 708, 737, 746, 787, 1016.

It has been suggested (975, 1017) that optically active epoxides can be obtained when optically active peroxy acids are used as oxidizing agents, but the excess of one optical isomer over the other is only about 1–7.5%. Asymmetric induction of optically active epoxides by direct epoxidation does not appear to be an efficient process; however, the chirality of the oxirane always correlates with that of the peroxy acid.

The stereochemical results of the epoxidation of four rigid methylene–cyclohexane systems have been studied by Carlson and Behn (208) with various peroxy acids and alkaline hydrogen peroxide–benzonitrile systems. The epoxidation of unhindered methylenecyclohexanes with peroxy acids gives mainly axial attack; epoxidation with the alkaline peroxide system gives mainly equatorial attack. The results are summarized in Table 25; nuclear magnetic resonance was used to determine product composition.

Table 25 shows that with all the peroxy acids and solvent systems the relatively unhindered olefins (A, B, C) in which there are two 1,3-position interactions with hydrogen for axial attack the predominant reaction product is that from axial attack. With the more hindered olefin (D) in which there is one 1,3-position interaction with hydrogen and one with the angular methyl group all of the peroxy acids give predominantly equatorial transfer of oxygen. Axial attack is somewhat solvent dependent, in agreement with Henbest (975), with methylene chloride and chloroform as solvents affording the least amount of axial attack. In contrast, the alkaline hydrogen peroxide system gave predominantly equatorial oxygen transfer, with all four olefins in the presence or absence of buffer or changing from potassium bicarbonate to carbonate.

TABLE 25.   PERCENTAGE OF AXIAL ATTACK IN OLEFIN
             EPOXIDATIONS

| Reagent | Olefin[a] | | | |
|---|---|---|---|---|
| | A | B | C | D |
| $m$-Chloroperoxybenzoic acid–$CH_2Cl_2$ | 59 | 69 | 65 | 2 |
| $m$-Chloroperoxybenzoic acid–$CH_3CN$ | 66 | — | — | — |
| $m$-Chloroperoxybenzoic acid–$Et_2O$ | 65 | — | — | — |
| $m$-Chloroperoxybenzoic acid–$CH_3OH$ | 68 | — | — | — |
| $m$-Chloroperoxybenzoic acid–$t$-BuOH | 64 | — | — | — |
| Monoperoxyphthalic acid–$Et_2O$ | 69 | 79 | 76 | 17 |
| Peroxyacetic acid–$CH_2Cl_2$ | 60 | 66 | 61 | 4 |
| $H_2O_2$–$C_6H_5CN$–$KHCO_3$–$CH_3OH$ | 14 | 32 | 33 | 2 |
| $H_2O_2$–$C_6H_5CN$–$KHCO_3 + PO_4^{3-}$–$CH_3OH$ | — | 30 | 29 | — |
| $H_2O_2$–$C_6H_5CN$–$K_2CO_3$ | — | 27 | — | — |

[a] A: $trans$-2-methylenedecahydronaphthalene; B: 4-$t$-butylmethylene cyclo-hexane; C: $trans$10-methyl-2-methylenedecahydronaphthalene; D: $trans$-9-methyl-2-methylenedecahydronaphthalene.

## D.  Mechanisms

Historically, the mechanism of peroxy acid epoxidation had been considered to involve either attack by positive hydroxyl ($^+OH$) or the addition of peroxy acid across the double bond, followed by ring closure, with elimination of carboxylic acid. Neither of these mechanisms is considered valid any longer.

The epoxidation of olefins with peroxy acids can be characterized as follows:

1. The reaction is first-order in olefin and in peroxy acid.

2. The primary product is the epoxide (oxirane) usually isolated in high to quantitative yields.

3. The reaction is often complete at or below room temperature in a few hours or less.

4. A wide variety of polar or nonpolar organic solvents can be used, as already discussed. Thus it is an extremely convenient and useful reaction for mechanistic study.

Since epoxidation is rapid in nonionizing solvents and there is little or no salt effect or catalysis by acids and the Hammett $\rho$ value is on the order of $-1$ (substituted stilbenes) or $+1$ (substituted peroxybenzoic acids), it has usually been assumed that ionic species or transition states cannot be involved. Furthermore, the high negative entropy of activation requires a highly ordered transition state.

The "molecular" mechanism first proposed by Bartlett (1018) and supported by others (756, 966, 1019) is consistent with the above facts, although there is no direct evidence for a "spiro" transition state:

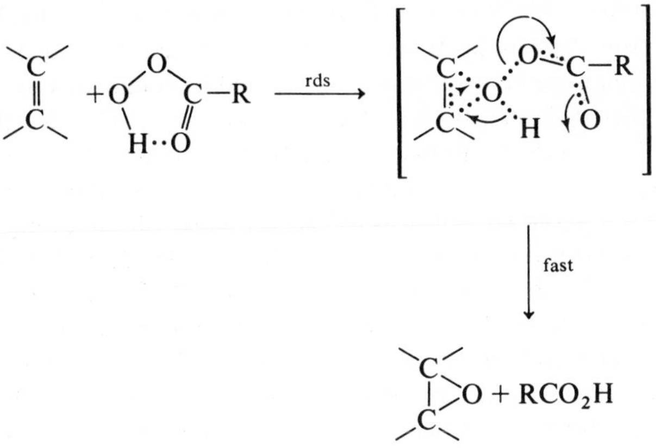

In this mechanism the transition state is nonionic and oxygen transfer occurs by a concerted intramolecular process via the spiro transition state shown. [Peroxy acids are known to have intramolecular hydrogen bonds 1020, 1021)].

An alternative mechanism involving 1,3-dipolar addition to the double bond as the rate-determining step (rds) has recently been suggested by Kwart and Hoffman (1022) on the basis of the similarity between the reactivity parameters and general kinetic characteristics of epoxidation and 1,3-dipolar addition reactions:

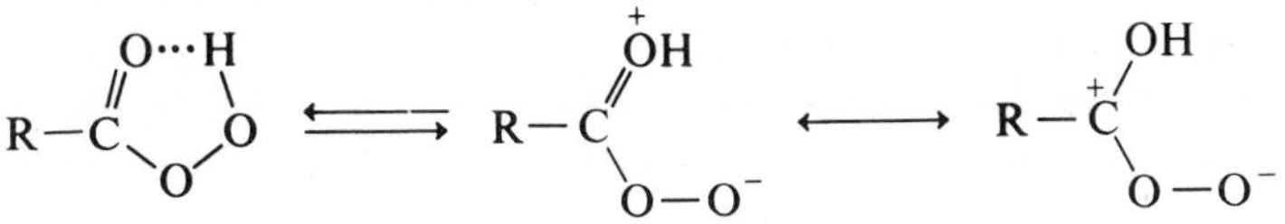

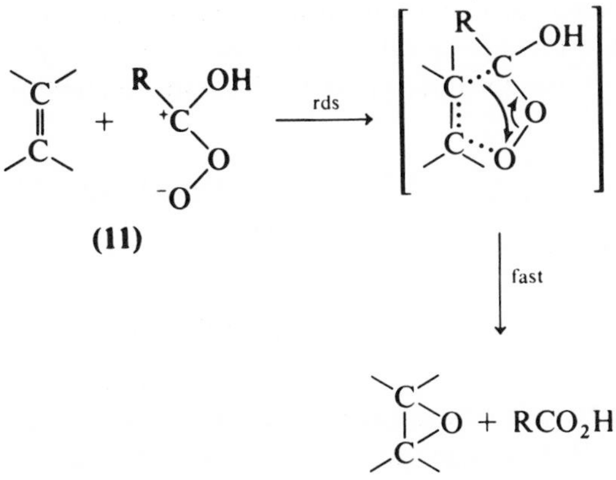

The rate-determining step involves 1,3-dipolar addition to the double bond of a hydroxycarbonyl oxide reagent (**11**) derived from the peroxy acid. Indirect evidence to support this proposal has been obtained from cases in which a carbonyl oxide, generated from a molozonide, can be displaced from its association with an indigenous carbonyl component by added olefin.

The 1,3-dipolar mechanism has been criticized by Bingham and co-workers (1019) and later defended by Kwart and co-workers (1023). Bingham et al. suggest that 1,3-dipolar addition in epoxidation is not rate determining because in comparing the rates of 1,3-dipolar addition of aromatic azides to norbornene and to cyclohexene the former reacts almost $10^4$ times more rapidly, yet in the epoxidation of the two olefins with peroxylauric acid in chloroform at 25° the second-order rate constants are essentially the same ($\sim 2 \times 10^{-2}$ l/mole-sec). In addition, 1,3-dipolar addition of nitrones is strongly accelerated by electron-withdrawing groups attached to the double bond, whereas such groups drastically reduce the epoxidation rate (1024).* Finally, the effect of alkyl substituents on epoxidation rates suggests a fairly symmetrical transition state, whereas the 1,3-dipolar mechanism would predict a heavily asymmetrical transition state (1024).*

Kwart et al. (1023) point out, however, that the rate differences observed in 1,3-dipolar additions of aromatic azides to norbornene and to cyclohexene may not be relevant to the hydroxycarbonyl oxide case, as it cannot be assumed that all 1,3-dipolar additions should proceed at vastly different rates with those olefins. Huisgen (1025–1027) has emphasized that a 1,3-dipolar reagent without a double bond in the sextet structure, as is the case with the assumed hydroxycarbonyl oxide, should not necessarily show rate differences in reactions with those cyclic olefins. Furthermore, Kwart et al. have heated the known 3-hydroxy-3,5,5-trimethyl-1,2-dioxacyclopentane, the product that would form from peroxyacetic acid and isobutylene by the dipolar mechanism, at 100° in deuteriochloroform and have shown that some isobutylene oxide is formed as a minor product. Although isobutylene oxide is not formed spontaneously and known dioxacyclopentanes (1,2-dioxolans) are more stable than the dipolar epoxidation mechanism requires, nevertheless the formation of the epoxide is possible from the required intermediate.

$$(CH_3)_2{-}C\underset{\overset{|}{H_2C}{-}\underset{\overset{|}{CH_3}}{C}\diagdown_{OH}}{\diagup^O}\diagdown_O \xrightarrow[100-125°]{\Delta} (CH_3)_2{-}\underset{O}{C{-}CH_2} + CH_3COHC\underset{O}{{-}C(CH_3)_2}$$

$$\qquad\qquad\qquad\qquad\qquad\qquad\qquad\qquad 1 \qquad\qquad\qquad\qquad 6$$

$$+ \ CH_3COCH{=}C(OCH_3)CH_3$$

$$9$$

To account for the low yield of isobutylene oxide on thermolysis, it was concluded that the 1,2-dioxolan never forms as such when the hydroxy-

* Also private communication from G. H. Whitham.

carbonyl oxide reagent attacks the olefin. A molecular-orbital and valence-bond representation has been employed to explain the epoxide-forming transition state in the dipolar mechanism (1023). The very act of approach to bond formation in the transition state produces rearrangement to epoxide by virtue of the conformational accommodation that produces O—O bond cleavage without actual formation of the 1,2-dioxolan (which would be the hypothetical nonplanar, stable product of 1,3-dipolar addition).

Azman, Borstnik, and Plesnicar (1028) have carried out calculations on the electronic structure of peroxy acids in the conventionally written intra-molecularly-hydrogen-bonded form and also in the hydroxycarbonyl oxide dipolar form, using the semiempirical Pariser–Parr–Pople SCF–MO method (1029, 1030). The energy difference between the intramolecularly hydrogen bonded and dipolar species is small (0.03 eV), and it is concluded that both epoxidation mechanisms are theoretically possible. In accord with the molecular mechanism (1018), the oxygen atom next to hydrogen in the peroxy-carboxyl group is the most likely place for attack by a nucleophile (the olefin) on the peroxy acid molecule. The $\pi$-electron distribution in the 1,3-dipolar form of the peroxy acid molecule, on the other hand, is also consistent with the 1,3-dipolar epoxidation mechanism (1022). The molecular mechanism is assumed to be more reasonable in nonpolar media and the dipolar mechanism in polar media. Additional work is required before a definitive conclusion can be reached.

The *cis*-directing effect in the epoxidation of allylic alcohols, already discussed, has been rationalized on the basis of the molecular mechanism of epoxidation (223, 242, 964, 965). Hydrogen bonding is assumed to occur between the hydroxy group and the peroxy acid, thereby effectively anchoring the latter on the hydroxy side of the molecule; the transition state is depicted in brackets:

## VI.　UNUSUAL AND MISCELLANEOUS EPOXIDATION STUDIES

### A.　Acetylenes

The expoxidation of acetylenes has long been of theoretical and practical interest; monoepoxidation should yield oxirenes directly and diepoxidation the dioxabicyclo derivatives:

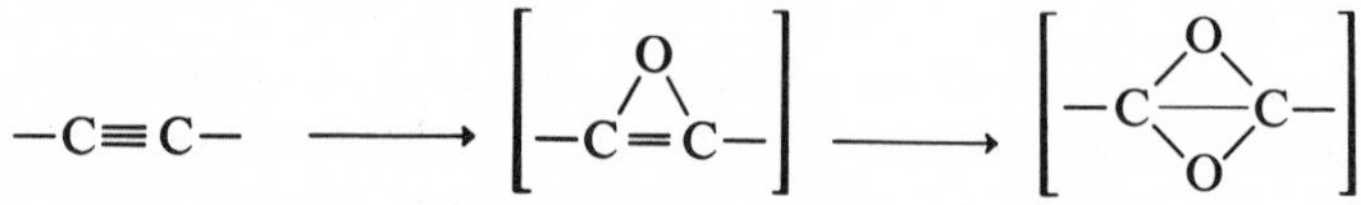

Neither of these classes of compounds has yet been prepared and isolated by any route, however. The reported preparation of 2-methyloxirene from the chromic anhydride oxidation of propyne has not been verified and is questionable in view of the vigorous reaction conditions (1031). The preparation of 2-chloro-3-phenyloxirene also seems unlikely (1032).

Oxirene, the simplest member of the oxirene series, may be considered to be $\pi$-isoelectronic with cyclobutadiene, just as furan is with benzene (1033). Simple molecular-orbital calculations predict zero delocalization energy and a singlet ground state for oxirene. An estimate of the strain energy of oxirene may be obtained by comparing it with other small-ring compounds. The strain energy of ethylene oxide, taken as the difference between the calculated and experimental heats of formation, is 13 kcal/mole. Assuming the same differential in strain energy holds between cyclopropane and cyclopropene ($\sim$27 kcal/mole), as in ethylene oxide and oxirene, the total strain energy of oxirene is about 40 kcal/mole. Since the strain energy of cyclopropane is about 12 kcal/mole more than that of ethylene oxide, oxirene might be expected to be more stable than cyclopropene, and therefore isolable, on this criterion.

In 1910 Prileschajew (1034) reported the first oxidation of an acetylene with a peroxy acid; he indicated that phenylacetylene yields methyl phenylacetate on oxidation with peroxybenzoic acid in ether. In contrast, Boeseken and Slooff (1035) reported in 1930 that peroxyacetic acid has no effect on phenylacetylene as well as on acetylene itself. Repetition of the oxidation of phenylacetylene by McDonald and Schwab (1033) with peroxybenzoic acid in chloroform that also contained methanol and ethanol yielded methyl phenylacetate (42.5%), benzoic acid (43.5%), benzaldehyde (11.1%), ethyl phenylacetate (8.1%), and methyl benzoate (3%). (The yield of benzoic acid is estimated to be at least 8% high owing to its presence in the original peroxybenzoic acid solution.)

The oxidation products of phenylacetylene can be rationalized on the assumption that the oxirene is the primary oxidation product and is either oxidized further to the unstable, easily rearranged, dioxabicyclo derivative or rearranges to phenylketene:

$$PhC\equiv C-H \xrightarrow{\text{PhCO}_3\text{H}} \left[ Ph-\underset{\underset{O}{\diagdown\diagup}}{C=C}-H \right]$$

$$\xrightarrow{[O]} \left[ PhC\overset{O}{\underset{O}{\diamond}}C-H \right] \longrightarrow \left[ Ph\overset{O}{\overset{\|}{C}}-\overset{O}{\overset{\|}{C}}-H \right]$$

$$\overset{O}{\overset{\|}{PhC}}-O-\overset{O}{\overset{\|}{C}}-H$$

$$\downarrow$$

$$PhCO_2H + HCO_2H$$

$$\xrightarrow{} [PhCH=C=O] \xrightarrow{\text{ROH}} PhCH_2CO_2R$$

With trifluoroperoxyacetic acid, phenylacetylene yields phenylacetic and benzoic acids; with peroxyacetic acid it yields benzyl acetate and acetylmandelic acid, in addition. The formation of the oxirene is assumed to be the first step, although in no case could it be isolated.

The oxidation of diphenylacetylene, a compound with no tautomerizable $\alpha$-hydrogens, has been studied by McDonald and Schwab (1033) and by Stille and Whitehurst (1036). The former used peroxybenzoic acid in chloroform, peroxyacetic acid in acetic acid–methylene chloride, and buffered trifluoroperoxyacetic acid in methylene chloride; the latter used peroxybenzoic and *m*-chloro- and *p*-nitroperoxybenzoic acids in various solvents. In no case is an oxirene or dioxabicyclo compound isolable, although the products obtained are best rationalized on the intermediacy of these small-ring species.

The oxidation of diphenylacetylene (1 mole) with trifluoroperoxyacetic acid (2 moles) yields benzil (76%) and benzoic acid (17%) (1033). Under the mild conditions and relatively short reaction times used by these investigators peroxyacetic acid and peroxybenzoic acid are unreactive (90% of diphenylacetylene recovered).

The polarity of the solvent and acidity of the system play an important role in the mechanistic fate of the diphenyloxirene intermediate (1036). In acetic acid the oxidation of diphenylacetylene with *m*-chloroperoxybenzoic acid yields benzoic acid mainly and smaller amounts of benzil and *O*-acetylbenzoin. In a benzene solvent with soldium carbonate present as an insoluble base no benzoic acid is formed; major products are benzophenone and *O*-benzoyl-*O*'- (*m*-chlorobenzoyl)dihydroxyphenylmethane. Benzoic acid and ethyl diphenylacetate are the main products in ethanol solution. Pathways for rationalizing the product distribution are given by Stille and Whitehurst (1036) on the assumption that an oxirene is formed first.

Acetylenes react considerably more slowly with peroxy acids than structurally similar olefins, as was first shown in a qualitative way by Boeseken and Slooff (1035) and quantitatively by Schlubach and Franzen in 1952 (1037). The latter compared the rate of oxidation of acetylenes and olefins with peroxybenzoic acid and showed that the former consume peroxy acid at one-thousandth the rate of the latter. Although Schlubach and Franzen claimed to have isolated oxirenes from the epoxidation of aliphatic acetylenes, later work showed that the product claimed to be the oxirene was in fact the isomeric α,β-unsaturated ketone (1038, 1039) and that by-products from ketene intermediates are also formed (1040). Over 2 moles of peroxy acid are consumed per mole of acetylenic compound and cleavage products predominate (1041). The products can be explained on the basis of oxirene formation as the primary oxidation step.

The low rate of reaction of the triple bond with peroxy acids permits the clean, selective epoxidation of double bonds in ene–yne compounds, as was first demonstrated by Malenok and co-workers from 1936 to 1941 (392, 609, 610, 613) and further explored from 1953 to 1958 (481, 494, 501, 612, 1042). Further studies on the selective epoxidation of ene–yne compounds and the structural types examined can be found in references 262, 546, 593, 1043–1048. (In some cases the α-glycol or hydroxyacyloxy compounds, not the oxiranes, are isolated.)

The Wolff rearrangement of α-diazoketones with ultraviolet irradiation has been suggested to involve ketocarbene and oxirene intermediates, although they were not detected or isolated but were deduced from the products of the reaction (1049).

$$CH_3 - {}^{13}\overset{\displaystyle O}{\underset{\displaystyle \|}{C}} - \overset{\displaystyle N_2}{\underset{\displaystyle \|}{C}} - CH_3$$

$$\Big\downarrow \; \underset{(-N_2)}{hv}$$

$$\left[ CH_3 - {}^{13}\overset{\displaystyle O}{\underset{\displaystyle \|}{C}} - \ddot{C} - CH_3 \right] \longrightarrow \left[ CH_3 - {}^{13}C \overset{\displaystyle O}{=} C - CH_3 \right]$$

$$\Big\downarrow$$

$$CO + {}^{13}CO + (CH_3)_2C: + (CH_3)_2 - {}^{13}C: \longleftarrow \left[ \begin{array}{l} (CH_3)_2C = {}^{13}C = O + \\ (CH_3)_2 - {}^{13}C = C = O \end{array} \right]$$

$$CH_3CH = CH_2 + CH_3 - {}^{13}CH = CH_2 + \underbrace{C_2H_6, C_2H_4, C_3H_8, CH_4}_{\text{minor products}}$$

High-resolution mass-spectrometry analysis was used to determine product composition.

## B.  Allenes

In his classical epoxidation studies Boeseken (1050) treated 1,1-dimethyl-allene with an excess of 30% peroxyacetic acid below 10°. The low temperature was required to avoid a violent reaction (compare with the low rate of oxidation of acetylenes). The principal product isolated was 3-acetoxy-3-methyl-2-butanone. Boeseken reported that, in spite of the long oxidation time, only one atom of oxygen was consumed.

$$(CH_3)_2C=C=CH_2 \xrightarrow[<10°]{CH_3CO_3H} (CH_3)_2 \underset{\underset{\underset{CH_3}{|}}{\underset{\underset{C=O}{|}}{\underset{O}{|}}}{C} - \underset{\underset{O}{\|}}{C} - CH_3$$

Later, Pansevich-Kolyada and Idelchik (1051, 1052) studied the oxidation of 2,3-dimethyl-2,3-hexadiene and 2-methyl-2,3-octadiene with peroxyacetic acid in ether solution. They also observed that allenes are rapidly oxidized, but the structures they reported for their reaction products have been questioned (1053, 1054).

$$R-CH=C=C(CH_3)_2 \xrightarrow[(C_2H_5)_2O]{CH_3CO_3H} \underset{\underset{OH}{|}}{R}CH - \overset{\overset{OCOCH_3}{|}}{C} \diagdown_O \diagup C(CH_3)_2$$

$$(R=C_2H_5, C_4H_9)$$

Peroxy acid epoxidation of the allene function is of considerable interest, as novel products are possible. An allene oxide (methyleneoxirane) should form on monoepoxidation and a dioxaspiroalkane on diepoxidation.

In 1966 Crandall and Machleder (1053) published a preliminary report and in 1968 a more complete account (1054) of the oxidation of tetramethylallene with peroxyacetic acid in methylene chloride solution in the presence of excess powdered sodium carbonate buffer. Although neither the mono-epoxidation nor diepoxidation products were isolated, the reaction products could be accounted for by a scheme that postulated their finite existence as unisolated reactive intermediates. The major products were 2-acetoxy-2,4-dimethyl-3-pentanone (12, 52%), as anticipated from Boeseken's early work (1050), and 2-acetoxy-4-hydroxy-2,4-dimethyl-3-pentanone (13, 39%); minor products were acetone,2,2,4,4-tetramethyl-3-oxetanone (14, 3%) and its corresponding lactone (15, 2%), and 4-hydroxy-2,4-dimethylpent-1-en-3-one (16, 4%). m-Chloroperoxybenzoic acid was substituted for peroxyacetic acid in one experiment; m-chlorobenzoate analogs of compounds 12 and 13 were obtained.

Oxidation of a fully methylated alkenylidenecyclopropane (17) with 1 equivalent of peroxyacetic acid in cold buffered methylene chloride yields two major products (18 and 19) and a third unidentified minor component (9%) (1055). Compound 18, obtained in 64% yield, is the acetate of a cyclopropyl ketone; compound 19, obtained in 27% yield, is 2,5,5,6-tetramethyl-6-hepten-3-yn-2-ol.

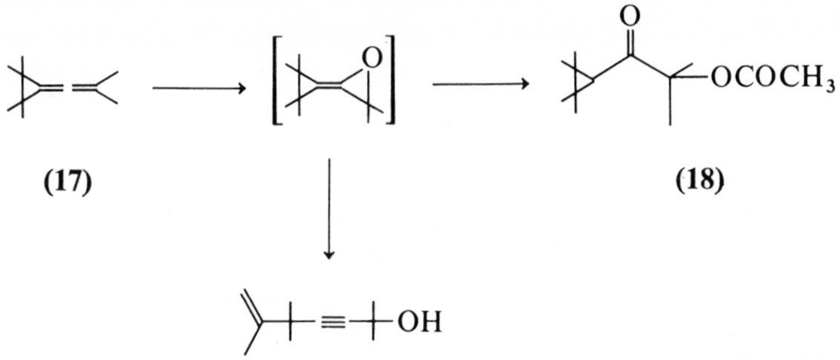

**(17)**                                                          **(18)**

**(19)**

The results are best rationalized by postulation of an allene oxide intermediate derived from selective attack of the peroxy acid at the allene double
bond remote from the cyclopropyl ring, which adds acetic acid and forms
compound **18** via the enol. The protonated allene oxide can fragment with
cleavage of both epoxide and cyclopropyl rings to establish the acetylenic
function in the open-chain cationic precursor to compound **19**.

In 1968 two reports simultaneously described the isolation of an allene
oxide and a dioxaspiroalkane. Crandall, Machleder, and Thomas (75)
isolated 5-*t*-butyl-2,2-dimethyl-1,4-dioxaspiro[2.2]pentane (**21**) by the epoxidation of 2,5,5-trimethyl-2,3-hexadiene (**20**) with 2 equivalents of buffered
peroxyacetic acid, and Camp and Greene (743) isolated 1,3-di-*t*-butylallene
oxide (**23**) by the epoxidation of excess 1,3-di-*t*-butylallene (**22**) with *m*-chloro-
peroxybenzoic acid.

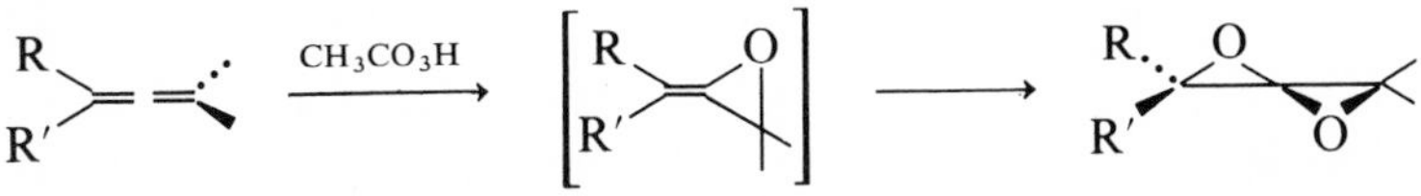

**(20)**                                                          **(21)**

(R = *t*-Bu, R′= H)

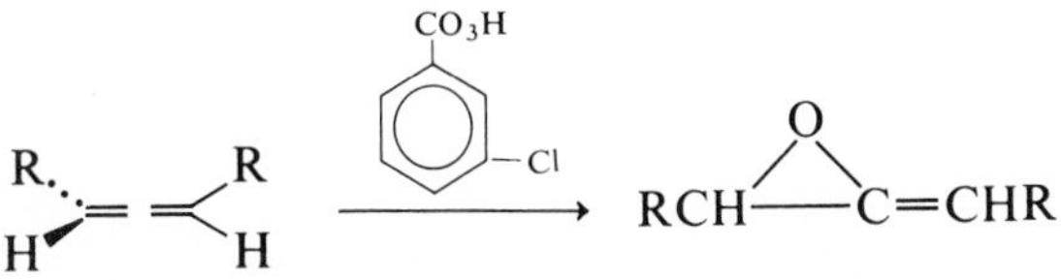

**(22)**                                **(23)**

Attempts by Crandall et al. (75) to isolate the allene oxide, the precursor
of compound **21**, by using only 1 equivalent of peroxyacetic acid failed; the
product consisted largely of unreacted compound **20** (50 %) and the dioxaspiro

compound **21** (40%). The illustrated stereochemistry of **21** is assigned on the assumption that the peroxy acid first attacks the more highly substituted double bond from the direction remote from the *t*-butyl group. The intermediate allene oxide apparently reacts with peroxy acid faster than the allene itself. This is attributed to the resonance interaction of the substituted oxygen atom, which increases the nucleophilicity of the remaining double bond.

Camp and Greene (743), however, were able to isolate an allene oxide (**23**) by using excess allene. The allene oxide (**23**) is readily isomerized to the cyclopropanone on heating to 100°. It is noteworthy that the epoxidation of 1,1-di-*t*-butylallene (**24**), a compound that might be expected to yield a more stable allene oxide owing to the shielding effect of two *t*-butyl groups, yields only 2,2-di-*t*-butylcyclopropanone (**25**), a stable compound, with 1 mole of buffered peroxyacetic acid (1056). The allene oxide is the presumed intermediate and undergoes an acid-catalyzed rearrangement.

**(24)**                              **(25)**

Even with excess peroxy acid, the cyclopropane (**25**) is obtained, not the dioxaspiro compound.

Finally, the selective epoxidation of unsaturated allenes of the structure $RCH=CHCH=C=CH_2$ ($R = H, CH_3$, or $C_2H_5$) with *p*-nitroperoxybenzoic acid yields epoxyallenes (791), thus suggesting that the ordinary olefinic double bond is more readily oxidized than the allenic system.

### C.   Enol Esters and Ethers

Vinyl esters, the simplest enol esters, epoxidize very slowly with peroxy acids (169, 171, 957, 969), and the resulting epoxides are unstable and rearrange and hydrolyze readily (169, 171). Some free-radical polymerization of the vinyl ester may also take place (171).

Specific rate constants for the epoxidation of vinyl laurate, vinyl oleate, and methyl oleate with peroxylauric acid in benzene at 25°, determined polarographically, are $12 \times 10^{-3}$, $270 \times 10^{-3}$, and $232 \times 10^{-3}$ l/mole-min, respectively (957). Thus the internal double bond in vinyl and methyl oleates is epoxidized about 25 times more rapidly than the vinyl double bond with peroxylauric acid. With peroxybenzoic acid as the oxidant, however, (171,

957), specific reaction-rate constants are higher, and the rate ratio for the expoxidation of internal to vinyl double bonds may be as high as 200. There is considerable uncertainty in the ratio of the rates of epoxidation of internal to vinyl double bonds because the latter value is so low that the ratio may be greatly in error. The difference is sufficiently large, however, for such compounds as vinyl oleate (171, 665), vinyl 4-pentenoate (969), vinyl undecylenate (969), and other similar ones (312, 328, 329, 969) to be cleanly and selectively epoxidized at the internal double bond, thereby providing a ready source of vinylepoxy monomers.

Vinyl acetate epoxide (**26**, R = CH$_3$) has been isolated in about 25% yield by the peroxybenzoic acid epoxidation of vinyl acetate for 6 days at 5° (169). The epoxide is readily hydrolyzed by aqueous acid or base to glycolaldehyde (**29**). The peroxyacetic acid epoxidation of vinyl laurate also yields glycolaldehyde (**29**), presumably via its acetate (**28**, R′ = CH$_3$), but the peroxylauric acid epoxidation of vinyl laurate or vinyl 2-ethylhexanoate yields the laurate ester of glycoladehyde (**28**, R′ = C$_{11}$H$_{23}$) (171). A pathway to account for these results has been suggested (171).

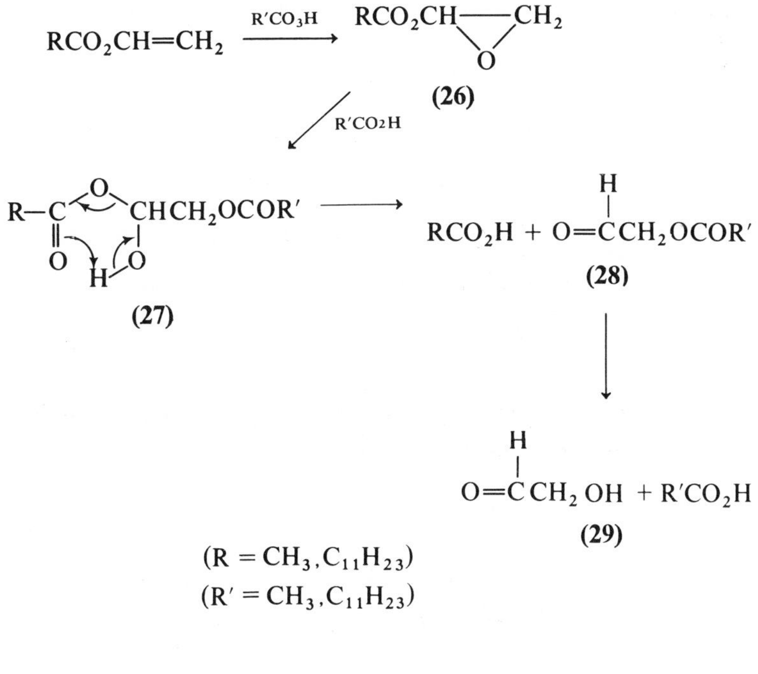

$$\text{RCO}_2\text{CH}=\text{CH}_2 \xrightarrow{\text{R'CO}_3\text{H}} \text{RCO}_2\text{CH}\!\!-\!\!\text{CH}_2$$

(**26**)

R′CO$_2$H

(**27**)

$$\longrightarrow \quad \text{RCO}_2\text{H} + \text{O}=\overset{\text{H}}{\underset{}{\text{C}}}\text{CH}_2\text{OCOR}'$$

(**28**)

$$\text{O}=\overset{\text{H}}{\underset{}{\text{C}}}\text{CH}_2\,\text{OH} + \text{R'CO}_2\text{H}$$

(**29**)

(R = CH$_3$, C$_{11}$H$_{23}$)

(R′ = CH$_3$, C$_{11}$H$_{23}$)

The proposed intermediate **27**, derived from the epoxide by ring opening by the carboxylic acid from the peroxy acid, cannot be isolated presumably because of its facile collapse to carboxylic acid and compound **28** by the six-center concerted mechanism shown.

The low rate of epoxidation of vinyl esters, coupled with the facile ring-opening and rearrangement reactions that take place, makes the resulting epoxy ethyl esters (**26**) rare compounds indeed.

Isopropenyl and other enol esters are epoxidized more rapidly by peroxy acids to yield labile epoxyacylates that undergo rapid intramolecular isomerization to $\alpha$-acyloxyketones when heated or chromatographed on silica gel or alumina (24, 722, 741, 763, 1057–1069). On low-temperature storage, however, some epoxyacylates have good stability. The $\alpha$-acyloxy-ketones are converted to $\alpha$-ketols on mild hydrolysis, thus providing a simple route from ketones to $\alpha$-ketols, as was first shown by Kritchevsky and Gallagher in the steroid field (1057, 1058).

The rearrangement of epoxyacylates to $\alpha$-acyloxyketones proceeds presumably by a concerted rearrangement (722).

In the older literature (1069) the peroxybenzoic acid epoxidation of 1-acetoxy-1-cyclohexene and 1-methyl-3-acetoxy-3-cyclohexene is reported to gives good yield (70 %) of the epoxides, isolated by distillation at temperatures in excess of 100°. Since heating above 100° is known to isomerize the epoxyacetate from 1-acetoxy-1-cyclohexene quantitatively to the α-ketoacetate (24, 722), the earlier report, which preceded the availability of instrumental methods for structure proof, must be considered erroneous.

Some typical enol ester oxidations with peroxy acids are shown below (reported yields are in parentheses below the compounds; the reference numbers are shown at right.) Enol esters with aryl substituents rearrange very readily, and intermediate epoxides are apparently not isolable.

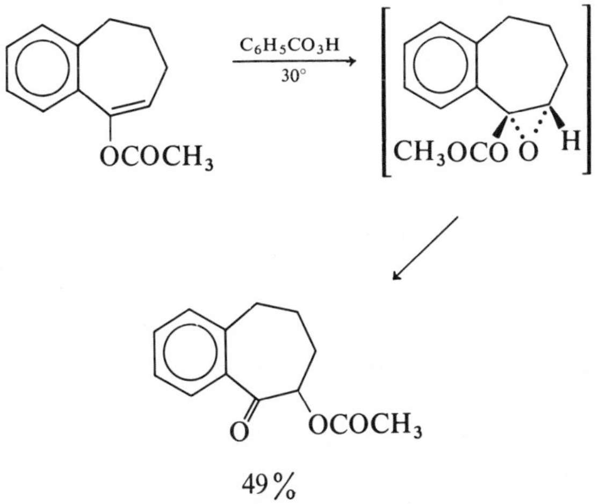

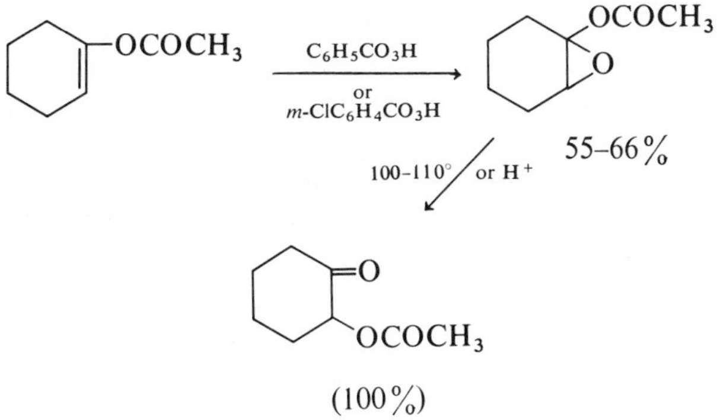

49 %

55–66%

(100%)

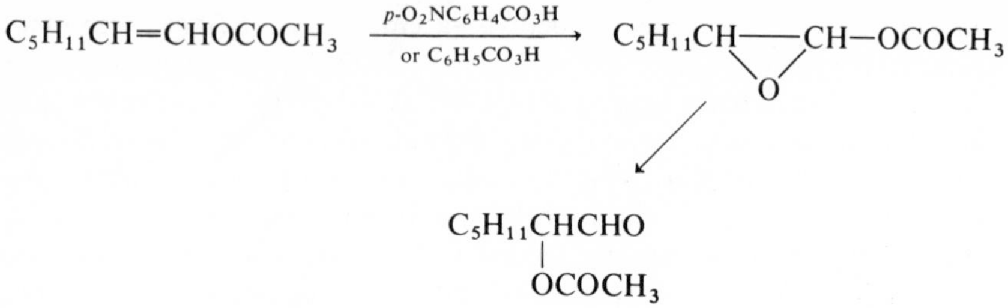

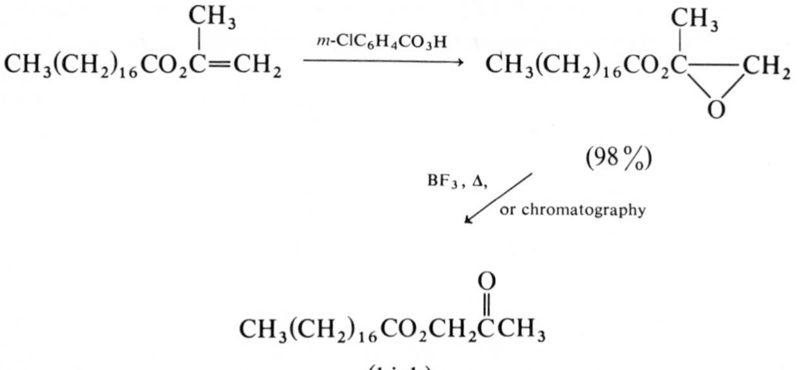

$$\text{C}_5\text{H}_{11}\text{CH}=\text{CHOCOCH}_3 \xrightarrow[\text{or C}_6\text{H}_5\text{CO}_3\text{H}]{p\text{-O}_2\text{NC}_6\text{H}_4\text{CO}_3\text{H}} \text{C}_5\text{H}_{11}\text{CH}\overset{\displaystyle\diagdown\!\!\diagup}{\underset{\text{O}}{}}\text{CH}-\text{OCOCH}_3$$

$$\text{C}_5\text{H}_{11}\text{CHCHO}$$
$$\overset{|}{\text{OCOCH}_3}$$

$$\text{CH}_3(\text{CH}_2)_{16}\text{CO}_2\overset{\overset{\text{CH}_3}{|}}{\text{C}}=\text{CH}_2 \xrightarrow{m\text{-ClC}_6\text{H}_4\text{CO}_3\text{H}} \text{CH}_3(\text{CH}_2)_{16}\text{CO}_2\overset{\overset{\text{CH}_3}{|}}{\text{C}}\overset{\displaystyle\diagdown\!\!\diagup}{\underset{\text{O}}{}}\text{CH}_2$$

(98%)

BF$_3$, Δ, or chromatography

$$\text{CH}_3(\text{CH}_2)_{16}\text{CO}_2\text{CH}_2\overset{\overset{\displaystyle\text{O}}{\|}}{\text{C}}\text{CH}_3$$

(high)

(Other peroxy acids besides those shown above can also be used.)

The epoxidation of enol ethers has received only limited study (137, 1069–1073). Although epoxidation is very rapid, owing to the high nucleophilicity of the double bond, acid-catalyzed epoxide-ring opening is also a facile reaction. It is doubtful, therefore, that early investigators had isolated the anticipated and reported epoxy ethers (1069, 1072, 1073).

Stevens and Tazuma (137) were able to synthesize an authentic epoxy ether in 90% yield by epoxidation of 2-methyl-1-ethoxy-1-phenylpropene with peroxybenzoic acid in ether at 0° for 30 sec, followed immediately by rapid passage (30 sec) of the solution through a column of activated alumina so as to separate benzoic acid from the labile epoxide. The structure of the epoxy ether was proved by comparison with an authentic specimen.

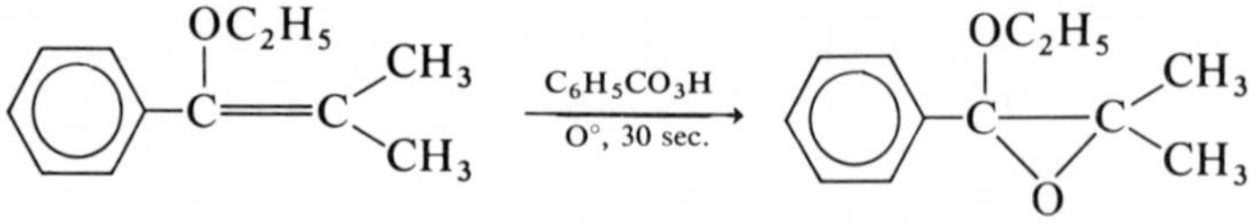

$$\text{C}_6\text{H}_5-\overset{\overset{\displaystyle\text{OC}_2\text{H}_5}{|}}{\text{C}}=\text{C}\overset{\nearrow\text{CH}_3}{\searrow\text{CH}_3} \xrightarrow[\text{0°, 30 sec.}]{\text{C}_6\text{H}_5\text{CO}_3\text{H}} \text{C}_6\text{H}_5-\overset{\overset{\displaystyle\text{OC}_2\text{H}_5}{|}}{\text{C}}\overset{\displaystyle\diagdown\!\!\diagup}{\underset{\text{O}}{}}\text{C}\overset{\nearrow\text{CH}_3}{\searrow\text{CH}_3}$$

(90%)

Identical reaction of 1-ethoxycyclohexene gave a 61% yield of the benzoate of α-hydroxycyclohexanone, presumably by way of the intermediate unstable epoxide (137).

Belleau and Gallagher (1071) reported that reaction of excess peroxybenzoic acid with $\Delta^{17}$- and $\Delta^{20}$-steroidal enol ethers yields ketosteroids and esters by cleavage of the intermediate epoxy ether. By treating a model epoxy ether, 1,2-epoxy-1-methoxy-2-methyl-1-(p-biphenyl)propane, with peroxybenzoic acid, Stevens and Tazuma (137) obtained a 27% yield of ester, thereby lending support to the pathway proposed by Belleau and Gallagher. The cleavage of enol ethers is paralleled by the cleavage of aromatic ethers to esters by peroxybenzoic acid (1074).

### D.  Other Epoxide Rearrangement Reactions

Acid-catalyzed or thermal rearrangement reactions of epoxides are well known even with relatively stable epoxides (1075). With some unsaturated compounds, epoxidation with peroxy acids forms intermediate epoxides that are so labile that they can be isolated, if at all, only by working at low temperatures for short reaction periods and by neutralizing the carboxylic acids derived from the peroxy acids. With acetylenes, as already noted, the low rate of epoxidation, coupled with facile ring-opening and rearrangement reactions, has made the intermediate oxirenes unisolable so far. Only specially structured allenes yield monoepoxides or diepoxides, which are readily converted to rearranged products even *in situ.* Enol esters and enol ethers can be converted to epoxides that are relatively stable at low temperatures, but they readily rearrange with acid, base, heat, or on chromatography.

Epoxide-ring opening is determined by the ease with which the carbon–oxygen bond is ruptured; this is always facilitated when there are electron-donating alkyl or aryl groups attached to the double bond that can stabilize intermediate positive ions (carbonium or oxocarbonium ions). The direction of epoxide-ring opening is governed usually by the relative stability of the two possible intermediate positive ions, but the rearranged product will depend on the relative migratory aptitude of the groups involved (1075). Epoxides can also rearrange under basic conditions, but this subject is not relevant here.

Byers and Hickinbottom (674) showed that the diisobutylenes, 2,4,4-trimethyl-2-pentene and 2,4,4-trimethyl-1-pentene, yield epoxides with peroxybenzoic acid in chloroform, but with peroxyacetic and peroxyformic acids a variety of products is obtained, including the epoxides and their rearranged products, unsaturated alcohol, ketone, aldehyde, and dioxane. The main course of formation of the abnormal products is via the labile epoxides

$$H_3C-\underset{\underset{CH_3}{|}}{\overset{\overset{CH_3}{|}}{C}}-CH_2-\underset{}{\overset{\overset{CH_3}{|}}{C}}=CH_2$$

$$\Big\downarrow\, HCO_3H \text{ or } CH_3CO_3H$$

$$H_3C-\underset{\underset{CH_3}{|}}{\overset{\overset{CH_3}{|}}{C}}-CH_2-\underset{O}{\overset{\overset{CH_3}{|}}{C}}\diagdown CH_2 \;+\; \text{1,2-glycol} + \text{aldehyde} + \text{ketone} + \text{dioxane}$$

1-Aryl- and 1-alkyl cyclopentenes and cyclohexenes give 2-substituted cycloalkanones in addition to, or instead of, epoxides (115, 492, 1076–1078). Earlier investigators who did not have infrared facilities may have overlooked the rearrangement owing to the similarity in physical properties between the epoxides and ketones. In the isolation of the epoxides by distillation the lowest distillation temperatures should be employed to minimize the rearrangement. For example, when ketone-free 1-methylcyclohexane epoxide is distilled at about 40° under reduced pressure, no rearrangement occurs (115).

Filler and co-workers (115) have summarized their results on the reaction of 1-substituted cyclohexenes with peroxybenzoic and monoperoxyphthalic acids (Table 26). More rearrangement occurs with peroxybenzoic acid; benzoic acid is the cause of this.

TABLE 26.   PRODUCTS FROM REACTION OF 1-SUBSTITUTED
            CYCLOHEXENES WITH PEROXY ACIDS[a]

| R | Peroxybenzoic Acid | | | Monoperoxyphthalic Acid | | |
|---|---|---|---|---|---|---|
| | Epoxide | Ketone | Diol | Epoxide | Ketone | Diol |
| $CH_3$ | 48 | 24 | 2 | 55 | 15 | 2 |
| $C_2H_5$ | 60 | 25 | — | 63 | 28 | — |
| $i\text{-}C_3H_7$ | 44 | 35 | — | 44 | 21 | — |
| $t\text{-}C_4H_9$ | — | 30 | — | — | 30 | — |
| $C_6H_5$ | — | 53 | — | — | — | 80 |

[a] Data from reference 115.

Oxidation of 1-alkyl-1-cyclopentenes (and 1-phenylcyclopentene) with excess formic acid and hydrogen peroxide (peroxyformic acid oxidant), followed by distillation of the oxidation products at atmospheric pressure, gives good yields of 2-alkylcyclopentanones (alkyl $= C_2H_5$, $n$-$C_3H_7$, $n$-$C_4H_9$) and 2-phenylcyclopentanone (1076–1078). When the distillation is conducted under vacuum, the products isolated are the *trans*-1,2-glycol monoformates formed by destruction of the intermediate oxiranes by ring opening, with the large excess of formic acid used during the epoxidation step. Similar results are obtained with 3-ethyl-2-pentene and ethylidene, propylidene, and butylidene cyclopentanes (1079).

When a halogen atom is attached to the double bond, facile epoxide-to-carbonyl rearrangement also occurs and the intermediate epoxides are not usually isolable (824, 1080–1082). Mousseron and Jacquier (1069), however, reported the preparation and isolation by distillation of epoxides in 75–80% yields by the peroxybenzoic acid epoxidation of 1-chlorocyclohexene, 1-chlorocyclopentene, 1-methyl-3-chloro-3-cyclohexene, and 2-chlorotetrahydronaphthalene. They observed that 1-chloro-1,2-epoxycyclohexane is relatively stable in an inert atmosphere but decomposes slowly on exposure to air to yield a small quantity of the rearranged product, 2-chlorocyclohexanone.

On the other hand, McDonald and Schwab (1081) reported that 1-chloro-1,2-epoxycyclohexane is difficult to isolate and readily rearranges at room temperature; it can be stored at $-10°$ for several months. 1-Chloro-4-methyl-1,2-epoxycyclohexane is more stable than the unmethylated compound but readily undergoes thermal or acid-catalyzed rearrangement with chlorine migration.

Chlorine migration in the epoxide-to-carbonyl rearrangement occurs with great facility (1080–1082). The epoxidation of *trans*-α-chlorostilbene with peroxybenzoic, peroxyacetic, or buffered peroxytrifluoroacetic acids yields the α-chloroketone desyl chloride in 63–76% yield, not the epoxide, although the latter is the intermediate.

$$C_6H_5CH{=}CClC_6H_5 \xrightarrow{\text{peroxy acids}} \overset{\overset{\displaystyle O}{\|}}{C_6H_5C}{-}\overset{\overset{\displaystyle Cl}{|}}{CH}C_6H_5$$
*(trans)*

That preferential chlorine migration occurs was shown by the peroxy acid epoxidation of *trans*-1-chloro-(p-tolyl)-2-phenylethylene (**30**) and *trans*-1-chloro-1-phenyl-2-(p-tolyl)ethylene (**31**). Only that α-chloroketone is obtained which involves chlorine migration.

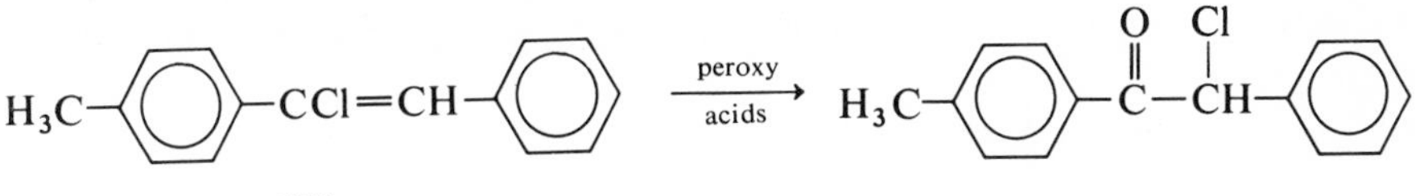

$$H_3C-\!\!\bigcirc\!\!-CCl\!=\!CH-\!\!\bigcirc \quad \xrightarrow[\text{acids}]{\text{peroxy}} \quad H_3C-\!\!\bigcirc\!\!-\overset{\overset{\displaystyle O}{\|}}{C}-\overset{\overset{\displaystyle Cl}{|}}{CH}-\!\!\bigcirc$$

**(30)**

$$\bigcirc\!\!-CCl\!=\!CH-\!\!\bigcirc\!\!-CH_3 \quad \longrightarrow \quad \bigcirc\!\!-\overset{\overset{\displaystyle O}{\|}}{C}-\overset{\overset{\displaystyle Cl}{|}}{CH}-\!\!\bigcirc\!\!-CH_3$$

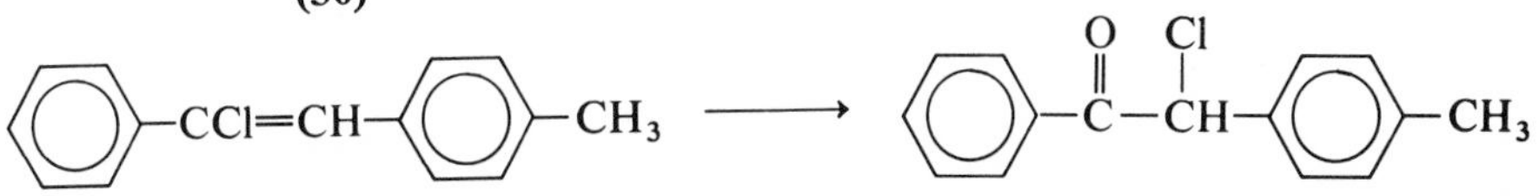

**(31)**

Hart and Lerner (824) have made use of the facile epoxide-to-carbonyl rearrangement to develop a single-step transformation of a number of alkenes to ketones with peroxytrifluoroacetic acid as the oxidant in the presence of boron fluoride etherate as the isomerization reagent. Reactions are complete within 15 min after mixing the reagents at 0–8° or on refluxing in methylene chloride. Table 27 lists the results obtained. The products result from hydrogen, methyl, acyl, chlorine, or bromine migration, as well as from ring contraction.

TABLE 27. SINGLE-STEP OXIDATION OF ALKENES TO KETONES WITH PEROXYTRIFLUOROACETIC ACID–BORON FLUORIDE[a]

| Alkene | Ketone | Yield (%) |
|---|---|---|
| 2,3-Dimethyl-2-butene | 3,3-Dimethyl-2-butanone | 75 |
| 2-Methyl-2-butene | 3-Methyl-2-butanone | 53 |
| cis-3-Methyl-2-pentene | 3-Methyl-2-pentanone | 63 |
| trans-3-Methyl-2-pentene | 3-Methyl-2-pentanone | 70 |
| 1-Methylcyclohexene | 2-Methylcyclohexanone | 41 |
| 1,2-Dimethylcyclohexene | 1-Acetyl-1-methylcyclopentane | 76 |
| $\Delta^{9,10}$-Octalin | Spiro[4.5]decan-6-one | 86 |
| 3-Chloro-2-methyl-2-butene | 3-Chloro-3-methyl-2-butanone | 77 |
| 3-Bromo-2-methyl-2-butene | 3-Bromo-3-methyl-2-butanone | 79 |
| 2,3-Dibromo-2-butene | 3,3-Dibromo-2-butanone | 61 |
| 3,4-Dimethyl-3-penten-2-one | 3,3-Dimethyl-2,4-pentanedione | 81 |
| 2-Cyclopentylidenecyclopentanone | Spiro[4.5]decane-6,10-dione | 43 |

[a] Data from Reference 824.

Cyclobutene and various methyl-substituted derivatives have been epoxidized with peroxybenzoic acid in methylene chloride or chloroform at 0 to −50° (61). In some cases epoxides are stable and can be isolated in yields up to 70% but usually 30–56%; both isolable and nonisolable intermediate epoxides rearrange, the former by the addition of protic or Lewis acids, the latter spontaneously. Results are summarized in the following equations (yields in parentheses below structure):

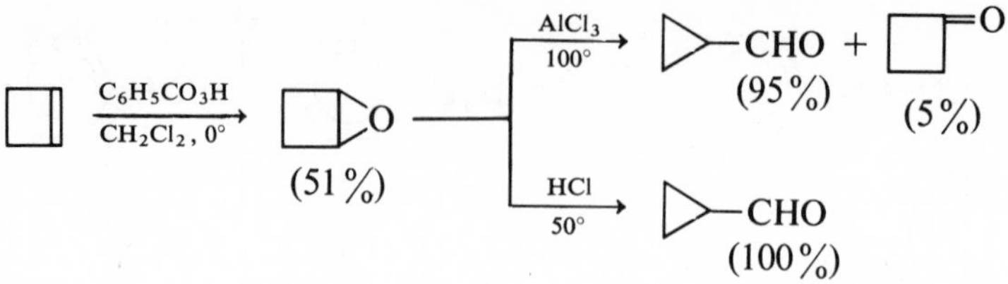

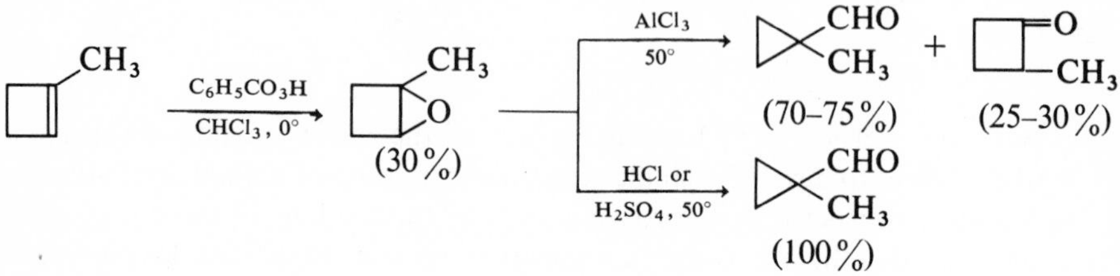

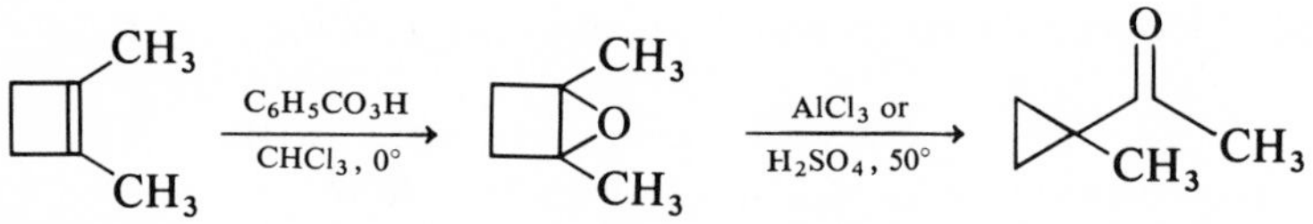

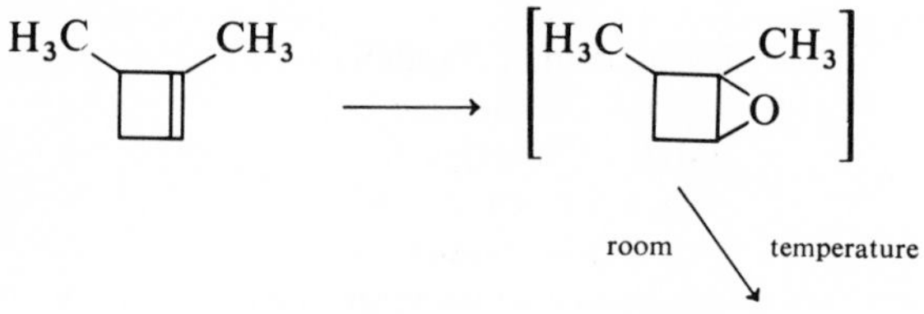

room　　temperature

rearranged products only

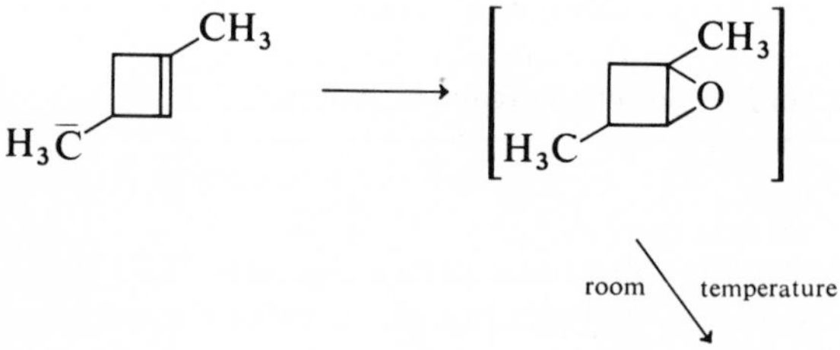

room　　temperature

rearranged products only

Epoxides of *o*-allylphenol and *o*-crotylphenol can be prepared by oxidation with peroxyacetic acid if mild reaction and isolation conditions are employed (336); *o*-propenylphenol does not yield an isolable epoxide. If the epoxides are heated or distilled, rearrangement occurs to coumaran or coumarone derivatives.

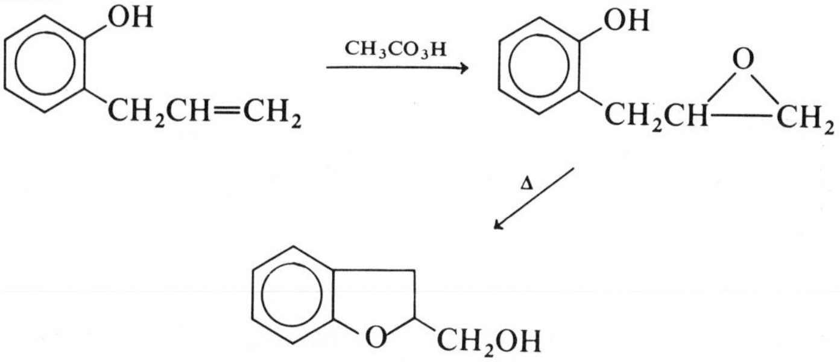

2-Hydroxymethylcoumaran

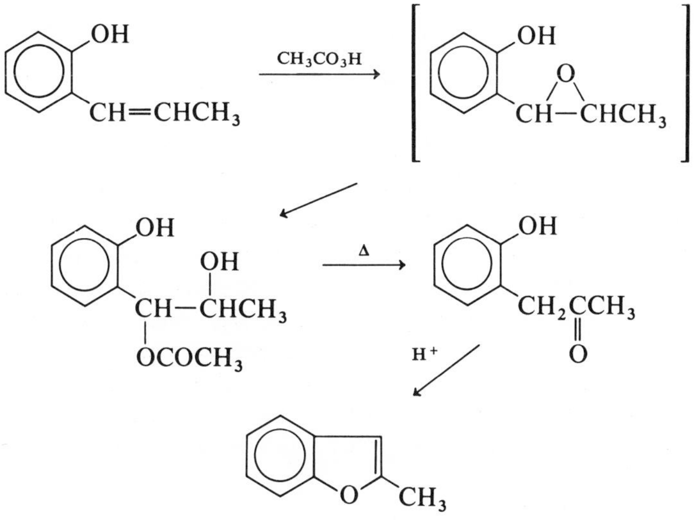

2- Methylcoumaran

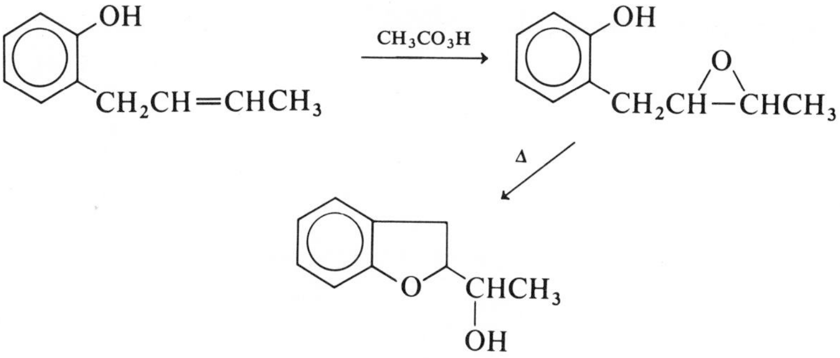

2-(1-Hydroxyethyl)coumaran

The epoxidation of cyclooctatetraene with peroxybenzoic acid is reported to yield a mixture of monoepoxide and a tricyclic rearranged product (73, 1083):

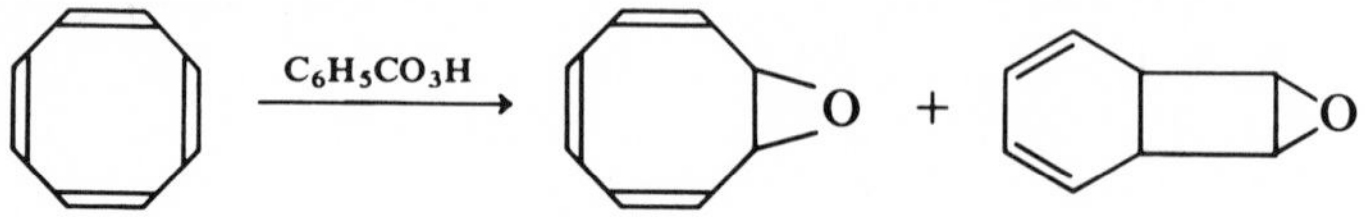

$$ \text{(octagon ring)} \xrightarrow{\text{C}_6\text{H}_5\text{CO}_3\text{H}} \text{(bicyclic epoxide)} \ \text{O} \ + \ \text{(tricyclic epoxide)} \ \text{O} $$

Octamethylcyclooctatetraene gives only the rearranged tricyclic epoxide (147).

Wheeler (1084) has suggested that the epoxidation of cyclooctatetraene proceeds by 1,4-addition to yield the cyclic ether:

The epoxidation of $\gamma,\delta$-unsaturated acids with *in situ* or preformed peroxy acids usually yields $\delta$-hydroxy-$\gamma$-lactones, rather than the anticipated epoxides (1085–1089). Although the epoxide is the intermediate, the molecule is suitably set up to undergo ring opening by the nucleophilic neighboring carboxy group. Increase in reaction temperature or the presence of strong acids increases the proportion of $\delta$- to $\gamma$-lactones (1089); ring opening appears to proceed stereospecifically *trans*. Results are summarized in the following equations from references 1085, 1088, and 1089, respectively (yields in parentheses below structure):

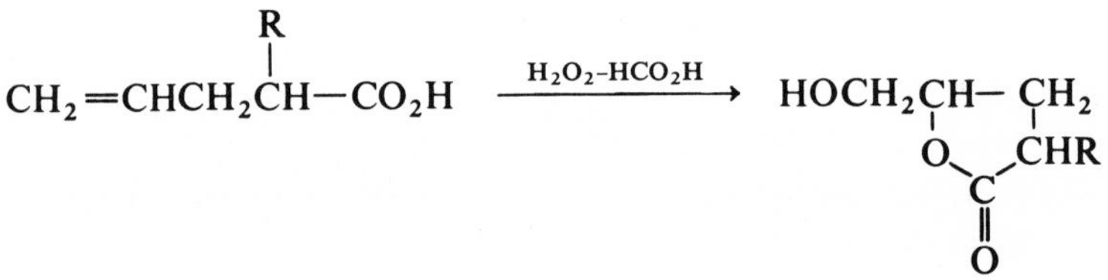

$$ \underset{}{\overset{\text{R}}{\underset{|}{\text{CH}_2\text{=CHCH}_2\text{CH}}}}\text{--CO}_2\text{H} \xrightarrow{\text{H}_2\text{O}_2\text{-HCO}_2\text{H}} \text{HOCH}_2\underset{\underset{\underset{\text{O}}{\overset{\|}{\text{C}}}}{\overset{|}{\text{O}}}}{\text{CH}}\text{--}\underset{\text{CHR}}{\overset{|}{\text{CH}_2}} $$

(80%)

(R = *n*- or *i*-C$_3$H$_7$, *n*- or *i*-C$_4$H$_9$, *n*- or *i*-C$_5$H$_{11}$

$$\underset{\begin{array}{cc}\text{Cl} & \text{R}\\ | & |\end{array}}{\text{H}_2\text{C}=\text{CCH}_2\text{CHCO}_2\text{H}} \xrightarrow{\text{H}_2\text{O}_2-\text{CH}_3\text{CO}_2\text{H}} \text{HOCH}_2-\overset{\text{Cl}}{\underset{}{\text{C}}}\underset{}{\text{---CH}_2}$$

$$(\text{R} = \text{C}_2\text{H}_5,\ \text{C}_3\text{H}_7,\ n\text{-Bu},\ i\text{-Bu})$$

$$(40\text{–}90\%)$$

$$(\text{R} = \text{CH}_3,\ \text{R}' = \text{H})$$
$$(\text{R} = \text{H},\ \text{R}' = \text{CH}_3)$$
$$(\text{R} = \text{C}_6\text{H}_5,\ \text{R}' = \text{H})$$
$$(\text{R} = \text{C}_6\text{H}_5,\ \text{R}' = \text{CH}_3)$$
$$[\text{R} = \text{H},\ \text{R}' = \text{C}_6\text{H}_5\ (cis\ \text{and}\ trans)]$$

2,3,3-Trimethyl-3-butenoic acid, a $\beta,\gamma$-unsaturated acid, yields a $\beta$-lactone on epoxidation with 40% peroxyacetic acid in methylene chloride at 15–20° (497). 3-Hexenoic acid, on the other hand, yields the anticipated epoxy acid (497). The epoxidation of ($\beta,1'$-epoxy-$\beta,10$-vinyl)*trans*-decalin, a vinyl ether, also yields a $\delta$-lactone on epoxidation with peroxybenzoic acid (1090). C-Nor-D-homosapogenins yield hydroxylactones with peroxybenzoic acid but epoxides with $m$-chloroperoxybenzoic acid (758).

Rearrangements also occur in the epoxidation of certain bicyclic compounds. Although norbornene gives an excellent yield of *exo-cis*-2,3-epoxynorbornane on epoxidation with various peroxy acids (96, 145, 259, 260, 780), norbornadiene yields an epoxide that easily rearranges and is difficult to isolate and maintain (144, 584, 1091). The bicyclic aldehyde bicyclo[3.1.0]-hex-2-ene-6-*endo*-carboxaldehyde is often the major product in spite of precautions to avoid rearrangement; other rearrangement products are also formed but in minor amounts (144, 584, 1091).

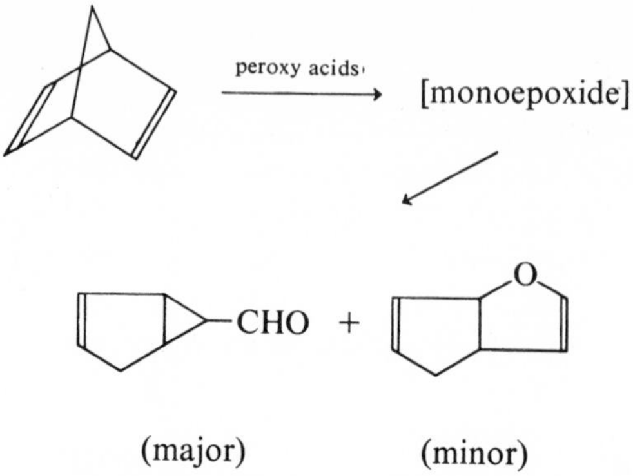

peroxy acids·    [monoepoxide]

—CHO  +

(major)              (minor)

Reaction of *endo*-dicyclopentadiene with peroxyformic acid yields a glycol monoformate (via the intermediate epoxide); hydrolysis yields a glycol that does not react with lead tetraacetate or periodic acid (1092). An *endo–exo* rearrangement that leads to a 1,3- or 1,4-glycol was suggested.

*exo*-Trimethylene-2-norbornene reacts with peroxyformic acid to give a rearrangement product on opening of the epoxide rings (1093). The glycol obtained by hydrolysis of the intermediate hydroxyformate is assigned the 2-*exo*-7-*syn* diol structure on the basis of physical (infrared) and chemical properties (formation of a *p*-nitrobenzylidene derivative and failure to react with periodic acid or to enhance the conductivity of boric acid). The trimethylene ring, which is *exo* in the starting material, is *endo* in the glycol.

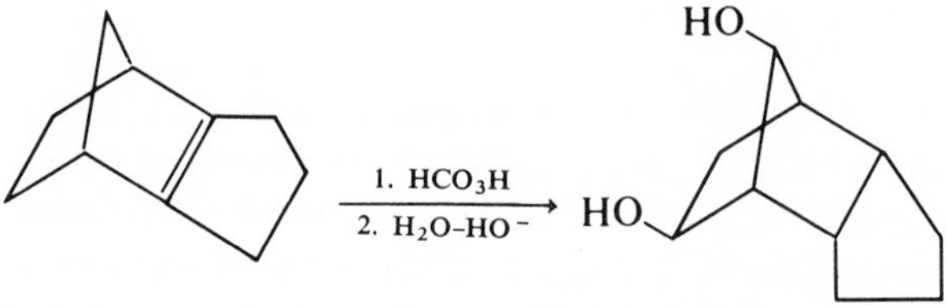

1. $HCO_3H$
2. $H_2O–HO^-$

*exo-cis*-Norbornene 2,3-epoxide, obtained from norbornene by epoxidation, also undergoes rearrangement when the epoxide ring is opened by formic or acetic acid, as shown by the isolation of the 2,7-diol after saponification (259).

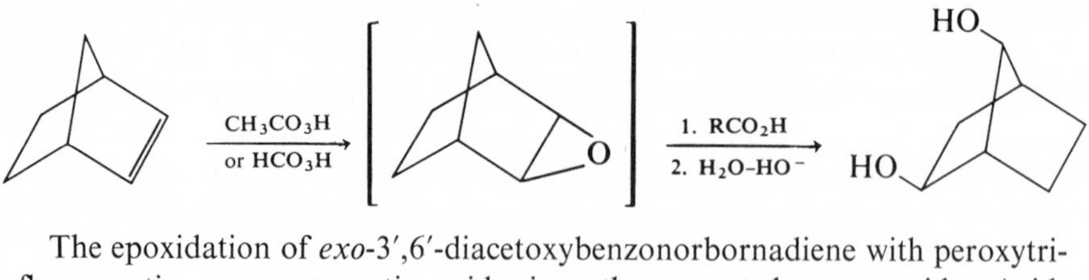

$CH_3CO_3H$
or $HCO_3H$

1. $RCO_2H$
2. $H_2O–HO^-$

The epoxidation of *exo*-3′,6′-diacetoxybenzonorbornadiene with peroxytrifluoroacetic or peroxyacetic acid gives the expected *exo*-epoxide. Acid-catalyzed acetolysis gives a hydroxyacetate formulated as *exo-anti-3',6'-*diacetoxybenzo-2-acetoxynorbornen-7-ol, the product of rearrangement (484). The rearrangement of the intermediate epoxide is also base catalyzed and is apparently without precedent. The reaction is rationalized on the assumption of $Ar_1$-3 participation, reported by Winstein and Baird (1094). *exo-3',6'-*Dimethoxybenzonorbornadiene behaves similarly only in the presence of peroxyacetic acid; peroxybenzoic and monoperoxyphthalic acids apparently fail to react.

An unusual epoxidation–ring opening–rearrangement reaction was reported by Kleinfelter and Von Schleyer (1095) in the peroxyformic acid epoxidation of some 2-arylnorbornenes. Reaction of 2-phenylnorbornene (R = $C_6H_5$) with a large excess of peroxyformic acid gives 2-endo-phenyl-2,3-*cis-exo-*norbornylene carbonate in maximum yields of 54–56%. The 2,3-*exo-cis-*diol corresponding to the epoxide gives a 76% yield of carbonate when treated similarly with peroxyformic acid. 2-*p*-Chlorophenylnorbornene also yields a carbonate (55%).

The formation of carbonates by the peroxyformic acid epoxidation of olefin is not a general reaction and is favored in molecules giving intermediates with eclipsed C—O bonds on adjacent carbon atoms.

1-Phenylcyclohexene does not yield a carbonate but a *cis*-diol, as do 1-phenylcycloheptene (1096) and indene (1097). Opening of the oxirane ring in a *cis* manner, a precedented reaction in other cases as well (984, 1095, 1098), is proposed as the primary step. A double inversion involving phenyl-group participation with protonated epoxide is viewed as the most likely mechanism (1095).

(R = aryl)

Oxidation of Criegee's hydrocarbon, octamethyl-*syn*-tricyclo[4.2.0.0$^{2,5}$]-octa-3,7-diene, with peroxytrifluoroacetic acid or acid-catalyzed rearrangement of its epoxide gives 1,3,4,5,6,7,8,9-octamethyl-2-oxatricyclo[5.2.0.0.$^{3,6}$]-nona-4,8-diene (1099) and not a structure assigned earlier.

The oxidation of substituted and unsubstituted tetrahydrochromans with excess peroxy acid is an excellent route to 6-ketononanolides (1100a, 1101). The primary products are epoxy ethers, which, as already discussed, undergo facile ring opening with excess peroxy acid (peroxybenzoic, *m*-chloroperoxybenzoic, and monoperoxyphthalic acids) to give hydroxy peroxyesters, which then fragment to 6-ketononanolides. Yields of 6-ketononanolides are in the range of 46–92% (1100, 1101).

(R = C$_6$H$_5$, *m*-ClC$_6$H$_4$, phthalic)

The oxidation technique has also been used to prepare 11- and 12-membered ring ketolactones (1100). If further oxidation of the carbonyl products by peroxy acids in a Baeyer–Villiger reaction is a problem, short reaction times must be used.

Reaction of 4-hydroxy-3,4-diphenyl-2-cyclopenten-1-one with an excess of *m*-chloroperoxybenzoic acid in methylene chloride yields not the expected epoxy carbinol but the ring-enlarged product, 3,5-dihydroxy-3,4-diphenyl-4,5-epoxypentanoic acid δ-lactone, quantitatively (1102).

The epoxidation of linalool and alcohols related to it does not yield oxiranes but rather hydroxytetrahydrofurans and hydroxytetrapyrans by intramolecular reaction of the appropriately placed hydroxy group with the oxirane function (1103–1106).

**(1104)**

Since the oxirane function can be introduced from either of two sides of the double bond, pairs of isomeric tetrahydropyrans and tetrahydrofurans are formed (four products).

The epoxidation of δ,ε-unsaturated ketones, followed by pyrolysis of the epoxide at 210°, yields [3.2.1.] bicyclic systems exclusively with a high degree of stereoselectivity (757). For example, when *cis*-6,7-epoxynonan-2-one, prepared by the epoxidation of the precursor *cis*-6-nonen-2-one with *m*-chloroperoxybenzoic acid, is heated to 210° in a base-washed sealed tube, nearly complete conversion (95%) takes place to yield a mixture of *exo*-6-ethyl-1-methyl-7,8-dioxabicyclo[3.2.1]octane (90%) and the corresponding *endo* isomer (10%) (757).

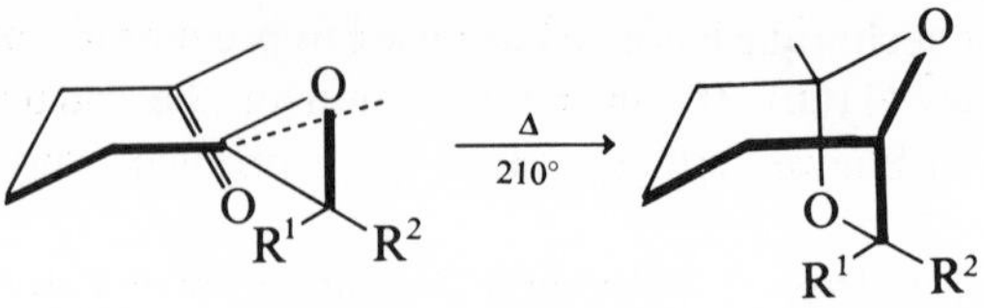

$$\xrightarrow[210°]{\Delta}$$

(95%)

$$(R^1 = R^2 = H)$$
$$(R^1 = C_2H_5, R^2 = H)$$
$$(R^1 = H, R^2 = C_2H_5)$$

The *exo* isomer is identical with breviocomin, the principal sex attractant of the western pine beetle *Dentroctonus brevicomis*. Similarly *trans*-6,7-epoxy-nonan-2-one yields nearly quantitatively *endo*-6-ethyl-1-methyl-7,8-diox-abicyclo[3.2.1]octene (epibrevicomin, 91%) and the *exo* isomer, brevicomin (9%). The reaction mechanism is not clear; other cyclizations of $\delta,\epsilon$-epoxy-ketones have also been reported but with acid catalysis (1107, 1108). An analogous thermal rearrangement (75% at 180°) is that of the epoxide of 1-allylcyclopropylacetate to 2,2-dimethylene-4-acetoxytetrahydrofuran via the intermediate orthoester, trioxabicyclo[3.2.1]octane (757).

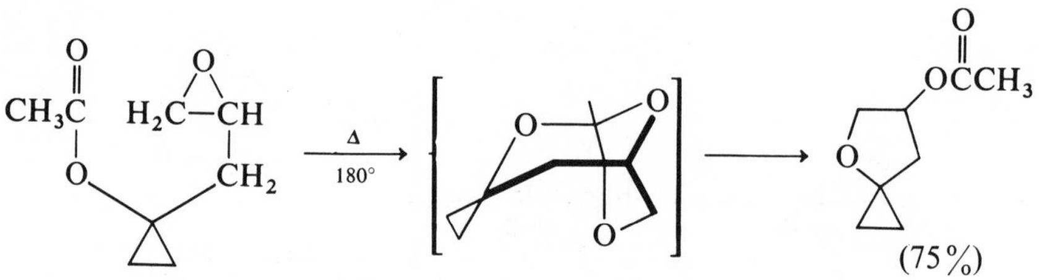

(75%)

In contrast to $\delta,\epsilon$-unsaturated ketones (757), the epoxidation of $\gamma,\delta$-unsaturated ketones with *m*-chloroperoxybenzoic acid does not yield the expected epoxides, but a mixture of products consisting of 2,7-dioxabicyclo-[2.2.1]heptanes and a 5-hydroxymethyldihydrofuran derivative (564). In one instance and under carefully controlled conditions the epoxidation of 6-methyl-5-hepten-2-one with peroxyacetic acid by the method of Crandall and Change (416) followed by rapid distillation yielded the thermally labile epoxide.

1-Vinylcyclopropanol undergoes an interesting ring enlargement on epoxidation with peroxybenzoic acid in ether to yield 2-hydroxymethyl-cyclobutanone (1109):

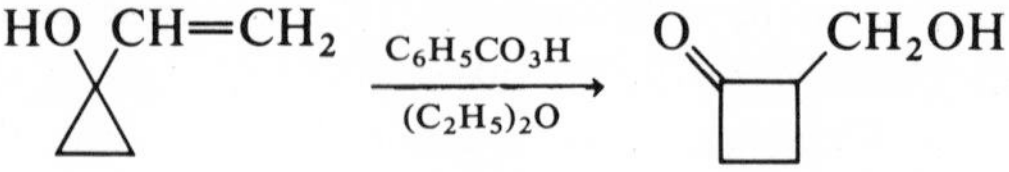

(32%)

In contrast, vinylcyclopropane yields the isolable oxirane on epoxidation with peroxyacetic acid in methylene chloride with excess sodium carbonate (664):

$$
\triangle\!\!-CH=CH_2 \xrightarrow[\text{Na}_2\text{CO}_3, \ 15\text{-}20°]{\text{CH}_3\text{CO}_3\text{H}, \ \text{CH}_2\text{Cl}_2} \triangle\!\!-CH\overset{O}{-\!\!\!-}CH_2
$$

$$(53\%)$$

The epoxidation of cyclic acetals of $\alpha,\beta$-unsaturated aldehydes (unsaturated dioxolanes) gives fair to good yields (28–67%) of epoxides with peroxyacetic acid in ethyl acetate solution (441):

$$
R^1\text{--CH}=\overset{\overset{\textstyle R^2}{|}}{C}\text{--}\overset{O\text{--}CH\text{--}R^3}{\underset{O\text{--}CH_2}{CH}} \xrightarrow[\text{CH}_3\text{CO}_2\text{C}_2\text{H}_5]{\text{CH}_3\text{CO}_3\text{H}} R^1\text{--CH}\overset{O}{-\!\!\!-}\overset{\overset{\textstyle R^2}{|}}{C}\text{--}\overset{O\text{--}CH\text{--}R^3}{\underset{O\text{--}CH_2}{CH}}
$$

$$(R^1 = CH_3, \ n\text{-}C_3H_7, \ \text{or} \ C_6H_5)$$
$$(R^2 = H, \ CH_3, \ \text{or} \ C_2H_5)$$
$$(R^3 = H \ \text{or} \ CH_3)$$

In contrast, when a noncyclic acetal, such as crotonaldehyde di-*n*-butylacetal, is similarly oxidized, a good yield of butyl crotonate (73%) is obtained (441). Sulfuric acid catalyzes the reaction.

$$
CH_3CH=CHCH(OC_4H_9)_2 \xrightarrow{\text{CH}_3\text{CO}_3\text{H}} CH_3CH=CH-CO_2C_4H_9
$$
$$(73\%)$$

The conversion of the acetal to the ester function also proceeds readily with saturated acetals to give 60–90% yields of esters (441).

$$
R\text{--}CH(OC_2H_5)_2 \xrightarrow[\text{H}^+]{\text{CH}_3\text{CO}_3\text{H}} R\text{--}CO_2C_2H_5 + C_2H_5OH + CH_3CO_2H
$$
$$(60\text{--}90\%)$$

$$(R = n\text{-}C_3H_7, \ C_6H_5, \ \text{or} \ C_6H_5CH(OC_2H_5)CH_2)$$

The epoxidation of unsaturated acylals proceeds normally (441).

$$
R_1\text{--CH}=\overset{\overset{\textstyle R^2}{|}}{C}\text{--}CH(OCOCH_3)_2 \xrightarrow{\text{CH}_3\text{CO}_3\text{H}} R^1\text{--CH}\overset{O}{-\!\!\!-}\overset{\overset{\textstyle R_2}{|}}{C}\text{--}CH(OCOCH_3)_2
$$
$$(55\text{--}75\%)$$

$$(R^1 = CH_3, \ C_2H_5, \ \text{or} \ C_3H_7)$$
$$(R^2 = C_2H_5)$$

The epoxidation of hexamethyl (Dewar benzene) with peroxyacetic acid–sodium carbonate or peroxybenzoic acid to give the monoepoxide (hexamethyl-5,6-epoxybicyclo[2.2.0]hex-2-ene) or diepoxide (hexamethyl-2,3:5,6-bisepoxy-bicyclo[2.2.0]hexane) is accompanied by a proton-catalyzed ring contraction, followed by valence isomerization to yield pentamethyl-5-acetylcyclopenta-diene (122).

Similar acid-catalyzed rearrangements to form three-membered rings have been reported for epoxycyclobutanes (61, 1110).

For other anomalous rearrangement reactions during epoxidation consult references 58, 966, 1111–1114.

# REFERENCES

1. D. Swern, *Chem. Rev.*, **45**, 1 (1949).
2. D. Swern, *Organic Reactions*, Vol. 7, Wiley, New York, 1953, Chapter 7.
3. W. R. Wragg, *Bull. Soc. Chim. France*, **1952**, 911.
4. M. S. Malinovskii, *Usp. Khim.*, **27**, 622 (1958).
5. S. Havel, *Chem. Listy*, **53**, 928 (1959).
6. B. Phillips and D. L. MacPeek, *Encyclopedia of Chemical Technology*, First Supplement, Interscience, New York, 1957, p. 622.
7. J. G. Wallace, *Hydrogen Peroxide in Organic Chemistry* E. I. du Pont de Nemours & Co., Wilmington, Del., 1960.
8. A. Rosowsky, *Heterocyclic Compounds with Three- and Four-Membered Rings*, Wiley, New York, 1964, Chapter 1.
9. J. G. Wallace, in *Encyclopedia of Chemical Technology*, 2nd ed., Vol. 8, Interscience, New York, 1965, p. 238.

10. FMC Corporation, *Preparation, Properties, Reactions, and Uses of Organic Peracids and Their Salts*, 1964.

11 D Swern, in *Encyclopedia of Polymer Science and Technology*, Vol 6, 1967, p. 83.

12. T. Ikawa, *Yuki Gosei Kagaku Kyokai Shi*, **26**, 506 (1968).

13. D. Swern, T. W. Findley, and J. T. Scanlan, *J. Amer. Chem. Soc.*, **66**, 1925 (1944).

14. D. Swern, T. W. Findley, and J. T. Scanlan, *J. Amer. Chem. Soc.*, **67**, 412 (1945).

15. D. Swern, *Chem. Rev.* **45**, 1 (1949).

16. N. Prileschajew, German Patent 230,723 (1908).

17. N. Prileschajew, *Ber.*, **42**, 4811 (1909).

18. N. Prileschajew, *J. Russ. Phys. Chem. Soc.*, **42**, 1387 (1910); *ibid.*, **43**, 609 (1911); *ibid.*, **44**, 613 (1912).

19. H. Batzer and E. Nikles, U.S. Patent 3,023,174 (1962); British Patent 865,340 (1961).

20. H. B. Henbest and B. Nicholls, *J. Chem. Soc.*, **1959**, 221.

21. S. Bernstein and R. Littell, *J. Org. Chem.*, **26**, 3610 (1961).

22. H. B. Henbest and R. A. L. Wilson, *J. Chem. Soc.*, **1957**, 1958.

23. J. Meinwald and B. C. Cadoff, *J. Org. Chem.*, **27**, 1539 (1962).

24. H. J. Shine and G. E. Hunt, *J. Amer. Chem. Soc.*, **80**, 2434 (1958).

25. H. B. Henbest and B. Nicholls, *J. Chem. Soc.*, **1957**, 4608.

26. R. Budziarek and F. S. Spring, *Chem. Ind.* (*London*), **1952**, 1102.

27. R. Budziarek, J. C. Hamlet, and F. S. Spring, *J. Chem. Soc.*, **1953**, 778.

28. R. B. Clayton, *J. Chem. Soc.*, **1953**, 2009.

29. E. M. Chamberlin, W. V. Ruyle, A. E. Erickson, J. M. Chemerda, L. M. Aliminosa, R. L. Erickson, G. E. Sita, and M. Tishler, *J. Amer. Chem. Soc.*, **75**, 3477, (1953).

30. R. Budziarek, G. T. Newbold, R. Stevenson, and F. S. Spring, *J. Chem. Soc.*, **1952**, 2892.

31. L. Ruzicka and P. Plattner, U.S. Patent 2,656,349 (1953).

32. P. Hofer, H. Linde, and K. Meyer, *Helv. Chim. Acta*, **45**, 1041 (1962).

33. O. Wintersteiner and M. Moore, *J. Org. Chem.*, **29**, 270 (1964).

34. M. I. Goryaev and F. S. Sharipova, *J. Gen. Chem. USSR*, **33**, 293 (1963).

35. O. N. Semikhatova-Arefova, *Uchenye Zapiski Rostov na Donu Univ.*, **25**, No. 7, 85 (1955).

36. Y. Ogata and I. Tabushi, *J. Amer. Chem. Soc.*, **83**, 3444 (1961).

37. E. A. Perfenov, G. A. Serebrennikova, and N. A. Preobrazhenskii, *Zh. Organ. Khim.*, **3**, 1766 (1967).

38. Y. V. Markova, K. K. Kuzmina, and M. N. Shchukina, *Zh. Obshch. Khim.*, **27**, 1270 (1957).

39. G. I. Poos and J. D. Rosenau, *J. Org. Chem.*, **28**, 665 (1963).

40. H. B. Henbest and B. Nicholls, *Proc. Chem. Soc.*, **1958**, 226.

41. P. G. Gassman and P. G. Pape, *J. Org. Chem.*, **29**, 160 (1964).

42. T. M. Medred and H. W. Christie, *J. Chem. Eng. Data*, **9**, 240 (1964).

43. D. W. Lewis, U.S. Patent 2,997,458 (1961).

44. R. Rambaud and M. Vessiere, *Bull. Soc. Chim. France*, **1960**, 1114.
45. T. Asahara, D. Saika, and M. Takahashi, *Seisan-Kenkyu*, **19**, 263 (1967).
46. J. Konoji, M. Iwamoto, and S. Yuguchi, Japanese Patent 13,457 (1968).
47. H. Faure and D. Gravel, *Can. J. Chem.*, **41**, 1452 (1963).
48. P. Hirsjärvi, P. Eenilä, H. Peltonen, L. Pirilä, and A. Päällysaho, *Suomen. Kem.*, **36B** 126 (1963).
49. G. V. Pigulevskii and S. A. Kozhin, *J. Gen. Chem. USSR*, **27**, 879 (1957).
50. M. I. Goryaev and G. A. Tolstikov, *Zh. Obshch. Khim.*, **31**, 664 (1961).
51. R. N. McDonald and T. E. Tabor, *J. Amer. Chem. Soc.*, **89**, 6573 (1967).
52. R. Villotti and A. Bowers, U.S. Patent 3,300,517 (1967).
53. A. Feldstein and C. A. Vanderwerf, *J. Amer. Chem. Soc.*, **76**, 1626 (1954).
54. E. M. Burgess, *J. Org. Chem.*, **27**, 1433 (1962).
55. D. Lavie, Y. Kashman, and E. Glotter, *Tetrahedron*, **22**, 1103 (1966).
56. G. Roberts, C. W. Shoppee, and R. J. Stephenson, *J. Chem. Soc.*, **1954**, 3178
57. A. Nickon and W. J. Mendelson, *J. Amer. Chem. Soc.*, **87**, 3921 (1965).
58. L. F. Fieser, K. Nakanishi, and W. Y. Huang, *J. Amer. Chem. Soc.*, **75**, 4719 (1953).
59. A. DaSettimo and M. F. Saettone, *J. Org. Chem.*, **29**, 3350 (1964).
60. T. Kauffmann, W. Jasching, and J. Schulz, *Ber.*, **99**, 1569 (1966).
61. J. L. Ripoll and J. M. Conia, *Tetrahedron Letters*, No. 15, 979 (1965); *Bull. Soc. Chim. France*, **1965**, 2755.
62. G. Wilke, U.S. Patent 3,014,928 (1961).
63. H. K. Wiese and S. B. Lippincott, U.S. Patent 2,978,464 (1961).
64. V. Prelog, K. Schenker, and H. H. Günthard, *Helv. Chim. Acta*, **35**, 1598 (1952).
65. J. J. Cahill, Jr., N. E. Wolff, and E. S. Wallis, *J. Org. Chem.*, **18**, 720 (1953).
66. L. N. Owen and G. S. Saharia, *J. Chem. Soc.*, **1953**, 2582.
67. T. W. Craig, G. R. Harvey, and G. A. Berchtold, *J. Org. Chem.*, **32**, 3743 (1967).
68. S. M. Nagvi, J. P. Horwitz, and R. Filler, *J. Amer. Chem. Soc.*, **79**, 6283 (1957).
69. B. M. Lynch and K. H. Pausacker, *J. Chem. Soc.*, **1955**, 1525.
70. W. J. Bailey and C. E. Knox, *J. Org. Chem.*, **25**, 511 (1960).
71. H. B. Henbest and R. L. Wilson, *Chem. Ind. (London)*, **1956**, 659.
72. G. Büchi and E. M. Burgess, *J. Amer. Chem. Soc.*, **84**, 3104 (1962).
73. J. P. Wibaut, *Verslag Gewone Vergade Afdeel Natuurk.*, **62**, 176 (1953).
74. A. C. Cope, P. T. Moore, and W. R. Moore, *J. Amer. Chem. Soc.*, **80**, 5505 (1958).
75. J. K. Crandall, W. H. Machleder, and M. J. Thomas, *J. Amer. Chem. Soc.*, **90**, 7346 (1968).
76. C. S. Dewey and R. A. Bafford, *J. Org. Chem.*, **30**, 491 (1965).
77. J. A. Franks, B. Tolbert, R. Steyn, and H. Z. Sable, *J. Org. Chem.*, **42**, 1440 (1965).
78. F. W. Kunstmann, H. Specht, and A. Stachowiak, German (East) Patent 47,127 (1966).
79. K. Takemoto, *Kogyo Kagaku Zasshi*, **63**, 186 (1960).

80. R. W. Rees, U.S. Patent 3,050,507 (1962).
81. P. D. Klimstra, *J. Med. Chem.*, **9**, 781 (1966).
82. R. B. Moffett and G. Slomp, Jr., *J. Amer. Chem. Soc.*, **76**, 3678 (1954).
83. N. S. Leeds, D. K. Fukushima, and T. F. Gallagher, *J. Amer. Chem. Soc.*, **76**, 2943 (1954).
84. S. Sasaki, *Sci. Repts. Tohoku Univ.*, first series, **42**, No. 1, 31 (1958).
85. D. C. Burke, J. H. Turnbull, and W. Wilson, *J. Chem. Soc.*, **1953**, 3237.
86. Syntex Corporation, British Patent 1,045,174 (1966).
87. H. C. Haas and N. W. Schuler, *J. Appl. Polymer Sci.*, **5**, 52 (1961).
88. F. Piozzi, P. Venturella, A. Bellino, and R. Mondelli, *Ric. Sci.*, **38**, 462 (1968).
89. L. M. Kindley, L. P. Glekas, and P. E. Ritt, *TAPPI*, **44**, 185A (1961).
90. M. Okada, A. Yamada, Y. Saito, and M. Hasunuma, *Chem. Pharm. Bull.*, **14**, 496 (1966).
91. E. E. van Tamelen, *J. Amer. Chem. Soc.*, **77**, 1704 (1955).
92. A. A. Durgaryan and S. A. Titanyan, *Izv. Akad. Nauk Armyan. SSR, Khim. Nauki*, **13**, No. 4, 263 (1960).
93. S. Hattori, *Yuki Gosei Kagaku Kyokai Shi*, **19**, 461 (1961).
94. G. R. Clemo and F. J. McQuillin, *J. Chem. Soc.*, **1952**, 3839.
95. A. S. Bawdekar, G. R. Kelkar, and S. C. Bhattacharya, *Tetrahedron Letters*, **1966**, 1225.
96. S. B. Soloway and S. J. Cristol, *J. Org. Chem.*, **25**, 327 (1960).
97. M. Vilkas, *Ann. Chim.*, **6**, 325 (1951).
98. P. G. Gassman and J. L. Marshall, *J. Amer. Chem. Soc.*, **88**, 2822 (1966).
99. A. Wende and A. Gesierich, *Plaste und Kautschuk*, **8**, 399 (1961).
100. H. H. Wasserman and N. E. Aubrey, *J. Amer. Chem. Soc.*, **77**, 590 (1966).
101. G. I. Poos, U.S. Patent 3,250,789 (1966).
102. V. F. Martynor and T. Chou, *Hua Hsueh Hsueh Pao*, **26**, 11 (1960).
103. A. Wende and A. Gesierich, *Plaste und Kautschuk*, **8**, 301 (1961).
104. G. Greber and L. Metzinger, *Makromol. Chem.*, **39**, 167 (1960).
105. F. L. Julietti, J. F. McGhie, B. L. Rao, W. A. Ross, and W. A. Cramp, *J. Chem. Soc.*, **1960**, 4514.
106. S. Wawzonek, P. D. Klimstra, R. E. Kallio, and J. E. Stewart, *J. Amer. Chem. Soc.*, **82**, 1421 (1960).
107. R. L. Rowland, A. Rodgman, and J. N. Schumacher, *J. Org. Chem.*, **29**, 16 (1964).
108. R. T. O'Connor, C. H. Mack, E. F. DuPré, and W. G. Bickford, *J. Org. Chem.*, **18**, 693 (1953).
109. W. G. Bickford, E. F. DuPré, C. H. Mack, and R. T. O'Connor, *J. Amer. Oil Chem. Soc.*, **30**, 376 (1953).
110. C. H. Mack and W. G. Bickford, U.S. Patent 2,865,931 (1958).
111. G. D. Meakins and M. W. Pemberton, *J. Chem. Soc.*, **1961**, 4676.
112. D. P. Popa and V. V. Titov, *Zh. Obshch. Khim.*, **1967**, 2459.
113. R. Budziarek, F. Johnson, and F. S. Spring, *J. Chem. Soc.*, **1952**, 3410.
114. H. Newman, *J. Org. Chem.*, **29**, 1461 (1964).
115. R. Filler, B. R. Camara, and S. M. Naqui, *J. Amer. Chem. Soc.*, **81**, 658 (1959).

116. G. D. Searle & Co., British Patent 885,782 (1961).
117. G. E. Arth, R. E. Beyler, and L. H. Sarett, U.S. Patent 3,004,994 (1961).
118. K. Schreiber, G. Schneider, and G. Sembdner, *Tetrahedron*, **22**, 1437 (1966).
119. R. V. Soler and J. V. Llados, Spanish Patent 241,871 (1958).
120. V. Mark, U.S. Patents 3,298,815 and 3,324,147 (1967).
121 B. Palameta and M. Prostenik, *Tetrahedron*, **19**, 1463 (1963).
122. W. Schäfer and H. Hellmann, *Angew. Chem., Intern., Ed.*, **6**, 518 (1967).
123. S. K. Rmaswami and S. C. Bhattacharya, *Tetrahedron*, **18**, 575 (1962).
124. N. P. Damodaran and S. Dev, *Tetrahedron*, **1968**, 4123.
125. P. A. Plattner, A. Fürst, F. Koller, and H. H. Kuhn, *Helv. Chim. Acta*, **37**, 258 (1954).
126. L. H. Briggs, R. C. Cambie, and P. S. Rutledge, *J. Chem. Soc.*, **1963**, 5374.
127. D. C. Umarani, K. G. Gore, and K. K. Chakravarti, *Tetrahedron Letters*, **1966**, 1255.
128. P. S. Kalsi, K. L. Bajaj, and I. S. Bhatia, *Perfum. Essent. Oil Rec.*, **58**, 864 (1967).
129. J. M. Tenlon and R. Wylde, *Compt. Rend.*, **C,262**, 1021 (1966).
130. P. A. Mayor and G. D. Meakins, *J. Chem. Soc.*, **1960**, 2792.
131. D. Lavie and E. C. Levy, *Tetrahedron Letters*, **1968**, 2097.
132. R. M. Bowman, A. Chambers, and W. R. Jackson, *J. Chem. Soc., C*, **1966**, 612.
133. L. F. Fieser, W. Y. Huang, and J. C. Babcock, *J. Amer. Chem. Soc.*, **75**, 116 (1953).
134. H. Heymann and L. F. Fieser, *J. Amer. Chem. Soc.*, **74**, 5938 (1952).
135. G. Defaye-Duchateau, *Compt. Rend.*, **251**, 1024 (1960).
136. P. A. Artamanov, *Zh. Obshch. Khim.*, **28**, 1355 (1958).
137. C. L. Stevens and J. Tazuma, *J. Amer. Chem. Soc.*, **76**, 715 (1954).
138. V. F. Martynor and I. M. Chou, *Hua Hsueh Hsueh Pao*, **26**, 53 (1960).
139. T. Colclough, J. I. Cuneen, and C. G. Moore, *Tetrahedron*, **15**, 187 (1961).
140. Y. Ogata and I. Tabushi, *J. Amer. Chem. Soc.*, **83**, 3440 (1961).
141. D. R. Campbell, J. O. Edwards, J. Maclachlan, and K. Polyar, *J. Amer. Chem. Soc.*, **80**, 5308 (1958).
142. G. Berti and F. Boltari, *Gazz. Chim. Ital.*, **89**, 2371 (1959).
143. G. B. Linden, R. E. Meyer, and C. R. Vanneman, U.S. Patent 3,316,278 (1967).
144. J. T. Lumb and G. H. Whitham, *J. Chem. Soc.*, **1964**, 1189.
145. J. K. Stille and D. R. Witherell, *J. Amer. Chem. Soc.*, **86**, 2188 (1964).
146. E. E. Smissman, T. L. Lemke, and O. Kristiansen, *J. Amer. Chem. Soc.*, **88**, 334 (1966).
147. G. Maier, *Ber.*, **96**, 2238 (1963).
148. S. D. Ross and E. R. Coburn, Jr., *J. Org. Chem.*, **26**, 323 (1961).
149. A. P. Gray, D. E. Heitmeier, and H. Kraus, *J. Amer. Chem. Soc.*, **84**, 89 (1962).
150. G. Berti, F. Bottari, B. Macchia, and F. Macchia, *Tetrahedron*, **21**, 3277 (1965).
151. A. Bhati, *Perfum. Essent. Oil Rec.*, **57**, 563 (1966).

152. G. Farges and A. Kergomard, *Bull. Soc. Chim. France*, **1963**, 51.

153. J. M. Coxon, E. Dansted, M. P. Hartshorn, and K. E. Richards, *Tetrahedron*, **24**, 1193 (1968)

154. C. E. Wheelock and P. J. Canterino, U.S. Patent 3,022,322 (1962).

155. Y. Urushibara, M. Chuman, and S. Wada, *Bull. Chem. Soc Japan*, **24**, 83 (1951).

156. M. E. Wall, H. A. Walens, and F. T. Tyson, *J. Org. Chem.*, **26**, 5054 (1961).

157. G. V. Pigulevskii and I. K. Mironova, *J. Gen. Chem. USSR*, **27**, 1184 (1951); *Zh. Obshch. Khim.*, **27**, 1101 (1957).

158. W. Reusch and C. K. Johnson, *J. Org. Chem.*, **28**, 2557 (1963).

159. W. Tagaki and T. Mitsui, *J. Org. Chem.*, **25**, 1476 (1960).

160. S. Szpilfogel, *Helv. Chim. Acta*, **34**, 843 (1951).

161. B. L. Van Duuren and F. O. Schmitt, *J. Org. Chem.* **25**, 1671 (1960).

162. T. Yamada, *Yukagaku*, **11**, 335 (1962).

163. D. Y. Curtin and D. B. Kellom, *J. Amer. Chem. Soc.*, **75**, 6011 (1953).

164. R. M. Bowman, A. Chambers, and W. R. Jackson, *J. Chem. Soc.*, C., **1966**, 1296.

165. H. Cristol, D. Dural, and G. Solladie, *Bull. Soc. Chim. France*, **1968**, 689.

166. F. D. Greene and W. Adam, *J. Org. Chem.*, **29**, 136 (1964).

167. J. B. Carr and A. C. Hintric, *J. Org. Chem.*, **29**, 2506 (1964).

168. A. Brenner and H. Schinz, *Helv. Chim. Acta*, **35**, 1615 (1952).

169. B. S. Gorton and J. A. Reeder, *J. Org. Chem.*, **27**, 2920 (1962).

170. R. E. Foster, U.S. Patent 2,687,406 (1954).

171. L. S. Silbert, Z. B. Jacobs, W. E. Palm, L. P. Witnauer, W. S. Port, and D. Swern, *J. Polymer Sci.*, **21**, 161 (1956).

172. L. V. Nozdrina, Y. I. Mindin, and K. A. Andrianov, *Izv. Akad. Nauk SSSR, Ser. Khim.*, **1967**, 2100.

173. H. Sakurai, T. Imoto, N. Hayashi, and M. Kumada, *J. Amer. Chem. Soc.*, **87**, 4001 (1965).

174. A. Wende and R. Bekker, Soviet Patent 138,623 (1961).

175. T. Chow, *Hua Hsueh Hsueh Pao*, **24**, 426 (1958).

176. E. Glotter, S. Greenfield, and D. Lavie, *J. Chem. Soc.*, **1968**, 1646.

177. T. Ikawa and T. Fukushima, *Can. J. Chem.*, **44**, 1817 (1966).

178. G. V. Pigulevskii, *J. Gen. Chem. USSR*, **4**, 616 (1934).

179. E. Raymond, *J. Chim. Phys.*, **28**, 480 (1931).

180. D. Swern, T. W. Findley, and J. T. Scanlan, *J. Amer. Chem. Soc.*, **66**, 1925 (1944).

181. E. Niki and Y. Kamiya, *Bull. Chem. Soc. Japan*, **40**, 583 (1967).

182. H. Böhme, *Ber.*, **70B**, 379 (1937).

183. H. Böhme and G. Steinke, *Ber.*, **70B**, 1709.

184. P. N. Chakravorty and R. H. Levin, *J. Amer. Chem. Soc.*, **64**, 2317 (1942).

185. A. Baeyer and V. Villiger, *Ber.*, **34**, 762 (1901).

186. M. B. Mueller, J. G. Iacoriello, and R. Rosenthal, U.S. Patent 3,155,638 (1964).

187. M. Halmos, J. Szabo, and A. Zoltai, *Acta Univ. Szeged. Acta Phys. Chem.*, **12**, 63 (1966).

188. E. M. Chamberlin, E. Tristram, T. Uttne, and J. M. Chemerda, *J. Amer. Chem. Soc.*, **79,** 456 (1957).

189. J. Julia and J. P. Lavaux, *Bull. Soc. Chim. France*, **1963,** 1223.

190. S. Julia and C. Montonnier, *Bull. Soc. Chim. France*, **1964,** 321.

191. I. Malunowicz and A. Mironowicz, *Bull. Acad. Pol. Sci., Ser. Sci. Chim.*, **16,** 579 (1968).

192. R. B. Clayton, A. Cranshaw, H. B. Henbest, E. R. H. Jones, B. J. Lovell, and G. W. Wood, *J. Chem. Soc.*, **1953,** 2009.

193. P. Bladon, H. B. Henbest, E. R. H. Jones, G. W. Wood, D. C. Eaton, and A. A. Wagland, *J. Chem. Soc.*, **1953,** 2916.

194. O. K. Sewell, J. H. Turnbull, and W. Wilson, *J. Chem. Soc.*, **1956,** 4689.

195. D. C. Burke, J. H. Turnbull, and W. Wilson, *J. Chem. Soc.*, **1953,** 3237.

196. M. V. Mijoric, W. Voser, H. Heusser, and O. Jeger, *Helv. Chim. Acta*, **35,** 964 (1952).

197. A. L. Draper, W. J Heilman, W. E. Schaefer, H. J. Shine, and J. N. Shoolery, *J. Org. Chem.* **27,** 2727 (1963).

198. J. Joska, J. Fajkos, and F. Sorm, *Coll. Czech. Chem. Commun.*, **31,** 298 (1966).

199. L. Ruzicka and P. Plattner, U.S. Patent 2,656, 349 (1953).

200. L. Sawlewicz, H. H. A. Horst, and K. Meyer, *Helv. Chim. Acta*, **1968,** 1353.

201. H. Walborsky and D. F. Loncrini, *J. Amer. Chem. Soc.*, **76,** 5396 (1954).

202. V. Matousek and V. Bazant, Czech Patent 92,652 (1959).

203. L. T. Eremenko and A. M. Korolev, *Izv. Akad. Nauk SSSR, Ser. Khim* , **1968,** 1125.

204. A. Chauvel, G. Clement, and J. C. Balaceanu, *Bull. Soc. Chim. France*, **1963,** 2025.

205. M. L. Sassiver and J. English, *J. Amer. Chem. Soc.*, **82,** 4891 (1960).

206. V. Bazant and V. Matousek, *Coll. Czech. Chem. Commun.*, **24,** 3758, (1959).

207. K. W. Wheeler, M. G. Van Campen, Jr., and R. S. Shelton, *J. Org. Chem.*, **25,** 1021 (1960).

208. R. G. Carlson and N. S. Behn, *J. Org. Chem.*, **32,** 1363 (1967).

209. R. Vlakhov, M. Holub, and V. Herout, *Coll. Czech. Chem. Commun.*, **32,** 822 (1967).

210. J. Streith, P. Pesnelle, and G. Ourisson, *Bull. Soc. Chim. France*, **1963,** 518.

211. L. Velluz, G. Amiard, and B. Goffinet, *Compt. Rend.*, **240,** 2076 (1955).

212. M. P. Kunstmann, D. S. Tarbell, and R. L. Autrey *J. Amer. Chem. Soc.*, **84,** 4115 (1962).

213. A. Aebi, D. H. R. Barton, and A. S. Lindsey, *Chem. Ind.* (*London*), **1953,** 487.

214. G. R. Ramage and R. Whitehead, *J. Chem. Soc.*, **1954,** 4336.

215. P. Teisseire, M. Plattier, W. Wojnarowski, and G. Ourisson, *Bull. Soc. Chim. France*, **1966,** 2749.

216. G. I. Stepanovic, I. Pejkovic-Tadic, B. Terecki and B. Jakovljevic-Simonovic, *Tetrahedron Letters*, **1966,** 6315.

217. A. Fürst and R. Scotoni, Jr. *Helv., Chim. Acta*, **36,** 1332 (1953).

218. *Coll. Czech. Chem. Commun.*, **32,** 3762 (1967).

219. L. I. Zakharkin, V. V. Korneva, and G. M. Kunitskaya, *Izv. Akad. Nauk USSR, Otd. Khim. Nauk*, **1961,** 1908.

220. A. C. Cope, J. K. Heeren, and J. Seeman, *J. Org. Chem.*, **28**, 516 (1963).
221. E E. Royals and L. L. Harrell, Jr., *J. Amer. Chem. Soc.*, **77**, 3405 (1955).
222. F. Bohlmann and H. Sim, *Ber.*, **88**, 1869 (1955).
223. M. Mousseron- Canet and J. C. Guilleux, *Bull. Soc. Chim. France*, **1966**, 3858.
224. Organon Laboratories, Ltd., British Patent 884,412 (1961).
225. E. L. Allred, C. L. Anderson, and R. L. Smith, *J. Org. Chem.*, **31**, 3493 (1966).
226. G. H. Alt and D. H. R. Barton, *J. Chem. Soc.*, **1954**, 1356.
227. S. O. Chan and E. J. Wells, *Can. J. Chem.*, **45**, 2123 (1967).
228. Dunlop Co., Ltd., British Patent, 1,098,884 (1968).
229. M. Mousseron, M. Mousseron-Canet, and G. Phillipe, *Compt. Rend.*, **258**, 3705 (1964).
230. N. S. Zefirov, A. F. Darydova, F. A. Abdulvaleeva, and Y. K Yurev, *Zh. Obshch, Khim.*, **36**, 197 (1966).
231. J. Iriarte, J. N. Shoolery, and C. Djerassi, *J. Org. Chem.*, **27**, 1139 (1962).
232. F. L. Julietti, J. F. McGhie, B. L. Rao, W. A. Ross, and W. A. Cramp, *J. Chem. Soc.*, **1960**, 4514.
233. K. Tsukida, *Vitamin*, **33**, 185 (1966).
234. P. Diassi and J. Fried, U.S. Patent 3,364,204 (1968).
235. A. D. Cross, E. Denot, R. Aceredo, R. Urquiza, and A. Bowers, *J. Org. Chem.*, **29**, 2195 (1964).
236 J. P. Ruelas, J. Iriarte, F. Kincl, and C. Djerassi, *J. Org. Chem.*, **23**, 1744 (1958).
237. K. Christen, M. Dünnenberger, C. B. Roth, H. Heusser, and O. Jeger, *Helv. Chim. Acta*, **35**, 1756 (1952).
238. D. S. Tarbell, R. M. Carman, D. D. Chapman, S. E. Cremer, A. D. Cross, K. R. Huffman, M. Kunstmann, N. J. McCorkindale, J. G. McNally, Jr., A. Rosowsky, F. H. L. Varino, and R. L. West, *J. Amer. Chem. Soc.*, **83**, 3096 (1961).
239. D. J. Goldsmith, *J. Amer. Chem. Soc.*, **84**, 3913 (1962).
240. F. Sorm, M. Streibl, J. Pliva, and V. Herout, *Chem. Listy*, **46**, 30 (1952).
241. M. Mousseron-Canet, B. Labeeuw, and J. C. Lanet, *Compt. Rend.*, **C262**, 1438 (1966).
242. M. Mousseron-Canet, B. Labeeuw, and J. C. Lanet, *Bull. Soc. Chim. France*, **1968**, 2125.
243. P. A. Plattner, A. Fürst, F. Koller, and H. H. Kuhn, *Helv. Chim. Acta*, **37**, 258 (1954).
244. N. Le Boulch, Y. Raoul, and G. Ourisson, *Bull. Soc. Chim. France*, **1967**, 2413.
245. D. Felix, A. Melera, J. Seibl, and E. Korats, *Helv. Chim. Acta*, **46**, 1513 (1963).
246. F. Hoffmann-La Roche and Co. A.G., Netherlands Patent Application 6,600,331 (1966).
247. M. Mousseron-Canet and J. C. Mani, *Bull. Soc. Chim. France*, **1965**, 481.
248. S. A. Julia and H. Heusser, *Helv. Chim. Acta*, **35**, 2080 (1952).
249. J. M. Coxon, M. P. Hartshorn, and D. N. Kirk, *J. Chem. Soc.*, **1964**, 2461.
250. H. B. S. Conacher and F. D. Gunstone, *Chem. Commun.*, **1968**, 281.
251. J. M. Coxon, M. P. Hartshorn, and D. N. Kirk, *Tetrahedron*, **23**, 3511 (1967).

252. M. Ferrari, E. L. Ghisalberti, U. M. Pagnoni, and F. Pelizzoni, *J. Amer. Oil Chem. Soc.*, **45**, 649 (1968).

253. M. Ferrari, E. L. Ghisalberti, U. M. Pagnoni, and F. Pelizzoni, *J. Amer. Oil Chem. Soc.*, **45**, 649 (1968).

254. A. K. Sen Gupta and H. Scharmann, *Fette, Seifen, Anstrichmittel*, **1968**, 265.

255. G. Berti, F. Bottari, and B. Macchia, *Gazz. Chim. Ital.*, **90**, 1763 (1960).

256. Y. Suhara, Y. Shibuya, and H. Otsuru, *Bull. Chem. Soc. Japan.* **40**, 1702 (1967).

257. Y. Suhara and T. Minami, *Bull. Chem. Soc. Japan*, **39**, 1968 (1966).

258. M. Mousseron-Canet and J. C. Mani, *Bull. Soc. Chim. France*, **1965**, 484.

259. H. Kwart and W. G. Vosburgh, *J. Amer. Chem. Soc.*, **76**, 5400 (1954).

260. W. Hueckel and O. Vogt, *Ann.*, **695**, 16 (1966).

261. C. R. Enzell and B. R. Thomas, *Tetrahedron Letters*, No. 4, 225 (1965).

262. A. Jenner and F. Everaerts, *Compt. Rend.*, **251**, 91 (1960).

263. C. R. Eddy, J. S. Showell, and T. E. Zell, *J. Amer. Oil Chem. Soc.*, **45**, 92 (1963).

264. M. Farooq and S. M. Osman, *Fette, Seifen, Anstrichmittel*, **61**, 636 (1959).

265. G. Schiemann and H. Martens, *Fette, Seifen, Anstrichmittel*, **67**, 1004 (1965).

266. E. F. Jenny and C. A. Grob, *Helv. Chim. Acta*, **36**, 1454 (1953).

267. A. C. Cope and J. K. Heeren, *J. Amer. Chem. Soc.*, **87**, 3125 (1965).

268. P. T. Lansbury and V. A. Pattison, *Tetrahedron Letters*, **1966**, 3073.

269. F. G. Bordwell and J. B. Biranowski, *J. Org. Chem.*, **32**, 629 (1967).

270. K. Meyer, German Patent, 1,111,399 (1961).

271. K. I. H. Williams, M. Smilowitz, and D. K. Fukushima, *J. Org. Chem.*, **28**, 2101 (1963).

272. J. Lange and C. Belzecki, *Acta Polon, Pharm.*, **18**, 177 (1961).

273. J. Moulines and R. Lalande, *Bull. Soc. Chim. France*, **1966**, 3387.

274. K. Tsukida, S. Yamane, and N. Yokota, *J. Vitaminol* (*Kyoto*), **1968**, 95.

275. A. A. Akhrem and I. G. Reshetova, *Izv. Akad. Nauk SSSR, Ser. Khim*, **1966**, 769.

276. B. C. Henshaw, D. W. Rome, and B. L. Johnson, *Tetrahedron Letters*, **1968**, 6049.

277. D. J. Goldsmith and C. J. Cheer, *J. Org. Chem.*, **30**, 2264 (1965).

278. A. Butenandt, E. Hecker, M. Hopp, and W. Koch, *Ann.*, **658**, 39 (1962).

279. A. Butenandt and E. Hecker, *Angew. Chem.*, **73**, 349 (1961).

280. J. J. Eisch and J. T. Trainor, *J. Org. Chem.*, **28**, 487 (1963).

281. G. N. Pandey and C. R. Mitra, *Tetrahedron Letters*, **1967**, 4683.

282. J. Böeseken, W. C. Smit, and Gaster, *Proc. Acad. Sci. Amsterdam*, **32**, 377 (1929).

283. W. C. Smit, *Rec. Trav. Chim.*, **49**, 675 (1930).

284. D. Swern, unpublished results, Eastern Regional Research Laboratory, U.S. Department of Agriculture.

285. V. A. Petrow, *J. Chem. Soc.*, **1939**, 998.

286. Y. R. Naves and G. Perrottet, *Helv. Chim. Acta*, **24**, 789 (1941).

287. G. King, *J. Chem. Soc.*, **1943**, 37.

288. E. M. Hicks, C. J. Berg, and E. S. Wallis, *J. Biol. Chem.*, **162**, 645 (1946).

289. F. S. Spring, *J. Chem. Soc.*, **1933**, 1345.
290. G. V. Pigulevskii and N. Simonova, *J. Gen. Chem. USSR*, **9**, 1928 (1939).
291. B. A. Arbuzow and B. M. Michailow, *J. Prakt. Chem.*, **127**, 1 (1930).
292. B. A. Arbuzow and B. M. Michailow, *J. Prakt. Chem.*, **127**, 92 (1930).
293. J. Böeseken and G. C. C. C. Schneider, *J. Prakt. Chem.*, **131**, 285 (1931).
294. G. V. Pigulevskii, *J. Gen. Chem. USSR*, **4**, 616 (1934).
295. S. Tanaka, *Mem. Coll. Sci., Kyoto Imp. Univ.*, **A22**, 97 (1930).
296. T. W. Findley, D. Swern, and J. T. Scanlan, *J. Amer. Chem. Soc.*, **67**, 412 (1945).
297. D. Swern, G. N. Billen, and J. T. Scanlan, *J. Amer. Chem. Soc.*, **68**, 1504 (1946).
298. B. Phillips, F. C. Frostick, Jr., and P. S. Starcher, *J. Amer. Chem. Soc.*, **79**, 5982 (1957).
299. B. Phillips and P. S. Starcher, U.S. Patent 2,785,185 (1957).
300. B. Phillips, F. C. Frostick, Jr., and P. S. Starcher, U.S. Patent 2,804,473 (1957); German Patent 1,043,315 (1958).
301. Allied Chemical Corp., *Chem. Week*, March 16, 1963, p. 63.
302. B. Phillips, P. S. Starcher, and B. D. Ash, *J. Org. Chem.*, **23**, 1823 (1958); German Patent 1,043,316 (1958).
303. B. Phillips and P. S. Starcher, U.S. Patent 2,814,641 (1957).
304. B. A. Arbuzov and A. G. Khismatullina, *Izv. Akad. Nauk SSSR, Otd. Khim. Nauk*, **1961**, 1280.
305. B. Phillips and D. L. Heywood, U.S. Patent 2,883,396 (1951).
306. B. Phillips and D. L. Heywood, U.S. Patent 2,917,521 (1959)
307. G. B. Payne, and C. W. Smith, U.S. Patent 2,776,301 (1957).
308. C. T. Yang, U.S. Patent 2,873,283 (1959).
309. T. Kataoka and H. Kakiuchi, *J. Japan Oil Chem. Soc.*, **13**, 642 (1964).
310. S. C. Johnson and Sons, Inc., British Patent 802,127 (1959).
311. L. Désalbres, B. Lahourcade, and J. Rache, *Bull. Soc. Chim. France*, **1956**, 761.
312. F. C. Frostick, Jr., B. Phillips, and P. S. Starcher, *J. Amer. Chem. Soc.*, **81**, 3350 (1955).
313. B. Phillips and P. S. Starcher, British Patent 735,974 (1955).
314. W. C. Fisher, S. M. Linder, R. L. Pelky, and H. P. Liao, French Patent 1,509,277 (1968).
315. S. W. Tinsley, P. S. Starcher, and J. P. Henry, U.S. Patent 2,946,808 (1960).
316. R. W. Martin and C. E. Pannell, U.S. Patent 2,965,607 (1960).
317. B. Phillips and P. S. Starcher, U. S. Patent 2,977,374 (1961).
318. B. Phillips and P. S. Starcher, British Patent 784,620 (1950).
319. Société Nationale des Petroles d'Aquitaine, Netherlands Patent Application 6,605,134 (1966).
320. P. S. Starcher, F. C. Frostick, Jr., and B. Phillips, *J. Org. Chem.*, **25**, 1420 (1960).
321. F. C. Frostick, Jr., and B. Phillips, U.S. Patent 2,935,556 (1960).
322. B. Phillips, P. S. Starcher, and D. L. MacPeek, U.S. Patent 2,969,377 (1961).
323. R. A. Sultanov, G. K. Bairamov, and S. I. Sadykh-Zade, *Zh. Organ. Khim.*, **1968**, 789.

324. Union Carbide Corp., British Patent 788,541 (1958).
325. B. Phillips and P. S. Starcher, U.S. Patent 2,794,812 (1957).
326. S. W. Tinsley, Jr. and E. Marcus, U.S. Patent 3,311,643 (1967).
327. Union Carbide Corp., British Patent 844,312 (1960).
328. Union Carbide Corp., British Patent 788,530 (1958).
329. Union Carbide Corp., British Patent 789,034 (1958).
330. B. Phillips, P. S. Starcher, and D. L. MacPeek, U.S. Patent 2,927,931 (1960).
331. J. E. Gill and J. Munro, *J. Chem. Soc.*, **1952**, 4630.
332. B. Phillips and F. C. Frostick, Jr., U.S. Patent 2,779,771 (1957).
333. I. V. Nicolescu and E. Angelescu, *Analele Univ. Bucaresti, Ser. Stiint, Nat.*, **12**, 149 (1963).
334. S. W. Tinsley, U.S. Patent 3,226,401 (1965).
335. S. A. Harrison and D. Aelony, *J. Org. Chem.*, **27**, 3310 (1962).
336. S. W. Tinsley, *J. Org. Chem.*, **24**, 1197 (1959).
337. General Mills, Inc., British Patent 841,589 (1960).
338. D. Aelony, *J. Appl. Polymer Sci.*, **4**, 141 (1960).
339. L. Goodman, S. Winstein, and R. Boschau, *J. Amer. Chem. Soc.*, **80**, 4312 (1958).
340. C. B. Havens and R. G. Brookens, U.S. Patent 3,068,193 (1962).
341. N. A. Belikova, K. V. Lebedeva, N. N. Melnikov, and A. F. Plate, *Zh. Obshch Khim.*, **35**, 1746 (1965).
342. R. R. Sauers, H. M. How, and H. Feilich, *Tetrahedron*, **21**, 983 (1965).
343. A. C. Cope and R. Gleason, *J. Amer. Chem. Soc.*, **84**, 1928 (1962).
344. Chemische Werke Hüls A.S., German Patent 1,178,847 (1964).
345. R. L. McConnell and H. W. Coover, Jr., U.S. Patent 3,297,724 (1967).
346. G. B. Payne and C. W. Smith, U.S. Patent 2,879,170 (1959).
347. R. M. Christenson and J. J. Jaruzelski, U.S. Patent 2,874,150 (1959).
348. D. J. Foster and D. E. Reed, Jr., U.S. Patent 3,347,875 (1967).
349. H. R. Guest, J. T. Adams, and B. W. Kiff, U.S. Patent 3,023,222 (1962).
350. Union Carbide Corp., British Patent 887,593 (1962).
351. R. W. Martin, and C. G. Schwarzer, U.S. Patent 2,965,608 (1960).
352. S. W. Tinsley, U.S. Patent 3,238,227 (1966).
353. B. Phillips and P. S. Starcher, U.S. Patent 3,056,806 (1962).
354. I. G. Tishchenko, *Geterogennye Reaktsii i Reakts. Sposobnost Sb.*, **1964**, 242.
355. J. W. Lynn, R. L. Roberts, and S. W. Tinsley, U. S. Patent 3,057,880 (1962).
356. B. Phillips, P. S. Starcher, C. W. McGary, and C. T. Patrick, U.S. Patent 2,917,491 (1959).
357. Société d'Electrochimie, French Patent 1,409,377 (1965).
358. J. W. Lynn, A. E. Brodhag, and S. W. Tinsley, *J. Chem. Eng. Data*, **8**, 445 (1963).
359. B. Phillips and P. S. Starcher, British Patent 858,793 (1961).
360. B. Phillips and P. S. Starcher, U.S. Patent 2,973,373 (1961).
361. Hitachi Chemical Industry Co., Ltd., Japanese Patent 19,693 (1966).
362. G. Bankwitz, German Patent 1,093,363 (1960).
363. J. K. Crandall and D. R. Paulson, *J. Org. Chem.*, **33**, 991 (1968).
364. W. V. McConnell and W. H. Moore, U.S. Patent 3,288,814 (1966).

365. Ciba, Ltd., Netherlands Patent Application 300,893 (1965).

366. H. Batzke and E. Nikle, U.S. Patent 2,988,554 (1961).

367. German Patent 1,087,594 (1960).

368. I. L. Kuranova, Y. D. Shenin, and G. V. Pigulevskii, *J. Gen. Chem. USSR*, **33**, 2916 (1963).

369. H. R. Guest and H. A. Stansbury, Jr., U.S. Patent 2,874,167 (1959).

370. Institut Tworzyw Sztucznych, Polish Patent 49,690 (1965).

371. M. Korach and W. E. Rideout, U.S. Patent 3,019,234 (1962).

372. M. Korach and W. H. Rideout, British Patent 846,534 (1960).

373. British Patent 852,097 (1960).

374. Société Nationale des Petroles d'Aquitaine, French Patent 1,468,814 (1967).

375. V. I. Pansevich-Kolyada and T. A. Galysheva, *Probl. Organ. Sinteza, Akad. Nauk SSSR, Otd. Obshch. I Tekhn. Khim.*, **1965**, 23.

376. G. B. Payne, *J. Amer. Chem. Soc.*, **27**, 3819 (1962).

377. E. M. Beavers and J. L. O'Brien, U.S. Patent 3,001,975 (1961).

378. B. Phillips, P. S. Starcher, and D. L. Heywood, U.S. Patent 2,806,860 (1957).

379. E. Marcus, D. L. MacPeek and S. W. Tinsley, U.S. Patent 3,373,214 (1968).

380. E. I. du Pont de Nemours & Co., British Patent 776,757 (1957).

381. Kyowa Fermentation Industry Co., Ltd., Japanese Patent 4706 (1966).

382. D. L. MacPeek, P. S. Starcher, and B. Phillips, *J. Amer. Chem. Soc.*, **81**, 680 (1959).

383. J. K. Crandall and L. H. C. Lin, *J. Org. Chem.*, **33**, 2375 (1968).

384. B. Phillips, P. S. Starcher, and D. L. MacPeek, British Patent 863, 446 (1961).

385. J. F. Rusling, G. R. Riser, M. E. Snook, and W. E. Scott, *J. Amer. Oil Chem. Soc.*, **45**, 760 (1968).

386. F. P. Greenspan and R. J. Gall, U.S. Patent 2,801,253 (1957).

387. F. P. Greenspan and R. J. Gall, U.S. Patent 2,692,271 (1954).

388. Food Machinery Corp., British Patent 739,609 (1955).

389. C. C. Price and D. D. Carmelite, *J. Amer. Chem. Soc.*, **88**, 4039 (1966).

390. D. F. O'Brien and J. W. Gates, Jr., *J. Org. Chem.*, **30**, 2593 (1965).

391. F. P. Greenspan and R. J. Gall, U.S. Patent 2,919,283 (1959).

392. R. J. Gall, F. P. Greenspan, and M. C. Daly, U.S. Patent 2,935,517 (1960).

393. L. I. Hansen and G. O. Sedgwick, U.S. Patent 3,051,729 (1962).

394. German Patent 1,081,462 (1960).

395. Japanese Patent 15,622 (1961).

396. V. I. Panevich-Kolyada and B. K. Bogush, *J. Gen. Chem. USSR*, **32**, 3485 (1962).

397. B. Phillips and P. S. Starcher, U.S. Patent 2,897,208 (1959).

398. H. P. Kaufmann, G. Hauschild, and R. Schickel, *Fette, Seifen, Anstrichmittel*, **63**, 239 (1961).

399. K. Gollnick and G. Schade, *Tetrahedron Letters*, **1966**, 2335.

400. Z. Chabudzinski and D. Sedzik, *Roczniki Chem.*, **40**, 1889 (1966).

401. G. Schiemann and W. Schneider, *Fette, Seifen, Anstrichmittel*, **69**, 749 (1967).

402. B. Phillips and P. S. Starcher, U.S. Patent 3,150,154 (1964).

403. R. L. Webb, U.S. Patent 2,902,495 (1959).

404. Dainippon Ink and Chemicals, Inc., Japanese Patent 24,373 (1965).

405. F. P. Greenspan, U.S. Patent 2,810,733 (1957).
406. Rohm and Haas Co., British Patent 790,063 (1958).
407. F. P. Greenspan and R. J. Gall, U.S. Patent 2,810,732 (1957).
408. E. M. Beavers and J. L. O'Brien, U.S. Patent 3,002,004 (1961).
409. Chemische Werke Hüls A.G., French Patent 1,490,471 (1967).
410. W. Stumpf and K. Rombusch, *Ann.*, **687**, 136 (1965).
411. R. A. Gray, U.S. Patent 2,997,483 (1961).
412. G. Wilke, U.S. Patent 3,014,928 (1961).
413. German Patent 1,058,987 (1959).
414. J. K. Crandall, D. B. Banks, R. A. Colyer, R. J. Watkins, and J. P. Arrington, *J. Org. Chem.*, **33**, 423 (1968).
415. A. C. Cope and W. N. Baxter, *J. Amer. Chem. Soc.*, **76**, 279 (1954).
416. J. K. Crandall and L. H. Chang, *J. Org. Chem.*, **32**, 435 (1967).
417. F. Scholnick, W. C. Ault, and W. S. Port, *J. Amer. Oil Chem. Soc.*, **40**, 229 (1963).
418. P. S. Starcher and S. W. Tinsley, U.S. Patent 2,999,867 (1961).
419. H. S. Bloch, U.S. Patent 3,054,807 (1962).
420. W. H. McKellin and S. M. Linder, British Patent 877,632 (1961).
421. H. Batzer, W. Fisch, and E. Nikles, U.S. Patents 3,211,750; 3,211,751, and 3,213,111 (1965).
422. F. C. Frostick, Jr., and B. Phillips, U.S. Patent 2,883,398 (1959).
423. B. Phillips and F. C. Frostick, Jr., U.S. Patent 2,999,865 (1961).
424. D. H. Mullins, B. Phillips, and F. C. Frostick, Jr., U.S. Patent 2,524,582 (1960).
425. F. C. Frostick, Jr., and B. Phillips, U.S. Patent 2,953,550 (1960).
426. P. S. Starcher, S. W. Tinsley, and D. H. Mullens, U.S. Patent 2,924,583 (1960).
427. German Patent 1,085,872 (1960).
428. B. Phillips and P. S. Starcher, U.S. Patent 2,794,029 (1957).
429. B. Phillips, F. C. Frostick, Jr., C. W. McGary, and C. T. Patrick, U.S. Patent 2,890,194 (1959).
430. B. Phillips, F. C. Frostick, Jr., C. W. McGary, and C. T. Patrick, U.S. Patent 2,890,209 (1959).
431. F. C. Frostick, Jr., and B. Phillips, U.S. Patent 2,716,123 (1955).
432. F. C. Frostick, Jr., and B. Phillips, U.S. Patent 2,786,067 (1957).
433. Farbwerke Höchst A.G., Belgian Patent 659,985 (1965).
434. F. C. Frostick, Jr., and B. Phillips, British Patent 793,150 (1958).
435. Knapsack A.G., German Patent 1,202,784 (1965).
436. B. L. Van Duuren, L. Orris, and N. Nelson, *J. Natl. Cancer Inst* , **35**, 707 (1965).
437. A. C. Cope, S. W. Fenton, and C. F. Spencer, *J. Amer. Chem. Soc.*, **74**, 5884 (1952).
438. M. Korach, D. B. Milsen, and W. H. Rideout, *J. Amer. Chem. Soc.*, **82**, 4328 (1960).
439. M Korach, U.S. Patent 2,916,402 (1959).
440. M. Korach, British Patent 855,199 (1960).

441. D. L. Heywood and B. Phillips, *J. Org. Chem.*, **25**, 1699 (1960); U.S. Patent 3,240,798 (1966).

442. R. B. Moffett and G. Slomp, Jr., *J. Amer. Chem. Soc.*, **76**, 3678 (1954).

443. H. V Anderson, E R. Garrett, and F. H. Lincoln, *J. Amer. Chem. Soc.*, **76**, 743 (1954).

444. D. Swern and H. B. Knight, U.S. Patent 3,049,504 (1962).

445. W. C. Ault and R. O. Feuge, U.S. Patent 3,050,481 (1961).

446. W. C. Ault and R. O. Feuge, British Patent 7,757, 326 (195J).

447. D. Swern and H. B. Knight, British Patent 775,714 (1957).

448. W. C. Ault and R. O. Feuge, U.S. Patent 895,966 (1952).

449. D. Swern and H. B. Knight, U.S. Patent 2,898,348 (1959).

450. E. P. Plueddemann and G. Fanger, *J. Amer. Chem. Soc.*, **81**, 2632 (1959).

451. G. B. Payne, C. W. Smith, and P. R. Van Ess, U.S. Patent 2,957,907 (1960).

452. R. F. Fischer, U.S. Patent 2,895,962 (1959).

453. C. W. Smith, G. B. Payne, and E C Shokal, U.S. Patent 2,856,369 (1959).

454. British Patent 862,588 (1961).

455. Henkel and Co., German Patent 1,082,263 (1960).

456. M. S. Malinovskii, V. G. Dryuk, A. G. Yudasina, and N. N. Maksimenko, *Zh. Organ. Khim.*, **2**, 2129 (1966).

457. R. Budziarek, R. Stevenson, and F. S. Spring, *J. Chem. Soc.*, **1952**, 4874.

458. B. Phillips, P. S. Starcher, and D. L. Heywood, U.S. Patent 2,999,868 (1961).

459. B. Phillips and P. S. Starcher, U.S. Patent 2,794,030 (1957).

460. B. Phillips, P. S. Starcher, and D. L. Heywood, British Patent 799,383 (1958).

461. S. C. Johnson and Sons, Ind., British Patent 611,126 (1959).

462. P. S. Starcher, D. L. MacPeek, and B. P. Phillips, U.S. Patent 2,861,084 (1958).

463. Union Carbide Corp., German Patent 1,043,312 (1958).

464. H. Y. Fan, W. E. Rader, M. Legator, and L. C. Larrick, U.S. Patent 3,215,708 (1965).

465. K. B. Wiberg and R. W. Ubersax, *Tetrahedron Letters*, **1968**, 3063.

466. B. Phillips and P. S. Starcher, U.S. Patent 2,894,959 (1959).

467. W. J. Middleton, *J. Org. Chem.*, **31**, 3731 (1966); U.S. Patent 3,194, 819 (1965).

468. E. C. Shokal, U.S. Patent 2,925,403 (1960).

469. J. W. Lynn, R. L. Roberts, and S. W. Tinsley, *J. Org. Chem.*, **26**, 3757 (1961).

470. German Patent 1,099,733 (1961).

471. B. Phillips, C. W. McGary, and C. T. Patrick, Jr., U.S. Patent 2,934,508 (1960).

472. P. Penczek, J. Vasvar, Z. Brojer, and K. Janicka, *Polimery*, **12**, 310 (1967).

473. S. R. Thorn, W. A. Dimter, Jr., and J. A. Gallagher, U.S. Patent 3,066,151 (1962).

474. A. W. Carlson, U.S. Patent 2,739,161 (1956).

475. S. P. Rowland and E. M. Beavers, U.S. Patent 2,963,490 (1960).

476. T. I. Temnikova and S. N. Semenova, *Zh. Organ Khim.*, **2**, 1171 (1966).

477. B. Phillips and P. S. Starcher, U.S. Patent 2,853,498 (1958).

478. B. Phillips and P. S. Starcher, U.S. Patent 2,745,847 (1956).

479. B. Phillips, P. S. Starcher, C. W. McGary, and C. T. Patrick, U.S. Patent 2,884,408 (1959).

480. B. Phillips and P. S. Starcher, British Patent 788,123 (1957).

481. N. M. Malenok, S. D. Kulkina, and Z. Y. Kortunenko, *Zh. Obshch. Khim.*, **28**, 428 (1958).

482. T. F. Mika and R. D. Sullivan, U.S. Patent 2,764,497 (1956).

483. Chemische Werke Hüls A.G., German Patent 1,210,889 (1966).

484. J. Meinwald and G. A. Wiley, *J. Amer. Chem. Soc.*, **80**, 3667 (1958); J. Meinwald, H. Nozaki, and G. A. Wiley, *J. Amer. Chem. Soc.*, **79**, 5579 (1957).

485. B. A. Arbuzov and V. A. Frinovskaya, *Dokl. Akad, Nauk SSSR*, **112**, 427 (1957).

486. I. F. Osejsenko and V. I. Pansevich-Kolyada, *Dokl. Akad. Nauk Belorusk. SSR*, **2**, 163 (1958).

487. V. F. Kucherov, A. L. Shabanov, and A. S. Onishchenko, *Izv. Akad. Nauk SSSR, Ser. Khim.*, **1966**, 689.

488. B. Phillips and P. S. Starcher, U.S. Patents 3,018,294 (1962); 3,151,130 (1964); 3,255,214 (1966); and 3,255,215 (1966).

489. V. I. Pansevich, *Zh. Obshch. Khim.*, **26**, 2161 (1956).

490. I. N. Nazarov, V. F. Kucherov, and V. G. Bukharov, *Izv. Akad. Nauk SSSR, Otd. Khim. Nauk* **1958**, 192.

491. I. N. Nazarov, V. F. Kucherov, and V. G. Bukharov, *Izv. Akad. Nauk SSSR, Otd. Khim. Nauk*, **1958**, 328.

492. M. M. Movsumzade, A. L. Shabanov, and F. K. Agaev, *Dokl. Akad. Nauk Azerb, SSR*, **23**, 15 (1967).

493. V. I. Pansevich-Kolyada, T. S. Strigalova, and Z. B. Idelchick, *Sbornik Statei Obshch. Khim.*, **2**, 1418 (1953).

494. N. M. Malenok and S. D. Kulkina, *Zh. Obshch. Khim.*, **24**, 1212, (1954); *ibid.*, 1837 (1954).

495. J. H. Bartlett, R. S. Brodkey, P. V. Smith, and H. K. Wiese, U.S. Patent 3,239,539 (1966).

496. W. V. McConnell and W. H. Moore, *J. Org. Chem.*, **30**, 3480 (1965).

497. J. Falbe, H. J. Schulze-Steinen, and F. Korte, *Ber.*, **97**, 1096 (1964).

498. C. G. Fitzgerald, A. J. Carr, M. Maienthal, and P. J. Franklin, *J. Soc., Plastics Eng.*, **13**, No. 1, (January 1957).

499. S. W. Tinsley, Jr. and D. L. MacPeek, U.S. Patent 3,337,587 (1967).

500. W. S. Port, L. L. Gelb, and W. C. Ault, U.S. Patent 2,975,149 (1961).

501. N. M. Malenok and S. D. Kulkina, *Zh. Obshch. Khim.*, **25**, 1462 (1955).

502. R. E. Buik and G. E. Inskeep, U.S. Patent 2,768,182 (1956).

503. E. P. Plueddemann, U.S. Patent 3,120,546 (1964).

504. S. P. Fore and G. Sumrell, *J. Amer. Oil Chem. Soc.*, **43**, 581 (1967).

505. I. N. Nazarov, V. F. Kucherov, and V. G. Bukharov, *Izv. Akad. Nauk SSSR, Otd. Khim Nauk*, **1958**, 192.

506. D. F. Ewing and C. Y. Hopkins, *Can. J. Chem.*, **42**, 1259 (1967).

507. O. L. Mageli, E. W. Patterson, and E. Y. Spencer, *Can. J. Chem.*, **31**, 23 (1953).

508. J. Dazzi, U.S. Patent 2,786,039 (1957).

509. B. Phillips and D. L. Heywood, U.S. Patent 2,891,969 (1959).
510. E. M. Beavers, S. P. Rowland, and R. G. White, U.S. Patent 2,997,484 (1961).
511. Rohm and Haas Co., British Patent 811,852 (1959).
512. V. F. Martynov and I. R. Belov, *J. Gen. Chem. USSR*, **32**, 2308 (1962).
513. T. C. Owen, C. L. Gladyss, and L. Field, *J. Chem. Soc.*, **1962**, 501.
514. L. Orthner and E. G. Fuelis, U.S. Patent 2,905,702 (1959).
515. D. Porret, W. Fisch, H. Batzer, and O. Ernst, U.S. Patent 3,147,279 (1964).
516. F. J. Sprules and H. C. Marks, U.S. Patent 2,802,800 (1957).
517. B. Phillips, P. S. Starcher, and D. L. MacPeek, U.S. Patent 3,235,570 (1966).
518. German Patent 1,083,797 (1960).
519. German Patent 1,078,106 (1960).
520. P. B. Subira and R. V. Soles, *Spanish Patent* 248,176 (1959).
521. M. S. Malinovskii, A. G. Yudasina, T. S. Skrodskaya, and V. G. Larionova, *Ukr. Khim. Zh.*, **33**, 598 (1967).
522. C. F. Krewson, G. R. Riser, and W. E. Scott, *J. Amer. Oil Chem. Soc.*, **43** 377 (1966).
523. G. V. Pigulevskii and A. E. Saprokluna, *Zh. Prikl. Khim.*, **30**, 1104 (1951).
524. E. E. van Tamelen and K. B. Sharpless, *Tetrahedron Letters*, **1967**, 2655.
525. C. R. Zanesco, *Helv. Chim. Acta*, **49**, 1002 (1966).
526. M. S. Malinovskii, V. G. Dryuk, and A. G. Yudasina, *Ukr. Khim. Zh.*, **1968**, 377.
527. V. I. Pansevich-Kolyada and A. E. Streltsov, *Probl. Organ. Sinteza, Akad Nauk SSSR, Otd. Obshch. i Tekhn. Khim.*, **1965**, 28.
528. S. H. Herzfield and M. Kleiman, U.S. Patent 2,583,569 (1952).
529. M. Kleiman, U.S. Patent 2,655,514 (1953).
530. B. L. Van Duuren, L. Langseth, B. M. Goldschmidt, and L. Orris, *J. Natl. Cancer Inst.*, **39**, 1217 (1967).
531. Archer-Daniels-Midland Co., Netherlands Patent Application 6,609,243 (1967).
532. E. Vogel, M. Biskup, A. Vogel, and H. Guenther, *Angew. Chem.*, **78**, 755 (1966).
533. E. Vogel, M. Biskup, A. Vogel, and H. Guenther, *Angew. Chem., Intern. Ed.*, **5**, 734 (1966).
534. B. K. Zeinalov, S. A. Iskenderova, and E. N. Burunova, *Azerb. Khim. Zh.*, **1968**, 81.
535. J. K. Crandall and D. R. Paulson, *J. Org. Chem.*, **33**, 3291 (1968).
536. B. Phillips and P. S. Starcher, U.S. Patent 2,857,402 (1958).
537. D. Porret and H. Batzer, U.S. Patent 3,127,417 (1964).
538. M. M. Azanovskaya and V. I. Pansevich-Kolyada, *Zh. Obshch. Khim.*, **27**, 384 (1957).
539. H. A. Suter and S. D. Cooley, U.S. Patent 2,903,465 (1959).
540. German Patent 1,015,782 (1957).
541. Z. Chabudzinski and M. Skwarek, *Roczniki Chem.*, **1968**, 283.
542. French Patent 1,252,694 (1961).
543. F. P. Greenspan and S. M. Linder, U.S. Patent 3,014,929 (1961).

544. F. P. Greenspan and S. M. Linder, British Patent 885,044 (1961).
545. E. E. Royals and J. C. Leffingwell, *J. Org. Chem.*, **31**, 1937 (1966).
546. J. M. Osbond, *J. Chem. Soc.*, **1961**, 5270.
547. D. Swern and W. E. Parker, *J. Org. Chem.*, **22**, 583 (1957).
548. D. Swern and G. B. Dickel, *J. Amer. Chem. Soc.*, **76**, 1957 (1954).
549. P. J. Gorrindo, *Chim. Ind.*, **85**, 907 (1961).
550. W. F. Neuhall, *J. Org. Chem.*, **24**, 1673 (1959).
551. G. B. Payne and P. H. Williams, *J. Org. Chem.*, **24**, 284 (1959).
552. I. G. Tishchenko, *Uchenyi Zopiski, Belorusk. Gosudarst. Univ. im. V. I. Lenina, Ser. Khim.*, **42**, 189 (1958).
553. D. M. Young, U.S. Patent 2,972, 619 (1961).
554. A. A. Akhram, *Izv. Akad. Nauk SSSR, Otd. Khim. Nauk*, **1959**, 750.
555. N. S. Drozdov and L. A. Grushetskaya, *Nauch Dokl. Vysshei Shkoly Khim. i Khim. Tekhnol.*, **1958**, No. 2, 339.
556. G. V. Pigulevskii and E. M. Rostomyan, *Zh. Obshch. Khim.*, **22**, 1987 (1952).
557. G. A. Razuvaev, V. S. Etlis, and E. P. Morozova, *Zh. Organ. Khim.*, **1**, 1567 (1965).
558. J. P. Schaefer, B. Horvath, and H. P. Klein, *J. Org. Chem.*, **33**, 2647 (1968).
559. H. Batzer and E. Nikles, U.S. Patent 2,991,293 (1961).
560. M. F. Sorokin and T. N. Fomicheva, *Tr. Mosk. Khim. Tekhnol. Inst.*, **1968**, 71, 74.
561. C. H. Mack and W. G. Bickford, *J. Org. Chem.*, **18**, 686 (1953).
562. G. V. Pigulevskii and A. E. Sokolova, *Zh. Obshch. Khim.*, **31**, 652 (1961).
563. B. H. Braun, M. Jacobson, M. Schwarz, P. E. Sonnet, N. Wakabayashi, and R. M. Waters, *J. Econ. Entomol.*, **1968**, 866.
564. Y. Gaoni, *J. Chem. Soc.*, C, **1968**, 2925, 2934.
565. R. L. Pruett, U.S. Patent 3,392,207 (1968).
566. G. V. Pigulevskii and I. L. Kuranova, *Zh. Prikl. Khim.*, **28**, 1353 (1955).
567. G. V. Pigulevskii and I. N. Naidenova, *Zh. Obshch. Khim.*, **28**, 234 (1958).
568. G. Maerker, E. T. Haeberer, and W. C. Ault, *J. Amer. Oil Chem. Soc.*, **43**, 100 (1960).
569. G. Maerker and E. T. Haeberer, *J. Amer. Oil Chem. Soc.*, **43**, 97 (1966).
570. G. Maerker, E. T. Haeberer, and S. F. Herb, *J. Amer. Oil Chem. Soc.*, **43**, 505 (1966).
571. R. Subbarao, G. V. Rao, and K. T. Achaya, *Tetrahedron Letters*, **1966**, 379.
572. J. L. O'Donnell, L. E. Gast, J. C. Cowan, W. J. de Jarlais, and G. E. McManna, *J. Amer. Oil Chem. Soc.*, **44**, 652 (1967).
573. E. Jungerman and P. E. Spoerri, *J. Amer. Chem. Soc.*, **75**, 4704 (1953).
574. G. V. Pigulevskii and I. L. Kuranova, *Dokl. Akad. Nauk SSSR*, **82**, 601 (1952).
575. German Patent 1,024,514 (1958).
576. G. B. Payne, U.S. Patent 2,980,706 (1961).
577. T. J. Logan, *J. Org. Chem.*, **26**, 3657 (1961).
578. V. I. Pansevich-Kolyada and L. I. Timoshek, *Zh. Obshch. Khim.*, **22**, 1392 (1952).
579. M. A. Nechaeva, L. A. Mukhamedova, R. A. Virobyants, and R. R. Shagidullin, *Khim. Geterotsikl. Soedin.*, **1967**, 1029.

580. G. V. Pigulevskii and N. A. Adrova, *Zh. Obshch. Khim.*, **27**, 136 (1957).
581. G. V. Pigulevskii and N. A. Adrova, *J. Gen. Chem. USSR*, **27**, 151 (1957).
582. A. A. Griswold and P. S. Starcher, *J. Org. Chem.*, **31**, 357 (1966).
583. J. Meinwald, S. S. Labana, L. L. Labana, and G. H. Wahl, *Tetrahedron Letters*, No. 23, 1789 (1965).
584. F. E. Franz, M. Dietrich, and A. Henshall, *Chem. Ind. (London)*, **1966**, 1177.
585. N. V. deBataafsche, British Patent 770,481 (1957).
586. H. M. Teeter, E. W. Bell, J. L. O'Donnell, M. J. Danzig, and J. C. Cowan, *J. Amer. Oil Chem. Soc.*, **35**, 238 (1958).
587. M. O. Farooq, S. M. Osman, and M. S. Siddiqui, *Rec. Trav. Chim.* **80**, 415 (1961).
588. I. L. Kuranova, Y. D. Shenin, and G. V. Pigulevskii, *J. Gen. Chem. USSR*, **33**, 2923 (1963).
589. Y. Suhara, *Tokyo Kogyo Shikensho Hokoku*, **53**, 217 (1958).
590. J. G. Wallace, U.S. Patent 2,833,813 (1958).
591. D. R. V. Golding, U.S. Patent 2,833,814 (1958).
592. L. A. Clarke and G. W. Eckert, U.S. Patent 3,000,823 (1961).
593. K. L. Mikolajczak, C. R. Smith, M. O. Bagley, and I. A. Wolff, *J. Org. Chem.*, **29**, 318 (1964).
594. V. F. Koutcherov and G. M. Segal, *Bull. Soc. Chim. France*, **1960**, 1412. (Consult for previous references of this group.)
595. J. B. Williamson, British Patent 820,461 (1959).
596. N. A. LeBel, R. N. Liesemer, and E. Mehmedbasich, *J. Org. Chem.*, **28**, 615 (1963).
597. J. D. C. Wilson, U.S. Patent 2,838,524 (1958).
598. German Patent 975,394 (1961).
599. G. H. Büchi, R. E. Erickson, N. W. Wakabayashi, and H. E. Eschinazi, U.S. Patent 2,968,660 (1961).
600. G. Büchi, W. D. McLeod, Jr., and J. Padilla, *J. Amer. Chem. Soc.*, **86**, 4438 (1964).
601. A. G. Wilbur, U.S. Patent 3,069,377 (1962).
602. J. Dazzi, U.S. Patent 2,889,338 (1959).
603. B. Phillips, P. S. Starcher, and D. L. MacPeek, U.S. Patent 3,141,027 (1964).
604. B. Phillips and F. C. Frostick, Jr., U.S. Patent 3,074,973 (1963).
605. G. V. Pigulevskii and A. E. Sokolova, *Zh. Obshch. Khim.*, **31**, 652 (1961).
606. V. I. Pansevich-Kolyada and B. K. Bogush, *Probl. Organ. Sinteza, Akad. Nauk SSRS, Otd. Obshch. i Tekhn. Khim.*, **1965**, 30.
607. G. B. Payne and P. H. Williams, *J. Org. Chem.*, **26**, 651 (1961).
608. V. I. Pansevich-Kolyada, and B. K. Bogush, *Vestsi Akad. Nauk Belorusk. SSR, Ser. Khim. Nauk*, **1967**, 55.
609. N. M. Malenok and I. Sologub, *J. Gen. Chem. USSR*, **II**, 983 (1941).
610. M. M. Malenok, *J. Gen. Chem. USSR*, **9**, 1947 (1939).
611. N. M. Malenok and I. V. Sologub, *J. Gen. Chem. USSR*, **6**, 1904 (1936).
612. N. M. Malenok, S. D. Kulkina, and Z. Y. Kovtunenko, *Zh. Obshch. Khim.*, **28**, 429 (1958).
613. N. M. Malenok and I. V. Sologub, *J. Gen. Chem. USSR*, **10**, 150 (1940).

614. S. Komori, Y. Shigeno, K. Yamamoto, and K. Umezawa, *Kogyo Kagaku Zasshi*, **64**, 1203 (1961).
615. F. P. Greenspan and R. E. Legat, Jr., U.S. Patent 2,829,130 (1958).
616. F. P. Greenspan and A. E. Pepe, U.S. Patent 2,829,131 (1958).
617. F. P. Greenspan, U.S. Patent 2,826,556 (1958).
618. F. P. Greenspan, U.S. Patent 2,829,135 (1958).
619. F. P. Greenspan and R. E. Legat, Jr., U.S. Patent 2,851,441 (1958).
620. C. E. Wheelock, U.S. Patent 2,914,490 (1959).
621. C. E. Wheelock, U.S. Patent 2,959,351 (1960).
622. F. P. Greenspan and R. E. Light, U.S. Patent 3,010,976 (1961).
623. C. E. Wheelock, U.S. Patent 3,025,273 (1962).
624. F. P. Greenspan and A. E. Pepe, U.S. Patent 3,030,336 (1962).
625. M. H. Reich and G. Nowlin, U.S. Patent 3,073,796 (1963).
626. M. H. Reich, G. Nowlin, and C. A. Heibergh, U.S. Patent 3,092,608 (1963).
627. Food Machinery Corp., German Patent 1,040,794 (1958).
628. F. P. Greenspan and A. E. Pepe, German Patent 1,111,398 (1961).
629. F. P. Greenspan and R. E. Light, Jr., U.S. Patent 2,875,178 (1959).
630. Food Machinery Corp., German Patent 1,038,283 (1958).
631. Hercules Powder Co., British Patent 892,361 (1962); J. D. Floyd, U.S. Patent 3,243,401 (1966).
632. W. P. Fitzgerald and P. V. Smith, U.S. Patent 2,842,513 (1958).
633. F. P. Greenspan and R. E. Light, Jr., U.S. Patent 2,932,627 (1960).
634. R. H. Reiff, U.S. Patent 2,928,805 (1960).
635. F. P. Greenspan and R. E. Light, U.S. Patent 2,927,934 (1960).
636. British Patent 787,293 (1957).
637. K. C. Tsou, H. E. Hoyt, and B. D. Halpern, *J. Polymer Sci.*, **A2**, 4425 (1964).
638. S. O. Greenlee and J. W. Pearce, U.S. Patent 3,023,178 (1962).
639. H. Krimm and H. Schnell, German Patent 974,868 (1960).
640. British Patent 744,831 (1956).
641. F. P. Greenspan, U.S. Patent 2,833,747 (1958).
642. British Patent 874,868 (1961).
643. P. E. Marlatt, A. R. Hanze, A. V. McIntosh, and R. H. Levin, U.S. Patent 2,592,586 (1952).
644. G. S. Myers and A. J. Verwijs, U.S. Patent 3,351,641 (1967).
645. J. Imamura, R. Ishioka, S. Sato, and N. Ohta, *Kogyo Kagaku Zasshi*, **69**, 1454 (1966); *ibid.*, 1459.
646. A. L. Stautzenberger and A. H. Richey, U.S. Patent 3,341,556 (1967).
647. A. W. Wahlroos, U.S. Patent 2,813,878 (1957).
648. S. P. Rowland and R. G. White, U.S. Patent 2,836,605 (1958).
649. Dutch Patent 90,753 (1959).
650. British Celanese, Ltd., Belgian Patent 650,764 (1964).
651. D. J. Reif and H. O. House, *Org. Syn.*, **38**, 83 (1958).
652. W. Dittmann and F. Stuerzenhofecker, *Ann.*, **699**, 177 (1966).
653. W. Dittmann, German Patent 1,246,757.
654. Z. Chabudzinski, *Roczniki Chem.*, **35**, 629 (1961).
655. D. Porret, W. Fisch, H. Batzer, and O. Ernst, U.S. Patent 3,210,375 (1965).
656. B. Phillips and P. S. Starcher, U.S. Patent 2,794,028 (1951).

657. Ciba Ltd., Netherlands Patent Application 6,603,747 (1966).

658. R. W. Gleason and J. T. Snow, *J. Org. Chem.*, **34**, 1963 (1969).

659. F. C. Frostick, Jr., and B. Phillips, U.S. Patent 3,018,259 (1962).

660. J. W. Lynn, R. L. Roberts, and S. W. Tinsley, U.S. Patent 3,122,568 (1964).

661. V. I. Pansevich-Kolyada, *Zh. Obshch. Khim.*, **30**, 3901 (1960).

662. L. A. Mukhamedova, T. M. Malyshko, R. R. Shagidullin, and N. V. Teptina, *Khim. Geterotsikl. Soedin*, **1968**, 3.

663. V. A. Ponomarenko, N. M. Khomutova, and A. P. Pavlov, *Khim. Geterotsikl. Soedin.*, **1967**, 387.

664. W. Kirmse and B. Kornrumpf, *Angew. Chem., Intern. Ed.*, **8**, 75 (1969).

665. D. Swern, U.S. Patent 3,020,292 (1962).

666. Imperial Chemical Industries Ltd., Netherlands Patent Application 6,504,592 (1965).

667. J. Imamura, N. Nagato, S. Sato, and N. Ohta, *Kogyo Kagaku Zasshi*, **69**, 1863 (1966).

668. R. V. Kucher, B. I. Chernyak, and A. N. Nikolaevskii, *Dopov. Akad. Nauk Ukr. SSR.*, Ser. B, **1968**, 550.

669. E. A. Blyumberg, P. I. Valov, Y. D. Norikov, and N. M. Emanuel, *Dokl. Akad. Nauk SSSR*, **167**, 579 (1966); P. I. Valov, E. A. Blyumberg, and N. M. Emanuel, *Izv. Akad. Nauk SSSR, Ser. Khim.*, **1966**, 1334.

670. Institute of Physical Chemistry, Academy of Sciences, USSR, British Patent 1,080,462 (1967).

671. Institut Français du Petrole, British Patent 1,058,328 (1967).

672. A. Greiner, German (East) Patent 54,359 (1967).

673. Institut Français du Petrole, des Carburants et Lubrifiants, French Patent 1,410,985 (1965).

674. A. Byers and W. J. Hickinbottom, *J. Chem. Soc.*, **1948**, 1328. See also *Nature*, **158**, 341 (1946); *ibid.*, **159**, 844 (1947); and *J. Chem. Soc.*, **1948**, 284, 1331.

675. W. Niederhauser and J. E. Koroly, U.S. Patent 2,485,160 (1949).

676. J. Dazzi, U.S. Patent 2,745,846 (1956); British Patent 728,258 (1955).

677. G. Stork, J. Romo, G. Rosenkranz, and C. Djerassi, *J. Amer. Chem. Soc.*, **73**, 3546 (1951).

678. C. Djerassi, O. Mancera, J. Romo, and G. Rosenkranz, *J. Amer. Chem. Soc.*, **75**, 3505 (1953).

679. C. Djerassi, O. Mancera, G. Stork, and G. Rosenkranz, *J. Amer. Chem. Soc.*, **73**, 4496 (1951).

680. H. Murata and F. Higo, U.S. Patent 3,112,325 (1963).

681. W. R. Sorenson, *J. Org. Chem.*, **24**, 1796 (1959).

682. W. Gündel and G. Dieckelmann, U.S. Patent 3,297,725 (1967).

683. Rohm and Haas Co., British Patent 791,285 (1958).

684. Sekisui Chemical Co. Ltd., Japanese Patent 22,930 (1965).

685. Société d'Electrochimie, French Patent 1,464,139 (1966).

686. J. A. Fioriti, A. P. Bentz, and R. J. Sims, *J. Amer. Oil Chem. Soc.*, **43**, 37 (1966).

687. H. Pachaly and O. Schlichting, U.S. Patent 2,845,442 (1958); German Patent 962,073 (1957).

688. British Patent 783,300 (1957).
689. E. G. E. Hawkins, *J. Chem. Soc.*, **1959**, 248.
690. O. Henberger and L. N. Owen, *J. Chem. Soc.*, **1952**, 910.
691. A. W. Ritter and S. P. Rowland, U.S. Patent 2,771,472 (1956).
692. J. Klinot, E. Ulehlova, and A. Vystrcil, *Coll. Czech. Chem. Commun.*, **32**, 2890 (1967).
693. Rohm and Haas Co., British Patent 815,301 (1959).
694. S. P. Rowland and R. G. White, German Patent 1,001,979 (1957).
695. W. C. G. Förster, U.S. Patent 2,890,228 (1959).
696. Rohm and Haas Co., German Patent 857,364 (1952).
697. F. P. Greenspan and R. J. Gall, U.S. Patent 2,774,774 (1956).
698. R. L. Wear, U.S. Patent 3,003,981 (1961).
699. W. F. Johns, *J. Org. Chem.* **32**, 4086 (1967).
700. R. L. Wear, U.S. Patent 2,944,035 (1960).
701. French Patent 1,216,374 (1960).
702. W. D. Emmons and A. F. Ferris, *J. Amer. Chem. Soc.*, **75**, 4623 (1953).
703. W. D. Emmons and A. S. Pagano, *J. Amer. Chem. Soc.*, **77**, 89 (1955).
704. British Patent 862,430 (1961).
705. W. D. Emmons, *J. Amer. Chem. Soc.*, **76**, 3468 (1954).
706. B. J. Calvert and J. D. Hobson, *J. Chem. Soc.*, **1964**, 5378.
707. L. H. Zalkow and S. K. Gabriel, *J. Org. Chem.*, **34**, 218 (1969).
708. G. I. Fray, R. J. Hilton, and J. M. Teire, *J. Chem. Soc.*, *C*, **1966**, 592.
709. M. S. Malinovskii, V. G. Dryuk, and V. I. Avramenko, *Zh. Organ. Khim.*, **1968**, 1725.
710. K. Hunger, *Ber.*, **1968**, 3530.
711. G. B. Payne, *J. Org. Chem.*, **26**, 663 (1961).
712. L. H. Zalkow, M. V. Kulkarni, and N. N. Girotra, *J. Org. Chem.*, **30**, 1679 (1965).
713. E. N. Zvonska, K. I. Eller, V. I. Tsetlin, B. I. Mitsner, and N. A. Preobrazhenskii, *Zh. Organ. Khim.*, **2**, 2184 (1966).
714. E. R. H. Jones, J. S. Stephenson, W. B. Turner, and M. C. Whiting, *J. Chem. Soc.*, **1963**, 2048.
715. E. J. Reist, I. G. Junga, M. E. Wain, O. P. Crews, L. Goodman, and B. R. Baker, *J. Org. Chem.*, **26**, 2139 (1961).
716. FMC Corporation (1962). *Chem. Week*, **92**, 55 (April 6, 1963).
717. Technical Bulletin, FMC Corporation (1963).
718. N. N. Schwartz and J. N. Blumbergs, *J. Org. Chem.*, **29**, 1976 (1964).
719. E. E. van Tamelen and D. Carty, *J. Amer. Chem. Soc.*, **89**, 3922 (1967).
720. A. Hasegawa and H. Z. Sable, *J. Org. Chem.*, **31**, 4149 (1966).
721. A. Hassner and P. Catsonlacos, *J. Org. Chem.*, **32**, 549 (1967).
722. K. L. Williamson, J. I. Coburn, and M. F. Herr, *J. Org. Chem.*, **32**, 3934 (1967).
723. Schering A.G., British Patent 1,049,988 (1966); German Patent 1,244, 744 (1967); French Patent 1,465,571 (1967).
724. W. S. Bowers, *Science*, **164**, 323 (1969).
725. B. Rickborn and J. Quartucci, *J. Org. Chem.*, **29**, 2476 (1964).

726. K. B. Wiberg, J. E. Hiatt, and G. Burgmaier, *Tetrahedron Letters*, **1968**, 5855.
727. L. A. Paquette, A. A. Youssef, and M. L. Wise, *J. Amer. Chem. Soc.*, **89**, 5246 (1967).
728. P. Catsoulacos, *Chim. Chron.*, **31**, 153 (1966).
729. D. J. Pasto and C. C. Cumbo, *J. Org. Chem.*, **30**, 1271 (1965).
730. F. Sweet and R. K. Brown, *Can. J. Chem.*, **46**, 707 (1968).
731. V. R. Valente and J. L. Wolfhagen, *J. Org. Chem*, **31**, 2509 (1966).
732. H. Smith, P. Wegfahrt, and H. Rapoport, *J. Amer. Chem. Soc.*, **90**, 1668 (1968).
733. H. C. Brown and A. Suzuki, *J. Amer. Chem. Soc.*, **89**, 1933 (1967).
734. R. L. Settine and C. McDaniel, *J. Org. Chem.*, **32**, 2910 (1967).
735. H. Inoue, T. Yoshida, and S. Tobita, *Tetrahedron Letters*, **1968**, 2945.
736. R. N. McDonald and R. N. Steppel, *J. Amer. Chem. Soc.*, **91**, 782 (1969).
737. D. B. Roll and A. C. Huitric, *J. Pharm. Sci.*, **55**, 942 (1966).
738. J. G. Traynham, G. R. Franzen, G. A. Knesel, and D. J. Northington, Jr., *J. Org. Chem.*, **32**, 3285 (1967).
739. R. M. Gerkin and B. Rickborn, *J. Amer. Chem. Soc.*, **89**, 5850 (1967).
740. C. R. Johnson, J. E. Keiser, and J. C. Sharp. *J. Org. Chem.*, **34**, 860 (1969).
741. J. A. Marshall and A. R. Hochstetler, *J. Amer. Chem. Soc.*, **91**, 648 (1969).
742. S. R. Sandler and F. R. Berg. *J. Chem. Eng. Data*, **12**, 447 (1967).
743. R. L. Camp and F. D. Greene, *J. Amer. Chem. Soc.*, **90**, 7349 (1968).
744. E. S. Rothman, *J. Amer. Oil Chem. Soc.*, **45**, 189 (1968).
745. J. Fried, J. W. Brown, and L. Borkenhagen, *Tetrahedron Letters*, **1965**, 2499.
746. D. K. Murphy, R. L. Alumbaugh, and B. Rickborn, *J. Amer. Chem. Soc.*, **69**, 2649 (1969).
747. B. Rickborn and S. Y. Lwo, *J. Org. Chem.*, **30**, 2212 (1965).
748. Schering, A.G., British Patent 1,095,958 (1967).
749. J. A. Marshall and A. R. Hochstetler, *J. Org. Chem.*, **33**, 2593 (1968).
750. A. T. Bottini and R. L. Van Etten, *J. Org. Chem.*, **30**, 2994 (1965).
751. M. Sprecher and D. Kost, *Tetrahedron Letters*, No. 9, 703 (1969).
752. E. Farkas and J. M. Owen, *J. Med. Chem.*, **9**, 510 (1966).
753. R. P. Stein, G. C. Buzby, Jr., and H. Smith, *Tetrahedron Letters*, **1966**, 5015.
754. H. Smith, R. Smith, and R. P. Stein, South African Patent 6704,741 (1968).
755. V. R. Valente and J. L. Wolfhagen, *J. Org. Chem.*, **31**, 2509 (1966).
756. R. G. Pews, *J. Amer. Chem. Soc.*, **89**, 5605 (1967).
757. H. H. Wasserman and E. H. Barber, *J. Amer. Chem. Soc.*, **91**, 3674 (1969).
758. W. F. Johns, *J. Org. Chem.*, **29**, 2545 (1964).
759. A. Padwa and R. Hartman, *J. Amer. Chem. Soc.*, **88**, 3759 (1966).
760. J. A. Marshall and J. J. Partridge, *J. Org. Chem.*, **33**, 4090 (1968).
761. C. J. Cheer and C. R. Johnson, *J. Org. Chem.*, **32**, 428 (1967).
762. C. J. Cheer and C. R. Johnson, *J. Amer. Chem. Soc.*, **90**, 178 (1968).
763. S. Serota and E. S. Rothman, *J. Amer. Oil Chem. Soc.*, **45**, 525 (1968).
764. W. Herz and R. C. Blackstone, *J. Org. Chem.*, **34**, 1257 (1969).
765. B. Rickborn and W. E. Lamke, *J. Org. Chem.*, **32**, 537 (1967).
766. K. H. Dahm, B. M. Trost, and H. Roeller, *J. Amer. Chem. Soc.*, **89**, 5292 (1967).

767. H. Schulz and I. Sprung, *Angew. Chem., Intern. Ed. Eng.*, **8**, 271 (1969).
768. C. W. J. Chang and S. W. Pelletier, *Tetrahedron Letters*, **1966**, 5483.
769. A. J. Lundeen and R. Van Hoozer, *J. Org. Chem.*, **32**, 3386 (1967).
770. Ciba Ltd., Netherlands Patent Application, 6,610,741 (1967).
771. J. A. Marshall and W. I. Fanta, *J. Org. Chem.*, **29**, 2501 (1964).
772. C. H. Heathcock, *J. Amer. Chem. Soc.*, **88**, 4110 (1966).
773. D. E. Bissing and A. J. Speziale, *J. Amer. Chem. Soc.*, **87**, 2683 (1965).
774. A. P. Gray and D. E. Heitmeier, *J. Org. Chem.*, **30**, 1226 (1965).
775. J. Meinwald and L. Hendry, *Tetrahedron Letters*, **1969**, 1657.
776. J. B. Lewis and G. W. Hedrick, *J. Org. Chem.*, **30**, 4271 (1965).
777. R. C. L. Chow and C. S. Marvel, *J. Polymer Sci.*, **6**, 1273 (1968).
778. S. Medvedev and O. Blokh, *J. Phys. Chem. USSR*, **4**, 721 (1933).
779. C. G. Overberger and R. W. Cummins, *J. Amer. Chem. Soc.*, **75**, 4250 (1953).
780. M. Vilkas, *Bull. Soc. Chim. France*, **1959**, 1401.
781. L. S. Silbert, E. Siegel, and D. Swern, *J. Org. Chem.*, **27**, 1336 (1962).
782. D. Swern and L. S. Silbert, unpublished results.
783. M. Dvolaitzky and J. Jacques, *Bull. Soc. Chim. France*, **1963**, 2793.
784. M. Mousseron-Canet and B. Labeeuw, *Bull. Soc. Chim. France*, **1968**, 4165.
785. G. Ponsinet, G. Ourisson, and G. Charles, *Bull. Soc. Chim. France*, **1967**, 4453.
786. E. Casaderall, C. Largeau, and P. Moreau, *Bull. Soc. Chim. France*, **1968**, 1514.
787. J. C. Richer and C. Freppel, *Can. J. Chem.*, **46**, 3709 (1968).
788. Y. Chretion-Bessiere and J. A. Retamar, *Bull. Soc., Chim. France*, **1963**, 884.
789. N. Langlois and B. Gastambide, *Bull. Soc. Chim. France*, **1965**, 2966.
790. N. LeBoulch, Y. Raoul, and G. Ourisson, *Bull. Soc. Chim. France*, **1965**, 646.
791. M. Bertrand and J. Grimaldi, *Compt. Rend.*, **C265**, 196 (1967).
792. J. Seyden-Penne, C. Gilbert, and B. Danree, *Compt. Rend.*, **C263**, 895 (1966).
793. D. A. Lightner and C. Djerassi, *Tetrahedron*, **21**, 583 (1965).
794. G. Berti, F. Bottari, P. L. Ferrarini, and B. Macchia, *J. Org. Chem.*, **30**, 4091 (1965).
795. G. Dieckelman, U.S. Patent 3,120,547 (1964).
796. J. Blum, *Compt. Rend.*, **248**, 2883 (1959).
797. German Patent 1,084,712 (1960).
798. A. G. Davies and J. E. Packer, *J. Chem. Soc.*, **1961**, 4390.
799. M. S. Crossley, A. C. Darby, H. B. Henbest, J. J. McCullough, B. Nicholls, and M. F. Stewart, *Tetrahedron Letters*, **1961**, 398.
800. H. B. Henbest, *Proc. Chem. Soc.*, **1963**, 159.
741. H. B. Henbest and J. J. McCullough, *Proc. Chem. Soc.*, **1962**, 74.
801. A. C. Darby, H. B. Henbest, and I. McClenoghan, *Chem. Ind. (London)*, **1962**, 462.
802. R. T. Aplin and L. Coles, *Chem. Commun.*, **1967**, 858.
803. R. W. White and W. D. Emmons, *Tetrahedron*, **17**, 31 (1962).
804. T. Suzuki, S. Hori, and S. Suzuki, *Kogyo Kagaku Zasshi*, **69**, 440 (1966).
805. M. S. Malinovskii, V. C. Dryuk, and S. P. Shamrovskaya, *Zh. Organ. Khim.*, **4**, 1887 (1968).
806. B. Radziszewski, *Ber.*, **17**, 1289 (1884).

807. K. B. Wiberg, *J. Amer. Chem. Soc.*, **75**, 3961 (1953).

808. G. B. Payne, P. H. Deming, and P. H. Williams, *J. Org. Chem.*, **26**, 659 (1961).

809. G. B. Payne and P. H. Deming, U.S. Patent 3,053,856 (1962).

810. G. B. Payne, *Tetrahedron*, **18**, 763 (1962)

811. R. J. Ferrier and N. Prasad, *J. Chem. Soc.*, C, **1969**, 575.

812. Y. Ogata and Y. Sawaki, *Tetrahedron*, **20**, 2065 (1964).

813. P. Radlick and S. Winstein, *J. Amer. Chem. Soc.*, **86**, 1866 (1964).

814. M. Mousseron, unpublished results, described in footnote 5 of reference 75.

815. P. H. Williams, U.S. Patent 3,113,951 (1963).

816. W. Stein, R. Brockmann, and G. Dieckelmann, U.S. Patent 3,141,896 (1964).

817. W. Stein, R. Brockmann, and G. Dieckelmann, German Patent 1,152,415 (1963).

818. B. A. Ershov, Z. G. Leus, and T. I. Temnikova, *Zh. Organ. Khim.*, **4**, 796 (1968).

819. G. S. Krishna Rao, S. Der, and P. C. Guha, *J. Indian Chem. Soc.*, **29**, 589 (1952).

820. H. B. Henbest, in *Organic Reaction Mechanisms*, Special Publication No. 19, The Chemical Society, London (1965), p. 83.

821. R. C. Ewins, H. B. Henbest, and M. A. McKervey, *Chem. Commun.*, **1967**, 1085.

822. R. M. Bowman, J. F. Collins, and M. F. Grunden, *Chem. Commun.*, **1967**, 1131.

823. M. S. Malinovskii, A. G. Yudasina, and N. G. Krivosheeva, *Zh. Obshch. Khim.*, **37**, 1666 (1967); *ibid.*, **38**, 1829 (1968).

824. H. Hart and L. R. Lerner, *J. Org. Chem.*, **32**, 2669 (1967).

825. H. Hibbert and P. Burt, in *Organic Syntheses*, Collective, Vol. I, Wiley, New York, 1941, p. 494.

826. D. Swern, T. W. Findley, and J. T. Scanlan, *J. Amer. Chem. Soc.*, **66**, 1925 (1944).

827. R. Pummerer and W. Reindel, *Ber.*, **66**, 335 (1933).

828. D. Swern, G. N. Billen, and J. T. Scanlan, *J. Amer. Chem. Soc.*, **68**, 1504 (1946).

829. T. W. Findley, D. Swern, and J. T. Scanlan, *J. Amer. Chem. Soc.*, **67**, 412 (1945).

830. W. R. Sorenson and T. W. Campbell, *Preparative Methods of Polymer Chemistry*, Interscience, New York, 1961, pp. 250 and 251.

831. L. F. Fieser and M. Fieser, *Reagents for Organic Synthesis*, Wiley, New York, 1967, p. 135.

832. United States Tariff Commission, *United States Production and Sales of Plasticizers*, February 1969.

833. F. Chevassus and R. de Broutelles, *The Stabilization of Polyvinyl Chloride*, St. Martin's Press, New York, 1963.

834. R. J. Gall and F. P. Greenspan, *Ind. Eng. Chem.*, **47**, 147 (1955).

835. F. P. Greenspan and R. J. Gall, *J. Amer. Oil Chem. Soc.*, **33**, 391 (1956).

836. E. I. duPont de Nemours & Co., Bulletins P-61-454 (1954), A-6282 (1955), and A-11387 (1960).

837. W. D. Niederhauser and J. E. Koroly, U.S. Patent 2,485,160 (1949).
838. T. W. Findley, D. Swern, and J. T. Scanlan, *J. Amer. Chem. Soc.*, **67**, 412 (1945).
839. D. Swern, *J. Amer. Chem. Soc.*, **69**, 1692 (1947).
840. J. G. Wallace, in *Encyclopedia of Chemical Technology*, Vol. 8, Interscience, New York, 1965, p. 238.
841. D. Swern, in *Encyclopedia of Polymer Science and Technology*, Vol. 6, 1967, p. 83.
842. Solvay Process Division, Allied Chemical Corporation, Bulletins HP-13, HP-20, and HP-23 (1959).
843. E. I. duPont de Nemours & Co., Bulletin A-313 (1958).
844. Becco Chemical Division, FMC Corporation, Bulletins 1S-854 (1954) and 69 (1955).
845. Shell Chemical Corporation, Bulletin SC:57-25 (1957).
846. R. J. Gall and F. P. Greenspan, *J. Amer. Oil Chem. Soc.*, **34**, 161 (1957).
847. R. J. Gall and F. P. Greenspan, U.S. Patent 2,919,283 (1959).
848. W. Wood and J. Termini, *J. Amer. Oil Chem. Soc.*, **35**, 331 (1958).
849. A. F. Chadwick, D. O. Barlow, A. A. D'Addieco, and J. G. Wallace, *J. Amer. Oil Chem. Soc.*, **35**, 355 (1958).
850. H. C. Wohlers, M. Sack, and H. P. Le Van, *Ind. Eng. Chem.*, **50**, 1685 (1958).
851. M. Sack and H. C. Wohlers, *J. Amer. Oil Chem. Soc.*, **36**, 623 (1959).
852. Solvay Process Division, Allied Chemical Corporation, Report 8, 58R (1958).
853. L. I. Hansen and A. G. Coutsicos, U.S. Patent 3,328,430 (1967).
854. D. Swern, *Paint and Varnish Production*, **46**, 27 (1956).
855. B. Phillips, F. C. Frostick, Jr., and P. S. Starcher, *J. Amer. Chem. Soc.*, **79**, 5982 (1957).
856. J. A. John and F. J. Weymouth, *Chem. Ind. (London)*, **1962**, 62.
857. B. Phillips, P. S. Starcher, and B. D. Ash, *J. Org. Chem.*, **23**, 1823 (1958).
858. H. Jones, paper presented at 33rd Fall Meeting of the American Oil Chemists' Society, Los Angeles, 1959.
859. F. P. Greenspan and S. M. Linder, U.S. Patent 3,130,207 (1964).
860. B. Phillips and D. L. MacPeek, in *Encyclopedia of Chemical Technology*, First Supplement, Interscience, New York, 1957, p. 622.
861. S. P. Rowland and E. M. Beavers, U.S. Patent 2,963,490 (1960).
862. F. P. Greenspan and R. J. Gall, U.S. Patent 2,810,732 (1957).
863. A. W. Wahlroos, U.S. Patent 2,813,878 (1957).
864. F. P. Greenspan and R. J. Gall, U.S. Patent 2,810,253 (1957).
865. A. A. D'Addieco, Canadian Patent 531,112 (1956).
866. D. F. Othmer, *Chem. and Met. Eng.*, **40**, 631 (1933); *ibid.*, **47**, 349 (1940).
867. E. I. duPont de Nemours & Co., Booklet A-114 (1955).
868. E. S. Shanley and J. R. Perrin, *Jet Propulsion*, **1958**, 382. Also issued as Bulletin 100, Becco Chemical Division, FMC Corporation.
869. Shell Chemical Corporation, Bulletin SC:59-44 (1959).
870. F. P. Greenspan and R. J. Gall, *Ind. Eng. Chem.*, **45**, 2722 (1953).
871. D. Swern, *Paint and Varnish Production*, **46**, 29, 35 (1956).
872. L. P. Witnauer, H. B. Knight, W. E. Palm, R. E. Koos, W. C. Ault, and D. Swern, *Ind. Eng. Chem.*, **47**, 2304 (1955).

873. J. T. Lutz, *Resin Review*, **18**, 14 (1968).

874. J. R. Darby, U.S. Patent 2,951,052 (1960).

875. Federal Register, **27**, 741 (1962).

876. P. L Larson, J. K. Finnegan, H. B. Haag, R. B. Smith, and G. R. Hennigar, *Toxical. Appl. Pharmacol.*, **2**, 649 (1960).

877. H. W. Chatfield, *Paint Manuf.*, **27**, 51 (1957).

878. H. W. Chatfield, *J. Oil and Colour Chem. Assoc.*, **41**, 301 (1958).

879. U.S. Patents 2,909,537 (1959) and 2,930,708 (1960); U.S. Patent 3,086,949 (1963).

880. A. Boake Roberts and Co. Bulletin, *Abrac A in Manufacture of Resins and Surface Coatings*, London, 1958.

881. C. W. Schroeder, U.S. Patents 2,872,427; 2,872,428; 2,886,473; and 2,903,381 (1959).

882. F. E. Condo and C. W. Schroeder, U.S. Patent 2,886,472 (1959); British Patent 783,740 (1957).

883. British Patent 758,728 (1956).

884. B. Phillips and F. C. Frostick, Jr., British Patent 788,530 (1958).

885. B. Phillips and F. C. Frostick, Jr., U.S. Patent 2,779,771 (1957).

886. L. S. Silbert and W. S. Port, *J. Amer. Oil Chem. Soc.*, **34**, 9 (1957).

887. British Patent 786,116 (1957).

888. J. W. Pearce and J. Kawa, *J. Amer. Oil Chem. Soc.*, **34**, 57 (1957).

889. S. O. Greenlee and J. W. Pearce, U.S. Patent 3,023,178 (1962).

890. F. Sprules and H. J. Marks, U.S. Patent 2,802,800 (1957).

891. British Patent 744,831 (1956).

892. A. W. Ritter and S. P. Rowland, U.S. Patent 2,771,472 (1956).

893. R. M. Brice and W. Budde, *Ind. Eng. Chem.*, **50**, 868 (1958).

894. F. C. Frostick, Jr., and B. Phillips, U.S. Patent 2,786,066 (1957).

895. R. VanCleve and D. H. Mullins, *Ind. Eng. Chem.*, **50**, 873 (1958).

896. F. C. Frostick, Jr., and B. Phillips, U.S. Patent 2,786,067 (1957).

897. E. I. duPont de Nemours & Co., Bulletins NC-58-1 and NC-58-1-A (1958).

898. M. Kleiman, U.S. Patent 2,655,514 (1953).

899. H. Bluestone, U.S. Patent 2,676,132 (1954).

900. G. Slomp, U.S. Patent 2,751,381 (1956).

901. G. Rosenkranz and C. Djerassi, U.S. Patent 2,773,887 (1956).

902. F. P. Greenspan, in *Chemical Reactions of Polymers*, edited by E. M. Fettes, Chapter IIE, Interscience, New York, 1964.

903. F. P Greenspan, C Johnston, and M. Reich, *Modern Plastics*, **37**, 142 (1959).

904. F. P. Greenspan, U.S. Patent 2,826,556 (1958).

905. F. P. Greenspan and R. E. Light, U.S. Patents 2,829,130 and 2,851,441 (1958); 2,875,178 (1959); 2,927,934 and 2,932,627 (1960); 3,010,976 (1961).

906. F. P. Greenspan and A. E. Pepe, U.S. Patents 2,829,131 (1958) and 3,030,336 (1962).

907. F. P. Greenspan, U.S. Patents 2,829,135 and 2,833,747 (1958).

908. British Patent 892,361 (1962); J. D. Floyd, U.S. Patent 3,243,401 (1966).

909. R. W. Rees, U.S. Patent 3,050,507 (1962).

910. F. P. Greenspan and R. E. Light, Canadian Patent 560,690 (1958).

911. K. L. Johnson, *J. Amer. Oil Chem. Soc.*, **41**, 191 (1964).

912. Anon., *Indian J. Technol.*, **1965**, 87.
913. H. Chotiner, *Oléagineux*, **16**, 451 (1961).
914. A. S. Danyushevskii, A. F. Vorobera, and A. L. Sergeeva, *Prasticeskie Massy*, **1960**, 20.
915. L. Kovacs, *Fette, Seifen, Anstrichmittel*, **63**, 251 (1961).
916. Y. Suhara, *J. Japan Oil Chem. Soc.*, **13**, 279 (1964).
917. H. A. Suter and S. D. Cooley, U.S. Patent 2, 903,465 (1959).
918. R. J. Gall, F. P. Greenspan, and M. C. Daly, U.S. Patent 2,935,517 (1960).
919. B. Werdelmann and G. Dieckelmann, U.S. Patent 3,248,404 (1966).
920. F. P. Greenspan and R. J. Gall, U.S. Patent 2,857,349 (1958).
921. S. P. Rowland and R. F. Conyne, U.S. Patent 2,822,368 (1968).
922. F. P. Greenspan, U.S. Patent 2,810,733 (1957).
923. S. P. Rowland and R. F. Conyne, U.S. Patent 2,963,455 (1960).
924. S. D. Douglas, U.S. Patent 3,008,911 (1961).
925. H. K. Latourette, H. M. Castrantas, R. J. Gall, and L. H. Dardoff, *J. Amer. Oil Chem. Soc.*, **37**, 559 (1960).
926. B. Phillips and P. S. Starcher, U.S. Patent 2,977,374 (1961).
927. H. K. Latourette and H. M. Castrantas, U.S. Patents 3,065,245 and 3,065,246 (1962).
928. W. C. Ault and R. O. Feuge, U.S. Patents 2,895,966 (1959) and 3,050,481 (1961).
929. D. Swern and H. B. Knight, U.S. Patents 2,898,348 (1959) and 3,049,504 (1962).
930. A. A. Manteuffel, G. R. Cook, and W. W. Cortiss, U.S. Patent 2,900,342 (1959).
931. G. W. Eckert, U.S. Patent 3,009, 878 (1961).
932. G. H. Tulk and T. A. Neuhaus, U.S. Patent 2,907,669 (1959).
933. T. W. Findley, U.S. Patent 2,964,454 (1960).
934. H. A. Newey, U.S. Patent 2,949,441 (1960).
935. T. W. Findley and J. L. Ohlson, U.S. Patent 2,964,484 (1960).
936. C. S. Ilardo, C. T. Bean, and P. Robitschek, U.S. Patent 2,992,196 (1961).
937. W. M. Budde and G. W. Matson, U.S. Patent 2,993,920 (1961).
938. K. R. Coney, R. B. Pyewell, and W. B. Webb, U.S. Patent 3,003,978 (1961).
939. M. H. Reich and G. Nowlin, U.S. Patent 3,073,796 (1963).
940. M. H. Reich, G. Nowlin, and C. A. Heiberger, U.S. Patent 3,092,608 (1963).
941. P. H. Rhodes, U.S. Patent 3,055,778 (1962).
942. V. Mark, U.S. Patents 3,298,815 and 3,324,147 (1967).
943. J. Boeseken and C. de Graff, *Rec. Trav. Chim.*, **41**, 199 (1922).
944. J. Boeseken and J. S. P. Blumberger, *Rec. Trav. Chim.*, **44**, 90 (1925).
945. J. Boeseken and G. Elsen, *Rec. Trav. Chim.*, **48**, 363 (1929).
946. W. C. Smit, *Rec. Trav. Chim.*, **49**, 686 (1930).
947. J. Boeseken and M. W. Maas Geesteranus, *Rec. Trav. Chim.*, **51**, 551 (1932).
948. J. Boeseken, *Rec. Trav. Chim.* **54**, 657 (1935).
949. J. Stuurmann, *Proc. Akad. Sci. Amsterdam*, **38**, 450 (1935).
950. J. Boeseken and J. Stuurmann, *Proc. Acad. Sci. Amsterdam*, **39**, 2 (1936).
951. J. Boeseken and J. Stuurmann, *Rec. Trav. Chim.*, **56**, 1034 (1937).

952. J. Boeseken and C. J. A. Hanegraaff, *Rec. Trav. Chim.*, **61**, 69 (1942).

953. H. Meerwein, A. Ogait, W. Prang, and A. Serini, *J. Prakt. Chem.* **113**, 9 (1926).

954. K. Bodendorf, *Arch. Pharm.*, **268**, 491 (1930).

955. P. Heinanen, *Ann. Acad. Sci. Fennicae*, **A59**, 3 (1943).

956. D. Swern, *J. Amer. Chem. Soc.*, **69**, 1692 (1947).

957. C. Ricciuti, L. S. Silbert, and W. S. Port, *J. Amer. Oil Chem. Soc.*, **34**, 134 (1957).

958. A. A. Durgaryan and S. A. Titanyan, *Izv. Akad. Nauk Armyan. SSR, Khim. Nauki*, **13**, 263 (1960).

959. A. Wende and A. Gesierich, *Plaste und Kautschuk*, **8**, 399 (1961).

960. Y. Nagai, *Yuki Gosei Kagaku Kyokaishi*, **19**, 527 (1961).

961. J. K. Stille and D. R. Witherell, *J. Amer. Chem. Soc.*, **86**, 2188 (1964).

962. W. Hueckel and O. Vogt, *Ann.*, **695**, 16 (1966).

963. W. Schneider, *Fette, Seifen, Anstrichmittel*, **69**, 421, 573 (1967).

964. H. B. Henbest and R. A. L. Wilson, *J. Chem. Soc.*, **1957**, 1958.

965. H. B. Henbest and R. A. L. Wilson, *Chem. Ind. (London)*, **1956**, 659.

966. D. R. Campbell, J. O. Edwards, J. Maclachlan, and K. Polgar, *J. Amer. Chem. Soc.*, **80**, 5308 (1958).

967. E. Giovanni and H. Wegmüller, *Helv. Chim. Acta*, **42**, 1142 (1959).

968. H. B. Henbest, and B. Nicholls, *J. Chem. Soc.*, **1959**, 221.

969. F. C. Frostick, Jr., B. Phillips, and P. S. Starcher, *J. Amer. Chem. Soc.*, **81**, 3350 (1959).

970. Y. Ogata and I. Tabushi. *J. Amer. Chem. Soc.*, **83**, 3440, 3444 (1961).

971. F. D. Gunstone and P. J. Sykes, *J. Chem. Soc.*, **1962**, 3063.

972. M. Mousseron-Canet, M. Mousseron, and C. Levallois, *Bull. Soc. Chim. France*, **1963**, 376.

973. F. Asinger, B. Fell, G. Hadik, and G. Steffan, *Ber.*, **97**, 1568 (1964).

974. F. D. Greene and W. Adam, *J. Org. Chem.*, **29**, 136 (1964).

975. H. B. Henbest, Chemical Society Special Publication 19, London, 1964, p. 23.

976. A. P. Gray, and D. E. Heitmeier, *J. Org. Chem.*, **30**, 1226 (1965).

977. B. Rickborn and S. Y. Lwo, *J. Org. Chem.*, **30**, 2212 (1965).

978. R. C. Ewins, H. B. Henbest, and M. A. McKervey, *Chem. Commun.*, **1967**, 1085.

979. F. Asinger, B. Fell, G. Hadik, and G. Stefan, *Ber.*, **97**, 1568 (1964).

980. A. N. Pudovik and B. E. Ivanov, *Zh. Obshch. Khim.*, **26**, 2771 (1956)., English translation, J. Gen. Chem. USSR. **26**, 3087 (1956).

981. I. M. Kolthoff and T. S. Lee, *J. Polymer Sci.*, **20**, 206 (1947).

982. A. Saffer and B. L. Johnson, *Ind. Eng. Chem.*, **40**, 538 (1948).

983. Y. Ishii and Y. Inamoto, *J. Chem. Soc. Japan, Ind. Chem. Sect.*, **63**, 765 (1960).

984. G. Berti and F. Bottari, *Gazz. Chim. Ital.*, **89**, 2380 (1959); *J. Org. Chem.*, **25**, 1286 (1960).

985. V. N. Sapunov and N. N. Lebedev, *Zh. Organ. Khim.*, **2**, 225 (1966).

986. P. Renolen and J. Ugelstad, *J. Chim. Phys.*, **57**, 634 (1960).

987. D. Swern, L. P. Witnauer, C. R. Eddy, and W. E. Parker, *J. Amer. Chem. Soc.*, **77**, 5537 (1955).

988. V. N. Sapunov and N. N. Lebedev, *Izv. Vysshikh Uchebn. Zavedenii, Khim. i Khim. Tekhnol.*, **8**, 771 (1965).

989. L. S. Silbert and D. A. Konen, paper presented at Middle Atlantic Regional Meeting, American Chemical Society, Philadelphia, February 1, 1968.

990. T. Yamada, S. Fukuda, and K. Hoshino, *J. Japan Oil Chem. Soc.*, **13**, 137 (1964).

991. G. V. Pigulevskii and P. A. Artamonov, *Zh. Obshch. Khim.*, **22**, 1140 (1952).

992. Y. Toyama, *Mem. Fac. Eng., Nagoya Univ.*, **4**, 121 (1952).

993. Y. Toyama and T. Yamamoto, *J. Chem. Soc. Japan, Ind. Chem. Sect.*, **55**, 164, 176 (1952).

994. A. K. Plisov and N. S. Skornyakova, *J. Gen. Chem. USSR* (English translation), **26**, 57 (1956).

995. A. K. Plisov and N. S. Skornyekova, *Tr. Odessk. Gosudarst. Univ. im I. I. Mechnikova, Ser. Khim. Nauk*, **146**, 33 (1956).

996. E.Giovanni and H. Wegmüller, *Helv. Chim. Acta*, **42, 1142** (1959).

997. K. Murai, F. Akazame and Y. Murakami, *J. Chem. Soc. Japan, Ind. Chem. Sect.*, **63**, 803 (1960).

998. Y. Suhara, *J. Japan Oil Chem. Soc.*, **9**, 607 (1960).

999. B. L. VanDuuren and F. L. Schmitt, *J. Org. Chem.*, **25**, 1761 (1960).

1000. A. K. Plisov and A. V. Bogatskii, *Zh. Obshch. Khim.*, **31**, 3324 (1961).

1001. P. I. Valov, E. A. Blyumberg, and T. V. Filippova, *Kinetica i Kataliz*, **8**, 760 (1967).

1002. J. S. Showell, J. R. Russell, and D. Swern, *J. Org. Chem.*, **27**, 853 (1962).

1003. E. de R. van Zuydewijn and J. Stuurmann, *Chem. Weekblad*, **33**, 540 (1936).

1004. M. S. Newman, N. Gill, and D. W. Thomson, *J. Amer. Chem. Soc.*, **89**, 2059 (1967).

1005. D. Swern, *J. Amer. Chem. Soc.*, **70**, 1235 (1948).

1006. L. P. Witnauer and D. Swern, *J. Amer. Chem. Soc.*, **72**, 3364 (1950).

1007. B. S. Mahl, M. S. Wadia, I. S. Bhatia, and P. S. Kalsi, *Perfum. Essent. Oil Record*, **1968**, 519.

1008. H. B. Henbest, B. Nicholls, W. R. Jackson, R. A. L. Wilson, N. S. Crossley, M. B. Meyers, and R. S. McElhinney, *Bull. Soc. Chim. France*, **1960**, 1365.

1009. H. Kwart and T. Takeshita, *J. Org. Chem.*, **28**, 670 (1963).

1010. H. B. Henbest and B. Nicholls, *Proc. Chem. Soc.*, **1958**, 225.

1011. H. B. Henbest, *Proc. Chem. Soc.*, **1963**, 159.

1012. N. A. LeBel and R. F. Czaja, *J. Org. Chem.*, **26**, 4768 (1961).

1013. E. J. Corey and R. L. Dawson, *J. Amer. Chem. Soc.*, **85**, 1782 (1963).

1014. J. C. Leffingwell and E. E. Royals, *Tetrahedron Letters*, **1965**, 3829.

1015. H. Kuczynski and K. Piatkowski, *Roczniki Chem.*, **33**, 299, 311 (1959).

1016. M. Veza, A. Kovendi, C. Schonberger, and D. Breazu, *Rev. Chim. (Bucharest)*, **1968**, 3.

1017. F. Montanari, I. Moretti, and G. Torre, *Boll. Sci. Fac. Chim. Ind. Bologna*, **1968**, 113.

1018. P. D. Bartlett, *Rec. Chem. Progr.*, **11**, 47 (1950).

1019. K. D. Bingham, G. D. Meakins, and G. H. Whitham, *Chem. Commun.*, **1966**, 445.
1020. P. A. Giguère and A. W. Olmos, *Can. J. Chem.*, **30**, 821 (1952).
1021. D. Swern, L. P. Witnauer, C. R. Eddy, and W. E. Parker, *J. Amer. Chem. Soc.*, **77**, 5537 (1955).
1022. H. Kwart and D. M. Hoffman, *J. Org. Chem.*, **31**, 419, (1966).
1023. H. Kwart, P. S. Starcher and S. W. Tinsley, *Chem. Commun.*, **1967**, 335.
1024. J. B. Lee and B. C. Uff, *Quart. Rev.*, **21**, 429 (1967).
1025. R. Huisgen, *Angew. Chem.*, **75**, 604, 742 (1963).
1026. R. Huisgen, *Proc. Chem. Soc.*, **1961**, 337.
1027. R. Huisgen, L. Mobius, G. Muller, H. Stangl, C. Szeimies, and J. M. Vernon, *Ber.*, **98**, 3992 (1965).
1028. A. Azman, B. Borstnik, and B. Plesnicar, *J. Org. Chem.*, **34**, 971 (1967).
1029. R. Pariser and R. G. Parr, *J. Chem. Phys.*, **21**, 466 (1953).
1030. J. A. Pople, *Trans. Faraday Soc.*, **49**, 1375 (1953).
1031. M. Berthelot, *Bull. Soc. Chim.*, **14**, 113 (1870).
1032. P. Muller and W. Schindler, U.S. Patent 2,432,118 (1947).
1033. R. N. McDonald and P. A. Schwab, *J. Amer. Chem. Soc.*, **86**, 4866 (1968).
1034. N. Prileschajew, *J. Russ. Phys. Chem. Soc.*, **42**, 1387 (1910).
1035. J. Boeseken and G. Slooff, *Rec. Trav. Chim.*, **49**, 95 (1930).
1036. J. K. Stille and D. D. Whitehurst, *J. Amer. Chem. Soc.*, **86**, 4871 (1964).
1037. H. H. Schlubach and V. Franzen, *Ann.*, **577**, 60 (1952).
1038. V. Franzen, *Ber.*, **87**, 1478 (1954).
1039. V. Franzen, *Ann.*, **614**, 31 (1958).
1040. V. Franzen, *Ber.*, **87**, 1219 (1954).
1041. H. H. Schlubach and V. Franzen, *Ann.*, **578**, 220 (1952).
1042. N. M. Malenok and I. V. Sologub, *Zh. Obshch. Khim.*, **23**, 1129 (1953).
1043. R. A. Raphael, *J. Chem. Soc.*, **1949**, 445.
1044. R. M. Evans, J. B. Fraser, and L. N. Owen, *J. Chem. Soc.*, **1949**, 248.
1045. V. A. Engelhardt and J. E. Castle, *J. Amer. Chem. Soc.*, **75**, 1734 (1953).
1046. F. D. Gunstone and M. A. McGee, *Chem. Ind.* (*London*), **1954**, 1112.
1047. F. Bohlmann and H. Sinn, *Ber.*, **88**, 1869 (1955).
1048. F. D. Gunstone and P. J. Sykes, *Chem. Ind* (*London*), **1960**, 1130.
1049. I. G. Csizmadia, J. Font, and O. P. Strausz, *J. Amer. Chem. Soc.*, **90**, 7360 (1968).
1050. J. Boeseken, *Rec. Trav. Chim.*, **54**, 657 (1935).
1051. V. I. Pansevich-Kolyada and Z. B. Idelchik, *Zh. Obshch. Khim.*, **24**, 1617 (1954).
1052. V. I. Pansevich-Kolyada and Z. B. Idelchik, *J. Gen. Chem.* USSR, **24**, 1601 (1954).
1053. J. K. Crandall and W. H. Machleder, *Tetrahedron Letters*, **1966**, 6037.
1054. J. K. Crandall and W. H. Machleder, *J. Amer. Chem. Soc.*, **90**, 7292 (1968).
1055. J. K. Crandall, D. R. Paulson, and C. A. Burnell, *Tetrahedron Letters*, **1968**, 5063.
1056. J. K. Crandall and W. H. Machleder, *J. Amer. Chem. Soc.*, **90**, 7347 (1968).

1057. T. H. Kritchevsky and T. F. Gallagher, *J. Biol. Chem.*, **179**, 507 (1949).

1058. T. H. Kritchevsky and T. F. Gallagher, *J. Amer. Chem. Soc.*, **73**, 184 (1951).

1059. T. H. Kritchevsky, D. L. Garmaise, and T. F. Gallagher, *J. Amer. Chem. Soc.*, **74**, 483 (1952).

1060. J. Pataki, G. Rosenkranz, and C. Djerassi, *J. Amer. Chem. Soc.*, **74**, 5615 (1952).

1061. L. F. Fieser, W. Y. Huang, and J. C. Babcock, *J. Amer. Chem. Soc.*, **75**, 116 (1953).

1062. D. Mancera, L. Miramontes, G. Rosenkranz, F. Sondheimer, and C. Djerassi, *J. Amer. Chem. Soc.*, **75**, 4428 (1953).

1063. A. H. Soloway, W. J. Considine, D. K. Fukushima, and T. F. Gallagher, *J. Amer. Chem. Soc.*, **76**, 2941 (1954).

1064. N. S. Leeds, D. K. Fukushima, and T. F. Gallagher, *J. Amer. Chem. Soc.*, **76**, 2943 (1954).

1065. P. D. Gardner, *J. Amer. Chem. Soc.*, **78**, 3421 (1956).

1066. J. J. Riehl, J. M. Lehn, and F. Hemmert, *Bull. Soc. Chim. France*, **1963**, 224.

1067. A. L. Draper, W. J. Heilman, W. E. Schaefer, H. J. Shine, and J. N. Shoolery, *J. Org. Chem.*, **27**, 2727 (1962).

1068. K. L. Williamson and W. S. Johnson, *J. Org. Chem.*, **26**, 4563 (1961).

1069. M. Mousseron and R. Jacquier, *Bull. Soc. Chim. France*, **1950**, 698.

1070. L. L. Smith, J. J. Goodman, H. Mendelsohn, J. P. Dusza, and S. Bernstein, *J. Org. Chem.*, **26**, 974 (1961).

1071. B. Belleau and T. F. Gallagher, *J. Amer. Chem. Soc.*, **74**, 2816 (1952).

1072. C. D. Hurd and O. E. Edwards, *J. Org. Chem.*, **14**, 680 (1949).

1073. M. Bergmann and A. Miekeley, *Ber.*, **54**, 2153 (1921); *ibid.*, **62**, 2297 (1929).

1074. H. Fernholz, *Ber.*, **84**, 110 (1951).

1075. R. E. Parker, and N. S. Isaacs, *Chem. Rev.*, **59**, 737 (1959).

1076. A. F. Plate and A. A. Melnikov, *Zh. Obshch. Khim.*, **30**, 935 (1960).

1077. A. F. Plate, A. A. Melnikov, T. A. Italinskaya, and R. A. Zelenko, *Zh. Obshch. Khim.*, **30**, 1250 (1960).

1078. J. P. Vila and R. Crespo, *Anales Real. Soc. Espan. Fis. y Quim.*, **48B**, 273 (1952).

1079. A. F. Plate, A. A. Melnikov, and A. A. Ovezova, *Zh. Obshch. Khim.*, **30**, 1256 (1960).

1080. R. N. McDonald and P. A. Schwab, *J. Amer. Chem. Soc.*, **85**, 820 (1963).

1081. R. N. McDonald and P. A. Schwab, *J. Amer. Chem. Soc.*, **85**, 4004 (1963).

1082. R. N. McDonald and P. A. Schwab, *J. Org. Chem.*, **29**, 2458 (1964).

1083. S. L. Friess and V. Boekelheide, *J. Amer. Chem. Soc.*, **71**, 4144 (1949).

1084. O. H. Wheeler, *J. Amer. Chem. Soc.*, **75**, 4858 (1953).

1085. M. T. Dangyan and J. V. Arakelyan, *Nauch. Tr. Erevan. Gosudarst. Univ., Ser. Khim. Nauk* **53**, 3 (1956).

1086. M. T. Dangyan and M. G. Zalinyan, *Nauch. Tr. Erevan. Gosudarst. Univ., Ser. Khim. Nauk*, **53**, 15 (1956); *ibid.*, **25**, 1956).

1087. M. T. Dangyan and G. M. Shakhnazaryan, *Zh. Obshch. Khim.*, **31**, 1643 (1961).

1088. M. T. Dangyan and E. G. Mesopyan, *Izv. Akad. Nauk Armyan. SSR, Khim. Nauki*, **14,** 147 (1961).
1089. G. Berti, *J. Org. Chem.*, **24,** 934 (1959).
1090. G. Baddeley and P. Hulme, *J. Chem. Soc.*, **1965,** 5148.
1091. J. Meinwald, S. S. Labana, and M. S. Chadha, *J. Amer. Chem. Soc.*, **85,** 582 (1963).
1092. M. Gates and S. P. Malchick, *J. Amer. Chem. Soc.*, **76,** 1378 (1954).
1093. L. Kaplan, H. Kwart, and P. von R. Schleyer, *J. Amer. Chem. Soc.*, **82,** 2341 (1960).
1094. S. Winstein and R. Baird, *J. Amer. Chem. Soc.*, **79,** 756 (1957).
1095. D. C. Kleinfelter and P. von R. Schleyer, *J. Amer. Chem. Soc.*, **83,** 2329 (1961).
1096. H. Riviere, *Bull. Soc. Chim. France*, **1964,** 97.
1097. W. E. Rosen, L. Dorfman, and M. P. Linfield, *J. Org. Chem.*, **29,** 1723 (1964).
1098. R. E. Parker and N. S. Isaacs, *Chem. Rev.*, **59,** 737 (1959).
1099. H. Hart and L. R. Lerner, *Tetrahedron*, **25,** 813 (1969).
1100. (a) I. J. Borowitz and G. J. Williams, *Tetrahedron Letters*, **1965,** 3813.
      (b) H. Immer and J. F. Bagli, J. Org. Chem. *33,* 2457 (1968) and Can. J. Chem. *46,* 3115 (1968).
1101. I. J. Borowitz, G. Gonis, R. Kelsey, R. Rapp, and G. J. Williams, *J. Org. Chem.*, **31,** 3032 (1966).
1102. A. Padwa and R. Hartman, *J. Amer. Chem. Soc.*, **88,** 3759 (1966).
1103. M. Mousseron-Canet, M. Mousseron, and C. Levallois, *Compt. Rend.*, **253** 1386 (1961).
1104. E. Klein, H. Farnow, and W. Rojahn *Ann.*, **675,** 73 (1964).
1105. M. Mousseron-Canet and C. Levallois, *Bull. Soc. Chim. France*, **1965,** 1339.
1106. M. Mousseron-Canet, C. Levallois, and H. Huerre, *Bull. Soc. Chim. France*, **1966,** 658.
1107. E. Demole and H. Wuest, *Helv. Chim. Acta*, **50,** 1314 (1967).
1108. U. Scheidegger, K. Schaffner, and O. Jeger, *Helv. Chim. Acta*, **45,** 400 (1962).
1109. H. H. Wasserman, R. E. Cochoy, and M. S. Baird, *J. Amer. Chem. Soc.*, **91,** 2375 (1969).
1110. R. Criegee and K. Noll, *Ann.* **627,** 1 (1959).
1111. F. L. Weisenborn and D. Taub, *J. Amer. Chem. Soc.*, **74,** 1329 (1952).
1112. D. Y. Curtin and A. Bradley, *J. Amer. Chem. Soc.*, **76,** 5777 (1954).
1113. L. H. Zalkow, E. J. Eisenbraun, and J. N. Shoolery, *J. Org. Chem.*, **26,** 981 (1961).
1114. J. Lhomme and G. Ourisson, *Bull. Soc. Chim. France*, **1964,** 1888.

*Chapter VI*

# Determination of Organic Peroxides by Physical, Chemical, and Colorimetric Methods

R. D. MAIR & R. T. HALL

**535**

---

## I.  INTRODUCTION

The detection and estimation of organic peroxides have always had important roles in the chemical technology of these compounds. Typically there may exist a need to detect and monitor traces of a particular peroxide compound, to resolve mixtures of several peroxide functions, or to confirm the reliability of an assay for purity. Coping with such problems at more than superficial depth soon reveals that the analysis of peroxides has a considerable complexity. Many pitfalls are present, along with numerous and challenging opportunities.

The complexity is rooted in the diversity of peroxide classes and also in the dual nature of these compounds: besides containing the characteristic —O—O— linkage, they also possess many of the physical and chemical properties of their nonperoxy analogs (e.g., hydroperoxides are related to alcohols, peroxy esters to esters).

Despite the discovery of numerous reactions for determining particular types of peroxides, however, and the proliferation of procedures that these reactions have spawned, there have yet to be devised any general methods applicable to all categories of peroxides. Such a goal has remained elusive and is probably unattainable. The use of a series of coordinated methods and techniques is more realistic.

The literature on peroxide analysis is not only extensive but cumbersome— a large body of knowledge arising from diverse fields of research that often seem to have little overlapping of basic interests. The authors have had recent occasion to cope with this problem while preparing a chapter, "Determination of Organic Peroxides," for the *Treatise on Analytical Chemistry* (1). The present treatment is an abridged version of that chapter. Other useful sources include books by Davies (2) and Hawkins (3), and reviews by Milas (4) and Martin (5).

In this chapter we attempt to summarize the analytical chemistry of organic peroxides as it is presently known. We have made an effort to satisfy both the reader interested in a broad treatment of the whole field of peroxide analysis and the chemist with a more specific objective, perhaps a recommended method for one peroxide type or a particular mixture of types, fresh ideas for the development of specific process-control methods, or a survey of methods for trace peroxide analysis in various systems. We do not discuss particular systems in detail. Rather our aim is to enable a chemist who is confronted with a specific analysis problem to select from these offerings a solution that will be appropriate and practical. We begin by discussing the relative analytical reactivities of the classes of peroxides, and their stabilities. This is followed by a section on the separation of peroxides from nonperoxides and from other peroxides, and on the conversion of peroxides to derivatives. Next comes the main body of the chapter, a discussion of the determination of organic peroxides by established or promising techniques. This material is presented under the three headings of "Physical and Instrumental Methods," "Chemical Reduction Methods," and "Colorimetric and Photometric Methods." The treatment of individual topics varies considerably in length, reflecting the relative significance of the method or technique and the literature references available. A final section outlines a selection of recommended procedures.

## II. REACTIVITY AND STABILITY OF ORGANIC PEROXIDES

### A.   Relative Analytical Reactivities

Analysts classify peroxides by reactivity toward analytical methods or reagents rather than by structure. From this viewpoint, we have arranged the peroxide classes in order of apparent decreasing ease of reduction. The sequence, based on peroxide analyses by the polarograph (6–8), by catalytic hydrogenation (9), and by reaction with iodide (9, 10), is as follows:

1. Peroxy acids
2. Diacyl peroxides
3. Hydroperoxides
4. Ozonides
5. Peroxy esters
6. Di-*n*-alkyl, di-*sec*-alkyl (and mixed) peroxides; olefin–oxygen copolymers; 1,4-epiperoxides; dimeric ketone peroxides
7. Di-*t*-aralkyl peroxides; endoperoxides
8. Trimeric ketone peroxides; di-*t*-alkyl peroxides

Our laboratory has recently proposed another simple classification (10) based on response to the widely used boiling sodium iodide–isopropyl alcohol reagent: *easily reduced* (stoichiometric iodine liberation within 5 min); *moderately stable* (partial reaction in 5 min); and *difficult-to-reduce* peroxides (inert to this reagent). This classification is the basis of our later discussion of iodine-liberation methods (Section IV-B-1-c, e, and f).

### B.   Stability

Among commercial peroxides, stability plays two key roles: it determines individual safety hazard ratings, and it defines the areas of technological utility.

The study of hazards from peroxides and of techniques for the safe handling of these materials in plant and laboratory has kept space with their commercial development. Several publications have recently reviewed the available information (11–16). Siemens categorized the hazardous properties of peroxides as flammability, thermal sensitivity, mechanical sensitivity, and explosiveness (16), a list to which Noller and Bolton have added physiological effects (13). Safe handling and storage of organic peroxides in the laboratory have been discussed by these authors (13) and by Sambale (15). Noller and associates (14) have proposed a relative hazard classification for organic peroxides based on specific test procedures; this summarizes the hazard potential for some 40 forms of the 26 peroxides (from seven classes or subclasses) that are in widest commercial use. Especially valuable in this subject area is a recent publication of ASTM Committee E-15, by Castrantas, Banerjee, and Noller (11).

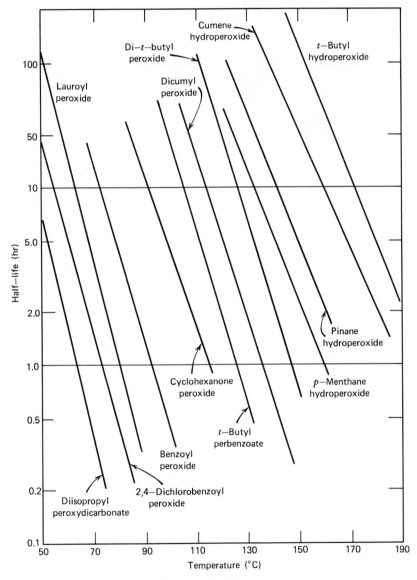

**Figure 1.** Relative thermal stabilities of various peroxides.

Chemical stability is important in determining areas of utility for hydroperoxides, and to some extent for peroxy acids and peroxy esters. The property of overriding significance to most commercial organic peroxides, however, is thermal stability, because end uses depend on their ability to dissociate into active free radicals at a technologically desired temperature. Such thermal decompositions give linear graphs of log half-time versus $1/T$ (°K). Over limited temperature ranges, a plot of log half-time versus $T$ (°C) is also substantially linear. We have used this arrangement in Figure 1 to illustrate the broad range of activity that is available.

It may seem surprising that alkyl hydroperoxides are the most thermally stable class, since they are chemically much less stable than dialkyl peroxides or peroxy esters. It is interesting also that thermal dissociation of hydroperoxides into radicals is greatly enhanced by the presence of di-$t$-butyl peroxide (17). For instance, a 1:1 mixture of the latter with cumene hydroperoxide or $t$-butyl hydroperoxide decreases the decomposition half-times of the hydroperoxides 60 and 450 times, respectively.

Doehnert and Mageli (18) list activation energies and frequency factors for many peroxides, and Guillet and co-workers (19) have shown how these parameters of the Arrhenius equation vary with molecular structure for the diacyl peroxides.

Molyneux (20) has recently published a careful evaluation of the literature data on Arrhenius perameters for dialkyl or aralkyl peroxides, based on the assumed homolytic, unimolecular scission:

$$ROOR' \longrightarrow RO\cdot + R'O\cdot$$

Theoretically the activation energy $E$ for this relatively weak bond should vary much less within this class than is indicated by the observed range of 31–38 kcal/mole between diethyl and di-$t$-butyl peroxides. Likewise, the frequency factors $A$ should all be close to $10^{13}$ sec$^{-1}$, whereas they are found to vary between $10^{13}$ and $10^{16}$ sec$^{-1}$. Furthermore, there is a linear "isokinetic" relationship between log $A$ and $E$ that implies a temperature (210°) at which all RO—RO' peroxides should have identical rate constants, $k = 0.1$ sec$^{-1}$. Attempts to rationalize the existing data have been unsuccessful (20).

### III.  SEPARATION AND CONVERSION OF ORGANIC PEROXIDES

The judicious and imaginative use of a variety of separation techniques can greatly simplify the determination of organic peroxides in complex systems. The simpler fractions thus produced may make an ambiguous situation clearcut, ready for straightforward analysis; or the separation procedure by itself may provide adequate analytical information. Alternatively, converting the peroxide or peroxides into other more readily determined compounds may achieve the desired analysis more readily. The intent of this section is not to present complete specific separation or conversion procedures, but rather to stimulate awareness of such possibilities by discussing a number of them briefly.

In analysis, situations requiring the separation of peroxides are widely varied and may include (a) separation at concentrations ranging from trace (e.g., removal of peroxidic contaminants from analytical solvents) to very high (e.g., removal of impurities from organic peroxides to be used as analytical standards; (b) separation of a specific class of peroxides from other peroxides and nonperoxides (e.g., isolation of peroxy acids from acid mixtures by selective extraction, or separation of dialkyl peroxides from hydroperoxides in a variety of ways); or (c) separation between members of the same class (e.g., resolution of the hydroperoxide mixtures produced in hydrocarbon autoxidation or the alkylidene peroxide mixtures produced by the reaction of hydrogen peroxide with aldehydes and ketones). In such separations one should utilize the chemical and physical properties of both the wanted and the unwanted components. One should be alert also to the possibility that two or more different approaches may be feasible and should consciously try to select that best suited to the situation.

All of the usual laboratory separation techniques can be used in organic peroxide analysis. Extraction (liquid–liquid partition) in separatory funnels may suffice, by virtue of its simplicity and predictability (even though the separations be less than optimal), when only limited effort can be justified. Laboratories possessing appropriate countercurrent distribution apparatus, such as the Craig–Post equipment or a Scheibel column, can obtain very effective separations of complex mixtures, including commercial peroxide preparations, for both analytical (referee) and preparative purposes. Precipitation-filtration will apply in the separation of $t$-hydroperoxides with concentrated alkali or the purification of a peroxide by recrystallization. Distillation is not widely employed except as a step in preparative work, having been generally replaced in analysis by gas chromatography. Chromatography is applicable in all its forms, and we consider it separately in Section IV-A-1.

### A. Removal of Peroxides from Solvents

Probably the earliest separation problem with peroxides was their removal from organic solvents, especially ethers, in which peroxides, formed by autoxidation, are contaminants and constitute a well-recognized and ever-present hazard. Chemical methods, summarized by Davies (2, Ch. 14) and Hawkins (3, Ch. 11), include treatment of the solvents with alkalies, sodium sulfite, acidic ferrous sulfate, stannous chloride, lead dioxide, zinc and acid, sodium and alcohol, lithium aluminum hydride, and triphenyl phosphine. Treatment with solid cerous hydroxide, as described by Ramsey and Aldridge (21), appears to be particularly effective for ethers. Shaking with copper-treated zinc dust (22) also has been recommended (23). Percolation through an active alumina column, introduced by Dasler and Bauer (24), is undoubtedly among the simplest, most effective, and most generally applicable solvent-purification techniques; for example, it removes not only peroxides but most other oxygenated impurities from hydrocarbons. Column purification on strongly basic ion-exchange resins (25) also has been successful for ethers, and peroxide reduction on resin columns in the sulfite form has been suggested (2, Ch. 14).

### B.  Separation of Peroxides from Mixtures by Means of Alkali

A widely used method for purifying hydroperoxides is to precipitate and isolate their sodium salts and then regenerate the hydroperoxide with a weak acid; this approach has been reviewed by Davies (2, Ch. 8). Primary and secondary hydroperoxides require low temperature and prompt regeneration in order to minimize their base-catalyzed decomposition to carbonyl compounds (2, Ch. 13; 26; 27). Tertiary hydroperoxides are more stable but require the use of substantial excess alkali to avoid formation of alcohol through slow elimination of oxygen (2, Chs. 8 and 13).

A variation in behavior is shown by $t$-butyl hydroperoxide. Although this compound dissolves without decomposition, neither its sodium nor its potassium salt precipitates from concentrated aqueous alkali solutions. An alcoholic medium produces opposite results, however: cooling hot solutions of $t$-butyl hydroperoxide and sodium or potassium butoxide in $t$-butyl alcohol (28) or diluting ice-cold solutions in concentrated ethanolic alkali with a large volume of cold acetone (29) causes precipitation of stable salts (containing alcohol of crystallization).

The concentration of the aqueous alkali has an effect on the consistency of the precipitated salts: with cumene hydroperoxide, concentrations below 20% sodium hydroxide produce slimy precipitates; with $p$-cymene hydroperoxides, to produce precipitates of good consistency we recommend 40–50% sodium hydroxide at ice temperatures. Our preferred procedure is to add a few milliters of 40% sodium hydroxide with vigorous stirring to a cold, dilute petroleum ether solution of the mixture. Sintered-stainless-steel funnels are better than sintered glass for filtrations involving such concentrated alkali.

The alkali-precipitation technique is a convenient way of isolating the impurities from concentrated hydroperoxides for further analysis. For instance, dicumyl peroxide and α-α,dimethylbenzyl alcohol can be separated almost quantitatively from 70% cumene hydroperoxide and each determined differentially in the presence of residual hydroperoxide by using a series of three different iodine-liberation procedures (10).

Partition between more dilute aqueous alkali and an immiscible organic solvent is useful in the analysis of autoxidised hydrocarbons, particularly aralkyl types, such as the disopropylbenzenes and triisopropylbenzenes. Typical distribution data on some pure components obtained in our laboratories are shown in Table 1. With these compounds no decomposition of hydroperoxide groups is observed in sodium hydroxide–hydrocarbon or sodium hydroxide–ether partition systems. Hydroperoxide loss occurs with sodium hydroxide–methyl isobutyl ketone, however.

TABLE 1. DISTRIBUTION COEFFICIENTS FOR SOME AUTOXIDIZED
*p*-DIISOPROPYLBENZENE COMPONENTS

| Solvent Pair | | Distribution Coefficients[a] for the Oxidate Components | | | | |
|---|---|---|---|---|---|---|
| Lower Phase | Upper Phase | Mono- alcohol | Monohydro- peroxide | Dialcohol | Alcohol Hydro- peroxide | Dihydro- peroxide |
| 7% Aqueous NaOH | Benzene | 0.0084 | 0.042 | 0.85 | 7.7 | 145 |
| 7% Aqueous NaOH | Methyl-ethyl ketone | — | — | — | 0.20 | 3.5 |
| 10% Aqueous NaOH | Ether | — | — | — | 0.6 | 15.1 |
| 20% Aqueous NaOH | Ether | — | — | — | 0.3 | 16.6 |
| 20% Aqueous NaOH | Methyl isobutyl ketone | — | — | — | <0.1 | 7.7 |

[a] Ratio of the concentration in the lower (alkali) phase to the concentration in the upper phase.

## C. Conversion of Peroxides to Derivatives

Peroxides may be converted into derivatives that are subsequently identified and determined in lieu of the parent peroxides. The most common approach, much used for proof of structure in studies of oxidation processes, is chemical reduction. There are many effective reagents (29–35), including thiophenol, sodium sulfite, lithium aluminum hydride, sodium borohydride, stannous chloride, zinc and acetic acid, zinc and hydrocloric acid, and sodium toluene-*p*-sulfinate and pyridine, as well as catalytic hydrogenation over nickel, platinum, or palladium. Trialkyl phosphines are especially useful, since they can be relied on to convert virtually every peroxide into the analog that corresponds to the plucking out of one of the oxygen atoms from its —O—O— bond.

Formation of derivatives from which the peroxide can be regenerated is frequently useful. In addition to salt formation with alkali, discussed above, this technique includes isolation of alkyl hydroperoxides as xanthhydryl or triphenylmethyl derivatives (36) and as derivatives of substituted cyclic immonium pseudo bases, salts, or ethers (37); for example, derivatives useful for hydroperoxide separation and characterization are formed with 1-methyl-6,8-dinitro-2-ethoxy-1,2-dihydroquinoline.

## IV.  DETERMINATION OF ORGANIC PEROXIDES

In this section we consider the many chemical reactions and physical techniques that are available for detecting and determining organic peroxides, including the many established methods as well as promising approaches that have not yet become widely used. These are listed in Table 2, in three sections. One contains physical or instrumental methods, based principally on physical (including spectral) properties, and the other two list chemical methods based on reaction with reducing agents or colorimetric reagents.

### A.  Physical and Instrumental Methods

Physical or instrumental methods for determining organic peroxides have been summarized in Table 2A.

### 1.  Chromatographic Techniques

(a)  GAS CHROMATOGRAPHY

Because instrumentation for it is so highly developed and widely available today, gas chromatography contends for the position of the preferred method for peroxide determination wherever it is applicable—that is, for the analysis of the peroxides that are sufficiently volatile and stable. Some successful applications are listed in Table 3; in many instances these include isolation and purification on a preparative scale. Bukata and co-workers (39) have summarized the earlier literature; Hahto and Beaulieu (104) list a more extensive bibliography.

Most separations have been on $t$-alkyl hydroperoxides and di-$t$-alkyl peroxides (38, 39), although primary and secondary hydroperoxides have been both analyzed and purified (106, 108, 111), and the dangerous diacetyl peroxide has been purified and characterized (104). The last two compounds in Table 3 probably approach the practical molecular-weight limit for the gas chromatography of peroxides. Bukata and co-workers (39) used a column temperature of 138° to separate this pair and reported relative standard deviations of 1%. The more volatile di-$t$-butyl peroxide had better precision, with a relative standard deviation of 0.3%.

TABLE 2. ESTABLISHED OR PROMISING METHODS FOR DETERMINING ORGANIC PEROXIDES

*A. Determination by Physical or Instrumental Means*

| Method or Technique | Applicable Peroxide Type | Applicable Concentration Range | Useful Features | Principal References |
|---|---|---|---|---|
| Gas chromatography | Relatively volatile and thermally stable | Trace to major component | Simple and reliable when applicable | 38, 39 |
| Liquid column chromatography | General | Trace to major component | Can purify standards, analyze mixtures | 40 |
| Thin-layer chromatography | All except relatively volatile | Trace to major component | Able to resolve mixtures of polyfunctional compounds | 41–45 |
| Paper chromatography | All except relatively volatile | Trace to major components | Able to resolve mixtures of polyfunctional compounds | 38, 46–50 |
| Pyrolysis–gas chromatography | Broad potential | Low to high | Quick assay for unreactive peroxides and relatively simple mixtures | 51 |
| Mass spectrometry | General—need slight volatility | Low to high | Where applicable can analyze mixtures directly | — |
| Infrared and near-infrared absorption | General, but accent on hydroperoxides | Moderately low to high | Can distinguish some ROOH's from analogous ROH's | 7, 52–59 |

TABLE 2—*continued*

| Method or Technique | Applicable Peroxide Type | Applicable Concentration Range | Useful Features | Principal References |
|---|---|---|---|---|
| Ultraviolet absorption | Aromatic peroxides | Low to high | Combines well with silicic acid column chromatography | 60 |
| Nuclear magnetic resonance | Broad potential | Above 10% | Structure determination; quick method for assay in crude autoxidation mixtures | 61–63 |
| Electron paramagnetic resonance | Accent on hydroperoxides | Trace to moderate | Quick determination, possible selectivity | 64 |
| Differential thermal analysis | All types | Moderately low to high | Quantitative analysis, sometimes of mixtures, convenient for determining thermal decomposition characteristics | 65 |
| Polarography (reduction with electrons) | All except di-t-alkyl and cyclic trimeric peroxides | Trace to major component | Convenient for following kinetics of peroxide formation and decomposition in chemical processing and autoxidation | 5, 6, 66–72 |

B. Determination by Chemical Reaction with Reducing Agents

| Reaction | Applicable Peroxide Type | Applicable Concentration Range | Useful Features | Principal References |
|---|---|---|---|---|
| Reduction with iodide | General | Trace to major component | Variety of procedures to fit many peroxide types | 3, Ch. 11; 5; 10 |
| Reduction with diphenyl sulfide | Peroxy acids only | Low to high | Specific reaction | 73, 74 |
| Reduction with triethyl arsine | Peroxy acids, diacyl peroxides, and alkyl hydroperoxides | Low to high | Differentiates diakyl peroxides | 73 |
| Reduction with tertiary phosphines | General | Low to high | Derivative preparation for gas chromatography | 32, 75 |
| Reduction with the stannous ion | Peroxy acids (?), diacyl peroxides, and alkyl hydroperoxides | Low to high | Avoids interference from organic sulfur compounds | 76, 77 |
| Reduction with the arsenite ion | Peroxy acids (?), diacyl peroxides, and alkyl hydroperoxides | Trace to high | Avoids interference from organic sulfur compounds | 78, 79 |
| Oxidation with $KMnO_4$ | Cyclohexanone hydroperoxides | Low to high | Active catalyst content | 80 |
| Reduction with hydrogen (catalytic) | General | Low to high | Derivative preparation | 2, Ch. 14; 9; 81; 82 |
| Reduction with $LiAlH_4$ | General | Low to high | Derivative preparation | 34, 83, 84 |
| Reduction with the ferrous ion[a] | — | — | — | |

*C. Determination by Reaction with Colorimetric Reagents*

| Reagent | Applicable Peroxide Type | Applicable Concentration Range | Useful Features | Principal References |
|---|---|---|---|---|
| Ferrous thiocyanate | Easily reduced | Trace to low | Rapid, precise, sensitive | 85–87 |
| Leuco chlorophenolindophenol[a] | Easily reduced | Trace to low | — | 86, 88, 89 |
| N,N′-diphenyl-p-phenylene-diamine | Easily reduced | Trace to low | Rapid, precise, sensitive | 90 |
| Leuco methylene blue | Easily reduced | Trace to low | Constant molar response, precise, very sensitive | 91–94 |
| N-benzoyl leuco methylene blue | Easily reduced | Trace to low | Reagent unaffected by air | 95 |
| sym-Diphenylcarbohydrazide | Hydroperoxides and diacyl peroxides | Trace to low | Rapid, precise, sensitive | 96 |
| Iodide | General | Trace to low | Constant molar response, precise, sensitive, widely applicable | 97–100 |
| Titanium(IV)[a] | Hydroperoxides and peroxy esters | Trace to low | — | 101–103 |

[a] Discussed but not recommended.

TABLE 3. SOME PEROXIDES DETERMINED BY GAS CHROMATOGRAPHY

| Compound | Reference |
|---|---|
| Diacetyl peroxide | 104 |
| Peroxyacetyl nitrate | 105 |
| Peroxypropionyl nitrate | 105 |
| Peroxybutyryl nitrate | 105 |
| Ethyl hydroperoxide | 106 |
| Isopropyl hydroperoxide | 106 |
| Allyl hydroperoxide | 107 |
| *n*-Butyl hydroperoxide | 108 |
| Isobutyl hydroperoxide | 61, 108 |
| *t*-Butyl hydroperoxide | 38, 39, 106 |
| *t*-Pentyl hydroperoxide | 38, 39 |
| *trans*-2-Methyl-3-pentenyl-2-hydroperoxide | 109 |
| 4-Methyl-3-pentenyl-2-hydroperoxide | 109 |
| *t*-Butyl peroxyacetate | 39 |
| *t*-Butyl peroxyisobutyrate | 39 |
| Diisopropyl peroxide | 110 |
| Di-*t*-butyl peroxide | 38, 39 |
| Di-*t*-pentyl peroxide | 38 |
| 2,5-Dimethyl-2,5-di(*t*-butylperoxy)hexane | 39 |
| 2,5-Dimethyl-2,5-di(*t*-butylperoxy)-3-hexyne | 39 |

A remarkably sensitive gas-chromatography method for peroxides is the determination of the peroxyacyl nitrates (along with many other nitrogen oxide derivatives) in air (105). These peroxides have been identified by Stephens and co-workers (112, 113) as primary eye irritants and phytotoxicants occurring in smog. Because of the extreme sensitivity of nitrate groups to the electron-capture mode of gas-chromatography detection, both peroxyacetyl nitrate and the even more damaging peroxypropionyl nitrate can be determined in a 2-ml sample of air at concentrations below 5 ppb.

Brill (109) used preparative and analytical gas chromatography as the basis for an important study in which he showed that the isomerization of hydroperoxides can occur readily in dilute nonpolar solutions. The compounds (*trans*-2-methyl-3-pentenyl-2-hydroperoxide, **1**, and 4-methyl-3-pentenyl-2-hydroperoxide, **2**) in the following equilibrium are readily interconverted:

$$
\underset{\textbf{(1)}}{\underset{\underset{OOH}{|}}{\overset{\overset{CH_3}{|}}{H_3C-C-CH=CH-CH_3}}} \;\;\rightleftharpoons\;\; \underset{\textbf{(2)}}{\underset{\underset{OOH}{|}}{\overset{\overset{CH_3}{|}}{H_3C-C=CH-CH-CH_3}}}
$$

The rate is inversely proportional to concentration, apparently an effect of hydrogen bonding. Brill's results cast doubt on the interpretation of some studies of the mechanism of autoxidation.

Hyden (51) has taken another approach to the use of gas chromatography in the analysis of peroxides, one that has many intriguing possibilities. A peroxide sample solution is injected directly into the *hot* injection port of a gas chromatograph, where the peroxide decomposes instantly; separation of its volatile decomposition products can then serve both to identify and quantitatively determine it. Hyden reports excellent results for di-*t*-butyl peroxide, which decomposes into ethane and acetone at 310°.

Apparently related is the method of Courtier (115), who used columns as hot as 200° and injection ports as hot as 250° for identifying both the peroxides and the diluting solvents in samples of commercial catalyst preparations for polyester polymerization.

Pyrolysis methods are not restricted to peroxides, of course, but are applicable to analysis in general. Characteristic features are the strong tendency of pyrogram patterns to vary with pyrolysis conditions and the consequent need for a pyrolysis technique that is both flexible and precise. The recent approach of Simon and Giacobbo (114), which employs high-frequency induction heating of samples as thin films on ferromagnetic conductors, is a long step toward such control; in their apparatus the Curie temperature of the conductor automatically provides a precisely reproducible pyrolysis temperature. Analytical chemists should consider broad development of the pyrolysis–gas chromatography approach, possibly coupled to liquid-column or thin-layer chromatography, and monitored when necessary by mass spectrometry, as has recently been described (116).

### (b) Liquid-Column Chromatography

Both adsorption and liquid–liquid partition chromatography methods are effective in the separation and analysis of peroxides. In adsorption chromatography separations are due to differences in the strength of adsorptive forces that hold the various components to the adsorbing substrate. Eggersglüss (40) has studied such factors extensively for peroxides. On the basis of 37 pure compounds from six peroxide classes, run on columns of seven individual adsorbents and eluted with three solvents, he concluded that adsorption is a function of the number and polarity of functional groups and that the relative strength of adsorption increases throughout the following series:

1. Dialkyl peroxide, R—OO—R
2. Hydroxyperoxide, R—C(OH)H—OO—R
3. Hydroperoxide, R—OOH

4. Peroxy acid, $R-C(O)-OOH$
5. Dihydroxyperoxide, $R-C(OH)H-OO-C(OH)H-R$
6. Hydroxyhydroperoxide, $R-C(OH)H-OOH$
7. Hydrogen peroxide, $HOOH$

The effect of moisture in the eluting solvent should be considered in peroxide separations on adsorption columns; too much water may decrease the adsorptive power of the substrate for specific compounds. Decomposition of peroxides on adsorbing substrates is also a drawback in this type of separation. Alkaline alumina is a prime offender, although it is said to be improved by pretreatment with acetic acid (2, Ch. 14). Davies has suggested (2, Ch. 14) that the ion-exchange resins used in ether purification (26) might separate hydroperoxide components of mixtures; gradient elution by a buffer of continuously varying pH should be considered in this approach.

In our laboratories, an effective scheme for the analysis of autoxidized isopropylbenzenes combines elution from silicic acid adsorption columns with identification and determination of the individual components (including nonperoxides) by ultraviolet absorption.*

The distribution coefficients in Table 1 suggest that liquid–liquid partition chromatography in aqueous alkali–organic systems should be effective in resolving peroxide mixtures, as has proved to be the case. Nonalkaline systems should also be operable, based on distribution coefficients in water–ether and water–petroleum ether systems tabulated by Eggersglüss for all six peroxide classes used in his adsorption study (40).

We have used a 20% aqueous sodium hydroxide–Celite column successfully in the analysis of isopropylbenzene oxidates.† To prepare a typical column we mix 10 ml of 20 wt % aqueous sodium hydroxide into 10 g of Celite, with a mortar and pestle, and pack it into a No. 2 chromatographic column; elution with hexane, diethyl ether, and ethyl–acetate acid in succession gives quantitative recoveries of monohydroperoxide, alcohol hydroperoxide, and dihydroperoxide from both *m*- and *p*-diisopropylbenzene oxidates.

In our experience with such ultraviolet-absorbing samples the silicic acid adsorption column coupled with ultraviolet analysis is more convenient and useful than partition systems. Particularly effective has been the UV Scanalyzer, an automatic gradient-elution column chromatograph developed in our laboratories by Kenyon and co-workers (60) and used here steadily for more than 10 years. Recently introduced detectors for liquid-column

---

*Private communication from A. A. Orr, Hercules Incorporated, Research Center, Wilmington, Delaware (1960).

†Private communication from A. J. Graupner, Hercules Incorporated, Wilmington, Delaware (1954).

chromatography based on differential refractometry (117), differential heats of adsorption (118), differential vapor osmometry (119), and continuous residue pyrolysis (120, 121), along with rapidly developing improvements in techniques and in commercially available hardware, promise broader utility for liquid-column separations in the analysis of peroxides and nonperoxides.

### (c) Paper Chromatography

A number of workers have developed applications of paper chromatography to peroxide analysis. The principal contributions are summarized in Table 4. Normal systems employ untreated paper (48) and paper coated with dimethyl-formamide (48) or ethylene glycol (46). In reversed-phase applications both silicone-oil-loaded (38, 46, 47) and acetylated (50) papers have been used. The success of these investigators in varying $R_f$ values widely by modifying the composition of stationary and mobile phases is a clear invitation to other peroxide analysts to try similar innovations.

Milas and Golubovic described a practical extension of this technique (49), wherein they scaled up a paper-chromatography separation (48) to a preparative scale on a cellulose powder–dimethylformamide column and eluted with pentane. They separated and determined the structures of six peroxides formed in the acid-catalyzed reaction of diethyl ketone and hydrogen peroxide.

$$\text{1.} \quad \underset{|}{\overset{R_2}{\phantom{|}}} \text{HOOC—OOH}$$

1.  $\overset{\displaystyle R_2}{\underset{\displaystyle |}{}}$
    HOOC—OOH

2.  $\overset{\displaystyle R_2}{\underset{\displaystyle |}{}}\quad\overset{\displaystyle R_2}{\underset{\displaystyle |}{}}$
    HO—C—OO—C—OOH

3.  $\overset{\displaystyle R_2}{\underset{\displaystyle |}{}}\quad\overset{\displaystyle R_2}{\underset{\displaystyle |}{}}$
    HOOC—OO—C—OOH

4.  $\overset{\displaystyle R_2}{\underset{\displaystyle |}{}}\quad\overset{\displaystyle R_2}{\underset{\displaystyle |}{}}\quad\overset{\displaystyle R_2}{\underset{\displaystyle |}{}}$
    HOO—C—OO—C—OO—C—OOH

5.  $\overset{\displaystyle R_2}{\underset{\displaystyle |}{}}\quad\overset{\displaystyle R_2}{\underset{\displaystyle |}{}}\quad\overset{\displaystyle R_2}{\underset{\displaystyle |}{}}\quad\overset{\displaystyle R_2}{\underset{\displaystyle |}{}}$
    HOO—C—OO—C—OO—C—OO—C—OOH

6.  A cyclic trimer peroxide

This approach deserves consideration, particularly for complex peroxide systems.

## (d) THIN-LAYER CHROMATOGRAPHY

Stemming mainly from the work of Stahl (122), thin-layer chromatography (TLC) increasingly competes with paper chromatography for a leading role in the separation of many compounds, including organic peroxides. The first significant work in this area was by Knappe and Peteri (44), who applied TLC to five important peroxide classes. They studied the separating ability of many mobile phases on silica gel and selected a pair that in combination usually gave unequivocal identification of a particular peroxide (confirmed by running a known). Still a third solvent mixture could resolve peroxides that had identical $R_f$ values with both of the standard mobile phases. These TLC systems and $R_f$ values obtained by means of them for 18 peroxides are listed in Table 5. Except for diacyl peroxides, which move relatively rapidly, the peroxides did not separate systematically by classes. Commercial ketone peroxides proved to be mixtures of several components.

Another pertinent study is that of Hayano and co-workers (43), who studied a series of seven compounds representing four classes of organic peroxides. These workers compared the performance of Knappe and Peteri's solvent systems with that of their own trio of three-component solvents, on silica gell. The latter systems and the $R_f$ values obtained by using them are presented in Table 6, and $R_f$ values obtained with the Knappe and Perteri solvents have been inserted in Table 5. Notice that in that case agreement between the two groups of workers is good for a pair of hydroperoxides, but discrepancies for three diacyl peroxides are sizable. Hayano and co-workers reported that the $R_f$ values in Table 6 are quite precise, having standard deviations in general below 0.05 $R_f$ units.

The most important contributions to the TLC or organic peroxides appear to be the comprehensive studies of Buzlanova and co-workers (41, 42), who reported on a total of 39 peroxides in five classes. Table 7 summarizes the solvent systems used by these workers [heptane–ether (15:1) and toluene–methanol (20:3)] and the $R_f$ values found on silica gel for some 26 peroxides, representing four major classes (41).

Data are presented also that were obtained with the toluene–methanol solvent on alumina; these results are similar to the data from silica gel, although the spots were more diffuse. Buzlanova and co-workers report partial decomposition on alumina for diacyl peroxides (41) and for ketone peroxides (42).

The peroxide spots were located by means of a spray reagent of $N,N$-dimethyl-$p$-phenylenediamine hydrochloride in methanol (128 ml methanol, 25 ml water, 1 ml glacial acetic acid, and 1.5 g of the amine hydrochloride). This reagent was said to be selective for detecting organic peroxides, giving a red-violet spot against a light background, and to have a sensitivity of 1 $\mu$g

TABLE 4.   SUMMARY OF. THE PRINCIPAL SYSTEMS FOR PAPERCHROMATOGRAPHY
SEPARATIONS OF ORGANIC PEROXIDES

| Type | Stationary Phase | Moving Solvent |
|---|---|---|
| Reversed phase | Whatman No. 4 paper, soaked in 5% silicone in cyclohexane, then dried | $H_2O$–EtOH–$CHCl_3$ (20:17:2) |
| Reversed phase | Partially acetylated Schleicher and Schüll 2043b paper | EtOAc–dioxane–$H_2O$ (2:4.5:4.6) |
| Normal | Whatman No. 1 paper: 1. Soaked in 50% dimethyl-formamide | Decalin |
| | 2. Soaked in 50% N-methyl-N-formamide (dried in each case) | Decalin |
| | 3. Untreated | n-BuOH–EtOH–$H_2O$ (45:5:50) |
| 1. Normal | Whatman No. 3 paper, untreated, between glass plates | Ether; or n-BuOH–ether–$H_2O$ (10:10:1); |
| 2. Normal | Whatman No. 3 paper, treated with (a) 5% and (b) 20% ethylene glycol in acetone, between silicone-treated glass plates | For (a) n-BuOH–petroleum ether (10:90); for (b) either $CHCl_3$–petroleum ether (1:1) or ether–petroleum ether (1:20) |
| 3. Reversed phase | Whatman No. 3 paper, treated with (a) 5% and (b) 20% silicone oil in petroleum ether, between silicone-treated glass plates | For (a) $H_2O$–EtOH–$CHCl_3$ (20:17:2); for (b) $H_2O$–MeOH (1:4) |
| Reversed phase | Whatman No. 3 paper, treated with 5% silicone oil solution, between silicone-treated glass plates | 1. 10% $H_2O$ in MeOH 2. 15% $H_2O$ in MeOH |

| Spray Reagent | Application and $R_F$ Range | Ref. |
|---|---|---|
| rous thiocyanate (freshly prepared) : to blution of NH₄SCN (5% w/v) and H₂SO₄ % v/v) add and dissolve eSO₄·(NH₄)₂SO₄·6H₂O (7% w/v) | $H_2O_2$ (0.84); aralkyl hydroperoxides (0.75–0.85); cycloalkyl hydroperoxides (0.5–0.6); trityl hydroperoxides and *t*-butyl peroxy esters (0.15–0.2); benzoyl peroxide (0.0) | 38 |
| *p*-Aminodimethylaniline hydrochloride nd 2% AcOH in 1:1 MeOH–H₂O KI in 1:1 MeOH–H₂O | $H_2O_2$ (0.83); aralkyl, cycloalkyl, and cyclic ether hydroperoxides (0.35–0.75); succinyl peroxide (0.8); benzoyl peroxide (0.2); lauroyl and stearoyl peroxides (0.0) | 50 |
| % *p*-Aminodimethylaniline hydrochloride )H–saturated aqueous KI-starch solution 3:2:5) (56%)–AcOH(10:90) | Some 22 pure peroxides of many types characterized by $R_F$ values from 0.0 to 1.0, which varied widely among the three solvent systems; also used to identify six complex peroxides produced by reacting diethyl ketone with $H_2O_2$ | 48 |
| rous thiocyanate (as above) nenylenediamine and acetaldehyde in queous AcOH (56%)–AcOH (10:90) | 1. $H_2O_2$ separated from hydroperoxides ($R_F$ values 0.1–0.5 and 0.95–1.0)<br><br>2. Excellent separations among many members of many peroxide classes —see original reference for example.<br><br>3. Dialkyl peroxides (0–0.5; H₂O₂, 0.7); hydroxyperoxides, cumene hydroperoxide (0.8–0.9), alkyl hydroperoxides (1.0). | 46 |
| rous thiocyanate (as above) 47 | Dialkyl peroxides (0.2–0.7); alkyl hydroperoxides (1.0) | 47 |

TABLE 5.  $R_f$ VALUES FOR SOME ORGANIC PEROXIDES BY THIN-LAYER
CHROMATOGRAPHY WITH THE SOLVENT SYSTEMS OF KNAPPE
AND PETERI[a]

| | Average $R_f$ Value (on Silica Gel) | | |
|---|---|---|---|
| Compound | I Toluene–CCl$_4$ (2:1) | II Toluene–HOAc (19:1) | III Petroleum Ether–EtOAc (49:1) |
| Lauroyl peroxide | 0.85 (0.71)[b] | 0.95 (0.92)[b] | |
| 2,4-Dichlorobenzoyl peroxide | 0.81 (0.66)[b] | 0.88 (0.92)[b] | |
| 4-Chlorobenzoyl peroxide | 0.74 | 0.94 | |
| Benzoyl peroxide | 0.55 (0.41)[b] | 0.70 (0.59)[b] | |
| t-Butyl peroxyoctoate | 0.28 | 0.55 | |
| Methyl isobutyl ketone peroxide, major component A | 0.25 | 0.55 | |
| t-Butyl peroxybenzoate | 0.24 | 0.47 | |
| Cyclohexanone peroxide, major component A | 0.21 | 0.38 | |
| Methyl ethyl ketone peroxide, major component A | 0.16 | 0.42 | |
| t-Butyl peroxyacetate | 0.12 | 0.32 | 0.18 |
| Cumene hydroperoxide | 0.11 (0.14)[b] | 0.33 (0.34)[b] | 0.09 |
| 2,2-Bis (t-Butyl peroxy) butene | 0.10 | 0.35 | |
| Methyl ethyl ketone peroxide, major component B | 0.10 | 0.10 | |
| t-Butyl hydroperoxide | 0.05 (0.06)[b] | 0.30 (0.34)[b] | |
| Methyl isobutyl ketone peroxide, major component B | 0.00 | 0.12 | |
| Cyclohexanone peroxide, major component B | 0.00 | 0.12 | |
| Cyclohexanone peroxide, minor component C | 0.00 | 0.10 | |
| Hydrogen peroxide | 0.00 | 0.00 | 0.00 |

[a] Reference 44.
[b] Data of Hayano, Ota, and Fukushima, (43).

peroxide. Color was said to develop at different rates depending on the type
of peroxide, with hydrogen peroxide and hydroperoxides showing up immediately and peroxy esters requiring heating at 70° for 10 min. Detection of
dialkyl peroxides required that the chromatogram be first sprayed with concentrated hydrochloric acid and then heated at 70° for 10–15 min, before

TABLE 6. $R_f$ VALUES FOR SOME ORGANIC PEROXIDES BY THIN-LAYER CHROMATOGRAPHY WITH THE SOLVENT SYSTEMS OF HAYANO, OTA, AND FUKUSHIMA[a]

| | Average $R_f$ Value (on Silica Gel)[b] | | |
|---|---|---|---|
| Compound | I<br>Benzene–<br>HOAc–MeOH<br>(10:1:1) | II<br>Chloroform–<br>HOAc–Water<br>(8:1:2) | III<br>Benzene–<br>HOAc–Water<br>(8:1:2) |
| Lauroyl peroxide | 0.90 | 0.97 | 0.82 |
| 2,4-Dichlorobenzoyl peroxide | 0.87 | 0.90 | 0.80 |
| Benzoyl peroxide | 0.80 | 0.87 | 0.63 |
| Dicumyl peroxide | 0.85 | 0.93 | 0.73 |
| Di-*t*-butyl peroxyphthalate | 0.73 | 0.66[c] | 0.24 |
| Cumene hydroperoxide | 0.54 | 0.59[c] | 0.27[c] |
| *t*-Butyl hydroperoxide | 0.34 | 0.37 | 0.13 |

[a] Reference 43.
[b] Standard deviation of the $R_f$ values is 0.01–0.04, except as noted.
[c] Standard deviation is 0.12.

developing with the peroxide spray reagent. Diaralkyl peroxides showed up much more readily than dialkyl peroxides, such as *t*-butyl-*t*-amyl peroxide. Di-*t*-butyl peroxide was too inert and too volatile to be detected. It was pointed out (41) that Knappe and Peteri (44) had apparently detected the presence not of di-*t*-butyl peroxide but rather of *t*-butyl hydroperoxide as an impurity.

It can be seen in Table 7 that all peroxides have lower mobilities in the heptane–ether solvent. This system is very effective for separating mixtures containing dialkyl and diacyl peroxides, which are relatively mobile, but has almost no effect on the more strongly adsorbed peroxy esters and hydroperoxides. On the other hand, the toluene–methanol system is effective in separating mixtures containing peroxy esters and hydroperoxides; in it all dialkyl and diacyl peroxides have $R_f$ values of about 1 and differ little from each other.

The increased mobility in toluene–methanol is attributed both to the greater polarity of the system and to the ability of methanol to displace peroxides from the adsorbent and form hydrogen bonds to the active sites (41). Increased concentrations of methanol in toluene produce linear increases in $R_f$; at 35% methanol hydrogen peroxide, which is notoriously difficult to move on a TLC plate, has an $R_f$ value above 0.5.

TABLE 7.   $R_f$ VALUES FOR SOME ORGANIC PEROXIDES BY THIN-LAYER
CHROMATOGRAPHY WITH THE SOLVENT SYSTEMS OF BUZLANOVA,
STEPANOVSKAYA, AND ANTONOVSKII[a]

| | $R_f$ Value | | |
| | Silica Gel | | Alumina |
| Compound | I<br>Heptane–<br>Et$_2$O<br>(15:1) | II<br>Toluene–<br>MeOH<br>(20:3) | II<br>Toluene–<br>MeOH<br>(20:3) |
|---|---|---|---|
| *Dialkyl Peroxides* | | | |
| Di-*t*-butyl peroxide | — | — | — |
| *t*-Butyl pinyl peroxide | 0.76 | 0.90 | — |
| *t*-Butyl-*t*-amyl peroxide | 0.74 | 0.95 | — |
| *t*-Butyl cumyl peroxide | 0.64 | 0.95 | 0.74 |
| Dicumyl peroxide | 0.55 | 0.95 | — |
| *Diacyl Peroxides* | | | |
| Stearoyl peroxide | 0.41 | 0.98 | 0.98 |
| Lauroyl peroxide | 0.37 | 0.98 | 0.98 |
| *p*-Chlorobenzoyl peroxide | 0.32 | 0.91 | — |
| Benzoyl peroxide | 0.21 | 0.74 | 0.95 |
| *Peroxy Esters* | | | |
| *t*-Butyl peroxy-(*o*-benzoyl)benzoate | 0.11 | 0.94 | — |
| *t*-Butyl peroxy-(*m*-nitro)benzoate | 0.08 | 0.90 | — |
| *t*-Butyl peroxycaproate | — | 0.88 | — |
| *t*-Butyl peroxypelargonate | — | 0.87 | — |
| 1,1′-Diperoxybenzoyldicyclohexyl peroxide | — | 0.83 | — |
| *t*-Amyl peroxyacetate | 0.14 | 0.80 | 0.87 |
| *t*-Butyl peroxybenzoate | 0.18 | 0.78 | 0.93 |
| 1,1-Diperoxybenzoylcyclohexane | — | 0.78 | — |
| *t*-Butyl peroxyacetate | 0.13 | 0.68 | 0.86 |
| *Hydroperoxides* | | | |
| 1,1-Diphenylethane hydroperoxide | 0.04 | 0.80 | — |
| *p*-Menthane hydroperoxide[b] | 0.07 | 0.66 | 0.51 |
| | | 0.50 | 0.39 |
| | | 0.25 | — |
| | | 0.18 | — |
| Cumene hydroperoxide | 0.05 | 0.54 | 0.37 |

TABLE 7 (*Continued*)

|  | $R_f$ Value | | |
|---|---|---|---|
|  | Silica Gel | | Alumina |
|  | I | II | II |
|  | Heptane– | Toluene– | Toluene– |
|  | Et$_2$O | MeOH | MeOH |
| Compound | (15:1) | (20:3) | (20:3) |
| Pinane hydroperoxide[c] | 0.20 | 0.45 | — |
|  | 0.09 | 0.35 | — |
|  |  | 0.23 | — |
| Tetralin hydroperoxide | 0.05 | 0.43 | 0.32 |
| *t*-Amyl hydroperoxide | 0.04 | 0.42 | — |
| *t*-Butyl hydroperoxide | 0.05 | 0.37 | 0.24 |
| *m*-Diisopropylbenzene dihydroperoxide | 0.00 | 0.30 | 0.11 |
| Hydrogen peroxide | 0.00 | 0.05 | 0.10 |

[a] Reference 41.
[b] A mixture of the oxidation products of *p*-menthane.
[c] A mixture of the oxidation products of pinane.

In a separate study of cyclic ketone peroxides (42) Buzlanova and co-workers chromatographed a group of 12 cyclohexanone and methylcyclohexanone peroxides on silica gel with toluene–methanol (20:3) as solvent. They found the interesting result that there were only three characteristic $R_f$ values (which were essentially independent of the presence or location of methyl substitution) as follows:

1. All 1,1'-dihydroxy peroxides had an $R_f$ of 0.05, identical with that of hydrogen peroxide, confirming the dissociation of these compounds into hydrogen peroxide and the original ketone.
2. The 1-hydroxy-1'-hydroperoxy peroxides had $R_f$ values of 0.12–0.14, less than those of the alkyl hydroperoxides, at 0.30–0.50.
3. The 1,1'-dihydroperoxy peroxides all had $R_f$ values of 0.74–0.75, which is consistent with intermolecular hydrogen bonding of the hydroperoxy group.

Both Buzlanova and co-workers and Hayano and co-workers found $R_f$ values to be identical for a compound whether alone or in a mixture.

Like paper chromatography, TLC is more frequently used as a qualitative than as a quantitative tool. With a running time of about 15 min and the ability to process many samples simultaneously on a single plate, TLC is valuable for preliminary surveys of new or complex peroxide samples and for monitoring column separations and conversion reactions. Studies of the autoxidation of fats by Rieche and co-workers (45) exemplify the practical value of TLC in such applications. TLC separations of dicumyl peroxide for quantitative estimation have been reported (123). Silica gel with pore sizes of 20–50 Å has been recommended for effective peroxide separations (124).

### 2. Infrared Absorption

The literature on the infrared absorption of organic peroxides is quite well summarized by Davies (2, Ch. 14) and Hawkins (3, Ch. 11), discussed by Bellamy (52), and reviewed by Karyakin (125).

An overemphasis of the limitations of infrared absorption in the analysis of organic peroxides has been based on the following two concepts:

1. The $-O-O-$ stretching mode is relatively symmetrical, with only slight vibrational changes in dipole moment, so that its absorption band can have only low intensity.

2. The force constants in the $-C-C-$, $-C-O-$, and $-O-O-$ bonds must all be similar, in view of their similar masses, so that the $-O-O-$ group should contribute principally to the structurally sensitive skeletal modes, without displaying a characteristic set of frequencies (126).

These theoretical predictions do not consider the fact that organic peroxides possess a skewed, modified hydrogen peroxide structure, with a characteristic dihedral angle of about 90–120° between the two molecular planes through the $-O-O-$ bond (8, 71, 127–130). Such a structure should have substantial dipole moment changes during vibrations and, provided its electronic interaction with the rest of the molecule is slight, could well have characteristic peroxide vibrational modes (perhaps more complex in form than a simple $-O-O-$ stretch), able to generate infrared absorptions at characteristic frequencies and with moderate intensities; this turns out to be the case for the hydroperoxide group, as will be seen in Section IV-A-2-b.

Infrared analysts (7, 52, 55, 57, 58, 131) generally agree that a characteristic $-O-O-$ stretching frequency does exist in the region of 850–890 cm$^{-1}$ for most peroxides, and that vibration of the dihedral angle in hydroperoxides produces another usually strong absorption in the 820–840 cm$^{-1}$ region. Because of accidental overlap by other (particularly skeletal) vibrations in the molecule, and by characteristic vibrations of alcohols, ethers, and oxiranes, however, analytical chemists must use these peroxide bands with caution in the qualitative examination of unknown systems and should always obtain independent confirmation.

(a) EMPIRICAL CORRELATIONS

Characteristic spectral features involving structural components other than the —O—O— bond have been established for some classes of peroxides. These empirical correlations have been made by examining the groups of peroxide spectra that have appeared from time to time, often in the wake of advances in synthesis techniques. Principal contributors have included Minkoff (58), who presented spectra of more than 30 peroxides from eight classes; Davison (53), who reported on the carbonyl absorptions of 24 symmetrical and asymmetrical diacyl peroxides and five peroxy esters; Swern and co-workers (132), who studied eight long-chain peroxy acids; and Williams and Mosher (59), who gave infrared spectral data for 17 primary and secondary alkyl hydroperoxides and the corresponding alcohols and with Welch tabulated spectra for 10 primary and secondary dialkyl peroxides (134). Hoffman (135) has reported spectral data for several allyl hydroperoxides, and Criegee and Paulig (111) have shown spectra for three cyclic peroxides.

Probably the most useful infrared spectra–structure correlations among peroxides are those of the peroxidic carbonyls, based mainly on Davison's work (53). For diacyl peroxides carbonyl absorption produces doublets split by about 25 cm$^{-1}$; they are centered in the range of 1769–1794 cm$^{-1}$ (5.65–5.57 $\mu$) for aromatic members of the class and in the slightly higher range of 1798–1808 cm$^{-1}$ (5.56–5.53 $\mu$) for aliphatic members. These bands are readily differentiated from the carbonyl doublets of anhydrides by the wider 65 cm$^{-1}$ frequency splitting of the latter. Peroxy esters have a single band in the range of 1758–1783 cm$^{-1}$ (5.69–5.61 $\mu$), higher in frequency that the singlet of its corresponding simple ester by an average of 36 cm$^{-1}$ (range of 1715–1765 cm$^{-1}$, or 5.83–5.67 $\mu$). Aliphatic peroxy acids have a similar band at 1747–1748 cm$^{-1}$ (5.72–5.71 $\mu$) (132) due to hydrogen-bonded peroxidic carbonyl, compared with 1710 cm$^{-1}$ (5.85 $\mu$) for hydrogen-bonded carboxylic acid carbonyl; one can positively identify peroxy acids by the failure of dilution to produce any unbonded carbony or hydroxyl absorption.

In other empirical correlations Williams and Mosher (59) noted that primary alkyl hydroperoxides have three bands at 1433, 1470, and 1488 cm$^{-1}$ (6.98, 6.80, and 6. 72 $\mu$, respectively), whereas the secondary hydroperoxides and all the corresponding primary and secondary alcohols have only a single absorption near 1470 cm$^{-1}$ (6.80 $\mu$). They found that secondary hydroperoxides have a double band with peaks at about 1340 and 1380 cm$^{-1}$ (7.46 and 7.25 $\mu$), whereas primary hydroperoxides and the alcohols show the 1380 cm$^{-1}$ (7.25 $\mu$) absorption only. Among *n*-alkyl hydroperoxides they observed a difference in behavior between homologs with even and odd numbers of carbon atoms: even-numbered compounds have a doublet at 807 and 826 cm$^{-1}$ (12.39 and 12.10 $\mu$, respectively), whereas the odd-numbered members have a single band at about 813 cm$^{-1}$ (12.30 $\mu$).

## (b) Hydroperoxide Group Frequencies

A relatively sophisticated treatment of the spectral properties of the hydroperoxide group is possible because of the following circumstance: the —C—O—O—H group is well enough isolated electronically from the rest of its molecule for it to be realistically treated as an independent four-atom molecule with a characteristic set of six vibrational frequencies and infrared-absorption bands (i.e., $3n - 6$ for an $n$-atom molecule). This study was carried out principally by Soviet workers (55, 57) and has been discussed by us in the *Treatise on Analytical Chemistry* (1) in considerable detail.

Kovner, Karyakin, and Efimov (57) calculated vibrational frequencies for COOH and COOD, based on a four-atom model whose bond distances and angles were derived from related molecules and whose force constants were estimated from frequencies assigned to these fundamental vibrations in cumene hydroperoxide and cumene deuterohydroperoxide. Agreement was excellent between calculated and observed frequencies. Gribov and Karyakin (55) took two additional steps. First, they computed the configurations of the normal vibrations of the COOH model, along with the relative position shifts of all atoms during the vibrations. (Of particular interest is the finding that the 880 cm$^{-1}$ (11.36 $\mu$) band previously assigned to a simple —O—O— stretching vibration is a more complex mode having important contributions also from COO bending and from flapping of the dihedral planes.) Next, they calculated the band intensities for the COOH absorptions; agreement was good with the observed intensities for cumene hydroperoxide.

For this concept of the hydroperoxide group as an electronically isolated molecular appendix with essentially independent fundamental frequencies, we derived added support from a close examination of the 17 spectra of alkyl hydroperoxides presented by Williams and Mosher (59).

Our assignments of the five fundamental COOH infrared-absorption bands accessible in these spectra were tabulated in *Treatise on Analytical Chemistry* (1) and agree well with the calculated frequencies. Karyakin has made a similar comparison (125).

## (c) Near-Infrared Analysis

The utility of near-infrared absorption for hydroperoxide analysis deserves wide recognition. The value of this technique is due to the combination of high resolution by modern instruments in the 3300–3700 cm$^{-1}$ (3–2.7 $\mu$) region of interest and the high molar absorptivities of the hydroperoxide bands occurring there. In our laboratories Goddu has studied the near-infrared analysis of hydroperoxides and derived alcohols and has summarized the pertinent information in a general treatment of near-infrared spectro-photometry (54). He found that the hydroxy bands of hydroperoxides at 3515–3550 cm$^{-1}$ (2.845–2.817 $\mu$) are well separated from those of alcohols at 3590–3640 cm$^{-1}$ (2.785–2.747 $\mu$) and phenols at 3610 cm$^{-1}$ (2.77 $\mu$), so that mixtures of such components can be analyzed. In carbon tetrachloride, alkyl and alicyclic hydroperoxides absorb at about 3550 cm$^{-1}$ (2.817 $\mu$), with molar absorptivities of 80–90 l/mole-cm. For aralkyl hydroperoxides this absorption is split into a doublet, typically at 3545 and 3520 cm$^{-1}$ (2.821 and 2.841 $\mu$), each element having an absorptivity value in the 30–40 l/mole-cm range. Electron-attracting substituents present in *p*-chlorocumyl, 3,4-dichlorocumyl, and *p*-nitrocumyl hydroperoxides withdraw electrons from the benzene ring and eliminate the splitting. The use of methylene chloride rather than carbon tetrachloride as solvent also removes this splitting from the unsubstituted aralkyl hydroperoxides and, in addition, lowers the resulting single frequency by about 20 cm$^{-1}$.

The near-infrared region should also be excellent for determining peroxy acids, in view of the data of Swern and co-workers (132): for eight long-chain peroxy acids the absorption frequency of the hydroxyl stretching mode was constant, at 3280 cm$^{-1}$ (3.048 $\mu$), over a concentration range in carbon tetrachloride from 0.006 to at least 0.3 M, and the molar absorptivities varied but little. The average for all peroxy acids over all concentrations was 5.16 l/mole-cm, with a maximum variation at any one concentration of only 3%. This behavior is due, of course, to the intramolecularly-hydrogen-bonded monomeric structure of peroxy acids in solution. Carboxylic acids, in contrast, show extremely variable absorptivities (54) because of their well-known equilibrium between the free monomer and the hydrogen-bonded dimer.

Near-infrared absorption is an excellent technique for following the decomposition and formation of hydroperoxides. It should also be valuable for studying the hydrolysis of peroxy esters and for identifying and determining the alcohol-reduction products of many peroxides. Other workers have described such uses (136).

### 3. Ultraviolet Absorption

Although ultraviolet spectra are much simpler and less specific than infrared ones, the absorption coefficients are usually much higher. We have found this spectral region to be useful in studying the autoxidation of isopropyl-benzenes,* particularly when coupled with automatic liquid-column chromatography on silica gel (60). Ultraviolet absorption is also used to determine excess reagent in a method based on chemical reduction by triphenyl phosphine (75); this is discussed later.

---

* Private communication from A. A. Orr, Hercules Incorporated, Research Center, Wilmington, Delaware (1960).

### 4. Nuclear Magnetic Resonance

Nuclear magnetic resonance (NMR), well established as a potent instrumental technique for elucidating molecular structures, is applicable to both the qualitative and the quantitative analysis of organic peroxides, although it has been relatively little used in this area. Pioneering studies on alkyl hydroperoxides by Fujiwara and co-workers (63) showed large differences in chemical shift among hydroperoxides, alcohols, and carboxylic acids not only for their oxygenated proton but also for the $\alpha$-methylene protons. Davies and co-workers studied the proton-magnetic-resonance (PMR) spectra of $t$-butyl and isobutyl groups in boron compounds (61) and have used PMR spectra to follow the autoxidation of several isomeric tributyl borons (62); they could readily differentiate $t$-butyl, $t$-butoxy, $t$-butylperoxy, isobutyl, isobutoxy, and isobutylperoxy groups attached to boron. Brill also used PMR to identify isomeric hydroperoxides in his isomerization studies (109), already mentioned in Section IV-A-1-a.

In the analysis of peroxides by PMR we are concerned with hydroperoxy protons and protons attached to carbons that are alpha ($\alpha$-CH) and beta ($\beta$-CH) to the functional group. The shielding of the hydroperoxy proton is appreciably less than that of the hydroxy proton (63), as indicated by the greater downfield shift of its resonance ($\delta = -9.1$ and $-4.4$ ppm, respectively, for $-$OOH and $-$OH), and is similar to that for the carboxy proton ($\delta = -9.8$ ppm). Available data comparing chemical shifts of $\alpha$-CH and of $\beta$-CH for alcohols and hydroperoxides are summarized in Table 8; shielding is again less in the hydroperoxides than in the alcohols, for both classes of proton. Resolution of $\alpha$-CH resonances in mixtures of hydroperoxides and alcohols should be straightforward, since there is a separation of about 0.4 ppm.

Recent data from our laboratory (138), in Table 8B and in Figure 2, show that the chemical shifts for $\beta$-CH in alcohols and the corresponding hydroperoxides differ by only 0.02–0.07 ppm. Such small differences can be resolved by currently available instruments, such as the Varian A-60A spectrometer. The NMR spectra in Figure 2 show the resolution of $\beta$-CH peaks for two three-component mixtures, each containing a dialkyl peroxide and the corresponding hydroperoxide and alcohol. The three peaks in Figure 2A represent, from left to right, $t$-butyl hydroperoxide, $t$-butyl alcohol, and di-$t$-butyl peroxide, in deuterochloroform; the first two are separated by 0.017 ppm. Figure 2B shows similar peaks for (from left to right) dicumyl peroxide, cumene hydroperoxide, and $\alpha,\alpha$-dimethylbenzyl alcohol in deuteroacetone; here the first pair are separated by only 0.013 ppm. (Concentrations are of the same order of magnitude, but unknown.)

Notice that the dialkyl peroxides in Figures 2A and B are at opposite ends of the triads. This variation illustrates the fact that the resonance position for the several components is frequently solvent dependent; in other runs

TABLE **8.** CHEMICAL SHIFT OF PROTONS ALPHA AND BETA TO HYDROXY AND HYDROPEROXIDE GROUPS, IN PMR SPECTRA

### A. Alpha Protons

| R | $(\delta_{CH_3} - \delta_{CH_2})_{ROH}$(ppm)[a, b] | $(\delta_{CH_3} - \delta_{CH_2})_{ROOH}$(ppm)[a, c] |
|---|---|---|
| *n*-Propyl | 2.53 | 2.93 |
| *n*-Butyl | 2.58 | 2.93 |
| *n*-Octyl | — | 2.93 |
| *n*-Nonyl | 2.58 | — |
| Isobutyl | 2.43 | 2.81 |

### B. Beta Protons

| R | $(\delta_{CH_3})_{ROH}$(ppm)[a, d] | $(\delta_{CH_3})_{ROOH}$(ppm)[a, d] |
|---|---|---|
| $(CH_3)_3C-$ | −1.22 | −1.24 |
| $H_3C-\!\!\langle\text{phenyl}\rangle\!\!-C(CH_3)_2-$ | −1.46 | −1.53 |
| $HC(CH_3)_2-\!\!\langle\text{phenyl}\rangle\!\!-C(CH_3)_2-$ | −1.53 | −1.57 |
| $Cl-\!\!\langle\text{phenyl}\rangle\!\!-C(CH_3)_2-$ | −1.47 | −1.50 |
| $Cl-\!\!\langle\text{dichlorophenyl}\rangle\!\!-C(CH_3)_2-$ | −1.50 | −1.52 |

[a] $\delta_{R'}$ = chemical shift of proton resonance in group R′ downfield from the proton resonance of tetramethylsilane.

[b] Data from reference 137.

[c] Calculated from the data of Fujiwara and co-workers (63).

[d] Data of Ward and Mair (138); samples were dissolved in deuterochloroform and run on a Varian A-60A spectrometer.

with the three cumene oxidate components, for example, there was superposition of alcohol and peroxide bands in deuterochloroform, whereas in the complexing solvents dimethyl sulfoxide and pyridine superposition of alcohol and hydroperoxide occurred. In some cases such variations are an asset, rather

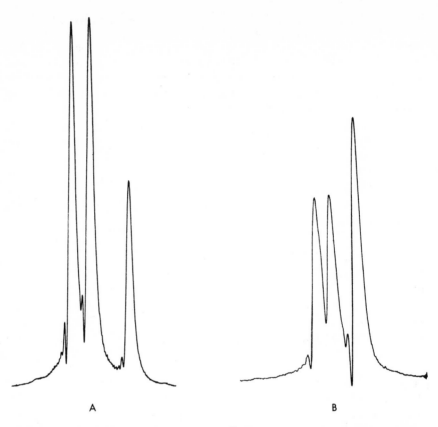

**Figure 2.** Nuclear magnetic resonances of β-CH protons in two peroxide systems, obtained with a Varian A-60A spectrometer. (*A*) From left to right, with chemical shifts from tetramethylsilane (in ppm): *t*-butyl hydroperoxide, $-1.248$; *t*-butyl alcohol, $-1.231$; and di-*t*-butyl peroxide, $-1.196$. Solvent: deuterochloroform. (*B*) From left to right, with chemical shifts from tetramethylsilane: dicumyl peroxide, $-1.534$; cumene hydroperoxide, $-1.520$; and cumyl(α,α-dimethylbenzyl) alcohol, $-1.493$. Solvent: deuteroacetone.

than a liability, in analytical studies. The resonance position can also be concentration dependent (63).

The NMR spectra of approximately 20 highly purified organic peroxides (hydroperoxides, peroxy acids, dialkyl peroxides, *t*-butylperoxy esters, and diacyl peroxides) and selected nonperoxidic analogs were recently reported by Swern and co-workers (139). The signal of the hydroperoxy proton is not only considerably downfield from that of the hydroxy proton but its position is also concentration dependent, although the upfield shift of the signal of the hydroperoxy proton with dilution is of much smaller magnitude than that in alcohols. The proton of the peroxycarboxylic acid group is even more deshielded than is the hydroperoxy proton, but its chemical shift is independent of concentration. A correlation is shown to exist between increasing acidity and the downfield shift of acidic protons in hydroperoxides and peroxy acids.

In conjunction with NMR spectroscopy, conversion to nonperoxides by reaction with triphenyl phosphine, as mentioned in Sections III-C and IV-B-4, should be particularly valuable in the analysis of peroxides, although use of such a combination has not been reported. The reaction is said to be quantitative, unequivocal, and rapid for almost every peroxide in chlorinated solvents and benzene (excellent for NMR work), and neither the reagent nor its oxidation product will interfere. Changes in band areas before and after treatment with the triphenyl phosphine reagent can be expected to give effective qualitative and quantitative analytical data.

### 5. *Electron Paramagnetic Resonance*

Electron paramagnetic resonance (EPR) has been studied by Klyueva and co-workers (64) for the quantitative determination of hydroperoxides. Their approach has two steps, the first of which is decomposition of hydroperoxide by means of variable-valence metal ions according to the mechanism

$$ROOH + M^n \longrightarrow RO\cdot + M^{n+1} + HO^-$$

$$ROOH + M^{n+1} \longrightarrow ROO\cdot + M^n + H^+$$

The second is interaction of the short-lived $RO\cdot$ and $ROO\cdot$ free radicals with a special reagent, a so-called reference compound, which is thereby converted into stable radicals that are quantitatively estimated in the EPR instrument. In order of decreasing activity metal salts useful for hydroperoxide decomposition are cobalt, copper, palladium, manganese, lead, silver, chromium, nickel, and iron naphthenates and stearates.

Initial work on benzene solutions of monohydroperoxides and dihydroperoxides of *m*- and *p*-diisopropylbenzenes employed diphenylamine (DPA) as the reference material and cobalt stearate as catalyst. Small amounts (0.07 ml) of benzene solutions of hydroperoxide containing a twofold excess of DPA were added to ampules containing some solid cobalt stearate on the bottom. These tubes were sealed, placed in hot water (85–90°) for 1 min, and then quenched in cold water.

Klyueva and co-workers (64) used *m*-diisopropylbenzene monohydroperoxide to study the quantitative relationship between hydroperoxide concentration and EPR signal intensity; for the latter they used the maximum span of the EPR spectrum (three approximately equal lines), since the shape of the spectrum did not change with hydroperoxide concentration. Response was linear over the hydroperoxide concentration range 0.01–0.2 wt % and then increased less rapidly at higher concentrations, probably because of an increase in the recombination rate of the hydroperoxide radicals. Sensitivity was 0.01 wt %, or $6 \times 10^{-4}$ $M$, corresponding to 8 $\mu$g of hydroperoxide in an ampule. The relative error was about 5%.

On trying to determine cumene hydroperoxide in a reaction mixture containing 30–35% acetone, 55–60% phenol, and 10% condensation products, they found that DPA was an unsuitable reference. A new reference, phenothiazine (thiodiphenylamine), proved to be acceptable. Stable free radicals readily formed during the reaction of cumene hydroperoxide and phenothiazine at room temperature without need for a catalyst. Again, response (a four-line signal) was linear to about 0.2 wt % of hydroperoxide and then fell off rapidly.

The use of EPR spectra for estimating organic peroxides is an interesting approach whose eventual utility remains to be seen. There would perhaps be possibilities for selectively estimating certain peroxides in mixtures, but considerably more work is required on pure model systems first. Kochi and Krusic (140) have recently reported the detection of radicals in photolysis of diacyl peroxides and the reaction of organophosphorus compounds with alkoxy radicals.

### 6. Thermal Analysis

Differential thermal analysis (DTA) can be of auxiliary service in the analysis and characterization of organic peroxides and compositions containing them. Melting and freezing peaks give information on identity, purity, and quantity for crystalline peroxides. Decomposition exotherms indicate relative thermal stabilities and can help define the role of specific peroxides in the curing of resin compositions.

A relatively new DTA technique known as differential scanning calorimetry (DSC) records the instantaneous rate of physical or chemical thermal processes as a function of temperature or time (65), thus giving peak areas directly in energy units. This development increases the applicability of thermal analysis to many types of compounds, including peroxides. Mathematical analysis of DSC peak shapes and areas can efficiently yield much information; from a single melting thermogram, for example, mole percent purity can be determined, and the kinetic parameters of thermal decompositions can be calculated from decomposition exotherms.

### 7. Polarography

Polarography has been used in determining organic peroxides for more than a quarter century (141–143). It is of great value in specific areas because of its high sensitivity, rapidity, and potential for qualitatively identifying and differentiating the components of mixtures. The status of this technique in peroxide analysis has been reviewed (5, 66, 144).

Martin (5) published an excellent review of the polarographic peroxide literature up to 1960. Johnson and Siddiqi (66) summarized the available information through 1964, with an extensive compilation of half-wave potentials. These are key references, along with a paper by Romantsev and Levin (145), who presented new information on the effect of solvent and supporting electrolyte. For details of the general theory and practice of polarography we refer the reader to Meites' recent textbook (146) and the *Treatise on Analytical Chemistry* (147, 148).

The volume of research on the polarographic characterization of peroxides is considerable. Several groups have studied water-soluble peroxides in pre- dominantly aqueous media (141, 143, 144, 149), an interest that arose mainly in researches on combustion chemistry. Other groups with a primary concern for autoxidation processes in fats and oils have devised and explored a general-purpose, nonaqueous polarographic system (6, 68, 72, 150) and have applied it broadly to studies of interest (67, 69, 151–154). Still others have compared aqueous and nonaqueous media and have examined the effects of such parameters as solvent polarity, nature of the supporting electrolyte, peroxide structure, molecular weight, and adsorption phenomena at the electrode on half-wave potential, diffusion-current constant, and polaro- graphic irreversibility (145, 149, 155, 156). A firm basis has recently been established, also, for the analysis of oxidized oils by oscillographic polaro- graphy (157). These matters are discussed further in this section.

Despite a substantial body of information on the polarography of organic peroxides, much remains to be done. In the following discussion our aim is to survey the present body of knowledge on peroxide polarography in such a way that one wishing to employ this technique in the solution of a specific problem can either select a suitable set of conditions from among the refer- enced studies or plan an appropriate methods-development program to work out such conditions for himself.

## (a) Experimental Factors in the Polarography of Peroxides

It appears to have been as difficult to settle on standard optimum proce- dures for polarographic analysis as on standard chemical methods, and for some of the same reasons—namely, that the interest of many workers has been limited to peroxides in only a few classes (e.g., water-soluble or water- insoluble members) and that for many applications there are many variations in procedure that work well or reasonably well.

A major factor in selecting the polarographic solvent medium is the solu- bility of the peroxide or the substrate containing it. Water is the usual starting point (144, 149) and methanol or ethanol the usual modifying solvent (145,

149, 158, 159), although dioxane has also been used (156) and ethylene glycol monoalkyl ethers would probably serve (160). Dimethylformamide is a very effective nonaqueous polarographic solvent (3, 145, 161). The most popular nonaqueous system has been 0.3 $N$ lithium chloride in benzene–methanol (1:1), a system devised by Lewis, Quackenbush, and DeVries (150); this has been an almost exclusive choice for polarographic studies of fat and oil peroxidation products (67, 69, 151–154, 162) and has been used for characterizing the nonaqueous behavior of many peroxides (6, 71, 72, 145).

Proske (163) took another approach to water-immiscible peroxides and substrates by solubilizing them with relatively high concentrations of polarographically suitable commercial wetting agents (specifically American Cyanamid's Aerosol® AY and MA, which were said to be diamyl and dihexyl sodium sulfosuccinates, respectively). With fats the detergent had to provide all of the supporting electrolyte; benzyl alcohol was used as a solubilizing aid.

Bernard (155) studied the polarographic behavior of more than 25 peroxides of many types and in 14 cases obtained comparative data between aqueous 0.1 $N$ sodium chloride, potassium chloride, or potassium sulfate and the nonaqueous system of 0.3 $N$ lithium chloride in benzene–methanol (1:1). In studying water-immiscible substrates with the aqueous system his technique was to extract water-soluble components into 0.1 $N$ potassium chloride. In general his results are compatible with those of other workers, although they have not been included in a recent compilation (66).

One unique feature of Bernard's data was the appearance of a second reduction wave at about $-1.1$ volts for the aqueous extracts of such diverse materials as ammonium persulfate, $t$-butyl peroxybenzoate, and peroxidized citral, tetralin, terpinolene, and dicyclopentadiene. In each case he attributed this wave to hydrogen peroxide, released by hydrolysis of the corresponding hydroperoxides; he based this identification on observations that the characteristic wave appeared in the extract only after some time and that it was enhanced by additions of hydrogen peroxide (155). It would appear worthwhile to confirm this identification by one of the techniques for removing hydrogen peroxide, either as a titanium (164) or a lanthanum (144) complex.

For the polarography of compounds that are difficult to reduce dimethylformamide saturated with tetramethylammonium chloride has been found* to be a particularly good system because its useful range of reduction potential extended to $-2.5$ volts (versus a mercury pool), well beyond the usual practical limit of $-2.0$ volts. In this system all dialkyl peroxides examined were successfully reduced. These compounds and their half-wave potentials are shown in Table 9. It is particularly noteworthy that di-$t$-butyl peroxide is reduced at an $E_{1/2}$ of $-2.10$ volts with a well-defined wave that is useful for

*Private communication from E. J. Forman and C. M. Wright, Hercules Incorporated, Research Center, Wilmington, Delaware (1959).

TABLE 9. HALF-WAVE POTENTIALS OF SOME DIALKYL PEROXIDES IN DIMETHYLFORMAMIDE[a]

| Compound | $E_{1/2}$ versus Hg Pool[b] (volts) |
|---|---|
| Di-*t*-butyl peroxide | −2.10 |
| *t*-Butyl-1,1,3,3-tetramethylbutyl peroxide | −1.93 |
| Dipinanyl peroxide | −1.88 |
| Di-*p*-menthyl peroxide | −1.75 |
| Bis(4-isopropylcumyl) peroxide | −1.75 |
| Bis(3,4-dichlorocumyl) peroxide | −1.75 |
| Dicumyl peroxide | −1.60 |
| Bis(2,4,6-tri-*t*-butylcyclohexadienone) peroxide | −1.37 |

[a] Private communication from E. J. Forman and C. M. Wright, Hercules Incorporated, Research Center, Wilmington, Delaware (1959).
[b] In dimethylformamide saturated with tetramethylammonium chloride.

quantitative analysis. We are not aware that the polarographic reduction of this peroxide has been previously reported.

Romantsev and Levin (145) found that proton availability in a solvent had a profound effect on the polarographic reduction wave for most types of peroxides. Decreasing protogenic character in the series water (or dilute ethanol), 95% ethanol, benzene–methanol (1:1), dimethylformamide was paralleled by increasingly irreversible, more elongated waveforms and substantial shifts in half-wave potential toward more negative voltages. An example is the shift in $E_{1/2}$ for dibenzoyl peroxide to −0.4 volt in dimethylformamide, the only system in which this peroxide has a reduction potential more negative than −0.1 volt.

This effect of solvent composition on irreversibility tends to be offset by adsorption of a peroxide on the mercury electrode, a phenomenon indicated by an increase in drop rate when the peroxide is present. Such adsorption was found for ethyl cumyl peroxide in 20% ethanol and for cumene hydroperoxide in 20 and 95% ethanol (145), and these compounds showed relatively sharp, reversible reduction waves. Bernard (155) and Skoog and Lauwzecha (149) found a similar dependence of wave characteristics on solvent composition.

A variety of salts and acids have been used as supporting electrolytes. Lithium chloride has been a frequent choice because lithium ions allow polarography in a more negative potential range than do sodium or potassium ions. Chloride has been the most widely used anion, although it would appear to be undesirable for aqueous studies of peroxy acids, peroxy esters, and diacyl peroxides because of its ease of oxidation at positive voltages. Both the lithium sulfate electrolyte used by Bruschweiler and Minkoff (144) and an acetate (68) appear to be better choices for the reductions of such peroxide classes. Alkaline systems should generally be avoided because they tend to decompose some peroxides. Many of the common buffers have been used (149, 156) but with little apparent benefit. If their use is contemplated, however, the analyst should be alert for effects on adsorption equilibria at the electrode in addition to the effects sought from the control of proton concentration.

Tetraalkylammonium salts as supporting electrolytes have significant adsorption effects that are well known (165). These quaternary cations exert a screening action that can markedly influence electrode reaction mechanisms involving adsorption and desorption. Romantsev and Levin (145) found $E_{1/2}$ values for peroxides to be consistently shifted toward the negative side whenever these electrolytes replaced alkali salts.

Many peroxides are highly reactive compounds and may react chemically with the components of the polarographic system. For example, aliphatic peroxy acids react with the methanol of the widely used benzene–methanol (1:1) solvent (68) (switching to a system of ammonium acetate in glacial acetic acid eliminated this difficulty); peroxy acids (144) and hydroperoxides (155) tend to hydrolyze in aqueous systems to produce hydrogen peroxide; and peroxide decomposition has been attributed to agar-agar, gelatin, sintered glass, and mercury (144), a special cell having been designed to minimize such effects (other workers have not reported such difficulties, however).

A pair of small, early waves occur in the 0.3 $N$ lithium chloride, benzene–methanol (1:1) system with fats and oils (67, 151, 152) and with peroxy acids and diacyl peroxides. Maack and Lück (166) attribute these waves to mercury ions produced when reactive peroxides are reduced by mercury metal. They reached this conclusion from cathode-ray polarographic studies of appropriate systems. (Three additional papers by these workers (157) establish oscillographic polarography as a useful new tool for studying the oxidation of fats and oils.) Since these waves have long lacked a definite explanation, this limited examination is a welcome lead that should be followed up. A study of the reaction of the various classes of peroxides with mercury, as a function of solvent composition and the presence of promoters or inhibitors, appears to be long overdue.

To avoid such reactions with mercury Roberts and Meek (167) developed a vibrating double electrode of bright platinum. It was so arranged that polarities could be reversed at periods in the range of 0.2–5 sec to help prevent the buildup of reaction products. Polarograms were much more stable than those obtained with single vibrating electrodes and were equivalent to those obtained with the dropping-mercury electrode. It has been suggested (5) that such electrodes should be of value for continuous process stream analysis.

## (b) QUANTITATIVE ANALYSIS

Although the reduction potential of a peroxide in a well-characterized polarographic system may be useful for qualitative identification, it is the diffusion-current constant $I$ that is important for absolute quantitative analysis. This is defined by

$$I = \frac{i_d}{Cm^{2/3}t^{1/6}}$$

where $i_d$ is the diffusion current, $m$ is the drop mass, $t$ is the drop period, and $C$ is the concentration. The value of $I$ determined for a compound in a particular medium by calibration with a particular capillary can be used to determine the compound quantitatively in the same medium from values of $i_d$ obtained on other polarographs with other capillaries. When pure specimens are not available for calibration, standards may be used whose concentration $C$ has been determined by appropriate chemical methods.

Diffusion-current constants vary both between and within classes of peroxides and are influenced by the composition of the media. Polarography hence must be considered to be ill-suited for the absolute quantitative analysis of the peroxide functional group; calibration with the peroxide to be determined is generally necessary, despite the data of Willits and co-workers (72) that show remarkably constant $I$ values for a series of diverse low-molecular-weight hydroperoxides. Martin (5) suggests that by using values for $I$ of 6.3 for aqueous solutions and 5.3 for benzene–methanol (1:1) one can obtain order-of-magnitude estimates of peroxide concentration probably accurate to within 30%, without calibration.

These statements do not imply that quantitative peroxide analyses are inherently inaccurate, but rather that calibration is necessary. Indeed, as precision is generally good, even at low concentrations, it needs only calibration with pure samples of the peroxide mixture in question, or with reliable alternate methods of analysis, for polarography to give accurate quantitative analyses at concentrations down to less than $10^{-4}$ M.

Most polarographers find it practical when possible to calibrate directly with individual pure peroxides or independently standardized samples (the comparative "standard solution" method described in Section V-B-4 of reference 147) rather than to use absolute diffusion-current constants determined independently or obtained from the literature; such direct calibration avoids the additional error from determining $m^{2/3}t^{1/6}$ for the currently used electrode capillary.

## (c)  THE EFFECTS OF PEROXIDE STRUCTURE

In peroxide analysis there is a close analogy between chemical and polarographic (electrochemical) methods: in the former reduction of the —O—O— bond is accomplished by chemical reagents, whereas in the latter the reduction is carried out directly by electrons. In each case, however, it is the strength of the —O—O— bond relative to the chemical or electrical reducing potential of the system that determines the response of a peroxide. This bond strength is related directly to peroxide structure; that is, to the nature of the one or two substituents by which the H—O—O—H progenitor has been transformed into a particular organic peroxide.

In polarographic systems the strength of the —O—O— bond in the peroxide will be the result of structural factors of two types, inductive and environmental. The inductive effects originate in a substituent group and should exert their influence independently of environment. Consider the expected behavior of organic peroxides relative to that of hydrogen peroxide. The latter is reduced with a half-wave potential $E_{1/2}$ of about $-1.0\pm0.1$ volt (versus a saturated calomel electrode) in a wide variety of media, as can be seen in Table 10, in which the available information on the effects of structure in peroxide polarography is summarized. Substituents that are electron donors should tend to stabilize the peroxide bond by inducing an increase in its electron density, thereby decreasing its electron affinity (71, 156). Alkyl groups are well-known electron donors, with the $t$-butyl substituent showing the greatest influence; it seems likely, therefore, that this factor is related to the general enhanced chemical stability of dialkyl peroxides and to their more negative half-wave reduction potentials relative to hydrogen peroxide (in anhydrous media). Electron-withdrawing groups, such as aromatic and carbonyl substituents, should have an opposite effect, and this may be exemplified by the large shift of $E_{1/2}$ for peroxy acids to the positive side of 0.0 volt and for diacyl peroxides to the $-0.1$ to $-0.4$-volt range. In the $t$-butyl peroxy esters, which reduce with $E_{1/2}$ values in the $-0.8$- to $-1.0$-volt range in anhydrous system, it would appear that the electron pushing and pulling effects are at a virtual standoff, although the peroxide bond in these compounds should be polarized.

In addition to these inductive effects there are environmental factors that act through the structure of a peroxide to influence its —O—O— bond strength. These are complex and have not been studied extensively. One is the effect of water content (apparent in Table 10), which suggests that protonation is important at some point in the reduction mechanism (145).

Another is the effect of adsorption processes at the dropping-mercury electrode (145, 156), with competition likely among a number of adsorption equilibria involving peroxides, their reduction products, solvent components, and supporting electrolyte. In many cases these environmental factors are dominant.

The effects of inductive and environmental factors may be seen in Table 10 for a series of alkyl hydroperoxides, dialkyl peroxides, and *t*-butyl peroxy esters. With hydroperoxides in a nonaqueous medium the nature of the R group in the $R-O-O-H$ structure apparently has a relatively small effect on $E_{1/2}$ values; these are generally displaced (from that of hydrogen peroxide) some 0.2–0.4 volt in the positive direction—opposite to the negative shift anticipated from electron induction by the alkyl group. Evidently an overriding environmental factor is operative; moreover, a persistent trend toward more positive reduction potentials with increasing size of the R-group shows that this environmental effect is being exerted through the substituent.

In the aqueous or substantially aqueous systems positive-shifting factors for the hydroperoxides are much stronger, as is also their direct relation to the size of the R-group. In these media, starting from hydrogen peroxide at $-0.9$ volt, there are sharp positive shifts of $E_{1/2}$ with increasing size of the *n*-alkyl substituent up through *n*-hexyl, which has an $E_{1/2}$ of $-0.1$ volt in 20% ethanol. Beyond this point the size effect tapers off quickly, with small, regular positive shifts up through *n*-nonyl hydroperoxide, for which $E_{1/2}$ in 20% ethanol is $-0.01$ volt versus the saturated calomel electrode.

It will be noted that in this system $E_{1/2}$ is more negative for *t*-butyl than for *n*-butyl hydroperoxide, which is in line with the greater inductive effect of the *t*-butyl group.

The effect of water content on half-wave potential is demonstrated throughout Table 10. Data from a series of three or four solvent compositions are listed for propyl, *t*-butyl, and cumyl hydroperoxides, ethyl cumyl peroxide, and *t*-butyl peroxybenzoate. In every case the higher the water content, the greater the positive $E_{1/2}$ shifts.

Only in anhydrous media are such environmental effects sometimes small enough not to overwhelm structural factors. For example, in benzene–methanol (1:1) dialkyl peroxides (but not hydroperoxides) can exhibit the negative shift relative to hydrogen peroxide reduction potentials anticipated through inductive effects of alkyl substitution.

In anhydrous media *t*-butylperoxy esters are remarkably like alkyl hydroperoxides in their polarographic behavior; they reduce in the same potential range and have a similar persistent small shift to more positive $E_{1/2}$ values with increasing size of the acyl group. In aqueous media there is a very strong positive shift, with *t*-butyl peroxybenzoate, for example, moving to the *positive* side of 0.0. volt. Such reduction behavior is more typical of

TABLE 10. VARIATIONS IN POLAROGRAPHIC HALF-WAVE POTENTIAL $E_{1/2}$ WITH PEROXIDE STRUCTURE AND MEDIA COMPOSITION FOR SEVERAL PEROXIDE CLASSES

$E_{1/2}$ (volts versus saturated calomel electrode) in Media of Decreasing Water Content

| Peroxide | Aqueous[a] | 20% Ethanol[b] | 95% Ethanol[c] | Benzene–Methanol (1:1)[d] | References |
|---|---|---|---|---|---|
| *Hydroperoxides* | | | | | |
| HOOH | −0.88, −0.93 | −1.07, 50% MeOH | −1.09 | −1.14, −1.16[e] | 144, 145, 155 |
| Methyl | −0.64, −0.79 | — | — | — | 144, 156 |
| Ethyl | −0.42, −0.44 | — | — | — | 144, 156 |
| Propyl | −0.35 | −0.49 | — | −0.98 | 145, 156 |
| n-Butyl | −0.20, −0.21 | — | — | — | 149 |
| sec-Butyl | −0.28 | — | — | — | 149 |
| t-Butyl | −0.35, −0.34, −0.27, −0.28 | −0.74, 50% MeOH | −1.04, 90% MeOH | −0.96, −1.10, −1.15[e] | 6, 72, 144, 149, 155 |
| n-Pentyl | −0.08, −0.19, −0.20 | −0.20 | — | −0.95 | 145, 149 |
| n-Hexyl | — | −0.08, −0.12 | — | −0.94 | 145, 149 |
| n-Heptyl | — | −0.03 | — | — | 149 |
| Heptenyl | — | −0.04 | — | −0.88 | 145 |
| n-Octyl | — | −0.02 | — | — | 149 |
| n-Nonyl | — | −0.01 | — | — | 149 |
| Cumyl | −0.10, −0.12 | +0.02 | −0.54 | −0.68, −0.82, −1.08[e] | 6, 72, 145 |

### Dialkyl Peroxides

|  |  |  |  |  |  |
|---|---|---|---|---|---|
| Diethyl | −0.5, −0.65, −0.63 | — | — | −1.63 | 144, 145, 155 |
| Ethylcumyl |  | −0.22 | −1.64 | −1.72 | 145 |
| Dicumyl |  |  | −1.35, 60% EtOH | −1.60, DMF[f,g] | — |
| Di-t-butyl |  |  | — | −2.10, DMF[f,g] | — |

### t-Butyl Peroxy Esters

|  |  |  |  |  |  |
|---|---|---|---|---|---|
| Peroxyacetate | −0.3 | — | — | −0.97, −1.02[e] | 5, 6 |
| Peroxybenzoate | Positive | −0.1 and −0.8, dioxane–water (80:20)[h] | — | −0.8, −0.82 | 5, 155, 156 |
| Peroxypelargonate |  | — | — | −0.96 | 71 |
| Peroxycaprate |  | — | — | −0.90 | 71 |
| Peroxylaurate |  | — | — | −0.87 | 71 |
| Peroxymyristate |  | — | — | −0.82 | 71 |

[a] Supporting electrolytes: 0.1 $M$ LiSO$_4$,; 0.01 $N$ HCl; 0.1 $N$ KCl and 0.04 $M$ Britten–Robinson buffer, pH 7.5; acetate buffer, pH 4; phosphate buffer, pH 9.

[b] Supporting electrolytes: 0.1 $N$ KCl, 0.1 $M$ H$_2$SO$_4$, 0.1 $N$ H$_2$SO$_4$.

[c] Supporting electrolytes: 0.3 $N$ LiCl, 0.1 $N$ KCl; 0.2 $N$ (CH$_3$)$_4$NOH in 60% EtOH.

[d] Supporting electrolyte: 0.3 $N$ LiCl.

[e] 0.03 $N$ LiCl + 0.01% ethyl cellulose.

[f] $E_{1/2}$ versus Hg pool; dimethylformamide (DMF) saturated with (CH$_3$)$_4$NCl; reported irreducible in other systems.

[g] Private communication from E. J. Forman and C. M. Wright.

[h] Ten substituted t-butyl peroxybenzoates show dual waves at −0.1 and −0.8 volt in dioxane–water; as water content increases, the first grows while the second diminishes (156).

peroxy acids than of hydroperoxides and suggests that in aqueous media peroxy ester reduction may proceed through a peroxy acid intermediate; this despite the generally accepted idea (28) that peroxy esters are derivatives of, and hydrolyze to, acids and hydroperoxides, rather than peroxy acids and alcohols.

In semi-aqueous solvent media peroxy esters show more complex behavior. Schulz and Schwarz (156), working with 0.25 $N$ ammonium nitrate in dioxane–water solutions found two waves for $t$-butylperoxy esters: one at $-0.85$ volt, which was diffusion controlled, and an earlier one at $-0.1$ volt, which had a kinetic character. As the water content of the medium was increased, this kinetic wave grew at the expense of the diffusion-controlled reduction, suggesting that the rate-controlling kinetic step giving rise to the early wave is the heterolytic polar cleavage of the $-O-O-$ bond in the field of the electrical double layer at the dropping-mercury electrode, promoted by the increased dielectric constant of the more aqueous media. They implied that the later, diffusion-limited wave resulted from a single-step homolytic cleavage. Other hypotheses can be advanced, however, involving hydrolysis, adsorption equilibria, and the role of protons in the reduction process. Certainly peroxy esters have a complex and perhaps a controversial polarographic behavior.

The situation is similarly complex for other peroxide classes, whose polarographic behavior can be altered significantly and systematically by such factors as the type of supporting electrolyte or the presence of polymeric additives. For instance, as already mentioned, the use of quaternary tetraalkylammonium electrolytes consistently shifts reduction potentials to more negative values (66, 145), presumably through a competitive adsorption process at the surface of the mercury drop. Again, the presence of a small amount of ethyl cellulose in benzene–methanol (1:1) will allow the resolution of a single wave due to fatty ester hydroperoxides into a double wave (6, 67) that has been attributed to monohydroperoxides and dihydroperoxides (162); this effect is maintained if the benzene is replaced by toluene or xylene but disappears with nonaromatic hydrocarbons and chloro compounds (6).

In view of the complex effects on both the qualitatively significant half-wave potential and the quantitatively significant diffusion-current constant by peroxide structure (both between and within classes), by composition of the solvent medium and selection of supporting electrolyte, and by the presence of modifying additives, there is a need for studies more extensive and penetrating in scope than those conducted heretofore. Such studies should take fuller advantage of the coupling of polarographic analysis with separation techniques (144, 145, 164) and chemical conversions; this should be a more effective approach with complex mixtures.

## (d) Quantitative Determination in a Galvanic Cell

A novel related technique for peroxide analysis is due to Kober (168). He has described a galvanic cell for the continuous and discontinuous quantitative determination of oxygen and inorganic and organic peroxides and hydroperoxides in liquids. The cell consists of a corrodible electrode made of lead, zinc, iron, tin, or cadmium tubing 1–2 meters long and 3–9 mm in diameter, coated on the outside with an inert plastic; this tube is separated by a concentric membrane sheath from an inert coaxial inner electrode of gold, platinum, silver, or copper wire; the cell is connected to a current-measuring apparatus and an integrator. The electrolyte is a 10–30% potassium hydroxide solution, which may contain auxiliary solvents. Oxygen and peroxides act as depolarizers in this system, and their content is calculated according to Faraday's law by the equation

$$m = \frac{M}{Fz} \int_0^t I\, dt,$$

where $m$ is the mass of the depolarizer, in grams, $M$ is its molecular weight, $F$ is the Faraday number, $z_0$ is the number of electrons transmitted in the reduction, and $\int_0^t I\, dt$ is the integral of the time changes of current strength, in ampere-seconds. Sensitivities down to $10^{-7}\%$ and an accuracy of $\pm 2\%$ were claimed. As examples, Kober cited the continuous and discontinuous analysis for cumene hydroperoxide at about 1% concentration, using lead and silver electrodes.

### B. Chemical Reduction Methods

Methods depending on quantitative reduction of peroxides by chemical reagents have been summarized in Table 2B. These are primarily wet-chemical volumetric procedures.

### 1. Iodide Methods

Reduction with iodide is the most widely used and the most generally useful of all the reactions for peroxide analysis; the variety of quantitative procedures based on this reduction is familiar to all who must analyze for these compounds. Perhaps because of this very variety, many workers seemingly unaware of available improvements have clung to old familiar methods, such as those of Wheeler (169) or Lea (170) for peroxides in oils, long after substantially improved approaches (171–173) had been developed.

The proliferation of methods has grown partly from the recurring needs for improvement in the accuracy, precision, speed, simplicity, or scope of peroxide methods, and partly from different opinions among chemists as to what in a procedure constitutes simplicity rather than complexity, and where the line should be drawn between the practical and impractical.

In this section we shall consider the basic factors affecting the quantitative determination of peroxides via reduction with iodide. Our intent is to assist analytical chemists in selecting or in tailoring iodide-reduction procedures to fit the requirements of their specific analytical situations. In each of the categories of iodide methods there are often acceptable alternatives in both apparatus and procedural details that are consistent with sound primary principles, but should also reflect such practical factors as the degree of analytical difficulty; the availability of equipment; realistically appraised requirements for speed, precision, and sensitivity; and the number and frequency of analyses to be performed. By taking sufficient pains, results of the highest quality will usually be obtainable. Judicious corner-cutting, on the other hand, can often produce simpler, quicker, and less expensive analyses that are entirely adequate for a particular application.

For judging the adequacy of a proposed procedure suitable criteria are necessary, unless its performance in the new situation can clearly be extrapolated from past experience. Our laboratory has consistently used the criteria (10) cited repeatedly by earlier workers (171, 174), namely, that results be substantially independent of sample size, reaction time, and reagent composition, within reasonable limits, and be unaffected by any nonperoxidic components that may be present in the sample.

## (a) Exclusion of Oxygen

Solvent systems proposed for iodide-reduction methods are nearly all based on acetic acid or isopropyl alcohol. Each of these has its advantages and disadvantages. On the one hand, alcohols are less active solvents, so that iodide reduces peroxides more slowly in alcohols than in acetic acid or chloroform–acetic acid blends, a disadvantage that is offset by the fact that alcohols prevent discernible interference through air oxidation of iodide (10, 173, 175). On the other hand, for the more active acetic acid–based solvents it has become firmly established (10, 171) that effective deaeration must be employed to avoid oxygen errors, and all acceptable procedures require this step.

A variety of techniques may be used for deaeration. Almost always this step involves purging with an inert gas from an appropriate source: nitrogen or carbon dioxide from a cylinder, or carbon dioxide generated in the reaction flask by adding small pieces of Dry Ice or a gram or two of sodium bicarbonate. We feel that the use of gas from a cylinder is more convenient once a distributing manifold and modified reaction flasks (10) have been constructed; in this preference we have substantial support (100, 171, 172). Direct addition of Dry Ice is widely used (176, 177) in room-temperature reductions and can function also at higher temperature (178) with assistance from a slit-rubber-tube (Bunsen) valve. In addition, the improvement by Wibaut and co-workers (133) of the sodium bicarbonate technique (179, 180) produces excellent results with peroxides that react at room temperature.

Complete deaeration without any purge gas can be achieved by allowing chloroform–acetic acid or ether–acetic acid mixtures to reflux, so that boiling solvent vapors physically exclude oxygen from the reaction zone (181). The fine performance of this clever adaptation is offset, however, by the specialized, somewhat ungainly glassware and a relatively complex perocedure.

The issue of specialized glassware is often present in iodometric peroxide methods, and almost always arises because of deaeration steps (10, 100, 133, 171, 181). We recommend that nonstandard apparatus be avoided wherever this is practical, because of the added costs in fabrication and maintenance, but that it be employed whenever it will significantly improve the operability or the results of a procedure.

Recommended deaeration procedures can range from vigorous and complex (100, 171, 181) to practical and casual (176, 177), with the latter approach giving surprisingly good results in appropriate circumstances. The primary consideration should be exclusion of oxygen during actual reduction of the peroxide (181), for it is then that reaction between free-radical intermediates and molecular oxygen can occur, to produce additional peroxide in proportion to the original peroxide content. Reasonable precautions also at other stages of the procedure, particularly during titration of the liberated iodine, should yield secondary improvements in accuracy and precision; such gains may be of crucial importance at low peroxide concentrations (100, 182).

(b)  ELIMINATION OF OLEFIN INTERFERENCE BY USE OF EXCESS IODIDE

The advantages of using a relatively large excess of iodide have been discussed recently (10). Besides increasing the rate of peroxide reduction this practice also prevents errors due to volatilization of iodine and to the addition of iodine to unsaturated materials (173). By maintaining the equilibrium

$$I_2 + I^- \rightleftharpoons I_3^-$$

far to the right, the excess iodide holds liberated iodine as triiodide ion, a form in which it differs from molecular iodine by being both nonvolatile and unable to undergo addition to double bonds.

Wager, Smith, and Peters (173) compared conditions corresponding to initial iodide concentrations of 0.26 and 0.95 $M$, and to final iodide–triiodide molar ratios of 10 and 44. They concluded that the amount of iodide is not critical provided a large excess is present. In our laboratory* an evaluation of iodide–isopropyl alcohol methods (173, 175) showed that liberation of iodine

*Private communication from R. F. Goddu, Hercules Incorporated, Research Center, Wilmington, Delaware (1952).

by hydroperoxides was stoichiometric for final iodide–triiodide ratios as low as 3–5, corresponding to consumption of 40–50% of the added iodide in the reaction,

$$ROOH + 3I^- + 2H^+ \longrightarrow ROH + H_2O + I_3^-$$

Results were some 30% low, however, for ratios of 0.3, in which 90% of the iodide was consumed.

The first substantial study of the possibility that interference may occur in peroxide analysis through iodine addition by unsaturated materials was made by Wagner, Smith, and Peters (173), who showed that whereas molecular iodine readily adds to the double bonds in a monoolefin (cyclohexene) and a conjugated diolefin (isoprene), there is no trace of such addition by triiodide. They considered the idea that triiodide addition might occur if catalyzed by peroxides, particularly in conjugated systems, and eliminated it by demonstrating that there is no iodine loss during reduction of benzoyl peroxide with sodium iodide in the presence of isoprene. Skellon and Wills (180) similarly reduced benzoyl peroxide in the presence of large excesses of oleic acid, without detecting iodine absorption; in addition, Banerjee and Budke studied the analysis for traces of peroxide in many solvents, saturated and unsaturated, and found no indication of interference from olefins (98). Wagner, Smith, and Peters (173) obtained further evidence on the non-reactivity of triiodide with olefins in the analysis of four autoxidized monolefins, diisobutylene, 2-pentene, cyclohexene, and tetralin, whose peroxidic components are the easily reduced alkyl hydroperoxides. In each case the resulting peroxide number was essentially independent of reaction time, sample size, and iodide concentration.

In these experiments with olefins the final iodide–triiodide molar ratios were all relatively high, ranging from 20 up to 1100, and above 50 in three-quarters of the cases. The data warrant the conclusion that at least in media based on isopropyl alcohol interferences from olefins of all types may be prevented in iodometric peroxide determinations by using sufficiently high proportions of excess iodide reagent. The same is probably true in the other useful media.

While interferences due to the volatility of iodine and its addition to olefinic double bonds can be eliminated by using a sufficient excess of alkali iodide reagent, this measure will not, of course, prevent interferences from reaction of iodine with acetone (175, 177, 183) or reduction with mercaptans and other sulfur compounds.

(c)  ANALYSIS OF EASILY REDUCED PEROXIDES

The classes of organic peroxides vary greatly in their ease of reduction by iodide, and a clear and seemingly natural distinction has been recognized between (a) the more easily reduced types that respond quickly and quantitatively to relatively mild reagents, and (b) those less easily reduced varieties that do not. It is on peroxides in the first category that the majority of iodometric analyses are performed; they are reduced by sodium or potassium iodide in acetic acid or an alcohol at room temperature (although the alcoholic reagents are customarily heated to boiling). It has been proposed that organic peroxides be classified operationally for iodometric analysis into three categories (10) based on response to a refluxing sodium iodide–isopropyl alcohol reagent. These categories, shown in Table 11, are as follows:

1. Easily reduced (rapid response).
2. Moderately stable (slow response).
3. Reduced with difficulty (no response).

The easily reduced subdivision includes principally the peroxy acids, the diacyl peroxides, and the alkyl hydroperoxides; its members are defined by the requirement for rapid quantitative reduction. A special case can be made for the *t*-butyl peroxy esters, which in effect, can be promoted from the moderately stable category by catalysis with a trace of ferric ion (184), as will be seen.

Two basic rules for sound quantitative iodometric peroxide analysis developed earlier in this section specify the exclusion of molecular oxygen except in alcoholic media, and the use of an adequate excess of iodide reagent. For the easily reduced peroxides, two other rules complete the basic

TABLE 11.  CLASSIFICATION OF ORGANIC PEROXIDES BY EASE OF REDUCTION WITH SODIUM IODIDE–ISOPROPYL ALCOHOL AT REFLUX

| Category | Membership | Response to Reagent[a] |
|---|---|---|
| Easily reduced | Peroxy acids, diacyl peroxides, all hydroperoxides, and certain other compounds | Liberate iodine stoichiometrically in 5 min or less |
| Moderately stable | Most peroxy esters, aldehyde peroxides, and ketone peroxides; di-*n*-alkyl peroxides; and mixed *n*-alkyl peroxides | Liberate iodine, but stoichiometric reduction requires more than 5 min |
| Difficult to reduce | Transannular peroxides, diaralkyl peroxides, di-*t*-alkyl peroxides, and mixed aralkyl–*t*–alkyl peroxides | Show no iodine liberation with the test reagent even on prolonged refluxing |

[a] Based on reaction of 0.5–1.0 millimole with 2 g NaI plus 2.5–5 ml acetic acid in 25–50 ml isopropyl alcohol boiling under reflux.

requirements: the reduction period must be adequate and there must be freedom from interference by other components of the sample. These components may include both substances that liberate iodine, such as the slower reacting, moderately stable peroxides or ferric iron (the latter interference can be controlled by complexing with a small amount of a fluoride salt), and those that consume it, such as acetone and mercaptans.

Some of the published iodide-reduction procedures that we consider to be satisfactory for quantitative analysis of the easily reduced peroxides are shown in Table 12. Iodometric procedures alone cannot differentiate mixtures of the several types of these reactive peroxides; there are more broadly based methods for doing so (73), as described in Section IV-B-2-c.

The potassium iodide–acetic acid methods in Table 12 specify brief reaction periods at room temperature. With chloroform also present, 1 hr is allowed for the complete reaction of lipid peroxide (Lea's cold method), probably an excessive period for peroxy acids and diacyl peroxides. The sodium iodide–isopropyl alcohol reagent mixtures are generally heated to shorten reaction times in this less active medium; reaction at room temperature is also reasonably fast, however. Buncel and Davies (185) used this reagent at room temperature to determine peroxysilanes, allowing the reaction to proceed overnight in the dark.

We call special attention to the method of Abrahamson and Linschitz (182), cited usually for the use of a dead-stop end point in the titration of liberated iodine in colored solutions. This method, in addition, has a remarkable combination of sensitivity, precision, and accuracy that places it in a class by itself for iodometric peroxide determination by a volumetric technique. For instance, 4 microequivalents of tetralin hydroperoxide were determined with better than 0.5% accuracy and precision, titrations being $16.60 \pm 0.03$ ml of 0.00026 N thiosulfate.

The modified reagent proposed by Hartman and White (183) deserves a broader evaluation than it has received; it was applied to only a few fats and oils and to benzoyl peroxide. This reagent consists of sodium iodide in $t$-butyl alcohol acidified with citric acid and modified with carbon tetrachloride. $t$-Butyl alcohol was selected because of its inertness to molecular iodine and its good solvent ability for fats; citric acid was used because it is a 50 times stronger acid than acetic acid, and hence a better source of hydrogen ions; carbon tetrachloride or chloroform kept the fat (and molecular iodine) in a bottom layer during titration.

This reagent has the following two noteworthy characteristics:

1. Its reducing power is appreciably enhanced over that of sodium iodide–isopropyl alcohol, acidified with acetic acid, and is approximately equivalent to that of sodium (or potassium) iodide–acetic acid–carbon tetrachloride (or

TABLE **12.** SOME SATISFACTORY IODOMETRIC METHODS FOR EASILY REDUCED PEROXIDES

| Reagent | Solvent | Temperature | Deaeration | Remarks | References |
|---|---|---|---|---|---|
| NaI | Isopropyl alcohol (acidified) | Boiling | None | Quantitative reaction in 5 min or less | 10, 173 |
| NaI or KI | Isopropyl alcohol (acidified) | Boiling | Continuous $N_2$ or $CO_2$ flushing during titration | Precise and accurate at trace peroxide levels, in colored solutions; uses electrometric dead-stop end point | 182 |
| KI | Acetic acid | Room | Several lumps of Dry Ice | Reaction period 5 min; results very precise | 176, 177 |
| KI | Acetic acid | Room | $NaHCO_3$ | Reaction period 10 or 15 min in the dark; very precise results | 133, 180 |
| KI | Acetic acid–chloroform (3 : 2) | Boiling | $N_2$ or $CO_2$ gas | Reaction period of 1 hr in the dark appears to be excessive; special apparatus | 171, 186 |
| KI | Acetic acid–chloroform (1 : 1) | Boiling | Chloroform vapor | Zero blank; reaction period 3–5 min | 181 |
| NaI | Acetic acid–chloroform (3 : 2) $4 \times 10^{-5}$ $M$ in $Fe^{3+}$ | Room | $N_2$ gas | Quantitative reduction of $t$-butylperoxy esters in 5–10 min | 184 |

chloroform); it can apparently yield quantitative results on easily reduced peroxides in 15 min at room temperature.

2. It consistently gives a zero blank, whether in the presence or absence of air. There appears to be a slight oxygen error, with results slightly higher in the presence of air because of induced peroxide formation (although not for the rapidly reacting benzoyl peroxide).

Areas needing investigation are the behavior of this reagent system on other classes of peroxides, with and without carbon tetrachloride, with and without heating, and with a more adequate amount of iodide, and the effect of substituting citric for acetic acid in the regular sodium iodide–isopropyl alcohol procedure (10, 173).

Silbert and Swern's $Fe^{3+}$-catalyzed modification (184) of Lea's cold method is included in Table 12 as a special case. The $t$-butylperoxy esters, for which this method is so effective, are categorized ordinarily as moderately stable peroxides, but are promoted to the easily reduced category by its use. This will be discussed further in the following subsection.

There have been two approaches to the continuous removal of iodine as it is formed during peroxide reduction in iodide–alcohol media (187, 188). In each case the primary aim is to prevent iodine loss through volatility and side reactions (188), or, specifically, through addition to conjugated diolefins (187). Despite the fact that fears of such loss appear to be generally unwarranted, the idea of continuously consuming liberated iodine is intriguing and deserves consideration in its own right.

One of the proposals, involving the continuous thiosulfate titration of liberated iodine by means of an automatic potentiometric titrator (188), appears to be too complex and fussy to offer advantage. The other approach (187) is more promising. Dahle and Holman (187) have incorporated in the reagent a known (excess) amount of a thioglycolate ester, whose thiol group consumes iodine as rapidly as it is produced; excess thioglycolate is determined by titration with a standard iodine solution. They used a solvent mixture of 50% absolute ethyl alcohol, 30% glacial acetic acid, and 20% chloroform, containing 0.2% (0.01 $M$) octyl thioglycolate. In a semimicro procedure up to 100 mg of lipid sample was dissolved in 1 ml of the solvent and reacted for 15 min after two drops of saturated aqueous potassium iodide had been added. Evaluation of this new procedure on lipids was clouded by use of the outmoded Wheeler method (169) as a standard of comparison. It would appear worthwhile to investigate the parameters of this method more extensively.

(d) CONTROL OF REDUCING POWER IN IODIDE REAGENTS

The more stable peroxides described in Table 11, which include essentially all types except peroxy acids, diacyl peroxides, and hydroperoxides, require more vigorous reaction conditions to bring about their quantitative reduction within a reasonable period of time. Several parameters that may be varied to increase and control the reducing power of iodide reagents have been discussed (10); they include the use of a catalyst, choice of solvent medium, choice of reaction temperature, addition of mineral acid, and, most importantly, the concentration of water. We shall now discuss these factors individually.

*Catalyst.* For a large, important class of organic peroxides—namely, the *t*-butylperoxy esters—use of a catalyst produces a dramatic increase in the rate of reduction by iodide (184), as we have noted; Silbert and Swern discovered that traces of $Fe^{3+}$ in an acetic acid–chloroform medium (protected from oxygen) produce a precise and stoichiometrically accurate liberation of iodine by these peroxides within 5 min at room temperature. In the absence of iron the reduction requires more than 4 hr, and even with stronger hydriodic acid reagents it takes nearly 1 hr (10). They proposed a five-step modified Haber–Weiss redox–free radical mechanism to account for the stoichiometry and products of reaction. According to this scheme, one-half the iodine is produced in a ferric–ferrous cycle wherein iodide reduces $Fe^{3+}$ to $Fe^{2+}$ and the latter is converted back to $Fe^{3+}$ by reducing the peroxy ester to an acid anion and a *t*-butoxy radical. The other half of the iodine is formed during decomposition of the *t*-butoxy radicals: these attack water or acetic acid to yield hydroxyl radicals, which oxidize iodide.

They found catalyst concentration to be very important, with an optimum of only 0.001% as ferric chloride hexahydrate, or $4 \times 10^{-5}$ M in $Fe^{3+}$. Higher catalyst levels gave low stoichiometry, perhaps because *t*-butoxy radicals would be present at higher concentrations and might decompose partially by alternative reactions.

The *t*-butylperoxy ester analysis is the only application that we have found in which reduction of a peroxide by iodide is significantly improved by a catalyst. The use of $Fe^{3+}$ with diacyl peroxides (189) and of cuprous chloride with aralkyl hydroperoxides (190) appears to be uncalled for in view of the inherent high reactivity of these peroxides.

*Choice of Solvent Medium.* Acetic acid has been virtually the exclusive choice as a reaction medium for iodide reduction of less reactive peroxides, as it is in this solvent that iodide shows its greatest reducing power. Alcohols (173, 175, 183) or acetic anhydride (191) are mainly useful with easily reduced peroxides. Acetone has been tried by several workers, but it is a most unsuitable solvent (175, 177, 183) because it reacts with part of the iodine.

Acetone-absorbed iodine is released slowly during titration with thiosulfate, causing the end point to fade quickly. We tried stepwise titration back to the end point on one occasion, and with persistence recovered substantially all of the missing iodine.

*Choice of Reaction Temperature.* The inherent reducing power of iodide reagent mixtures can be varied widely by modifying their composition. In addition, the strength of any such reagent system can be modified by varying its temperature, which would seem to allow great flexibility in tailoring the reducing power of a peroxide reagent. There are some practical considerations in varying temperature, however. Room temperature is the lowest that has been found useful and is the easiest to use. The reflux temperature of the reagent is also convenient to employ and is the maximum that should normally be considered. A pitfall, however, is that the presence of volatile components may lower the reagent's refluxing temperature significantly. For example, acetic acid solutions of sodium iodide boil at nearly 120°; but the addition of chloroform or carbon tetrachloride as auxiliary solvent or the dilution of peroxide samples with benzene or cyclohexane markedly lowers the boiling temperature of such a reagent, thereby slowing the peroxide reduction.

In the range between ambient and reflux reaction temperatures can be controlled directly with heating baths, a device used in several well-known methods: the "hot" modification of the Lea method (171) employs a boiling-water bath and also one maintained at 77°, and the hydriodic acid method of Vaughan and Rust (192, 193) carries out reduction of di-*t*-butyl peroxide at 60°. Such methods of heating have seemed to us to be unnecessary and undesirable complications in most instances, to be considered only when they are more conveniently available than boiling under reflux. This view would require revision if future studies should show reaction temperature to be an important parameter in tailoring reagent activity.

*The Roles of Mineral Acid and Water Concentration.* In the chapter on peroxide analysis in the *Treatise on Analytical Chemistry* (1) we have discussed at some length the manner in which the reducing power of sodium iodide–acetic acid reagents changes when water or mineral acids are added as additional components. The addition of such acids as hydriodic, hydrochloric, phosphoric, or sulfuric sharply increases the reducing potency of these iodide reagents, whereas the addition of water brings about a sharp decrease.

The general increase in reducing power when mineral acids are added can be attributed to higher concentrations of hydriodic acid (a strong reductant) produced by reaction with the iodide salt. In fact the original reducing power of iodide–acetic acid must be largely due to the presence of small but significant concentrations of hydriodic acid formed from the dissolved iodide.

The general decrease in the reducing power of iodide–acetic acid or hydriodic acid–acetic acid solutions when water is added can be attributed to hydration of the hydriodic acid through a series of equilibria that may be represented by the equation

$$HI + nH_2O \; \rightleftharpoons \; HI \cdot H_2O + (n-1)H_2O \; \rightleftharpoons$$

$$HI \cdot 2H_2O + (n-2)H_2O \; \rightleftharpoons \; HI \cdot 3H_2O + (n-3)H_2O \; \rightleftharpoons$$

$$\text{higher complexes}$$

These complexes are probably solvated as well. It is our hypothesis that the reducing power of a hydriodic acid molecule is inversely related to the number of water molecules in its coordination sphere. This means that the unhydrated or lightly hydrated hydriodic acid species at the left of the above series of equilibria should be the active reducing agents in the system. Addition of water decreases their concentration.

This apparently straightforward effect of water is moderated by a competing hydration reaction that ties up some of the water and thus regulates the amount available to participate in the above equilibria. This reaction is the association of acetic acid and water:

$$(HOAc)_2 + 2H_2O \; \rightleftharpoons \; 2HOAc \cdot H_2O$$

Since acetic acid is the principal component of the reagent system, its effect on the extent of hydriodic acid hydration can be large. In fact it is realistic to look on the acetic acid–water ratio as the parameter that controls the reducing power of the hydriodic acid that may be present.

The credibility of the above equilibria was established through two studies (10), one on the formation of hydriodic acid *in situ* by addition of concentrated (37%) hydrochloric acid, and the other on the direct use of blends of constant-boiling (57%) hydriodic acid and acetic acid. (Both these mineral acids contain substantial amounts of water: the hydrochloric, 3.4 moles/mole, and the hydriodic, 5.3 moles/mole; hence adding them to glacial acetic acid very materially decreases the acetic acid–water ratio.)

The first study is summarized in Figure 3, which shows the extent of reduction of dicumyl peroxide in 10 and 30 min at room temperature by various combinations of acetic acid, hydrochloric acid, iodide, and water. The controlling parameter is clearly water content; the extent of reduction increases linearly with decreasing water content in the reactants from about 15% down at least as far as 1%.

As a practical demonstration of this water effect, one gets a significantly more potent reagent mixture by adding 2 ml of concentrated hydrochloric acid to 50 ml of an iodide–acetic acid solution than by adding 5 ml, because the smaller portion of acid yields a drier reagent.

Figure 4 summarizes the second study, in which the reducing powers of a number of acetic acid–hydriodic acid blends are compared, mostly at reflux

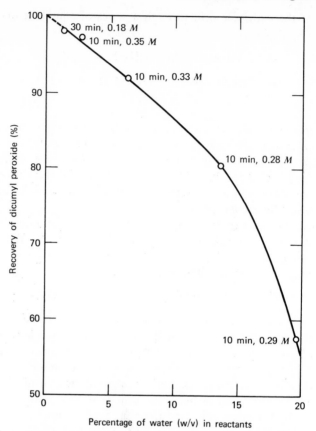

**Figure 3.** Effect of water on HI methods at room temperature. Various proportions of acetic acid, 37% HCl, water, and NaI or KI. Reaction time: 10 and 30 min; molar concentration of iodide: 0.18, 0.35, 0.33, etc.

temperature. To provide moderate reaction rates the substrate is 7-cumyl alcohol, a compound whose reduction by hydriodic acid presents an interference in some peroxide methods. Again, as in Figure 3, there is a smooth reciprocal relation between reducing power and water content. Notice that the results are almost independent of iodide concentration and reaction time. Notice also that the results for the reagent produced from sodium iodide and hydrochloric acid fall on the same curve.

The reagent at the right of Figure 4, containing 37% water, is the well-known reagent of Vaughan and Rust (192, 193). It consists of equal parts by volume of acetic and 57% hydriodic acids, and was employed at the recommended 60°.

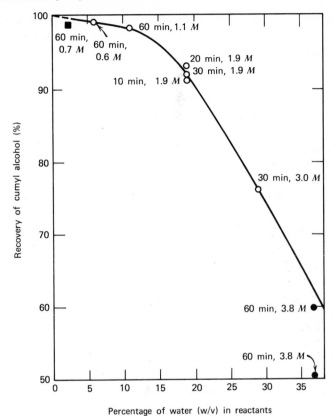

**Figure 4.** Effect of water on HI methods at reflux. ○ Acetic acid–57% HI, refluxed at about 120°: ● acetic acid–57% HI, reacted at 60°, as in reference 53; ■ acetic acid–NaI–37° HCl, reflux at about 120°; reaction time: 10, 20, 30, and 60, min; molar concentration of iodide: 0.7, 1.1, 3.0, etc.

Some water must be present to reduce the volatility of anhydrous hydrogen iodide and to mitigate its extremely potent reducing power. The latter can be illustrated by an experience in which a sodium iodide–acetic acid solution was acidified with concentrated sulfuric acid, so as to provide a solution of anhydrous hydrogen iodide. The result was an immediate rapid reduction to hydrogen sulfide, with copious liberation of iodine.

As might be anticipated, quite satisfactory reagents can be obtained by acidifying sodium iodide–acetic acid solutions with small amounts of concentrated hydriodic, rather than concentrated hydrochloric, acid (10). The discovery that concentrated hydriodic acid can be freed of iodine by simply shaking with mercury before use (5, 194) has made this substitution practical.

Water contents are equivalent, as the two acids contain equal concentrations of water (75%) on a volume basis. Data on blanks give evidence that the use of hydriodic, rather than hydrochloric, acid produces a slightly stronger reagent (1).

### (e) ANALYSIS OF MODERATELY STABLE PEROXIDES

Above the level of reducing power adequate for easily reduced peroxides, two additional useful levels have been defined, one with intermediate and one with strong reducing properties. Examples are described in Table 13. The methods with intermediate potency are a compromise between the ability to reduce a large group of moderately stable peroxides (Table 11) and the need to be relatively free from interference by nonperoxidic components. Methods in the strong category, however, are more seriously affected by such interference; reduction of nonperoxides is apt to be severe and unavoidable without destroying the effectiveness of the reagent in the intended peroxide reduction.

Reagents for the two intermediate-level methods in Table 13 are closely related, both stemming from a solution of six grams of sodium iodide in 50 ml of acetic acid. In the first case 2 ml of 37% hydrochloric acid is added and the reagent is used at room temperature. In the second case the solution is used at the boiling point after adding 3 ml of water to suppress the reduction of nonperoxides.

We have used these methods extensively for the determination of dicumyl peroxide in autoxidized cumene and for similar analyses in related aralkyl systems. Frequently the samples contain potentially reducible by-products, typified by cumyl alcohol and α-methylstyrene in autoxidized cumene. Our experience has shown that the second method, the boiling sodium iodide–acetic acid–water reagent, is more reliable for such samples than is the first, the room-temperature hydriodic acid method. With the latter, although neither the alcohol nor olefin by-products show the slightest trace of iodine liberation in the absence of dicumyl peroxide, they are both partially reduced in its presence, however, causing peroxide analyses to be proportionately high.

This interference appears to have been specifically due to peroxide reversal of Markownikoff's rule during addition of hydriodic acid to the side-chain double bond of α-methylstyrene (10); the resulting primary organic iodide group should be reduced readily by excess hydriodic acid, unlike the normally expected tertiary iodide. Such reversed addition in the presence of peroxides is well known for hydrobromic acid, although it does not occur for hydrochloric acid and had not been observed previously for hydriodic acid; Royals had predicted such a possibility, however (195).

The preferred method (No. 2 in Table 13) required two modifications of the room-temperature method. The first was to drastically reduce the hydriodic acid concentration (from 0.5 to $10^{-4}$ or $10^{-5}$ $M$) by eliminating the addition of mineral acid while countering the attendant reduced reactivity by heating the reagent to boiling. It was found that this reagent continued to reduce cumyl alcohol and $\alpha$-methylstyrene, now in the absence of peroxides as well as in their presence.

The second modification was to add an appropriate small amount of water. This has an effect summarized in Figure 5. The shaded areas indicate that

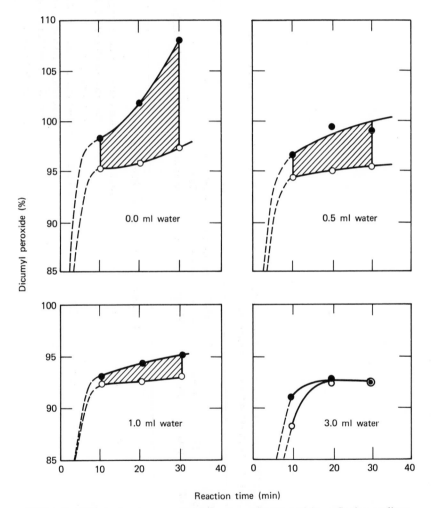

**Figure 5.** Effect of water on reduction of cumyl alcohol by refluxing sodium iodide–acetic Acid. ○ 75 mg of dicumyl peroxide; ● 75 mg of dicumyl peroxide + 250 mg of cumyl alcohol; used 3 g of NaI (generous measure) in 50 ml of acetic acid, reaction at reflux.

TABLE 13.   COMPOSITION OF USEFUL IODOMETRIC REAGENTS FOR MODERATELY STABLE[a] AND DIFFICULT-TO-REDUCE[a] PEROXIDES

| Reagent[b]: Acetic Acid Solution of | HI Molarity | Total I⁻ Molarity | H₂O Concentration | | Reaction Temperature | Remarks | Reference |
|---|---|---|---|---|---|---|---|
| | | | Molarity | % (w/v) | | | |
| *A. Methods of Intermediate Reducing Power* | | | | | | | |
| 1. NaI–HCl[c] | 0.5 | 0.75 | 1.7 | 3.1 | Room | Quantitative reaction of dicumyl peroxide in 10 min; di-*t*-butyl peroxide inert | 10 |
| 2. NaI–H₂O[d] | ~10⁻⁵ | 0.75 | 3.1 | 5.7 | Boiling[e] | Quantitative reduction of dicumyl peroxide in 20 min; di-*t*-butyl peroxide inert; little interference from nonperoxides | 10 |

### B. Methods of Strong Reducing Power

| | | | | | | | |
|---|---|---|---|---|---|---|---|
| 1. NaI–HCl[c, f] | 0.5 | 0.75 | 1.7 | 3.1 | Boiling[e] | Quantitative reduction of di-*t*-butyl peroxide in 5 min, frequent interference from nonperoxides, e.g., quantitative reduction of cumyl alcohol | 10 |
| 2. KI–HCl[c] | 0.17 | 0.17 | 6.8 | 12.3 | Boiling[e] | Weaker reagent than preceding entry; only 90% reduction of di-*t*-butyl peroxide and no attack on cumyl alcohol | 73 |

[a] As defined in Table 12.
[b] All reagents deaerated with $N_2$ or $CO_2$ gas.
[c] Forms HI.
[d] Method II in reference 10.
[e] Approximately 120°.
[f] Method III in reference 10.

the addition of increasing amounts of water causes the interference to decrease progressively. Notice that when interference is eliminated (3.0 ml of water added), the necessary reaction period for dicumyl peroxide reduction had increased from 10 to 20 min.

Notice also the stoichiometric deficiency in iodine liberation by the dicumyl peroxide, which was a pure specimen. As has been reported (10), this behavior is not well understood, but it is remarkably precise. The following are typical reproducible recoveries obtained for several pure bis(aralkyl) peroxides by the recommended procedure (30-min reaction at reflux under inert gas with 6 g of sodium iodide and 3 ml of water in 50 ml of acetic acid):

| Peroxide | Recovery (%) |
|---|---|
| Dicumyl peroxide | 92.5 |
| Di-(p-cymene)peroxide | 87.5 |
| Di-(p-isopropylbenzene)peroxide | 89.0 |

When results are corrected for these conversions, they appear to be both accurate and precise. This method is applicable to other similar peroxides after calibration with pure specimens. Such analyses have been found to be independent of sample size and reaction time, within broad limits (10).

A successful application of this method to the determination of 0.25–0.5% dicumyl peroxide in fire-resistant polystyrene materials (containing other peroxides) recently has been described by Brammer, Frost, and Reid (123); the dicumyl peroxide was recovered by acetone extraction, separated from other peroxides by thin-layer chromatography, by the method of Knappe and Peteri (44), and determined directly on the silica gel substrate of the TLC spot by a tenfold scaledown of the original iodometric procedure (10). Only about 1-mg portions of dicumyl peroxide were used; at the 0.5% level the standard deviation was 0.03% absolute, corresponding to 95% confidence limits of ±12% relative for single determinations.

An alternative and much simpler technique for the final analysis in this case would be ultraviolet spectrophotometry, on an extract of the TLC spot.

Whether controlled additions of water could have similarly suppressed the previously discussed interferences in the room-temperature hydriodic acid method (2 ml of 37% hydrochloric acid per 50 ml of iodide solution) has not been investigated. There is a good possibility that such would prove to be the case, although reaction periods might have to be substantially lengthened.

(f)  ANALYSIS OF DIFFICULT-TO-REDUCE PEROXIDES

The final two methods in Table 13 bracket the practical upper and lower limits for reducing power of reagents that can determine the peroxides that are difficult to reduce. The first, our so-called method III, has the most powerful reagent for peroxide analysis that has been described (10). The second is the method used by Horner and Jürgens (73) to determine dialkyl peroxides when resolving mixtures of peroxide classes by the group of procedural combinations described in Section IV-B-3. The reasons why this pair of methods should differ in reducing power are clear in Table 13, which shows that the second has four times the water content and less than one-fifth the hydriodic acid concentration of the first.

TABLE 14.   COMPARISON OF TWO METHODS FOR DIFFICULT-TO-REDUCE PEROXIDES

| Sample | Reaction Time (min) | Apparent Percentage of Peroxide | |
|---|---|---|---|
| | | Mair and Graupner Method III[a] | Horner and Jürgens ROOR Method[b] |
| Dicumyl peroxide (commercial product) | 50 | 271.5 (90.5% of 3 moles $I_2$ per mole) | 86.5 |
| | 30 | — | 89.0, 90.2 |
| Di-*t*-butyl peroxide (nominally 97% pure) | 50 | 96.0 | 87.0 |
| | 30 | — | 86.5 |

[a] Procedure as in reference 10; reagent is 6 g NaI + 2 ml concentrated HCl in 50 ml acetic acid.

[b] Reagent as in reference 73: 1 g KI (as 1 ml saturated aqueous solution) + 5 ml concentrated HCl in 30 ml acetic acid; apparatus and procedural steps the same as for method III in reference 10.

A limited comparison that was made of these two methods is summarized in Table 14. The Horner and Jürgens method apparently reduced both dicumyl and di-*t*-butyl peroxides quantitatively; the much stronger method III performed as previously reported (10), quantitatively reducing di-*t*-butyl peroxide and producing 3 moles rather than 1 mole of iodine per mole of dicumyl peroxide. Relatively low results for di-*t*-butyl peroxide by the Horner and Jürgens method probably indicate that significant losses occurred due to sample volatility under the conditions of the analysis. This loss does not occur in method III, apparently because reduction is much more rapid and is complete in less than 5 min.

In view of the effectiveness of water concentration in controlling the reducing power of a sodium iodide–acetic acid system, as clearly shown in Figure 5, the few results in Table 14 by the Horner and Jürgens reagent (composition as detailed in Table 13) suggest that water is similarly effective in hydriodic acid–acetic systems. We recommend that this lead be evaluated systematically and that the effects of using intermediate reaction temperatures be explored also, in tailoring iodometric reagents for the difficult-to-reduce peroxides.

As indicated above in Section IV-B-1-d, the practical upper limit on reducing power of hydriodic acid reagents is set by the minimum water content necessary to prevent evolution of gaseous hydrogen iodide from the boiling reaction mixture. The reagent for the third method in Table 13, the so-called method III, is about at this limit. This reagent consists of 6 g of sodium iodide and 2 ml of 37% hydrochloric acid per 50 ml of acetic acid (10). Because this solution is employed at the boiling point, an important step in the procedure is the avoidance of even a modest inert-gas purge of the reaction flask during boiling in order to prevent what can otherwise be substantial losses of hydrogen iodide and—in the case of volatile compounds, such as di-t-butyl peroxide—substantial losses of the sample also.

It is probable that the methods in Table 13B will satisfactorily reduce the other family of relatively inert peroxides, the cyclic trimeric ketone peroxides; we have not, however, tested this point.

A modified approach to the analysis of peroxides that are reduced only with difficulty is the procedure of Braithwaite and Penketh, for the determination of butadiene polyperoxide (196). They have used a boiling reagent solution of lithium iodide in isooctanol, acidified with phosphoric acid. This interesting combination allows the use of an alcohol at a high enough temperature to promote rapid reaction, along with good iodide solubility. This new system has been applied to no other peroxides; it would appear to deserve a broader evaluation.

### (g)  Guidelines on Potential Interference by Hydriodic Acid Reduction of Nonperoxides

A study of reactivity by several classes of potentially reducible nonperoxides in method III (10) has turned up two apparent generalizations: it is clear that an alcohol and the olefin resulting from its dehydration are equally reducible; it seems also that to be reducible, nonperoxides must contain a cyclic structure.

Among terpenes and related compounds, there is a clean division of the monocyclics into reactive and unreactive types. Thus all difunctional monocyclics (i.e., menthadienes, menthenols, and menthadiols) liberate exactly 1 mole of iodine within 15 min, whereas the corresponding monofunctional menthenes or menthols are virtually inert even after prolonged reaction periods. Bicyclic terpenes are also virtually inert unless they can isomerize to bifunctional monocyclics.

Rates of reduction of exocyclic double bonds conjugated with aromatic rings, and the equivalent alcohols, depend on the levels of alkyl substitution on the α-carbon. For example, α-methylstyrene and cumyl (α,α-dimethyl-benzyl) alcohol are reduced promptly and quantitatively; styrene and α-methylbenzyl alcohol, on the other hand, are reduced more slowly and incompletely.

A double bond conjugated with two benzene rings, as in *trans*-stilbene, is almost inert. So is the nonaromatic conjugated double bond in 1-phenyl-cyclohexene.

## 2. Organic Sulfide and Arsine Methods

Diphenyl sulfide rapidly and quantitatively reduces peroxy acids, with no significant attack on other peroxides, including diacyl peroxides and alkyl hydroperoxides (73). Another thioether, *p*-nitrothioanisole, has the same selectivity. It should be recognized, however, that this behavior of these two reagents is not the rule for thioethers but rather a useful exception; only these two among the organic sulfides react so slowly with diacyl peroxides and hydroperoxides. Thiophane and diethyl sulfide (74), in contrast, reduce dibenzoyl peroxide quantitatively within 15 min at room temperature. Table 15 shows how reactivity toward a typical diacyl peroxide varies with the structure of the thioether.

TABLE 15. RELATIVE EFFECTIVENESS OF THIOETHERS IN
DECOMPOSING DIBENZOYL PEROXIDE[a]

| Thioether | Reaction Half-time (min) |
| --- | --- |
| Thiophane | 4 |
| Diethyl sulfide | 5 |
| Dibutyl sulfide | 24 |
| *p*-Methoxythioanisole | 50 |
| Ethyl *p*-tolyl sulfide | 85 |
| *p*-Methylthioanisole | 122 |
| Dibenzyl sulfide | 122 |
| Diisopropyl sulfide | 150 |
| Thioanisole | 185 |
| *p*-Nitrothioanisole | 7 days |
| Diphenyl sulfide | 7 days |

[a] Date of Horner and Jürgens (74).

Triethyl arsine quantitatively reduces peroxy acids, diacyl peroxides, and alkyl hydroperoxides, producing, respectively, 1 and 2 moles of carboxylic acid and 1 mole of alcohol (73). Reduction with triphenyl arsine is equally quantitative, but slower.

### 3.  Resolution of Peroxide Classes by Combined Use of Iodide, Alkali, Sulfide, and Arsine Methods

Horner and Jürgens have coupled reduction with diphenyl sulfide and triethyl arsine to two iodide-reduction procedures (one mild and one more vigorous) and an acidimetric titration to obtain a flexible scheme for quantitatively resolving, without a prior separation, mixtures of any two or all of the four peroxide classes: peroxy acids, diacyl peroxides, alkyl hydroperoxides, and dialkyl peroxides, along with organic acids (73). Such discrimination should make this approach a powerful tool for studying peroxide syntheses and reactions, two particularly promising applications being the alkylidene peroxide mixtures formed by the reaction of hydrogen peroxide or alkyl hydroperoxides with aldehydes or ketones and certain other compounds, and the mixtures resulting from extensive autoxidation of fats and oils.

The analytical scheme proposed by Horner and Jürgens consists of a group of eleven procedures; these have been summarized in Table 16. Each procedure is designed to analyze one specific peroxide mixture out of all possible combinations of the four classes; nonperoxidic organic acids are determined as well. Specificity has been obtained primarily by the use of two reagents: diphenyl sulfide, which rapidly and specifically converts peroxy acids to carboxylic acids while itself being converted to diphenyl sulfoxide; and triethyl arsine, which is inert to dialkyl peroxides but reacts quantitatively with the three other peroxide types according to the equations:

$$R-\overset{\overset{\displaystyle O}{\|}}{C}-OOH + Et_3As \xrightarrow{H_2O} R-\overset{\overset{\displaystyle O}{\|}}{C}-OH + Et_3AsO \cdot H_2O$$

$$R-\overset{\overset{\displaystyle O}{\|}}{C}-OO-\overset{\overset{\displaystyle O}{\|}}{C}-R + Et_3As \xrightarrow{H_2O} 2R-\overset{\overset{\displaystyle O}{\|}}{C}-OH + Et_3AsO \cdot H_2O$$

$$R-OOH + Et_3As \xrightarrow{H_2O} R-OH + Et_3AsO \cdot H_2O$$

Triethyl arsine is used in two modes: one as a standard solution whose consumption is a measure of the total easily reduced peroxides, and the other as a conversion agent that specifically reveals diacyl peroxides through the additional acidity (2 moles) produced by their reaction.

The methods of Horner and Jürgens appear to be aimed primarily at the analysis of mixtures whose qualitative compositions are known, at least as to peroxide classes. In some cases this will represent the analytical situation realistically. Where there is no positive information about the nature of the sample, however, the final procedure in Table 16 has been provided; it covers the possible presence of all four peroxide classes.

The general utility of this scheme for the resolution of peroxide mixtures into their component classes would, it seems, be greatly enhanced if there were a way to include peroxy esters. We have found no indication in the literature that this feat has been achieved, and we have no personal experience in this particular area. We encourage a study of the response of peroxy esters to organic sulfides and arsines.

### 4. Organic Phosphine Methods

According to Horner and Jurgeleit (32), virtually every type of peroxide reacts with tertiary phosphines to give the corresponding tertiary phosphine oxide and the analogous nonperoxy product corresponding to the reaction:

$$R^1-O-O-R^2 + PR^3 \longrightarrow R^1-O-R^2 + OPR^3$$

Reactions are rapid at room temperature except for dialkyl peroxides, which require heating. They cited numerous examples of such conversions, in high yields, to the specific products indicated by the above equation, in which $R^1$ and $R^2$ can be H, acyl, alkyl, or aralkyl. Hydroxyalkyl and dihydroxyalkyl hydroperoxides are converted to aldehydes and ketones, respectively; ozonides, depending on structure, produce aldehydes or ketones. The only exception to the consistent specificity is with diacyl peroxides, which in the presence of alcohol yield some ester as well as anhydride; this is the circumstance that led Horner and Jürgens (83) to select tertiary arsines over phosphines as their preferred reagent.

This study of phosphines made clear the possibility of differentiating alkyl hydroperoxides from diakyl peroxides through the much more rapid reaction of the hydroperoxides (32). Dulog and Burg developed this idea into a seemingly sound and effective method for determining organic peroxides of every class (except a few that are outstandingly unreactive) and for differentiating quantitatively between the easily reduced peroxides (hydroperoxides, peroxy acids, diacyl peroxides, and ketone peroxides) and the more sluggish dialkyl peroxides.

In all cases a weighed (excess) quantity of the stable and easily purified triphenyl phosphine was reacted with the peroxide in an aromatic hydrocarbon medium in an inert atmosphere. The easily reduced peroxides reacted quantitatively within 10 min at room temperature, whereas the dialkyl peroxides required refluxing in xylene under nitrogen for 2 hr (some peroxides of intermediate reactivity, such as *t*-butyl peroxybenzoate, were refluxed in benzene for 1 hr).

The importance of an inert atmosphere cannot be overemphasized. In current evaluations we have found that results can be high and variable unless oxygen is rigorously excluded from *dissolved* triphenyl phosphine. This reagent is considerably more demanding of good anaerobic technique than are

TABLE 16. RESOLUTION OF PEROXIDE CLASSES—QUANTITATIVE DETERMINATION OF MIXTURES CONTAINING UP TO FOUR CLASSES: PEROXY ACIDS, DIACYL PEROXIDES, ALKYL HYDROPEROXIDES, AND DIALKYL PEROXIDES[a]

| Components Present | Procedure Outline | Calculations[b] |
|---|---|---|
| A. Organic acids, peroxy acids, diacyl peroxides | 1. *Determine total peroxide content with KI*: react an aliquot part of the sample mixture in 2–5 ml benzene or chloroform with 1 ml saturated aqueous KI in 30 ml glacial acetic acid under $CO_2$ for 10–15 min at room temperature; dilute with 30–50 ml water and titrate liberated iodine with 0.1 $N$ sodium thiosulfate. Consumption: $a$ ml 0.1 $N$ thiosulfate. | $a \times 0.1 = meq_{ox}$, peroxy acid + diacyl peroxide |
| | 2. *Determine diacyl peroxide with KI after destroying peroxy acid with $Ph_2S$*: add a second aliquot of the sample mixture to diphenyl sulfide slightly in excess of its total peroxide content, allow to react for 10 min and then determine the diacyl peroxide iodometrically as above. Consumption: $b$ ml 0.1 $N$ thiosulfate. | $b \times 0.1 = meq_{ox}$, diacyl peroxide $c \times 0.1 = meq_{ac}$, acid + peroxy acid $(a - b) \times 0.1 = meq_{ox}$, peroxy acid |
| | 3. *Determine total acid content with NaOH*: titrate a third aliquot of the sample mixture with 0.1 $N$ NaOH to a phenolphthalein end point. Consumption: $c$ ml 0.1 $N$ alkali. | $\left(c - \dfrac{a - b}{2}\right) \times 0.1 = meq_{ac}$, organic acid |

B. Organic acids, peroxy acids, alkyl hydroperoxides

1. *Determine total peroxide content with KI*: as in A.1. Consumption: $a$ ml 0.1 $N$ thiosulfate.
2. *Determine alkyl hydroperoxide with KI after destroying peroxy acid with* $Ph_2S$: as in A.2. Consumption: $b$ ml 0.1 $N$ thiosulfate.
3. *Determine total acid content with NaOH*: as in A.3. Consumption: $c$ ml 0.1 $N$ alkali.

As in A, simply inserting alkyl hydroperoxide for diacyl peroxide

C. Diacyl peroxides, alkyl hydroperoxides

1. *Determine total peroxide content with KI*: as in A.1. Consumption: $a$ ml 0.1 $N$ thiosulfate.
2. *Determine diacyl peroxide with NaOH after conversion to acid with* $Et_3As$: to a second aliquot part of the sample mixture under nitrogen add 300–400 mg triethyl arsine and let stand 30 min; dilute with 30 ml water and titrate the resulting acids with 0.1 $N$ NaOH to a phenolphthalein end point; determine an acidity blank on the triethyl arsine reagent and subtract from the sample titration. Consumption (net): $b$ ml 0.1 $N$ alkali.

$a \times 0.1 = \text{meq}_{ox}$, diacyl peroxide + alkyl hydroperoxide

$b \times 0.1 = \text{meq}_{ac} = \text{meq}_{ox}$, diacyl peroxide

$(a - b) \times 0.1 = \text{meq}_{ox}$, alkyl hydroperoxide

D. Organic acids, peroxy acids, dialkyl peroxides

1. *Determine total peroxide content with HI*: react an aliquot of the sample mixture in 2–5 ml benzene or chloroform with 1 ml saturated aqueous KI and 5 ml concentrated HCl in 30 ml glacial acetic acid under $CO_2$ or $N_2$ by boiling for 30 min; after cooling, dilute with 30–50 ml water and titrate liberated iodine with 0.1 $N$ sodium thiosulfate. Consumption: $a$ ml 0.1 $N$ thiosulfate.

$a \times 0.1 = \text{meq}_{ox}$, peroxy acid + dialkyl peroxide

$b \times 0.1 = \text{meq}_{ox}$, peroxy acid

$(a - b) \times 0.1 = \text{meq}_{ox}$, dialkyl peroxide

$\left(c - \dfrac{b}{2}\right) \times 0.1 = \text{meq}_{ac}$, organic acid

TABLE 16.—Continued

| Components Present | Procedure Outline | Calculations[b] |
|---|---|---|
| | 2. *Determine peroxy acid with standard* Et$_3$As: to a second aliquot of the sample mixture under CO$_2$ add an excess of 0.1 $N$ (0.05 $M$) solution of triethyl arsine in benzene and allow to stand for 15 min; then back-titrate the unconsumed triethyl arsine with an 0.1 $N$ iodine solution. Consumption: $b$ ml 0.1 $N$ triethyl arsine. | $a \times 0.1 = $ meq$_{ox}$, diacyl peroxide $+$ dialkyl peroxide |
| | 3. *Determine total acid content with NaOH*: as in A.3. Consumption: $c$ ml 0.1 $N$ alkali. | |
| E. Diacyl peroxides, dialkyl peroxides | 1. *Determine total peroxide content with HI*: as in D.1. Consumption: $a$ ml 0.1 $N$ thiosulfate. | $b \times 0.1 = $ meq$_{ox}$, diacyl peroxide |
| | 2. *Determine diacyl peroxide with standard* Et$_3$As: as in D.2, but with the reaction period extended to 45 min. Consumption: $b$ ml 0.1 $N$ triethyl arsine. | $(a - b) \times 0.1 = $ meq$_{ox}$, dialkyl peroxide |
| F. Alkyl hydroperoxides, dialkyl peroxides | 1. *Determine the total peroxide content with HI*: as in D.1. Consumption: $a$ ml 0.1 $N$ thiosulfate. | As in E, simply substituting alkyl hydroperoxide for diacyl peroxide |
| | 2. *Determine alkyl hydroperoxide with standard* Et$_3$As: as in E.2. Consumption: $b$ ml 0.1 $N$ triethyl arsine. | |
| G. Organic acids, peroxy acids, diacyl peroxides, alkyl hydroperoxides | 1. *Determine the total peroxide content with KI*: as in A.1. Consumption: $a$ ml 0.1 $N$ thiosulfate: $\Sigma$ peroxy acid, diacyl peroxide, alkyl hydroperoxide. | $(a - b) \times 0.1 = $ meq$_{ox}$, peroxy acid $(d - c) \times 0.1 = $ meq$_{ac}$ = meq$_{ox}$, diacyl peroxide |

2. *Determine diacyl peroxide + alkyl hydro-peroxide with KI, after destroying peroxy acid with* Ph₂S: react a second aliquot portion of the sample mixture with about 300 mg diphenyl sulfide for 10 min. Determine remaining peroxides iodometrically as in A.1. Consumption: $b$ ml 0.1 $N$ thiosulfate: Σ diacyl peroxide, alkyl hydroperoxide.

$$\left(c - \frac{a-b}{2}\right) \times 0.1 = \text{meq}_{ac}, \text{ organic acid}$$

$$[b - (d-c)] \times 0.1 = \text{meq}_{ox}, \text{ alkyl hydro-peroxide}$$

3. *Determine initial total acid content with NaOH*: as in A.3. Consumption: $c$ ml 0.1 $N$ alkali: Σ acid, peroxy acid.

4. *Determine total acid content with NaOH after converting diacyl peroxide to acid with* Et₃As: as in C.2 (30-min reaction); include blank correction. Consumption (net): $d$ ml 0.1 $N$ alkali: Σ acid, peroxy acid, diacyl peroxide.

H. Organic acids, peroxy acids, diacyl peroxides, dialkyl peroxides,

1. *Determine total peroxide content with* HI: as in D.1. Consumption: $a$ ml 0.1 $N$ thiosulfate: Σ peroxy acid, diacyl peroxide, dialkyl peroxide.

$$(a - b) \times 0.1 = \text{meq}_{ox}, \text{ dialkyl peroxide}$$

$$c \times 0.1 = \text{meq}_{ox}, \text{ diacyl peroxide}$$

$$(b - c) \times 0.1 = \text{meq}_{ox}, \text{ peroxy acid}$$

2. *Determine peroxy acid + diacyl peroxide with standard* Et₃As: as in E.2 (45 min reaction period). Consumption: $b$ ml 0.1 $N$ triethyl arsine: Σ peroxy acid, diacyl peroxide.

$$\left(d - \frac{b-c}{2}\right) \times 0.1 = \text{meq}_{ac}, \text{ organic acid}$$

3. *Determine diacyl peroxide with standard* Et₃As, *after destroying peroxy acid with* Ph₂S: react a third aliquot of the sample mixture with about 200–300 mg diphenyl sulfide for 10 min; then determine diacyl peroxide on this reaction mixture according to H.2 (or E.2). Consumption: $c$ ml 0.1 $N$ triethyl arsine: diacyl peroxide.

TABLE **16.**—Continued

| Components Present | Procedure Outline | Calculations[b] |
|---|---|---|
| | 4. *Determine organic acid + peroxy acid with* NaOH: as in A.3. Consumption: *d* ml 0.1 *N* alkali: Σ acid, peroxy acid. | |
| I. Organic acids, peroxy acids, alkyl hydroperoxides, dialkyl peroxides | 1. *Apply steps* H.1, H.2, H.3, *and* H.4: substitute alkyl hydroperoxide for diacyl peroxide whenever the latter occurs in section H. | Calculate as in Section H, substituting alkyl hydroperoxide for diacyl peroxide. |
| J. Organic acids, diacyl peroxides alkyl hydroperoxides, dialkyl peroxides | 1. *Determine total peroxide content with* HI: as in D.1. Consumption: *a* ml 0.1 *N* thiosulfate: Σ diacyl peroxide, alkyl hydroperoxide, dialkyl peroxide | $(a - b) \times 0.1 = $ meq$_{os}$, dialkyl peroxide $(d - c) \times 0.1 = $ meq$_{ac}$ = meq$_{os}$, diacyl peroxide |
| | 2. *Determine diacyl peroxide + alkyl hydroperoxide with standard* Et$_3$As. Consumption: *b* ml 0.1 *N* triethyl arsine: Σ diacyl peroxide, alkyl hydroperoxide. | $c \times 0.1 = $ meq$_{ac}$, organic acid $[b - (d - c)] \times 0.1 = $ meq$_{os}$, alkyl hydroperoxide |
| | 3. *Determine organic acid with* NaOH: as in A.3. Consumption: *c* ml 0.1 *N* alkali: acid. | |
| | 4. *Determine total acid content with* NaOH, *after converting diacyl peroxide to acid with* Et$_3$As: as in G.4. Consumption: *d* ml 0.1 *N* alkali: Σ acid, diacyl peroxide. | |

K. Organic acids, peroxy acids, diacyl peroxides, alkyl hydroperoxides, dialkyl peroxides

1. *Determine total peroxide content with HI:* as in D.1. Consumption: $a$ ml 0.1 $N$ thiosulfate: Σ peroxy acid, diacyl peroxide, alkyl hydroperoxide, dialkyl peroxide.

$(a-b) \times 0.1 = meq_{ox}$, dialkyl peroxide
$(b-c) \times 0.1 = meq_{ox}$, peroxy acid
$(e-d) \times 0.1 = meq_{ox}$, diacyl peroxide
$[c-(e-d)] \times 0.1 = meq_{ox}$, alkyl hydroperoxide

2. *Determine peroxy acid + diacyl peroxide + alkyl hydroperoxide with standard Et₃As:* as in E.2. Consumption: $b$ ml 0.1 $N$ triethyl arsine: Σ peroxy acid, diacyl peroxide, alkyl hydroperoxide.

$\left(d - \dfrac{b-c}{2}\right) \times 0.1 = meq_{ac}$, organic acid

3. *Determine diacyl peroxide + alkyl hydroperoxide with standard Et₃As, after destroying peroxy acid with Ph₂S:* as in H.3. Consumption: $c$ ml 0.1 $N$ triethyl arsine: Σ diacyl peroxide, alkyl hydroperoxide.

4. *Determine organic acid + peroxy acid with NaOH:* as in A.3. Consumption: $d$ ml 0.1 $N$ alkali: Σ acid, peroxy acid.

5. *Determine total acid content with NaOH, after converting diacyl peroxide to acid with Et₃As:* as in G.4. Consumption: $e$ ml 0.1 $N$ alkai: Σ acid, peroxy acid, diacyl peroxide.

---

[a] Reference 73.
[b] In the calculations a milliequivalent of peroxide based on reduction of the —O—O— bond ($meq_{ox}$) equals $\frac{1}{2}$ millimole, whereas a milliequivalent of an organic acid ($meq_{ac}$) equals 1 millimole.

the iodide–acetic acid reagents. Three alternatives—a gravimetric, a spectro-photometric, and a titrimetric procedure—were provided for estimating the amount of excess phosphine reagent, which is the final step in the quantitative procedure. The first two were based on conversion of triphenyl phosphine into the water-soluble $\alpha$-hydroxymethyltriphenylphosphonium chloride by reaction with formaldehyde and hydrochloric acid, as described by Hoffman (197):

$$P(C_6H_5)_3 + HCHO \underset{NaOH}{\overset{HCl}{\rightleftharpoons}} [(C_6H_5)_3PCH_2OH]^+Cl^-$$

In the gravimetric method excess phosphine is separated by extraction into water as the hydroxymethyl complex, then converted back to phosphine with alkali, and extracted directly into peroxide-free ether for evaporation and final weighing. This finish-up procedure seems to be particularly suited for occasional analyses; although it requires more manipulation than is desirable in routine determinations, it requires no special reagent preparation or standardization.

In the spectrophotometric approach excess phosphine is estimated from absorption at 275 m$\mu$ by the aqueous solution of the hydroxymethyl complex after an appropriate dilution; absorbance is independent of both formalde-hyde and hydrochloric acid concentration. We think the procedure could be made quicker and easier by simply extracting the reaction mixture with a measured volume of acidified formaldehyde solution and then withdrawing an aliquot from the resulting aqueous layer for an appropriate dilution and photometry; this would avoid the washing steps necessary for transfer of the aqueous layer, as specified in the published procedure (75).

For the titrimetric finish-up a different approach is taken, in which the excess phosphine is reacted with a known (excess) amount of hydrogen peroxide, and the excess of the latter then titrated with standard ceric sulfate. This is a straightforward, simple procedure, whose only apparent disadvan-tage appears to be the additional 30 min required for the phosphine–hydrogen peroxide reaction.

Dulog and Burg (75) report consistently good agreement between these triphenyl phosphine methods and iodometric methods. In addition, they quantitatively resolved mixtures of cumene hydroperoxide and dicumyl peroxide, and of methyl linoleate hydroperoxide and 2,2-bis($t$-butylperoxy) butane, both in methyl linoleate.

The quantitative determination of ozonides by reaction with triphenyl phosphine was recently described by Lorenz (198), who found that this peroxide class will liberate iodine stoichiometrically from iodide only when the reduction yields ketones rather than aldehydes. By reaction with a weighed excess of the phosphine reagent in ethanol or in heptane–phenol solution, 9 out of 10 purified monomeric ozonides prepared from a group of olefins were estimated to be more than 95% pure. Excess reagent was titrated with standard aqueous iodine solution. Reaction at room temperature required up to 3 days for completion, but probably could be safely speeded up by using a somewhat higher temperature or a more reactive reagent, such as triethyl phosphine (32). Air was excluded to prevent direct oxidation of reagent or autoxidation of aldehydes, both of which could cause high results.

We propose yet another approach to the use of tertiary phosphines in the resolution of peroxide mixtures, suggested by their ability to achieve clear-cut usually quantitative, and broadly general conversions of organic peroxides into specific nonperoxide analogs. In this scheme total peroxidic function would be determined from the consumption of phosphine reagent and individual peroxides would be estimated from their analogs via established chemical or instrumental methods. This approach should be particularly valuable for studying the formation and reactions of organic peroxides in labile systems.

Phosphine conversion of the readily available α-hydroperoxy olefins and diacyl peroxides (189) into α-hydroxy olefins and unsymmetrical acid anhydrides has been suggested as a useful organic preparative reaction.

### 5. Stannous Methods

Reduction with the stannous ion avoids interference in the iodide methods by sulfur compounds (76) that consume liberated iodine. Barnard and Hargrave (76) have described earlier work with the stannous chloride reagent and studied the effects of many parameters. They devised a method that is highly precise and accurate for hydroperoxides, although rather long and complex; diacyl peroxides and presumably peroxy acids are also determined. Egerton and co-workers (77) found reaction to be much more rapid in alkaline media, without apparent hydroperoxide loss despite the known instability of primary and secondary hydroperoxides to alkali (2, Ch. 13; 26, 27).

### 6. Arsenite Methods

Reduction with the arsenite ion is another reaction that has been found useful where there are interferences with iodide methods. Walker and Conway (79) have modified Siggia's original proposal (78) into a remarkably accurate, precise, and sensitive method for determining hydroperoxide in petroleum products down to concentration levels of less than 0.1 meq/l (peroxide number 0.1). Gasolines and fuel oils, but not refined mineral oil, were found to contain components that interfere seriously in iodide-reduction methods (173, 175) at the typically low hydroperoxide concentrations, by consuming liberated iodine.

This interference is probably due to the sulfur compounds present in most petroleum products. It might be argued that iodine absorption by olefins in the sample is perhaps also a contributing factor. We tend to discount this possibility, being convinced that the triiodide ion is unable in general to react with olefins, as discussed above in Section IV-B-1-b. In view of the extremely high ratio of olefin to peroxide (or triiodide) that must have existed in certain of Walker and Conway's samples, however, we cannot with certainty rule out the chance that some of the iodine was absorbed by the olefins.

The Walker and Conway method appears to be quite practical for product-control testing in a peroleum-refinery laboratory, the application for which it was devised. We do not recommend it for general use, however, because it is slow.

Apparently no one has described any attempt to apply either stannous or arsenite reagents to peroxy esters or dialkyl peroxides. We might expect peroxy esters to reduce readily in an alkaline medium, and primary and secondary dialkyl peroxides might also respond. The scope of these two reagents should be surveyed more broadly, since useful specificity may exist.

### 7. Potassium Permanganate—Discrimination of Hydroperoxide and Peroxide Groups in Cyclohexanone Peroxide Systems

Titration with potassium permanganate oxidizes hydrogen peroxide to oxygen and water, and serves to determine it quantitatively. Szobor and Back (80) have studied the possibility of utilizing this titration, along with differences in rates of conversion to hydrogen peroxide by hydrolysis, to resolve the mixtures of peroxy compounds that constitute commercial cyclohexanone peroxide. They came up with some unexpected and useful results.

The principal cyclohexanone peroxide components active as polymerization catalysts are 1-hydroxy-1'-hydroperoxycyclohexyl peroxide and the 1,1'-di-hydroperoxy derivative. Szobor and Back found that when these compounds were dissolved in glacial acetic acid, diluted with three volumes of water, acidified with sulfuric acid, and allowed to stand for 24 hr at 25°, titration of the reaction mixture with 0.1 $N$ potassium permanganate was accompanied by vigorous evolution of oxygen gas and consumed titrant equivalent to the total active oxygen content of the starting peroxides; this was the expected result of quantitative hydrolysis to hydrogen peroxide. On the other hand, when the same diluted solutions were titrated with permanganate immediately, there was no evolution of oxygen and the consumption of titrant corresponded to the active oxygen from the hydroperoxy groups alone. Evidence was presented that an $Mn^{2+}$–peroxide complex had been formed. Szobor and Back propose that such a titration be used to determine the total active hydroperoxy content of cyclohexanone peroxide mixtures.

1,1'-Dihydroxycyclohexyl peroxide will interfere with the proposed hydroperoxide titration. This component on immediate titration consumed titrant equivalent to its active oxygen content, and simultaneously evolved oxygen gas (80). Such behavior indicates that the dihydroxy derivative undergoes rapid hydrolysis, a fact noted also by Buzlanova and co-workers (42) in a TLC evaluation of the peroxide mixture, discussed in Section IV-A-1-d. Fortunately the evolution of oxygen will signal that interference by this possible component is occurring.

### 8. Catalytic Hydrogenation

Catalytic hydrogenation has been relatively little used, although recommended by Davies (2, Ch. 14). Its chief utility in peroxide analysis has been for derivative preparations in proofs of structure. Kinetic studies of the hydrogenation of various peroxide and nonperoxide groups on several types of catalyst (81, 82) showed numerous examples of selectivity, although the order of reactivity for various peroxides varied considerably with the type of catalyst employed (e.g., nickel, palladium, or platinum).

Criegee and co-workers (9) used rates of hydrogenation on palladium black in acetic acid and rates of iodine liberation from sodium iodide–acetic acid reagent to differentiate between and characterize a group of peroxides. The two reduction methods gave closely similar results: hydroperoxides reacted quantitatively in a few minutes, whereas peroxy esters and dialkyl peroxides were substantially less reactive, and dimeric and trimeric acetone peroxides still less so.

### 9. Lithium Aluminum Hydride

Lithium aluminum hydride has been relatively little used with peroxides. There is the drawback that lithium aluminum hydride is not specific in its action, reacting generally with all compounds containing an active hydrogen, and quantitatively reducing ketones and esters (199). It may be a useful conversion agent, however, whenever peroxide reduction should produce characteristic alcohols or when there is other pertinent information about a sample. Studies of *p*-cymene oxidation (27) provide an excellent example of such an application.

Lithium aluminum hydride reduces organic hydroperoxides smoothly at room temperature to the corresponding alcohols (84) with the evolution of exactly 2 moles of hydrogen per mole of hydroperoxide. Other peroxide classes form 1 mole of hydrogen per mole, dibenzoyl peroxide reacting with almost explosive vigor, methyl $\alpha$-tetralyl peroxide and ascaridole at moderate rates, and di-*t*-butyl peroxide very slowly. The stoichiometry corresponds to fission of the $-O-O-$ bond and direct formation of lithium aluminum alkoxide. This reaction has been used to determine methyl linoleate hydroperoxide quantitatively (75) and to determine hydroperoxide impurities in isopropyl- and isopropenylacetylene (200).

Use of excess hydride reagent converts peroxides to the corresponding alcohols quantitatively. With excess peroxide, however, it is clear that only two of the four hydrogens in lithium aluminum hydride are reactive (34) and that the resulting dialkoxide will not reduce peroxides:

$$ROOR + LiAlH_4 \longrightarrow LiAl(OR)_2H_2 + H_2$$

$$ROOR + LiAl(OR)_2H_2 \longrightarrow \text{no reaction}$$

There is parallel behavior in the reduction of nitriles by this reagent at $35°$ or lower (83).

An analogous but milder reagent, sodium borohydride, is particularly useful for the reduction of triglyceride peroxides, since it does not reduce ester groups and so leaves the basic fat structure intact (33).

### 10.  Ferrous Methods

In the past chemists have made extensive use of reduction by the ferrous ion for the quantitative determination of easily reduced organic peroxides, particularly the alkyl hydroperoxides formed in the development of rancidity and in other cases of autoxidation. These methods have long been controversial. On the one hand, results under a given set of conditions are consistently precise for a given peroxide or autoxidized substrate, and hence proportional to results by reliable iodometric procedures. On the other hand, the proportionality factor varies with the nature of the reaction solvent, has unsystematically different values for different peroxidized materials, and for a given peroxide and a given solvent can vary by a factor of as much as 5 or 6 with the extent of aeration of the reaction mixture—results being higher than iodometric values in the presence of oxygen and lower than iodometric ones in its absence.

Many careful studies have been undertaken in efforts to understand these phenomena (76, 86, 201–204), but have failed in varying degrees because the chemistry involved was not comprehended. A substantial understanding of the ferrous methods has come about, however, through the definitive work of Kolthoff and Medalia (85, 205), who drew on basic studies of the reaction mechanisms to elucidate the analytical difficulties.

Their treatment recognizes both positive and negative error factors. Positive errors occur only when air is present during peroxide reduction; this situation induces the formation of additional hydroperoxide from the dissolved oxygen by a free-radical-chain reaction, tending to produce excess ferric ion in proportion to the original peroxide concentration and the oxygen

content. Negative errors arise from several chain reactions in which RO·
radicals can be consumed without a corresponding oxidation of ferrous ions
and therefore without production of the theoretical 2 moles of ferric ion
per mole of peroxide. The nature of the solvent has a very large effect on
these negative factors, as various types of solvent molecules participate in
the several chain reactions to differing degrees. Alcohols are doubly harmful
because they can convert a portion of the ferric ion product back to ferrous,
by a reaction in which alcohol free radicals are oxidized to carbonyl compounds. Other workers have confirmed obtaining lower results with alcohol
than with acetone-based reagents (86, 201).

As both positive and negative errors can occur in the presence of air, it is
apparent that these may sometimes offset one another to yield approximately
correct results fortuitously (203). When air has been rigorously excluded,
however, only negative errors are possible, and unless they too are eliminated, results by anaerobic ferrous methods must be low. Kolthoff and
Medalia's procedures (85) based on deaerated aqueous acetone are the soundest
attempt in this direction of which we are aware. Chapman and Mackay have
used a similar anaerobic acetone reagent (201).

In view of the extremely complex reaction system in ferrous methods, with
so many significant variables to be controlled (85), we recommend that
volumetric ferrous *macro* methods be rejected in favor of others based on
more tractable reagents. The use of ferrous methods continues to be widespread in trace colorimetric analyses (85–87, 201) because of speed, simplicity,
and sensitivity and because it may be satisfactory in many cases to obtain
relative, rather than absolute, peroxide concentrations.

## C. Colorimetric and Photometric Methods

Although one might choose to view the colorimetric estimation of peroxides as both instrumental and chemical in nature, their treatment as a fully
independent class can be justified by the way in which they are normally
used. Colorimetric methods are usually called for when one must determine
peroxides at trace levels in groups of samples, especially in limited portions.
Such situations typically involve monitoring autoxidation in natural products
or the estimation of peroxide catalysts and residual cross-linking agents in
commercial processes.

Methods for determining peroxides by colorimetric and photometric procedures are summarized in Table 2C.

### 1. Colorimetric Ferrous Methods

About colorimetric ferrous methods there are two opposing schools of
thought. One advocates striving to eliminate errors as thoroughly as possible.
As we have seen in the preceding section, gross errors of opposite sign can

occur simultaneously with these methods. Kolthoff and Medalia (85) feel that this is inherently undesirable and that such errors should be systematically eliminated: positive errors by using only anaerobic systems, and negative errors by finding better solvents than acetone and trying additives—such as the bromide ion, and maleic or fumaric acids—that are known to suppress free-radical oxidations (206). There would appear to be distinct possibilities of improving stoichiometry to the point where more than a few types of autoxidized substrates would yield precise and fairly accurate peroxide analyses.

The other position, taken by Lea (86), is that there is "... little prospect of so complicated a reaction being made truly stoichiometric particularly when applied to fats of variable fatty acid composition oxidized under a variety of conditions." He feels that nothing is to be gained by excluding atmospheric oxygen, because the method is then less sensitive and less simple. Furthermore, the characteristically good precision of the reactions under many empirically fixed situations, including the presence of air, should make it possible to get accurate results when necessary, for a particular substrate, by calibrating the selected colorimetric ferrous procedure with an iodometric or other reliable method.

A procedure representing each position is described in Section V. That selected for use in the presence of oxygen is Smith's adaptation (87) of a method by Loftus-Hills and Thiel (207). Another adaptation of the latter method is due to Driver, Koch, and Salwin (208). By changing from methanol–benzene (30:70) to an ethanol–benzene (80:20) as the solvent system, these workers found that color developed satisfactorily at room temperature, so that heating was unnecessary. This circumstance permitted a considerably simplified procedure, with color development directly in a colorimeter cur-vette. The method has been evaluated by Hamm and Hammond (209).

### 2. *Leuco Chlorophenolindophenol*

Two careful evaluations (86, 88) of Hartmann and Glavind's method (89) based on 3,5-dichloro-4-,4′-dihydroxydiphenylamine (leuco dichlorophenol-indophenol) have shown it to behave similarly to the ferrous method, with results severely high in the presence and low in the absence of oxygen. Precision was much inferior to that of preferred aerobic ferrous methods (86), and sensitivity only one-fourth as great, with respective molar absorptivities of 7300 versus 28,200 l/mole-cm (87, 88). The leuco chlorophenolindophenol method accordingly is not recommended for general use.

### 3. N,N′-*Diphenyl-*p-*phenylenediamine*

The best of the aromatic diamine reagents appears to be *N,N′*-diphenyl-*p*-phenylenediamine, which has been recommended by Ryland (90). The first thorough evaluation of this type of reagent was by Ueberreiter and Sorge (94), who developed a procedure based on the oxidation of 4,4'-diaminodiphenylamine to the indamine dye, phenylene blue (**3**).

(**3**)

They found this method inferior to their well-known leuco methylene blue method because rate of formation of indamine and fading due to solvolysis both depended strongly on the effective acid strength of the solvent medium. Molar absorptivity (at 640 m$\mu$) varied similarly, increasing with fluorenone peroxide, for example, from 28,000 to 42,000 l/mole-cm as acetic acid was modified with 5% trichloroacetic acid, and to 54,000 l/mole-cm with the further addition of 1% water.

With *N,N'*-diphenyl-*p*-phenylenediamine, Ryland (90) found good sensitivity, good reagent stability, and low blanks under her preferred conditions (given in Table 17). Color development varied with solvent composition, however; for example, *t*-butyl hydroperoxide gave the following molar absorptivities in media consisting of benzene modified with 65 vol % of the indicated solvents:

| Auxiliary Solvent | Molar Absorptivity (l/mole-cm) |
|---|---|
| Cyclohexane | 25,000 |
| Methyl alcohol | 21,000 |
| Acetone | 10,500 |
| None (all benzene) | 24,000 |

In benzene all of the peroxides studied obeyed Beer's law.

Table 17 summarizes data for other peroxides in the benzene system. It is clear that the stoichiometry of the color development, a one-electron oxidation of the diamine to the acid-stable semiquinone free radical (90, 210), is inexact in this system and variable among the peroxides, although constant for a given peroxide. In particular, three of the peroxides in the table formed more than the theoretical 2 moles of semiquinone per mole of peroxide, thus giving apparent molar absorptivities above the nominal limit of 27,400 l/mole-cm. This behavior is similar to that observed with ferrous thiocyanate (85, 86) and leuco dichlorophenolindophenol (86, 88) in the presence of air, and probably has a similar cause—namely, induced formation of hydroperoxide by dissolved oxygen during the free-radical-chain reactions that are part of the peroxide-reduction process.

TABLE 17.   COLORIMETRIC DETERMINATION OF PEROXIDES WITH
            $N,N'$-DIPHENYL-$p$-PHENYLENEDIAMINE[a]

| Peroxide | Rate of Color Development[b] | Molar Absorptivity[c, d] (l/mole-cm) |
|---|---|---|
| Succinyl peroxide | Rapid[e] | 17,600 |
| Benzoyl peroxide | Rapid[e] | 24,400 |
| $t$-Butyl hydroperoxide | Rapid[e] | 24,400 |
| Cumene hydroperoxide | Slower[f] | 24,400 |
| Lauroyl peroxide | Slower[f] | 32,600 |
| $t$-Butyl peroxybenzoate | Slower[f] | 35,500 |
| $t$-Butyl peroxyacetate | Rapid[e] | 54,000 |
| Di-$t$-Butyl peroxide | No reaction | — |

[a] Data and procedure of Ryland (90).

[b] A 2-ml volume of 0.05 $M$ diamine and 2 ml of 0.18 $M$ trichloroacetic acid in benzene, plus the peroxide sample in benzene, mixed at ambient temperature in a 25-ml volumetric flask (wrapped in aluminum foil) and diluted to the mark with benzene.

[c] At 700m$\mu$; there is another band at 385 m$\mu$ that has some 50% greater sensitivity.

[d] Theoretical value is 27,400 l/mole-cm based on the figure of 13,700 l/mole-cm for one-electron oxidation of the diamine to the semiquinone (210).

[e] Almost instantaneous, color fully developed in 5 min.

[f] A little slower, but full color developed within 30 min.

In order to appraise realistically the role of aromatic diamine reagents in trace peroxide analysis, analytical chemists need to study the effects of deaeration, which previously had been overlooked (90, 94, 211). If color development were found to be grossly low in the absence of oxygen, paralleling the behavior of ferrous thiocyanate and leuco dichlorophenolindophenol reagents, then one should rate $N,N'$-diphenyl-$p$-phenylenediamine (the best of the aromatic diamines) as approximately equivalent to, but not significantly better than, the best ferrous thiocyanate procedure (86, 87), whether in speed, simplicity, convenience, precision, or sensitivity.

Ryland evaluated several other diamine reagents (90). $N,N'$-dimethyl- and $N,N,N',N'$-tetramethyl-$p$-phenylenediamines were relatively rapidly oxidized by atmospheric oxygen, and the latter was much slower in color development than the $N,N'$-diphenyl derivative; neither offered any advantage. Methylene-4,4'-bis($N,N$-dimethylaniline), also known as tetramethyl base and described as a spot-test reagent for peroxides (212), showed no color development in solution under conditions similar to those used with the other diamines. If the reaction mixture was evaporated on filter paper, however, the characteristic blue color developed very rapidly, as in the original spot test.

### 4. Leuco Methylene Blue

The colorimetric determination of trace concentrations of organic peroxides by reduction with leuco methylene blue, developed by Ueberreiter and Sorge (93, 94), has two characteristics highly to be prized in such methods. One is a relatively high sensitivity, about three times better than that of the ferrous thiocyanate (87), aromatic diamine (90, 94), and iodide colorimetric methods (99, 100). The second is an essentially constant molar response for all reactive peroxide types, a property documented by statistical analysis of a mass of data (93) obtained over about a tenfold range of concentrations on seven peroxides that included four ketone peroxides, two diacyl peroxides, and an alkyl hydroperoxide. We calculated that the average molar absorptivity in this group had been 77,200 l/mole-cm, with a 95% confidence interval of ±3900 l/mole-cm. Among other important colorimetric peroxide methods, only the iodometric procedures (to be discussed in a subsequent section) have a similarly constant absolute response. This behavior is valuable because it allows a general calibration to be obtained from any conveniently available known peroxide.

Along with this sensitivity and absolute accuracy is a precision quite adequate for a trace method. The pooled relative standard deviation of Ueberreiter and Sorge's data (93), with 43 degrees of freedom, is 1.85%. This indicates at the 95% confidence level that duplicate determinations should differ by not more than 5.2% for peroxide numbers down to 0.025. In addition, the procedure is rapid and simple, and can be carried out in the presence of air. It is applicable to benzene-soluble samples, including polymers, and color development for more slowly reacting peroxides can be accelerated safely by heating briefly (92, 93). There is some additional comment on the leuco methylene blue method in the next section.

An excellent description of this method by Martin (5) includes an improved procedure for the phenylhydrazine reduction of methylene blue (213) (the process by which the reagent is prepared) and a careful presentation of Ueberreiter and Sorge's original procedure.

Despite all the foregoing advantages of this method, many analytical chemists in need of reliable trace peroxide analyses have turned away from it because of difficulties in preparing suitably pure leuco methylene blue reagent and in storing its dilute benzene solution. For instance, although the presence of oxygen seems to be harmless during color development, both oxygen and water vapor cause deterioration of the dilute reagent solution during storage and must be rigorously excluded; the leuco methylene blue must be prepared, recrystallized, and dried under a similarly inert atmosphere (93).

It is our view that this method has too high a potential value to warrant being dismissed because of the cited difficulties. It would seem a better course, rather, to try developing preparation and handling techniques that would render these difficulties tolerable. Some steps in this direction have been taken. Thus Martin (5) and Banderet and co-workers (91) describe nearly identical methylene blue reduction processes that improve yields of the leuco product by precipitating its ammonium salt. In manipulation of the pure product and its benzene solutions the latter workers made extensive use of flasks and ampules fitted with break-seals. In this way, although with considerable effort, they prepared very satisfactory reagent solutions.

A more practical answer to the problem of reagent supply is for producers of specialty analytical reagents to prepare and market a colorimetric-reagent grade of leuco methylene blue, perhaps in sealed glass ampules containing enough of the material to prepare a liter of reagent.

Another approach to the problem of storing benzene solutions of leuco methylene blue in a suitably decolorized state has been described by Dulog (92). Following a suggestion by Ueberreiter and Sorge (93), he stored the reagent solution in the reservoir of an automatic buret over platinized asbestos in an atmosphere of hydrogen rather than nitrogen. Filter paper fastened over the inlet prevented passage of catalyst to the buret. With this technique the use of pure, light-yellow leuco methylene blue was not necessary; greenish, surface-oxidized preparations served just as well, since the solution was reduced before use. It was necessary to keep the solution well protected from light.

### 5.  N-*Benzoyl Leuco Methylene Blue*

In the *Treatise on Analytical Chemistry* we have carefully compared the general behavior of leuco methylene blue with that reported for *N*-benzoyl leuco methylene blue by Eiss and Giesecke (95). The latter reagent was obtained commercially in a suitably pure grade, and its benzene solutions were unaffected by air so that they could be stored satisfactorily in brown bottles at room temperature. The *N*-benzoyl leuco derivative was thus apparently free from the two chief drawbacks of leuco methylene blue. It had an important defect of its own, however, in that reaction with peroxides was relatively slow.

Attempts to accelerate this reaction by heating produced erratic results, even at 30°. Zirconium naphthenate was an effective promoter, however, reducing the reaction period for several hydroperoxides from about 40 hr to less than 1 hr. For benzoyl peroxide the time decrease was from 120 to 30 hr; this is still an unreasonably slow determination for such a reactive peroxide. Most other metallic naphthenates had no appreciable affect. These selectively variable reaction periods appreciably diminish the general effectiveness of this reagent.

The benzoyl leuco methylene blue method was also very sensitive to the effects of light, apparently more so than the leuco methylene blue method; for the latter, no protection from light was specified originally (5, 93, 94), although later workers took general precautions (91, 92).

An important difference between these two reagents lies in their relative sensitivities for determining peroxide, the sensitivity of the benzoyl derivative being about twice as great as that of leuco methylene blue. This can be seen in Table 18, where the absorbance data of Eiss and Giesecke have been expressed as molar absorptivities; the average for the peroxides studied is 155,000 l/mole-cm, compared with 77,200 l/mole-cm for leuco methylene blue (93). As with leuco methylene blue, the response of the benzoyl derivative is largely independent of peroxide structure, and precision appears to be highly satisfactory.

TABLE 18.   MOLAR ABSORPTIVITIES AT 662 m$\mu$ FOR PEROXIDES REACTED
WITH $N$-BENZOYL LEUCO METHYLENE BLUE[a]

| Peroxide | Period for Complete Color Development | Molar Absorptivity[b] (l/mole-cm) |
|---|---|---|
| Cumene hydroperoxide | 40 min (38 hr)[c] | 148,000 ± 4000[d] |
| p-Menthane hydroperoxide | 2 hr | 153,000 ± 2500[d] |
| t-Butyl hydroperoxide | 30 min (36 hr)[c] | ~ 144,000[e] |
| Benzoyl peroxide | 30 hr (120 hr)[c] | 150,000 ± 3000[d] |
| Lauroyl peroxide | 5 hr | 183,000 ± 4000[d] |
| | | Average = 155,000 |

[a] Calculated from the data of Eiss and Giesecke (95).
[b] At 622 m$\mu$. Comparable value for leuco methylene blue (93), based on seven peroxides, was 77,200 ± 3900 l/mole-cm.
[c] Without use of zirconium naphthenate.
[d] Ninety-five percent confidence interval.
[e] Did not obey Beer's law; required working curve.

Methylene blue cation should be the color body produced in common by both reagents on reaction with peroxides. Its molar absorptivity as determined from pure methylene blue trichloroacetate (93) is 88,100 l/mole-cm. This then should be theoretically the maximum molar response toward peroxides by either of the leuco reagents, much lower than the average value of 155,000 l/mole-cm cited above for the benzoyl derivative.

The explanation developed in the *Treatise on Analytical Chemistry* (1) is that *N*-benzoyl leuco methylene blue does produce almost exactly 1 mole of methylene blue cation per mole of peroxide, as required in the straight-forward oxidation–reduction reaction; the apparently high results (175 % of theory) are attributed to additional hydroperoxide formed from dissolved oxygen during reaction of the original peroxide, the well-known "oxygen error."

Unsubstituted leuco methylene blue, on the other hand, is believed to generate the cation color body much less efficiently, due to a complex scheme of equilibria among a variety of oxidized reagent forms that was recognized by Ueberreiter and Sorge (93); they found a maximum molar absorptivity of only 42,700 l/mole-cm for peroxide-oxidized leuco methylene blue and some significant spectral differences. The higher effective absorptivity of 77,200 l/mole-cm for this reagent, based on peroxide content, is 180 % of their maximum based on the reagent, and seems due, logically, to an oxygen error paralleling that proposed for *N*-benzoyl leuco methylene blue.

As in the case of the aromatic diamine reagents, it appears that no one has examined the implied oxygen effect for leuco methylene blue and the *N*-benzoyl derivative by studying their behavior as peroxide reagents in oxygen-free systems. A systematic study should be undetaken so that the relative merits of ferrous thiocyanate, aromatic diamines, and leuco methy-lene blue as colorimetric reagents for determining traces of peroxides can be assessed objectively.

### 6.   sym-*Diphenylcarbohydrazide*

Earlier work on the use of this reagent in the colorimetric detection and estimation of peroxides has been summarized by Hamm and co-workers (96). Stamm (214, 215, 216) originally proposed the use of diphenylcarbohydrazide for detecting rancidity in oils, suggesting that the red color resulted from reaction with acids, aldehydes, and ketones. Hartmann and Glavind (89) proposed, however, that the basis of the Stamm test was the oxidation of diphenylcarbohydrazide to diphenylcarbazone by hydroperoxide:

PhNH—NHCONHNHPh + ROOH $\longrightarrow$

$$PhN=NCONHNHPh + ROH + H_2O,$$

since its results paralleled their indophenol test.

Hamm and co-workers have come up with a sensitive and apparently soundly based quantitative method for determining peroxides in fats and oils by use of the diphenylcarbohydrazide reagent (96); their method should be seriously considered by anyone trying to select a suitable colorimetric peroxide method. The reaction is carried out in tetrachloroethane–acetic acid (80:20) containing 0.1 % of purified 1,5-diphenylcarbohydrazide; a 3-min heating period in boiling water is specified to develop the color. Absorbance is measured at 565 m$\mu$.

It has been shown that diphenylcarbazone is the reaction product in this modified Stamm test and that in the absence of oxygen each mole of peroxide (lauroyl peroxide) produces 1 mole of this derivative. When oxygen is not excluded, each mole of peroxide oxidizes 2.631 moles of diphenylcarbohydrazide. When this factor was included in the calculation, agreement was excellent between analyses obtained by the colorimetric method and by an iodometric method, on oxidized soybean oil (96).

The aborbance is quite temperature sensitive, decreasing about 1.5% for a temperature increase of 1°. Color intensity should be read within 30 min of heating.

Hamm and Hammond (209) have made a comparison of an iodometric procedure, the modified Stamm method, and the colorimetric iron method of Driver and co-workers (208), calibrating the latter two with lauroyl peroxide (these produced, respectively, 2.63 and 1.46 moles of oxidized reagent per mole of peroxide). For the analysis of milk fat, corrected values were in excellent agreement by the three methods. The iron method could determine peroxide numbers down to 0.009, whereas the modified Stamm test could go down to 0.003.

### 7.  Colorimetric Iodide Methods

The range of peroxide analyses via the well-established iodide-reduction methods can be extended to concentration levels well below the practical limits of volumetric procedures by estimating the liberated iodine colorimetrically, either directly (97, 100, 217, 218) or indirectly (99). Most widely used (97, 100) has been reaction at room temperature with potassium iodide in acetic acid–chloroform (2:1), a reagent system similar to that commonly employed for the volumetric determination of peroxides in fats and oils (101, 171), and with the same limitations as to types of peroxides that may be determined; for example, there is no discernible reaction by dicumyl or di-$t$-butyl peroxide (97).

The absorbing species in these procedures is the triiodide ion (100), the form in which iodine exists quantitatively at iodide–iodine ratios of 5 and larger (100). This ion has a molar absorptivity of 22,700 l/mole-cm in the acetic acid–chloroform solvent, 0.05 $M$ in potassium iodide, with a maximum at 362 m$\mu$.

In practice none of the colorimetric iodine methods uses measurements at the absorption peak, partly because borosilicate glass has a slight absorption that would necessitate careful cell matching (a difficulty readily eliminated by the use of fused-quartz or other high-silica cells) and partly because absorbances would usually be too high. Heaton and Uri's procedure (100) was to carry out their photometry at a wavelength between 380 and 500 m$\mu$ such that the triiodide solutions from their samples had absorbances in a desirable range, and then convert these values by means of a wavelength–absorptivity relation (1, 100) into equivalent absorbances at 400 m$\mu$, where the calibration had been established. Calibrations based on standards of either iodine or linoleic acid hydroperoxide were equivalent, and Beer's law was followed up to iodine concentrations of $5 \times 10^{-4}$ $M$. To eliminate detectable blank values a special cell was used in which the operations of deaeration, sample reduction, and triiodide spectrophotometry were carried out.

Banerjee and Budke (97) in modifying this method chose to set up two procedures covering the 0–1 and 1–10 ranges of peroxide number, the first calibrated with triiodide absorption at 410 m$\mu$, the second at 470 m$\mu$. For the lower range they used a specially constructed cell related to that of Heaton and Uri, and maintained positive nitrogen pressure throughout the reaction, but for the higher range they used ordinary volumetric flasks and measured the absorbance (rapidly) in regular 1-cm cells. Evaluation of the method showed satisfactory recoveries for 17 commercial peroxides representing four easily reduced types.

Heaton and Uri's careful work demonstrated in both the volumetric assay and trace colorimetric ranges that scrupulous elimination of atmospheric oxygen can reduce iodometric blank values to the vanishing point, at least in methods for easily reduced peroxides. To bubble nitrogen through a solution for an hour in order to effect deaeration, as these workers did, appears to be excessively cautious, however. Banerjee and Budke obtained acceptable results even at the lowest concentration levels using a total deaeration period of only 6 min. Their blanks corresponded to less than 2 $\mu$g of active oxygen, or a peroxide number of less than 0.05 on a 5-ml sample (97).

As with the volumetric methods, it is evident that analytical chemists using colorimetric iodide-reduction procedures for determining peroxides have much leeway in devising variants that best fit their requirements of speed, simplicity, precision, accuracy, and type of sample. An interesting example is Anton's adaptation for determining hydroperoxides in nylon yarn (217); sample and potassium iodide–acetic acid reagent were reacted in tetrafluoropropanol at room temperature and triiodide absorbance was measured at 400 m$\mu$; relative standard deviation was $\pm 5\%$, and the sensitivity was 1 micromole/gram.

There have been no published colorimetric versions of the methods for difficult-to-reduce peroxides. If such methods are needed, however, their development should be straightforward. It would be important to eliminate the traces of reducing impurities in acetic acid that are believed to be responsible for most of the iodine liberation in blanks with adequately vigorous reagents (10); refluxing over powdered dichromate, as described by Lea (86), might be suitable. If there should be significant interference from highly colored sample substrates, this might be eliminated by an appropriate extraction with an immiscible solvent, after the peroxide reduction.

Dubouloz and co-workers in determining lipid peroxides, have used extensively a colorimetric method that employs indirect estimation of the liberated iodine. In its modern version (99) this method appears quite practical: the reaction medium, two parts of 10% benzoic acid in chloroform and one part of methanol, contains a standard (excess) amount of an intensely colored compound, thiofluorescein, that is quantitatively decolorized by iodine; in a 5-min reaction at 70° in nitrogen-purged centrifuge tubes, peroxides are reduced by potassium iodide, a proportionate amount of thiofluorescein is consumed, and the remaining color then extracted into 10.0 ml of 10% aqueous sodium carbonate containing cyanide or "Complexon III" and estimated by absorbance at 590 m$\mu$ after centrifuging. Peroxide numbers are obtained from the absorbance decrease relative to a blank run.

This indirect determination, although effective, has no real advantage over direct triiodide spectrophotometry. Sensitivities are about equal, with thiofluorescein decoloration by iodine corresponding to a molar absorptivity of about 25,000 l/mole-cm and triiodide absorption at 362 m$\mu$ to about 23,000 l/mole-cm. Supposed advantages, based on relative freedom from iodine losses due to either volatility or reabsorption by olefinic substrates, can be discounted because of the excellent quantitative record of the direct methods (97, 100) and the results of a study by Banerjee and Budke (98) of trace peroxide analyses in olefins and other unsaturated solvents.

In this connection we wish to emphasize our conviction (discussed in Section IV-B-1-b) that loss of iodine through volatility or reabsorption by olefins does not occur when an adequate excess of iodide is present, although many have considered such errors to be potential hazards (75, 94, 99, 173, 219).

### 8. Other Colorimetric Methods

Formation of a yellow-colored complex between titanium(IV) and hydrogen peroxide has long been known as a moderately sensitive method for the colorimetric estimation of either the metal ion or the peroxide (220). It can also determine those organic peroxides that first decompose to hydrogen peroxide by hydrolysis (101–103), principally hydroperoxides and peroxy esters.

Martin has discussed a number of other colorimetric methods (5), but these have relatively little utility. They include the oxidation of phenolphthalin to phenolphthalein, by hydrogen peroxide,* and color development from a reagent consisting of vanillin in 70% sulfuric acid (221).

---

*Unpublished work by T. O. Norris and A. L. Ryland, E. I. duPont de Nemours & Co., Wilmington, Delaware.

A recent method for determining hydroperoxide groups in oxidized polyethylene, by Mitchell and Perkins (222), differs from conventional colorimetric methods in using absorption of infrared rather than visible radiation as the final quantitative measuring step. Reaction of oxidized polyethylene films with gaseous sulfur dioxide for as little as 10 min at room temperature gave apparently quantitative conversion of hydroperoxide groups (which absorb at 3520 $cm^{-1}$) into dialkyl sulfate groups (with a principal absorption at 1195 $cm^{-1}$). Presumably these groups arise through formation of an alkyl hydrogen sulfate intermediate by the reaction (222)

$$ROOH + SO_2 \longrightarrow ROSO_3H$$

followed by esterification with hydroxyl groups present in the oxidized polymer film. This method was calibrated with dilauryl sulfate. A sensitivity of 2 ppm $-OOH$ was observed, with a potential limit of 0.1 ppm by use of thicker specimens. This is due to the relatively high absorptivity of the 1195 $cm^{-1}$ sulfate band, which was consistently 14 times stronger than that of the 3520 $cm^{-1}$ hydroperoxide band. Extension to other polyolefins was suggested (223).

Purcell and Cohen (224) have studied the microdetermination of peroxides by kinetic colorimetry, a technique wherein the rate of color development aids in identifying the peroxides present in a sample. They were concerned particularly with the classes of oxidants found in polluted atmospheres typical of smog conditions. From a group of procedures evaluated for applicability to atmospheric oxidants (225), they selected three reagents for their kinetic studies: ferric thiocyanate, neutral potassium iodide, and molybdate-catalyzed potassium iodide. All three agents gave immediate maximum color development with ozone and peroxyacetic acid, and more gradual color development with "slow oxidants"—acetyl peroxide, nitrogen dioxide, alkyl hydroperoxides, and peroxyacyl nitrates. These slow oxidants were evaluated by plotting the undeveloped absorbance ($A\infty - A$) versus time; single-component systems gave the straight line of a pseudo-first-order reaction, with a characteristic half-life. Hydrogen peroxide gave slow color development only with neutral potassium iodide; addition of the molybdate catalyst caused immediate full reaction.

## V.  SUMMARY OF RECOMMENDED METHODS AND PROCEDURES

The many approaches to determining organic peroxides and the variety of methods presently available for their estimation have been reviewed in preceding sections of this chapter. Procedures for carrying out the cited analyses are generally available in the original references. The *Treatise on Analytical Chemistry* (1) presents detailed instructions for carrying out a dozen of these methods comprising a group that we consider to be most generally useful. We conclude the present chapter by summarizing these selected procedures briefly and again calling attention to several additional promising methods.

### A. Gas Chromatography

Gas chromatography is often the preferred method when applicable; that is, when organic peroxides have sufficient volatility and thermal stability (39, 104, 109). Use of a glass column and on-column sample loading minimizes decomposition. Thermal-conductivity detectors are generally specified; spectacular sensitivities have been reported on peroxyacyl nitrates in smog through the use of an electron-capture detector, however (105).

Deliberate on-column pyrolysis of peroxides (51, 225) or pyrolysis followed by gas chromatography are alternative approaches that may be selectively useful. Control of pyrolysis conditions to give reproducible cracking-product distributions may require refined instrumentation, however.

### B. Polarography

Its broad applicability makes polarography especially useful in peroxide analysis. Virtually all organic peroxides can be reduced and quantitatively determined by this method, down to relatively low concentrations, when there is no interference from reducible nonperoxides; and such interferences can generally be eliminated by appropriate preliminary separations by thin-layer or column chromatography.

Conventional polarographic equipment and techniques should be used (1, 147, 148). Most analytical situations can be covered by three standard media—namely, an aqueous system with lithium chloride or lithium sulfate as supporting electrolyte, for water-soluble systems (144, 149); a benzene–methanol (1:1) solvent with lithium chloride electrolyte, for water-immiscible substrates, such as fats and oils (150); *N,N*-dimethylformamide (DMF) saturated with tetramethylammonium chloride, for the more difficult-to-reduce peroxides,* including di-*t*-butyl peroxide. Commercially available DMF may be purified by fractional distillation (1) if necessary.

A variant worth considering for continuous analysis is a galvanic cell (168) containing a corrodible-metal-tube electrode with an inert coaxial core. Peroxides depolarize the system, and their concentration is proportional to the current.

*Private communication from E. J. Forman and C. M. Wright, Hercules Incorporated, Research Center, Wilmington, Delaware (1959).

## C.  Iodide-Reduction Methods

Iodide-reduction methods are by far the most widely used of the chemical reduction methods, and procedures have proliferated. In the *Treatise on Analytical Chemistry* (1) we have given detailed procedures for four such recommended methods (1).

One is *for determining easily reduced peroxides* by reacting them with sodium iodide in boiling, acidified isopropyl alcohol (10), essentially the procedure of Wagner, Smith, and Peters (173). In this procedure exclusion of atmospheric oxygen is not necessary. Other acceptable iodide procedures for easily reduced peroxides are those of Wibaut and co-workers (133), Lea (86, 171), Paschke and Wheeler (172), and Cass (117).

*For determining moderately stable peroxides*, specifically dicumyl and related bis(aralkyl) peroxides, we gave a procedure developed in our laboratory (10) in which these compounds are reacted with sodium iodide in boiling acetic acid modified with 6% water. This reagent suppresses interference from the related *t*-alcohols or their α-methylstyrene dehydration products. It is essentially without effect on di-*t*-alkyl peroxides. Oxygen is excluded by blanketing with inert gas in a special reaction flask; a convenient multiunit setup was described.

*For determining difficult-to-reduce peroxides* we recommend a similar procedure based on a related reagent proposed by Horner and Jürgens (73), namely, 1 part saturated aqueous potassium iodide, 5 parts concentrated hydrochloric acid, and 30 parts glacial acetic acid, which can also quite properly be viewed as hydriodic acid in acetic acid modified with 12% water. We boil samples for 30 min or more in the equipment utilized in the preceding procedure. Although our experience with it is limited, this reagent, like the previous one, appears to reduce bis(aralkyl) peroxides quantitatively without interference from their associated nonperoxides, and in addition to reduce di-*t*-alkyl peroxides.

A similar but more vigorous reagent, one that approaches the practical upper limit in reducing power, has been extensively evaluated (10). It consists of 6 parts of solid sodium iodide and 3 parts of 37% hydrochloric acid in 50 parts of glacial acetic acid, and is also employed at the boiling point. In comparison with the Horner and Jürgens reagent, as can be seen in Table 13, this stronger version contains five times as much hydriodic acid and only one-quarter as much water to moderate activity. Although it reduces quantitatively the initial reduction products of many peroxides, and many nonperoxides, its general utility appears to be limited.

Specifically *for determining t-butylperoxy esters* we recommend the iron-catalyzed procedure of Silbert and Swern (184). In this procedure the sample in chloroform is reacted in the dark at room temperature with sodium iodide in glacial acetic acid containing sufficient ferric chloride to make the system $4 \times 10^{-5}$ $M$ in $Fe^{3+}$. Reaction is complete in 5–10 min.

Deaeration, accomplished by flushing with nitrogen, does not appear to be complete, since the iron-containing acetic acid solvent seems to be used without removal of dissolved oxygen. It should be determined whether careful deaeration modifies the results.

### D.  Mixtures of Peroxides from Several Classes

Systems containing peroxides from a number of classes can be analyzed quantitatively as to class without prior separation, by means of a group of chemical methods described by Horner and Jürgens (73). These procedures are given in Table 16. Appropriate reductions with diphenyl sulfide, triethyl arsine, potassium iodide, and hydriodic acid, along with acidimetry, allow quantitative determination in the same sample of any or all of the chemical classes: carboxylic acid, peroxy acid, diacyl peroxide, alkyl hydroperoxide, and dialkyl peroxide. Peroxy esters are not included in the scheme.

### E.  Reduction with Sodium Arsenite

For determining traces of peroxides in petroleum products we recommend the procedure of Walker and Conway (79), which employs reduction with a sodium arsenite reagent. This method avoids the interference from sulfur compounds present in most gasolines and fuel oils that causes low results by iodide reduction methods. Since large samples can be used, one can accurately determine very low hydroperoxide concentrations in petroleum products.

### F.  Colorimetric Methods

In the *Treatise on Analytical Chemistry* (1) we have detailed four colorimetric procedures for determining trace amounts of easily reduced organic peroxides, two based on reduction with ferrous salts, one with iodide, and one with leuco methylene blue.

The first is the method of Smith (87), cited by Lea (86) as the most precise and sensitive of the ferrous thiocyanate methods for colorimetric determination of traces of hydroperoxide in lipids. Reduction is carried out in benzene–methanol (7:3), without removal of oxygen. Although peroxide estimates are known to be high by a factor of almost 2, one can obtain good absolute accuracy when necessary by calibrating with an independent method. Substrates must be soluble in the benzene–methanol medium.

The second colorimetric iron method is the procedure of Kolthoff and Medalia (85). It utilizes reduction with ferrous perchlorate in acetone, in the absence of oxygen. For cumene hydroperoxide the method gives results that are accurate (2 moles of ferric ion per mole) and reasonably precise. Precision is satisfactory for lipid and fatty acid hydroperoxides, also, although in these cases the stoichiometry appears to be severely and unpredictably low.

The recommended colorimetric iodide procedure is that of Banerjee and Budke (91). It is simple, with adequate precision and accuracy, and is broadly applicable to estimating hydroperoxides, particularly in organic solvents, in the concentration range down to 10 ppm of active oxygen. (There is a more sensitive variation for the range of 1–10 ppm.) The method may be calibrated with a standard iodine solution, hydrogen peroxide, or a known hydroperoxide. Color development is carried out in small volumetric flasks purged with nitrogen. The reaction medium is acetic acid–chloroform (2:1), 80% of which can be replaced by isopropyl alcohol where necessary to obtain miscibility of samples.

The recommended leuco methylene blue procedure is Dulog's modification (92) of the original Ueberreiter and Sorge procedure (93). Many problems in preparing and storing leuco methylene blue solutions have been avoided by maintaining the reagent solution in a dispensing buret over platinized asbestos, where it may be reduced with hydrogen before use to maintain its initial high analytical quality.

Color development is carried out by refluxing for up to 15 min in a benzene medium acidified with trichloroacetic acid. Oxygen is not excluded, but strong light should be avoided.

The leuco methylene blue method has the greatest sensitivity of the useful colorimetric peroxide methods, outperforming the ferrous thiocyanate and iodide methods in this regard by a factor of about 3. In addition calibration can be carried out with any convenient reactive peroxide, since the response of leuco methylene blue to peroxides appears to be constant on a molar basis and independent of their structure.

# REFERENCES

1. R. D. Mair and R. T. Hall, "Determination of Organic Peroxides," *Treatise on Analytical Chemistry*, Part II, Vol. 14, edited by I. M. Kolthoff and P. J. Elving, Interscience, New York, *in preparation*.

2. A. G. Davies, *Organic Peroxides*, Butterworths, London, 1961.

3. E. G. E. Hawkins, *Organic Peroxides*, Van Nostrand, Princeton, N.J., 1961.

4. N. A. Milas, "Organic Peroxides and Peroxy Compounds," in *Encyclopedia of Chemical Technology*, Vol. 10, edited by R. E. Kirk and D. F. Othmer, Interscience, New York, 1953, p. 58.

5. A. J. Martin, "Determination of Organic Peroxides," in *Organic Analysis*, Vol. IV, edited by John Mitchell, Jr., Interscience, New York, 1960, pp. 1–64.

6. E. J. Kuta and F. W. Quackenbush, *Anal. Chem.*, **32**, 1069 (1960).

7. L. S. Silbert, *J. Amer. Oil Chem. Soc.*, **39**, 480 (1962).

8. D. Swern and L. S. Silbert, *Anal. Chem.*, **35**, 880 (1963).

9. R. Criegee, W. Schorrenberg, and J. Becke, *Ann. Chem.*, **565**, 7 (1949).

10. R. D. Mair and A. J. Graupner, *Anal. Chem.*, **36**, 194 (1964).

11. H. M. Castrantas, D. K. Banerjee, and D. C. Noller, *Fire and Explosion Hazards of Peroxy Compounds*, ASTM STP 394, American Society for Testing Materials, 1965.

12. J. E. Guillet and M. F. Meyer, *Ind. Eng. Chem., Prod. Res. Develop.*, **1**, 226 (1962).

13. D. C. Noller and D. J. Bolton, *Anal. Chem.*, **35**, 887 (1963).

14. D. C. Noller, S. J. Mazurowski, G. F. Linden, F. J. G. DeLeeuw, and O. L. Mageli, *Ind. Eng. Chem.*, **56**, No. 12, 18 (1964).

15. E. Sambale, *Plaste Kautschuk*, **10** (1), 33 (1963).

16. A. M. E. Siemens, *Brit. Plastics*, **35**, 357 (1962).

17. O. L. Mageli, S. D. Stengel, and D. F. Doehnert, *Mod. Plastics*, **36**, (No. 7), 135 (1959).

18. D. F. Doehnert and O. L. Mageli, *Mod. Plastics*, **36** (No. 6), 142 (1959).

19. J. E. Guillet, T. R. Walker, M. F. Meyer, J. P. Hawk, and E. B. Towne, *Ind. Eng. Chem., Prod. Res. Develop.*, **3**, 257 (1964).

20. P. Molyneux, *Tetrahedron*, **22**, 2929 (1966).

21. J. B. Ramsey and F. T. Aldridge, *J. Amer. Chem. Soc.*, **77**, 2561 (1955).

22. H. E. Fierz-David, *Chimia*, **1**, 246 (1947).

23. A. Weissberger, E. S. Proskauer, J. A. Riddick, and E. E. Toops, Jr., *Organic Solvents*, 2nd ed., Interscience, New York, 1955.

24. W. Dasler and C. D. Bauer, *Ind. Eng. Chem., Anal. Ed.*, **18**, 52 (1946).

25. R. N. Feinstein, *J. Org. Chem.*, **24**, 1172 (1959).

26. N. Kornblum and H. E. De La Mare, *J. Amer. Chem. Soc.*, **73**, 880 (1951).

27. G. S. Serif, C. F. Hunt, and A. N. Bourns, *Can. J. Chem.*, **31**, 1229 (1953).

28. N. A. Milas and D. M. Surgenor, *J. Amer. Chem. Soc.*, **68**, 642 (1946).

29. N. Kornblum and H. E. De La Mare, *J. Amer. Chem. Soc.*, **74**, 3079 (1952).

30. A. G. Davies, *J. Chem. Soc.*, **1958**, 3474.

31. A. G. Davies and R. Feld, *J. Chem. Soc.*, **1956**, 665.

32. L. Horner and W. Jurgeleit, *Ann. Chem.*, **591**, 138 (1955).

33. M. Matic and D. A. Sutton, *Chem. Ind. (London)*, **1953**, 666.

34. M. Matic and D. A. Sutton, *J. Chem. Soc.*, **1952**, 2679.

35. H. R. Williams and H. S. Mosher, *J. Amer. Chem. Soc.*, **76**, 3495 (1954).

36. A. G. Davies, R. V. Foster, and R. Nery, *J. Chem. Soc.*, **1954**, 2204.

37. A. Rieche, E. Schmitz, and P. Dietrich, *Chem. Ber.*, **92**, 2239 (1959).

38. M. H. Abraham, A. G. Davies, D. R. Llewellyn, and E. M. Thain, *Anal. Chim. Acta*, **17**, 499 (1957).

39. S. W. Bukata, L. L. Zabrocki, and M. F. McLaughlin, *Anal. Chem.*, **35**, 885 (1963).

40. W. Eggersglüss, *Organische Peroxyde*, Verlag Chemie, Weinheim, 1951.

41. M. M. Buzlanova, V. F. Stepanovskaya, and V. L. Antonovskii, *J. Anal. Chem. USSR*, (English translation), **21**, 1324 (1966).

42. M. M. Buzlanova, V. F. Stepanovskaya, A. F. Nesterov, and V. L. Antonovskii, *J. Anal. Chem. USSR*, (English translation), **21**, 454 (1966).

43. S. Hayano, T. Ota, and Y. Fukushima, *Bunseki Kagaku*, **15**, 365 (1966); through *Chem. Abstr.*, **67**, 39941r (1967).

44. E. Knappe and D. Peteri, *Z. Anal. Chem.*, **190**, 386 (1962).

45. A. Rieche, M. Schulz, H. E. Seyfarth, and G. Gottschalk, *Fette, Seifen, Anstrichmittel,* 64, 198 (1962).
46. J. Cartlidge and C. F. H. Tipper, *Anal. Chim. Acta,* 22, 106 (1960).
47. G. Dobson and G. Hughes, *J. Chromatog.,* 16, 416 (1964).
48. N. Milas and I. Belic, *J. Amer. Chem. Soc.,* 81, 3358 (1959).
49. N. Milas and A. Golubovic, *J. Amer. Chem. Soc.,* 81, 3361 (1959).
50. A. Rieche and M. Schulz, *Angew. Chem.,* 70, 694 (1958).
51. S. Hyden, *Anal. Chem.,* 35, 113 (1963).
52. L. Bellamy, *The Infrared Spectra of Complex Molecules,* 2nd ed., Wiley, New York, 1958, Chapters 7 and 8.
53. W. H. T. Davison, *J. Chem. Soc.,* 1951, 2456.
54. R. F. Goddu, "Near-Infrared Spectrophotometry," in *Advances in Analytical Chemistry and Instrumentation,* Vol. 1, edited by C. N. Reilley, Interscience, New York, 1963, p. 406.
55. L. A. Gribov and A. V. Karyakin, *Opt. Spectr. (USSR)* (English translation), 9, 350 (1960).
56. R. T. Holman, C. Nickell, O. S. Privett, and P. R. Edmondson, *J. Amer. Oil Chem. Soc.,* 35, 422 (1958).
57. M. A. Kovner, A. V. Karyakin, and A. P. Efimov, *Opt. Spectr. (USSR)* (English translation), 8, 64 (1960).
58. G. J. Minkoff, *Proc. Roy. Soc. (London),* A228, 287 (1951).
59. H. R. Williams and H. S. Mosher, *Anal. Chem.,* 27, 517 (1955).
60. W. C. Kenyon, J. E. McCarley, E. G. Boucher, A. E. Robinson, and A. K. Wiebe, *Anal. Chem.,* 27, 1888 (1955).
61. A. G. Davies, D. G. Hare, and R. F. M. White, *J. Chem. Soc.,* 1960, 1040.
62. A. G. Davies, D. G. Hare, and R. F. M. White, *J. Chem. Soc.,* 1961, 341.
63. S. Fujiwara, M. Katayama, and S. Kamio, *Bull. Chem. Soc. Japan,* 32, 657 (1959).
64. N. D. Klyueva, G. E. Muratova, A. I. Kashlinskii, and A. V. Sokolov, *J. Anal. Chem. USSR* (English Translation), 22, 253 (1967).
65. E. S. Watson, M. J. O'Neill, J. Justin, and N. Brenner, *Anal. Chem.,* 36, 1233 (1964).
66. R. M. Johnson and I. W. Siddiqi, *J. Polarog. Soc.,* 11, 72 (1965).
67. E. J. Kuta and F. W. Quackenbush, *J. Amer. Oil Chem. Soc.,* 37, 148 (1960).
68. W. E. Parker, C. Ricciuti, C. L. Ogg, and D. Swern, *J. Amer. Chem. Soc.,* 77, 4037 (1955).
69. C. Ricciuti, C. O. Willits, C. L. Ogg, S. G. Morris, and R. W. Riemenschneider, *J. Amer. Oil Chem. Soc.,* 31, 456 (1954).
70. M. Schulz and K. H. Schwarz, *Z. Chemie,* 7, 176 (1967).
71. L. S. Silbert, L. P. Witnauer, D. Swern, and C. Ricciuti, *J. Amer. Chem. Soc.,* 81, 3244 (1959).
72. C. O. Willits, C. Ricciuti, H. B. Knight, and D. Swern, *Anal. Chem.,* 24, 785 (1952).
73. L. Horner and E. Jürgens, *Angew. Chem.,* 70, 266 (1958).
74. L. Horner and E. Jürgens, *Ann. Chem.,* 602, 135 (1957).

75. L. Dulog and K. H. Burg, *Z. Anal. Chem.*, **203**, 184 (1964).
76. D. Barnard and K. R. Hargrave, *Anal. Chim. Acta*, **5**, 476 (1951).
77. A. C. Egerton, A. J. Everett, G. J. Minkoff, S. Rudrakanchana, and K. C. Salooja, *Anal. Chim. Acta*, **10**, 422 (1954).
78. S. Siggia, *Anal. Chem.*, **19**, 872 (1947).
79. D. C. Walker and H. S. Conway, *Anal. Chem.*, **25**, 923 (1953).
80. A. Szobor and I. Back, *Tetrahedron Letters*, **1966**, 3985.
81. A. A. Balandin, L. Kh. Freidlin, and N. V. Nikiforova, *Bull. Acad. Sci. USSR, Div. Chem. Sci.* (English translation), **1958**, 126.
82. A. A. Balandin, L. Kh. Freidlin, and N. V. Nikiforova, *Bull. Acad. Sci. USSR, Div. Chem. Sci.* (English translation), **1959**, 1138.
83. L. H. Amundsen and L. S. Nelson, *J. Amer. Chem. Soc.*, **73**, 242 (1951).
84. D. A. Sutton, *Chem. Ind.* (*London*), **1951**, 272.
85. I. M. Kolthoff and A. I. Medalia, *Anal. Chem.*, **23**, 595 (1951).
86. C. H. Lea, *J. Sci. Food Agric.*, **3**, 586 (1952).
87. G. H. Smith, *J. Sci. Food Agric.*, **3**, 26 (1952).
88. L. Hartman and M. D. L. White, *J. Sci. Food Agric.*, **3**, 112 (1952).
89. S. Hartmann and J. Glavind, *Acta Chem. Scand.*, **3**, 954 (1949).
90. A. L. Ryland, paper presented at the 142nd Meeting of the American Chemical Society (Division of Analytical Chemistry), Atlantic City, N.J., September 1962.
91. A. Banderet, M. Brendle, and G. Riess, *Bull. Soc. Chim. France*, **1965**, 626.
92. L. Dulog, *Z. Anal. Chem.*, **202**, 192 (1964).
93. G. Sorge and K. Ueberreiter, *Angew. Chem.*, **68**, 486 (1956).
94. K. Ueberreiter and G. Sorge, *Angew. Chem.*, **68**, 352 (1956).
95. M. I. Eiss and P. Giesecke, *Anal. Chem.*, **31**, 1558 (1959).
96. D. L. Hamm, E. G. Hammond, V. Parvanah, and H. E. Snyder, *J. Amer. Oil Chem. Soc.*, **42**, 920 (1965).
97. D. K. Banerjee and C. C. Budke, *Anal. Chem.*, **36**, 792 (1964).
98. D. K. Banerjee and C. C. Budke, *Anal. Chem.*, **36**, 2367 (1964).
99. P. Dubouloz, J. Fondarai, J. Laurent, and R. Marville, *Anal. Chim. Acta*, **15**, 84 (1956).
100. F. W. Heaton and N. Uri, *J. Sci. Food Agric.*, **9**, 781 (1958).
101. H. Pobiner, *Anal. Chem.*, **33**, 1423 (1961).
102. C. N. Satterfield and A. H. Bonnell, *Anal. Chem.*, **27**, 1174 (1955).
103. W. C. Wolfe, *Anal. Chem.*, **34**, 1328 (1962).
104. M. P. Hahto and J. E. Beaulieu, *J. Chromatog.*, **25**, 472 (1966).
105. E. F. Darley, K. A. Kettner, and E. R. Stephens, *Anal. Chem.*, **35**, 589 (1963).
106. A. D. Kirk and J. H. Knox, *Trans. Faraday Soc.*, **56**, 1296 (1960).
107. S. Dykstra and H. S. Mosher, *J. Amer. Chem. Soc.*, **79**, 3474 (1957).
108. C. F. Wurster, Jr., L. J. Durham, and H. S. Mosher, *J. Amer. Chem. Soc.*, **80**, 327 (1958).
109. W. F. Brill, *J. Amer. Chem. Soc.*, **87**, 3286 (1965).
110. G. R. McMillan, *J. Amer. Chem. Soc.*, **83**, 3018 (1961).
111. R. Criegee and G. Paulig, *Chem. Ber.*, **88**, 712 (1955).

112.  J. T. Middleton, *Ann. Rev. Plant Physiol.*, **12**, 431 (1961).
113.  E. R. Stephens, E. F. Darley, O. C. Taylor, and W. E. Scott, *Intern. J. Air Pollution*, **4**, 79 (1961); through *Chem. Abstr.*, **55**, 19082h (1961).
114.  W. Simon and H. Giacobbo, *Chem. Ingr.-Tech.*, **37**, 709 (1965).
115.  J. C. Courtier, *Methodes Phys. Anal.*, **1966**, 23; through *Chem. Abstr.*, **65**, 17125g (1966).
116.  J. Voellmin, P. Kriemler, I. Omura, J. Seibl, and W. Simon, *Microchem. J.*, **11**, 73 (1966).
117.  Waters Associates, Inc., Framingham, Mass., trade literature on model R-4 liquid chromatograph.
118.  Japan Electron Optics Laboratory Co., Ltd., (JEOL), trade literature on model JLC-2A "Universal" automatic recording liquid chromatograph.
119.  W. Simon, J. T. Clerc, and R. E. Dohner, *Microchem. J.*, **10**, 495 (1966).
120.  J. E. Stouffer, *Barber-Colman Chromatogram*, Vol. 5, No. 1 (1965).
121.  J. E. Stouffer, T. E. Kersten, and P. M. Krueger, *Biochim. Biophys. Acta*, **93**, 191 (1964).
122.  E. Stahl, *Chemiker-Ztg.*, **82**, 323 (1958).
123.  J. A. Brammer, S. Frost, and V. W. Reid, *Analyst*, **92**, 91 (1967).
124.  M. Schulz, H. Seeboth, and W. Wieker, *Z. Chem.*, **2**, 279 (1962).
125.  A. V. Karyakin, *Russian Chem. Rev.* (English translation), **30**, 460 (1961).
126.  N. Sheppard, *Discussions Faraday Soc.*, **9**, 322 (1950).
127.  W. Lobunez, J. R. Rittenhouse, and J. G. Miller, *J. Amer. Chem. Soc.*, **80**, 3505 (1958).
128.  R. L. Redington, W. B. Olson, and P. C. Cross, *J. Chem. Phys.*, **36**, 1311 (1962).
129.  M. T. Rogers and T. W. Campbell, *J. Amer. Chem. Soc.*, **74**, 4742 (1952).
130.  F. D. Verderame and J. G. Miller, *J. Phys. Chem.*, **66**, 2185 (1962).
131.  G. J. Minkoff, *Discussions Faraday Soc.*, **9**, 320 (1950).
132.  D. Swern, L. P. Witnauer, C. R. Eddy, and W. E. Parker, *J. Amer. Chem. Soc.*, **77**, 5537 (1955).
133.  J. P. Wibaut, H. B. van Leeuwen, and B. van der Wal, *Rec. Trav. Chim.*, **73**, 1033 (1954).
134.  F. J. Welch, H. R. Williams, and H. S. Mosher, *J. Amer. Chem. Soc.*, **77**, 551 (1955).
135.  J. Hoffman, *J. Org. Chem.*, **22**, 1747 (1957).
136.  V. A. Terent'ev, N. Kh. Shtivel, and V. L. Antonovskii, *Zh. Prikl. Spektrosk.*, **5**, 463 (1966); through *Chem. Abstr.*, **66**, 52055d (1967).
137.  American Physics Institute, NMR Spectral Data Catalog (A.P.I. Project No. 44).
138.  G. A. Ward and R. D. Mair, *Anal. Chem.*, **41**, 538 (1969).
139.  D. Swern, A. H. Clements and T. M. Luong, *Anal. Chem.*, **41**, 412 (1969).
140.  J. K. Kochi and P. J. Krusic, *J. Amer. Chem. Soc.*, **91**, 3940, 3944 (1969).
141.  A. Dobrinskaja and M. Neumann, *Acta Physicochim. URSS*, **10**, 297 (1939).
142.  B. A. Gosman, *Coll. Czech. Chem. Commun.*, **7**, 467 (1935).
143.  V. Stern and S. Polak, *Acta Physicochim. URSS*, **11**, 797 (1939).
144.  H. Brüschweiler and G. J. Minkoff, *Anal. Chim. Acta*, **12**, 186 (1955).

145. M. F. Romantsev and E. S. Levin, *J. Anal. Chem. USSR* (English Translation), **18**, 957 (1963).

146. L. Meites, *Polarographic Techniques*, 2nd ed., Interscience, New York, 1965, p. 396.

147. L. Meites, "Voltammetry at the Dropping Mercury Electrode (Polarography)," in *Treatise on Analytical Chemistry*, Part I, Vol. 4, I. M. Kolthoff and P. J. Elving, Interscience, New York, 1963, Chapter 46.

148. C. N. Reilley and R. W. Murray, "Introduction to Electrochemical Techniques," in *Treatise on Analytical Chemistry*, Part I, Vol. 4, edited by I.M. Kolthoff and P. J. Elving, Interscience, New York, 1963, Chapter 43.

149. D. A. Skoog and A. B. H. Lauwzecha, *Anal. Chem.*, **28**, 825 (1956).

150. W. R. Lewis, F. W. Quackenbush, and T. DeVries, *Anal. Chem.*, **21**, 762 (1949).

151. S. S. Kalbag, K. A. Narayan, S. S. Chang, and F. A. Kummerow, *J. Amer. Oil Chem. Soc.*, **32**, 271 (1955).

152. W. R. Lewis and F. W. Quackenbush, *J. Amer. Oil Chem. Soc.*, **26**, 53 (1949).

153. D. Swern, J. E. Coleman, H. G. Knight, C. Ricciuti, C. O. Willits, and C. R. Eddy, *J. Amer. Chem. Soc.*, **75**, 3135 (1953).

154. C. O. Willits, C. Ricciuti, C. L. Ogg, S. G. Morris, and R. W. Riemenschneider, *J. Amer. Oil Chem. Soc.*, **30**, 420 (1953).

155. M. L. J. Bernard, *Ann. Chim. (Paris)*, **10**, 315 (1955).

156. M. Schulz and K. H. Schwarz, *Monatsber. Deut. Akad. Wiss. Berlin*, **6**, (7), 515 (1964).

157. H. Lueck and R. Maack, *Fette, Seifen, Anstrichmittel*, **68**, 81, 305, 609 (1966).

158. P. A. Giguère and D. Lamontagne, *Can. J. Chem.*, **29**, 54 (1951).

159. F. Reimers, *Sbornik Mezinarod., Polarog. Sjezdu. Praze, 1st Congr., 1951*, Part 1, 203; through *Chem. Abstr.*, **46**, 6998d (1952).

160. T. D. Parks and K. A. Hansen, *Anal. Chem.*, **22**, 1268 (1950).

161. S. Wawzonek, E. W. Blaha, R. Berkey, and M. E. Runner, *J. Electrochem. Soc.*, **102**, 235 (1955).

162. F. W. Quackenbush and E. J. Kuta, paper presented at the 142nd Meeting of the American Chemical Society (Division of Analytical Chemistry), Atlantic City, N.J., September 1962.

163. G. Proske, *Anal. Chem.*, **24**, 1834 (1952).

164. W. M. MacNevin and P. F. Urone, *Anal. Chem.*, **25**, 1760 (1953).

165. R. Parsons, "The Structure of the Electrical Double Layer and Its Influence on the Rates of Electrode Reactions," in *Advances in Electrochemistry and Electrochemical Engineering*, Vol. 1, edited by P. Delahay, Interscience, New York, 1961, p. 1.

166. R. Maack and H. Lück, *Experientia*, **19**, 466 (1963).

167. E. R. Roberts and J. S. Meek, *Analyst*, **77**, 43 (1952).

168. R. Kober, German (East) Patent 53,354 (1967); through *Chem. Abstr.* **67**, 122070b (1967).

169. D. H. Wheeler, *Oil & Soap*, **9**, 89 (1932).

170. C. H. Lea, *Proc. Roy. Soc. (London)*, **B108**, 175 (1931).

171. C. H. Lea, *J. Soc. Chem. Ind. (London)*, **65**, 286 (1946).

172.  R. F. Paschke and D. H. Wheeler, *Oil & Soap*, **21**, 52 (1944).
173.  C. D. Wagner, R. H. Smith, and E. D. Peters, *Anal. Chem.*, **19**, 976 (1947).
174.  M. E. Stansby, *Ind. Eng. Chem.*, *Anal. Ed.*, **13**, 627 (1941).
175.  V. R. Kokatnur and M. Jelling, *J. Amer. Chem. Soc.*, **63**, 1432 (1941).
176.  P. D. Bartlett and R. Altschul, *J. Amer. Chem. Soc.*, **67**, 812 (1945).
177.  W. E. Cass, *J. Amer. Chem. Soc.*, **68**, 1976 (1946).
178.  W. E. Cass, *J. Amer. Chem. Soc.*, **72**, 4915 (1950).
179.  H. A. Liebhafsky and W. H. Sharkey, *J. Amer. Chem. Soc.*, **62**, 190 (1940).
180.  J. H. Skellon and E. D. Wills, *Analyst*, **73**, 78 (1948).
181.  B. T. D. Sully, *Analyst*, **79**, 86 (1954).
182.  E. W. Abrahamson and H. Linschitz, *Anal. Chem.*, **24**, 1355 (1952).
183.  L. Harman and M. D. L. White, *Anal. Chem.*, **24**, 527 (1952).
184.  L. S. Silbert and D. Swern, *Anal. Chem.*, **30**, 385 (1958).
185.  E. Buncel and A. G. Davies. *J. Chem. Soc.*, **1958**, 1550.
186.  J. H. Skellon and M. H. Thurston, *Analyst*, **73**, 97 (1948).
187.  L. K. Dahle and R. T. Holman, *Anal. Chem.*, **33**, 1960 (1961).
188.  J. S. Matthews and J. F. Patchan, *Anal. Chem.*, **31**, 1003 (1959).
189.  L. S. Silbert and D. Swern, *J. Amer. Chem. Soc.*, **81**, 2364 (1959).
190.  H. Hock and H. Kropf, *Chem. Ber.*, **92**, 1115 (1959).
191.  K. Nozaki, *Ind. Eng. Chem.*, *Anal. Ed.*, **18**, 583 (1946).
192.  F. H. Dickey, J. H. Raley, F. F. Rust, R. S. Treseder, and W. E. Vaughan, *Ind. Eng. Chem.*, **41**, 1673 (1949).
193.  W. E. Vaughan and F. F. Rust, U.S. Patent 2,403,771 (1946) (to Shell Development Co.).
194.  C. T. Handy and H. S. Rothrock, *J. Amer. Chem. Soc.*, **80**, 5306 (1958).
195.  E. E. Royals, *Advanced Organic Chemistry*, Prentice-Hall, Englewood Cliffs, N.J., 1954, p. 369.
196.  B. Braithwaite and G. E. Penketh, *Anal. Chem.*, **39**, 1470 (1967).
197.  H. Hoffmann, *Angew. Chem.*, **72**, 77 (1960).
198.  O. Lorenz, *Anal. Chem.*, **37**, 101 (1965).
199.  F. A. Hochstein, *J. Amer. Chem. Soc.*, **71**, 305 (1949).
200.  A. P. Terent'ev, G. C. Larikova, and E. A. Bondarevskaya, *J. Anal. Chem. USSR* (English translation), **21**, 312 (1966).
201.  R. A. Chapman and K. Mackay, *J. Amer. Oil Chem. Soc.*, **26**, 360 (1949).
202.  C. H. Lea, *J. Soc. Chem. Ind.* (*London*), **64**, 106 (1945).
203.  C. D. Wagner, H. L. Clever, and E. D. Peters, *Anal. Chem.*, **19**, 980 (1947).
204.  C. D. Wagner, R. H. Smith, and E. D. Peters, *Anal. Chem.*, **19**, 982 (1947).
205.  I. M. Kolthoff and A. I. Medalia, *J. Amer. Chem. Soc.*, **71**, 3777, 3784, 3789 (1949).
206.  J. H. Merz and W. A. Waters, *J. Chem. Soc.*, **1949**, S15.
207.  G. Loftus-Hills and C. C. Thiel, *J. Dairy Res.*, **14**, 340 (1946).
208.  M. G. Driver, R. B. Koch, and H. Salwin, *J. Amer. Oil Chem. Soc.*, **40**, 504 (1963).
209.  D. L. Hamm and E. G. Hammond, *J. Dairy Sci.*, **50**, 1166 (1967).
210.  H. Linschitz, J. Rennert, and T. M. Korn, *J. Amer. Chem. Soc.*, **76**, 5839 (1954).

211. P. R. Dugan, *Anal. Chem.*, **33**, 696, 1630 (1961).
212. G. H. Foxley, *Analyst*, **86**, 348 (1961).
213. P. Landauer and H. Weil, *Ber.*, **43**, 198 (1910).
214. J. Stamm, *Bull. Soc. Pharm. Esthonia*, **5**, 181 (1925).
215. J. Stamm, *J. Pharm. Chim.*, **118**, 214 (1926).
216. J. Stamm, *Analyst*, **51**, 416 (1926).
217. A. Anton, *J. Appl. Polymer Sci.*, **9**, 1631 (1965).
218. A. M. Siddiqi and A. L. Tappel, *Chemist-Analyst*, **44**, 52 (1955).
219. N. N. Dastur and C. H. Lea, *Analyst*, **66**, 90 (1941).
220. I. M. Kolthoff and E. B. Sandell. *Textbook of Quantitative Inorganic Analysis*, 3rd ed., Macmillan, New York, 1952, pp. 574, 600, 705.
221. S. Arrhenius, *Acta Chem Scand.*, **9**, 715 (1955).
222. J. Mitchell, Jr., and D. M. Smith, *Aquametry*, Interscience, New York, 1948, p. 142.
223. J. Mitchell, Jr., and L. R. Perkins, *Symposium on Weatherability of Plastic Materials*, Manufacturing Chemists Association, National Bureau of Standards, Gaithersburg, Md., February 8, 1967.
224. T. C. Purcell and I. R. Cohen, *Environ. Sci. Technol.*, **1**, 431 (1967).
225. I. R. Cohen, T. C. Purcell, and A. P. Altshuller, *Environ. Sci. Technol.*, **1**, 247 (1967).

*Chapter VII*

# Physical Properties of Organic Peroxides

LEONARD S. SILBERT

---

## I. STRUCTURE

Hydrogen peroxide, the parent member in the peroxide family, is a structural model for most organic peroxides. In a century of effort toward elucidating its structure all of the important classical physical properties were acquired. The library of information extending to 1955 had fortunately been comprehensively reviewed in an elegant monograph (1). Since this book's publication, references to subsequent structural studies of the hydrogen peroxide molecule are limited and incomplete (2,3).

Until 1934 the structure of hydrogen peroxide was generally conceived in terms of simple extensions of the water molecule, illustrated in Figure 1 as linear, planar, or trigonal configurations. The inadequacy of these planar concepts stimulated an evaluation by dipole moment (4) and a theoretical treatment by quantum mechanics that together have supported a skewed conformation as an appropriate description of its structure.* Spectroscopic and X-ray investigations (see Section A) subsequently confirmed the skewed structure.

The skewed molecule with symmetry $C_2$ (Figure 1d) is describable by four parameters: the O—O distance, the O—O—H distance, the O—O—H angle,

---

* Reference 1 contains a detailed account of the investigations attempting to relate dipole moments to these classical configurations.

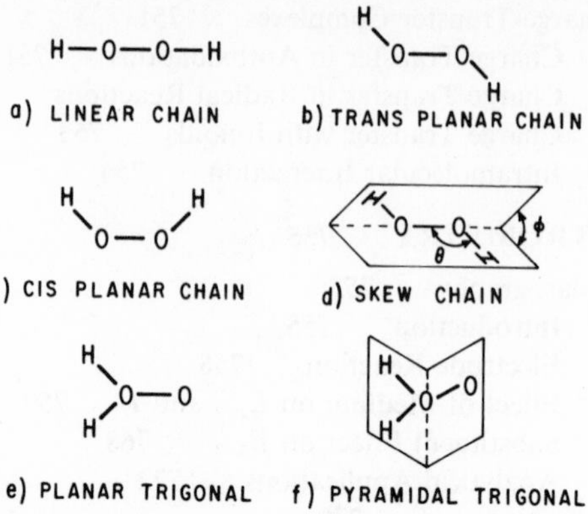

**Figure 1.** Reasonable configurations of the hydrogen peroxide molecule. (Reference 1, page 311).

and the azimuthal or dihedral angle between the two planes each defined by the O—O—H grouping. Of these properties, the dihedral angle is the structural feature of principal chemical interest. Its evaluation in hydrogen peroxide and in a variety of hydrogen peroxide complexes has contributed invaluably to an understanding of the compound's chemical and physical properties that stimulated similar structural determinations of its organic derivatives.

## A. The Dihedral Angle

### 1. Hydrogen Peroxide

Penney and Sutherland (5,6) undertook the theoretical treatment of the dihedral angle and derived the value of 90° by including the repulsion of unshared electron pairs occupying $sp$-hybrid orbitals (Figure 2). They further suggested that repulsion between the hydrogen atoms would enlarge this angle to 100°, a value later shown to be consistent with the angle computed from the dipole moment (7). Amako and Giguère (8) in their subsequent LCAO-SCF treatment arrived at the larger azimuthal angle of 120° on the assumption of $sp^2$ hybridization, application of approximate integrals, and exclusion of $1s$ electrons on the oxygen atoms. A more exact calculation by Kaldor and Shavitt (3) resulted in an equilibrium dihedral angle of 114.5° in comparison with 111.5° that Hunt et al. (9,10) measured from the far-infrared spectrum on hydrogen peroxide and deuterium peroxide vapor. With respect to dihedral values derived from the infrared spectrum, Giguere and Bain (11) a decade earlier had derived the very low value of $82 \pm 20°$,

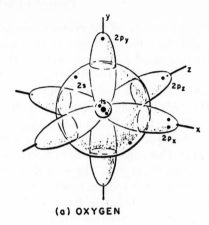

(a) OXYGEN

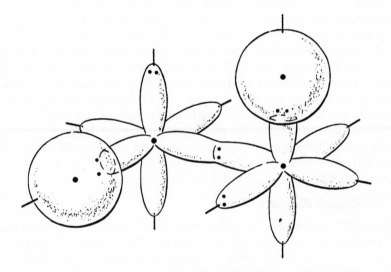

(b)  HYDROGEN  PEROXIDE

**Figure 2.**  Quantum mechanical representation of the electronic structure of atomic oxygen and hydrogen peroxide. (Reference 1, page 317).

but Redington, Olson, and Cross (12), finding this value too low, reported $119.8 \pm 3°$ from a careful analysis of the infrared rotational-vibrational bands obtained for hydrogen peroxide vapor.

Crystalline hydrogen peroxide and several of its crystalline complexes were examined by X-ray analysis. In 1951 Abrahams, Collin, and Lipscomb (13) reported short intermolecular $O \cdots O$ distances (2.78 Å) in the structure of crystalline hydrogen peroxide on the presumption that these were hydrogen-bonded atoms, although the hydrogen-atom positions were not actually determined. They calculated a dihedral angle of 93.8° on the assumption of linearity for the triatom (O—H···O) distance. The azimuthal angle in a variety of crystalline hydrogen peroxide complexes varies in accordance with the nature of the coordinating compound, as illustrated in such peroxy-hydrates as $CO(NH_2)_2 \cdot H_2O_2$ (106°) (14), $H_2O_2 \cdot H_2O$ (138.8°) (15), potassium oxalate monoperhydrate (101.6°) (16), and rubidium oxalate monoper-hydrate (103.4°) (16). This angle is clearly subject to the effects of strong polar and steric forces within the crystal. The most striking example is provided by the peroxide conformation in sodium oxalate monoperhydrate, which Pedersen and Pedersen (17) concluded to be *trans* planar on the basis of their X-ray and proton-magnetic-resonance studies.

Busing and Levy (2) submitted a single crystal of solid hydrogen peroxide to neutron-diffraction analysis. Their method provides accurate positions for both oxygen and hydrogen atoms and, accordingly, an accurate geometric description of the molecule in the crystal. The dihedral angle (90.2 ± 0.6°) in the crystalline state differs significantly from the accurately recorded angle (119.8°) in the vapor state (12). This discrepancy between precise values measured for two different physical states of hydrogen peroxide, as well as the differences shown in a comparison of these results with and among those for crystalline hydrogen peroxide–containing complexes, indicates the sensitivity of the dihedral angle to the molecular environment. If the dihedral angle in the simple hydrogen peroxide molecule is so readily influenced by the molecular environment, similar effects would be expected in the more complicated structures of organic peroxide derivatives.

## 2. *Organic Peroxides*

Prior to 1952 the concept of a Penney–Sutherland model for organic peroxides had not been seriously considered or pursued. Descriptions of organic peroxides as linear or planar molecules persisted for nearly two decades after Penney and Sutherland had published their classic paper. The views prevailing during this period on organic peroxide structures are illustrated in the following examples covering a range of peroxide types that had been considered and studied by a variety of techniques.

1.  Peroxy acids were the first peroxides recognized to differ structurally from their nonperoxy analogs, the carboxylic acids. The basic clue was uncovered in the period 1910–1915 by D'Ans and co-workers (18,19), who showed that the boiling points of peroxy acids were below those of the corresponding carboxylic acids despite their higher molecular weights. Four decades passed before a partially correct interpretation was suggested by Davison (20) and further elaborated by Giguère and Olmos (21). On the basis of carbonyl

and hydroxyl absorptions in the infrared region, they correctly interpreted the peroxy acid molecule as existing exclusively in a hydrogen-bonded, intra-molecularly chelated monomeric ring in the liquid and vapor states, as depicted by structure **1** (21).

**(1)**            **(2)**

Strong monomeric chelation was also proposed by Emanuel and co-workers (22) on the basis of chemical and infrared evidence on lower aliphatic peroxy acids to account for the lower combustion temperatures that chelation renders to these substances. Chelated monomeric molecules vaporize more readily than dimers and hence are more combustible in the vapor phase than in the liquid phase.

Swern and colleagues (23) adapted the chelate interpretation of peroxy acid structures for molecules in the vapor and liquid state to molecules in the solid state but extended the chelate concept to include intermolecular as well as intramolecular hydrogen-bonded rings (structure **2**).

In each of these investigations the conformation of the hydrogen-bonded ring was considered to be either planar or only slightly puckered. Giguère and Olmos had proposed bond distances and angles in conformity with their planar model. A calculation of the dipole moment in accordance with this model gave the value $1.83D$, nearly $0.5D$ less than the actual value of $2.32D$ (24). No reasonable changes in the bond moments for the O—H and C=O links in bond angles are able to bring the moment of the planar model close to $2.32D$.

2. Minkoff (25) in a preliminary communication "provisionally" suggested an O—O stretching band at 830–890 cm$^{-1}$ for a variety of peroxide types. These results could only have been reported with the knowledge that the vibrational frequency in the infrared region confirms the presence of a dipole moment and that such a conformation is in contradiction to a linear or planar structure.

Sheppard (26), however, rebutted attempts to correlate infrared absorptions with the peroxide grouping from the theoretical prediction that the vibration of an O—O bond in dialkyl peroxides, because of symmetry, would not be associated with a very great change in dipole moment. He acknowledged that different properties may be shown by the hydroperoxide group owing to its attachment at the end of the molecule, which removes the restriction of a symmetrical conformation.

3. Aroyl peroxides would seem to be good molecular candidates for a planar conformation. An X-ray study of the dibenzoyl peroxide structure

reported by Kasatochkin, Perlina, and Ablesova (27) characterized the crystal structure as a projection of two molecules in an elementary cell in the form of elongated molecules along the *c*-axis (Figure 3). They indicated the molecules of benzoyl peroxide to be stacked in the unit cell in a flat planar form.

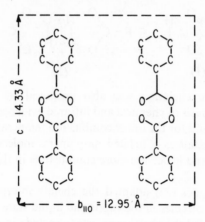

**Figure 3.** Crystal structure dibenzoyl peroxide: projection of two molecules in the [110] plane (ring hydrogen atoms omitted). (Reference 27).

4.   In his treatment of bond energies in peroxides Walsh (28) suggested the polarity of hydroperoxides to be the result of a skewed conformation. In contradistinction, he regarded the molecules of di-*t*-butyl peroxide to conform closely to a *trans* structure.

5.   Oesper and Smyth (29) assumed that the insertion of an additional oxygen atom into benzoic anhydride in forming benzoyl peroxide would allow comparatively greater freedom between the carbonyl groups. They were surprised to find moments for the peroxide ($1.58 \times 10^{18} D$) that were less than half those of the anhydride ($4.15 \times 10^{18} D$). The significance of these values was not further considered. Dipole moments for *t*-butyl peroxyphthalate and di-*t*-butyl diperoxyphthalate were measured by Voltz (30), but no attempt was made to relate this property to structure.

Although Davison (20) had first suggested a skewed structure for peroxides and the Penney–Sutherland model for the hydrogen peroxide molecule had received extensive experimental support and acceptance, a skewed conformation for organic peroxides was not experimentally established until Rogers and Campbell (31) reported the electric moments for ascaridole, *t*-butyl hydroperoxide, and di-*t*-butyl peroxide and computed the dihedral angle for the latter two compounds. Their evidence provided the necessary impetus for exploring the physical characteristics of the peroxide bond as it is found in diverse peroxide classes.

### (a)  Hydroperoxides and Dialkyl Peroxides

Rogers and Campbell (31) derived the dihedral angle $\varphi$ for *t*-butyl hydroperoxide ($\varphi = 100°$) and di-*t*-butyl peroxide ($\varphi = 123°$) from their electric-moment measurements. These values had shown conclusively that the angle of twist in the parent hydrogen peroxide molecule is maintained on derivatization and in addition had suggested a dependency of this angle on the molecular environment. Shortly afterward Miller and his co-workers (32) measured the electric moments for a representative group of peroxides. Their observed moments for hydroperoxides and dialkyl peroxides agreed with the calculated values on the basis of a fixed skew conformation about the peroxide grouping. A dihedral angle was calculated for *t*-butyl hydroperoxide ($\varphi = 100°$) and *n*-butyl–*t*-butyl peroxide ($\varphi = 105°$) based on the fixed skew, in general accordance with this angle in hydrogen peroxide. Their dihedral angle ($\varphi = 123°$) between the C—O—O planes in di-*t*-butyl peroxide agreed with the Rogers and Campbell value (31). Although it is significantly larger than the angle in hydrogen peroxide or in the few organic peroxides that had been measured, its magnitude was explained as a steric effect of the two large butyl groups imposing a high potential barrier to rotation about the O—O bond.

### (b)  Diacyl Peroxides

The electric moments of aliphatic and aromatic peroxides are comparable to the moments of hydroperoxides and dialkyl peroxides (32). The additional modes of intramolecular rotation and the effects of resonance by carbonyl and phenyl groups complicate a precise interpretation of the results. Nevertheless it had been concluded that the pair of carbonyl groups facing inwardly toward the O—O group (Figure 4*a*) as opposed to an outward orientation (Figure 4*b*) accounts for the observed dihedral angle (105°).

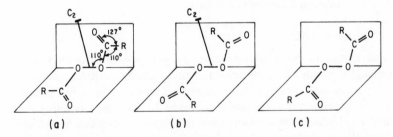

**Figure 4.** Configurations of the dibenzoyl peroxide molecule: (a) Inward orientation of acyl groups, symmetry $C_2$; (b) Outward orientation, Symmetry $C_2$; (c) Mixed orientation, Symmetry $C_1$. (Reference 33).

An indirect estimation of this angle was derived from X-ray powder data obtained for an alternating series of even and odd chain members of aliphatic diacyl peroxides and their nonperoxy analogs, the acid anhydrides (34). The approximate value, 123°, was calculated by compensating for the odd–even chain differences between the two classes of compounds and by assuming the inward orientation of the carbonyl functions.

Finally, definitive single-crystal studies were instituted on a few selected examples of acyl peroxides (35–37) and peroxy acids (38, 39). The dihedral angle in crystalline dibenzoyl peroxide (91 ± 2°) (36) and in crystalline 4,4′-dichlorodibenzoyl peroxide (81.1 ± 3.4°) (37) confirms the skewed structure for this class of peroxides. The values for these peroxides in the solid state are in essential agreement with the solution values derived from the electric moments of acyl and aroyl peroxides. The smaller angle displayed in the chloroperoxide demonstrates the influence that an altered molecular environment may exercise on peroxide structures; this had been noted earlier for hydrogen peroxide and its complexes.

Single-crystal analysis (36) confirmed the inward carbonyl orientation deduced earlier from dielectric-constant data (32). Fayat (33) also favored this conformation from an analysis of infrared carbonyl absorption bands that he obtained for benzoyl peroxide in four solvents and in the solid film. He eliminated a third possible conformation of symmetry $C_1$ (Figure 4c) and two linear conformations of symmetry $C_{2h}$ (Figure 5a) and $C_s$ (Figure 5b).

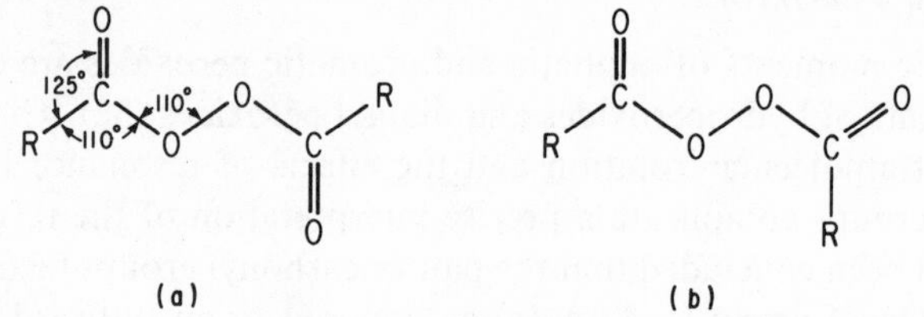

(a)                                    (b)

**Figure 5.** Planar-linear configurations of dibenzoyl peroxide: (a) Inward orientation of carbonyl groups, Symmetry $C_{2h}$; (b) Mixed orientation, Symmetry $C_s$. (Reference 33).

## (c) PEROXY ACIDS

Confirmation of the skewed structure in hydroperoxides, dialkyl peroxides, and diacyl peroxides by dielectric-moment studies prompted a reassessment of the peroxy acid structure (24). A revised model from a flat planar to a puckered five-membered chelate ring linked by hydrogen bonding between the hydroxy hydrogen and the carbonyl oxygen was concluded from dielectric-constant determinations on long-chain aliphatic peroxy acids. The structure is depicted as an "open book" model in which the dihedral angle between opposed pages is 72° (Figure 6). The angle, although significantly smaller

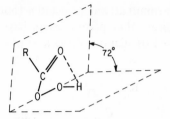

**Figure 6.** A peroxyacid molecule intramolecularly hydrogen
bonded in a skewed configuration.
(Reference 40).

than in most other organic peroxides, is a reasonable result of strong hydrogen
bonding on the barrier to rotation. It is more stable than the planar conformation for which lone-pair electrons in hybridized orbitals on each oxygen atom
are *cis* and repulsive interactions are at a maximum.

This conclusion disagrees with the quantum-mechanical results reported
by Yonezawa, Kato, and Yamamoto (41), who have calculated the *cis* form
to be 1.8 kcal/mole more stable than a skew form with a dihedral angle of
100°. A review of this paper shows that they have chosen classical values for
the bond lengths and bond angles that may not be correct for these compounds (see Figure 13 for these values in crystalline peroxypelargonic acid)
and have compared their *cis* form to an incorrect skew model. Their conclusions are not in accord with the experimental observations but may be in
harmony with the skew model they have chosen.

Experimental confirmation of a skewed conformation for peroxy acids in
the liquid or solution state brought into question the validity of the dimeric
dichelate planar structure deduced for these compounds in the crystalline
state (structure **2**). Single-crystal X-ray analysis instituted on peroxypelargonic
and *o*-nitroperoxybenzoic acids clarified the nature of their structures. A most
important feature contradicting the earlier concept is the intermolecular
hydrogen bonding uniting the molecules in infinite spirals along the twofold
screw axis. The molecules pack in extensions of infinite chains because of the
intermolecular-hydrogen-bond linkings. The intermolecular hydrogen bond
largely determines the value of the dihedral angle, which measures 133° in
peroxypelargonic acid and 146° in *o*-nitroperoxybenzoic acid. These values
are the largest aximuthal angles that have been reported for organic peroxides.

### (d) PEROXY ESTERS

Verderame and Miller (42) determined the electric moments of *t*-butyl
peroxy esters of aliphatic acids and of *t*-butyl peroxybenzoate. A detailed
analysis of the molecular moment as a function of the bond moments, bond

angles, and dihedral (azimuthal) angles $\omega$ and $\varphi$ (both shown in the reference model **3** at zero) indicate that peroxy esters have the Penney–Sutherland structure with restricted oscillation of the acyl group out of plane with the peroxide grouping.

(3)

The dihedral angle $\varphi$ was computed to fall within the range of 100–150°, for which the oscillating twist ($\omega$-libration) is centered in the vicinity of 35°. In this structure the carbonyl function is oriented toward the peroxide oxygen; that is, the carbonyl is *cis* to the *t*-butyl group. Yonezawa, Kato, and Yamamoto (41) are in agreement with this structure on the basis of energy requirements and molecular-orbital calculations.

A comparison of peroxy esters with ordinary esters is of interest. In general dipole-moment values for the latter indicate a *cis* orientation of the carbonyl and alkoxy groups (43, 44), whereas electron-diffraction studies show $\omega$ values of 25–40° for methyl formate and methyl acetate (45, 46).

## B.  Crystal Structures

In the last decade efficient synthetic methods for preparing most classes of peroxides in high yield and high purity have been developed. These methods ensure easy access to a variety of peroxide derivatives for single-crystal three-dimensional analyses. One unfortunate limitation of organic peroxides is their low resistance to decomposition under prolonged X-ray bombardment, but the thermal effect of exposure has been largely offset by enclosing the instrument containing the crystal at reduced temperatures ($-15$ to $-30°$) (38, 39). However, severe crystal disintegration may result by chemical action of the absorbed radiation. This effect necessitates reducing the time of exposure, which also reduces accuracy by restricting the observed spectra (39).

Pertinent crystallographic data for peroxides that have been exposed to single-crystal analysis are listed in Table 1, which also includes preliminary data for a few peroxides whose structures were not completely analyzed.

### 1.  Diacyl Peroxides

(a)  Dibenzoyl Peroxide (36)

The dibenzoyl peroxide molecule consists of two nearly planar benzoyloxy groups linked at the singly bonded oxygen atoms forming the peroxide bond

# TABLE 1. CRYSTAL DATA OF PEROXIDES

| Compound | System and Space Group | Cell Dimensions (Å) | Angle (degrees) | O–O Bond Length (Å) | Number of Molecules per Unit Cell | Density (g/cc)[a] and Volume (Å)³ | References |
|---|---|---|---|---|---|---|---|
| **Hydrogen Peroxide** $H_2O_2$ $H_2O_2 \cdot 2H_2O$ | Tetragonal $P4_12_12$ $C_2/c$ | $a=4.06$ $b=8.00$ $a=9.400$ $b=9.479$ $c=4.51$ | $\phi=89$ $\beta=121.33$ $\phi=130$ | 1.481 | 4 4 | $D_m=1.71$[b] $V=131.9$ $D_x=1.36$[c] | 1, p. 192; 2; 13;47 15 |
| *Acyl Peroxides* | | | | | | | |
| Benzoyl peroxide $Ph-C-O-O-C-Ph$ (C=O) | Orthorhombic $P2_12_12_1$ | $a=8.95$ $b=14.24$ $c=9.40$ | $\phi=91$ | 1.46 | 4 | $D_m=1.33$ $D_x=1.34$ $V=1210$ | 36 |
| 4,4'-Dichlorobenzoyl peroxide $ClC_6H_4-C-O-O-C-C_6H_4Cl$ (C=O) | Monoclinic $P2_1/a(C^5{}_{2h})$ | $a=25.47$ $b=7.80$ $c=6.85$ | $\beta=98.60$ $\phi=81.1$ | 1.48 | 4 | $D_m=1.53$ $D_x=1.535$ | 37 |
| 4,4'-Dibromobenzoyl peroxide $BrC_6H_4-C-O-O-C-C_6H_4Br$ (C=O) | Monoclinic $P2_1/a(C^5{}_{2h})$ | $a=25.94$ $b=7.91$ $c=6.83$ | $\beta=96.83$ $\phi$ not reported | | 4 | $D_m=1.90$ $D_x=1.909$ | 37 |
| *Peroxy Acids* | | | | | | | |
| Peroxypelargonic acid $Me(CH_2)_7CO_3H$ | Monoclinic $P2_1/C$ | $a=23.49$ $b=4.80$ $c=9.64$ | $\beta=106.0$ $\phi=133$ | 1.44 | 4 | $D_m=1.11$ $D_x=1.11$ $V=1044$ | 39 |
| o-Nitroperoxybenzoic acid (structure: benzene ring with $-CO_3H$ and $NO_2$) | Monoclinic $P2_1/C$ | $a=13.84$[d], $13.75$[e] $b=8.03$[d], $7.95$[e] $c=7.51$[d], $7.47$[e] | $\beta=112$[f] $\phi=146$ | 1.478 | 4 | $D_m=1.576$[d], $1.614$[e] $D_x=1.572$[d], $1.614$[e] $V=773.8$[d], $753.5$[e] | 38 |

TABLE 1. (Continued)

| Compound | Cell Dimensions | | | O—O Bond Length (Å) | Number of Molecules per Unit Cell | Density (g/cc)[a] and Volume (Å³) | References |
|---|---|---|---|---|---|---|---|
| | System and Space Group | (Å) | Angle (degrees) | | | | |
| *Peroxy Acids (continued)* | | | | | | | |
| p-Nitroperoxybenzoic acid $O_2NC_6H_4CO_3H$ | Orthorhombic $P2_12_12_1$ | | | | | | 35 |
| *Cyclic Peroxides* | | | | | | | |
| 3,6-Diphenyl-1,2,4,5-tetraoxacyclohexane | Monoclinic $P2_1/C$ | $a = 6.09$ $b = 7.85$ $c = 12.37$ | $\beta = 93.9$ $\phi = 88.2$ | 1.48 | 2 | $D_m = 1.36$ $D_x = 1.38$ | 48 |
| 3,3,6,6-Tetramethyl-1,2,4,5-tetraoxacyclohexane (dimeric acetone peroxide) | Undetermined twinning | | | | 4 | $V = 800.9$ | 49 |
| 3,3,6,6-Tetrakis-bromomethyl-1,2,4,5-tetraoxacyclohexane | Monoclinic $P2_1/n$ | $a = 10.53$ $b = 5.38$ $c = 10.65$ | $\beta = 98$ $\phi = 60.2$ | 1.45 | 2 | $D_m = 2.45$ $D_x = 2.57$ | 50 |
| Dimeric cyclohexanone peroxide | Triclinic $P_1$ | $a = 5.827$ $b = 6.061$ $c = 9.489$ | $\alpha = 85.37$ $\beta = 88.58$ $\gamma = 62.68$ | 1.48 | 1 | $D_m = 1.27$ $D_x = 1.28$ | 49 |

# TABLE 1. (Continued)

| Compound | System and Space Group | Cell Dimensions ($\text{Å}$) | Angle (degrees) | O–O Bond Length ($\text{Å}$) | Number of Molecules per Unit Cell | Density (g/cc)[a] and Volume ($\text{Å}^3$) | References |
|---|---|---|---|---|---|---|---|
| *Cyclic Peroxides (continued)* | | | | | | | |
| Dimeric cycloheptanone peroxide | Monoclinic $P2_1/C$ | $a = 9.399$ $b = 6.396$ $c = 11.682$ | $\beta = 102.4$ | 1.472 | 2 | $D_m = 1.25$ $D_x = 1.24$ | 51 |
| Dimeric cyclooctanone peroxide | Monoclinic $P2_1/C$ | $a = 9.791$ $b = 7.334$ $c = 11.779$ | $\beta = 114.2$ | 1.474 | 2 | $D_m = 1.21$ $D_x = 1.22$ | 52 |
| Dimeric cyclododecanone peroxide | Triclinic $P\bar{1}$ | $a = 5.62$ $d(010) = 6.97$ $d(001) = 14.22$ | $\alpha = 85.0$ | | 1 | | 49 |

Structural diagrams of dimeric cycloheptanone peroxide, dimeric cyclooctanone peroxide, and dimeric cyclododecanone peroxide.

TABLE 1. (Continued)

| Compound | System and Space Group | Cell Dimensions | | Angle (degrees) | O—O Bond Length (Å) | Number of Molecules per Unit Cell | Density (g/cc)[a] and Volume (Å³) | References |
|---|---|---|---|---|---|---|---|---|
| | | (Å) | | | | | | |
| *Cyclic Peroxides (continued)* | | | | | | | | |
| Trimeric acetone peroxide | Monoclinic $P2_1/C$ | $a = 13.925$ $b = 10.790$ $c = 7.970$ | | $\beta = 91.6$ | 1.483 | 4 | $D_m = 1.22$ $D_x = 1.23$ | 49, 53 |
| Trimeric cyclopentanone peroxide | $P6_3 22$ | $a = 5.45$ $c = 9.90$ | | | | 2/3 | | 49 |

TABLE 1. (Continued)

| Compound | System and Space Group | Cell Dimensions | | O–O Bond Length Å | Number of Molecules per Unit Cell | Density (g/cc)[a] and Volume (Å³) | References |
|---|---|---|---|---|---|---|---|
| | | (Å) | Angle (degrees) | | | | |
| *Cyclic Peroxides (continued)* | | | | | | | |
| Trimeric cyclohexanone peroxide | Orthorhombic $P_{bca}$ | $a = 9.95$ $b = 11.78$ $c = 31.93$ | | | 8 | | 49 |
| *Ozonide* | | | | | | | |
| 3-Carbomethoxy-5-anisyl-1,2,4-trioxacyclopentane | Orthorhombic | $a = 25.39$ $b = 10.96$ $c = 8.17$ | | | 8 | | 49 |

[a] $D_m$ = density experimentally measured; $D_x$ = density calculated from crystal data.
[b] At $-20°$ temperature.
[c] At $-190°$ temperature.
[d] At $25°$ temperature.
[e] At $-15°$ temperature.
[f] At $-15°$ temperature $\beta = 112°40'$.

(Figure 7). The dihedral angle at the peroxide bond is 91°. The two phenyl groups, each containing coplanar ring atoms, intersect in planes at 84°, along a line nearly parallel with the peroxide bond. Neither the O-16—C-7—O-15—O-17 nor the O-18—C-8—O-17—O-15 grouping is planar, but they exhibit a slight torsional displacement of 2.5 and 4.2°, respectively, about the C-7—O-15 and the C-8—O-17 bonds. The molecular packing, the dihedral angle at the peroxide bond, and the two dihedral angles between the phenyl groups and the bonded carboxy groups are illustrated in Figure 8. The crystal may be enantiomorphic or it may contain both D and L molecules, in which case the crystal may be composed of microdomains, each containing molecules of the same sense (35, 36).

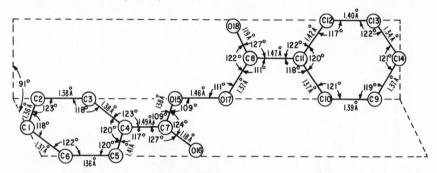

**Figure 7.** Skewed configuration of dibenzoyl peroxide in the crystalline state with interatomic distances and angles. (Reference 36).

### (b)  HALOGEN-SUBSTITUTED DIBENZOYL PEROXIDES (37)

4,4′-Dichlorobenzoyl and 4,4′-dibromobenzoyl peroxides are isostructural. These halogenated peroxides crystallize in the monoclinic system, rather than in the orthorhombic symmetry of the parent dibenzoyl peroxide. Further evidence of a difference in molecular packing is the 10° diminution in the dihedral angle to 81°. The molecule resembles dibenzoyl peroxide shown in Figure 7.

Caticha-Ellis and Abrahams (37) stated their preliminary findings that difluorodibenzoyl and diiododibenzoyl peroxides are nonisostructural. Apparently chlorine and bromine are more nearly balanced in size and electronegativity to give crystal structures that are more nearly identical.

### 2.  Peroxy Acids

### (a)  o-NITROPEROXYBENZOIC ACID (38)

A dihedral angle of 146° separates the C—O—O and O—O—H planes in o-nitroperoxybenzoic acid (Figure 9). The small torsional angle of 5° about the C—O bond, which is the dihedral angle between the O—C—O and

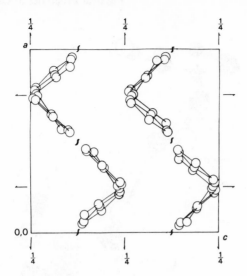

**Figure 8.** Projection of dibenzoyl peroxide molecules in the [010] plane depicting the molecular packing, the dihedral angle at the peroxide bond, and the angle of torsion at the C—C bond between the phenyl and carboxyl groups. (Reference 36).

C—O—O planes, displaces the hydrogen atom and the carboxy oxygen atom on opposite sides of the C—O—O plane. The latter plane imparts an angle of 58° with the plane of the ring. The nitrogroup is twisted 28° out of the ring plane. Five of the atoms in the benzene ring are coplanar. The carbon atom attached to the $NO_2$—group is displaced 0.025 Å from this plane, whereas the carbon atom of the O—C—O—O—H group is displaced 0.074 Å on the opposite side of the plane. The crystal structure is of the racemate type (35, 38).

(b) PEROXYPELARGONIC ACID (39)

Crystals of peroxypelargonic acid belong to the monoclinic system. The direction of the chain axis is approximately [102] (Figures 10 and 11), projecting a long spacing ($\alpha \sin \beta$) of 22.55 Å. The chain packing can be referred to the orthorhombic $O' \parallel$ subcell with $P2_122$ symmetry. Carbon atoms from C-3 to C-9 are coplanar. The chain twists at C-3, at which juncture the C-3—C-2 and C-2—C-1 bonds form angles of 4 and 8°, respectively, in opposite directions to the plane of the main chain. This stereochemistry is visualized in Figure 12. The five atoms of the peroxycarboxy group having the stereochemistry shown in Figure 13 (O-1, O-3 *cis* with respect to the bond O-1—O-2) are coplanar within the accuracy of the analysis.

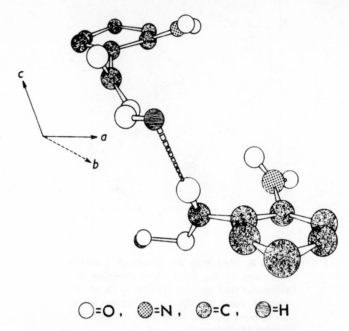

$$\bigcirc = O, \; \circledcirc = N, \; \textcircled{\tiny{}} = C, \; \ominus = H$$

**Figure 9.** Crystalline structure of o-nitroperoxybenzoic acid: a perspective view of glide related molecules intermolecularly hydrogen bonded. Ring hydrogens omitted for clarity. (Reference 38).

The molecules are hydrogen bonded in infinite spirals about the screw axis. The intermolecular O-1···O-3' hydrogen-bonded distance between screw-related molecules is 2.74 Å, making the angle of 120° with O-1—O-2. The peroxy acid dihedral angle between the C-1—O-2—O-1 and O-2—O-1—O-3' planes is 133°. The crystal is racemic and thereby composed of equal numbers of left- and right-handed spirals (35, 39).

### 3. *Cyclic Peroxides*

#### (a) PHTHALOYL PEROXIDE (54)

The crystalline structure of phthaloyl peroxide has not been experimentally determined. Its enhanced stability compared with that of benzoyl peroxide, a fifty-seven-fold difference in the rate of decomposition, attracted attention to its cylic structure. Greene (54) calculated the dihedral angle to be 38°6', assuming classical bond distances and internal angles of 120° for his calculations. He gave a tentative explanation for the greater stability of phthaloyl peroxide over acyclic peroxides by asserting that oxygen–oxygen fission in the latter takes place by means of simple stretching of the oxygen–oxygen

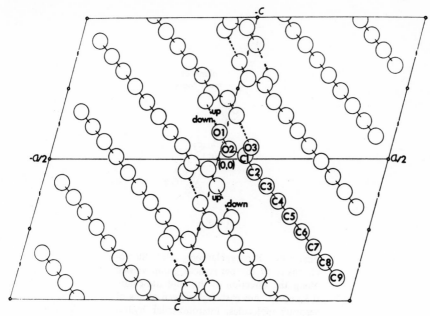

**Figure 10.** Crystalline structure of peroxypelargonic acid. Projection along the *b*-axis; direction of chain axis approximately [102]. (Reference 39).

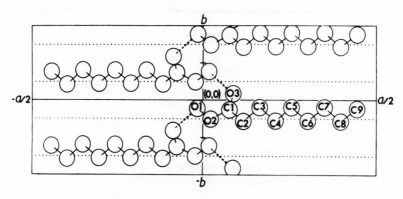

**Figure 11.** Crystalline structure of peroxypelargonic acid. Projection along [c] of screw-axis related molecules. (Reference 39).

bond, whereas the cyclic peroxide achieves fission only by twisting the bond between the carbonyl carbon and the ring carbon.

(b) HYDRATED SODIUM PEROXOBORATE (55)

Hydrated sodium peroxoborate was the first crystalline cyclic peroxide to be analyzed for its structure. The description of this crystalline inorganic compound is included among the organic cyclic peroxides to provide an

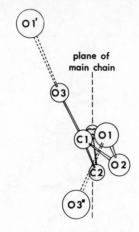

**Figure 12.** Peroxypelargonic acid: Stereo-chemistry of the peroxyacid group viewed along the direction of the aliphatic chain. (03 and 03″ are carbonyl oxygen atoms of separate molecules, intermolecular hydrogen bonds shown. (Reference 39).

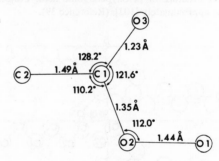

**Figure 13.** Bond distances and angles in the coplanar peroxy carboxylic acid group of peroxypelargonic acid. (Reference 39).

interesting comparison with some aspects of their structures, such as the dihedral angle.

The triclinic unit cell (space group $P\bar{1}$) contains one formula unit $Na_2B_2(O_2)_2(OH)_4 \cdot 6H_2O$. The shape of the $B_2(O_2)_2(OH)_4^{2-}$ ion and the dihedral angle of 64° in the double peroxo bridge are schematically diagrammed in Figure 14a and b, respectively.

The sodium ions are at symmetry centers of oxygen octahedra that share opposite edges. The arrangement forms octahedral chains, shown in part in

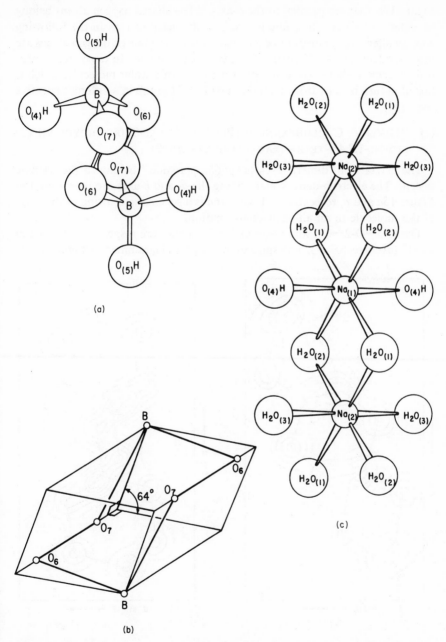

**Figure 14.** Crystal structure of hydrated sodium peroxoborate: (a) Shape of the $B_2(O_2)_2(OH)_4{}^{2-}$ ion; (b) the double peroxo bridge interconnecting the two boron atoms; (c) a part of the hydrated sodium chains constituting the octahedra. (Reference 55).

Figure 14c, that are parallel to the b-axis. "The shared oxygen atoms belong to water molecules. The sodium ions link the negative ions in the following way. In alternate octahedra in each chain, the two unshared oxygen atoms are incorporated in O-4—H groups from two negative ions. In every other octahedron each unshared oxygen atom forms part of a water molecule, which is linked in turn by a hydrogen bond to an O-5—H group in a different negative ion" (55).

(c)  "DIMERIC CYCLOHEXANONE PEROXIDE" (3,6-SPIRO-DICYCLOHEXY-
LIDENE-1,2,4,5-TETRAOXACYCLOHEXANE) (49)

The crystals of "dimeric cyclohexanone peroxide" belong to the triclinic system. The electron-density maps along the a- and b-axes are reproduced in Figure 15a and b, respectively. Figure 16a is a schematic drawing of the rings of the molecule in an all-chair conformation.

The O-3—O-2—C-1 and O-4—O-1—C-1 angles are respectively 107.5 and 107.1° (average 107.3°). The spiro carbon atom diagrammed in Figure 16b is

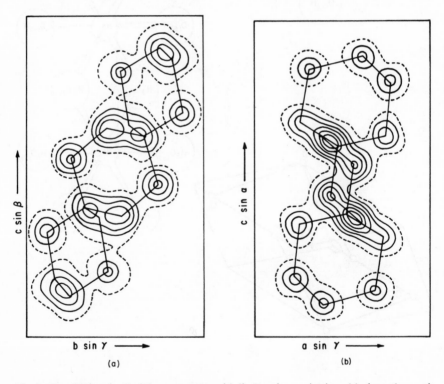

(a)                                                     (b)

**Figure 15.**  "Dimeric Cyclohexanone Peroxide": Fourier projections (a) along the a-axis and (b) along the b-axis. (Reference 49).

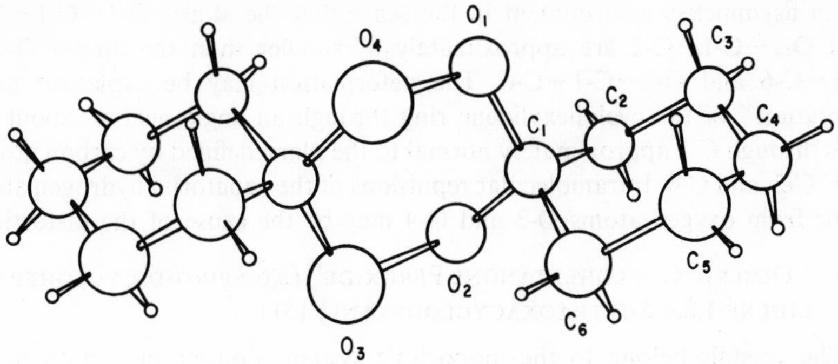

(a)

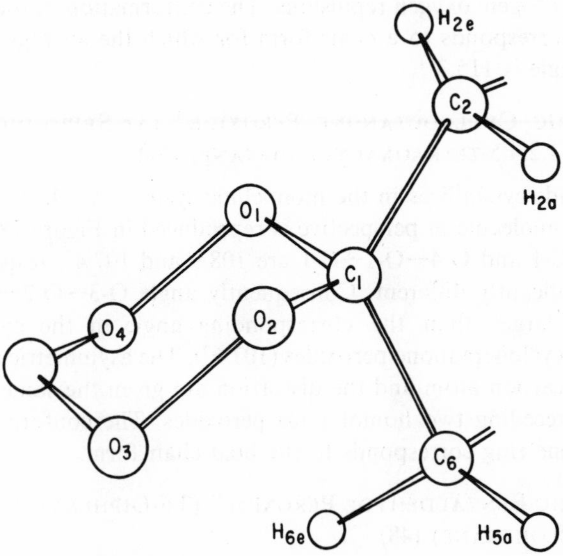

(b)

**Figure 16.** Schematic drawing of the "dimeric cyclohexanone peroxide" molecule: (a) view along [100]; (b) representation of the environment of the spiro carbon atom showing proximity of the equatorial hydrogens to peroxide oxygen. (Reference 49).

in an asymmetric environment in the sense that the angles O-1—C-1—C-2 and O-2—C-1—C-2 are approximately 8° smaller than the angles O-1—C-1—C-6 and O-2—C-1—C-6. The deformation may be explained as a "rotation" of the cyclohexylidene ring through an angle near 4° about an axis through C-1 approximately normal to the plane defined by carbon atoms C-1, C-2, and C-6. Intramolecular repulsions of the equatorial hydrogen atom H-6$e$ from oxygen atoms O-3 and O-4 may be the cause of the distortion.

(d)  "DIMERIC CYCLOHEPTANONE PEROXIDE" (3,6-SPIRO-DICYCLOHEPTY-
    LIDENE-1,2,4,5-TETRAOXACYCLOHEXANE) (51)

The crystals belong to the monoclinic system. Fourier projection maps along the $b$- and $c$-axes and a schematic representation of the molecules are reproduced in Figure 17$a$, $b$, and $c$, respectively.

The O-3—O-2—C-1 and O-4—O-1—C-1 angles are 107.6 and 108.3°, respectively. The environment of the spiro carbon is asymmetric of the kind described for "dimeric cyclohexanone peroxide," that is, as a "rotation" of approximately 4° about an axis through C-1 normal to the plane defined by C-1, C-2, and C-7. The distortion may similarly be explained by intramolecular hydrogen–oxygen repulsions. The conformation of the cycloheptylidene ring corresponds to a chair form for which the average value of the C—C—C angle is 115.3°.

(e)  "DIMERIC CYCLOOCTANONE PEROXIDE" (3,6-SPIRO-DICYCLOOCTY-
    LIDENE-1,2,4,5-TETRAOXACYCLOHEXANE) (52)

The peroxide cyrstallizes in the monoclinic system. A schematic representation of the molecule in perspective is reproduced in Figure 18. The angles O-3—O-2—C-1 and O-4—O-1—C-1 are 108.9 and 107.4°, respectively, and may be significantly different. Consequently angle O-3—O-2—C-1 may be significantly larger than the corresponding angle in the cyclohexanone (107.5°) and cycloheptanone peroxides (107.6°). The asymmetric environment of the spiro carbon atom and the distortion are given the same explanation as for the preceding two homologous peroxides. The conformation of the cyclooctylidene ring corresponds to the boat-chair form.

(f)  "DIMERIC BENZALDEHYDE PEROXIDE" (3,6-DIPHENYL-1,2,4,5-TETRA-
    OXACYCLOHEXANE) (48)

This aromatic peroxide crystallizes in the monoclinic system. A schematic representation of the molecule in perspective is reproduced in Figure 19. Angles O-3—O-2—C-7 and O-4—O-1—C-7 are each 105.5°. The angle between the planes defined by oxygen atoms O-1, O-2, O-3, and O-4 and the benzene ring is 88.2°. The benzene ring is geometrically displaced normal to the plane defined by the four oxygen atoms of the tetraoxacyclohexane ring.

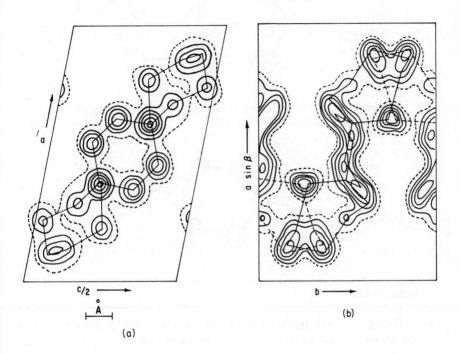

(a)

(b)

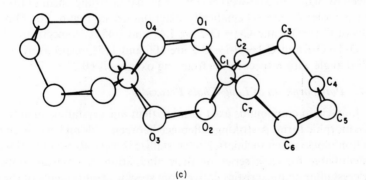

(c)

**Figure 17.** Structure of "dimeric cycloheptanone peroxide": Fourier projection (a) along the *b*-axis and (b) along the *c*-axis; (c) Schematic representation of the molecule. (Reference 51).

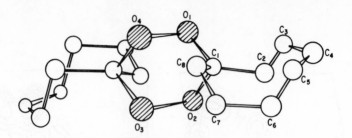

**Figure 18.** Schematic drawing of the "dimeric cyclooctanone peroxide" molecule. (Reference 52).

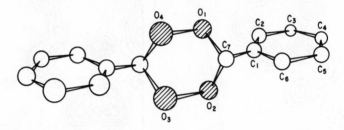

**Figure 19.** Schematic drawing of the "dimeric benzaldehyde peroxide" molecule. (Reference 48).

(g) "DIMERIC 1,3-DIBROMOACETONE PEROXIDE" (3,3,6,6-TETRAKISBRO-
MOMETHYL-1,2,4,5-TETRAOXACYCLOHEXANE (50)

This peroxide crystallizes in the monoclinic system. Projections of the molecules and their arrangement in the unit cell along the *a*- and *b*-axes are provided by Schulz and co-workers (50). The spatial arrangement of the atoms within a molecule are schematically diagrammed in Figure 20. The bond lengths of C-1—O-1 and C-1—O-2 are 1.47 and 1.45 Å, respectively. Angles C-1—O-2—O-1 and C-1—O-1—O-2 are 106 and 115°, respectively, and the dihedral angle with respect to the four ring oxygens is 60.2°.

### 4. Polymorphism of Long-Chain Peroxides

Most long-chain compounds exist in more than one crystalline modification, or polymorphic form. A striking difference between odd- and even-numbered chain homologs is their inclination to crystallize in separate modifications that are identifiable for each series by their alternation in physical properties. Their crystalline characteristics derive from specific arrangements of the long, flexible chains. Since the general phenomenon of polymorphism of long-chain compounds has already been treated in detail (56–58), a brief description with updated information will suffice to outline the major features.

The carbon atoms comprising the $CH_2$ groups in long aliphatic chains are

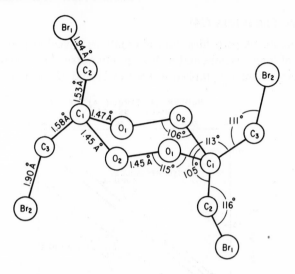

**Figure 20.** Schematic representation of the "dimeric 1,3-dibromoacetone peroxide" molecule with bond distances and bond angles. (Reference 50).

arranged in the crystal in a flat zigzag pattern at an average angle of 112° (59) and a repeat period (distance between alternate atoms) of 2.54 Å, a configuration allowing the overall length to be maximal and cylindrical in contour. The long rodlike chains bundle into layers in arrays determined by closest packing requirements and by a tendency of the symmetry of molecular layering to be as high as possible—that is, with the chain axes arranged parallel and the terminal methyl group lying in a plane to the "heads" and "tails" of molecules structuring the upper and lower sheets. This simple form may be distorted by substituents that translate the methylene groups on a spiral rather than a plane.

Alternation of properties in homologous series is associated with variation in the angle of tilt—that is, the inclination of the chain axis with respect to the plane through the end groups of the molecule. However, it is not generally stated or known that homologs possessing the same angle of tilt are not necessarily isomorphous, because they may differ in the nature of their end or side packing; for example, the alternation of melting points is attributed to differences in terminal packing density at the layer interface (60). Homologs may have the same angle of tilt, but the van der Waals interactions at this interface may not be equivalent at different orientations, and arrangements of the end groups because of variations in "head" to "tail" contacts. In view of their paraffinic nature peroxide derivatives of long-chain aliphatic acids are expected to retain the general physical characteristics generally associated with polymorphic behavior.

## (a)  DIACYL PEROXIDES (34)

Long-chain diacyl peroxides show alternations typical of paraffinic derivatives. Plots of long spacings and melting points versus the number of carbon atoms per acyl chain are graphed in Figures 21 and 22, respectively, and the

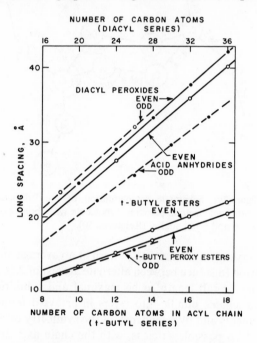

**Figure 21.** Long spacings vs. carbon number in long chain diacyl peroxides, *t*-butyl peroxy esters, and their non-peroxide analogs. (Reference 34).

values are recorded in Table 2. Acid anhydrides, the nonperoxide analogs, are included in the graphs for comparison.

The angle of tilt is nearly the same for the even and odd diacyl peroxide series, indicating essentially equivalent lateral chain packing. Although the acyl-chain moieties are even or odd numbered, the total carbon number per molecule is even, suggesting the likelihood of a single series. On the other hand, any observable differences between the series pertains to the nature of the end packing. The odd-numbered chain series shows the larger long spacing per carbon and consequently the larger end-packing value. The lower-melting-point relationship is in accord with these long-spacing results, which confirm the odd-carbon chain series as representing the higher energy modification.

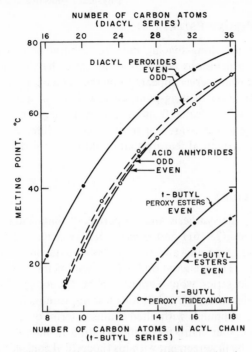

**Figure 22.** Melting point vs. carbon number in long chain diacyl peroxides, *t*-butyl peroxyesters, and their non-peroxide analogs. (Reference 34).

TABLE 2.  MELTING-POINT AND LONG-SPACING DATA OF LONG-CHAIN PEROXIDES[a]

| Number of Carbons in Chain | Parent Acid | Diacyl Peroxides | | *t*-Butyl Peroxy Esters | |
|---|---|---|---|---|---|
| | | mp (°C) | Long Spacing (Å)[b] | mp (°C) | Long Spacing (Å)[c] |
| Even series: | | | | | |
| 10 | Capric | 40.5–41.0 | 24.60 | −6.5 | 13.40 |
| 12 | Lauric | 54.7–55.0 | 29.05 | 8.0–8.6 | 15.25 |
| 14 | Myristic | 63.9–64.4 | 33.40 | 20.5–21.0 | 17.00 |
| 16 | Palmitic | 71.4–71.9 | 37.75 | 30.3–30.6 | 18.80 |
| 18 | Stearic | 76.5–76.9 | 42.20 | 38.7–39.3 | 20.55 |
| Odd series: | | | | | |
| 9 | Pelargonic | 13.0–13.5 | 12.30 | — | 12.45 |
| 13 | Tridecanoic | 48.3–48.8 | 32.15 | 10.0–10.4 | 15.73 |

[a] Data from reference 34.
[b] Tilt, $\beta$: even, 57°13′; odd, 57°53′.
[c] Tilt, $\beta$: even, 43°7′.

## (b)  $t$-Butyl Peroxy Esters (34)

Melting-point and long-spacing data for $t$-butyl peroxy esters result in contraindicated conclusions concerning the series having the higher energy form. The odd-carbon series appears to be the higher energy modification on the basis of their lower melting-point curve (Figure 21 and Table 2), but the lower energy form is indicated from their long-spacing-versus-carbon-number relationship. Side spacings are apparently making different contributions in the two series, a structural feature that could be clarified by a single-crystal examination.

## (c)  Peroxy Acids

Early work on the synthesis and properties of long-chain aliphatic monobasic peroxy acids (23, 61) and dibasic peroxy acids (62) failed to reveal alternation by long-spacing and melting-point data on the even- and odd-carbon homologs. The even and odd peroxy acid members in both the monobasic and dibasic series would appear to be homologously isomorphous.

Larsson (60) discussed the conditions leading to nonalternation in homologous series of compounds with tilted chains in terms of end-group structure. He accepted an earlier claim for nonalternation of monobasic peroxy acids (61), stating it to be in agreement with his theoretical considerations; from the data on single-crystal studies on peroxypelargonic acid (40) he deduced the subcell index $O' \parallel$ (2,0,1) for this chain packing to be in accordance with his index conditions for nonalternation. It appears to be overlooked, however, that alternation in these compounds may still exist through subtle differences in the terminal and lateral stacking of the layers.

Silbert and co-workers (63) remeasured the long spacings and melting points on carefully prepared and purified odd and even monobasic peroxy acids (Table 3). A statistical least-squares analysis* on the separate series resulted in two linear and nearly parallel lines having the following relationships between long spacing $L$ and carbon number $n$:

Odd series:
$$L_{\text{odd}} = 2.123n + 3.409$$
Standard deviation of slope = 0.0055
Standard deviation of intercept = 0.041
Even series:
$$L_{\text{even}} = 2.134n + 3.303$$
Standard deviation of slope = 0.0048
Standard deviation of intercept = 0.070

---

* L. S. Silbert, E. Siegel, and D. A. Lutz, unpublished data.

TABLE 3. MELTING POINTS AND LONG-SPACINGS OF MONOBASIC PEROXY ACIDS

| Number of Carbon Atoms | Peroxy Acid | mp (°C) | Long Spacing (Å) |
|---|---|---|---|
| Even series: | | | |
| 20 | Peroxyeicosanoic | 70.5–70.8 | 45.97 |
| 18 | Peroxystearic | 64.9–65.4 | 41.78 |
| 16 | Peroxypalmitic | 60.3–60.8 | 37.43 |
| 14 | Peroxymyristic | 55.0–55.3 | 33.17 |
| 12 | Peroxylauric | 50.0–50.5 | 28.84 |
| 10 | Peroxycapric | 40.5–40.8 | 24.66 |
| 8 | Peroxycaprylic | 29.8–30.3 | 20.41 |
| Odd series: | | | |
| 19 | Peroxynonadecanoic | 68.3–68.5 | 38.83, 43.73 |
| 17 | Peroxyheptadecanoic | 62.0–62.5 | 39.53 |
| 15 | Peroxypentadecanoic | 56.8–57.0 | 35.25 |
| 13 | Peroxytridecanoic | 51.2–51.5 | 31.00 |
| 11 | Peroxyhendecanoic | 44.0–44.2 | 26.79 |
| 9 | Peroxypelargonic | 34.6–34.8 | 22.50 |

Although the slopes of both lines are virtually identical (i.e., their difference is statistically insignificant), the odd and even intercept values were significantly different by the "Student's *t* test." The test was significant at the 97.5% level, indicating separate populations. An additional observation indicating the presence of two polymorphs was provided by long-spacing measurements on peroxynonadecanoic acid.

Separate and distinct melting-point curves (Figure 23) were obtained, although the difference is subtle; the odd series comprises the class with the lower melting curve. The odd series is seen to be the less stable modification by having the lower melting curve and the larger end packing. Although the curves merge at C-18, these conclusions are not invalidated, because it is established that alternation becomes less pronounced as the chain length increases (64).

Dibasic peroxy acids obtained by careful preparation and purification (63) gave separate melting-point curves for the odd- and even-numbered members (Table 4A, Figure 24). The corresponding curves for the parent dibasic acids are shown for comparison. Alternation is not evident from the long spacings but is revealed instead by the molar volumes showing a change in the side packing for these compounds as the determining factor.

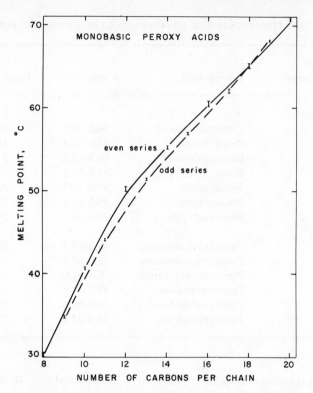

**Figure 23.** Melting point variation with carbon number in the even- and odd-membered series of long chain aliphatic monobasic peroxy acids. (Reference 63).

## 5.  Coordinated Ligand Peroxides

Several heavy-metal-complex salts have been oxidized to contain the $-O_2-$ group bonded to the metal ion. Their precise structures have intrigued investigators since Fremy first prepared dinuclear cobaltic $\mu$-peroxo coordination compounds in 1852 (65). These peroxidized complexes may be obtained at different metal oxidation states in coordination with ligands of varying complexity and are characterized by intense colors. Several of their structures had been rationalized or deduced indirectly, but the uncertainties of such methods have necessitated recourse to definitive analyses through single-crystal X-ray diffraction methods.

Peroxide complexes of cobalt (65–67), chromium (68, 69), molybdenum (70), tungsten (71), niobium (72), iridium (73–75), and platinum (76) have all yielded their structures to single-crystal analysis and provided in the process an improved understanding of their chemical nature. Since detailed data and

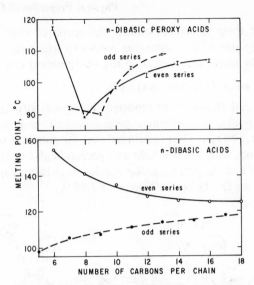

**Figure 24.** Melting point variation with carbon number in the dibasic peroxyacid and parent diacid series: Upper graph, dibasic peroxyacids (Reference 63); lower graph, dibasic acids.

TABLE 4A. PHYSICAL PROPERTIES OF DIBASIC PEROXY ACIDS

| | | Long Spacing | | | Molar Volume[a] | |
|---|---|---|---|---|---|---|
| $n$-Diperoxy Acid | mp (°C) | Å | $\Delta L/\Delta CH_2$ | $d^{30}$ | $cm^3$ | $\Delta V/\Delta CH_2$ |
| 1,16-Diperoxyhexa-decanedioic ($C_{16}$) | 106.0–106.5 | 17.53 | | 1.163 | 273.8 | — |
| 1,14-Diperoxytetra-decanedioic ($C_{14}$) | 105.2–106.2 | 15.64[b], 16.73[c] | — | 1.219, 1.51[c] | 238.2, 252.3[c] | 35.6, 21.5[c] |
| 1,13-Diperoxytri-decanedioic ($C_{13}$) | 108.5–108.9 | 15.64 | 1.09 | 1.233 | 224.1 | 14.1, 28.2 |
| 1,12-Diperoxydo-decanedioic ($C_{12}$) | 101.3–104.0 | 14.61 | 1.03 | 1.240 | 211.5 | 12.6 |
| 1,11-Diperoxyhen-decanedioic ($C_{11}$) | 104.0–105.0 | 13.56 | 1.05 | 1.275 | 194.7 | 16.8 |
| Diperoxysebacic ($C_{10}$) | 98.0–98.5 | 12.49 | 1.07 | 1.277 | 183.4 | 11.3 |
| Diperoxyazelaic ($C_9$) | 89.8–90.2 | 11.43 | 1.06 | 1.316 | 167.3 | 16.1 |
| Diperoxysuberic ($C_8$) | 89.0–89.2 | 10.36 | 1.07 | 1.328 | 155.3 | 12.0 |
| Diperoxypimelic ($C_7$) | 91.5–92.5 | 9.29 | 1.07 | 1.414 | 135.9 | 19.4 |
| Diperoxyadipic ($C_6$) | 116.5–117.2 | 6.85 | 2.44 | — | — | — |

[a] Molar volume = molecular weight/$d_4^{30}$.
[b] Crystallized from 97% alcohol.
[c] Crystallized from chloroform.

descriptions of their structures exceed the scope of this chapter, a brief presentation of some of the interesting modes by which the $-O_2-$ function is bonded to the heavy-metal atom in the coordination complex will suffice.

### (a) EXAMPLE OF A PEROXOCOBALTATE

Vannerberg and Brosset (65) reported the crystal structure of the paramagnetic compound decammine:$\mu$:peroxodicobalt pentanitrate, $[(NH_3)_5$-$CoO_2Co(NH_3)_5](NO_3)_5$. They concluded that the cobalt atoms are linked by a peroxide group, with the peroxide axis perpendicular to the cobalt–cobalt axis (Figure 25) and the peroxo group bonded to cobalt atoms by $d$-$\pi$ bonds. They obtained an O—O bond distance of 1.45 Å.

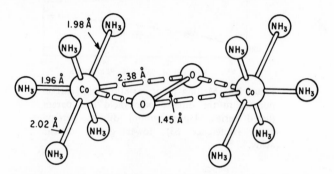

**Figure 25.**   Geometric configuration of the ion
$[(NH_3)_5CoO_2Co(NH_3)_5]^{5+}$.

Broken lines indicate $\pi$-bonding in a complex by cobaltic ions with the bridged peroxide group. (Reference 65).

Schaefer and Marsh (64) reexamined the geometry of the cation in single crystals containing sulfate in place of nitrate as anions in the formula $(NH_3)_5CoO_2Co(NH_3)_5SO_4(HSO_4)_3$. In their structure (Figure 26) the bridged peroxo group is neither perpendicular to the cobalt–cobalt axis nor bonded to the metal atoms by $d$-$\pi$ bonds. Each oxygen atom of the peroxo group is $\sigma$-bonded to a cobalt atom in a zigzag, nearly coplanar, arrangement of the four atoms, with only a small dihedral angle. This structure is in contrast to the Vannerberg–Brosset and the hydrogen peroxide structures. The oxygen–oxygen bond distances measured 1.31 Å, a value significantly shorter than the 1.45–1.48 Å found in peroxides and larger than the 1.28 Å found in alkali-metal superoxides (77). On the assumption that the nature of the anion would have no important effect on the cation's geometry, Schaefer and Marsh concluded that the Vannerberg–Brosset structure is incorrect. The absence of an 880 cm$^{-1}$ band in the infrared region (65) is in accord with this staggered configuration for which there would be an absence of a dipole moment.

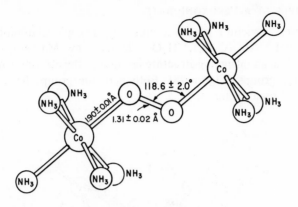

**Figure 26.** Geometric configuration of the ion
$$[(NH_3)_5CoO_2Co(NH_3)_5]^{5+}$$
with the structure of two cobaltic ions $\sigma$-bonded to the peroxide atoms. (Reference 66).

## (b) EXAMPLE OF A PEROXOCHROMATE

Stomberg (69) examined the crystal and molecular structure of several peroxochromates. Oxodiperoxopyridinechromium(VI), $[CrO(O_2)_2Py]$, where Py = pyridine, is chosen for illustration in Figure 27 (68). The pair of peroxide groups form two three-membered $CrO_2$ rings approximately disposed at right angles to one another. The O—O bond of 1.404 Å is shorter than the usual peroxide bond, indicating increased bond order and showing the $\pi$-orbitals of the peroxo group as contributing more to the O—O bonding than the $\pi^*$ orbital contributes to the antibonding orbital. A short, strong band at 880 cm$^{-1}$ has been identified for the O—O stretching vibration (78).

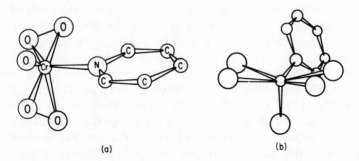

**Figure 27.** Geometric configuration of the $[CrO(O_2)py]$ molecule: (a) projection approximately perpendicular to the base plane; (b) projection approximately parallel to the base plane. (Reference 69).

## (c) EXAMPLE OF A PEROXONIOBATE

The crystal structure of ammonium diperoxodioxalatoniobate mono-hydrate, $(NH_4)_3NbO(C_2O_4)_3 \cdot H_2O$, obtained by Mathern, Weiss, and Rohmer (72), is an intriguing structure because of the *cis* conformation of the two peroxide groups (Figure 28). Both peroxide groups have the normal 1.48 Å O—O bond distance.

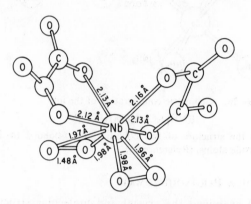

**Figure 28.** Geometric configuration of the diperoxodioxalatoniobate ion showing the *cis* arrangement of the two peroxide groups. (Reference 72).

## (d) EXAMPLES OF PLATINUM-FAMILY OXYGEN CARRIERS

Iridium (79) and platinum (76, 80) chelates form reversible 1:1 oxygen complexes. These synthetic molecular oxygen carriers offer advantages in their simplicity and variety as models for the natural oxygen carriers hemo-globin and hemocyanin, for gaining information about the mode of attach-ment of molecular oxygen and for studying the phenomenon of oxygen reversibility. An intriguing property of the iridium molecular oxygen carrier is the change in function from reversibility in the chlorocomplex to irreversi-bility in the iodocomplex (81). Perspective representations of the molecules $IrO_2Cl(CO)(PPh_3)_2$ (Figure 29), $IrO_2I(CO)(PPh_3)_2$ (Figure 30), and $[Pt(O_2)(PPh_3)_2]$, 1.5 $C_6H_6$ (Figure 31), where Ph = phenyl, illustrate two modes by which the oxygen molecule is bonded to the coordination sphere.

The O—O distance of 1.30 Å in the iridium chlorocomplex in which oxygen is bound reversibly is larger than this distance in molecular oxygen (1.21 Å) but significantly lower than the 1.45–1.49 Å bond distance in typical per-oxides. They are also equidistant from the iridium atom; this is consistent with Griffith's model of the π-bonding of molecular oxygen in hemoglobin

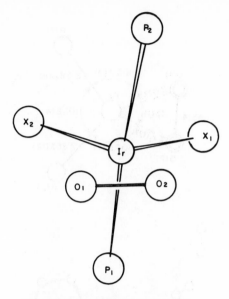

**Figure 29.** A skeletal drawing of the
$IrO_2Cl(CO)[P(C_6H_5)_3]_2$

molecule in perspective. Phenyl rings
attached to $P_1$ and $P_2$ are omitted; $X_1$ and
$X_2$ are the disordered Cl—CO positions.
(Reference 73a).

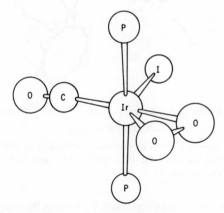

**Figure 30.** Geometric structure of the
$IrIO_2(CO)[P(C_6H_5)]_2$

molecule. Phenyl groups attached to the
phosphorus atoms are omitted. (Reference
82).

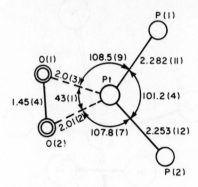

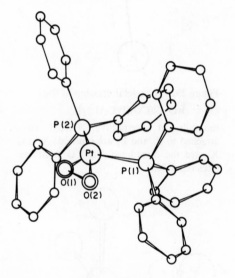

**Figure 31.** Geometry of the
{Pt(O₂)[P(C₆H₅)₃]₂}
molecule showing the near coplanarity of the
five atoms — Pt, P(1), P(2), O(1), and O(2).
(Reference 76).

(82). The iridium iodocomplex contrasts by having its oxygen bound irreversibly for the normal O—O distance of 1.47 Å (82).

The properties of the platinum compound appear to be a compromise of the properties of the two iridium halogen complexes. The —O₂— bonding in the oxygen-reversible platinum complex differs from the iridium chlorocomplex oxygen carrier in having a normal O—O distance (1.45 Å) and a nearly coplanar arrangement with platinum and the two phosphorus atoms.

One additional complex of the iridium family has shown a startling feature. In the iridium complex $\{Ir(O_2)[(C_6H_5)_2PCH_2CH_2P(C_6H_5)_2]_2\}[PF_6]$ the addition of oxygen is irreversible. The strong metal-to-oxygen bond is accompanied by a weak O—O bond having the unusually long O—O distance of 1.625 Å (83). In the nearly isostructural rhodium complex, where addition of oxygen is reversible, the weaker metal-to-oxygen bond is accompanied by a stronger O—O bond, whose length is 1.418 Å (83).

## C. Recapitulation

The skewed conformation of the parent hydrogen peroxide molecule is maintained in most peroxide derivatives. However, the structure of the peroxide and its physical environment can modify the basic angular value (90°) of the O—O group, leading to conformations in which this angle is enlarged or compressed. Certain peroxides—such as hydroperoxides, primary and secondary dialkyl peroxides, and diacyl peroxides—show only minor deviation from the basic angle. The steric effects of bulky substituents, however, as in di-*t*-butyl peroxide, in which this angle is enlarged, give a more nearly *trans* planar structure.

The associative force of hydrogen bonding is an important factor in the modification of structure. Peroxy acids in the vapor or liquid states favor intramolecular hydrogen bonding, with compression of the dihedral angle to approximately 70° (i.e., toward a *cis* conformation), whereas in the crystal state they favor intermolecular hydrogen bonding that expands the angle toward 140° to give a nearly *trans* conformation.

The variety of peroxide structures is best shown by the metalloperoxide derivatives in which a wide range of conformations and configurations is possible. The noble metals are of particular interest because of their capacity to accommodate the peroxide grouping in either a $\sigma$-bonded catenated structure or a loosely held association maintained by $d$-$\pi$ bonding. The nature of the organic ligands in these complexes determines which of the structures is preferred as well as the reversibility or irreversibility of the metalloperoxide complexes.

The importance of explicit structural determinations of peroxides is to be emphasized since their physical and chemical properties are partly dependent on structure. Certain peroxide types, such as di-*t*-butyl peroxide, peroxy esters, and certain perfluoro peroxides, require further structural analysis, which may help to clarify the factors involved in their relatively high stabilities.

## II.  ULTRAVIOLET ABSORPTION

### A.  Hydrogen Peroxide and Organic Peroxides

The ultraviolet absorption spectrum of hydrogen peroxide has been measured in the region extending from the edge of the visible near 4000 Å to the edge of the Schumann near 1950 Å (Figure 32*a*) (1, p. 287). The spectrum is a rising continuum of parabolic shape with no maximum and no evidence of bands in the structure. The spectrum illustrated in the figure is constructed from measurements carried out on hydrogen peroxide in the neat phase, in aqueous solutions, and in the vapor phase. The curve indicates the the presence of an inflection point at each end of the investigated wavelength region. The rising curvature at the low wavelength cut-off suggests the possible presence of a maximum for wavelengths in the Schumann ultraviolet (84a).

A systematic and comprehensive spectral examination of organic peroxides has not been carried out in the ultraviolet. This is not surprising since the lack of distinctive absorptions for peroxides has not encouraged such studies.

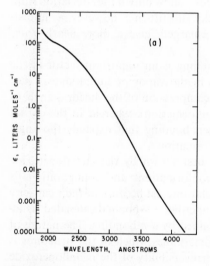
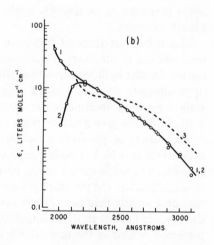

**Figure 32.**

(a)  Ultraviolet absorption spectrum of hydrogen peroxide: Composite curve based on measurements on neat liquid, aqueous solutions, and vapor; $\varepsilon = \log (I_0/I)C^{-1}L^{-1}$, 1. mole$^{-1}$ cm$^{-1}$. (Reference 1, page 287).

(b)  Ultraviolet absorption of dialkyl peroxides:

    (1)  dimethyl peroxide (Reference 84d).

    (2)  dimethyl peroxide (Reference 84e).

    (3)  di-t-butyl peroxide (Reference 84b, page 443).

However, the growing interest in the photolytic decomposition of peroxides (84b) appears to be attracting renewed interest in the examination of peroxide derivatives.

The absorption spectral curves of simple alkyl peroxides (84c–e, 85a) (Figure 32*b*) bear a resemblance to the hydrogen peroxide spectrum. The spectrum of dimethyl peroxide by Toth and Johnston (84d) agrees with the results of Takezaki et al. (84e) for the region extending from the visible to 2150 Å. Below 2150 Å the results by the two groups of investigators are in disagreement since the spectral curve by the former group shows an increasing continuum, whereas the curve by the latter reaches a maximum peak and then abruptly descends. The spectrum obtained in this region by Takezaki et al., was inferred to have been derived from measurements extended beyond the reliable range of their spectrometer (84d).

Dialkyl peroxides (84c–e, 85b) (Figure 32*b*) and alkyl hydroperoxides (84c, 85a, 85b) absorb at much longer wavelengths than do ethers and alcohols (84b), the absorptions by the latter class of compounds appearing as banded structures below 2000 Å. The overlapping lone-pair orbitals of the two adjacent peroxide oxygen atoms lead to low-energy bonding and high-energy antibonding lone-pair molecular orbitals that are correspondingly filled with electrons (84b). The transitions are assumed to involve two closely adjacent levels $\pi_{p_{x,y}} \to \sigma_z^*$ that may account for the absorption shift to longer wavelengths.

Carbonyl derivatives undergo the $n \to \pi^*$ transition in the accessible ultraviolet (86). The attachment to carbonyl carbon of an electronegative substituent bearing a lone-pair of electrons will cause a hypsochromic (blue) shift relative to acetaldehyde ($\lambda_{max} = 2930$ Å, $\varepsilon = 12$) (86, chapt. 9). Carboxylic acids ($\varepsilon = 41$) and esters ($\varepsilon = 60$) show this absorption at 2040 Å suggesting a similar absorption for peroxycarboxyl derivatives in the vicinity of this wavelength. However, peroxyacetic acid and its anion (Figure 32*c*) (87a) and saturated acyl peroxides (solutions in hexane or cyclohexane, $\varepsilon \cong 47$ to 62 at 2537 Å) (87b, c) display a rising continuum toward shorter wavelengths without forming a maximum in the region examined. *t*-Butyl peroxylaurate in comparison exhibits a pair of weak absorptions at 2170 Å ($\varepsilon \cong 80$) and 2700 Å ($\varepsilon \cong 50$), although band assignments were not made (87d).

Aromatic peroxides (Table 4B) show bands that are typical of benzene and its derivatives (86, chapt. 12). The characteristic secondary band of benzene is centred around 2560 Å ($^1B_{2\mu} \leftarrow {}^1A_{1g}$ transition) and the two shorter wavelength bands are found at 2030 Å (primary band; $^1B_{1\mu} \leftarrow {}^1A_{1g}$ transition) and 1850 Å ($^1E_{2\mu} \leftarrow {}^1A_{1g}$ transition). Substitution in benzene shifts the latter two bands bathochromically (red shift) by about equal amounts. On the basis of these comparisons, the secondary band may be assigned to dicumyl peroxide, cumyl hydroperoxide, and tetralin hydroperoxide (doublet) (Table 4B).

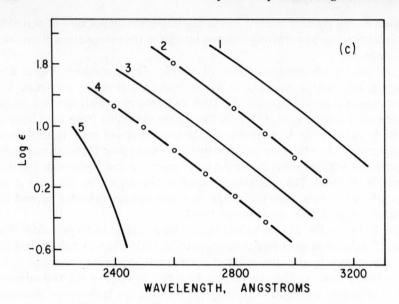

**Figure 32**
(c)   Ultraviolet absorption of peroxyacetic acid and peroxyacetate
      solutions: (1) perhydroxyl ion; (2) peroxyacetate ion; (3)
      hydrogen peroxide; (4) peroxyacetic acid; (5) acetic acid.
      (Reference 87a).

Aroyl peroxides exhibit one or more absorption bands (Table 4B). The benzoyl peroxide spectrum in Figure 32d (88a) is typical of this class of compounds and further illustrates the effect of solvent on the absorption bands. The spectra of aroyl peroxides and their corresponding acid anhydrides are similar (88a) and show primary and secondary bands at wavelengths close to those for benzoic acid (2300 Å, 2730 Å) and benzoate anion (2240 Å, 2680 Å). The agreement of the ratio $\lambda_{sec}/\lambda_{prim} = 1.19$ for benzoyl peroxide with the near constancy of 1.19–1.28 for monosubstituted benzenes lends support to the band assignments. Unfortunately, these spectra have not been extended into the Schumann to allow an assessment of the carbonyl $n \to \pi^*$ transition, although this transition is expected to lie submerged under the high-intensity primary band.

Because of strong absorption in the ultraviolet, hydrogen peroxide may be analyzed at concentrations as low as 8 ppm, even though Beer's law is not applicable (1b, p. 563). Ultraviolet absorption spectra of aroyl peroxides and their corresponding acid anhydrides (88a) are too similar for general analytical applications. In one useful case, the close similarity in ultraviolet absorption

TABLE 4B. PRIMARY AND SECONDARY BANDS AND EXTINCTION COEFFICIENTS FOR AROMATIC PEROXIDES

| Peroxide[a] | Primary Band | | Secondary Band | | Ref. |
|---|---|---|---|---|---|
| | $\lambda_{max}$, Å | $\varepsilon_{max}$ | $\lambda_{max}$, Å | $\varepsilon_{max}$ | |
| Dicumyl Peroxide[b] | — | — | 2570 (structured) <br> <2400[c] | 415 | 88b |
| Hydroperoxides | | | | | |
| Cumene hydroperoxide[d] | — | — | 2580 (structured) <br> <2300[c] | 200 | 88b |
| Tetralin hydroperoxide | — | — | 2740 <br> 2670 | 518 <br> 530 | 88a |
| Aroyl Peroxides | | | | | |
| Benzoyl peroxide | | | | | |
| in benzene | 2310 | $3.98 \times 10^4$ | 2740 | $2.16 \times 10^3$ | 88a |
| in chloroform | — | — | 2740 | $2.34 \times 10^3$ | 88a |
| o-Toluoyl peroxide | 2340 | $2.75 \times 10^4$ | 2800 | $3.55 \times 10^3$ | 88a |
| o,o'-Dichlorobenzoyl peroxide | 2330 | $2.47 \times 10^4$ | 2810 | $2.89 \times 10^3$ | 88a |
| p-Toluoyl peroxide | 2400 | $4.2 \times 10^4$ | — | — | 88a |
| p,p'-Dichlorobenzoyl peroxide | 2470 | $4.17 \times 10^4$ | — | — | 88a |
| Cinnamoyl peroxide | 2850 | $5.76 \times 10^4$ | — | — | 88a |

[a] In benzene solution except where indicated.
[b] In hexane and carbon tetrachloride solution.
[c] Continuum to below this wavelength may be attributed to the peroxide absorption.
[d] In carbon tetrachloride solution.

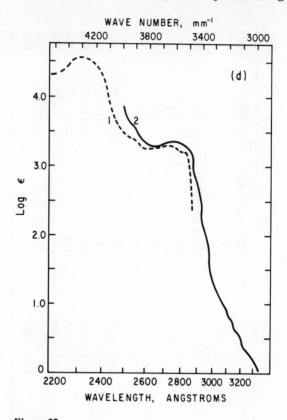

**Figure 32**

(d)  Ultraviolet absorption of dibenzoyl peroxide:
(1) in benzene solution; (2) in chloroform
solution. (Reference 88a).

of phthaloyl peroxide to phthalic anhydride has been quoted as evidence in
support of structure (4) rather than (5) (54). Applications that have been carried
out in the ultraviolet region are unimportant for the peroxide group and are
primarily concerned with the remainder of the molecule. By way of illustration, conjugation of the isolated double bonds of methyl linoleate during
autoxidation (89, 90) and the formation of α,β-unsaturated ketones as
products of methyl oleate autoxidation (91, 92) have been determined by
ultraviolet spectroscopy.

The availability of new spectrophotometers capable of recording spectra to the far (vacuum) region of 170 m$\mu$ or even lower wavelengths (93, 94) and the newly instituted studies in this region on chromophores (95) may arouse new interest for extensions to peroxide chemistry.

## B. Hydrogen Sesquioxide

Czapski and Bielski (96) reported the production of hydrogen sesquioxide, $H_2O_3$, by electron-beam radiolysis of an air-saturated perchloric acid solution. The compound is interesting because of its existence as the third member of the homologous series, beginning with water and hydrogen peroxide. Hydrogen sesquioxide absorbs in the ultraviolet region below 280 m$\mu$ to give an absorption spectrum similar to that of hydrogen peroxide but of higher intensity (97).

After an extended controversy over the existence of a fourth member of the series, the existence of hydrogen superoxide ($H_2O_4$) has been confirmed (98, 99) and its heat of decomposition reported (100). No spectral data have been obtained on these compounds other than the infrared-absorption spectra of a peroxide radical condensate that Soviet workers have reported on their preparation and identification of higher hydrogen peroxides (101).

## C. Peroxy Radicals

The absorption spectrum of the perhydroxyl radical, HOO·, was investigated in connection with the absorption spectra and kinetics of hydrogen sesquioxide (97). The absorption of the perhydroxy radical is slightly dependent on acid concentration, forming a maximum at approximately 235 m$\mu$ ($\varepsilon_{max} \simeq 1.1 \times 10^3$ l/mole-cm).

The hydroperoxy radical is structurally interesting by favoring an isosceles-triangle configuration (102). This structure was determined from a molecular-orbital study of its geometry, using LCAO–MO–SCF calculations on closed-shell ions to give a value of 9.5 eV for the ionization potential and 4.4 eV for the dissociation energy (102).

McCarthy and MacLachlan (103, 104) have recorded the ultraviolet spectrum of the cyclohexylperoxy radical from 260 to 300 m$\mu$; it has maxima at 275 m$\mu$ ($\varepsilon_{max} \simeq 2 \times 10^3$ l/mole-cm) and 297 m$\mu$ ($\varepsilon_{max} \simeq 1 \times 10^3$ l/mole-cm).

## III. INFRARED ABSORPTION

The study of the structure of molecules by infrared spectroscopy is based on the observed characteristic frequencies of vibration of functional groups and larger structural units (105). The frequencies of these vibrations are not only dependent on the nature of the particular bonds themselves but are also affected by the entire molecule and its environment. Resonance and inductive effects of substituents may shift the group frequencies. Frequency changes are sometimes a function of the sum of the electronegativities of the substituents (106, p. 391; 107). Also, a frequency change in one part of a spectrum is frequently related to a frequency change elsewhere, a relationship that is helpful for specific identifications (106).

Raman spectra furnish information on normal vibrations, which are frequently weak or inactive in the infrared region (and vice versa). The two techniques are complementary by providing information on the complete vibrational spectrum of the molecule (108).

A variety of peroxides have been examined by infrared spectroscopy in search of structural correlations with frequencies that would be useful in identification and analysis, but only a few peroxides have been examined by Raman spectroscopy. Szymanski (109) has reexamined the group frequency assignments for typical compounds containing C—O, O—O, and O—H functions. It is his opinion that the frequencies of peroxides and hydroperoxides do not differ from those of related alcohols, ethers, acids, esters, and phenols except for modification by resonance and inductive effects. Identification of group frequencies in these classes of compounds is difficult since other bands often appear in the same region from 1400 to 800 cm$^{-1}$. Nonetheless, bands that do appear in the region of C—O, O—O, and O—H group frequencies can give useful information for identification if both the concept of group frequencies and the spectra of model compounds are utilized (109). Raman spectra, when obtained, have been instrumental in the identification of frequencies and their assignments (109).

It is generally helpful to have infrared spectra on specific peroxides available for reference since it is often impractical for each laboratory to obtain these curves for all possible examples. The task of locating publications containing critical infrared spectra has been simplified by Hershenson (110), who compiled a reference index to 16,000 compounds. The compilation includes approximately 95 citations specific to peroxides.

## A.  Hydrogen Peroxide. Infrared Fundamentals

Infrared, Raman, and diffraction methods have provided information on the molecular dimensions and angles of hydrogen peroxide. Identification of the experimental absorption bands with the vibrational motions of hydrogen peroxide has been carried out mainly by Simon and Fehér (111–113), Bailey and Gordon (114), and Giguère (115, 116). These studies have been summarized by Giguère (116) and by Schumb, Satterfield, and Wentworth (1, p. 326 ff.). The general conclusions that have been reached on the fundamental frequencies are updated by recent data and presented in this section in capsule form.

The skewed structure of hydrogen peroxide, model $d$ of Figure 1, has 12 modes of motion: three translational modes that do not enter our spectroscopic discussion, three rotational modes, and six vibrational modes, which are displayed in Figure 33. The model has a twofold axis of symmetry, which classifies it as a member of the $C_2$ point group.

Because of its permanent dipole moment, hydrogen peroxide will have an

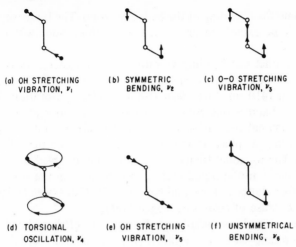

(a) OH STRETCHING VIBRATION, $\nu_1$

(b) SYMMETRIC BENDING, $\nu_2$

(c) O—O STRETCHING VIBRATION, $\nu_3$

(d) TORSIONAL OSCILLATION, $\nu_4$

(e) OH STRETCHING VIBRATION, $\nu_5$

(f) UNSYMMETRICAL BENDING, $\nu_6$

**Figure 33.** Normal modes of vibrational motion for the hydrogen peroxide molecule. (Reference 116).

absorption spectrum arising from rotational motions; an infrared-absorption band will arise by vibrational motions associated with a change in the dipole moment; and a band in the Raman spectrum will arise by each vibrational motion involving a change in molecular polarizability. Overtones, the raising of the vibration energy by more than one quantum at a time, and combination frequencies due to raising the energy of two vibrations simultaneously will be observed. The order and numbering of the fundamental vibrations, $\nu_1$, $\nu_2$, etc., is in accordance with the frequency and symmetry of the vibration involved (116). A brief summary of their correlations is presented.

1. The O—O stretching vibration $\nu_3$: this vibration is established as the absorption at 877 cm$^{-1}$ for the liquid and solid phases, and remains unchanged in both hydrogen and deuterium peroxides on changing the mass of the hydrogen atom. Bain and Giguère (117) failed to observe any indication of this band for hydrogen peroxide vapor; this emphasizes the need for further investigations by means of laser Raman spectroscopy.

2. In-phase and out-of-phase O—H stretching vibrations $\nu_1$ and $\nu_5$: these vibrations are identified with the absorption band near 3600 cm$^{-1}$. The frequency of this band changes by the factor $1^{-1/2}$ on replacement of hydrogen by deuterium, to establish this as a hydrogen motion. The gas-phase $\nu_1$ vibration band has never been isolated, but $\nu_1$ and $\nu_5$ are believed to be degenerate because of "loose coupling of the two vibrating OH bars through the O—O bond" (12).

3. The asymmetrical bending vibration $\nu_6$: the infrared bands for vapor (1266 cm$^{-1}$), liquid (1353 cm$^{-1}$), and solid (1378 cm$^{-1}$) phases are attributed

to the asymmetrical bending of the hydroxy group. The frequency change for the deuterated peroxide confirms association of this motion with the hydrogen atom.

4. The symmetrical bending vibration $v_2$: the assignment is speculative. The calculated value for this vibration is 1380 cm$^{-1}$ (12). In the crystalline solid a band at 1410 cm$^{-1}$ has been assigned to the $v_2$ vibration (118), but no corresponding band is observed in the liquid or gaseous phases (117).

5. The torsional oscillation or hindered internal rotation $v_4$ (12, 119): this mode in the gas phase is assigned to 314 cm$^{-1}$ and the overtone to 575 cm$^{-1}$ (119). The torsional frequency is more than doubled in the condensed phases, 635 cm$^{-1}$ in the liquid (120) and 695 cm$^{-1}$ in the solid state (118). More recently a doublet at 690 and 650 cm$^{-1}$ (121) has been assigned to $v_4$ in the crystal, because of strong hydrogen bonds.

6. Overtones and combinations: the majority of these bands have been evaluated in recent years with certainty (1, p. 326 ff.; 12, 116, 117, 121). Near-infrared bands in the vicinity of 7040, 4715, and 3600 cm$^{-1}$ (1.42, 2.12, and 2.8 $\mu$, respectively) have analytical interest (see Section B-1-a).

## B.  Empirical Correlations of Organic Peroxides

Until 1951 there were few papers on infrared or Raman spectra of inorganic or organic peroxides. Frequencies of 870–880 cm$^{-1}$ were ascribed to the stretching vibration of the —O—O— link in hydrogen and deuterium peroxides (11, 117) and similarly in the range of 830–890 cm$^{-1}$ to the peroxide-group vibration in alkyl and acyl peroxides, hydroperoxides, peroxy acids, and peroxy esters (20, 25, 122–124).

X-ray analysis and dipole-moment studies on hydrogen peroxide and organic peroxides show the atoms of the HOOH and COOC chains to be nonlinear and nonplanar (Figure 1d). The —O—O— link is twisted to form a dihedral angle at 70–150° with respect to hydrogen or carbon, the exact angle depending on the structure of the neighbouring group attached to each oxygen atom. The presence of a light hydrogen atom at the end of the COOH chain in hydroperoxides and the more asymmetrical substitution structure it forms in comparison with COOC must increase the probability of development of a characteristic band.

In a theoretical calculation of the vibration in a model system XOOX with X of different masses (ranging from mass 15 for methyl to infinite mass) Minkoff (125) showed that the X—O and O—O vibrational frequencies vary within the comparatively narrow limits 819–985 cm$^{-1}$ for O—O, 975–1082 cm$^{-1}$ for $v_s$ (X—O), and 824–1044 cm$^{-1}$ for $v_a$ (X—O). Kovner, Karyakin, and Epimov (126) calculated the theoretical vibrational modes of the free COOH and COOD groups assuming a 90° dihedral peroxide configuration. A set of four nonlinear atoms requires six vibrational modes ($3n - 6$), which were experimentally observed by extension of the spectral absorptions into the far-infrared region. Good agreement was obtained between calculated and observed frequencies for these groups in isopropylbenzene hydroperoxide (Table 5). The intense band at 840 cm$^{-1}$ was assigned to oscillations of the dihedral angle $\varphi$ between the C—O—O and O—O—H planes.

TABLE 5. CALCULATED AND OBSERVED VIBRATIONAL FREQUENCIES OF THE COOH AND COOD GROUPS IN HYDROPEROXIDES[a]

| | | COOH | | | | COOD | |
| | | Vibrational Frequency (cm$^{-1}$) | | | | Vibrational Frequency (cm$^{-1}$) | |
| Bonds and Angles[b] | | Calculated | Observed | Bonds and Angles[b] | | Calculated | Observed |
|---|---|---|---|---|---|---|---|
| COO | *d* | 586 | 585 | COOD | *d* | 549 | — |
| COOH | $\varphi$ | 834 | 840 | COOD | $\varphi$ | 680 | — |
| O—O | *s* | 886 | 880 | O—O | *s* | 873 | 855 |
| C—O | *s* | 1156 | 1155 | C—O | *s* | 1156 | 1155 |
| OOH | *d* | 1321 | 1325 | OOD | *d* | 997 | 995 |
| O—H | *s* | 3450 | 3450 | O—D | *s* | 2539 | 2550 |

[a] Data from reference 126.
[b] $s$ = stretching modes; $d$ = deformation modes; $\varphi$ = dihedral out-of-plane deformation.

In recent years infrared examination of an extended range of aliphatic and aromatic peroxide classes supports this band region for an O—O vibrational frequency (40, 127, 128). Unfortunately absorption in the region of 880 cm$^{-1}$ is not an unambiguous indication of the presence of an O—O group since molecular skeletal vibrations of the *t*-butoxy (129, 130) and epoxide (93, 94, 131) groups, C—C stretch, and CH$_3$ rocking vibrations (132) fall in this region. Symmetrical vibrations of the COOC chain do not produce a great change in the dipole moment, so that the corresponding infrared bands will be weak. The masses and the force constants for the O—O group in this chain are very similar to the values for C—O and C—C groups; hence the appearance of an unequivocal band specific for the O—O group is improbable in most compounds because of vibrational coupling. However, the presence of a characteristic O—O band will depend on the angle between successive bonds in the molecular chain. Vibrational coupling will not occur if this angle is about 90°, and hence in this case a characteristic O—O band will be clearly defined (133).

The bands of O—O, C—O, and O—H vibrations are of primary interest for distinguishing peroxides. Absorptions of other groups are generally less characteristic for distinguishing peroxides from similarly absorbing components except in specific cases.

## 1. O—O Stretching

### (a) HYDROPEROXIDES

The O—O stretching frequency for hydroperoxides has been calculated to be 883 cm$^{-1}$ (126, 127, 134). A weak absorption is experimentally found at 877–847 cm$^{-1}$, but this is a region in which the corresponding alcohols also absorb; this prohibits the use of this band for hydroperoxide characterization (124, 135).

### (b) DIALKYL PEROXIDES

Dialkyl peroxides have a calculated O—O stretch at 985–819 cm$^{-1}$ (125, 126, 134). Only a few low-molecular-weight simple aliphatic members of dialkyl peroxides have been examined. Leadbeater (122) determined Raman and infrared spectra for diethyl ether (intense infrared band at 882 cm$^{-1}$ and intense Raman bands at 831, 882, 910, and 919 cm$^{-1}$), di-α-dihydroxydiethyl peroxide (intense Raman bands at 831, 883, and 911 cm$^{-1}$), and α-mono-hydroxyethyl peroxide (intense Raman bands at 800, 830, 881, and 912 cm$^{-1}$). Minkoff (125) found that two bands (862 and 847 cm$^{-1}$) for diethyl ether were replaced by one band (877 cm$^{-1}$) for diethyl peroxide; the latter band was assigned provisionally to the O—O stretch.

Di-$t$-butyl peroxide has a strong band at 874 cm$^{-1}$ compared with $t$-butyl hydroperoxide's absorptions, a medium band at this frequency, and a more intense one at 847 cm$^{-1}$ (124, 129, 136). The large dihedral angle for di-$t$-butyl peroxide constrains the molecule to a high degree of planarity, whereby the intensity of the O—O stretching vibration is diminished. An infrared and Raman investigation on the vibrational spectra and assignments of di-$t$-butyl peroxide in the gas, liquid, and solid states led to the conclusion that the O—O stretching motion is associated with the pair of Raman frequencies at 904 and 862 cm$^{-1}$ (130). Infrared spectra for a series of mixed $t$-butyl peroxides have been published and may aid in their characterization (137).

### (c) ACETAL AND KETAL PEROXIDES

Minkoff (125) reported the spectra of low-molecular-weight monohydroxy-alkyl hydroperoxides, dihydroxydialkyl peroxides, and monohydroxydialkyl peroxides. Weak to medium bands in the 910–830 cm$^{-1}$ region were observed for which a positive O—O assignment could not be made.

Kharasch and Sosnovsky (138) published spectra for the following peroxides as a reference source for this class: 1,1-dihydroxydicyclohexyl peroxide, 1-hydroxy-2-chlorocyclohexyl hydroperoxide, 1-hydroxy-2-bromocyclohexyl hydroperoxide, 1,1-di-*t*-butylperoxycyclohexane, dicyclohexylidene diperoxide, bis(1-hydroxybenzyl) peroxide, dibenzaldiperoxide, and bis(hydroxymethyl) peroxide. The spectra of bis(hydroxyalkyl) peroxides were also reported by Fritsch and Deatherage (139).

### (d) PEROXY ACIDS

Spectra for peroxyformic and peroxyacetic acids were reported by Giguère and Olmos (21). Peroxyformic acid in the vapor state showed three weak bands, at 872, 859, and 842 cm$^{-1}$, and in the liquid state a single weak band at 855 cm$^{-1}$. Peroxyacetic acid showed three very strong bands at 868, 861, and 853 cm$^{-1}$ in the vapor state and a strong band at 858 cm$^{-1}$ in the liquid state. The 855 and 853 cm$^{-1}$ bands were assigned to O—O stretch. Spectra for peroxypropionic and peroxybutyric acids also reveal a weak-to-medium band in the vicinity of 870 cm$^{-1}$ (140). Peroxypalmitic acid in solution (23) has a medium-intensity band at 865 cm$^{-1}$ (11.56 $\mu$) that is absent in the corresponding parent carboxylic acids. Solid peroxypalmitic acid (23) has two medium-intensity bands in this region, at 879 and 890 cm$^{-1}$, but a drawn-out band with a weak absorption at 891 cm$^{-1}$ for solid palmitic acid makes a positive assignment difficult.

In the aromatic series peroxybenzoic acid has a band at 880 cm$^{-1}$ (141). Bands at 846–873 cm$^{-1}$ for a series of *o*-, *m*-, and *p*-substituted peroxybenzoic acids (—Cl, —Br, —F, and —NO$_2$) were assigned to O—O stretch. They show a small shift toward lower frequencies after deuteration and toward higher frequencies on dissolution.

### (e) ACYL PEROXIDES

Long-chain diacyl peroxides have a medium semibroad band at 892 cm$^{-1}$, in contrast to a weak, very broad band at 925 cm$^{-1}$ for anhydrides (25, 40, 124). Benzoyl peroxide has a Raman band at 883 cm$^{-1}$ (122) that does not appear in the infrared region (124).

### (f) PEROXY ESTERS

The spectra of several aromatic (5, 136, 142) and aliphatic (40, 136, 142) *t*-butyl peroxy esters have been reported. In general a medium 845–870 cm$^{-1}$ band and a weak, broad 915–930 cm$^{-1}$ band are observed. *t*-Butyl esters also show a similar pair of bands in these regions, weak and broad at 845 cm$^{-1}$ and medium at 948 cm$^{-1}$. *t*-Butyl esters and peroxy esters are distinguishable from normal alkyl esters, which show only a trace of a doublet in this region. The absorption band at 850 cm$^{-1}$ has been attributed to the butoxy rather than to the peroxide group (129, 136), but a more detailed study is needed in view of the 904 and 862 cm$^{-1}$ O—O stretching assignments observed in the Raman for di-$t$-butyl peroxide (130).

### 2.  C=O Stretching

#### (a)  DIACYL PEROXIDES

Peroxides containing the structure —C(O)OOC(O)— show a doublet in the carbonyl absorption, resembling anhydrides (143). The splitting in the carbonyl band is attributed to coupling. The pair of carbonyl bands differs in relative intensities, a useful feature that aids in characterization. In acyclic aliphatic anhydrides and benzoic anhydride the band at the higher carbonyl frequency has a higher intensity, whereas in cyclic anhydrides acyclic aliphatic peroxides, and dibenzoyl peroxide the band at the lower carbonyl frequency is more intense. The splitting in anhydrides is about 60 cm$^{-1}$.

Davison (20) examined a series of acyl, aroyl, and asymmetrical peroxides. Aliphatic diacyl peroxides have a pair of carbonyl bands in the regions 1820–1811 and 1796–1784 cm$^{-1}$, and an average split of 25 cm$^{-1}$. The larger frequency split shown by anhydrides is the result of its shorter coupling link (143). The introduction of two aroyl groups in the peroxide lowers the band frequencies and diminishes the average split to 22 cm$^{-1}$. Carbonyl absorptions for aryl peroxides are in the range 1805–1780 and 1783–1758 cm$^{-1}$. Asymmetrical peroxides with an aryl and alkyl group attached to opposing carbonyls show a mean carbonyl splitting larger than normal (ranging from 24–33 cm$^{-1}$) because of the effect of aryl conjugation on the adjacent carbonyl group.

In crystalline dibenzoyl peroxide the preferred structural conformation of the carbonyl groups is an orientation directed inward toward the peroxide grouping (Figure 4$a$). Crystalline forces impose conformations that may not exist in the liquid phase because of free rotation. Fayat (33) considered this problem for benzoyl peroxide in the liquid phase to determine its preference from among the six possible conformations and symmetries. His infrared analysis of the doublet carbonyl frequencies and their relative intensities in several solvents and in solid films led to the conclusion of the inward orientation for the compound in solution as well as in the crystalline state.

#### (b)  PEROXY ACIDS

Peroxy acids have a single carbonyl absorption in the range of 1760–1710 cm$^{-1}$ in the vapor, liquid, or solution. This band was determined by Giguère and Olmos (21) in peroxyformic (1739 cm$^{-1}$) and peroxyacetic (1760 cm$^{-1}$) acids. Long-chain peroxy acids in either the solution or solid phase absorb at 1747–1748 cm$^{-1}$ (23). Carbonyl absorptions in aromatic peroxy acids are doublets in the solid and singlets in solution (141). The doubling is believed to be due to correlation field splitting since it disappears in solution. The single carbonyl absorption observed in solution remains unchanged on dilution because of intramolecular hydrogen bonding. In comparison carboxylic acids exhibit monomeric and dimeric bands in solution and a single band for the dimer in the solid state (23).

### (c) PEROXY ESTERS

Aliphatic series of *t*-butyl peroxy esters absorb at 1783–1750 cm$^{-1}$ (20, 142); the aromatic series absorb at 1771–1758 cm$^{-1}$ (20). Minkoff (125) published the spectra of several peroxy esters. Aliphatic series of *t*-butyl esters absorb at the distinctly lower frequency range of 1735–1712 cm$^{-1}$ (128, 136).

### 3. *O—H Stretching*

### (a) HYDROPEROXIDES

Holman, Edmondson, and co-workers (144, 145) correlated hydrogenic and other functional-group absorptions in the near-infrared region [10,000–3300 cm$^{-1}$ (1–3 $\mu$)] for fatty acids, their esters and oxidized products, and hydroperoxides. Specific vibration bands for the —OOH grouping were observed at 3550, 4530, and 6850 cm$^{-1}$ (2.82, 2.07, and 1.46 $\mu$, respectively). The 3550 cm$^{-1}$ band is difficult to differentiate from similar bands of the hydroxy, aldehyde, ketone, acid, and ester groups, all of which absorb in this region, but the 6850 and 4530 cm$^{-1}$ bands are readily differentiated from neighboring hydroxy absorptions by the use of quartz optics. The maximum at 4530 cm$^{-1}$ is probably a combination band, and the maximum at 6850 cm$^{-1}$ is a harmonic of the 3550 cm$^{-1}$ fundamental. Their low intensity necessitates use of either high concentrations or long absorption cells.

In other regions of the infrared spectrum hydroperoxides do not exhibit specific O—H vibrations that distinguish them from other hydroxy-containing compounds (127), but the observed bands are sufficiently different in frequency from alcohol and phenol OH bands to be useful in analyses of their mixtures (146). Monomeric hydroperoxides have the O—H stretching vibration band at approximately 3580–3530 cm$^{-1}$ (124, 125, 127, 134, 135, 146). [Hydroperoxide O—H frequencies reported by Williams and Mosher (135) are approximately 100 cm$^{-1}$ lower than this range, which is in disagreement with the results of other investigators.] Hydrogen bonding reduces this frequency range to 3500–3000 cm$^{-1}$. The bands of monomeric hydroperoxides are narrower than the bands of the hydrogen-bonded polymeric forms. Hydrogen bonding in $\alpha$-acetylenic hydroperoxides appears to be stronger than in corresponding $\alpha$-acetylenic alcohols by showing a larger shift in the absorption range of the O—H stretching vibration (147).

The O—H stretch of hydroperoxides is strongly influenced by deuteration and solvent. The stretching frequency of OOD shifts to about 2500 cm$^{-1}$. For instance, the O—H band of cumene hydroperoxide, located at 3402 cm$^{-1}$, shifts to 2513 cm$^{-1}$ on deuteration (148). A doublet hydroxy band is observed in aralkyl hydroperoxides, such as cumene and p-cymene hydroperoxides; it is caused by intramolecular hydrogen bonding with the π-electrons of the aromatic ring (149). Figure 34 compares the absorption of cumene hydroperoxide with that of an aliphatic hydroperoxide. The intensities of the doublet are approximately equal to the intensity of the single peak of t-butyl hydroperoxide.

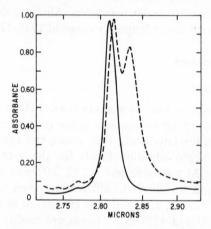

**Figure 34.** Infrared absorption spectra of t-butyl hydroperoxide (solid line) and cumene hydroperoxide (dashed line) in carbon tetrachloride. (Reference 146).

Substituents in the aromatic ring affect the splitting. In derivatives of cumene hydroperoxide containing electron-withdrawing groups (—Cl and —NO$_2$) the splitting is eliminated since the ring is depleted of sufficient electron density for intramolecular complexation. The splitting may also be removed by the use of methylene chloride as solvent in the place of carbon tetrachloride; methylene chloride shifts the fundamental to approximately 20 cm$^{-1}$ lower frequencies (146). This effect arises by hydrogen bonding with this solvent, which disrupts the weaker intramolecular hydrogen bonding of low energy [0.90 kcal/mole (150)]. The absorptivity of hydroperoxides is nearly constant over a concentration range from 0.2 to 20 mM (146).

Alkyl hydroperoxides reveal an interesting characteristic alternating pattern in the region 830–800 cm$^{-1}$ (135). Hydroperoxides with an even number of carbons in the alkyl chain (butyl to decyl) show two maxima at about 825 and

805 cm$^{-1}$, whereas the odd-numbered members show a single maximum at about 813 cm$^{-1}$. Liquid alcohols do not mimic this characteristic.

Absorption bands at approximately 1100 cm$^{-1}$ are to be expected for the C—O stretching mode and the O—H deformation mode. This region is evidently not uniquely characteristic (105). Bands in the region 880–700 cm$^{-1}$ due to the nonplanar deformational vibrations of the O—H group have been observed only for hydrogen peroxide (118) and isopropylbenzene hydroperoxide (126).

## (b) PEROXY ACIDS

A sharp O—H stretching band at approximately 3300 cm$^{-1}$ distinguishes peroxy acids from carboxylic acids in the near-infrared region. Swern and co-workers (23) reported the constancy of this band at 3280 cm$^{-1}$ in carbon tetrachloride solutions for eight long-chain peroxy acids over a concentration range from 0.006 to at least 0.3 $M$, with little variation in molar absorptivities. The average for all peroxy acids and concentrations was 51.6 l/mole-cm, with a maximum variation at any concentration of only 3%. A strong absorption for liquid peroxyformic acid (3378 cm$^{-1}$) and liquid peroxyacetic acid (3310 cm$^{-1}$) is in contrast to their very weak absorptions in this region in the vapor state (21). In the solid state the hydroxy band of long-chain peroxy acids broadens and shifts to lower frequencies (3200 cm$^{-1}$).

The OO—H bands at 6850 and 4830 cm$^{-1}$ that advantageously distinguish hydroperoxides from other functional absorptions in the near-infrared region are very weak for peroxyacetic acid (145). These bands have no analytical utility for peroxyacetic acid (other peroxy acids have not been examined in this region), since acetic acid also exhibits low-intensity bands at these frequencies (145).

Peroxybenzoic acids (141) have narrower bands and absorb at higher frequencies (3270–3250 cm$^{-1}$) in solution than they do in the solid state (3260–3232 cm$^{-1}$). $o$-Nitroperoxybenzoic acid (solution, 3295 cm$^{-1}$; solid, 3305 cm$^{-1}$) and $p$-fluoroperoxybenzoic acid (solution, 3250 cm$^{-1}$; solid, 3290 cm$^{-1}$) are exceptions by having the reverse frequency order for these two states.

The single sharp hydroxy band at approximately 3300 cm$^{-1}$ arises from the intramolecular-hydrogen-bonded monomeric structure **1** (22, 24, 127). Intermolecularly bonded carboxylic acids, by comparison, have a broadened band for O—H stretch in the 2900 cm$^{-1}$ region and have variable absorptivities because of an equilibrium between free monomers and hydrogen-bonded dimers (106, p. 170).

Peroxy acids have a strong band in the region 1450–1400 cm$^{-1}$ (21, 141), which shifts to 1130–1100 cm$^{-1}$ on deuteration. This band is attributed to the O—H in-plane deformation vibration.

A strong band in the vicinity of 1200 cm$^{-1}$, generally accompanied by one or two weaker bands on the higher frequency side, appears for dissolved or solid peroxy acids. Deuteration does not significantly alter the intensity or frequency of the 1200 cm$^{-1}$ band for peroxy acids in solution, but relative intensity changes are observed for the solid and a frequency change of 8–10 cm$^{-1}$ is noted in the case of $p$-chloroperoxybenzoic acid. A similar band in the spectra of the parent carboxylix acids is about 50 cm$^{-1}$ higher in frequency. The band is assigned to the C—O stretching vibration coupled to the O—H deformation.

### 4.  Fluoroperoxides

When directly attached to the peroxide group, a fluorine atom has a dramatically strong effect on the structure and vibrational spectrum of the peroxide. Fuorine peroxide, FOOF, has the same $C_2$ symmetry as hydrogen peroxide but the O—O bond (1.217 Å) is much shorter than in hydrogen peroxide (1.48 Å), being of about the same length as in the oxygen molecule (151). The O—F bonds (1,575 Å) are longer than they are in OF$_2$ (1.409 Å). The O—O bond order is nearly the same as in the oxygen molecule, and the O—F bonds are weaker than they are in OF$_2$. The O—O stretching frequency in O$_2$F$_2$ is 1306 cm$^{-1}$, but in $^{18}$O$_2$F$_2$ this band disappears, to be replaced by a 1239 cm$^{-1}$ band. Because of the similar masses of oxygen and fluorine, the O—O and O—F modes are mixed.

The strength of the O—O bond is attributed to the interactions of fluorine atoms with the unpaired $\pi^*$ orbitals on the oxygen atoms (151, 152). A three-center O—O—F grouping is formed with a molecular orbital that is antibonding with respect to O—O and bonding with respect to O—F. In this electronic arrangement the bond order is nearly unchanged from that in oxygen, and the O—F bond is weakened by forming part of a three-centered bonding system rather than a two-centered, two-electron bond system.

Bis(monofluorocarbonyl) peroxide, FC(O)OOC(O)F, has the O—O stretching mode of medium intensity at 912 cm$^{-1}$ (153). The frequency is higher than it is for organic peroxides and considerably lower than the frequency for O$_2$F$_2$. This may be attributed to inteıcalation of the carbonyl carbon between fluorine and peroxide oxygen, which prevents direct electronic interaction and attenuates the electronegative effect. The $\pi$-electrons of carbonyl could serve as an electronic funnel for fluorine interacting with oxygen; this would account for the observed shift of the O—O stretch to higher frequency. The carbonyl stretch (1899, 1905, and 1934 cm$^{-1}$) is split (32 cm$^{-1}$) by coupling of the C=O modes, as expected for acyl peroxides, and the high-frequency

branch is additionally split by interaction with the $O=C\overset{\displaystyle F}{\underset{\displaystyle O}{\diagup}}$ groupings.

Bis(trifluoromethyl) peroxide, $CF_3OOCF_3$, has the O—O stretch of medium intensity in the infrared region (890 $cm^{-1}$), which in the Raman (886 $cm^{-1}$) is the most intense band (153, 154). The O—O stretching mode has essentially the frequency found for organic and hydrogen peroxides. In view of the fact that the inductive effects of $CF_3$ and F are nearly the same (155), some shift to higher frequency would be expected. Evidently fluorine's strong effect on oxygen is transmitted largely through direct electronic interaction, as in $O_2F_2$, rather than by induction; the intercalation of atoms by $\sigma$-bonding prevents this interaction. Negative hyperconjugation of $CF_3$ (156) may have the effect of increasing the strength of the peroxide bond by diminishing electron repulsion between the nonbonding $p$-orbitals of the contiguous oxygens shown in structure **6**.

$$F^-$$
$$+$$
$$F—C=O—O—CF_3$$
$$|$$
$$F$$

**(6)**

Pentafluorosulfur peroxofluoroformate, $SF_5OOC(O)F$, is of comparative interest. Two intense absorptions at 889 and 937 $cm^{-1}$ were tentatively assigned to $F_5S$—O stretch rather than to the O—O bond (157).

A variety of new fluoroperoxides is being prepared and described with increasing frequency (158). A few illustrations of these unique compounds are bis(difluorofluoroxymethyl) peroxide ($FOCF_2OOCF_2OF$) (159), fluorocarbonyl trifluoromethyl peroxide [$FC(O)OOC(O)CF_3$] (160), bis(trifluoromethyl) trioxide ($CF_3OOOCF_3$) (161, 162), and $C_3H_7OOF$ (163). Infrared assignments have been tentatively made for the sharp absorption bands in each case.

### 5. Ozonides

The ozonide absorption band was initially but incorrectly assigned to the carbonyl region by Briner and co-workers (164). The erroneous assignment was traced to the presence of carbonyl compounds formed by ozonide decomposition (165) and accordingly corrected (166–168). Criegee et al. (168) examined the ozonides of 18 compounds of varied structures, observing for 13 of these a medium-strong absorption band in the range 1064–1042 $cm^{-1}$. This absorption range 1100–1000 $cm^{-1}$ as characteristic for ozonides was confirmed for the ozonides formed from olefinic hydrocarbons—in particular from short-chain alkenes by Garvin and Schubert (169), for di-$t$-butylethylene by Schröder (170), for $n$-alk-1-enes (1110 and 1062 $cm^{-1}$) by Menyailo et al. (171), and for butene-2 by Loan et al. (172). No Raman band corresponding to the ozonide band in the 1100–1000 $cm^{-1}$ region was found (173). The infrared correlation is of limited value owing to the C—O stretching absorption in this region and has, in fact, been attributed to the C—O bonds in the ozonides (171). Greenwood and Haske (174, 175) made a comparison of the spectra for hex-3-ene ozonide and di-$n$-propyl peroxide and found several coincident bands close to 950, 1380, and 1460 cm$^{-1}$.

Ozonides prepared from olefinic fatty acid esters were separated and analyzed. Riezebos et al. (176) isolated six geometrical isomeric ozonides from ozonized methyl oleate. These compounds proved to be a triad of $cis$–$trans$ pairs (7, 8, and 9), all of which absorbed at 1110 cm$^{-1}$ along with bands at 830 cm$^{-1}$ for the $cis$ isomer and 1320 cm$^{-1}$ for the $trans$ isomer, in correspondence with the products predicted on the basis of Criegee's theory (177, 178).

$$
\underset{\textbf{(7)}}{Me_3(CH_2)_7CH\overset{O-O}{\underset{O}{\diagup\diagdown}}CH(CH_2)_7COOMe}
\qquad
\underset{\textbf{(8)}}{Me(CH_2)_7CH\overset{O-O}{\underset{O}{\diagup\diagdown}}CH(CH_2)_7Me}
$$

$$
\underset{\textbf{(9)}}{MeOOC(CH_2)_7CH\overset{O-O}{\underset{O}{\diagup\diagdown}}CH(CH_2)_7COOMe}
$$

Privett and Nickell (179) reported similar products and absorption bands from the ozonolysis of methyl oleate and elaidate.

## C. Analytical Applications

Several peroxides have been determined analytically by their specific absorption bands. Infrared spectroscopy has been fruitful in studying the mechanism of peroxide formation in photooxidation (127). In kinetic measurements of peroxide decompositions Raley and co-workers (180) made use of the 873 cm$^{-1}$ (11.46 cm$^{-1}$) absorption band to measure di-$t$-butyl peroxide decompositions, and Bartlett and Hiatt (181) applied the carbonyl region to $t$-butyl peroxy esters. Benzoyl peroxide (997, 1223, 1768, and 1792 cm$^{-1}$) and aliphatic diacyl peroxides (890, 1060, 1782, and 1812 cm$^{-1}$) follow Beer's law at the designated frequencies in quantitative determinations ($\pm 3\%$ relative error) of their presence in styrene or polystyrene (182). The sensitivity of the carbonyl absorption at 1724 cm$^{-1}$ is in fact sufficient to determine acyl end groups in a polystyrene prepared and labeled with a diacyl peroxide of azobenzene-4-carboxylic acid as initiator (183).

Infrared spectroscopy appears to have some analytical advantages over other methods for the determination of hydroperoxides. Kohler's peroxide (184) and Julian's peroxide (185) were shown by infrared analysis to be hydroperoxides rather than cyclic peroxides as originally assumed. The region of 3400–3600 cm$^{-1}$, corresponding to the vibrations of the "free" O—H group of the hydroperoxides and alcohols, was used for the determinations of *t*-butyl hydroperoxide and *t*-butanol in *t*-butyl peroxyacetate, peroxybenzoate, or di-*t*-butyl peroxide and for cumene hydroperoxide and α,α-dimethylbenzyl alcohol in cumyl peroxide (186). Even the decomposition products of α,α-dimethylbenzyl hydroperoxide could be assessed by making absorption measurements at appropriate wave numbers (187). Frankel, Evans, and Cowan (188) used infrared-absorption spectra in their investigations of the products from thermal dimerization of fatty ester hydroperoxides. Bands in the near-infrared region (6800 and 4530 cm$^{-1}$) are specific for hydroperoxides whose formations in the autoxidations of fatty oils could be followed in this spectral region (144, 145). The 3700–3330 cm$^{-1}$ region has been studied for its application to an examination of autoxidation products of fats (189). O'Connor's reviews (190, 191) on the application of infrared-absorption spectra to lipid chemistry are invaluable reference sources with regard to oxidation and autoxidation studies.

Some peroxides are difficult to measure accurately by chemical methods. Ascaridole, a cyclic peroxide constituent of chenopodium oil, is a common illustration of a peroxide that has resisted reliably accurate and quantitative determinations by reduction of the peroxide function. It does not react stoichiometrically with iodide ion, but it is determinable in the infrared region at 935.5 cm$^{-1}$ (10.69 $\mu$) (error limit 1–2%), encountering no interference at this wavelength (192, 193).

Peroxyacyl nitrates, $RC(O)OONO_2$, a series of unstable peroxides that have their origin via photochemical reactions in polluted air, were discovered by laboratory tests with a long-path infrared spectrophotometer (194, 195). Studies on their physiological properties are necessary due to their lachrymatory effects on man and their toxicity to plant life. The absorptivities of peroxyacetyl, peroxypropionyl, and peroxybutyryl nitrates at seven bands between 1835 and 793 cm$^{-1}$ have been reported (196).

A final example is given to show that spectral data alone may not always permit reliable interpretations. Milas and Arzoumanidis (197) prepared a peroxide they considered to be di-*t*-butyl trioxide, $Me_3COOOCMe_3$, by reacting triphenyl phosphine dichloride and triphenyl phosphine oxide with *t*-butyl hydroperoxide. Their interpretation of the compound's structure was based on infrared, Raman, nuclear magnetic resonance (NMR), dipole moment, and elemental analysis. Bartlett and Günther (198) on further examination have indicated this compound to be the peroxyketal 2,2-di-*t*-butylperoxypropane, $Me_2C(OO—t-Bu)_2$.

## IV. NUCLEAR MAGNETIC RESONANCE

Two initial studies were published simultaneously in 1959 on the use of nuclear magnetic resonance (NMR) spectroscopy for characterizing peroxides. Fujiwara, Katayama, and Kamio (199) acquired data on a few hydroperoxides, whereas Davies, Hare, and White (200–202) applied NMR spectroscopy to a series of isobutyl and *t*-butyl, isobutoxy, and *t*-butoxy, and isobutylperoxy and *t*-butylperoxy boron compounds. With respect to the latter investigation, Davies et al. attempted to follow the path of the autoxidation of isomeric tributyl boron compounds and the isomerization of the intermediates to products. In the succeeding decade minor interest appeared in the application of NMR to peroxide chemistry. Attention was directed mainly to solution of specialized problems on dialkyl peroxide and hydroperoxides (203–208). Swern, Clements, and Luong (209), recognizing the need to fill this void, obtained the NMR characteristics of a representative range of peroxides. The technique was formerly used by Brill (210) to identify isomeric hydroperoxides qualitatively, and the analytical application was extended by Ward and Mair (211) to the quantitative analysis of peroxides. The power of the NMR technique as a probe of structure is clearly evident from the results of such studies on ozonides and perfluoroperoxides. It has recently proven to be an elegant tool for probing free-radical reactions.

### A. Hydrocarbon Peroxides

#### 1. Hydroperoxides

The peak area and chemical shifts of the labile proton in the —OOH and —OH functions are dependent on sample concentration (209, 211) and solvent composition (211) because of hydrogen bonding, chemical exchange, and solvation effects. The concentration dependency of the OO—H signal is recorded for *t*-butyl hydroperoxide in Table 6; the shift to high field that occurs on dilution is presumed to result from a decrease in intermolecular hydrogen bonding (209). In $CHCl_3$ or $CDCl_3$ solutions of hydroperoxides and alcohols at corresponding concentrations the OO—H gives a signal about 8 ppm downfield from tetramethylsilane, whereas the corresponding OH of alcohols resonates at about 1.6–2.0 ppm. Despite the appearance of aromatic proton absorptions at the same downfield region as OO—H, interference with their identifications may be eliminated by dilution to shift the OO—H frequency. Proton exchange with $D_2O$ is also useful since quantitative measurements can be effected by calculating the relative peak areas in this region before and after exchange.

A mechanism that may explain the large deshielding effect in hydroperoxides and the shift to highfield on dilution was not offered by previous

TABLE 6. CONCENTRATION DEPENDENCE OF
OO—H RESONANCE OF *t*-BUTYL
HYDROPEROXIDE IN CHLOROFORM
SOLUTIONS[a]

| *t*-BuOOH Concentration ($M$) | —OO—H Chemical Shift (ppm)[b] |
|---|---|
| 1.51 | 8.15 |
| 1.00 | 7.97 |
| 0.50 | 7.78 |
| 0.01 | 7.33 |
| 0.00 | 7.1–7.2 |

[a] Data from reference 209.
[b] Downfield from tetramethylsilane, internal reference.

investigators, but a rationale based on Reid and Connor's model for alcohols (212) seems to be applicable. The proton signal of the hydroxylic hydrogen in alcohols is a weighted average of the monomeric molecule and its various hydrogen-bonded species, consisting of a cyclic trimer and other polymeric forms. The cyclic trimer, a six-membered-ring structure with three mutually reinforcing hydrogen bonds, was suggested to be a more stable association than the cyclic dimer (213, 214). Reid and Connor recognized the cyclic trimeric model of associated alcohol molecules (structure **10**) as a "stable" structure for which delocalization of σ-bonds about the six hydrogen-bonded atoms could occur within the ring.

(10)          (11)          (12)

A "circulating current" loop may be set up within this system, and the protons may undergo a shift to low field relative to their frequency in the usual hydrogen-bonded state. Hydroperoxides are reported (215) to associate in structures that closely fit this model. Aside from some association as linear polymeric species, they give rise to two types of six-membered-ring associations, a dimer as the main species (structure **11**) and a trimer (structure **12**) in lesser amount. Both cyclic structures are appropriate for setting up a "circulating current" loop and deshielding the ring protons as in the alcohol trimer. A more pronounced deshielding in hydroperoxides may have its origin in the larger electronegativity of two oxygen atoms in the peroxide, which would attract electron density from the protons.

Since the peroxide functionality and the alcohol group would not have identical electronegativities, a difference in their inductive effects on $\alpha$-methylene protons is not surprising. The $\alpha$-methylene protons in alkyl hydroperoxides are shifted about 20 Hz down-field from the resonance line of the group in the corresponding alcohols (199), a difference Ward and Mair (211) confirmed between benzyl hydroperoxide ($-290$ Hz) and benzyl alcohol ($-271$ Hz) in carbon tetrachloride solution, with tetramethylsilane as internal reference. The $\alpha$-proton peak is split into multiplets by $\beta$-proton coupling; the vicinal coupling constants are in the 6–9 Hz range.

The resonance line of $\beta$-protons is affected by solvent and shows a variation with solvent of 1–4 Hz for hydroperoxides, alcohols, and dialkyl peroxides (Table 7) (211). In very favorable cases with adequate resolution these peaks

TABLE 7.  EFFECT OF SOLVENT ON THE $\beta$-PROTON
CHEMICAL SHIFT IN HYDROPEROXIDES AND
DIALKYL PEROXIDES[a]

| | | Chemical Shift ($H_z$)[b] | | |
|---|---|---|---|---|
| R | Solvent | ROOR | ROOH | ROH |
| $Me_3C$ | CDCl$_3$ | $-73.0$ | $-76.3$ | $-75.2$ |
| | DMSO | $-70.3$ | $-69.5$ | $-68.3$ |
| | Pyridine | $-74.2$ | $-81.0$ | $-82.2$ |
| Me<br>\|<br>Ph$-$C$-$<br>\|<br>Me | CDCl$_3$ | $-91.4$ | $-92.2$ | $-91.4$ |
| | DMSO | $-89.4$ | $-91.5$ | $-91.5$ |
| | Pyridine | $-95.5$ | $-102.4$ | $-102.4$ |
| | Acetone | $-92.1$ | $-91.3$ | $-89.7$ |

[a] Data from reference 211.
[b] Relative to tetramethylsilane as internal indicator.

may serve in the qualitative and quantitative analysis of these compounds as components of a mixture. Quantitative measurements make use of the peak area, although Ward and Mair (211) obtained satisfactory results by using the peak height as a measure of the relative concentration of the components. Table 8 contains additional data on alcohols and peroxides in carbon tetrachloride. Some of the resonances reported may be useful in analyses.

TABLE 8. NMR SPECTRAL DATA OF HYDROPEROXIDES AND ALCOHOLS[a]

| | | Chemical Shift of Indicated Proton (ppm)[b] | | | | |
|---|---|---|---|---|---|---|
| | | Hydroperoxide | | Alcohol | | |
| Hydroperoxide | Concentration (M) | —OO—H | β-CH₃ | —O—H | β-CH₃ | Reference |
| t-Butyl hydroperoxide | 1.0 | 7.97 | 1.25 | 1.62* | 1.23* | 209 |
| | | | 1.27 | | 1.25 | 211[c] |
| 2,5-Dimethyl-2,5-dihydroperoxyhexane | 0.62 | 8.12 | 1.20 | | | 209 |
| 4-Hydroperoxy-2,6-di-t-butyl-4-methylcyclohexa-2,5-dien-1-one | | 7.58 | 1.40 | 2.01* | 2.42* | 209 |
| 1-Hydroxy-1'-hydroperoxydicyclohexyl peroxide | 0.41 | 9.16 | | | | 209 |
| PhC(Me₂)COOH | | 7.87 | 1.55 | | | 209 |
| | | | 1.54* | | 1.52* | 211[c] |
| 4-Me—C₆H₄—C(Me₂)OOH | | | 1.53* | | 1.46* | 211[c] |
| 4-Me₂CH—C₆H₄—C(Me₂)OOH | | | 1.57* | | 1.53* | 211[c] |
| 4-Cl—C₆H₄—C(Me₂)OOH | | | 1.50* | | 1.47* | 213 |
| 3,4-Cl₂—C₆H₃—C(Me)₂OOH | | | 1.52* | | 1.50* | 213 |

[a] In CCl₄ solution; the asterisk indicates CDCl₃.
[b] Downfield from tetramethylsilane as internal reference.
[c] Data from reference 211 converted from hertz to parts per million for comparison.

## 2. Peroxy Acids (209)

Nuclear-magnetic-resonance spectral data obtained for the hydroxy proton and α-methylene protons of several peroxy acids and their nonperoxidic analogs are summarized in Table 9. The acidic proton resonance of peroxy acids appears as a broad singlet in the range of 11.90–11.75 ppm (tetramethylsilane = 0) and is shifted upfield relative to carboxylic acids.

TABLE 9. NMR SPECTRAL DATA FOR PEROXY ACIDS AND CARBOXYLIC ACIDS[a,b]

| | | Chemical Shift of Indicated Proton (ppm)[c] | | | | |
|---|---|---|---|---|---|---|
| | | Peroxy Acid | | Carboxylic Acid | | |
| Acid | Concentration (M) | —OO—H | α-CH₂ | —O—H | α-CH₂ (or C—H) | |
| Peroxybenzoic | | 11.75 | | 13.15 | | (Benzoic) |
| Peroxypelargonic | 0.67 | 11.20 | 2.40 | 11.84 | 2.35 | (Pelargonic) |
| | 0.067 | 11.37 | | 10.70 | | |
| Peroxylauric | 0.50 | 10.90 | 2.40 | 11.84 | 2.34 | (Lauric) |
| Cyclohexane-peroxycarboxylic | | 11.40* | 2.30* | 11.54* | 2.40* | (Cyclohexane-carboxylic) |

[a] Data from reference 209.
[b] In CCl₄ solution; the asterisk indicates CDCl₃.
[c] Downfield from tetramethylsilane as internal reference.

The Reid–Connor model (see preceding subsection) may be used to explain the chemical shift of the acidic proton in carboxylic acids and their cyclic dimers (212). Peroxy acids differ from carboxylic acids by their preference for intramolecular hydrogen bonding. The acidic proton, because of the skewed conformation, resides in the zone of positive shielding of the carbonyl group.

Dilution of peroxy acids has little effect on the chemical shift (Table 9). This result is consistent with its intramolecularly-hydrogen-bonded structure. On the other hand, the hydroxy-proton resonance of carboxylic acids shifts upfield as the concentration is reduced beyond mole fraction 0.25, this result indicating dimer dissociation. Below 2 mole % acid the dissociation is extremely sensitive to concentration (216, 217).

The chemical shifts of the acidic proton of typical carboxylic acids, peroxy acids, hydroperoxides, and alcohols are compared with the $pK_a$ values (Table 10) (209). The comparison shows a downfield shift of the hydrogen-resonance

TABLE 10.   CHEMICAL SHIFTS OF ACIDIC PROTONS AND $pK$ VALUES[a]

| Compounds | | Chemical Shift (ppm)[b] | $pK_a{}^c$ |
|---|---|---|---|
| Carboxylic acids | $(RCO_2H)$ | 11–13.5[d] | 3–6 |
| Peroxycarboxylic acids | $(RCO_3H)$ | 10.9–11.8 | 7–8 |
| Hydroperoxides | $(ROOH)$ | 7.6–9.2[d] | 11.6–12.8 |
| Alcohols | $(ROH)$ | 0.5–5.5[d] | ~16 |

[a] Data from reference 209.
[b] Downfield from tetramethylsilane as internal reference. Values obtained for approximately 10% solutions in $CCl_4$ or $CDCl_3$.
[c] Typical values.
[d] Resonances shift upfield on dilution due to changes in the equilibrium from dimer to monomer. Alcohol chemical-shift values are very sensitive to concentration changes.

signal with increasing acidity. The clear separation of proton resonances among these hydroxylic compounds establishes their diagnostic utility.

The α-methylene protons of aliphatic peroxy acids ($\delta = 2.40$ ppm) give rise to typical triplets. They are deshielded only slightly (about $-0.05$ Hz) in comparison with α-methylene protons in the corresponding carboxylic acids.

### 3. Dialkyl Peroxides

β-Methyl chemical shifts observed among several dialkyl peroxides (Table 11) are not greatly different excepting for a small deshielding effect shown by the aromatic ring in cumyl derivatives (209). The effect of solvent on the chemical shift of protons at the β-carbon has been already referred to (see Section A-1) as having some analytical value.

TABLE 11. PROTON CHEMICAL SHIFTS IN DIALKYL PEROXIDES, ROOR[a]

| | Proton Chemical Shift (ppm)[b] | |
| --- | --- | --- |
| Peroxide | Me<br>$\mid$<br>$-O-O-C-Me$<br>$\mid$<br>Me | $\mid$<br>$-O-O-C-Me$<br>$\mid$ |
| Di-*t*-butyl Peroxide | 1.17<br>1.18[c] | 1.12[c] |
| 2,5-Dimethyl-2,5-di(*t*-butylperoxy) hexane | 1.17 | 1.15 |
| 4-*t*-Butylperoxy-2,6-di-*t*-butyl-4-methylcyclohexa-2,5-dien-1-one | 1.19* | 1.32* |
| Dicumyl peroxide | | 1.53 |
| Ascaridole | | 1.31 |

[a] Data from reference 209.
[b] Downfield from tetramethylsilane as internal reference. Values obtained from CCl₄ solution; the asterisk indicates CDCl₃.
[c] Values for the nonperoxidic analog *t*-butyl isopropyl ether, obtained in CDCl₃.

The spectrum of ascaridole (13) is consistent with its established structure (209).

(13)

The peroxy group is believed to be exerting a small long-range deshielding effect on the vinyl protons (a doublet of doublets centered at 6.4 ppm for two hydrogens) on the basis that the nonperoxidic analog [2.2.2]bicyclooctene has this similar resonance centered at 6.26 ppm. Splitting of the ascaridole vinyl proton resonance arises from coupling with coupling constants of $J_{\alpha\beta}$ and with $J = 1$ Hz for the long-range effect of either $J_{\alpha\delta}$ or $J_{\beta\gamma}$.

### 4. t-Butyl Peroxy Esters (209)

The resonance line of $t$-butyl protons in $t$-butyl peroxyacetate (1.28 ppm) and $t$-butyl peroxybenzoate (1.36 ppm) shows a small upfield shift from the corresponding frequency in dialkyl ($t$-butyl type) peroxides and in $t$-butyl acetate (1.95 ppm). This is a useful characteristic for checking the chemical purity of these compounds. Methyl protons in acetate and peroxyacetate resonate at the same position (2 ppm).

### 5. Diacyl Peroxides (209)

The spectra of dibenzoyl peroxide and dilauroyl peroxide are identical with those of their nonperoxidic analogs, the anhydrides, with no unusual features in evidence. The $\alpha$-methylene protons of dilauroyl peroxide resonate at the same position as those in peroxylauric acid.

### 6. Oxyperoxides

Peroxides that are hydrogen peroxide adducts of aldehydes and ketones and other structurally related derivatives are conveniently grouped under this classification.

Ketal hydroperoxides, such as 1'-hydroxy-1'-hydroperoxydicyclohexyl peroxide (Table 8) undergo facile equilibrium in solution with other species, the formation of which complicates determination of the concentration dependency of OO—H and O—H protons (209). Compounds of this class have been reported to hydrogen bond intramolecularly in a five-membered configuration (206, 218, 219). The intramolecular hydrogen bond deshields the proton relative to the non-hydrogen-bonded form, causing it to suffer further deshielding effects by constraining the proton in the proximate region of the peroxide and hydroxy groupings. The corresponding nonhydroperoxidic derivative, 1,1'-dihydroxycyclohexyl peroxide, in deuterochloroform exhibits a weak, broad singlet at 9.78 ppm that is tentatively assigned to the hydroxy protons.

Seebach (205) prepared several oxyperoxides and a few analogous azaperoxides (206) by reacting their respective antecedents, 2,5-dimethylfuran (equation 1) and pentamethylpyrrole (equation 2) with hydrogen peroxide or $t$-butyl hydroperoxide. Several of these structures, a few examples of which are shown below (a–e), were deduced from their proton resonances.

(a, d, e: $H_2O_2$ reaction; b: *t*-BuOOH reaction; c: from a by benzoylation)

In a related manner, Schulz and co-workers (208) were able to elucidate the five- and six-membered-ring structures of sugar 1,2-peroxy esters by NMR analysis.

### 7. Cyclic Peroxides

Peroxides occur in two conformations in equimolar amounts of compounds **14** and **15**, which are optical antipodes.

Their rate of interconversion depends on the barrier to rotation about the O—O bond. The effects of this equilibrium were measured on three cyclic peroxides: 3,3,6,6-tetramethyl-1,2-dioxane (203, 204); 3,3,5,5-tetramethyl-1,2-dioxacyclopentane (204); and acetone diperoxide (1,1,4,4-tetramethyl-2,3,5,6-tetraoxacyclohexane) (207).

The interconversion between the isomers of tetramethyldioxane (**16** and **17**), studied by NMR, has been treated as a typical rate process.

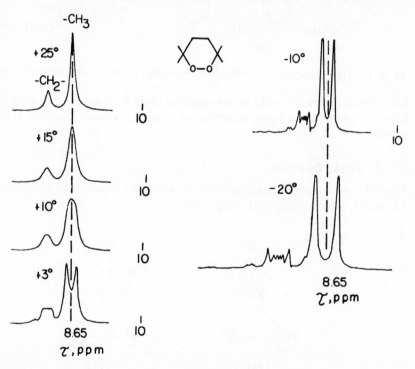

**(16)**                              **(17)**

At room temperature the NMR spectrum of tetramethyldioxane in carbon disulfide (Figure 35) (203) consists of two peaks, the rapid interconversion of the molecules being responsible for single methyl and single methylene proton peaks. A single peak results for the methyl groups since these are not coupled to other hydrogens. At lower temperatures the rate of the chair-to-chair interconversion is diminished, leading to proton resonances that are a doublet for

**Figure 35.** Proton magnetic resonance spectra at 60 Mc. sec$^{-1}$ of 3,3,6,6-tetramethyl-1,2-dioxane at various temperatures. (Reference 203).

methyl and a sextuplet for methylene. The relative chemical shift $(v_a - v_e)$ arises from the chemical nonequivalence of axial and equatorial methylene and methyl groups.

A simpler spectrum is obtained from acetone diperoxide (Figure 36) (204). Since this molecule contains only methyl groups, the two conformers result in

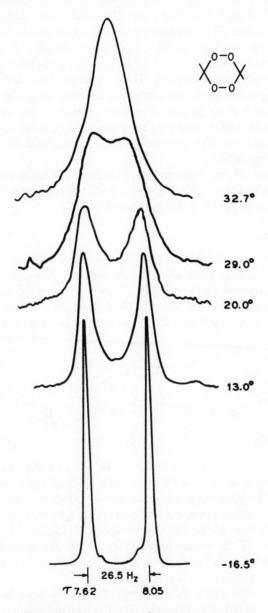

**Figure 36.** Proton magnetic resonance spectra at 60 Mc. sec$^{-1}$ of acetone diperoxide (1,1,4,4-tetramethyl-2,3,5,6-tetraoxacyclohexane) at various temperatures. (Reference 207).

two sharp absorptions, with a chemical-shift difference between axial and equatorial methyl groups of 0.44 ppm. The temperature dependence of the separation of the methyl resonances allows the activation energy for the inter-conversion to be calculated as 18.5 kcal/mole (frequency factor $4.0 \times 10^{16}$ $sec^{-1}$) for tetramethyldioxane and 12.3 kcal/mole (frequency factor $3.95 \times 10^{10}$ $sec^{-1}$) for acetone diperoxide. These values and the derivable thermo-dynamic parameters suggest that tetramethyldioxane molecules pass through a flexible transition state comprised of five coplanar atoms, whereas acetone diperoxide traverses an intermediate form possessing a greater degree of order than the starting form. The latter compound is considered to pass through a twist-boat intermediate.

The spectrum of the five-membered-ring tetramethyldioxacyclopentane shows no methyl nonequivalence down to $-40°$; this indicates that the barrier height is less than that for a six-membered ring.

### 8. *Ozonides*

For many years the primary ozonide obtained by the addition of ozone to a carbon–carbon double bond eluded isolation and structure confirmation.* The products expected for the primary addition of ozone are the molozonides of structures **18** (1,2,3-trioxolane) or **19** (Staudinger's molozonide) or a $\pi$-complex of structure **20** (220–222). The observation of glycol formation on

|     (18)     |     (19)     |     (20)     |

reduction (223, 224) eliminates structure **20** from consideration (222). Evi-dence establishing compound **18** as a stable product was reported a decade earlier (223–225). Recently its existence has gained support through an NMR examination of *cis*- and *trans*-molozonides in $CCl_2F_2$ solution at $-130°$, the *trans* isomer existing as the more stable species (226, 227). These results are shown in Table 12. At temperatures above $-130°$ the unstable molozonide decomposes to the usual ozonolysis products of normal ozonide, aldehyde, and oligomers, as shown in Table 12.

According to the widely held Criegee theory, the unstable molozonides dissociate to carbonyl and carbonyl oxide (Criegee zwitterion) fragments (**21**).

---

* Discussions on the mechanism of ozonolysis are available in reviews by Bailey (220) and Menyailo and Pospelov (221).

TABLE 12. NMR SPECTRAL DATA FOR *trans*- AND *cis*- ALKENE OZONATION MIXTURE[a,b]

| Alkene | Aldehyde CHO | Ozonide CH | Molozonide CH | Propanal CH₂ | Ethanal CH₃ | Methylene | Methyl |
|---|---|---|---|---|---|---|---|
| | | | | | | | |
| *trans:* | | | | | | | |
| | | | **−130° Spectra** | | | | |
| 2-Butene | 9.52 (d) | 5.00 (q), 4.90 (q)[d] | 4.12 (q) | | 1.98 (d) | 1.52 (m) | 1.15 (d) |
| 2-Pentene | 9.62 (s) | 4.95 (u) | 4.08 (m) | 2.38 (u) | 2.05 (d) | 1.60 (m) | 1.27 (d), 0.85 (t) |
| 3-Hexene | | | 4.07 (t) | | | | 0.90 (t) |
| | | | **−70° Spectra** | | | | |
| 2-Butene | 9.55 (d) | 5.07 (q) 4.98 (q)[e] | | | 1.97 (d) | 1.47 (u) | 1.13 (d) |
| 2-Pentene | 9.57 (s) | 4.98 (m)[e] | | 2.27 (q) | 1.95 (d) | 1.48 (m) | 1.22 (d), 0.82 (t) |
| 3-Hexene | 9.60 (s) | 4.93 (t), 4.87 (t)[e] | | 2.27 (q) | | | 0.80 (t) |
| | | | | | | | |
| *cis:* | | | | | | | |
| | | | **−130° Spectra** | | | | |
| 2-Butene | 9.50 (u) | 5.05 (q), 4.97 (q)[d] | 4.52 (u) | | 1.80 (d) | | 1.17 (d), 1.08 (d) |
| 2-Pentene | 9.65–9.56 (m) | 4.93 (m)[d] | 4.35 (m) | | | Overlapping multiplets 2.02–0.42 | |
| 3-Hexene | | 4.95 (t), 4.87 (t)[d] | 4.53 (u) | | | 1.48 (u) | 0.80 (u) |
| | | | **−50° Spectra** | | | | |
| 2-Butene | | 5.17 (q), 5.10 (q)[e] | | | 1.93 (d) | | 1.23 (d) |
| 2-Pentene | | 5.70–4.80 (m)[e] | | | | 2.27–1.30 (m) | 1.27 (d), 0.87 (t) |
| 3-Hexene | | 5.05 (t), 5.00 (t)[e] | | | | 1.58 (m) | 0.88 (t) |

[a] Data from references 226 and 227.

[b] Authentic ozonides of 2-butene and 3-hexene, authentic 3-hexene ozonation oligomer, and aldehydes were used for peak assignments.

[c] All δ values related to internal tetramethylsilane = 0.00. CHCl₂F was used as an internal standard. Abbreviations: d, doublet; m, multiplet; q, quartet; s, singlet; t, triplet; u, unresolved.

[d] Oligomer evidenced by weak absorption on low-field side of peak.

[e] Marked absorption, which disappeared on recooling the mixture to −130°, on the low-field side of the peak indicated considerable oligomer.

The zwitterions recombine to produce the normal ozonide (1,2,4-trioxolane, **22**) as the stable isolatable product.

$$15 \longrightarrow \begin{array}{c} R^1 \\ \phantom{} \\ R^3 \end{array}\!\!\overset{+}{C}-O-\overset{-}{O} \; + \; \begin{array}{c} R^2 \\ \phantom{} \\ R^4 \end{array}\!\!C{=}O \longrightarrow \begin{array}{c} R^1 \cdots \\ \phantom{} \\ R^3 \end{array}\!\!C\!\!\begin{array}{c} O-O \\ \phantom{} \\ O \end{array}\!\!C\!\!\begin{array}{c} ..R^2 \\ \phantom{} \\ R^4 \end{array}$$

$$(21) \hspace{5cm} (22)$$

In hydroxylic solvents other products are formed, but these are not germane to this discussion.

Riezebos and co-workers (176) examined the stable *cis*- and *trans*-ozonides isolated from methyl oleate ozonizations. They observed no difference in the spectra of the methine protons in the two isomeric forms (chemical shift of 5.1 on the $\delta$-scale). Subsequent workers (174, 228, 229), on examination of this crucial region in the spectra of ozonides prepared from olefins, observed overlapping multiplets between $\tau = 4.6$ and $\tau = 5.2$ ppm. These observations are assembled in Table 13.

Fliszár et al. (230) observed the methine resonance in *cis*- and *trans*-stilbene ozonides at $\tau = 3.79$ and $3.83$, respectively. The assignment was based on the correlation of a downfield resonance for *cis* found in aliphatic ozonides. Bishop et al. (229) reported that the relative order of these assignments for the stilbene ozonides were in error, correcting the order by a reversal of assignments, as reported in Table 13.

The low-temperature technique was used in proving that ozonization of dimethyl anthracene yields 9,10-transannular ozonide (231). Kolsaker (232) isolated the *cis*- and *trans*-ozonides of methyl *p*-methoxy cinnamates (**23**), which contain an electron-withdrawing group attached directly to the ozonide ring.

$$\text{MeO} - \underset{\underset{H \quad H}{}}{\overset{\overset{H \quad H}{}}{\bigcirc}} - \text{HC}_\beta \underset{O-O}{\overset{O}{\diagdown}} \text{CH}_\alpha\text{COOMe}$$

$$(23)$$

No spin–spin interaction between the two ring protons is observed. This characteristic was interpreted as a difference in conjugative power of the normal ozonide ring relative to the epoxide ring and to that postulated for the molozonide. The difference in chemical shifts of the $\beta$-protons between isomeric ozonides was used to determine the configuration of the isomers.

TABLE **13.** METHINE-PROTON SHIFTS IN NORMAL OZONIDES[a]

| Carbon Number | Ozonide | $\tau$ (Multiplicity)[b] | | References |
|---|---|---|---|---|
| 4 | (ozonide structure) | *trans* | 4.75 (q) | 174, 228 |
| | | *cis* | 4.70 (q) | |
| 5 | (ozonide structure) | *trans* | 4.7–5.15 (m) | 174, 228 |
| | | *cis* | 4.79 (q), 4.95 (t) | |
| 6 | (ozonide structure) | *trans* | 5.04 (t) | 174, 176, 228 |
| | | *cis* | 5.02 (t) | |
| 6 | (ozonide structure) | *trans* | 4.68–5.05 (m) | 228 |
| | | *cis* | 4.68–5.16 (m) | |
| 8 | (ozonide structure) | *trans* | 5.25 (d) | 228 |
| | | *cis* | 5.23 (d) | |
| 7 | (ozonide structure) | *trans* | 4.9 (q), 5.3 (s) | 228 |
| | | *cis* | 4.79, (q) 5.32 (s) | |
| 8 | (ozonide structure) | *trans* | 5.08 (t), 5.35 (s) | 228 |
| | | *cis* | 5.01 (t), 5.35 (s) | |
| 10 | (ozonide structure) | *trans* | 5.37 (s) | 228 |
| | | *cis* | 5.3 (s) | |
| 6 | (ozonide structure) | *trans* | 4.87 (q), 5.19 (d) | 228 |
| | | *cis* | 4.78 (q), 5.20 (d) | |
| 14 | 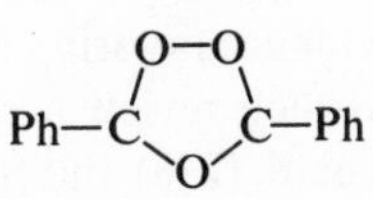 Ph—C(O—O / O)C—Ph | *trans* | 3.79 | 229, 230[c] |
| | | *cis* | 3.83 | |

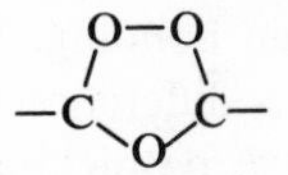

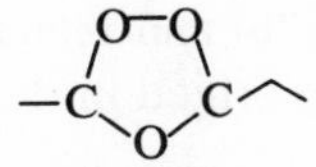

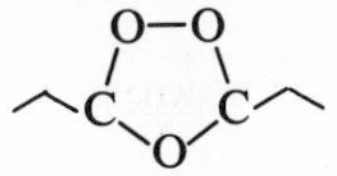

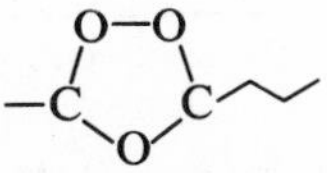

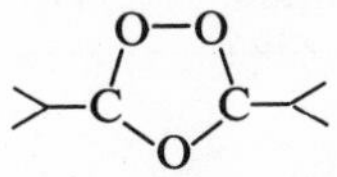

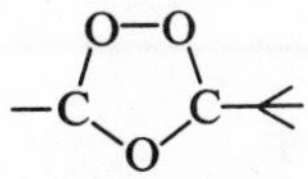

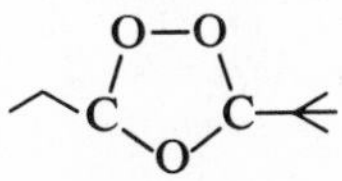

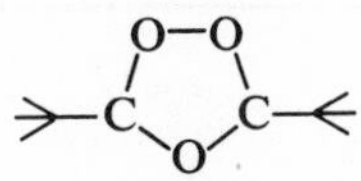

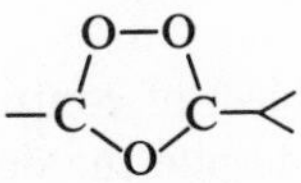

[a] See reference 228 for assignments of other protons.
[b] Abbreviations are as follows: q, quartet; t, triplet; d, doublet; s, singlet; m, multiplet.
[c] Fliszár et al. (230) report reversed assignments (see text), which were corrected by Story et al. (229).

## B. Peroxyfluorides and Fluorocarbon Peroxides

The NMR spectral characteristics of some oxygen fluorides of the general formula $O_xF_2$ ($x = 1$–$6$) (158, 233) are given in Table 14. The $^{19}F$ chemical shifts are externally referenced to trichlorofluoromethane. An intriguing characteristic of these spectra is the large shift to low field. The large shift in the fluoroperoxides, as clearly noted in $O_2F_2$ (234) as compared with $OF_2$, indicates a great difference in the fluorine–oxygen bonding. Nebgen, Metz, and Rose (235) suggested an explanation for this difference in $O_2F_2$. They propose that in this compound there is extensive deshielding, which they associate with the observed long O—F bond (236) and relate to overlap of the fluorine atom $p$-orbital with the antibonding orbital in the oxygen molecule.

TABLE 14.   $^{19}F$ AND $^{17}O$ NMR DATA FOR OXYGEN FLUORIDES AND COMPARATIVE SUBSTANCES

| Compound | $^{19}F$ Chemical Shift Relative to $CCl_3F$ (ppm)[a] | $^{17}O$ Position | $^{17}O$ Chemical Shift Relative to $H_2{}^{17}O$ (ppm)[b] |
|---|---|---|---|
| $F_2$ (77°K) | −422 | | |
| $OF_2$ (77°K) | −249 | $^{17}OF_2$ | −830[a] |
| $(O_2F)_n$ | | ? | −971 |
| | | | −1512 |
| $O_2F_2$ (145°K) | −865 | $^{17}O_2F_2$ | −647[c] |
| $O_3$ | | $O-^{17}O-O$ | −1032 |
| | | $^{17}O-O-^{17}O$ | −1598 |
| $O_3F_2$ ( <145°K) | −877 | | |

[a] Reference 235.
[b] Reference 237.
[c] Triplet.

The structure of "$O_3F_2$" is currently the subject of controversy. Its existence as a molecular entity remains uncertain despite the demonstration of definable physical properties. Arguments and evidence favoring any of several possible structures will not be repeated here, but their review is recommended in the detailed expositions presented by Nebgen et al. (238) and Solomon et al. (237). Solomon et al. (237) had also investigated the $^{17}O$ resonance lines from "$O_3F_2$" and from several reference materials in order to probe the controversial structure (Table 14).

Fluorine-magnetic-resonance spectral data for a variety of unusual fluorocarbon peroxides are assembled in Table 15 (239, 240). In the perfluorodioxide derivative the $CF_3$ group resonance is split by fluorine nuclei across

TABLE 15.   [19]F NMR SPECTRAL DATA FOR FLUORINATED PEROXIDES

| Compound | Chemical Shift)[a] (Multiplet) | Coupling Constants (Hz) |
|---|---|---|
| I[b,c]  $\overset{O}{\underset{\parallel}{FC}}-O-O-\overset{O}{\underset{\parallel}{CF}}$ | 34.1 (s) | |
| II[b]  $\underset{\alpha}{F_3C}-O-O-\overset{O}{\underset{\parallel}{C}}\underset{\beta}{F}$ | $\alpha = 69.5$ (d) $\beta = 32.4$ (q) | $J_{\alpha\beta} = 3.4$ |
| III[b,d]  $F_3C-O-O-CF_3$ | 69.0 (s) | |
| IV[b]  $(F_3C-O-O-)_2CO$ | 69.6 (s) | |
| V[b]  $F_3C-O-O-H$ | 70.5 (s) | |
| VI[b]  $\underset{\alpha}{F_3C}-O-O-\underset{\beta}{CF_2}-O-\underset{\gamma}{F}$ | $\alpha = \phantom{0}69.0$ (d, t) $\beta = \phantom{0}80.6$ (d, q) $\gamma = -156.7$ (t, u) | $J_{\alpha\beta} = 3.4$ $J_{\alpha\gamma} = 1.4$ $J_{\beta\gamma} = 35.1$ |
| VII[b,e]  $(\underset{\alpha}{F_3C}-O-O-)_2\underset{\beta}{CF}-O-\underset{\gamma}{F}$ | $\alpha = \phantom{0}68.7$ $\beta = \phantom{0}90.6$ $\gamma = -168.0$ | $J_{\alpha\beta} = 3.5$ $J_{\alpha\gamma} < 3$ $J_{\beta\gamma} = 2.5$ |
| VIII[f]  $\underset{\alpha}{F_3C}-O-O-O-\underset{\beta}{CF_2}-\underset{\gamma}{CF_3}$ | $\alpha = 68.7$ (s) $\beta = 96.4$ (q) $\gamma = 83.8$ (t) | $J_{\alpha\beta} < 1$ $J_{\beta\gamma} = 1.5$ |
| IX[f]  $\underset{\alpha}{F_3C}-O-O-\underset{\beta}{CF_2}-\underset{\gamma}{CF_3}$ | $\alpha = 68.7$ (t) $\beta = 95.7$ (q,q) $\gamma = 83.2$ (t) | $J_{\alpha\beta} = 4.3$ $J_{\beta\gamma} = 1.5$ |
| X[f]  $\underset{\alpha}{F_3C}-O-\underset{\beta}{CF_2}-\underset{\gamma}{CF_3}$ | $\alpha = 56.2$ (t) $\beta = 91.2$ (q, q) $\gamma = 87.6$ (t) | $J_{\alpha\beta} = 9.2$ $J_{\beta\gamma} \sim 2.2$ |

[a] In $\phi^*$ units, defined as $10^6(H_S - H_{CCl_3F})/H_{CCl_3F}$, using $CCl_3F$ as solvent and reference in accordance with the standard technique of Filipovich and Tiers (241).
[b] Reference 240.
[c] Fox and Franz (242) report $\phi^* = 34.4$.
[d] Fluorine absorptions of the $CF_3OO$ group in a series of perfluoroalkyl trifluoromethyl peroxides and trioxides are observed in the region of $\phi^* = 68–69$. See Table 16 and reference 239.
[e] All peaks were overall doublets with unresolved fine structure.
[f] Reference 239.

an intervening pair of oxygen atoms, but the effect is not transmitted across three intervening oxygen atoms in $CF_3OOOCF_2CF_3$. The perfluoroether derivative serves as a reference model for these compounds. Data on numerous ethers, peroxides, and trioxides are summarized in Table 16 (239). Asym-

TABLE 16.   $^{19}$F NMR STRUCTURE DETERMINATION OF $R_FO_{1-3}R_F'^{a,b}$

| Compounds | Chemical Shift[c] | | | (Hz) |
| | $CF_3O-$ | $C\underline{F}_3CF_2O-$ | $CF_3C\underline{F}_2O-$ | $\|\leftarrow J\rightarrow\|$ $-CF_2O_nCF_2$ |
|---|---|---|---|---|
| Ethers    $(O_n = 1)$ | 55–58 | 87–88 | 88–91 | 9–10 |
| Peroxides $(O_n = 2)$ | 68–69 | 83–84 | 96–98 | ~4 |
| Trioxides $(O_n = 3)$ | 68–69 | 83–84 | 96–98 | <1 |

[a] Data from reference 239.
[b] $R_F$ refers to a perfluoroalkyl substituent.
[c] In $\phi^*$ units defined as $10^6(H_S - H_{CCl_3F})/H_{CCl_3F}$, using $CCl_3F$ as solvent and reference in accordance with the standard technique of Filipovich and Tiers (241).

metrical derivatives like $CF_3O_{1-3}C_2F_5$ are readily distinguished by the appropriate F—F' coupling constants, as shown in column 5. The coupling across the oxygen atoms is probably 0.3 Hz for trioxides and smaller still for tetroxides. Thompson (239) recommends this approach for determining the number of oxygens between perfluoroalkyl groups as an alternative to analysis of splitting patterns of the weak $^{13}$C satellites of the symmetrical homologs (e.g., $CF_3O_{1-3}CF_3$). In such cases the F—F' coupling constants are not directly observable due to the inherent symmetry of the molecule.

## C.   Chemically Induced Dynamic Nuclear Polarization

Nuclear-magnetic-resonance spectroscopy is an elegant instrumental technique for penetrating the chemical problems of free radicals and their reactions. These include the electronic structures of free radicals, their reactivities, their rate processes involving zero activation energy, and their specific interactions (243). A review of the pertinent publications appropriate to the subject (243) emphasizes investigations carried out on stable radicals. An intriguing recent development is the use of NMR for probing the reactions of radicals generated from organic peroxides and azo compounds by thermal and photochemical methods (244–247) or from the reaction of alkyl halides with lithium alkyls (248–250). The phenomenon is called "chemically induced dynamic nuclear polarization" (CIDNP) (247). The initial experimental investigations on this phenomenon were instituted by Ward and Lawler (248–250) and

Fischer and Bargon (244–246). The latter pair of investigators have also published a convenient introductory review of their own work (247).

During rapid free-radical reactions emission and/or enhanced absorption lines appear in $^1$H NMR spectra, becoming normal absorption lines at the end of the reaction. A simple example characteristic of the emission phenomenon is displayed in the spectra for dibenzoyl peroxide decompositions (Figure 37) (247). The protons responsible for the abnormal resonances show the effects of polarization, which are associated only with the *products* of the reaction.

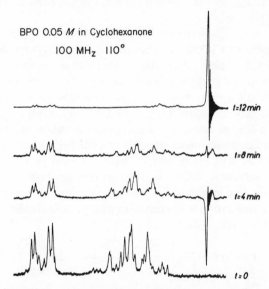

**Figure 37.** Proton magnetic resonance spectra during thermal decomposition of dibenzoyl peroxide in illustration of the chemically induced dynamic nuclear polarization phenomenon. (Reference 247).

A brief and qualitative explanation of CIDNP will suffice, because the literature may be referred to for the formal mathematical arguments (244–250). In the free radical there are two dynamically coupled spin systems, electronic and nuclear (proton). During the free-radical reaction a change in the spin state of one system produces a change in the spin state of the other system. The decomposition of a radical source generates a pair of radicals with antiparallel electron spins. Since the radicals are produced with their spin states equally populated, spin-state transitions $(+\frac{1}{2} \rightarrow -\frac{1}{2})$ will occur, producing a set of populations governed by the usual (electron spin) Boltzmann distribution. Since the electronic and nuclear spins are coupled, these transitions will produce corresponding transitions in the nuclear spin system. The type of coupling dictates the subsequent behavior of the nuclear spins.

For scalar coupling the electronic spin transition $(+\frac{1}{2} \rightarrow -\frac{1}{2})$ induces a nuclear spin transition $(-\frac{1}{2} \rightarrow +\frac{1}{2})$. The population of the lower nuclear spin state and hence the intensity of the resulting NMR signal are thereby enhanced.

For dipolar coupling the electronic spin transition $(+\frac{1}{2} \rightarrow -\frac{1}{2})$ induces primarily the nuclear spin transition $(+\frac{1}{2} \rightarrow -\frac{1}{2})$. In this case the nuclear spin state of higher energy becomes more populated than the nuclear spin state of lower energy and stimulated emission of the product molecules takes place at the resonance frequency.

Several acyl peroxide decompositions have been investigated as examples of the application of CIDNP. Fischer and Bargon (247) studied the decomposition of dibenzoyl peroxide at 100° in cyclohexanone solvent. Solutions of the peroxide prepared at room temperature were transferred to the preheated probe of the NMR spectrometer, and the spectra were repeatedly recorded. Before decomposition (Figure 37) $(t = 0)$ a normal spectrum is observed for dibenzoyl peroxide protons. As decomposition proceeds $(t \simeq 4 \text{ min})$, the emission line attains maximum intensity; on further reaction $(t \simeq 7 \text{ min})$ the intensity decreases, and the resonance eventually reappears as an absorption line of normal intensity. On the basis of its line position and by analysis of the products, the line is assigned to benzene protons. In accordance with the decomposition in equation 4, the transient phenyl radicals abstract a hydrogen atom from the solvent:

where X represents the solvent.

The introduction of paramagnetic ions reduces the effect of dynamic nuclear polarization by increasing the rate of relaxation of the free-radical electron spins, the emission decreasing with increasing ion concentration (247). In this manner Fischer and Bargon (247) were able to follow the absorption amplitude of benzene derived from dibenzoyl peroxide decompositions without the complications of emission. They determined an experimental enhancement factor ($V_{exp}$) for the CIDNP:

$$V_{exp} = \frac{A^* - A^0}{A^0}$$

where $A^0$ is the absorption amplitude (more correctly, intensities) of the nonemitting benzene at the observed time for the emission maximum and $A^*$ is the emission-amplitude maximum. Their values of $V_{exp}$ determined at various operating frequencies of the NMR spectrometer are $-2.9$ at 100 MHz, $-8.5$ at 56.4 MHz, and $-20$ at 40 MHz. Similar results were obtained for chlorobenzene from chlorobenzoyl peroxide decompositions.

Diacetyl peroxide decomposed in dimethyl phthalate at 100° to give a single emission line (247, 251). Higher acyl peroxides showed resonances with emission and enhanced absorption character, arising in some cases from olefins derived from disproportionation reactions of the transient alkyl radicals:

$$2RCH_2CH_2\cdot \longrightarrow RCH{=}CH_2 + RCH_2CH_3 \tag{5}$$

Radicals generated from benzoyl peroxide and lauroyl peroxide decompositions in *o*-dichlorobenzene solutions show evidence of CIDNP during their iodine-atom abstraction reactions with isopropyl iodide (252). For benzoyl peroxide decompositions polarization is observed in the *ortho* protons of iodobenzene and the methyl protons of propane. The methane septuplet of isopropyl iodide ($\delta = 4.34$) shows both emission and enhanced absorption. The vinyl methylene protons of propene ($\delta = 5.14$) give rise to only enhanced absorption, and the vinyl methane protons ($\delta = 6.0$), only emission.

The undecyl radical from lauroyl peroxide abstracts the iodine atom from isopropyl iodide. CIDNP is observed in the methylene protons of the methylene iodide group of *n*-undecyl iodide ($\delta = 3.2$), the methine proton of isopropyl iodide ($\delta = 4.34$), and in the vinyl methylene ($\delta = 5.16$) and methine protons ($\delta = 6.0$) of the product 1-undecene. CIDNP thus furnishes direct evidence for an iodine-atom transfer between identical alkyl moieties; that is, the transfer of an iodine atom from isopropyl iodide to the isopropyl radical.

It is to be noted and emphasized that the developments rapidly taking place in photochemistry will be assisted even more effectively by the observation of strong nuclear polarizations resulting from photochemically initiated reactions proceeding through triplet-state intermediates (253).

## V. ELECTRON SPIN RESONANCE

Peroxides are a convenient source of free radicals, which are generally nonperoxidic in nature and thus distinguishable from intermediate peroxy radicals obtained by autoxidation processes. Methods capable of distinguishing among different radical types may then provide detailed data about molecular structure and kinetics. Since free radicals are characterized by the presence of an unpaired electron, they are detectable by a magnetic-resonance technique, called electron-spin-resonance (ESR) or electron-paramagnetic-resonance (EPR) spectroscopy, which allows investigations of their interactions with the molecular environment. A brief review of EST theory is presented to provide useful background information to the discussion on peroxy- and peroxide-generated free radicals.

## A.  Brief Introduction to ESR Theory (254, 255)

An electron possesses a spin and in association with the spin a magnetic moment. In a uniform magnetic field the electron spin is allowed only parallel or antiparallel orientations to the field. The energy separation $\Delta E$ of the corresponding energy levels is equal to $g\beta H$, where $g$ is the proportionality constant characteristic of the spin system, $\beta$ in the Bohr magneton, and $H$ is the strength of the external magnetic field. Transitions between the energy levels are induced by electromagnetic radiation at the frequency $v$ for the resonance condition

$$\Delta E = hv = g\beta H$$

where $h$ is Planck's constant. The signal for a free electron is a single line of energy absorption, corresponding to a change in direction of the electron magnetic moment.

The electron spin in the presence of magnetic nuclei is perturbed by interaction with the nuclear spins. This gives rise to hyperfine splitting (hfs) of the electron-spin energy levels. (The hyperfine-splitting constant is usually designated by $a$ or $A$, in gauss or oersteds.) Since electron magnetic moments are approximately 1000 times larger than nuclear magnetic moments, the hyperfine splitting is a small fraction of the total electron-spin energy. The unpaired electron spin is allowed two orientations in a magnetic field ($M_s = \pm\frac{1}{2}$), and each nuclear spin is allowed $2I + 1$ orientations [$M_s$ and $M_I$ are the quantum numbers for the $z$-components of the electron spin and nuclear spin ($I$), respectively]. The coupling between them gives rise to $2I + 1$ resonance lines of equal intensity. The value of $I$ is $\frac{1}{2}$ for $^1H$, $^{13}C$, $^{15}N$, and $^{19}F$, and 1 for $^2H$ and $^{14}N$. The simple example of an unpaired electron in association with a nucleus ($I = +1$) is illustrated in Figure 38.

A more complicated spectrum is obtained for the unpaired electron in the presence of several nuclei because of interaction with each nucleus. When the coupling nuclei are all equivalent, $n$ nuclei contribute a total of $2nI + 1$ lines. When more than one group of equivalent nuclei are present, the total number of lines is given by the product of the number of lines appropriate to each group of equivalent nuclei—that is, $(2n_1I_1 + 1)(2n_2I_2 + 1)$, where $n_1$ is the

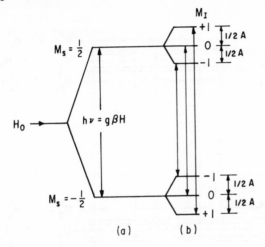

**Figure 38.** Splitting of electron energy levels by a magnetic field $H_0$ for electron spin resonance. (a) Effect of electron spin only : (b) effect of adding a nucleus with spin $I = 1/2$. Arrows indicate the allowed transitions; A is the hyperfine splitting constant.

number of equivalent nuclei with spin $I_1$, $n_2$ is the number of equivalent nuclei with spin $I_2$, and so on. Figure 39 shows the energy states and correlation diagram for an electron-spin interaction with one nitrogen ($^{14}N$) nucleus and one hydrogen ($^1H$) nucleus, giving rise to six hyperfine-splitting components. The intensities of the lines from interaction with one nucleus are all equal. The intensities from interaction with $n$ equivalent spin $\frac{1}{2}$ nuclei are obtained from Pascal's triangle—that is, the coefficients in the expansion of $(1 + x)^n$. The hyperfine-splitting components are disposed symmetrically about the center of the spectrum.

Free radicals in ESR are identified by their characteristic hyperfine splittings, their $g$-factor values, and their relaxation times. The nature of the medium and the temperature affect these characteristics. The hyperfine splittings of free radicals provide the most information in ESR spectra. Line shapes, which may be Gaussian or Lorentzian, and line widths are determined by the nature and extent of the interactions between the spin system and the environment. The $g$-factor, a dimensionless constant, is the next most important parameter. It is particularly useful for free radicals that have no hyperfine structure, such as peroxy free radicals ($RO_2\cdot$). It is calculated from the resonance frequency as given by the center of the ESR spectrum, using the equation $h\nu = g\beta H$. Its value for the free electron is 2.002319, which is the usual point of reference for organic free radicals. The $g$-factors for free radicals are generally in the

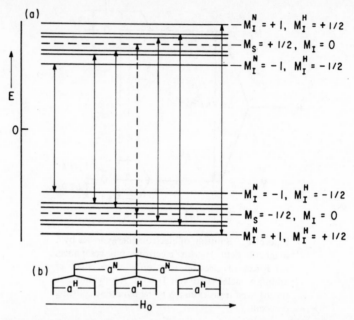

**Figure 39.** (a) Splitting of electron energy levels by a magnetic field $H_0$ for electron spin resonance showing for example the effect of adding two nuclei $^{14}N$ and $^{1}H$. Energy transitions are allowed for $\Delta M = 0$, $\Delta M_s = 1$; hyperfine splitting constant ($a^N$, $a^H$) g-value, and line width are defined. (b) A simplified correlation diagram of the transitions. (Patterned after Reference 256).

vicinity of 2.0. Differences in the structure and charge of the free radical are reflected in differences in $g$-factors.

Anisotropic interactions (interactions that depend on the orientation of the radical) are encountered when the crystalline electric field lacks spherical symmetry. In this case the interaction of the unpaired electron with the external magnetic field and with magnetic nuclei in the molecule are described by tensor relationships. The $g$-factor tensors and their corresponding hyperfine-splitting-constant tensors are conveniently described by their three principal values. For axial symmetry with $z$ taken as the axis of symmetry,

$$g_z = g_{\parallel}; \qquad g_x = g_y = g_{\perp}$$
$$A_z = A_{\parallel}; \qquad A_x = A_y = A_{\perp}$$

Rapid tumbling in liquids, in gases, and in the crystals, depending on the temperature, averages out all anisotropic interactions and leads to the isotropic or average $g$-factor, with sharp line spectra often being the result. The isotropic values are derived by substituting the above values into the relationships

$$g_{iso} = \frac{g_\| + 2g_\perp}{3}; \qquad A_{iso} = \frac{A_\| + 2A_\perp}{3}$$

For anisotropic interactions the hyperfine-splitting constant of the radical will depend on its orientation to the magnetic field. In this case the ESR transition frequencies for systems involving a single nucleus are a function of the angle between the magnetic field and the symmetry axis.

## B.  $HO_2\cdot$ and $HO\cdot$ Radicals

### 1.  *Genesis of Radicals by Irradiation*

Free radicals produced by ultraviolet (257, 258), electron beam (259), or nuclear (260) irradiation of hydrogen peroxide solutions can be stabilized and detected when generated in a solid matrix at low temperatures (258). The type of radicals produced, whether $HO_2\cdot$ or $HO\cdot$, and their characteristics have attracted a variety of ESR investigations since the initial studies by Ingram and co-workers (257).

The perhydroxy free radical ($HO_2\cdot$) is detected in a matrix when formed by irradiation of hydrogen peroxide in water below 145°K (259) or in argon at 4.2°K (259, 261). The ESR spectrum of the $HO_2\cdot$ radical in a glassy hydrogen peroxide–water matrix is illustrated in Figure 40; good resolution is obtained

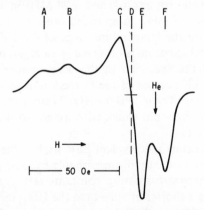

**Figure 40.** The first derivative of the ESR absorption spectrum of $HO_2\cdot$ in a glassy $H_2O_2$—$H_2O$ sample, recorded at a microwave frequency of 9164.0 $MH_z$. $H_e$ is the magnetic field strength of the free electron; the magnetic field strength H increases in the direction from left to right. g values and hyperfine splittings were measured at values of the magnetic field indicated at positions A–F. (Reference 259).

on careful annealing after irradiation at 77°K. The parameters for the $g$-tensor and the hyperfine-splitting tensor $A$ are recorded in Table 17. Similar principal

TABLE 17.    PRINCIPAL COMPONENTS OF $g$-TENSOR AND HYPERFINE-
SPLITTING TENSOR FOR $HO_2\cdot$ AND $DO_2\cdot$[a]

| Radical | $g_1$ | $g_2$ | $g_3$ | $|A_1|/h$ | $|A_2|/h$ | $|A_3|/h$ |
|---------|-------|-------|-------|-----------|-----------|-----------|
| | | | | MHz[b] | | |
| $HO_2\cdot$ in $H_2O$ (77°K) | 2.0353 | 2.0086 | 2.0042 | $-39.2$ | $-10.0$ | $-43.9$ |
| $DO_2\cdot$ in $D_2O$ (77°K) | 2.0344 | 2.0086 | 2.0031 | $-6.0$ | $-1.5$ | $-6.5$ |
| | $g_\| = g_1$ | $g_\perp = \frac{1}{2}(g_2 + g_3)$ | | | | |
| $HO_2\cdot$ in argon (4.2°K) | 2.0393 | 2.0044 | | $-37.8$ | $-7.0$ | $-40.9$ |

[a] Data from reference 259.
[b] Negative sign for hyperfine splitting is a theoretical assignment.

$g$-factors for the $HO_2\cdot$ radical were also obtained after radical formation by gamma or ultraviolet irradiation of a single crystal of hydrogen peroxide–urea addition complex (262). These values imply that 29 % of the unpaired-electron density is located on the central oxygen atom. The isotropic hyperfine-splitting constant of the $HO_2\cdot$ radical is $-31$ MHz, as compared with $-75$ MHz for the $HO\cdot$ radical in water vapor (263) and $\pm 74$ MHz in water ice (264).

The same spectrum of $HO_2\cdot$ in annealed samples of hydrogen peroxide–water is obtained over the temperature range of 4.2–145°K (259). A similar but not quite identical spectrum is observed in an argon matrix at 4.2°K (259, 261). The shape of the spectrum in the water matrix indicates restricted rotation of the radicals because of the formation of hydrogen bonds, which are not present in argon. The $g_1$ axis for $HO_2\cdot$ is along the O—O axis, whereas $g_3$, which is near the free-spin value, corresponds to the axis perpendicular to the molecular plane.

On the basis of separate independent ESR studies, the $HO\cdot$ (263–265) and the $HO_2\cdot$ radicals are evidently distinguishable by their $g$-factors (Tables 17 and 18) isotropic hyperfine-splitting constants (see preceding subsection). The $HO\cdot$ radical has a shorter lifetime than the $HO_2\cdot$ radical (266, 267). The ESR spectrum of the $HO_2\cdot$ radical disappears above 145°K in water (259) and nearly vanishes at 200°K when the radical is embedded in the urea complex (262).

Despite the $HO_2\cdot$ radical's fleeting existence at higher temperatures, Livingston and Zeldes (268) were able to measure the radical in a flow system patterned on Dixon and Norman's method (269). By this technique they obtained

TABLE **18.** EXPERIMENTAL AND THEORETICAL SPIN-HAMILTONIAN
PARAMETERS FOR THE HYDROXY RADICAL

| Parameter | HO· in Ice[a] | HO· in Vapor[b] | Theoretical[c] |
|---|---|---|---|
| $g_{\parallel}$ | $2.0615 \pm 0.010$ | | 2.0631 |
| $g_{\perp}$ | $2.0095 \pm 0.005$ | | 2.0077 |
| $A_z$ | $7 \pm 5$ Oe | $5 \pm 0.3$ Oe | 5.019 Oe |
| $A_x = A_y$ | $-43 \pm 5$ Oe | $-42.6$ Oe | $-38.87$ Oe |
| $A_{iso}$ | $-26.3 \pm 5$ Oe | $-26.7$ Oe | $-24.24$ Oe |

[a] Data from reference 264.
[b] Data from reference 263.
[c] Data from reference 265.

a singlet ($g$ value 2.015) for $HO_2\cdot$ in hydrogen peroxide solution under ultra-violet irradiation. The resonance was attributed to both $HO_2\cdot$ and to the ionic $O^{2-}\cdot$ radical, whose formation depends on the acidity of the medium. Although formation of HO· radical pairs is the primary reaction in solution, their high reactivity contraindicates their presence and only $HO_2\cdot$ radicals are observed. The reaction is formulated as follows:

$$HO\cdot + H_2O_2 \longrightarrow H_2O + HO_2\cdot \qquad (6)$$

The HO· radical is formed, but it is not measured by ESR in alcohol solutions purged free of dissolved oxygen; the only free radicals observed are those formed by the abstraction of a hydrogen alpha to the hydroxy group:

$$HO\cdot + RCH_2OH \longrightarrow H_2O + R\dot{C}HOH \qquad (7)$$

In analogous experiments with oxygen-saturated solutions the paramagnetic species that are found are attributed to the organic peroxy radicals generated by the addition of oxygen to the hydroxyalkyl radical normally present in oxygen-free media:

$$R\dot{C}HOH + O_2 \longrightarrow \begin{array}{c} R-CH-OH \\ | \\ O_2\cdot \end{array} \qquad (8)$$

### 2. *Radicals from Oxidation–Reduction Reactions*

Electron-spin-resonance spectra of short-lived free radicals derived by heavy-metal reduction of hydrogen peroxide can be obtained in a flow system on passage of the sample through the cavity cell of the spectrometer (269). These radicals may abstract hydrogen from suitable substrates to generate transient radicals of lower stability (266). Several studies on systems involving titanium (266, 269–276), iron (266, 275, 276), and cerium (275, 277) have been described.

The $Ti^{3+}$–hydrogen peroxide system has been closely examined by two groups (266, 274). Florin, Sicilio, and Wall (274) examined the ESR spectra of paramagnetic species in a flow system, relating the observed kinetics to generation and disappearance schemes. Their conception of the most likely free-radical structure favors the titanium peroxy radical complex $Ti-O-O\cdot$, proposed by Turkevich et al. (278) and Fischer (279). The radical is basically an ionized $HO_2\cdot$ radical in a complex form. Takakura and Rånby (266) succeeded in obtaining spectra with two well-resolved peaks, one with $g = 2.0140$ (low field) and the second with $g = 2.0125$ (2.5 gauss higher), attributed to the $HO_2\cdot$ and $HO\cdot$ radicals, respectively. The relative ratio of the two peaks depends on acidity and the $H_2O_2:Ti^{3+}$ ratio in the reaction system. The evidence indicates that their coordination with $Ti^{4+}$ ensures their survival for detection by ESR. These radicals decay less rapidly than the secondary radicals derived by their hydrogen abstractions from organic substrates when present in the system. Mechanistically the birth and demise of the $HO\cdot$ radical are analogous to the Haber–Weiss scheme (280, 281) for Fenton-reagent reactions:

$$Ti^{3+} + H_2O_2 \longrightarrow Ti^{4+} + HO^- + HO\cdot \tag{9}$$

$$HO\cdot + Ti^{3+} \longrightarrow HO^- + Ti^{4+} \tag{10}$$

At appropriate acidities and $H_2O_2:Ti^{3+}$ ratios the formation of $HO_2\cdot$ comes about by hydrogen-atom transfer to the $HO\cdot$ radical:

$$HO\cdot + H_2O_2 \longrightarrow H_2O + HO_2\cdot \tag{11}$$

Saito and Bielski (277) examined the formation of the $HO_2\cdot$ radical in aqueous solution, using ceric sulfate:

$$Ce^{4+} + H_2O_2 \longrightarrow Ce^{3+} + H^+ + HO_2\cdot \tag{12}$$

$$Ce^{4+} + HO_2\cdot \longrightarrow Ce^{3+} + H^+ + O_2 \tag{13}$$

The $g$-value for this radical species is 2.016; the resonance peak has a line width of 27 gauss. A similar ESR resonance for the $Ce^{4+}$–$H_2O_2$ system was observed by Bains et al. (275) with a line width of about 24 gauss, a $g$-value of 2.016, and a relative signal height of approximately 0.1 times the corresponding signal height of the $Ti^{4+}$–radical complex. They could not detect the presence of the $HO\cdot$ radical unless $Ti^{4+}$ ions were included in the system. Evidently the $Ce^{3+}$ ions do not form stable complexes with the radicals.

## C. RO$_2$· and RO· Radicals

### 1. Genesis of Radicals by Irradiation

Radical formation by fission of the O—O bond is supported by evidence from bond-energy considerations and product analysis (282). Detection of alkoxy radicals RO· by ESR would appear to be not difficult, but they have been more difficult to detect than the HO· radical. Only recently has ESR evidence for the detection of *t*-butoxy radicals been reported (283).

Piette and Landgraf (284) obtained ESR spectra of radicals derived from photolytically decomposed hydroperoxides and concluded these to be alkoxy radicals. However, Ingold and Morton (285) and Maguire and Pink (286) were unable to detect the formation of *t*-BuO· from *t*-BuOOH nor any radical species from either pure *t*-Bu—OO—*t*-Bu at 77°K or in solution matrices at 195°K. Instead the detectable species were *t*-butyl peroxy radicals (t-BuO$_2$·). Ingold and Morton (285) related the origin of *t*-BuO$_2$· radicals to the presence of *t*-butyl hydroperoxide as an impurity in *t*-Bu—OO—t-Bu. The primary formation of the *t*-BuO· radical was recognized, but the high reactivity of this radical accounts for its short life. Its attack on the hydrogen of the hydroperoxide grouping leaves the *t*-butyl peroxy radical as the detectable component. Maguire and Pink (286) in a further detailed study of the reaction found that alkyl peroxides, such as di-*t*-butyl peroxide and cumyl peroxide, are not influenced by their structures in carrying out their role as radical sources for hydrogen abstractions. In hydrocarbon solution these peroxides attack the solvent to generate the same solvent radical, which then reacts with dissolved oxygen in formation of the corresponding solvent peroxy radical. When di-*t*-butyl peroxide is dispersed by deposition on silica gel and photolyzed at low temperature (77°K), the radicals formed are the decomposition products of the undetected *t*-butoxy radical: the methyl radical (quartet) and a radical believed to be $CH_3COCH_2$· (triplet). *t*-Butyl hydroperoxide gave high radical concentrations formed by the reaction sequence

$$t\text{-BuOOH} \longrightarrow t\text{-BuO·} + \text{HO·} \tag{14}$$

$$t\text{-BuOOH} + \text{HO·} \longrightarrow t\text{-BuO}_2\text{·} + \text{H}_2\text{O} \tag{15}$$

$$t\text{-BuOOH} + t\text{-BuO·} \longrightarrow t\text{-BuO}_2\text{·} + t\text{-BuOH} \tag{16}$$

Attempts by other investigators working with di-*t*-butyl peroxide (287), cumene hydroperoxide (288), and cumene peroxide (288) failed to detect alkoxy radicals. They may nonetheless be used effectively since the expected similarity of alkoxy radicals to HO· radicals allows reactions in nonpolar solvents in order to avoid the protonation effects and solubility difficulties encountered in aqueous media (287).

It has been suggested by Bersohn and Thomas (287) that $g$ values for RO·
are to be found below 2.0100 and alkyl peroxy $g$ values would fall within the
range 2.0140–2.0190. Weiner and Hammond (283) have presented ESR
evidence for the presence of the $t$-BuO· radical. For solid $t$-Bu—OO—$t$-Bu at
$-70°$ they observed a one-line spectrum, with $g = 2.004 \pm 0.004$, with peak-
to-peak width of 12 gauss, in contrast to $g = 2.015 \pm 0.004$ of a one-line
spectrum for $t$-butyl peroxy radicals. Decay of the signal with $g = 2.004$ at
room temperature was followed, and the termination constant $10^9 \ M^{-1} \sec^{-1}$
for $t$-BuO· was compared with $10^4 \ M^{-1} \sec^{-1}$ for the $t$-BuOO· radical. The
$g$ value for $t$-BuO· is predicted to be larger than the free-electron value of
2.0023 because of a greater spin-orbit coupling constant for oxygen than for
carbon.

Thomas (290) has pointed out a unique characteristic of $t$-peroxy radicals
($t$-butyl peroxy, cumyl peroxy, and diphenyl ethyl peroxy) that has not been
shown by other organic radicals. The line widths of the resonances of the
latter radicals normally increase with increasing viscosity and decreasing
temperature. Peroxy radicals show the opposite behavior of the line width,
which decreases with increasing viscosity and decreasing temperature. For
example, the line width of the $t$-butyl peroxy radical signal drops with increasing viscosity from approximately 35 gauss in pentane to a limiting value of
about 14 gauss, all at 25°. The line width as a function of temperature drops
from approximately 20 gauss at 25° to a limiting value of approximately 3
gauss at about $-70°$. The behavior is tentatively attributed to additional
spin-orbital relaxation mechanisms.

### 2.  γ-Irradiation of Di-t-butyl Peroxide

Di-$t$-butyl peroxide in rigid solvent matrices (ethers, amines, hydrocarbons)
at $-196°$ decomposes under $\gamma$-irradiation in a manner that is different from
that observed in photolytic processes (291). Excitation is not operative as in
the photolytic reaction scheme,

$$t\text{-Bu}—OO—t\text{-Bu} \xrightarrow{h\nu \, (\lambda \, < \, 313 \, m\mu)} t\text{-BuO·}^* + t\text{-BuO·} \qquad (17)$$

$$t\text{-BuO·}^* \longrightarrow \text{·CH}_3 + \text{CH}_3\text{COCH}_3 \qquad (18)$$

$$t\text{-BuO·} + \text{SH} \longrightarrow t\text{-BuOH} + \text{S·} \qquad (19)$$

whereby the primary radical is the excited species (292). In the $\gamma$-irradiation
process peroxide captures electrons emanating from the ionization of matrix
molecules to form the molecular anion of the peroxide. The anion, which is
deep blue, absorbs at about 560 m$\mu$. It does not decompose to methyl radicals
until the $\gamma$-irradiated sample is subsequently illuminated with visible light.
The following sequence of reactions in contrast to photolysis (equations

17–19) accounts for the features of radiolysis followed by reaction 18 (the matrix molecule is symbolized by M):

$$M \xrightarrow{\;\gamma\;} (M^+) \text{ mobile} + \varepsilon^- \tag{20}$$

$$(M^+) \text{ mobile} + t\text{-Bu}-OO-t\text{-Bu} \longrightarrow t\text{-Bu}-OO-t\text{-Bu}^+ + M \tag{21}$$

$$\varepsilon^- + t\text{-Bu}-OO-t\text{-Bu} \longrightarrow t\text{-Bu}-OO-t\text{-Bu}^- \tag{22}$$

$$t\text{-Bu}-OO-t\text{-Bu}^- \xrightarrow{hv \text{ (visible)}} t\text{-Bu}-OO-t\text{-Bu} + \varepsilon^- \tag{23}$$

$$\varepsilon^- + t\text{-Bu}-OO-t\text{-Bu}^+ \longrightarrow t\text{-BuO} \cdot + t\text{-BuO} \cdot^* \tag{24}$$

### 3. Radicals from Oxidation–Reduction Reactions

$t$-Peroxy radicals from $t$-butyl and cumene hydroperoxides may be generated in nonaqueous solvents by metal decomposition catalysts (293). The metals are soluble in the organic solvents in the form of their naphthenates, soaps, and acetylacetonates. The stability of the radical depends on the type of heavy metal employed and qualitatively follows the order Mn < Co < V. The ESR signal of the peroxy radical in the presence of vanadium is still evident after 60 hr at room temperature. The stability of the radical suggests an association with the metal ion formerly observed for the system $H_2O_2-Ti^{3+}-Ti^{4+}$ (see Section B-2).

### D. Radicals from Fluorinated Compounds

Lontz (294) found that the radical $CF_3\dot{C}FCONH_2$ in the solid state is oxidized by oxygen to give a free radical assumed to be $CF_3C(O_2\cdot)FCONH_2$. The peroxy radical's ESR spectrum exhibited no hyperfine splitting and gave $g$ values measured along the three principal axes of 2.0102, 2.0193, and 2.0269. Toriyama and Iwasaki (295) carried out detailed ESR investigations on the analogous peroxy radical $\cdot O_2CF_2CONH_2$, trapped in a $\gamma$-irradiated single crystal of trifluoroacetamide. Peroxy radicals of poly(tetrafluoroethylene) have also been investigated (296, 297).

    Vanderkooi and Fox (298) photolyzed samples of $CF_3OF$ and $CF_3OOCF_3$ in diamagnetic solvents at $-196°$. The irradiations were carried out in liquids at low temperature (1:1 $CF_4-CF_3Cl$ mixtures or $NF_3$, which remain liquid at $-196°$) since their isotropic ESR spectra are highly resolved and more convenient to interpret than the broad spectra recorded for their glassy or polycrystalline states. Identical six-line ESR spectra were derived from both compounds. Vanderkooi and Fox believed the identifiable species to be $CFOO\cdot$ radicals. The origin of the peroxy radical $CFOO\cdot$ was attributed to the initial formation of $CF_3O\cdot$ (still unidentified spectroscopically), followed by oxidation in the presence of traces of oxygen, which are experimentally difficult to remove completely. Fessenden's work (299) subsequently refuted the $CF_3OO\cdot$ radical as the active species and claimed the $CF_3OOO\cdot$ radical as the correct species. He carried out similar experiments with $CF_3OOCF_3$ at $-196°$ in the presence of small amounts of oxygen enriched in $^{17}O$ so that the $^{17}O$ hyperfine structure could be observed.

Oxygen fluorides $(O_xF_2)$ are an intriguing family of compounds, whose structures and decomposition mechanisms may be elucidated by ESR (300, 301). Compounds up to tetraoxygen difluoride, the highest member currently submitted to ESR examination (301), yield the same $FOO\cdot$ radical. The spectrum of undiluted, solid $O_4F_2$ at 77°K yields a single broad-band signal, with a peak-to-peak line width of 52 gauss and a $g$ value of 2.0035. The spectrum of the compound in dilute solutions of carbon tetrafluoride($CF_4$) at 77°K gives a sharp doublet for the $FOO\cdot$ radical. The same doublet is also derived by several approaches: electron irradiation of carbon tetrafluoride in the presence of oxygen enriched in $^{17}O$ (302), photolysis of oxygen difluoride $(OF_2)$ (303, 304), decomposition of dioxygen difluoride $(O_2F_2)$ (300), and either decomposition of trioxygen difluoride $(O_3F_2)$ or a thermal equilibrium of the following form (300):

$$2(O_3F_2) \quad \rightleftharpoons \quad O_2F_2 + 2(FOO\cdot) \qquad (25)$$

Fessenden and Schuler (302) point to their $g$ value (2.0038) for $FO_2\cdot$ as being considerably smaller than it is in alkylperoxy radicals [2.0148–2.0155 (289, 305)] and in $HO_2\cdot$ [2.016 (277); 2.0132 (306)], this result indicating that large negative contributions to the $g$-factor arise from the presence of the fluorine atom. The unpaired electron in $F-O-O\cdot$ is promoted into a relatively low lying antibonding $O-F$ orbital (302). An alternative representation by Kasai and Kirshenbaum (300) is that the fluorine nucleus exists at a relatively long distance from the $O_2$ moiety in the molecule, with the unpaired electron present mainly in the $O_2$ antibonding $\pi$-orbital.

### E.  Applications

A particularly important application of peroxides relates to their use as free-radical sources for generating transient radical species from a variety of suitable compounds. A few examples are provided by $\alpha$-hydrogen abstraction from carboxylic acids (270) and alcohols (268), addition to olefins (307), and oxidation of amines (308) and oximes (309) to nitroxide radicals of the types $R_2NO\cdot$ and $R_3CNHO\cdot$, respectively.

Specificity has been noted in the hydrogen-abstraction characteristics of radicals generated by the oxidation–reduction process (276). The type of radical and its reactivity in radical–metal complexes depend on the metal and nature of the complex. For example, the ESR spectrum of the tetrahydro-furanyl radical shows whether tetrahydrofuran is attacked at the $\alpha$- or $\beta$-

position. Oxidation by $Ti^{3+}/H_2O_2$ (electrophilic) or $Fe^{2+}/H_2O_2$ (nucleophilic) takes place at the $\alpha$-position (structure **24**) or $\beta$-position (structure **25**), respectively.

$$
\begin{array}{cc}
\text{H}_2\text{C}-\text{CH}_2 & \text{H}\text{C}-\text{CH}_2 \\
\text{C}\cdot\quad\text{CH}_2 & \text{H}_2\text{C}\quad\text{CH}_2 \\
\text{H}\quad\text{O} & \text{O} \\
(24) & (25)
\end{array}
$$

Two attractive approaches to the preparation of specific alkyl and perfluoroalkyl free radicals and related metastable species have been developed using peroxides as the radical-generating reagent. Diacyl peroxides

$$(RCO_2)_2 \xrightarrow{h\nu} 2R\cdot + CO_2 \tag{26}$$

or peroxy esters

$$RCO_2OCMe_3 \xrightarrow{h\nu} R\cdot + CO_2 + \cdot OCMe_3 \tag{27}$$

photolyzed (2500–3500 Å) at low temperatures directly in the cavity of the spectrometer give an intense ESR spectrum of the desired alkyl radical (310). The alternative method employs a reaction between an organometallic compound of the Group III elements [organoboron (311–313), aluminum (311, 312), gallium (311, 312), and phosphorus compounds (311, 312)] and a readily available, reactive free radical (311–313). These elements bear a vacant orbital, allowing a homolytic free-radical replacement of the alkyl grouping. For example, trimethyl boron reacts with photochemically generated *t*-butoxy radicals in the ESR cavity, giving the ESR spectrum of the generated radical:

$$Me_3B + t\text{-BuO}\cdot \longrightarrow Me\cdot + Me_2B-O-t\text{-Bu} \tag{28}$$

A commonly used standard substance in ESR measurements is the stable radical 1,1-diphenyl-2-picrylhydrazyl (DPPH), which displays a quintet in its spectrum. As a result of an investigation on the effect of oxidizing impurities on the radical activity of DPPH (314), Ueda uncovered one of the few analytical applications of ESR (315). *t*-Hydroperoxides form a complex with DPPH, reducing the quintet to a triplet that is further resolved to 17 lines after degassing. No effect on the ESR spectrum was observed on addition of hydrogen peroxide (314) to the DPPH solution, but addition of peroxyacetic or *m*-chloroperoxybenzoic acid decomposes DPPH exponentially with time (315). Formation of the complex and its associated triplet in ESR is the basis of an analysis for hydroperoxides. The reaction of DPPH with *t*-butyl hydroperoxide is faster than that with cumene hydroperoxide.

Electron-spin-resonance spectroscopy is invaluable for investigating oxidation processes in which peroxy radicals are the active intermediates. In addition to the peroxy radicals that have been discussed, ESR may be applied advantageously to the study of peroxy radicals in the polymers of hydrocarbons (316, 317) and perfluorocarbons (296, 297). The ESR technique has recently found a thermodynamic application in the measurement of the heat of reaction between a hydrocarbon radical and oxygen to give peroxy radicals as product. The simplicity of the technique is bound to be important for other systems as well. The illustration provided by Janzen, Ayers, and Johnston (318) is the decomposition of crystalline triphenylmethyl halide or triphenylmethylacetic acid to the triphenylmethyl radical ($g = 2.0024$ and line width = 10.7 gauss). Admission of oxygen to the cavity cell traps the radicals to give the triphenyl peroxy radicals $Ph_3C-O-O\cdot$ ($g = 2.014$ and line width = 7.4 gauss). The reaction is a reversible equilibrium

$$Ph_3C\cdot + O_2 \quad \rightleftharpoons \quad Ph_3C-O-O\cdot \tag{29}$$

that may be measured by alternately heating and cooling the closed system. A value of 9.0 kcal/mole for the heat of reaction was obtained. The direct thermochemical measurement of a reversible free-radical oxygenation may now be applied by this technique to systems of biochemical interest, such as the autoxidation of vitamin $B_{12r}$ (319), which has been observed to be a reversible system.

## VI.  MASS SPECTROMETRY

The mass spectrometer gives a pattern of positive ions produced by electron bombardment of the molecules. The technique is important for obtaining ionization and appearance potentials, molecular weights, structural data, and invaluable insights on the rearrangement processes of reactive ionic species (320, 321). Mass-spectra-studies on organic peroxides are sparse, and considerable additional study is warranted.

### A.  Hydrogen Peroxide

Foner and Hudson (322) and Robertson (323) obtained direct experimental proof of the $HO_2\cdot$ free radical in a mass spectrograph by generating the radicals in a bimolecular reaction between hydrogen atoms and oxygen molecules. Foner and Hudson (324) later extended their work by this technique to derive important thermodynamic information on the intermediate species. The facile formation of $HO_2\cdot$ radicals in the mass spectrograph allowed the measurement of the ionization potential $I$ for the reaction

$$HO_2\cdot + e \quad \longrightarrow \quad HO_2^+ + 2e \qquad I_{HO_2\cdot} = 11.53 \pm 0.02 \text{ eV}$$

The appearance potential $A$ for the $HO_2^+$ ion was obtained from hydrogen peroxide for the reaction

$$H_2O_2 + e \longrightarrow HO_2^+ + H\cdot + e \qquad A_{HO_2^+} = 15.36 \pm 0.05 \text{ eV}$$

From these experimental values and appropriate thermochemical relationships, the following bond energies were derived:

At 0°K:

$$D_{0 \ (H-OOH)} = 88.4 \pm 2 \text{ kcal/mole}$$
$$D_{0 \ (H-O_2\cdot)} = 45.9 \pm 2 \text{ kcal/mole}$$
$$\Delta H_0^\circ{}_{(HO_2\cdot)} = 5.7 \pm 2 \text{ kcal/mole}$$

At 298°C:

$$D_{H-OOH} = 89.6 \pm 2 \text{ kcal/mole}$$
$$D_{H-O_2\cdot} = 47.1 \pm 2 \text{ kcal/mole}$$
$$\Delta H_{298}^\circ{}_{(HO_2\cdot)} = 5.0 \pm 2 \text{ kcal/mole}$$

### B. Di-*t*-butyl Peroxide

Lavrovskaya et al. (325) used the method of ion-charge transfer to detect free radicals produced from di-*t*-butyl peroxide by pyrolysis at 900°. Samples were bombarded with 10.23-eV $NH_3^+$ ions issuing from an auxiliary mass spectrometer. The complex pattern produced from the thermal decomposition contained the ions at $m/e$ 73, which corresponds to $Me_3CO^+$ and were derived from the *t*-butoxy radical.

### C. Hydroperoxides

Fragmentation spectra by electron impact on a series of isomeric pentyl, hexyl, and heptyl hydroperoxides were obtained by Burgess and co-workers (326). The main features of their mass spectra were inferred by analogy to the fragmentation of alcohols. The initial ionization

$$
\begin{array}{ccc}
\overset{\displaystyle R^1}{\underset{\displaystyle R^2}{\overset{|}{\underset{|}{H-C-\ddot{O}-\ddot{O}-H}}}}
& \longrightarrow &
\overset{\displaystyle R^1}{\underset{\displaystyle R^2}{\overset{|}{\underset{|}{H-C-\overset{.+}{\ddot{O}}-\ddot{O}-H}}}}
\end{array}
\qquad (30)
$$

is conceived as a process of removing an electron from the nonbonding orbital of one of the oxygen atoms. From the mass-spectral data the C—C bond in the peroxides is more susceptible to fragmentation than in the corresponding alcohols. This may be explained by the greater electronegativity of

—OOH as compared with —OH. The general results obtained indicate the relative order of stability of the molecular ions (M) to be tertiary > secondary > primary and pentyl > hexyl > heptyl. Alkyl ions arise from the molecular ion by elimination of $HO_2 \cdot$, since the formation of $R^+$ is of a lower ionization potential than that of $HO_2^+$.

A cyclic transition state is invoked to explain the rearrangements. (A full arrow symbolizes the transfer of an electron pair, whereas a "fishhook" indicates the transfer of a single electron.) Formation of the (M-18) peak is important for 3-pentyl hydroperoxide (26), which is postulated in the following equation to arise in a manner analogous to alcohol fragmentation (326):

$$
\begin{array}{ccc}
\overset{\cdot\cdot+}{HO} \cdots \cdots H & & \overset{\cdot+}{:O} \!-\! CH_2 \\
| \quad\quad | & \longrightarrow \ H_2O \ + & | \quad\quad | \\
O \cdots \cdots CH_2 & & MeCH_2CH \!-\! CH_2 \\
| \quad\quad | & & \\
MeCH_2CH \!-\! CH_2 & &
\end{array}
$$

(26)                                    ($m/e$ 86)

An (M-46) peak suggests elimination of ethylene and water:

$$
\begin{array}{cc}
\overset{\cdot\cdot}{HO:}\overset{H}{\curvearrowright} CH_2 & \overset{+O\cdot}{\|} \\
| \quad\quad | & \longrightarrow \ MeCH_2CH + C_2H_4 + H_2O \\
\overset{+}{:O:} \ CH_2 & \\
MeCH_2HC &
\end{array}
$$

(26)                                    ($m/e$ 58)

Hydrogen peroxide is eliminated in a transition state analogous to water elimination to yield products from 1-heptyl hydroperoxide, as shown by the following equations:

$$
\begin{array}{cc}
H \cdots \cdots \overset{+\cdot\cdot}{:O}OH & \left( R\!-\!HC \!-\!\! CH_2 \right)^+ \\
R\!-\!HC \cdots \cdots CH_2 \ \longrightarrow & \quad\quad\quad \cdot \ + H_2O_2 \\
(CH_2)_n & (CH_2)_n
\end{array}
$$

(M)$^+$                                    (M-34)$^+$

$$
\begin{array}{c}
\overset{\cdot\cdot\cdot}{\underset{}{+}}OOH \\
H \} \ CH_2 \\
C_3H_7\!-\!HC \curvearrowleft CH_2 \quad \longrightarrow \ C_5H_{10}{}^{\cdot+} + C_2H_4 + H_2O_2 \\
H_2C
\end{array}
$$

($m/e$ = 70)

Hydrocarbon fragments of the general formula $C_nH_{2n-1}$ (i.e., 27, 41, 55, 69, etc.) and $C_nH_{2n+1}$ (i.e., 15, 29, 43, 57, 71, etc.) are produced, as may be expected from the aliphatic chain of these hydroperoxides, which are typical hydrocarbon derivatives.

### D. Cyclic Peroxides

Bertrand, Fliszár, and Rousseau (327) examined the behavior of a series of tetraoxacyclohexane derivatives toward electron impact to assess the relative stabilities of phenyl, cyclohexyl, and tetraoxacyclohexyl rings. The spectrum of the related ketal peroxide, 1,1′-dihydroxydicyclohexyl peroxide, was included in their study for comparison. The latter compound did not give a molecular ion radical (27) at 230°. A low-intensity peak at $m/e$ 196, the highest in the spectrum, indicates the fragment as a precursor of the $m/e$ 98 ion:

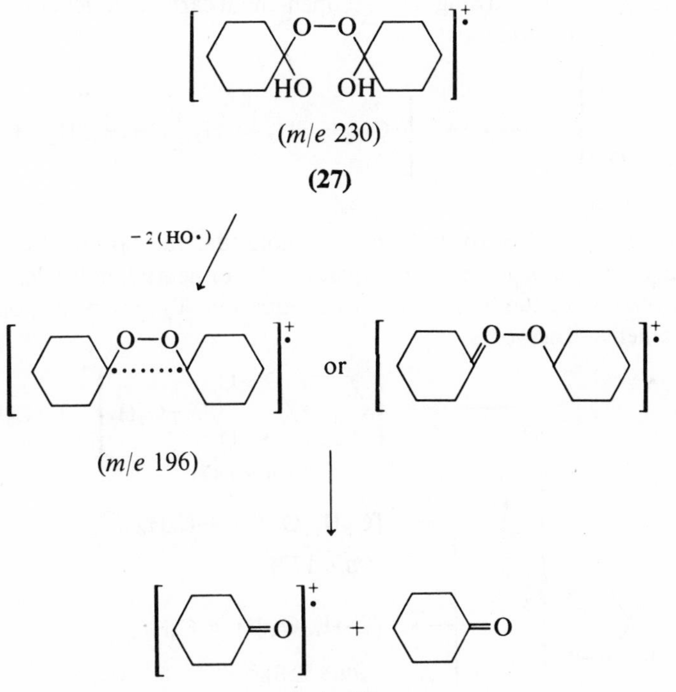

*(m/e 230)*

**(27)**

$-2(\text{HO}\cdot)$

*(m/e 196)*　　　or

*(m/e 98)*

Rupture of the peroxide bond leads to the fragment $m/e$ 115 at a very low intensity, but this mode of bond breaking suggests a second route to the $m/e$ 98 ion fragment:

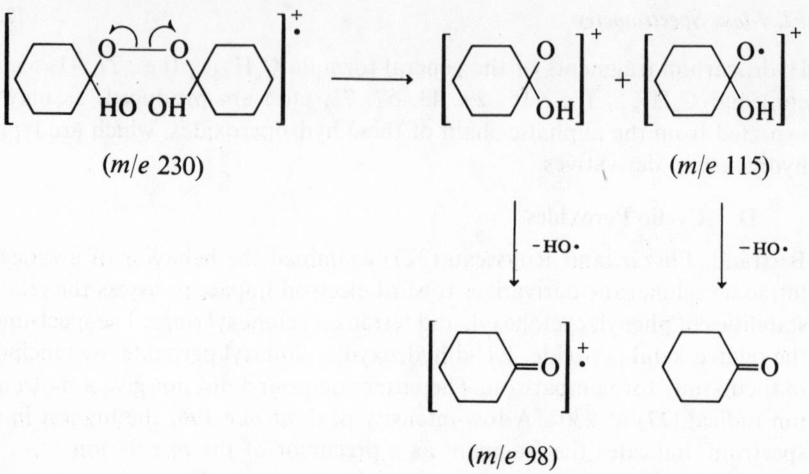

$$\left[\text{(m/e 230)}\right]^{+\cdot} \longrightarrow \left[\text{(m/e 115)}\right]^{+} + \left[\text{(m/e 115)}\right]^{+}$$

(m/e 230)

(m/e 115)

$-HO\cdot$      $-HO\cdot$

(m/e 98)

Subfragments may be derived from these precursors; for example, the α-cleavage of $m/e$ 115 rearranges to the open-chain carboxylic acid ion, followed by a series of degradations:

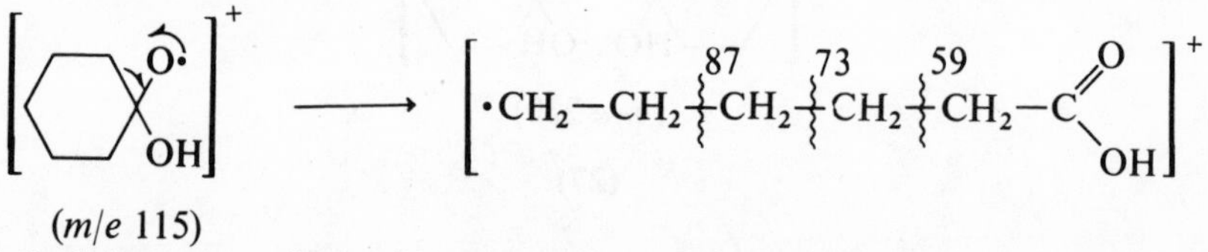

$$\left[\text{(m/e 115)}\right]^{+} \longrightarrow \left[\cdot CH_2-CH_2 \overset{87}{\{}CH_2 \overset{73}{\}}CH_2 \overset{59}{\{}CH_2-C\underset{OH}{\overset{O}{\diagdown}}\right]^{+}$$

(m/e 115)

Cyclohexanone diperoxide forms the molecular radical ion $m/e$ 228. Loss of an allyl or cyclopropyl species (equation 38), or neutral molecules (equation 39), or scission of the heterocyclic ring (equation 40) may be responsible for the fragments observed.

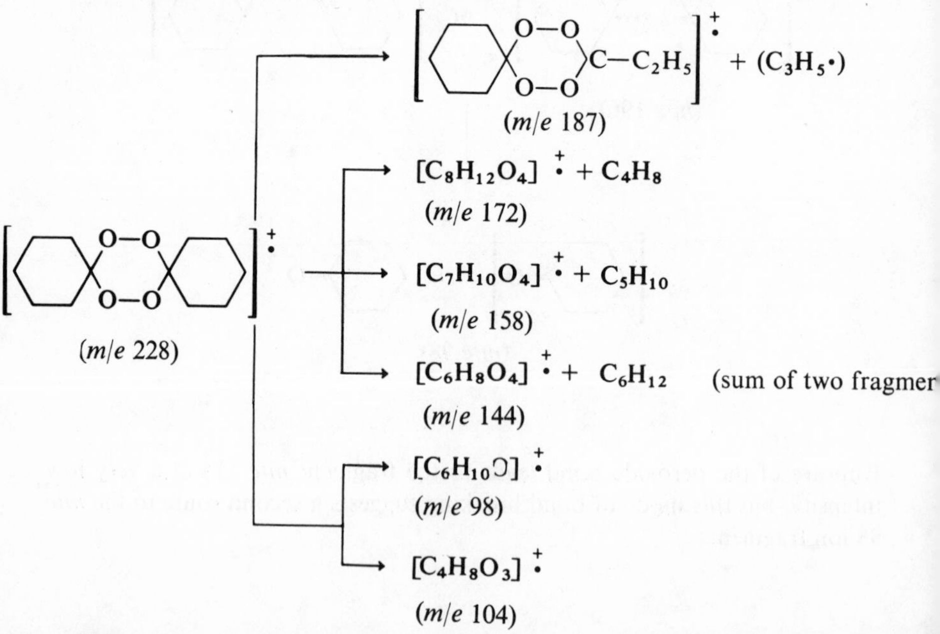

$$\left[\underset{O-O}{\overset{O-O}{\diagup}}C-C_2H_5\right]^{+\cdot} + (C_3H_5\cdot)$$

(m/e 187)

$[C_8H_{12}O_4]^{+\cdot} + C_4H_8$

(m/e 172)

$[C_7H_{10}O_4]^{+\cdot} + C_5H_{10}$

(m/e 158)

$[C_6H_8O_4]^{+\cdot} + C_6H_{12}$     (sum of two fragmer

(m/e 144)

$[C_6H_{10}O]^{+\cdot}$

(m/e 98)

$[C_4H_8O_3]^{+\cdot}$

(m/e 104)

(m/e 228)

734

The phenyl derivatives, such as benzophenone diperoxide, yield $O_2^+$ (*m/e* 32), along with neutral and positively charged radical ketone fragments:

$$\begin{bmatrix} Ph & O{-}O & Ph \\ & \diagdown C \diagup \quad \diagdown C \diagup & \\ Ph & O{-}O & Ph \end{bmatrix}^{+ \cdot} \longrightarrow \begin{array}{c} Ph \\ \diagdown \\ \diagup \\ Ph \end{array} C{=}O \; + \; O_2 \; + \; \begin{array}{c} Ph \\ \diagdown \\ \diagup \\ Ph \end{array} C{=}O^{\cdot +}$$

(*m/e* 396)                    (*m/e* 182)

Bertrand, Fliszár, and Rousseau (327) concluded from these and additional mass-spectral data on several ketone diperoxide derivatives that the order of ring stability is cyclohexyl < tetraoxacyclohexyl < phenyl. Ledaal (328) disputed this relative order between cyclohexyl and tetraoxacyclohexyl rings. He obtained spectra for cyclohexanone diperoxide (Figure 41) by two different methods of introducing samples into a high-resolution mass spectrometer.

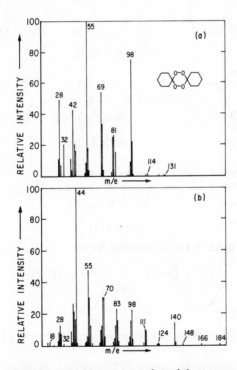

**Figure 41.** Mass spectra of cyclohexanone diperoxide. (a) Direct introduction to ionizing chamber; (b) indirect introduction (thermolysis at 190°C) (Reference 328).

One spectrum (Figure 41*a*) was obtained by direct introduction to the ionization chamber of the sample below its decomposition temperature. A second spectrum (Figure 41*b*) was obtained thermolytically by indirect introduction at 190°. The molecular ion in either case was not detected, in contrast to the result of Bertrand et al. (327).

Direct introduction of sample produced oxygen, along with the *m/e* 98 radical ion of cyclohexanone and the peaks corresponding to *m/e* 114, 130, and 131 (equations 42, 43, 44, and 45, respectively and the spectrum in Figure 41).

$$(m/e\ 228) \qquad\qquad (m/e\ 98)$$

$$(m/e\ 114)$$

$$(m/e\ 130)$$

$$(m/e\ 131)$$

Ion *m/e* 114 may exist in the alternative forms

Ion *m/e* 131, the highest observed, may arise by transfer of a hydrogen radical (equation 45).

Thermolysis resulted in an intense $CO_2$ peak (*m/e* 44) in place of the oxygen peak, which is virtually absent. The highest fragment, *m/e* 184, and the most important subfragment, *m/e* 140, correspond to 11-undecanolactone ($C_{11}H_{20}O_2$) and cyclodecane ($C_{10}H_{20}$), respectively. The formation of this lactone by photolytic or thermal decompositions of cyclohexanone diperoxide was reported by Story et al. (329).

In accordance with these results, Ledaal (328) emphasized three important considerations in obtaining mass spectra of peroxides:

1. Use of high resolution to ensure identification of the fragments.
2. Judicious choice of temperature in relation to the compound's thermal stability.
3. The derivation of spectra by using techniques of both direct and indirect sample introduction to differentiate between the effects of electron impact and thermal decomposition.

### E. Ozonides

Rousseau and co-workers (330) submitted a series of ozonides to mass-spectra analysis to determine the characteristic features of the fragmentation patterns. Their investigation on the fragmentation processes was aided by isotopically labeling the ozonides at the ether linkage with oxygen-18.

Scission occurs at the heterocyclic ring by two rupture paths (28), involving the O—O bond (path 1) and the C—O bond (path 2).

(28)

The important fragment ions appearing in the spectra correspond to $(M\text{-}32)^+$ (29), $R^1R^2CO_2^+$ (30 and 31), and $R^1R^2CO\cdot^+$, which are characteristic of these ozonides.

(29)          (30)          (31)

Cleavage of the O—O bond is the main fragmentation mode, whereas cleavage of the C—O bond is less frequent. Both paths result in ions having the same $m/e$ ratio, excepting path 2, which additionally furnishes the $(M\text{-}32)^+$ peak by loss of oxygen. Oxygen-18 labeling at the ether oxygen was used to distinguish between forms 30 and 31, which have the same $m/e$ ratio.

The decomposition of the $[M\text{-}32]^+$ and $[R^1R^2CO_2]^+$ ions, which were described for the specific examples, depends on the nature of the R substituents. The decomposition of $R^1R^2CO^+$ ions is well established for many species (321, p. 101):

$$\underset{R^1}{\overset{R^2}{\diagdown}}C\!\!=\!\!\overset{+}{O} \xrightarrow{-R^2\cdot} R\!\!-\!\!\overset{+}{C}\!\!\equiv\!\!O \xrightarrow{-CO} [R]^+$$

An intense $m/e$ 105 peak corresponding to $PhC\!\equiv\!O^+$ is present when phenyl is attached directly to the ozonide ring. The aromatic group confers enhanced stability on the ionic species, so that most of the ionic current is due to this type of carbonyl fragment. Ozonides containing the $XCH_2$ grouping (where $X = MeCH_2-$, $Ph-$, or $Cl-$) gives rise to fragments that are stabilized to a lesser degree, so that the ionic current is due mainly to ions of secondary decompositions.

The cyclic ozonide (32) included among the ozonides studied is not comparable to the acyclic types.

(32)

Although O—O cleavage is the primary site of fission, C—C cleavage is a competitive decomposition path.

## VII. MOLECULAR ASSOCIATIONS

### A. Hydrogen Bonding

Hydrogen bonding is a characteristic interaction of hydroxylic compounds with strong influences on their physical properties and many of their chemical reactions (331). Its importance in the kinetics, mechanisms, or applications of peroxide decompositions had been recognized by several groups of workers largely from the evidence of qualitative spectral data (124, 332–337). In recent years the need to clarify the role of hydrogen bonding in peroxide chemistry has found expression in a few quantitative measurements of the properties and strength of hydrogen bonds.

In hydroxylic compounds the electronegative character of oxygen, the small size of the proton, the polarity of the O—H link, and configurational requirements for interactions of the proton with an electron-donating source contribute to the formation and stability of the hydrogen bond. The hydrogen-bonding characteristics of the hydroxy group are modified by its attachment to the electronegative oxygen atom in the hydroperoxy group. The modified properties derive from an altered polarity in the O—H link, the availability of two oxygen atoms of different basicities that compete as hydrogen-bonding acceptors, and, in comparison with their nonperoxide analogs, the conformational intramolecular interactions permitted by the additionally linked oxygen atom with appropriately positioned groups.

### 1. Hydrogen Peroxide

Hydrogen peroxide, like its analog water, is essentially unassociated in the vapor phase (1, p. 344 ff.), but physical properties—such as melting point, boiling point, and dielectric constant—indicate a high degree of association in the liquid phase (338). Hydrogen peroxide's high dielectric constant ($74D$ at $20°$) approximates water's higher value ($80D$ at $20°$). The high dielectric value is attributed to intermolecular associations linking the molecules in linear chains (338), although the possible presence of cyclic structures may not be disregarded (1, p. 344 ff.).

Hydrogen peroxide is weakly acidic in its relation to water because it protonates the more basic oxygen atom of water to a small degree (1, p. 392 ff.):

$$H_2O_2 + H_2O \rightleftharpoons HO_2^- + H_3O^+$$

The heat of dissociation for this reaction, 8.2 kcal/mole, was determined from equilibria measured at several temperatures. The weaker basicity of hydrogen peroxide (1, p. 329 ff.) allows protonations by strong acids to remain "strong" in this medium (338). The proton appears to have a normal mobility in hydrogen peroxide, in contrast to the abnormal conductance of the proton and the hydroxy ion in water (338). The conductance has been explained as a straightforward ionic migration in a chain-structured matrix rather than electron transference from oxygen to oxygen.

The strong forces of association are clearly evident for hydrogen peroxide in the crystalline state (1, p. 344 ff.). The crystals are dense, and the molecules are compactly packed by hydrogen bonding as the principal intermolecular force. The O—H—O distance between molecules, 2.78 Å, is in correspondence with this distance in water ice. Because of the dihedral angle at the O—O bond, the molecules are rigidly maintained by the hydrogen bonds in a geometric disposition prescribing an infinite helix along the fourfold screw axes of the crystal. The strong intermolecular hydrogen-bonding forces within the hydrogen peroxide crystal and in the crystalline dihydrate, $H_2O_2 \cdot 2H_2O$, have been evaluated in further detail from the infrared vibrational frequencies (117, 118).

### 2.   *Hydroperoxides*

The association behavior of *t*-butyl hydroperoxide (339, 340) and isomeric heptane hydroperoxides (341) has been studied by infrared spectroscopy. Partition experiments with the former (340) and viscosity determinations on the latter (341) supplement the spectral data by providing additional insight on their associations.

In carbon tetrachloride solutions hydroperoxides show a double infrared-absorption peak in the O—H stretching region. The sharp absorption of *t*-butyl hydroperoxide at 3554 cm$^{-1}$ (340) [3540 (339) and 3551 cm$^{-1}$ (342) have been reported by other investigators] belongs to the fundamental vibration of an OH group free of hydrogen bonding, whereas the broad band with a maximum at about 3400 cm$^{-1}$ belongs to vibrations of the hydrogen-bonded OH group (Figure 42). The frequency of the associated hydroxy

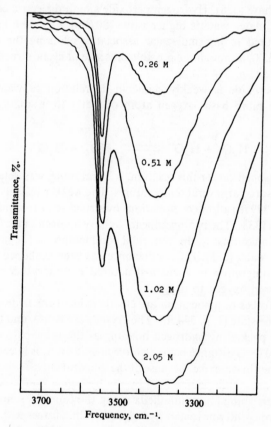

**Figure 42.** Effect of concentration on the infrared absorption of free and hydrogen-bonded *t*-butyl hydroperoxide in carbon tetrachloride. (Reference 340).

absorption band, in contrast to that of alcohols and phenols (106, p. 96) is characteristically unaffected by changes in concentration. Geiseler and Zimmermann (341) chose to apply the 6900 cm$^{-1}$ (1.46-$\mu$) band, the first harmonic vibration of the free hydroxy group, to measure hydrogen bonding in isomeric heptane hydroperoxides; the band shows a concentration dependency of the extinction coefficient. The hydrogen-bonded structures are considered to be cyclic dimers (33), accompanied by higher chain polymers, but supplementary viscosity measurements have supported the presence of cyclic trimers (34) as well.

(33)          (34)

The dimerization constants $K_{\text{dimer}}$ for the equilibrium 2P $\rightleftharpoons$ PP at two temperatures in pure carbon tetrachloride were measured from the decrease in absorbance of the free hydroxy group and used to obtain the thermodynamic parameters recorded in Table 19. Despite disagreement in this table

TABLE 19. HYDROGEN-BONDED COMPLEXES OF *t*-BUTYL HYDROPEROXIDE IN CARBON TETRACHLORIDE SOLUTION: THERMODYNAMIC PROPERTIES[a]

| Complex | $K$(l/mole) | | $-\Delta H$ (kcal/mole) | $-\Delta S$ (e.u.) |
|---|---|---|---|---|
| | 70° | 30° | | |
| *t*-Butyl hydroperoxide (TBHP): | | | | |
| TBHP dimer | 0.60 | 1.90 | 5.95 | 18.4 |
| TBHP dimer[b] | 0.46[c] | 1.03[d] | 6.3 | — |
| TBHP–styrene | 0.17 | 0.27 | 2.39 | 10.5 |
| TBHP–benzene | 0.11 | 0.15 | 1.58 | 9.0 |
| TBHP–*o*-dichlorobenzene | 0.13 | 0.15 | 0.74 | 6.2 |
| TBHP–chlorobenzene | 0.10 | — | — | — |
| Deuterated *t*-butyl hydroperoxide (TBDP): | | | | |
| TBDP dimer | 0.64 | 1.65 | 4.9 | 1.5 |
| TBDP–styrene | 0.50 | 0.84 | 2.7 | 9 |
| TBDP–benzene | 0.25 | — | — | — |

[a] Except where otherwise noted, data are from reference 340.
[b] Data from reference 339.
[c] At 40.5°.
[d] At 17.4°.

between the two sets of $K_{dimer}$ values reported by two groups, both sets of data result in the same dimerization energy of 6 kcal/mole. Zharkov and Rudnevskii (339) derived the polymerization constant for higher degrees of association [i.e., $K' = 3.95$ (17.4°) and 2.75 (40.5°)], from which they obtain the value of 2.8 kcal/mole as the hydrogen-bonding energy for each subsequent attachment. The equilibrium constant is higher in dimerization than in subsequent stages of association because the entropy loss is greater (339), which indicates that the equilibrium is largely in favor of the formation of cyclic dimers. Since the degree of association is $\bar{n} = 3.2$ (339) at 1.1 ml/l (17.4°) and the presence of trimers was indicated (341), the cyclic structures assumed by hydroperoxide associations is in contrast to the open-linked chains (dimers and polymers) of alcohols, for which $\bar{n} = 3$–4 (343).

The position of the hydroperoxide grouping in an aliphatic chain influences the strength of the association (Table 20). The heats of association for the four

TABLE 20.    HYDROGEN-BONDING EQUILIBRIA OF MONOMERIC WITH
DIMERIC AND TRIMERIC HEPTANE HYDROPEROXIDE ISOMERS:
THERMODYNAMIC PARAMETERS AT 30°[a]

| Hydroperoxide Position | $\Delta H$ (kcal/mole) | | $\Delta F$ (cal/mole) | | $\Delta S$ (e.u.) | |
|---|---|---|---|---|---|---|
| | $\Delta H_{12}$ | $\Delta H_{13}$ | $\Delta F_{12}$ | $\Delta F_{13}$ | $\Delta S_{12}$ | $\Delta S_{13}$ |
| 1 | −4.9 | −4.2 | −680 | 0 | −13.9 | −13.9 |
| 2 | −4.7 | −5.3 | +210 | −550 | −16.2 | −15.7 |
| 3 | −4.4 | −5.4 | + 64 | −680 | −14.7 | −15.6 |
| 4 | −4.5 | −5.9 | −160 | −790 | −14.3 | −16.2 |

[a] Data from reference 215.

isomeric heptane hydroperoxides were calculated from equilibrium measurements to be in the range of 4.4–4.9 kcal/mole for the dimer and 4.2–5.9 kcal/mole for the trimer. Primary hydroperoxides are most strongly associated, as is the case for alcohols (215). Among secondary hydroperoxides, the strength of the associations increases as the hydroperoxide grouping is moved toward the central carbon position in the chain.

In carbon tetrachloride solutions of $t$-butyl hydroperoxide containing aromatic solvents Walling and Heaton (340) observed the appearance of a low-frequency shoulder on the fundamental OH absorption. In solutions of pure aromatic solvents the free OH peak of $t$-butyl hydroperoxide disappears completely and a single peak of lower frequency appears as a broader and more intense band. They recognized the change as intermolecular hydrogen bonding with $\pi$-electrons of the aromatic ring, with the shifts paralleling solvent basicity. Ring substituents in the aromatic solvent may also complex with hydroperoxides. For example, anisole shows a two-peak spectrum indicative of competitive bonding at the ether oxygen. Benzotrifluoride is also cited as forming a double peak. Although this observation was not further amplified, a possible explanation may be found in the hyperconjugation of the trifluoromethyl group (156), which would make available a region of high electron density about this group for competitive hydrogen bonding. The association constants $K$ in carbon tetrachloride solution for the association of *t*-butyl hydroperoxide with aromatic solvents were determined for the equilibrium $A + B \rightleftharpoons AB$ from the decrease in absorbance at 3554 cm$^{-1}$ (340).

Deuteration of the hydroperoxide group shifts the absorption frequency of *t*-butyl hydroperoxide to 2630 cm$^{-1}$ (340). The shift is accompanied by a corresponding change in the equilibrium constant and in the thermodynamic parameters of the dimer (Table 19). A stronger hydrogen bond forms with deuterium than with hydrogen, and the former accordingly decreases the tendency toward dimerization by increasing the association with aromatic $\pi$-electrons. The energy and entropy values of hydrogen bonding for *t*-butyl hydroperoxide and its deuterated (ROOD) derivative are given in Table 19.

Zharkov and Rudnevskii (150) observed intramolecular hydrogen bonding in cumene (isopropylbenzene) hydroperoxide. They obtained 0.9 kcal/mole as the energy of this bond. An independent qualitative observation of intramolecular hydrogen bonding for this compound was reported by Barnard, Hargrave, and Higgins (149) and for phenyl cyclohexyl hydroperoxides by Kwart and Keen (344). Shabalin and co-workers (345) later carried out a detailed extension of Zharkov and Rudnevskii's studies in corroboration and amplification of the general results.

The infrared spectrum of neat cumene hydroperoxide in the O—H fundamental vibrational region contains a broad asymmetric absorption band, with the maximum at about 3430 cm$^{-1}$ (345). The absorption band, typical of associated hydroxy groups, diminishes on dilution in carbon tetrachloride, and a pair of narrow overlapping absorption bands appears simultaneously at 3551 and 3519 cm$^{-1}$. The higher frequency band is assigned to free OH and the lower band to intramolecular hydrogen bonding with aromatic ring $\pi$-electrons.

The intramolecular hydrogen bond of cumene hydroperoxide is readily broken by the introduction of aromatic or oxygenated solvents but is replaced by intermolecular hydrogen bonds. The spectral changes accompanying addition of a complexing solvent at increasing concentrations in carbon tetrachloride solutions of the hydroperoxide are illustrated in Figure 43*a*. In dilute solutions a third band due to intermolecular hydrogen bonding forms at the expense of the double peaks; for example, the frequencies of the intermolecular and intramolecular hydrogen bonds are widely disparate with dioxane as the complexing component (Figure 43*a*) and are sufficiently resolved to

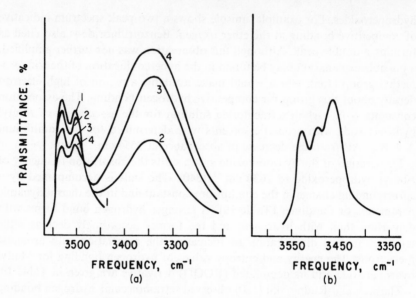

**Figure 43.** Effect of electron-donor complexing solvents on the infrared absorption of cumene hydroperoxide ($\sim$0.03M) in $CCl_4$ solutions. (a) Addition of dioxane; 1-cumene hydroperoxide only; 2-0.063M dioxane; 3-0.200M dioxane; 4-0.50M dioxane. (b) Addition of pentamethylbenzene (0.3M). (Reference 345).

appear as peaks on a broad envelope with pentamethylbenzene (Figure 43b), but complexation with benzene results in a single unresolved peak because of the small frequency difference.

The free hydrogens and intramolecular hydrogen bonds characteristically diminish together in the same ratio as the intermolecular hydrogen bonding increases with increasing concentration of the complexing solvent. It would seem *a priori* that relatively higher concentrations of free hydroxy species ought to develop as the intramolecularly-hydrogen-bonded species are depleted by their conversion to intermolecular hydrogen bonds. Instead the remaining concentration of hydroxy species reequilibrates almost instantaneously to reestablish the ratio of free to intramolecularly-hydrogen-bonded hydroxy species. The equilibrium constant for this conversion is 1.8 at 20°, showing that approximately one-third of the molecules are intramolecularly hydrogen bonded. (345)

The associations between cumene hydroperoxide and solvents of increasing basicity are reflected by a shift of the OH band to lower frequency, an increase in the half-width of the band (Table 21), and an increase in the equilibrium constant (Table 22). An estimate of the change in energy of intermolecular hydrogen bonds in a series of aromatic solvents falls in the narrow energy

TABLE 21. INTERMOLECULAR ASSOCIATION OF HYDROPEROXIDES WITH
SOLVENTS: FREQUENCY, SHIFT, HALF-WIDTH, AND
INTEGRATED INTENSITY DATA OF THE OH ABSORPTION
BANDS

| Compound and Solvent | $\nu$ (cm$^{-1}$) | $\nu_{OH}$ (cm$^{-1}$) | $\Delta\nu_{1/2}$ (cm$^{-1}$) | $A \times 10^{-3}$ (l/cm$^2$-mole) |
|---|---|---|---|---|
| | *Cumene hydroperoxide*[a] | | | |
| Carbon tetrachloride | 3551 | — | 30 | 7.7 |
| Chloroform | 3530 | 21 | 58 | 9.7 |
| o-Dichlorobenzene | 3522 | 29 | 58 | 11.7 |
| Methylene chloride | 3521 | 30 | 55 | 11.8 |
| Chlorobenzene | 3510 | 41 | 54 | 11.9 |
| Iodobenzene | 3504 | 47 | 72 | 14.5 |
| Benzene and homologs | 3503–3476 | 48–75 | 43–55 | 13.2–14.9 |
| Nitrobenzene | 3490 | 61 | 92 | 21.9 |
| Nitromethane | 3485 | 66 | 98 | 22.3 |
| Anisole | 3435 | 116 | 90 | 25.9 |
| Acetophenone | 3432 | 119 | — | — |
| Benzoquinone | 3430 | 121 | — | — |
| Benzophenone | 3425 | 126 | — | — |
| Acetonitrile | 3417 | 134 | 113 | 30.5 |
| t-Butyl peroxide | 3400 | 151 | 120 | 35.3 |
| Acetone | 3375 | 176 | — | — |
| Cyclohexanone | 3365 | 186 | — | — |
| Dioxane | 3357 | 194 | 133 | 53.2 |
| Diethyl ether | 3335 | 216 | 135 | 60.0 |
| Diisopropyl ether | 3326 | 225 | 122 | 51.2 |
| | *t-Butyl hydroperoxide*[b] | | | |
| Hexane | 3561 | | 29 | |
| Cyclohexane | 3556 | | 29 | |
| Carbon tetrachloride | 3554 | | 30 | |
| o-Dichlorobenzene | 3530 | | 46 | |
| m-Dichlorobenzene | 3530 | | 46 | |
| Chlorobenzene | 3523 | | 53 | |
| Benzene | 3509 | | 58 | |
| t-Butylbenzene | 3504 | | 64 | |
| Styrene | 3500 | | 66 | |
| Anisole | 3480, 3418 | | 73, 114 | |

[a] Data from reference 342.
[b] Data from reference 340.

TABLE 22.  INTERMOLECULAR ASSOCIATION
EQUILIBRIUM CONSTANTS OF CUMENE
HYDROPEROXIDE[a]

| Acceptor | $K$ (l/mole) | |
| --- | --- | --- |
|  | Hydroperoxide | Phenol |
| Pseudocumene | 0.17 | — |
| Pentamethylbenzene | 0.35 | 0.23 |
| Anisole | 0.46 | 1.01 |
| Benzoquinone | 1.25 | — |
| $t$-Butyl peroxide | 0.65 | 0.97 |
| Benzophenone | 1.67 | 7.82 |
| Acetone | 1.92 | 12.31 |
| Diisopropyl ether | 1.78 | 9.51 |
| Dioxane | 2.20 | 8.7 |
| Acetophenone | 2.28 | 8.32 |
| Cyclohexanone | 2.50 | 16.83 |

[a] Data from reference 342.

range of 0.95 kcal/mole for benzene to 1.51 kcal/mole for pentamethylbenzene
(345). Oxygenated solvents, which have higher basicities than aromatic solvents, form stronger hydrogen-bond associations with correspondingly pronounced spectral differences (Table 21).

Shabalin and Kiva (342) and Richardson and Smith (219) from their
respective studies on intermolecular and intramolecular hydrogen-bond associations were able to cast some light on the relative differences between
peroxide oxygen (oxygen attached to carbon) and hydroxy oxygen as the
proton acceptor in hydroperoxides. Hydroperoxides are known to be stronger
acids than alcohols but weaker acids than phenols (85a, 149). Shabalin and
Kiva (342) found support for this intermediate position from a comparison of
band shifts relative to free OH for associations by hydroxylated compounds
with diethyl ether and dioxane; for example, cumene hydroperoxide shows
frequency shifts with these two solvents of 216 and 194 cm$^{-1}$, respectively;
$\alpha,\alpha$-dimethylbenzyl alcohol, the nonperoxide analog, shows shifts of 142 and
131 cm$^{-1}$, respectively; and phenol, 275 and 241 cm$^{-1}$, respectively. It was
then necessary to determine the relative basicity of the two types of oxygen in
hydroperoxides. The peroxide oxygen in cumene hydroperoxide is expected
to approximate more closely the nature and basicity of oxygen in di-$t$-butyl
peroxide than the hydroxy oxygen. Since association between these two peroxides develops an absorption band at 3400 cm$^{-1}$ corresponding to the shift
$\Delta v_{OH} = 151$ cm$^{-1}$, the basicity of the peroxide oxygen is lower than the

basicity of the ether oxygen (Table 21), yet sufficiently high to compare with that of acetonitrile.

Richardson and Smith (219) also concluded that the peroxide oxygen has a lower basicity. They compared the relative capability of an ether oxygen and a peroxide group to associate intramolecularly with an alcohol. Comparative data for two of the compounds examined, 2-$t$-butylperoxy-2-methylpropanol (**35**, $\Delta v = 45$ cm$^{-1}$, $\Delta H_{ib} = -0.95$ kcal/mole) and 2-methyl-2-neopentoxy-1-propanol (**36**, $\Delta v = 54$ cm$^{-1}$, $\Delta H_{ib} = -1.07$ kcal/mole), shows the ether compound with the higher frequency shift between free and intramolecularly-hydrogen-bonded species and the higher enthalpy for this bonding in a 1,5-interaction. A 1,5-interaction that forms in preference to a 1,6-interaction has also been reported for cyclohexanone peroxides of the type shown in structure **37** (218).

(35)          (36)          (37)

The degree of proton acceptability by the hydroxy oxygen in hydroperoxides is more difficult to assess, but Shabalin and Kiva (342) cite as an indication of this oxygen's lower basicity the fact that hydroperoxides form cyclic dimers. It seems to this writer that the two oxygens in hydroperoxides differ in basicity by their attachment to atoms of different inductions; that is, the peroxide oxygen is more basic than the oxygen attached to hydrogen because of its union to the electron-releasing methylene group. Similarly the alcohol oxygen is more basic than the hydroxy oxygen of hydroperoxides because of its attachment to carbon, compared with the latter by its attachment to the electronegative oxygen.

The dimerization constants allow a comparison of the tendencies of cumene hydroperoxide and its nonperoxide analog, dimethylbenzyl alcohol, for self-association. The dimerization equilibrium constant at 20° is 0.62 l/mole for cumene hydroperoxide, 0.80 l/mole for $\alpha,\alpha$-dimethylbenzyl alcohol, and 2.4 l/mole for phenol. The constant for cumene hydroperoxide and the value of 1.03 l/mole for $t$-butyl hydroperoxide (339) are closer to the constant that is characteristic of alcohols than of phenols.

A possible equilibrium between $\alpha$-ketohydroperoxides and perepoxide (equation 47) was examined by Richardson and Steed (346). The cyclic structure **39** that had been proposed for aromatic derivatives by Kohler (347) was proved to be the α-ketohydroperoxide **38** by Rigaudy (184) and the latter work confirmed by Fuson and Jackson (348).

$$
\begin{array}{ccc}
\underset{\displaystyle \overset{\displaystyle O}{\underset{\displaystyle C}{\parallel}}}{} & & \\
R^1 \overset{\diagup}{\underset{\displaystyle HOO}{\diagdown}} CR^2R^3 & \rightleftharpoons & R^1\!\!-\!\underset{\displaystyle O\!-\!O}{\overset{\displaystyle OH}{C}}\!-\!CR^2R^3
\end{array}
$$

(38)                                 (39)

Richardson and Steed found no evidence in the infrared and NMR spectra for the presence of a perepoxide. They did establish the presence of inter-molecular and intramolecular hydrogen bonding, the latter **40** arising via the equilibrium:

$$
Me_2CH \overset{\displaystyle \overset{\displaystyle O}{\parallel}}{\underset{\displaystyle OOH}{C}} CMe_2 \rightleftharpoons Me_2CH \overset{\diagup}{\diagdown}
$$

(40)

A monomer–dimer enthalpy value of − 1.45 kcal/mole and an intramolecular-hydrogen-bonding enthalpy value of − 1.86 kcal/mole were obtained from these equilibria.

### 3. Peroxy Acids

Some of the hydrogen-bonding characteristics of peroxy acids were discussed in connection with their structures and infrared-spectra properties. In the liquid and vapor phases the molecules exist as intramolecularly-hydrogen-bonded monomeric species, which accounts for their observed properties. Molecular-weight determinations (23, 24) and infrared analysis (20, 21) estab-lished the monomeric state assumed by peroxy acids in solution, in contrast to the dimeric association of carboxylic acids (106, p. 161; 349; 350). Other properties shown by peroxy acids are in contrast to those of the parent acids; for example, their higher vapor pressures (351, 352) and higher p$K$ values (lower acidity) (23, 85a, 353) also reflect their internal hydrogen bonding.

A normal coordinate analysis of the force constants for bonds in the per-oxycarboxyl grouping had been carried out by Brooks and Haas (354), who assumed a planar cyclic structure in their calculations. Their force-constant and frequency values for peroxyformic and peroxyacetic acids are summarized in Table 23. The calculated O—H⋯O (hydrogen bond), O—H, C—O, and C=O stretching constants are surprisingly appropriate for molecules having

TABLE **23.** FREQUENCIES AND APPROXIMATE FORCE-CONSTANT VALUES FOR PEROXY ACIDS[a]

| | | Frequency $(cm^{-1})$ | | | |
| | Force | Peroxyformic Acid | | Peroxyacetic Acid | |
| Assignment | Constant[b] | Obsd[c] | Calcd | Obsd[c] | Calcd |
|---|---|---|---|---|---|
| O—H stretch | 5.8 | 3367 | 3367 | 3310 | 3310 |
| C—H stretch | 4.8 | 2987 | 2987 | | |
| C=O stretch | 9.2 | 1739 | 1739 | 1760 | 1760 |
| O—O—H bend | 0.44 | 1453 | 1453 | 1450 | 1450 |
| H—C bend | 0.62 | 1340 | 1340 | | |
| C—O stretch | 5.3 | 1243 | 1243 | 1248 | 1249 |
| O—O stretch | 3.5 | 859 | 859 | 868 | 868 |
| C—C stretch | 3.9 | | | 861 | 861 |
| O=C—O bend | 1.9 | 810 | 810 | 656 | 656 |
| Me—C bend | 1.7 | | | | 513 |
| Hydrogen-bond stretch | 0.6 | | 437 | | 411 |

[a] Data from reference 354.
[b] Units are millidynes per angstrom for stretching constants and millidynes per angstrom per square radian for bending constants.
[c] Data from reference 21.

strong intramolecular hydrogen bonding, in view of the incorrectness of their basic assumption of a planar model. Experimentally the correct structure of the grouping is a skewed configuration with a 72° dihedral angle.

The physical state of a compound may bring about important changes in the nature of its hydrogen bonding. With peroxy acids, crystallization brings into play a change from intramolecular to intermolecular hydrogen bonding of the molecules, leading to a corresponding helical arrangement of the peroxycarboxyl chains. See Sections I A2(c) and I B2. The dihedral angle imposed by the arrangement of molecules in the solid state also seems to depend on the nature of the alkyl or aryl group of the peroxy acid. A change in the nature of hydrogen bonding in the solid state is indicated in the infrared spectra by a shift and broadening in $v_{OH}$ and a broadening of the $v_{C=O}$ bands.

Although the energy of intramolecular hydrogen bonding in peroxy acids has been determined, its precise value remains uncertain. Giguère and Olmos (21) have estimated the value 6 kcal/mole for peroxyformic acid and 7 kcal/mole for peroxyacetic acid from their infrared spectra. A comparable estimate was subsequently determined from the heats of combustion of peroxy acids, *t*-butyl peroxy esters and their nonperoxide analogs (355) by means of the following equations

$$\underset{\substack{\text{O} \\ \|}}{} \text{RCOH(g)} + \tfrac{1}{2}\text{O}_2(g) \longrightarrow \underset{\substack{\text{O}\cdots\text{H} \\ \| \;\;\; |}}{} \text{RCOO(g)}; \qquad \varDelta H° = 20.0 \text{ kcal/mole} \qquad (49)$$

$$\underset{\substack{\text{O} \\ \|}}{} \text{RCO}-t = \text{Bu(g)} + \tfrac{1}{2}\text{O}_2(g) \longrightarrow \underset{\substack{\text{O} \\ \|}}{} \text{RCOO}-t = \text{Bu};$$
$$\varDelta H°. = 15 \text{ kcal/mole} \qquad (50)$$

and application of the Benson–Buss additivity law (356), whereby the hydrogen-bonding energy $\varDelta H_{hb}$ is computed from the relationship

$$\varDelta H_{hb} = \varDelta H_{peracid} - \varDelta H_{perester} = [\text{O}-(\text{CO})(\text{C})] - [\text{O}-(\text{CO})(\text{H})] +$$
$$[\text{O}-(\text{O})(\text{H})] - [\text{O}-(\text{O})(\text{C})]$$

The energy of the intramolecular hydrogen bond is calculated to be $5 \pm 3$ kcal/mole. Unfortunately the relatively high estimate of uncertainty associated with the absolute value mars its reliability. Benson, in disagreement with this large value, suggests about 1 or 2 kcal/mole.* Evidently an accurate redetermination of this bond energy would be desirable.

Hydrogen bonding performs an important role in the reactions of peroxy acids. A well-known illustration is the essential function of intramolecular hydrogen bonding in the nonionic mechanism of olefin epoxidations (357). The cyclic hydrogen-bonded configuration coordinates with the olefin in a transition state (**41**) to effect *cis* addition.

(**41**)

Epoxidation rates are found to be lower in oxygenated solvents than in neutral hydrocarbon media. The cause of the solvent effect has been attributed to the rupturing of the intramolecular hydrogen bond and to the formation of a stronger intermolecular hydrogen bond with the more basic oxygenated

* Private communication from S. W. Benson.

solvent (358). Consequently intermolecular associations with peroxy acids lead to a reduction in the effective concentration of the more active cyclic species and to an increase in the entropy of activation.

## B. Charge-Transfer Complexes

Molecular associations termed charge-transfer (CT) or $\pi$-complexes play an important directive role in many chemical reactions (359–361). Despite extensive experimental work in this area in the last two decades, the participation of peroxides in these associations has been largely neglected. Only two papers represent attempts at quantifying the data; it is to be hoped that this will be rectified as interest develops.

Generally charge-transfer associations involve two molecules, one as the electron donor, for which the ionization potential is a measure of donor ability, and the recipient as the electron acceptor, which is related to electron affinity and reduction potential (359). The complexations are accompanied by changes in physical properties, most often measured spectroscopically. The spectra arise from the promotion of an electron in one molecule to an excited level in another which is stabilized by resonance between a no-bond structure and a polar electronic structure.

### 1. Charge Transfer in Autoxidations

The solution of oxygen in various solvents produces optical absorption bands unobservable in either component alone and without the formation of a stable complex (362). This type of charge transfer, called contact charge transfer, is "collisional" in origin. It has been measured spectroscopically in aromatic hydrocarbons (363, 364), the CT spectra appearing on the edge of the hydrocarbon absorption (about 280 m$\mu$ for cumene and 290 m$\mu$ for *p*-cymene). The property is analytically useful for the continuous measurement of oxygen in the autoxidation studies of reactive hydrocarbons (364).

A reversible CT absorbing species is formed between oxygen and ethers (Figure 44) because the absorbance is reduced on displacement of $O_2$ by nitrogen sparging and returns on reoxygenation (365). Ethers react with the oxygen on irradiation to give ethyl acetate as the major product. The CT complex is describable as a hybrid of a nonbonded structure (**42**) and a charge-separated ("dative") bond structure (**43**).

$$O_2$$

$$R \overset{\displaystyle O}{\diagup \diagdown} R$$

**(42)**

$$O_2^-$$

$$R \overset{\displaystyle O}{\diagup \overset{+}{} \diagdown} R$$

**(43)**

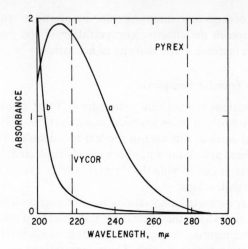

**Figure 44.** Ultraviolet spectra of oxygen-diethyl ether (curve a) with nitrogen-ether as a reference compared to nitrogen-ether (curve b) with empty cell as a reference. Spectra were obtained for samples in cells of 1-cm path length; cutoff regions of vycor and pyrex are indicated. (Reference 365).

The excited state undergoes chemical reactions that are proposed to proceed in accordance with the following equations:

$$\text{MeCH}_2\text{OCH}_2\text{Me} + \text{O}_2 \rightleftharpoons \overset{\overset{\displaystyle \text{O}_2}{\displaystyle \cdot\cdot}}{\text{MeCH}_2\text{OCH}_2\text{Me}} \xrightarrow{hv} \overset{\overset{\displaystyle \text{O}_2^{-}}{|}}{\underset{+}{\text{MeCH}_2\text{OCH}_2\text{Me}}}$$

$$\overset{\overset{\displaystyle \text{O}^{-}}{\text{O}^{\diagup}}}{\underset{+}{\text{MeCH}_2\text{OCH}_2\text{Me}}} \longrightarrow \overset{\overset{\displaystyle \text{OH}}{\text{O}^{\diagup}}}{\underset{+}{\text{MeCH}_2\text{O}-\overset{-}{\text{C}}\text{HMe}}} \longrightarrow \overset{\overset{\displaystyle \text{OH}}{\text{O}^{\diagup}}}{\text{MeCH}_2\text{OCHMe}}$$

$$\overset{\overset{\displaystyle \text{O}}{\|}}{\text{MeCH}_2\text{OCMe}} + \text{H}_2\text{O}$$

### 2. Charge Transfer in Radical Reactions

The problem of π-complexes involving a radical RO₂· was initiated by Boozer and Hammond (366, 367) to explain the kinetics of the inhibition of aromatic hydrocarbon oxidation by amine and phenol inhibitors. They proposed a mechanism involving a π-complex between peroxy radicals and inhibitors (InH):

$$RO_2^{\boldsymbol{\cdot}} + InH \; \rightleftharpoons \; [RO_2^{-\boldsymbol{\cdot}}\cdots \overset{+}{InH}]$$

$$[RO_2^{-\boldsymbol{\cdot}}\cdots \overset{+}{InH}] + RO_2^{\boldsymbol{\cdot}} \; \longrightarrow \; ROOH + RO_2In \qquad (54)$$

Thomas and Tolman (368) confirmed the kinetics for which Thomas (369) sought evidence of a stable π-complex, using pyridine (Py) as a typical amine electron donor for the equilibrium

$$RO_2^{\boldsymbol{\cdot}} + Py \; \rightleftharpoons \; [RO_2^{-\boldsymbol{\cdot}}\cdots \overset{+}{py}]K$$

Direct experiments by ESR to determine radical concentrations did not give positive results in support of the equilibrium ($K = 0.4$ 1/mole) but rather a displacement primarily in the direction of the free radical. Buchachenko and Sukhanova (336) in their review of the subject asserted the difficulty of reaching an unambiguous confirmation for the existence of π-complexes between $RO_2^{\boldsymbol{\cdot}}$ and aromatic compounds. In accordance with Tokumaru and Simamura's (370) theoretical considerations, the reactivity of a free radical depends primarily on its electron-donating property. Certainly peroxy radicals, circumscribed as they are by a high electron density, should show high electron-donating properties. The reason for the low stability of Thomas' cumylperoxy radical–pyridine complex may be found in this assessment and attributed to repulsive interaction between two electron-donor species.

Tokumaru and Simamura (370) presented a theoretical treatment of the features leading to free-radical-induced decompositions of peroxides. Aroyl peroxides and peroxy acids contain a vacant antibonding orbital $2p\sigma_\mu$ about the O—O bond available for association with the free radical's unpaired electron. They consider the free radicals as having electron-donating properties and the carbonyl grouping as contributing a subordinate role in the associations. Electron-attracting substituents in benzoyl peroxide lower the energy level of the $2p\sigma_\mu$ orbital about the O—O bond, which is accompanied by an increase in the strength of the CT interaction between the attacking radical and the peroxide and a lowering of the activation energy for induced decomposition.

The concept of charge transfer was invoked as a qualitative explanation of the mechanism of benzoyl peroxide decomposition in chlorobenzene and bromobenzene solutions (371) and for the decarboxylative decompositions of aliphatic and aromatic carboxylic acids in solutions of iodine and peroxides to the corresponding alkyl and aryl iodides (372). The latter reaction involves the participation of hydrogen bonding in conjunction with charge transfer.

### 3. Charge Transfer with Ionoids

Anionoid reagents (Lewis bases) engage in CT interactions by electron donation to the vacant orbital in aroyl peroxides and peroxy acids (370). Tokumaru and Simamura cite the following examples as reactions that may

be understood in CT terms: the rapid decomposition of benzoyl peroxide by
N,N-dimethylanilines (373, 374); the reaction of substituted benzoyl peroxides
with alkali iodides (370) or with phosphines (375–377); the reactions of peroxybenzoic acids with olefins (378), sulfides (379, 380), and sulfoxides (370);
and the oxidation of sulfides with hydrogen peroxide (381, 382).

A quantitative determination of a CT complex was obtained by Nandi and
Nandi (383) for the interaction between benzoyl peroxide and Rhodamine
6GX Color Base. A color change is effected from yellow (480–484 and 450–
454 m$\mu$) before interaction to pink (515 m$\mu$) for the CT associate to colorless
for the final products.

Peroxides possessing an orbital of the donor type may interact with catio-
noid reagents (Lewis acids) (370). This type of interaction has been measured
between di-$t$-butyl peroxide and iodine (384). Although the oxygen atoms in
this peroxide do not lend themselves to ultraviolet measurement, a slight shift
was observed for the $I_2$ band, indicating a weakly interacting CT effect.

### 4.  Intramolecular Interaction

Homolysis of a peroxide bond is anchimerically accelerated by an iodine,
sulfur, or olefinic substituent in the *ortho* position of aroyl peroxides (385–
387) or $t$-butyl peroxy esters (388–394) or in the $\beta$-position of acyl peroxides
(395). Relative decomposition rates of the *ortho*-substituted peroxides to the
unsubstituted derivatives are as high as 10,000 times greater. These substituents
are known to be effective charge-transfer agents that may allow this type of
reactivity to take place intramolecularly when situated in appropriate loca-
tions in the molecule. In the illustration for an $o$-thio-substituted peroxy ester
(388, 389)

(44)                                              (45)

the decomposition has been postulated to pass through a singlet transition state represented by canonical forms **44** and **45**. Although this class of compound was not spectroscopically examined for evidence of charge-transfer, the canonical structures are indicative of the presence of charge transfer in the ground state.

## VIII.  ELECTROMETRY

### A.  Polarography

#### 1.  Introduction

Accounts of the general theory and practice of polarography are available in selected textbooks and monographs (396–400). Specialized reviews on the nonaqueous polarography of organic compounds have been published by Gutmann and Schöber (401), LeGuillanton (402), and Wawzonek (403), and on the polarography of peroxides by Martin in 1960 (404), Johnson and Siddiqi in 1965 (405), and Mair and Hall in 1969 (406). These reviews are supplemented by Wawzonek's (407) and Pietrzyk's (408) general reviews in *Analytical Chemistry*, which together provide a convenient compendium of research in this field.

The polarographic method is based on the study of electrolytes with the use of a mercury capillary electrode, the potential of which is measured against that of a macroelectrode. The macroelectrode is usually an aqueous saturated calomel half-cell, connected by means of a salt bridge, but a mercury pool or silver electrode may also be employed.

A polarographic curve, or polarogram, is obtained for an electroreducible or electrooxidizable substance by measuring the variation of current with a continuously increasing voltage (Figure 45). The potential at the point of

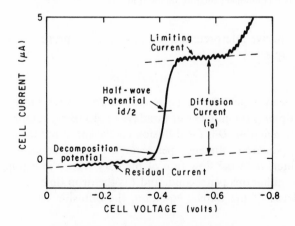

**Figure 45.**  Typical normal polarogram.

inflection of a polarogram is the "half-wave potential," which is closely related to the standard potential of the electrode reaction. Improvement in resolving power and in the measurement of half-wave potentials may be obtained by derivative polarograms, which are often necessary to overcome the difficulties in the resolution of polarographic steps. These are obtained by measuring the rate of change of cell current with applied voltage ($di/dE$) (Figure 46) to obtain a maximum at the half-wave potential. Derivative polarograms are recorded routinely by the newer instruments. A digital voltmeter may be incorporated into the system to provide a visual display of continuously generated potentials and hence the direct observation of the half-wave potential.

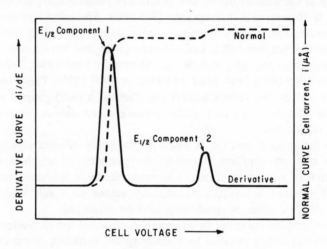

**Figure 46.** Idealized normal and derivative polarogram of a two component mixture in solution.

The quantitative interpretation of polarographic processes is expressed by the Ilkovic equation:

$$i_d = 607nD^{1/2}Cm^{2/3}t^{1/6} \tag{56}$$

where $i_d$ is the average diffusion current in microamperes during the life of the mercury drop, $n$ is the number of faradays of electricity required per mole of the electrode reaction, $D$ is the diffusion coefficient of reducible or oxidizable substance in square centimeters per second, $C$ is the concentration in millimoles per liter, $m$ is the flow rate of mercury from the dropping electrode capillary in milligrams per second, and $t$ is the drop time in seconds.

On the basis of the Ilkovic equation, the diffusion current constant $I$ is defined as

$$I = \frac{i_d}{Cm^{2/3}t^{1/6}} \tag{57}$$

This constant is nearly characteristic of a particular class of compounds with the same functional group. The half-wave potential and the direct proportionality between concentration and limiting diffusion current provide the two basic facts for the use of polarography in qualitative and quantitative analysis.

Electrochemical reductions may be reversible or irreversible. These terms reflect the rates of electrode processes. The irreversibility of an electrode reaction means that the reaction proceeds at a rate too slow to maintain equilibrium at the electrode surface, but chemically irreversible steps are also electrochemically irreversible (400). The equations for an irreversible wave are

$$E_{1/2} = E^0 + \frac{2.3RT}{n\alpha F} \log 0.87 \, k_f^0 \sqrt{t/D} \tag{58}$$

and the more practical relation

$$E = E_{1/2} - \frac{2.3RT}{n\alpha F} \log \frac{i}{i_d - i} \tag{59}$$

where $E_{1/2}$ is the half-wave potential, $E^0$ is the standard potential, $R$ is the universal gas constant, $T$ is the absolute temperature, $n$ is the number of electrons per molecule participating in the slow stage of reduction, $\alpha$ is the transfer coefficient, $F$ is the faraday constant, $k_f^0$ is the velocity constant of the heterogeneous electrode process, $t$ is the drop time of the capillary, $D$ is the diffusion coefficient, $i$ is the current for any given potential $E$, and $i_d$ is the diffusion-current constant given by the Ilkovic equation. A plot of $(FT/F) \log [i/(i_d - i)]$ against $-E$ is linear with slope $\alpha n$ and is referred to as a "log plot." If it is assumed that $n = 1$, then $\alpha$ is equal to $0.15 - 0.30$ (409).

The electrochemical behavior of hydrogen peroxide has been extensively investigated since Thenard's initial studies in 1818, which ultimately established a basis for its manufacture by electrolytic processes (1, Ch. 1). Since Heyrovsky's introduction of polarography as an electrochemical technique it has been known that dissolved oxygen in electrolyte solutions is reducible at the dropping-mercury electrode to produce a polarogram consisting of two waves: the initial wave representing reduction of oxygen to hydrogen peroxide and the subsequent wave the reduction of hydrogen peroxide (410). Heyrovsky's initial work was shortly followed by his determination of the reduction potentials of hydrogen peroxide in acid and alkaline solutions (411).

Organic peroxides were first detected polarographically by Gosman (412) in ether autoxidation products. The polarographic analysis of known individual organic peroxides (methyl hydroperoxide and diethyl peroxide) was subsequently described by Dobrinskaja and Neumann (413) and Stern and

Polak (414). In the years following these studies on peroxides the technique has proved to be invaluable in the detection and analysis of oxidation products in fats and oils, and the combustion products of fuels (see Subsection 5). The polarographic methods—because of their high sensitivity, rapidity, and inherent capability for qualitatively identifying and differentiating components of a mixture—complement spectroscopic methods of analysis.

### 2. Electrode Reaction

The reductions of peroxides are irreversible but, in some cases the waves approach those for a two-electron reversible reduction (415). The electrochemical reduction of hydrogen peroxide cleaves the O—O linkage by the addition of two electrons, forming the hydroxy ion, which is neutralized to water in proton-containing media (396, pp. 552–558).

$$\text{HO}-\text{OH} + 2e + 2\text{H}^+ \longrightarrow 2\text{H}_2\text{O} \quad \text{(in acid medium)}$$

$$\text{HO}-\text{OH} + 2e \longrightarrow 2\text{HO}^- \quad \text{(in alkaline medium)}$$

Organic peroxides behave at the dropping-mercury electrode in a manner similar to that of hydrogen peroxide and suggestive of a related mechanism of reduction for yielding the corresponding hydroxy compounds (34, 128, 396, 409, 415):

$$\text{ROOR}' + 2e + 2\text{H}^+ \longrightarrow \text{ROH} + \text{R}'\text{OH}$$

Schulz and Schwarz (416–418) investigated the mechanism of reduction in more detail by examining the effect of substitution of benzoyl peroxide and the effect of increasing solvent polarity (416, 417). The thermochemical rupture of a peroxide bond (RO—OR′) may occur *heterolytically*, an ionic dissociation ranging from complete dissociation $\text{RO}^- + \text{R}'\text{O}^+$ to slight polarization $\text{RO}^{\delta-}\cdots^{\delta+}\text{OR}'$, or *homolytically*, a symmetrical rupture leading to a pair of free radicals (RO· + R′O·) (292, pp. 57–59). The mode of cleavage of the peroxide bond depends on the nature of the solvent and the substituents. Heterolytic cleavage is favored in a highly dipolar O—O bond induced by substituents of electron-attracting power. This effect is enhanced in media of high dielectric constant. In a similar manner cleavage of a peroxide bond in polarographic reductions is influenced by the solvent medium to take place heterolytically, in which case two electrons are added in sequence, or homolytically by a two-electron addition in a single process leading to a pair of anions. The two mechanisms are illustrated in Figure 47 for the reduction of substituted benzoyl peroxides. A plot of half-wave potentials versus the Hammett substituent values for several substituted benzoyl preoxides results in two intersecting regression lines (Figure 48). These lines break in slope at $\sigma$ for the *p*-chloro substituent, indicating a change from a radical (homolytic) to a polar (heterolytic) reduction.

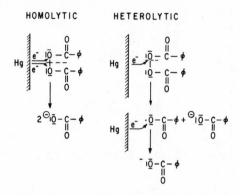

**Figure 47.** Electrode-reduction mechanism of the O—O bond in dibenzoyl peroxide: 2-electron reduction (homolytic) and 2-step 1-electron reductions (heterolytic). Reference 418).

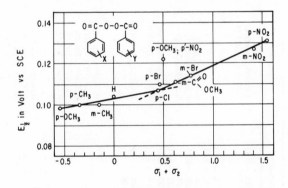

**Figure 48.** Half-wave potential-Hammett $\sigma$-substituent curve for substituted dibenzoyl peroxides. Supporting electrolyte 0.25N $NH_4NO_3$ in dioxane water (4:1); peroxide concentration, $5 \times 10^{-4}$ M. (See Table 24) (Reference 418).

### 3. Effect of Medium on $E_{1/2}$ and I

Half-wave potentials and diffusion-current constants of various peroxides in aqueous and nonaqueous solutions are compiled in Table 24. The half-wave potential and diffusion-current constant of hydroperoxides in 5–40% aqueous alcohol are both independent of concentration (415). The half-wave potentials of hydroperoxides in nonaqueous solvents (methanol, ethanol, isopropanol, benzene–methanol, benzene–isopropanol, nitromethane, and acetic acid) are

TABLE 24. POLAROGRAPHIC DATA FOR PEROXIDES

| Peroxide | $E^{1/2}$ (volts versus saturated calomel electrode) | | | Diffusion-Current Constant $I^c$,[d] | References |
| | Aqueous[a] | Alcohol | Benzene–Methanol (1:1)[b] | | |
| --- | --- | --- | --- | --- | --- |
| Hydrogen peroxide | -0.88, -0.93 / -1.03[e] | -1.07 (50% MeOH)[f] / -1.09 (93% EtOH)[g] | -1.14 / -1.16[h] | 4.4 (A) | 409, 421, 422 |
| Saturated hydroperoxides: | | | | | |
| Methyl | -0.64, -0.79 | | | | |
| Ethyl | -0.42, -0.44 | | | 3.3 / 3.0 | 416–418, 423 / 416–418, 423 |
| Propyl | -0.35 | -0.49 (20% EtOH) | -0.98 | 4.63 (B–M) | 409, 416–418 |
| 1-Butyl | -0.20, -0.21 | -0.26 (5% EtOH) | | 3.2 (5% EtOH) | 415 |
| 2-Butyl | -0.28 | -0.28 (5% EtOH) | | 3.1 (5% EtOH) | 415 |
| t-Butyl | -0.33, -0.34, / -0.35, -0.27, / -0.28 | -0.74 (50% MeOH) / -1.04 (90% MeOH) | -0.96, -1.10, -1.04, / -1.15[h] | 5.8 (B–M) / 3.6 (A) | 415, 418, 420 / 423, 424, 425 / 426 |
| 1-Pentyl | -0.08, -0.19, / -0.20 | -0.20 (20% EtOH) | -0.95 | 3.1 (5% EtOH) / 3.17 (A) / 2.8 (20% EtOH) / 3.06 (5% EtOH) | 409, 415 |
| 2-Pentyl | | -0.24 (20% EtOH) | | 4.27 (B–M) / 2.6 | 415 |
| 3-Pentyl | | -0.22 (20% EtOH) | | 2.5 | 415 |
| 3-Methyl-1-Butyl | | -0.23 (20% EtOH) | | 2.6 | 415 |
| Cyclopentyl | | -0.25 (20% EtOH) | | 2.6 | 415 |
| 1-Hexyl | | -0.12 (20% EtOH) | -0.94 | 2.62 (20% EtOH) / 4.23 (B–M) / 4.27 (B–M)[i] | 409, 415 |
| 2-Hexyl | | -0.16 (20% EtOH) | | 2.5 | 415 |
| 3-Hexyl | | -0.16 (20% EtOH) | | 2.5 | 415 |
| Cyclohexyl | | -0.14 (20% EtOH) | | 2.5 | 415 |
| 1-Heptyl | | -0.14 (20% EtOH) | | | 415 |
| 2-Heptyl | | -0.12 (20% EtOH) | | 2.4 | 415 |
| 1-Octyl | | -0.02 (20% EtOH) | | 2.3 | 415 |
| 2-Octyl | | -0.80 (20% EtOH) | | 2.3 | 415 |
| 1-Nonyl | | -0.01 (20% EtOH) | | 2.2 | 415 |
| p-Menthane | | | -0.76 | 5.9 | 420 |
| Pinane | | | -0.80 | 5.9 | 420 |

| Compound | | | | | Ref. |
|---|---|---|---|---|---|
| Tetralin | | | −0.73 | 5.8 | 420 |
| Phenylcyclohexane | | | −0.66,h −1.08h | | 425 |
| t-Butylisopropylphenyl | | | −1.06 | | |
| Unsaturated hydroperoxides: | | | | | |
| 3-Hydroperoxy-heptene-1 | | −0.04 (20% EtOH) | −0.88 | 2.88 (20% EtOH)<br>4.45 (B–M) | 409 |
| Cyclohexene | −0.05 | | −0.77 | 5.9 (B–M) | 420, 427 |
| Cumene | −0.10, −0.12 | +0.20 (20% EtOH) | −0.68, −0.82,<br>−1.08h<br>−0.879 | 5.8 (B–M) | 409, 420, 425 |
| α-Pinene | 0.0 | | −0.82, −0.845 | 5.9 (B–M)<br>5.8 (B–M) | 426<br>420, 424 |
| Methyl oleate | | −0.54 (95% EtOH) | −0.61 | 2.66 (20% EtOH)<br>2.74 (95% EtOH)<br>4.40 (B–M) | 420 |
| Methyl ethyl ketone hydroperoxide | | | −0.949 | | 426 |
| Steroid hydroperoxides: | | | | | |
| 21-Acetoxy-3,3-<br>bishydroperoxypregnan-20-one | | −1.3 (pH 5.6)<br>−1.4 (pH 7.6) | | 2.85 (pH 2.5–3.5)<br>3.55 (pH 5.6)<br>3.33 (pH 7.6) | 428 |
| 21-Acetoxy-3,3-<br>bishydroperoxypregnan-11,20-dione | | −0.7 (pH 5.6)<br>−0.5 (pH 7.6) | | 3.21 (pH 5.6 and 7.6)<br>3.07 (pH 3.5)<br>3.27 (pH 2.5) | 428 |
| 3,3-Bishydroperoxy-Δ⁴-pregnen-20-one | | −1.32 (pH 5.6)<br>−1.4 (pH 7.6) | | 3.92 (pH 5.6)<br>3.46 (pH 7.6) | 428 |
| 21-Acetoxy-3,3-bishydroperoxy-Δ⁴-pregnen-20-one | | −1.3 (pH 5.6 and 7.6) | | 3.00 (pH 5.6)<br>2.62 (pH 7.6) | 428 |
| 3α-Acetoxy-17,17-bishydroperoxyetiocholan-11-one | | −0.7 (pH 5.6 and 7.6) | | 3.52 (pH 5.6)<br>2.81 (pH 7.6)<br>3.6 (pH 2.5)<br>3.27 (pH 3.5) | 428 |
| 3α-Acetoxy-20,20-bishydroperoxypregnane | | −1.3 (pH 5.6 and 7.6) | | 3.05 (pH 5.6)<br>2.79 (pH 7.6) | 428 |

TABLE 24. POLAROGRAPHIC DATA FOR PEROXIDES (Continued)

| Peroxide | $E^{1/2}$ (volts versus saturated calomel electrode) | | | Diffusion-Current Constant $I$[c,d] | References |
|---|---|---|---|---|---|
| | Aqueous[a] | Alcohol | Benzene–Methanol (1:1)[b] | | |
| 3α-Acetoxy-20,20-bishydroperoxypregnan-11-one | | −0.95 (pH 5.6), −1.5 (pH 7.6) | | 3.12 (pH 5.6), 3.20 (pH 7.6) | 428 |
| Dialkyl Peroxides: | | | | | |
| Di(hydroxymethyl) | −0.4 | | | | 414 |
| Diethyl | −0.5, −0.65, −0.64 | | −1.63 | 3.0 (A), 3.93 (B–M), 3.80 (B–M)[j] | 409, 414, 423 |
| Bis(1-hydroxyheptyl) | | | 0.0[h] and −1.20[h] | | 425 |
| Methyl ethyl ketone peroxide | | | −0.60[h] and −1.26[h] | | 425 |
| Dioxane | | | −0.88 | | 424 |
| Di-t-butyl | 0.0 | | −2.10 (DMF)[k] | | 406 |
| Di-cumyl | | | −1.60 (DMF)[k] | | 406 |
| Ethyl cumyl | | −1.35 (60% EtOH), −0.22 (20% EtOH), −1.64 (95% EtOH) | −1.72 | 1.7 (20% EtOH), 1.97 (95%EtOH) | 409 |
| Peroxy acids: | | | | | |
| Peroxyformic | +0.20 | | | 5.2 (A) | 423 |
| Peroxyacetic | +0.20 | | 0.0 | 4.6 (A) | 423, 425 |
| Peroxypropionic | +0.20 | | | 4.0 (A) | 423 |
| $C_{17}$ to $C_{18}$ aliphatic | 0.00 to −0.06[i] | | 0.0 to −0.06 | | 61 |
| Peroxy esters | | | | | |
| t-Butyl peroxyacetate | −0.03 | | −0.97, −1.02 | 4.7 | 404, 425 |
| t-Butyl peroxynonanoate | | | −0.96 | 4.8 | 34 |
| t-Butyl peroxydecanoate | | | −0.90 | 4.4 | 34 |
| t-Butyl peroxydodecanoate | | | −0.87 | | 34 |
| t-Butyl peroxytetradecanoate | | | −0.82 | 4.6 | 34 |
| t-Butyl peroxybenzoate | Positive | | | | 424 |
| t-Butyl-substituted peroxybenzoates (10 compounds) | −0.094 and −0.846[m,n] | | −0.8, −0.82 | | 404, 416–418 |
| t-Butyl peroxyphthalate | | | −0.70[h] and −1.05[h] | | 425 |
| t-Butyl peroxytosylate | | 0.0 (100% MeOH) | | | 429 |
| Diacyl peroxides: | | | | | |
| Dinonanoyl | | | −0.10 | 4.0 | 34 |
| Didecanoyl | | | −0.10 | 4.4 | 34 |
| Didodecanoyl | | | −0.09 | 4.2 | 34 |
| Ditetradecanoyl | | | −0.12 | 3.6 | 34 |
| Dihexadecanoyl | | | −0.10 | 3.4 | 34 |

| Compound | | | | |
|---|---|---|---|---|
| Dioctadecanoyl | | | | |
| Succinic acid peroxide | −0.01 (95% EtOH) | −0.08 | 2.8 | 34 |
| Dibenzoyl | | −0.19 | 3.1 (B–M) | 420 |
| | | 0.0, | 3.0 (B–M) | 409, 420, 430 |
| | | −0.4 (DMF), | 3.58 (95% EtOH) | |
| | | −0.24 to +0.29 | 1.85 (DMF | 416–418 |
| X = Y = | | | | |
| p-OCH$_3$ | | +0.098 | | |
| p-CH$_3$ | | +0.100 | | |
| m-CH$_3$ | | +0.100 | | |
| H | | +0.104 | | |
| p-Cl | | +1.107 | | |
| p-Br | | +0.110 | | |
| m-Br | | +0.114 | | |
| m-COOCH$_3$ | | +0.106 | | |
| m-NO$_2$ | | +0.127 | | |
| p-NO$_2$ | | +0.131 | | |
| X = p-NO$_2$;   Y = p-OCH$_3$ | | +0.122 | | |
| Bis(2,4-dichlorobenzoyl) | | 0.00[h] | | 425 |
| Transannular peroxide: | | | | |
| Ascaridole | −0.72 | −1.22[h] | | 425 |
| Polyalkylidene peroxides: | | | | |
| Dioxymethyl | −0.42 (2.5% EtOH),[o] | | | 414 |
| Acetone diperoxide | −1.03 (2.5% EtOH)[o] | | | 431 |
| | −1.42 (100% EtOH)[o] | | | |

TABLE 24. POLAROGRAPHIC DATA FOR PEROXIDES (Continued)

| Peroxide | $E^{1/2}$ (volts versus saturated calomel electrode) | | | Diffusion-Current Constant I[c,d] | References |
| --- | --- | --- | --- | --- | --- |
| | Aqueous[b] | Alcohol | Benzene–Methano l(1:1)[b] | | |
| Acetone triperoxide | | -1.18 (95% EtOH)[o,p] | -1.46 (95% EtOH)[j] | 6.9 (95% EtOH)[o] | 409 |
| (Peroxides) | | -1.36 (95% EtOH)[q] | -1.93 (DMF)[q] | 8.4 (95% EtOH)[q] | |
| | | -1.82 (95% EtOH)[g] | | 3.2 (DMF)[q] | |
| Cyclohexanoner | -0.0 | | | | |
| Bis(1-hydroxycyclohexyl) | | | -0.92 | | 424 |
| 1-Hydroxy 1′-hydroperoxy-dicyclohexyl | | | -1.28 (B–M, 3:7) | | 432 |
| Bis(1-hydroperoxy-cyclohexyl) | | | -0.90, -1.28 | | 432 |
| | | | -0.40, -0.90, -1.28 | | 432 |

[a] Supporting electrolytes: 0.1 M LiSO₄.; 0.01 N HCl, 0.1 N KCl, and 0.04 M Bitter–Robinson buffer, pH 7.5; acetate buffer, pH 4; phosphate buffer, pH 9.

[b] Supporting electrolyte: 0.3 N LiCl.

[c] $I = i_d/Cm^{2/3}t^{1/16}$. The diffusion-current constants reported by Skoog and Lauwzecha (415) are designated $i_d/c$. Since this term does not include the capillary constant, values listed in this table were accordingly adjusted by the appropriate capillary constant.

[d] Designation of abbreviations: A, aqueous; Ac, glacial acetic acid; B–M, benzene–methanol (1:1); B–M, 3:7, benzene–methanol (3:7); D, dioxane; DMF, dimethylformamide.

[e] Supporting electrolyte: 0.01 N LiOH.

[f] Supporting electrolyte: 0.1 N KCl; 0.1 M H₂SO₄; 0.1 F H₂SO₄.

[g] Supporting electrolyte: 0.1 N KCl; 0.1 N LiCl; 0.2 N (CH₃)₄ NOH in 60% EtOH.

[h] Supporting electrolyte: 0.3 N LiCl + 0.01% ethyl cellulose.

[i] The alcoholic solution for all steroid hydroperoxides was 85% ethanol in water.

[j] Supporting electrolyte: 0.1 N (CH₃)₄NCl.

[k] $E_{1/2}$ versus mercury pool; DMF saturated with (CH₃)₄NCl; irreducible in other systems.

[l] In glacial acetic acid.

[m] In 80% dioxane in water.

[n] First wave grows and second wave diminishes with increase in water content (416–418).

[o] $E_{1/2}$ dependent on alcohol content.

[p] Supporting electrolyte: 0.3 N CH₃COONa.

[q] Supporting electrolyte: 0.1 N (C₄H₉)₄NCl.

[r] The structure of cyclohexanone peroxide was not fully defined but appears to be 1-hydroxycyclohexyl hydroperoxide.

dependent on concentration (419). The diffusion-current constant of hydroperoxides in benzene–methanol (1:1) is independent of concentration, but the waves at high concentration of hydroperoxide are distorted by the formation of maxima, which complicate the analysis (420). The effect of concentration on half-wave potentials and diffusion-current constant has not been closely examined for other peroxide classes.

Solvents used in polarography are selected not only for solubilization of the peroxide but also for their strong influence on the half-wave potential and diffusion coefficient. Water has the advantages of good solubility for electrolyte and low electrical resistance, but it dissolves only low-molecular-weight peroxides (415, 423). Proske (421) enhanced the solvation properties of water for water-immiscible organic peroxides by solubilizing them with the commercial wetting agents Aerosol® AY and AM (American Cyanamide Co.), which are stated to be dihexyl and diamyl sodium sulfosuccinates, respectively. The wetting agents also performed as supporting electrolytes in the analysis of fat peroxides. Addition of benzyl alcohol improves both shape and height of the wave, apparently by increasing mutual solubility of the soap micelles and water. Most peroxides may be solubilized by the addition of water-soluble organic solvents, such as methanol or ethanol (409, 415, 430, 433) or ethylene glycol monoethyl ether (434). Such cosolvents as dioxane, acetone, and acetonitrile (416, 417) and tetrahydrofuran (435) may also serve, but they have been less commonly used for peroxides.

Nonaqueous solvents are highly effective in polarographic studies on organic peroxides. Lewis, Quackenbush, and DeVries (436) devised benzene–methanol (1:1) containing lithium chloride (0.3 $N$) as electrolyte. This medium has been the most widely used system for the polarographic investigations of peroxides in fats and oils (422, 424, 437–448) and for determining the polarographic characteristics of a variety of peroxides in a nonaqueous system (34, 409, 420, 425). Aside from a few nonaqueous media—benzene–methanol, ethanol, and glacial acetic acid for peroxy acid determinations (61, 62, 449) and dimethylformamide (406, 409), the latter having decided advantages for peroxides—the utility of other organic solvents has not attracted much interest.

The half-wave potential of peroxides is independent of pH (415, 419), but Skoog and Lauwzecha (415) reported a very slight positive shift with increasing pH. Schulz and Schwarz (416–418) also found the reductions of hydroperoxides to be independent of pH in the pH range 3–9 but to increase below pH 3 and above pH 9. The protogenic character of the solvent strongly affects the electrochemical reduction of peroxides (409). Decreasing protogenic character of solvent in the series water or dilute aqueous ethanol, 95% ethanol, benzene–methanol (1:1), dimethylformamide is paralleled by a systematic negative shift in half-wave potential for reduction of the peroxide (Table 24) and by a wave elongation, indicating increasing irreversibility. In aprotic solvents, such as dimethylformamide (with tetraethylammonium chloride as supporting electrolyte), the half-wave potential becomes more negative. For instance, benzoyl peroxide shifts from $-0.1$ volt in benzene–methanol (1:1) to $-0.4$ volt in dimethylformamide, the only solvent in which this peroxide is more negative (409). It is to be noted that the polarographic reduction of di-$t$-butyl peroxide ($-2.1$ volts) has been reported for the first time by application of a solvent in which the negative potential range is extended to $-2.5$ volts (406).

Organic solvents have a marked effect on wave shape; this is evident for the reduction of cumyl hydroperoxide (Figure 49$a$) and ethyl cumyl peroxide (Figure 49$b$) (409). The waves are elongated more in nonaqueous solvents than in water or in dilute ethanol. Reductions in such solvents often follow the equation of an irreversible wave (equations 58 or 59). Deviations from linearity for [log $i(i_d - i)$] versus $-E$ are presumably determined by the absorption of peroxide on the mercury electrode. This was experimentally detected for an electrode in the presence of a peroxide, which showed a decrease in drop time. The absorbability increases with increasing polarity of the solvent and is greater for hydroperoxides than for dialkyl peroxides (409, 416–418). The influence of solvent composition on the wave characteristics of peroxide reductions has also been observed by other investigators (415, 424).

Supporting electrolytes—such as potassium, lithium, or ammonium chloride, methyl acid sulfate, lithium methoxide, lithium sulfate and lithium hydroxide—have been used in water, aqueous alcohol and benzene–methanol solutions (405, 409, 423, 436). Salts of sodium or potassium allow reductions at potentials more positive than about $-1.8$ volts; with lithium salts the negative range is extended 0.1 volt (397, p. 62). Chloride as an anion is undesirable in the anodic region because of its ease of oxidation, the preferred electrolyte being lithium sulfate (409, 423), lithium acetate (61), or ammonium nitrate (418). Alkaline systems are to be avoided because of hydrolysis or decomposition of peroxides in such media and because of their adverse effect of elongating the polarographic wave (415–418). The use of buffers, although they are not of major importance in peroxide determinations, permits satisfactory polarograms (415).

The negative potential range can be extended to $-2.5$ volts versus the saturated calomel electrode as reference base) by using salts of quaternary amines, such as tetraethylammonium chloride or bromide (397, p. 62; 409). A variety of these salts is commercially available for this application.

The negative shift in the half-wave potential on changing from lithium to tetraalkylammonium salts is attributed to the screening effect of the quaternary amine salts on the electrode surface, which influences electrode reaction mechanisms involving adsorption and desorption (397, p. 68; 450). In such solvents as acetonitrile and dimethylformamide, in which halides may interfere

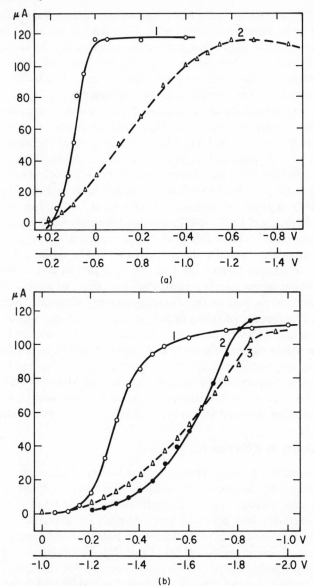

**Figure 49.** Polarization curves showing effect of organic solvent on wave shape.

(a) Cumyl hydroperoxide:
1—in 20% ethanol, origin +0.20V; 2—in 95% ethanol, origin −0.20V.
(b) Ethyl cumyl peroxide:
1—in 20% ethanol, origin 0.0V; 2—in 95% ethanol, origin −1.0V;
3—in benzene/methanol (1:1), origin −1.0V. (Reference 409).

by functioning as strong bases (451), quaternary ammonium halides may be replaced by tetraethyl fluoborate (452). This salt has the added advantage of being less absorbed than other quaternary ammonium salts at the cathode because of its small size. Should the need arise for electroreductions at higher negative potentials, the range may be extended to $-3$ volts by means of tetrabutylammonium perchlorate in glyme solution.*†

Experimental difficulties are encountered with reactive peroxides in their reduction at the mercury electrode. High concentrations of peroxide react with mercury to deposit a calomel film on the surface of the mercury pool; this interferes with potential measurements (427). A small early wave is observed for oxidized fats and oils between 0 and $-2.3$ volts (422, 436–438, 442), which is the region of reduction of peroxy acids and diacyl peroxides (34, 425, 449) in benzene–methanol (1:1) with 0.3 $N$ lithium chloride as electrolyte. Maack and Lück (454) investigated the early wave with a cathode-ray polarograph and related this peak to the reduction of the mercury ion. Reactive peroxides, especially diacyl peroxides (448), oxidize the mercury metal of the electrode to mercury ions. In light of this result this class of peroxides should be intensively investigated in a variety of solvent systems on an inert electrode as well as the dropping-mercury electrode. Roberts and Meek (455) developed a vibrating bright platinum electrode in order to aviod the complications of such reactions with mercury. Their electrode, used in conjunction with a calomel electrode and agar-salt bridge, gave polarograms from ethyl hydroperoxide reductions equivalent to those obtained by conventional dropping-mercury electrodes. Brüschweiler and Minkoff (423) observed that peroxides may be catalytically decomposed by agar, gelatin, or sintered glass, necessitating the need for an external calomel reference electrode.

### 4.  Substituent Effect on $E_{1/2}$

Peroxides are reduced either chemically by a variety of chemical reagents or electrochemically by electrons as the reagent. The ease of reducing a peroxide depends in large measure on the strength of the O—O bond. The relation between reducibility and peroxide bond strength for aliphatic peroxides is suggested by a correlation between half-wave potential and energy of activation for thermal decomposition (128) (Table 25). This relation is similar to the order of peroxide reactivity with the iodide ion, for example, di-$t$-butyl peroxide < trimeric ketone peroxides < dialkyl peroxides and dimeric ketone

---

* The original reference to this report was misplaced and was not relocated in the literature. The use of tetrabutylammonium perchlorate in dimethyl sulfoxide extends the range to $-2.7$ volts (453).

† The inherent dangers of handling perchlorates, especially in organic solvents, are to be recognized and the necessary safety precautions implemented.

TABLE 25. HALF-WAVE POTENTIAL AND ACTIVATION ENERGY OF DECOMPOSITION[a]

| Parameter | Di-*t*-butyl Peroxide | Dialkyl Peroxide | *t*-Butyl Peroxy Ester | Hydro-peroxide | Diacyl Peroxides | Peroxy Acids |
|---|---|---|---|---|---|---|
| Half-wave potential (volts versus saturated calomel electrode) | −2 | −1 | −0.8 to −1.0 | −0.6 to −0.9 | −0.10 | 0.00 to −0.06 |
| Activation energy (kcal/mole) | 38 to 40 | 36 to 37 | 35 to 36 | 27 to 32 | 30 | 24 |

(a) Data from reference 128.

peroxides < *t*-butyl peroxy esters < hydroperoxides < diacyl peroxides $\gtrsim$ peroxy acids (34, 128, 425). Chemical reactions with other reagents show a similar relationship (456).

The strength of the peroxide bond depends on the structure (the nature of the substituents) and on the environment (the nature of the solvent, electrolyte, concentration, pH). Electrophilic substituents weaken the O—O bond by diminishing its electron density; the effect is a displacement of the half-wave potential to positive (toward anodic) values. Correspondingly, electrophobic substituents strengthen the peroxide bond by increasing electron density at this site causing a shift to negative (toward cathodic) values.

The electron-donating effect of alkyl branching is evident in a series of dialkyl peroxides and hydroperoxides listed in Tables 24 and 25. Dialkyl peroxides that contain an electron-donating alkyl group on each of the oxygen atoms are thermochemically more stable and reducible at potentials more negative than hydroperoxides that contain a single alkyl substituent. This effect of the alkyl group is further evident in comparing peroxides containing the electron-withdrawing group attached to the peroxide oxygen; that is, peroxy esters, diacyl peroxides, and peroxy acids. Tertiary and secondary alkyl hydroperoxides reduce at potentials more negative than *n*-alkyl hydroperoxides because of the electron-donating effect of alkyl branching.

Di-*t*-butyl peroxide is thermochemically the most stable peroxide in the aliphatic series and is accordingly reduced at the most negative potential. Both of these characteristics may be attributed to the strong electron-donating induction of the *t*-butyl group, which places the largest electron density about the peroxide oxygen atoms. However, steric hindrance due to the bulky *t*-butyl group may contribute to the cathodic shift by impeding the orientation of the peroxy group at the electrode surface, since an essential step in electroreduction (457) must be overlap of the bulky electron source with the antibonding orbital of the reducible group. The energy required to add an electron to the lowest unoccupied molecular orbital is greater for a compound that is sterically hindered than for one less hindered.

Reduction of a single bond in comparison with a $\pi$-bond requires a more negative potential for addition of an electron to the lowest empty molecular orbital (457, 458). This explains the facility of reducing at less negative potentials those peroxides that bear a carbonyl function on the peroxide group. It should be noted that it is the peroxycarbonyl group that is the electroreducible site in these compounds rather than the peroxide bond itself. In *t*-butyl peroxy esters the electron-donating alkyl group and the electron-withdrawing carbonyl function approximately cancel their mutual push-pull effects, so that these compounds are reduced at near the potential range for dialkyl peroxides and hydroperoxides.

*A priori*, diacyl peroxides would be expected to reduce at the lowest potentials because of the attachment of two electronegative acyl groups. Peroxy acids, instead, claim this distinction. This has been attributed to intramolecular hydrogen bonding (**1**)

$$R-C\overset{\displaystyle O\cdots}{\underset{\displaystyle O-O}{\diagup}}\overset{\displaystyle H}{\diagup}$$

**(1)**

in which electrons in the cyclic configuration are withdrawn from the peroxide group toward the hydrogen bond through the polarization of the carbonyl function (405). The five-membered chelate ring is a structure of lower energy than the open-chain form (355). Cyclization appears to be equivalent in effect to a second electron-withdrawing acyl grouping, since peroxy acids and acyl peroxides reduce in the same potential region. It is of interest that the intramolecularly-hydrogen-bonded structure may also account for the peroxy acids, peroxyformic and peroxyacetic, having a higher diffusion rate than hydrogen peroxide (423). The ease of reducing aliphatic peroxy acids (0.0 volt) suggests that the reduction of aromatic peroxy acids may take place in the anodic potential region similar to that of aroyl peroxides.

The influence of environmental factors on half-wave potentials was already discussed in a general way. The nature of the medium influences the strength of the peroxide bond for electroreductive cleavage either by operating directly on the peroxy group or indirectly through its effect on the skeletal structure. In a series of solvents dialkyl peroxides, hydroperoxides, peroxy esters, and aliphatic acyl peroxides manifest positive shifts in their half-wave potentials with increasing chain length. The shift appears to be most evident for the lowest members of the series. Peroxy acids do not show this trend, although refined measurements may suggest a slight effect. Positive shifts in half-wave potential are largest on moving from nonaqueous to aqueous media. It is clearly evident from the data in Table 24 that the medium is exerting its influence through the chain, the magnitude approaching a limit in each peroxide series. The limit in the hydroperoxide series in 20% ethanol is approached at $n$-hexyl ($-0.03$ volt) and continues with slight positive increments up to $n$-nonyl ($-0.01$ volt). The chain-length limit is not known for each series because of insufficient extensions of the peroxide classes with respect to chain length and variety of solvents.

Aliphatic $t$-butyl peroxy esters resemble hydroperoxides in their polarographic behavior by reducing in the adjacent potential range and by showing corresponding positive shifts in $E_{1/2}$ with increasing alkyl-chain size. $t$-Butyl peroxyacetate has been reduced in aqueous solution [$-0.3$ volt versus saturated calomel electrode (SCE)] and in nonaqueous media such as benzene–methanol (1:1) ($-1.02$ volts versus SCE) (425). $t$-Butyl peroxybenzoate and its substituted derivatives give polarograms in aqueous dioxane solution consisting of two waves, each depending on solvent composition (416–418). The first wave with a kinetic character occurs at $-0.094$ volt (versus SCE) and the second wave, which is diffusion controlled, forms at $-0.85$ volt (versus SCE). The first wave increases at the expense of the second wave in solutions of increasing water content, suggesting the rate-controlling step to be the heterolytic polar wave cleaving the O—O bond at the dropping-mercury electrode. The second wave was implied to be the single-step homolytic cleavage.

The mechanism that is described by Given and Peover (459) to explain the reductions of polynuclear aromatic hydrocarbons in solvents of low and high proton availability and that is probably applicable to aromatic ketenes and quinones provides a rationale for the electroreduction of peroxybenzoates. In the first wave one electron is reversibly added and the free-radical ion produced either diffuses away from the electrode or is protonated on the electrode. If the latter is true, it must be reduced at the same potential as the first ion. [Given and Peover cite Hoijtink's (460) quantum-mechanical conclusions that the reduction potential of a radical RH· must always be less negative than that of a hydrocarbon molecule from which it is formed. This conclusion is assumed to be applicable to peroxybenzoate reductions.] Under these circumstances only one wave of two-electron height will be observed. If the mononegative ion is not protonated on the electrode, it may diffuse and protonate away from the electrode and a further wave will be observed when the potential is sufficiently negative. Thus in aprotic solvents and solvents of low proton availability two one-electron waves will be observed. This mechanism is offered as a tentative explanation of the results reported by Schulz and Schwarz (418).

In completely aqueous media $t$-butyl peroxybenzoate moves to the positive side of 0.0 volt. Mair and Hall (406) offer the suggestion that peroxy ester reductions may proceed through a peroxy acid intermediate; this is contrary to the accepted view (35, 461) that peroxy esters are derivatives of, and hydrolyze to, acids and hydroperoxides rather than to peroxy acids and alcohols.

Benzoyl peroxide and its substituted derivatives reduce in dioxane–water (0.25 $N$ $NH_4NO_3$) on the positive side of 0.0 volt (418). Inductive and resonance effects of the substituents influence the half-wave potential in the expected manner—that is, toward more positive potentials by electrophilic groups and toward less positive potentials by electrophobic groups. However, the Hammett-substituent effect on the half-wave potential of dibenzoyl peroxides is slight.

### 5.  *Analytical Applications*

Electrochemical reductions of peroxides have not been adequately examined for applications other than for analysis. Knowledge of the reaction mechanism of peroxides at the electrode surface is scanty, and an understanding of the process is limited by the absence of controlled-potential electrolytic studies, which are essential for a detailed description of intermediate and final reduction products. In special organic-synthesis applications controlled-potential studies may be of value in the selective reduction of two or more peroxide groups in a molecule or of reducible nonperoxide groups in a molecule containing a peroxide grouping.

Polarography is a valuable technique for peroxide determinations. It is rapid and capable of high sensitivity ($10^{-6}$–$10^{-7}$ M) but may be less precise than some conventional procedures. An accuracy of 1–3% is general in the range of $10^{-4}$–$10^{-5}$ M and of about 10% at $10^{-6}$ M (444, 462, 463). A statistical comparison of the polarographic method with the Wheeler iodide and stannous chloride methods for determining hydroperoxides gave results that were not significantly different and in support of the polarographic method for its reliability as well as its specificity (462).

Quantitative analysis depends on the diffusion-current constant $I$, expressed by equation 57. The value of this constant for the determinable compound in a selected medium is obtained by calibration with a particular capillary. The constant may be applied in the *absolute method* (equation 57) for computation of the unknown concentration of the compound from the observed $i_d$ and the measured value of $m$ and $t$ for the dropping-mercury electrode used. Alternatively any of the following three techniques may be applied (396, pp. 375–385):

1. *Empirical calibration curves* of the analyzable compound may be constructed showing diffusion current or waveheight as a function of concentration from which the concentration of an unknown is assessed.

2. In the *method of standard addition* the polarogram of the unknown solution is first recorded, then a known value of a standard solution of the analyzable compound is added to this solution, and a second polarogram is recorded. The concentration is determined from the increase in the diffusion current.

3. The *internal standard method* is based on the fact that relative diffusion-current constants of various compounds at equal concentrations are independent of the particular capillary used provided all other factors are kept constant. Given the accurate diffusion-current constant for these compounds, one compound may be used as an internal standard to calibrate the capillary used for determination of the other. Pure compounds or standards of known concentration analyzed by chemical methods may be used for calibrations.

Because diffusion-current constants vary between and within classes of peroxides and are affected by the nature of the medium, a limitation is placed on the accuracy in quantitative analysis. Willits et al. (420) found $I$ to be nearly equivalent ($5.85 \pm 0.05$) for a variety of low-molecular-weight hydroperoxides. This value cannot be used as a constant in high-accuracy determinations of unknown peroxide mixtures, since the diffusion coefficient of other peroxide types (e.g., diacyl peroxides) are not equivalent and vary with chain length.

Martin (404) suggests, for order-of-magnitude estimates within 30%, the use of $I$ values of 6.3 for aqueous solutions and 5.3 for benzene–methanol (1:1) solutions.

Silbert et al. (34) sought a relation between the diffusion coefficient $I$ and the molecular weight $M$ that would take advantage of the dependence of $I$ on chain length. The relations using either $i_d M^{2/3}/C$ or $i_d M^{1/2}/C$ equals a constant are comparable ($C$ is concentration and the capillary constant $m^{2/3}t^{1/6}$ is assumed a constant in each series and may be useful for determining the approximate molecular weight of the unknown peroxide). The general applicability of these relations has not been examined.

The half-wave potential affords detection of specific peroxide types as well as other reducible functions in oxidized products. The special capabilities of polarography provide the incentive to investigate complex mixtures in autoxidized solvents, fuels, and fats in greater detail.

It is imperative in polarographic analysis to first remove dissolved oxygen from solution in the polarographic cell by its displacement with helium, nitrogen, or carbon dioxide. Otherwise the current resulting by reductions at potentials more negative than about 0.1 volt (versus SCE) may mask the waves of the reducible components. To prevent substrate-concentration changes in volatile solvents, the purging gas must be saturated with the same solvent or solvent mixture prior to its passage through the cell. Typical polarograms of oxygen reductions in aqueous media are shown in Figure 50.

The pronounced maximum in curve 1 is a common phenomenon, also exhibited by other compounds. The maxima may be eliminated by addition of small amounts of certain high-molecular-weight materials, such as dyestuffs, gelatin, or ethyl cellulose (396, pp. 156–188; 425). The reduction of oxygen in the aprotic solvent dimethyl sulfoxide has been shown to proceed through two steps at the dropping-mercury electrode (464). The first wave is due to the reaction $O_2 + e = O_2^-$ and the second wave of lower height to the reaction $O_2^- + e + \text{cation} = \text{peroxide product}$. An excellent detailed review is available on the electrochemical reductions of oxygen (465). The polarographic method has been usefully applied to the determination of oxygen in various media (396, pp. 375–385; 466–470), fats (471), petrol (472), and soil (473).

Reimers (474) examined ether peroxides by removing hydrogen peroxide with ether-saturated 0.1 $N$ lithium hydroxide. Extraction of hydrogen peroxide eliminated the complicated nature of its wave, which was 23% higher than the wave from an equivalent concentration of ether peroxide. He determined acetaldehyde and ether peroxide simultaneously by directly analyzing the ether solution in ethanol, using tetramethylammonium hydroxide as electrolyte. Autoxidized isopropyl ether and $n$-butyl ether gave distinctly different polarograms, illustrating the nonidentity and differentiation of the unidentified peroxides in these two solvents (436).

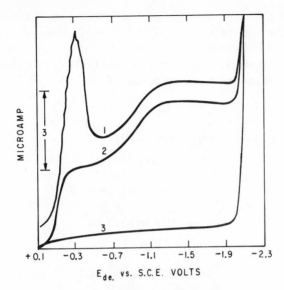

**Figure 50.** Polarographic curve of oxygen.
(1) 0.05N potassium chloride solution saturated with air.
(2) After addition of a maximum suppressor (trace of methyl red).
(3) Residual current after removal of oxygen by nitrogen sparging. (Reference 396, page 553).

Simultaneous analysis of organic peroxides and aldehydes has been of interest to several other investigators (413, 414, 423, 475, 476). Hydrogen peroxide may be removed by its precipitation with lanthanum acetate in alkaline solution (423) or with titanium tetrachloride in 1 *N* hydrochloric acid, followed by addition of ammonium hydroxide (477). A three-step technique was subsequently developed in which hydrogen peroxide is determined in an acid buffer solution without interference by aldehydes; formaldehyde is determined by titanium salt precipitation of hydrogen peroxide in alkaline solution; and acetaldehyde is estimated in the presence of dimedone, which reactively removes formaldehyde (478). In one study of acetaldehyde oxidation the cool-flame condensate obtained was shown to consist of acetaldehyde, formaldehyde, and diethyl peroxide by means of their polarograms (413).

Ketone peroxides have been conveniently analyzed polarographically. Acetone diperoxide decompositions have been followed polarographically in 2.5 and 30% alcohol solutions (431). Reductions in aqueous solutions containing up to 5% ethanol were characterized by a two-step wave. With solutions containing more than 10% ethanol, the polarograms consisted of one wave independent of pH. Conditions for the polarographic determination of bis(1-hydroxycyclohexyl) peroxide (**46**), 1-hydroperoxycyclohexyl 1-hydro-xycyclohexyl peroxide (**47**), and bis(1-hydroperoxycyclohexyl) peroxide (**48**)* were established and a method was described for the polarographic reduction of a mixture of **46** and **47** in benzene–methanol (3:7) against a background of 0.2 $N$ acetic acid and 0.2 $N$ sodium acetate (432).

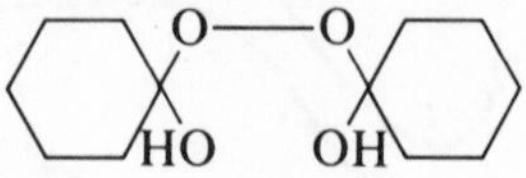

(**46**)

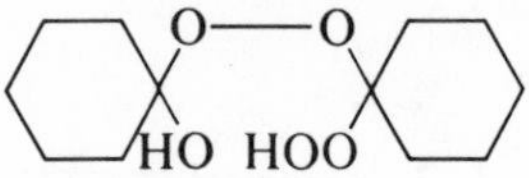

(**47**)

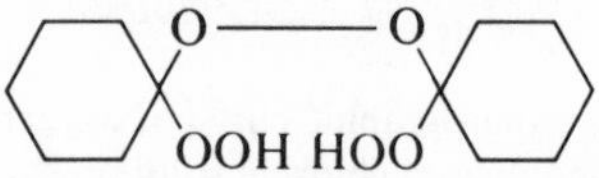

(**48**)

Steroid bishydroperoxides (428) containing the $\diagup\!\!\!\!C(OOH)$ function were examined in water–ethanol (1:20) with sodium $p$-toluene sulfonate as electrolyte between pH 2 and 8. These compounds reduced at four half-wave potential ranges (0 to 0.2; 0 to $-0.05$; $-0.25$ to $-0.45$, and $-1.15$ to $-1.25$ volts versus SCE).

Gasoline-autoxidation products were qualitatively examined polarographically (406, 479). Hydroperoxides in motor gasoline have been measured by the polarographic and Ampot methods, and the results agreed excellently with chemical analysis (426). The Ampot (Sargent Scientific Company) measures diffusion currents in a modification of the polarographic method. The authors conclude that the agreement among the three analytical techniques and among the half-wave potentials for a variety of cracked gasolines treated by thermal, catalytic, hydrogenation, and ultraviolet processes justifies dispensing with the polarogram for these measurements and simplifying the analytical procedure by relying entirely on two diffusion-current measurements obtained with any simple device capable of polarometric measurements.

The autoxidation of fats and oils produces a complicated mixture of oxidized products, whose identifications, although still incompletely resolved, have been partially aided by polarography. It was shown in the autoxidation

* Shown in different form as structure **37**.

of fats (lard, oleate, and linoleate esters) that a direct proportionality exists between peroxide values measured polarographically and chemically (421, 436, 437, 480). Lewis and Quackenbush (437) presented the first polarographic evidence showing that the iodometric method for determining the peroxide value includes more than one type of peroxide and that all three waves in lard disappear when it is treated with potassium iodide. Subsequently several groups of workers, particularly Willits et al. (441, 442) and Swern et al. (440) demonstrated that all of the peroxides formed during autoxidation could not be related to the hydroperoxide content alone, although this is the major constituent in the initial autoxidation stages. Lück and Maack (447) applied oscillographic polarography in a quantitative differentiation between the two techniques of polarographic and chemical analyses, the polarographic values amounting to less than the chemical values by approximately 50 % for fats and by 10–28 % for methyl esters of unsaturated fatty acids. Swern (440) had attributed this difference to the formation of peroxides that he designated "cyclic peroxides."

The isolation of peroxides other than hydroperoxides was accomplished by Khan (481), who successfully crystallized solid peroxides from autoxidized methyl oleate and methyl stearolate. The polarograms of these peroxides were similar to those of ketone peroxides. Autoxidations of soybean oil at 60° by Kalbag and co-workers (422) indicated the presence of peroxides in larger proportions in soybean oil than in methyl esters of similar composition. This material was associated with the polymerized fraction and reduced at a more positive potential than did hydroperoxides. Peroxides produced in parallel oxidations of soybean oil in the presence of 0.1 % cobalt catalyst were qualitatively different from those produced in the absence of the drier. These peroxides were considered to have the cyclic structure on the basis of their polarographic characteristics and from the studies of Knight and co-workers (482).

Although hydroperoxides and cyclic peroxides of the aldehyde and ketone types are among the products of oxidized fats, their precise identification and structures must await isolation of the pure compounds by sophisticated separation techniques. It is noteworthy in several of the foregoing studies (422, 438–442) that the identity of an observed preliminary wave between 0 and $-0.2$ volt that had been attributed to an unidentified peroxide type appears to have been resolved with a cathode-ray polarograph. Maack and Lück (454) traced the cause of this peak to a reduction of mercury ions, which were derived from the mercury metal of the electrode by peroxide oxidation to mercury ions.

Polarography has been applied to the structural determination of some polymeric peroxides (483–485) and to the analysis of residual peroxide initiators in polymers (486–489). Bovey and Kolthoff (483) examined polystyrene peroxide in a benzene–ethanol–water medium. The peroxide concentrations were calculated on the basis of the —C(Ph)HCH$_2$OO— unit. The polarograms with half-wave potential close to $-1.0$ volt (versus SCE) resembled those described by Stern and Polak (414) for diethyl peroxide. It was suggested that the dip in the diffusion current for the polymer is due to desorption of the compound at potentials more negative than $-1.5$ volts. Barnes et al. (484) studied the peroxide polymers of methyl methacrylate and vinyl acetate in acetone as solvent and 0.1 $N$ tetramethylammonium bromide as electrolyte. The autoxidation products of methacrylates, peroxide, and pyruvate were determined in benzene–methanol (1:1) and 0.3 $M$ lithium chloride as electrolyte. Residual initiator concentrations of benzoyl peroxide in polymethyl methacrylate (487) and of cumene hydroperoxide in butadiene–styrene latex (489) were readily analyzed polarographically.

Ascaridole, a bicyclic terpene peroxide constituent of oil of chenopodium, cannot be correctly analyzed iodometrically, since the amount of iodine liberated is not in stoichiometric proportion (490). The results are generally too high (491–493) because of partial reaction at the double bond, but an accurate calculation is derived by using a correction formula (491, 492). The polarographic method is preferable to the iodometric procedure for the analysis of this peroxide because of its specificity, rapidity, and accuracy. Ascaridole reductions are independent of pH (494) and have been reported at $-0.89$ volt (versus SCE) in 50% ethanol containing 0.2 $N$ lithium sulfate (494), $-1.20$ volts (versus SCE) in benzene–ethanol (3:7) containing 0.1 $N$ lithium chloride (495), and $-0.86$ volt (versus SCE) in 50% ethanol containing 0.1 $N$ lithium chloride buffered at pH 3 (495).

The use of polarographic analysis in following reaction kinetics has been reviewed by Semerano (496), Page (497), and Zuman (498), and additional examples were outlined by Perrin (400, p. 301). Examples of its use with peroxides are the kinetics of peroxylauric acid expoxidation of unsaturated fatty acid esters (449), the decomposition of $t$-butyl pertosylate (429), the autoxidation of colza oil (rapeseed oil) (499), and the kinetics of alkaline hydrolysis of $t$-butyl peroxybenzoate (500).

### B.  Potentiometry

Peroxy acids, hydroperoxides, and certain peroxide derivatives of aldehydes or ketones containing the OOH group are weak acids, titratable with base. Peroxy acids, which have a p$K_a$ of about 3.5 units greater than the parent acid, may be titrated potentiometrically in water as weak acids with 0.1 $N$ sodium hydroxide solution (85a, 353). The potentiometric titration curve and first-derivative curve obtained on a typical water-soluble peroxy acid when titrated with dilute alkali is illustrated in Figure 51. The first step corresponds to the parent carboxylic acid and the second to the peroxy acid. Carbon

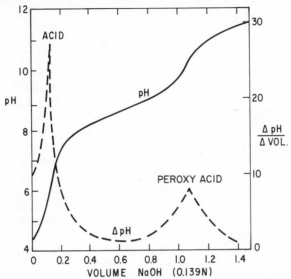

**Figure 51.** Potentiometric titration and first derivative curve of a carboxylic acid-peroxyacid mixture in aqueous solution. (Reference 85a).

dioxide interferes in the titration of peroxyformic acid but not with the higher members of the series—peroxyacetic, peroxypropionic, or *n*-peroxybutyric acids. The $pK_a$ of the peroxy acids is estimated from the pH at the half-neutralization point.

Hydrogen peroxide is a stronger acid than hydroperoxides. The dissociation constant of hydrogen peroxide, initially determined by Joyner (501), was redetermined by Evans and Uri (502) with regard to the effect of ionic strength. The dissociation constant at 20° for zero ionic strength was found to be $1.78 \times 10^{-12}$ (i.e., pH 11.75). The measurements at other temperatures gave the following results: at 15°, $K = 1.39 \times 10^{-12}$; 20°, $1.78 \times 10^{-12}$; 25°, $2.24 \times 10^{-12}$; 30°, $3.55 \times 10^{-12}$. The heat of dissociation calculated from these results is 8.2 kcal/mole.

Everett and Minkoff (85a) compared their dissociation constant of hydrogen peroxide with those of various hydroperoxides (Table 26), using a spectroscopic technique for the determination of the pH at the neutralization point. The $pK_a$ values range from 11.6 to 12.8. The hydroperoxides can be titrated potentiometrically in strongly basic nonaqueous media; Martin (503) chose ethylenediamine with sodium aminoethoxide as titrant and antimony electrodes. An inverse relationship is suggested between the potential at the half-neutralization point in ethylenediamine and the $pK_a$ value of the hydroperoxide in aqueous solution (Table 26). The hydroperoxide values indicate a

TABLE **26.**   EFFECT OF ACIDITY AND HYDROPEROXIDE
STRUCTURE ON POTENTIAL AT
HALF-NEUTRALIZATION

| Hydroperoxide | $pK_a$ (20°)[a] | Potential at Half-Neutralization (volt)[b] |
|---|---|---|
| Hydrogen peroxide | 11.6, 11.75[c] | 0.58 |
| Methyl | 11.5 | — |
| Ethyl | 11.8 | 0.55 |
| Isopropyl | 12.1 | — |
| Cyclohexyl | — | 0.54 |
| Cumene | 12.6 | No inflection |
| 2-Isobutyl propyl | 12.8 | — |
| t-Butyl | 12.8 | No inflection |

[a] Data from reference 85a.
[b] Data from reference 503.
[c] Data from reference 502.

relationship between structure and potential at half-neutralization. The acid strength of a hydroperoxide decreases in the order primary > secondary > tertiary.

Peroxy acids were also stated to be titratable in ethylenediamine (503). This is surprising in view of known peroxy acid oxidations of amines (504). Later work by Lefort and Sorba (505) showed the incorrectness of the claim by proving the titratable species to be the parent acid.

## IX.   MOLAR REFRACTION

Molar refraction (molecular refractivity), an aspect of molar volume, is an additive and constitutive property (506–508). Although molar refraction has contributed to the classical elucidation of structure of many organic compounds, there have been no important structural analyses of peroxides by this technique. This may have resulted from the general attraction for the more elegant spectroscopic techniques that were developed during the same period, during which various peroxide classes were being synthesized.

Despite the simplicity of density and refractive-index measurements, reliable data from which molar refractions can be derived on pure compounds are limited. Molecular refractivity is nevertheless an important measure of the polarizability of the outer-shell electrons in a molecule and in this regard is necessary for many physical (509) and chemical (508) considerations (refractive indices, electric dipole moments, term values of nonpenetrating orbits, van der Waals forces, electronic mechanisms).

Polarizability is expressed by the relation

$$R_M = \tfrac{4}{3}\pi N \alpha_M$$

where $R_M$ is molar refraction, $N$ is Avogadro's number, and $\alpha_M$ is the molecular polarizability. The most important of several relations used to measure molar refraction is the Lorentz–Lorenz relation, expressed by

$$R_M = \frac{M}{d}\frac{n^2 - 1}{n^2 + 2}$$

where $M$ is the molecular weight, $d$ is the density, and $n$ is the refractive index.

In an early publication Rieche (84c) reported two different values for the molar refraction of the peroxide group; that is, 4.04 for dialkyl peroxides and 3.7 for hydroperoxides. Milas (510) recognized the inconvenience of having different group values for different peroxide classes. He proposed a single value of 2.19 for the peroxidic oxygen (O\*). This value was derived as the difference in molar refractivity between hydrogen peroxide and water. Justification for the single value is illustrated from Rieche's examples (84c). By subtracting 1.643, the atomic refractivity of ether oxygen, from 4.04 and similarly 1.525, the atomic refractivity of alcohol oxygen, from 3.7, values of 2.40 and 2.17, respectively, are obtained, in agreement with the atomic refraction of peroxidic oxygen. Milas indicated the near identity of this value to the atomic refractivity of the carbonyl oxygen (2.211). He correctly rejected the double-bonded oxooxide structure for peroxides

$$\text{R}-\overset{\displaystyle \|}{\underset{\displaystyle \text{O}}{\text{O}}}-\text{R} \qquad\qquad (\text{R} = \text{H or alkyl})$$

on the basis of the then available evidence for the catenated single-bonded structure R—O—O—R, which is now well established.

Compounds with deviations from additivity (exaltations) may be explained as a constitutive peculiarity, an indication of the tightening or loosening of electrons due to the polarizing influence of the surrounding field. A few selected examples of the molar refractions of peroxides and their nonperoxidic analogs are listed in Table 27. The data in Table 27 taken from Milas' compilation (510) include a few examples of this writer's unpublished results.

In Table 27 the calculated molar refractions of peroxides are derived by addition of the peroxidic oxygen value 2.19 to the observed or calculated molar refraction of the nonperoxidic compound. The exaltations appear as an excess to any exaltation that may reside in the nonperoxidic compound.

TABLE 27. COMPARISON OF THE MOLAR REFRACTIVITY OF SELECTED LIQUID PEROXIDES AND OF THE CORRESPONDING NONPEROXIDIC ANALOGS[a]

| Peroxide | Molar Refractivity $MR_D$ | | | Exaltation |
|---|---|---|---|---|
| | Observed | Calculated | Observed[b] | |
| | *Hydroperoxides* | | *Alcohols* | |
| $C_2H_5OOH$ | 15.17 | 15.08 | 12.89 | 0.09 |
| $i\text{-}C_4H_9OOH$ | 24.42 | 24.41 | 22.22 | 0.01 |
| $t\text{-}C_4H_9OOH$ | 28.95 | 28.91 | 26.72 | 0.04 |
| $\overset{\displaystyle Me}{\underset{\displaystyle \vert}{Ph-CHOOH}}$ | 39.51 | 39.27 | 37.08 | 0.24 |
| OOH (indanyl) | 42.31 | 41.68 | 39.49* | 0.63 |
| | *Dialkyl Peroxides* | | *Ethers* | |
| $C_2H_5OOC_2H_5$ | 24.74 | 24.66 | 22.47 | 0.08 |
| $t\text{-}C_4H_9OO-t\text{-}C_4H_9$ | 43.36 | 42.87 | 40.68 | 0.49 |
| OOMe (indanyl) | 47.35 | 46.41 | 44.22 | 0.94 |
| | *Hydroxyalkyl Peroxides* | | *Hemiacetals* | |
| $MeCH(OH)OOMe$ | 21.35 | 21.41 | 19.22* | -0.06 |
| $C_2H_5OOCH_2OH$ | 21.56 | 21.41 | 19.22* | 0.15 |
| | t-*Butyl Peroxy Esters* | | t-*Butyl Esters* | |
| $Me(CH_2)_7C(O)OO-t\text{-}Bu$ | 66.70[c] | 66.18[c] | 63.00*[c] | 0.52 |
| $C_6H_5C(O)OO-t\text{-}Bu$ | 54.84 | 54.04 | 51.86 | 0.79 |
| $MeCH{=}CHC(O)OO-t\text{-}Bu$ | 43.85 | 42.51 | 40.33* | 1.34 |
| | *Diacyl Peroxides*[c] | | *Anhydrides*[c] | |
| $[Me(CH_2)_6C(O)O]_2$ | 80.78 | 80.07 | 77.88 | 0.71 |
| $[Me(CH_2)_7C(O)O]_2$ | 90.04 | 89.38 | 87.19 | 0.66 |

[a] Data from reference 510 except where indicated.
[b] Calculated values are indicated by an asterisk.
[c] Unpublished data of L. S. Silbert, E. Siegel, and D. A. Lutz.

Simple aliphatic alkyl hydroperoxides, dialkyl peroxides, and hydroxyalkyl-peroxides show no significant exaltations.

Real exaltations, however, appear in alkyl hydroperoxides and dialkyl peroxides containing the aromatic nucleus or a carbon–carbon double bond attached to the molecule in the position beta to the peroxide oxygen. They also appear in both aliphatic and aromatic peroxy esters and in diacyl peroxides, indicating the perturbation effect of the attached carbonyl.

Milas (510) explained the large exaltation in aromatic peroxy esters as a consequence of the resonance structures

$$
\ddot{\underset{..}{O}} \qquad\qquad \ddot{\underset{}{O}}
$$
$$
R:C:\ddot{\underset{..}{O}}:\ddot{\underset{..}{O}}:C(CH_3)_3 \longleftrightarrow R:\ddot{C}: \quad :\ddot{\underset{..}{O}}:C(CH_3)_3
$$
$$
\underset{..}{\ddot{O}}.
$$

in which a canonical structure with two oxygen atoms attached to the same carbon are equivalent. In terms of current knowledge on the structure of peroxy compounds a more likely explanation may be found for these acyl, aromatic, and olefinic derivatives in the skewed conformation, whereby $\pi$-electrons may interact by perturbation with the $p$-orbital electrons of the peroxidic oxygens.

## REFERENCES

1. W. C. Schumb, C. N. Satterfield, and R. L. Wentworth, *Hydrogen Peroxide*, Reinhold, New York, 1955.
2. W. R. Busing and H. A. Levy, *J. Chem. Phys.*, **42**, 3054 (1965).
3. U. Kaldor and I. Shavitt, *J. Chem. Phys.*, **44**, 1823 (1966).
4. E. P. Linton and O. Maass, *Can. J. Res.*, **7**, 81 (1932).
5. W. G. Penney and G. B. B. M. Sutherland, *Trans. Faraday Soc.*, **30**, 898 (1934).
6. W. G. Penney and G. B. B. M. Sutherland, *J. Chem. Phys.*, **2**, 492 (1934).
7. M. T. Rogers and T. W. Campbell, *J. Amer. Chem. Soc.*, **74**, 4742 (1952).
8. Y. Amako and P. A. Giguère, *Can. J. Chem.*, **40**, 765 (1962).
9. R. H. Hunt, R. A. Leacock, C. W. Peters, and K. T. Hecht, *J. Chem. Phys.*, **42**, 1931 (1965).
10. R. H. Hunt and R. A. Leacock, *J. Chem. Phys.*, **45**, 3141 (1966).
11. P. A. Giguère and O. Bain, *J. Phys. Chem.*, **56**, 340 (1952).
12. R. L. Redington, W. B. Olson, and P. C. Cross, *J. Chem. Phys.*, **36**, 1311 (1962).
13. S. C. Abrahams, R. L. Collin, and W. N. Lipscomb, *Acta Cryst.*, **4**, 15 (1951).
14. C. Lu, E. W. Hughes, and P. A. Giguère, *J. Amer. Chem. Soc.*, **63**, 1507 (1941).
15. I. Olovsson and D. H. Templeton, *Acta Chem. Scand.*, **14**, 1325 (1960).
16. B. F. Pedersen, *Acta Chem. Scand.*, **21**, 779 (1967).

17.  B. F. Pedersen and B. Pedersen, *Acta Chem. Scand.*, **18**, 1454 (1964).

18.  J. D'Ans and A. Kneip, *Ber.*, **48**, 1136 (1915).

19.  D. Swern, *Chem. Rev.*, **45**, 1 (1949).

20.  W. H. T. Davison, *J. Chem. Soc.*, **1951**, 2456.

21.  P. A. Giguère and A. W. Olmos, *Can. J. Chem.*, **30**, 821 (1952).

22.  Z. K. Maïzus, G. Ya. Timofeeva and N. M. Emanuel, *Dokl. Akad. Nauk SSSR*, **70**, 655 (1950); Z. K. Maïzus, V. M. Cherednichenko, and N. M. Emanuel, *Dokl. Akad. Nauk, SSSR*, **70**, 855 (1950); B. S. Neporent, T. E. Pavlovskaya, N. M. Emanuel, and N. G. Yaroslavkiĭ, *Dokl. Akad. Nauk SSSR*, **70**, 1025 (1950); D. G. Knorre and N. M. Emanuel, *Zh. Fiz. Khim.*, **26**, 425 (1952).

23.  D. Swern, L. P. Witnauer, C. R. Eddy, and W. E. Parker, *J. Amer. Chem. Soc.*, **77**, 5537 (1955).

24.  J. R. Rittenhouse, W. Lobunez, D. Swern, and J. G. Miller, *J. Amer. Chem. Soc.*, **80**, 4850 (1958).

25.  G. J. Minkoff, *Discussions Faraday Soc.*, **9**, 320 (1950).

26.  N. Sheppard, *Discussions Faraday Soc.*, **9**, 322 (1950).

27.  V. Kasatochkin, S. Perlina, and K. Ablesova, *Compt. Rend. Acad. Sci. URSS*, **47**, 36 (1945).

28.  A. D. Walsh, *J. Chem. Soc.*, **1948**, 331.

29.  P. F. Oesper and C. P. Smyth, *J. Amer. Chem. Soc.*, **64**, 768 (1942).

30.  S. E. Voltz, *J. Amer. Chem. Soc.*, **76**, 1025 (1954).

31.  M. T. Rogers and T. W. Campbell, *J. Amer. Chem. Soc.*, **74**, 4742 (1952).

32.  W. Lobunez, J. R. Rittenhouse, and J. G. Miller, *J. Amer. Chem. Soc.*, **80**, 3505 (1958).

33.  C. Fayat, *Compt. Rend.*, **C265**, 1406 (1967).

34.  L. S. Silbert, L. P. Witnauer, D. Swern, and C. Ricciuti, *J. Amer. Chem. Soc.*, **81**, 3244 (1959).

35.  G. A. Jeffrey, R. K. McMullan, and M. Sax, *J. Amer. Chem. Soc.*, **86**, 949 (1964).

36.  M. Sax and R. K. McMullan, *Acta Cryst.*, **22**, 281 (1967).

37.  S. Caticha-Ellis and S. C. Abrahams, *Acta Cryst.*, **B24**, 277 (1968).

38.  M. Sax, P. Beurskens, and S. Chu, *Acta Cryst.*, **18**, 252 (1965).

39.  D. Belitskus and G. A. Jeffrey, *Acta Cryst.*, **18**, 458 (1965).

40.  D. Swern and L. S. Silbert, *Anal. Chem.*, **35**, 880 (1963).

41.  T. Yonezawa, H. Kato, and O. Yamamoto, *Bull. Chem. Soc. Japan*, **40**, 307 (1967).

42.  F. D. Verderame and J. G. Miller, *J. Phys. Chem.*, **66**, 2185 (1962).

43.  G. W. Wheland, *Resonance in Organic Chemistry*, Wiley, New York, 1955, p. 235.

44.  R. J. B. Marsden and L. E. Sutton *J. Chem. Soc.*, **1936**, 1383.

45.  C. P. Smyth, *Dielectric Behavior and Structure*, McGraw-Hill, New York, 1955, p. 306.

46.  P. W. Allen and L. E. Sutton, *Acta Cryst.*, **3**, 46 (1950).

47.  K.-H. Linke and A. Klaeren, *Acta Cryst.*, **B24**, 1619 (1968).

48.  P. Groth, *Acta Chem. Scand.*, **21**, 2711 (1967).

49. P. Groth, *Acta Chem. Scand.*, **18**, 1301 (1964); *ibid.*, **21**, 2608 (1967).
50. M. Schulz, K. Kirschke, and E. Höhne, *Ber.*, **100**, 2242 (1967).
51. P. Groth, *Acta Chem. Scand.*, **18**, 1801 (1964); *ibid.*, **21**, 2631 (1967).
52. P. Groth, *Acta Chem. Scand.*, **19** 1497 (1965); *ibid.*, **21**, 2695 (1967).
53. P. Groth, *Acta Chem. Scand.*, **23**, 1311 (1969).
54. F. D. Greene, *J. Amer. Chem. Soc.*, **81**, 1503 (1959).
55. A. Hansson, *Acta Chem. Scand.*, **15**, 934 (1961).
56. T. Malkin, in *Progress in the Chemistry of Fats*, Vol. I, edited by R. T. Holman, W. O. Lundberg, and T. Malkin, Academic Press, New York, 1952, pp. 1–17.
57. R. T. O'Connor, in *Fatty Acids* (*Their Chemistry, Properties, Production, and Uses*), 2nd ed., Part 1, edited by K. S. Markley, Interscience, New York, 1960, p. 285.
58. L. Ivanovszky, *J. Polymer Sci.*, **58**, 249 (1962).
59. P. W. Teare, *Acta Cryst.*, **12**, 294 (1959).
60. K. Larsson, *J. Amer. Oil Chem. Soc.*, **43**, 559 (1966).
61. W. E. Parker, C. Ricciuti, C. L. Ogg, and D. Swern, *J. Amer. Chem. Soc.*, **77**, 4037 (1955).
62. W. E. Parker, L. P. Witnauer, and D. Swern, *J. Amer. Chem. Soc.*, **79**, 1929 (1957).
63. L. S. Silbert, E. Siegel, and D. A. Lutz, *Abstracts of the 146th Meeting of the American Chemical Society, Denver, Colorado, January 1964*, p. 22-D.
64. E. von Sydow, *Arkiv Kemi*, **9**, 231 (1956).
65. N.-G. Vannerberg and C. Brosset, *Acta Cryst.*, **16**, 247 (1963).
66. W. P. Schaefer and R. E. Marsh, *J. Amer. Chem. Soc.*, **88**, 178 (1966).
67. M. Calligaris, G. Nardin, and L. Randaccio, *Chem. Comm.*, **1969**, 763.
68. R. Stomberg, *Arkiv. för Kemi*, **22**, 29 (1964).
69. R. Stomberg, *Arkiv för Kemi*, **24**, 283 (1965).
70. D. Grandjean and R. Weiss, *Bull. Soc. Chim. France*, **1967**, 3040, 3044, 3049, 3054, 3058.
71. F. W. B. Einstein and B. R. Penfold, *Acta Cryst.*, **17**, 1127 (1964).
72. G. Mathern, R. Weiss, and R. Rohmer, *Chem. Comm.*, **1969**, 70.
73. (a) J. A. Ibers and S. J. LaPlaca, *Science*, **145**, 920 (1964); (b) S. J. LaPlaca and J. A. Ibers, *J. Am. Chem. Soc.*, **87**, 2581 (1965).
74. J. A. McGinnety, R. J. Doedens, and J. A. Ibers, *Inorg. Chem.*, **6**, 2243 (1967).
75. J. A. McGinnety and J. A. Ibers, *Chem. Comm.*, **1968**, 235.
76. T. Kashiwagi, N. Yasuoka, N. Kasai, M. Kakudo, S. Takahashi, and N. Hagihara, *Chem. Comm.*, **1969**, 743.
77. L. Pauling, *The Nature of the Chemical Bond*, Cornell University Press, Ithaca, New York, N.Y., 1969, p. 352.
78. W. P. Griffith, *J. Chem. Soc.*, **1962**, 3948.
79. L. Vaska, *Science*, **140**, 809 (1963).
80. S. Takahashi, K. Sonogashira, and N. Hagihara, *J. Chem. Soc. Japan*, **87**, 610 (1966).
81. J. S. Griffith, *Proc. Roy. Soc.* (*London*), **A235**, 23 (1956).
82. J. A. McGinnety, R. J. Doedens, and J. A. Ibers, *Science*, **155**, 709 (1967).

83.   J. A. McGinnety, N. C. Payne, and J. A. Ibers, *J. Am. Chem. Soc.*, **91**, 6301 (1969).

84a.  R. B. Holt, C. K. McLane, and O. Oldenberg, *J. Chem. Phys.*, **16**, 225 (1948).

84b.  J. G. Calvert and J. N. Pitts, Jr., *Photochemistry*, Wiley, New York, 1967, pp. 202, 441–450, 196.

84c.  A. Rieche, *Alkylperoxyde und Ozonide*, Steinkopff, Dresden, 1931.

84d.  L. M. Toth and H. S. Johnston, *J. Am. Chem. Soc.*, **91**, 1276 (1969).

84e.  Y. Takezaki, T. Miyazaki, and N. Nakahara, *J. Chem. Phys.*, **22**, 536 (1954).

85a.  A. J. Everett and G. J. Minkoff, *Trans. Faraday Soc.*, **49**, 410 (1953).

85b.  A. Egerton, E. J. Harris, and G. H. S. Young, *Trans. Faraday Soc.*, **44**, 745 (1948).

86.   H. H. Jaffé and M. Orchin, *Theory and Applications of Ultraviolet Spectroscopy*, Wiley, New York, 1962, Chapter 9 and Chapter 12.

87a.  P. A. Giguère and A. W. Olmos, *Can. J. Chem.*, **34**, 689 (1956).

87b.  O. J. Walker and G. L. E. Wild, *J. Chem. Soc.*, **1937**, 1132.

87c.  R. A. Sheldon and J. K. Kochi, *J. Am. Chem. Soc.*, **92**, 4395 (1970).

87d.  W. H. Simpson and J. G. Miller, *J. Am. Chem. Soc.*, 90, 4093 (1968).

88a.  J. W. Breitenbach and J. Derkosch, *Monat. Chem.*, **81**, 530 (1950).

88b.  R. G. Norrish and M. H. Searby, *Proc. Roy. Soc. (London)*, **A237**, 464 (1956).

89.   K. T. Zilch, H. J. Dutton, and J. C. Cowan, *J. Amer. Oil Chem. Soc.*, **29**, 244 (1952).

90.   H. H. Sephton and D. A. Sutton, *Chem. Ind. (London)*, **1953**, 667.

91.   C. E. Swift, F. G. Dollear, and R. T. O'Connor, *Oil & Soap*, **23**, 355 (1946).

92.   C. E. Swift, F. G. Dollear, L. E. Brown, and R. T. O'Connor, *J. Amer. Oil Chem. Soc.*, **25**, 39 (1948).

93.   W. I. Kaye, *Appl. Spectrosc.*, **15**, 89 (1961); *ibid.*, **15**, 130 (1961).

94.   W. I. Kaye, *Anal. Chem.*, **34**, 287 (1962).

95.   O. V. Agashkin and A. E. Lyuts, *Russian Chem. Rev.* (English Translation), **36**, 427 (1967).

96.   G. Czapski and B. H. J. Bielski, *J. Phys. Chem.*, **67**, 2180 (1963).

97.   B. H. J. Bielski and H. A. Schwarz, *J. Phys. Chem.*, **72**, 3836 (1968).

98.   N. I. Kobozev, I. I. Skorokhodov, L. I. Nekrasov, and E. I. Makarova, *Zh. Fiz. Khim.*, **31**, 1843 (1957).

99.   J. A. Wojtowicz, F. Martinez, and J. A. Zaslowsky, *J. Phys. Chem.*, **67**, 849 (1963).

100.  D. A. Csejka, F. Martinez, J. A. Wojtowicz, and J. A. Zaslowsky, *J. Phys. Chem.*, **68**, 3878 (1964).

101.  T. V. Yagodovskaya and L. I. Nekrasov, *Zh. Fiz. Khim.*, **41**, 211 (1967); *ibid.*, **40**, 1304 (1966).

102.  M. E. Boyd, *J. Chem. Phys.*, **37**, 1317 (1962).

103.  R. L. McCarthy and A. MacLachlan, *Trans. Faraday Soc.*, **57**, 1107 (1961).

104.  R. L. McCarthy and A. Maclachlan, *J. Chem. Phys.*, **35**, 1625 (1961).

105.  C. N. R. Rao, *Chemical Applications of Infrared Spectroscopy*, Academic Press, New York, 1963.

106.  L. J. Bellamy, *The Infrared Spectra of Complex Molecules*, 2nd ed., Wiley, New York, 1958.

107. L. J. Bellamy, *Chem. Ind.* (*London*), **1957**, 26.
108. H. A. Szymanski, Ed., *Raman Spectroscopy, Theory and Practice*, Plenum, New York, 1967.
109. H. A. Szymanski, Ed., *Progress in Infrared Spectroscopy*, Vol. III, Plenum, New York, 1967.
110. H. M. Hershenson, *Infrared Absorption Spectra—Index for 1945–1957*, Academic Press, New York, 1959.
111. A. Simon and F. Fehér, *Z. Elektrochem.*, **41**, 290 (1935).
112. A. Simon, *Z. Angew. Chem.*, **51**, 794 (1938).
113. F. Fehér, *Ber.*, **72B**, 1778 (1939).
114. C. R. Bailey and R. R. Gordon, *Trans. Faraday Soc.*, **34**, 1133 (1938).
115. P. A. Giguère, *Can. J. Res.*, **28B**, 485 (1950).
116. P. A. Giguère, *J. Chem. Phys.*, **18**, 88 (1950).
117. O. Bain and P. A. Giguère, *Can. J. Chem.*, **33**, 527 (1955).
118. P. A. Giguère and K. B. Harvey, *J. Mol. Spectrosc.*, **3**, 36 (1959).
119. D. Chin and P. A. Giguère, *J. Chem. Phys.*, **34**, 690 (1961).
120. R. C. Taylor, *J. Chem. Phys.*, **18**, 898 (1950).
121. R. L. Miller and D. F. Hornig, *J. Chem. Phys.*, **34**, 265 (1961).
122. R. Leadbeater, *Compt. Rend.*, **230**, 829 (1950).
123. M. D. Magat, *Discussions Faraday Soc.*, **10**, 330 (1951).
124. O. D. Shreve, M. R. Heether, H. B. Knight, and D. Swern, *Anal. Chem.*, **23**, 282 (1951).
125. G. J. Minkoff, *Proc. Roy. Soc.*, (*London*), **A224**, 176 (1954).
126. M. A. Kovner, A. V. Karyakin, and A. P. Efimov, *Optics and Spectroscopy*, **8**, 64 (1960).
127. A. V. Karyakin, *Russian Chem. Rev.* (English translation), **30**, 460 (1961).
128. L. S. Silbert, *J. Amer. Oil Chem. Soc.*, **39**, 480 (1962).
129. A. R. Philpotts and W. Thain, *Anal. Chem.*, **24**, 638 (1952).
130. D. C. McKean, J. L. Duncan, and R. K. M. Hay, *Spectrochim. Acta*, **23A**, 605 (1967).
131. O. D. Shreve, M. R. Heether, H. B. Knight, and D. Swern, *Anal. Chem.*, **23**, 277 (1951).
132. D. Chapman, *J. Amer. Oil Chem. Soc.*, **42**, 353 (1965).
133. G. Herzberg, *Infra-red and Raman Spectra of Polyatomic Molecules*, Van Nostrand, New York, 1945, p. 199.
134. L. A. Gribov and A. V. Karyakin, *Optics and Spectroscopy*, **9**, 350 (1960).
135. H. R. Williams, and H. S. Mosher, *Anal. Chem.*, **27**, 517 (1955).
136. H. A. Ory, *Anal. Chem.*, **32**, 509 (1960).
137. T. Tanaka, *Sci. Rept. Saitama Univ.*, **A3**, 225 (1960).
138. M. S. Kharasch and G. Sosnovsky, *J. Org. Chem.*, **23**, 1322 (1958).
139. C. W. Fritsch and F. E. Deatherage, *J. Amer. Oil Chem. Soc.*, **33**, 109 (1956).
140. E. R. Stephens, P. L. Hanst, and R. C. Doerr, *Anal. Chem.*, **29**, 776 (1957).
141. R. Kavčič, B. Plesničar, and D. Hadži, *Spectrochim. Acta*, **23A**, 2483 (1967).
142. Yu. N. Anisimov, S. S. Ivanchev, and A. I. Yurzhenko, *Zh. Prikl. Spektrosk.*, **6**, 239 (1967).

143. L. J. Bellamy, B. R. Connelly, A. R. Philpotts, and R. L. Williams, *Z. Elektrochem.*, **64**, 563 (1960).

144. R. T. Holman and P. R. Edmondson, *Anal. Chem.*, **28**, 1533 (1956).

145. R. T. Holman, C. Nickell, O. S. Privett, and P. R. Edmondson, *J. Amer. Oil Chem. Soc.*, **35**, 422 (1958).

146. R. F. Goddu, in *Advances in Analytical Chemistry and Instrumentation*, Vol. I, edited by C. N. Reilley, Interscience, New York, 1960, p. 347.

147. N. A. Milas and O. L. Mageli, *J. Amer. Chem. Soc.*, **75**, 5970 (1953).

148. A. Simon, B. Jentzsch, and I. Menzel, *Ber.*, **90**, 1023 (1957).

149. D. Barnard, K. R. Hargrave, and G. M. C. Higgins, *J. Chem. Soc.*, **1956**, 2845.

150. V. V. Zharkov and N. K. Rudnevskii, *Optics and Spectroscopy*, **7**, 497 (1959).

151. K. R. Loos, C. T. Goetschel, and V. A. Campanile, *Chem. Commun.*, **1968**, 1633.

152. R. D. Spratley and G. C. Pimentel, *J. Amer. Chem. Soc.*, **88**, 2394 (1966).

153. A. J. Arvía and P. J. Aymonino, *Spectrochim. Acta*, **18**, 1299 (1962).

154. J. R. Durig and D. W. Wertz, *J. Mol. Spectrosc.*, **25**, 465 (1968).

155. W. A. Sheppard, *J. Amer. Chem. Soc.*, **85**, 1314 (1963).

156. W. A. Sheppard, *J. Amer. Chem. Soc.*, **87**, 2410 (1965).

157. R. Czerepinski and G. H. Cady, *J. Amer. Chem. Soc.*, **90**, 3854 (1968).

158. J. J. Turner, *Endeavor*, **27**, 42 (1968).

159. M. Lustig and J. K. Ruff, *Chem. Commun.*, **1967**, 870.

160. R. L. Cauble and G. H. Cady, *J. Org. Chem.*, **33**, 2099 (1968).

161. L. R. Anderson and W. B. Fox, *J. Amer. Chem. Soc.*, **89**, 4313 (1967).

162. P. G. Thompson, *J. Amer. Chem. Soc.*, **89**, 4316 (1967).

163. I. J. Solomon, A. J. Kacmarek, J. N. Keith, and J. K. Raney, *J. Amer. Chem. Soc.*, **90**, 6557 (1968).

164. E. Briner, B. Susz, and E. Dallwigk, *Helv. Chim. Acta*, **35**, 340, 345, 353 (1952); *Compt. Rend.*, **234**, 1932 (1952).

165. R. Criegee, G. Blust, and H. Zinke, *Ber.*, **87**, 766 (1954).

166. E. Briner and E. Dallwigk, *Helv. Chim. Acta*, **39**, 1446, 1826 (1956).

167. E. Briner and E. Dallwigk, *Compt. Rend.*, **243**, 630 (1956).

168. R. Criegee, A. Kerchow, and H. Zinke, *Ber.*, **88**, 1878 (1955).

169. D. Garvin and C. Schubert, *J. Phys. Chem.*, **60**, 807 (1956).

170. G. Schröder, *Ber.*, **95**, 733 (1962).

171. A. T. Menyailo, M. V. Pospelov, and A. I. Shekuteva, *Neftekhim.*, **4**, 894 (1964).

172. L. D. Loan, R. W. Murray, and P. R. Storey, *J. Amer. Chem. Soc.*, **87**, 737 (1965).

173. E. Briner, C. Christol, H. Christol, S. Fliszár, and G. Rossetti, *Helv. Chim. Acta*, **46**, 2249 (1963).

174. F. J. Greenwood and B. J. Haske, *Tetrahedron Letters*, No. 11, 631 (1965).

175. F. L. Greenwood and B. J. Haske, *J. Org. Chem.*, **30**, 1276 (1965).

176. G. Riezebos, J. C. Grimmelikhuysen, and D. A. Van Dorp. *Rec. Trav. Chim.*, **82**, 1234 (1963).

177. R. Criegee and G. Wenner, *Ann.*, **564**, 9 (1949).

178. R. Criegee and G. Lohaus, *Ann.*, **583**, 6 (1953).

179. O. S. Privett and E. C. Nickell, *J. Lipid Res.*, **4**, 208 (1963).
180. J. H. Raley, F. F. Rust, and W. E. Vaughan, *J. Amer. Chem. Soc.*, **70**, 1336 (1948).
181. P. D. Bartlett and R. Hiatt, *J. Amer. Chem. Soc.*, **80**, 1398 (1958).
182. Yu. N. Anisimov, S. S. Ivanchev, and A. I. Yurzhenko, *Zh. Anal. Khim.*, **21**, 113 (1966).
183. H. Kämmerer, F. Rocaboy, K.-G. Steinfort, and W. Kern, *Makromol. Chem.*, **53**, 80 (1962).
184. J. Rigaudy, *Compt. Rend.*, **226**, 1993 (1948).
185. C. Dufraisse, A. Etienne, and J. Rigaudy, *Compt. Rend.*, **226**, 1773 (1948).
186. V. A. Terent'ev, N. Kh. Shtivel' and V. L. Antonovskiĭ, *Zh. Prikl. Spectrosk.*, **5**, 463 (1966).
187. I. I. Shabalin and V. A. Simanov, *Neftekhim.*, **5**, 251 (1965); through *Anal. Abstr.*, **14**, 2032 (1967).
188. E. N. Frankel, C. D. Evans, and J. C. Cowan, *J. Amer. Oil Chem. Soc.*, **37**, 418 (1960).
189. H. T. Slover and L. R. Dugan Jr., *J. Amer. Oil Chem. Soc.*, **35**, 350 (1958).
190. R. T. O'Connor, *J. Amer. Oil Chem. Soc.*, **33**, 1 (1956).
191. R. T. O'Connor, *J. Amer. Oil Chem. Soc.*, **38**, 648 (1961).
192. M. Maruyama, *J. Pharm. Japan*, **72**, 927 (1952), through *Chem. Abstr.*, **46**, 9784d (1952).
193. M. Maruyama, *Pharmazie*, **8**, 595 (1953); through *Chem. Abstr.*, **50**, 15024i (1956).
194. E. R. Stephens, in *Proceedings of an International Symposium on Chemical Reactions in the Lower and Upper Atmosphere*, Interscience, New York, 1961.
195. E. R. Stephens, P. L. Hanst, R. C. Doerr, and W. E. Scott, *Ind. Eng. Chem.*, **48**, 1498 (1956).
196. E. R. Stephens, *Anal. Chem.*, **36**, 928 (1964).
197. N. A. Milas and G. G. Arzoumanidis, *Chem. Ind. (London)*, **1966**, 66.
198. P. D. Bartlett and P. Günther, *J. Amer. Chem. Soc.*, **88**, 3288 (1966).
199. S. Fujiwara, M. Katayama, and S. Kamio, *Bull. Chem. Soc. Japan*, **32**, 657 (1959).
200. A. G. Davies, D. G. Hare, and R. F. M. White, *J. Chem. Soc.*, **1960**, 1040.
201. A. G. Davies, D. G. Hare, and R. F. M. White, *Chem. Ind. (London)*, **1959**, 1315.
202. A. G. Davies, D. G. Hare, and R. F. M. White, *J. Chem. Soc.*, **1961**, 341.
203. G. Claeson, G. Androes, and M. Calvin, *J. Amer. Chem. Soc.*, **83**, 4357 (1961).
204. H. Friebolin and W. Maier, *Z. Naturforsch.*, **16a**, 640 (1961).
205. D. Seebach, *Ber.*, **96**, 2712 (1963).
206. D. Seebach, *Ber.*, **96**, 2723 (1963).
207. R. W. Murray, P. R. Story, and M. L. Kaplan, *J. Amer. Chem. Soc.*, **88**, 526 (1966).
208. M. Schulz, H.-F. Boeden, P. Berlin, and W. R. Bley, *Ann.*, **715**, 172 (1968).
209. D. Swern, A. H. Clements, and T. M. Luong, *Anal. Chem.*, **41**, 412 (1969).
210. W. F. Brill, *J. Amer. Chem. Soc.*, **87**, 3286 (1965).
211. G. A. Ward and R. D. Mair, *Anal. Chem.*, **41**, 538 (1969).

212. C. Reid and T. M. Connor, *Nature*, **180**, 1192 (1957).
213. R. Mecke, *Discussions Faraday Soc.*, **9**, 161 (1950).
214. A. D. Cohen and C. Reid, *J. Chem. Phys.*, **25**, 790 (1956).
215. G. Geiseler and H. Zimmermann, *Z. Physik. Chem. (Frankfurt)*, **43**, 84 (1964).
216. L. W. Reeves, *Trans. Faraday Soc.*, **55**, 1684 (1959).
217. J. W. Emsley, J. Feeney, and L. H. Sutcliffe, *High Resolution Nuclear Magnetic Resonance Spectroscopy*, Vol. 1, Pergamon Press, Oxford, 1965, p. 543.
218. A. H. M. Cosijn and M. G. J. Ossewold, *Rec. Trav. Chim.*, **87**, 1264 (1968).
219. W. H. Richardson and R. S. Smith, *J. Org. Chem.*, **33**, 3882 (1968).
220. P. S. Bailey, *Chem. Rev.*, **58**, 925 (1958).
221. A. T. Menyailo and M. V. Pospelov, *Russian Chem. Rev.* (English translation), **36**, 284 (1967).
222. P. S. Bailey, J. A. Thompson, and B. A. Shoulders, *J. Amer. Chem. Soc.*, **88**, 4098 (1966).
223. P. Criegee and G. Schröder, *Ber.*, **93**, 689 (1960).
224. F. L. Greenwood, *J. Org. Chem.*, **29**, 1321 (1964).
225. F. L. Greenwood, *J. Org. Chem.*, **30**, 3108 (1965).
226. L. J. Durham and F. L. Greenwood, *J. Org. Chem.*, **33**, 1629 (1968).
227. L. J. Durham and F. L. Greenwood, *Chem. Commun.*, **1967**, 843; *ibid.*, **1968**, 24.
228. R. W. Murray, R. D. Youssefyah, and P. R. Story, *J. Amer. Chem. Soc.*, **89**, 2429 (1967).
229. C. E. Bishop, D. D. Denson, and P. R. Story, *Tetrahedron Letters*, No. 55, 5739 (1968).
230. S. Fliszár, J. Carles, and J. Renard, *J. Amer. Chem. Soc.*, **90**, 1364 (1968).
231. R. E. Erickson, P. S. Bailey, and J. C. Davis, Jr., *Tetrahedron*, **18**, 389 (1962).
232. P. Kolsaker, *Acta Chem. Scand.*, **19**, 223 (1965).
233. A. G. Streng and A. V. Grosse, *J. Amer. Chem. Soc.*, **88**, 169 (1966).
234. N. J. Lawrence, J. S. Ogden, and J. J. Turner, *Chem. Commun.*, **4**, 102 (1966).
235. J. W. Nebgen, F. I. Metz, and W. B. Rose, *J. Amer. Chem. Soc.*, **89**, 3118 (1967).
236. R. H. Jackson, *J. Chem. Soc.*, **1962**, 4585.
237. I. J. Solomon, J. N. Keith, A. J. Kacmarek, and J. K. Raney, *J. Amer. Chem. Soc.*, **90**, 5408 (1968).
238. J. W. Nebgen, W. B. Rose, and F. I. Metz, *J. Mol. Spectrosc.*, **20**, 72 (1966).
239. P. G. Thompson, *J. Amer. Chem. Soc.*, **89**, 4316 (1967).
240. R. L. Talbott, *J. Org. Chem.*, **33**, 2095 (1968).
241. G. Filipovich and G. V. D. Tiers, *J. Phys. Chem.*, **63**, 761 (1959).
242. W. B. Fox and G. Franz, *Inorg. Chem.*, **5**, 946 (1966).
243. A. L. Buchachenko and N. A. Sysoeva, *Russian Chem. Rev.* (English translation), **37**, 798 (1968).
244. J. Bargon, H. Fischer and U. Johnsen, *Z. Naturforsch.*, **22a**, 1551 (1967).
245. J. Bargon and H. Fischer, *Z. Naturforsch.*, **22a**, 1556 (1967).
246. J. Bargon and H. Fischer, *Z. Naturforsch.*, **23a**, 2109 (1968).
247. H. Fischer and J. Bargon, *Accts. Chem. Res.*, **2**, 110 (1969).

248. H. R. Ward, *J. Amer. Chem. Soc.*, **89**, 5517 (1967).

249. H. R. Ward and R. G. Lawler, *J. Amer. Chem. Soc.*, **89**, 5518 (1967).

250. R. G. Lawler, *J. Amer. Chem. Soc.*, **89**, 5519 (1967).

251. R. Kaptein, *Chem. Phys. Letters*, **2**, 261 (1968).

252. H. R. Ward, R. G. Lawler, and R. A. Cooper, *Tetrahedron Letters*, No. 7, 527 (1969).

253. G. L. Closs and L. E. Closs, *J. Amer. Chem. Soc.*, **91**, 4549 (1969) and succeeding papers.

254. R. S. Alger, *Electron Paramagnetic Resonance*, Interscience Publishers, New York, 1968.

255. M. C. R. Symons, in *Advances in Physical Organic Chemistry*, edited by V. Gold, Academic Press, 1963, p. 283.

256. G. A. Russell, *Science*, **161**, 423 (1968).

257. D. J. E. Ingram, W. G. Hodgson, C. A. Parker, and W. T. Rees, *Nature*, **176**, 1227 (1955).

258. R. C. Smith and S. I. Wyard, *Nature*, **186**, 226 (1960).

259. S. J. Wyard, R. C. Smith, and F. J. Adrian, *J. Chem. Phys.*, **49**, 2780 (1968).

260. J. Kroh, B. C. Green, and J. W. T. Spinks, *J. Amer. Chem. Soc.*, **83**, 2201 (1961).

261. F. J. Adrian, E. L. Cochran, and V. A. Bowers, *J. Chem. Phys.*, **47**, 5441 (1967).

262. T. Ichikawa, M. Iwasaki, and K. Kuwata, *J. Chem. Phys.*, **44**, 2979 (1966).

263. H. E. Radford, *Phys. Rev.*, **126**, 1035 (1962).

264. T. E. Gunter, *J. Chem. Phys.*, **46**, 3818 (1967).

265. K. Kayama, *J. Chem. Phys.*, **39**, 1507 (1963).

266. K. Takakura and B. Rånby, *J. Phys. Chem.*, **72**, 164 (1968).

267. E. J. Hart, *Rec. Chem. Progr.*, **28**, 25 (1967).

268. R. Livingston and H. Zeldes, *J. Chem. Phys.*, **44**, 1245 (1966).

269. W. T. Dixon and R. O. C. Norman, *J. Chem. Soc.*, **1963**, 3119.

270. W. T. Dixon, R. O. C. Norman, and A. L. Buley, *J. Chem. Soc.*, **1964**, 3625.

271. H. Yoshida and B. Rånby, *Acta Chem. Scand.*, **19**, 1495 (1965).

272. J. R. Steven and J. C. Ward, *J. Phys. Chem.*, **71**, 2367 (1967).

273. P. J. Baugh, O. Hinojosa, and J. C. Arthur, Jr., *J. Phys. Chem.*, **71**, 1135 (1967).

274. R. E. Florin, F. Sicilio, and L. A. Wall, *J. Phys. Chem.*, **72**, 3154 (1968).

275. M. S. Bains, J. C. Arthur, Jr., and O. Hinojosa, *J. Phys. Chem.*, **72**, 2250 (1968).

276. T. Shiga, A. Boukhors, and P. Douzou, *J. Phys. Chem.*, **71**, 4264 (1967).

277. E. Saito and B. H. J. Bielski, *J. Amer. Chem. Soc.*, **83**, 4467 (1961).

278. Y. S. Chiang, J. Craddock, D. Mickewich, and J. Turkevich, *J. Phys. Chem.*, **70**, 3509 (1966).

279. H. Fischer, *Ber. Bunsenges, Phys. Chem.*, **71**, 685 (1967).

280. F. Haber and J. Weiss, *Naturwiss.*, **20**, 948 (1932).

281. F. Haber and J. Weiss, *Proc. Roy. Soc. (London)*, **A147**, 332 (1934).

282. C. Walling, *Free Radicals in Solution*, Wiley, New York, 1957.

283. S. Weiner and G. S. Hammond, *J. Amer. Chem. Soc.*, **91**, 2182 (1969).

284.  L. H. Piette and W. C. Landgraf, *J. Chem. Phys.*, **32**, 1107 (1960).
285.  K. U. Ingold and J. R. Morton, *J. Amer. Chem. Soc.*, **86**, 3400 (1964).
286.  W. J. Maguire and R. C. Pink, *Trans. Faraday Soc.*, **63**, 1097 (1967).
287.  J. Q. Adams, *J. Amer. Chem. Soc.*, **90**, 5363 (1968).
288.  J. J. Zwolenik, *J. Phys. Chem.*, **71**, 2464 (1967).
289.  M. Bersohn and J. R. Thomas, *J. Amer. Chem. Soc.*, **86**, 959 (1964).
290.  J. R. Thomas, *J. Amer. Chem. Soc.*, **88**, 2064 (1966).
291.  T. Shida, *J. Phys. Chem.*, **72**, 723 (1968).
292.  A. V. Tobolsky, and R. B. Mesrobian, *Organic Peroxides, Their Chemistry, Decomposition and Role in Polymerization*, Interscience, New York, 1954.
293.  R. W. Brandon and C. S. Elliott, *Tetrahedron Letters*, No. 44, 4375 (1967).
294.  R. Lontz, *Bull. Amer. Phys. Soc.*, **8**, 328 (1963).
295.  K. Toriyama and M. Iwasaki, *J. Phys. Chem.*, **73**, 2663 (1969).
296.  M. Iwasaki and Y. Sakai, *J. Polymer Sci.*, **A2**, 265 (1968).
297.  H. Tanaka, A. Matsumoto, and N. Goto, *Bull. Chem. Soc. Japan*, **37**, 1128 (1964).
298.  N. Vanderkooi, Jr., and W. B. Fox, *J. Chem. Phys.*, **47**, 3634 (1967).
299.  R. W. Fessenden, *J. Chem. Phys.*, **48**, 3725 (1968).
300.  P. H. Kasai and A. D. Kirshenbaum, *J. Amer. Chem. Soc.*, **87**, 3069 (1965).
301.  A. D. Kirshenbaum and A. G. Streng, *J. Amer. Chem. Soc.*, **88**, 2434 (1966).
302.  R. W. Fessenden and R. H. Schuler, *J. Chem. Phys.*, **44**, 434 (1966).
303.  A. Arkell, *J. Amer. Chem. Soc.*, **87**, 4057 (1965).
304.  F. Neumayr and N. Vanderkooi, Jr., *Inorg. Chem.*, **4**, 1234 (1965).
305.  R. W. Fessenden and R. H. Schuler, *J. Chem. Phys.*, **39**, 2147 (1963).
306.  L. H. Piette and G. Bulow, in *Preprints of the Symposium on the Use of Electron Spin Resonance in the Elucidation of Reaction Mechanisms*, 147th Annual Meeting, American Chemical Society Meeting, Philadelphia, 1964.
307.  E. E. Beasley and R. S. Anderson, *J. Chem. Phys.*, **40**, 2565 (1964).
308.  G. M. Coppinger and J. D. Swalen, *J. Amer. Chem. Soc.*, **83**, 4900 (1961).
309.  J. Q. Adams, *J. Amer. Chem. Soc.*, **89**, 6022 (1967).
310.  J. K. Kochi and P. J. Krusic, *J. Amer. Chem. Soc.*, **91**, 3940 (1969).
311.  P. J. Krusic and J. K. Kochi, *J. Amer. Chem. Soc.*, **91**, 3942 (1969).
312.  J. K. Kochi and P. J. Krusic, *J. Amer. Chem. Soc.*, **91**, 3944 (1969).
313.  A. G. Davies and B. P. Roberts, *Chem. Commun.*, **1969**, 699.
314.  H. Ueda, Z. Kuri, and S. Shida, *J. Chem. Phys.*, **36**, 1676 (1962).
315.  H. Ueda, *Anal. Chem.*, **35**, 2213 (1963).
316.  C. H. Bamford and J. C. Ward, *Trans. Faraday Soc.*, **58**, 971 (1962).
317.  H. Fischer, K.-H. Hellwege, and P. Neudörfl, *J. Polymer Sci.*, **A1**, 2109 (1963).
318.  C. L. Ayers, E. G. Janzen, and F. J. Johnston, *J. Amer. Chem. Soc.*, **88**, 2610 (1966).
319.  J. H. Bayston, N. K. King, F. D. Looney, and M. E. Winfield, *J. Amer. Chem. Soc.*, **91**, 2775 (1969).
320.  H. Budzikiewicz, C. Djerassi, and D. H. Williams, *Mass Spectrometry of Organic Compounds*, Holden-Day, San Francisco, 1967.
321.  K. Biemann, *Mass Spectrometry, Organic Chemical Application*, McGraw-Hill, New York, 1962.

322. S. N. Foner and R. L. Hudson, *J. Chem. Phys.*, **21**, 1608 (1953).

323. A. J. B. Robertson, *Applied Mass Spectrometry*, Institute of Petroleum, London, 1954, p. 112.

324. S. N. Foner and R. L. Hudson, *J. Chem. Phys.*, **36**, 2681 (1962).

325. G. K. Lavrovskaya, M. I. Markin, and V. L. Tal'roze, *Tr. Komiss. Anal. Khim., Akad. Nauk SSSR*, **13**, 474 (1963).

326. A. R. Burgess, R. D. G. Lane, and D. K. Sen Sharma, *J. Chem. Soc., B*, **1969**, 341.

327. M. Bertrand, S. Fliszár, and Y. Rousseau, *J. Org. Chem.*, **33**, 1931 (1968).

328. T. Ledaal, *Tetrahedron Letters*, No. 41, 3661 (1969).

329. P. R. Story, D. D. Denson, C. E. Bishop, B. C. Clark, Jr., and J.-C. Farine, *J. Amer. Chem. Soc.*, **90**, 817 (1968).

330. J. Castonguay, M. Bertrand, J. Carles, S. Fliszár, and Y. Rousseau, *Can. J. Chem.*, **47**, 919 (1969).

331. G. C. Pimentel and A. L. McClellan, *The Hydrogen Bond*, Freeman, San Francisco, 1960.

332. L. Bateman and H. Hughes, *J. Chem. Soc.*, **1952**, 4594.

333. A. V. Tobolsky and L. R. Matlack, *J. Polymer Sci.*, **55**, 49 (1961).

334. H. E. De La Mare, *J. Org. Chem.*, **25**, 2114 (1960).

335. J. O. Edwards, in *Peroxide Reaction Mechanisms*, edited by J. O. Edwards, Interscience, New York, 1962, pp. 70, 84 ff.

336. A. L. Buchachenko and O. P. Sukhanova, *Russian Chem. Rev.* (English translation), **36**, 192 (1967).

337. R. West and R. H. Baney, *J. Phys. Chem.*, **64**, 822 (1960).

338. Edward S. Shanley, in *Peroxide Reaction Mechanisms*, edited by J. O. Edwards, Interscience, New York, 1962, p. 129.

339. V. V. Zharkov and N. K. Rudnevskii, *Optics and Spectroscopy*, **12**, 264 (1962).

340. C. Walling and L. Heaton, *J. Amer. Chem. Soc.*, **87**, 48 (1965).

341. G. Geiseler and H. Zimmermann, *Naturwiss*, **51**, 106 (1964).

342. I. I. Shabalin and E. A. Kiva, *Optics and Spectroscopy*, **24**, 377 (1968).

343. A. Ens and F. E. Murray, *Can. J. Chem.*, **35**, 170 (1957).

344. H. Kwart and R. T. Keen, *J. Amer. Chem. Soc.*, **81**, 943 (1959).

345. I. I. Shabalin, M. A. Klimchuk, and E. A. Kiva, *Optics and Spectroscopy*, **24**, 294 (1968).

346. W. H. Richardson and R. F. Steed, *J. Org. Chem.*, **32**, 771 (1967).

347. E. P. Kohler, *Amer. Chem. J.*, **36**, 185 (1906).

348. R. C. Fuson and H. L. Jackson, *J. Amer. Chem. Soc.*, **72**, 1637 (1950).

349. G. A. Jeffrey and M. Sax, *Acta Cryst.*, **16**, 430 (1963).

350. J. H. Robertson, *Acta Cryst.*, **17**, 316 (1964).

351. A. C. Egerton, W. Emte, and G. J. Minkoff, *Discussions Faraday Soc.*, **10**, 278 (1951).

352. W. S. Singleton, in *Fatty Acids* (*Their Chemistry, Properties, Production, and Uses*), 2nd ed., Part 1, edited by K. S. Markley, Interscience, New York, 1960, p. 499.

353. R. Wolf, *Bull. Soc. Chim. France*, **1954**, 644.

354. W. V. F. Brooks and C. M. Haas, *J. Phys. Chem.*, **71**, 650 (1967).

355.  H. A. Swain, Jr., L. S. Silbert, and J. G. Miller, *J. Amer. Chem. Soc.*, **86,** 2562 (1964).

356.  S. W. Benson and J. H. Buss, *J. Chem. Phys.*, **29,** 546 (1958).

357.  P. D. Bartlett, *Rec. Chem. Progr.*, **11,** 47 (1950).

358.  P. Renolen and J. Ugelstad, *J. Chim. Phys.*, **57,** 634 (1960).

359.  E. M. Kosower, in *Progress in Physical Organic Chemistry*, Vol. 3, edited by S. G. Cohen, A. Streitwieser, Jr., and R. W. Taft, Interscience, New York, 1965, pp. 81–163.

360.  G. Briegleb, *Elektronen-Donator-Acceptor-Komplexe*, Springer-Verlag, Berlin, 1961.

361.  L. J. Andrews and R. M. Keefer, *Molecular Complexes in Organic Chemistry*, Holden-Day, San Francisco, 1964.

362.  H. Bradley, Jr., and A. D. King, Jr., *J. Chem. Phys.*, **47,** 1189 (1967).

363.  J. Betts and J. C. Robb, *Nature*, **215,** 274 (1967).

364.  J. Betts and J. C. Robb, *Trans. Faraday Soc.*, **64,** 2402 (1968).

365.  V. I. Stenberg, R. D. Olson, C. T. Wang, and N. Kulevsky, *J. Org. Chem.*, **32,** 3227 (1967).

366.  C. E. Boozer and G. S. Hammond, *J. Amer. Chem. Soc.*, **76,** 3861 (1954).

367.  C. E. Boozer, G. S. Hammond, C. E. Hamilton, and J. N. Sen, *J. Amer. Chem. Soc.*, **77,** 3233 (1955).

368.  J. R. Thomas and C. A. Tolman, *J. Amer. Chem. Soc.*, **84,** 2930 (1962).

369.  J. R. Thomas, *J. Amer. Chem. Soc.*, **85,** 591 (1963).

370.  K. Tokumaru and O. Simmaura, *Bull. Chem. Soc. Japan*, **36,** 333 (1963).

371.  G. B. Gill and G. H. Williams, *J. Chem. Soc.*, **1965,** 7127.

372.  L. S. Silbert, *J. Amer. Oil Chem. Soc.*, **46,** 615 (1969).

373.  M. Imoto and S. Choe, *J. Polymer Sci.*, **15,** 485 (1955).

374.  L. Horner and K. Sherf, *Ann.*, **573,** 35 (1951).

375.  L. Horner and W. Jurgeleit, *Ann.*, **591,** 138 (1955).

376.  M. A. Greenbaum, D. B. Denney, and A. K. Hoffmann, *J. Amer. Chem. Soc.*, **78,** 2563 (1956).

377.  D. B. Denney and M. A. Greenbaum, *J. Amer. Chem. Soc.*, **79,** 979 (1957).

378.  Y. Nagai, *Yuki Gosei Kagaku Kyokaishi*, **19,** 537 (1961).

379.  C. G. Overberger and R. W. Cummins, *J. Amer. Chem. Soc.*, **75,** 4250 (1953).

380.  A. Cerniani and G. Modena, *Gazz. Chim. Ital.*, **89,** 843 (1959).

381.  G. Modena and L. Maioli, *Gazz. Chim. Ital.*, **87,** 1306 (1957).

382.  G. Modena, *Gazz. Chim. Ital.*, **89,** 834 (1959).

383.  P. K. Nandi and U. S. Nandi, *J. Phys. Chem.*, **69,** 4071 (1965).

384.  H. Tsubomura and R. P. Lang, *J. Amer. Chem. Soc.*, **83,** 2085 (1961).

385.  J. E. Leffler, R. D. Faulkner, and C. C. Petropoulos, *J. Amer. Chem. Soc.*, **80,** 5435 (1958).

386.  J. E. Leffler and A. F. Wilson, *J. Org. Chem.*, **25,** 424 (1960).

387.  W. Honsberg and J. E. Leffler, *J. Org. Chem.*, **26,** 733 (1961).

388.  J. C. Martin and W. G. Bentrude, *Chem. Ind.* (*London*), **1959,** 192.

389.  J. C. Martin, D. L. Tuleen, and W. G. Bentrude, *Tetrahedron Letters*, No. 6, 229 (1962).

390.  W. G. Bentrude and J. C. Martin, *J. Amer. Chem. Soc.*, **84,** 1561 (1962).

391. D. L. Tuleen, W. G. Bentrude, and J. C. Martin, *J. Amer. Chem. Soc.*, **85**, 1938 (1963).

392. T. W. Koenig and J. C. Martin, *J. Org. Chem.*, **29**, 1520 (1964).

393. J. C. Martin and T. W. Koenig, *J. Amer. Chem. Soc.*, **86**, 1771 (1964).

394. T. H. Fisher and J. C. Martin, *J. Amer. Chem. Soc.*, **88**, 3382 (1966).

395. J. E. Leffler and J. S. West, *J. Org. Chem.*, **27**, 4191 (1962).

396. I. M. Kolthoff and J. J. Lingane, *Polarography*, Vols. I and II, Interscience, New York, 1952.

397. P. Zuman, *Organic Polarographic Analysis*, Macmillan, New York, 1964.

398. L. Meites, *Polarographic Techniques*, 2nd ed., Interscience, New York, 1965.

399. O. H. Müller, in *Physical Methods of Organic Chemistry*, 3rd ed., Vol. I, Part IV, edited by A. Weissberger, Interscience, New York, 1960, pp. 3155–3279.

400. C. L. Perrin, in *Progress in Physical Organic Chemistry*, Vol. 3, edited by S. G. Cohen, A. Streitwieser, Jr., and R. W. Taft, Interscience, New York, 1965, pp. 165–316.

401. V. Gutmann and G. Schöber, *Angew. Chem.*, **70**, 98 (1958).

402. G. LeGuillanton, *Bull. Soc. Chim. France*, **1963**, 2359.

403. S. Wawzonek, *Talanta*, **12**, 1229 (1965).

404. A. J. Martin, in *Organic Analysis*, Vol. IV, edited by J. Mitchell, Jr., I. M. Kolthoff, E. S. Proskauer, and A. Weissberger, Interscience, New York, 1960, pp. 1–60.

405. R. M. Johnson and I. W. Siddiqi, *J. Polarog. Soc.*, **11**, 72 (1965).

406. R. D. Mair and R. T. Hall, in *Organic Analysis*, Vol. 14, Interscience, New York, in press.

407. S. Wawzonek, *Anal. Chem.*, **21**, 61 (1949); *ibid.*, **22**, 30 (1950); *ibid.*, **26**, 65 (1954); *ibid.*, **28**, 638 (1956); *ibid.*, **30**, 661 (1958); *ibid.*, **32**, 1445 (1960); *ibid.*, **34**, 182R (1962); *ibid.*, **36**, 200R, 220R (1964).

408. D. J. Pietrzyk, *Anal. Chem.*, **38**, 285R (1966); *ibid.*, **40**, 205R (1968).

409. M. F. Romantsev and E. S. Levin, *J. Anal. Chem. USSR* (English translation), **18**, 957 (1963).

410. J. Heyrovsky, *Trans. Faraday Soc.*, **19**, 818 (1924).

411. J. Heyrovsky, *Casopia Lekaru, Ceskych.*, **7**, 242 (1927).

412. B. A. Gosman, *Coll. Czech. Chem. Commun.*, **7**, 467 (1935).

413. A. A. Dobrinskaja and M. B. Neumann, *Acta Physicochim. URSS*, **10**, 297 (1939).

414. V. Stern and S. Polak, *Acta Physicochim. URSS*, **11**, 797 (1939).

415. D. A. Skoog and A. B. H. Lauwzecha, *Anal. Chem.*, **28**, 825 (1956).

416. M. Schulz and K. H. Schwarz, *Monatsber. Deut. Akad. Wiss. Berlin*, **6**, (7), 515 (1964).

417. M. Schulz and K. H. Schwarz, *Abhandl. Deut. Akad. Wiss. Berlin, Kl. Chem., Geol. Biol.*, **1964** (1), 119.

418. M. Schulz and K. H. Schwarz, *Z. Chemie*, **7**, 176 (1967).

419. I. A. Korshunov and A. I. Kalinin, *Khim. Perekisnykh Soedin., Akad. Nauk SSSR, Inst. Obshch. i Neorgan. Khim.*, **1963**, 279, through *Chem. Abstr.*, **60**, 15149e (1964).

420. C. O. Willits, C. Ricciuti, H. B. Knight, and D. Swern, *Anal. Chem.*, **24**, 785 (1952).

421. G. E. O. Proske, *Anal. Chem.*, **24**, 1834 (1952).

422. S. S. Kalbag, K. A. Narayan, S. S. Chang, and F. A. Kummerow, *J. Amer. Oil Chem. Soc.*, **32**, 271 (1955).

423. H. Brüschweiler and G. J. Minkoff, *Anal. Chim. Acta*, **12**, 186 (1955).

424. M. Bernard, *Ann. Chim. (Paris)*, **10**, 315 (1955); *Compt. Rend.*, **236**, 2412 (1953).

425. E. J. Kuta and F. W. Quackenbush, *Anal. Chem.*, **32**, 1069 (1960).

426. M. L. Whisman and B. H. Eccleston, *Anal. Chem.*, **30**, 1638 (1958).

427. M. B. Neïman and M. I. Gerber, *Zh. Anal. Khim.*, **1**, 211 (1946); through *Chem. Abstr.*, **41**, 4096a (1947).

428. M. Legrand, V. Delaroff, and M. Bimbad, *Rec. Trav. Chim.*, **77**, 1034 (1958).

429. P. D. Bartlett and T. G. Traylor, *J. Amer. Chem. Soc.*, **83**, 856 (1961).

430. P. A. Giguère and D. Lamontagne, *Can. J. Chem.*, **29**, 54 (1951).

431. B. N. Moryganov, A. I. Kalinin, and L. N. Mikhotova, *J. Gen. Chem. USSR* (English translation), **32**, 3414 (1962).

432. V. L. Antonovskii and Z. S. Frolova, *J. Anal. Chem.* (English translation), **19**, 696 (1964).

433. F. Reimers, *Quart. J.*, *Pharm. Pharmacol.*, **19**, 473 (1946).

434. T. D. Parks and K. A. Hansen, *Anal. Chem.*, **22**, 1268 (1950).

435. J. Hakl, *Chem. Listy*, **61**, (4), 536 (1967); through *Anal. Abstr.*, **15**, 3974 (1968).

436. W. R. Lewis, F. W. Quackenbush, and T. DeVries, *Anal. Chem.*, **21**, 762 (1949).

437. W. R. Lewis and F. W. Quackenbush, *J. Amer. Oil Chem. Soc.*, **26**, 53 (1949).

438. E. J. Kuta and F. W. Quackenbush, *J. Amer. Oil Chem. Soc.*, **37**, 148 (1960).

439. C. Paquot and J. Mercier, *Compt. Rend.*, **236**, 1802 (1953).

440. D. Swern, J. E. Coleman, H. B. Knight, C. Ricciuti, C. O. Willits, and C. R. Eddy, *J. Amer. Chem. Soc.*, **75**, 3135 (1953).

441. C. O. Willits, C. Ricciuti, C. L. Ogg, S. G. Morris, and R. W. Riemenschneider, *J. Amer. Oil Chem. Soc.*, **30**, 420 (1953).

442. C. Ricciuti, C. O. Willits, C. L. Ogg, S. G. Morris, and R. W. Riemenschneider, *J. Amer. Oil Chem., Soc.*, **31**, 456 (1954).

443. E. Becker and A. Niederstebruch, *Fette, Seifen, Anstrichmittel*, **68**, 135 (1966).

444. A. Niederstebruch and I. Hinsch, *Fette, Seifen, Anstrichmittel*, **69**, 637 (1967).

445. H. Lück and R. Maack, *Fette, Seifen, Anstrichmittel*, **68**, 81 (1966).

446. H. Lück and R. Maack, *Fette, Seifen, Anstrichmittel*, **68**, 305 (1966).

447. H. Lück and R. Maack, *Fette, Seifen, Anstrichmittel*, **68**, 609 (1966).

448. R. Maack and H. Lück, *Fette, Seifen, Anstrichmittel*, **70**, 245 (1968).

449. C. Ricciuti, L. S. Silbert, and W. S. Port, *J. Amer. Oil Chem. Soc.*, **34**, 134 (1957).

450. S. R. Missan, E. I. Becker, and L. Meites, *J. Amer. Chem. Soc.*, **83**, 58 (1961).

451. A. J. Parker, *Quart. Rev. (London)*, **16**, 163 (1962).

452. N. S. Moe, *Acta Chem. Scand.*, **19**, 1023 (1965).

453. R. C. Buchta and D. H. Evans, *Anal. Chem.*, **40**, 2181 (1968).

454. R. Maack and H. Lück, *Experientia*, **19**, 466 (1963).

455. E. R. Roberts and J. S. Meek, *Analyst*, **77**, 43 (1952).

456. A. J. Martin, *J. Chem. Ed.*, **35**, 10 (1958).

457. F. L. Lambert and K. Kobayashi, *J. Amer. Chem. Soc.*, **82**, 5324 (1960).

458. P. J. Elving and B. Pullman, in *Advances in Chemical Physics*, Vol. 3, edited by I. Prigogine, Interscience, New York, 1961, p. 1 ff.

459. P. H. Given and M. E. Peover, *Advances in Polarography*, Vol. 3, (Second International Congress of Polarography, Cambridge, 1959), Pergamon, 1961, p. 948 ff.

460. G. J. Hoijtink, J. Van Schooten, E. De Boer, and W. I. Aalbersberg, *Rec. Trav. Chim.*, **73**, 355 (1954).

461. N. A. Milas and D. M. Surgenor, *J. Amer. Chem. Soc.*, **68**, 642 (1946).

462. C. Ricciuti, J. E. Coleman, and C. O. Willits, *Anal. Chem.*, **27**, 405 (1955).

463. L. Dulog, *Z. Anal. Chem.*, **202**, 258 (1964).

464. E. L. Johnson, K. H. Pool, and R. E. Hamm, *Anal. Chem.*, **38**, 183 (1966).

465. V. S. Bagotskii, L. N. Nekrasov, and N. A. Shumilova, *Russian Chem. Rev.* (English translation), **34**, 717 (1965).

466. V. S. Griffiths and M. I. Jackson, *Talanta*, **9**, 205 (1962); *ibid.*, **9**, 871 (1962).

467. J. E. A. Van Gheluwe, A. Buday, and A. L. Stock, *Proc. Amer. Soc. Brew. Chem.*, **1963**, 58; through *Anal. Abstr.*, **11**, 3473 (1964).

468. R. Briggs and W. H. Mason, British Patent 936,724.

469. D. T. Sawyer and J. L. Roberts, Jr., *J. Electroanal. Chem.*, **12**, 90 (1966).

470. F. Rallo and L. Rampayzo, *J. Electroanal. Chem. Interfacial Electrochim.*, **16**, 61 (1968).

471. K. Täufel and F. Linow, *Nährung.*, **7**, 41 (1963).

472. G. L. Woodroffe, *Talanta*, **11**, 967 (1964).

473. H. A. P. Ingram, *Nature*, **202**, 1312 (1964).

474. F. Reimers, *Sbornik Mezinarod. Polarog. Sjezdu Praze*, 1st Congr., Part 1, 203 (1951); through *Chem. Abstr.*, **46**, 6998d (1952).

475. M. I. Gerber, A. A. Dobrinskaya, and M. B. Neĭman, *Tr. Vsesoyuz. Konf. Anal. Khim.*, **2**, 585 (1943); through *Chem. Abstr.*, **39**, 3760 (1945).

476. M. N. Mikhaĭlova and M. B. Neiman, *Zavodskaya Lab.*, **9**, 166 (1940); through *Chem. Abstr.*, **34**, 5788 (1940).

477. W. M. MacNevin and P. F. Urone, *Anal. Chem.*, **25**, 1760 (1953).

478. S. Sandler and Y.-H. Chung, *Anal. Chem.*, **30**, 1252 (1958).

479. P. Urone, *Diss. Abstr.*, **20**, 2595 (1960).

480. J. Mercier, *Oléagineux*, **13**, 165 (1958).

481. N. A. Khan, *Pakistan J. Sci. Ind. Res.*, **2**, 287 (1960).

482. H. B. Knight, E. F. Jordan, Jr., R. E. Koos, and D. Swern, *J. Amer. Oil Chem. Soc.*, **31**, 93 (1954).

483. F. A. Bovey and I. M. Kolthoff, *J. Amer. Chem. Soc.*, **69**, 2143 (1947).

484. D. E. Barnes, R. M. Elofson, and G. D. Jones, *J. Amer. Chem. Soc.*, **72**, 210 (1950).

485. A. I. Kalinin, E. M. Perepletchikova, I. A. Korshunov, and E. N. Zil'berman, *Zh. Obshch. Khim.*, **36**, 1563 (1966); through *Chem. Abstr.*, **66**, 61277e (1967).

486. M. Bohdanecký and J. Exner, *Chem. Listy.*, **48**, 1506 (1954); through *Anal. Abstr.*, **3**, 465 (1956).

487. T. Takeuchi, N. Yokouchi, and Y. Takayama, *Japan Analyst*, **4**, 234 (1955); through *Anal. Abstr.*, **3**, 164 (1956).

488. V. N. Dmitrieva, O. V. Meshkova, and V. D. Bezuglyi, *Zh. Anal. Khim.*, **19**, 389 (1964); through *Chem. Abstr.*, **61**, 1276c (1964).

489. J. Paściak and Z. Wójcik, *Chemia Anal.*, **10**, 355 (1965); through *Anal. Abstr.*, **13**, 4987 (1966).

490. A. Osol and G. E. Farrer, Eds., *The Dispensatory of the United States of America*, Philadelphia, 1947, p. 252.

491. A. H. Beckett and G. O. Jolliffe, *J. Pharm. Pharmacol.*, **5**, 869 (1953).

492. M. Maruyama, *Ann. Rept. Takamine Lab.*, **7**, 103 (1955); through *Chem. Abstr.*, **50**, 16042c (1956).

493. G. Dusinsky and M. Tyllova, *Chem. Zvesti*, **16**, 701 (1962); through *Chem. Abstr.*, **58**, 8233a (1963).

494. B. Bitter, *Sbornik Mezinarod. Polarog. Sjezdu Praze*, 1st Congr., Part 1, 771 (1951); through *Chem. Abstr.*, **46**, 11582h (1952).

495. M. Maruyama, *Ann. Rept. Takamine Lab.*, **5**, 65 (1953); through *Chem. Abstr.*, **49**, 15177g (1955).

496. G. Semerano, *Proc. First Intern. Congr. Polarog. Prague*, **1**, 300 (1951).

497. J. E. Page, *Quart. Rev. (London)*, **6**, 262 (1952).

498. P. Zuman, *Proc. Intern. Symp. Microchem.* (Birmingham, 1958), Pergamon, New York, 1960, p. 294.

499. H. Niewiadomski, M. Ploszynski, J. Sawicki, and W. Zwierzykowski, *Oléagineux*, **16**, 175 (1961).

500. V. L. Antonovskii, M. N. Buzlanova, and Z. S. Frovola, *Kinetika i Kataliz*, **8**, 671 (1967); through *Chem. Abstr.*, **67**, 901904 (1967).

501. R. A. Joyner, *Z. Anorg. Chem.*, **77**, 103 (1912).

502. M. G. Evans and N. Uri, *Trans. Faraday Soc.*, **45**, 224 (1949).

503. A. J. Martin, *Anal. Chem.*, **29**, 79 (1957).

504. D. Swern, *Chem. Rev.*, **45**, 34 (1949).

505. D. Lefort and J. Sorba, *Oléagineux*, **16**, 13 (1961).

506. N. Bauer, K. Fajans, and S. Z. Lewis, in *Physical Methods of Organic Chemistry*, 3rd ed., Vol. 1, Part 2, edited by A. Weissberger, Interscience, New York, 1960, pp. 1138–1281.

507. S. S. Batsanov, *Refractometry and Chemical Structure*, translated from Russian by P. P. Sutton, Consultants Bureau, New York, 1961.

508. A. E. Remîck, *Electronic Interpretations of Organic Chemistry*, Wiley, New York, 1943.

509. L. Pauling and E. B. Wilson, Jr., *Introduction to Quantum Mechanics, with Applications to Chemistry*, McGraw-Hill, New York, 1935, p. 226.

510. N. A. Milas, D. M. Surgenor, and L. H. Perry, *J. Amer. Chem. Soc.*, **68**, 1617 (1946).

*Chapter VIII*

# Acyl Peroxides

## R. HIATT

## I. INTRODUCTION

Acyl peroxides may be defined as substances of the type $RC(O)OOC(O)R'$, where R and R' are either alkyl or aryl. Compounds in which R and R' are alkoxy groups, though properly termed peroxydicarbonates, have traditionally been included in the acyl peroxide category, and are so in this chapter. Also included are some compounds, of mixed parentage: $ROC(O)OOC(O)R'$ and $RSO_2OOC(O)R'$.

Acyl peroxides have been one of the most frequently used sources of free radicals, and interest in their various modes of decomposition has been keen. Rather surprisingly, however, no full and comprehensive account of their characteristics and reactions has appeared in the last 30 years. Reviews of the early work are cited in references 85, 216, and 225c. More recent brief summaries can be found in books and articles by Boguslavskaya (47), A. G. Davies (91), Hawkins (210), Karnojitsky (269), Mageli and Sheppard (350), and Tobolsky and Mesrobian (519), and more detailed treatments of some particular aspect can be found in references 541 (free-radical reactions); 510 (concerted homolyses); 139 and 140 (nucleophilic displacements on $-O-O-$); 9, 112, and 559 (homolytic aromatic substitution); 148 and 423 (polymerization); 351 and 479 (quantitative analysis); 265 and 464 (polarography); and 512 (infrared spectra).

The development of highly accurate analytical techniques of the last 10–15 years has resulted in such a wealth of material that seemingly, with the aid of computer analysis, almost any problem in acyl peroxide chemistry could be solved now [e.g., De Tar (113) on the thermal decomposition of benzoyl peroxide in benzene]. The flow of information shows no sign of diminishing; judging by the output of the last 6 months (January–June 1969), the next year should produce at least 50 articles worth citing in a review such as this.

This chapter attempts no computer analysis but does provide a fairly complete index to the relevant literature. To cite every instance in which an acyl peroxide was used to initiate a free-radical-chain reaction would be nearly impossible and hardly pertinent. Rather, the points of concern include methods of synthesis and analysis, characterization of new peroxides, modes and rates of decomposition, and what has happened to the pieces.

## II.  SYNTHESIS OF ACYL PEROXIDES

Most symmetrical acyl peroxides can be readily prepared by the reaction of an acyl chloride or anhydride with either a metal peroxide or hydrogen peroxide plus a base. Asymmetrical peroxides, RC(O)OOC(O)R', are synthesized by using the sodium salt of a peroxy acid together with an acyl chloride or anhydride; the peroxy acid is frequently synthesized *in situ* by the autoxidation of an aldehyde. Preparations utilizing carbodiimides or imidazolides have been developed.

### A.  Syntheses with Sodium Peroxide

The earliest synthesis of acyl peroxides (acetyl, butyryl, isovaleryl, and benzoyl) by Brodie (54) in 1864 utilized barium peroxide, a reagent that still seems preferred on occasion [e.g., in the preparation of $(CF_3CO_2)_2$ from $(CF_3CO)_2O$ (370)]. However, sodium peroxide soon became (529) and has continued to be the most commonly used metallic peroxide. The original procedure of Vanino (529) involved a homogeneous solution of sodium peroxide and acyl chloride in aqueous acetone, but higher yields are usually obtainable with a two-phase system (417), adding a solution of the acid chloride in benzene (418), toluene (418, 507a), or cyclohexane (462, 507a) to a dilute aqueous solution of the peroxides.

$$2RC\overset{\overset{\displaystyle O}{\|}}{-}Cl + Na_2O_2 \longrightarrow RC\overset{\overset{\displaystyle O}{\|}}{-}O-O-\overset{\overset{\displaystyle O}{\|}}{C}-R + 2NaCl \qquad (1)$$

The acyl peroxide thus formed precipitates from solution, thereby being protected from hydrolysis, so that yields are usually 80% or better. A variant is to add small amounts of water to an acetone (27) or ether (484) solution of the acyl chloride or anhydride containing solid sodium peroxide in suspension. This appears to be the best method for making $Ac_2O_2$ (419, 484). Notably explicit directions for the preparation of aroyl peroxides can be found in references 417, 418, and 507a; of acyl peroxides in reference 150; and of thenoyl peroxides in references 462 and 463. Peroxy dicarbonates are usually synthesized similarly from sodium peroxide and a chloroformate (360, 499, 500, 553).

$$2RO\overset{\overset{\displaystyle O}{\|}}{-}CCl + Na_2O_2 \longrightarrow R-O-\overset{\overset{\displaystyle O}{\|}}{C}-O-O-\overset{\overset{\displaystyle O}{\|}}{C}-O-R \qquad (2)$$

Reaction temperatures of 0–5° are sufficiently low for most acyl peroxide syntheses, but the more thermally labile members, such as $(CF_3CO_2)_2$ (370), $(CCl_3CO_2)_2$ (370, 567), and $(PhCH_2CO_2)_2$ (23), require temperatures of $-15$ to $-20°$, obtained by loading the aqueous sodium peroxide solution with sodium chloride. Even under these conditions the peroxide from hydratropoyl chloride could not be isolated, only its decomposition products having been found (177).

## B.   Syntheses with Hydrogen Peroxide

The obvious advantages of hydrogen peroxide over aqueous sodium peroxide are the more ready control of alkalinity and the greater convenience of preparing more concentrated or nonaqueous reagent mixtures. Instances utilizing dilute aqueous hydrogen peroxide and sodium hydroxide (194, 225b, 247, 411, 557) presumably have resulted from more ready availability or requirements for a patentable process. However peroxides containing base-sensitive groups, such as $(CF_3C_6H_4CO_2)_2$ (340), $(CHOC_6H_4CO_2)_2$ (194), or $(CH_3CO_2C_6H_4-CO_2)_2$ (80), are best prepared from 30% hydrogen peroxide and sodium bicarbonate; acyl peroxides of long-chain fatty acids can be synthesized in high yield with 50–60% hydrogen peroxide and pyridine (480). Kochi and Mocadlo (296) give excellent directions for using 30% hydrogen peroxide and pyridine to prepare peroxides from series of branched-chain or small-ring-containing acids in nearly quantitative yields.

Some success has been reported in the use of hydrogen peroxide–urea complexes to synthesize acyl peroxides highly sensitive to hydrolysis. Thus De Tar and Carpino (114) were able to prepare an impure sample of *trans*-2,3-diphenylacrylyl peroxide, though neither the *cis* isomer nor *o,o'*-dibenzoyl-benzoyl peroxide could be synthesized by this means. The method has worked much better in the preparation of several $\omega$-iodo acyl peroxides (561). Formation of the acyl peroxide as a urea inclusion complex may itself tend to stabilize against hydrolysis, as shown by Heslinga and Schwaiger (218) for the series $AcO_2C(O)(CH_2)_nC(O)O_2Ac$, $n = 2–7$.

## C.   Acyl Peroxides from Peroxy Acids

The first synthesis of acetyl benzoyl peroxide is attributed to Baeyer and Villiger (12b), the general method using the sodium salt of a peroxy acid being further developed by Wieland et al. (556, 557).

$$Ac_2O + BzO_2Na \longrightarrow BzO_2Ac \qquad (3)$$

Good yields are obtainable under homogeneous conditions [aqueous acetone (556, 557)], or by suspending the sodium salt in a cyclohexane solution of the acid chloride (65, 90). The peroxy acid itself, together with an acyl chloride and pyridine, may be used (437). A number of unusual peroxides have been prepared in this fashion, including nitro acyl peroxides [e.g., $(NO_2)_3CCH_2$-$CH_2C(O)O_2Ac$] (477), polymers with built-in peroxide groups (from polyacrylyl chloride and peroxybenzoic acid) (485), acyl peroxycarbonates $(RC(O)OOC(O)OR')$ (429, 437), and acyl peroxysulfonates [$RC(O)OOSOOR'$] (341).

An alternative procedure exploited extensively by Ol'dekop and co-workers (399, 401, 403), though developed much earlier (12a) and the subject of several patents (6, 69, 253), employs the autoxidation of an aldehyde in the presence of an acid anhydride. The process, which requires sodium acetate (399, 401, 403) or lithium acetate (253) as a catalyst, appears suited only to the more thermally stable peroxides since 30–40° temperatures are required to autoxidize the aldehydes.

$$RCHO \xrightarrow{O_2} \overset{\overset{O}{\parallel}}{RCO_2H} \xrightarrow[NaOAc]{(R'CO)_2O} \overset{\overset{O}{\parallel}\ \overset{O}{\parallel}}{RCO_2CR'} \qquad (4)$$

Pseudo-first-order rate constants for acyl peroxide formation from peroxyacetic (394) and peroxybenzoic (393) acids in mixtures of acetic and sulfuric acids have been measured by Ogata and co-workers. The reaction followed $H_0$, conforming to the postulated mechanism:

$$AcOH \underset{}{\overset{H^+}{\rightleftharpoons}} CH_3\overset{+}{C}(OH)_2 \underset{}{\overset{AcO_2H}{\rightleftharpoons}} CH_3\underset{\underset{OH}{|}\ \underset{H}{|}}{\overset{\overset{OH}{|}}{C}-\overset{+}{O}-OAc} \rightleftharpoons$$

$$Ac_2O_2 + H_2O \qquad (5)$$

The rate of reverse reaction was also measured (394), but under somewhat different conditions (see Section IV-D).


### D.  Diimide Syntheses

Both symmetrical and asymmetrical acyl peroxides can be synthesized without the intermediacy of the acyl chloride or anhydride, using dicyclohexyldiimide as a condensing agent (182).

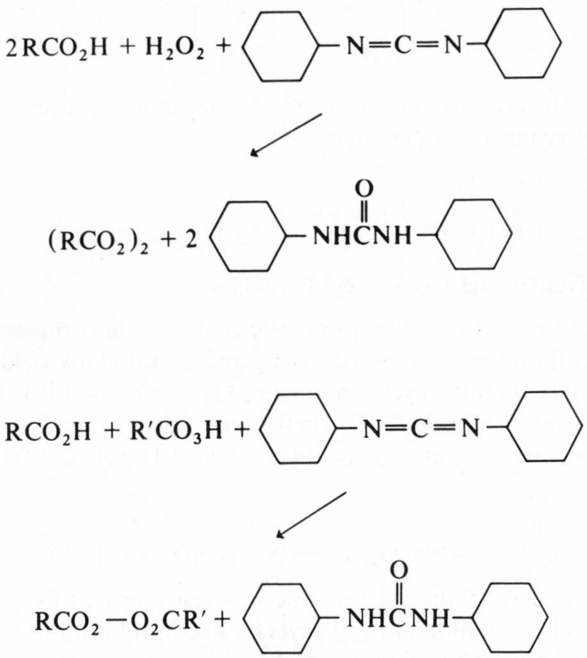

$$2\,RCO_2H + H_2O_2 + \bigcirc\!\!-\!N\!=\!C\!=\!N\!-\!\bigcirc$$

$$(RCO_2)_2 + 2\,\bigcirc\!\!-\!\overset{\displaystyle O}{\overset{\displaystyle \|}{NHCNH}}\!-\!\bigcirc$$

(6)

$$RCO_2H + R'CO_3H + \bigcirc\!\!-\!N\!=\!C\!=\!N\!-\!\bigcirc$$

$$RCO_2\!-\!O_2CR' + \bigcirc\!\!-\!\overset{\displaystyle O}{\overset{\displaystyle \|}{NHCNH}}\!-\!\bigcirc$$

The reaction, which can be pictured as proceeding through an acid–diimide complex (121, 182), has given 75–90% yields of benzoyl peroxide, palmitoyl peroxide, and *trans-t*-butylcyclohexanoyl peroxide, etc., and a 40% yield of cyclic monomeric phthaloyl peroxide (182).

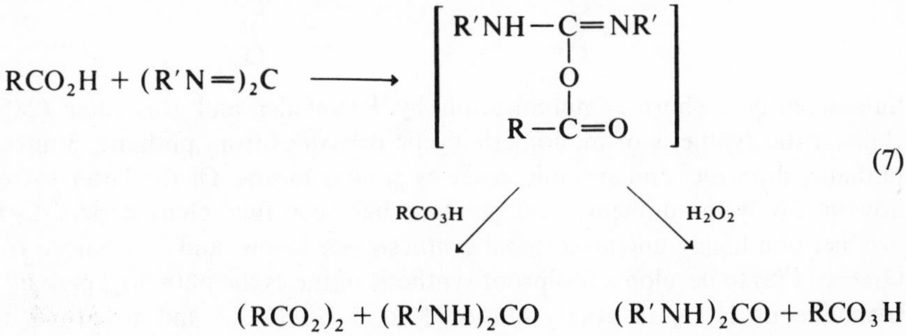

$$RCO_2H + (R'N\!=\!)_2C \longrightarrow \begin{bmatrix} R'NH\!-\!C\!=\!NR' \\ | \\ O \\ | \\ R\!-\!C\!=\!O \end{bmatrix}$$

(7)

$$RCO_3H \swarrow \qquad\qquad \searrow H_2O_2$$

$$(RCO_2)_2 + (R'NH)_2CO \qquad (R'NH)_2CO + RCO_3H$$

Staab and co-workers (496) have found *N,N*-dicarbonyldiimidazole to be an equally effective condensing reagent. Its utility in preparing peroxides that are difficult to obtain by other means has been amply demonstrated by Singer and Kong's (483) synthesis of

$$\left(\begin{array}{c} Ph \\ \diagdown \\ H \end{array} C=C \begin{array}{c} Me \\ \diagup \\ \diagdown CO_2 \end{array}\right)_2 \tag{8}$$

Imidazole itself has been shown to effect a facile interconversion of asymmetrical and symmetrical peroxides (496c).

$$2RCO_2CR' \underset{\text{imidazole}}{\rightleftharpoons} (RCO_2)_2 + (R'CO_2)_2 \tag{9}$$

### E. Polymeric and Cyclic Acyl Peroxides

Treatment of the acyl chlorides of dibasic acids [succinic, fumaric, phthalic, etc., (410, 528)] with sodium or hydrogen peroxide ordinarily yields polymeric peroxides rather than the cyclic monomers (12c, 367, 468). It is not clear in the case of oxalyl chloride (367) whether the viscous, highly-water-soluble product is polymeric or merely the diperoxy acid HOOC(O)C(O)OOH.

$$Cl-\overset{O}{\overset{\|}{C}}-(CH_2)_nC-\overset{O}{\overset{\|}{C}}-Cl + Na_2O_2 \longrightarrow HO[O-\overset{O}{\overset{\|}{C}}(CH_2)_n\overset{O}{\overset{\|}{C}}-O]_mOH \tag{10}$$

Cyclic monomeric phthaloyl peroxide was first prepared by K. E. Russell (456) with a dilute solution of the diacyl chloride in chloroform together with dilute buffered aqueous sodium peroxide.

$$\tag{11}$$

Subsequently a short communication by Kleinfeller and Rastädter (285) claimed the synthesis of monomeric cyclic peroxides from phthalic, 3-nitrophthalic, diphenic, and succinic acids by similar means. Of the latter three, however, only the diphenic acid peroxide has been fully characterized as a product of a highly unconventional synthesis (see below) and it remained for Greene (178) to develop a foolproof synthesis of the cyclic phthaloyl peroxide (nonaqueous $H_2O_2$ in ether solution + solid $NaHCO_3$) and to prove its structure unambiguously.

### F. Unusual Syntheses

A novel method (425, 426) involves the ozonolysis of trialkyl phosphite–1,2-diketone adducts. The reaction appears to be particularly suited to the synthesis of cyclic peroxides, an 80% yield of diphenic acid peroxide being obtained, as compared with 18% for benzoyl or acetyl peroxide, from benzil and biacetyl, respectively (425). Its general utility is as yet unknown, however.

$$\text{(12)}$$

The oxidation of potassium salts of carboxylic acids, either by Kolbe electrolysis (519) or by elemental fluorine (143), has led to acyl peroxides, though the procedure has been used rarely.

$$FC(NO_2)_2CH_2CH_2CO_2^- K^+ \xrightarrow[\;H_2O\;]{F_2} [FC(NO_2)_2CH_2CH_2CO_2]_2 \qquad (13)$$

Fluoroformyl peroxide has been isolated from the reaction of carbon monoxide, fluorine, and oxygen at room temperature (7) and has also been prepared by the photolysis of oxalyl fluoride in the presence of oxygen (87).

$$\underset{F-C-C-F}{\overset{O\quad O}{\underset{\parallel\quad\parallel}{}}} + O_2 \xrightarrow{h\nu} \underset{F-C-O-O-C-F}{\overset{O\qquad\quad O}{\underset{\parallel\qquad\quad\parallel}{}}} \qquad (50\%) \quad (14)$$

The latter is reminiscent of Wieland's isolation, in low yield, of ethyl peroxy-dicarbonate from the thermolysis of $Ph_3CN=NC(O)OEt$ in the presence of oxygen (553) and of one postulated termination reaction in acetaldehyde autoxidation (363).

$$2\underset{MeC-O_2\cdot}{\overset{O}{\underset{\parallel}{}}} \longrightarrow Ac_2O_2 + O_2 \qquad (15)$$

Finally, many benzoyl peroxides are stable enough to undergo ring nitration without decomposition. Gelissen and Hermans (167) prepared *m,m'*-dinitrobenzoyl peroxide from benzoyl peroxide and a mixture of nitric and sulfuric acids. Ol'dekop (399a) and Shreibert (477b) and their co-workers have likewise synthesized numerous asymmetrical nitrobenzoyl-acyl peroxides.

### G.  Purification of Acyl Peroxides

Acyl peroxides, with some exceptions (see Appendix), are solids at room temperature and usually can be recrystallized readily from mixtures of a solvent, such as benzene or chloroform, with a nonsolvent (methanol, ethanol, or petroleum ether) (80, 182, 418, 507b). Frequently, however, if the starting acyl chloride is pure, the peroxide obtained directly from the reaction medium will be more than 98% pure (by iodometric titration) (150, 191, 480). For peroxides that are oils, pure starting materials are the only way to ensure a pure product (191), except for those few (e.g., acetyl, propionyl, and butyryl) that can be distilled under high vacuum (445a, b) (*CAUTION*).

*Solid acyl peroxides should be handled with extreme caution*; most are shock sensitive (350), and instances are reported of peroxides that, after being handled safely on several occasions, suddenly detonated at the touch of a spatula (225b, 331). Acetyl peroxide deserves special mention. The material crystallizes from the ether solution in which it is prepared after standing overnight at $-78°$ (419, 484) and can be filtered and evacuated to dryness (mechanically *and behind a screen*). At this point it is safest to add a solvent of choice. Slagle and Shine (484) say the crystals can be kept a week at $-78°$ but recommend against it. Significantly, they do not report the melting point of the pure material. Crystalline acetyl peroxide melting at "about 30°" may be obtained by distillation at reduced pressure of the commercially available (454) 25% solution in dimethyl phthalate (470). This material, which can be touched—but not tapped—without detonation, appears, however, still to contain about 20% of dimethyl phthalate.

It will be apparent that melting points are not very good criteria of purity for acyl peroxides. Many melt with decomposition, and the decomposition or detonation point may vary with the rate of heating.

### III.  PHYSICAL CHARACTERISTICS

### A.  Structure of Acyl Peroxides

Dipole moments of acyl peroxides in benzene solution (Table 1) show that the compounds have a skewed structure; calculations showed these data to be consistent with a dihedral angle around the O—O bond of 100° and the

TABLE 1. DIPOLE MOMENTS OF ACYL PEROXIDES IN
BENZENE[a]

| Peroxide | Temperature (°C) | $\mu$ (debyes) |
|---|---|---|
| $(C_6H_5CO_2)_2$ | 30 | 1.60 |
| $(p\text{-}ClC_6H_4CO_2)_2$ | 30 | 1.33 |
| $(p\text{-}ClC_6H_4CO_2)_2$ | 20 | 1.36 |
| $[Me(CH_2)_{10}CO_2]_2$ | 20 | 1.32 |

[a] Data from reference 344.

carbonyls of the carboxy groups facing inward (344). Such a picture agreed
with the powder X-ray data of Silbert, Swern, and their colleagues (509). More
recent X-ray studies of benzoyl peroxide crystals have given dihedral angles of
93 (264) and 91° (460), with an O—O distance of $1.46 \pm 0.015$ Å (460).

$$91\text{-}93° \quad (16)$$

Similar studies of p-Cl and p-Br benzoyl peroxides have indicated a rather
smaller dihedral angle ($81.1 \pm 3.4°$) but an identical O—O bond length (71).
The phenyl groups in the crystal occupy planes intersecting at an angle of 84°
along a line nearly parallel to the O—O bond (460).

## B. Bond Energies

The —O—O— bond-dissociation energy in peroxides may be taken to equal
the enthalpy of activation for thermal homolysis (32b):

$$(RCO_2)_2 \longrightarrow 2RCO_2\cdot \qquad (17)$$

Typical values of $D_{O-O}$ for acyl peroxides adopted by Jaffe et al. (262) are
$Ac_2O_2$, 29.5; $(EtCO_2)_2$, 30.0; $(PrCO_2)_2$, 30.0; and $Bz_2O_2$, 31.0 kcal/mole.
The cleavage of the O—O bond for acyl peroxides is thus considerably easier
than it is for dialkyl peroxides and peroxy esters [$D_{O-O} = 35\text{-}39$ kcal/mole
(414, 509a) or for hydroperoxides ($D_{O-O} = 40\text{-}43$ kcal/mole) (33)].

Jaffe, Prosen, and Szwarc (262) have also computed the heats of formation
for the above peroxides from combustion data, using the information together
with data on O—O bond energies to calculate values of $\Delta H_f$ for $RCO_2\cdot$ and

to estimate enthalpies for reaction 18. For R = Me, Et, Pr, and Ph, $\Delta H$ is $-17$, $-14$, $-13$, and $-3$ kcal/mole, respectively.

$$RCO_2\cdot \quad \longrightarrow \quad R\cdot + CO_2 \tag{18}$$

### C.  Infrared Spectra

Davison (96) has published an extensive compilation of carbonyl stretching frequencies for stright-chain aliphatic peroxides and $p$-substituted benzoyl peroxides, and Ivanchev and co-workers (258) give a complete analysis of the spectra of the peroxides $[Me(CH_2)_nCO_2]_2$, where $n = 2, 3, 4, 6, 8, 10,$ 14, and 16, as well as for the acids themselves and several of their anhydrides. Sevchenko and Zyat'kov (467) have reported the spectra of a number of asymmetric peroxides, $AcO_2COR$ and $AcO_2COAr$. A representative sample is shown in Table 2.

TABLE 2.  CARBONYL STRETCHING FREQUENCIES FOR
$(RCO_2)_2$ AND $(RCO)_2O$

| R | Carbonyl Stretching Frequency $(cm^{-1})$[a] | | | |
| | Peroxide[b,c] | | Anhydride[d] | |
|---|---|---|---|---|
| Me | 1820 | 1796 | 1824[e] | 1748[e] |
| $Me(CH_2)_2$ | 1816 | 1784 | 1821 | 1752 |
| $Me(CH_2)_6$ | 1812 | 1786 | 1819 | 1751 |
| $Me(CH_2)_{10}$ | 1811 | 1786 | | |
| $Me(CH_2)_{16}$ | 1811 | 1786 | 1818 | 1751 |
| $p$-$MeOC_6H_4$ | 1780 | 1758 | | |
| $p$-$MeC_6H_4$ | 1785 | 1763 | | |
| $C_6H_5$ | 1789 | 1767 | 1780[e] | 1725[e] |
| $p$-$BrC_6H_4$ | 1793 | 1771 | | |
| $p$-$NO_2C_6H_4$ | 1799 | 1779 | | |

[a] Solutions in carbon tetrachloride, except for the substituted benzoyl peroxides, which were in ethylene dichloride.
[b] Data from reference 96.
[c] Ivanchev's values (258) are consistently higher by two to three wave numbers.
[d] Data from reference 258 unless otherwise indicated.
[e] Data from reference 31a.

Several points are notable:

1.  The carbonyl absorption occurs at about the same position as in the corresponding anhydride, but at sufficiently higher frequencies than in the carboxylic acid $[\nu_{C=O} = 1708 \pm 3$ cm$^{-1}$ for $RCO_2H$ in $CCl_4$ (258)] or the

corresponding ester [$\nu_{C=O} = 1742 \pm 7$ cm$^{-1}$ for RCO$_2$R (31a)], to permit a ready distinction of the peroxide from these, its typical decomposition products (512).

2. The carbonyl absorption, like that of the anhydrides, is split due to coupling between the two carbonyl groups (31, 96, 258); the average distance between the two maxima for peroxides (25–30 cm$^{-1}$) is less than that for anhydrides ($\sim 68$ cm$^{-1}$), the difference being ascribed (31b, 96, 258) to the fact that in the anhydrides the direct carbonyl-to-carbonyl distance is less by 1.4 Å than that for peroxides (509b). Calculations and full discussions are contained in articles by Bellamy (31b), Popov (416), and their co-workers.

3. Electron-withdrawing groups shift the carbonyl absorption to higher frequencies. Table 2 illustrates the effect for some ring-substituted benzoyl peroxides; Davison (96) has shown that for both *meta-* and *para*-substituted compounds the shift correlates rather nicely with the Hammett $\sigma$ values, even though the total spread of $\nu_{C=O}$ frequencies is small.

Aliphatic acyl peroxides also exhibit absorptions at 890–905, 1060, and 1125–1130 cm$^{-1}$, assigned respectively to O—O, $\overset{\overset{\text{O}}{\|}}{\text{C}}$—O—O—$\overset{\overset{\text{O}}{\|}}{\text{C}}$, and C—O stretching modes (258, 512). Benzoyl peroxide shows only the C—O (1224 cm$^{-1}$) and the $\overset{\overset{\text{O}}{\|}}{\text{C}}$—O—O—$\overset{\overset{\text{O}}{\|}}{\text{C}}$ (997 cm$^{-1}$) absorptions (258). Ivanchev et al. (258) have calculated O—O force constants from these data (4.65, 3.86, and 3.74 × 10$^5$ dynes/cm for Bz$_2$O$_2$, [Me(CH$_2$)$_{2-4}$CO$_2$]$_2$, and [Me(CH$_2$)$_{6-16}$CO$_2$]$_2$, respectively), and remarked that these conform qualitatively at least to the thermal stabilities of the peroxides.

## D. Ultraviolet Spectra

Electronic spectra of acyl peroxides in the normal ultraviolet–visible range are not very interesting and few have been published. Walker and Wild (539) have reported that acetyl peroxide in ethanol exhibits a continuous spectrum starting at 2800 Å and reaching an extinction coefficient of 100 at about 2300 Å. They note its similarity in shape to that of hydrogen peroxide, which occurs at slightly lower wavelength. In fact, however, the spectrum of acetyl peroxide is equally similar to that of acetic acid. Breitenbach and Derkosch (52a) have observed that ultraviolet spectra of ring-substituted benzoyl peroxides are virtually indistinguishable from those of the corresponding anhy-

drides. Spectra of some asymmetrical peroxides (**1**)

**(1)**

have been recorded by Ol'dekop and co-workers (403a).

## IV.  CHEMICAL CHARACTERISTICS

### A.  Thermal Decomposition

One might say of any peroxide, as of Cawdor, "Nothing in his life became him like the leaving of it" (469). For acyl peroxides the phrase is particularly apt; their usefulness derives from their decomposition, and research has been overwhelmingly devoted to unraveling the complexities. To summarize the more detailed discussion in the following section, thermal decomposition is comprised of three main elements:

1.  Homolysis of the O—O bond

$$
\underset{\displaystyle \overset{\displaystyle O}{\|}}{R-C}-O-O-\underset{\displaystyle \overset{\displaystyle O}{\|}}{C-R} \xrightarrow{\Delta} 2R\overset{\displaystyle O}{\underset{\displaystyle \|}{C}}-O\cdot \tag{19}
$$

2.  Radical-induced decomposition

$$
R-\overset{O}{\underset{\|}{C}}-O-O-\overset{O}{\underset{\|}{C}}-R + R'\cdot \longrightarrow R-\overset{O}{\underset{\|}{C}}-O-R' + R-\overset{O}{\underset{\|}{C}}-O\cdot \tag{20}
$$

3.  "Polar" decomposition via carboxy inversion.

$$
R-\overset{O}{\underset{\|}{C}}-O-O-\overset{O}{\underset{\|}{C}}-R \longrightarrow R-O-\overset{O}{\underset{\|}{C}}-O-\overset{O}{\underset{\|}{C}}-R
$$

$$
\longrightarrow CO_2 + R-O-\overset{O}{\underset{\|}{C}}-R \tag{21}
$$

Aroyl peroxides may undergo radical-induced decomposition via attack on the ring as well. In cases where R· is a particularly stable radical O—O homolysis and C—C homolysis are probably synchronous;

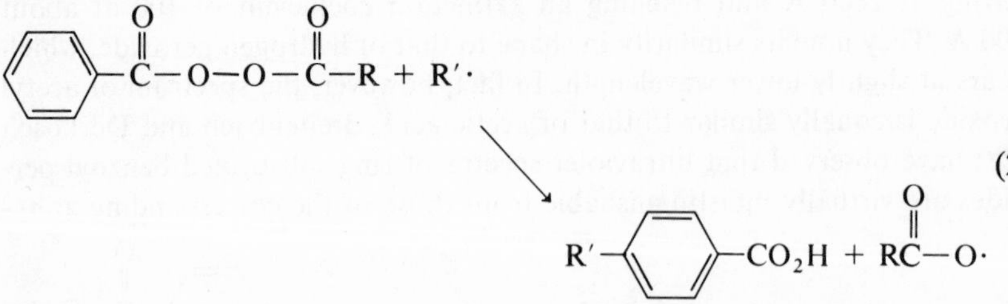

$$
\tag{22}
$$

where it is not, competition between $CO_2$ loss from $RCO_2\cdot$ and diffusion strongly influence the rates and products of decomposition.

$$( RCO_2)_2 \longrightarrow R\cdot + CO_2 + \cdot O_2CR \quad (\text{or } 2R\cdot + 2\ CO_2) \tag{23}$$

$$(RCO_2)_2 \rightleftharpoons [RCO_2\cdot \ \cdot O_2CR] \begin{cases} 2\ RCO_2\cdot \\ RCO_2\cdot + CO_2 + R\cdot \\ RCO_2R + CO_2 \\ R{-}R + 2\ CO_2 \\ 2\ R\cdot + 2\ CO_2 \end{cases} \tag{24}$$

It is not easy to determine how much each of the several modes contributes to the overall process of thermal decomposition of any peroxide. Generally speaking, aroyl peroxides are more subject to free-radical attack at O—O than the aliphatic analogs; carboxy inversion is facilitated by any factors (including the polarity of the solvent) that tend to stabilize $RCO_2^-$, and by branching at the $\alpha$-carbon.

Thermal stability then is a composite quantity that depends on the concentration of the peroxide (higher concentrations promote induced decomposition), the solvent [which may facilitate either carboxy inversion or induced decomposition, or attack the peroxide directly (see Section IX-A)], and the structural features of the peroxide itself. Thermal stability of $(RCO_2)_2$ decreases as $R\cdot$ is progressively made a more stable radical. Noteworthy compilations of data concerning the effect of structure on the *gross rate* of thermal decomposition in dilute solution have been made by Boozer (49), Guillet (189–191), Thynne (186), and their co-workers. Those of Boozer et al. (49), the most extensive, are shown in Table 3.

The products of thermal decomposition are chiefly useful insofar as they are free radicals, serving to initiate polymerization (541) and telomerization chains (138, 541), or for studies of the inherent reactivity of species with an unpaired electron [see, for example, Szwarc et al. on "methyl affinities" (57, 187, 214, 338, 487); Greene (181), Hart (324), Schwetlick (466), and their co-workers on alkyl radicals; and Cadogan, Hey, Williams et al. (55b; 67; 77; 93; 95a, b; 157; 168; 188; 197; 198; 219–225a; 339), Lynch, Pausacker, and co-workers (129, 346, 347, 372, 408, 409), Eliel (133, 141, 142) and others (28, 374, 524) on aryl radicals].

The use of acyl peroxides in organic syntheses has recently been reviewed by Karnojitsky (269), but instances in which strictly thermal decomposition has synthetic utility apart from simply providing free radicals are uncommon. Most notable perhaps is the synthesis by Fieser, Leffler, and co-workers (149, 150) of some 200 potential antimalarials via the following reaction:

$$+ \ (R'CO_2)_2 \quad \xrightarrow[\text{90-95}^\circ]{\text{AcOH}} \quad \tag{25}$$

(30–60%)

TABLE 3. THERMAL STABILITY OF ACYL PEROXIDES $[(RCO_2)_2]$ IN SOLUTION[a]

| R | Relative Rate[b] of Thermal Decomposition at 40° |
|---|---|
| PhCHMe | $>10^6$ |
| t-Butyl | $10^6$ |
| $PhCH_2$ | 1695 |
| $Et_2CH$ | 1680 |
| $Me_2CH$ | 1490 |
| $CH_2Cl$ | 1120 |
| $PhCH_2CHMe$ | 780 |
| ⟨cyclohexyl⟩ | 36 |
| $MeCH=CH$ | 5 |
| $Ph(CH_2)_4$ | 4 |
| Pr | 2.2 |
| Et | 2.0 |
| Me | 1.0 |
| Ph | 0.3 |

[a] Data from reference 49.
[b] In PhCl; first-order decomposition rates measured by $CO_2$ evolution; relative to $Ac_2O_2$, for which $k_{1(40°)} = 2.2 \times 10^{-7}$ sec$^{-1}$.

Aroyl peroxides may serve as a reasonably convenient source of biaryls (377, 559), and both aryl and alkyl iodides (482) and bromides (8) can be obtained by thermal decompositions in the presence of iodine or carbon tetrabromide, respectively. The decomposition of benzoyl peroxide in the presences of iodine plus an olefin yields mixtures of the 1,2-dibenzoate and iodobenzoate (41, 163, 164, 412).

$$(ArCO_2)_2 \underbrace{\begin{array}{c} \xrightarrow{\text{Ar'H}} \text{Ar Ar'} \\ \xrightarrow{\text{I}_2} \text{ArI} \end{array}} \tag{26}$$

## B. Thermal Decomposition of Pure Peroxides—Explosion

Low-molecular-weight acyl peroxides and peroxydicarbonates in the pure state [e.g., $(MeCO_2)_2$, $(CCl_3CO_2)_2$, $(CF_3CO_2)_2$, $(ROCO_2)_2$, where R = Et, *i*-Pr, *sec*-Bu] detonate on standing at room temperature, and attempted pyrolyses of milligram quantities may result in violent explosions (353). Even the relatively stable benzoyl peroxide explodes on heating a few degrees above its melting point (153, 519). Less sensitive peroxides are of some practical interest, and patents claim considerable thermal and shock stability for peroxydicarbonates with large R groups, such as 4-*t*-butyl or 4-*t*-pentyl, cyclohexyl (383b), and acyl peroxides of the type $(RO(CH_2)_nCO_2)_2$ (540), $RCO_2O_2C(CH_2)_2$-$CO_2O_2CR$ (190c), or $(RCH=CR'CO_2)_2$ (190a). [The latter peroxides are sensitive to shock, however, if R' = H (190a).]

The thermal decomposition of neat benzoyl peroxide at temperatures below its detonation point is reported (144) to yield mainly carbon dioxide and biphenyl, together with smaller amounts of phenyl benzoate and benzene. A recent careful study by Fine and Gray (153) of the explosive decomposition of the solid at 105–140° has shown it to be largely thermal in origin; that is, temperature increase due to inefficient dissipation of the heat of decomposition. Isothermal decompositions at these temperatures *in vacuo*, where no explosion took place, proceeded four to five times as fast as in dilute solution, due mainly to induced decomposition (153).

## C. Photolytic Decomposition

Direct photolysis of acyl peroxides, which requires light of 2500–3000 Å* (294, 539), is a convenient way to produce alkyl (260, 294, 295, 317, 568) or aryl (76) radicals at low temperatures. Electron-spin-resonance (ESR) signals of irradiated solutions or glasses even at −140 to −170° arise only from R· (or its reaction products), not from $RCO_2$· (76, 261, 294, 295, 568).

Sensitizers that appear to facilitate photodecomposition by direct energy transfer† (377, 531, 545) include benzophenone (377, 545), acetophenone (545), 2-acenaphthone (545), anthracene and other polynuclear aromatics (545, 345) and even benzene (377) or toluene (502, 531). Sensitization by benzophenone and other ketones, which is readily quenched by oxygen (377), probably occurs via excited triplet states of the sensitizer (377, 545), whereas that produced by

---

* Excepting, of course, those constructed with built-in chromophores—such as bis(phenylazo) benzoyl peroxides (193, 266) or dinicotinoyl peroxides (368). The former may be used to photoinitiate polymerizations (193).

† Photosensitization by benzophenone in the presence of benzhydrol, however, appears to involve a ketyl-radical chain (489):

$$Ph_2C=O* + Ph_2CHOH \longrightarrow 2Ph_2\dot{C}OH$$
$$Ph_2\dot{C}OH + Bz_2O_2 \longrightarrow BzOH + BzO·$$
$$BzO· + Ph_2CHOH \longrightarrow BzOH + Ph_2\dot{C}OH$$

benzene or toluene is unaffected by oxygen (377, 502) and may involve excited singlet states (377). For ketones there appears to be a threshold $n \rightarrow \pi^*$ transition energy of about 59 kcal/mole; those having transitions of lower energy to an excited triplet state, such as fluorenone, benzil, or 2,3-pentanedione, do not photosensitize the decomposition of acyl peroxides (527, 545), although they do act as photosensitizers for the decomposition of hydroperoxides (527) via a ketyl-radical mechanism.

The products from photolytic decomposition, reported in considerable detail for acetyl peroxide (539), benzoyl peroxide (377, 545), *trans*-4-cyclohexanecarbonyl peroxide (545), and benzoyl phenylacetyl peroxide (502), appear to rise exclusively from homolysis followed by typical reactions of free radicals (377, 502, 545). Carboxy-inversion products, which form a large part of the products of the thermal decomposition of *trans*-4-cyclohexanecarbonyl (181, 184, 324, 548) and benzoylphenylacetyl (502, 504) peroxides are absent.

Even for peroxides not prone to give carboxy inversion, such as benzoyl or cinnamoyl, there are subtle differences between benzene-sensitized photolysis and thermal decomposition (377, 520). These can be partly explained by the difference in temperature (25 versus 78–80°); that is, the higher yields of esters from solvent obtained in the photolyses result from less rapid decarboxylation of the acyloxy radical (377).

$$ArCO_2\cdot \left\{ \begin{array}{l} \longrightarrow Ar\cdot + CO_2 \\ \overset{H}{\diagup} \diagdown \quad \xrightarrow{\ O_2\ } ArCO_2Ph \\ \xrightarrow{PhH} ArCO_2 \end{array} \right. \qquad (27)$$

However, the benzene-sensitized photolysis of benzoyl peroxide also yields more ester from cage recombination than either thermal decomposition (80°) or benzophenone-sensitized photolysis (25°) (13 versus $\sim 3\%$) (377). A possible explanation is that the excited benzene singlet may transfer sufficient energy to produce Ph·, $PhCO_2\cdot$, and $CO_2$ directly, thus eliminating the competition between $CO_2$ loss from $PhCO_2\cdot$ and diffusion, which governs ester-from-cage* in the other cases (113, 377).

$$Bz_2O_2 \left\{ \begin{array}{l} \xrightarrow[PhH]{hv} [Ph\cdot\ CO_2\ PhCO_2\cdot] \longrightarrow PhCO_2Ph + CO_2 \\ \qquad\qquad\qquad\qquad\qquad\searrow \\ \qquad\qquad\qquad\qquad Ph\cdot + CO_2 + PhCO_2\cdot \\ \xrightarrow[\text{or } hv + Ph_2CO]{\Delta} [PhCO_2\cdot\ PhCO_2\cdot] \longrightarrow 2\ PhCO_2\cdot \text{ etc.} \end{array} \right. \qquad (28)$$

\* Ester-from-cage is readily distinguished from ester-from-solvent by reacting either $(ArCO_2)_2$ in benzene or $Bz_2O_2$ in ArH (377).

The photolysis of acetyl peroxide with high-energy sensitizers has similarly been proposed (375) to yield $CH_3 \cdot$ directly.

Some novel photolytic syntheses from fluoroformyl peroxide have been reported by Talbott (514). When this compound is photolyzed alone, a 50% yield of $CF_3O_2CF_3$ is obtained, whereas in the presence of $CF_2N_2$ the main product is a peroxy ester, $CF_3O_2C(O)F$.

## D. Hydrolysis and Alcoholysis

Hydrolysis of an acyl peroxide yields one equivalent each of acid and peroxy acid (78) and affords a practical synthesis for peroxybenzoic acid (50, 410, 480b). Quantitative hydrolysis of $Ac_2O_2$ to $H_2O_2 + 2AcOH$ may be achieved with 2 $M$ $HClO_4$ (352).

$$(RCO_2)_2 + H_2O \longrightarrow R\overset{\overset{\displaystyle O}{\|}}{C}-OH + R\overset{\overset{\displaystyle O}{\|}}{C}-O-OH \qquad (29)$$

The rates of hydrolysis are said to parallel those for analogous anhydrides (46a), but the only modern data are those of Ogata et al. for $Ac_2O_2$ (394) and $AcO_2C(O)C_6H_4-X$ (393) in mixtures of acetic and sulfuric acids. They have shown that there is a mobile equilibrium (for which $k_-$ for reaction 29 follows $H_0$ and exhibits a Hammett $\rho$-factor of $-0.37$) and propose the following mechanism:

$$AcOH \underset{fast}{\overset{H^+}{\rightleftharpoons}} Me\overset{+}{C}(OH)_2 \underset{slow}{\overset{ArCO_3H}{\rightleftharpoons}} \left[ \begin{array}{c} HO \quad H \quad O \\ \backslash \quad / \quad \| \\ MeCO-O-CAr \\ | + \\ OH \end{array} \right]^+$$

$$\qquad (30)$$

$$Ar\overset{\overset{\displaystyle O}{\|}}{C}-O-O-Ac + H_3O^+$$

At 45° and 3.6 $N$ sulfuric acid the equilibrium constant for $AcO_2Bz$ was found to be approximately 0.04 over a twelvefold $H_2O$ concentration range (393). Interestingly, under these conditions the equilibrium mixture contained no appreciable benzoyl peroxide, acetyl peroxide, or peroxyacetic acid.

$$K = \frac{[AcO_2Bz][H_2O]}{[BzO_2H][HOAc]} \qquad (31)$$

Base-catalyzed hydrolysis of $(FCO_2)_2$ yields $F^-$, $CO_3^{2-}$, $O_2$, and $CO_2$, probably via the intermediate acid and peroxy acid (7).

The reaction of acetyl peroxide with the methoxide ion in methanol has been used to determine its carbonyl $^{18}O$ content (516).

$$\underset{\|}{\overset{^{18}O}{\underset{\|}{\phantom{O}}}}\quad \underset{\|}{\overset{^{18}O}{\phantom{O}}} \qquad\qquad\qquad \underset{\|}{\overset{^{18}O}{\phantom{O}}}$$

$$Me-C-O-O-C-Me + MeO^- \longrightarrow Me-C-O-Me \quad (32)$$

Similarly the phenoxide ion in chloroform produces phenyl benzoate from benzoyl peroxide (546). However, there appears to have been little attempt to study the straight alcoholysis of acyl peroxides or to sort it out from the rapid radical-induced decomposition that occurs in alcoholic solvents (167, 384). The decomposition of acetyl peroxide in $C_3$ or $C_4$ alcohols at 90° yields some of the acetyl ester, but the main products ($CH_4$, $CO_2$, $MeCO_2Me$, alcohol derivatives, and oxidation products) are clearly homolytic in nature (283). With phenols, attack is predominantly on the peroxide linkage (546)*.

### E. Reduction of Acyl Peroxides

Reduction methods are not highly developed, there being little call for their use in acyl peroxide chemistry of interest in the last several decades. Reaction with $I^-$ is rapid and quantitative; it serves mainly to determine peroxide concentration (see Section V-A) but has been used by Schwartz and Leffler (465) to analyze mixtures of $p\text{-}ClC_6H_4CO_2O_2CC_6H_4I$ and $(p\text{-}IC_6H_4CO_2)_2$, produced by the partial decomposition of the latter in carbon tetrachloride.

$$\text{NaI} \quad | \quad \text{AcMe-H}_2\text{O} \qquad\qquad\qquad\qquad (33)$$

Quaternary ammonium borohydrides reduce benzoyl peroxide fairly cleanly, the yield of recovered benzoic acid being 65–100%, depending on the solvent and temperature (506). Benzoyl peroxide reacts with aluminum isopropoxide as well, but not readily, and the product mixture is complex (438); 3 hr at 70–75° gives 80% decomposition but only a 30% yield of benzoic acid (92). Sodium hydride does not react with benzoyl peroxide at moderate tempera-

---

* However, the evidence cited by Singer (Vol. I, Chapter V, p. 299) to the effect that peroxyesters can be synthesized by nucleophilic displacement by alcohols or phenols on the peroxide oxygen is quite tenuous and probably should be discounted entirely. (Note added in proof.) R. H.

tures and has been used to precipitate benzoic acid, produced via homolytic reactions, as the sodium salt (104).

The reduction of acyl peroxides to anhydrides under mild conditions and in good yield can be achieved with phosphines or phosphites (73, 230, 232, 237) (see Section IX-A).

Virtually quantitative reduction of $(RCO_2)_2$ to RH, $CO_2$, and $RCO_2H$ can be achieved with chromous perchlorate in aqueous ethanol (296b). The reaction takes the course

$$(RCO_2)_2 + Cr^{2+} \longrightarrow RCO_2^- Cr^{3+} + RCO_2 \cdot \qquad (34)$$

$$RCO_2 \cdot \longrightarrow CO_2 + R \cdot \xrightarrow{Cr^{2+} + H_2O} RH + Cr^{3+}OH^- \qquad (35)$$

Most R groups, including cyclopropyl and cyclobutyl, yield unrearranged

RH, but $\left( \triangleright\!\!-CH_2-CO_2 \right)_2$ gives only 1-butene (296b).

### F. Other Reactions

To be discussed in subsequent sections are reactions involving electrophilic attack on the O—O bond by $H^+$ and Lewis acids; nucleophilic attack by phenols, amines, phosphines, sulfides, carbanions, and olefins; and catalytic decompositions by metal ions.

## V. ANALYSIS OF ACYL PEROXIDES*

Iodometric titration and infrared spectroscopy have been the most frequently used analytical techniques in the chemistry of acyl peroxides. Polarographic and chromatographic methods are also available. All of these methods have been reviewed by Martin (351) and by Siggia (479), and have been considered in some detail with respect to the analysis of hydroperoxides in Chapter I of this volume.

### A. Iodometric Titration

Liberation of iodine from an acidified iodide solution has served as an almost universal criterion of acyl peroxide purity.†

$$(RCO_2)_2 + 2I^- \xrightarrow{H^+} I_2 + 2RCO_2^- \qquad (36)$$

* See Chapter VI for a detailed discussion of analytical methods.

† o-Iodobenzoyl peroxide (332) is an unusual case for which iodine liberation does not distinguish peroxide from its rearrangement product, the iodosobenzoate

Complete reaction with potassium iodide in aqueous acetic acid at room temperature takes 10–15 min (384, 507b) and requires blanketing with an inert atmosphere to avoid concomitant oxidation of $I^-$ by oxygen. Reaction can be accelerated considerably by adding a trace of ferric chloride (407, 481) or by using sodium iodide in acetone acidified by carbon dioxide. An equally popular procedure developed by Wagner, Smith, and Peters (537) is to reflux the sample with sodium iodide in isopropanol–acetic acid for several minutes. Atmospheric oxygen is expelled by the boiling, and blank titrations are virtually negligible (349).

These procedures are also suitable for the analysis of peroxydicarbonates, which appear to react somewhat faster than acyl peroxides (500).

Normally the iodine produced is titrated with thiosulfate (351, 479), but it may be estimated equally well spectroscopically (14).

Measuring the rate of color formation under precisely controlled conditions has been used to distinguish between different types of peroxides (422). At 27° in a neutral aqueous potassium iodide solution color development followed first-order kinetics with $t_{1/2}$ for peroxyacetic acid, hydrogen peroxide, acetyl peroxide, and methyl hydroperoxide being 0, 14, 17, and 33 min, respectively.

Acetyl peroxide in mixtures containing both peroxyacetic acid and hydrogen peroxide may be estimated quantitatively by first adding an excess of diethanol sulfide, then potassium iodide after the hydrogen peroxide and peroxy acid have been reduced (86).

## B.  Other Chemical Methods

The reduction of acyl peroxides to anhydrides by triphenyl phosphine is quantitative (236, 237) and rapid [10–20 min in benzene (132) or isopropanol (498) at room temperature].

$$(RCO_2)_2 + Ph_3P \longrightarrow (RCO)_2O + Ph_3PO \qquad (37)$$

The excess phosphine can be separated readily and weighed (132) or determined spectrophotometrically (498).

Colorometric methods based on the oxidation of amines include the formation of methylene blue from the leuco base (493) and the reaction with $N,N$-dimethyl-$p$-phenylenediamine sulfate (131). Benzoyl peroxide has been shown to give quantitative results with these reagents, but the methods have been little used.

The reaction of aniline with benzoyl or lauroyl peroxide has been used, with results comparable in precision to those obtained by iodometric methods (413). Careful control of conditions is required, however, to achieve reproducible stoichiometry (413). Extremely small amounts of benzoyl peroxide can be detected by its reaction with hexamethylene tetramine to yield formaldehyde (147).

## C. Polarographic Analysis

Excellent reviews and discussions on polarographic analysis of peroxides have been published by Johnson and Siddiqui (265) and Schulz and Schwarz (464), whereas Donner-Maack and Lück (128) have applied oscillopolarographic techniques to the analysis of benzoyl, lauroyl, and succinyl peroxides.

Half-wave potentials for acyl peroxides fall in the region of $-0.1$ to $+0.4$ volt, depending on the solvent–electrolyte combination (265), of which typical examples are methanol–benzene–lithium chloride (316, 450), dioxane–water–ammonium nitrate (464), 95% ethanol–tetrabutylammonium chloride (265, 450), and dimethylformamide–tetrabutylammonium chloride (265, 450). Acyl peroxides are thus more readily reduced than hydroperoxides ($E_{1/2} = -0.6$ to $-1.1$ volts), peroxy esters ($E_{1/2} \simeq -1.0$ volt), and dialkyl peroxides ($E_{1/2} > -1.5$ volts), but at about the same half-wave potential as peroxy acids ($E_{1/2} \simeq 0.0$ volt) (265, 316, 450, 509a), and peroxydicarbonates ($E_{1/2} \simeq 0.0$ volt) (126).

For substituted benzoyl peroxides a plot of $E_{1/2}$ versus Hammett $\sigma$ values is roughly linear (464, 523). Schulz and Schwarz (464) have detected a slight break in the plot occurring at bis($p$-chlorobenzoyl) peroxide and suggest that the peroxides with substituents more electron donating than Cl cleave homolytically at the dropping-mercury electrode, whereas those with electron-withdrawing substituents cleave heterolytically. (Actually a much better defined break occurs in $\sigma$-plots of the rates of thermal decomposition and is discussed in Section VI.)

Diffusion currents increase linearly with peroxide concentration and can be used for accurate quantitative analysis of $10^{-3}$–$10^{-4}$ $M$ solutions (126, 128, 316, 351, 464).

## D. Chromatographic Analysis

Acyl peroxides can be separated from other peroxidic materials by paper (366, 448) or thin-layer chromatography on silica gel or alumina (35, 63, 152, 211, 286, 400). Eluants are generally mixed solvents, such as dimethylforma-mide–decalin (366), ethanol–heptane (63), toluene–carbon tetrachloride (286), toluene–acetic acid (211, 286), and spots are developed with either acidified potassium iodide (35, 366) or $p$-aminodimethylaniline hydrochloride (152, 286, 366, 448) sprays. The technique has been particularly useful for following the hydrolysis of acyl peroxides to peroxy acids (393, 394) or the isomerization of asymmetrical to symmetrical peroxides under basic conditions (400).

$$2AcO_2Bz \xrightarrow[H_2O]{Na_2CO_3} Ac_2O_2 + Bz_2O_2 \tag{38}$$

Mixtures of acyl peroxides [e.g., lauroyl, benzoyl, bis(p-chlorobenzoyl), and bis(2,4-dichlorobenzoyl)] can be readily separated with the less polar eluant mixtures (63, 152, 211, 286).

The combination of relatively high boiling points and thermal instability renders gas-chromatography techniques impractical for the analysis of most acyl peroxides. Hahto and Beaulieu (195) have, however, chromatographed acetyl peroxide on didecyl phthalate–FluoroPak-80, recovering material not extensively decomposed in a Dry Ice trap.

Identification of acyl peroxides by their gas–liquid chromatography pyrolysis patterns at 200° has been explored (83) but not extensively developed.

### E.   Measurement of Rates of Decomposition of Acyl Peroxides

Although unsuited to determining absolute concentrations, infrared spectrometry offers a convenient and reasonably precise method of monitoring relative concentrations and is frequently used in rate studies. The method is particularly advantageous where other types of peroxides may be formed during the reaction, as in the pyrolysis of acyl peroxides in polymer films (192c). Most workers have used the strong carbonyl absorption at 1790–1810 cm$^{-1}$ for this purpose (for typical examples see references 184, 204–206, 323b, 333, 334), but the 1223 cm$^{-1}$ (4) and 1002 cm$^{-1}$ (170) bands for benzoyl peroxide and the 1060 cm$^{-1}$ (4) band for lauroyl peroxide may be used with equal precision (4).

Carbon dioxide evolution has been used to measure rates of decomposition (49, 445, 486, 487), although the data require careful interpretation since the yield of carbon dioxide is seldom if ever quantitative (445a, 486). This offers no problem, of course, if only a measure of overall thermal stability is desired (49).

Thermal-decomposition rates for several acyl peroxides and peroxydicarbonates have been measured by differential thermal analysis (18, 427).

### VI.   THERMAL DECOMPOSITION OF AROYL PEROXIDES

In the absence of special structural factors aroyl peroxides undergo radical-induced decomposition more readily and carboxy inversion less readily than aliphatic acyl peroxides. Loss of carbon dioxide from $ArCO_2$· is slower than it is from $RCO_2$·. Both types of peroxide exhibit solvent-cage [  ] reactions. Thus the decomposition of benzoyl peroxide in a solvent (XH) can be at least partly described by the following schematic:

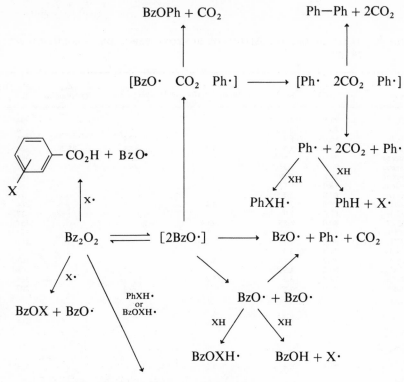

## A. Rates of Decomposition of Benzoyl Peroxide

Rates of thermal decomposition of benzoyl strongly depend on the solvent. Illustrative peroxide rate constants from the work of Nozaki and Bartlett (24, 384), Barnett and Vaughan (17), and Hartman, Sellers, and Turnbull (207) are shown in Table 4, but more dramatic are the percentage decompositions of 0.2 $M$ benzoyl peroxide in 1 hr at 79.8° in various media (384): highly halogenated solvents, 13–14%; aromatic hydrocarbons, 15–20%; most alkanes, 20–50%; ethers and alcohols, 80–100%. Decompositions in phenylacetylenes are also very rapid (15, 16). Furthermore, in most solvents the rates show higher than first-order dependence on peroxide concentration (17, 207, 362, 384).

Nozaki and Bartlett (384) were the first to demonstrate conclusively that these phenomena were due to concurrent homolysis and radical-induced decomposition

$$-\frac{\partial [Bz_2O_2]}{dt} = k_1[Bz_2O_2] + k_i[Bz_2O_2]^n \tag{39}$$

TABLE **4.** RATES OF DECOMPOSITION OF BENZOYL PEROXIDE IN SOLUTION AT 78–80°

| Solvent | $T$ (°C) | $k_1 \times 10^5$ (sec$^{-1}$) | $10^5 k_i$[(l/mole)$^{1/2}$ sec$^{-1}$] | $E_{a1}$ (kcal/mole) | $E_{ai}$ (kcal/mole) | Ref |
|---|---|---|---|---|---|---|
| CCl$_4$ | 79.8 | 2.1 | 3.7 | | | 384 |
| CCl$_4$ | 80.0 | 3.3–3.8[a] | | | | 4 |
| CHCl$_3$ | 80.0 | 1.8[a] | | 27.4[a] | | 505 |
| Benzene | 79.8 | 3.3 | 4.3 | 33.3 | 25.2 | 384 |
| Benzene | 78.0 | 3.8[a,b] | | 29.9 | | 362 |
| Benzene | 80.0 | 3.3[a] | | 29.9 | | 207 |
| Benzene | 78.0 | 1.88 | 3.74 | | | 168a |
| Benzene | 78.0 | 1.67[c] | | | | 168a |
| Benzene | 80.0 | 2.7 | | | | 17 |
| PhCH$_3$ | 79.8 | 3.3 | 4.3 | | | 384 |
| PhC$_2$H$_5$ | 80.0 | 3.3 | | | | 17 |
| PhC$_2$H$_5$ | 80.0 | 1.85 | 9.63 | | | 157 |
| PhCH(CH$_3$)$_2$ | 80.0 | 1.95 | 9.50 | | | 157 |
| $p$-(CH$_3$)$_2$C$_6$H$_4$ | 80.0 | 1.75 | 7.25 | | | 157 |
| PhNO$_2$ | 79.8 | 3.3 | 4.3 | | | 384 |
| PhNO$_2$ | 80.2 | 3.56 | 5.96 | | | 168d |
| PhNO$_2$ | 80.2 | 1.93[c] | | | | 168d |
| $t$-BuPh | 79.8 | 3.3 | 15.3 | | | 384 |
| $t$-BuPh | 80.0 | 4.5[a] | | 30.4 | | 207 |
| 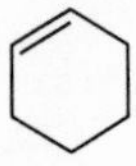 (cyclohexene) | 79.8 | 1.9 | 4.6 | | | 384 |
| 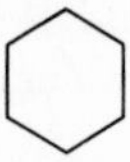 (cyclohexane) | 79.8 | 6.4 | 3.2 | | | 384 |
| 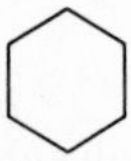 (cyclohexane) | 80.0 | 10.3[a] | | 28.2 | | 207 |
| 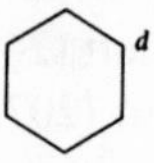 (cyclohexane)[d] | 80.0[d] | 5.3[d] | | | | 507b |
| $n$-C$_6$H$_{14}$ | 80 | 3.8[a] | | 29.0 | | 207 |
| Ac$_2$O | 79.8 | 7.5 | 3.0 | 30.7 | 25.0 | 384 |
| AcOH | 79.8 | 8.1 | 51 | | | 384 |
| AcPh | 80.0 | 4.3[a] | | 30.2 | | 43 |
| Dioxane + 3,4-dichloro-styrene | 80.0 | 4.2[a] | | | | 507a |
| Dioxane | 80.0 | 67[a,b] | | | | 17 |
| $n$-BuOH | 80.0 | 61[a] | | | | 17 |

[a] "First-order" rate constant for (Bz$_2$O$_2$)$_0$ = 0.05 $M$ or less.
[b] In air.
[c] In the presence of galvinoxyl.
[d] With 0.015 $M$ styrene, extrapolated from 91.1° by assuming $E_a \simeq$ 30 kcal/mole.

and to provide a kinetic analysis in mechanistic terms. Application of the steady-state approximation to the chain reaction (in its simplest form) leads to expression 39, where $n$, the order in $[Bz_2O_2]$ of the induced term, equals $\frac{3}{2}$ if termination is between like radicals and equals 1 if unlike radicals terminate. The term $k_i$ is, of course, a complex constant including $k$ for reactions 41, 42, and 43.

$$Bz_2O_2 \longrightarrow 2BzO\cdot \longrightarrow 2Ph\cdot + 2CO_2 \tag{40}$$

$$BzO\cdot \ (or\ Ph\cdot) + XH \longrightarrow BzOH\ (or\ PhH) + X\cdot \tag{41}$$

$$X\cdot + Bz_2O_2 \longrightarrow BzOX + BzO\cdot \tag{42}$$

$$X\cdot + X\cdot \ (or\ BzO\cdot\ or\ Ph\cdot) \longrightarrow termination \tag{43}$$

Generally speaking, for hydrocarbons and halogenated solvents $n = \frac{3}{2}$, and rate constants for both unimolecular homolysis and radical attack on peroxide can be obtained from the kinetic analysis (384) (Table 4). However, in ethers and alcohols $n = 1$ [decompositions are unimolecular for about one half-life, after which there is autoretardation due to buildup of products that inhibit induced decomposition (24)] and homolysis rates can be measured only in the presence of additives that suppress the induced reaction (24, 507a).

Scavengers that effectively suppress induced decomposition (Table 4) may be of two types:

1.  Substances that react with Ph· or BzO· to produce new radicals that terminate faster than they attack the peroxide, such as styrenes (507), acetophenone (43, 359), oxygen (24, 455, 505), or iodine (199, 507).

2.  Relatively stable free radicals, such as diphenylpicrylhydrazyl (27, 160, 474, 475) and galvinoxyl (168a), that do not attack the peroxide but react with active radicals to form nonradical products.

The most careful study of a scavenger of the first type is that of Swain, Schaad, and Kresge (507b) on benzoyl peroxide decomposition in cyclohexane. Addition of 0.015 $M$ styrene to 0.003 $M$ benzoyl peroxide in this solvent is sufficient to suppress induced decomposition almost completely. Higher styrene concentrations (up to 0.15 $M$) affect the rate only very slightly, and the products only insofar as the yield of benzene is somewhat reduced (507b).

Gill and Williams (168) have reported a series of elegant studies on the decomposition of benzoyl peroxide in benzene, nitrobenzene, and other solvents with and without added galvinoxyl. Most interestingly, they showed that $k_1$ in the presence of galvinoxyl was slightly lower than the $k_1$ evaluated by the $1 + \frac{3}{2}$ analysis in the absence of scavenger* (Table 4); this indicates

---

* An earlier similar finding by Swain et al. (507b) with styrene as a radical scavenger cannot be discounted. However, the issue was somewhat confused since $Bz_2O_2$ in $CCl_4$ or $Ac_2O + 1\ M$ styrene gave *larger* $k_1$ values than those obtained by $1 + \frac{3}{2}$ order analysis of decompositions in these solvents without styrene.

the presence of a small first-order induced component as well as the more readily observable $\frac{3}{2}$ order reaction.* In benzene at a peroxide concentraton. of 0.01 $M$ it was estimated to account for about 20% of the decomposition (168a).

Despite the complexity of the situation, it is apparent that thermal decompositions of dilute solutions of benzoyl peroxide ($[Bz_2O_2]_0 \leqslant 0.05$ $M$) will frequently *appear* to be unimolecular (Table 4).

Clearly, however, the rates for the *noninduced* part of the decomposition, sorted out by whatever means, are rather insensitive to solvent structure or polarity (Table 4), in accordance with the view that this reaction is indeed a homolysis. (However, see Section VII C.)

## B. Rates of Decomposition of Other Aroyl Peroxides

Rates of thermal decomposition of a wide spectrum of ring-substituted benzoyl peroxides have been determined by Swain, Stockmayer, and Clarke (507a), Blomquist and Buselli (43), O'Driscoll and White (392b), and Cooper (80), either by using inhibitors to suppress induced decomposition (43, 507a) or by measuring the rate at which the peroxide initiated polymerization† (80, 392b). Selected values from the more than 50 peroxides investigated are shown in Table 5. As first pointed out by Swain et al. (507a), electron-withdrawing substituents in the *meta* and *para* positions generally retard decomposition, whereas the reverse is true of electron-donating groups. The rate constants fit a Hammett $\rho\sigma$ plot fairly well (using for $\sigma$ the sum of the two groups) with $\rho = -0.38$ (43, 80, 507a, 562). This was rationalized by Swain and his co-workers (507a) in terms of an inductive effect removing (or adding to) the excess of electron density on the peroxidic oxygens, thereby stabilizing (or destabilizing) it with respect to cleavage.

$$\text{X} \overbrace{\phantom{xxx}} \overset{\overset{O}{\parallel}}{\underset{\leftarrow}{C}} - O - O - \overset{\overset{O}{\parallel}}{\underset{\rightarrow}{C}} \overbrace{\phantom{xxx}} \text{X} \qquad (44)$$

---

* This difference between rates determined iodometrically and determined by the disappearance of scavenger must be in part due to solvent-cage reactions. However, from Swain, Schaad, and Kresge's studies of decompositions in cyclohexane (507b) it appears that the total products from nonscavengeable radicals amount to only 2–3%.

† Cooper's experimentally observed quantity was $(fk_d/k_t)\frac{1}{2}k_p$, which for comparison with the decomposition-rate studies required good values for propagation and termination constants, as well as an assumption about the initiating efficiency $f$ of the peroxide. The difference between Cooper's decomposition-rate constants and those of Swain and of Blomquist (Table 5) is too great to be explained by the 10° difference in temperature and mainly results from his assumption that $f = 1.0$. The more recent and elegant analysis of O'Driscoll and White (392b) permits evaluation of both $k_d$ and $f$ (which decreases regularly from 0.8 for $p$-MeO to 0.27 for $p$-NO$_2$) as well as $k_i$ (here the rate constant for the attack of the polystyryl radical on peroxides).

TABLE 5.  RATES OF THERMAL DECOMPOSITION OF $\left( \underset{X}{\underbrace{\phantom{xxx}}} - \overset{\displaystyle O}{\overset{\displaystyle \|}{C}} - O \right)_2$

IN SOLUTION

| X | $E_a{}^a$(kcal/mole) | $80°^a$ | $80°^b$ | $90°^{c,d}$ | $70°^{c,e}$ |
|---|---|---|---|---|---|
| | | $k_1 \times 10^5$ (sec$^{-1}$) | | | |
| p-MeO | 28.7 | 15.6 | 11.8 | 32.9 | 1.85 |
| m-MeO | 28.9 | 6.41 | 5.72 | | 0.90 |
| p-Me | 29.9 | 5.92 | 6.12 | 19.3 | 1.09 |
| m-Me | 30.2 | 4.70 | 4.40 | | 0.60 |
| H | 30.2 | 4.32 | 4.20 | 13.3 | 0.70 |
| p-Cl | 30.4 | 3.84 | 3.62 | | 0.54 |
| m-Cl | 30.7 | 2.85 | 2.53 | | 0.36 |
| p-CN | 31.2 | 2.43 | 2.03 | 9.7 | 0.36 |
| m-CN | | | 1.70 | | |
| p-NO$_2$ | 30.3 | 4.34 | | 8.7 | |
| m-NO$_2$ | 30.2 | 3.80 | | | |
| 3,5-(NO$_2$)$_2$ | 31.2 | 1.87 | | | |
| o-MeO | 27.2 | 215.0 | | | 13.8 |
| o-Me | 30.2 | 31.3 | 10.0$^f$ | | 2.8 |
| o-Cl | 29.4 | 39.0 | | | 2.8 |
| o-NO$_2$ | 28.6 | 134.0 | | | |
| o-Ph | | 50.0 | 25.0$^f$ | | |
| o-OPh | 29.0 | | | | |

$^a$ In acetophone (43).
$^b$ In dioxane$+2M$ 3,4-dichlorostyrene (507a).
$^c$ Calculated from the rate of initiation of styrene polymerization.
$^d$ Data from reference 392b.
$^e$ Data from reference 80.
$^f$ In carbon tetrachloride plus 1 $M$ styrene (185).

Nitrobenzoyl peroxides do not fit the Hammett plot (43, 392b). Their anomalously fast rates of decomposition may be due in part to some unsuppressed induced decomposition. O'Driscoll and White (392b) have shown that p-nitrobenzoyl peroxide is attacked by polystyryl radicals at least 50 times more readily than is benzoyl peroxide. O'Driscoll and White's value for homolysis is at least lower than that for benzoyl peroxide (Table 5). However, it may be that p-nitrobenzoyl peroxide and the p-cyano as well, both of which have low initiating efficiencies (392b), partly decompose by a carboxy-inversion mechanism.

All *ortho*-substituted peroxides appear to decompose considerably faster than other benzoyl peroxides (43, 80, 185). The reason is probably steric for the examples shown in Table 5; the entropies of activation are greater than for the *meta* and *para* analogs (43). In other cases, to be discussed later, anchimeric assistance to the cleavage of the O—O bond has been demonstrated.

2-Thenoyl peroxides substituted in positions 4 and 5 decompose in carbon tetrachloride + 3,4-dichlorostyrene at rates that are very similar to those of benzoyl peroxides and obey a $\rho\sigma$ relationship with $\rho = -0.44$ (462, 463). Again, the 5-nitro compound decomposes anomalously fast, as does the 5-phenyl, rather surprisingly.

### C. Mechanisms of Induced Decomposition

Benzoyl peroxide refluxed in ethyl ether yields 0.9 mole of $\alpha$-ethoxyethyl benzoate per mole of decomposed peroxide (70a). Decomposition in *t*-butyl methyl ether or *t*-butyl methyl sulfide produces substantial yields of *t*-BuOCH$_2$OBz or *t*-BuSCH$_2$OBz (213), whereas in cyclohexene some cyclohexenyl benzoate is formed (127). In the presence of hexaphenylethane, triphenylmethyl benzoate is produced (201, 556), and heating with trialkyl tin hydrides gives R$_3$SnOBz + BzOH quantitatively (380, 381). In each case [i.e., ethers (104, 130), cyclohexene (127), triphenylmethyl (127), and alkyl tin hydrides (380)] isotopic-labeling experiments have demonstrated that radical attack occurs at the peroxidic oxygen.

$$\underset{\text{Ph}-\overset{\overset{\displaystyle ^{18}\text{O}}{\|}}{\text{C}}-\text{O}-\text{O}-\overset{\overset{\displaystyle ^{18}\text{O}}{\|}}{\text{C}}-\text{Ph} + \text{EtO}-\overset{\displaystyle \cdot}{\text{C}}\text{HMe}}{}$$

$$\longrightarrow \quad \text{Ph}-\overset{\overset{\displaystyle ^{18}\text{O}}{\|}}{\text{C}}-\text{O}-\overset{\overset{\displaystyle \text{OEt}}{|}}{\text{C}}\text{HMe} + \text{PhCO}_2\cdot \quad (45)$$

Thus the radical-induced chain in these cases is composed of displacement on O—O alternating with* hydrogen abstraction

$$\text{PhCO}_2\cdot + \text{EtOCH}_2\text{Me} \longrightarrow \text{EtO}\overset{\displaystyle \cdot}{\text{C}}\text{HMe} + \text{PhCO}_2\text{H} \quad (46)$$

Other examples of this type of induced decomposition are provided by the polystyryl radical (397, 420), benzaldehyde (457, 550), and some intramolecular displacements (21, 44).†

---

* Yields of carbon dioxide are relatively low in decompositions in which induced decomposition predominates (70a, 201, 380), as is reasonable in the presence of ethers or tin hydrides with readily abstractable hydrogens.

† Reactions 47, 48, and 49 are from references 457 and 550, 44, and 21, respectively.

$$BzH \xrightarrow{\ BzO \cdot \ } Bz \cdot \xrightarrow{\ Bz_2O_2 \ } Bz_2O + BzO \cdot \tag{47}$$

$$(48)$$

$$(49)$$

Induced decomposition by triphenylmethyl is not a chain reaction in the same sense. Two trityl radicals are used up for each molecule of peroxide destroyed, forming a basis for quantitative analysis of $Ph_3C \cdot$ (200), though the relative yields of products $Ph_3COBz$, $BzOH$, and $Ph_3CAr$ [from solvent ArH (32, 201, 555, 556)] vary with initial concentrations of the reactants (201). The negligible yields of carbon dioxide (201, 556) and biphenyl (201), as well as the approximately 15% yield of $Ph_3C-{}^{18}O-C(O)Ph$ from $[PhC({}^{18}O)O]_2$ (127), suggest the following mechanism (32, 110):

$$\longrightarrow Ph_3COBz + BzO \cdot \tag{50}$$

The rate constant $k$ for reaction 50 has been determined by Suehiro and co-workers (503) for several substituted benzoyl peroxides in benzene and other solvents (Table 6) and shown to have a $\rho$ value (in benzene) of 1.45, which increases with increasing solvent polarity.

TABLE 6.  RATE CONSTANTS FOR RADICAL DISPLACEMENT ON THE PEROXIDE LINKAGE OF BENZOYL PEROXIDE[a]

| Radical | Solvent | $T\,(°C)$ | $k_i$ (l/mole-sec)[a] | $\Delta H\dagger$ (kcal/mole) | $\Delta S\dagger$ (e.u.) | Ref. |
|---------|---------|-----------|-----------------------|-------------------------------|--------------------------|------|
| Ph$_3$C· | Benzene | 25 | 1.16 | 10.4 | −24 | 503 |
| Ph$_3$C· | Chlorobenzene | 25 | 2.40 | 9.3 | −26 | 503 |
| Ph$_3$C· | Anisole | 25 | 2.26 | 9.4 | −26 | 503 |
| Ph$_3$C· | Nitrobenzene | 25 | 6.58 | 5.6 | −36 | 503 |
| Ph$_3$C· | Benzene | 60 | 7–14[b] | | | |
| Polystyryl | Styrene | 60 | 8.0 | | | 420 |

[a] $R· + Bz_2O_2 \xrightarrow{k_i} ROBz + BzO.$
[b] Calculated from $k_i$ values given in reference 503 for 15, 25, and 35°.

Evidently the transition state has considerable ionic character in which the attacking radical acts as an electron donor (503).

$$
\begin{array}{ccccc}
\text{Bz—O} \;\cdot\text{OBz} & \longleftrightarrow & \text{BzO—OBz} & \longleftrightarrow & \text{BzO·} \quad ^-\text{OBz} \\
\quad | & & & & \\
\quad \text{CPh}_3 & & \cdot\text{CPh}_3 & & ^+\text{CPh}_3
\end{array}
\qquad (52)
$$

Similar conclusions have been drawn by Rübsamen and co-workers (453) and Walling and Azar (542b), who showed that attack by Et$_3$Sn· (453) or R$\overset{.}{C}$HOR′ (542b) on asymmetrical peroxides occurs predominantly at the more electropositive carbon.

$$
\underset{\text{(70\%)}}{\text{AcOSnEt}_3} + \underset{\text{(30\%)}}{\text{Et}_3\text{SnO}-\overset{\displaystyle O}{\overset{\|}{C}}\!\!-\!\!\!\bigcirc\!\!\!-\text{OEt}}
$$

(53)

Chain-transfer constants $C_t$ for substituted benzoyl peroxides in polymerizing styrene show a regular increase as the substituent groups are made more electron donating (81), and $\rho$ for $k_i$, the rate constant for polystyryl-radical attack on peroxide, is 1.7 (392b) [though here again nitrobenzoyl and cyanobenzoyl peroxides react anomalously fast (392b)]. From analysis of the products (397), and the very similar rate constants (357, 420) (Table 6) and $\rho$ values for polystyryl versus triphenylmethyl attack, it is apparent that polystyryl also performs a radical displacement on the peroxidic oxygen.

$$\text{polystyryl} \cdot \begin{array}{l} \xrightarrow[k_i]{\text{Bz}_2\text{O}_2} \text{BzO} \cdot + \text{polymer} \qquad (54) \\[2em] \xrightarrow[k_p]{\text{PhCH: CH}_2} \text{polystyryl} \cdot \end{array}$$

$$(55)$$

The speed at which *different* radicals effect displacement on the O—O bond has been shown by Tokumaru and Simamura to depend qualitatively on their electron-donating ability: $R\overset{\cdot}{C}HOR > R\cdot \gg CCl_3\cdot$ (522). The relatively slower induced decomposition in noncyclic as compared with cyclic ethers (384) has been suggested by Walling and Azar (542b) to result from a competitive cleavage of $R\overset{\cdot}{C}HOR$ to RCHO and the less active $R\cdot$. In the cyclic case the active $\alpha$-alkoxy radical is readily regenerated.

$$(56)$$

A variation in the mechanism of diplacement on the O—O bond occurs with radicals that are effective hydrogen-atom donors, such as ketyls (489), or electron donors, such as ketyl-radical anions (165). Radical-induced decompositions in aliphatic alcohols may well involve direct hydrogen-atom donation (24, 384), judging from the substantial yields of aldehydes or ketones produced (283, 395, 542b).

$$\text{Ph}_2\overset{\cdot}{C}-\text{O}-\text{H} \quad \underset{\underset{\text{Bz}}{|}}{\text{O}}-\text{O}-\text{Bz} \longrightarrow \text{Ph}_2\text{CO} + \text{BzOH} + \text{BzO} \cdot \quad (57)$$

$$\text{Ph}_2\overset{\cdot}{C}-\overset{-}{\text{O}}\overset{+}{\text{M}} + \text{Bz}_2\text{CO}_2 \longrightarrow \text{Ph}_2\text{CO} + \text{Bz}\overset{-}{\text{O}}\overset{+}{\text{M}} + \text{BzO} \cdot \quad (58)$$

DeTar and co-workers (113, 120) have shown the major pathway for induced decomposition in benzene to be hydrogen-atom donation by cyclohexadienyl radicals.

$$\text{(cyclohexadienyl-Ph,H)} + Bz_2O_2 \longrightarrow BzOH + BzO\cdot + Ph-Ph$$

## D. Radical Attack on the Aromatic Ring

A distinctly different mechanism for induced decomposition operates with radicals that are neither good electron donors nor good hydrogen-atom donors. Decompositions of aroyl peroxides in carbon tetrachloride (46b, 224a, 273, 550), chloroform (67a, 224a), bromotrichloromethane (67a), methylene bromide (67b), cyclohexane (550), or acetic acid (169) produce *ortho-* and *para-*substituted acids among other products.*

$$Bz_2O_2 \xrightarrow{CCl_4} CCl_3-\langle\rangle-CO_2H + \underset{[CCl_3 \text{ ortho}]}{\langle\rangle-CO_2H} \tag{60}$$

(1 mole)                          (0.27 mole)                    [0.01 mole (224a)]

$$Bz_2O_2 \xrightarrow{AcOH} \underset{CH_2CO_2H}{\langle\rangle-CO_2H} + \underset{CH_2CO_2H \text{ ortho}}{\langle\rangle-CO_2H} + \underset{Ph}{\langle\rangle-CO_2H} \tag{61}$$

(1 mole)                 (0.09 mole)              (0.07 mole)            (0.11 mole)

It seems unlikely that these arise from radical–radical coupling, although Schwartz and Leffler (465) have shown that the radical produced from bis(*p*-iodobenzoyl) peroxide is stable enough to abstract from carbon tetrachloride.

$$\left(\underset{I}{\langle\rangle-CO_2}\right)_2 \xrightarrow{CCl_3\cdot} \underset{\cdot}{\langle\rangle-CO_2} -O_2C-\underset{I}{\langle\rangle} \xrightarrow{CCl_4} \underset{Cl}{\langle\rangle-CO_2} -O_2C-\underset{I}{\langle\rangle} \tag{62}$$

Walling and his co-workers (544, 550), observing that none of the correspondingly substituted benzene was ever isolated from these decompositions, have proposed that the intermediate is an α-lactone that undergoes intramolecular hydrogen transfer.

---

* These are not found, however, when bromoform is the solvent (224a).

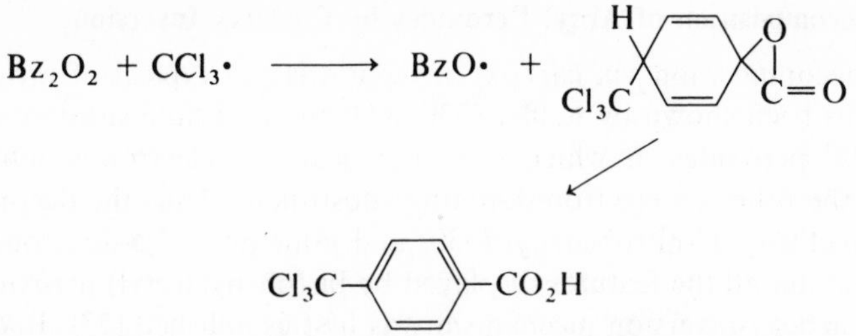

$$Bz_2O_2 + CCl_3\cdot \longrightarrow BzO\cdot + \text{(structure)}$$

(63)

$$Cl_3C\!\!-\!\!\text{(ring)}\!\!-\!\!CO_2H$$

Most recently he and Cekovic (544) have reported an elaborate study on the thermal and photolytic decompositions of acetylbenzoyl and acetylchloro-benzoyl peroxides, which from the yields of the various toluic acids produced appears to confirm the postulated mechanism.*

Cadogan, Hey, and Hibbert (67a), however, have argued for a kind of concerted displacement, which Walling (550) discounted on the basis of the absence of any deuterium-isotope effect.

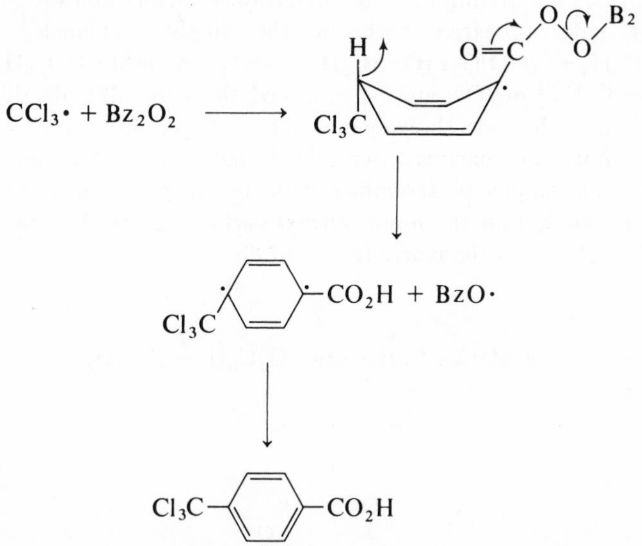

$$CCl_3\cdot + Bz_2O_2 \longrightarrow Cl_3C\text{(structure)}$$

$$Cl_3C\text{(ring)}\!\!-\!\!CO_2H + BzO\cdot$$

(64)

$$Cl_3C\!\!-\!\!\text{(ring)}\!\!-\!\!CO_2H$$

Its absence, say Cadogan et al. (67a) can be explained if the first step, formation of the cyclohexadienyl radical, is rate determining.

The decomposition of benzoyl peroxide in benzene does not yield any phenylbenzoic acid (120). It must be supposed the phenyl radicals are less active in this solvent than in acetic acid due to complexing with the aromatic ring.

---

* Recently Leffler and Zepp (J. Amer. Chem. Soc., **92**, 3713 (1970)) in a thorough reexamination of α-naphthoyl peroxide decomposition have suggested that the naphthyl naphthoate in the products is generated in this fashion. (Note added in proof.) R. H.

## E.  Decomposition of Aroyl Peroxides by Carboxy Inversion

Although decomposition via carboxy inversion is not typical of aroyl peroxides, it has been shown by Leffler (330, 334) to contribute substantially in
asymmetrical peroxides in which one ring bears an electron-withdrawing
group and the other an electron-donating substituent. Thus the decompositions of $p$-methoxy–$p'$-nitrobenzoyl (330) and $p$-methoxy–3,5-dinitrobenzoyl
peroxides exhibit all the features displayed by bis(phenylacetyl) peroxide, for
which the carboxy-inversion mechanism was first established (23). Rates are
sensitive to solvent polarity and are increased by added acid, in conformity
with the Brønsted catalysis law; initiation of vinyl polymerization is inefficient, particularly in polar media (330, 334).* The products consist largely of
$p$-methoxyphenol and the nitrobenzoic acids, whereas in thionyl chloride as
solvent the mixed carbonic anhydride, $p$-MeO—$C_6H_4$—$OC(O)OC(O)$—
$C_6H_4$—$p$-$NO_2$ can be isolated (330).

Leffler (330) hypothesizes that the push–pull effect of the dissimilar substituents provides the driving force for heterolysis. Denny and his co-workers
(99, 105) have prepared both of the singly $^{18}O$-labeled isomers,
$p$-MeO—$C_6H_4$—$C(^{18}O)O_2C(O)$—$C_6H_4$—$p$-$NO_2$ and $p$-MeO—$C_6H_4$—C(O)-
$O_2C(^{18}O)$—$C_6H_4$—$p$-$NO_2$, and decomposed them in sulfur dioxide to the
mixed carbonic anhydride, showing that whereas the carbonyl oxygen attached
to the $p$-methoxy side remains unscrambled, that on the $p$-nitro side becomes
partly, but not completely, scrambled with the ether oxygen. Denny is thus
forced to conclude that the $p$-nitrophenyl carboxy anion does not become
completely free during the rearrangement (99).

$$p\text{-}MeO - C_6H_4CO_2 - O_2C_6H_4 - p\text{-}NO_2$$

(65)

---

* This claim has been challenged recently by Imoto et al. (563a), who found that the
initiating efficiency of $p$-methoxy-$p'$-nitrobenzoyl peroxide for the polymerization of styrene
or acrylonitrile, though low, was identical in the three solvents benzene, acetone, and
dimethylformamide.

## F. Products of Decomposition of Aroyl Peroxides

### 1. Cleavage of Arylcarboxy Radicals and Cage Recombinations

Apart from radical-induced decompositions, the main determinant of product distribution from decomposing aroyl peroxides is the competition between the loss of $CO_2$ from $ArCO_2\cdot$ and its reaction, by abstraction from or addition to, other species.

$$(ArCO_2)_2 \longrightarrow \longrightarrow ArCO_2\cdot \begin{cases} \longrightarrow Ar\cdot + CO_2 \\ \xrightarrow{\underset{/}{\overset{\backslash}{C}}=\overset{/}{\underset{\backslash}{C}}} ArCO_2-\overset{|}{\underset{|}{C}}-\overset{|}{\underset{|}{C}}\cdot \quad (66) \\ \xrightarrow{RH} ArCO_2H + R\cdot \end{cases}$$

As shown some years ago by Hammond and Soffer (202), the benzoyl radical can be trapped quantitatively with an efficient scavenger. The intermediate is presumed to be a hypoiodite from the products in the absence of water (199).

$$Bz_2O_2 \longrightarrow 2BzO\cdot \xrightarrow{I_2+H_2O} 2BzOH \qquad (67)$$

The decomposition of benzoyl peroxide in vinyl monomers yields little carbon dioxide; most of the initiator fragments are found in the polymer as BzO— rather than as Ph— (20, 40, 436). Bevington and his co-workers (36–40), using isotopically labeled benzoyl peroxides in monomer–benzene solutions, have evaluated $k_c/k_a$ (where the subscripts $c$ and $a$ represent cleavage and addition, respectively) in a number of systems; in for example, styrene–benzene at 60° $k_c/k_a = 0.4$ mole/l, $E_{a_c} - E_{a_a} = 6.6$ kcal/mole (36, 40); in methyl methacrylate–benzene $k_c/k_a$ at 60° $= 3.3$ moles/l (38). Their estimates (40) for the absolute value of $k_c$ in benzene are $7.4 \times 10^3$ sec$^{-1}$ at 60° and $2.4 \times 10^4$ sec$^{-1}$ at 80°.

For cleavage of the benzoyloxy radical versus hydrogen abstraction ($k_a'$), Walling (542a) has made a start (Table 7) on a study analogous to that done for alkoxy radicals (see Chapter I of this volume for leading references). Evidently, as for $t$-BuO$\cdot$, the competition depends not only on the substrate but on the solvent as well (542a) (Table 7).

Beyond this, it is perhaps worth noting a few examples of decompositions in which noncleavage of aroyloxy radicals leads to interesting products. Chief among these are cases in which intramolecular displacement leads to lactones (95c, 106, 115, 272).

$$(68)$$

$$(20\text{–}30\%)$$

TABLE 7.  CLEAVAGE VERSUS HYDROGEN ABSTRACTION BY THE
           BENZOYLOXY RADICAL[a]

| Substrate | Solvent | $k_c/k_a'$ (moles/l)[b] | $E_{a_c} - E_{a_a}'$ (kcal/mole)[c] |
|-----------|---------|------------------------|--------------------------------------|
| Cyclohexane | Carbon tetrachloride | 0.098 | 4.7 |
| Cyclohexane | Benzene | 1.44 | 5.0 |
| Cyclohexane | Acetone | 0.076 | 0.0 |
| Cyclohexane | Chlorobenzene | 0.058 | −0.3 |
| Ethyl ether | Carbon tetrachloride | 1.96 | |
| Benzaldehyde | Benzene | 0.5 | |
| Isopropanol | Benzene | 0.14 | |
| Anisole | Benzene | 0.12 | |
| Cumene | Benzene | [d] | |

[a] Data from reference 542a.
[b] Ratio of the rate constant for cleavage and the rate constant for abstraction from substrate at 70°.
[c] Difference in activation energies for the two processes.
[d] Too small to measure.

Denny and Klemchuck (106) have shown, using bis($o$-2-deuterophenyl benzoyl) peroxide, that lactone formation proceeds with an isotope effect of 1.38. Bis($o$-2-biphenylyl benzoyl) peroxide yields both triphenylene and the lactone, the ratio depending on the solvent (115).

$$(69)$$

$$(29\%) \qquad\qquad (9\%)$$

Bis($o$-phenoxybenzoyl) peroxide refluxed in benzene yields less than 1% carbon dioxide (117), the major identified product (25%) being phenyl salicylate, formed presumably by a radical displacement (117).

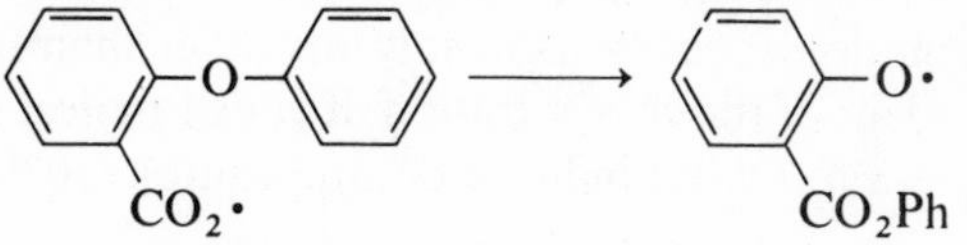

$$(70)$$

The decomposition of benzoyl peroxide in napthalene (94) and other poly-nuclear aromatic solvents (88, 449) produces substantial yields of benzoates, probably via benzoyloxy addition (119), though the contribution from induced decomposition has not been evaluated.

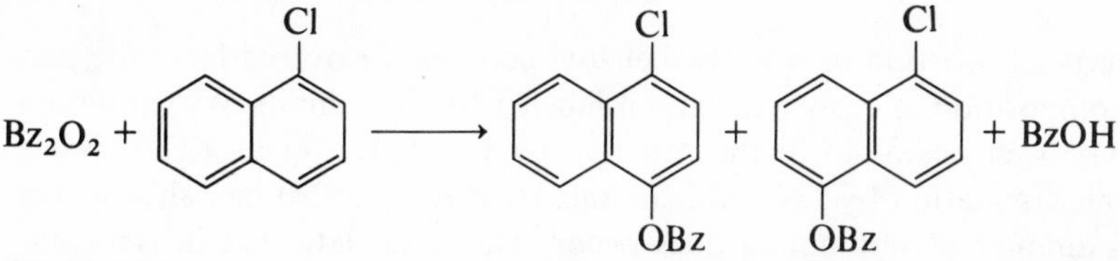

$$(71^*)$$

(1 mole)      (0.4 mole)      (0.2 mole)    (0.95 mole)

Yields of simple aryl benzoates from the decompositions of benzoyl per-oxide in mononuclear aromatics are low, a few percent or less (346, 377) but are increased by the presence of oxygen (19, 377). Simamura and co-workers (377) have suggested reversible formation of a $\sigma$-complex.

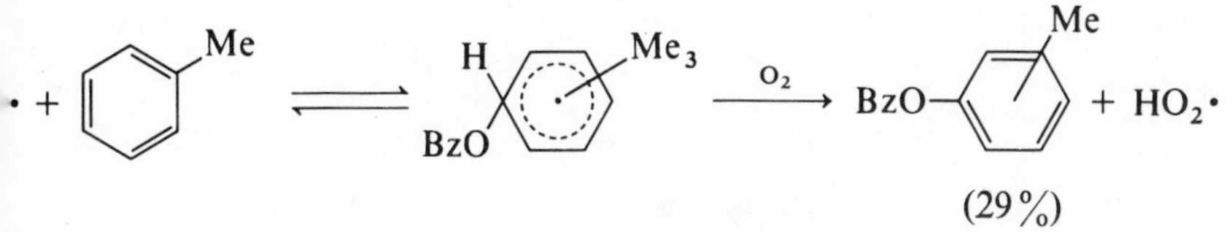

$$(72)$$

(29%)

There is considerable evidence to suggest that cage recombination of the primary and secondary radicals competes with carbon dioxide cleavage from benzoyloxy radicals and with diffusion.

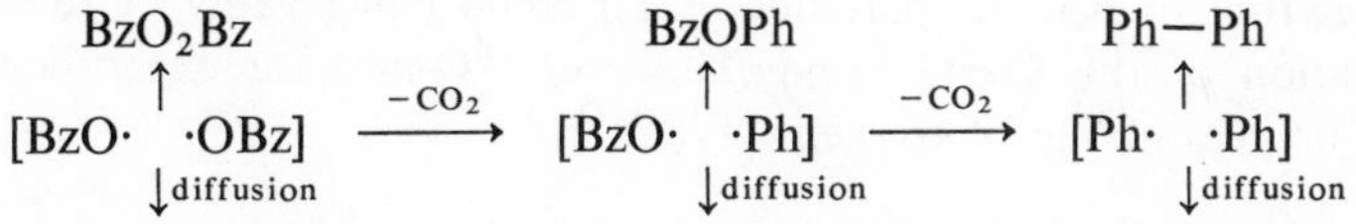

$$(73)$$

The rates of decomposition and the efficiencies of radical production decrease in polymerizing solutions of increasing viscosity (170, 171, 255, 257,

---

* Yields from reference 88.

259, 365), reaching their lowest point when reactions are carried out in polymer films (52b, 192, 423, 424); rates measured by the consumption of radical scavengers, such as galvinoxyl, are slightly lower than those determined by the titration of residual peroxide (168a, 544). Biphenyl and phenyl benzoate are produced from decompositions in solvents other than benzene (507b), in amounts independent of the concentration of added radical scavenger so long as enough is present to inhibit induced decomposition (507b):

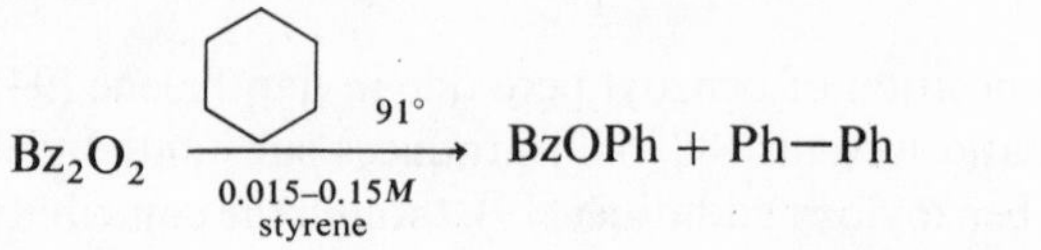

$$\text{Bz}_2\text{O}_2 \xrightarrow[\substack{0.015-0.15M \\ \text{styrene}}]{\text{91°}} \text{BzOPh} + \text{Ph—Ph} \tag{74}$$

(1 mole)                          (0.020 mole) (0.0015 mole)

The crucial question of whether benzoyl peroxide recovered from the partial decomposition of carbonyl-oxygen-labeled $\text{Bz}_2\text{O}_2$ exhibits oxygen scrambling has been answered in the negative by Kobayashi et al. (287). But a recent reexamination by J. C. Martin and H. Hargis (353b) has shown that a small amount of scrambling does occur. They calculate that in isooctane at 80° some 4.6% of the caged benzoyloxy radicals reform peroxide, whereas in the more viscous mineral oil the return is 17.8%.

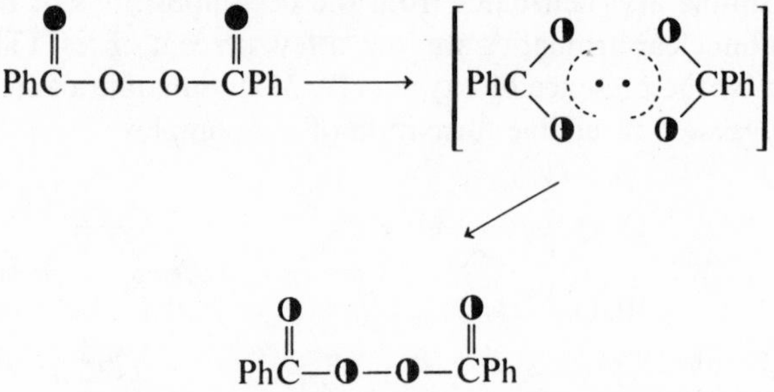

$$\text{PhC—O—O—CPh} \longrightarrow \left[ \text{PhC} \overset{O}{\underset{O}{\diagdown}} (\cdot\cdot) \overset{O}{\underset{O}{\diagup}} \text{CPh} \right] \tag{75}$$

$$\downarrow$$

$$\text{PhC—O—O—CPh}$$

● indicates $^{18}\text{O}$ labelled oxygen and ◐,
oxygen containing half of the original label.

The extraordinary thermal stability of the cyclic phthaloyl peroxide [about 50 times that of benzoyl peroxide (178)] could conceivably be due to easy reformation of the O—O bond. However $^{18}\text{O}$-labeling experiments have shown no scrambling of oxygens (178c).

### 2. Phenyl Radicals

Two points seem paramount here:

1.   Aroyl peroxides do indeed yield free aryl radicals, shown by the similarity in products from the decomposition of benzoyl peroxide and other

sources of phenyl radicals, triphenyl methylazobenzene and *N*-nitrosoacetani-lide* (55, 188, 219, 224, 347). Equally illustrative are the results of the decompositions of the optically active compound **2** in benzene, carbon tetra-chloride, and bromotrichloromethane (116).

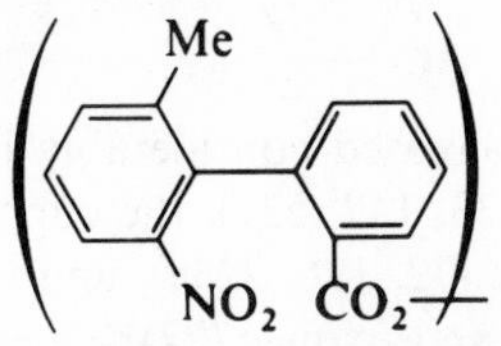

It may be inferred that the aroyloxy radical does not racemize, due to hindered rotation, since the peroxide itself is stable with respect to racemization.

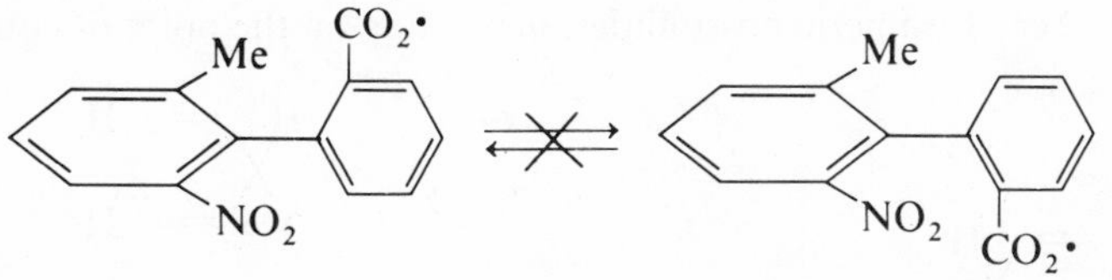

$$(76)$$

However, the products (**3**) are partially to largely racemic, depending on the rate of rotation of compound **4** versus its rate of abstraction.

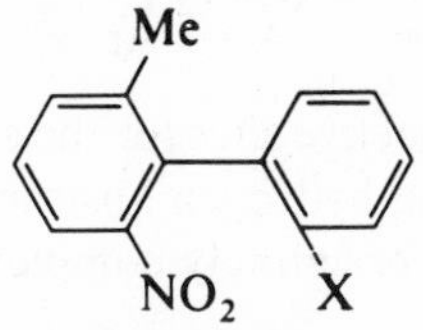

$(X = H, Cl, Br)$

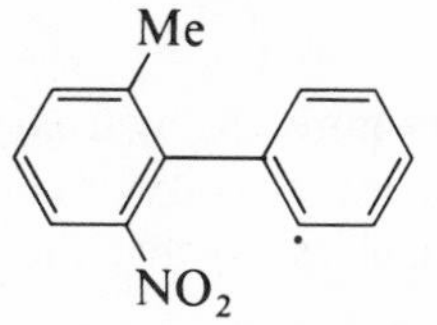

**(3)**

Razuvaev and Zateev (443b) have demonstrated that the phenyl radical itself does not isomerize.

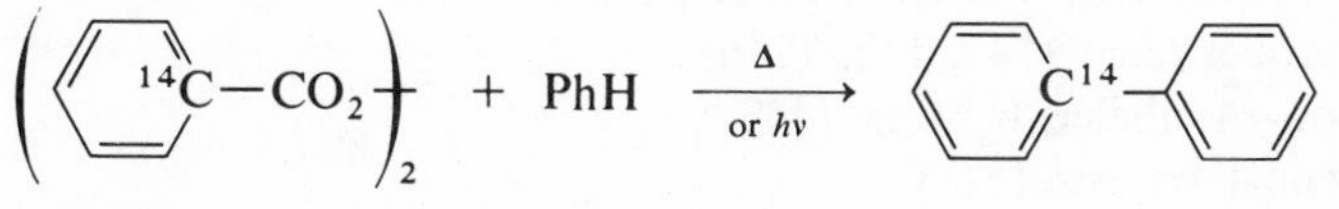

$$(77)$$

* This similarity has been disputed by Eliel and co-workers (141), who showed that in $C_6H_5D$, benzoyl peroxide decompositions differed from those of the nitroso and azo compounds in yields of dihydrobiphenyls and biaryls containing two deuterium atoms and in the deuterium-isotope effect displayed in the formation of biphenyl versus deuterobiphenyl. They concluded that triphenylmethylazobenzene and *N*-nitrosoacetanilide yielded "hot" (i.e., not thermodynamically equilibrated) phenyl radicals. Considering the complexity of the processes leading to these products (113, 374a), however, the proposition cannot be said to be proved. Gragarov and Turkina (174) have also reported different yields of isotopic products from deuterobenzene and different Ph· sources.

2.  Arylation of an aromatic solvent is a two-stage process, addition of the aryl radical followed by hydrogen abstraction (119, 120, 347, 559).

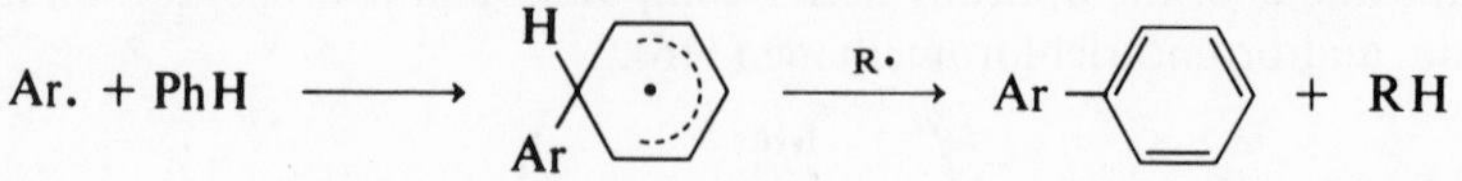

$$\text{Ar.} + \text{PhH} \longrightarrow \quad \longrightarrow^{R\cdot} \quad \text{Ar}\!\!-\!\!\bigcirc + \text{RH} \quad (78)$$

Yields of biaryls are increased considerably by the presence of oxidizing agents, such as oxygen (133, 142, 521), the cupric ion (221), chloranil (196), iodine (196), nitrobenzene (72, 196, 223a), and other nitro and nitroso compounds, hydroxylamines, and oximes (72). These latter are thought to form nitroxyl radicals as the active dehydrogenating species (72).

Competing with such oxidative reactions is dimerization or disproportionation of the cyclohexadienyl radicals (119, 120, 408). Obviously equation 79 holds a number of isomeric possibilities, depending on the point of coupling.

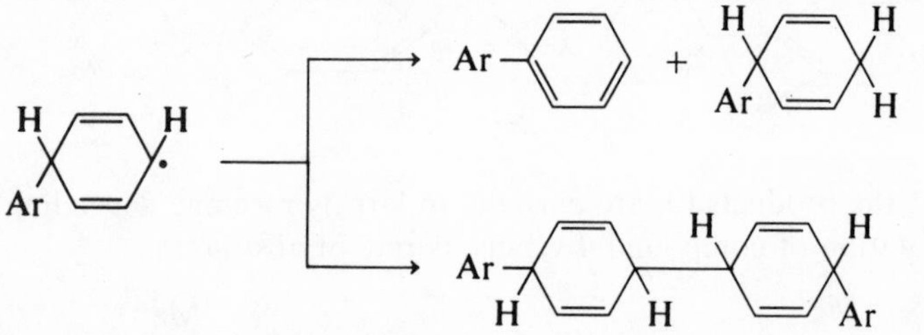

$$(79)$$

Furthermore, the hydroaryls appear to provide attractive sites for the addition of aryl and aroyloxy radicals as well as undergoing hydrogen abstraction by these species, so that the total product mixture is exeedingly complex (120).

Careful studies of the products of the decomposition of benzoyl and substituted benzoyl peroxides in aromatic and nonaromatic solvents may be cited as follows:

Benzene (61, 119, 120, 167, 168a, 223b, 347a, 369, 408).
Deuterobenzenes (141, 142, 174b).
Carbon-14-labeled benzene (443c).
Benzene + oxygen (521).
Benzene + nitric oxide (168b).
Toluene (89, 198, 222, 223c).
Ethylbenzene (89, 157).
Cumene (89, 157).
t-Butylbenzene (223c, 374a).
Cyclohexylbenzene (443c).
Xylenes (157).
Naphthalene (95b, 347b).

Fluorobenzene (223a).
Penta- and hexafluorobenzene (61, 77, 339).
Chlorobenzene (95a, 197, 222, 223b, 347b).
Bromobenzene (93, 374a).
Iodobenzene (55, 60).
Anisole (346, 374a).
Nitrobenzene (168d, 197, 222, 347b, 374a).
Methyl benzoate (223d).
Pyridine (129, 220, 223c, 404, 409).
Pyrazoles (129).
Thiazoles (129, 533).
Carbon tetrachloride (224a, 524).
Chloroform (67a, 224a).
Bromoform (67b, 224a).
Methylene dibromide (67b).
Bromotrichloromethane (67a, 225a).
Carbon tetrachloride + cyclohexene (304).
Ethyl acetate (224b).
Acetic acid (169, 278, 471).
Water (27b).

Three of these, more or less typical, are summarized in Table 8.

Recently, MacKay (348) has reported an elaborate study of the thermal decomposition products of 3-thenoyl and 5-nitro-2-thenoyl peroxides in benzene (and other aromatic solvents). Most interestingly, a small amount of biphenyl was produced, a product not previously found when the peroxide itself did not supply a phenyl radical. MacKay suggests a complicated process based in part on the increased yield of biphenyl in the presence of added acid (348).

$$RCO_2\cdot + PhH \longrightarrow$$

(80)

Earlier studies of products from thenoyl and nicotinoyl peroxides are due to Ford and MacKay (155).

A few samples of products of decomposition of benzoyl peroxide in the presence of other reactive substrates may be mentioned here.

TABLE 8. PRODUCTS OF DECOMPOSITION OF 0.2 $M$ BENZOYL
PEROXIDES IN SOLVENTS AT 80°

| | Solvent | | |
|---|---|---|---|
| Product | Cumene[a] | Nitrobenzene[b] | Chloroform[c] |
| PhH | 0.51 | | 1.71 |
| PhCl | | | 0.01 |
| PhCO$_2$H | 0.52 | 0.852 | Nil |
| S—S | 0.30[d] | | 0.57[e] |
| Ph—S | 0.22[f] | 0.797[g] | |
| Ph—S—S—Ph | 0.22[h] | 0.146[h] | |
| Esters | 0.13 | 0.08 | |
| Cl$_3$C⟨benzene ring⟩—CO$_2$H | | | 0.17 |

[a] Data from reference 157.
[b] Data from reference 186d.
[c] Data from reference 67a.
[d] Bicumyl.
[e] C$_2$Cl$_6$.
[f] Isopropylbiphenyls.
[g] Nitrobiphenyls.
[h] Tetrahydroquaterphenyls.

Kaplan (268) has shown that heating benzoyl peroxide in trimethylene di-iodide yields cyclopropane and iodobenzene almost quantitatively. Only traces of 3-phenylpropyl iodide are found. When the solvent is a mixture of methylene iodide and 1,1-diphenylethylene, a good yield of 1,1-diphenylcyclopropane is obtained (268).

The decomposition of benzoyl peroxide in trichloroethylene gives mainly chlorobutenes and butadienes, but about 12% of the phenyl groups end up as 1,2,3,4-tetrachloronaphthalene (244).

The decompositions of benzoyl peroxide in the presence of isocyanides (476), carbodiimides (103), and diazo compounds (342, 564) yield rather disappointingly commonplace products, explicable on the basis of BzO· or Ph· attack.

### G.   The Decomposition of Benzoyl Peroxide in Benzene

The various factors relating to the homolytic decomposition of aroyl peroxides have been discussed; it seems fitting to close with a description of

DeTar's tour de force (113), application of computer analysis to the hundred-year-old problem: What happens when benzoyl peroxide decomposes in benzene?

DeTar started with some 300 possible reactions and the detailed product studies for decompositions at different initial concentrations provided by his own group (120) and by Gill and Williams (168a) (Figure 1). The equations

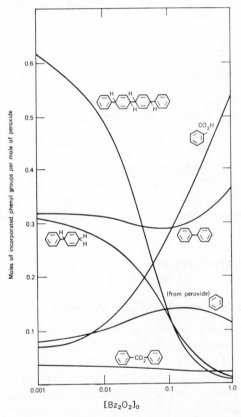

**Figure 1.** Products [calculated (113)] from the decomposition of benzoyl peroxide in benzene at 80°.

were reduced to 30 types for which approximate rate constants could be assigned, with the net result that at $[Bz_2O_2]_0 = 0.3\ M$, 101 reactions were found to contribute $0.05\%$ or more to product formation. DeTar gives tables of calculated versus observed products for $[Bz_2O_2]_0 = 0.001, 0.01, 0.1$, and $0.3\ M$ at 10 and $90\%$ decomposition, and observes that there are two discrepancies:

1. The mechanism gives the correct amount of biphenyl + dihydrobiphenyl, but their ratio is too low. This is explained (113) as an experimental difficulty in analyzing for dihydrobiphenyl.

2. The total amount of induced decomposition is too low. DeTar's mechanism requires the major routes to be

$$Bz_2O_2 + \cdot \underset{H}{\overset{Ph}{\bigcirc}} \longrightarrow BzOH + BzO\cdot + Ph-Ph \qquad (81)$$

$$Bz_2O_2 + \underset{H \;\; OBz}{\overset{Ar}{\bigcirc}} \longrightarrow BzOH + BzO\cdot + \underset{OBz}{\overset{Ar}{\bigcirc}} \qquad (82)$$

Increasing the rate constants for these reactions causes the yield of benzoic acid to be too high. Increasing the contribution from

$$Bz_2O_2 + Ar\cdot \longrightarrow ArOBz + BzO\cdot \qquad (83)$$

would make the ester yields too high.

There are still problems to be solved, but DeTar did not have at that time Martin's recent results on the rate of recombination of benzoyloxy radicals, (353b) though he does speculate that this might provide the key to a correct solution (113).

## VII.  DECOMPOSITIONS OF ALIPHATIC ACYL* AND AROYL–ACYL* PEROXIDES

### A.  Induced Decomposition of Acyl Peroxides

Acetyl peroxide is not as readily attacked by radicals as benzoyl peroxide is; its chain-transfer constant is polymerizing styrene is too small to measure (81). First-order constants for its thermal decompositions are insensitive to initial peroxide concentration (as long as $[P]_0 < 0.1 \; M$) and to most solvents (336, 518), though decompositions are considerably more rapid in alcohols (451, 518). Decomposition of acetyl peroxide in diisopropyl ether produces no $\alpha$-alkoxy ester, $AcOCMe_2O-i$-Pr (276), typical of the induced decomposition of aroyl peroxides in ethers. Equal quantities of acetone and $i$-PrOH are produced, together with isopropyl acetate equal to the sum of these, suggesting that $AcO\cdot$ or $Me\cdot$ produced by homolysis attacks the ether (276).

Acetyl peroxide appears to be attacked by the triethyl tin radical (381); in the presence of $Et_3SnH$ a 58% yield of $Et_3SnOAc$ is produced, although the

---

* In this section the term "acyl" is used to indicate peroxides without an aryl group directly attached to the carbonyl group.

temperature required (45°) is somewhat higher than that for benzoyl peroxide (30°) (381) and the reaction is not as clean.

$$Et_3Sn\cdot + Ac_2O_2 \longrightarrow Et_3SnAc + AcO\cdot \xrightarrow{Et_3SnH} AcOH + Et_3Sn\cdot$$

$$(84)$$

Some hydrogen is formed by the reaction of the acetic acid formed with $Et_3SnH$, and carbon dioxide and methane account for about half the acetyl peroxide. In fact it is not completely clear (to this author) that reaction 84 is required to explain Rübsamen's observations.

Acyl peroxides with longer alkyl chains undergo induced decomposition fairly readily. DeTar and Lamb (118) showed that $\delta$-phenylpentanoyl peroxide decomposes about 12 % faster at an initial concentration of 0.12 $M$ than at 0.05 $M$ in carbon tetrachloride at 77.1°. Rates fit a $1 + \frac{3}{2}$ order plot, with $k_1 = 7.00 \times 10^5$ and $k_i = 3.21 \times 10^5$ l/mole-sec. Induced decomposition is evidEnt for propionyl and butyryl peroxides in isocotane, $n$-hexane, or dioxane, but negligible in benzene, toluene, and acetic acid (486). Cooper (81) has found that octanoyl or lauroyl peroxides have chain-transfer constants with the polystyryl radical equal to that of benzoyl peroxide, and acyl peroxides with unsaturated chains show greater values (Table 9). The variation

TABLE 9.   CHAIN-TRANSFER CONSTANTS FOR
ACYL PEROXIDES $(RCO_2)_2$ IN
POLYMERIZING STYRENE AT $70°$[a]

| R | $C_p$[b] |
|---|---|
| Methyl | $\sim 0$ |
| $n\text{-}C_3H_7$ | 0.018 |
| $n\text{-}C_7H_{15}$ | 0.098 |
| $n\text{-}C_{13}H_{27}$ | 0.116 |
| $MeCH=CH_2$ | 0.146 |
| $PhCH=CH_2$ | 1.10 |
| Phenyl | 0.075 |

[a] Data from reference 81.
[b] $C_p$ = ratio of the rate constant for attack of the polystyryl radical on peroxide to the rate constant for polystyryl addition to monomer.

in transfer ability with the nature of the alkyl group makes it tempting to speculate that initial attack is not at the $O—O$ bond. In fact this type of induced decomposition has recently been demonstrated in a few cases (44, 376, 561).

$$\left( \underset{\text{Br(CH}_2)_3\overset{\text{O}}{\overset{\|}{\text{C}}}-\text{O}}{} \right)_2 \xrightarrow{\text{R}\cdot} \text{RBr} + \text{H}_2\text{C} \overset{\text{H}_2\text{C}-\text{CH}_2}{\underset{\text{O}}{\diagdown}} \overset{\diagup}{\underset{\diagdown}{\text{C}=\text{O}}} \qquad (85)$$

$$\overset{\text{H}_2\text{C}-\text{CH}_2}{\text{H}_2\text{C}\diagdown_{\text{O}}\diagup\text{C}=\text{O}} + \text{RCO}_2\cdot \qquad (86)$$

$$\text{Br(CH}_2)_3\text{CO}_2\cdot \longrightarrow \text{CO}_2 + \text{Br(CH}_2)_3\cdot$$

Thermal decomposition of $\gamma$-bromobutyryl peroxide was shown by Blomstrom et al. (44) to yield lactone, carbon dioxide, and trimethylene bromide quantitatively; similar products were formed from the corresponding iodo compound. Woolford and Gedye (561) have found similar reactivity for $[\text{I(CH}_2)_n\text{C(O)O}]_2$, where $n = 4$ or 5. Muramoto et al. (376) have shown that cinnamoyl and propioloyl peroxides and peroxy esters are attacked by radicals at the site of unsaturation exclusively.

$$(\text{PhCH}=\text{CHCO}_2)_2 \quad + \quad \bigcirc$$

$$\downarrow \text{C}_6\text{H}_{12} \,\, 80°$$

$$\text{PhCH}-\text{CH}-\overset{\text{O}}{\overset{\|}{\text{C}}}-\text{O} \quad \text{O}-\overset{\text{O}}{\overset{\|}{\text{C}}}-\text{CH}=\text{CHPh} \qquad (87)$$

$$\downarrow$$

$$\text{PhCH}=\text{CH} \;+\; \text{CO}_2 \;+\; \text{PhCH}=\text{CHCO}_2\cdot$$

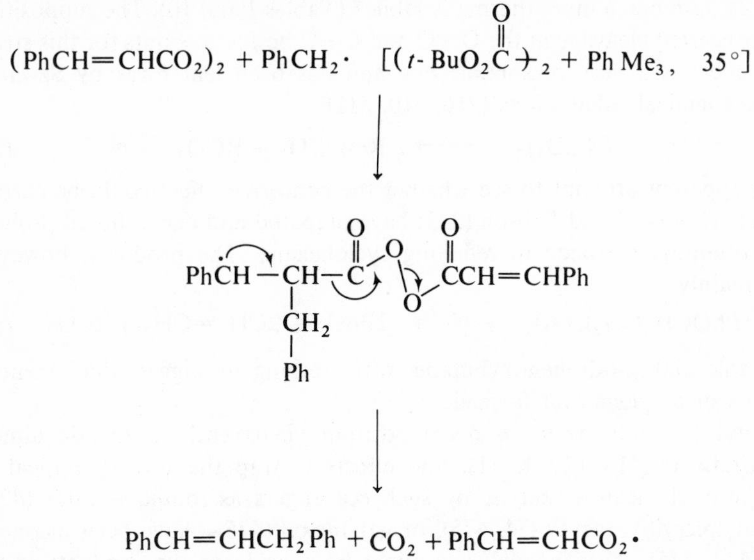

$$(PhCH=CHCO_2)_2 + PhCH_2 \cdot \quad [(t\text{-}BuO_2\overset{\overset{\displaystyle O}{\|}}{C}\,)_2 + Ph\,Me_3, \quad 35°]$$

$$PhCH=CHCH_2Ph + CO_2 + PhCH=CHCO_2 \cdot$$

(88)

However, $(PhCH_2CH_2CO_2)_2$, $(Ph_3CCH_2CO_2)_2$, and $(\triangleright\!-\!CO_2)_2$ are attacked very readily by the triphenylmethyl radical, 1 hr in benzene solution at 25° sufficing for the complete decomposition of the peroxide.* Furthermore, Cass (70b) found that lauroyl peroxide decomposed very readily in ether, producing 88% of 1-ethoxyethyl laurate, which can be explained only by the normal type of radical displacement on the O—O bond.

The asymmetrical acetyl benzoyl peroxides undergo induced decomposition quite readily, either at the O—O bond by α-alkoxyalkyl (542b) or trialkyl tin (453) radicals, or on the benzene ring by $CH_3\cdot$ or $CCl_3\cdot$ (544).

## B. Effects of Structure on the Rate of Thermal Decomposition—Concerted Cleavage at O—O and C—C

Recent interest in acyl peroxides has focused on two areas: cage reactions versus concerted decomposition, and decomposition via the carboxy-inversion route, of which the first is dealt with here.

In solution acetyl peroxide decomposes thermally at about the same rate as benzoyl peroxide (Table 10), but other acyl peroxides, $(RCO_2)_2$, where the R groups as radicals exhibit resonance or inductive stabilization, such as benzyl (23, 566), allyl (190, 191), trichloromethyl (333), and trifluoromethyl

---

* These peroxides yield little or no $Ph_4C$, the concomitant product of $Ph_3C\cdot$ decomposition of benzoyl peroxide in benzene, and the other products were incompletely identified. The point of $Ph_3C\cdot$ attack is not really clear (110).

(58, 125), are much more thermally labile* (Tables 1 and 10). The supposition that concerted cleavage at the O—O and C—C bonds accounts for this structural effect is a very reasonable one and has been lent force by Szwarc's thermochemical calculations (216, 510, 511).

$$(RCO_2)_2 \longrightarrow R\cdot + CO_2 + RCO_2\cdot \qquad (89)$$

In an apparent attempt to see whether the concerted effect could be carried further, Horspool and Pauson (243) have prepared and decomposed $\beta$-phenoxypropionoyl peroxide in refluxing cyclohexane. The products, however, were mainly

$$(PhOCH_2CH_2CO_2)_2 \xrightarrow{\quad\times\quad} 2PhO\cdot + 2CH_2{=}CH_2 + 2CO_2 \qquad (90)$$

phenetole and 1,4-diphenoxybutane, with nothing to suggest that phenoxy radicals or ethylene ever formed.

Acetyl peroxide itself on decomposition yields carbon dioxide almost quantitatively (215) (Table 11), and efforts to trap the acyloxy radical by conventional means—that is, by such scavengers as iodine + water (472), diphenylpicrylhydrazyl (474, 475), or galvinoxyl (475)—have been inconclusive (474, 475). Also inconclusive might be considered the products of the decomposition of acetyl peroxide in $^{14}CH_3CO_2H$ (159), which were interpreted as a small amount of exchange between acetoxy radicals and acetic acid.† The rates of the decomposition of $(CD_3CO_2)_2$ are experimentally indistinguishable from those of $(CH_3CO_2)_2$ (299), but the abundance of $^{13}CO_2$ and $C^{18}O_2$ produced from acetyl peroxide compared with natural abundances shows a small but incontrovertible isotope rate effect (173), $k_{12C}/k_{13C} = 1.023 \pm 0.003$; $k_{16O}/k_{16O} = 1.023 \pm 0.007$, the latter leading to the assertion that cleavage of the C—C bond must accompany cleavage of the O—O bond (173).

However, Martin and his co-workers (516, 352) have ended any controversy§, demonstrating by isotopic labeling that in solvents, at least, two

---

* Nitro substitution on the alkyl group appears to stabilize the peroxide: for R = Pr—, $(NO_2)_3C(CH_2)_2—$, $(NO_2)_2CH(CH_2)_3—$, respectively, $k_1$ at 95° in $CCl_4$ = 1.58, 0.018, and 0.068 min$^{-1}$ (478). It appears to be a nice example of Swain's thesis (507a) that electron-withdrawing groups stabilize the O—O bond.

† Goldstein's words are more pointed (173): "Only two of the many reports (159, 472–475)‡ of successful acetoxyl trapping, where induced decomposition was believed absent, have survived subsequent discoveries (475, 535)‡ and even these (159, 475)‡ need not implicate more than a small fraction of reactant molecules."

‡ Reference numbers changed to those of this chapter.

§ The controversy has been reopened by Goldstein and coworkers, (J. Amer. Chem. Soc., 92, 4119, 4120, 4122 (1970)) who have (1), shown that in many respects labelled oxygen scrambling in acyl peroxides is better explained by sigmatropic shifts than by cage recombination of acyloxy radicals and (2) adduced isotope rate effects for $AC_2O_2$ decomposition considerably larger than hitherto claimed and more in line with a concerted homolysis. The discussion following must be considered in the light of these new data.

TABLE 10. RATES OF DECOMPOSITION OF SOME ACYL PEROXIDES $(RCO_2)_2$

| R | Solvent | $T(°C)$ | $k_1 \times 10^5$ (sec$^{-1}$) | $E_a$ (kcal/mole) | Ref. |
|---|---|---|---|---|---|
| Me | Vapor | 85.2 | 22.4 | 29.5 | 445a |
| Me | Benzene | 85.2 | 16.6 | 32.3 | 336 |
| Me | Toluene | 85.0 | 17.1 | 33 | 539, 451 |
| Me | Isooctane | 85.2 | 14.9 | 32.2 | 336 |
| Me | Cyclohexane | 85.2 | 12.7 | 31.4 | 336 |
| Me | Carbon tetrachloride | 85.2 | 11.7 | 33.4 | 138 |
| Me | Acetic Acid | 85.2 | 13.0 | 30.2 | 336 |
| Me | Benzene | 80.0 | 8.7 | — | 475 |
| Me | 1-Hexane | 80.0 | 8.73 | 31.7 | 475 |
| Me | Cyclohexene | 80.0 | 7.0 | 31.9 | 475 |
| Et | Vapor | 85.0 | 16.0 | 30.0 | 445b |
| Et | Benzene | 85.0 | 24.0 | 30.9 | 486 |
| n-Pr | Vapor | 85.0 | 20.0 | 29.6 | 445b |
| n-Pr | Benzene | 85.0 | 33.1 | 31.4 | 486 |
| i-Pr | Isooctane | 85.0 | 680.0 | 27.3 | 488 |
| (cyclopropyl) | p-Cymene | 70.0 | 1.0 | 29.1 | 186 |
| (cyclopropyl) | Carbon tetrachloride | 70.0 | 0.44 | — | 206 |
| (cyclopentyl) | Carbon tetrachloride | 70.0 | 31.7 | — | 206 |
| (cyclohexyl) | Carbon tetrachloride | 70.0 | 71.0 | — | 206 |
| (cyclobutyl-CH₂—) | Carbon tetrachloride | 70.0 | 2.8 | — | 206 |
| (cyclohexyl-CH₂—) | Carbon tetrachloride | 70.0 | 2.6 | 28.6 | 206 |
| (cyclopropyl-CH₂—) | Carbon tetrachloride | 45.0 | 15.3 | 23.8 | 320 |
| (cyclohexyl) | Carbon tetrachloride | 44.0 | 5.0 | 26.3 | 205 |
| (cyclohexyl) | Carbon tetrachloride | 45.0 | 20.0 | 26.0 | 320 |
| t-Bu | Carbon tetrachloride | 50.0 | 71.4 | 15.9 | 324 |
| PhCH₂ | Toluene | 0.0 | 2.5 | 23.0 | 23 |
| CCl₃ | Benzene | 0.0 | ~100 | — | 333 |
| Apocamphyl | Benzene | 80.0 | 23.0 | — | 22 |
| Triptycyl | Benzene | 80.0 | 14.2 | — | 22 |

**TABLE 11.** PRODUCTS OF DECOMPOSITION OF SOME ACYL PEROXIDES, $(RCO_2)_2$, PERTAINING TO CAGE RECOMBINATION OF RADICALS

| R | Percentage of Theoretically Possible Product | | | | | Conditions | | | |
|---|---|---|---|---|---|---|---|---|---|
| | $CO_2$ | R–R | RH + R–2H | $RCO_2R$ | SX[a,b] | [P]$_0$[c] | Solvent | T (°C) | Ref. |
| Me | 103 | 96.0 | 3.1 | | | | Vapor | 240 | 445a |
| Me | 88 | 4.6 | 70 | | | | Isooctane | 80 | 355 |
| Me | | 3.4 | | | | | Hexane | 60 | 51 |
| n-Pr | | | | 13 | 15[a], 9[b] | 1.3 | $CCl_4$ | 80 | 279 |
| i-Pr | | | | | 13[a], 9[b] | 1.3 | $CCl_4$ | 80 | 279 |
| i-Pr | 100 | | 22[d] | | | 1.3 | Isooctane | 55 | 488 |
| i-Pr | | 12[e] | 21 | 20[f] | 35[a] | 0.1 | $CCl_4$ | 50 | 320 |
| △ | 74–88 | | | 16 | 62–70[b] | 0.1 | $CCl_4$ | 70 | 206 |
| △—CH₂— | 66 | | | 56 | ~8[b] | 0.05 | $CCl_4$ | 78 | 205 |
| n-C₅H₁₁ | | | | | 43[b] | 0.2 | $CH_2Br_2$ | 80 | 67b |
| n-C₆H₁₃ | | 20 | 2.6 | 10–15 | | 0.1 | $CCl_4$ | 74 | 124 |
| n-C₇H₁₅ | | | | | 84[b] | 0.2 | $CH_2Br_2$ | 80 | 67b |
| ⬡ | | 6[g] | 8[h] | 47 | 24[a] | 0.1 | $CCl_4$ | 50 | 320 |
| ⬡ (methyl) | 93 | | | 22–24 | 75–78[b] | 0.1 | $CCl_4$ | 70 | 206 |

| Structure | | | | | | | | | |
|---|---|---|---|---|---|---|---|---|---|
| cyclohexyl–CH$_2$– | 88 | | 30 | 66[b] | 0.1 | CCl$_4$ | 70 | 206 |
| 4-$t$-Bu–cyclohexyl | 57 | | 90 | 1.3[b] | | C$_2$H$_2$Br$_4$ | 50 | 324 |
| $t$-Bu–cyclohexyl | | | | 20[b] | 0.1 | CCl$_4$ | 80 | 181, 184 |
| bicyclic | | 9 | 50 | 36[b] | | CCl$_4$ | 80 | 274 |
| Me–CH(Et)– | | | 30–40 | | | PhCCl$_3$ | 80 | 280 |

TABLE 11. (Continued)

| R | Percentage of Theoretically Possible Product | | | | | Conditions | | | |
|---|---|---|---|---|---|---|---|---|---|
| | $CO_2$ | R — R | RH + R — 2H | $RCO_2R$ | $SX^{a,b}$ | $[P]_0^c$ | Solvent | $T(°C)$ | Ref. |
| PhCH$_2$ | | 12 | | 24 | | 0.3 | CCl$_4$ | 0 | 23 |
| PhCHMe | | 8-8 | | 20 | | | Et$_2$O | 0 | 177 |
| PhCH$_2$CHMe | 70 | — | 4$^h$ | 60 | 20$^a$ | | CCl$_4$ | 77 | 123 |
| Ph(CH$_2$)$_4$ | 84 | 24.6 | 6$^h$ | 19.5 | 41$^a$ | 0.001 | CCl$_4$ | 77 | 122b |
| Ph, Ph, Me (structure) | | 38$^i$ | | | | | Benzene | 80 | 538 |
| Triptycyl | 50 | | 50$^j$ | 20$^k$ | | | Benzene | 80 | 22 |

a Solvent dimer.
b R· abstraction product.
c Initial concentration of peroxide in moles per liter.
d In the presence of added benzoquinone.
e Added scavenger (BDPA) cuts this to 6%.
f Plus 27% of acyl carbonic anhydride.
g Added scavenger (BDPA) had no effect.
h Alkene + RCO$_2$H.
i Open-chain dimer; added I$_2$ cuts this to zero.
j Triphenylene.
k Includes equal amounts of RCO$_2$H + ROH.

acyloxy radicals are the first products and diffusion and reformation of acetyl peroxide are competing reactions. They have further shown by deuterium labeling that in isooctane all the ethane and methyl acetate produced result from cage recombinations of methyl and acetoxy radicals (516b).

$$
\overset{\overset{18}{\bullet}}{\underset{}{\|}} \quad \overset{\overset{18}{\bullet}}{\underset{}{\|}}
$$

$$
\text{Me}\overset{\|}{\text{C}}\!-\!\text{O}\!-\!\text{O}\!-\!\overset{\|}{\text{C}}\!-\!\text{Me} \xrightarrow[80^\circ]{\text{isooctane}} \left[\,2\ \text{MeC}\overset{\mathbf{O}}{\underset{\mathbf{O}}{\cdots}}\,\right] \xrightarrow[\text{diffusion}]{62\%}
$$

$$
(91)
$$

$$
\overset{38\%}{\searrow}
$$

$$
(\text{MeC}\!\overset{\mathbf{O}}{\!-\!}\mathbf{O}\,)_{\!\overline{2}}
$$

This had been concluded earlier by Braun, Rajenback, and Eirich (51), who found that the yields (EtH, 3.4%, AcOMe, 13% in *n*-hexane at 60°) increased with increased viscosity of the solvent in proportion to $(\eta)^{\frac{1}{2}}$ (51, 421). The data yield a rate constant for the decarboxylation of the acetoxy radical of $1.6 \times 10^9\ \text{sec}^{-1}$ (60°), with an activation energy of 6.6 kcal/mole (51). Szwarc and his co-workers (215, 337, 445c) came to similar conclusions from the invariance of ethane and methyl acetate yields in the presence of added free-radical scavengers.* The course of acetyl peroxide decompositions can thus be interpreted in terms of the following equation:

$$
\begin{array}{ccccc}
\text{Ac}_2\text{O}_2 & & \text{AcOMe} & & \text{EtH} \\
\downarrow\uparrow & \xrightarrow{k_c} & \uparrow & \longrightarrow & \uparrow \\
[2\text{AcO}\!\cdot] & & [\text{AcO}\!\cdot\ \ \text{CO}_2\ \ \cdot\text{Me}] & & [\text{Me}\!\cdot\ \ 2\text{CO}_2\ \ \cdot\text{Me}] \quad (92) \\
\downarrow\text{diffusion} & & \downarrow\text{diffusion} & & \downarrow\text{diffusion}
\end{array}
$$

---

\* Vogt and Hamil (535) have shown that in 3-methylpentane at 80° the yields of EtH and MeOAc are sensitive to the concentration of added iodine and applied pressure (1–6000 atm). The former observation is explicable in terms of Hammond and Waits' (203) calculations, which show that even at low concentrations of scavenger there is a reasonable probability that a radical pair will be " born " with a scavenger within its solvent cage. Vogt's findings that methyl acetate was *not* among the products at 1-atm pressure even in the absence of iodine is in direct conflict with those of Herk, Feld, and Szwarc (215), Braun et al. (51), and Taylor and Martin (516b), and is rather mysterious.

Lastly, Koenig and his group in a series of elegant and closely reasoned papers (298–302) have shown that the isotope rate effects for decomposition, mentioned above, derive mainly from changes in the rate of decarboxylation of acyloxy radicals, thus affecting their availability for cage recombination to peroxide. For $CH_3CO_2\cdot \longrightarrow CH_3\cdot + CO_2$ at 75° $k_{3H}/k_{3D} = 1.09$. The acetoxy radical, though decarboxylating very rapidly, lives long enough to exchange partly with the radioactive acetate ion (302).

$$CH_3CO_2\cdot + {}^{14}CH_3CO_2^- \xrightleftharpoons{\text{DMF}} {}^{14}CH_3CO_2\cdot + CH_3CO_2^- \qquad (93)$$

In most cases the products derived from solvents in which acetyl peroxide has been decomposed are those to be expected from methyl radicals released in the system. In the 1940s Kharasch and his co-workers investigated most of the possibilities for solvent dimers: decomposition in acetic acid or in iso-butyric acid yields 50 % succinic or 60 % tetramethylsuccinic acid, respectively (276); esters (281a) and acyl chlorides (419) also yield $\alpha,\alpha'$-dimers; in ketones 1,4-diketones are produced (282); alcohols appear to give some glycol, though the product mixture is very complex, containing esters and oxidation products as well (283). Decompositions in benzene produce toluene, but in alkyl-benzenes—such as toluene, ethylbenzene, or cumene—the predominant products arise from coupling of the benzyl-type radicals (281b, 559). More recently Schwetlick and Helm (466) have looked at the coupling products of mixtures of toluene with acetone, acetonitrile and chloroform, obtaining total amounts of $RR + R'R + R'R'$ on the order of 30–50 %.

Herk and Szwarc (214) have shown that methyl radicals from acetyl peroxide are not "hot" since they yield the same H/D abstraction ratio as methyl radicals from other sources.

The products of the decomposition of acetyl peroxide in cyclohexene, however, have caused considerable concern (173, 353a, 355, 473, 475) owing to the fact that some 10–15 % (353a, 355, 473) of the acetoxy radicals appeared to be trapped, appearing as cyclohexyl acetate. The rate of decomposition was no faster than in other solvents (475) (Table 10), ruling out any direct attack of cyclohexene on the peroxide, but the very rapid decarboxylation of AcO· seemed to preclude attack of free acetoxy radicals on the solvent. Again the problem has been solved by Martin et al. (355) in scrupulous detail, using peroxides labeled with ${}^{18}O$ in the C=O group and conducting decompositions in the presence and absence of galvinoxyl.

It appears that the acetoxy radicals unite with olefin very rapidly to yield a $\pi$-complex. This complex, though still essentially in the solvent cage, may react with methyl or another acetoxy radical to produce compounds 5 and 6 (yields not affected by galvinoxyl) or diffuse into solution, where its collapse to cyclohexyl acetate depends on the availability of a hydrogen donor (355).

(5)                    (6)

Thus, although the thermal decomposition of acetyl peroxide is fairly well understood, that of other acyl peroxides is more difficult to assess. For most, decarboxylation of the acyloxy radical appears to be rapid, if not concerted (206, 488). [Cinnamoyl (520) and cyclopropane carbonyl peroxides are exceptions, decarboxylation leading to the unstable vinyl or cyclopropyl radicals.] Yields of R—R [from $(RCO_2)_2$] are relatively high from decompositions in solution (Table 11), as well as in the pure state (216). The formation of R—R and of $RCO_2R$ is generally considered to result from radical–radical combination in the solvent cage (Table 11, references therein), as has in some cases (122b, 123, 320) the formation of $R'CH_2CH_3 + R'CH=CH_2$ or $R'CH=CH_2 + R'CH_2CH_2CO_2H$ by radical–radical disproportionation. Admittedly the sum of these yields a maximum possible value for the cage recombination. Likewise, a minimum value for the number of radicals escaping the cage can be obtained by assessing the amount of RCl or $C_2Cl_6$ produced when the peroxides are decomposed in carbon tetrachloride (Table 11).

A less qualitative approach has been through the use of radical scavengers, measuring both their effect on the yield of "cage" products and their rate of disappearance compared with that of titratable peroxide. The method has its pitfalls* but appears to give consistent results with the three stable free radicals galvinoxyl, diphenylpicrylhydrazyl, and α,α-bis(diphenylene)-β-phenylallyl (319, 355, 544). With their use, Lamb et al. (319, 320) have demonstrated that cyclohexanecarbonyl peroxide in carbon tetrachloride produces only 20% "free" radicals at 35° and about 30% at 85°. The yield from isobutyryl peroxide is about 50%, insensitive to changes in temperature (320).

---

* Investigating the effect of iodine, alkenes, and mercaptans, alone or in combination, on the carbon dioxide yield from 5-phenylpentanoyl peroxide, DeTar and Lamb (118) found many cases in which the rate of decomposition as well as the products were affected by the presence of scavenger. Denisov (98) has used α-naphthol to assess the yield of free radicals from butyryl and other acyl peroxides, claiming that the rate of disappearance of naphthol is independent of its concentration. However, his rate of naphthol disappearance is 10–15 times as fast as the rate of decomposition of butyryl peroxide as determined by Szwarc et al. (486). His approximately 50% radical-formation efficiencies (based on two naphthol molecules for one peroxide molecule) are very much in line with a 1–1 naphthol–peroxide nucleophilic displacement, well established for other phenols with peroxides (see the next section).

Products from the former peroxide (45° in benzene with an excess of galvin-oxyl) are as follows (in moles per mole of peroxide):

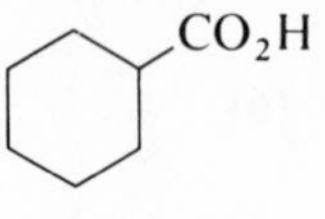

$CO_2H$

(0.12)

$CO_2$

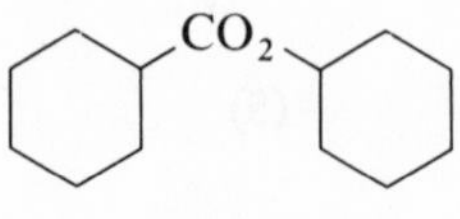

(0.40)

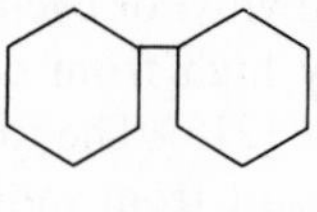

(0.04) 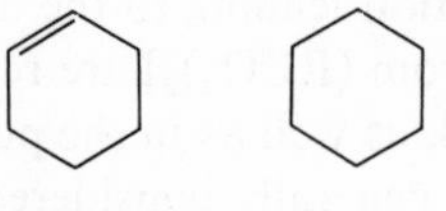            (0.28)

## C.  Carboxy Inversion in Acyl Peroxides

The main trouble in interpreting the scavenger data—in terms of percentage of cage recombination, induced decomposition, and enhancement of the decomposition rate due to concerted decomposition—is contribution from the carboxy-inversion reaction, appreciation of whose importance has been steadily growing over the last several years (107, 184, 320, 385, 504, 548). Once considered rather a curiosity, to be manipulated by push–pull inductive groups on the peroxide nucleus (330, 334), it is now found to contribute substantially to the decomposition of many acyl peroxides, in nonpolar solvents, in the absence of added acid, and particularly for those peroxides whose rapid decompositions had been interpreted as concerted to yield stable types of alkyl radicals.

There appears to be three clues to the likelihood of an acyl peroxide's decomposition partly by carboxy inversion:

1.  Sensitivity of the rate of decomposition to solvent polarity or added acid, shown with isobutyryl peroxide (320) and cyclopropylacetyl peroxide (205), as well as with phenylacetyl peroxide (23).

2.  Increases in the rate of decomposition with applied pressure, as expected for the volume decrease in going to the transition state (543, 548, 549), shown with *trans*-4-*t*-butylcyclohexanecarbonyl peroxide (548) and isobutyryl peroxide (320).

3.  High yields of ester, which in all cases investigated have shown a high degree of retention of the original peroxide's configuration.

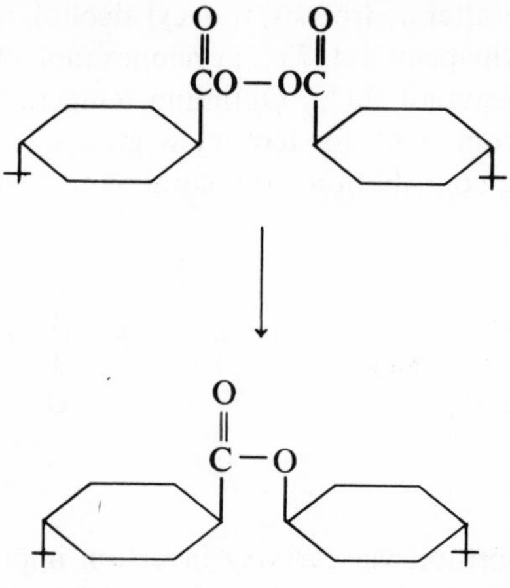

(94)

[90.0% yield, complete retention (324)]

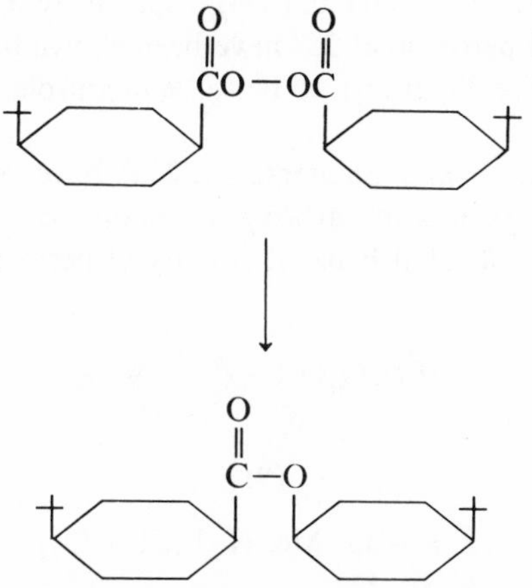

(95)

[90.4% yield, complete retention (324)]

Originally shown by Kharasch et al. (280) for L-(+)-2-methylbutanoyl peroxide, it has since been shown for hydratropoyl peroxide (177), 3-phenyl-2-methylpropanoyl peroxide (123, 385), and *trans*-4-*t*-butylcyclohexanecarbonyl peroxide (184, 324). From the sterically constrained apocamphoyl (224) and triptycene carbonyl (22) peroxides, the high yields of ester (Table 11) also strongly suggest a carboxy inversion route.

Denny and Sherman (107) have seriously proposed carboxy inversion of mixed acyl–aroyl peroxides as a synthetic method for converting carboxylic acids to alcohols with one less carbon, obtaining from compound **7** with the appropriate R group (after hydrolysis) tridecyl alcohol, 58%; 1-octanol, 58%; 1-hexanol, 66%; 2-ethylpentanol, 73%; cyclohexanol, 69%; $t$-butanol, 40%; and 2,4,4-trimethylpentanol, 50%. Optimum temperatures required for the conversion ranged from $-30°$ for tertiary R groups to 60–100° for primary R groups, suggesting contribution from compound **8** in the transition state.

$$R-\overset{\overset{\displaystyle O}{\|}}{C}-O_2-\overset{\overset{\displaystyle O}{\|}}{C}-\langle\!\!\langle\ \rangle\!\!\rangle-NO_2$$

**(7)**

$$R^+ \quad \overset{\overset{\displaystyle O}{\|}}{\underset{\underset{\displaystyle O}{\|}}{C}} \quad {}^-O-\overset{\overset{\displaystyle O}{\|}}{C}-\langle\!\!\langle\ \rangle\!\!\rangle-NO_2$$

**(8)**

Extensive decomposition via carboxy inversion implies, of course, a low yield of scavengeable radicals, and a low or negative entropy of activation for decomposition. Values for the latter for the decomposition of cyclopropylacetyl peroxide in carbon tetrachloride and phenylacetylbenzoyl peroxide in benzene are 3.1 (205) and $-5.0$ e.u. (504) respectively. Free-radical yields for phenylacetylbenzoyl peroxide at 25° have been shown by Suehiro et al. (504), who used diphenylpicrylhydrazyl, to be 4.0% in anisole solution and 6.6% in benzene.

Quite recently Lamb and Sanderson (323a) have measured free-radical yields (1–15%) and extents of carboxy inversion (55–75%) for the thermal decomposition of substituted benzoyl isobutyryl peroxides **(9)**.

$$i\text{-}PrCO_2O_2C-\langle\!\!\langle\ \rangle\!\!\rangle-X$$

**(9)**

$(X = t\text{-}Bu, Me, H, F, Cl, NO_2)$

Analysis of the data in terms of competing reactions showed, as might be expected that $k_1'$ exhibited a positive $\rho$-factor ($+1.0$).

$$\text{Peroxide} \xrightarrow{k_1} [2R\cdot] \longrightarrow 2R\cdot \tag{96}$$
$$\Big\downarrow k_1' \qquad\qquad \downarrow$$
$$\qquad\qquad\qquad \text{cage products}$$
$$\text{Inversion}$$

The startling result is that $\rho$ for $k_1$ was also positive ($+0.7$), the direct opposite of what is found for diaroyl peroxides (Section VI-B) and contrary to the accepted rationale for stability of these peroxides towards homolysis (Section VI-B).

The question of just how much the acyl carbonic anhydride figures in the carboxy-inversion reaction is open. The decomposition of phenylacetyl-*p*-bromobenzoyl peroxide at 30° in benzene (only 2% of radicals scavengeable by galvinoxyl) has yielded (184) 30% of compound **10** together with 45% of phenylacetyl *p*-bromobenzoate (and 10% of diphenylmethane).

$$Br-\overset{\displaystyle}{\bigcirc}-\overset{\overset{\displaystyle O}{\|}}{C}-O-\overset{\overset{\displaystyle O}{\|}}{C}-O-CH_2Ph$$

**(10)**

The carbonic anhydride was shown to be too thermally stable to be the precursor of the ester, and since the $^{18}O$ of the peroxide $C{=}O$ had become almost completely scrambled in the recovered phenylacetyl *p*-bromobenzoate, Greene (184) prefers to think that this compound does after all result from the cage recombination of radicals. However, since Denny (99b) has shown that the carbonic anhydride (from compound **11**)

also has scrambled oxygens, this cannot be taken as a serious objection to a largely polar transition state (107) leading to both carbonic anhydride and to ester.*

$$MeO-\bigcirc-\overset{\overset{\displaystyle O}{\|}}{C}-O_2\overset{\overset{\displaystyle O}{\|}}{C}-\bigcirc-NO_2$$

$$\left[\;\;\underset{\underset{+}{\overset{\displaystyle\setminus}{C}}}{\overset{\overset{\displaystyle O}{\|}}{RC}}\quad^-O-\overset{\overset{\displaystyle O}{\|}}{C}-R\;\;\longleftrightarrow\;\;R^+\;\;\underset{\underset{\displaystyle O}{\|}}{\overset{\displaystyle O}{\diagdown}}{\overset{}{C}}O-\overset{\overset{\displaystyle O}{\|}}{C}-R\;\;\longleftrightarrow\;\;R-O\overset{\overset{+}{C{=}O}}{\underset{}{}}\quad^-\overline{O}-\overset{\overset{\displaystyle O}{\|}}{C}R\;\;\right]$$

$$(97)$$

$$\overset{\overset{\displaystyle O}{\|}}{RO C}-O-\overset{\overset{\displaystyle O}{\|}}{C}-R \qquad\qquad \overset{\overset{\displaystyle O}{\|}}{RO C}R + CO_2$$

* Walling and coworkers (J. Amer. Chem. Soc., **92**, 4927 (1970)) have recently proposed this sort of transition state augmented by contributions from singlet diradical structures to be the pathway for *all* acyl peroxide decompositions. Polar versus radical-type products are governed by the medium and the rate of spin interconversion. The hypothesis is supported by new data for isobutyryl peroxide decompositions as well as selected examples from the literature, including the well established but poorly understood phenomenon of "molecularly assisted" homolysis. The most telling point is that accelerated decompositions of peroxides caused by a polar environment *give faster rates of free radical production* (than the slower decomposition in nonpolar media) even though the efficiency of radical production may be very low, and the bulk of the products of the polar type.

## D.  Anchimeric Assistance in Acyl and Aroyl Peroxide Decomposition

Certain acyl or aroyl peroxides with double bonds $\gamma,\delta$ to the carbonyl group (301, 321–323b) or with iodo substituents strategically placed (332, 335) undergo thermal decomposition at a considerably faster rate than their saturated analogs (322, 323b), the noniodo-substituted compound (332, 335) or benzoyl peroxide (301) (Tables 3, 10, and 12). It is apparent that in these cases the O—O bond undergoes assisted cleavage in the manner more fully documented for similarly constituted peroxy esters by Martin and co-workers (34b, 154, 354). Reactions appear to be uncomplicated by radical-induced decomposition, the effect of radical scavengers (301, 321, 323b) or of changes in initial concentration (335) being nil. Rates do increase as solvents are made more polar (301, 323b, 332), but with the example most likely proceed via a carboxy-inversion route, o-iodobenzoyl-p-nitrobenzoyl peroxide, Honsberg and Leffler (227) have shown that added trichloroacetic acid has little effect. The polar character of the transition state has further been demonstrated by Lamb et al. (323b) for a series (12), where a $\rho\sigma^+$ plot gave $\rho = -1.16$.

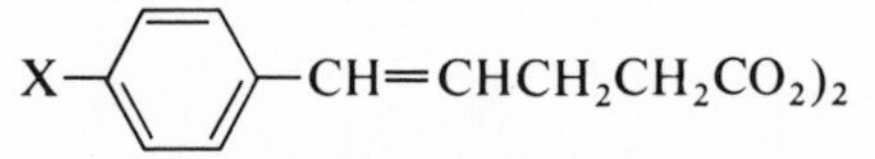

$$X-\langle\bigcirc\rangle-CH{=}CHCH_2CH_2CO_2)_2$$

**(12)**

$$(X = MeO, Me, H, Cl, F)$$

Product studies show the nature of the assistance quite clearly, the o-iodobenzoyl peroxides producing iodoso carboxylates almost quantitatively (332).

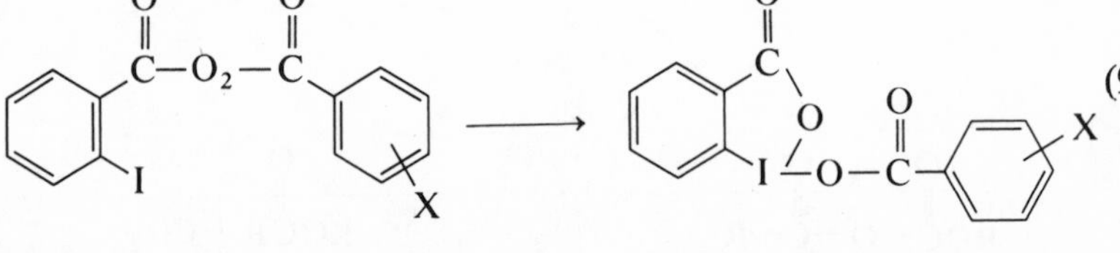

(98)

Products of the o-vinylbenzoyl benzoyl peroxides are more complex but include a high proportion of phthalides and coumarins (301).

   Both exo- and endo-norbornene-5-carbonyl peroxides appear not to undergo any similar internally assisted cleavage, the rates of decomposition being about the same as those of the saturated analogs (204) (Table 12). The endo isomer does exhibit an instance of internal radical-induced decomposition, however,

TABLE 12. ANCHIMERIC ASSISTANCE IN ACYL PEROXIDE DECOMPOSITION

| Peroxide | Solvent | $T$ (°C) | $k_1 \times 10^5$ (sec$^{-1}$) | $\Delta H\ddagger$ (kcal/mole) | $\Delta S\ddagger$ (e.u.) | Ref. |
|---|---|---|---|---|---|---|
| [PhCH=CHCH$_2$CH$_2$CO$_2$)$_2$ | Benzene | 60 | 6.68 | 21.8 | −12.4 | 323b |
| (Ph(CH$_2$)$_4$CO$_2$]$_2$ | Benzene | 60 | 1.0 | 30.5 | 10.3 | 323b |
| | Chlorobenzene | 70 | 30.2 | 20.2 | −8.7 | 301 |
| | Chloroform | 22 | 186 | — | — | 332 |
| | Benzene | 62.5 | 23 | — | — | 335 |
| | Carbon tetrachloride | 65.9 | 81 | 28.5 | 17 | 204 |
| | Carbon tetrachloride | 65.9 | 84 | — | — | 204 |

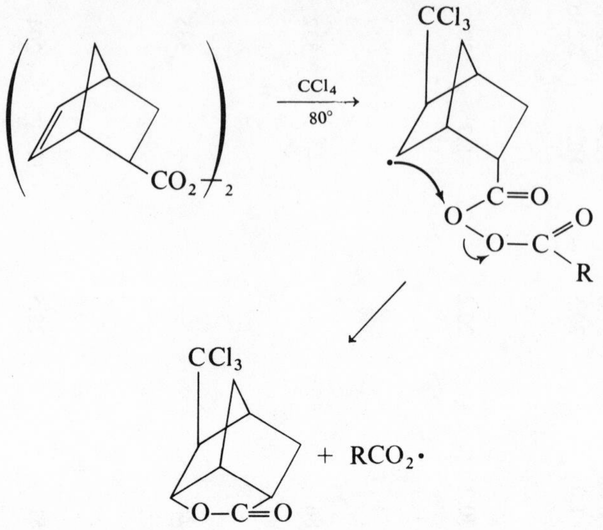

(99)

whereas more interestingly, the *exo* compound produces a similar product via radical rearrangement (204).

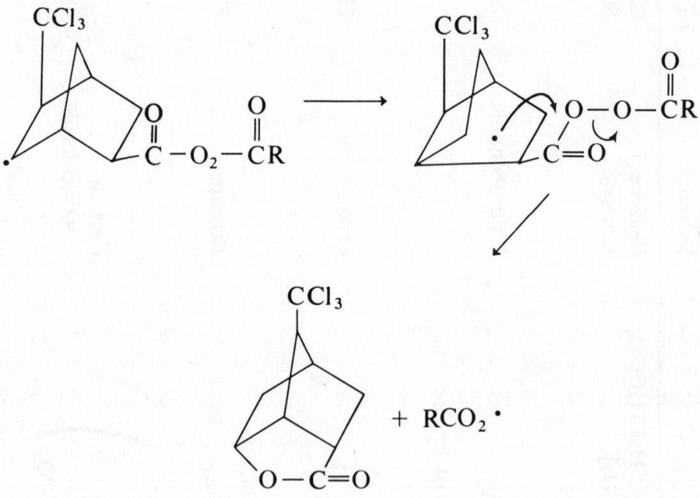

(100)

*o*-Azidobenzoyl peroxide is not considered by Leffler (333) to exhibit anchimerically assisted decomposition, although the rate in benzene at 80° ($k_1 = 94 \times 10^{-5}$ sec$^{-1}$) is more than 10 times that of benzoyl peroxide. The products, for the most part, leave the azido group intact.

(1 mole)  (1.02 mole)  (0.18 mole)  (101)

(0.20 mole)  (0.20 mole)  (0.10 mole)

Triphenylmethylacetyl peroxide undergoes phenyl migration on decomposition, yielding largely triphenylethylene and rearranged ester (102, 447). The formation of ester, at least, has been proposed to involve concertion (102), since its ether oxygen contained 65% of the $^{18}O$ originally present as carbonyl in the peroxide. None of the normal type of ester, $Ph_3CCH_2C(O)CH_2CPh_3$, could be found.

(25%)

## VIII. PEROXYDICARBONATES AND OTHER DIVERSE TYPES

Although peroxydicarbonates are useful low-temperature initiators for vinyl polymerization (499b, 501) [rates of decomposition at 50° are similar to those of benzoyl peroxide at 80° (Tables 2 and 13)], little has been published about the characteristics of their thermal decomposition. The rates of decomposition were determined for several of these compounds ([ROC(O)O]$_2$, R = Et, $i$-Pr, PhCH$_2$, NO$_2$CMe$_2$CH$_2$ (500); R = $i$-Pr (79)) in the early 1950s. Razuvaev and Terman have since added others to the list [R = Me, Bu, $i$-Bu, $t$-Bu, amyl, and cyclohexyl (439c), and Ph (441)] (Table 13).

There is a belief that peroxydicarbonates are particularly sensitive to radical-induced decomposition (79, 500, 501). This is undoubtedly true for the pure substances (500, 501), as might be expected for any thermally labile substance at such high concentration. Addition of 1% of iodine to pure diisopropyl peroxydicarbonate reduces the rate of decomposition by a factor of 60 (500).

TABLE 13. RATES OF THERMAL DECOMPOSITION OF SOME
PEROXYDICARBONATES AND BENZOYL PERSULFONATES

| R | $[P]_0$[a] | Solvent | $T$ (°C) | $k_1 \times 10^5$ (sec$^{-1}$) | $E_a$ (kcal/mole) | Ref. |
|---|---|---|---|---|---|---|
| Me | — | Benzene | 45 | 2.72 | — | 439c |
| Et | 2.2 | t-Butanol | 45 | 1.47 | 32 | 158 |
| Et | 1.1 | t-Butanol | 45 | 1.25 | 33 | 158 |
| Et | 0.21 | [b] | 45 | 1.4 | 30.4 | 500 |
| i-Pr | 0.15 | [b] | 50 | 2.28 | 28.1 | 500 |
| i-Pr | 0.14 | Toluene | 50 | 3.03 | — | 500 |
| i-Pr | 0.02 | Benzene–styrene | 54.3 | 5.0 | — | 79 |
| ⬡ (cyclohexyl) | — | Benzene | 45 | 4.15 | 29.7 | 439c |
| PhCH$_2$ | — | Benzene | 45 | 1.1 | 31.8 | 439c |
| PhCH$_2$ | 0.053 | Toluene | 50 | 2.92 | — | 500 |

$$R-\overset{\displaystyle O}{\underset{\displaystyle O}{S}}-O_2Bz$$

| R | | Solvent | $T$ (°C) | $k_1 \times 10^5$ (sec$^{-1}$) | $E_a$ (kcal/mole) | Ref. |
|---|---|---|---|---|---|---|
| Et | — | Carbon tetrachloride | 30 | 1.6 | 25.4 | 432b |
| Et | — | Benzene | 30 | 2.5 | 26.0 | 432b |
| Et | — | Isopropanol | 30 | 44.1 | 29.2 | 432b |

[a] Initial concentration of peroxide in moles per liter.
[b] 2,2'-Oxydiethylene bis(allyl carbonate).

However, referring to Table 13, it is apparent that at the concentrations in solution at which rates of decomposition are ordinarily measured, the peroxydicarbonates are considerably less sensitive to induced decomposition than benzoyl peroxide. Decompositions are uniformly first order (79, 158, 439, 500) even at 2.2 $M$ in t-butanol (158), and rate constants show very little dependence on the solvent or on initial concentration over a wide range (Table 13). Cohen and Sparrow (79) found that addition of styrene to the benzene solvent gave no change in the rate of decomposition but were so influenced by the belief as to call their results "apparent first-order rate constants."

This is not to say that induced decomposition is not a factor. Terman and co-workers (312) found cyclohexyl peroxydicarbonate to have a 70% efficiency of initiation for the autoxidation of cumene. Cohen and Sparrow (79) found only 21 % of *meso*-2,3-diphenylbutane, the solvent-coupling product, from the isopropyl compound in ethylbenzene at 54°, but also obtained an unidentified oil, which if calculated as *dl*-2,3-diphenylbutane, raises solvent coupling to 70 %. McBay and Tucker (360) have examined the products from the decomposition of methyl, ethyl, isopropyl, and butyl peroxydicarbonates in cumene, cymene, and toluene at 110–150°. Solvent-coupling products ranged from 100 to 10 % at the low and high extremes of initial peroxide concentration, and carbonate esters were found to replace solvent-coupling products at the high initial concentration; this is suggestive of the following induced reaction:

$$PhCH_2\cdot + (R\overset{\overset{\displaystyle O}{\|}}{O}CO)_2 \longrightarrow R\overset{\overset{\displaystyle O}{\|}}{O}COCH_2Ph + R\overset{\overset{\displaystyle O}{\|}}{O}CO\cdot \qquad (103)$$

Recently Van Sickle (529) has reported rate constants for the unimolecular homolysis ($k_1$) and the $\frac{3}{2}$ order induced decomposition ($k_i$) for cyclohexyl peroxydicarbonate in benzene at 50° as $3.9 \times 10^{-5}$ sec$^{-1}$ and $1.39 \times 10^{-4}$ (l/mole)$^{1/2}$ sec$^{-1}$, respectively.

We ourselves recently became interested in peroxydicarbonates as a possible low-temperature source of alkoxy radicals, although the results of previous workers (79, 360, 439) indicated that the decarboxylation of the alkyl carbonate radical was slow compared with abstraction from solvent (79, 360, 361, 439a) or addition to vinyl monomers (439b). This is confirmed by the products of the decomposition of *sec*-butyl peroxydicarbonate in toluene (Table 14). A little 2-butanone is formed, even at low temperatures, but the butanone and acetaldehyde, clues to the presence of *sec*-butoxy radicals in these media (226), are not produced in substantial amounts below 100°.

$$\textit{sec}\text{-BuO}\cdot \left\{ \begin{array}{l} \xrightarrow{\text{S}\cdot} Me\overset{\overset{\displaystyle O}{\|}}{C}CH_2Me + SH \\[2ex] \longrightarrow AcH + Et\cdot \end{array} \right. \qquad (104)$$

More particularly, decarboxylation does not compete with diffusion from the solvent cage. No di-*sec*-butyl peroxide is formed.

$$(\textit{sec}\text{-BuO}\overset{\overset{\displaystyle O}{\|}}{C}\text{-O})_2 \longrightarrow 2[\textit{sec}\text{-BuO}\overset{\overset{\displaystyle O}{\|}}{C}\text{-O}\cdot] \overset{\times}{\longrightarrow} 2[\textit{sec}\text{-BuO}\cdot] + 2CO_2$$

$$\downarrow \text{diffusion}$$

TABLE 14. PRODUCTS OF DECOMPOSITION OF PEROXYDICARBONATES, $(R-OCO-)_2$ with the structure having $O=$ on each carbonate carbon

| R | $[P]_0$[b] | Solvent | $T$ (°C) | Products[a] | | | | Ref. |
|---|---|---|---|---|---|---|---|---|
| | | | | Aldehyde | Ketone | Alcohol | $(PhCH_2)_2$ | |
| sec-Butyl | 0.061 | Toluene | 35 | 0[c] | 0.11[d] | 1.75[e] | 0.72 | 226 |
| sec-Butyl | 0.078 | Toluene | 46 | Trace[c] | 0.28[d] | 1.71[e] | 0.65 | 226 |
| sec-Butyl | 0.078 | Toluene | 110 | 0.03[c] | 0.11[d] | 1.57[e] | 0.54 | 226 |
| Cyclohexyl | 0.05 | Benzene | 50 | 0 | 0.86[f] | 1.02[g] | — | 529 |
| Cyclohexyl | 0.05 | Toluene | 50 | 0 | 0.15[f] | 1.73[g] | 0.58 | 529 |
| Cyclohexyl | 0.05 | Styrene | 50 | 0 | 0[f] | 0.73[g] | — | 529 |
| Cyclohexyl | 0.05 | Cyclohexane | 50 | 0 | 0.05[f] | 1.52[g] | — | 529 |

[a] In moles per mole of decomposed peroxide.
[b] Acetaldehyde.
[c] Initial concentration of peroxide, in moles per liter.
[d] Methyl ethyl ketone.
[e] 2-Butanol.
[f] Cyclohexanone.
[g] Cyclohexanol.

Van Sickle (529) has come to similar conclusions recently for cyclohexyl peroxydicarbonate based both on the absence of products from the cyclohexyloxy radical cleavage and on the drastic diminution of $CO_2$ yield when decompositions are done in solvents, such as styrene, that trap radicals by addition.

Phenyl peroxydicarbonate (441) may prove to act differently since decarboxylation yields the resonance-stabilized phenoxy radical. This peroxide is said to be more labile than other peroxydicarbonates (441a) and inhibits rather than initiates polymerization (441b).

Razuvaev and his co-workers have prepared a number of benzoyl (429) and acetyl (437) peroxycarbonates [$RC(O)O_2C(O)OR'$, $R = Ph$ or $Me$; $R' = Me$, Ph, and cyclohexyl]. These appear to be very sensitive to radical-induced decomposition, judging from the low energies of activation reported (24–25 kcal/mole) (429b) and the products, which, in chloroform or carbon tetrachloride, include *o*- and *p*-trichloromethylbenzoic acids (429c, 429d).

Also investigated by Razuvaev et al. (432) are benzoyl persulfonates ($BzO_2SO_2R$, $R = Me$, Et, Pr, and *i*-Pr). These tend to decompose by a carboxy-inversion process. The rates are very sensitive to solvent polarity (432b) or to added acids (432a) (Table 13), and the products even in carbon tetrachloride include 45% of the carbonic sulfonic anhydride $PhOC(O)OSO_2R$.

## IX. REACTIONS OF ACYL PEROXIDES WITH NUCLEOPHILES

The acyl peroxy linkage is readily attacked by many nucleophiles, including amines, phosphines, sulfides, carbanions, halide ions, enamines, and, in some cases, simple olefins. The first-formed ion pair may be stable, as when N: is a carbanion, or decompose routinely ($N: = I^-$ in $H_2O$), leading to useful synthetic or analytical schemes. Or its decomposition may proceed by diverse pathways, yielding a multitude of products and a field-day for mechanists.

$$
\begin{array}{c} O \\ \parallel \\ C-R \end{array} \quad N: O \quad \begin{array}{c} O \\ \parallel \\ C-R \end{array} \longrightarrow \left[ \overset{+}{N}-O-\overset{\overset{\displaystyle O}{\parallel}}{C}-R \quad \overset{-}{O}-\overset{\overset{\displaystyle O}{\parallel}}{C}-R \right] \longrightarrow \text{products} \tag{106}
$$

Table 15 shows rate data for a number of such reactions. Reactants and conditions are too diverse for close comparisons, though in general it is

TABLE 15.  RATES OF NUCLEOPHILIC ATTACK ON ACYL PEROXIDES

| Peroxide | Nucleophile | Solvent | $T$ (°C) | $k_2 \times 10^5$ (l/mole-sec)[a] | $E_a$ ($\Delta H\ddagger$)[b] (kcal/mole) | $A$ ($\Delta S\ddagger$)[c] (e.u.) | Ref. |
|---|---|---|---|---|---|---|---|
| Benzoyl peroxide | PhNMe$_2$ | Toluene | 0 | 1.5[d] | 14.8 | 10[7.57] | 175 |
| | PhNMe$_2$ | Toluene | 0 | 10.7[e] | — | — | 175 |
| | PhNMe$_2$ | Benzene–styrene | 0 | 10 | — | — | 547 |
| | PhNMe$_2$ | Styrene | 0 | 28 | (11.0) | (−9.9) | 547 |
| | (MeO)$_3$P | Toluene | 30 | 140 | — | — | 530 |
| | $n$-Bu$_2$S | Chloroform | 20 | 360 | 14.8[f] | 10[8.6f] | 235 |
| | PhSMe | Chloroform | 20 | 41 | — | — | 235 |
| | $p$-NO$_2$C$_6$H$_4$S-Me | Chloroform | 20 | 0.9 | — | — | 235 |
| | PhSMe | Chloroform | 20 | ≥1,700[f] | — | — | 232 |
| | Cyclohexene | Carbon tetrachloride | 80 | 683 | — | — | 183a |
| | $trans$-Stilbene | Carbon tetrachloride | 80 | 1,370 | — | — | 183b |
| | $trans$-4,4′-Dimethoxystilbene | Carbon tetrachloride | 80 | 53,200 | (11.6) | (−27.3) | 183a |
| | Ph—C≡C—Ph | Carbon tetrachloride | 80 | 4.4 | — | — | 183c |

| | | | | | | | |
|---|---|---|---|---|---|---|---|
| $(m\text{-}BrC_6H_4CO_2)_2$ | trans-4,4'-Dimethoxystilbene | Benzene | 44.8 | 105 | — | — | 180 |
| | $Me_2C{=}CMe_2$ | Benzene | 44.8 | 4.7 | 17.7f | 107.9f | 179 |
| $(p\text{-}NO_2C_6H_4CO_2)_2$ | trans-4,4'-Dimethoxystilbene | Benzene | 44.8 | 958 | — | — | 180 |
| Benzoyl Peroxide | m-Cresol | Benzene | 20 | 0.11 | (17.4) | (−26.7) | 546 |
| | $p\text{-}MeOC_6H_4OH$ | Benzene | 20 | 33.1 | (15.3) | (−22.3) | 546 |
| | $I^-(Na^+)$ | 95% EtOH | 20.5 | 5,600 | (16.4) | (−8.6) | 525b |
| | $I^-(Rb^+)$ | 95% EtOH | 20.5 | 16,700 | (13.9) | (−14.7) | 525b |
| | $Cl^-(Li^+)$ | Toluene–AcOH (~1:1) | 65 | 9.8 | — | — | 293 |
| | $Cl^-(Li^+)$ | Dimethylformamide | 60 | ~900f | 19.4 | $10^{11.7f}$ | 13 |

a $k_2 = -\dfrac{[\partial(\text{peroxide})/\partial t]}{[\text{peroxide}][\text{nucleophile}]}$

b Values for $\Delta H^{\ddagger}$ are shown in parentheses.

c Values for $\Delta S^{\ddagger}$ are shown in parentheses.

d,e Analyzed as a chain reaction: $-\dfrac{\partial[Bz_2O_2]}{\partial t} = 3k_2[Bz_2O_2][DMA] + k_b[Bz_2O_2]^{3/2}[DMA]^{1/2}$;

d $k_2$; e $3k_2 + k_b$ (see text).

f Calculated from data in reference.

clear that activation energies are low and $A$-factors or entropies are indicative of considerable steric restriction in the transition state.

### A. Reactions with Amines

No nucleophilic displacement on peroxidic oxygen has received more attention than that by amines (29), and of peroxides attacked, none more than the acyl variety. Extensive product studies for the reaction with acyl peroxides are due to Horner (228, 230, 231, 233, 234, 238–241), Edward (134, 136), Hawkins (68), and their co-workers, following on the early work of Gambarjan (161, 162). The amine–peroxide combination as an initiator for vinyl polymerization was investigated by Meltzer and Tobolsky (364), Lal and Green (318), and Imoto et al. (252) in the early 1950s, and at about the same time the work of Bartlett and Nozaki (24) and Walling and Indictor (541, 547) laid a groundwork for mechanistic analysis. More recent work is as follows:

*Products*: Henbest (56, 212), Huisgen (248, 249, 251), and Swan (146, 452, 508).

*Efficiencies of free-radical production*: O'Driscoll (386–391) and Imoto (459, 563b).

*Kinetics and rates*: Antonovskii (5), Chaltykyan and Beileryan (30, 74, 458, 490, 491) and other Soviet workers, (1, 497, 532, 534), Favini (145), Okoda (396), and Nandi (378).

*Mechanism*: Denny (100), Graham (175), Kashino (270, 271), and Hrabak (245, 329).

Free radicals in the reacting medium have been observed by ESR spectroscopy and tentatively identified (358, 492, 497). The reactions of amides with acyl peroxides (551) and of peroxydicarbonates with amines (84, 440) have been studied.

The reaction of amines with acyl peroxides is much more rapid than the thermal decomposition of the peroxides alone (24). For example, benzoyl peroxide with dimethylaniline at $0°$ in styrene or chloroform exhibits an apparent second-order rate constant of $2$–$3 \times 10^{-4}$ $sec^{-1}$ (547). Isopropyl peroxydicarbonate reacts with similar rapidity (84), but acetyl (233) and lauroyl (145, 396) peroxides react somewhat slower.

The types of products depend on whether the amine is primary, secondary, or tertiary and are rather similar to those produced by the action of bromine on amines (111).

Under an air atmosphere, dimethylaniline with benzoyl peroxide yields benzoic acid, $N$-methylaniline, and formaldehyde almost quantitatively, together with small amounts of $p$-benzoyldimethylaniline and a substance once thought to be 2,2'-bisdimethylaminobiphenyl (240) or 4,4'-bisdimethylaminodiphenylmethane (23), but later found to be $N$-$p$-dimethylaminobenzylaniline (146).

Under nitrogen, the yield of this latter substance is increased (146) and a number of polymeric and cyclic structures of similar constitution are found (452a), such as the following:

$$Me_2N-\!\!\left\langle\;\right\rangle\!\!\left[CH_2-\underset{\underset{Me}{|}}{N}-\!\!\left\langle\;\right\rangle\right]_n\!\!CH_2-\underset{\underset{Me}{|}}{N}-Ph,\;(n\!=\!0,1,2,)$$

The yield of formaldehyde is drastically reduced, and the main product (38% in one instance) is tetramethylbenzidine (175). In the presence of *N*-phenyl-maleimide a tetrahydroquinoline adduct is formed (508).

$$Bz_2O_2 + PhNMe_2 + \overset{O}{\underset{O}{\parallel}} N-Ph \longrightarrow \qquad (107)$$

Triethylamine with benzoyl peroxide in air yields diethylamine plus acetaldehyde (238a), together with the enamine diethylvinylamine (547).

Some tertiary amines, such as quinuclidines, form stable amine oxides (251).

$$\xrightarrow{\;Bz_2O_2\;} \qquad (108)$$

(80%)

Secondary amines in solution with benzoyl peroxide do not absorb oxygen (68, 238b) [though they do take up nitric oxide (238b)] and yield hydroxyamine benzoates which, if the amine is aryl, rearrange to *o*-benzamidophenols (68, 134a, 136a, 238b).

$$\text{(structure: aniline with H, Me on N)} + Bz_2O_2 \xrightarrow[\text{benzene, 20 min}]{0°} \text{(structure: BzN, Me, OH on benzene)} \qquad (109)$$

[30% (134a)]

Yields of the benzamidophenols can be increased to 60–70% by the presence of sodium hydroxide (452b).

N-Alkylanilines also yield a little of the corresponding N,N-dialkylhydrazo-benzene, pointing to the intermediacy of $Ph-\overset{\bullet}{N}-R$ (452b) [also inferred by the reactivity with nitric oxide, but not with oxygen (238b)].

Primary amines (i.e., anilines) yield complex mixtures, including benzoic acid, benzanilides, benzamidophenols, and azobenzenes (68, 143, 238b, 452b). Again, the yield of phenols is increased by the presence of sodium hydroxide (452b).

$$\text{(aniline-NH}_2\text{)} + Bz_2O_2 \xrightarrow[\substack{\text{benzene} \\ + \text{NaOH}}]{5°} \text{(HO, NHBz structure)} + \text{(NHBz structure)} + PhN{=}NPh \qquad (110)$$

$$(2\%) \qquad\qquad (15\%) \qquad\qquad (7\%)$$

The combination of benzoyl peroxide with a tertiary amine, but not with primary and secondary amines, initiates vinyl polymerization (68, 238b). The rates of initiation are first order in each component (252, 364, 547) or nearly so (391). [O'Driscoll and Schmidt (391) found the order to vary with temperature, being $([Bz_2O_2][PhNMe_2])^{0.994}$ at 30° and $([Bz_2O_2][PhNMe_2])^{0.836}$ at 60° for styrene polymerization]. The efficiencies of initiation are quite low, however; the rate of disappearance of peroxide being 5–10 times greater than the rate of production of radicals (387, 390, 391, 547) and decreases with increasing temperature. The polymers obtained contain little nitrogen, implicating benzoyloxy as the responsible radical (240, 547).

Mechanistically speaking, the first stage of the amine–peroxide reaction is, unquestionably, nucleophilic attack on the O—O bond (547). The rates of reaction are enhanced by electron-withdrawing groups on the phenyl nucleus of benzoyl peroxide (234, 239, 387, 390) ($\rho = 1.6$–2.7 for reactions with dimethylaniline or diethylaniline, depending on solvent and temperature) and decreased by similar groups on the amine (234, 239, 387) [$\rho$ for $Et_2N-C_6H_4-X + Bz_2O_2$ is $-2.7$ (387)]. The ostensible first product of the reaction has been identified as the quaternary ammonium salt (56, 166), though there is spectral evidence that this is preceded by formation of an amine–peroxide complex (386, 389, 497). The quaternary salt can be obtained equally well by the reaction of the amine oxide with a carboxylic anhydride; it has been shown to yield the same products from these precursors (45, 248, 249) and to initiate polymerization with equal facility (459).

$$Ac_2O_2 + Me_3N \longrightarrow [Me_3\overset{+}{N}-OAc \quad {}^-OAc]$$

$$\xrightarrow{\text{HBr}} [Me_3\overset{+}{N}-OAc]Br^- \quad (111)$$

Beyond this point, however, there is much less certainty. Walling and Indictor (547) proposed that for dimethylaniline–benzoyl peroxide the decomposition of the ion pair was largely ionic, mainly on the basis of inefficient initiation of polymerization. Graham and Mesrobian (175) have since shown, however, that in this system, the low efficiency arises from the fact that most of the peroxide disappears by induced decomposition, that the "second-order" reactions of amine plus peroxide obtained by earlier workers (252, 364, 386–391, 547) were largely fortuitous, and that the true rate expression contains two terms:*

$$\frac{-\partial[Bz_2O_2]}{\partial t} = k_a[Bz_2O_2][DMA] + k_b[Bz_2O_2]^{\frac{3}{2}}[DMA]^{\frac{1}{2}} \quad (112)$$

The expression is the familiar one for radical-induced decomposition, but Graham (175) was forced to conclude from analysis of the products and their effect on the rate of reaction that the species responsible was not phenyl or the simple imino radical, but tetramethylbenzidine or its ion radical.

The presence of induced decomposition partly explains the complex kinetics observed by Sogomonyan and Chaltykyan (490, 491) for the reaction of ethanolamines with benzoyl peroxide and suggests why reactions of dimethylaniline with peroxides should be slower in styrene than in benzene (84, 547).

---

* Graham and Mesrobian (175) were able to derive an expression of this form using steady-state approximations, in which $k_a$ was equal to three times the bimolecular rate constant for initiation by amine plus peroxide;

$$DMA + Bz_2O_2 \xrightarrow{k_2} 2 \text{ free radicals}$$

Their value for $k_2$ ($0.5 \times 10^{-5}$ l/mole-sec at $0°$ in toluene, $E_a = 14.8$ kcal/mole) agreed precisely with that obtained from the autoxidation of DMA initiated by DMA $+ Bz_2O_2$. As shown in Table 15, the sum $k_a(= 3k_2) + k_b$ is nearly identical to the "bimolecular" rate constant observed by Walling and Hodgdon (546), measured simply as $-\partial[Bz_2O_2]/\partial t =$ "$k_2$" $[Bz_2O_2][DMA]$. The clear presumption is that in the latter work as well as the many other reports of simple bimolecular kinetics the concentrations of reactants have been varied over an insufficient range; the reported constants are composites and have a tenuous connection to the actual rate of amine–peroxide interaction.

But the generally complex solvent effects (547) or differences between one tertiary amine and another (84, 490, 491) cannot yet be said to be clearly understood.

The reaction of primary (5, 396) or secondary (30) amines with benzoyl peroxide also is claimed to be first order in each reactant, though Kashino et al. (270) have found that a better fit is obtained if the amine concentration is corrected for that complexed with benzoic acid formed as the reaction proceeds. There seems to be little attention paid to the overall stoichiometry in these kinetic studies, but that found by Antonovskii (5), 0.4 mole of α-naphthylamine used per mole of lauroyl peroxide destroyed, bodes ill for any simple interpretation.

The reaction of benzoyl peroxide labeled with $^{18}O$ in the C=O group with dibenzylamine in ether has been shown by Denny (100) to produce 94% of N,N-dibenzylbenzamide, with the labeled oxygen still in the carbonyl group. In the benzamidophenol produced when the reaction is done in carbon tetrachloride the oxygens are almost completely scrambled. Denny (100) interprets both reactions as ionic, the first as a straight displacement, the second as a migration involving a free carboxylate anion.

$$(113)$$

$$(114)$$

But ESR measurements show the presence of free radicals in the reaction of benzoyl peroxide with both primary (358, 492) and secondary (270) amines. In the one case in which the structure of the radical has been clearly identified as diaryl nitroxyl (492) one aryl group comes from each species.

$$\left( Me-\!\!\!\bigcirc\!\!\!-CO_2 \right)_2 + PhNH_2 \longrightarrow \qquad (115)$$

## B. Reactions with Phosphines, Phosphites, and Arsines

Acyl peroxides are reduced readily at room or slightly elevated temperatures by compounds of the type $R_3P$ or $(RO)_3P$, where R can be either alkyl or aryl (29, 62, 66, 73, 105, 230, 232). For aroyl peroxides, at least, the reaction is quantitative (236, 498); the resulting anhydride plus phosphine oxide or phosphate can be recovered in more than 80–95% yield (62, 105b, 232, 237), depending on the ease of isolation (62). Acetyl peroxide with triphenylphosphine at 20° yields a small amount (0.7%) of carbon dioxide (230), perhaps resulting from competing thermal homolysis, and aroyl peroxides also yield free-radical type of products if the reactants are combined in the absence of a solvent and the considerable exothermicity of the reaction is allowed to elevate the temperature (62). However, in a solvent at temperatures of 40° or below the combination does not initiate either vinyl polymerization or autoxidation (230, 237).

$$Bz_2O_2 + (EtO)_3P \xrightarrow[\text{36°, 1.5 hr}]{\text{ether}} Bz_2O + (EtO)_3PO \qquad (116)$$

$$(81\%) \quad [97\% \,(62)]$$

On long standing or heating the products can react further (62, 494, 495).

$$(EtO)_3PO + Bz_2O \longrightarrow \left[ (EtO)_2 \overset{O}{\underset{\uparrow}{P}} \right]_3 CPh \qquad (117)$$

$$Bz_2O_2 + Ph_2PCl \xrightarrow[25°]{\text{PhCl}} Bz_2O + Ph_2\overset{O}{\underset{\uparrow}{P}}Cl \longrightarrow BzOPPh_2 + BzCl \quad (118)$$

The mechanism of the initial reaction has been firmly established to be nucleophilic attack of the phosphorus compound on the O—O linkage (29, 66). The oxygen-labeling experiments of Denny and co-workers (105b, 176) provide strong evidence for an intermediate ion pair

$$\text{PhC}-\text{O}_2\text{CPh} + \text{Ph}_3\text{P} \longrightarrow \left[ \text{PhC}-\text{O} \quad - \quad \text{C}-\text{Ph} \right]$$

$$\text{(119)}$$

$$\text{Ph}_3\text{PO} + \text{PhC}-\text{O}-\text{C}-\text{Ph}$$

in which the carboxylate ion is free enough both to randomize its oxygens and to undergo exchange with added foreign carboxylate (176). If the O—O bond is polarized by means of a nitro substituent on one of the phenyl rings of the peroxide, the phosphine attacks the more electropositive oxygen almost exclusively (105b).

$$\text{Bz}_2\text{O}_2 + \text{Ph}_3\text{P} \xrightarrow{p-\text{NO}_2\text{C}_6\text{H}_4\text{CO}_2^-} \text{Ph}_3\text{PO} + p\text{-NO}_2\text{C}_6\text{H}_4\text{CO}_2\text{Bz} \quad (120)$$

Asymmetrical phosphines (229, 242) or phosphites (530) yield oxides of partially racemized configuration, suggested to result from inversion in the ion-pair stage (229, 242). The extent of racemization depends on solvent polarity: 49% in petroleum ether, 78% in methanol, 88% in MeCN for MePhPrP + Bz$_2$O$_2$* (242).

---

* Such a mechanism for inversion conflicts, of course, with Denny's claim (105b) that the oxygen in Ph$_3$PO from Ph$_3$P + $p$-NO$_2$C$_6$H$_5$CO$_2$O$_2$CC$_6$H$_4$—$p$-Me in pentane solution is exclusively from the nitro-substituted side. Any internal displacement in the ion pair should yield the more stable $p$-nitrobenzoate anion. However, racemization has not yet been demonstrated for an asymmetrical triarylphosphine. Furthermore, the MePhPrP–Bz$_2$O$_2$ reaction is not as clean as could be desired, yielding a considerable amount of benzoic acid as well as anhydride and phosphine oxide (242).

$$Bz_2O_2 \; + \; \overset{Me}{\underset{Pr}{P{-}Ph}}$$

$$\left[ \underset{Pr}{BzO{-}\overset{+}{\overset{Me}{P}}{-}Ph} \; \bar{O}Bz \right] \longrightarrow \left[ Bz\bar{O} \; \underset{Pr}{Ph{-}\overset{Me}{P}{-}OBz} \right] \qquad (121)$$

$$\underset{Pr}{O{\leftarrow}\overset{Me}{P}{-}Ph} \qquad\qquad \underset{Pr}{Ph{-}\overset{Me}{P}{\rightarrow}O}$$

The rate of reaction of acyl peroxides with phosphines or phosphites appears to be rather insensitive to solvent polarity; the reaction proceeds with more or less equal facility in pentane (176), benzene (495), methylene dichloride (105b), ether (62, 237), and isopropanol (498). The analogous reaction, $Ph_3P + BzO_2{-}t\text{-Bu}$, for which second-order rate constants have been established (105a), is affected very little by changes in the solvent.

The kinetics of benzoyl peroxides reactions with phosphites are said to show similar characteristics (530), but the data are not yet readily available.

Benzoyl peroxide reacts quite readily at room temperature with ylides, presumably via a polar mechanism since the presence of galvinoxyl or other inhibitors of free-radical reactions has no influence (108). Denny and Valega (108) have proposed the following mechanism to account for the products:

$$Ph_3P{=}CH{-}Bz \; + \; Bz_2O_2$$

$$\left[ \underset{OBz}{BzCH{-}\overset{+}{P}Ph_3} \; \bar{O}Bz \right] \longrightarrow \underset{OBz}{PhC}\overset{O^-}{\underset{}{{-}}}\underset{OBz}{CH{-}\overset{+}{P}Ph_3}$$

$$(122)$$

$$\underset{OBz \quad OBz}{PhC}\overset{O}{\overset{\diagdown\diagup}{{-}}}CH \xrightarrow{\;O\;} \underset{OBz}{PhCH}{-}\overset{O}{\overset{\|}{C}}{-}OBz + Ph_3P \xrightarrow{Bz_2O_2} Ph_3PO \; + \; Bz_2O$$

The reaction of triarylarsines with acyl peroxides has received much less attention than the phosphine reaction, but apparently it follows a similar course (230, 236). At any rate, the reduction is rapid (230) and can be used to estimate acyl peroxide concentration quantitatively (236).

### C.  Reactions with Organic Sulfides

Like hydroperoxides and peroxy acids, acyl peroxides are decomposed fairly rapidly by organic sulfides (213, 230, 235). The products, however, include a mixture of sulfoxide, anhydride, carboxylic acid, and $\alpha$-acyloxy thioethers, as if nucleophilic attack, analogous to the phosphine reaction, were overlaid by a competing radical-induced decomposition as occurs with ethers (235). Cyclic phthaloyl peroxide is unique in that it yields only sulfoxide plus anhydride in a reaction about 50 times as fast as that of $Bz_2O_2$. The sulfoxide is further oxidized to the sulfone (232).

$$Bz_2O_2 + t\text{-}Bu\text{--}S\text{--}Me \xrightarrow{\ 18\,°C\ }$$

$$\longrightarrow \quad BzOH + t\text{-}Bu\text{--}S\text{--}CH_2OBz$$
$$48\% \qquad\qquad 21\%$$

$$\qquad\qquad\qquad\qquad\qquad\qquad\qquad O \qquad\qquad (123)$$
$$\qquad\qquad\qquad\qquad\qquad\qquad\qquad \uparrow$$
$$\longrightarrow \quad Bz_2O \quad + \quad t\text{-}Bu\text{--}S\text{--}Me$$
$$(no$$
$$quantitative \qquad 7\% \ (213)$$
$$estimate)$$

The most extensive exploration is due to Horner and Jurgens (235), who reacted acetyl and benzoyl peroxides with 12 different sulfides. In chloroform solution at 0–20° benzoyl peroxide reacts about 10 times as fast as acetyl peroxide; reaction with aryl sulfides is much slower than with alkyl sulfides, and with nitroaryl sulfides is slower still (Table 15). The reaction of benzoyl peroxide with sulfides is first order in each component and gives the same $k_2$ whether determined by the disappearance of benzoyl peroxide, the appearance of RS($\rightarrow$O)R or RS—C(OBz)H—R' whether the 1% ethanol in the chloroform solvent is present or removed, though this affects the RS($\rightarrow$O)R/ RSC(OBz)H—R' product ratio. Thus the benzoyloxy sulfide appears not to arise through a radical-induced pathway after all but via the same ion pair that precedes sulfoxide formation (235).

$$2O_2 + RSCH_2R' \longleftarrow \left[ \begin{array}{c} \underset{\underset{\overset{|}{\underset{C}{\underset{|}{Ph}}}}{\overset{O}{\overset{||}{O}}}}{\underset{\overset{+}{O}}{\overset{BzO}{\underset{|}{R-S-CH_2R'}}}} \longleftrightarrow \underset{\overset{|}{Ph}}{\overset{Ph}{R-S-CHR'}} \end{array} \right] \quad (124)$$

BzOH + RS'—CHR'
|
OBz

## D. Reactions with Olefins, Enamines, and Phenols

Reactions with olefins, enamines, and phenols can all be considered mechanistically as nucleophilic displacements* on the O—O bond by the $\pi$-electrons of a carbon–carbon double bond, assisted in the latter cases by conjugated nonbonded $p$-electrons.

$$R-\overset{O}{\overset{||}{C}}-O-O \quad \overset{}{\underset{\overset{|}{C=O}}{\underset{|}{R}}} \quad \overset{\cdots}{C=C-X} \longrightarrow R-CO_2^- + R\overset{O}{\overset{||}{C}}-O-\overset{}{C}-\overset{X^+}{\overset{|}{C}} \quad (125)$$

Where such assistance is not available (i.e., simple olefins) clear examples are rare. In acetyl peroxide decompositions, as mentioned in Section VII-B, it has been shown that olefin intervention occurs only after initial O—O homolysis; likewise the rates of benzoyl peroxide decomposition in olefinic solvents are similar to those in alkanes or aromatic hydrocarbons (384) (Table 15).

* It is a moot question whether simple C=C assisted decomposition occurs by heterolysis or homolysis. Several examples given earlier in Section VII–D might equally well have been cited here, as might internal assistance by the iodide ion. For hydroperoxides, generally speaking, any substance that moderately accelerates the rate of decomposition also increases the rate of free-radical production, though the two increases are not always parallel. Similar data are mostly lacking for acyl peroxides but, where extant (see below), suggest that both modes occur simultaneously.

However, cyclic phthaloyl peroxide decomposes very rapidly in the presence of alkenes (178, 183) (Table 15) to give predominantly a mixture of cyclic phthalate and lactonic orthoester.

$$(25\%) \qquad\qquad (75\%) \qquad (126)$$

Some allylic monophthalate is produced from olefins with an allylic hydrogen atom [the main product from tetramethylethylene (179), 45% from cyclohexene (183a)]. The rates of reaction in an inert solvent (CCl$_4$) are first order in peroxide and in olefin (183) and parallel the reactivities of peroxy acids rather than methyl or trichloromethyl affinities (183a); for $p,p'$-substituted *trans*-stilbenes $\rho\sigma^+$ plots were linear, with $\rho = -1.65$.

$$(127)$$

Greene and Rees (183a) have concluded that their reaction is polar. Going further, Greene, Adam, and Cantril (180) have shown that with the most reactive olefins (mono- and dimethoxystilbenes) both bis(*m*-bromobenzoyl) and bis(*p*-nitrobenzoyl) peroxides give similar reactions. In both there is a small (10–25%) but significant* free-radical component that can be suppressed by added galvinoxyl (180). The rates for the polar reaction are 200–2000 times less fast than those for phthaloyl peroxide (179, 180) (Table 15) but yield $\rho$-factors of $+1.2$ for substitution on the peroxide nucleus and $-1.0$ for stilbene substitution (180).

Bis(*m*-bromobenzoyl) peroxide epoxidizes tetramethylethylene in high yield (179); $^{18}O$ labeling shows that the epoxide oxygen derives solely from peroxidic oxygens; insensitivity to added galvinoxyl rules out a radical-chain mechanism (179). It will be noted that epoxide + anhydride is an obvious alternative pathway from the intermediates in equation 127. Epoxide has been a suspected minor by-product in some phthaloyl peroxide–olefin reactions (183b).

$$(m\text{-}BrC_6H_4CO_2)_2 + Me_2C{=}CMe_2$$

$$\Big\downarrow \; 45° \; \diagup \; \text{benzene}$$

$$\overset{\displaystyle O}{\overset{\displaystyle \diagup\backslash}{Me_2C{-}CMe_2}} + (m\text{-}BrC_6H_4CO)_2O \tag{128}$$

$$(72\%) \qquad\qquad (75\%)$$

Benzoyl peroxide has been used to benzoylate a number of enamines in 20–80% yield (10, 48, 263, 325). Although the focus has been on synthesis rather than on mechanisms, the reaction is reasonably formulated as nucleophilic attack on the O—O bond (10, 325).

---

* Greene et al. (180) propose assisted homolysis:

$$(ArCO_2)_2 + \overset{\backslash}{\underset{\diagup}{C}}{=}\overset{\diagup}{\underset{\backslash}{C}} \longrightarrow ArCO_2{\cdot} + \overset{ArCO_2}{\overset{|}{\underset{\diagup}{\overset{\backslash}{C}}{-}\overset{\cdot}{\underset{\backslash}{C}}\diagup}}$$

$$\text{\textit{trans}}$$

$$\Big\downarrow \; (ArCO_2)_2 \diagup$$

$$\overset{ArCO_2 \;\; O_2CAr}{\overset{\backslash|\quad|\diagup}{\underset{\diagup\quad\backslash}{C{-}C}}} + ArCO_2{\cdot}$$

$$\text{\textit{meso} + dl}$$

The reaction of phthaloyl peroxide with norbornene in carbon tetrachloride yields a small amount of carbon tetrachloride addition product, which also clearly derives from a competing radical-chain mechanism (183b).

$$\text{(129)} \qquad [78\% \ (325)]$$

In a few cases the intermediate benzoyl enamines have been isolated and converted to oxazoles (263).

$$\text{(130)}$$

$$
\underset{\text{MeC}=\text{CH}-\overset{\displaystyle \text{O}}{\overset{\|}{\text{C}}}-\text{Me}}{\overset{\text{NH}_2}{|}} \xrightarrow{\text{Bz}_2\text{O}_2} \underset{\text{MeC}=\text{C}-\overset{\displaystyle \text{CMe}}{\underset{\|}{\text{O}}}}{\overset{\text{H}_2\text{N} \quad \text{OBz}}{|\qquad|}} \xrightarrow[\text{H}^+]{\Delta}
$$

In general yields appear to be better at low temperatures, to vary with the amine used in no readily explicable fashion, and are not always easy to reproduce (10, 263). Studies on the mechanism of reaction are said to be in progress (263).

The decomposition rates of aroyl peroxides are enhanced by the presence of phenols (24). The rates are directly proportional to phenol concentration (25a, 546), are accelerated by electron-donating substituents (25b, 546) and retarded by large *ortho* groups on the phenol (25b, 546) (Table 15), and exhibit an isotope effect for ArOH/ArOD of about 1.3 (546). The products do not include carbon dioxide and consist mainly of *ortho* aroyloxy phenols (42, 82, 100, 136a), in which the integrity of the aroyl carbonyl group is largely maintained (100).

$$
(\text{Ph}\overset{{}^{18}\text{O}}{\overset{\|}{\text{C}}}-\text{O})_2 + \text{Me}-\!\!\!\left\langle\!\!\!\bigcirc\!\!\!\right\rangle\!\!\!-\text{OH}
$$

$$\text{(131)}$$

$$
\text{Me}\!\!\left\langle\!\!\!\bigcirc\!\!\!\right\rangle\!\!-\underset{{}^{18}\text{OBz}}{\text{OH}} \ + \ \text{Me}-\!\!\!\left\langle\!\!\!\bigcirc\!\!\!\right\rangle\!\!\!-\underset{\underset{{}^{18}\text{O}}{\|}}{\overset{}{\text{O}-\text{C}-\text{Ph}}}\!\!\!-\text{OH}
$$

$$(13\%) \qquad\qquad\qquad (87\%)$$

The latter finding militates against the mechanism of Walling and Hodgdon (546), namely, a series of four-center reactions culminating in a Claisen-type rearrangement that transformed carbonyl oxygen to ether oxygen. Though explaining the presence of some $^{18}O$ in the phenolic OH still requires some gymnastics (100), the evidence points to an initial nucleophilic displacement by carbon (42, 100).

$$(132)$$

With $\beta$-naphthol the process is repeated to yield 30–40% of the 1,1-dibenzoyl ketone (42). In general, however, phenols with both *ortho* positions blocked react very slowly (546); the products are explicable in terms of phenoxy–phenoxy and phenoxy–benzoyloxy radical coupling (82, 254).

### E.  Reactions with Carbanions

Explorations of the synthetic utility of displacement by Grignard reagents (328), sodiomalonic esters (325a, b), and sodium derivatives of $\beta$-keto esters and nitriles (326) on benzoyl peroxide as well as on peroxy esters and peroxy acids (325c) are mainly due to Lawesson and co-workers. The reactions of sodiomalonates, etc., appear to be best carried out at 0° in benzene solution and, from the high yields of products [70–90% (325b)] coupled with $^{18}O$ labeling data (326), are concluded to proceed via carbanion displacement on O—O, uncomplicated by any free-radical component.

$$(133)$$

Grignard reagents, however, generally yield only 20–40% of the expected ester (137, 328).

$$Bz_2O_2 + PhMgBr \xrightarrow[\text{Et}_2\text{O}]{0-5°} BzOPh + BzOH + PhBr \quad (134)*$$

$Et_2$ (1 mole)          (0.35 mole)    (1.27 moles)    (0.21 mole)

The products are explained by Lawesson and Yang (328) in terms of disproportionation of RMgBr to $R_2Mg + MgBr_2$, followed by competing reactions.

---

\* Yields from reference 328.

$$R_2Mg + Bz_2O_2 \longrightarrow 2ROBz + Mg(OBz)_2 \tag{135}$$

$$MgBr_2 + Bz_2O_2 \longrightarrow Br_2 + Mg(OBz)_2 \tag{136}$$

$$RMgBr + Br_2 \longrightarrow RBr + MgBr_2 \tag{137}$$

The reaction with phenyl lithium suffered no such disability but gave a 61 % yield of triphenylcarbinol, based on phenyl lithium (328).

## F.  Reactions with Halide Ions

The reduction of acyl peroxides by the iodide ion, though long a standard method of analysis (Section V-A) has been studied kinetically only a few years ago (525). The reaction in either 100 or 95 % ethanol is first order in $I^-$ and in benzoyl peroxide; $k_2$ values for 15 *meta*- and *para*-substituted benzoyl peroxides give a linear $\rho\sigma$ plot, with $\rho = +0.8$, in accord with a rate-determining displacement on O—O by $I^-$ (525a).

$$I^- \overset{\frown}{\underset{Bz}{O-OBz}} \xrightarrow[\text{slow}]{} BzO^- + BzO-I \xrightarrow[\text{fast}]{I^-} BzO^- + I_2 \tag{138}$$

Rate constants also depend on the associated cation (Table 15), increasing tenfold over the series LiI < NaI < KI < CsI < RbI (525b), in line with a reasonable assumption as to the cation's ability to neutralize the developing negative charge on oxygen (525b).

The decompositions of benzoyl peroxide by $Br^-$ (293, 297) or $Cl^-$ (13, 293, 297) are slower (Table 15) but still substantially faster than thermal homolysis. The combination of benzoyl peroxide, ArH, and lithium bromide or lithium chloride in acetic acid yields 2BzOH, ArX, HX, and in some cases unreacted $X_2$ (293); in dimethylformamide the formation of some $HC(O)NMeCH_2OBz$ suggests the intermediacy of benzoyloxy radicals (13), although neither styrene nor acrylonitrile undergoes polymerization (13).

Bimolecular rate constants have been obtained for the initial stages of the $Cl^--Bz_2O_2$ reaction of benzoyl peroxide with the chloride ion (13, 293) (Table 15), but these are so sensitive to the nature and concentration of a chlorinatable substrate as to suggest that the rate-determining reaction is, in fact, attack on benzoyl hypochlorite (293). The similarity of products to those from the Hunsdiecker reaction also suggests the intermediacy of benzoyl hypochlorite (59).

$$Cl^- + Bz_2O_2 \underset{}{\overset{\text{fast}}{\rightleftharpoons}} BzO^- + BzOCl \xrightarrow[\text{slow}]{ArH} BzOH + ArCl \tag{139}$$

The thermal homolysis of valeryl peroxide competes with nucleophilic attack by $Cl^-$; the products include *n*-BuCl, thought to arise via a chain reaction (297). Lithium bromide, however, decomposes this peroxide rapidly at 0° to yield $2BuCO_2Li + Br_2$ quantitatively (297).

$$(BuCO_2)_2 + LiCl \xrightarrow[\text{AcOH}]{55°} BuCO_2Li + BuCO_2Cl \qquad (140)$$

$$BuCO_2Cl + Bu\cdot \longrightarrow BuCl + BuCO_2\cdot \longrightarrow Bu\cdot + CO_2 \quad (141)$$

## X. REACTIONS OF ACYL PEROXIDES WITH ELECTROPHILES

### A. Reactions with Proton or Lewis Acids

Carboxylic acids, as mentioned in Section VI-E, catalyze the carboxy inversion of acyl peroxides prone to decompose in that fashion (23, 205, 330, 334). Benzoyl peroxide, though little affected by these substances (23), undergoes carboxy inversion quite readily in the presence of boron trifluoride [3 hr at 0° (250)], aluminum chloride [24 hr at 0° (135)], or antimony pentachloride [4 hr at 25° (101, 109)], yielding, under optimal conditions, 90% of phenyl benzoate (101, 109, 250, 446). Electron-donating substituents on the phenyl nucleus accelerate the rearrangement, and electron-withdrawing groups retard it (250).

$$(142)$$

Asymmetrically substituted peroxides yield a mixture of esters in which that derived from the more stable carboxylate ion predominates (135, 250).

$$(143)$$

Though seemingly a reasonable possibility, Lewis acid–catalyzed decompositions of benzoyl peroxide in aromatic solvents yield little (109) or no (135, 250) benzoxylated aromatics. Such products are found resulting from the aluminum chloride–catalyzed decomposition of bis(p-nitrobenzoyl) peroxide (135, 109).

$$(NO_2\text{-}\langle\rangle\text{-}CO_2)_2 + AlCl_3 \xrightarrow[25°]{anisole} NO_2\text{-}\langle\rangle\text{-}\overset{O}{\overset{\|}{C}}\text{-}O\text{-}\langle\rangle\text{-}OMe$$

$$(28\%)$$

$$(144)$$

However, other products indicate free-radical mechanisms at work, which could explain the formation of aryl nitrobenzoates equally well (109).

Hydroxylations of toluene, however, achieved through peroxydicarbonate decompositions catalyzed by aluminum chloride (305, 310, 430b) or ferric chloride (307, 430a), occur in *ortho–meta–para* proportions that are strongly indicative of electrophilic substitution (305, 307, 310) (Table 16).

$$(145)$$

TABLE 16. DECOMPOSITION OF ISOPROPYL
PEROXYDICARBONATE BY ALUMINUM CHLORIDE

| Substrate | Yield of ArOH (%)[a] | | | | Ref. |
|---|---|---|---|---|---|
| | Total | *ortho* | *meta* | *para* | |
| Anisole | 76 | 20 | — | 80 | 309 |
| Toluene | 50 | 34 | 11 | 55 | 310 |
| Diphenyl ether | 39 | 43 | — | 57 | 309 |
| Diphenyl | 20 | 38 | — | 62 | 309 |
| Chlorobenzene | 2 | + | + | + | 309 |

[a] Peroxide–AlCl$_3$–ArH(1:2:20) at 0°.

Similar reactions utilizing other aromatic compounds (Table 16) tend to corroborate this interpretation, though, as amply demonstrated by Kovacic and co-workers (305, 307, 309, 310), total yields of phenols depend on several factors, such as the proportions of catalyst, peroxydicarbonate, and substrate, and temperature.

### B. Reactions with Mercury, Aluminum, Tin, and Lead as Free Metals, Salts, and Alkyls

A number of product studies reported mainly by Razuvaev, Ol'dekop, and their co-workers are grouped here, more for convenience than for any homogeneity of mechanism. The reactions are those of acyl peroxides (usually Bz$_2$O$_2$, Ac$_2$O$_2$, or AcO$_2$Bz) with Hg (435), Hg$_2$(O$_2$CR)$_2$ (402c, d, e), Hg(O$_2$CR)$_2$ (402a, b, f), HgR$_2$ (433b, 442c, 444), Sn or SnCl$_2$ (434), Ph$_2$Sn (311), R$_4$Sn (403d, 442a, b, d), R$_3$SnX (536), Al (517), Al($i$-PrO)$_3$ (438a, b), EtAl(OEt)$_2$ (438c), R$_3$Al (371, 433a, 438d), Ph$_4$Pb (403d), and Pt or Ni (428, 431, 517); a sampling is shown in Table 17.

From the products it is apparent that, generally speaking, free-radical reactions play a part, ranging from cases such as Hg (435) or Sn (434), where initiation probably occurs by simple thermal homolysis, to those of Ph$_2$Sn (311), Et$_3$Sn(OEt) (536b), and Et$_3$Al (433a, 438d), which cause fairly rapid decomposition at room temperature and are not unreasonably suggested to involve prior complexation (433a, b; 536a).

TABLE 17. SOME PRODUCTS FROM DECOMPOSITIONS OF BENZOYL PEROXIDE IN THE PRESENCE OF METALS OR ORGANOMETALLIC COMPOUNDS[a]

| $M(X)_n$ | $T$ (°C) | Yield of Product (mole per mole of $Bz_2O_2$) | | | | | Ref. |
|---|---|---|---|---|---|---|---|
| | | $M(X)_m(OBz)_{n-m}$ | $M(X)_m(Ph)_{n-m}$ | BzOH | $CO_2$ | $XH$[b] | |
| Hg | 80 | — | 0.31 (PhHgOBz) | — | — | — | 435 |
| $HgEt_2$ | 80 | 0.72 ($m=1$), 0.01 ($m=0$) | 0 | 0.34 | 0.1 | 0.32c | 442c |
| $Hg(OAc)_2$ | 80 | 0.70 (MeHgOAc) | 0.1 ($m=0$) | — | Present | —d | 402a |
| $Sn$[e] | 80 | 0.44 [$Sn(OH)_2(OBz)_2$] | — | — | — | — | 434 |
| $SnCl_2$ | 25 | 0.34 [$SnCl_2(OBz)_2$] | — | — | Trace | —f | 434 |
| $SnPr_4$[g] | 80–100 | 0.75 ($m=3$), 0.23 ($m=2$) | — | 0.53 | 0.22 | 0.75 | 536a |
| $SbCl_3$ | 80 | 1.0 [$SbCl_3(OBz)_2$] | — | — | 0 | — | 434 |
| $AlEt_3$ | 25 | 0.75 ($m=2$) | — | — | 0 | — | 433a |
| $Al(Et)(OEt)_2$ | 25 | —[$Al(OEt_2)(OBz)$] | — | — | | —h | 438c |

a Solvent is benzene, unless otherwise noted.
b Plus X—2H.
c Plus $EtC_6H_4CO_2HgEt$ (0.13), $EtC_6H_4CO_2H$ (0.27), EtPh (0.14), and traces of Hg, $C_4H_{10}$, BzOEt, and BzOHg.
d Plus traces of methane and ethane.
e In benzene–water.
f Plus some $SnCl_4$.
g In tetrapropyl tin.
h Plus EtOBz (0.18).

$$Bz_2O_2 + R_3Al \;\rightleftharpoons\;
\begin{array}{c}
\underset{R}{\overset{R}{\diagdown}}\overset{\delta^-}{Al}\diagup R \\
\\
\underset{\delta^+}{}\;\;O \cdots\; R \;\; O \\
\;\;\; \| \qquad | \quad \| \\
Ph-C\diagdown\; O \diagup O-C-Ph \\
\;\;\;\;\; O
\end{array}
\qquad (146)$$

$$R_2AlOBz + ROBz \qquad \text{radicals}$$

$$Bz_2O_2 + Ph_2Sn \xrightarrow[25°]{16\ hr} \overset{BzO\;\;\;\;OBz}{\underset{\;}{Ph_2Sn-SnPh_2}} \qquad (147)$$

$$[81\% \, (311)]$$

The trialkyl aluminum–acyl peroxide combination is potentially the most interesting, occupying as it does presently a place of prominence in the patent literature as a low-temperature initiator for vinyl polymerization, equalled formerly only by benzoyl peroxide + dimethylaniline. So far the only mechanistic study reported appears to be that of Milovskaya and Zhuravleva (371), who showed that in vinyl acetate radical production takes the form

$$\frac{\partial(\text{radicals})}{dt} = k_2[Et_3Al][Bz_2O_2] \qquad (148)$$

where $k_2 = 10^{9.56}\, e^{-15.8/RT}$ ($= 5.6 \times 10^{-3}$ l/mole-sec at 20°) (371a). Both ethyl and benzoyloxy radicals are implicated as the initiating species by investigation of the products of reaction in inert solvents (371b) (Table 17); the polymer itself contained too many groups incorporated via chain transfer for any definite conclusions to be drawn (371b).

By using methyl violet as an indicator $Et_3Al$ was shown to be "neutralized" very rapidly by benzoyl peroxide in a Lewis acid–Lewis base type of reaction (371b). The stoichiometry suggested the formation of a complex $(Et_3Al)_2\cdots(Bz_2O_2)$. In the presence of vinyl monomers, however, the complex is thought to be $Et_3Al\cdots M\cdots Bz_2O_2$.

## XI. DECOMPOSITIONS OF ACYL PEROXIDES BY METAL IONS OF VARIABLE VALENCE

Since the discovery of the $Cu^+$–$Cu^{2+}$ catalyzed reactions of benzoyl peroxide with organic substrates by Kharasch and Fono (275), there have been numerous examples of their synthetic utility (26, 53, 221, 288, 296, 313–315, 515).

The combination has served principally to substitute BzO— or Ph— on olefins:*

$$Bz_2O_2 + MeCH=CHMe \xrightarrow[\text{MeCN}]{Cu^+ - Cu^{2+}} BzOH + \underset{\underset{\text{(80–90\%)}}{}}{BzO-\overset{\overset{Me}{|}}{C}HCH=CH_2} \quad (149)$$

$$\underset{\text{(80–95\%)}}{}$$

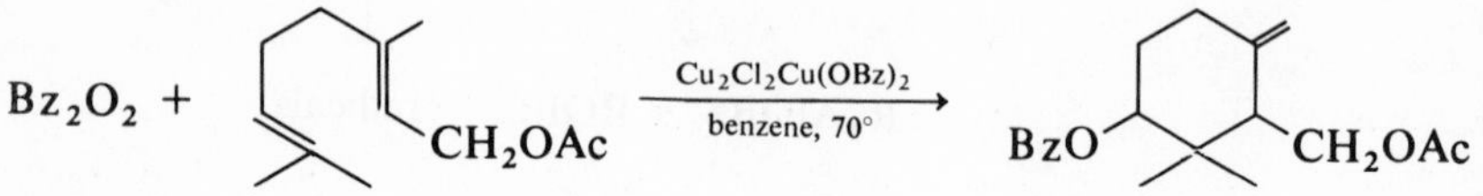

(150)

(55–60%)

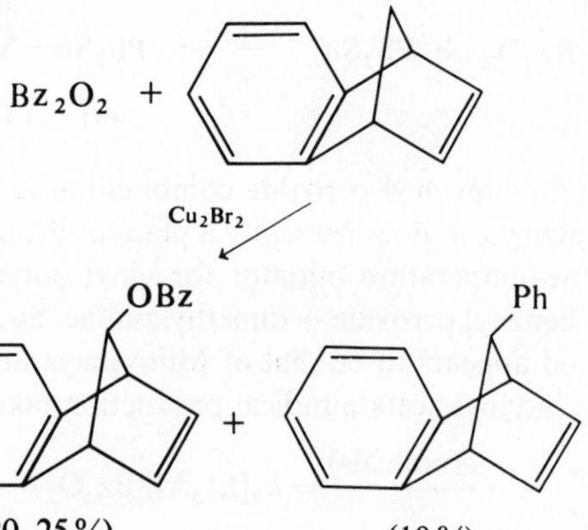

(151)

(20–25%)    (10%)

In an unreactive solvent and with a large excess of $Cu_2Br_2$, the peroxide is reduced to benzoic acid (27), whereas in aromatic solvents good yields of substitution products are obtained (221, 292, 306, 308, 313–315).

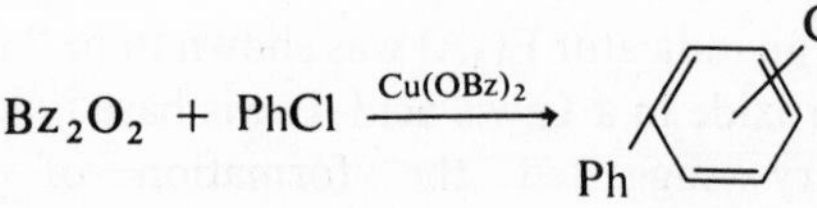

(152)†

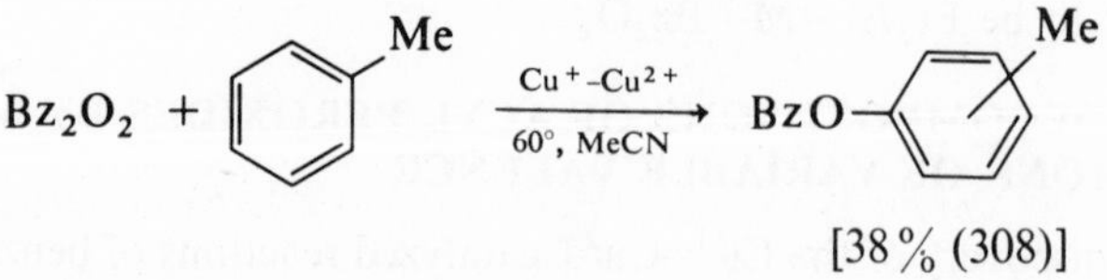

(153)

[38% (308)]

---

* Reactions 149, 150, and 151 are from references 288, 53, and 26, respectively.
† From reference 221.

The decompositions of acyl (as opposed to aroyl) peroxides in an unreactive solvent leads to carbon dioxide, carboxylic acid, and alkene in virtually quantitative amounts (172, 289–292). In the presence of high olefin concentration products of attack by R·, but not by $RCO_2·$, are observed (296a).

$$[Me(CH_2)_6CO_2]_2 \tag{154}$$

$$\xrightarrow[\text{80°, MeCN}]{\text{Cu(OAc)}_2} \quad CO_2 \quad + \quad Me(CH_2)_6CO_2H \quad + \quad Me(CH_2)_4CH=CH_2$$

$$\text{(0.96 mole)} \qquad \text{(0.92 mole)} \qquad \text{(0.94 mole)}$$

The patterns of mechanism have been amply clarified by Kochi and co-workers (288–291, 296, 297).* The initial reduction by $Cu^+$ is followed by a number of competitions in which loss of $CO_2$ from $RCO_2·$ always wins unless R is aryl or methyl (296a).

$$(RCO_2)_2 + Cu^+ \longrightarrow RCO_2^- + RCO_2· + Cu^{2+} \tag{155}$$

$$\text{(156)}$$

The alkyl radicals, whether the initial R· or R'· from the addition of R· to substrate, reduce $Cu^{2+}$ back to $Cu^+$, becoming product, R—H, or in some cases R—X via ligand transfer.

$$\text{(157)}$$

Kochi (291) has demonstrated that in one case at least, ligand-transfer product is preceded by a carbonium ion, but that the loss of a proton to form olefin is

---

* Ref. 296a includes an excellent summary and unified picture for all types of peroxides with $Cu^I$—$Cu^{II}$.

synchronous (eq. 158) of the two products, acetate and olefin, only the acetate has equilibrated labels.

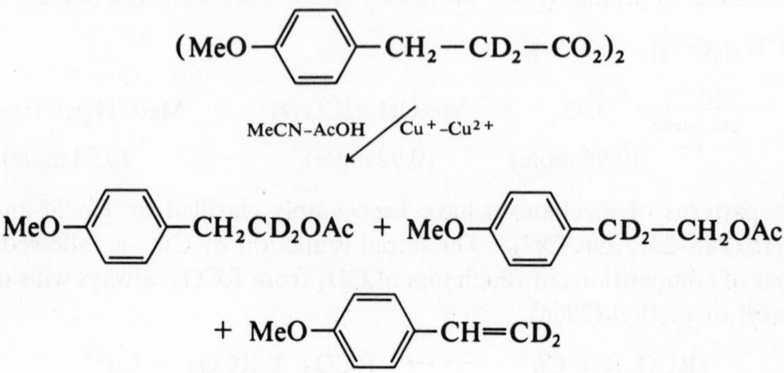
(MeO–C6H4–CH2–CD2–CO2)2
MeCN-AcOH
Cu+-Cu2+
MeO–C6H4–CH2CD2OAc + MeO–C6H4–CD2–CH2OAc
+ MeO–C6H4–CH=CD2

(158)

However, the rearrangement encountered in the benzoyl peroxide substitution on norbornadiene strongly suggests that carbonium-ion intermediates can be involved in olefin formation as well.

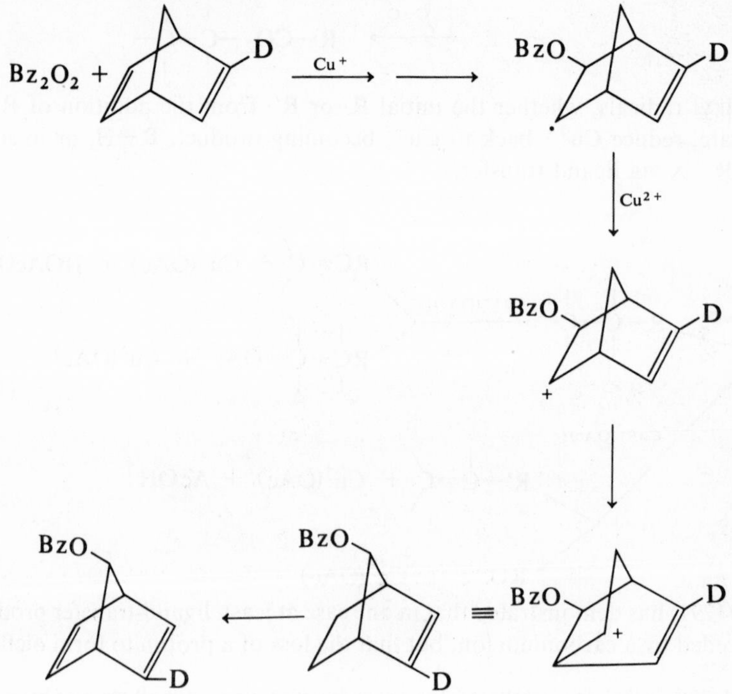
Bz2O2 +
D
Cu+
BzO
D
Cu2+
BzO
D
+
BzO
D
≡
BzO
D
BzO
D

(159)

Kochi and Bemis (290) have recently obtained stable solutions of cuprous acetate in acetonitrile–acetic acid mixtures and measured a rate constant for the reduction of valeryl peroxides by $Cu^+$ (equation 155); $k_{25.5°} = 0.08$ l/mole-sec. The oxygenation of aromatics by benzoyl peroxide (308, 314) or by peroxydicarbonates, which give higher yields perhaps because of less rapid decarboxylation (306, 313, 315), and copper salts can be rationalized on the same basis (308, 314, 315). It is a bit disconcerting, however, that this sup- posedly free-radical substitution gives essentially the same *ortho–meta–para* ratio as that from the aluminum chloride–catalyzed reaction (Table 17) on which basis an ionic mechanism was postulated.

$$Bz_2O_2 \xrightarrow{Cu^+} BzO^- + BzO\cdot \xrightarrow{PhMe} \quad \quad (160)$$

[38%; $o:m:p = 56:18:26$ (314)]

$$(i\text{-}PrO{-}CO_2)_2 \xrightarrow{Cu^+} \xrightarrow{PhMe} \quad \quad (161)$$

[92%; $o:m:p = 57:15:28$ (306)]

It should be noted further that subsequent to the demonstration that $Cl^-$ and $Br^-$, themselves, cause acyl peroxides to decompose (Section IX-F) no decompositions catalyzed by copper halides should be rationalized solely on the action of $Cu^+$ or $Cu^{2+}$. Kochi and Subramanian (297) found that addi- tion of cupric chloride to a lithium chloride–catalyzed decomposition of valeryl peroxide did not increase the rate of reaction, though the yields of valeric acid and butyl chloride were increased somewhat. The copper-catalyzed decomposition is retarded through the oxidation of $Cu^+$ to $Cu^{2+}$ by $RCO_2X$ or $X_2$ but can be observed if these substances are removed by the addition of styrene, 1-octene, mesitylene, etc. (297).

The trio of metal ions that figure so heavily in the decompositions of hydroperoxides (Chapter 1 and 2 of this volume), $Fe^{2+}$, $Co^{2+}$, $Mn^{2+}$, have seen little use in combination with acyl peroxides. Though all three, as well as $Cu^+$, have been used to promote low-temperature acyl peroxide–catalyzed vinyl polymerization (303) and manganous acetylacetonate has been shown to accelerate vastly the benzoyl peroxide–initiated addition of sodium bisulfite or thiophenol to olefins (552), the reactions are slow; 3 days at room temperature are required to obtain a 30% yield of bisulfite addition product from 1-octene (552). The difficulty is, of course, that, unlike hydroperoxides, acyl peroxides, having no proton attached to oxygen, cannot reduce the metal ion from its higher valence state directly.

The ferrous-ion-catalyzed decomposition of benzoyl peroxide in ethanol has been studied in some detail by Hasegawa and co-workers (208, 209). The cycle, which requires reduction of $Fe^{3+}$ by solvent-derived radicals, yields a steady-state concentration of $Fe^{2+}$ after a few minutes, shown spectroscopically to be proportional to the initial concentration of the ferrous ion (208). The second-order rate constant for reaction 162 was found to be 8.4 l/mole-sec at 25°, with an activation energy of 14.2 kcal/mole and an $A$-factor of $2.2 \times 10^{11}$.

$$Bz_2O_2 + Fe^{2+} \longrightarrow BzO^- + BzO\cdot + Fe^{3+} \tag{162}$$

$$BzO\cdot + EtOH \longrightarrow BzOH + Me\overset{\bullet}{C}HOH \tag{163}$$

$$Me\overset{\bullet}{C}HOH + Fe^{3+} \longrightarrow Fe^{2+} + AcH + H^+ \tag{164}$$

Rather surprisingly, methoxy-substituted benzoyl peroxides react faster, and nitro-substituted ones slower, with $Fe^{2+}$ than benzoyl peroxide itself ($\rho \simeq -0.3$) (209). This is the opposite direction from either radical or nucleophilic attack on the O—O bond and is interpreted by Hasegawa (209) as evidence for initial complexation in which peroxide acts as the donor.

# APPENDIX

This appendix is a compilation of some acyl peroxides that have been characterized, their melting points (where given), and references to their products and rates of thermal decomposition in solution.

TABLE A-1. SYMMETRICAL DIACYL PEROXIDES, $(RCO_2)_2$

| | | | References | |
|---|---|---|---|---|
| $C_n$ | R | mp[a] | Products | Rate |
| 0 | F | −42.5 (bp, 15.9) (7) | 7 | |
| 1 | $CH_3$ | —(419, 470, 484) | See text (VII B) | See text (VII B) |
| | $CD_3$ | —(299, 516) | 516 | 299, 300 |
| | $CH_3$ (carbonyl $^{14}C$) | —(516) | 516 | |
| | $CH_3$ (carbonyl $^{18}O$) | −(353, 355) | 355 | |
| | $CH_2Cl$ | 85 (528b, 417) | | |
| | $CCl_3$ | —(333, 370, 528b, 567) | | 333 |
| | $CF_3$ | —(370) | | |
| 2 | $MeCH_2$ | Oil (78, 262, 445b) | 486 | 445b, 486 |
| | $ICH_2CH_2$ | 68–73[b] (335) | 335 | 335 |
| | $CF_3CCl_2$ | —(125) | | |
| | $CF_3CF_2$ | —(565) | | |
| 3 | $MeCH_2CH_2$ | —(262, 445b) | 486 | 80, 445b, 486 |
| | $(Me)_2CH$ | Oil (279, 320) | 320, 488 | 80, 320, 488 |
| | | 80–80.5 (296b) | 206 | 186 |
| | | Oil (206) | | |
| | $HO_2CCH_2CH_2$ | 132–133[b] (343) | | 427 |
| | $BrCH_2CHBrCH_2$ | 76–78 (80) | | 80 |
| | $BrCH_2CH_2CH_2$ | Oil (44) | 44 | |
| | $ICH_2CH_2CH_2$ | 30–31 (44) | 44 | |
| | $CF_3CF_2CF_2$ | −15 (565) | | |
| | $(NO_2)_3CCH_2CH_2$ | 114–115 (284b) | 478 | 478 |
| | $FC(NO_2)_2CH_2CH_2$ | —(143) | 478 | 478 |
| 4 | $Me(CH_2)_3$ | Oil (289) | 289 | 297, 323b |
| | $Me_2CHCH_2$ | Oil (54, 289, 296b) | 289 | |
| | $EtCHMe$ | Oil (289, 296b) | 184 | |
| | L-(+)-$EtCHMe$ | Oil (280) | 280 | |
| | $t$-Bu | —(296b) | | |
| | $CH_2:CHCH_2CH_2$ | —(296a) | | 323b |
| | $EtCH:CH$ | Oil (190b) | | 190b |
| | ◇— | Oil (206, 296b) | | 206 |
| | ▷—$CH_2$— | 5 (290b) | | |
| | | —(205, 206) | 205 | 205 |

| $C_n$ | R | mp[a] | Products | Rate |
|---|---|---|---|---|
| | MeO$_2$C(CH$_2$)$_2$ | 62 (172) | | |
| | trans-MeO$_2$CCH:CH | —(172) | | |
| | I(CH$_2$)$_4$ | 47.5–8.3 (561) | 561 | |
| | (NO$_2$)$_2$CH(CH$_2$)$_3$ | 88–89 (284b) | 478 | 478 |
| | (NO$_2$)$_2$CCl(CH$_2$)$_3$ | 85–89 (284a) | | |
| | (NO$_2$)$_2$Br) CH$_2$)$_3$ | 69–71 (284a) | | |
| | CF$_3$CF$_2$CF$_2$CF$_2$ | —(565) | | |
| 5 | Me(CH$_2$)$_4$ | | | 80, 566 |
| | t-BuCH$_2$ | 43.5–44.3 (289) | | |
| | MeCH:CHCH:CH | 113–114 (80) | | 80 |
| | 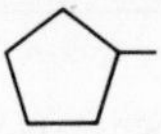 (cyclopentyl) | Oil (206) | | 206 |
| | 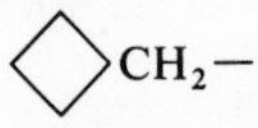 (cyclobutyl)CH$_2$— | Oil (206) | | 206 |
| | I(CH$_2$)$_5$ | 39–40 (561) | 561 | |
| | (NO$_2$)$_2$CH(CH$_2$)$_4$ | 88–89 (284a) | | |
| 6 | Me(CH$_2$)$_5$ | —(124) | 67, 124 | 566 |
| | MeO$_2$C(CH$_2$)$_4$ | 36–37 (172) | | |
| | 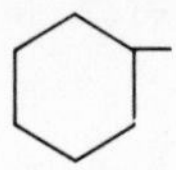 (cyclohexyl) | Oil (206, 319) | 206, 319 | 80, 206, 319 |
| | 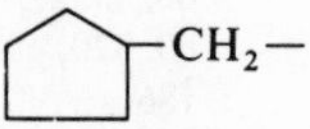 (cyclopentyl)CH$_2$— | Oil (206) | | 206 |
| 7 | Me(CH$_2$)$_6$ | 21.8–22.4 (480a) 23 (80); 29 (172) | | 80, 191, 427 |
| | Me(CH$_2$)$_3$CHEt | Oil (191) | | 191 |
| | MeCH$_2$CH$_2$CH:CEt | Oil (191) | | 191 |
| | CHMe$_2$CH:CEt | Oil (191) | | 191 |
| | 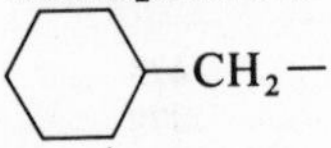 (cyclohexyl)—CH$_2$— | Oil (206) | 206 | 205 |
| | (norbornyl) | 87 (204) 55–56 (204) | 204 204 | 204 204 |
| | (norbornenyl) | 49.5–50.5 (204) 45–47 (204) | 204 204 | 204 204 |
| | 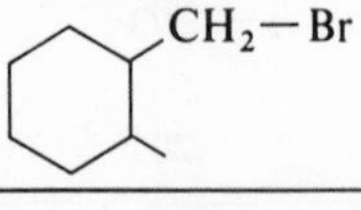 (cyclohexyl)CH$_2$—Br | 65[b] (44) | 44 | |

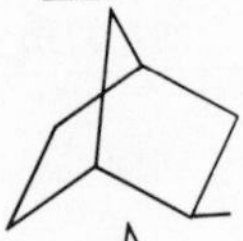

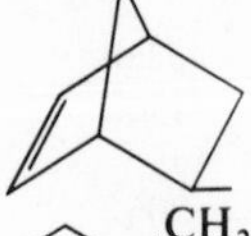

| $C_n$ | R | References | | |
|---|---|---|---|---|
| | | mp[a] | Products | Rate |
| | PhCH$_2$ | 41 (528) | 23 | 23 |
| | | —(23, 225b) | | |
| | 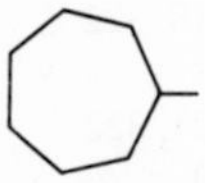 (cycloheptyl) | Oil (206) | | 206 |
| 8 | Me(CH$_2$)$_7$ | $n = _D^{25}$ 1.4410 (67b); | | 191 |
| | | 13.0–13.5 (480a) | 67b | |
| | CMe$_3$CH$_2$CHMeCH$_2$ | Oil, $n_D^{20}$ = 1.4382 (405) | | |
| | Me(CH$_2$)$_5$CH:CH | Oil (190b) | | 191 |
| | Me(CH$_2$)$_4$CH:CHCH$_2$ | Oil (191) | | 191 |
| | MeCH$_2$CH$_2$CMe$_2$CH:CH | Oil (190b) | | 190b |
| | PhCH$_2$CH$_2$ | 38[b] (558) | | 80, 505 |
| | PhCD$_2$CH$_2$ | —(291) | | |
| | PhCH:CH | 133 (557); 134 (496c) | 520, 557 | 80, 505 |
| | $p$-NO$_2$C$_6$H$_4$CH:CH | 140 (80) | | 80 |
| | PhC:C | 94 (376) | 376 | 376 |
| | PhOCH$_2$CH$_2$ | 93–94 (243) | 243 | |
| | Me(CH$_2$)$_5$OCH$_2$CH$_2$ | Oil (540) | | 540 |
| 9 | Me(CH$_2$)$_8$ | 44–45 (172, 382); | | 191 |
| | | 40–41 (480a) | | |
| | Me(CH$_2$)$_3$CHEtCH:CH | Oil (190b) | | 191 |
| | MeO$_2$C(CH$_2$)$_7$ | 41–43 (172) | | |
| | PhCH$_2$CHMe | (123) | 123, 385 | |
| | PhCHMeCH$_2$ | ~0 (447) | 447 | |
| | $p$-MeC$_6$H$_4$CH$_2$CH$_2$ | —(290b) | | |
| | $p$-MeOC$_6$H$_4$CH$_2$CH$_2$ | —(290b) | | |
| | $p$-MeOC$_6$H$_4$CH$_2$CD$_2$ | —(291) | | |
| | $p$-MeOC$_6$H$_4$CH:CH | 93 (80) | | 80 |
| | $trans$-PhCH:CMe | 72–74[b] (483) | 483 | |
| | PhOCH$_2$CH$_2$CH$_2$ | Oil (540) | | 540 |
| | 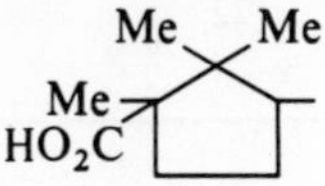 (structure: Me Me / Me / HO$_2$C cyclopentane) | 142[b] (367b) | | |
| | 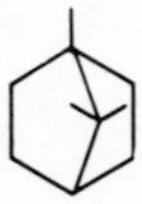 (bicyclic structure) | —(274) | 274 | 22 |
| 0 | MeO$_2$C(CH$_2$)$_8$ | 50–51 (172) | | |
| | $t$-Bu—(cyclohexyl)— ($trans$ and $cis$) | 88–89 (181, 324) | 181, 324 | 324, 548 |
| | | —(324) | 324 | 324 |

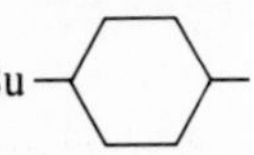

TABLE A-1.   (Continued)

| $C_n$ | R | mp[a] | References Products | References Rate |
|---|---|---|---|---|
| | $Ph(CH_2)_4$ | 33–34 (122a, 292) | 122 | 118, 322 |
| | $PhCMe_2CH_2$ | 40 (447) | 447 | |
| | trans-$PhCH:CHCH_2CH_2$ | 60 (322) | 322 | 321, 322 |
| | $PhCH:CHCH:CH$ | 108–110 (80) | | 80 |
| | trans-$p$-$ClC_6H_4CH:CHCH_2CH_2$ | 93–95 (323b) | | 323b |
| | trans-$p$-$FC_6H_4CH:CHCH_2CH_2$ | 59–60 (323b) | | 323b |
| | $I(CH_2)_{10}$ | 75.5–76.5 (561) | 561 | |
| 11 | $Me(CH_2)_{10}$ | 54.7–55 (480a); 55.8–56.3 (496c) | | 80, 191, 566 |
| | $Ph(CH_2)_5$ | 27–28 (292) | 292 | 292 |
| | $p$-$MeOC_6H_4(CH_2)_4$ | 61–62.5 (323b) | | 323b |
| | trans-$p$-$MeC_6H_4CH:CHCH_2CH_2$ | 93–93.5 (323b) | | 323b |
| | trans-$p$-$MeOC_6H_4CH:CHCH_2CH_2$ | —(323b) | | 323b |
| 12 | $Me(CH_2)_{11}$ | 48.3–48.8 (480a) | | |
| | $Me(CH_2)_9OCH_2CH_2$ | Oil (540) | | 540 |
| 13 | $Me(CH_2)_{12}$ | 59 (80); 63.9–64.4 (480a) | | 80 |
| 14 | $Ph_2CHCH_2$ | 100[b] (447) | 447 | |
| | trans-$PhCH:CPh$ | —(114) | | |
| 15 | $Me(CH_2)_{14}$ | 67–68 (80) 70.5–71.5 (182); 71.4–71.9 (480a) | | 80, 566 |
| 16 | 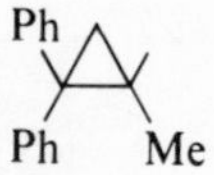 Ph, Ph, Me (cyclopropane structure) [ + (R)] | 129[b] (538) | 538 | |
| 17 | $Me(CH_2)_{16}$ | 76.5–76.9 (480a) | | 566 |
| 20 | $Ph_3CCH_2$ | 133.5[b] (102); 138[b] (447) | 102, 447 | |
| | $(p$-$NO_2C_6H_4)_3CCH_2$ | 165–170[b] (447) | 447 | |
| | triptyl | 294–296 (22) | 22 | 22 |
| 21 | $Ph_3CCH_2CH_2$ | 116[b] (560) | 560 | |

[a] The reference number is shown in parentheses.
[b] Decomposes.

TABLE **A-2.** SYMMETRICAL DIAROYL PEROXIDES

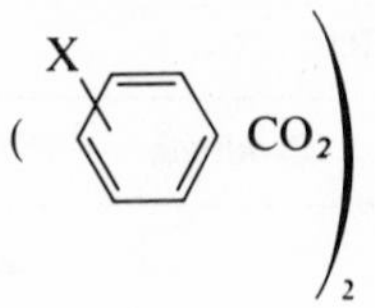

| X | mp[a] | References Products | References Rate |
|---|---|---|---|
| | | References | |
| | | Products | Rate |
| H | 106–107 (161a, 54, 78) | See text (VI F,G) | See text (VI A) |
| *p*-T | (39, 507b) | 507b | |
| H (carbonyl[13]C) | 106–107 (11) | | |
| H (carbonyl[18]O) | —(100, 127) | | |
| *o*-Br | 108 (80) | 55a, 222 | 80 |
| *m*-Br | 131[b] (507a) | 55a | 80, 392b, 507a |
| *p*-Br | 143[b] (182); 144[b] (418); 151–152[b] (507a) | 55a | 507 |
| *o*-Cl | 95 (43) | 43, 80 | |
| *m*-Cl | 122–123[b] (43, 507a) | | 43, 80, 507a |
| *p*-Cl | 137–138[b] (167); 140[b] (43) (496c, 507a) | 55a, 60, 67, 198 | 43, 80, 392b, 507d |
| *o*-F | 66–67 (80) | | 80 |
| *m*-F | 87–88 (80) | | 80 |
| *p*-F | 97–98 (503); 93 (80); 94–95[b] (369) | 369 | 80 |
| *o*-I | —(332) | 332 | 80, 332 |
| *m*-I | 128 (80) | | 80 |
| *p*-I | 157 (80, 415) | 55a, 465 | 80, 465 |
| *o*-NO$_2$ | 145 (43); 147[b] (225b) | 222, 332 | 43, 80, 332 |
| *m*-NO$_2$ | 136–137[b] (418); 137 (43, 507) | 408 | 43 |
| *p*-NO$_2$ | 155–156[b] (418); 157–158[b] (43, 507) | 67, 198, 408 | 43, 392b |
| *o*-N$_3$ | 108.5–109 (333) | 333 | 333 |
| 6-Cl-3-NO$_2$ | 146 (80) | | 80 |
| 3,5-(NO$_2$)$_2$ | 190–192 (284a); 158 (80); 161–162[b] (43) | | 43, 80 |
| 3,4-Cl$_2$ | —(392b) | | 392b |
| 2,3,4,5-(Cl)$_4$ | 154 (80) | | 80 |
| 2,3,4,5,6-(F)$_5$ | 72 (61) | 61 | |
| *o*-Me | 54 (80); 52.5–53.5 (43) | 185 | 43, 80, 185 |
| *m*-Me | 54 (43, 503, 507a) | | 43, 80, 507a |
| *p*-Me | 136[b] (43, 197, 496c, 507a, 555) | 55a, 67, 198 | 43, 80, 392b, 507a |
| *o*-MeO | 80 (80); 85–86 (43) | | 43, 80 |

TABLE A-2. (Continued)

| X | mp[a] | References Products | Rate |
|---|---|---|---|
| | | Products | Rate |
| m-MeO | 82–83 (43, 507a) | | 43, 80, 507a |
| p-MeO | 126–127 (418); 128[b] (43, 503, 507a) | 408 | 43, 392b, 507a |
| o-HO$_2$C | 125[b] (225b) | | |
| m-CN | 163–164[b] (507a) | | 507a |
| p-CN | 165 (80); 176[b] (507a) | | 80, 392b, 507a |
| 3,4-OCH$_2$O | 131 (80) | | 80 |
| p-CH$_2$Cl | 148.5 (194) | | |
| p-CHO | —(194) | | |
| o-CF$_3$ | 70–72 (340) | | |
| m-CF$_3$ | 59.5–61 (340) | | |
| p-CF$_3$ | 125–126[b] (340) | | |
| 3-CF$_3$—5-NO$_2$ | 158–160[b] (340) | | |
| p-Et | 58–59 (80) | | 80 |
| p-CH$_2$:CH | | | 80 |
| o-EtO | | | 80 |
| p-EtO | 131–132 (80) | | 80 |
| o-MeCO$_2$ | —(418) | | |
| p-MeCO$_2$ | 120–122 (80) | | 80 |
| o-MeO$_2$C | —(225b) | | 80 |
| p-MeO$_2$C | 154 (80) | | 80 |
| 3,5-(Me)$_2$ | 118–120 (80) | | 80 |
| 2,3-(MeO)$_2$ | 97 (80) | | 80 |
| 3,4-(MeO)$_2$ | 119–120 (80) | | 80 |
| 3,5-CF$_3$)$_2$ | 98–99.5 (340) | | |
| p-i-Pr | 98 (80) | | 80 |
| p-EtCHMe (dl) | 49–50 (356) | | |
| p-Et-CHMe (l) | 45.5–47 (356) | | |
| p-t-Bu | 142–143 (80) | | 80, 507a |
| o-Ph | 101.5–102.5[b] (185); 105–106 (292) | 185 | 185 |
| p-Ph | 171[b] (557) | 557 | |
| o-(2-D—C$_6$H$_4$) | 107–109 (106, 272) | 106, 272 | |
| o-(3-NO$_2$—C$_6$H$_4$) | 123 (272) | 272 | |
| o-(3-CN—C$_6$H$_4$) | 118–120 (272) | 272 | |
| o-PhO | 66–67 (43, 117) | 117 | 43 |
| m-Ph$_4$N=N | 102–103[b] (267) | | 397 |
| p-Ph$_4$N=N | 145–146 (193); 148–150[b] (267) | | |
| o-(2-Me—6-NO$_2$—C$_6$H$_3$) | 137[b] (116) | 116 | |
| o-PhCH$_2$ | 72.5–73 (185); 74–75 (292) | 185 | 185 |

TABLE A-2. (Continued)

| X | mp[a] | References Products | Rate |
|---|---|---|---|
| ɔ-PhCH$_2$ | 87–88 (185) | 185 | 185 |
| ɔ-PhCH$_2$O | 102 (80) | | 80 |
| ɔ-Me(CH$_2$)$_7$O | 61 (80) | | 80 |
| ɔ-(Me(CH$_2$)$_3$CHEtCH$_2$O$_2$C) | Oil, $n_D^{20}=1.5159$ (383a) | | |
| ɔ-(1-Naphthyl) | 108–115 (115) | 115 | |
| ɔ-(1-menthyl—O$_2$C—) | 117–118 (356) | | 356 |
| ɔ-(2-Biphenylyl) | —(115) | 115 | 115 |
| (phthaloyl peroxide structure) | 126–127[b] (178a) | 178a | 178a |
| (diphenyl diacyl peroxide structure) | 70[c] (425, 426) | | |
| (bis(1-naphthoyl) peroxide structure) | 98.2 (225b, 273) | 95b, 273 | |
| (bis(2-naphthoyl) peroxide structure) | 138 (273); 140 (225b) | 95b, 273 | |

[a] The reference number is shown in parentheses.
[b] Decomposes.
[c] Explodes.

TABLE A-3.  ASYMMETRICAL DIACYL PEROXIDES

$$RCO_2O_2CR'$$

| R | R' | Refractive Index or mp | Reference |
|---|---|---|---|
| Me | Et | $n_D^{20} = 1.4069$ | 403c |
| Me | Pr | $n_D^{20} = 1.4123$ | 403b |
| Me | MeCH:CH | $n_D^{20} = 1.4451$ | 253, 401a |
| Me | HO$_2$CCH$_2$CH$_2$ | 84 | 461a |
| Me | $n$-Bu | $n_D^{20} = 1.4164$ | 403c |
| Me | $i$-Bu | $n_D^{20} = 1.4145$ | 403b |
| Me | Me(CH$_2$)$_4$ | $n_D^{20} = 1.4250$ | 403c |
| Me | ⬡— | 42 | 399b |
| Me | Me$_2$CHCH$_2$CHMeCH$_2$ | — | 253 |
| Et | $n$-Pr | $n_D^{20} = 1.4160$ | 403c |
| Et | MeCH:CH | $n_D^{20} = 1.4438$ | 401a |
| Et | HO$_2$CCH$_2$CH$_2$ | 51 | 461a |
| Et | $n$-Bu | $n_D^{20} = 1.4202$ | 403c |
| Et | ⬡— | Oil | 399b |
| Et | Me$_2$CHCH$_2$CHMeCH$_2$ | — | 253 |
| Pr | MeCH:CH | $n_D^{20} = 1.4462$ | 401a |
| Pr | HO$_2$CCH$_2$CH$_2$ | 71 | 461a |
| Pr | ⬡— | Oil | 399b |
| Bu | Me(CH$_2$)$_5$ | $n_D^{20} = 1.4399$ | 401b |

TABLE A-4. ASYMMETRICAL ACYL AROYL PEROXIDES

$$BzOO_2CR$$

| | References | | |
|---|---|---|---|
| R | mp[a] | Products | Rate |
| $CH_2Cl$ | 33–33.5 (399b, 250, 453) | | 453 |
| Et | $n_D^{20} = 1.5097$ (403b, 453) | | 453 |
| Pr | $n_D^{20} = 1.5040$ (403b) | | |
| MeCH:CH | 38.3 (401a) | | |
| *n*-Bu | $n_D^{20} = 1.4991$ (399e) | | |
| *i*-Bu | 30–31.5 (399b) | | |
| $(NO_2)_3C(CH_2)_3$ | 100–102 (284a) | | |
| $(NO_2)_2CH(CH_2)_4$ | 80–82 (284a) | | |
| | 56[b] (558) | | |
| $PhCH_2$ | 33.5–34[b] (502); 35[b] (556) | 504 | 504 |
| $4-ClC_6H_4CH_2$ | 48–49[b] (502) | | 502 |
| $4-BrC_6H_4CH_2$ | | 184 | |
| $3,4-Cl_2C_6H_3CH_2$ | 65–66[b] (182) | | |
| $PhCH_2CH_2$ | 46[b] (558, 556) | | |
| PhCH:CH | 92–93 (557) | 557 | |
| *t*-Bu—⬡— (*trans*) | 81[b] (182) | | |
| $Me(CH_2)_{10}$ | 29 (453) | | 453 |
| $Me(CH_2)_{14}$ | 46–47 (453, 496c) | | 453 |
| $Me(CH_2)_{16}$ | 78–79 (80) | | 80 |

| | References | | |
|---|---|---|---|
| X | mp[a] | Products | Rate |
| H | 37–39 (379, 453, 544) | 542b, 544 | 453, 544 |
| *o*-Br | 28.5–29.5 (399b) | | 398 |
| *m*-Br | 58.5–59 (399b) | | 398 |
| *p*-Br | 66.5 (399b) | | 398 |
| *o*-Cl | 13–13.5 (399b) | | 398 |
| *m*-Cl | 53–54 (403a,b; 453) | | 453 |
| *p*-Cl | 49.5 (403a,b; 453; 544) | 542b, 544 | 398, 453, 544 |

| X | mp[a] | References Products | Rate |
|---|---|---|---|
| | | | |
| o-F | 9.5 (399f) | | |
| o-OH | 32 (399b) | | |
| p-I | 5.5 (399f) | | |
| m-NO$_2$ | 66.5–67 (399a) | | 398 |
| p-NO$_2$ | 105 (399c); 102–103 (453) | | 398, 453 |
| 2,6-Cl$_2$ | 54.5 (544) | 544 | 544 |
| 4-Br-3-NO$_2$ | 92.6 (399a) | | |
| 4-Cl-3-NO$_2$ | 95–96 (399a) | | |
| 2,4,6-Cl$_3$ | 71 (544) | 544 | 544 |
| o-Me | $n_D^{20} = 1.5126$ (403b) | | 398 |
| m-Me | 32–32.5 (403b) | | |
| p-Me | 65–65.5 (403a,b) | | |
| o-MeO | 25.5–26 (399b) | | 398 |
| p-MeO | 59.5 (403a,b; 453) | | 453 |
| p-CN | 87–88 (453) | | 453 |
| o-(CO$_2$H) | 95 (461a) | | |
| 4-Me-3-NO$_2$ | 75.8–76.8 (399a) | | |
| p-EtO | 72–73 (453) | | 453 |
| o-AcO | 32–32.5 (399b) | | 398 |
| 2,4-Me$_2$ | 24–24.8 (399c) | | |
| i-Pr | $n_D^{20} = 1.5110$ (399f) | | |
| 2,4,6-Me$_3$ | 53.5–54 (399c) | | |
| p-Ph | 107 (399b) | | 398 |

$$R-CO_2O_2C-\!\!\!\left\langle\!\!\!\bigcirc\!\!\!\right\rangle\!\!-X$$

| R | X | mp[a] | References Products | Rate |
|---|---|---|---|---|
| | | | | |
| CH$_2$Br | p-Cl | 36 (67b) | 67b | |
| CH$_2$Cl | p-MeO | 48 (453) | | 453 |
| Et | m-Cl | $n_D^{20} = 1.5170$ (403b) | | |
| Et | p-NO$_2$ | 56 (461a) | | |
| Et | o-(CO$_2$H) | 85 (461a) | | |
| CH$_3$(CH$_2$)$_4$ | o-(CO$_2$H) | — (461a) | | |
| CH$_3$(CH$_2$)$_6$ | m-Cl | Oil (107) | | |

[a] The reference number is shown in parentheses.
[b] Decomposes.

**904**

TABLE A-5. ASYMMETRICAL DIAROYL PEROXIDES

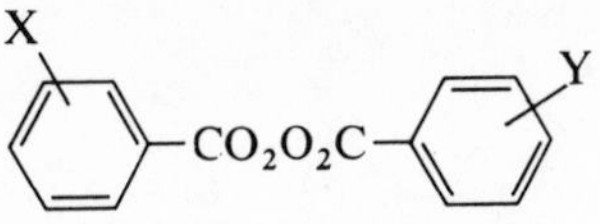

| | | References | | |
|---|---|---|---|---|
| X | Y | mp[a] | Products | Rate |
| H | o-Br | 81 (399e) | | |
| H | p-Br | 93.4–94.5 (182) | | |
| H | o-Cl | 54.5–55 (399d); 49–50 (558) | | |
| H | m-Cl | 88.5–89.5 (399d) | | |
| H | p-Cl | 72.5–73 (399d); 84 (496c, 557) | 557 | |
| H | o-NO$_2$ | —(250) | | |
| H | m-NO$_2$ | 101 (80); 135[b] (399a) | | 80 |
| H | p-NO$_2$ | 113–114 (80, 496c, 557) | 557 | 80 |
| H | o-Me | 34.5–3.5 (496c) | | |
| H | p-Me | 89.5–90.5 (496c) | | |
| H | o-MeO | —(250) | | |
| H | m-MeO | 57–59 (507a) | | 507a |
| H | p-MeO | 68–74 (507a); 74 (557) | 557 | 507a |
| H | m-CN | 101–102[b] (507a) | | 507a |
| H | p-i-Pr | 86.5 (399f) | | |
| H | p-Ph | 138–141 (105b) | | |
| p-Br | m-NO$_2$ | 130–130.5[b] (399a) | | |
| o-I | p-NO$_2$ | 214[b] (227) | 227 | 227 |
| p-Me | m-Br | 90–92 (507a) | | 507a |
| p-Me | p-NO$_2$ | —(105b) | | |
| p-MeO | p-NO$_2$ | 107–108[b] (182, 330) | 330 | 330 |
| p-MeO | p-NO$_2$ (carbonyl $^{18}$O) | 108–108.5 (99) | 99 | |
| p-MeO | 3,5-(NO$_2$)$_2$ | 97[b] (334) | | 334 |
| H | 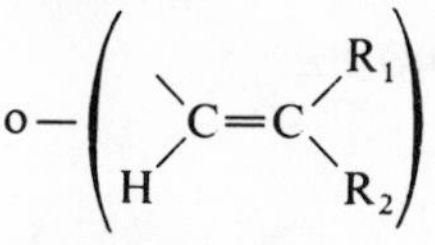 $o-\left(\begin{array}{c} \text{C}=\text{C} \\ \text{H} \quad \text{R}_1 \\ \quad\quad \text{R}_2 \end{array}\right)$ | | | |
| H | R$^1$ = H, R$^2$ = Ph | 58–59 (301) | | 301 |
| H | R$^1$ = H, R$^2$ = m-NO$_2$Ph | 114[b] (301) | | 301 |
| H | R$^1$ = H, R$^2$ = p-NO$_2$Ph | 114[b] (301) | 301 | 301 |
| H | R$^1$ = p-NO$_2$Ph, R$^2$ = p-NO$_2$Ph | 114[b] (301) | | 301 |
| p-NO$_2$ | R$^1$ = H, R$^2$ = Ph | 95[b] (301) | | 301 |
| p-NO$_2$ | R$^1$ = H, R$^2$ = p-NO$_2$Ph | 114[b] (301) | | 301 |
| p-MeO | R$^1$ = H, R$^2$ = p-NO$_2$Ph | 110–112 (301) | | 301 |

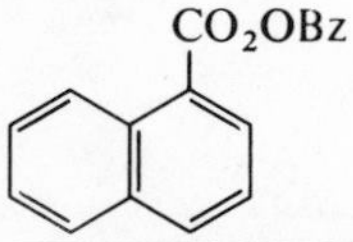

67 (556)

[a] The reference number is shown in parentheses.
[b] Decomposes.

TABLE A-6. HETEROCYCLIC PEROXIDES

| R | | References | |
|---|---|---|---|
| | mp[a] | Products | Rate |

| | $(RCO_2)_2$ | | |
|---|---|---|---|
| | 88–89 (368) | 155 | |
| | 166[b] (80) | | |
| | 86–87[b] (367b) | | |
| | 104[b] (367b) | | |

| | $RC(O)O_2Bz$ | | |
|---|---|---|---|
| | 57.5–58 (65) | 65 | |
| | 92–92.5 (65) | | |
| | 102–102.3 (65) | | |
| | 49.5–50 (63) | | |

TABLE A-6. (Continued)

| | References | | |
|---|---|---|---|
| R | mp[a] | Products | Rate |

| H | 92–93[b] (52c); 100–101 (463) | | 463 |
| 4-Br | 126 (463) | | 463 |
| 5-Br | 130 (463) | | 463 |
| 5-Cl | 110 (463) | | 463 |
| 5-NO₂ | 121–121.5 (348, 463) | 348 | 463 |
| 3-Me | 92–93 (462) | | 462 |
| 4-Me | 113 (463) | | 463 |
| 5-Me | 104 (463) | | 463 |
| 4-Br-5-Et | 98–102 (462) | | 462 |
| 5-*t*-Bu | 93 (463) | | 463 |
| 5-Ph | 81[b] (462) | | 462 |

| H | 121–122[b] (348, 363) | 348 | 363 |
| 5-Br | 102–103 (362) | | 362 |
| 5-Cl | 72–73 (362) | | 362 |
| 5-NO₂ | 153–154 (362) | | 362 |
| 4,5-Br₂ | 158–159[b] (362) | | 362 |
| 2,5-Cl₂ | 92–93 (362) | | 362 |
| 2,5-Me₂ | — (362) | | 362 |

[a] The reference number is shown in parentheses.
[b] Decomposes.

TABLE A-7. PEROXYDICARBONATES
$(ROCO_2)_2$

| R | Refractive Index or mp[a] | Products | Rate |
|---|---|---|---|
| | | References | |
| Me | — (360, 500) | 360 | 439c |
| Et | $n_D^{20} = 1.4017$ (500, 360, 553) | 360 | 500 |
| ClCH$_2$CH$_2$ | $n_D^{20} = 1.4582$ (500, 360) | 360 | |
| MeOCH$_2$CH$_2$ | $n_D^{20} = 1.4250$ (500) | | |
| MeCHCO$_2$Et | $n_D^{20} = 1.4266$ (500) | | |
| n-Pr | — (360, 500) | 79, 360 | 79 |
| i-Pr | $n_D^{20} = 1.4034$ (500, 360) | 79, 360 | 79, 500, 501 |
| CH$_2$:CHCH$_2$ | $n_D^{20} = 1.434$ (500) | | |
| n-Bu | — (360) | 360 | 439c |
| i-Bu | $n_D^{20} = 1.4148$ (500, 360) | 360 | 439c |
| t-Bu | — (439c) | | 439c |
| MeCH$_2$CH(NO$_2$)CH$_2$ | 50 (500) | | |
| Me$_2$C(NO$_2$)CH$_2$ | 100–101 (500) | | 500 |
| t-BuCH$_2$ | 45 (500) | | |
| 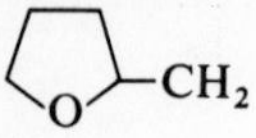 [tetrahydrofuranyl]—CH$_2$ | — (500) | | |
| 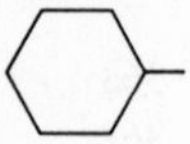 [cyclohexyl] | 46 (500) | 439a, 529 | 439a, 529 |
| 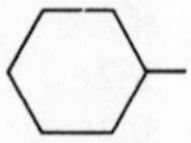 [4-methylcyclohexyl] | — (439b) | | |
| Ph | 82–83[b] (441a) | 441a | 441a, b |
| PhCH$_2$ | 101–102 (500) | 439a | 439a, c; 500 |
| t-Bu—[cyclohexyl] | — (383b) | | |
| EtCMe$_2$—[cyclohexyl] | — (383b) | | |
| 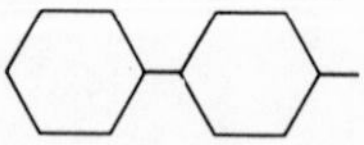 [bicyclohexyl] | — (383b) | | |

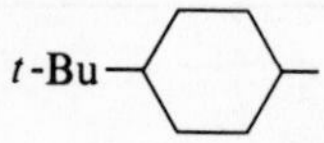

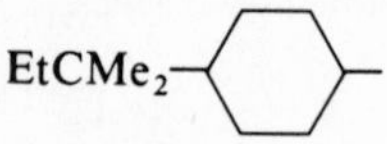

[a] The reference number is shown in parentheses.
[b] Carbonyl $^{14}$C.

TABLE **A-8.** ACYL PEROXYCARBONATES

$$\begin{matrix} O & O \\ \parallel & \parallel \\ RC-O_2COR' \end{matrix}$$

| R | R' | References | | |
|---|---|---|---|---|
| | | mp[a] | Products | Rate |
| Me | Me | — (437) | 437 | |
| Me | ⬡ | — (437) | 437 | |
| Me | Me(CH₂)₁₀ | 18 (406) | | |
| Ph | Me | −3 (429a) | 429c | 429b |
| Ph | ⬡ | 38 (429a) | 429d | 429b |
| Ph | Ph | 50 (429a) | | 429b |

[a] The reference number is shown in parentheses.

TABLE **A-9.** BENZOYL PEROXYSULFONATES

$$\begin{matrix} O \\ \parallel \\ BzO-OS-R \\ \parallel \\ O \end{matrix}$$

| R | References | | |
|---|---|---|---|
| | mp[a] | Products | Rate |
| Me | 54–55 (341) | 432a, b | 432b |
| Et | 46.5–47.5 (341) | 432a, b | 432b |
| *n*-Pr | 24–25 (341) | | 432b |
| *i*-Pr | 49–50 (341) | | 432b |
| PhCH₂ | — (341) | | |

[a] The reference number is shown in parentheses.

TABLE A-10. MULTI- AND POLYACYL PEROXIDES

$$RCO_2O_2C(CH_2)_nCO_2O_2CR$$

| | | References | | |
|---|---|---|---|---|
| R | $n$ | mp[a] | Products | Rate |
| Me | 2 | 119–120 (461b, 218) | | |
| Me | 4 | 61–62 (218) | | |
| Me | 7 | — (218) | | |
| Et | 2 | 56–57 (461b) | | |
| Ph | 2 | 104 (461b) | | |
| $(PhCO_2O_2CCH_2CH_2 +)_2$ | | 93–94 (80) | | 80 |
| CO₂O₂Bz / CO₂O₂Bz (ortho-benzene) | | 119 (80, 461b) | | 80 |

$$\begin{array}{c} CH_2 \\ \parallel \\ RCO_2O_2CCH_2-C-CO_2O_2R \end{array}$$

| R | mp[a] | Rate |
|---|---|---|
| Ph | 54–55 (256) | 256 |
| $n\text{-}C_7H_{15}$ | 9–11 ($n_D^{20} = 1.4463$) (256) | 256 |

$$RCO_2O_2CCH_2CHClCO_2O_2CR$$

| R | mp[a] | Rate |
|---|---|---|
| Ph | 75–76 (256) | 256 |
| $n\text{-}C_7H_{15}$ | 14–15 ($n_D^{20} = 1.4350$) (256) | 256 |

| | | |
|---|---|---|
| Polysuccinyl | (528a) | |
| Polyfumaryl | (528a) | |
| Polysebacoyl | (526a) | 526 |
| Polyterephthaloyl | (528a) | |

[a] The reference number is shown in parentheses.

# REFERENCES

1. E. I. Ablyakimov and K. A. Makarov, *Reakts. Sposobnost. Org. Soedin.*, **4**, 191 (1967); *Chem. Abstr.*, **69**, 76310.
2. G. R. Allen and M. J. Weiss, *J. Org. Chem.*, **27**, 4681 (1962).
3. J. K. Allen and J. C. Bevington, *Trans. Faraday Soc.*, **56**, 1762 (1960).
4. Yu. N. Anisimov, S. S. Ivanchev, and A. I. Yurzhenko, *Zh. Anal. Khim.*, **21**, 113 (1966).
5. V. C. Antonovskii and L. D. Bezborodova, *Zh. Fiz. Khim.*, **42**, 351 (1968).
6. H. R. Appel, U.S. Patent 3,397,245 (1968); *Chem. Abstr.*, **69**, 76954.
7. A. J. Arvia, P. J. Aymonino, and H. J. Schumacher, *Z. Anorg. Allgem. Chem.*, **316**, 327 (1962).
8. T. Asahara, D. Saika, and Y. Nakashima, *Yukagaku*, **14**, 288 (1965); *Chem. Abstr.*, **63**, 17870.
9. D. R. Augood and G. H. Williams, *Chem. Rev.*, **57**, 123 (1957).
10. R. L. Augustine, *J. Org. Chem.*, **28**, 581 (1962).
11. G. Ayrey and C. G. Moore, *J. Chem. Soc.*, **1956**, 1356.
12. (a) A. Baeyer and V. Villiger, *Ber.*, **33**, 1569 (1900); (b) *ibid.*, **33**, 1581 (1900); (c) *ibid.*, **34**, 762 (1901).
13. C. H. Bamford and E. F. T. White, *J. Chem. Soc.*, **1960**, 4490.
14. D. K. Bannerjee and C. C. Budke, *Anal. Chem.*, **36**, 792, 2367 (1964).
15. I. M. Barkalov, A. A. Berlin, V. I. Gol'danskii, and M.-K. Kuo, *Vysokomolekul. Soedin*, **5**, 368 (1963); *Chem. Abstr.*, **58**, 763.
16. I. M. Barkalov, V. I. Gol'danskii, and M.-K. Kuo, *Dokl. Akad. Nauk SSSR*, **151**, 1123 (1963).
17. B. Barnett and W. E. Vaughan, *J. Phys. Chem.*, **51**, 926, 942 (1947).
18. K. E. J. Barrett, *J. Appl. Polymer Sci.*, **11**, 1617 (1967).
19. P. D. Bartlett and A. Altschul, *J. Amer. Chem. Soc.*, **67**, 812 (1945).
20. P. D. Bartlett and S. G. Cohen, *J. Amer. Chem. Soc.*, **65**, 543 (1943).
21. P. D. Bartlett and L. B. Gortler, *J. Amer. Chem. Soc.*, **85**, 1864 (1963).
22. P. D. Bartlett and F. D. Greene, *J. Amer. Chem. Soc.*, **76**, 1088 (1954).
23. P. D. Bartlett and J. E. Leffler, *J. Amer. Chem. Soc.*, **72**, 3030 (1950).
24. P. D. Bartlett and K. Nozaki, *J. Amer. Chem. Soc.*, **69**, 2299 (1947).
25. (a) J. J. Batten and M. F. R. Mulcahy, *J. Chem. Soc.*, **1956**, 2949; (b) *ibid.*, **1956**, 2959.
26. M. A. Battiste and M. E. Brennan, *Chem. Ind. (London)*, **1966**, 1494.
27. C. E. H. Bawn and S. F. Mellish, *Trans. Faraday Soc.*, **47**, 1216 (1951).
28. (a) A. L. J. Beckwith and G. W. Evans, *Proc. Chem. Soc.*, **1962**, 63; (b) A. L. J. Beckwith and R. O. C. Norman, *J. Chem. Soc.*, **1969**, 400.
29. E. J. Behrman and J. O. Edwards, *Progress in Phys. Org. Chem.*, **4**, 93 (1967).
30. (a) N. M. Beileryan, F. O. Karapetyan, and O. A. Chaltykyan, *Dokl. Akad. Nauk Arm. SSR*, **43**, 108 (1966); *Chem. Abstr.*, **66**, 37179; (b) N. M. Beileryan, F. O. Karapetyan, and O. A. Chaltykyan, *Arm. Khim. Zh.*, **19**, 128 (1966); *Chem. Abstr.*, **65**, 3683; (c) *ibid.*, **19**, 828 (1966); *Chem. Abstr.*, **67**, 43136;

(d) N. M. Beileryan, B. M. Sogomonyan, and O. A. Chaltykyan, *Dokl. Akad. Nauk Arm. SSR*, **45**, 13 (1967); *Chem. Abstr.*, **68**, 68224.

31. (a) L. J. Bellamy, *Infra-red Spectra of Complex Molecules*, 2nd ed., Methuen. London (1958), pp. 127–130; (b) L. J. Bellamy, B. R. Connelly, A. R. Philpotts, and R. L. Williams, *Z. Elektrochem.*, **64**, 563 (1960).

32. R. A. Benkeser and W. Schroeder, *J. Amer. Chem. Soc.*, **80**, 3314 (1958).

33. S. W. Benson, in *Organic Peroxides*, Vol. I, edited by D. Swern, Interscience, New York, 1970.

34. (a) W. G. Bentrude and K. R. Darnall, *J. Amer. Chem. Soc.*, **90**, 3588 (1968); (b) W. G. Bentrude and J. C. Martin, *J. Amer. Chem. Soc.*, **84**, 1561 (1962).

35. D. Bernhardt, *Z. Chemie*, **8**, 237 (1968).

36. J. C. Bevington, *Proc. Roy. Soc. (London)*, **A239**, 420 (1957).

37. J. C. Bevington and M. Johnson, *Makromol. Chem.*, **102**, 73 (1967).

38. J. C. Bevington and B. W. Malpass, *J. Polymer Sci.*, **A2**, 1893 (1964).

39. J. C. Bevington and G. Sinatti, *J. Polymer Sci.*, **A4**, 7 (1966).

40. J. C. Bevington and J. Toole, *J. Polymer. Sci.*, **28**, 413 (1958).

41. T. E. Benzmenova and R. A. Dorofeeva, *Khim. Geterotsikl. Soedin.*, **1967**, 1116; *Chem. Abstr.*, **69**, 77037.

42. V. P. Bhatia and K. B. L. Mathur, *Tetrahedron Letters*, **1966**, 4057.

43. A. T. Blomquist and A. J. Buselli, *J. Amer. Chem. Soc.*, **73**, 3883 (1951).

44. D. C. Blomstrom, K. Herbig, and H. E. Simmons, *J. Org. Chem.*, **30**, 959 (1965).

45. J. Boekelheide and D. L. Harrington, *Chem. Ind. (London)*, **1955**, 1423.

46. (a) J. Boëseken and H. Gelissen, *Verslag Akad. Wetenschappen Amsterdam*, **34**, 456 (1925); (b) *Rec. Trav. Chim.*, **43**, 869 (1924).

47. L. S. Boguslavskaya, *Usp. Khim.*, **34**, 1199 (1965).

48. F. Bohlmann and H. Peter, *Ber.*, **99**, 3362 (1966).

49. C. E. Boozer, C. B. Love, J. M. Motes, R. Turner, J. Toney, B. G. Maxey, W. W. Christian, and W. W. Stevens, U.S. Department of Commerce, Office of Technical Services, P. B. Rept. 154949, 1959 15 pp; *Chem. Abstr.*, **59**, 2609.

50. C. Braun, *Organic Syntheses*, collective Vol. I, Wiley, New York, 1932, p. 431.

51. W. Braun, L. Rajenbach and F. R. Eirich, *J. Phys. Chem.*, **66**, 1591 (1962).

52. (a) J. W. Bretenbach and J. Derkosch, *Monatsch.*, **81**, 530 (1950); (b) J. W. Breitenbach and H. Frittum, *J. Polymer Sci.*, **29**, 565 (1958); (c) J. W. Breitenbach and H. Karlinger, *Monatsch.*, **80**, 739 (1949).

53. R. Breslow, J. T. Groves and S. S. Olin, *Tetrahedron Letters*, **1966**, 4717.

54. B. C. Brodie, *J. Chem. Soc.*, **17**, 266 (1864).

55. (a) D. L. Brydon and J. I. G. Cadogan, *Chem. Commun.*, **1966**, 744; (b) *J. Chem. Soc., C*, **1968**, 819.

56. D. Buckley, S. Dunstan, and H. B. Henbest, *J. Chem. Soc., C*, **1957**, 4901.

57. R. P. Buckley and M. Szwarc, *Proc. Roy. Soc. (London)*, **A240**, 396 (1957).

58. O. H. Bullitt, U.S. Patent 2,559,630 (1951).

59. N. J. Bunce and D. D. Tanner, *J. Amer. Chem. Soc.*, **91**, 6096 (1969).

60. J. F. Bunnett and C. C. Wamser, *J. Amer. Chem. Soc.*, **88**, 5534 (1966).

61. J. Burdon, J. G. Campbell, and J. C. Tatlow, *J. Chem. Soc., C*, **1969**, 822.

62. A. J. Burn, J. I. G. Cadogan, and P. J. Bunyan, *J. Chem. Soc.*, **1963**, 1527.

63. M. M. Buzlavova, V. F. Stepanovskaya, and V. L. Antonovskii, *Zh. Anal. Khim.*, **21**, 1491 (1966).

64. G. S. Bylina, A. P. El'nitiskii, and Yu. A. Ol'dekop, *Vysokomol. Soed.*, **8**, 1386 (1966).

65. D. R. Byrne, F. M. Gruen, D. Priddy, and R. D. Schuetz, *J. Heterocyclic Chem.*, **3**, 369 (1966).

66. J. I. G. Cadogan, *Quart. Rev. (London)*, **16**, 208 (1962).

67. (a) J. I. G. Cadogan, D. H. Hey, and P. G. Hibbert, *J. Chem. Soc.*, **1965**, 3939; (b) *ibid.*, **1965**, 3950.

68. C. W. Capp and E. G. E. Hawkins, *J. Chem. Soc.*, **1953**, 4106.

69. T. F. Carruthers, U.S. Patent 1,985,886 (1935); *Chem. Abstr.*, **29**, 1104.

70. (a) W. E. Cass, *J. Amer. Chem. Soc.*, **69**, 500 (1947); (b) *ibid.*, **72**, 4915 (1950).

71. S. Catischa-Ellis and S. C. Abrahams, *Acta Cryst.*, **B24**, 277 (1698).

72. G. R. Chalfout, D. H. Hey, K. S. Y. Liang, and M. J. Perkins, *Chem. Commun.*, **1967**, 367.

73. F. Challenger and V. K. Wilson, *J. Chem. Soc.*, **1927**, 213.

74. O. A. Chaltykyan, N. M. Beileryan, and E. R. Sarukhanyan, *Izv. Akad. Nauk SSSR, Khim. Nauki*, **17**, 21 (1964).

75. E. A. Chandross and F. I. Sonntag, *J. Amer. Chem. Soc.*, **86**, 3179 (1964).

76. I. I. Chkheidze, V. I. Irofimov, and A. T. Kroitskii, *Kinetika i Kataliz*, **8**, 453 (1967).

77. P. A. Claret, J. Coulson, and G. H. Williams, *J. Chem. Soc.*, *C*, **1968**, 341.

78. A. M. Clover and G. F. Richmond, *Amer. Chem. J.*, **29**, 179 (1903).

79. S. G. Cohen and D. B. Sparrow, *J. Amer. Chem. Soc.*, **72**, 611 (1950).

80. W. Cooper, *J. Chem. Soc.*, **1951**, 3106.

81. W. Cooper, *J. Chem. Soc.*, **1952**, 2408.

82. S. L. Cosgrove and W. A. Waters, *J. Chem. Soc.*, **1949**, 3189; *ibid.*, **1951**, 388.

83. J. C. Courtier, *Methodes Phys. Anal.*, **1966**, 23.

84. J. C. Crano, *J. Org. Chem.*, **31**, 3615 (1966).

85. R. Criegee, in *Methoden der organischen Chemie*, Vol. VIII, Part III, edited by D. Müller, Thieme, Stuttgart, 1952, p. 38.

86. F. W. Czech and M. J. McCarthy, *Chemist-Analyst*, **55**, 11 (1966).

87. R. Czerepinski and G. H. Cady, *Inorg. Chem.*, **7**, 169 (1968).

88. R. L. Dannley and M. Gippen, *J. Amer. Chem. Soc.*, **74**, 332 (1952).

89. R. L. Dannley and B. Zaremsky, *J. Amer. Chem. Soc.*, **77**, 1588 (1955).

90. J. D'Aness and H. Gould, *Ber.*, **92**, 2559 (1959).

91. A. G. Davies, *Organic Peroxides*, Butterworths, London, 1961, pp. 63–71.

92. A. G. Davies and C. D. Hall, *J. Chem. Soc.*, **1963**, 1192.

93. D. I. Davies, D. H. Hey, and M. Tiecco, *J. Chem. Soc.*, **1965**, 7062.

94. D. I. Davies, D. H. Hey, and G. H. Williams, *J. Chem. Soc.*, **1958**, 1878.

95. (a) D. I. Davies, D. H. Hey, and H. G. Williams, *J. Chem. Soc.*, **1961**, 562; (b) *ibid.*, **1961**, 3116; (c) D. I. Davies and C. Waring, *J. Chem. Soc.*, *C*, **1967**, 1639.

96. W. H. T. Davison, *J. Chem. Soc.*, **1951**, 2456.

97. E. T. Denisov, *Izv. Akad. Nauk SSSR, Ser. Khim.*, **1963**, 2037.

98. E. T. Denisov, S. S. Ivanchev, L. A. Zborschchik, and V. N. Zolotova, *Izv. Akad. Nauk SSSR, Ser. Khim.*, **1968**, (7), 1500.

99. (a) D. B. Denny, *J. Amer. Chem. Soc.*, **78**, 590 (1956); (b) D. B. Denny and D. G. Denny, *J. Amer. Chem. Soc.*, **79**, 4806 (1957).

100. D. B. Denny and D. Z. Denny, *J. Amer. Chem. Soc.*, **82**, 1389 (1960).

101. D. B. Denny and D. Z. Denny, *J. Amer. Chem. Soc.*, **84**, 2455 (1962).

102. D. B. Denny, R. L. Ellsworth, and D. Z. Denny, *J. Amer. Chem. Soc.*, **86**, 1116 (1964).

103. D. B. Denny and D. Feig, *J. Amer. Chem. Soc.*, **81**, 225 (1959).

104. D. B. Denny and D. Feig, *J. Amer. Chem. Soc.*, **81**, 5322 (1959).

105. (a) D. B. Denny, W. F. Goodyear, and B. Goldstein, *J. Amer. Chem. Soc.*, **82**, 1393 (1960); (b) D. B. Denny and M. A. Greenbaum, *J. Amer. Chem. Soc.*, **79**, 979 (1957).

106. D. B. Denny and P. P. Klemchuck, *J. Amer. Chem. Soc.*, **80**, 3289 (1958).

107. D. B. Denny and N. Sherman, *J. Org. Chem.*, **30**, 3760 (1965).

108. D. B. Denny and T. M. Valega, *J. Org. Chem.*, **29**, 440 (1964).

109. D. Z. Denny, T. M. Valega, and D. B. Denny, *J. Amer. Chem. Soc.*, **86**, 4 (1964).

110. D. B. Denny and H. M. Weiss, *J. Org. Chem.*, **28**, 1415 (1963).

111. N. C. Deno and R. E. Fruit, *J. Amer. Chem. Soc.*, **90**, 3502 (1968).

112. O. C. Dermer and M. T. Edmisson, *Chem. Rev.*, **57**, 77 (1957).

113. D. F. DeTar, *J. Amer. Chem. Soc.*, **89**, 4058 (1967).

114. D. F. DeTar and L. A. Carpino, *J. Amer. Chem. Soc.*, **77**, 6370 (1955).

115. D. F. DeTar and C. C. Chu, *J. Amer. Chem. Soc.*, **82**, 4969 (1960).

116. D. F. DeTar and J. C. Howard, *J. Amer. Chem. Soc.*, **77**, 4393 (1955).

117. D. F. DeTar and A. Klyusky, *J. Amer. Chem. Soc.*, **77**, 4411 (1955).

118. D. F. DeTar and R. C. Lamb, *J. Amer. Chem. Soc.*, **81**, 122 (1959).

119. D. F. DeTar and R. A. J. Long, *J. Amer. Chem. Soc.*, **80**, 4742 (1958).

120. D. F. DeTar, R. A. J. Long, J. Randleman, J. Bradley, and P. Duncan, *J. Amer. Chem. Soc.*, **89**, 4051 (1967).

121. D. F. DeTar and R. Silverstein, *J. Amer. Chem. Soc.*, **88**, 1013 (1966).

122. (a) D. F. DeTar and C. Weis, *J. Amer. Chem. Soc.*, **78**, 4296 (1956); (b) *ibid.*, **79**, 3041 (1957).

123. D. F. DeTar and C. Weis, *J. Amer. Chem. Soc.*, **79**, 3045 (1957).

124. D. F. DeTar and D. V. Wells, *J. Amer. Chem. Soc.*, **82**, 5839 (1960).

125. A. L. Dittman and J. M. Wrightson, U.S. Patent 2,705,706 (1955).

126. V. N. Dmitrieva, O. V. Merkhova, and V. D. Bezuglyi, *Zh. Anal. Khim.*, **19**, 389 (1964).

127. W. E. Doering, K. Okamoto, and H. Krauch, *J. Amer. Chem. Soc.*, **82**, 3579 (1960).

128. D. R. Donner-Maack, and D. H. Lück, *Fette, Seifen, Anstrichmittel*, **70**, 245 (1968).

129. H. J. M. Dou and B. M. Lynch, *Tetrahedron Letters*, **1965**, 897.

130. E. H. Drew and J. C. Martin, *Chem. Ind.* (*London*), **1959**, 925.

131. (a) P. R. Dugan. *Anal. Chem.*, **33**, 696, 1630 (1961); (b) P. R. Dugan and R. D. O'Neill, *Anal. Chem.*, **35**, 414 (1963).

132. L. Dulog and K. H. Berg, *Z. Anal. Chem.*, **203**, 189 (1964).

133. M. Eberhardt and E. J. Eliel, *J. Org. Chem.*, **27**, 2289 (1962).

134. (a) J. T. Edward, *J. Chem. Soc.*, **1954**, 1464; (b) *ibid.*, **1965**, 222.

135. J. T. Edward, H. S. Chang, and S. A. Samad, *Can. J. Chem.*, **40**, 804 (1962).

136. (a) J. T. Edward and S. A. Samad, *Can. J. Chem.*, **41**, 1027 (1963); (b) *ibid.*, **41**, 1638 (1963).

137. J. T. Edward and S. A. Samad, *Pakistan J. Sci. Ind. Res.*, **7**, 200 (1964).

138. F. G. Edwards and F. R. Mayo, *J. Amer. Chem. Soc.*, **72**, 1265 (1950).

139. J. O. Edwards, in *Peroxide Reaction Mechanisms*, edited by J. O. Edwards, Wiley, New York, 1962, p. 67.

140. J. O. Edwards and R. Curci, *Kirk–Othmer Encyclopedia of Chemical Technology* Vol. 14, 2nd ed., Wiley, New York, 1967, pp. 820–839.

141. E. L. Eliel, M. Eberhardt, O. Simamura, and S. Meyerson, *Tetrahedron Letters*, **1962**, 749.

142. E. L. Eliel, S. Meyerson, Z. Welvant, and S. H. Wilen, *J. Amer. Chem. Soc.*, **82**, 2936 (1960).

143. L. T. Eremenko and F. Ya. Natsibullin. *Izv. Akad. Nauk SSSR Ser. Khim.*, **1968**, 686.

144. H. Erlenmeyer and W. Schoenaur, *Helv. Chim. Acta*, **19**, 338 (1936).

145. G. Favini, *Gazz. Chim., Ital.*, **89**, 2121 (1959).

146. (a) J. M. Fayadh, D. W. Jessop, and G. A. Swan, *Proc. Chem. Soc.*, **1964**, 236; (b) *J. Chem. Soc.*, C, **1966**, 1605.

147. F. Feigl, D. Goldstein, and D. Haguenauer-Castro, *Z. Anal. Chem.*, **178**, 419 (1961).

148. E. Ferraris, *Materie Plastiche*, **29**, 163 (1963).

149. L. F. Fieser, M. T. Leffler, et al., *J. Amer. Chem. Soc.*, **70**, 3151 (1948).

150. L. F. Fieser and A. E. Oxford, *J. Amer. Chem. Soc.*, **64**, 2060 (1942).

151. L. F. Fieser and R. B. Turner, *J. Amer. Chem. Soc.*, **69**, 2338 (1947).

152. P. Fijolka and R. Gnauck, *Plaste und Kautschuk*, **13**, 343 (1966).

153. D. H. Fine and P. Gray, *Combust. & Flame*, **11**, 71 (1967).

154. T. H. Fisher and J. C. Martin, *J. Amer. Chem. Soc.*, **88**, 3382 (1966).

155. (a) M. C. Ford and D. MacKay, *J. Chem. Soc.*, **1957**, 4620; (b) *ibid.*, **1958**, 1294.

156. M. C. Ford and W. H. Waters, *J. Chem. Soc.*, **1951**, 824.

157. W. R. Foster and G. H. Williams, *J. Chem. Soc.*, **1962**, 2862.

158. H. N. Friedlander, *J. Polymer Sci.*, **58**, 455 (1962).

159. A. J. Fry, B. M. Tolbert, and M. Calvin, *Trans. Faraday Soc.*, **49**, 1444 (1953).

160. V. I. Galibei, S. S. Ivanchev, and A. I. Yurzhenko, *Vysokomolekul. Soedin.*, **7**, 1746 (1965); *Chem. Abstr.*, **64**, 3689.

161. (a) S. Gambarjan, *Ber.*, **42**, 4003 (1909); (b) *ibid.*, **58B**, 1775 (1925).

162. S. Gambarjan and O. A. Chaltykyan, *Ber.*, **60B**, 390 (1927).

163. M. O. G. Garcia, *Revista de la Facultad de Ciencies, Oviedo*, **3**, 28 (1962).

164. M. O. G. Garcia and J. M. Pertierra, *An. Real. Soc. Espan. Fiz-Quim.*, Ser. B, **63**, 435 (1967).

165. J. F. Garst, D. Walmsley, and W. R. Richards, *J. Org. Chem.*, **27**, 2924 (1962).

166. W. B. Geiger, *J. Org. Chem.*, **23**, 298 (1958).
167. (a) H. Gelissen and P. H. Hermans, *Ber.*, **58**, 285, 476, 764, 984, 2396 (1925); (b) *ibid.*, **59B**, 63 (1926).
168. (a) G. B. Gill and G. H. Williams, *J. Chem. Soc.*, **1965**, 995; (b) *ibid.*, 5756; (c) *ibid.*, 7127; (d) *ibid.*, *B*, **1966**, 880.
169. M. T. Gladstone, *J. Amer. Chem. Soc.*, **76**, 1581 (1954).
170. G. P. Gladyshev, P. E. Meiserle, T. T. Omarov, and S. R. Rafikov, *Dokl. Akad. Nauk SSSR*, **168**, 1093 (1966).
171. R. H. Gobran, M. B. Bernbaum, and A. V. Tobolsky, *J. Polymer Sci.*, **46**, 431 (1960).
172. S. Goldschmidt, H. Spaeth, and L. Beer, *Ann.*, **649**, 1 (1961).
173. M. J. Goldstein, *Tetrahedron Letters*, **1964**, 1601.
174. (a) I. P. Gragarov and M. Ya. Turkina, *Dokl. Akad. Nauk SSSR*, **140**, 1317 (1961); (b) *Zh. Obshch. Khim.*, **33**, 1894 (1963).
175. D. M. Graham and R. B. Mesrobian, *Can. J. Chem.*, **41**, 2938, 2945 (1963).
176. M. A. Greenbaum, D. B. Denny, and A. K. Hoffman, *J. Amer. Chem. Soc.*, **78**, 2563 (1956).
177. F. D. Greene, *J. Amer. Chem. Soc.*, **77**, 4869 (1955).
178. (a) F. D. Greene, *J. Amer. Chem. Soc.*, **78**, 2246 (1956); (b) *ibid.*, 2250; (c) *ibid.*, **81**, 1503 (1959).
179. F. D. Greene and W. Adam, *J. Org. Chem.*, **29**, 136 (1964).
180. F. D. Greene, W. Adam, and J. E. Cantrill, *J. Amer. Chem. Soc.*, **83**, 3461 (1961).
181. F. D. Greene, C. C. Chu, and J. Walia, *J. Org. Chem.*, **29**, 1285 (1964).
182. F. D. Greene and J. Kazan, *J. Org. Chem.*, **28**, 2168 (1963).
183. (a) F. D. Greene and W. W. Rees, *J. Amer. Chem. Soc.*, **80**, 3432 (1958); (b) *ibid.*, **82**, 890 (1960); (c) *ibid.*, **82**, 893 (1960).
184. F. D. Greene, H. P. Stein, C. C. Chu, and M. Vane, *J. Amer. Chem. Soc.*, **86**, 2080 (1964).
185. F. D. Greene, G. R. Van Norman, J. C. Cantrill, and R. D. Gilliam, *J. Org. Chem.*, **25**, 1790 (1960).
186. G. Greig and J. C. J. Thynne, *Trans. Faraday Soc.*, **63**, 2196 (1967).
187. J. Gresser, A. J. Rajenbach and M. Szwarc, *J. Amer. Chem. Soc.*, **83**, 3005 (1961).
188. W. S. Grieve and D. H. Hey, *J. Chem. Soc.*, **1934**, 1797.
189. J. E. Guillet, U.S. Patent 3,232,921 (1966); *Chem. Abstr.*, **64**, 12832.
190. (a) J. E. Guillet, J. P. Hawk, and E. B. Towne, German Patent 1,131,407 (1962); *Chem. Abstr.*, **57**, 8734; (b) French Patent 1,345, 330 (1963); *Chem. Abstr.*, **61**, 5805; (c) J. E. Guillet, Canadian Patent 725,694 (1966).
191. J. E. Guillet, T. R. Walker, M. F. Meyer, J. P. Hawk, and E. B. Towne, *Ind. Eng. Chem. Prod. Res. Develop.*, **3**, 257 (1964).
192. (a) H. C. Haas, *J. Polymer Sci.*, **39**, 493 (1959); (b) *ibid.*, **54**, 287 (1961); (c) *ibid.*, **55**, 33 (1961).
193. H. C. Haas and E. M. Idelson, U.S. Patent 3,271,384 (1966); *Chem. Abstr.*, **65**, 15280.
194. H. C. Haas, N. W. Schuler and H. S. Kolesinski, *J. Polymer Sci.*, Part A-1, **5**, 2964 (1967).

195. M. P. Hahto and J. E. Beaulieu, *J. Chromatog.*, **25**, 472 (1966).

196. C. D. Hall, *Chem. Ind. (London)*, **62**, 384 (1965).

197. J. K. Hambling, D. H. Hey, S. Orman, and G. H. Williams, *J. Chem. Soc.*, **1961**, 3108.

198. J. K. Hambling, D. H. Hey, and G. H. Williams, *J. Chem. Soc.*, **1962**, 487.

199. G. S. Hammond, *J. Amer. Chem. Soc.*, **72**, 3737 (1950).

200. G. S. Hammond, A. Ravve, and J. F. Modic, *Anal. Chem.*, **24**, 1373 (1952).

201. G. S. Hammond, J. T. Rudesill, and F. J. Modic, *J. Amer. Chem. Soc.*, **73**, 3929 (1951).

202. G. S. Hammond and L. M. Soffer, *J. Amer. Chem. Soc.*, **72**, 4711 (1950).

203. G. S. Hammond and H. P. Waits, *J. Amer. Chem. Soc.*, **86**, 1911 (1964).

204. H. Hart and G. J. Chloupek, *J. Amer. Chem. Soc.*, **85**, 1155 (1963).

205. H. Hart and R. A. Cipriani, *J. Amer. Chem. Soc.*, **84**, 3697 (1962).

206. H. Hart and D. P. Wyman, *J. Amer. Chem. Soc.*, **81**, 4891 (1959).

207. P. F. Hartman, H. G. Sellers, and D. Turnbull, *J. Amer. Chem. Soc.*, **69**, 2416 (1947).

208. S. Hasegawa and N. Nishimura, *Bull. Chem. Soc. Japan*, **33**, 775 (1960).

209. S. Hasegawa, N. Nishimura, S. Mitsumoto, and K. Yokoyama, *Bull. Chem. Soc., Japan*, **36**, 522 (1963).

210. E. G. E. Hawkins, *Organic Peroxides*, Van Nostrand, Princeton, N.J., 1961, pp. 300–329.

211. S. Hayano, T. Ota, and Y. Fukushima, *Bunseki Kagaku*, **15**, 365 (1966).

212. H. B. Henbest and R. Patton, *J. Chem. Soc.*, **1960**, 3557.

213. (a) H. B. Henbest, J. A. W. Reid, and C. J. M. Stirling, *J. Chem. Soc.*, **1964**, 1217; (b) *ibid.*, **1964**, 1220.

214. L. Herk and M. Szwarc, *J. Amer. Chem. Soc.*, **82**, 3558 (1960).

215. L. Herk, M. Feld, and M. Szwarc, *J. Amer. Chem. Soc.*, **83**, 2998 (1961).

216. P. H. Hermans, *Rec. Trav. Chim.*, **54**, 760 (1935).

217. P. H. Hermans and J. Van Eyk, *J. Polymer Sci.*, **1**, 407 (1946).

218. L. Heslinga and W. Schwaiger, *Rec. Trav. Chim.*, **85**, 75 (1966).

219. D. H. Hey, *J. Chem. Soc.*, **1934**, 1966.

220. D. H. Hey, K. S. Y. Liang, and M. J. Perkins, *J. Chem. Soc.*, C, **1967**, 2679.

221. D. H. Hey, K. S. Y. Liang, and M. J. Perkins, *Tetrahedron Letters*, **1967**, 1477.

222. D. H. Hey, H. N. Moulden, and G. H. Williams, *J. Chem., Soc.*, **1960**, 3769.

223. (a) D. H. Hey, M. J. Perkins, and H. G. Williams, *Chem. Ind. (London)*, **1963**, 83; (b) *J. Chem. Soc.*, **1963**, 5604; (c) D. H. Hey, D. A. Singleton, and G. H. Williams, *J. Chem. Soc.*, **1963**, 5612; (d) D. H. Hey, F. C. Saunders, and G. H. Williams, *J. Chem. Soc.*, **1964**, 3409; (e) D. H. Hey, M. J. Perkins, and G. H. Williams, *J. Chem. Soc.*, **1964**, 3412.

224. (a) D. H. Hey and J. Peters, *J. Chem. Soc.*, **1960**, 79; (b) *ibid.*, **1960**, 88.

225. (a) D. H. Hey, and R. Tewfik, *J. Chem. Soc.*, **1965**, 2402; (b) D. H. Hey and E. W. Walker, *J. Chem. Soc.*, **1948**, 2213; (c) D. H. Hey and W. A. Waters, *Chem. Rev.*, **21**, 169 (1937).

226. R. Hiatt and S. Szilagyi, *Can. J. Chem.*, **48**, 615 (1970).

227. W. Honsberg and J. E. Leffler, *J. Org. Chem.*, **26**, 733 (1961).

228. L. Horner, *J. Polymer Sci.*, **18**, 438 (1955).

229. L. Horner, *Organo-Phosphorus Compounds*, International Symposium, Heidelberg, 1964 IUPAC, Butterworths, London, 1964.

230. L. Horner and B. Anders, *Ber.*, **95**, 2470 (1962).

231. L. Horner and C. Betzel, *Ann.*, **579**, 175 (1953).

232. L. Horner and M. Brüggemann, *Ann.*, **635**, 27 (1960).

233. L. Horner, H. Brüggemann, and K. H. Knappe, *Ann.*, **626**, 1 (1959).

234. L. Horner and H. Junkermann, *Ann.*, **591**, 53 (1955).

235. L. Horner and E. Jürgens, *Ann.*, **602**, 135 (1957).

236. L. Horner and E. Jürgens, *Angew. Chem.*, **70**, 268 (1958).

237. L. Horner and W. Jurgeleit, *Ann.*, **591**, 138 (1955).

238. (a) L. Horner and W. Kirmse, *Ann.*, **597**, 48 (1958); (b) *ibid.*, **597**, 66 (1958).

239. L. Horner and K. Scherf, *Ann.*, **573**, 35 (1951).

240. L. Horner and E. Schwenk, *Ann.*, **566**, 69 (1950).

241. L. Horner and H. Steppan, *Ann.*, **606**, 47 (1957).

242. L. Horner and H. Winkler, *Tetrahedron Letters*, **1964**, 3275.

243. W. M. Horspool and L. Pauson, *Monatsh Chem.*, **98**, 1256 (1967).

244. W. L. Howard and R. E. Gilbert, *J. Org. Chem.*, **27**, 2685 (1962).

245. F. Hrabak and M. Vacek, *Coll. Czech. Chem. Commun.*, **30**, 573 (1965).

246. H. H. Huang and P. K. K. Lim, *J. Chem. Soc.*, C, **1967**, 2432.

247. R. Huegel, P. L. Moudelini, and G. Sugri, Italian Patent 640,244 (1962); *Chem. Abstr.*, **60**, 2840.

248. R. Huisgen and F. Bayerlein, *Ann.*, **630**, 138 (1959).

249. R. Huisgen, F. Bayerlein, and W. Heydkamp, *Ber.*, **92**, 3223 (1959).

250. R. Huisgen and W. Edl, *Angew. Chem.*, **74**, 588 (1962).

251. R. Huisgen and W. Kolbeck, *Tetrahedron Letters*, **1965**, 783.

252. (a) M. Imoto and S. Choe, *J. Polymer Sci.*, **15**, 485 (1955); (b) M. Imoto, T. Otsu, and K. Kimura, *J. Polymer Sci.*, **15**, 475 (1955); (c) M. Imoto, T. Otsu, and T. Ota, *Makromol. Chem.*, **16**, 10 (1955); (d) M. Imoto and K. Takemoto, *J. Polymer Sci.*, **18**, 377 (1955).

253. Imperial Chemical Industries, Ltd., French Patent 1,337,986 (1963); *Chem. Abstr.*, **60**, 2787.

254. H. Inoue, O. Simamura, and K. Takamizawa, *Bull. Chem. Soc. Japan*, **35**, 1958 (1962).

255. S. S. Ivanchev and S. G. Erigova, *Dokl. Akad. Nauk SSSR*, **183**, 602 (1968).

256. S. S. Ivanchev and A. I. Prisyazhnyuk, *Dokl. Akad. Nauk SSSR*, **179**, 858 (1968).

257. S. S. Ivanchev, L. V. Skubilina, and E. T. Denisov, *Vysokomolekul. Soedin.*, **B9**, 706 (1967); *Chem. Abstr.*, **67**, 117495.

258. S. S. Ivanchev, A. I. Yurzhenko, and Yu. N. Anisimov, *Zh. Fiz. Khim.*, **39**, 1900 (1965).

259. S. S. Ivanchev, A. I. Yurzhenko, and V. I. Galibei, *Dokl. Akad. Nauk SSSR*, **152**, 1159 (1963).

260. S. S. Ivanchev, A. I. Yurzhenko, A. F. Lukovnikov, U. N. Gak, and S. M. Kvaska, *Teor. Eksp. Khim.*, **4**, 780 (1968).

261. S. S. Ivanchev, A. I. Yurzhenko, A. F. Lukovnikov, S. I. Peredereeva, and Yu. V. Gak, *Dokl. Akad. Nauk SSSR*, **171**, 894 (1966).

262. I. Jaffe, E. J. Prosen, and M. Szwarc. *J. Chem. Phys.*, **27**, 416 (1957).
263. H. J. Jakobsen, E. H. Larsen, P. Madsen, and S. O. Lawesson, *Arkiv Kemi*, **24**, 519 (1965).
264. G. A. Jeffrey, R. K. McMullan, and M. Sax, *J. Amer. Chem. Soc.*, **86**, 949 (1964).
265. R. M. Johnson and I. W. Siddiqui, *J. Polarog. Soc.*, **11**, 72 (1965).
266. (a) H. Kaemmerer, F. Rocaboy, and W. Kern, *Makromol. Chem.*, **51**, 220 (1962); (b) *ibid.*, **51**, 224 (1962); (c) H. Kaememerer, F. Rocaboy, K. G. Steinfort, and W. Kern, *Makromol. Chem.*, **53**, 80 (1962).
267. H. Kaemmerer, K. G. Steinfort, and F. Rocaboy, *Makromol Chem.*, **63**, 214 (1963).
268. L. Kaplan, *J. Amer. Chem. Soc.*, **89**, 1753 (1967).
269. V. J. Karnojitsky, *Chim. Ind.*, **98**, 363 (1967).
270. S. Kashino, Y. Mugino, and S. Hasegawa, *Bull. Chem. Soc. Japan.*, **40**, 2004 (1967).
271. S. Kashino, N. Nishimura, K. Ino, and S. Hasegawa, *Bull. Chem. Soc. Japan*, **40**, 33 (1967).
272. G. W. Kenner, M. A. Murray, and C. M. B. Tylor, *Tetrahedron*, **1**, 259 (1957).
273. M. S. Kharasch and R. C. Dannley, *J. Org. Chem.*, **10**, 406 (1945).
274. M. S. Kharasch, F. Engelmann, and W. Urry, *J. Amer. Chem. Soc.*, **65**, 2428 (1943)
275. (a) M. S. Kharasch and A. Fono, *J. Org. Chem.*, **23**, 324 (1958); (b) *ibid.*, **24**, 606 (1959).
276. M. S. Kharasch, N. H. Friedlander, and W. H. Urry, *J. Org. Chem.*, **16**, 531 (1951).
277. M. S. Kharasch and M. T. Gladstone, *J. Amer. Chem. Soc.*, **65**, 15 (1943).
278. M. S. Kharasch, E. V. Jensen, and W. Urry, *J. Org. Chem.*, **10**, 386 (1945).
279. M. S. Kharasch, S. S. Kane, and H. C. Brown, *J. Amer. Chem. Soc.*, **63**, 526 (1941).
280. M. S. Kharasch, J. Kuderna, and W. Nudenberg, *J. Org. Chem.*, **19**, 1283 (1955).
281. (a) M. S. Kharasch, H. C. McBay, and W. H. Urry, *J. Org. Chem.*, **10**, 394 (1945); (b) *ibid.*, **10**, 401 (1945).
282. M. S. Kharasch, H. C. McBay, and W. H. Urry, *J. Amer. Chem. Soc.*, **70**, 1269 (1948).
283. M. S. Kharasch, J. L. Rowe, and W. H. Urry, *J. Org. Chem.*, **16**, 905 (1951).
284. (a) R. I. Kibal'nikova, A. M. Kurdyukov, and A. I. Shreibert, *Mater. Nauch. Konf. Sovnarkhoz, Nixhnevolzhsk, Ekon. Raiona, Volgograd Politekh. Inst.*, No. 2, 97 (1965); *Chem. Abstr.*, **66**, 65243; (b) *ibid.*, 100 (1965); *Chem. Abstr.* **67**, 2699.
285. H. Kleinfeller and K. Rastädter, *Angew. Chem.*, **65**, 543 (1953).
286. E. Knappe and D. Peterli, *Z. Anal. Chem.*, **190**, 386 (1962).
287. M. Kobayashi, H. Minato, and Y. Ogi, *Bull. Chem. Soc. Japan*, **41**, 2822 (1968).
288. J. K. Kochi, *J. Amer. Chem. Soc.*, **84**, 1572 (1962).
289. J. K. Kochi, *J. Amer. Chem. Soc.*, **85**, 1958 (1963).

290. (a) J. K. Kochi and A. Bemis, *Tetrahedron*, **24**, 5099 (1968); (b) *J. Amer. Chem. Soc.*, **90**, 4038 (1968).

291. J. K. Kochi, A. Bemis, and C. L. Jenkins, *J. Amer. Chem. Soc.*, **90**, 4616 (1968).

292. J. K. Kochi and R. D. Gilliom, *J. Amer. Chem. Soc.*, **86**, 5251 (1964).

293. J. K. Kochi, B. M. Graybill, and M. Kurz, *J. Amer. Chem. Soc.*, **86**, 5257 (1964).

294. J. K. Kochi and P. J. Krusic, *J. Amer. Chem. Soc.*, **91**, 3940 (1969).

295. J. K. Kochi, P. J. Krusic, and D. R. Eaton, *J. Amer. Chem. Soc.*, **91**, 1877 (1969).

296. (a) J. K. Kochi and H. E. Mains, *J. Org. Chem.*, **30**, 1862 (1965); (b) J. K. Kochi and P. E. Mocadlo, *J. Org. Chem.*, **30**, 1134 (1965).

297. J. K. Kochi and R. V. Subramanian, *J. Amer. Chem. Soc.*, **87**, 1508 (1965).

298. T. W. Koenig, *J. Amer. Chem. Soc.*, **91**, 2558 (1969).

299. T. W. Koenig and W. D. Brewer, *Tetrahedron Letters*, **1965**, 2773.

300. T. W. Koenig and R. Cruthoff, *J. Amer. Chem. Soc.*, **91**, 2562 (1969).

301. T. W. Koenig and J. C. Martin, *J. Org. Chem.*, **29**, 1520 (1964).

302. T. W. Koenig and R. Wielesek, *J. Amer. Chem. Soc.*, **91**, 2551 (1969).

303. A. Konishi and K. Nambu, *J. Polymer Sci.*, **54**, 209 (1961).

304. E. C. Kooyman and E. Farenhorst, *Rec. Trav. Chim.*, **70**, 867 (1951).

305. P. Kovacic and M. E. Kurz, *J. Amer. Chem. Soc.*, **87**, 4811 (1965).

306. P. Kovacic and M. E. Kurz, *J. Amer. Chem. Soc.*, **88**, 2068 (1966).

307. P. Kovacic and M. E. Kurz, *Chem. Commun.*, **1966**, 321.

308. P. Kovacic and M. E. Kurz, *Tetrahedron Letters*, **1966**, 2689.

309. P. Kovacic and M. E. Kurz, *J. Org. Chem.*, **31**, 2011 (1966).

310. P. Kovacic and S. T. Morneweck, *J. Amer. Chem. Soc.*, **87**, 1566 (1965).

311. H. G. Kuivila and E. R. Jakusik, *J. Org. Chem.*, **26**, 1430 (1961).

312. Z. I. Kulitski, L. M. Terman, V. F. Tsepalov, and V. Ya. Shlyapintokh, *Izv. Akad. Nauk SSSR, Otd. Khim. Nauk*, **1963**, 253.

313. M. E. Kurz and P. Kovacic, *J. Amer. Chem. Soc.*, **89**, 4960 (1967).

314. M. E. Kurz and P. Kovacic, *J. Org. Chem.*, **33**, 1950 (1968).

315. M. E. Kurz, P. Kovacic, A. K. Bose, and I. Kugajevsky, *J. Amer. Chem. Soc.*, **90**, 1818 (1968).

316. E. J. Kuta and F. W. Quackenbush, *Anal. Chem.*, **32**, 1069 (1960).

317. M. G. Kuz'min and I. V. Berezin, *Dokl. Akad. Nauk SSSR*, **148**, 377 (1963).

318. J. Lal and R. Green, *J. Polymer Sci.*, **17**, 403 (1955).

319. R. C. Lamb and J. G. Pacifici, *J. Amer. Chem. Soc.*, **86**, 914 (1964).

320. R. C. Lamb, J. G. Pacifici and P. W. Ayers, *J. Amer. Chem. Soc.*, **87**, 3928 (1965).

321. R. C. Lamb, J. G. Pacifici, and L. P. Spadafino, *J. Org. Chem.*, **30**, 3102 (1965).

322. R. C. Lamb, F. F. Rogers, G. D. Dean and F. W. Voigt, *J. Amer. Chem. Soc.*, **84**, 2635 ( (1962).

323. (a) R. C. Lamb and J. R. Sanderson, *J. Amer. Chem. Soc.*, **91**, 4034 (1969); (b) R. C. Lamb, L. P. Spadafino, R. G. Webb, E. B. Smith, W. E. McNew, and J. G. Pacifini, *J. Org. Chem.*, **31**, 147 (1966).

324. H. H. Lau and H. Hart, *J. Amer. Chem. Soc.*, **81**, 4897 (1959).

325. (a) S. O. Lawesson and T. Busch, *Acta Chem. Scand.*, **13**, 1716 (1959); (b) S. O. Lawesson, T. Busch, and C. Berglund, *Acta Chem. Scand.*, **15**, 260 (1961); (c) S. O. Lawesson and C. Frisell, *Arkiv Kemi*, **17**, 409 (1961).

326. S. O. Lawesson, C. Frisell, D. Z. Denny, and D. B. Denny, *Tetrahedron*, **19**, 1229 (1963).

327. S. O. Lawesson, H. J. Jakobsen, and E. H. Larsen, *Acta Chem. Scand.*, **17**, 1188 (1963).

328. S. O. Lawesson and N. C. Yang, *J. Amer. Chem. Soc.*, **81**, 4230 (1959).

329. J. Lokaj and F. Hrabak, *Makromol. Chem.*, **119**, 23 (1968).

330. J. E. Leffler, *J. Amer. Chem. Soc.*, **72**, 67 (1950).

331. J. E. Leffler, *Chem. Eng. News*, **41**, (48), 45, (1963).

332. J. E. Leffler, R. D. Faulkner, and C. C. Petropoulis, *J. Amer. Chem. Soc.*, **80**, 5435 (1958).

333. J. E. Leffler and H. H. Gibson, *J. Amer. Chem. Soc.*, **90**, 4117 (1968).

334. J. E. Leffler and C. C. Petropoulis, *J. Amer. Chem. Soc.*, **79**, 3068 (1957).

335. J. E. Leffler and J. S. West, *J. Org. Chem.*, **27**, 4191 (1962).

336. M. Levy, M. Steinberg, and M. Szwarc, *J. Amer. Chem. Soc.*, **76**, 5978 (1954).

337. M. Levy and M. Szwarc, *J. Amer. Chem. Soc.*, **76**, 5981 (1954).

338. M. Levy and M. Szwarc, *J. Amer. Chem. Soc.*, **77**, 1949 (155).

339. P. Lewis and G. H. Williams, *J. Chem. Soc.*, B, **1969**, 120.

340. J. Lichtenberger and F. Weiss, *Bull. Chim. Soc. France*, **1962**, 915.

341. V. R. Likhterov, V. S. Etlis, G. A. Razuvaev, and A. V. Gorelik, *Vysokomolekul. Soedin.*, **4**, 357 (1962); *Chem. Abstr.*, **57**, 12691.

342. H. Lind and E. Fahr, *Tetrahedron Letters*, **1966**, 4505.

343. W. Lobunez, J. R. Rittenhouse, and J. S. Miller, *J. Amer. Chem. Soc.*, **80**, 3505 (1958).

344. R. Lombard and G. Schroeder, *Bull. Soc., Chim. France*, **1963**, 2800.

345. C. Luner and M. Szwarc, *J. Chem. Phys.*, **23**, 1978 (1955).

346. B. M. Lynch and R. B. Moore, *Can. J. Chem.*, **40**, 1461 (1962).

347. (a) B. M. Lynch and K. H. Pausacker, *Austral. J. Chem.*, **10**, 40 (1957); (b) *ibid.*, **10**, 164 (1957).

348. D. MacKay, *Can. J. Chem.*, **44**, 2881 (1966).

349. R. D. Mair and A. J. Graupner, *Anal. Chem.*, **36**, 194 (1964).

350. O. L. Mageli and C. S. Sheppard, in *Organic Peroxides*, Vol. I, edited by D. Swern, Interscience, New York, 1970.

351. A. J. Martin, *Organic Analysis*, Vol. IV, Interscience, New York, 1960, pp. 1–64.

352. J. C. Martin and S. Dombchik, *Oxidation of Organic Compounds—I, Advances in Chemistry Series*, No. 75, American Chemical Society, Washington, D.C., 1968, p. 269.

353. (a) J. C. Martin and E. H. Drew, *J. Amer. Chem. Soc.*, **83**, 1232 (1961); (b) J. C. Martin and H. Hargis, *ibid.*, **91**, 5399 (1969).

354. J. C. Martin and T. W. Koenig, *J. Amer. Chem. Soc.*, **86**, 1771 (1964).

355. J. C. Martin, J. W. Taylor, and E. H. Drew, *J. Amer. Chem. Soc.*, **89**, 129 (1967).

356. C. S. Marvel, R. L. Frank, and E. Prill, *J. Amer. Chem. Soc.*, **65**, 1647 (1943).

357.  F. R. Mayo, R. A. Gregg, and M. S. Matheson, *J. Amer. Chem. Soc.*, **73,** 1691 (1951).

358.  (a) F. N. Mazitova and Yu. M. Ryzhmanov, *Dokl. Akad. Nauk SSSR*, **161,** 1346 (1965); (b) F. N. Mazitova, Yu. M. Ryzhmanov, R. R. Shagidullin, and I. A. Lamanova, *Neftekhim.*, **5,** 904 (1965); (c) F. N. Maxitova, Yu. M. Ryshmanov, Yu. V. Yabloka, and O. S. Durova, *Dokl. Akad. Nauk SSSR*, **153,** 354 (1963).

359.  H. C. McBay and S. M. McBay, *Amer. Chem. Soc.*, Div. Paint, Plastics, Printing Ink Chem., *Preprints*, **19,** No. 1, 1 (1959).

360.  H. C. McBay and O. Tucker, *J. Org. Chem.*, **19,** 869 (1954).

361.  H. C. McBay, O. Tucker, and P. T. Groves, *J. Org. Chem.*, **24,** 536 (1959).

362.  J. H. McClure, R. E. Robertson, and A. C. Cuthbertson, *Can. J. Res.*, **20B,** 103 (1942).

363.  C. A. McDowell and L. K. Sharples, *Can. J. Chem.*, **36,** 251 (1958).

364.  T. H. Melzer and A. V. Tobolsky, *J. Amer. Chem. Soc.*, **76,** 5178 (1954).

365.  P. E. Messerle, N. M. Zenit, and G. P. Gladyshev, *Vysokomolekul. Soedin.*, **B9,** 608 (1967); *Chem. Abstr.*, **67,** 100424.

366.  N. A. Milas and I. Belic, *J. Amer. Chem. Soc.*, **81,** 3358 (1959).

367.  (a) N. A. Milas and P. C. Panagiotakos, *J. Amer. Chem. Soc.*, **68,** 534 (1946); (b) N. A. Milas and A. McAlevy, *J. Amer. Chem. Soc.*, **55,** 349 (1933); **56** 1219 (1934).

368.  N. A. Milas and P. C. Panagiotakos, *J. Amer. Chem. Soc.*, **62,** 1878 (1940).

369.  P. Miles and H. Suschitzky, *Tetrahedron*, **19,** 385 (1963).

370.  W. T. Miller, A. L. Dittman, and S. K. Reed, U.S. Patent 2,580,358 (1951).

371.  (a) E. B. Milovskaya and T. G. Zhuravleva, *Vysokomolekul. Soedin.*, **6,** 1035 (1964); *Chem. Abstr.*, **61,** 10777; (b) E. B. Milovskaya, T. G. Zhuravleva, and L. V. Zamoyskaya, *J. Polymer Sci.*, C-2, 899 (1967).

372.  N. Mingin and K. H. Pausacker, *Austral. J. Chem.*, **18,** 821 (1965).

373.  S. L. Mkhitaryan, N. M. Beileryan, and O. A. Chaltykyan, *Izv. Akad. Nauk Arm. SSR, Khim. Nauki*, **16,** 527 (1963); *Chem. Abstr.*, **61,** 724.

374.  (a) R. T. Morrison, J. Cazes, N. Samkoff, and C. A. Howe, *J. Amer. Chem. Soc.*, **84,** 4152 (1962); (b) R. T. Morrison, *Report AFOSR*, U. S. Clearinghouse Fed. Sci. Tech. Inform., AD 652250, 1967; *Chem. Abstr.*, **68,** 58887.

375.  F. G. Moses, *Diss. Abstr.*, **B28,** 845 (1967).

376.  N. Muramoto, T. Ochiai, O. Simamura, and M. Yoshica, *Chem. Commun.*, **1968,** 717.

377.  T. Nakata, K. Tokumaru, and O. Sinamura, *Tetrahedron Letters*, **1967,** 3303.

378.  P. K. Nandi and U. S. Nandi, *J. Phys. Chem.*, **69,** 4071 (1965).

379.  J. U. Nef, *Ann.*, **298,** 274 (1897).

380.  W. P. Neumann and K. Rübsamen, *Ber.*, **100,** 1621 (1967).

381.  W. Neumann, K. Rübsamen, and R. Sommer, *Ber.*, **100,** 1063 (1967).

382.  W. W. Nielson, R. L. Friedman, and W. E. Laurence, U.S. Patent 3,367,951 (1968); *Chem. Abstr.*, **69,** 18621.

383.  (a) M. Noury and van der Lande, Netherlands Patent Application 6,606,159 (1967); *Chem. Abstr.*, **68,** 69705; (b) Netherlands Patent Application 6,700,636 (1967); *Chem. Abstr.*, **68,** 2634.

384. K. Nozaki and P. D. Bartlett, *J. Amer. Chem. Soc.*, **68**, 1686 (1946).

385. S. Oae, T. Kashiwagi and S. Kozuka, *Chem. Ind. (London)*, **1965**, 1694.

386. K. F. O'Driscoll, T. P. Konen, and K. M. Connolly, *J. Polymer Sci.*, A-1, **5**, 1789 (1967).

387. K. F. O'Driscoll, P. F. Lyons, and R. Patsiga, *J. Polymer Sci.*, A, **3**, 1567 (1965).

388. K. F. O'Driscoll and S. McArdle, *J. Polymer Sci.*, **40**, 557 (1959).

389. K. F. O'Driscoll and E. N. Richezza, *J. Polymer Sci.*, **46**, 211 (1960).

390. K. F. O'Driscoll and E. N. Richezza, *Makromol. Chem.*, **47**, 15 (1961).

391. K. F. O'Driscoll and J. F. Schmidt, *J. Polymer Sci.*, **45**, 189 (1960).

392. (a) K. F. O'Driscoll and P. J. White, *J. Polymer Sci.*, B, **1**, 597 (1963); (b) *ibid.*, A **3**, 283 (1965).

393. Y. Ogata, Y. Furuya, and K. Aoki, *Bull. Chem. Soc. Japan*, **38**, 838 (1965).

394. Y. Ogata, Y. Furuya, and J. Maekawa, K. Okano, *J. Amer. Chem. Soc.*, **85**, 961 (1963).

395. K. Ohkita, H. Kagahara, M. Iida, Y. Watanabe, and S. Suzuki, *Nippon Gomu Kyokaishi*, **35**, 22 (1962); *Chem. Abstr.*, **58**, 4454.

396. Y. Okoda, *Kogyo Kagaku Zasshi*, **65**, 1085 (1962); *Chem. Abstr.*, **58**, 6664.

397. (a) O. F. Olaj, J. W. Breitenbach, and I. Hofreiter, *Makromol. Chem.*, **91**, 264 (1966); (b) *Monatsh.*, **98**, 997 (1967).

398. Yu. A. Ol'dekop and G. S. Bylina, *Dokl. Akad. Nauk Belorusk. SSR*, **8**, 316 (1964); *Chem. Abstr.*, **61**, 10778.

399. (a) Yu. A. Ol'dekop and A. P. Elnitskii, *Zh. Organ. Khim.*, **1**, 876 (1965); (b) Yu. A. Ol'dekop, G. S. Bylina, L. K. Grakovich, Zh. I. Buloichik, and Zh. D. Teif, *Zh. Organ. Khim.*, **1**, 82 (1965); (c) Yu. A. Ol'dekop and A. P. El'nitskii, *Zh. Organ. Khim.*, **2**, 1257 (1966); (d) Yu. A. Ol'dekop, G. S. Bylina, I. K. Burykina, and G. S. Kislyak, *Zh. Organ. Khim*, **2**, 2175 (1966); (e) Yu. A. Ol'dekop, G. S. Bylina, and S. F. Petrashkevich, *Akad. Nauk SSSR, Otd. Obshch. Tekh. Khim.*, **1967**, 152; *Chem. Abstr.*, **68**, 39292; (f) Yu. A. Ol'dekop, G. S. Bylina, and L. K. Burykina, *Izv. Akad. Nauk Belorusk. SSR, Ser. Khim. Nauk*, **1968**, 118; *Chem. Abstr.*, **70**, 96350.

400. Yu. A. Ol'dekop and A. P. El'nitskii, *Zh. Obshch. Khim.*, **34**, 3478 (1964).

401. (a) Yu. A. Ol'dekop, A. P. El'nitskii, S. F. Petrashkevich, and A. A. Karaban, *Izv. Akad. Nauk Belorusk. SSR, Khim. Nauk*, **1968**, 80; *Chem. Abstr.*, **70**, 11057; (b) Yu. A. Ol'dekop, A. P. El'nitskii, and S. I. Budai, *ibid.*, **1968**, 117; *Chem. Abstr.*, **70**, 11026.

402. (a) Yu. A. Ol'dekop and N. A. Maier, *Dokl. Akad. Nauk Belorusk. SSR*, **4**, 288 (1960); *Chem. Abstr.*, **55**, 10377; (b) *Izv. Akad. Nauk Belorusk SSR, Ser. Fiz. Tekh. Nauk*, **1960**, No. 2, 37; *Chem. Abstr.*, **57**, 16385. (c) Yu A. Ol'dekop, N. A. Maier, and V. N. Pshenichnyi. *Zh. Obshch. Khim.*, **34**, 317 (1964); (d) *ibid.*, **35**, 9047 (1965); (e) *ibid.*, **36**, 1408 (1966); (f) *ibid.*, **38**, 1441 (1968).

403. (a) Yu. A. Ol'dekop, A. N. Sevchenko, I. P. Zvat'kov, G. S. Bylina, and A. P. El'nitskii, *Dokl. Akad. Nauk SSSR*, **128**, 1201 (1959); (b) *Zh. Obshch. Khim.*, **31**, 2904 (1961); (c) Yu. A. Ol'dekop A. N. Sevchenko, I. P. Zyat'kov, and A. P. El'nitskii, *Zh. Obshch. Khim.*, **33**, 2771 (1963); (d) Yu. A. Ol'dekop and R. F. Sokolova, *Zh. Obsch. Khim.*, **23**, 1159 (1953).

404. J. Overhoff and G. Tilman, *Rec. Trav. Chim.*, **48**, 993 (1929).

405. A. Pajaczkowski, British Patent 849,028 (1960); *Chem. Abstr.*, **55**, 6928.

406. A. Pajaczkowski and J. M. Turner, U.S. Patent 3,108,093 (1963); *Chem. Abstr.*, **56**, 4952.

407. V. Patek, *Chem. Promysl*, **14**, 375 (1964).

408. K. H. Pausacker, *Austral. J. Chem.*, **10**, 49 (1957).

409. K. H. Pausacker, *Austral. J. Chem.*, **11**, 200 (1958).

410. H. von Pechmann and L. Vanino, *Ber.*, **27**, 1510 (1894).

411. C. A. Peri, B. Domenicali, and P. L. Mondelini, German Patent 1,192,181 (1965); *Chem. Abstr.*, **63**, 8205.

412. A. Perret and A. Perrot, *Helv. Chim. Acta*, **28**, 558 (1945).

413. L. N. Petrova, E. N. Novikova, and A. B. Skvortsova, *Zh. Anal. Khim.*, **21**, 494 (1966).

414. K. S. Pitzer, *J. Amer. Chem. Soc.*, **70**, 2140 (1948).

415. S. Polowinski and W. Reimschussel, *Zeszyty Nauk Politech. Lodz. Chem.*, No. 14, 87 (1964); *Chem. Abstr.*, **63**, 11778.

416. E. M. Popov, A. Kh. Khomenko, and P. P. Shorygin, *Izv. Akad. Nauk SSSR, Ser. Khim.*, **1965**, 51.

417. C. C. Price, R. W. Kell, and E. Krebs, *J. Amer. Chem. Soc.*, **64**, 1103 (1942).

418. C. C. Price and E. Krebs, *Organic Syntheses*, Collective Vol. III, Wiley, New York, 1955, p. 649.

419. C. C. Price and H. Moritz, *J. Amer. Chem. Soc.*, **75**, 3686 (1953).

420. W. A. Pryor and T. L. Pickering, *J. Amer. Chem. Soc.*, **84**, 2705 (1962).

421. W. Pryor and K. Smith, *J. Amer. Chem. Soc.*, **89**, 1741 (1967).

422. T. C. Purcell and I. R. Cohen, *Environmental Science and Technology*, **1**, 431 (1967).

423. R. Rado, *Chem. Listy*, **61**, 785 (1967).

424. (a) R. Rado and M. Lazar, *J. Polymer Sci.*, **45**, 257 (1960); (b) *ibid.*, **62**, 167 (1962).

425. F. Ramirez, S. B. Bhatia, R. B. Mitra, Z. Hamlet, and N. B. Desai, *J. Amer. Chem. Soc.*, **86**, 4394 (1964).

426. F. Ramirez, N. B. Desai, and R. B. Mitra, *J. Amer. Chem. Soc.*, **83**, 492 (1961).

427. N. H. Ray, *J. Chem. Soc.*, **1960**, 4023.

428. G. A. Razuvaev, *Tetrahedron Supplement No. 3*, 224 (1959).

429. (a) G. A. Razuvaev, V. A. Dodonov, and V. S. Etlis, *Izv. Akad. Nauk SSSR, Ser. Khim.*, **1964**, 426; (b) G. A. Razuvaev, V. A. Dodonov, and B. N. Moryganov, *Izv. Akad. Nauk SSSR, Ser. Khim.*, **1964**, 430; (c) G. A. Razuvaev, V. A. Dodonov, and T. I. Starostina, *Zh. Organ. Khim.*, **2**, 857 (1966); (d) G. A. Razuvaev, V. A. Dodonov, B. N. Moryganov, and T. I. Starostina, *Zh. Organ. Khim.*, **3**, 1602 (1967).

430. (a) G. A. Razuvaev, N. A. Kartashova, and L. S. Boguslavskaya, *Zh. Obshch. Khim.*, **34**, 2093 (1964); (b) *Zh. Org. Khim.*, **2**, 1372 (1966).

431. G. A. Razuvaev and V. N. Latyeava, *Zh. Obshch. Khim.*, **28**, 2233 (1958).

432. (a) G. A. Razuvaev, V. R. Likhterov, and V. S. Etlis, *Tetrahedron Letters*, **1961**, 527; (b) *Zh. Obsch. Khim.*, **32**, 2033 (1962).

433. (a) G. A. Razuvaev, E. V. Mitrofanova, and G. G. Petukhov, *Zh. Obshch. Khim.*, **31**, 2340 (1961); (b) G. A. Razuvaev, E. V. Mitrofanova, and N. S. Vyazankin, *Dokl. Akad. Nauk SSSR*, **144**, 132 (1962).

434. G. A. Razuvaev, B. N. Moryganov, E. P. Dlin, and Yu. A. Ol'dekop, *Zh. Obshch. Khim.*, **24**, 262 (1954).

435. G. A. Razuvaev, Yu. A. Ol'dekop, and L. N. Grobov, *Dokl. Akad. Nauk SSSR*, **88**, 77 (1953).

436. G. A. Razuvaev, G. G. Petukhov, and V. A. Dodonov, *Tr. po Khim. i Khim. Tekhnol.*, **3**, 193 (1960).

437. G. A. Razuvaev, T. I. Starostina, V. A. Dodonov, and A. A. Golubev, *Zh. Organ. Khim.*, **4**, 1030 (1968).

438. (a) G. A. Razuvaev, L. P. Stepovik, and E. V. Mitrofanova, *Izv. Akad. Nauk SSSR, Ser. Khim.*, **1964**, 162; (b) *Zh. Obshch. Khim.*, **35**, 1095 (1965); (c) G. A. Razuvaev and L. P. Stepovik, *Zh. Obshch. Khim.*, **35**, 1672, (1965); (d) G. A. Razuvaev, L. P. Stepovik, V. A. Dodonov, and G. V. Nesterov, *Zh. Obshch. Khim.*, **39**, 123 (1969).

439. (a) G. A. Razuvaev and L. M. Terman, *Zh. Obshch. Khim.*, **30**, 2387 (1960); (b) G. A. Razuvaev, L. M. Terman, and D. M. Yanovskii, *Dokl. Akad. Nauk SSSR*, **161**, 614 (1965); (c) *Vysokomolekul. Soedin.*, **B9**, 208 (1967).

440. G. A. Razuvaev, L. M. Terman, L. N. Mikhotova, and D. M. Yanovskii, *Zh. Organ. Khim.*, **1**, 79 (1965).

441. (a) G. A. Razuvaev, L. M. Terman, and D. M. Yanovskii, *Zh. Organ. Khim.*, **1**, 274 (1965); (b) G. A. Razuvaev, L. M. Terman, and K. F. Bol'shakova, *Vysokomolekul. Soedin.*, **B10**, 38 (1968); *Chem. Abstr.*, **68**, 69359.

442. (a) G. A. Razuvaev and N. S. Vyazankin, *Khim. Perekisnykh Soedin., Akad. Nauk SSSR, Inst. Obshch. i Neorgan. Khim.*, **1963**, 283; *Chem. Abstr.*, **60**, 15708; (b) G. A. Razuvaev, N. S. Vyzankin, and O. S. D'yachkovskaya, *Zh. Obshch. Khim.*, **32**, 2161 (1962); (c) G. A. Razuvaev, N. S. Vyazankin, and E. V. Mitrofanova, *Zh. Obshch. Khim.*, **34**, 675 (1964); (d) G. A. Razuvaev, N. S. Vyazankin, and O. A. Shchepetkova, *Tetrahedron*, **18**, 667 (1962).

443. (a) G. A. Razuvaev, B. G. Zateev, and G. G. Petukhov, *Sbornik Nauch. Rabot. Inst. Fiz.-Org. Khim., Akad. Nauk Belorusk. SSR*, **1960**, 41; *Chem. Abstr.*, **56**, 3375; (b) G. A. Razuveav and B. G. Zateev, *Dokl. Akad. Nauk SSSR*, **148**, 863 (1963); (c) *Zh. Obshch. Khim.*, **33**, 673 (1963).

444. G. A. Razuvaev, S. F. Zhil'tsov, O. N. Druzhkov, and G. G. Petukov, *Zh. Obshch. Khim.*, **36**, 258 (1966).

445. (a) A. Rembaum and M. Szwarc, *J. Amer. Chem. Soc.*, **76**, 5975 (1954); (b) *J. Chem. Phys.*, **23**, 909 (1955); (c) *J. Amer. Chem. Soc.*, **77**, 3486 (1955).

446. A. F. A. Reynhart, *Rec. Trav. Chim.*, **46**, 54, 62 (1927).

447. W. Rickatson and T. S. Stevens, *J. Chem. Soc.*, **1963**, 3960.

448. A. Rieche and M. Schulz, *Angew. Chem.*, **70**, 694 (1958).

449. I. M. Roitt and W. A. Waters, *J. Chem. Soc.*, **1952**, 2695.

450. M. F. Romantsev and E. S. Levin, *Zh. Anal. Khim.*, **18**, 1109 (1963).

451. S. D. Ross and M. A. Fineman, *J. Amer. Chem. Soc.*, **73**, 2176 (1951).

452. (a) R. B. Roy and G. A. Swan, *Chem. Commun.*, **1966**, 427; (b) *J. Chem. Soc., C*, **1968**, 80; (c) *ibid.*, **1968**, 83.

453. K. Rübsamen, W. P. Neumann, R. Sommer, and U. Frommer, *Ber.*, **102**, 1290 (1969).

454. H. A. Rudolph and R. L. McEwen, U.S. Patent 2,458,207 (1949); *Chem. Abstr.*, **43**, 3444.

455.  G. A. Russell, *J. Amer. Chem. Soc.*, **78**, 1044 (1956).

456.  K. E. Russell, *J. Amer. Chem. Soc.*, **77**, 4814 (1955).

457.  F. F. Rust, F. H. Seubold, and W. E. Vaughan, *J. Amer. Chem. Soc.*, **70**, 3256 (1948).

458.  E. R. Sarukhanyan, N. M. Beileryan, and O. A. Chaltykyan, *Dokl. Akad. Nauk Arm. SSR*, **38**, 285 (1964); *Chem. Abstr.* **61**, 13158.

459.  T. Sato, K. Takemoto, and M. Imoto, *Makromol. Chem.*, **104**, 297 (1967).

460.  M. Sax and R. K. McMullan, *Acta Cryst.*, **22**, 281 (1967).

461.  (a) G. Schroeder and R. Lombard, *Bull. Soc. Chim. France*, **1964**, 542; (b) *ibid.*, **1964**, 1603.

462.  R. D. Schuetz, F. M. Gruen, D. R. Byrne, and R. L. Brennan, *J. Heterocyclic Chem.*, **3**, 184 (1966).

463.  R. D. Shuetz and D. M. Teller, *J. Org. Chem.*, **27**, 410 (1962).

464.  (a) M. Schulz and K. H. Schwarz, *Z. Chemie*, **7**, 176 (1967); (b) *Monatsber. Deut. Akad. Wiss., Berlin*, **6**, 515 (1964).

465.  M. M. Schwartz and J. E. Leffler, *J. Amer. Chem. Soc.*, **90**, 1368 (1968).

466.  (a) K. Schwetlick, *Tetrahedron*, **22**, 785 (1966); (b) K. Schwetlick and S. Helm, *Tetrahedron*, **22**, 793 (1966).

467.  A. N. Sevchenko and I. P. Zyat'kov, *Dokl. Akad. Nauk Belorusk. SSR*, **6**, 697 (1962); *Chem. Abstr.*, **58**, 6344.

468.  H. A. Shah, F. Leonard, and A. V. Tobolsky, *J. Polymer Sci.*, **7**, 537 (1951).

469.  W. S. Shakespeare, *Macbeth*, Act I, Scene 4.

470.  E. S. Shanley, *J. Amer. Chem. Soc.*, **72**, 1419 (1950).

471.  C. Shih, D. H. Hey, and G. H. Williams, *J. Chem. Soc.*, **1958**, 1885.

472.  H. J. Shine and D. M. Hoffman, *J. Amer. Chem. Soc.*, **83**, 2782 (1961).

473.  H. J. Shine and J. R. Slagle, *J. Amer. Chem. Soc.*, **81**, 6309 (1959).

474.  H. J. Shine, R. E. Spillet, and R. D'Hollander, *Annual Report, The Robert A. Welch Foundation*, 1964–1965 p. 59.

475.  H. J. Shine, J. A. Waters, and D. M. Hoffman, *J. Amer. Chem. Soc.*, **85**, 3613 (1963).

476.  T. Shono, M. Kimura, Y. Ito, K. Nishida, and R. Oda, *Bull. Chem. Soc. Japan*. **37**, 635 (1964).

477.  (a) A. I. Shreibert, N. V. Elsakov, A. P. Khardin, and V. I. Ermarchenko, *Zh. Organ. Khim.*, **3**, 1755 (1967); (b) A. I. Shriebert, A. P. Khardin, A. M. Kurdyukov, E. N. Elsakov, V. I. Ermarchenko, R. I. Kibal'nikova, and E. G. Tikhonova, *Zh. Organ. Khim.*, **4**, 1699 (1968).

478.  A. I. Shriebert, A. P. Khardin, and S. Yu. Sizov, *Zh. Organ. Khim.*, **4**, 182 (1968).

479.  E. Siggia, *Quantitative Organic Analysis via Functional Groups*, Wiley, New York, 1963, pp. 255–295.

480.  (a) L. S. Silbert and D. Swern, *J. Amer. Chem. Soc.*, **81**, 2364 (1959); (b) *J. Org. Chem.*, **27**, 1336 (1962).

481.  L. S. Silbert and D. Swern, *Anal. Chem.*, **30**, 385 (1958).

482.  L. S. Silbert, D. Swern, and T. Asahara, *J. Org. Chem.*, **33**, 3670 (1968).

483.  L. A. Singer and N. P. Kong, *J. Amer. Chem. Soc.*, **88**, 5213 (1966).

484.  J. R. Slagle and H. J. Shine, *J. Org. Chem.*, **24**, 107 (1959).

485. G. Smets and W. van Rillaer, *J. Polymer Sci.*, A, **2**, 2417 (1964).
486. J. Smid, A. Rembaum, and M. Szwarc, *J. Amer. Chem. Soc.*, **78**, 3315 (1956).
487. J. Smid and M. Szwarc, *J. Amer. Chem. Soc.*, **78**, 3322 (1956).
488. J. Smid and M. Szwarc, *J. Chem. Phys.*, **29**, 432 (1958).
489. (a) W. Smith, *Tetrahedron*, **25**, 2071 (1969); (b) W. Smith and B. W. Rossiter, *Tetrahedron*, **25**, 2059 (1969).
490. (a) B. M. Sogomonyan, N. M. Beileryan, and O. A. Chaltykyan, *Dokl. Akad. Nauk Arm. SSR*, **34**, 201 (1962); *Chem. Abstr.*, **58**, 52; (b) *Armyansk. Khim. Zh.*, **19**, 123 (1966); *Chem. Abstr.*, **65**, 3716; (c) *ibid.*, **19**, 391 (1966); *Chem. Abstr.*, **66**, 2017.
491. B. M. Sogomonyan and O. A. Chaltykyan, *Khim. Perekisnykh Soedin.*, *Akad. Nauk SSSR, Inst. Obshch. i Neorgan. Khim.*, **1963**, 270; *Chem. Abstr.*, **61**, 1725.
492. K. Someno and O. Kikuchi, *Kogyo Kagaku Zasshi*, **68**, 1527 (1965); *Chem. Abstr.*, **63**, 16180.
493. G. Sorge and K. Überreiter, *Angew. Chem.*, **68**, 486 (1956).
494. G. Sosnovsky and D. J. Rawlinson, *Chem. Ind.* (*London*), **1967**, 120.
495. G. Sosnovsky and D. J. Rawlinson, *J. Org. Chem.*, **33**, 2325 (1968).
496. (a) H. A. Staab, *Angew. Chem., Intern. Ed.*, **1**, 351 (1962); (b) H. A. Staab, F. Graf, and W. Rohr, *Ber.*, **98**, 1128 (1965); (c) *ibid.*, **98**, 1122 (1964).
497. S. D. Stavrova, G. V. Peregudov, and M. F. Margaritova, *Dokl. Akad. Nauk SSSR*, **157**, 636 (1964).
498. R. A. Stein and V. Slawson, *Anal. Chem.*, **35**, 1008 (1963).
499. (a) F. Strain, U.S. Patent 2,370,588 (1945); *Chem. Abstr.*, **39**, 4891; (b) U.S. Patent 2,464,062 (1949); *Chem. Abstr.*, **43**, 5641.
500. F. Strain, W. E. Bissinger, W. R. Dial, H. Rudolph, B. J. DeWitt, H. C. Stevens, and J. H. Langston, *J. Amer. Chem. Soc.*, **72**, 1254 (1950).
501. W. A. Strong, *Ind. Eng. Chem.*, **56**, 33 (1964).
502. T. Suehiro, S. Hibino, and T. Saito, *Bull. Chem. Soc. Japan*, **41**, 1707 (1968).
503. T. Suehiro, A. Kanoya, H. Hara, T. Nakahama, M. Omori, and T. Komori, *Bull. Chem. Soc. Japan*, **40**, 668 (1967).
504. T. Suehiro, H. Tsuruta, and S. Hibino, *Bull. Chem. Soc. Japan*, **40**, 674 (1967).
505. I. V. Sukmanskaya and A. I. Yurzhenko, *Zh. Obshch. Khim.*, **30**, 2108 (1960).
506. E. A. Sullivan and A. A. Hinckley, *J. Org. Chem.*, **27**, 3731 (1962).
507. (a) C. G. Swain, W. T. Stockmayer, and J. T. Clarke, *J. Amer. Chem. Soc.*, **72**, 5426 (1950); (b) C. G. Swain, L. J. Schaad, and A. J. Kresge, *J. Amer. Chem. Soc.*, **80**, 5313 (1958).
508. G. A. Swan, *Chem. Commun.*, **1969**, 20A.
509. (a) D. Swern and L. S. Silbert, *Anal. Chem.*, **35**, 880 (1963); (b) L. S. Silbert, L. P. Witnauer, D. Swern, and C. Ricciuti, *J. Amer. Chem. Soc.*, **81**, 3244 (1959).
510. M. Szwarc, Reference 139, pp. 153–174.
511. M. Szwarc and L. Herk, *J. Chem. Phys.*, **29**, 438 (1958).
512. H. A. Szymanski, *Prog. Infrared Spectrosc.*, **3**, 139–152 (1967).
513. H. Takeuchi, T. Nagai, and N. Tokura, *Tetrahedron*, **23**, 1783 (1967).
514. R. C. Talbott, *J. Org. Chem.*, **33**, 2095 (1968).

515. H. Tanida and T. Tsuji, *J. Org. Chem.*, **29**, 849 (1964).

516. (a) J. W. Taylor and J. C. Martin, *J. Amer. Chem. Soc.*, **88**, 3650 (1966); (b) *ibid.*, **89**, 6904 (1967).

517. L. M. Terman and G. A. Razuvaev, *Zh. Obschch. Khim.*, **31**, 3132 (1961).

518. W. M. Thomas and M. T. O'Shaughnessy, *J. Polymer Sci.*, **11**, 455 (1953).

519. A. V. Tobolsky and R. B. Mesrobian, *Organic Peroxides*, Interscience, New York, 1954.

520. K. Tokumaru, *Chem. Ind. (London)*, **1969**, 297.

521. K. Tokumaru, K. Horie, and O. Simamura, *Tetrahedron*, **21**, 867 (1965).

522. (a) K. Tokumaru and O. Simamura, *Bull. Chem. Soc. Japan*, **36**, 76 (1963); (b) *ibid.*, **36** 333 (1963).

523. M. Tonaka, M. Nishimura, and S. Hasegawa, *Rept. Res. Lab. Surface Sci., Fac. Sci. Okayama Univ.*, **2**, 289 (1966); *Chem. Abstr.* **66**, 85358.

524. E. A. Trosman and Kh. S. Bagdasar'yan, *Zh. Fiz. Khim.*, **38**, 141 (1964).

525. (a) G. Tsuchihashi, S. Miyajima, T. Otsu, and O. Simamura, *Tetrahedron*, **21**, 1039 (1965); (b) G. Tsuchihashi, M. Matsuchima, S. Miyajima, and O. Simamura, *Tetrahedron*, **21**, 1049 (1965).

526. N. S. Tsveltkov and R. F. Markovskaya, *Vysokomolekul. Soedin.*, **8**, 1299 (1966); *Chem. Abstr.*, **65**, 15506.

527. K. Überreiter and W. Bruns, *Makromol. Chem.*, **68**, 24 (1963).

528. (a) L. Vanino and E. Thiele, *Ber.*, **29**, 1727 (1896); (b) L. Vanino and E. Uhlfelder, *Ber.*, **33**, 1043 (1900).

529. D. E. Van Sickle, *J. Org. Chem.*, **34**, 3446 (1969).

530. S. Varga, Thesis, Rutgers University, 1968; *Diss. Abstr.*, **29B**, 2370 (1969).

531. I. N. Vasilev and V. A. Krongauz, *Kinetika i Kataliz*, **4**, 204 (1963).

532. T. T. Vasil'eva and R. Kh. Freidlina, *Izv. Akad. Nauk SSSR, Ser. Khim.*, **1968**, 1093.

533. (a) G. Vernin, H. J. M. Dou, and J. Metzger, *Compt. Rend. Acad. Sci. Paris*, **C264**, 336 (1967); (b) G. Vernin and J. Metzger, *Bull. Soc. Chim. France*, **1963**, 2504.

534. E. D. Vilenskaya, K. I. Ivanov, and M. M. Razarenova, *Zh. Organ. Khim.*, **4**, 1043 (1968).

535. T. C. Vogt and W. H. Hamil, *J. Phys. Chem.*, **67**, 292 (1963).

536. (a) N. S. Vyazankin, G. A. Razuvaev, and T. N. Brevnova, *Zh. Obshch. Khim.*, **34**, 1005 (1964); (b) N. S. Vyazankin, G. A. Razuvaev, O. S. D'yachkovskaya, and O. A. Shchepetkova, *Dokl. Akad. Nauk SSSR*, **143**, 1348 (1962); (c) *Khim. Perekisnykh Soedin. Akad. Nauk SSSR, Inst. Obshch. i Neorgan. Khim.*, **1963**, 298; *Chem. Abstr.*, **60**, 12040.

537. C. D. Wagner, R. H. Smith, and E. D. Peters, *Ind. Eng. Chem., Anal. Ed.*, **19**, 976 (1947).

538. H. M. Walborsky, C. J. Chen, and J. L. Webb, *Tetrahedron Letters*, **1964**, 3551.

539. O. J. Walker and G. L. Wild, *J. Chem. Soc.*, **1937**, 1132.

540. Wallace and Tiernan, Netherlands Patent Application 6,604,977 (1966); *Chem. Abstr.*, **66**, 76469.

541. C. Walling, *Free Radicals in Solution*, Wiley, New York, 1958.

542. (a) C. Walling and J. C. Azar, *J. Org. Chem.*, **33**, 3885 (1968); (b) *ibid.*, **33**, 3888 (1968).

543. C. Walling and L. Bollyky, *J. Amer. Chem. Soc.*, **86**, 3750 (1964).

544. C. Walling and Z. Cekovic, *J. Amer. Chem. Soc.*, **89**, 6681 (1967).

545. C. Walling and M. J. Gibian, *J. Amer. Chem. Soc.*, **87**, 3413 (1965).

546. C. Walling and R. B. Hodgdon, *J. Amer. Chem. Soc.*, **80**, 228 (1958).

547. C. Walling and N. Indictor, *J. Amer. Chem. Soc.*, **80**, 5814 (1958).

548. C. Walling, H. N. Moulden, J. H. Waters, and R. C. Neuman, *J. Amer. Chem. Soc.*, **87**, 518 (1965).

549. C. Walling and J. Pellon, *J. Amer. Chem. Soc.*, **79**, 4786 (1957).

550. C. Walling and E. S. Savas, *J. Amer. Chem. Soc.*, **82**, 1738 (1960).

551. W. Walter, M. Steffen, and C. Heyns, *Ber.*, **99**, 3204 (1966).

552. C. H. Wang, R. McNair, and P. Levins, *J. Org. Chem.*, **30**, 3817 (1965).

553. H. Wieland, H. von Hove, and K. Börner, *Ann.*, **446**, 31 (1926).

554. H. Wieland and J. Maier, *Ber.*, **64**, 1205 (1931).

555. H. Wieland and A. Meyer, *Ann.*, **551**, 249 (1942).

556. H. Wieland, T. Ploetz, and K. Indest, *Ann.*, **532**, 179 (1937).

557. H. Wieland and G. A. Razuvaev, *Ann.*, **480**, 157 (1930).

558. H. Wieland, S. Schapiro, and H. Metzger, *Ann.*, **513**, 93 (1934).

559. G. H. Williams, *Homolytic Aromatic Substitution*, Pergamon, London, 1960.

560. J. W. Wilt and J. A. Lundquist, *J. Org. Chem.*, **29**, 921 (1964).

561. R. G. Woolford and R. N. Gedye, *Can. J. Chem.*, **45**, 291 (1967).

562. T. Yamamoto and T. Azumi, *Himeji Kogyo Daigaku Kenkyu Hokoku*, No. 13, 77 (1961); *Chem. Abstr.*, **56**, 359 (1961).

563. (a) H. Yano, K. Takemoto, and M. Imoto, *J. Makromol. Sci. Chem.*, **2**, 81 (1968); (b) *ibid.*, **2**, 739 (1968).

564. M. Yoshida, O. Simamura, K. Miyaka, and L. V. Tam, *Chem. Ind. (London)*, **1966**, 2060.

565. D. M. Young and W. M. Stoops, U.S. Patent 2,792,423 (1957); *Chem. Abstr.*, **51**, 15583.

566. A. I. Yurzhenko, S. S. Ivanchev, and V. I. Galibei, *Dokl. Akad. Nauk SSSR*, **140**, 1348 (1961).

567. C. Zimmerman, U.S. Patent 2,580,373 (1951); *Chem. Abstr.* **46**, 6668.

568. A. V. Zubkov, A. T. Koritskii, and Ya. S. Lebedev, *Dokl. Akad. Nauk SSSR*, **180**, 1150 (1968).

# Index